Third Edition

Essentials *of* Biology

Sylvia S. Mader

Michael Windelspecht
Appalachian State University

With contributions by

Lynn Preston
Tarrant County College

Penn Foster Schools

McGraw Hill Education

5 6 7 8 9 0 RMN RMN 15 14

ISBN-13: 978-0-07-813535-4
ISBN-10: 0-07-813535-4

Learning Solutions Consultant: Heidi Freund Degheri
Project Manager: Lynn Nagel

Brief Contents

About the Authors

Dr. Sylvia S. Mader has authored several nationally recognized biology texts published by McGraw-Hill. Educated at Bryn Mawr College, Harvard University, Tufts University, and Nova Southeastern University, she holds degrees in both Biology and Education. Over the years she has taught at University of Massachusetts, Lowell, Massachusetts Bay Community College, Suffolk University, and Nathan Mathew Seminars. Her ability to reach out to science-shy students led to the writing of her first text, *Inquiry into Life,* that is now in its thirteenth edition. Highly acclaimed for her crisp and entertaining writing style, her books have become models for others who write in the field of biology.

Although her writing schedule is always quite demanding, Dr. Mader enjoys taking time to visit and explore the various ecosystems of the biosphere. Her several trips to the Florida Everglades and Caribbean coral reefs resulted in talks she has given to various groups around the country. She has visited the tundra in Alaska, the taiga in the Canadian Rockies, the Sonoran Desert in Arizona, and tropical rain forests in South America and Australia. A photo safari to the Serengeti in Kenya resulted in a number of photographs for her texts. She was thrilled to think of walking in Darwin's steps when she journeyed to the Galápagos Islands with a group of biology educators. Dr. Mader was also a member of a group of biology educators who traveled to China to meet with their Chinese counterparts and exchange ideas about the teaching of modern-day biology.

Dr. Michael Windelspecht serves as the Introductory Biology Coordinator at Appalachian State University, in Boone NC, where he directs a program that enrolls over 4,500 non-science majors annually. He was educated at the University of Maryland, Michigan State University, and the University of South Florida. As an educator, Dr. Windelspecht teaches not only introductory biology for non-majors, but also biology for science majors, genetics, and human genetics. In addition to his teaching assignments, Dr. Windelspecht is active in promoting the scientific literacy of secondary school educators. He has led multiple workshops on integrating water quality research into the science curriculum, and has spent several summers teaching Pakistani middle school teachers.

As an author, Dr. Windelspecht has published five reference textbooks, and multiple print and online lab manuals. He served as the series editor for a ten-volume work on the human body. For years Dr. Windelspecht has been active in the development of multimedia resources for the online and hybrid science classrooms. Along with his wife, Sandra, he owns a multimedia production company that actively develops and assesses the use of new technologies in the classroom.

Preface

Essentials of Biology was written to not only engage non-major students in the science of biology, but also to encourage them to integrate scientific concepts into their lives. While the textbook follows a traditional organization, it contains a variety of print and multimedia assets that allow the instructor to promote scientific literacy and appreciation of our scientific world.

The authors identified several goals that guided the revision of *Essentials of Biology,* Third Edition.

- **Evolution** coverage increased
- **Genetics of human disease** coverage expanded
- **Applications** added to enhance the relevancy of content for students
- **Media** assets integrated in textbook and in Connect™ Biology

Evolution

- A new evolutionary tree is presented for Kingdom fungi.
- **Chapter 19: Both Water and Land: Animals** has been extensively reworked to reflect the modern understanding of the relationships between the major groups of animals. New evolutionary trees have been added that address the new relationships among the protostome invertebrates.
- The section on human evolution in chapter 19 has been updated to include the influence of *Homo floresiensis* and the sequencing of the Neandertal genome on *Homo sapiens* evolutionary heritage.
- Many of the Connecting the Concepts boxes at the end of the each section direct the student to sections of the textbook that contain additional information on evolutionary topics.

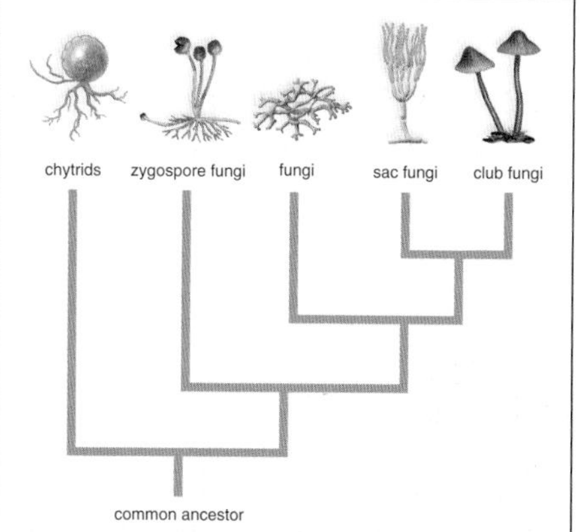

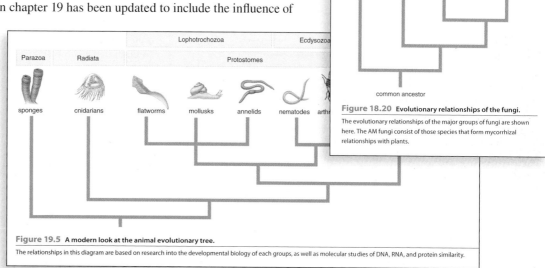

Figure 18.20 Evolutionary relationships of the fungi.
The evolutionary relationships of the major groups of fungi are shown here. The AM fungi consist of those species that form mycorrhizal relationships with plants.

chytrids   zygospore fungi   fungi   sac fungi   club fungi

common ancestor

Figure 19.5 A modern look at the animal evolutionary tree.
The relationships in this diagram are based on research into the developmental biology of each groups, as well as molecular studies of DNA, RNA, and protein similarity.

Lophotrochozoa   Ecdysozoa
Parazoa   Radiata   Protostomes
sponges   cnidarians   flatworms   mollusks   annelids   nematodes   arthr

Genetics of Human Disease

Throughout the book, the genetic basis of human diseases (such as Down syndrome, cystic fibrosis, and Huntington disease) are identified and discussed.

Connections and Misconceptions

What causes cystic fibrosis?

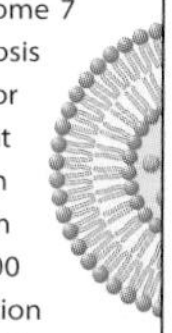

In 1989, scientists determined that defects in a gene on chromosome 7 causes cystic fibrosis (CF). This gene, called *CFTR* (short for cystic fibrosis conductance regulator), codes for a protein that is responsible for the movement of chloride ions across the membranes of cells that produce mucus, sweat, and saliva. Defects in this gene cause an improper water/salt balance in the excretions of these cells, which in turn leads to the symptoms of CF. Currently, there are over 1,400 known mutations in the CF gene. This tremendous amount of variation accounts for the differences in the severity of the symptoms in CF patients. By knowing the precise gene that causes the disease, scientists have been able to develop new treatment options for people with CF. At one time, people with CF rarely saw their 20th birthday; now it is routine for people to live into their 30s and 40s. New treatments, such as gene therapy, are being explored for sufferers of CF.

Applications

Many new and/or improved applications, such as the Connections and Misconceptions readings, have been added to the text to help students understand the relevancy of biological concepts to their lives. Refer to the inside back cover for a complete listing of application readings.

Connections and Misconceptions

Why can't you drink seawater?

Seawater is hypertonic to our cells. It contains approximately 3.5% salt, whereas our cells contain 0.9%. Once the salt had entered your blood, your cells would shrivel up and die as they lost water trying to dilute the excess salt. Your kidneys can only produce urine that is slightly less salty than seawater, so you would dehydrate providing the amount of water necessary to rid your body of the salt. In addition, salt water contains high levels of magnesium ions, which cause diarrhea and further dehydration.

 Animation Osmosis

Connections and Misconceptions

How is a paternity test done, and why are the results always a percentage instead of a certainty?

In a paternity test, a sample from the mother, child, and possible father are taken and amplified by PCR; then the samples are electrophoresed and the banding patterns compared. In a typical paternity test, 13–16 human identity markers (HID) are examined. These markers, or genes, are found in all humans, but in differing lengths of repeats that are inherited from parent to child.

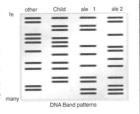

DNA Band patterns

When a certain amount of these markers match in the child and possible father, the result is a percentage of certainty that the man is the father (normally, 99.0% or higher). Only by comparing entire genomes, which co___ ___ ___ ___ of the lions of base pairs and approximately 20,000–25,000 gen___ ___ result in 100% certainty.

Connections and Misconceptions

What is RNA interference?

RNA interference, or RNAi, is an experimental procedure in which small pieces of RNA are used to "silence" the expression of specific alleles. These RNA sequences are designed to be complementary to the mRNA transcribed by a gene of interest. Once the complementary RNA sequences enter the cell, they bind with the target RNA, producing double-stranded RNA molecules. These double-stranded RNA molecules are then broken down by a series of enzymes within the cell. First discovered in worms, RNAi is believed to have evolved in eukaryotic organisms as a protection against certain types of viruses. Research into developing RNAi treatments for a number of human diseases, including cancer and hepatitis, are currently underway.

 Animation RNA Interference   **Video** Halting Hepatitis

Connections and Misconceptions

What has happened to Darwin's finches since the nineteenth century?

The finches of the Galápagos Islands have continued to provide a wealth of information on evolutionary processes. Starting in 1973, a team of researchers led by Drs. Peter and Rosemary Grant began a 30-year study of a species of groundfinch on the island of Daphne Major in the Galápagos. Through detailed measurements and observations of over 19,000 birds, the Grants were able to document the evolutionary change of a species in response to changes in the environment in action. The details of their study were reported in the Pulitzer Prize–winning book *The Beak of the Finches: A Story of Evolution in* **Video** Finches Natural Selection

Connections and Misconceptions

How do drugs that regulate depression and anxiety work?

In general, pharmaceutical drugs that regulate behavior work by regulating the amount of neurotransmitter in the synapse. For example, drugs such as Xanax and Valium increase the levels of gamma-aminobutyric acid. These medications are used for 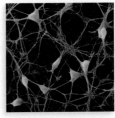 panic attacks and anxiety. Reduced levels of norepinephrine and serotonin are linked to depression. Drugs such as Prozac, Paxil, and Cymbalta allow norepinephrine and/or serotonin to accumulate at the synapse, usually by blocking their reabsorption. By increasing the levels of these chemicals, the postsynaptic cell receives a more constant chemical message, which explains their effectiveness as antidepressants.

Media Integration

A significant new feature of this edition is the integration of content and animation, video, and audio assets. Virtually every section of the textbook is now linked to MP3 files, animations of biological processes, National Geographic videos, or ScienCentral videos.

 MP3 files. These three-to-five minute audio files not only serve as a review of the material in the chapter, but also assist the student in the pronunciation of scientific terms.

 Animations. Drawing on McGraw-Hill's vast library of animations, the authors have selected animations that will enhance the student's understanding of complex biological processes.

 Videos. Two different types of movies are integrated into this edition of the text. The ScienCentral videos are short news clips on recent advances in the sciences. The National Geographic videos provide the student with a glimpse of the complexity of life that normally would not be possible in the classroom.

 Virtual labs. Simulated experiments allow students to explore the topics covered in some chapters.

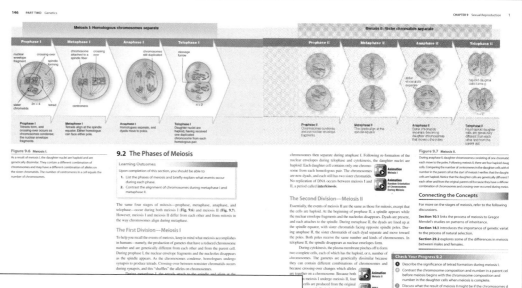

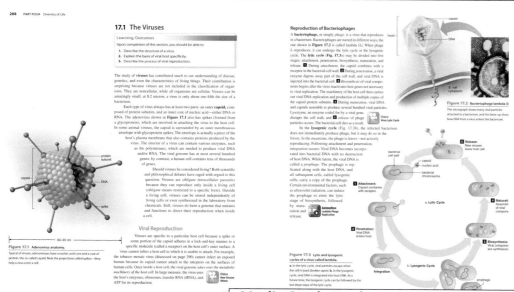

Media Study Tools

www.mhhe.com/maderessentials3

Enhance your study of this chapter with study tools and practice tests. Also ask your instructor about the resources available through ConnectPlus, including the media-rich eBook, interactive learning tools, and animations.

DNA and Genes

The virtual lab "DNA and Genes" provides an interactive look at how mutations influence the amino acid sequence of a protein.

A Student's Guide to Using This Textbook

Chapter Outline
Lists the major sections that will be discussed in the chapter.

NEW Before You Begin
Links the content of the chapter with material from earlier in the text. The questions designate important topics that you should understand before proceeding into the chapter.

NEW Learning Outcomes
Provide you with an overview of what you are to know. Your instructor can assign activities through Connect™ to help you achieve these outcomes.

NEW Media Integration
Enhances your study of biology with media. Go to www.mhhe.com/maderessentials3 to access the animations, videos, and MP3 files referenced throughout this book. Ask your instructor about related quizzes that are available through Connect™ Biology.

7

Energy for Cells

OUTLINE

BEFORE YOU BEGIN

Before beginning this chapter, take a few moments to review the following discussions.

Section 3.2 What are the roles of carbohydrates in living organisms?

Section 5.2 What is the ATP cycle?

Section 6.1 What do the terms *oxidation* and *reduction* mean?

Not All Diets Are Created Equal

An obesity epidemic rages unfettered in the United States. Not surprisingly, an entire industry has emerged, hawking weight-loss supplements, "miracle" fat burners, and fad diets. Many diets may help you initially lose weight, but have you ever considered the science behind how these diets claim to work? When you do, you will see that many of these claims simply can't be verified. Many fad diets rob the body of the essential nutrients it needs to stay healthy, yet millions of people spend a fortune, hoping for a miracle, and many of them suffer health problems as a result.

Your cells break down glucose in a process called cellular respiration to extract the energy from your food to create the ATP they need. To lose weight, it all comes down to regulating the input of energy while providing adequate levels of important vitamins and minerals for the cells to convert these nutrients to ATP. When cells have excess energy, or a deficiency in vitamins, the excess is converted to fat. In this chapter, you will learn how cells use glucose, fat, and protein to produce ATP. You will also learn how the body uses metabolites to build glucose, protein, and fat. Understanding these processes will give you the information you need to make informed decisions about weight loss, diets, and nutritional supplements.

17.1 The Viruses

Learning Outcomes

Upon completion of this section, you should be able to

1. Describe the structure of a virus.
2. Explain the basis of viral host specificity.
3. Describe the process of viral reproduction.

The study of **viruses** has contributed much to our understanding of disease, genetics, and even the characteristics of living things. Their contribution is surprising because viruses are not included in the classification of organisms. They are noncellular, while all organisms are cellular. Viruses can be amazingly small; at 0.2 micron, a virus is only about one-fifth the size of a bacterium.

Each type of virus always has at least two parts: an outer **capsid**, composed of protein subunits, and an inner core of nucleic acid—either DNA or RNA. The adenovirus shown in **Figure 17.1** also has spikes (formed from a glycoprotein), which are involved in attaching the virus to the host cell. In some animal viruses, the capsid is surrounded by an outer membranous

---

the agent, he or sh...
some antiviral drugs are available. Most antiviral compounds, such as ribavirin and acyclovir, are structurally similar to nucleotides; therefore, they interfere with viral genome synthesis. Compounds related to acyclovir are commonly used to suppress herpes outbreaks. HIV is treated with antiviral compounds specific to a retrovirus. The well-publicized drug AZT and others block reverse transcriptase. And HIV protease inhibitors block the enzymes required for the maturation of viral proteins.

  Video Flu Jail Cell   Animation Antiviral Agents

Connecting the Concepts

For more on viruses, refer to the following discussions.

Section 13.3 explores how viruses may be used in gene therapy.

Section 26.5 provides additional information on the HIV virus and the disease AIDS.

Section 29.3 examines some of the sexually transmitted diseases (STDs) that are caused by viruses.

Check Your Progress 17.1

❶ Describe the structure of a virus.

❷ Categorize the types of viral reproduction.

❸ Summarize the ways viruses can emerge as an infectious disease.

NEW Connecting the Concepts
Directs you to areas of the textbook that provide additional information on a topic. Many of the Connecting the Concepts have been designed to enhance your understanding of evolutionary biology.

Check Your Progress
Questions at the end of each section help you assess and/or apply your understanding of the material in the section. The questions progress in difficulty (red, yellow, green) to ensure you are going beyond memorization of content.

Media Study Tools

Provide a link to the *Essentials of Biology* website, that contains practice tests, animations, and videos organized and integrated by chapter to help you succeed in your study of biology. The ConnectPlus™ platform provides a media-rich eBook, interactive learning tools, and access to the LearnSmart™ system for enhanced student performance.

NEW Virtual Labs

Allow you to investigate topics associated with the content of the chapter from a scientific perspective.

Summary

Provides an excellent overview of the chapter concepts using concise, bulleted summaries, summary tables, and key illustrations.

Key Terms

Page-referenced to help you master the scientific vocabulary.

Media Study Tools

www.mhhe.com/maderessentials3

Enhance your study of this chapter with study tools and practice tests. Also ask your instructor about the resources available through ConnectPlus, including the media-rich eBook, interactive learning tools, and animations.

Enzyme-Controlled Reactions
The virtual lab "Enzyme-Controlled Reactions" provides an interactive investigation of how environmental conditions regulate enzyme activity.

The Chapter in Review

Summary

5.1 What Is Energy?

Potential energy can be converted to kinetic energy, and vice versa. Solar and chemical energy are forms of potential energy; mechanical energy is a form of kinetic energy. Two energy laws hold true universally:

- Energy cannot be created or destroyed, but it can be transferred or transformed.
- Energy can be converted from one form to another, but some is lost as unusable forms. Therefore, the entropy of the universe is increasing, and only a constant input of energy maintains the organization of living things.

5.2 ATP: Energy for Cells

Energy flows from the sun through chloroplasts and mitochondria, which produce ATP. Because ATP has three adjoining negative phosphate groups, it is a high-energy molecule that tends to break down to ADP + P, releasing energy. ATP breakdown is coupled to various energy-requiring cellular reactions, including protein synthesis, active transport, and muscle contraction. Cellular respiration provides the energy for the production of ATP. The following diagram summarizes the ATP cycle:

5.3 Metabolic Pathways and Enzymes

A metabolic pathway is a series of reactions that proceed in an orderly, step-by-step manner. Each reaction requires an enzyme that is specific to its substrate. Enzymes bring substrates together at an enzyme's active site and speed reactions by lowering the energy of activation. The activity of most enzymes and metabolic pathways is regulated by feedback inhibition.

5.4 Cell Transport

The plasma membrane is semipermeable; some substances can freely cross the membrane, and some must be assisted across if they are to enter the cell.

Passive transport requires no metabolic energy and moves substances from a higher to a lower concentration.

- In simple diffusion, molecules move from an area of higher concentration to the area of lower concentration until the concentration of molecules is the same at both sites. Some molecules, such as dissolved gases, cross plasma membranes by simple diffusion.
- In facilitated diffusion, molecules diffuse across a plasma membrane through a channel protein (aquaporins are channel proteins for water) or with the assistance of carrier proteins.
- Osmosis is the simple diffusion of water through aquaporins across a membrane toward the area of lower water concentration (higher solute concentration).

Active transport requires metabolic energy (ATP) and moves substances from a lower to a higher concentration across a membrane.

- A transport protein acts as a pump that causes a substance to move against its concentration gradient. For example, the sodium-potassium pump carries Na^+ to the outside of the cell and K^+ to the inside of the cell.

Thinking Scientifically

1. Some coupled reactions in cells, including many involved in protein synthesis, use the nucleotide GTP as an energy source instead of ATP. What would be the advantage of using GTP instead of ATP as an energy source for these cellular reactions?
2. Cystic fibrosis is a genetic disorder caused by a defective membrane transport protein. The defective protein closes chloride channels in membranes, preventing chloride from being exported out of cells. This results in the development of a thick mucus on the outer surfaces of cells. This mucus clogs the ducts that carry digestive enzymes from the pancreas to the small intestine, clogs the airways in the lungs, and promotes lung infections. How do you think the defective protein results in a thick, sticky mucus outside the cells, instead of a loose, fluid covering?
3. Some prokaryotes, especially archaea, are capable of living in extreme environments, such as deep-sea heat vents, where temperatures can reach 80°C (176°F). Few organisms can survive at this temperature. What adaptations might archaea possess that allow them to survive in such extreme heat?

Bioethical Issue

Arsenic in Drinking Water

Arsenic is chemically similar to phosphorus, and it may form arsenate ions (AsO_4^{3-}) within the body that can substitute for phosphate (PO_4^{3-}) in many enzymatic reactions. Arsenate blocks some of the processes in the cell that are responsible for producing ATP, resulting in cell death. Metabolically active cells are particularly vulnerable to this toxin.

Arsenic occurs naturally in the water and soil; thus, arsenic levels in drinking water are a concern in many parts of the world. In 2000, the Clinton administration directed the Environmental Protection Agency (EPA) to reset the standard for maximal allowable levels in water to 10 parts per billion (ppb) from the previous level of 50 ppb set in 1942. These rules were suspended by the Bush administration in 2001, citing the high cost of compliance by utilities. Supporters asserted that the costs of compliance would be significant and that smaller utilities serving rural communities would be particularly hard hit. Other critics objected on the basis that the cost-to-benefit ratio of the new standards did not justify the new rules. But many health and environmental organizations and many members of Congress blasted the decision to rescind the new rules, citing studies demonstrating health problems associated with arsenic

---

Thinking Scientifically

Questions give you an opportunity to reason as a scientist. (See Appendix A for answers.)

Bioethical Issue

Readings discuss a variety of controversial topics that confront our society so you can begin to understand those issues in the context of biology.

Testing Yourself

Questions help you review material and prepare for tests. (See Appendix A for answers.)

Key Terms

active site 84	induced fit model 84
active transport 88	isotonic solution 87
aquaporin 86	kilocalorie 78
calorie 78	kinetic energy 78
coupled reaction 81	metabolic pathway 83
endocytosis 89	osmosis 87
energy 78	phagocytosis 89
energy laws 79	pinocytosis 89
energy of activation (E_a) 85	plasmolysis 88
entropy 79	potential energy 78
enzyme 83	receptor-mediated
enzyme inhibition 84	endocytosis 89
exocytosis 89	simple diffusion 86
facilitated diffusion 86	sodium-potassium pump 88
feedback inhibition 84	solute 86
heat 79	solution 86
hypertonic solution 88	solvent 86
hypotonic solution 88	substrate 84

Testing Yourself

Choose the best answer for each question.

1. Which of the following is not a fundamental law of energy?
   a. Energy cannot be created or destroyed.
   b. Energy can be changed from one form to another.
   c. Energy is lost when it is converted from one form to another.
   d. Potential energy cannot be converted to kinetic energy.

4. ATP is a good source
   a. it is versatile—ab
   b. its breakdown is c reactions.
   c. it provides just th reactions.
   d. All of these are co

5. Compared with carbo
   a. is less organized.
   b. has less potential

For questions 6–9, match th answer may include more th

Key:
   a. oxygen
   b. carbon dioxide
   c. water
   d. carbohydrates
6. Consumed by cellula
7. Produced by cellular
8. Consumed by photos
9. Produced by photosy
10. In the following figu required for a reacti the energy of activati

11. In the induced fit model of enzyme catalysis,
    a. the substrate exactly fits multiple enzymes.
    b. many substrates fit every enzyme.
    c. the enzyme changes shape slightly to accommodate the substrate.
    d. the enzyme doesn't fit the substrate until it is induced by another substrate.

A brilliant visuals program brings biology to life!

Multi-Level Perspective

Illustrations depicting complex structures show macroscopic and microscopic views to help you see the relationships between increasingly detailed drawings.

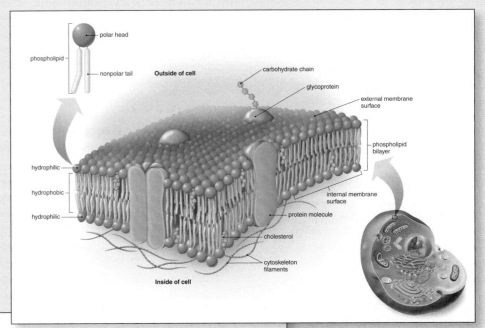

polar head
phospholipid
nonpolar tail
Outside of cell
carbohydrate chain
glycoprotein
external membrane surface
phospholipid bilayer
hydrophilic
hydrophobic
hydrophilic
internal membrane surface
protein molecule
cholesterol
cytoskeleton filaments
Inside of cell

cuticle
upper epidermis
palisade mesophyll
leaf vein
spongy mesophyll
lower epidermis

Water and minerals enter leaf through xylem.
Sugar exits leaf through phloem.

×94

Combination Art

Drawings of structures are paired with micrographs to give you the best of both perspectives: the realism of photos and the explanatory clarity of line drawings.

nucleus
mitochondrion
central vacuole
chloroplast

Leaf cell

guard cell
O$_2$ and H$_2$O exit leaf through stoma.
CO$_2$ enters leaf through stoma.
epidermal cell

Stoma and guard cells

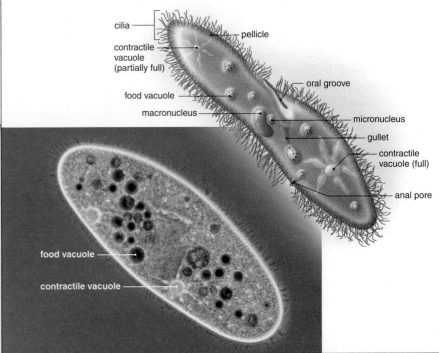

cilia
contractile vacuole (partially full)
food vacuole
macronucleus
pellicle
oral groove
micronucleus
gullet
contractile vacuole (full)
anal pore

food vacuole
contractile vacuole

Process Figures

Complex processes are broken down into a series of smaller steps that are easy to follow. Numbers guide you through the process.

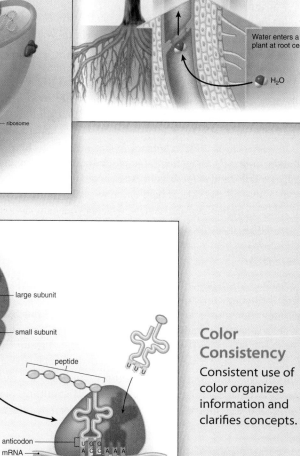

Evaporation of water (transpiration) creates tension that pulls the water column from the roots to the leaves.

H_2O

Cohesion and adhesion of water molecules keeps the water column intact within xylem.

Water enters a plant at root cells.

H_2O

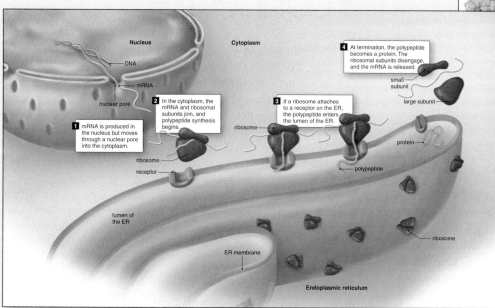

Nucleus

DNA

mRNA

nuclear pore

Cytoplasm

1 mRNA is produced in the nucleus but moves through a nuclear pore into the cytoplasm.

2 In the cytoplasm, the mRNA and ribosomal subunits join, and polypeptide synthesis begins.

ribosome

receptor

3 If a ribosome attaches to a receptor on the ER, the polypeptide enters the lumen of the ER.

ribosome

4 At termination, the polypeptide becomes a protein. The ribosomal subunits disengage, and the mRNA is released.

small subunit

large subunit

protein

polypeptide

lumen of the ER

ribosome

ER membrane

Endoplasmic reticulum

amino acid

complementary base pairing

template DNA strand

RNA polyme

Transc the nu joined polyme comple of DNA

U G G

anticodon

tRNA binding sites

P site   A site

large subunit

mRNA binding site

small subunit

b. Ribosome

peptide

anticodon

mRNA

5′

ribosome

U G G
A C C A A A

3′

c. tRNA–amino acid at ribosome

amino acid methionine

initiator tRNA

mRNA

small ribosomal subunit

start codon

P site

Color Consistency

Consistent use of color organizes information and clarifies concepts.

Teaching and Learning Tools

McGraw-Hill Higher Education and Blackboard® have teamed up.

Do More

Blackboard, the Web-based course-management system, has partnered with McGraw-Hill to better allow students and faculty to use online materials and activities to complement face-to-face teaching. Blackboard features exciting social learning and teaching tools that foster more logical, visually impactful, and active learning opportunities for students. You'll transform your closed-door classrooms into communities where students remain connected to their educational experience 24 hours a day.

This partnership allows you and your students access to McGraw-Hill's Connect™ and Create™ right from within your Blackboard course—all with one single sign-on.

Not only do you get single sign-on with Connect™ and Create™, you also get deep integration of McGraw-Hill content and content engines right in Blackboard. Whether you're choosing a book for your course or building Connect™ assignments, all the tools you need are right where you want them—inside of Blackboard.

Gradebooks are now seamless. When a student completes an integrated Connect™ assignment, the grade for that assignment automatically (and instantly) feeds your Blackboard grade center.

McGraw-Hill and Blackboard can now offer you easy access to industry leading technology and content, whether your campus hosts it, or we do. Be sure to ask your local McGraw-Hill representative for details.

LearnSmart™

LearnSmart™ is available as an integrated feature of McGraw-Hill Connect™ Biology and provides students with a GPS (**G**uided **P**ath to **S**uccess) for your course. Using artificial intelligence, **LearnSmart™** intelligently assesses a student's knowledge of course content through a series of adaptive questions. It pinpoints concepts the student does not understand and maps out a personalized study plan for success. This innovative study tool also has features that allow instructors to see exactly what students have accomplished, and a built-in assessment tool for graded assignments. Visit the site below for a demonstration.

www.mhlearnsmart.com

McGraw-Hill Connect™ Biology

www.mhhe.com/maderessentials3

McGraw-Hill Connect™ Biology provides online presentation, assignment, and assessment solutions. It connects your students with the tools and resources they'll need to achieve success.

With Connect™ Biology you can deliver assignments, quizzes, and tests online. A robust set of questions and activities are presented and aligned with the textbook's learning outcomes. As an instructor, you can edit existing questions and author entirely new problems. Track individual student performance—by question, assignment, or in relation to the class overall—with detailed grade reports. Integrate grade reports easily with Learning Management Systems (LMS), such as WebCT and Blackboard. And much more.

ConnectPlus™ Biology provides students with all the advantages of Connect™ Biology, plus 24/7 online access to an eBook. This media-rich version of the book is available through the McGraw-Hill Connect™ platform and allows seamless integration of text, media, and assessments.

To learn more, visit

www.mcgrawhillconnect.com

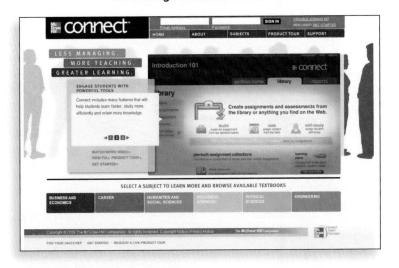

My Lectures—Tegrity

Tegrity Campus™ records and distributes your class lecture, with just a click of a button. Students can view anytime/anywhere via computer, iPod, or mobile device. It indexes as it records your PowerPoint® presentations and anything shown on your computer so students can use keywords to find exactly what they want to study. Tegrity is available as an integrated feature of McGraw-Hill Connect™ Biology or as standalone.

Animations for a New Generation

Dynamic, 3D animations of key biological processes bring an unprecedented level of control to the classroom. Innovative features keep the emphasis on teaching rather than entertaining.

- An options menu lets you control the animation's level of detail, speed, length, and appearance, so you can create the experience you want.
- Draw on the animation using the white board pen to highlight important areas.
- The scroll bar lets you fast forward and rewind while seeing what happens in the animation, so you can start at the exact moment you want.
- A scene menu lets you instantly jump to a specific point in the animation.
- Pop ups add detail at important points and help students relate the animation back to concepts from lecture and the textbook.
- A complete visual summary at the end of the animation reminds students of the big picture.
- Animation topics include: Cellular Respiration, Photosynthesis, Molecular Biology of the Gene, DNA Replication, Cell Cycle and Mitosis, Membrane Transport, and Plant Transport.

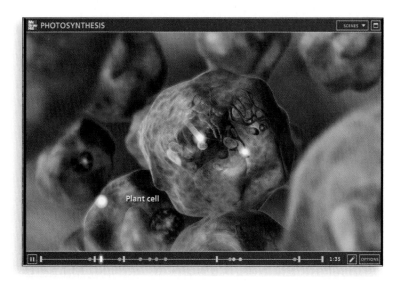

Create

With McGraw-Hill Create™, **www.mcgrawhillcreate.com,** you can easily rearrange chapters, combine material from other content sources, and quickly upload content you have written like your course syllabus or teaching notes. Find the content you need in Create by searching through thousands of leading McGraw-Hill textbooks. Arrange your book to fit your teaching style. Create even allows you to personalize your book's appearance by selecting the cover and adding your name, school, and course information. Order a Create book and you'll receive a complimentary print review copy in 3–5 business days or a complimentary electronic review copy (eComp) via email in minutes. Go to **www.mcgrawhillcreate.com** today and register to experience how McGraw-Hill Create™ empowers you to teach your students your way.

Traditional Resources

As always, McGraw-Hill offers the following resources for instructors.

Presentation Tools

Everything you need for outstanding presentations in one place!

www.mhhe.com/maderessentials3

- *FlexArt Image PowerPoints*—including every piece of art that has been sized and cropped specifically for superior presentations, as well as labels that can be edited and flexible art that can be picked up and moved on key figures. Also included are tables, photographs, and unlabeled art pieces.
- *Lecture PowerPoints with Animations*—animations illustrating important processes are embedded in the lecture material.
- *Animation PowerPoints*—animations only are provided in PowerPoint.
- *Labeled JPEG Images*—Full-color digital files of all illustrations that can be readily incorporated into presentations, exams, or custom-made classroom materials.
- *Base Art Image Files*—unlabeled digital files of all illustrations.

Presentation Center

In addition to the images from your book, this online digital library contains photos, artwork, animations, and other media from an array of McGraw-Hill textbooks.

Computerized Test Bank

A comprehensive bank of test questions is provided within a computerized test bank powered by McGraw-Hill's flexible electronic testing program, EZ Test Online. A new tagging scheme allows you to sort questions by Bloom's difficulty level, learning outcome, topic, and section. With EZ Test Online, instructors can select questions from multiple McGraw-Hill test banks or author their own, and then either print the test for paper distribution or give it online.

Instructor's Manual

The instructor's manual contains chapter outlines, lecture enrichment ideas, and discussion questions.

Laboratory Manual

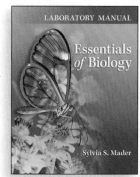

The *Essentials of Biology Laboratory Manual* is written by Dr. Sylvia Mader. Every laboratory has been written to help students learn the fundamental concepts of biology and the specific content of the chapter to which the lab relates, as well as gain a better understanding of the scientific method.

Companion Website

www.mhhe.com/maderessentials3

The *Essentials of Biology* companion website allows students to access a variety of free digital learning tools that include:

- Chapter-level quizzing
- Animations and videos
- Vocabulary flashcards
- Virtual labs

Overview of Content Changes to *Essentials of Biology,* Third Edition

Chapter 1   *A View of Life*

This chapter previews the text by discussing the characteristics of life, principles of evolution, organization of the biosphere, and the scientific process. A new application feature has been added to this chapter that addresses the number of species present on the planet.

Part I   *The Cell*

Chapter 2: The Chemical Basis of Life has had new applications added that address the source of elements, irradiation of food, the ability of insects to walk on water, and the acidity of the stomach. In **Chapter 3: The Organic Molecules of Life,** new applications have been added on organic produce, high-fructose corn syrup, soluble and insoluble fiber, omega-3 fatty acids, and the effect of relaxers on hair. In addition, the virtual lab on nutrition has been referenced at the end of the chapter. **Chapter 4: Inside the Cell** has been enhanced with new SEM micrograph images of spiderlings. Applications on the size of the largest cell, *E. coli,* the endosymbiotic theory, and the uses of collagen are now located in the chapter. A new table on the differences between prokaryotic and eukaryotic cells is now a component of the Chapter in Review material. **Chapter 5: The Dynamic Cell** includes a reworked description of osmosis (fig. 5.12), and the addition of application material covering the BMR, use of creatine phosphate supplements, the problems of drinking seawater, the genetic basis of cystic fibrosis, and a virtual lab on enzyme-controlled reactions. **Chapter 6: Energy for Life** features new applications on the origin of oxygen in the atmosphere, use of grow-lights in greenhouses, forests and carbon sequestration, and the evolution of C4 plants. In **Chapter 7: Energy for Cells,** figure 7.2 has been enhanced to illustrate the inputs and outputs of each stage of glucose breakdown in a cell, and figure 7.8 now illustrates the interaction of $FADH_2$ with the electron transport chain. Applications in this chapter include the use of chromium supplements, and a discussion of the influence of lactic acid on muscle soreness.

Part II   *Genetics*

Chapter 8: Cellular Reproduction includes new application material on the topics of chromosome number, the difference between apoptosis and necrosis, and the carcinogenic nature of cigarette smoke. A virtual lab, The Cell Cycle and Cancer, is also referenced at the end of the chapter. In **Chapter 9: Sexual Reproduction,** new application readings on meiosis and gamete production, female age and the incidence of Down syndrome, and Jacobs syndrome and male aggressiveness, have been added. For **Chapter 10: Patterns of Inheritance,** several new application readings have been introduced, including material on Mendel's experimental approach, the meaning of dominant and recessive traits, the genetic basis of skin color, and the difference between the dominance and prevalence of a trait. A reference to a virtual lab on Punnett Squares and Sex-Linked traits is included at the end of the chapter. **Chapter 11: DNA Biology and Technology** contains new application readings on the length of time required to copy the DNA of a human cell, and paternity testing, and reference is made to a virtual lab on DNA and Genes at the end of the chapter. **Chapter 12: Gene Regulation and Cancer** has been enhanced through the inclusion of application readings on the nutritional value of cloned animals, stem cells and human disease, X-inactivation, and transposons. The virtual lab, Cell Cycle and Cancer, is referenced at the end of the chapter. **Chapter 13: Genetic Counseling** has been reorganized to include sections on Counseling for Genetic Disorders and Testing for Genetic Disorders. New applications, such as the use of FISH for karyotype analysis, the genetic basis of Down syndrome, azoospermia, OTC genetic tests, and RNA interference, have been added.

Part III   *Evolution*

Chapter 14: Darwin and Evolution contains new application readings on the Grant's study of the Galapagos finches and vestigial organs in humans. A new figure (14.11) is included to indicate Wallace's contribution to the study of biogeography. The content of **Chapter 15: Evolution on a Small Scale** has been reorganized to place the discussion of natural selection between the content on microevolution. New material on sexual selection is now featured in section 15.1 and a new application reading on evolutionary change in a population is included. In **Chapter 16: Evolution on a Large Scale,** section 16.1 places a greater emphasis on speciation and table 16.1 now includes the probable causes of each of the mass extinction events. New application readings focus on how scientists identify new species, polyploidy in animals, the importance of the Burgess Shale, and the current mass extinction crisis. Coverage of the differences between the five-kingdom and three-domain system of classification has also been expanded.

Part IV   *Diversity of Life*

Several of the illustrations in **Chapter 17: The Microorganisms: Viruses, Bacteria, and Protists,** have been updated. Figure 17.6 now includes information on the H1N1 and H5N1 viruses, and figure 17.10 has been enhanced to illustrate internal and external structures of a prokaryote. Figure 17.19 has been revised to reflect the modern interpretation of the events in the endosymbiotic theory. Application readings on the H1N1 virus, prions and Alzheimer disease, and giardiasis are located throughout the chapter. In **Chapter 18: Land Environment: Plants and Fungi,** a distinction between the microphyll and megaphyll plants has been added to figure 18.2. A new diagram (fig. 18.20) illustrates the evolutionary relationship among the fungi. In addition, subheadings are provided in the text for the major phyla of fungi. New application readings for this chapter include the use of horsetail supplements, carnivorous plants, and the metabolic consequences of eating death cap mushrooms. The content of **Chapter 19: Both Water and Land: the Animals,** has been extensively revised to reflect the modern understanding of the evolutionary relationships in the animals. The terms eumetazoan and parazoan are now introduced to distinguish the sponges from other forms of animals. Figure 19.5 now includes a distinction between the lophotrochozoan and ecdysozoan invertebrates and the labels in figure 19.29 have been updated to reflect the major events in the evolution of the vertebrates. New application readings for this chapter include discussions on the loss of the coral reefs, the importance of *Homo floresiensis*, and the genetic similarity of *H. sapiens* and the Neandertals. The virtual lab, Classifying Arthropods, is referenced at the end of the chapter.

Part V *Plant Structure and Function*

Chapter 20: Plant Anatomy and Growth includes new application readings on the world's tallest tree, paper manufacturing, cork production, and how to read information on fertilizer bags. The virtual lab, Plant Transpiration, is referenced at the end of the chapter. In **Chapter 21: Plant Responses and Reproduction,** new application readings include the identification of a tomato as a fruit or a vegetable, and an examination as to whether transgenic crops are organic.

Part VI *Animal Structure and Function*

Chapter 22: Being Organized and Steady contains new application readings on the topics of epithelial tissue renewal and the speed at which neurons communicate. **Chapter 23: The Transport Systems** includes application readings on heart murmurs, varicose veins, and blood doping. The virtual lab, Blood Pressure, is referenced at the end of the chapter. In **Chapter 24: The Maintenance Systems,** new application readings on the topics of mucous production, kidney transplants, and hunger are now included. **Chapter 25: Human Nutrition** includes new application readings on trans-fatty acids, sodium in the diet, the 10,000 step program, and the health benefits of green tea. The virtual lab, Nutrition, is referenced at the end of the chapter. **Chapter 26: Defenses Against Disease** includes new discussions of clonal selection, the chemical secretions of neutrophil molecules, the RV 144 HIV vaccine trials in Thailand, as well as an enhanced coverage of the definitions of "self" and "non-self". Applications in this chapter include coverage of the use of biopsies for detecting cancer, the mode of action of antihistamines, the role of a fever in the immune response, and the controversy concerning MMR vaccines and autism. In **Chapter 27: The Control Systems,** application readings on the pharmacology of antidepressants, medical marijuana, the induction of labor, and gestational diabetes are included in the chapter. In addition, new material on the chemical

nature of methamphetamines, and their effects on the body, has been introduced. **Chapter 28: Sensory Input and Motor Output** contains new application readings on LASIK surgery, the mode of action of aspirin, and the use of rigor mortis to estimate the time of death. The virtual lab, Muscle Stimulation, is referenced at the end of the chapter. **Chapter 29: Reproduction and Development** has been reorganized, and a new section introduces contraception, assisted reproductive technologies, and sexually transmitted diseases. The chapter now contains application readings on the topics of the gestation periods of viviparous animals, hormone replacement therapy, assisted reproductive technologies, and the concept of females as the "default" gender in human development.

Part VII *Ecology*

Chapter 30: Ecology and Populations has been reorganized so the discussion of ecology precludes the coverage of population biology. The material on age structure diagrams has been revised to place greater focus on replacement reproduction. Statistics on human population growth rates (fig. 30.4) have also been updated. This chapter contains new application readings on global birth and death rates, global population densities, and the human influence on extinction rates. The virtual lab, Population Biology, is referenced at the end of the chapter. **Chapter 31: Communities and Ecosystems** contains new application readings, including examples of mutualistic relationships between humans and other species, chemoautotrophs and deep sea thermal vents, greenhouse gases, and the role of estuaries. The virtual lab, Model Ecosystems, is referenced at the end of the chapter. In **Chapter 32: Human Impact on the Biosphere,** new application readings have been added that cover the amount of water required to produce food, methylmercury accumulation in the food chain, and personal actions that conserve energy and/or water.

Acknowledgments

Dr. Sylvia Mader represents one of the icons of science education. Her dedication to her students, coupled to her clear, concise, writing style, has benefited the education of thousands of students over the past three decades. It is an honor to continue her legacy, and to bring her message to the next generation of students.

Over the years, several instructors have lent their talent to ensuring the success of *Essentials of Biology*. For this edition, I want to thank Lynn Preston of Tarrant County Community College for her work on several chapters and the end of chapter material. Throughout each chapter, I have strove to ensure that the material was written and illustrated in the familiar Mader style.

Many dedicated and talented individuals assisted in the development of *Essentials of Biology*. I am very grateful for the help of so many professionals at McGraw-Hill who were involved in bringing this book to fruition. In particular, let me thank Rose Koos, the developmental editor who lent her exemplary talents, advice, and patience to all those who worked on this text. The publisher and editor, Michael Hackett, provided invaluable assistance and support during the development of this edition. The senior project manager, Jayne Klein, faithfully and carefully steered the book through the publication process. Tamara Maury, the marketing manager, tirelessly promoted the text and educated the sales reps on its message. Eric Weber, the digital product manager, played an integral role in developing the Connect™ resources for this text.

The design of the book is the result of the creative talents of Laurie Janssen and many others who assisted in deciding the appearance of each element in the text. Electronic Publishing Services followed their guidelines as they created and reworked each illustration, emphasizing pedagogy and beauty to arrive at the best presentation on the page. Lori Hancock and Evelyn Jo Johnson did a superb job of finding just the right photographs and micrographs.

Who I am, as an educator and an author, is a direct reflection of what I have learned from my students. Education is a two-way street, and it is my honest opinion that both my professional and personal life have been enriched by my interactions with my students. They have encouraged me to learn more, teach better, and never stop questioning the world around me. I would also like to acknowledge my wife Sandra. She has never waivered in her patience and support of my endeavors.

The third edition of *Essentials of Biology* would not have the level of excellence associated with a Mader textbook had it not been for the input of the many contributors and reviewers listed below.

McGraw-Hill's 360° Development Process is an ongoing, never-ending, market-oriented approach to building accurate and innovative print and digital products. It is dedicated to continual large-scale and incremental improvement driven by multiple customer feedback loops and checkpoints. This is initiated during the early planning stages of our new products, and intensifies during the development and production stages, then begins again upon publication in anticipation of the next edition. This process is designed to provide a broad, comprehensive spectrum of feedback for refinement and innovation of our learning tools, for both student and instructor. The 360° Development Process includes market research, content reviews, course- and product-specific symposia, accuracy checks, and art reviews. We appreciate the expertise of the many individuals involved in this process.

Ancillary Authors

Instructor's Manual
Farshad Tamari, *Kean University*

Practice Tests
Josh Dobkins and Denise Slayback-Barry, *Keiser University*

Test Bank
Joy Brookshire, *Kennesaw State University*
Dave Cox, *Lincoln Land Community College*

Lecture Outlines/Image PowerPoints
Lynn Preston, *Tarrant County Community College*

FlexArt Manuscript
Sharon Thoma, *University of Wisconsin–Madison*

eBook Quizzes
Dan Matusiak, *St. Charles Community College*

Connect Question Bank
Krissy Johnson, *Appalachian State University*
Alex James, *Appalachian State University*

Third Edition Reviewers

Joseph Arruda, *Pittsburg State University*
Lisa Boggs, *Southwestern Oklahoma State University*
Julio Budde, *Youngstown State University*
Peter Chabora, *Queens College of CUNY*
Maitreyee Chandra, *Diablo Valley College*
Glenn Cohen, *Troy University*
Josh Dobkins, *Keiser University*

Emma Castro, *Victor Valley College*
Carl Estrella, *Merced College*
Karl Fath, *Queens College of CUNY*
Victor Fet, *Marshall University*
Lynn Ruth Firestone, *Brigham Young University—Idaho*
Teresa Fischer, *Indian River State College*
Cynthia Fluharty, *Drexel University*

Cynthia Ford, *Pittsburg State University*
Brandon Foster, *Wake Technical Community College*
Steven Gorsich , *Central Michigan University*
Thomas Hemmerly, *Middle Tennessee State University*
Rebecca Hoffman, *Drexel University*
Andrea Holgado, *Southwestern Oklahoma State University*

Robert Holmes, *Hutchinson Community College*
Sandra Johnson, *Middle Tennessee State University*
David Jones, *Dixie State College of Utah*
Hinrich Kaiser, *Victor Valley College*
Arnold Karpoff, *University of Louisville*
Ruhul Kuddus, *Utah Valley University*
Brenda Leady, *University of Toledo*
Tammy Liles, *Bluegrass Community and Technical College*
Delia Lister, *Pittsburg State University*
Lisa Maranto, *Prince George's Community College*

Shawn Meagher, *Western Illinois University*
Susan Meiers, *Western Illinois University*
Christine Morin, *Prince George's Community College*
Esther Muehlbauer, *Queens College of CUNY*
Alam Nur-E-Kamal, *Medgar Evers College–CUNY*
Tricia Paramore, *Hutchinson Community College*
Mary Perry, *Keiser University*
Angela Porta, *Kean University*
Rongsun Pu, *Kean University*

Michael Rutledge, *Middle Tennessee State University*
Michael Scott, *Lincoln University*
Jennifer Stanford, *Drexel University*
Marshall Sundberg, *Emporia State University*
Muatasem Ubeidat, *Southwestern Oklahoma State University*
Michael Woller, *University of Wisconsin–Whitewater*
Aimee Wurst, *Lincoln University*
Calvin Young, *Fullerton College*

Second Edition Reviewers

Eugene L. Bass, *University of Maryland Eastern Shore*
Bill Radley Bassman, *Stern College*
Loren A. Bertocci, *Salem International University*
Barbara Boss, *Keiser University*
Patty Bostwick Taylor, *Florence Darlington Technical College*
Anthony Botyrius, *York College of Pennsylvania*
Ben F. Brammell, *Morehead State University*
Mimi Bres, *Prince George's Community College*
Carol Britson, *University of Mississippi*
Nikolle L. Brown, *State Center Community College District, Madera Center*
Matthew Burnham, *Jones County Junior College*
Claire M. Carpenter, *Yakima Valley Community College*
Steven Christenson, *Brigham Young University Idaho*
Julie Clements, *Keiser University*
Anthony Cooper, *Albany State University*
Laurie-Ann Crawford, *Hawkeye Community College*
Peter A. Daempfle, *SUNY College of Technology at Delhi*
Debbie Dalrymple, *Thomas Nelson Community College*
Chris Davison, *Long Beach City College*
Jean DeSaix, *University of North Carolina at Chapel Hill*
Josh Dobkins, *Keiser University*
Li'anne Drysdale, *Ozarks Technical Community College, Richwood Valley Facility*
Pamela Faber, D.A., *Heidelberg College*
Farshad Tamari, *Kean University*

Jeffry Fasick, *Kean University*
Lynn Firestone, *Brigham Young University, Idaho*
Diane Fritz, *Northern Kentucky University*
Wendy Garrison, *University of Mississippi*
Norma Golovoy, *Schoolcraft College*
David M. Gordman, *Pittsburg State*
Tamar Goulet, *University of Mississippi*
Colleen Hatfield, *California State University, Chico*
Jennifer A. Herzog, *Herkimer County Community College*
Elizabeth Hodgson, *York College of Pennsylvania*
Mary Ruth Hood, *Tomball College*
George A. Jacob, *Xavier University*
Ashok Jain, *Albany State University*
Scott Johnson, *Wake Technical Community College*
Gregory A. Jones, *Santa Fe Community College*
Mary King Kananen, *Penn State Altoona*
Brenda Kelley, *Alpena Community College*
David Knowles, *East Carolina University*
Eliot Krause, *Seton Hall University*
Ruhul H. Kuddus, *Utah Valley University*
Brenda Leady, *University of Toledo*
Debabrata Majumdar, *Norfolk State University*
Lisa Maranto, *Prince George's Community College*
Rob McCandless, *Methodist University*
Michael McCarthy, *Eastern Arizona College*
Quintece M. McCrary, *University of Maryland–Eastern Shore*
John Mishler, *Delaware Valley College*
Sarah E. Myer, *Oral Roberts University*
Joanna Padolina, *Virginia Commonwealth University*
Kathleen Pelkki, *Saginaw Valley State University*

Mary Perry, *Keiser University*
Patrick K. Pfaffle, *Carthage College*
Megan Pickering Stringer, *Jones County Junior College*
Rongsun Pu, *Kean University*
Denice Robertson, *Northern Kentucky University*
Carolina Ross, *Laramie County Community College*
Barbara J. Salvo, *Carthage College*
Susan Sawyer, *Kellogg Community College*
John Richard Schrock, *Emporia State University*
James M. Schwartz, *Front Range Community College*
Marilyn Shopper, *Johnson County Community College*
William Simcik, *Tomball College, NHMCCD*
Robert Singer, *New Jersey City University*
Denise Slayback-Barry, *Keiser University*
Kathryn Sloan Ponnock, *Delaware Valley College*
Bo Sosnicki, *Piedmont Virginia Community College*
Chebium Subrahmanyam, *Florida A & M University*
Marshall D. Sundberg, *Emporia State University*
Karen S. Teitelbaum, *State University of NY College of Technology at Delhi*
Kip R. Thompson, *Ozarks Technical Community College*
Kevin R. Toal, *St. Louis Community College at Florissant Valley*
Kate Warpeha, *University of Illinois at Chicago*
Stan Willenbring, *Dabney S Lancaster Community College*
Heather Wilson-Ashworth, *Utah Valley University*
W. Brooke Yeager III, *Luzerne County Community College*

First Edition Reviewers

Sylvester Allred, *Northern Arizona University*
Paul E. Arriola, *Elmhurst College*
Tammy Atchison, *Pitt Community College*
James S. Backer, *Daytona Beach Community College*
Gail F. Baker, *LaGuardia Community College*

Sirakaya Beatrice, *Pennsylvania State University*
Carla Bundrick Benejam, *California State University–Monterey Bay*
Charles L. Biles, *East Central University*
Donna H. Bivans, *Pitt Community College*

Steven G. Brumbaugh, *Greenriver Community College*
Neil Buckley, *SUNY–Plattsburgh*
Nancy Butler, *Kutztown University*
Michelle Cawthorn, *Georgia Southern University*
Van D. Christman, *Brigham Young University*

Genevieve C. Chung, *Broward Community College*
Kimberly Cline-Brown, *University of Northern Iowa*
Mary C. Colavito, *Santa Monica College*
Mark A. Coykendall, *College of Lake County*
Don C. Dailey, *Austin Peay State University*
Cathy A. Davison, *Empire State College*
Bonnie L. Dean, *West Virginia State University*
William R. DeMott, *Indiana-Purdue University–Fort Wayne*
Amy Stinnett Dewald, *Eureka College*
Lee C. Drickamer, *Northern Arizona University*
Marie D. Dugan, *Broward Community College*
James W. DuMond, Jr., *Texas State University*
Kathryn A. Durham, *Lorain County Community College*
Andrew R. Dyer, *University of South Carolina–Aiken*
Steven E. Fields, *Winthrop University*
Lynn Firestone, *Brigham Young University–Idaho*
Susan Fisher, *Ohio State University*
Edison R. Fowlks, *Hampton University*
Dennis W. Fulbright, *Michigan State University*
Ron Gaines, *Cameron University*
John R. Geiser, *Western Michigan University*
Beatriz Gonzalez, *Santa Fe Community College*

Andrew Goyke, *Northland College*
Richard Gringer, *Augusta State University*
Lonnie J. Guralnick, *Western Oregon University*
William F. Hanna, *Massasoit Community College*
Lisa K. Johansen, *University of Colorado at Denver*
Ragupathy Kannan, *University of Arkansas–Fort Smith*
Arnold Karpoff, *University of Louisville*
Darla E. Kelly, *Orange Coast College*
Elaine B. Kent, *California State University–Sacramento*
Scott L. Kight, *Montclair State University*
Kristin Lenertz, *Black Hawk College*
Melanie Loo, *California State University–Sacramento*
Michelle Malott, *Minnesota State University–Moorhead*
Paul Mangum, *Midland College*
Mara Manis, *Hillsborough Community College*
Karen Benn Marshall, *Montgomery College–Takoma Park*
Cynthia Conaway Mauroidis, *Northwest State Community College*
Elizabeth McPartlan, *De Anza College*
Dwight Meyer, *Queensborough Community College*

Rod Nelson, *University of Arkansas–Fort Smith*
Donald J. Padgett, *Bridgewater State College*
Tricia L. Paramore, *Hutchinson Community College*
Brian K. Paulson, *California University of Pennsylvania*
Debra K. Pearce, *Northern Kentucky University*
Lisa Rapp, *Springfield Technical Community College*
Jill Raymond, *Rock Valley Community College*
Cara Shillington, *Eastern Michigan University*
Lee Sola, *Glendale Community College*
John D. Sollinger, *Southern Oregon University*
Andrew Storfer, *Washington State University*
Janis G. Thompson, *Lorain County Community College*
Briana Timmerman, *University of South Carolina*
James R. Triplett, *Pittsburg State University*
Paul Twigg, *University of Nebraska–Kearney*
Garland Rudolph Upchurch, Jr., *Texas State University, San Marcos*
James A. Wallis II, *St. Petersburg College, Tarpon Springs Campus*
Cosima B. Wiese, *College Misericordia*
Melissa Zwick, *Longwood University*

General Biology Symposia

Every year McGraw-Hill conducts several General Biology Symposia, which are attended by instructors from across the country. These events are an opportunity for editors from McGraw-Hill to gather information about the needs and challenges of instructors teaching non-majors level biology courses. They also offer a forum for the attendees to exchange ideas and experiences with colleagues they might not have otherwise met. The feedback we have received has been invaluable, and has contributed to the development of *Essentials of Biology*, Third Edition and its ancillaries.

Norris Armstrong, *University of Georgia*
David Bachoon, *Georgia College and State University*
Sarah Bales, *Moraine Valley Community College*
Lisa Bellows, *North Central Texas College*
Joressia Beyer, *John Tyler Community College*
James Bidlack, *University of Central Oklahoma*
Mark Bloom, *Texas Christian University*
Paul Bologna, *Montclair University*
Linda Brandt, *Henry Ford Community College*
Marguerite Brickman, *University of Georgia*
Bradford Boyer, *Suffolk County Community College*
Art Buikema, *Virginia Polytechnic Institute*
Sharon Bullock, *Virginia Commonwealth University*
Raymond Burton, *Germanna Community College*
Nancy Butler, *Kutztown University of Pennsylvania*
Jane Caldwell, *West Virginia University*
Carol Carr, *John Tyler Community College*
Kelly Cartwright, *College of Lake County*
Rex Cates, *Brigham Young University*
Sandra Caudle, *Calhoun Community College*
Genevieve Chung, *Broward Community College*

Jan Coles, *Kansas State University*
Marian Wilson Comer, *Chicago State University*
Renee Dawson, *University of Utah*
Lewis Deaton, *University of Louisiana at Lafayette*
Jody DeCamilo, *St. Louis Community College*
Jean DeSaix, *University of North Carolina at Chapel Hill*
JodyLee Estrada-Duek, *Pima Community College, Desert Vista*
Laurie Faber-Foster, *Grand Rapids Community College*
Susan Finazzo, *Broward Community College*
Theresa Fischer, *Indian River Community College*
Theresa Fulcher, *Pellissippi State Technical College*
Dennis Fulbright, *Michigan State University*
Steven Gabrey, *Northwestern State University*
Cheryl Garett, *Henry Ford Community College*
Farooka Gauhari, *University of Nebraska–Omaha*
John Geiser, *Western Michigan University*
Cindy Ghent, *Towson University*
Julie Gibbs, *College of DuPage*
William Glider, *University of Nebraska, Lincoln*

Christopher Gregg, *Louisiana State University*
Carla Guthridge, *Cameron University*
Bob Harms, *St. Louis Community College, Meramec*
Wendy Hartman, *Palm Beach Community College*
Tina Hartney, *California State Polytechnic University*
Kelly Hogan, *University of North Carolina, Chapel Hill*
Eva Horne, *Kansas State University*
David Huffman, *Texas State University, San Marcos*
Shelley Jansky, *University of Wisconsin, Stevens Point*
Sandra Johnson, *Middle Tennessee State University*
Tina Jones, *Shelton State Community College*
Arnold Karpoff, *University of Louisville*
Jeff Kaufmann, *Irvine Valley College*
Kyoungtae Kim, *Missouri State University*
Michael Koban, *Morgan State University*
Todd Kostman, *University of Wisconsin Oshkosh*
Steven Kudravi, *Georgia State University*
Nicki Locascio, *Marshall University*
Dave Loring, *Johnson County Community College*

Janice Lynn, *Alabama State University*

Phil Mathis, *Middle Tennessee State University*

Mary Victoria McDonald , *University of Central Arkansas*

Susan Meiers , *Western Illinois University*

Daryl Miller, *Broward Community College, South Campus*

Marjorie Miller, *Greenville Technical College*

Meredith Norris, *University of North Carolina at Charlotte*

Mured Odeh, *South Texas College*

Nathan Olia, *Auburn University, Montgomery*

Rodney Olsen, *Fresno City College*

Alexander Olvido, *Virginia State University*

Clark Ovrebo, *University of Central Oklahoma*

Forrest Payne, *University of Arkansas at Little Rock*

Nancy Pencoe, *University of West Georgia*

Murray P. "Pat" Pendarvis, *Southeastern Louisiana University*

Jennie Plunkett, *San Jacinto College*

Scott Porteous, *Fresno City College*

David Pylant, *Wallace State Community College*

Fiona Qualls, *Jones County Junior College*

Eric Rabitoy, *Citrus College*

Karen Raines, *Colorado State University*

Kirsten Raines, *San Jacinto College*

Jill Reid, *Virginia Commonwealth University*

Darryl Ritter, *Okaloosa-Walton College*

Chris Robinson , *Bronx Community College*

Robin Robison, *Northwest Mississippi Community College*

Vickie Roettger , *Missouri Southern State University*

Bill Rogers, *Ball State University*

Vicki Rosen, *Utah State University*

Kim Sadler, *Middle Tennessee State University*

Cara Shillington, *Eastern Michigan University*

Greg Sievert, *Emporia State University*

Jimmie Sorrels, *Itawamba Community College*

Judy Stewart, *Community College of Southern Nevada*

Julie Sutherland, *College of DuPage*

Bill Trayler, *California State University–Fresno*

Linda Tyson, *Santa Fe Community College*

Eileen Underwood, *Bowling Green State University*

Heather Vance-Chalcraft, *East Carolina University*

Marty Vaughan, *IUPUI, Indianapolis*

Paul Verrell, *Washington State University*

Thomas Vogel, *Western Illinois University*

Brian Wainscott, *Community College of Southern Nevada*

Jennifer Warner, *University of North Carolina, Charlotte*

Scott Wells, *Missouri Southern State University*

Lan Xu, *South Dakota State University*

Robin Whitekiller, *University of Central Arkansas*

Allison Wiedemeier, *University of Illinois, Columbia*

Michael Windelspecht, *Appalachian State University*

Mary Wisgirda , *San Jacinto College, South Campus*

Tom Worcester, *Mount Hood Community College*

Frank Zhang, *Kean University*

Michelle Zjhra, *Georgia Southern University*

Victoria Zusman, *Miami Dade College*

Contents

PART VI   Animal Structure and Function

CHAPTER **22**

Being Organized and Steady   412

CHAPTER **23**

The Transport Systems   430

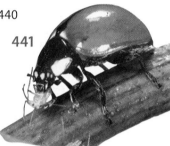

CHAPTER **24**

The Maintenance Systems   450

CHAPTER **25**

Human Nutrition   475

CHAPTER **26**

Defenses Against Disease   496

CHAPTER **27**

The Control Systems   515

CHAPTER **28**

Sensory Input and Motor Output   538

CHAPTER **29**

Reproduction and Development   558

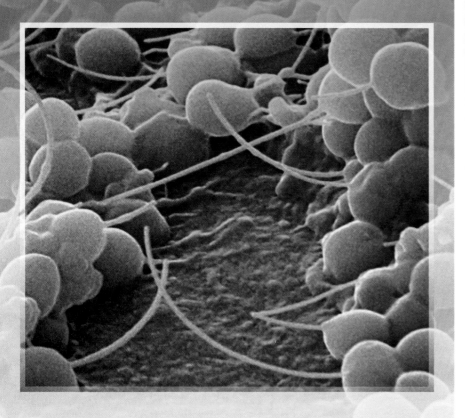

1

A View of Life

Your Skin Is Their World

It is hard to believe that the number of bacterial organisms living on your skin right now is greater than the number of human organisms living on this planet. How can more than 6 billion individual organisms live on the surface of your skin? The answer is that bacterial cells are extremely small—so small that we tend not to think of them. Because most people cringe at the thought of "germs," a stroll through just about any aisle of your local grocery store will reveal a slew of products that claim to be antibacterial. Advertisers tend to make us think that all bacteria are harmful. But only a handful of bacterial species are actually dangerous to humans; the overwhelming majority are beneficial. The billions of bacteria that live on your skin are termed normal flora (or resident bacteria), and they essentially occupy all the available space on your skin surface. Should a detrimental bacterium come along, it will literally have no room to settle and multiply. Because the normal flora protect us in this way, we certainly don't want to do away with them.

The bacterial cells occupying your skin have a lot in common with the cells that make up your body. This chapter introduces the common characteristics, the diversity, and the organization of living things, as well as explaining how scientists perform their studies.

1.1 The Characteristics of Life

Learning Outcomes

Upon completion of this section, you should be able to

1. Explain the basic characteristics that are common to all living things.
2. Describe the levels of organization of life.
3. Summarize how the terms *homeostasis, metabolism,* and *adaptation* all relate to living organisms.
4. Recognize the special relationship between life and evolution.

The science of biology is the study of life—from huge, menacing sharks to miniature, exotic orchids. Life is diverse. Living **organisms** may be found everywhere on our planet, from thermal vents at the bottom of the ocean to the coldest reaches of Antarctica. Yet, despite life's diversity, all living organisms share certain characteristics. These include levels of organization, the ability to acquire materials and energy, the ability to maintain homeostasis, the ability to respond to stimuli, the ability to reproduce and develop, and the ability to adapt and evolve to changing conditions. By studying these characteristics, we gain insight into the complex nature of life, which helps us distinguish living organisms from nonliving things.

Video
Coral Reef Ecosystems

Living Things Are Organized

The complex organization of living things begins with small molecules that join to form larger molecules within a **cell,** the smallest, most basic unit of life. Although a cell is alive, it is made from nonliving molecules (**Fig. 1.1**).

The majority of the organisms on the planet are single-celled. Plants, fungi, and animals are **multicellular** and are composed of many types of cells.

Figure 1.1 Levels of biological organization.

The cell is the smallest unit of life. The biosphere includes all the life on the planet.

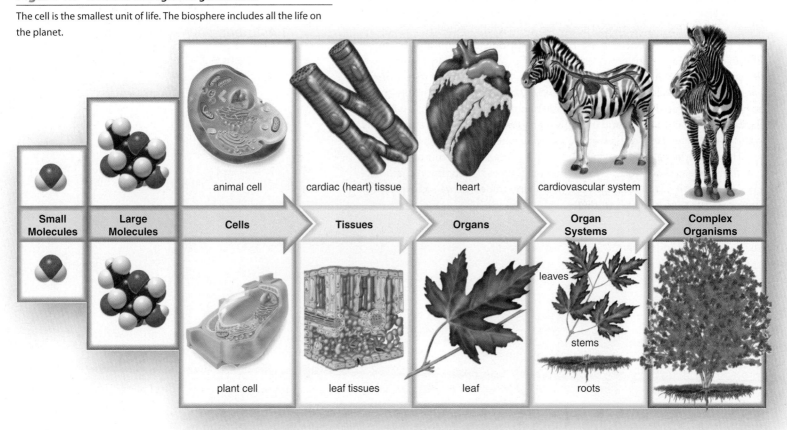

Small Molecules	Large Molecules	Cells	Tissues	Organs	Organ Systems	Complex Organisms

animal cell — cardiac (heart) tissue — heart — cardiovascular system

plant cell — leaf tissues — leaf — leaves / stems / roots

In multicellular organisms, similar cells combine to form **tissues.** Tissues make up **organs,** as when various tissues combine to form a heart or a leaf. Organs work together in **organ systems;** for example, the heart and blood vessels form the cardiovascular system. Various organ systems work together within complex organisms.

Later in this chapter, we will see that levels of biological organization extend beyond the individual organism. For example, all the members of one species in a particular area belong to a population. Zebras form one of the many populations on an African plain. All the populations in a given area make up a community. The community of populations interacts with the physical environment to form an ecosystem. Finally, all the Earth's ecosystems make up the biosphere.

Connections and Misconceptions

How many species are there?

The truth is, we really don't know. While most scientists believe that there are probably as many as 15 million species on the planet, some estimate that there may be over 100 million different species. So far, fewer than 2 million species have been identified, and most of those are insects. However, new species are being discovered all the time—a recent discovery of a new species of spiny pocket mouse in 2010 in the mountains of Venezuela is one example.

Living Things Acquire Materials and Energy

Living things cannot maintain their organization or carry on life's activities without an outside source of materials and energy (**Fig. 1.2**). Food provides nutrient molecules, which are used as building blocks or energy sources. **Energy** is the capacity to do work, and it takes work to maintain the organization of the cell and the organism. When cells use nutrient molecules to make their parts and products, they carry out a sequence of chemical reactions. The term **metabolism** encompasses all the chemical reactions that occur in a cell.

The ultimate source of energy for nearly all life on Earth is the sun. Plants and certain other organisms are able to capture solar energy and carry on **photosynthesis,** a process that transforms solar energy into the chemical energy of nutrient molecules. Animals and plants get energy by metabolizing nutrient molecules made by photosynthesizers.

Living Things Maintain an Internal Environment

For metabolic processes to continue, living things need to keep themselves stable in temperature, moisture level, acidity, and other factors critical to maintaining life. This is called **homeostasis**—the maintenance of internal conditions within certain boundaries.

a.

b.

Figure 1.2 Acquiring nutrient materials and energy.

a. Osprey, a type of hawk, eating a fish. **b.** Plants acquire energy from the sun. Humans harvest those plants for food.

Many organisms depend on behavior to regulate their internal environment. A chilly lizard may raise its internal temperature by basking in the sun on a hot rock. When it starts to overheat, it scurries for cool shade. Other organisms have control mechanisms that do not require any conscious activity. When a student is so engrossed in her textbook that she forgets to eat lunch, her liver releases stored sugar to keep the blood sugar level within normal limits. Hormones regulate sugar storage and release, but in other instances the nervous system is involved in maintaining homeostasis.

Living Things Respond

Living things find energy and/or nutrients by interacting with their surroundings. Even unicellular organisms can respond to their environment. The beating of microscopic hairs or the snapping of whip-like tails moves them toward or away from light or chemicals. Multicellular organisms can manage more complex responses. A monarch butterfly can sense the approach of fall and begin its flight south, where resources are still abundant. A vulture can smell meat a mile away and soar toward dinner.

The ability to respond often results in movement: The leaves of a plant turn toward the sun, and animals dart toward safety. Appropriate responses help ensure survival of the organism and allow it to carry on its daily activities. Altogether, we call these activities the *behavior* of the organism.

Living Things Reproduce and Develop

Life comes only from life. Every type of living thing can **reproduce,** or make another organism like itself. Bacteria and other types of unicellular organisms simply split in two. In multicellular organisms, the reproductive process usually begins with the pairing of a sperm from one partner and an egg from the other partner. The union of sperm and egg, followed by many cell divisions, results in an immature individual, which grows and develops through various stages to become an adult.

An embryo develops into a whale or a yellow daffodil or a human being because of the specific set of genes inherited from its parents (**Fig. 1.3**). In all organisms, the **genes** are made of long DNA (deoxyribonucleic acid) molecules, but even so the genes are different between species and individuals. Genetic comparisons are the basis of paternity testing. Although an individual's overall genetic makeup is unique, it consists of DNA from both parents. Thus, the DNA profiles of a child and his or her biological parents are measurably similar. DNA provides the blueprint or instructions for the organization and metabolism of the particular organism. All cells in a multicellular organism contain the same set of genes, but only certain ones are turned on in each type of specialized cell.

Living Things Have Adaptations

Adaptations are modifications that make organisms suited to their way of life. Some hawks have the ability to catch fish (see Fig. 1.2); others are best at catching rabbits. Hawks can fly, in part, because they have hollow bones to

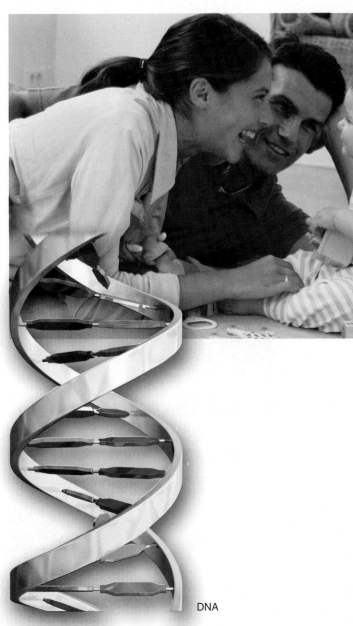

DNA

Figure 1.3 A human family.

Whether they are unicellular or multicellular, all organisms reproduce. Offspring receive a copy of their parents' DNA and therefore a copy of their genes.

reduce their weight and flight muscles to depress and elevate their wings. When a hawk dives, its strong feet take the first shock of the landing, and its long, sharp claws reach out and hold onto the prey. Hawks have exceptionally keen vision, which enables them not only to spot prey from great heights but also to estimate distance and speed.

Penguins look very different from hawks, although they are both birds. Penguins are adapted to an aquatic existence in the Antarctic. While birds such as hawks have forelimbs proportioned for flying, penguins have stubby, flattened wings suitable for swimming. Their feet and tails serve as rudders in the water, but their flat feet also allow them to walk on land. Penguins live where it is cold; an extra layer of downy feathers is covered by short, thick feathers to form a waterproof coat. Layers of blubber also keep the birds warm in cold water.

The manner in which species become adapted to their environment is discussed in the next section of this chapter. The members of a species have a similar structure and, if sexually reproducing, are capable of interbreeding.

Check Your Progress 1.1

1. Identify the levels of biological organization and the basic characteristics of life.
2. Summarize homeostasis and explain why it is important in a living organism.
3. Hypothesize how some of the characteristics of life lead to evolution.

1.2 Evolution: The Core Concept of Biology

Learning Outcomes

Upon completion of this section, you should be able to

1. Distinguish among the three domains of life.
2. Distinguish among the major kingdoms.
3. Explain the process of natural selection.
4. Explain the relationship between natural selection and evolution.

Connecting the Concepts

To better understand how biologists study the characteristics of life, refer to the following discussions.

Section 6.1 explores how plants acquire energy from the sun to manufacture sugars.

Section 21.2 examines how plants respond to stimuli such as light and gravity.

Section 29.4 investigates how humans develop from a single-celled embryo to an adult.

From bacteria to bats, toadstools to trees, whip-poor-wills to whales—there is a tremendous wealth of diversity in the living world, yet all organisms are united by a common heritage, which began during the early years of our planet. Just as you have a family history that goes back generations, all species can trace their history to the original living organism. **Evolution,** the process by which species have changed and diversified since life first arose, explains the unity—all living things share the same characteristics—and the diversity of life. If you have ever studied a family tree, then you know it is a diagram used to show how individuals are related to one another and descended from common ancestors. An evolutionary tree is very much like a family tree, only it depicts relationships between groups of organisms. The evolutionary tree in **Figure 1.4** illustrates the lineages of the major groups of living things. The tree summarizes the history of life on Earth over the approximately 3.5 billion years that have passed since the first, ancestral cell. All the different groups of organisms on our planet are related to one another and are represented by branches on the same "family tree of life."

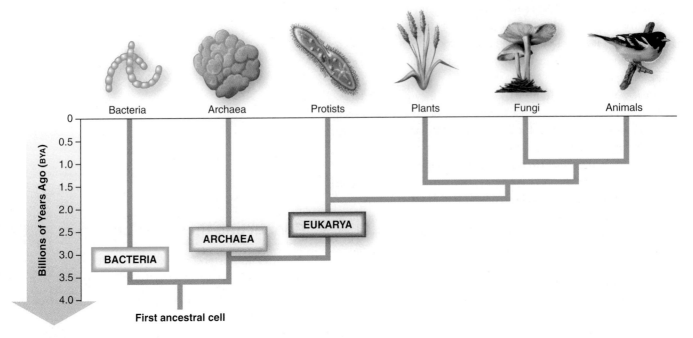

Figure 1.4 An evolutionary tree.

Organisms grouped on the same branch of the tree have a common ancestor located at the base of the branch. Organisms grouped on the same branch (e.g., fungi and animals) are more closely related to one another (i.e., have a more recent common ancestor) than organisms on different branches (e.g., animals and plants). The base of the tree itself represents the common ancestor of all living things.

Table 1.1 Levels of Biological Organization

Category	Human	Corn
Domain	Eukarya	Eukarya
Kingdom	Animalia	Plantae
Phylum	Chordata	Anthophyta
Class	Mammalia	Liliopsida
Order	Primates	Commelinales
Family	Hominidae	Poaceae
Genus	Homo	Zea
*Species**	*H. sapiens*	*Z. mays*

*To specify an organism, you must use the full binomial name, such as *Homo sapiens*.

The Diversity of Life

Think of an enormous department store, offering thousands of different items for sale. The various items are grouped in departments—electronics, apparel, furniture, and so on—to make them easy for customers to find. Because life is so diverse, it is helpful to have a system that groups organisms into categories. Two areas of biology help us group organisms into categories: **Taxonomy** is the discipline of identifying and naming organisms according to certain rules, and **systematics** makes sense out of the bewildering variety of life on Earth by classifying organisms according to their presumed evolutionary relationships. As systematicists learn more about evolutionary relationships between species,the taxonomy of a given organism may change. Systematicists are even now making observations and performing experiments that will one day bring about changes in the classification system adopted by this text.

Categories of Classification

The classification categories, from least inclusive to most inclusive, are **species, genus, family, order, class, phylum, kingdom,** and **domain (Table 1.1).** Each successive category above species contains more types of organisms than the preceding one. Species placed within one genus share many specific characteristics and are the most closely related, while species placed in the same domain share only general characteristics. For example, all species in the genus *Pisum* look pretty much the same—that is, like pea plants—but species in the plant kingdom can be quite varied, as is evident when we compare grasses with trees. By the same token, only modern humans are in the genus *Homo*, but many types of species, from tiny hydras to huge whales, are members of the animal kingdom. Species placed in different domains are the most distantly related.

Scientific Naming

Biologists give each living thing a two-part scientific name called a **binomial name.** For example, the scientific name for the garden pea is *Pisum sativum*; our own species is *Homo sapiens*. The first word is the genus, and the second word is the specific epithet of a species within a genus. The genus may be

×20,000

Figure 1.5 Domain Archaea.

Archaea are capable of living in extreme environments. This is *Methanosarcina mazei,* a methane-generating prokaryote.

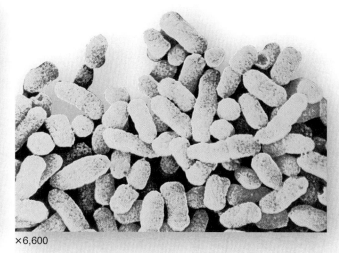

×6,600

Figure 1.6 Domain Bacteria.

Bacteria are structurally simple but metabolically diverse. This is *Escherichia coli.* a common prokaryote found in our intestinal tracts.

abbreviated, such as *P. sativum* or *H. sapiens*. Scientific names are universally used by biologists to avoid confusion. Common names tend to overlap, and often they are in the languages of the people who use those names. But scientific names are based on Latin, a universal language that not too long ago was well known by most scholars.

Domains

Biochemical evidence (obtained from the study of DNA and proteins) suggests that there are only three domains of life: **domain Bacteria, domain Archaea,** and **domain Eukarya.** Both domain Archaea (**Fig. 1.5**) and domain Bacteria (**Fig. 1.6**) contain prokaryotes. Prokaryotes are unicellular, and they lack the membrane-bounded nucleus found in the eukaryotes of domain Eukarya.

 **Animation** Three Domains

Prokaryotes are structurally simple but metabolically complex. Archaea live in aquatic environments that lack oxygen or are too salty, too hot, or too acidic for most other organisms. Perhaps these environments are similar to those of the primitive Earth and archaea are representative of the first cells that evolved. Bacteria are found almost everywhere—in the water, soil, and atmosphere, as well as on our skin and in our mouths and large intestines. Although some bacteria cause diseases, others perform many services, both environmental and commercial. For example, they are used to conduct genetic research in our laboratories (the *E.coli* in Fig. 1.6 is one example), to produce innumerable products in our factories, and to purify water in our sewage treatment plants.

Kingdoms

Systematicists are just beginning to understand how to categorize archaea and bacteria into kingdoms. Domain Eukarya, on the other hand, has four distinct kingdoms with which you may be familiar. **Protists** (kingdom Protista) range from unicellular to a few multicellular organisms (**Fig. 1.7**). Some are photosynthesizers; others must ingest their food. Among the **fungi** (kingdom Fungi) are the familiar molds and mushrooms that, along with many types of bacteria, help decompose dead organisms (**Fig. 1.8**). **Plants** (kingdom Plantae) are

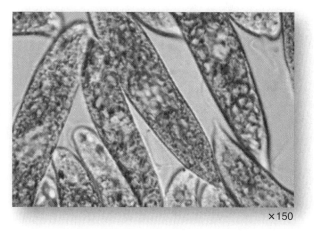

×150

Figure 1.7 Domain Eukarya, kingdom Protista.

Many protists are unicellular. This is *Euglena,* an interesting organism with both plant and animal-like characteristics.

Figure 1.8 Domain Eukarya, kingdom Fungi.

Fungi are multicellular and break down organic debris. This is an *Amanita* mushroom.

Figure 1.9 **Domain Eukarya, kingdom Plantae.**

Plants are multicellular photosynthesizers. This is the bristlecone pine, *Pinus longaeva*, one of the oldest organisms on the planet.

Figure 1.10 **Domain Eukarya, kingdom Animalia.**

Animals are multicellular and ingest their food. This is the bobcat, *Lynx rufus*.

well known as multicellular photosynthesizers (**Fig. 1.9**). **Animals** (kingdom Animalia) are multicellular organisms that ingest their food (**Fig. 1.10**). The three domains and the four kingdoms within the domain Eukarya are depicted in the evolutionary tree in Figure 1.4. This tree, which is largely based on the DNA of organisms, says that the domain Bacteria was the first to arise in the history of life, followed by the domain Archaea and finally the domain Eukarya. The domain Archaea is more closely related to the domain Eukarya than either is to the domain Bacteria. Among the Eukarya, the protists gave rise to the kingdoms of plants, fungi, and animals. Later in this text, we will return to this evolutionary tree, and the evolution of each kingdom will be discussed separately. Now that we know how scientists group organisms, let's return to our discussion of evolution and the process by which diversity arises.

Natural Selection

In the nineteenth century, two naturalists—Charles Darwin and Alfred Russell Wallace—came to the conclusion independently that evolution occurs by means of a process called **natural selection.** Charles Darwin is the more famous of the two because he wrote a book called *On the Origin of Species*, which presented much data to substantiate the occurrence of evolution by natural selection. Since that time, evolution has become the core concept of biology because the theory explains so many different types of observations in every field of biology. The study of evolution encompasses all levels of biological organization. Indeed, much of today's evolution research is carried out at the molecular level, comparing the DNA of different groups of organisms to determine how they are related. Looking at how life has changed over time, from its origin to the current day, helps us understand why there are so many different kinds of organisms and why they have the characteristics they do. Our growing knowledge of evolution by natural selection also has practical applications, including the prevention and treatment of disease.

Natural selection is a process in which the first three steps result in a population becoming adapted to the environment (in step 4):

1. The members of a population have heritable variations: differences in their DNA that can be passed from one generation to the next.
2. The population produces more offspring than the resources of an environment can support.
3. There is competition for resources; as a result, the more adapted individuals survive and reproduce to a greater extent than those that lack the adaptations.
4. Across generations, a larger proportion of the population becomes adapted to the environment.

One example that Darwin used concerned giraffes, which are browsers feeding on tree leaves. Imagine an early population of giraffes in which the neck length varied from short to long. Giraffes with long necks had an advantage over those with short necks because they could reach more food. The ability to capture more food allowed long-neck giraffes to have more offspring than short-neck giraffes. Therefore, slowly over many generations giraffes became exclusively long-necked animals.

Today, we know that, because of selection, resistance to antibiotic drugs has become increasingly common in a number of bacterial species, including those that cause tuberculosis, gonorrhea, and staph infections. Antibiotic drugs such as penicillin kill susceptible bacteria. However, some bacteria in the body of a patient undergoing antibiotic treatment may be unharmed by the drug. Bacteria can survive antibiotic drugs in many different ways. For example, certain bacteria can endure treatment with penicillin because they break down the drug, rendering it harmless. If even one bacterial cell lives because it is antibiotic-resistant, then its descendants will inherit this drug-defeating ability. The more antibiotic drugs are used, the more natural selection favors resistant bacteria, and the more often antibiotic-resistant infections will occur.

Descent with Modification

Darwin said that evolution represents descent with modification. One species can be a common ancestor to several species, each adapted to a particular set of environmental conditions. Specific adaptations allow species to play particular roles in their environment. The diversity of life-forms is best understood in terms of the many different ways in which organisms carry on their life functions within the environment where they live, acquire energy, and reproduce.

The Hawaiian honeycreepers are a remarkable example of evolution (**Fig. 1.11**). The 50+ species of honeycreepers all evolved from 1 species of finch, which likely originated in North America and arrived in the Hawaiian islands between 3 and 5 million years ago. Modern honeycreepers have an assortment of bill shapes adapted to different types of food. Some honeycreeper species have curved, elongated bills used for drinking flower nectar. Others have strong, hooked bills suited to digging in tree bark and seizing wood-boring insects or short, straight, finch-like bills for feeding on small seeds and fruits. Even with such dramatic differences in feeding habits and bill shapes, honeycreepers still share certain characteristics, which stem from their common finch ancestor. The various honeycreeper species are similar in body shape and size, as well as mating and nesting behavior.

Connecting the Concepts

Evolution is the unifying theme in the study of biology. For more information, refer to the following discussions.

Section 14.1 provides more background information on Darwin and natural selection.

Section 15.2 examines how small changes in genetic material can produce evolutionary change.

Section 16.1 explores how evolutionary processes influence species over long periods of time.

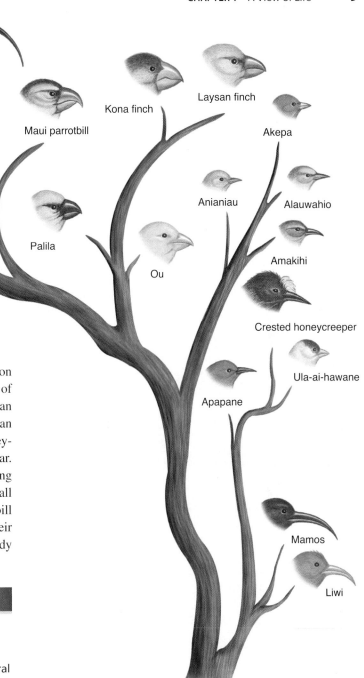

Figure 1.11   Evolution of Hawaiian honeycreepers.

Hawaiian honeycreepers, descendants of a single ancestral species, display an amazing diversity of bill shapes and sizes.

Check Your Progress 1.2

1. List the eight classification categories, from least to most inclusive.

2. Describe the process of natural selection and explain its importance in the study of biology.

3. Explain why the concept of descent with modification is important in understanding evolutionary change.

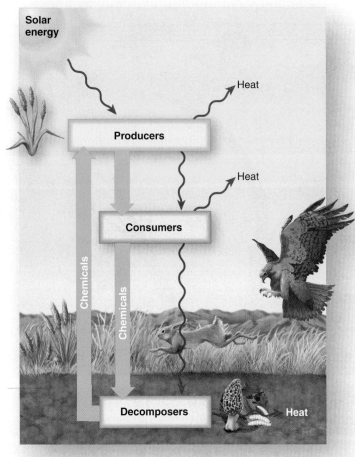

Figure 1.12 A grassland, a terrestrial ecosystem.

In an ecosystem, chemical cycling (aqua arrows) and energy flow (red arrows) begin when plants use solar energy and inorganic nutrients to produce their own food. Chemicals and energy are passed from one population to another in a food chain. Eventually, energy dissipates as heat. With the death and decomposition of organisms, chemicals are returned to living plants once more.

Connecting the Concepts

For more on the functions of ecosystems, and the human threats to both ecosystems and the biosphere, refer to the following discussions.

Section 31.2 takes a more detailed look at the movement of energy and nutrients in an ecosystem.

Section 32.1 examines how human activity can influence the balance of an ecosystem.

Check Your Progress 1.3

1. Contrast a community with an ecosystem.
2. Differentiate between energy flow and chemical cycling in an environment.
3. Create a pro/con list of how humans impact the environment; include at least one pro and one con.

1.3 Ecosystems and the Biosphere

Learning Outcomes

Upon completion of this section, you should be able to

1. Distinguish among populations, communities, and ecosystems.
2. Contrast chemical and energy recycling within an ecosystem.

The organization of life extends beyond the individual to the **biosphere,** the zone of air, land, and water at the surface of the Earth where living organisms are found. Individual organisms belong to a **population,** all the members of a species within a particular area. The populations within a **community** interact among themselves and with the physical environment (soil, atmosphere, etc.), thereby forming an **ecosystem.**

Energy flow and nutrient cycling in an ecosystem climate largely determine where different ecosystems are found in the biosphere. For example, deserts exist in areas of minimal rain, while forests require much rain. The two most biologically diverse ecosystems—tropical rain forests and coral reefs—occur where solar energy is most abundant. One example of an ecosystem in North America is the grasslands, which are inhabited by populations of rabbits, hawks, and various types of grasses, among many others. These populations interact with each other by forming food chains in which one population feeds on another. For example, rabbits feed on grasses, while hawks feed on rabbits and other organisms.

As **Figure 1.12** shows, ecosystems are characterized by chemical cycling and energy flow, both of which begin when photosynthesizers, such as grasses, take in solar energy and inorganic nutrients to produce food (organic nutrients) by photosynthesis. Chemical cycling (aqua arrows) occurs as chemicals move from one population to another in a food chain, until death and decomposition allow inorganic nutrients to be returned to the photosynthesizers once again. Energy (red arrows), on the other hand, flows from the sun through plants and the other members of the food chain as they feed on one another. The energy gradually dissipates and returns to the atmosphere as heat. Because energy does not cycle, ecosystems could not stay in existence without solar energy and the ability of photosynthesizers to absorb it.

Human Influence on Ecosystems

The human population tends to modify ecosystems for its own purposes. Humans clear forests or grasslands in order to grow crops; later, they build houses on what was once farmland; finally, they convert small towns into cities. As coasts are developed, humans send sediments, sewage, and other pollutants into the sea. In the process, they disrupt the normal flow of energy and nutrients in the ecosystem (Fig. 1.12).

Tropical rain forests, coral reefs, and most of the other ecosystems are severely threatened as the human population increases in size. Some coral reefs are 50 million years old, and yet, in just a few decades human activities have destroyed 10% of all reefs and seriously degraded another 30%. At this rate, nearly three-quarters could be destroyed within 50 years. Similar statistics are available for tropical rain forests.

Video Coral Reef Ecosystems

It has long been clear that human beings depend on healthy ecosystems for food, medicines, and various raw materials. We are only now beginning to realize that we depend on them even more for the services they provide. The workings of ecosystems ensure that environmental conditions are suitable for the continued existence of humans.

1.4 Science: A Way of Knowing

Biology, with its numerous branches (**Fig. 1.13**), is the scientific study of life. Religion, aesthetics, ethics, and science are all ways that human beings have of finding order in the natural world. Science differs from the other fields by its process, which often involves the use of the scientific method (**Fig. 1.14**).

Observation

The scientific method begins with **observation.** We can observe with our noses that dinner is almost ready, observe with our fingertips that a surface is smooth and cold, and observe with our ears that a piano needs tuning. Scientists also extend the ability of their senses by using instruments; for example, the microscope enables them to see objects that could never be seen by the naked eye. Finally, scientists may expand their understanding even further by taking advantage of the knowledge and experiences of other scientists. For instance, they may look up past studies on the Internet or at the library, or they may write or speak to others who are researching similar topics.

Hypothesis

After making observations and gathering knowledge about a phenomenon, a scientist uses inductive reasoning. **Inductive reasoning** occurs whenever a person uses creative thinking to combine isolated facts into a cohesive whole. Chance alone can help a scientist arrive at an idea. The most famous case pertains to the antibiotic penicillin, which was discovered in 1928. While examining a petri dish of bacteria that had accidentally become contaminated with the mold *Penicillium*, Alexander Fleming observed an area around the mold that was free of bacteria. Fleming had long been interested in finding cures for human diseases caused by bacteria, and he was very knowledgeable about antibacterial substances. So when Fleming saw the dramatic effect of *Penicillium* mold on bacteria, he reasoned that the mold might be producing an antibacterial substance. We call such a possible explanation for a natural event a **hypothesis.** A hypothesis is based on existing knowledge, so it is much more informed than a mere guess. Fleming's hypothesis was supported by further study. Sometimes a hypothesis is not supported and must be either modified and subjected to additional study or rejected.

All of a scientist's past experiences, no matter what they might be, may influence the formation of a hypothesis. But a scientist considers only hypotheses that can be tested by experiments or further observations. Moral and religious beliefs, while very important to our lives, differ among cultures and through time and are not always testable.

Figure 1.13 Biologists.

Biologists work in a variety of settings. For example, (**a**) some botanists work in greenhouses; (**b**) some biologists, such as this biochemist, work in laboratories; and (**c**) many ecologists and environmentalists collect data in the field.

a. Botanist

b. Biochemist

c. Ecologists

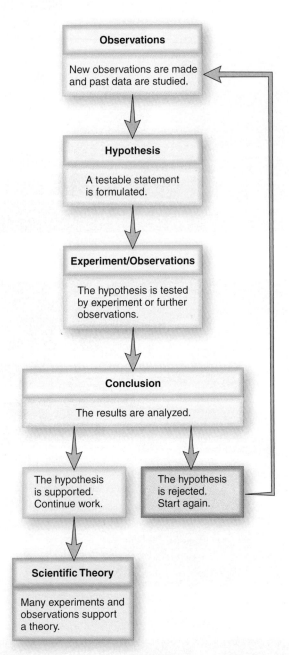

Figure 1.14 Flow diagram for the scientific method.

On the basis of new and/or previous observations, a scientist formulates a hypothesis. The hypothesis is tested by further experiments and/or observations, and new data either support or do not support the hypothesis. Following an experiment, a scientist often chooses to retest the same hypothesis or to test a related hypothesis. Conclusions from many different but related experiments may lead to the development of a scientific theory. For example, studies pertaining to development, anatomy, and fossil remains all support the theory of evolution.

Experiments/Further Observations

Scientists often perform an **experiment,** which is a series of procedures designed to test a specific hypothesis. The manner in which a scientist intends to conduct an experiment is called the **experimental design.** A good experimental design ensures that scientists are testing what they want to test and that their results will be meaningful. When an experiment is done in a laboratory, all conditions can be kept constant, except for an **experimental variable,** which is deliberately changed. One or more **test groups** are exposed to the experimental variable, but one other group, called the **control group,** is not. If, by chance, the control group shows the same results as the test group, the experimenter knows the results do not support the hypothesis.

Scientists often use **model** organisms and model systems to test a hypothesis. Models allow the scientist to control specific variables and environmental conditions in a way that may not be possible in the natural environment. For example, ecologists may use computer programs to model how human activities will affect the climate of a specific ecosystem. Cell biologists may use mice for modeling the effects of a new drug. While models provide useful information, they do not always answer the original question completely. For example, medicine that is effective in mice should ideally be tested in humans, and ecological experiments that are conducted using computer simulations need to be verified by actual field experiments. Biologists, and all other scientists, continuously design and revise their experiments to better understand how different factors may influence their original observation.

Data

The results of an experiment are referred to as the **data.** Mathematical data are often displayed in the form of a graph or table. Sometimes studies rely on statistical data. Statistical analysis allows a scientist to detect relationships in the data that may not be obvious on the surface. Let's say an investigator wants to know if eating onions can prevent women from getting osteoporosis (weak bones). The scientist conducts a survey asking women about their onion-eating habits and then correlates these data with the condition of their bones. Other scientists critiquing this study would want to know: How many women were surveyed? How old were the women? What were their exercise habits? What criteria were used to determine the condition of their bones? And what is the probability that the data are in error? The greater the variance in the data, the greater the probability of error. In any case, even if the data do suggest a correlation, scientists would want to know if there is a specific ingredient in onions that has a direct biochemical or physiological effect on bones. After all, correlation does not necessarily mean causation. It could be that women who eat onions eat lots of vegetables and have healthier diets overall than women who do not eat onions. In this way, scientists are skeptics who always pressure one another to keep investigating.

Conclusion

Scientists must analyze the data in order to reach a **conclusion** about whether a hypothesis is supported or not. Because science progresses, the conclusion of one experiment can lead to the hypothesis for another experiment (Fig. 1.14). In other words, results that do not support one hypothesis can often help a scientist formulate another hypothesis to be tested. Scientists report their findings in scientific journals, so that their methodology and data are available to other scientists. Experiments and observations must be *repeatable*—that is, the reporting scientist and any scientist who repeats the experiment must get the same results, or else the data are suspect.

Scientific Theory

The ultimate goal of science is to understand the natural world in terms of **scientific theories,** which are accepted explanations for how the world works. Some of the basic theories of biology are the cell theory, which says that all organisms are composed of cells; the gene theory, which says that inherited information dictates the form, function, and behavior of organisms; and the theory of evolution, which says that all organisms have a common ancestor and that each organism is adapted to a particular way of life.

The theory of evolution is considered the unifying concept of biology because it pertains to many different aspects of organisms. For example, the theory of evolution enables scientists to understand the history of life, the variety of organisms, and the anatomy, physiology, and development of organisms. The theory of evolution has been a very fruitful scientific theory, meaning that it has helped scientists generate new testable hypotheses. Because this theory has been supported by so many observations and experiments for over 100 years, some biologists refer to the theory of evolution as the **principle** of evolution, a term sometimes used for theories that are generally accepted by an overwhelming number of scientists. Others prefer the term **law** instead of *principle.*

How to Do a Controlled Study

We now know that most stomach and intestinal ulcers (open sores) are caused by the bacterium *Helicobacter pylori*. Let's say investigators want to determine which of two antibiotics is best for the treatment of an ulcer. When clinicians do an experiment, they try to vary just the experimental variables—in this case, the medications being tested. A control group is not given the medications, but one or more test groups are given the medications. If by chance, the control group shows the same results as a test group, the investigators immediately know the results of their study are invalid because it would mean the medications may have nothing to do with the results. The study depicted in **Figure 1.15** shows how investigators may study this hypothesis:

Hypothesis: Newly discovered antibiotic B is a better treatment for ulcers than antibiotic A, which is in current use.

Next, the investigators might decide to use three experimental groups: one control group and two test groups. It is important to reduce the number of possible variables (differences), such as sex, weight, and other illnesses, among the groups. Therefore, the investigators *randomly* divide a very large group of volunteers equally into the three groups. The hope is that any differences will be distributed evenly among the three groups. This is possible only if the investigators have a large number of volunteers.

The three groups are to be treated like this:

Control group: Subjects with ulcers are not treated with either antibiotic.
Test group 1: Subjects with ulcers are treated with antibiotic A.
Test group 2: Subjects with ulcers are treated with antibiotic B.

Figure 1.15 A controlled laboratory experiment to test the effectiveness of a medication in humans.

In this study, a large number of people were divided into three groups. The control group received a placebo and no medication. One of the test groups received medication A, and the other test group received medication B. The results are depicted in a graph, and it shows that medication B was a more effective treatment than medication A for the treatment of ulcers.

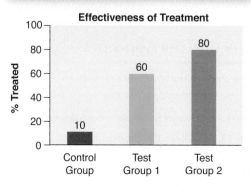

State Hypothesis:
Antibiotic B is a better treatment for ulcers than antibiotic A

Perform Experiment:
Groups were treated the same except as noted

Control group: received placebo

Test group 1: received antibiotic A

Test group 2: received antibiotic B

Collect Data:
Each subject was examined for the presence of ulcers

Effectiveness of Treatment

After the investigators have determined that all volunteers do have ulcers, they will want the subjects to think they are all receiving the *same* treatment. This is an additional way to protect the results from any influence other than the medication. To achieve this end, the subjects in the control group can receive a **placebo,** a treatment that appears to be the same as that administered to the other two groups but actually contains no medication. In this study, the use of a placebo would help ensure the same dedication by all subjects to the study.

The Results

After two weeks of administering the same amount of medication (or placebo) in the same way, the stomach and intestinal lining of each subject are examined to determine if ulcers are still present. Endoscopy, a procedure depicted in the photograph in Figure 1.15, is one way to examine a patient for the presence of ulcers. This procedure is performed under sedation and involves inserting an endoscope—a small, flexible tube with a tiny camera on the end—down the throat and into the stomach and the upper part of the intestine. Now, the doctor can see the lining of these organs and can check for ulcers. Tests performed during an endoscopy can also determine if *Helicobacter pylori* is present.

Because endoscopy is somewhat subjective, it is probably best if the examiner is not aware of which group the subject is in; otherwise, the prejudice of the examiner may influence the examination. When neither the patient nor the technician is aware of the specific treatment, it is called a *double-blind* study.

In this study, the investigators may decide to determine the effectiveness of the medication by the percentage of people who no longer have ulcers. So, if 20 people out of 100 still have ulcers, the medication is 80% effective. The difference in effectiveness is easily read in the graph portion of Figure 1.15.

Conclusion: On the basis of their data, the investigators conclude that their hypothesis has been supported.

Publication of Scientific Studies

Scientific studies are customarily published in a scientific journal (**Fig. 1.16**), so that all aspects of a study are available to the scientific community. Before information is published in scientific journals, it is typically reviewed by experts, who ensure that the research is credible, accurate, unbiased, and well executed. Another scientist should be able to read about an experiment in a scientific journal, repeat the experiment in a different location, and get the same (or very similar) results. Some articles are rejected for publication by reviewers when they believe there is something questionable about the design or the manner in which an experiment was conducted. Scientific magazines (Fig. 1.16), such as *Scientific American,* are involved with reporting scientific findings to the general public. The information in these articles is usually obtained from articles first published in scientific journals.

As mentioned previously, the conclusion of one experiment often leads to another experiment. The need for scientists to expand on findings explains why science changes and the findings of yesterday may be improved upon tomorrow.

Figure 1.16 Scientific publications.

Scientific journals, such as *Science,* are scholarly journals in which researchers share their findings with other scientists. Scientific magazines such as *Scientific American, New Scientist,* and *Science News,* contain articles that are usually written by reporters for a broader audience.

Connecting the Concepts

For more information on the central theories in biology, refer to the following discussions.

Section 4.2 examines the basic principles of the cell theory.

Section 14.1 explores the development of the theory of evolution.

Check Your Progress 1.4

1. Identify, in order, the steps of the scientific method.
2. Devise a controlled study and explain why it is an important research process.
3. Explain why publishing scientific studies is important.

1.5 Science and Bioethical Issues

Learning Outcomes

Upon completion of this section, you should be able to

1. Explain the role of bioethics in scientific studies.
2. Explain the significance of biodiversity.

Many scientists work in the field or the laboratory, collecting data and coming to conclusions that sometimes seem remote from our everyday lives. Other scientists are interested in using the findings of past and present scientists to produce a product or develop a technique that does affect our lives. The application of scientific knowledge for a practical purpose is called **technology.**

Who should decide how, and even whether, a technology is put to use? Making value judgments is not a part of science. Ethical and moral decisions must be made by all people. Therefore, the responsibility for how to use the fruits of science must reside with people from all walks of life, not with scientists alone. **Bioethics** is the branch of ethics that is concerned with the development and consequences of biological technology. Scientists should provide the public with as much information as possible, but all citizens, including scientists, should make decisions about the use of technologies.

Biodiversity is perhaps the single most significant bioethical issue that we face today. Biodiversity may be defined as variation in life on Earth, and it usually refers to the number of different species. The present number of species has been estimated to be as high as 15 million, but so far fewer than 2 million have been identified, named, and classified. The biodiversity that surrounds us is the result of approximately 3.5 billion years of evolution. Unfortunately, technology-enabled human activities present a major threat to biodiversity. As the human population grows, more and more of the biosphere is exploited for human needs, such as food and housing.

Extinction is the death of a species or a larger taxonomic group. Although it is natural for species extinctions to occur over time, human activities have increased the extinction rate by a factor of 100 to 1,000. It is estimated that as many as 400 species per day are lost because of human activities. Many biologists are alarmed about the present rate of extinction and the resulting loss of biodiversity worldwide.

Our own species may be in peril if biodiversity loss continues, because we are interconnected with all of life on Earth in ways that we do not yet fully understand. We depend on other living things to supply us with food, clothing, medicines, and construction materials. We need other species to help clean up our pollution, preserve soil fertility, and maintain the balance of gases in the atmosphere. We can protect biodiversity through the preservation of ecosystems. Just as a native fisherman who assists in overfishing a reef is doing away with his own food source, we as a society are contributing to the destruction of our home, the biosphere, when we destroy ecosystems. If instead we adopt a conservation ethic that preserves ecosystems, we are helping ensure the continued existence of all species, including our own.

Connecting the Concepts

For more information on the loss of biodiversity, refer to the following discussions.

Section 16.2 explores some of the causes of mass extinctions in the history of life.

Section 32.2 examines how human activity and population growth are influencing the levels of biodiversity on the planet.

Check Your Progress 1.5

1. Describe bioethics and state why it is important to all people, not just scientists.
2. Explain why biodiversity is important to life on Earth.
3. Predict what type of impact the extinction of one species would have on the Earth.

Media Study Tools

The Chapter in Review

Summary

1.1 The Characteristics of Life

Living things, often called organisms, share several common characteristics. Organisms

- are organized (have levels of organization).
- acquire materials and energy.
- respond to external stimuli.
- reproduce and develop.
- have adaptations.

1.2 Evolution: The Core Concept of Biology

Evolution explains the unity and diversity of life. Descent from a common ancestor explains why organisms share some characteristics, and adaptation to various ways of life explains the diversity of life-forms. An evolutionary tree diagram may be used to describe how groups of organisms are related to one another.

According to the rules of taxonomy, species belong to the following categories (from the most inclusive to the least inclusive):

Domains:
- Archaea: prokaryotes
- Bacteria: prokaryotes
- Eukarya: eukaryotes

Kingdoms in domain Eukarya:
- Protista: unicellular to multicellular organisms with various modes of nutrition
- Fungi: live on debris; molds and mushrooms
- Plantae: multicellular photosynthesizers
- Animalia: multicellular organisms that ingest food

To classify a particular organism, scientists use a consistent sequence of categories. A binomial name consists of the genus and the specific epithet.

1.3 Ecosystems and the Biosphere

Populations within a community interact with one another and with the physical environment, forming an ecosystem. With an ecosystem, chemicals cycle, but energy flows unidirectionally and eventually becomes heat.

Human activity can have a negative influence on the flow of energy and nutrients in an ecosystem.

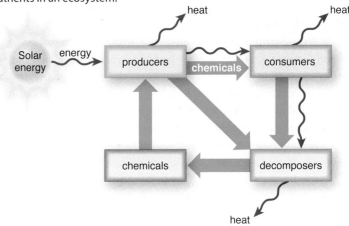

1.4 Science: A Way of Knowing

The scientific process includes a series of systematic steps known as the scientific method:

- Observations, which use the senses and may also include studies done by others
- A hypothesis (a statement to be tested)
- New observations and experiments
- A conclusion reached by analyzing data to determine whether the results support or do not support the hypothesis

A hypothesis confirmed by many different studies becomes known as a theory. Examples are the cell theory, the gene theory, and the theory of evolution, which is the unifying concept of biology.

1.5 Science and Bioethical Issues

Scientific findings often lead to the development of a technology that can be of service to human beings. Technologies have both benefits and drawbacks. Bioethics is concerned with the consequences of using biological technology. Every member of society needs to be prepared to participate in making bioethical decisions. Human activities, many of which are made possible by advances in technology, are damaging ecosystems and causing a worldwide loss of biodiversity; this is arguably the most important bioethical dilemma of our time.

Key Terms

adaptation   4
animal   8
binomial name   6
biodiversity   15
bioethics   15
biology   11
biosphere   10
cell   2
class   6
community   10
conclusion   12
control group   12
data   12
domain   6
domain Archaea   7
domain Bacteria   7
domain Eukarya   7
ecosystem   10
energy   3
evolution   5
experiment   12
experimental design   12
experimental variable   12
extinction   15
family   6
fungus   7
gene   4
genus   6
homeostasis   3

hypothesis   11
inductive reasoning   11
kingdom   6
law   13
metabolism   3
model   12
multicellular   2
natural selection   8
observation   11
order   6
organ   3
organism   2
organ system   3
photosynthesis   3
phylum   6
placebo   14
plant   7
population   10
principle   13
protist   7
reproduce   4
scientific theory   13
species   6
systematics   6
taxonomy   6
technology   15
test group   12
tissue   3

Testing Yourself

Choose the best answer for each question.

1. The region of the Earth's surface where all organisms are found is called the
   a. ecosystem.
   c. community.
   b. population.
   d. biosphere.

2. In the following diagram, label the levels of biological organization:

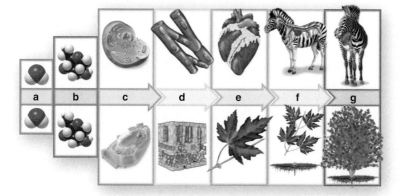

3. All of the chemical reactions that occur in a cell are called
   a. homeostasis.
   c. heterostasis.
   b. metabolism.
   d. cytoplasm.

4. Which of the following statements is not true?
   a. Darwin described evolution as descent with modification.
   b. Evolution is an explanation for both the unity and the diversity of life.
   c. Evolution occurs through a process called natural selection.
   d. Evolution is just one man's idea.

5. A modification that helps equip organisms for their way of life is a(n)
   a. homeostasis.
   c. adaptation.
   b. natural selection.
   d. extinction.

6. The process of turning solar energy into chemical energy is called
   a. work.
   c. photosynthesis.
   b. metabolism.
   d. respiration.

7. Place the steps in the process of natural selection in the correct order:
   (1) Over generations, a larger proportion of the population becomes adapted to the environment.
   (2) The population produces more offspring than the environment can support.
   (3) Members of a population have heritable variations.
   (4) The more adapted individuals survive and reproduce to a greater extent than those individuals lacking adaptations.
   a. 2, 4, 3, 1
   d. 2, 3, 1, 4
   b. 4, 1, 3, 2
   e. 3, 2, 4, 1
   c. 3, 1, 4, 2
   f. 1, 2, 4, 3

For questions 8–12, choose one answer from the following key.

Key:
   a. unicellular organism
   b. multicellular organism
   c. neither
   d. both

8. Capable of responding to the environment.

9. Reproduces by uniting an egg with a sperm.

10. Contains tissues.

11. Contains genes.

12. Creates energy.

13. A group of interbreeding individuals is a(n)
   a. ecosystem.
   c. community.
   b. species.
   d. phylum.

14. Information collected from a scientific experiment is known as
   a. a scientific theory.
   c. data.
   b. a hypothesis.
   d. a conclusion.

15. Taxonomy is a biological discipline in which organisms are grouped according to their
   a. geographic location.
   c. common ancestries.
   b. size.
   d. gene number.

16. Identify the correct statement.
   a. Order is more inclusive than kingdom in taxonomy.
   b. Taxonomical organization includes four domains.
   c. Systematics deals with evolutionary relationships between species.
   d. Systematics gave us the current system of scientific naming of organisms.

17. A representation of a natural phenomenon studied by scientists when the real object is impossible to study is a(n)

   a. model.
   b. experimental variable.
   c. control group.
   d. scientific theory.

18. A hypothesis cannot be formed without which of the following?

   a. experimentation
   b. observation
   c. data
   d. theory

Thinking Scientifically

1. Based on the evolutionary tree below, which prokaryotic domain gave rise to the domain Eukarya? Which kingdom in domain Eukarya gave rise to plants, animals, and fungi?

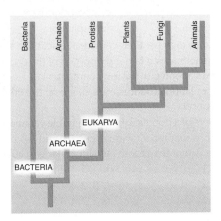

2. Several years before Alexander Fleming discovered penicillin, he made another exciting discovery. Fleming had a bad cold but kept working in his lab. He decided to put some of his nasal mucus in a petri dish of bacteria to see what would happen. Fleming observed that, over time, the mucus-covered area of the petri dish became clear of bacteria. What hypothesis might he have formulated when he saw the effect of his nasal mucus on the bacteria? Fleming soon became notorious for squeezing lemon juice into people's eyes to make them produce tears. He would then place the tears in petri dishes of bacteria. What hypothesis do you think he was testing?

Bioethical Issue

Placebo Use in Clinical Drug Trials

Clinical trials on drug efficacy are typically carried out on two groups of patients. The test group receives the drug, while the control group receives a placebo. The control group determines the extent to which psychology affects people who believe they are taking the new drug.

Suppose a new cancer drug has been developed. Many people sign up to participate in a clinical trial to test the drug's ability to slow the progression of the disease. These people know that they may receive a placebo in place of the real drug, but they participate because they have run out of options. Is it right to give the placebo to some of the participants, knowing that their disease is going to progress during the course of the trial? What if the drug were one that alleviated pain in chronic pain sufferers? In this case, no disease progression would occur during the course of the trial, but participants taking the placebo would have to endure serious pain.

2

The Chemical Basis of Life

Radiation Causes Chemical Damage to Cells

You are probably aware that the effects of exposure to extreme levels of radiation in humans include sunburns, hair loss, cancer, and maybe even death. The former spy Alexander Litvinenko died of poisoning by the radioactive element polonium-210, for example. But what is it about exposure to radioactivity that causes such damage to living cells? When we refer to a "radioactive atom," we really mean that the atom is not stable and has a tendency to release subatomic particles. If subatomic particles are released from an atom, they can cause major damage. When organisms are exposed to radioactivity, loose subatomic particles penetrate their cells and cause damage to cellular molecules, such as the genetic material (DNA). When DNA is damaged, a mutation may result. Many mutations may cause cancer; thus, exposure to high levels of radioactivity often causes cancer in humans. Radioactivity can also cause so much damage that many cells (and the individual) die.

Chemistry is the very essence of all organisms. You may have never thought of your body as simply a pile of chemicals, how the chemistry of your body changes when you eat or drink certain foods, or the changes that occur in cells when they are exposed to radiation. But more and more we see the boundaries between chemistry and biology becoming blurred. Scientists have made much progress in medicine, agriculture, and ecology by understanding the chemical reactions that occur in cells. Knowledge of the basic principles of chemistry will greatly enhance your ability to understand biology and will provide a foundation on which to build further competency.

OUTLINE

BEFORE YOU BEGIN

Before beginning this chapter, take a few moments to review the following discussions.

Section 1.1 What are the basic characteristics of all living organisms?

Figure 1.2 How do molecules relate to cells in the levels of biological organization?

2.1 The Nature of Matter

Learning Outcomes

Upon completion of this section, you should be able to

1. Distinguish among the types, location, and charge of subatomic particles.
2. Explain how isotopes are useful in the study of biology.
3. Relate how the arrangement of electrons determines an element's reactivity.
4. Contrast ionic and covalent bonds.
5. Relate how bonding leads to the formation of molecules and compounds.

When you kiss your sweetheart, pat your dog, catch a bus, or mow your lawn—everything you are touching is matter. **Matter** refers to anything that takes up space and has mass. Matter can exist as a solid, a liquid, or a gas. Not only are we made of matter, but so are the water we drink and the air we breathe.

All matter, both nonliving and living, is composed of elements. Formally speaking, an **element** is a substance that cannot be broken down into another substance by ordinary chemical means. There are only 92 naturally occurring elements, and each of these differs from the others in its properties. (A *property* is a physical or chemical characteristic, such as density, solubility, melting point, and reactivity.)

Both the Earth's crust and all organisms are made up of elements, but these elements occur in different proportions (**Fig. 2.1**). Six elements—carbon, hydrogen, nitrogen, oxygen, phosphorus, and sulfur—are of special significance to us. They make up about 98% of the body weight of most organisms. The acronym CHNOPS helps us remember these six elements, whose properties are essential to the uniqueness of living things, from simple one-celled life-forms to complex multicellular organisms.

oxygen (O) 65%

carbon (C) 18%

hydrogen (H) 10%

nitrogen (N) 5%

calcium (Ca) 2%

phosphorus (P) 1.1%

other elements, including sulfur 0.9%

Figure 2.1 Elements in living organisms.

If analyzed at the level of atoms, human beings are mostly composed of oxygen, carbon, hydrogen, nitrogen, calcium, and phosphorus.

Connections and Misconceptions

Where do elements come from?

We are all familiar with elements. Iron, sodium, oxygen, and carbon are all common in our lives, but where do they originate from?

Normal chemical reactions do not produce elements. The majority of the heavier elements, such as iron, are produced only by the intense chemical and physical reactions within stars. When these stars reach the end of their life, they explode, producing a supernova. Supernovas scatter the heavier elements into space, where they eventually are involved in the formation of planets.

The late astronomer and philosopher Carl Sagan (1934–1996) frequently referred to humans as "star stuff." In many ways we, and all other living organisms, are formed from elements that originated within the stars.

Atomic Structure

The *atomic theory* states that elements consist of tiny particles called **atoms.** Because each element consists of only one kind of atom, the same name is given to an element and its atoms. This name is represented by one or two letters, called the **atomic symbol.** For example, the symbol H stands for a hydrogen atom, and the symbol Na (for *natrium* in Latin) stands for a sodium atom.

If we could look inside a single atom, we would see that it is made mostly of three types of subatomic particles: **neutrons,** which have no electrical charge; **protons,** which have a positive charge; and **electrons,** which have a negative charge. Protons and neutrons are located within the center of an atom, which is called its **nucleus,** while electrons move about the nucleus.

Figure 2.2 shows the arrangement of the subatomic particles in a helium atom, which has only two electrons. In Figure 2.2*a*, the stippling shows the probable location of electrons, and in Figure 2.2*b*, the circle represents the approximate location of electrons. Most of an atom is empty space. In fact, if we could draw an atom the size of a baseball stadium, the nucleus would be like a gumball in the center of the stadium, and the electrons would be tiny specks whirling about in the upper stands. Usually, we can only indicate where the electrons are expected to be. In our analogy, the electrons might very well stray outside the stadium at times.

From our discussion of elements, you might expect each atom to have a certain mass. In effect, the **atomic mass** of an atom is just about equal to the sum of its protons and neutrons. Protons and neutrons are assigned one atomic mass unit each. Electrons are so small that their mass is assumed to be zero in most calculations. The term *atomic mass* is used, rather than *atomic weight,* because mass is constant but weight changes according to the gravitational force of a planet.

All atoms of an element have the same number of protons. This is called the atom's **atomic number.** The number of protons makes an atom unique and may be used to identify which element the atom belongs to.

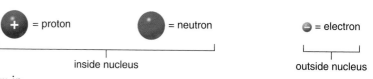
inside nucleus   = proton   = neutron   = electron   outside nucleus

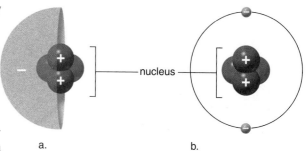

a.   b.

Figure 2.2   **Two models of helium (He).**

Atoms contain subatomic particles, which are located as shown in these two simplified models of helium. Protons are positively charged, neutrons have no charge, and electrons are negatively charged. Protons and neutrons are within the nucleus, and electrons are outside the nucleus. **a.** This model shows electrons as a negatively charged cloud around the nucleus. **b.** In this model, the average location of electrons is sometimes represented by a circle.

The Periodic Table

Once chemists discovered a number of the elements, they began to realize that the elements' chemical and physical characteristics recur in a predictable manner. The periodic table (**Fig. 2.3**) was developed as a way to display the elements, and therefore the atoms, according to these characteristics. In a periodic table, the atomic number is written above the atomic symbol. The mass number is written below the atomic symbol. For example, the carbon atom is shown in this way:

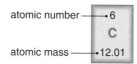
atomic number ⟶ 6
C
atomic mass ⟶ 12.01

Every atom is in a particular period (the horizontal rows) and in a particular group (the vertical columns). The atoms in group 8 are called the *noble gases* because they are gases that rarely react with another atom, for reasons we will discuss later. Notice that helium (He) and neon (Ne) are noble gases.

The atomic number tells you the number of positively charged protons, as well as the number of negatively charged electrons if the atom is electrically neutral. To determine the usual number of neutrons, subtract the number of protons from the mass number, written below the atomic symbol, and take the closest whole number.

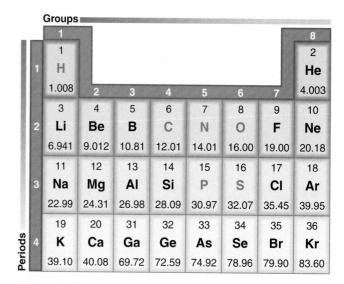

Figure 2.3   **A portion of the periodic table.**

In the periodic table, the elements, and therefore the atoms that compose them, are in the order of their atomic numbers but arranged in periods (horizontal rows) and groups (vertical columns). All the atoms in a particular group have certain chemical characteristics in common. The six elements previously mentioned (CHNOPS) are highlighted in red.

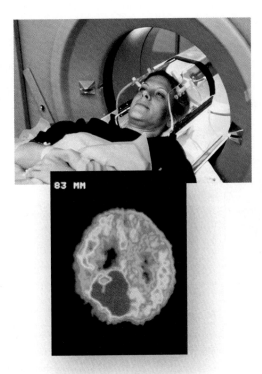

Figure 2.4 **PET scan.**

In a PET scan, red indicates areas of greatest metabolic activity, and blue means areas of least activity. Computers analyze the data from different sections of an organ—in this case, the human brain.

Isotopes

Isotopes are atoms of the same element that differ in the number of neutrons. In other words, isotopes have the same number of protons, but they have different mass numbers. In some cases, a nucleus with excess neutrons is unstable and may decay and emit radiation. These isotopes are said to be radioactive. However, not all isotopes are radioactive. The radiation given off by radioactive isotopes can be detected in various ways. Most people are familiar with the use of a Geiger counter to detect radiation. However, other methods to detect radiation exist that are useful in medicine and science.

Animation
Half-Life

Uses of Radioactive Isotopes   The importance of chemistry to biology and medicine is nowhere more evident than in the many uses of radioactive isotopes. For example, radioactive isotopes can be used as tracers to detect molecular changes or to destroy abnormal or infectious cells. Since both radioactive isotopes and stable isotopes contain the same number of electrons and protons, they essentially behave the same in chemical reactions. This means that you can put a small amount of radioactive isotope in a sample and it becomes a **tracer** by which to detect molecular changes because it behaves as the stable isotope. Positron-emission tomography (PET) utilizes tracers to determine the comparative activity of tissues. A radioactively labeled glucose tracer that emits a subatomic particle, known as a positron, is injected into the body. The radiation given off is detected by sensors and analyzed by a computer. The result is a color image that shows which tissues took up glucose and are thus metabolically active (**Fig. 2.4**). Although not shown in Figure 2.4, a PET scan can help diagnose a malfunctioning thyroid, a brain tumor, Alzheimer disease, epilepsy, or a stroke.

Animation
Nuclear Medicine

Radioactive substances in the environment can cause harmful chemical changes in cells, damage DNA, and cause cancer. The release of radioactive particles following a nuclear power plant accident can have far-reaching and long-lasting effects on human health. Even relatively small doses of radioactivity can be fatal—former KGB spy Alexander Litvinenko was fatally poisoned with the radioactive element polonium-210 (**Fig. 2.5***a*). But the effects of radiation can also be put to good use. Packets of radioactive isotopes can be placed in the body, so that the subatomic particles emitted destroy only cancer cells, with little risk to the rest of the body (**Fig. 2.5***b*).

Radiation from radioactive isotopes has been used for many years to sterilize medical and dental equipment. Since the terror attacks in 2001, mail that

Figure 2.5 **High levels of radiation.**

a. High levels of radiation can cause damage to the body. Former KGB spy Alexander Litvinenko was fatally poisoned with a high dose of radioactivity. **b.** Radiation can also be used to heal. Physicians treat patients with high levels of radioactive isotopes to kill cancer cells.

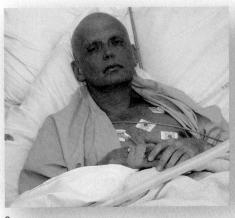

a.

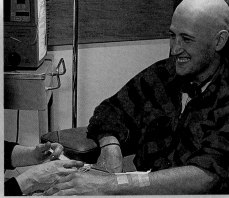

b.

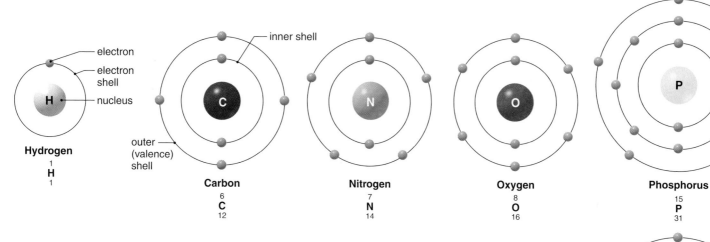

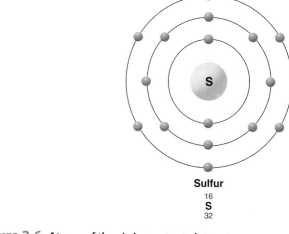

Figure 2.6   Atoms of the six important elements.

Electrons orbit the nucleus at particular energy levels (electron shells): The first shell contains up to two electrons, and each shell thereafter can contain up to eight electrons (until we consider atoms with an atomic number above 20). Each shell is to be filled before electrons are placed in the next shell.

is destined for the White House and congressional offices in Washington, DC, is irradiated to protect against dangerous biological agents, such as anthrax.

Connections and Misconceptions

Does irradiation add radioactive particles to food?

No, the process of food irradiation exposes certain types of foods to a form of radiation called ionizing radiation. While ionizing radiation is very useful in killing bacteria on the foods, it does not accumulate in or on the food. If you shine a light on a wall, the wall will not accumulate the light, nor will the wall emit light when you turn your light out. The same is true of ionizing radiation and the food irradiation process. Rather, food irradiation helps protect our food supply against disease-causing bacteria, such as *Salmonella* and *Escherichia coli* 0157:H7.

Arrangement of Electrons in an Atom

Electrons in atoms are much like the blades of a ceiling fan. When the fan is moving, it is difficult to see the individual blades, and all you see is a whirling blur. When the fan is stopped, each blade has a specific location and can be seen. Likewise, the electrons of an atom are constantly moving. Although it is not possible to determine the precise location of an individual electron at any given moment, it is useful to construct models of atoms that show electrons at discrete energy levels about the nucleus (**Fig. 2.6**). It seems reasonable to suggest that negatively charged electrons are attracted to the positively charged nucleus, and therefore it takes an increasing amount of energy to push them farther away from the nucleus. Electrons in outer energy levels, therefore, contain more energy than those in inner energy levels.

Each energy level contains a certain number of electrons. In the models shown in Figure 2.6, the energy levels (also called **electron shells**) are drawn as concentric rings about the nucleus. These shells represent the energy of the electrons, not necessarily their physical location. The first shell closest to the nucleus can contain two electrons; thereafter, each additional shell can contain eight electrons. For these atoms, each lower level is filled with electrons before the next higher level contains any electrons.

The sulfur atom, with an atomic number of 16, has two electrons in the first shell, eight electrons in the second shell, and six electrons in the third, or

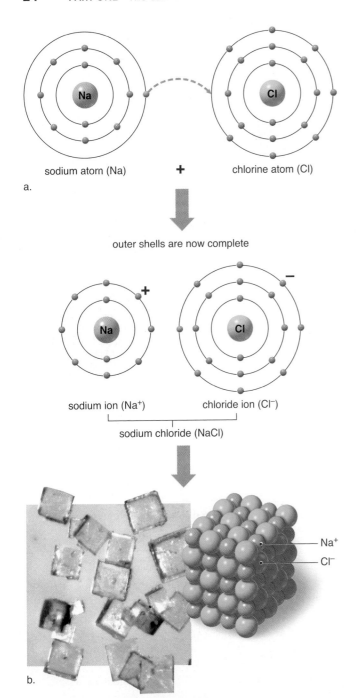

a.

outer shells are now complete

sodium atom (Na)   +   chlorine atom (Cl)

sodium ion (Na⁺)   chloride ion (Cl⁻)

sodium chloride (NaCl)

Na⁺
Cl⁻

b.

Figure 2.7 Formation of sodium chloride.

a. During the formation of sodium chloride, an electron is transferred from the sodium atom to the chlorine atom. At the completion of the reaction, each atom has eight electrons in the outer shell, but each also carries a charge as shown. **b.** In a sodium chloride crystal, attraction between the Na⁺ and Cl⁻ ions causes the atoms to assume a three-dimensional lattice shape.

outer, shell. Revisit the periodic table (see Fig. 2.3), and note that sulfur is in the third period. In other words, the period tells you how many shells an atom has. Also note that sulfur is in group 6. The group tells you how many electrons an atom has in its outer shell.

If an atom has only one shell, the outer shell is complete when it has two electrons. If an atom has two or more shells, the outer shell is most stable when it has eight electrons; this is called the **octet rule.** As mentioned previously, atoms in group 8 of the periodic table are called the noble gases because they do not ordinarily undergo reactions. Atoms with fewer than eight electrons in the outer shell react with other atoms in such a way that each has a completed outer shell after the reaction. Atoms can give up, accept, or share electrons in order to have eight electrons in the outer shell. In other words, the number of electrons in an atom's outer shell, called the **valence shell,** determines its chemical reactivity. The size of an atom is also important. Both carbon (C) and silicon (Si) atoms are in group 4, and therefore they have four electrons in their valence shells. This means they can bond with as many as four other atoms in order to achieve eight electrons in their outer shells. But carbon in period 2 has two shells, and silicon in period 3 has three shells. The smaller atom, carbon, can bond to other carbon atoms and form long-chained molecules, while the larger silicon atom is unable to bond to other silicon atoms. Still, the chemical properties of atoms—that is, the ways in which they react—are largely determined by the arrangement of their electrons.

Types of Chemical Bonds

A group of atoms bonded together is called a **molecule.** When a molecule contains atoms of more than one element, it can be called a **compound.** Compounds and molecules contain two types of chemical bonds: ionic and covalent. The type of bond that forms depends on whether two bonded atoms share electrons or whether one has given electrons to the other. For example, in hydrogen gas (H_2) the two hydrogen atoms are sharing electrons in order to fill the valence shells of both atoms. When sodium chloride (NaCl) forms, however, the sodium atom (Na) gives an electron to the chlorine (Cl) atom, and in that way each atom has eight electrons in the outer shell.

Ionic Bonding

An **ionic bond** forms when two atoms are held together by the attraction between opposite charges. The reaction between sodium and chlorine atoms is an example of how an ionic bond is formed. Consider that sodium (Na), with only one electron in its third shell, usually gives up an electron (**Fig. 2.7a**). Once it does so, the second shell, with eight electrons, becomes its outer shell. Chlorine (Cl), on the other hand, tends to take on an electron, because its outer shell has seven electrons. If chlorine gets one more electron, it has a completed outer shell. So, when a sodium atom and a chlorine atom react, an electron is transferred from sodium to chlorine. Now both atoms have eight electrons in their outer shells.

This electron transfer causes these atoms to become **ions,** or charged atoms. The sodium ion has one more proton than it has electrons; therefore, it has a net charge of +1 (symbolized by Na⁺). The chloride ion has one more electron than it has protons; therefore, it has a net charge of −1 (symbolized by Cl⁻). Negatively charged ions often have names that end in "ide," and thus Cl⁻ is called a chloride ion. In the periodic table, atoms in groups 1 and 2 and groups 6 and 7 become ions when they react with other atoms. Atoms in groups 2 and 6 always transfer two electrons. For example, calcium becomes Ca^{2+}, while oxygen becomes O^{2-}.

Animation
Ionic Bonds

Ionic compounds are often found as **salts.** Sodium chloride is table salt. Salts can exist as a dry solid or in a dissociated (ionized) form. A sodium chloride crystal illustrates the solid form of a salt (Fig. 2.7*b*). When salts are placed in water, they release ions as they dissolve. NaCl separates into Na^+ and Cl^-. Ionic compounds are most commonly found in this dissociated (ionized) form in biological systems because these systems are 70–90% water.

Covalent Bonding

A **covalent bond** results when two atoms share electrons in order to have a completed outer shell. In a hydrogen atom, the outer shell is complete when it contains two electrons. If hydrogen is in the presence of a strong electron acceptor, it gives up its electron to become a hydrogen ion (H^+). But if this is not possible, hydrogen can share with another atom, and thereby have a completed outer shell. For example, one hydrogen atom can share with another hydrogen atom. In this case, the two orbitals overlap and the electrons are shared between them—that is, you count the electrons as belonging to both atoms:

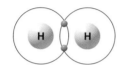

Hydrogen gas (H_2)

Rather than drawing an orbital model like the one above, scientists often use simpler ways to indicate molecules. A *structural formula* uses straight lines, as in H—H. The straight line is used to indicate a pair of shared electrons. A *molecular formula* omits the lines that indicate bonds and simply shows the number of atoms involved, as in H_2.

Sometimes, atoms share more than two electrons to complete their octets. A double covalent bond occurs when two atoms share two pairs of electrons, as in this molecule of oxygen gas:

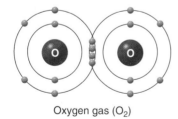

Oxygen gas (O_2)

In order to show that oxygen gas (O_2) contains a double bond, the structural formula is written as O=O to indicate that two pairs of electrons are shared between the oxygen atoms.

It is also possible for atoms to form triple covalent bonds, as in nitrogen gas (N_2), which can be written as N≡N. Single covalent bonds between atoms are quite strong, but double and triple bonds are even stronger.

A single atom may form bonds with more than one other atom. For example, the molecule methane results when carbon binds to four hydrogen atoms (**Fig. 2.8***a*). In methane, each bond actually points to one corner of a four-sided structure called a tetrahedron (Fig. 2.8*b*). The best model to show this arrangement is a ball-and-stick model (Fig. 2.8*c*). Space-filling models (Fig. 2.8*d*) come closest to showing the actual shape of a molecule. The shapes of molecules help dictate the functional roles they play in organisms.

Methane (CH₄)

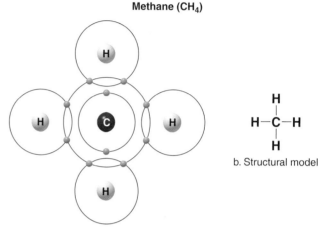

a. Electron model showing covalent bonds

b. Structural model

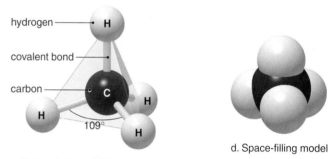

c. Ball-and-stick model

d. Space-filling model

Figure 2.8   Shapes of covalently bonded molecules.

An electron model (**a**) and a structural model (**b**) show that methane (CH₄) contains one carbon atom bonded to four hydrogen atoms. **c.** The ball-and-stick model shows that, when carbon bonds to four other atoms, as in methane, each bond actually points to one corner of a tetrahedron. **d.** The space-filling model is a three-dimensional representation of the molecule.

Chemical Reactions

Chemical reactions are very important to organisms. For example, we have already noted that the process of photosynthesis enables plants to make molecular energy available to themselves and other organisms. An overall equation for photosynthesis indicates that some bonds are broken and others are formed:

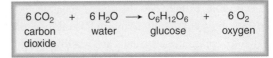

$$6\ CO_2 \quad + \quad 6\ H_2O \quad \longrightarrow \quad C_6H_{12}O_6 \quad + \quad 6\ O_2$$

carbon dioxide   water   glucose   oxygen

This equation says that six molecules of carbon dioxide react with six molecules of water to form one glucose molecule and six molecules of oxygen. The **reactants** (molecules that participate in the reaction) are shown on the left of the arrow, and the **products** (molecules formed by the reaction) are shown on the right. Notice that the equation is "balanced"—that is, the same number of each type of atom occurs on both sides of the arrow.

Note the glucose molecule in the equation above. Glucose is more complex than any molecule discussed so far. The arrangement of atoms in a glucose molecule is shown below:

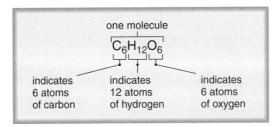

one molecule
$$C_6H_{12}O_6$$

indicates   indicates   indicates
6 atoms   12 atoms   6 atoms
of carbon   of hydrogen   of oxygen

MP3
Chemical Bonding

Animation
Ionic and Covalent Bonding

Connecting the Concepts

Cells use a variety of chemical reactions to maintain the characteristics of life. For a more detailed look at some of these processes, refer to the following discussions.

Section 5.3 examines how linked chemical reactions, called metabolic pathways, can be used to assemble complex compounds in cells.

Section 6.1 explores how photosynthetic organisms can manufacture glucose from carbon dioxide, sunlight, and water.

Section 7.1 investigates the chemical reactions that allow cells to extract the energy in glucose.

Check Your Progress 2.1

1. Describe the structure of an atom, including the charge of each subatomic particle.
2. Identify the differences between an atom and an isotope, as well as the beneficial uses of radioactive isotopes.
3. Compare and contrast covalent and ionic bonds.
4. Summarize the octet rule and explain how it affects an element's reactivity.
5. Distinguish between reactants and products and explain why it is important to "balance" a chemical reaction.

2.2 Water's Importance to Life

Life began in water, and it is the single most important molecule on Earth. All organisms are 70–90% water; their cells consist of membranous compartments enclosing aqueous solutions. Water has unique properties, which make it a life-supporting substance. These properties of water stem from the structure of the molecule.

The Structure of Water

The electrons shared between two atoms in a covalent bond are not always shared equally. Atoms differ in their **electronegativity**—that is, their affinity for electrons in a covalent bond. The electronegativity of an atom is related to its size; larger atoms tend to have a higher affinity, and electronegativity, for electrons than do smaller atoms. In addition, atoms that are more electronegative tend to hold shared electrons more tightly than do those that are not. This causes the bond to become **polar,** meaning that the atoms on both sides of the bond are partially charged, even though the overall molecule itself bears no net charge. For example, in water, oxygen shares electrons with two hydrogen atoms. Oxygen is more electronegative than hydrogen, so the two bonds are polar. The shared electrons spend more time orbiting the oxygen nucleus than the hydrogen nuclei, and this unequal sharing of electrons makes water a polar molecule. The covalent bonds are angled, and the molecule is bent roughly into a Λ shape. The point of the Λ (oxygen) is the negative (−) end, and the two hydrogens are the positive (+) end (**Fig. 2.9***a*).

The polarity of water molecules causes them to be attracted to one another. The positive hydrogen atoms in one molecule are attracted to the negative oxygen atoms in other water molecules. This attraction is called a **hydrogen (H) bond,** and each water molecule can engage in as many as four H bonds (Fig. 2.9*b*). The covalent bond is much stronger than an H bond, but the large number of H bonds in water makes the H bond strong overall. The properties of water are due to its polarity and its ability to form hydrogen bonds.

Properties of Water

Water is so familiar that we take it for granted, but without water, life as we know it would not exist. The properties of water that support life are solvency, cohesion and adhesion, high surface tension, high heat capacity, high heat of vaporization, and varying density.

Animation
Water Properties

Figure 2.9 **The structure of water.**

a. The space-filling model shows the ∧ shape of a water molecule. Oxygen attracts the shared electrons more than hydrogen atoms do, and this causes the molecule to be polar: The oxygen carries a slightly negative charge and the hydrogens carry a slightly positive charge. **b.** The positive hydrogens form hydrogen bonds with the negative oxygen in nearby molecules. Each water molecule can be joined to other water molecules by as many as four hydrogen bonds.

Water is a solvent   Because of its polarity and H-bonding ability, water dissolves a great number of substances. Molecules that are attracted to water are said to be **hydrophilic** (*hydro*, water; *phil*, love). Polar and ionized molecules are usually hydrophilic. Nonionized and nonpolar molecules that are not attracted to water are said to be **hydrophobic** (*hydro*, water; *phob*, fear).

When a salt such as sodium chloride (NaCl) is put into water, the negative ends of the water molecules are attracted to the sodium ions, and the positive ends of the water molecules are attracted to the chloride ions. This attraction causes the sodium ions and the chloride ions to break up, or dissociate, in water:

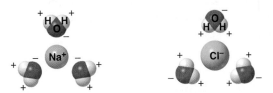

The salt NaCl dissociates in water.

Water dissolves many polar nonionic substances, such as long chains of glucose, by forming H bonds with them. When ions and molecules disperse in water, they move about and collide, allowing reactions to occur. As mentioned, cellular fluids are aqueous solutions of various substances, and so are the oceans. Without the dissolving power of water, aquatic organisms could not take up the substances they need from the water.

Water molecules are cohesive and adhesive   *Cohesion* refers to the ability of water molecules to cling to each other due to hydrogen bonding. Because of cohesion, water exists as a liquid under ordinary conditions of temperature and pressure. The strong cohesion of water molecules is apparent because water flows freely, yet the molecules do not separate from each other. *Adhesion* refers to the ability of water molecules to cling to other polar surfaces. This is because of water's polarity. The positive and negative poles of water molecules cause them to adhere to other polar surfaces.

Due to cohesion and adhesion, liquid water is an excellent transport system. Both within and outside the cell, water assists in the transport of nutrient and waste materials. Many multicellular animals contain internal vessels in which water assists the transport of materials. The liquid portion of blood, which transports dissolved and suspended substances about the body, is 90% water. The cohesion and adhesion of water molecules allow blood to fill the tubular vessels of the cardiovascular system, making transport possible. Cohesion and adhesion also contribute to the transport of water in plants. Plants have their roots anchored in the soil, where they absorb water, but the leaves are uplifted and exposed to solar energy. Water evaporating from the leaves is immediately replaced with water molecules from transport vessels that extend from the roots to the leaves (**Fig. 2.10**). Because water molecules are cohesive, a tension is created that pulls a water column up from the roots. Adhesion of water to the walls of the vessels also helps prevent the water column from breaking apart.

Water has a high surface tension   Because the water molecules at the surface are more strongly attracted to each other than to the air above, water molecules at the surface cling tightly to each other. Thus, we say that water

Figure 2.10   Cohesion and adhesion of water molecules.

How does water rise to the top of all trees? Water-filled vessels extend from the roots to the leaves. When water evaporates from the leaves, this water column is pulled upward due to the cohesion of water molecules with one another and the adhesion of water molecules to the sides of the vessel.

(labels on figure)
Water evaporates, pulling the water column from the roots to the leaves.

Water molecules cling together and adhere to sides of vessels in stems and tree trunks.

Water enters a plant at root cells.

Water in water column

exhibits surface tension. The stronger the force between molecules in a liquid, the greater the surface tension. Hydrogen bonding is the main force that causes water to have a high surface tension. If you slowly fill a glass with water, you may notice that the level of the water forms a small dome above the top of the glass. This is due to the surface tension of water.

Connections and Misconceptions

How do some insects walk on water?

Anyone who has visited a pond or stream has witnessed insects walking on the surface of the water. These insects, commonly called water striders, have evolved this ability by adapting to two properties of water—its surface tension and the fact that water is a polar molecule. By trapping small air bubbles in the hairs on their legs, water striders are able remain buoyant and not break the surface tension of the water molecules. Many water striders also secrete a nonpolar wax, which further repels the water molecules and keeps the insect afloat.

Water has a high heat capacity The many hydrogen bonds that link water molecules allow water to absorb heat without greatly changing in temperature. Water's high heat capacity is important not only for aquatic organisms but for all organisms. Because the temperature of water rises and falls slowly, terrestrial organisms are better able to maintain their normal internal temperatures and are protected from rapid temperature changes.

Water also has a high heat of vaporization: It takes a great deal of heat to break the hydrogen bonds in water so that it becomes gaseous and evaporates into the environment. If the heat given off by our metabolic activities were to go directly into raising our body temperature, death would follow. Instead, the heat is dispelled as sweat evaporates (**Fig. 2.11**).

Because of water's high heat capacity and high heat of vaporization, temperatures along the Earth's coasts are moderate. During the summer, the ocean absorbs and stores solar heat; during the winter, the ocean releases it slowly. In contrast, the interior regions of the continents experience abrupt changes in temperature.

Water is less dense as ice Unlike other substances, water expands as it freezes, which explains why cans of soda burst when placed in a freezer and how roads in northern climates become bumpy because of "frost heaves" in the winter. Since

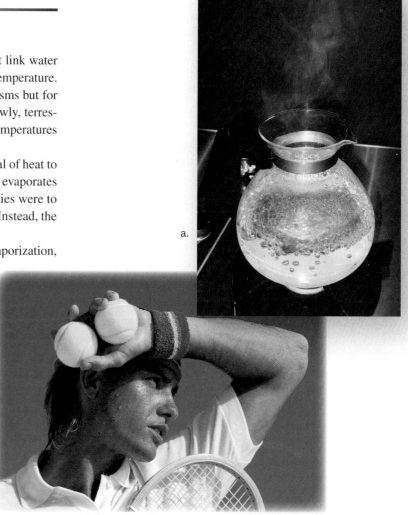

Figure 2.11 Heat of vaporization.

At room temperature, water is a liquid. **a.** Water takes a large amount of heat to vaporize at 100°C. **b.** It takes much body heat to vaporize sweat, which is mostly liquid water, and this helps keep our bodies cool when the temperature rises.

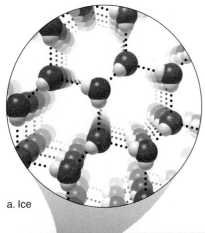

a. Ice

ice layer

b. Pond

Figure 2.12 Properties of ice.
a. The geometric requirements of hydrogen bonding of water molecules cause ice to be less dense than liquid water. **b.** Therefore, bodies of water freeze from the top down, and organisms in ponds and lakes are protected during the winter.

water expands as it freezes, it is less dense than liquid water, and therefore ice floats on liquid water (**Fig. 2.12**).

If ice were more dense than water, it would sink, and ponds, lakes, and perhaps even the ocean would freeze solid, making life impossible in the water as well as on land. Instead, bodies of water always freeze from the top down. When a body of water freezes on the surface, the ice acts as an insulator to prevent the water below it from freezing. This protects aquatic organisms, so that they can survive the winter. As ice melts in the spring, it draws heat from the environment, helping prevent a sudden change in temperature that might be harmful to life.

Connecting the Concepts

For more on how living organisms are adapted to handling water, refer to the following discussions.

Section 5.4 explores how water moves across the membranes of cells.

Section 20.7 examines how cohesion moves water in plants.

Section 23.3 examines the fluid components of blood in animals.

Check Your Progress 2.2

1. Identify the special properties of water.
2. Describe the structure of a water molecule and explain how the structure allows it to have special properties.
3. Describe how some of the special properties of water make it essential to life. Give specific examples for a plant, a human, and a fish.

2.3 Acids and Bases

Learning Outcomes

Upon completion of this section, you should be able to

1. Distinguish between an acid and a base.
2. Interpret the pH scale.
3. Explain the purpose of a buffer.

Figure 2.13 shows that, when water dissociates (breaks apart), it releases an equal number of **hydrogen ions (H^+)** and **hydroxide ions (OH^-),** as in the following reaction:

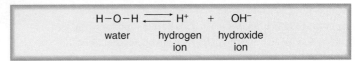

$$H-O-H \rightleftharpoons H^+ \quad + \quad OH^-$$
water   hydrogen   hydroxide
              ion   ion

Acidic Solutions (High H⁺ Concentration)

Lemon juice, vinegar, tomato juice, and coffee are all acidic solutions. What do they have in common? Acidic solutions have a sharp or sour taste, and therefore we sometimes associate them with indigestion. To a chemist, **acids** are substances that dissociate in water, releasing hydrogen ions (H⁺). For example, an important acid is hydrochloric acid (HCl), which dissociates in this manner:

$$HCl \rightarrow H^+ + Cl^-$$

The acidity of a substance depends on how fully it dissociates in water. HCl dissociates almost completely; therefore, it is called a strong acid. If hydrochloric acid is added to a beaker of water, the number of hydrogen ions (H⁺) increases greatly (**Fig. 2.14**).

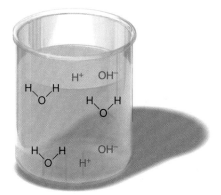

Figure 2.13 Dissociation of water molecules.

Dissociation of water molecules produces an equal number of hydrogen ions (H⁺) and hydroxide ions (OH⁻). (These illustrations are not meant to be mathematically accurate.)

Connections and Misconceptions

How strong is the acid in your stomach?

Within the gastric juice of the stomach is hydrochloric acid (HCl), which has a pH value between 1.0 and 2.0. This makes the contents of your stomach around 1 million times more acidic that water and 100 times more acidic than vinegar. While theoretically the gastric juice in your stomach is able  to dissolve metals, such as steel, in reality the contents of your stomach are exposed to these extreme pH values for only a short period of time before moving into the remainder of the intestinal tract, where the acid levels are quickly neutralized.

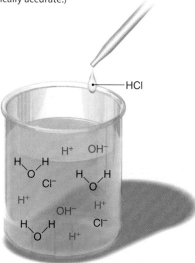

Figure 2.14 Action of a strong acid in water.

Hydrochloric acid (HCl) releases hydrogen ions (H⁺) as it dissociates. The addition of HCl to water results in a solution that is acidic (more H⁺ than OH⁻).

Basic Solutions (Low H⁺ Concentration)

Milk of magnesia and ammonia are common basic (alkaline) solutions that most people are familiar with. Basic solutions have a bitter taste and feel slippery when in water. To a chemist, **bases** are substances that either take up hydrogen ions (H⁺) or release hydroxide ions (OH⁻). For example, an important base is sodium hydroxide (NaOH), which dissociates in this manner:

$$NaOH \rightarrow Na^+ + OH^-$$

Like acids, the strength of a base is determined by how fully it dissociates. Dissociation of sodium hydroxide is almost complete; therefore, it is called a strong base. If sodium hydroxide is added to a beaker of water, the number of hydroxide ions increases (**Fig. 2.15**).

Many strong bases, such as ammonia, are useful household cleansers. Ammonia has a poison symbol and carries a strong warning not to ingest the product. Neither acids nor bases should be tasted, because they are quite destructive to cells.

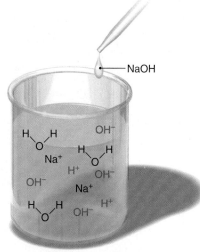

Figure 2.15 Action of a strong base in water.

Sodium hydroxide (NaOH) releases OH⁻ as it dissociates. The addition of NaOH to water results in a solution that is basic (more OH⁻ than H⁺).

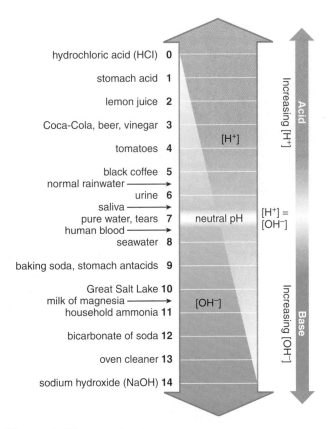

Figure 2.16   The pH scale.

The proportionate amount of hydrogen ions to hydroxide ions is indicated by the diagonal line. Any solution with a pH above 7 is basic, while any solution with a pH below 7 is acidic.

a.

Figure 2.17   Effects of acid deposition.

The burning of fossil fuels, such as gasoline and coal, leads to acid deposition, which causes (**a**) tree deaths and (**b**) fish kills in lakes and ponds.   b.

pH and the pH Scale

pH[1] is a mathematical way of indicating the number of hydrogen ions in a solution. The **pH scale** is used to indicate the acidity or basicity of a solution. The pH scale ranges from 0 to 14 (**Fig. 2.16**). A pH of 7 represents a neutral state in which the hydrogen ion and hydroxide ion concentrations are equal, as in pure water. A pH below 7 is acidic because the hydrogen ion concentration, commonly expressed in brackets as [H^+], is greater than the hydroxide concentration, [OH^-]. A pH above 7 is basic because [OH^-] is greater than [H^+]. Further, as we move down the pH scale from pH 7 to pH 0, each unit has 10 times the acidity [H^+] of the previous unit. As we move up the scale from 7 to 14, each unit has 10 times the basicity [OH^-] of the previous unit.

MP3
Water and pH

The pH scale was devised to eliminate the use of cumbersome numbers. For example, the hydrogen ion concentrations of these solutions are on the left, and the pH is on the right:

	[H^+] (moles per liter)		pH
0.000001	=	1×10^{-6}	6 (acid)
0.0000001	=	1×10^{-7}	7 (neutral)
0.00000001	=	1×10^{-8}	8 (base)

The effect of pH on organisms is dramatically illustrated by the phenomenon known as acid deposition. When fossil fuels are burned, sulfur dioxide and nitrogen oxides are produced, and they combine with water in the atmosphere to form acids. These acids then come in contact with organisms and objects, leading to damage or even death (**Fig. 2.17**).

MP3
Acid-Base Balance

Buffers and pH

In the human body, pH needs to be kept within a narrow range in order to maintain homeostasis. A **buffer** is a chemical or a combination of chemicals that keeps pH within normal limits. Buffers resist pH changes because they can take up excess hydrogen ions (H^+) or hydroxide ions (OH^-). Many commercial products, such as some aspirins, shampoos, and deodorants, are buffered as an added incentive for us to buy them.

Connecting the Concepts

For more on how organisms have evolved mechanisms to adjust pH, refer to the following discussions.

Section 24.1 explores how the respiratory system adjusts to changes in pH caused by the presence of carbon dioxide gas.

Section 24.2 examines how both the excretory system of insects and the urinary system of humans regulate pH levels.

Check Your Progress 2.3

1. Explain the use of the pH scale and how it is arranged.
2. Generalize what information is given by the pH of a solution.
3. Summarize how buffers are used to regulate the pH of the human body.

[1] pH is defined as the negative log of the hydrogen ion concentration [H^+]. A log is the power to which 10 must be raised to produce a given number.

Media Study Tools

www.mhhe.com/maderessentials3

Enhance your study of this chapter with study tools and practice tests. Also ask your instructor about the resources available through ConnectPlus, including the media-rich eBook, interactive learning tools, and animations.

The Chapter in Review

Summary

2.1 The Nature of Matter

Definitions to remember:

- **Matter:** Takes up space, has mass, and is composed of elements.
- **Element:** A fundamental constituent of matter that cannot be broken down into another substance; composed of atoms.
- **Nucleus:** The center of an atom; contains protons (+) and neutrons.
- **Atomic mass:** Protons plus neutrons.
- **Isotopes:** Atoms of same type that differ in the number of neutrons.
- **Atomic number:** Determined by the number of protons in an atom; equals number of electrons when atom is neutral.
- **Electrons** (–)

  First shell: Complete with two electrons.

  Other shells: Complete with eight electrons, assuming an atomic number of 20 or below.

  Valence shell: Number of electrons in outer shell determines reactivity of atom.
- **Octet rule:** Atoms react with one another in order to have a completed outer shell with eight electrons.

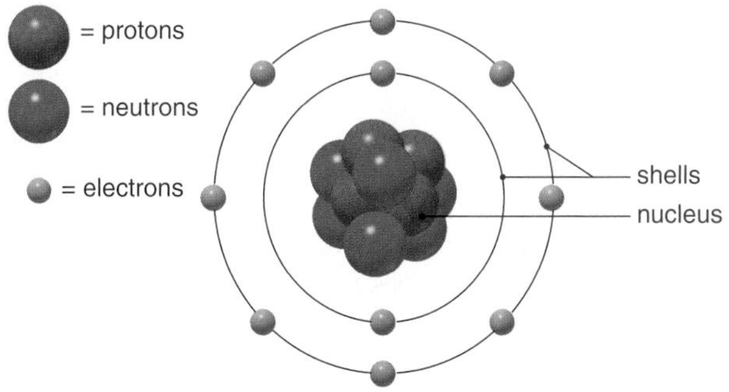

- **Ionic bond:** Attraction between oppositely charged ions. Ions form when atoms lose or gain one or more electrons to achieve a completed outer shell.
- **Covalent bond:** Sharing of electrons between two atoms. There are single covalent bonds (sharing one pair of electrons), double (sharing two pairs of electrons), and triple (sharing three pairs of electrons).

2.2 Water's Importance to Life

The two covalent O—H bonds of water are polar because the electrons are not shared equally between oxygen and hydrogen. The positively and negatively charged ends of the molecules are attracted to each other to form hydrogen bonds. The polarity and hydrogen bonding in water account for its unique properties, which can be summarized as follows:

Properties	Chemical Reason(s)	Effect
Water is a solvent.	Polarity	Water facilitates chemical reactions.
Water is cohesive and adhesive.	Hydrogen bonding; polarity	Water serves as a transport medium.
Water has a high surface tension.	Hydrogen bonding	The surface tension of water is hard to break.
Water has a high heat capacity.	Hydrogen bonding	Water protects organisms from rapid changes in temperature.
Water has a high heat of vaporization.	Hydrogen bonding	Water helps organisms resist overheating.
Water is less dense as ice.	Hydrogen bonding	Ice floats on liquid water.

2.3 Acids and Bases

Water dissociates to produce an equal number of hydrogen ions and hydroxide ions. This is neutral pH. This illustration shows the range of acidic and basic solutions:

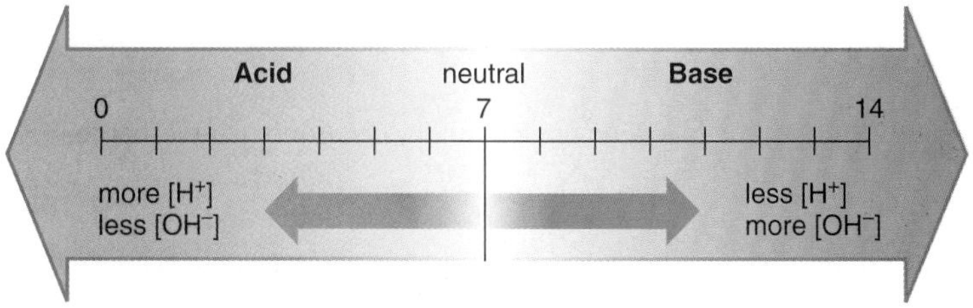

Cells are sensitive to pH changes. Biological systems contain buffers that help keep the pH within a normal range.

Key Terms

acid 31
atom 21
atomic mass 21
atomic number 21
atomic symbol 21
base 31
buffer 32
compound 24
covalent bond 25
electron 21
electronegativity 27
electron shell 23
element 20
hydrogen (H) bond 27
hydrogen ion (H^+) 30
hydrophilic 28
hydrophobic 28
hydroxide ion (OH^-) 30
ion 24
ionic bond 24
isotope 22
matter 20

molecule   24
neutron   21
nucleus   21
octet rule   24
pH   32
pH scale   32
polar   27

product   26
proton   21
reactant   26
salt   25
tracer   22
valence shell   24

Testing Yourself

Choose the best answer for each question.

1. The atomic mass of an atom depends primarily on the number of
   a. protons and neutrons.
   b. positrons.
   c. neutrons and electrons.
   d. protons and electrons.

2. The most abundant element by weight in the human body is
   a. carbon.
   b. hydrogen.
   c. oxygen.
   d. nitrogen.

3. Which of the following is true referring to the octet rule?
   a. The octet rule deals with the electrons in the valence shell only.
   b. The octet rule deals with the stability of atoms in chemical bonds.
   c. The octet rule can pertain to the noble gases.
   d. Both a and b are true.

4. How many more electrons does nitrogen require to fill its outer shell?
   a. 0
   b. 4
   c. 2
   d. 3

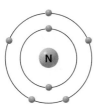

Nitrogen

5. Which of the following has a naturally occurring neutral charge?
   a. neutron
   b. atom
   c. proton
   d. both a and b

6. In the following equation, name the reactants: NaCl → Na + Cl
   a. Na
   b. Cl
   c. NaCl
   d. both a and b

7. A covalent bond in which electrons are not shared equally is called
   a. polar.
   b. normal.
   c. nonpolar.
   d. neutral.

8. Water flows freely but does not separate into individual molecules because water is
   a. cohesive.
   b. hydrophilic.
   c. hydrophobic.
   d. adhesive.

9. Compounds having an affinity for water are said to be
   a. cohesive.
   b. hydrophilic.
   c. hydrophobic.
   d. adhesive.

10. Water freezes from the top down because
    a. water has a high surface tension.
    b. ice has a high heat capacity.
    c. ice is less dense than water.
    d. ice is less cohesive than water.

11. Water can absorb a large amount of heat without much change in temperature because it has a high
    a. surface tension.
    b. heat capacity.
    c. hydrogen ion (H^+) concentration.
    d. hydroxide ion (OH^-) concentration.

12. A solution has a large amount of (H^+) dissociated into it; which of the following will act as a buffer to this solution?
    a. NaOH
    b. HCl
    c. OH^-
    d. H^+

Thinking Scientifically

1. If you watch a bird dive into water, you will see that the surface of the water is smooth and continuous as the bird enters. However, when the bird flies back into the air, many drops of water fly into the air as well (see Fig. 2.9). Explain this observation based on the properties of water.

2. Like carbon, silicon has four electrons in its outer shell, yet life evolved to be carbon-based. What is there about silicon's structure that might prevent it from sharing with four other elements and prevent it from forming the many varied shapes of carbon molecules?

Bioethical Issue

Acid Precipitation

Acid precipitation is produced when atmospheric water is polluted by sulfur dioxide and nitrous oxide emissions. These compounds are converted to sulfuric and nitric acids, are absorbed into water droplets in the atmosphere, and then fall back to Earth as acid precipitation. These emissions are mostly produced by the burning of fossil fuels, particularly coal. Because the emissions are airborne, the acid precipitation may occur far from where the fossil fuels are being burned. For example, most of the acid precipitation in Scandinavian countries comes from emissions produced in Great Britain, central Europe, and Russia. Similarly, much of the acid precipitation in southeastern Canada results from industrial activity in the midwestern and eastern United States.

Acid precipitation harms aquatic ecosystems and affects the balance of species within ecosystems on land. It also corrodes metals and some types of building material, such as limestone and marble. However, not all scientists are concerned. The National Acidic Precipitation Assessment Program (NAPAP) in 1990 concluded that, although acid precipitation has had a noticeable impact on many ecosystems as well as human activities, the damage from acid precipitation was not as widespread as had been feared. However, the program was reauthorized and continues to monitor the situation, as do other environmental groups. Although the emissions responsible for acid precipitation have been linked to respiratory problems in humans, no studies to date have shown acid precipitation to be harmful to human health.

To address these problems, many industrialized countries, including the United States and Canada, signed the Convention on Long-Range Transboundary Air Pollution (CLRTAP) between 1979 and 1983. While this and other agreements are a positive first step in addressing the problem, acid precipitation persists, and it has become a large problem in other parts of the world, particularly in India and southwest China.

Do you think acid precipitation is a significant environmental problem? Why or why not? Do you think the countries producing pollution should compensate their neighbors for the damage being done? Are countries at least obligated to minimize the effects of their air pollutants on other countries? And since many of the emissions that cause acid precipitation are due to the burning of coal for electricity, would you support the use of nuclear energy to reduce the use of coal-burning plants?

3

The Organic Molecules of Life

Not All Cholesterol Is Bad

Without cholesterol, your body would not function well. What is it about cholesterol that can cause problems in your body? Although blood tests now distinguish between "good" cholesterol and "bad" cholesterol, in reality there is only one molecule called cholesterol. Cholesterol is essential to the normal functioning of cells. It is a part of the cell's outer membrane, and it serves as a precursor to hormones, such as the sex hormones testosterone and estrogen. Without cholesterol, your cells would be in major trouble.

So when does cholesterol become "bad"? Your body packages cholesterol in high-density lipoproteins (HDLs) or low-density lipoproteins (LDLs), depending on such factors as genetics, your diet, your activity level, and whether you smoke. Cholesterol packaged as HDL is primarily on its way from the tissues to the liver for recycling. This reduces the likelihood of the formation of deposits called plaques in the arteries; thus, HDL cholesterol is considered "good." LDL is carrying cholesterol to the tissues; high levels of LDL cholesterol contribute to the development of plaque, which can result in heart disease, making LDL the "bad" form. When cholesterol levels are measured, the ratio of LDL to HDL is determined. Exercising, improving your diet, and not smoking can all lower LDL levels. In addition, a variety of prescription medications can help lower LDL levels and raise HDL levels. These lipoproteins are just one example of the organic molecules that our body needs to function correctly.

In this chapter, you will learn about the structure and function of the major classes of organic molecules, including carbohydrates, lipids, proteins, and nucleic acids.

OUTLINE

BEFORE YOU BEGIN

Before beginning this chapter, take a few moments to review the following discussions.

Section 2.1 What is a covalent bond?

Section 2.2 What is the difference between hydrophobic and hydrophilic molecules?

Figure 2.8 How are molecules represented in a diagram?

3.1 Organic Molecules

Learning Outcomes

Upon completion of this section, you should be able to

1. Distinguish between organic and inorganic molecules.
2. Recognize the importance of functional groups in determining the chemical properties of an organic molecule.

The study of chemistry can be divided into two major categories: **organic chemistry** and **inorganic chemistry**. The difference between the two is relatively simple. **Organic chemistry** is the study of organic molecules. An **organic** molecule is one that contains atoms of carbon and hydrogen. Organic molecules make up portions of cells, tissues, and organs **(Fig. 3.1)**. An inorganic molecule does not contain carbon. Water (H_2O) and table salt (NaCl) are examples of inorganic molecules. This chapter focuses on the diversity and functions of organic molecules.

The Carbon Atom

A bacterial cell can contain over 5,000 different organic molecules, and a plant or animal cell has twice that number. What is there about carbon that makes organic molecules so diverse and so complex? Carbon, with a total of six electrons, has four electrons in the outer shell. In order to acquire four electrons to complete its outer shell, a carbon atom almost always shares

Figure 3.1 Organic molecules as structural materials.

Some organic molecules are purely structural in nature. Related organic molecules (**a**) help hold a plant stem erect, (**b**) make the shell of a crab hard, and (**c**) support the wall of a bacterium.

b.

c. ×38,000 a.

electrons with CHNOPS, the elements that make up most of the weight of living things (see page 20).

Because carbon is small and needs to acquire four electrons, carbon can bond with as many as four other elements. Carbon atoms most often share electrons with other carbon atoms. The C—C bond is stable, and the result is that carbon chains can be quite long. Hydrocarbons are chains of carbon atoms that are bonded only to hydrogen atoms. Any carbon atom of a hydrocarbon molecule can start a branch chain, and a hydrocarbon can turn back on itself to form a ring compound (**Fig. 3.2**). Carbon can form double bonds with itself and other atoms. **Isomers** are molecules that have the same number and kinds of atoms but different chemical properties because the atoms occur in different arrangements. It is apparent, then, that the chemistry of carbon leads to a huge structural diversity of organic molecules. Since structure dictates function, it also means these molecules have a wide range of diverse functions.

Connections and Misconceptions

Is there a connection between organic molecules and organic produce?

As you just learned, organic molecules are those that contain carbon and hydrogen. Vegetables and fruits are all organic in that they contain these elements. The use of the term *organic* when related to farming means that there are certain production standards in place when growing the food. Normally, it means no pesticides or herbicides with harsh chemicals were used and the crop was grown as naturally as possible. So, as you can see, there are two different ways that the term *organic* may be used.

The Carbon Skeleton and Functional Groups

The carbon chain of an organic molecule is called its skeleton, or backbone. This terminology is appropriate because, just as your skeleton accounts for your shape, so does the carbon skeleton of an organic molecule account for its shape. The reactivity of an organic molecule is largely dependent on the attached functional groups (**Fig. 3.3**). A **functional group** is a specific combination of bonded atoms that always has the same chemical properties and therefore always reacts in the same way, regardless of the particular carbon skeleton to which it is attached. As in Figure 3.3, we often use an *R* to stand for the rest of the molecule to save space, because only the functional group is involved in the reaction.

The functional groups of an organic molecule therefore help determine its chemical properties. For example, sugars contain a carbon skeleton with many attached polar —OH groups. Thus, although hydrocarbons are nonpolar and **hydrophobic** (not soluble in water), glucose with several —OH groups is actually **hydrophilic** (soluble in water; see Chapter 2). Because cells are composed mainly of water, the ability to interact with and be soluble in water profoundly affects the activity of organic molecules in cells.

Organic molecules containing carboxyl groups (—COOH) are both polar (hydrophilic) and weakly acidic. They partially ionize and release hydrogen ions in solution:

$$—COOH \rightarrow —COO^- + H^+$$

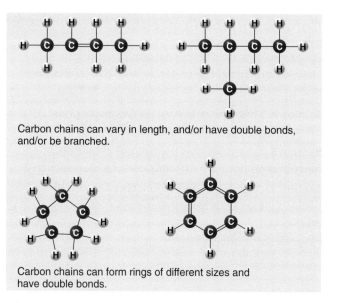

Carbon chains can vary in length, and/or have double bonds, and/or be branched.

Carbon chains can form rings of different sizes and have double bonds.

Figure 3.2 **Hydrocarbons are highly versatile.**

Hydrocarbons contain only hydrogen and carbon. Even so, they can be quite varied, according to the number of carbons, the placement of any double bonds, possible branching, and possible ring formation.

Functional Groups		
Group	**Structure**	**Found in**
Hydroxyl	$R—O—H$	Alcohols, sugars
Carboxyl	$R—C(=O)—O—H$	Amino acids, fatty acids
Amino	$R—N(H)(H)$	Amino acids, proteins
Sulfhydryl	$R—S—H$	Amino acid cysteine, proteins
Phosphate	$R—O—P(=O)(O)—O—H$	ATP nucleic acids
R = remainder of molecule		

Figure 3.3 **Functional groups.**

Molecules with the same carbon skeleton can still differ according to the type of functional group attached. Many of these functional groups are polar, helping make the molecule soluble in water. In this illustration, the remainder of the molecule, the hydrocarbon chain, is represented by an *R*.

Connecting the Concepts

For examples of how functional groups contribute to the chemical properties of a molecule, refer to the following discussions.

Section 4.3 demonstrates how the functional groups of a phospholipid help form the plasma membrane.

Section 5.2 examines how the removal of phosphate functional groups from ATP can power cellular reactions.

Section 11.1 explains how the functional groups of nucleotides are linked together to form DNA and RNA chains.

Check Your Progress 3.1

1. Identify what makes a molecule organic, and clarify the importance of the distinction.
2. Summarize the structure, function, and importance of functional groups.
3. Describe the attributes of a carbon atom that allow it to form varied molecules, and outline the importance of this characteristic.

3.2 The Biological Molecules of Cells

Learning Outcomes

Upon completion of this section, you should be able to

1. Summarize the categories of carbohydrates and provide examples of their diverse biological functions.
2. Summarize the categories of lipids and provide examples of their diverse biological functions.
3. Summarize the variety of protein types and provide examples of their diverse biological functions.
4. Summarize the two categories of nucleic acids and describe their biological functions.

Despite their great diversity, organic molecules in cells, also called **biological molecules,** are grouped into only four categories: carbohydrates, lipids, proteins, and nucleic acids. You are very familiar with these molecules because certain foods are known to be rich in carbohydrates, lipids, and proteins, as illustrated in **Figures 3.4, 3.5,** and **3.6.** When you digest these foods, they break down into

Figure 3.4 Carbohydrates.

The familiar carbohydrates, sugar and starch, are present in the foods shown. A diet loaded with these carbohydrates may make you prone to type 2 diabetes and other illnesses. In contrast, moderate amounts of multigrain foods are better for you and provide fiber to keep you regular.

Figure 3.5 Lipid foods.

A diet rich in fat can indeed make you fat. However, you do need some fat in your diet. Choose vegetable oils over the animal fat in lard, butter, and other dairy products. The fats in vegetable oils lower your risk of cardiovascular disease if they haven't been hydrogenated, as they are in processed foods.

Figure 3.6 Protein foods.

It's surprising how little protein is needed in the diet. Just 6 ounces a day will provide you with all the amino acids you need. Some forms of protein, such as beef, contain high amounts of fat. Fish, however, contains beneficial oils that lower the incidence of cardiovascular disease. Also, the diet can include an egg a day, usually with no ill effects.

subunit molecules. Your body then takes these subunits and builds from them the large macromolecules that make up your cells. Many different foods also contain nucleic acids, the type of molecule that forms your genetic material.

Certain biological molecules in cells are composed of a large number of the same type of subunits, called **monomers.** When many monomers join, the result is a **polymer.** A protein can contain hundreds of amino acid monomers, and a nucleic acid can contain hundreds of nucleotide monomers. How can polymers get so large? Just as a train increases in length when boxcars are hitched together one by one, so a polymer gets longer as monomers bond to one another.

A cell uses the same type of reaction to synthesize any type of biological molecule. It is called a **dehydration synthesis reaction,** because the equivalent of a water molecule—that is, an —OH (hydroxyl group) and an —H (hydrogen atom)—is removed as the reaction occurs (**Fig. 3.7***a*). To break down a biological molecule, a cell will use an opposite type of reaction: During a **hydrolysis** (*hydro*, water; *lysis*, break) **reaction,** an —OH group from water attaches to one monomer, and an —H from water attaches to the other monomer (Fig. 3.7*b*). In other words, water is used to break the bond holding monomers together.

Carbohydrates

In living organisms, **carbohydrates** are almost universally used as an immediate energy source. However, for many organisms, such as plants and fungi, they also have structural functions. Carbohydrates may exist either as saccharide (sugar) monomers or as polymers of saccharides. Typically, the sugar glucose is a common monomer of carbohydrate polymers. The term *carbohydrate* may refer to a single sugar molecule (monosaccharide), two bonded sugar molecules (disaccharide), or many sugar molecules bonded together (polysaccharide).

Monosaccharides: Ready Energy

Because **monosaccharides** have only a single sugar molecule, they are also known as simple sugars. A simple sugar can have a carbon backbone consisting of three to seven carbons. The word *carbohydrate* might make you think that every carbon atom is bonded to an H and an —OH. This is not strictly correct, as you can see by examining the structural formula for glucose (**Fig. 3.8**). Still, sugars do have many polar —OH groups, which make them soluble in water.

Glucose, with six carbon atoms, has a molecular formula of $C_6H_{12}O_6$. Glucose has two important isomers, called *fructose* and *galactose*, but even so, we usually think of glucose when we see the formula $C_6H_{12}O_6$. That's because glucose has a special place in the chemistry of organisms. Photosynthetic organisms, such as plants and bacteria, manufacture glucose using energy from the sun. Glucose is also the preferred immediate source of energy for nearly all types of organisms. In other words, glucose has a central role in the energy-reactions of cells.

Ribose and **deoxyribose,** with five carbon atoms, are significant because they are found in the nucleic acids RNA and DNA, respectively. RNA and DNA are discussed later in this chapter.

Disaccharides: Varied Uses

A **disaccharide** (*di*, two; *saccharide*, sugar) contains two monosaccharides bonded together. Disaccharides are the form in which sugars are usually

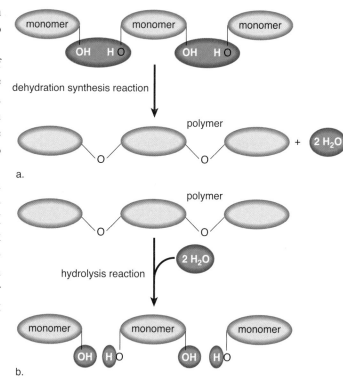

Figure 3.7 Synthesis and breakdown of polymers.

a. In cells, synthesis often occurs when monomers bond during a dehydration synthesis reaction (removal of H_2O). **b.** Breakdown occurs when the monomers in a polymer separate because of a hydrolysis reaction (the addition of H_2O).

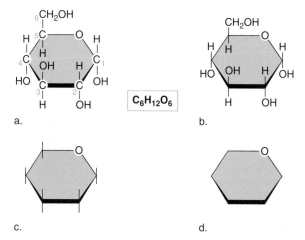

$C_6H_{12}O_6$

Figure 3.8 Glucose.

Each of these structural formulas represents glucose. **a.** The carbon skeleton (with carbon atoms numbered) and all attached groups are shown. **b.** The carbon skeleton is omitted. **c.** The carbon skeleton and attached groups are omitted. **d.** Only the ring shape of the molecule remains.

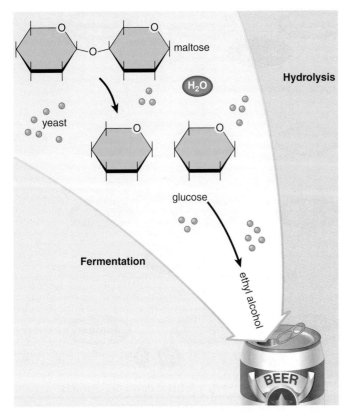

Figure 3.9 **Breakdown of maltose, a disaccharide.**

Maltose is the energy source for yeast during the production of beer. Yeasts differ as to the amount of maltose they convert to alcohol, so selection of the type of yeast is important for the correct result.

Connections and Misconceptions

What is high-fructose corn syrup?

Many beverages made commercially, including many cola drinks, contain high-fructose corn syrup (HFCS). In the 1980s, a commercial method was developed for converting the glucose in corn syrup to the much sweeter-tasting fructose. Nutritionists are not in favor of eating highly *processed*  foods that are rich in sucrose, HFCS, and white starches. They say these foods provide "empty" calories, meaning that, although they supply energy, they don't supply any of the vitamins, minerals, and fiber needed in the diet. In contrast, minimally processed foods provide glucose, starch, and many other types of nutritious molecules.

transported. For example, the brewing of beer relies on maltose, a disaccharide usually derived from barley. During the production of beer, yeast breaks down maltose to two units of glucose and then uses the glucose as an energy source in a process called *fermentation.* A waste product of this reaction is ethyl alcohol (**Fig. 3.9**).

Sucrose, a disaccharide acquired from sugar beets and sugarcane, is of special interest because we use it at the table and in baking as a sweetener. Our body digests sucrose into its two monomers, glucose and fructose. Later, the fructose is changed to glucose, our usual energy source. If the body doesn't need more energy at the moment, the glucose can be metabolized to fat. While glucose is the energy source of choice for animal cells, fat is the body's primary energy storage form. That's why eating lots of sugary desserts can make you gain weight.

Lactose is a disaccharide commonly found in milk. Lactose contains a glucose molecule combined with a galactose molecule. Individuals who are lactose intolerant are not able to break down the lactose disaccharide. The disaccharide then moves through the intestinal tract undigested, where the normal intestinal bacteria use it as an energy source. Symptoms of lactose intolerance include abdominal pain, gas, bloating, and diarrhea.

Polysaccharides as Energy Storage Molecules

Polysaccharides are polymers of monosaccharides. Some types of polysaccharides function as short-term energy storage molecules because they are much larger than a sugar and are relatively insoluble. Polysaccharides cannot easily pass through the plasma membrane and are kept (stored) within the cell.

Plants store glucose as **starch.** The cells of a potato contain granules where starch resides during winter until energy is needed for growth in the spring. Notice in **Figure 3.10***a* that starch exists in two forms—one is non-branched and the other is branched. Animals store glucose as **glycogen,** which is more highly branched (Fig. 3.10*b*). Branching subjects a polysaccharide to more attacks by hydrolytic enzymes; therefore, branching makes a polysaccharide easier to break down. Spiraling makes these polysaccharides more compact.

The storage and release of glucose from liver cells are controlled by hormones. After we eat, the pancreas releases the hormone insulin, which promotes the storage of glucose as glycogen. When cells need glucose for energy, the hormone glucagon can hydrolyze the glucose monomers from glycogen.

Polysaccharides as Molecules

Some types of polysaccharides function as structural components of cells. **Cellulose** is the most abundant of all the carbohydrates, which in turn are the most abundant of all the organic molecules on Earth. Plant cell walls contain cellulose, and therefore cellulose is plentiful in the wood of tree trunks. And the seeds of cotton plants have long fibers composed mostly of cellulose.

The bonds joining the glucose subunits in cellulose are different from those found in starch and glycogen (Fig. 3.10*c*). As a result, the molecule does not spiral or have branches. The long glucose chains are held parallel to each other by hydrogen bonding to form strong microfibrils and then fibers. The fibers crisscross within plant cell walls for even more strength.

The digestive juices of animals can't hydrolyze cellulose, but some microorganisms can digest it. Cows and other ruminants (cud-chewing animals) have an internal pouch, where microorganisms break down cellulose to glucose. In humans, cellulose has the benefit of serving as dietary fiber, which maintains regular elimination and digestive system health.

Connections and Misconceptions

What is the difference between soluble and insoluble fiber?

Fiber, also called roughage, is composed mainly of the undigested carbohydrates that pass through the digestive system. Most fiber is derived from the structural carbohydrates of plants. This includes such material as cellulose, pectins, and lignin. Fiber is not truly a nutrient, since we do not use it directly for energy or cell building, but it is an extremely important component of our diet. Fiber not only adds bulk to material in the intestines, keeping the colon functioning normally, but also binds many types of harmful chemicals in the diet, including cholesterol, and prevents them from being absorbed. There are two basic types of fiber—insoluble and soluble. Soluble fiber dissolves in water and acts in the binding of cholesterol. Soluble fiber is found in many fruits, as well as oat grains. Insoluble fiber provides bulk to the fecal material and is found in bran, nuts, seeds, and whole-wheat foods.

Chitin, which is found in the exoskeleton of crabs and related animals, such as lobsters and insects, is also a polymer of glucose. However, each glucose subunit has an amino group ($-NH_2$) attached to it. Since the functional groups attached to organic molecules determine their properties, chitin is chemically different from cellulose, even though the linkage between the glucose molecules is like that found in cellulose. Even though chitin, like cellulose, is not digestible by humans, it still has many good uses. Seeds are coated with chitin, and this protects them from attack by soil fungi. Because chitin also has antibacterial and antiviral properties, it is processed and used in medicine as a wound dressing and suture material. Chitin is even useful during the production of cosmetics and various foods.

MP3
Carbohydrates

Lipids

Although molecules classified as **lipids** are quite varied, they have one characteristic in common: They are all insoluble in water due to their long, nonpolar hydrocarbon chains and their relative lack of hydrophilic functional groups. It is said that oil and water do not mix. For example, salad dressings are rich in vegetable oils. Even after shaking, the vegetable oil will separate out from the water.

Lipids are very diverse and they have varied structures and functions. **Fats** (such as bacon fat, lard, and butter) and **oils** (such as corn oil, olive oil, and coconut oil) are some well-known lipids. In animals, fats are used for both insulation and long-term energy storage. They are used to insulate

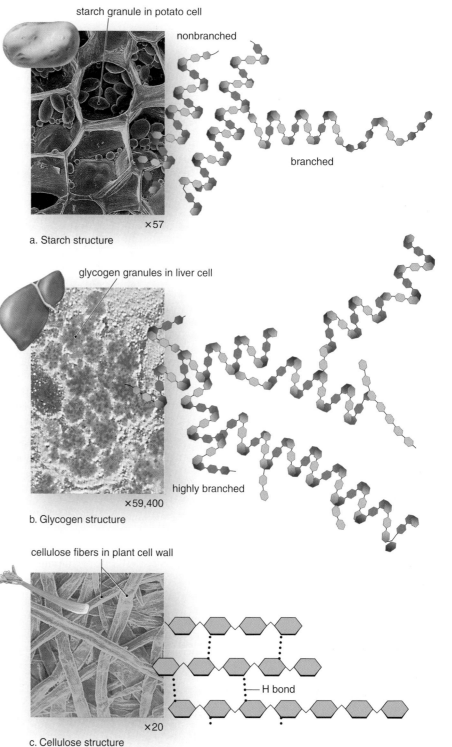

a. Starch structure

b. Glycogen structure

c. Cellulose structure

Figure 3.10   Starch and glycogen structure and function.

a. Glucose is stored in plants as starch. The electron micrograph shows the location of starch in plant cells. Starch is a chain of glucose molecules that can be nonbranched or branched. **b.** Glucose is stored in animals as glycogen. The electron micrograph shows glycogen deposits in a portion of a liver cell. Glycogen is a highly branched polymer of glucose molecules. **c.** In plant cell walls, each cellulose fiber contains several microfibrils. Each microfibril contains many polymers of glucose hydrogen-bonded together. Three such polymers are shown.

marine mammals from cold arctic waters and to protect our internal organs from damage. Instead of fats, plants use oils for long-term energy storage. In animals, the secretions of oil glands help waterproof skin, hair, and feathers (**Fig. 3.11**).

Fats and Oils: Long-Term Energy Storage

Fats and oils contain two types of subunit molecules: glycerol and fatty acids (**Fig. 3.12**). **Glycerol** contains three —OH groups. The —OH groups are polar; therefore, glycerol is soluble in water. A **fatty acid** has a long chain of carbon atoms bonded only to hydrogen, with a carboxyl group at one end. A fat or an oil forms when the carboxyl portions of three fatty acids react with the —OH groups of glycerol. This is a dehydration synthesis reaction because, in addition to a fat molecule, three molecules of water result. Fats and oils are degraded during a hydrolysis reaction, in which water is added to the molecule (Fig. 3.12). Because three long fatty acids are attached to the glycerol molecule, fats and oils are called **triglycerides.** This structure can pack a lot of energy into one molecule. Thus, it is logical that fats and oils are the body's primary long-term energy storage molecules.

Fatty acids are the primary components of fats and oils. Most of the fatty acids in cells contain 16 or 18 carbon atoms per molecule, although smaller or larger ones are also found. Fatty acids are either saturated or unsaturated (**Fig. 3.13**). **Unsaturated fatty acids** have double bonds in the carbon chain wherever the number of hydrogens is less than two per carbon atom (Fig. 3.13*a*). **Saturated fatty acids** have no double bonds between the carbon atoms (Fig. 3.13*b*). The carbon chain is saturated, so to speak, with all the hydrogens it can hold. Saturation or unsaturation of a fatty acid determines its chemical and physical properties.

In general, oils are liquids at room temperature because they contain unsaturated fatty acids. Notice in Figure 3.13*a* that the double bond creates a bend in the fatty acid chain. Such kinks prevent close packing between the

Figure 3.11

Preening in birds.

As a bird preens, it transfers oil to its feathers from a gland at the base of its tail. Because the oil repels water, the feathers become waterproof.

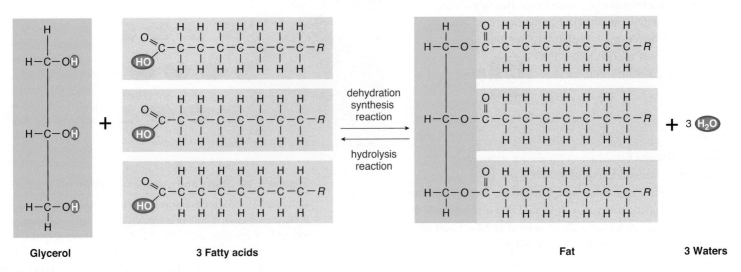

Glycerol   3 Fatty acids   Fat   3 Waters

Figure 3.12 Synthesis and breakdown of fat.

Following a dehydration synthesis reaction, glycerol is bonded to three fatty acid molecules, and water is given off. Following a hydrolysis reaction, the bonds are broken due to the addition of water. *R* represents the remainder of the molecule, which in this case is a continuation of the hydrocarbon chain, composed of 16 or 18 carbons.

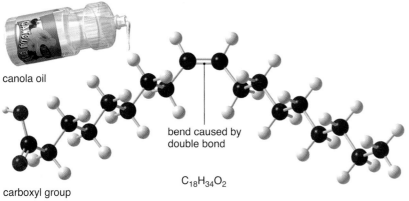

canola oil

bend caused by
double bond

$C_{18}H_{34}O_2$

carboxyl group

a. Oleic acid, a monounsaturated fatty acid (one double bond) found in canola oil.

butter

carboxyl group $C_{18}H_{36}O_2$

b. Stearic acid, a saturated fatty acid (no double bonds) found in butter.

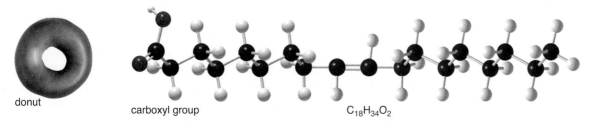

donut   carboxyl group $C_{18}H_{34}O_2$

c. Elaidic acid, a trans fatty acid (one double bond) found in many snack foods.

Figure 3.13 Fatty acids.

A fatty acid has a carboxyl group attached to a long hydrocarbon chain. **a.** If there is one double bond between adjacent carbon atoms in the chain, the fatty acid is monounsaturated. **b.** If there are no double bonds, the fatty acid is saturated. A diet high in saturated fats appears to contribute to diseases of the heart and blood vessels. **c.** Hydrogenation of oils is not always complete. Further, at some remaining double bonds, the attached hydrogens are in the trans configuration. If so, the fatty acid is a trans fatty acid. Trans fats are also linked to cardiovascular disease.

hydrocarbon chains and account for the fluidity of oils. On the other hand, butter, which contains mostly saturated fatty acids, is a solid at room temperature. The saturated fatty acid chains can pack together more tightly because they have no kinks (Fig. 3.13b).

Trans fats contain fatty acids that have been partially hydrogenated (combined with hydrogen) to make them more saturated, and thus more solid (Fig. 3.13c). Trans fats are found in many processed foods—margarine, baked goods, and fried foods in particular. Saturated fats and trans fats have been linked to a cardiovascular disease called atherosclerosis, in which lipid material, called plaque, accumulates inside blood vessels. Plaque contributes to high blood pressure and heart attacks. Unsaturated oils, particularly *monounsaturated* (one double bond) oils but also *polyunsaturated* (many double bonds) oils, have been found to be protective against atherosclerosis. These healthy oils are found in abundance in olive oil, canola oil, and certain fish.

Phospholipids: Membrane Components

Phospholipids, as implied by their name, contain a phosphate functional group. Essentially, a phospholipid is constructed like a triglyceride, except that, in place of the third fatty acid attached to glycerol, there is a charged phosphate group. The phosphate group is usually bonded to another polar functional group, indicated

Connections and Misconceptions

What are omega-3 fatty acids?

Not all fats are bad. In fact, some of them are essential to our health. A special class of unsaturated fatty acids, the omega-3 fatty acids, are considered both an essential and a developmentally important nutrient. The name omega-3 (also called n-3 fatty acids) is derived from the location of the double bond in the carbon chain. The three important omega-3 fatty acids are linolenic acid (ALA), docosahexaenoic acid (DHA), and eicosapentaenoic acid (EPA). Omega-3 fatty acids are a major component of the fatty acids in the brain, and adequate amounts of them appear to be important in children and young adults. A diet that is rich in these fatty acids also offers protection against cardiovascular disease, and research is ongoing with regard to other health benefits. DHA may reduce the risk of Alzheimer disease. DHA and EPA may be manufactured from APA in small amounts within our body. Some of the best sources of omega-3 fatty acids are cold-water fish, such as salmon and sardines. Flax oil, also called linseed oil, is an excellent plant-based source of omega-3 fatty acids.

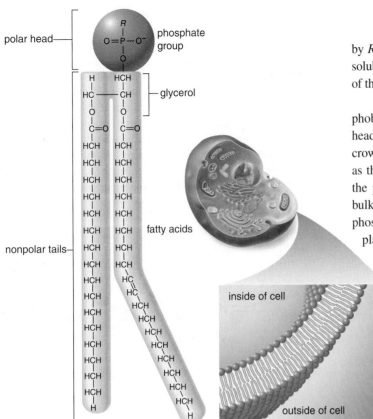

polar head

R

O=P—O⁻ | phosphate group
O

HCH | glycerol
HC — CH
O O
C=O C=O
HCH HCH
HCH HCH
HCH HCH
HCH HCH
HCH HCH
HCH HCH — fatty acids
HCH HCH
HCH HC
HCH HC
HCH HCH
HCH HCH
HCH HCH
HCH HCH
HCH HCH
HCH HCH
HCH HCH
H H

nonpolar tails

a. Phospholipid structure

inside of cell

outside of cell

b. Plasma membrane of a cell

Figure 3.14 **Phospholipids from membranes.**

a. Phospholipids are constructed like fats, except that, in place of the third fatty acid, they have a charged phosphate group. The hydrophilic (polar) head group is soluble in water, whereas the two hydrophobic (nonpolar) tail groups are not. **b.** This causes the molecules to arrange themselves as a bilayer in the plasma membrane that surrounds a cell.

by *R* in **Figure 3.14***a*. Thus, one end of the molecule is hydrophilic and water soluble. This portion of the molecule is the polar head. The hydrocarbon chains of the fatty acids, or the nonpolar tails, are hydrophobic and not water soluble.

Because phospholipids have both hydrophilic (polar) heads and hydrophobic (nonpolar) tails, they tend to arrange themselves so that only the polar heads interact with the watery environment outside and the nonpolar tails crowd inward away from the water. Between two compartments of water, such as the outside and inside of a cell, phospholipids become a bilayer in which the polar heads project outward and the nonpolar tails project inward. The bulk of the plasma membrane that surrounds cells consists of a fairly fluid phospholipid bilayer, as do all the other membranes in the cell (Fig. 3.14*b*). A plasma membrane is essential to the structure and function of a cell, and thus phospholipids are vital to humans and other organisms.

Steroids: Four Fused Rings

Steroids are lipids that possess a unique carbon skeleton made of four fused rings, shown in red in **Figure 3.15.** Unlike other lipids, steroids do not contain fatty acids, but they are similar to other lipids because they are insoluble in water. Steroids are also very diverse. The types of steroids differ primarily in the types of functional groups attached to their carbon skeleton.

Cholesterol (Fig. 3.15*a*) is a component of an animal cell's plasma membrane, and it is the precursor of other steroids, such as the sex hormones testosterone and estrogen. The male sex hormone, testosterone, is formed primarily in the testes, and the female sex hormone, estrogen, is formed primarily in the ovaries (Fig. 3.15*b,c*). Testosterone and estrogen differ only by the functional groups attached to the same carbon skeleton, yet they have a profound effect on the body and the sexuality of humans and other animals.

Anabolic steroids, such as synthetic testosterone, can be used to increase muscle mass. The result is usually unfortunate, however. The presence of the steroid in the body upsets the normal hormonal balance: The testes atrophy (shrink and weaken), and males may develop breasts; females tend to grow facial hair and lose hair on their head. Because steroid use gives athletes an unfair advantage and destroys their health—heart, kidney, liver, and psychological disorders are common—they are banned by professional athletic associations.

MP3 Lipids

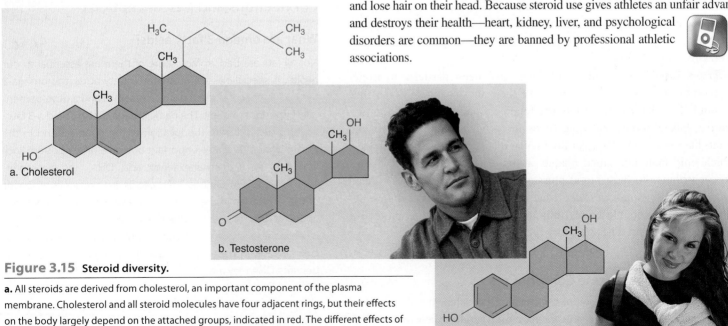

a. Cholesterol

b. Testosterone

c. Estrogen

Figure 3.15 **Steroid diversity.**

a. All steroids are derived from cholesterol, an important component of the plasma membrane. Cholesterol and all steroid molecules have four adjacent rings, but their effects on the body largely depend on the attached groups, indicated in red. The different effects of **(b)** testosterone and **(c)** estrogen on the body are due to different groups attached to the same carbon skeleton.

Proteins

Proteins are of primary importance in the structure and function of cells. Here are some of their many functions:

Support Some proteins are structural proteins. Examples include the protein in spiderwebs; keratin, the protein that contributes to hair and fingernails; and collagen, the protein that lends support to skin, ligaments, and tendons (**Fig. 3.16***a*).

Metabolism Many proteins are **enzymes.** They bring reactants together and thereby act as catalysts, speeding up chemical reactions in cells. Enzymes are specific for particular types of reactions and can function at body temperature.

Transport Channel and carrier proteins in the plasma membrane allow substances to enter and exit cells. Other proteins transport molecules in the blood of animals—for example, **hemoglobin,** found in red blood cells, is a complex protein that transports oxygen (Fig. 3.16*b*).

Defense Some proteins, called antibodies, combine with disease-causing agents to prevent those agents from destroying cells and causing diseases and disorders.

Regulation Hormones are regulatory proteins. They serve as intercellular messengers that influence the metabolism of cells. For example, the hormone insulin regulates the concentration of glucose in the blood, while human growth hormone (hGH) contributes to determining the height of an individual.

Motion The contractile proteins actin and myosin allow parts of cells to move and cause muscles to contract (Fig. 3.16*c*). Muscle contraction enables animals to move from place to place and substances to move through the body. It also regulates body temperature.

The structures and functions of cells differ according to the type of protein they contain. Muscle cells contain actin and myosin; red blood cells contain hemoglobin; support cells produce the collagen they secrete.

a. Structural proteins

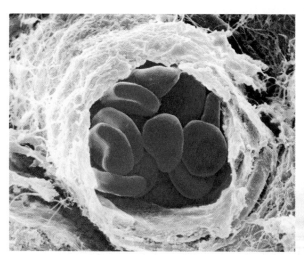

b. Transport proteins

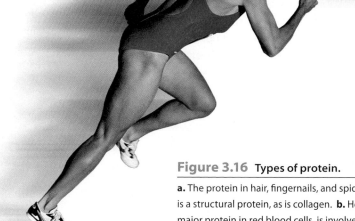

c. Contractile proteins

Figure 3.16 Types of protein.

a. The protein in hair, fingernails, and spiderwebs (keratin) is a structural protein, as is collagen. **b.** Hemoglobin, a major protein in red blood cells, is involved in transporting oxygen. **c.** Contractile proteins, actin and myosin, cause muscles to move.

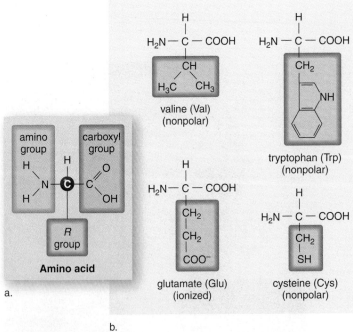

Figure 3.17 Amino acids.

a. Structure of an amino acid. **b.** Proteins contain arrangements of 20 different kinds of amino acids, four of which are shown here. Amino acids differ by the particular R group (blue) attached to the central carbon. Some R groups are nonpolar and hydrophobic, some are polar and hydrophilic, and some are ionized and hydrophilic.

Connections and Misconceptions

How do perms and relaxers work on hair?

Hair, composed of proteins, can be altered chemically to be curly or straight. For instance, a perm contains chemicals that break and reform the bonds between the functional groups in the amino acids to form disulfide bonds, resulting in spirals and making curls. A relaxer will break the bonds between disulfide bond regions to straighten the proteins. Neither are permanent, because they do not alter the genes controlling the original shape of the proteins that make the hair.

Amino Acids: Subunits of Proteins

Proteins are polymers, and their monomers are called **amino acids.** Amino acids have a unique carbon skeleton, in which a central carbon atom bonds to a hydrogen atom, two functional groups, and a side chain, or R group (**Fig. 3.17**a). The name *amino acid* is appropriate because one of the two functional groups is an —NH_2 (amino group) and another is a —COOH (an acid group, as discussed on page 37).

There are 20 different amino acids, which vary according to their particular R group. The R groups range in complexity from a single hydrogen atom to a complicated ring compound. The unique chemical properties of an amino acid depend on the chemical properties of the R group. For example, some R groups are polar, some are charged, and some are hydrophobic. The amino acid cysteine, for instance, has an R group that ends with a sulfhydryl (—SH) group, which can connect one chain of amino acids to another by a disulfide bond, —S—S—. Four amino acids commonly found in cells are shown in Figure 3.17b.

Peptides

Figure 3.18 shows how two amino acids join by a dehydration synthesis reaction between the carboxyl group of one and the amino group of another. The resulting covalent bond between two amino acids is called a **peptide bond.** The atoms associated with the peptide bond share the electrons unevenly because oxygen is more electronegative than nitrogen.

Therefore, the peptide bond is polar, and hydrogen bonding is possible between the C=O of one amino acid and the N—H of another amino acid in a polypeptide.

A **peptide** is two or more amino acids covalently bonded together, and a **polypeptide** is a chain of many amino acids joined by peptide bonds. Proteins are polypeptide chains that have folded into complex shapes (see the following section). A protein may contain one or more polypeptide chains. While some proteins are small, others are composed of a very large number of amino acids. Ribonuclease, a small protein that breaks down RNA, contains barely more than 100 amino acids. But some proteins are very large, as is titin, which contains over 33,000 amino acids! Titin is an integral part of the structure of your muscles; without it, your muscles would not function properly.

Figure 3.18 Synthesis and degradation of peptide.

Following a dehydration synthesis reaction, a peptide bond joins two amino acids, and water is given off. Following a hydrolysis reaction, the bond is broken due to the addition of water.

The amino acid sequence determines a protein's final three-dimensional shape, and thus its function. Each polypeptide has its own normal sequence. Proteins that have an abnormal sequence of amino acids often have the wrong shape and cannot function properly.

Animation
Protein
Denaturation

Shape of Proteins

All proteins have multiple levels of structure. These levels are called the primary, secondary, tertiary, and quaternary structures (**Fig. 3.19**). A protein's sequence of amino acids is called its *primary structure*. Consider that an almost infinite number of words can be constructed by changing the number and sequence of the 26 letters in our alphabet. In the same way, many different proteins can result by varying the number and sequence of just 20 amino acids.

The *secondary structure* of a protein occurs when portions of the amino acid chain take on a certain orientation in space, depending on the number and identity of the amino acids present in the chain. Portions of the polypeptide can have the spiral shape called an alpha helix, or it can turn back on itself, like an accordion, a shape called a beta pleated sheet. Hydrogen bonds between nearby peptide bonds maintain the secondary structure of a protein. Any polypeptide may contain one or more secondary structure regions within the same chain.

Proteins also have a *tertiary structure*. The tertiary structure of a protein is its overall three-dimensional shape that results from the folding and twisting of its secondary structure. The tertiary structure is held in place by interactions

Figure 3.19 Levels of protein organization.

All proteins have a primary structure. Fibrous proteins have a secondary structure; they are either helices (e.g., keratin, collagen) or pleated sheets (e.g., silk). Globular proteins always have a tertiary structure, and most have a quaternary structure (e.g., hemoglobin, enzymes).

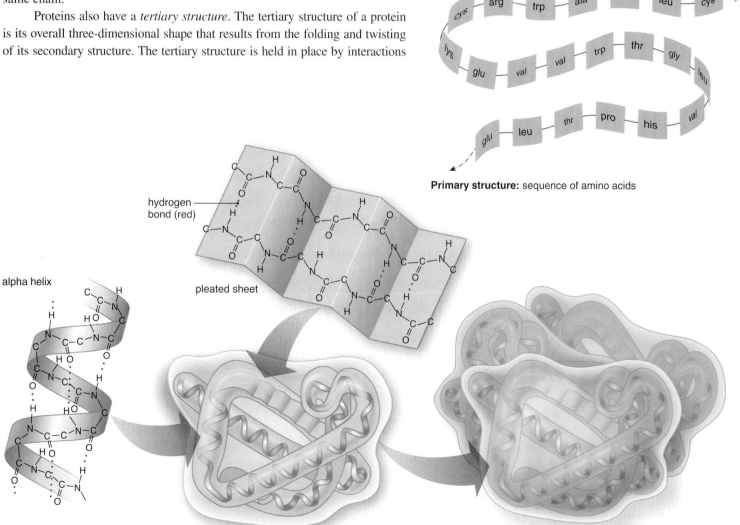

Primary structure: sequence of amino acids

Secondary structure: alpha helix and pleated sheet

Tertiary structure: globular shape

Quaternary structure: more than one polypeptide

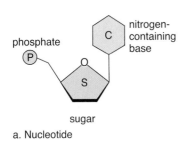

phosphate

nitrogen-
containing
base

sugar

a. Nucleotide

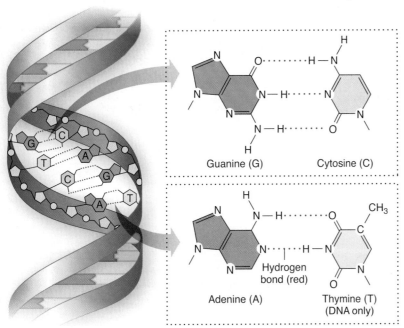

Guanine (G)   Cytosine (C)

Hydrogen
bond (red)

Adenine (A)   Thymine (T)
(DNA only)

b. DNA structure with base pairs: G with C and A with T

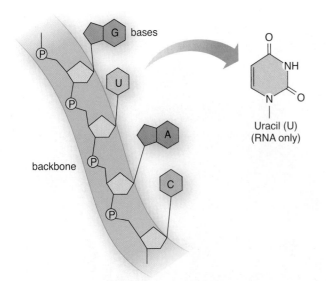

bases

Uracil (U)
(RNA only)

backbone

c. RNA structure with bases G, U, A, C

Figure 3.20 DNA and RNA structure.

a. Structure of a nucleotide. **b.** Structure of DNA and its bases.
c. Structure of RNA in which uracil replaces thymine.

between the *R* groups of amino acids making up the helices and beta pleated sheets within the polypeptide. Several types of interactions are possible. For example, ionic bonds, hydrogen bonds, and even covalent bonds can occur between *R* groups. Also, a tight packing of side chains can occur when hydrophobic *R* groups come together to avoid contact with water.

The tertiary structure of a protein determines its function, and this structure can be affected by the environmental conditions, such as pH and temperature. Acids and bases may interrupt interactions between *R* groups and affect a protein's structure. Likewise, a rise in temperature can disrupt the shape of an enzyme. For example, frying an egg disrupts the tertiary structure of the proteins in it, causing the protein albumin to become solid and change color. The protein has been **denatured** (broken down and inactivated) and has lost its function.

Some proteins, such as hemoglobin and insulin, have a *quaternary structure* because they contain more than one polypeptide chain. Each polypeptide chain in the protein has its own primary, secondary, and tertiary structures. The quaternary structure is determined by how the individual polypeptide chains interact. The quaternary structure can also affect a protein's function. If hemoglobin's quaternary structure is disrupted, it can no longer carry oxygen in the blood.

The overall shapes of many proteins may be classified as fibrous or globular. **Fibrous proteins** adopt a rod-like structure. *Keratin,* the fibrous protein in hair, fingernails, horns, reptilian scales, and feathers, has many helical regions. *Collagen,* the protein that gives shape to the skin, tendons, ligaments, cartilage, and bones of animals, also contains many helical secondary structures and adopts a rod-like shape. Thus, most fibrous proteins have structural roles. **Globular proteins** have a rounded or irregular three-dimensional tertiary shape. Enzymes are globular proteins, as is hemoglobin.

MP3
Proteins

Nucleic Acids

DNA (*deoxyribonucleic acid*) and **RNA** (*ribonucleic acid*) are the **nucleic acids** in cells. Early investigators called them nucleic acids because they were first detected in the nucleus. As already stated, DNA stores genetic information. Each DNA molecule contains many genes, and genes specify the sequence of the amino acids in proteins. RNA is the molecule that aids in transcribing and translating DNA into a protein.

Nucleic acids are polymers in which the monomer is called a **nucleotide.** All nucleotides, whether they are in DNA or RNA, have three parts: a phosphate, a 5-carbon sugar, and a nitrogen-containing base (**Fig. 3.20***a*). The phosphate is simply —PO_4. The sugar is deoxyribose in DNA and ribose in RNA, which accounts for their names. (Deoxyribose has one less oxygen than does ribose.) Each nucleotide of DNA contains one of four bases: adenine (A), guanine (G), cytosine (C), or thymine (T).

As you can see in Figure 3.20*b*, DNA is a double helix, meaning that two strands spiral around one another. In each strand, the backbone of the molecule is composed of phosphates bonded to sugars, and the bases project to the inside. Interestingly, the base guanine (G) is always paired with cytosine (C), and the base adenine (A) is always paired with thymine (T). This is called *complementary base pairing.* Complementary base pairing holds the two strands together and is very important when DNA makes a copy of itself, a process called replication.

RNA differs from DNA not only by its sugar but also because it uses the base uracil (U) instead of thymine. Whereas DNA is double-stranded, RNA is single-stranded (Fig. 3.20c). Complementary base pairing also allows DNA to pass genetic information to RNA. The information is stored in the sequence of bases. DNA has a triplet code called codons, and every three bases stands for one of the 20 amino acids in cells. Once you know the sequence of bases in a gene, you know the sequence of amino acids in a polypeptide.

Many scientists around the world were involved in the Human Genome Project. Not long ago, members of this project determined the sequence of all the base pairs in all of our 20,500 genes. They hope this information will lead to cures for many of the human illnesses that are caused by proteins that don't function as they should.

ATP: An Energy Molecule

In addition to being one of the subunits of nucleic acids, a derivative of the adenine nucleotide has a metabolic function that is very important to most cells. When adenosine (adenine plus ribose) is modified by the addition of three phosphate groups, it becomes adenosine triphosphate (ATP), which acts as an energy carrier in cells. ATP will be discussed in more detail in Chapter 5.

MP3
Nucleic Acids

Relationship Between Proteins and Nucleic Acids

We have learned that the functional group of an amino acid determines its behavior, and that the order of the amino acids within a polypeptide determines its shape. The shape of a protein determines its function. The structure and function of cells are determined by the types of proteins they contain. The same is true for organisms—that is, they differ with regard to their proteins. Next, we learned that DNA, the genetic material, bears instructions for the sequence of amino acids in polypeptides. The proteins of organisms differ because their genes differ.

Sometimes very small changes in a gene cause large changes in the protein encoded by the gene, resulting in illness. For example, in sickle cell disease, the affected individual's red blood cells are sickle-shaped because, in one location on the protein, the amino acid valine (Val) appears where the amino acid glutamate (Glu) should be (**Fig. 3.21**). This substitution makes red blood cells

Figure 3.21 **Sickle cell disease.**

One of the amino acid chains in hemoglobin is 146 amino acids long. Sickle cell disease, characterized by sickled red blood cells, results when valine occurs instead of glutamate at the sixth amino acid position.

H_2N— Val — His — Leu — Thr — Pro — Glu — Glu —
Normal hemoglobin

H_2N— Val — His — Leu — Thr — Pro — Val — Glu —
Sickle cell hemoglobin

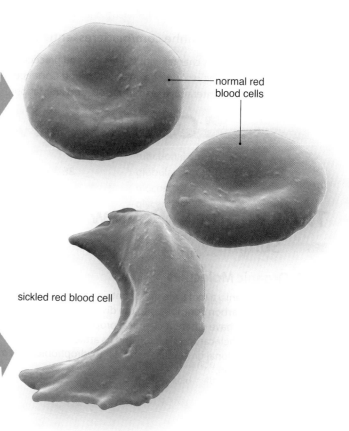

normal red blood cells

sickled red blood cell

4

Inside the Cell

OUTLINE

BEFORE YOU BEGIN

Before beginning this chapter, take a few moments to review the following discussions.

Section 1.1 What are the basic characteristics of life?

Section 2.2 What properties of water enable it to support life?

Section 3.2 What are the basic roles of carbohydrates, fats, proteins, and nucleic acids in living organisms?

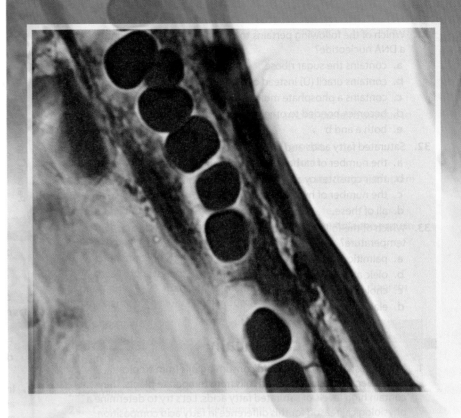

The Diversity and Unity of Cells

Red blood cells are the most abundant cell type in your body, and they perform a critical job—delivering oxygen throughout your body and picking up carbon dioxide waste from its cells. Along with other types of blood cells, new red blood cells are continually produced from precursor cells, known as stem cells, located in the marrow of your bones. Red blood cells do not resemble a typical animal cell in structure because they are chock full of hemoglobin, the pigment that makes them red and enables them to carry oxygen. Human red blood cells also lack a nucleus and DNA. Because of this, they live only about 120 days before being destroyed by the liver and the spleen. Thus, the body continually faces the daunting task of replacing the 3 million red blood cells it loses every day. If this process of replacing the red blood cells is impaired, as occurs in a condition called aplastic anemia, a person can quickly die.

Cells are the fundamental units of all living things. Although the diversity of organisms is incredible, the cells of all organisms share many similarities. In fact, there are only a few structural differences between most types of cells. In this chapter, you will learn about the two main evolutionary classes of cells and how they are similar as well as different. Learning about the structure of cells will later help you understand their complex functions and interactions.

4.2 The Two Main Types of Cells

Learning Outcomes

Upon completion of this section, you should be able to

1. Distinguish between prokaryotic and eukaryotic cells.
2. Identify the structures of a prokaryotic cell.

All cells have an outer membrane called a plasma membrane. The plasma membrane encloses a semifluid substance called the **cytoplasm** and the cell's genetic material. The plasma membrane regulates what enters and exits cells; the contents of the cytoplasm carry on chemical reactions; and the genetic material provides the information needed for growth and reproduction. When a cell reproduces, it produces more cells. The concepts that all organisms are composed of cells and that cells come only from preexisting cells are the two central tenets of the **cell theory.**

Cells are divided into two main types, according to the way their genetic material is organized. **Prokaryotic** (Greek; *pro*, before, and *karyon*, kernel or nucleus) **cells** are so named because they lack a membrane-bounded nucleus. Their DNA is located in a region of the cytoplasm called the nucleoid. The other type of cell, **eukaryotic cell** (Greek; *eu*, true), has a nucleus that houses its DNA. Eukaryotic cell structure is discussed in Section 4.4.

Prokaryotic Cells

Organisms from the domains Archaea and Bacteria are prokaryotic cells (see Figs. 1.5 and 1.6) whose evolutionary history dates back to the first cells on Earth (around 3.5 billion years ago). Prokaryotic cells are generally much smaller in size and simpler in structure than eukaryotic cells (**Fig. 4.3**). Their small size and simple structure allow them to reproduce very quickly and effectively; they exist in great numbers in the air, in bodies of water, in the soil, and even on you. Prokaryotes are an extremely successful group of organisms.

Bacteria are well known because they cause some serious diseases, including tuberculosis, throat infections, and gonorrhea. Even so, the biosphere would not long continue without bacteria. Many bacteria decompose dead remains and contribute to ecological cycles. Bacteria also assist humans in another way—we use them to manufacture all sorts of products, from industrial chemicals to food-stuffs and drugs. The active cultures found in a container of yogurt are bacteria that are beneficial to us. There are also billions of bacteria teeming within your intestines! We are dependent on them for many things, including the synthesis of some vitamins we can't make ourselves.

Our knowledge about how DNA specifies the sequence of amino acids in proteins was much advanced by doing experiments utilizing *E. coli*, a bacterium that lives in the human large intestine. The fact that this information applies to all prokaryotes and eukaryotes, even ourselves, reveals the remarkable unity of living things and gives evidence that all organisms share a common descent.

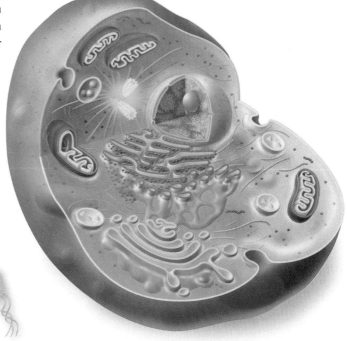

cell

flagella

Prokaryotic cell: simple internal structure

Eukaryotic cell: complex internal structure

Figure 4.3 Comparison of prokaryotic cells and eukaryotic cells.

Prokaryotic cells are much smaller, and less complex, than eukaryotic cells.

Connections and Misconceptions

Doesn't *E. coli* cause intestinal problems?

The majority of *E. coli* in our body not only do not cause disease but actually benefit us by providing small amounts of vitamin K and some B vitamins. These *E. coli* even defend your intestinal tract against infection by other bacteria. However, one strain of *E. coli* (called O157:H7) produces a toxin (called Shiga toxin) that causes the intestines to become inflamed and secrete fluids, causing diarrhea. It typically is obtained from eating undercooked meat products. Still, even friendly strains of *E. coli* can cause problems in humans if they enter the urinary or genital tract.

Video
E. coli Wars

Bacterial Structure

In bacteria, the cytoplasm is surrounded by a plasma membrane. Most bacteria possess a cell wall, and sometimes also a capsule (**Fig. 4.4**). The cytoplasm contains a variety of enzymes. Enzymes are organic catalysts that speed up the many types of chemical reactions that are required to maintain an organism. The plasma membranes of prokaryotes and eukaryotes have a similar structure (see page 60). The plasma membrane regulates the movement of materials into and out of the cell. The **cell wall** maintains the shape of the cell, even if the cytoplasm should happen to take up an abundance of water. The **capsule** is a protective layer of polysaccharides lying outside the cell wall.

The DNA of a bacterium is located in a single circular, coiled chromosome that resides in a region called the **nucleoid.** The many proteins specified by bacterial DNA are synthesized on tiny structures called **ribosomes.** A bacterial cell contains thousands of ribosomes.

The appendages of a bacterium—namely, the flagella, fimbriae, and conjugation pili—are made of protein. Some bacteria have **flagella** (sing., flagellum), which are tail-like appendages that allow bacteria to propel them-

Figure 4.4 A prokaryotic cell.

The structures in this diagram are characteristic of most prokaryotic cells.

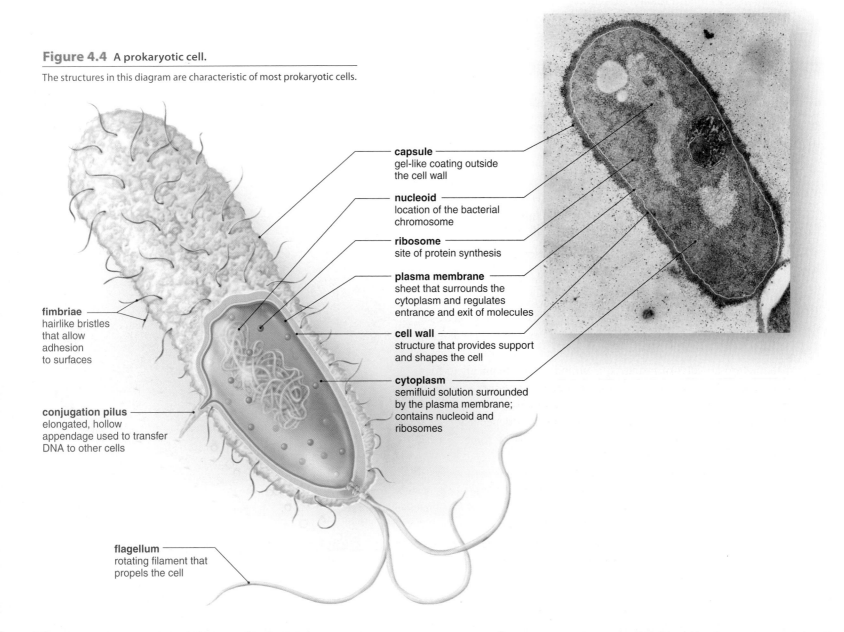

fimbriae
hairlike bristles
that allow
adhesion
to surfaces

conjugation pilus
elongated, hollow
appendage used to transfer
DNA to other cells

flagellum
rotating filament that
propels the cell

capsule
gel-like coating outside
the cell wall

nucleoid
location of the bacterial
chromosome

ribosome
site of protein synthesis

plasma membrane
sheet that surrounds the
cytoplasm and regulates
entrance and exit of molecules

cell wall
structure that provides support
and shapes the cell

cytoplasm
semifluid solution surrounded
by the plasma membrane;
contains nucleoid and
ribosomes

selves. The bacterial flagellum has a filament, a hook, and a basal body, which is a series of rings anchored in the cell wall and plasma membrane. The hook rotates 360° within the basal body, and this rotary motion propels bacteria—the bacterial flagellum does not move back and forth like a whip. Sometimes flagella occur only at the ends of a cell, and other times they are dispersed randomly over the surface.

Fimbriae are small, bristle-like fibers that sprout from the cell surface. They don't have anything to do with motility, but they do help bacteria attach to a surface, such as a tabletop or even your skin! **Conjugation pili** are rigid, tubular structures that bacteria use to pass DNA from cell to cell. Bacteria can also take up DNA directly from the external environment or by way of viruses. All three of these methods allow bacteria to pass information, including genes for drug resistance, from cell to cell.

Connecting the Concepts

For more information on prokaryotic organisms and the bacteria, refer to the following discussions.

Section 11.3 describes how bacteria may be genetically altered to benefit humans.

Section 17.3 explores the evolutionary history of the prokaryotes and their relationships to the eukaryotes.

Section 26.2 examines how bacteria work both with, and against, our immune system.

Check Your Progress 4.2

1. Identify the major distinctions between a prokaryotic cell and a eukaryotic cell.
2. Outline how bacteria can be both harmful and beneficial to humans.
3. Distinguish among the three types of bacterial appendages and describe what they have in common.

4.3 The Plasma Membrane

Learning Outcomes

Upon completion of this section, you should be able to

1. Recognize the key components of the cell plasma membrane.
2. Explain how the plasma membrane regulates the passage of molecules into and out of the cell.
3. Describe the diverse functions of the proteins embedded in the plasma membrane.

The **plasma membrane** marks the boundary between the outside and inside of a cell. Its integrity and function are vital to a cell because it acts much as a gatekeeper, regulating the passage of molecules and ions into and out of the cell.

In both prokaryotes and eukaryotes, the plasma membrane is a phospholipid bilayer (see Fig. 3.14*b*) with embedded proteins. In this bilayer, the polar (hydrophilic) heads of the phospholipids are orientated in two directions

(**Fig. 4.5**). In the outer layer of the membrane, the phospholipid heads face toward the external environment. On the interior layer of the membrane, the heads of phospholipid molecules are directed toward the interior cytoplasm of the cell. The nonpolar (hydrophobic) tails of the phospholipids face inward toward each other, where there is no water. Cholesterol molecules, present in the plasma membrane of eukaryotic cells, lend support to the membrane, giving it the general consistency of olive oil.

The fluid-mosaic model states that the protein molecules embedded in the membrane have a pattern (a mosaic) within the phospholipid bilayer (Fig. 4.5). The pattern varies according to the type of cell, but it may also vary within the same membrane at different times. When you consider that the plasma membrane of a red blood cell contains over 50 different types of proteins, you can see why the membrane is said to be a mosaic. Short chains of sugars are attached to the outer surface of some of these proteins, forming glycoproteins. The sugar chain helps a protein perform its particular function. Many glycoproteins are involved in establishing the identity of the cell. They often play an important role in the immune response against disease-causing agents entering the body.

MP3
Membrane
Structure

Functions of Membrane Proteins

The proteins that occur within a membrane have varied, specific functions.

Channel Proteins

Channel proteins form a tunnel across the entire membrane, allowing only one or a few types of specific molecules to simply move across the membrane (**Fig. 4.6**a). For example, aquaporins are channel proteins that allow water to enter or exit a cell. Without aquaporins in the kidneys, your body would soon dehydrate.

Figure 4.5   **A model of the plasma membrane.**

The plasma membrane is composed of a phospholipid bilayer. The polar heads of the phospholipids are at the surfaces of the membrane; the nonpolar tails make up the interior of the membrane. Proteins embedded in the membrane have various functions (see Fig. 4.6).

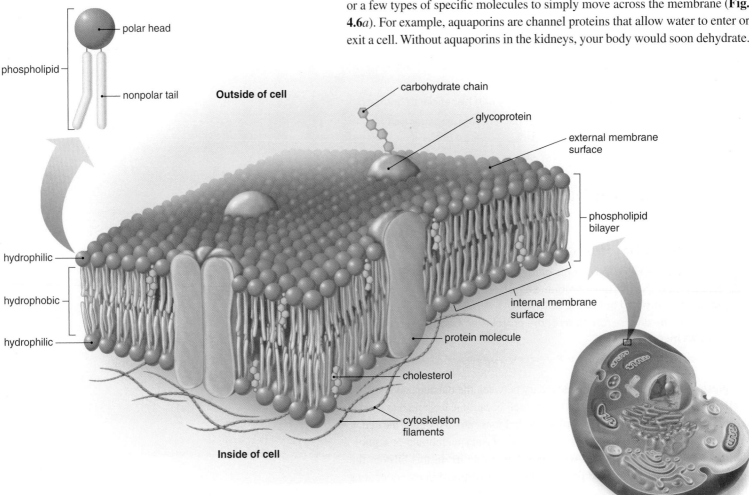

Transport Proteins

Transport proteins are also involved in the passage of molecules and ions through the membrane. They often combine with a substance and help it move across the membrane, with an input of energy (Fig. 4.6*b*). For example, a transport protein conveys sodium and potassium ions across a nerve cell membrane. Without this transport protein, nerve conduction would be impossible.

Cell Recognition Proteins

Cell recognition proteins are glycoproteins (Fig. 4.6*c*). Among other functions, these proteins enable our body to distinguish between our own cells and the cells of other organisms. Without this distinction, pathogens would be able to freely invade the body.

Receptor Proteins

A receptor protein has a shape that allows a specific molecule, called a signal molecule, to bind to it (Fig. 4.6*d*). The binding of a signal molecule causes the receptor protein to change its shape and thereby bring about a cellular response. For example, the hormone insulin binds to a receptor protein in liver cells, and thereafter these cells store glucose.

Enzymatic Proteins

Some plasma membrane proteins are enzymatic proteins that directly participate in metabolic reactions (Fig. 4.6*e*). Without enzymes, some of which are attached to the various membranes of the cell, a cell would never be able to perform the degradative and synthetic reactions that are important to its function.

Junction Proteins

As discussed on page 73, proteins are also involved in forming various types of junctions between cells (Fig. 4.6*f*). The junctions assist cell-to-cell adhesion and communication. The adhesion junctions in your bladder keep the cells bound together as the bladder swells with urine.

Connecting the Concepts

For more information on the role of the plasma membrane proteins, refer to the following discussions.

Section 5.4 describes how aquaporins and other membrane proteins move molecules across the membrane.

Section 27.1 illustrates how channel proteins are involved in the function of the nervous system.

Section 28.1 examines how special photoreceptor proteins located in the cells of the eyes aid in vision.

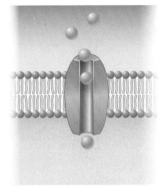
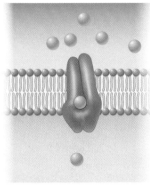

a. Channel protein   b. Transport protein

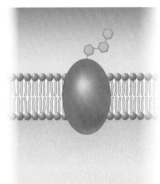
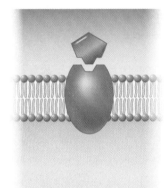

c. Cell recognition protein   d. Receptor protein

e. Enzymatic protein   f. Junction proteins

Figure 4.6 Membrane protein diversity.

These are some of the types of proteins embedded in the plasma membrane.

Check Your Progress 4.3

1. Identify the basic components that make up the structure of the plasma membrane.

2. Explain why the plasma membrane is described as a fluid-mosaic model.

3. Distinguish among the six types of membrane proteins by function.

4.4 Eukaryotic Cells

As discussed in Chapter 1, protists, fungi, plants, and animals are all composed of eukaryotic cells. Unlike prokaryotic cells, eukaryotic cells have a membrane-bounded **nucleus** that houses their DNA. The DNA forms the genes, which are located on chromosomes.

Eukaryotic cells are much larger than prokaryotic cells, and as mentioned they often have projections that increase their surface area. They are also compartmentalized and contain small structures, called **organelles,** that differ in structure and function. All the molecules necessary to perform a particular function are concentrated inside the organelle.

In this chapter, the organelles are divided into four categories:

1. ***The nucleus and the ribosomes.*** The nucleus communicates with the ribosomes, which are the small particles that synthesize proteins coded for by DNA located within the chromosomes.

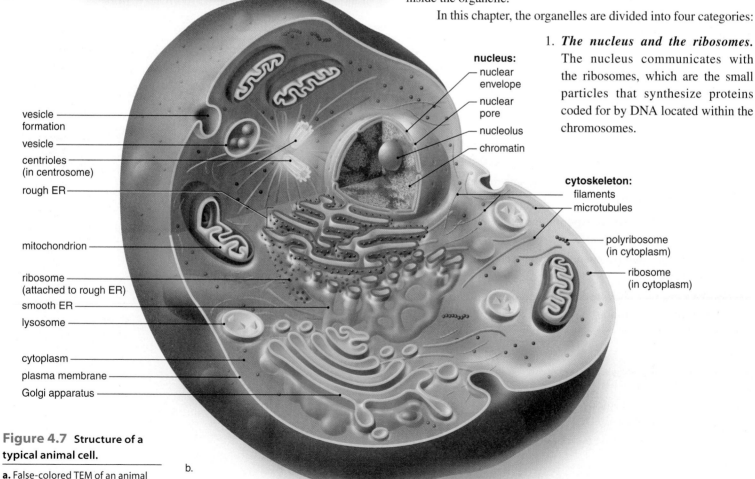

plasma membrane

nuclear envelope

nucleolus

chromatin

endoplasmic reticulum

a.

×10,000

vesicle formation

vesicle

centrioles (in centrosome)

rough ER

mitochondrion

ribosome (attached to rough ER)

smooth ER

lysosome

cytoplasm

plasma membrane

Golgi apparatus

b.

nucleus:
— nuclear envelope
— nuclear pore
— nucleolus
— chromatin

cytoskeleton:
filaments
microtubules

polyribosome (in cytoplasm)

ribosome (in cytoplasm)

Figure 4.7 Structure of a typical animal cell.

a. False-colored TEM of an animal cell. **b.** Generalized drawing.

2. ***Organelles of the endomembrane system (ES).*** Each of these organelles has its own particular set of enzymes and produces its own products. The products are carried between the ES organelles by **transport vesicles,** small, membranous sacs that isolate the products from the cytoplasm.

3. ***The energy-related organelles.*** The energy-related organelles—chloroplasts in plant cells and mitochondria in both plant and animal cells—are more self-contained than those of the ES. They even contain DNA and possess ribosomes that resemble those of prokaryotic cells.

4. ***The cytoskeleton.*** The **cytoskeleton** is a lattice of protein filaments and tubules that maintains the shape of the cell and assists in the movement of organelles. Without an efficient means of moving organelles and their products, eukaryotic cells could not exist.

Each structure in the typical animal and plant cells shown in **Figures 4.7** and **4.8** has been given a particular color, and this color will be used to represent the same structure throughout this text.

MP3
Cellular Organelles

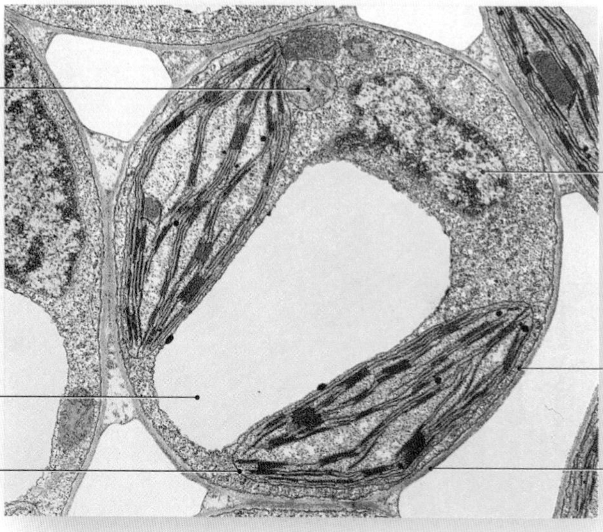

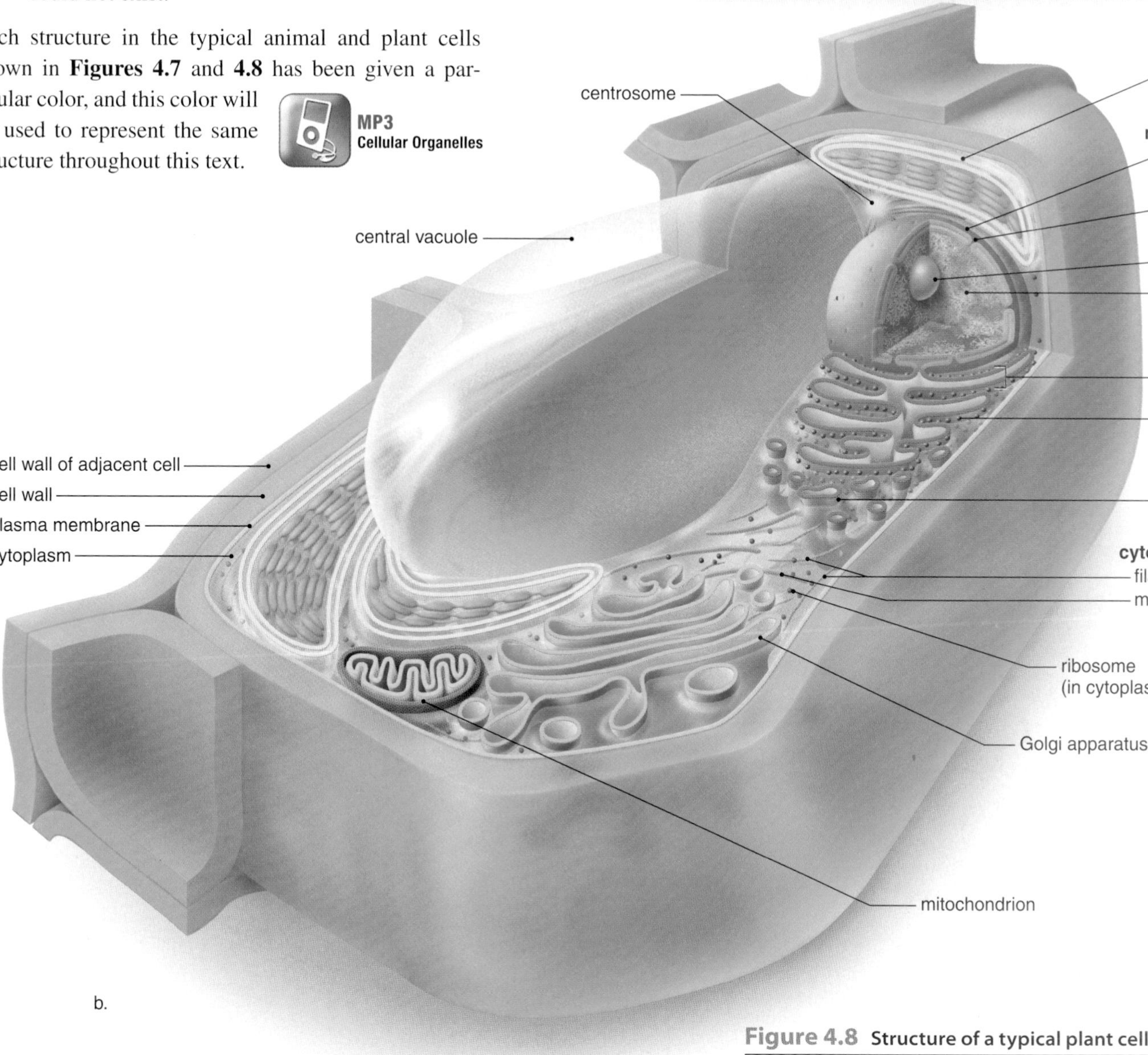

×7,700

a.

b.

Figure 4.8 Structure of a typical plant cell.

a. False-colored TEM of a plant cell. **b.** Generalized drawing.

Nucleus and Ribosomes

The nucleus stores genetic information, and the ribosomes in the cytoplasm use this information to carry out protein synthesis.

The Nucleus

Because of its large size, the nucleus is one of the most noticeable structures in the eukaryotic cell (**Fig. 4.9**). The nucleus contains **chromatin** within a semifluid nucleoplasm. Chromatin looks grainy, but actually it is a network of strands. Just before the cell divides, the chromatin condenses and coils into rod-like structures called **chromosomes.** All the cells of an organism contain the same number of chromosomes, except for the egg and sperm, which usually have half this number.

Chromatin (and therefore chromosomes) is composed of DNA, protein, and some RNA. The DNA is organized into genes, each of which has a specific sequence of nucleotides that codes for a polypeptide. The coded information is relayed from the nucleus to the ribosomes by a type of RNA called messenger RNA (mRNA). This mRNA has a sequence of bases that mirrors the sequence of bases in a gene. This sequence specifies the order of amino acids that is to occur in a particular polypeptide. At the ribosome, the information in the mRNA is translated into a polypeptide chain. A protein may contain one or more polypeptides. Because the proteins of a cell determine its structure and functions, the nucleus may be thought of as the command center of the cell.

Within the nucleus is a dark structure called a **nucleolus.** In

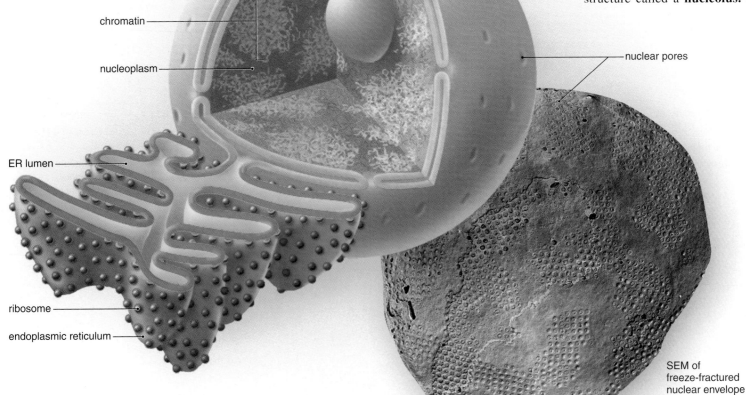

nuclear envelope — outer membrane
inner membrane

nucleolus

chromatin

nucleoplasm

ER lumen

ribosome

endoplasmic reticulum

nuclear pores

SEM of
freeze-fractured
nuclear envelope

Figure 4.9   **Structure of the nucleus.**

The nuclear envelope contains pores that allow substances to pass from the nucleus to the cytoplasm.

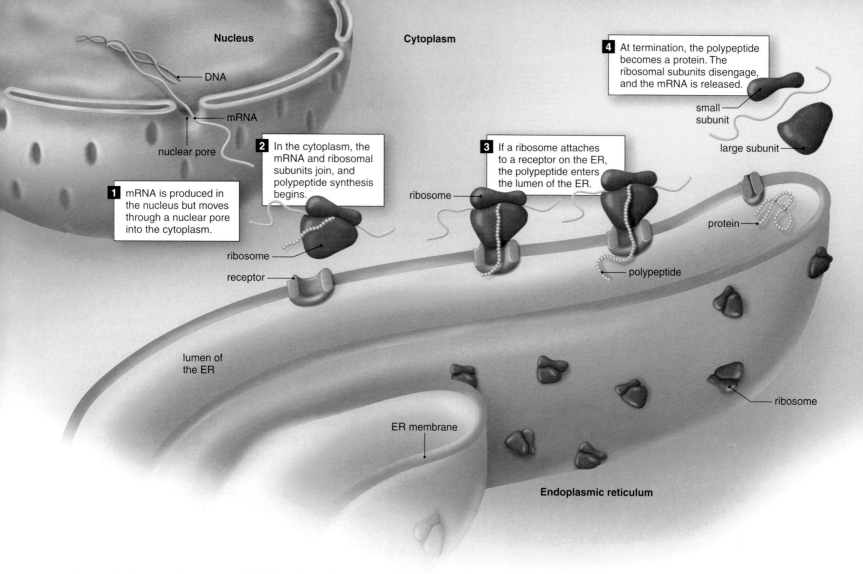

Nucleus

DNA

mRNA

nuclear pore

Cytoplasm

1 mRNA is produced in the nucleus but moves through a nuclear pore into the cytoplasm.

2 In the cytoplasm, the mRNA and ribosomal subunits join, and polypeptide synthesis begins.

ribosome

3 If a ribosome attaches to a receptor on the ER, the polypeptide enters the lumen of the ER.

ribosome

4 At termination, the polypeptide becomes a protein. The ribosomal subunits disengage, and the mRNA is released.

small subunit

large subunit

protein

polypeptide

receptor

lumen of the ER

ER membrane

ribosome

Endoplasmic reticulum

the nucleolus, another type of RNA, called ribosomal RNA (rRNA), is produced. Proteins join with rRNA to form the subunits of ribosomes. The assembled ribosomal subunits are then sent out of the nucleus into the cytoplasm, where they join and assume their role in protein synthesis.

The nucleus is separated from the cytoplasm by a double membrane of phospholipids known as the **nuclear envelope.** Located throughout the nuclear envelope are nuclear pores that allow the nucleus to communicate with activities in the cytoplasm. The **nuclear pores** are of sufficient size (100 nm) to permit the passage of ribosomal subunits and mRNA out of the nucleus into the cytoplasm, as well as the passage of proteins from the cytoplasm into the nucleus.

Ribosomes

Ribosomes are found in both prokaryotes and eukaryotes. In both types of cells, ribosomes are composed of two subunits, one large and one small. Each subunit has its own mix of proteins and rRNA. As mentioned, ribosomes are sites of protein synthesis. It is here that the instructions carried by the mRNA molecule are decoded with the help of another type of RNA, transfer RNA (tRNA). Using the ribosome as a workbench, the tRNA translates the information in the mRNA into a polypeptide chain. Proteins may contain one or more polypeptide chains.

In eukaryotic cells, some ribosomes occur freely within the cytoplasm. Other ribosomes are attached to the **endoplasmic reticulum (ER),** an organelle of the endomembrane system. After the ribosome binds to a receptor at the ER, the polypeptide being synthesized enters the lumen (interior) of the ER, where it may be further modified (**Fig. 4.10**) and assume its final shape.

Figure 4.10 The nucleus, ribosomes, and endoplasmic reticulum (ER).

After mRNA leaves the nucleus, it attaches itself to a ribosome, and polypeptide synthesis begins. When a ribosome combines with a receptor at the endoplasmic reticulum (ER), the polypeptide enters the lumen of the ER through a channel in the receptor. Exterior to the ER, the ribosome splits, releasing the mRNA while a protein takes shape inside the ER lumen.

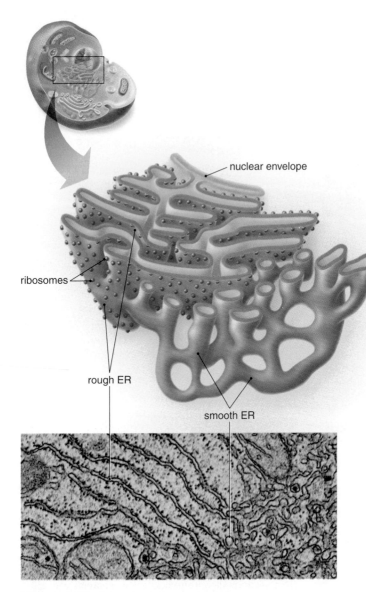

nuclear envelope

ribosomes

rough ER

smooth ER

Figure 4.11 **Endoplasmic reticulum (ER).**

The rough ER consists of flattened saccules and has ribosomes present on its surface. The ribosomes synthesize polypeptides, which then enter the rough ER for modification. The smooth ER lacks ribosomes and is more tubular in structure. Lipids are synthesized by the smooth ER, which can have other functions as well.

Endomembrane System

The **endomembrane system** consists of the nuclear envelope, the membranes of the endoplasmic reticulum, the Golgi apparatus, and many small, membranous sacs called **vesicles.** This system helps compartmentalize the cell, so that particular enzymatic reactions are restricted to specific regions. Transport vesicles carry molecules from one part of the system to another.

Endoplasmic Reticulum

The endoplasmic reticulum (ER) consists of a complicated system of membranous channels and saccules (flattened vesicles). It is physically continuous with the outer membrane of the nuclear envelope (**Fig. 4.11**).

Rough ER is studded with ribosomes on the side of the membrane that faces the cytoplasm; therefore, rough ER is able to synthesize polypeptides. It also modifies the polypeptides after they have entered the central enclosed region of the ER, called the lumen, where a protein takes shape. The rough ER forms transport vesicles, which take proteins to other parts of the cell. Often, transport vesicles are on their way to the plasma membrane or the Golgi apparatus (described below).

Smooth ER, which is continuous with rough ER, does not have attached ribosomes. Smooth ER synthesizes lipids, such as phospholipids and steroids. The functions of smooth ER are dependent on the particular cell. In the testes, it produces testosterone and, in the liver, it helps detoxify drugs. Regardless of any specialized function, smooth ER also forms transport vesicles that carry molecules to other parts of the cell, notably the Golgi apparatus.

Golgi Apparatus

The **Golgi apparatus,** named for its discoverer, Camillo Golgi, consists of a stack of 3 to 20 slightly curved, flattened saccules resembling pancakes (**Fig. 4.12**). The Golgi apparatus may be thought of as a transfer station. First, it receives transport vesicles sent to it by rough and smooth ER. The molecules within the vesicles are modified as they move between saccules. For example, sugars may be added to or removed from proteins within the saccules of the Golgi. Finally, the Golgi apparatus sorts the modified molecules and packages them into new transport vesicles according to their particular destinations. Outgoing transport vesicles may return to the ER or proceed to the plasma membrane, where they discharge their contents during **secretion.** In animal cells, some of the vesicles that leave the Golgi are lysosomes, which are discussed next.

Lysosomes

Lysosomes are vesicles, produced by the Golgi apparatus, that digest molecules and even portions of the cell itself. Sometimes, after engulfing molecules outside the cell, a vesicle formed at the plasma membrane fuses with a lysosome. Lysosomal enzymes then digest the contents of the vesicle. In Tay-Sachs disease, a genetic disorder, lysosomes in nerve cells are missing an enzyme for a particular lipid molecule. The cells become so full of storage lipids that they lose their ability to function. In all cases, the individual dies, usually in childhood. Research is currently underway on using advances in medicine, such as gene therapy, to provide the missing enzyme for these children.

Animation
Lysosomes

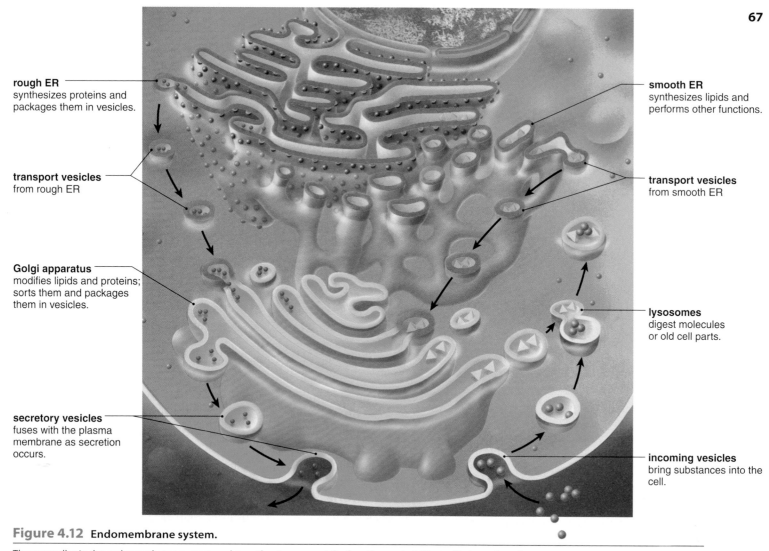

rough ER synthesizes proteins and packages them in vesicles.

transport vesicles from rough ER

Golgi apparatus modifies lipids and proteins; sorts them and packages them in vesicles.

secretory vesicles fuses with the plasma membrane as secretion occurs.

smooth ER synthesizes lipids and performs other functions.

transport vesicles from smooth ER

lysosomes digest molecules or old cell parts.

incoming vesicles bring substances into the cell.

Figure 4.12 Endomembrane system.

The organelles in the endomembrane system work together to carry out the functions noted. Plant cells do not have lysosomes.

Vacuoles

Vacuoles, like vesicles, are membranous sacs, but vacuoles are larger than vesicles. The vacuoles of some protists are quite specialized; they include contractile vacuoles for ridding the cell of excess water and digestive vacuoles for breaking down nutrients (**Fig. 4.13***a*). Vacuoles usually store substances, such as nutrients or ions. Plant vacuoles contain not only water, sugars, and salts but also pigments and toxic molecules (Fig. 4.13*b*). The pigments are responsible for many of the red, blue, and purple colors of flowers and some leaves. The toxic substances help protect a plant from herbivorous animals. Few animal cells contain vacuoles, but fat cells, or adipocytes, contain a very large lipid-engorged vacuole that takes up nearly two-thirds of the volume of the cell (Fig. 4.13*c*)!

Energy-Related Organelles

Chloroplasts and mitochondria are the two eukaryotic membranous organelles that specialize in energy conversion. **Chloroplasts** use solar energy to synthesize carbohydrates. **Mitochondria** (sing., mitochondrion) break down carbohydrates

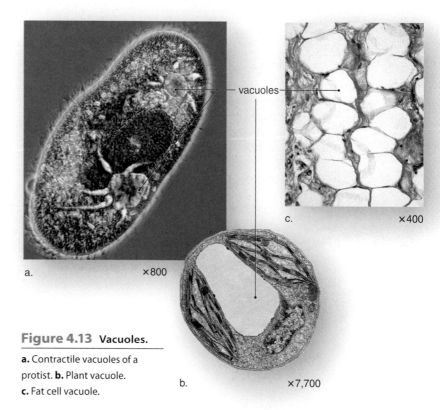

vacuoles

a. ×800

c. ×400

b. ×7,700

Figure 4.13 Vacuoles.

a. Contractile vacuoles of a protist. **b.** Plant vacuole. **c.** Fat cell vacuole.

to produce **adenosine triphosphate (ATP)** molecules. The production of ATP is of great importance because ATP serves as a carrier of energy in cells. The energy of ATP is used whenever a cell synthesizes molecules, transports molecules, or carries out a special function, such as muscle contraction or nerve conduction. Without a constant supply of ATP, no cell could exist for long.

Chloroplasts

The chloroplast, an organelle found in plants and algae, is the location where carbon dioxide gas, water, and energy from the sun are used to produce carbohydrates by the process of photosynthesis. The chloroplast is quite large, being twice the width and as much as five times the length of a mitochondrion. Chloroplasts have a three-membrane system. They are bounded by a double membrane, which includes an outer membrane and an inner membrane. The large inner space, called the **stroma,** contains a concentrated mixture of enzymes and disc-like sacs called thylakoids. The **thylakoids** are formed from the third membrane. A stack of thylakoids is called a **granum.** The lumens of thylakoid sacs form a large internal compartment called the thylakoid space (**Fig. 4.14**). The pigments that capture solar energy are located in the thylakoid membrane, and the enzymes that synthesize carbohydrates are in the stroma. The carbohydrates produced by chloroplasts serve as organic nutrient molecules for plants and, indeed, for all living things on Earth.

The discovery that chloroplasts have their own DNA and ribosomes supports an accepted theory that chloroplasts are derived from photosynthetic bacteria that entered a eukaryotic cell in the distant past. As shown in Figure 4.8, plant cells contain *both* mitochondria and chloroplasts.

Mitochondria

Mitochondria are much smaller than chloroplasts, and they are usually visible only under an electron microscope. We think of mitochondria as having a shape like that shown in **Figure 4.15,** but actually they often change shape, becoming longer and thinner or shorter and broader. Mitochondria can form long, moving chains, or they can remain fixed in one location—often where energy is most needed. For example, they are packed between the contractile elements of cardiac cells and wrapped around the interior of a sperm's flagellum.

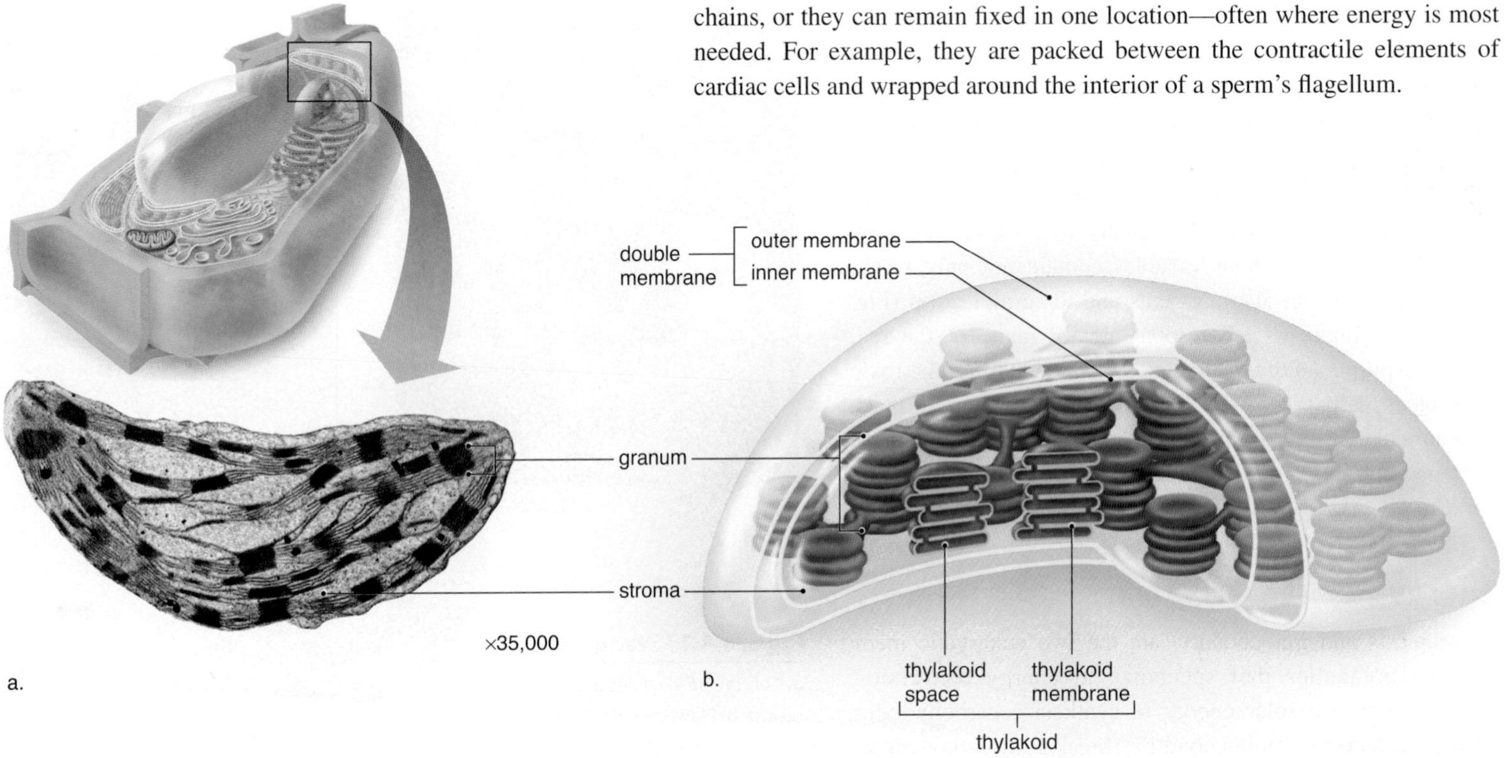

Figure 4.14 **Chloroplast structure.**

a. Electron micrograph of a chloroplast, the organelle that carries out photosynthesis. **b.** Generalized drawing that reveals the grana, each of which is composed of a stack of membranous sacs called thylakoids. In some grana, but not all, the thylakoid spaces are interconnected.

×35,000

a.

b.

Like chloroplasts, mitochondria are bounded by a double membrane. The inner membrane is highly convoluted into folds, called **cristae,** that project into the matrix. Cristae increase the surface area of the inner membrane so much that, in a liver cell, they account for about one-third of the total membrane in the cell.

Mitochondria are often called the powerhouses of the cell because they produce most of the ATP the cell utilizes. The inner membrane encloses the **matrix,** which contains a highly concentrated mixture of enzymes that assists the breakdown of carbohydrates and other nutrient molecules. These reactions supply the chemical energy that permits ATP synthesis to take place on the cristae. The complete breakdown of carbohydrates, which also involves the cytoplasm, is called **cellular respiration** because oxygen is needed and carbon dioxide is given off.

The matrix also contains mitochondrial DNA and ribosomes. The presence of mitochondrial DNA and ribosomes is evidence that mitochondria and chloroplasts have similar origins and are derived from bacteria that took up residence in an early eukaryotic cell. Today, with a few rare exceptions, all eukaryotic cells have mitochondria.

The Cytoskeleton and Motor Proteins

The **cytoskeleton** is a network of interconnected protein filaments and tubules that extends from the nucleus to the plasma membrane in eukaryotic cells. Much as bones and muscles give an animal structure and produce movement, the elements of the cytoskeleton maintain cell shape and, along with **motor proteins,** allow the cell and its organelles to move. But unlike an animal's skeleton, the cytoskeleton is highly dynamic—its elements can be quickly assembled and disassembled as appropriate. The cytoskeleton includes microtubules, intermediate filaments, and actin filaments. The major motor proteins are myosin, kinesin, and dynein.

Connections and Misconceptions

How do we know that mitochondria and chloroplasts were once bacteria?

The DNA found in these organelles is structured differently than that found in the nucleus. Both mitochondria and chloroplasts contain a single, circular chromosome—similar to those found in prokaryotes. The genes located on this chromosome are very closely related to prokaryotic genes in both structure and function. In addition, both mitochondria and chloroplasts reproduce in a manner very similar to the process of binary fission in bacteria. These observations, coupled with detailed analyses of mitochondrial and chloroplast DNA, have strongly suggested that both of these organelles arose from an ancient symbiotic event (also called endosymbiosis) that played an important role in the evolution of the eukaryotes.

Animation
Endosymbiosis

Figure 4.15 **Mitochondrion structure.**

a. Generalized drawing that reveals the cristae within a mitochondrion.
b. Electron micrograph of a mitochondrion, the organelle that is involved in cellular respiration.

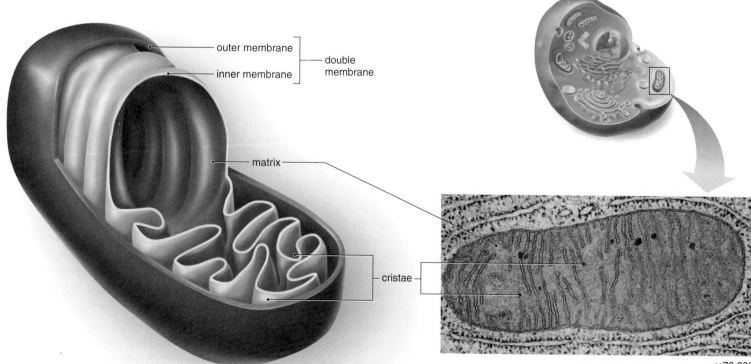

outer membrane
inner membrane
double membrane

matrix

cristae

cristae

×70,000

a.

b.

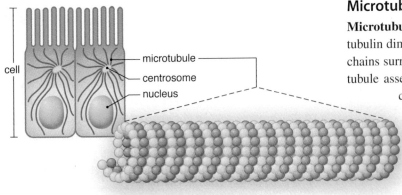

Figure 4.16 Microtubules.

Microtubules are located throughout the cell and radiate outward from the centrosome.

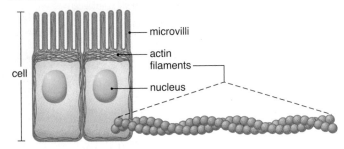

Figure 4.17 Actin filaments.

Actin filaments are organized into bundles or networks just under the plasma membrane, where they lend support to the shape of a cell.

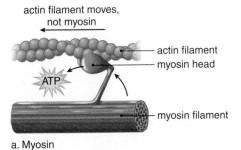

a. Myosin

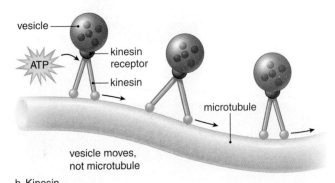

b. Kinesin

Figure 4.18 Motor proteins.

a. Myosin creates movement by detaching and reattaching to actin to pull the actin filament along the myosin filament. **b.** The motor molecule kinesin carries organelles along microtubule tracks. One end binds to an organelle, and the other end attaches, detaches, and reattaches to the microtubule. ATP supplies the energy for both myosin and kinesin.

Microtubules

Microtubules are small, hollow cylinders composed of 13 long chains of tubulin dimers (two tubulin molecules at a time). In electron micrographs, the chains surround what appears to be an empty central core (**Fig. 4.16**). Microtubule assembly is controlled by the **centrosome,** a microtubule organizing center, which lies near the nucleus. Microtubules radiating from the centrosome help maintain the shape of the cell and act as tracks along which organelles and other materials can move.

Intermediate Filaments

Intermediate filaments are intermediate in size between actin filaments and microtubules. They are rope-like assemblies of proteins that typically run between the nuclear envelope and the plasma membrane. The network they form supports both the nucleus and the plasma membrane.

The protein making up intermediate filaments differs according to the cell type. Intermediate filaments made of the protein keratin give great mechanical strength to skin cells.

Actin Filaments

Each **actin filament** consists of two chains of globular actin monomers twisted about one another in a helical manner to form a long filament. Actin filaments support the cell, forming a dense, complex web just under the plasma membrane (**Fig. 4.17**). Actin filaments also support projections of the plasma membrane, such as microvilli.

Motor Proteins

Motor proteins associated with the cytoskeleton are instrumental in allowing cellular movements. The major motor proteins are myosin, kinesin, and dynein.

Myosin often interacts with actin filaments when movement occurs. For example, myosin is interacting with actin filaments when cells move in an amoeboid fashion and/or engulf large particles. During animal cell division, actin, in conjunction with myosin, pinches the original cell into two new cells. When a muscle cell contracts, myosin pulls actin filaments toward the middle of the cell (**Fig. 4.18***a*).

Kinesin and dynein move along microtubules much as a car travels along a highway. First, an organelle, perhaps a vesicle, combines with the motor molecule, and then the motor molecule attaches, detaches, and reattaches farther along the microtubule. In this way, the organelle moves from one place to another in the cell (Fig. 4.18*b*). Kinesin and dynein are acting similarly when transport vesicles take materials from the Golgi apparatus to their final destinations.

Centrioles

Centrioles are short cylinders with a 9 + 0 pattern of microtubule triplets—that is, nine sets of triplets occur in a ring, and none are in the middle of the cylinder (**Fig. 4.19**). In animal cells and most protists, two centrioles lie at right angles to one another in the middle of a centrosome. A centrosome, as mentioned previously, is the major microtubule organizing center for the cell. Therefore, it is possible that centrioles are also involved in the process by which microtubules assemble and disassemble.

Before an animal cell divides, the centrioles replicate, and the members of each pair are at right angles to one another. Then each pair becomes part of a separate centrosome. During cell division, the centrosomes move apart and most

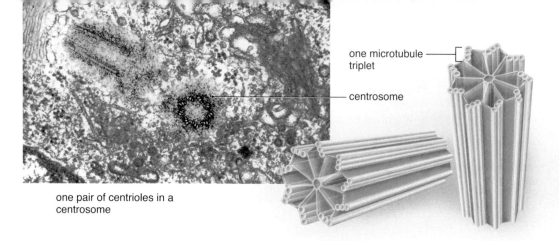

one pair of centrioles in a centrosome

one microtubule triplet

centrosome

Figure 4.19 Centrioles.

A pair of centrioles lies to one side of the nucleus in an animal cell.

likely play a role in organizing a microtubular apparatus (the spindle), which assists the movement of chromosomes. In any case, each new cell eventually has its own centrosome and pair of centrioles.

Plant and fungal cells have the equivalent of a centrosome, but this structure does not contain centrioles, suggesting that centrioles are not necessary for the assembly of cytoplasmic microtubules.

Cilia and Flagella

In eukaryotes, **cilia** and **flagella** are hair-like projections that can either move stiffly, like an oar, or undulate, like a whip (**Fig. 4.20***a*). Cells that have these organelles are capable of movement. Cilia are much shorter than flagella, but they have a similar construction. Both are membrane-bounded cylinders with a basal body lying in the cytoplasm. They have a matrix that contains nine microtubule doublets arranged in a circle around two central microtubules. This is called the 9 + 2 pattern of microtubules (Fig. 4.20*b*). Cilia and flagella move when the microtubule doublets slide past one another.

Connecting the Concepts

The function of many of the organelles and structures in this section will be expanded upon later in the text. For some examples, refer to the following discussions.

Section 6.1 explores how the chloroplast is involved in photosynthesis.

Section 8.3 describes how the centrosomes and microtubules play a role in eukaryotic cell division.

Section 17.3 examines the evolutionary challenges that faced prokaryotic organisms on the early Earth.

Check Your Progress 4.4

1. Identify the components of the nucleus and ribosomes and give a function for each.

2. Describe the components of the endomembrane system and list the function of each component.

3. Summarize the special functions of vacuoles and hypothesize what might happen if they are not present in a cell.

4. Compare and contrast the structure and function of the two energy-related organelles of a eukaryotic cell.

5. Distinguish between the cytoskeleton proteins and motor proteins by structure and function.

cilia in bronchial wall of lungs

flagella of sperm

a.

Flagellum

Flagellum cross section

central microtubules

microtubule doublet

dynein side arms

TEM ×350,000

plasma membrane

TEM ×101,000

triplets

Basal body

Basal body cross section

b.

Figure 4.20 Cilia and flagella.

a. Cilia in the bronchial wall and the flagella of sperm are organelles capable of movement. The cilia in the bronchi of our lungs sweep mucus and debris back up into the throat, where it can be swallowed or ejected. The flagella of sperm allow them to swim to the egg. **b.** Cilia and flagella have a distinct pattern of microtubules bounded by a plasma membrane.

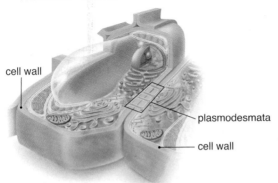

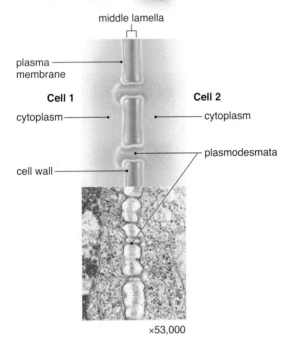

Figure 4.21 Plasmodesmata.

Plant cells are joined by membrane-lined channels, called plasmodesmata, that contain cytoplasm. Through these channels, water and small molecules can pass from cell to cell.

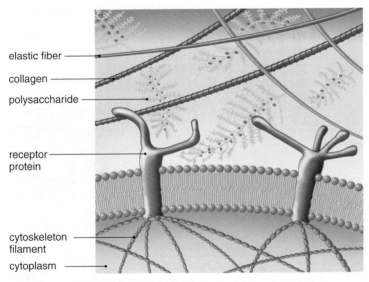

Figure 4.22 Animal cell extracellular matrix.

The extracellular matrix supports an animal cell and affects its behavior.

4.5 Outside the Eukaryotic Cell

Learning Outcomes

Upon completion of this section, you should be able to

1. Describe the structure of a plant cell wall.
2. State the purpose of the extracellular matrix.
3. Distinguish among the types of junctions present between eukaryotic cells.

Now that we have completed our discussion of the plasma membrane and the cell contents, you might think we have finished our tour of the cell. However, most cells also have extracellular structures formed from materials the cell produces and transports across its plasma membrane.

Plant Cell Walls

All plant cells have a **cell wall.** A *primary cell wall* contains cellulose fibrils and noncellulose substances, and these allow the wall to stretch when the cell is growing. Adhesive substances are abundant outside the cell wall in the middle lamella, a layer that holds two plant cells together.

For added strength, some plant cells have a *secondary cell wall* that forms inside the primary cell wall. The secondary wall has a greater quantity of cellulose fibrils, which are laid down at right angles to one another. Lignin, a substance that adds strength, is a common ingredient of secondary cell walls.

In a plant, living cells are connected by **plasmodesmata** (sing., plasmodesma), numerous narrow, membrane-lined channels that pass through the cell wall (**Fig. 4.21**). Cytoplasmic strands within these channels allow the direct exchange of some materials between adjacent plant cells and eventually among all the cells of a plant. The plasmodesmata allow only water and small solutes to pass freely from cell to cell.

Exterior Cell Surfaces in Animals

Animal cells do not have a cell wall, but they do have two other exterior surface features of interest: (1) An extracellular matrix exists outside the cell, and (2) various junctions occur between some cell types.

Extracellular Matrix

An **extracellular matrix (ECM)** is a meshwork of fibrous proteins and polysaccharides in close association with the cell that produced them (**Fig. 4.22**). Collagen and elastin are two well-known proteins in the extracellular matrix. Collagen resists stretching and elastin gives it resilience. The polysaccharides play a dynamic role by directing the migration of cells along collagen fibers during development. Other ECM proteins bind to receptors in a cell's plasma membrane, permitting communication between the extracellular matrix and the cytoskeleton within the cytoplasm of the cell.

The extracellular matrices of tissues vary greatly. They may be quite flexible, as in cartilage, or rock solid, as in bone. The rigidity of the extracellular matrix is influenced mainly by the number and types of protein fibers present and how they are arranged. The extracellular matrix of bone is very hard because, in addition to the components mentioned, mineral salts—notably, calcium salts—are deposited outside the cell.

Junctions Between Cells

Three types of junctions are found between certain cells: adhesion junctions, tight junctions, and gap junctions. The type of junction found between two cells depends on whether or not the cells need to be able to exchange materials and if the cells need to be joined together very tightly.

In **adhesion junctions,** internal cytoplasmic plaques, firmly attached to the cytoskeleton within each cell, are joined by intercellular filaments (**Fig. 4.23***a*). The result is a sturdy but flexible sheet of cells. In some organs—such as the heart, stomach, and bladder, where tissues must stretch—adhesion junctions hold the cells together.

Adjacent cells are even more closely joined by **tight junctions,** in which plasma membrane proteins actually attach to each other, producing a zipper-like fastening (Fig. 4.23*b*). The cells of tissues that serve as barriers are held together by tight junctions; for example, urine stays within kidney tubules because the cells of the tubules are joined by tight junctions.

A **gap junction** allows cells to communicate. A gap junction is formed when two identical plasma membrane channels join (Fig. 4.23*c*). The channel of each cell is lined by six plasma membrane proteins that allow the junction to open and close. A gap junction lends strength to the cells, but it also allows small molecules and ions to pass between them. Gap junctions are important in heart muscle and smooth muscle because they permit the flow of ions that is required for the cells to contract as a unit.

Connections and Misconceptions

Do collagen creams remove wrinkles?

Not exactly. Collagen is a protein that is active in the middle layers of the skin, called the dermis. Collagen placed on the surface of the skin can't pass through the tough outer layers to where it is needed. However, the moisturizers that are present in the cream prevent water loss and make the skin more elastic, until the cream wears off.

Connecting the Concepts

For more on the structures described in this section, refer to the following discussions.

Section 20.2 examines how differences in the composition of the cell walls in plants help identify the type of plant tissue.

Section 22.1 describes the role of collagen in the formation of connective tissue in animals.

Check Your Progress 4.5

1. Contrast the primary cell wall with the secondary cell wall in plants.
2. Describe the composition of the extracellular matrix of an animal cell.
3. Compare the structure and function of adhesion, tight, and gap junctions.

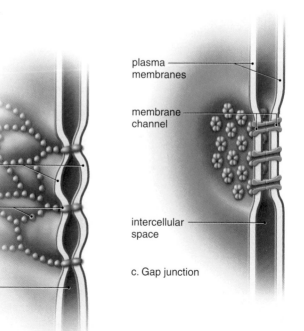

Figure 4.23 Junctions between cells of the intestinal wall.
a. In adhesion junctions, intercellular filaments run between two cells. **b.** Tight junctions between cells form an impermeable barrier because their adjacent plasma membranes are joined. **c.** Gap junctions allow communication between two cells because adjacent plasma membrane channels are joined.

a. Adhesion junction

plasma membranes

filaments of cytoskeleton

intercellular filaments

intercellular space

plasma membranes

tight junction proteins

intercellular space

b. Tight junction

plasma membranes

membrane channel

intercellular space

c. Gap junction

Media Study Tools

 plus+

|BIOLOGY

The Chapter in Review

Summary

4.1 Cells Under the Microscope

- Cells are microscopic in size. Although a light microscope allows you to see cells, it cannot make out the detail that an electron microscope can.
- Cells must remain small in order to have an adequate amount of surface area per cell volume.

4.2 The Two Main Types of Cells

- All cells have a plasma membrane, a cytoplasm, and genetic material.
- Some of the differences between prokaryotic and eukaryotic cells are presented in the following table:

Differences between Prokaryotic and Eukaryotic Cells	Prokaryotic Cells	Eukaryotic Cells
Size	1–5 μm	Typically > 50 μm
DNA Location	Nucleoid region	Nucleus
Organelles	No	Yes
Chromosomes	One, circular	Multiple, linear

4.3 The Plasma Membrane

The plasma membrane of both prokaryotes and eukaryotes is a phospholipid bilayer.

- The phospholipid bilayer regulates the passage of molecules and ions into and out of the cell.
- The fluid-mosaic model of membrane structure shows that the embedded proteins form a mosaic (varying) pattern.
- The types of embedded proteins are channel, transport, cell recognition, receptor, and enzymatic proteins.

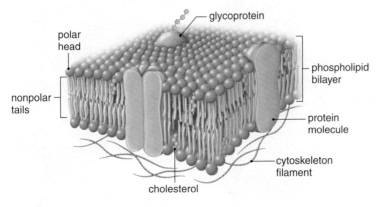

4.4 Eukaryotic Cells

Eukaryotic cells, which are much larger than prokaryotic cells, contain organelles, such as the ones in the following diagram:

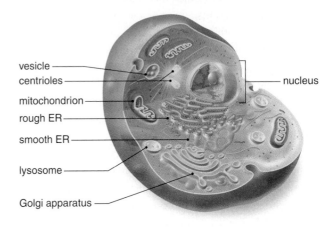

Nucleus and Ribosomes

- The nucleus houses chromatin, which contains DNA, the genetic material. During division, chromatin becomes condensed into chromosomes.
- The nucleolus is an area within the nucleus where ribosomal RNA is produced. Proteins are combined with the rRNA to form the subunits of ribosomes, which exit the nucleus through nuclear pores.
- Ribosomes in the cytoplasm synthesize polypeptides, which later become proteins with the help of mRNA, a DNA intermediary.

Endomembrane System

- Rough ER has ribosomes, which produce polypeptides, on its surface. These polypeptides enter the ER, are modified, and become proteins, which are then packaged in transport vesicles.
- Smooth ER synthesizes lipids, but it also has various metabolic functions, depending on the cell type. It can also form transport vesicles.
- The Golgi apparatus is a transfer station that receives transport vesicles and modifies, sorts, and repackages proteins into transport vesicles that fuse with the plasma membrane as secretion occurs.
- Lysosomes are produced by the Golgi apparatus. They contain enzymes that carry out intracellular digestion.

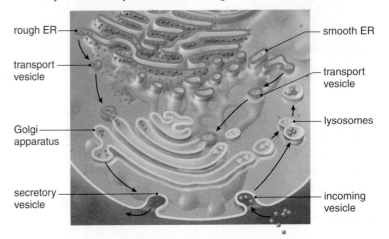

Vacuoles and Vesicles

- Vacuoles are large, membranous sacs specialized for storage, contraction, digestion, and other functions.
- Vesicles are small, membranous sacs.

Energy-Related Organelles

- Chloroplasts capture the energy of the sun and carry on photosynthesis, which produces carbohydrates.
- Mitochondria are the site of cellular respiration. They break down carbohydrate-derived products to ATP.

The Cytoskeleton and Motor Proteins

- The cytoskeleton maintains cell shape and allows the cell and the organelles to move.
- Microtubules radiate from the centrosome and are present in cytoplasm. They also occur in centrioles, cilia, and flagella.
- Intermediate filaments support the nuclear envelope and the plasma membrane. They give mechanical strength to skin cells, for example.
- Actin filaments are long, thin, helical filaments that support the plasma membrane and projections of the cell.
- Motor proteins allow cellular movements to occur and move vesicles and organelles within the cell. The motor protein myosin is often interacting with actin when movement occurs, and kinesin interacts with microtubules to move vesicles and other organelles from place to place.

Centrioles

- Centrioles are present in animal cells, but not plant cells.
- In animal cells and those of most protists, two centrioles lie at right angles just outside the nucleus.

Cilia and Flagella

- Cilia and flagella are hair-like projections that allow some cells to move.
- Both have a cylindrical construction with an internal 9 + 2 pattern of microtubules.
- Their basal bodies resemble centrioles.

4.5 Outside the Eukaryotic Cell

Plant cells have a freely permeable cell wall, with cellulose as its main component.

- Some plant cells have a secondary cell wall containing lignin and cellulose, which give them rigidity.
- Small, membrane-lined channels called plasmodesmata span the cell wall and contain strands of cytoplasm, which allow materials to pass from one cell to another.
- The extracellular matrix contains proteins and polysaccharides produced by the cell that helps support cells and aids in communication between cells.
- Adhesion junctions and tight junctions, if present, help hold cells together.
- Gap junctions allow the passage of small molecules between cells.

Key Terms

actin filament 70
adenosine triphosphate (ATP) 68
adhesion junction 73
capsule 58
cell 55
cell theory 57
cellular respiration 69
cell wall 58, 72
centriole 70
centrosome 70
chloroplast 67
chromatin 64
chromosome 64
cilium (pl., cilia) 71
conjugation pili 59
cristae 69
cytoplasm 57
cytoskeleton 63, 69
endomembrane system 66
endoplasmic reticulum (ER) 65
eukaryotic cell 57
extracellular matrix (ECM) 72
fimbria (pl., fimbriae) 59
flagellum (pl., flagella) 58, 71
gap junction 73
Golgi apparatus 66
granum 68
lysosome 66
matrix 69
microtubule 70
mitochondrion 67
motor protein 69
nuclear envelope 65
nuclear pore 65
nucleoid 58
nucleolus 64
nucleus 62
organelle 62
plasma membrane 59
plasmodesmata 72
prokaryotic cell 57
ribosome 58
rough ER 66
secretion 66
smooth ER 66
stroma 68
surface-area-to-volume ratio 56
thylakoid 68
tight junction 73
transport vesicle 63
vacuole 67
vesicle 66

Testing Yourself

Choose the best answer for each question.

1. Which of the following is not found in a prokaryotic cell?
   a. cytoplasm
   b. ribosome
   c. plasma membrane
   d. mitochondrion
   e. nucleoid

2. Which of the following can be viewed only with an electron microscope?
   a. virus
   b. chloroplast
   c. bacterium
   d. human egg

3. Identify the incorrect statement.
   a. A bacterial cell can contain thousands of ribosomes.
   b. A light microscope allows us to see cells and their inner complexity.
   c. In both eukaryotes and prokaryotes, the plasma membrane is a phospholipid bilayer.
   d. A receptor protein has a special shape that allows a signal molecule to bind to it.

4. The plasma membrane is called a fluid mosaic because it contains
   a. waxes suspended within a mosaic of phospholipids.
   b. a mosaic of proteins suspended within a phospholipid bilayer.
   c. a polysaccharide mosaic suspended within a protein bilayer.
   d. a mosaic of phospholipids suspended within a protein bilayer.

5. Eukaryotic cells compensate for a low surface-to-volume ratio by
   a. taking up materials from the environment more efficiently.
   b. lowering their rate of metabolism.
   c. compartmentalizing their activities into organelles.
   d. reducing the number of activities in each cell.

6. Ribosomal subunits are produced in which organelle?
   a. mitochondria
   b. endoplasmic reticulum
   c. cytoplasm
   d. nucleolus

For questions 7–10, match the items to those in the key.

Key:

    a. channel proteins
    b. transport proteins
    c. cell recognition proteins
    d. receptor proteins
    e. enzymatic proteins
    f. tight junction proteins

  7. Assist with cell-to-cell communication.
  8. Combine with molecules to carry them across plasma membranes.
  9. Combine with signaling molecules.
 10. Participate in metabolic reactions.

For questions 11–14, match the items to those in the key.

Key:

    a. single membrane
    b. double membrane
    c. no membrane

 11. Ribosome
 12. Nucleus
 13. Lysosome
 14. Mitochondrion
 15. The organelle that sorts proteins and sends them to their final destinations is the
    a. ribosome.   **c.** Golgi apparatus.
    b. vacuole.   **d.** lysosome.
 16. Plant vacuoles may contain
    a. flower color pigments.
    b. toxins that protect plants against herbivorous animals.
    c. sugars.
    d. all of these.
 17. The interior of a mitochondrion is called the
    a. cristae.   **c.** matrix.
    b. thylakoid.   **d.** granum.
 18. Label the parts of the following cell.

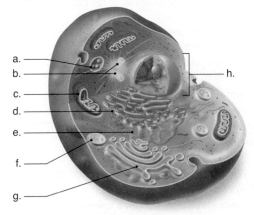

 19. Which of these structures is involved in protein synthesis?
    a. ribosomes   **c.** mitochondria
    b. plasma membrane   **d.** microtubules

Thinking Scientifically

1. The 1992 movie *Lorenzo's Oil* recounts the story of six-year-old Lorenzo Odone and his battle with adrenoleukodystrophy (ALD), which is caused by a malfunctioning organelle called a peroxisome. A peroxisome is an organelle that ordinarily contains enzymes capable of breaking down long-chain fatty acids. In ALD, the enzyme is missing, and when long-chain fatty acids accumulate, muscle weakness and loss of muscle control result.
    a. How might a dietitian be able to help a person with ALD?
    b. What organelle discussed in this chapter is most similar in structure and function to a peroxisome?

2. Protists of the phylum Apicomplexa cause malaria and contribute to infections associated with AIDS. These parasites are unusual because they contain plastids. (Chloroplasts are a type of plastid, but most plastids store materials.) Scientists have discovered that an effective antibiotic that kills the parasite inhibits the functioning of the plastid. Can you conclude that the plastid is necessary to the life of the parasite? Why or why not?

3. Scientists reason that a cell could not evolve until it had an outer plasma membrane. Why does that seem logical?

Bioethical Issue

Human Cloning

Cloning animals, and the specter of cloning humans, is an issue that has caused much concern among the scientific community, governments, and the general public. The use of a fertilized egg, from which the nucleus has been removed, is in disfavor because it requires the destruction of a human embryo.

    Recent technological advances may sidestep this concern but will create others. In 2007, scientists in the United Kingdom applied for licenses to create human-nonhuman cell lines, and this avoids the current prohibition on using human embryos in scientific research. Instead of using a human egg, the scientists proposed using fertilized eggs from other animal species, such as rabbits, and adding a human cell nucleus to it to produce a human-animal hybrid cell, which could then be implanted into a surrogate mother.

    Proponents of this measure point out that this technology avoids the destruction of human embryos. However, there is still the moral issue of whether it is acceptable to clone individual people. Some insist that such technology could be abused to create clones of unsuspecting individuals. While those who oppose the measure insist that cloning human beings is still wrong, and that even the destruction of animal embryos is questionable, the procedure does satisfy the letter of the law prohibiting the destruction of human embryos for such research. Consider that the nucleus, and thus the DNA, of the resulting clone is definitely human, but the other contents of the cell—for example, mitochondria, endoplasmic reticulum, and cytoskeleton—are that of another species. Since mitochondria contain DNA, not all of the DNA of the resulting cell would be human.

    Do you believe that it is acceptable or unacceptable to clone human beings, even if human embryos are not destroyed to do so? Are the fears of unauthorized use of the technology real or unfounded? And should resulting clones be considered to be fully human? Would you be willing for taxpayer funds to be available for this type of research? Why or why not?

5

The Dynamic Cell

Can Water Be Poisonous?

Although we normally think of all the wonderful things water does for our body, drinking an excessive amount of it during a short period of time can be toxic or even fatal. In 2006, a young mother in California died after drinking several gallons of water in an attempt to win a video game system. Drinking too much water has also been reported in people who are under the influence of drugs such as Ecstasy.

In humans, water intake typically equals water output, primarily by the kidneys. When huge amounts of water are consumed very quickly, the kidneys sometimes can't keep up and the excess water accumulates around the cells of the body. This causes them to swell. This swelling, especially in the brain, causes symptoms similar to those associated with intoxication, and the condition is called *water intoxication,* or *hyponatremia.*

What makes cells swell? As you will learn in this chapter, the process of osmosis causes water to move from the side of a cell containing more water to the side containing less water. In the case of water intoxication, first there is much more water outside the cells than inside the cells. Then, when this water moves into the cells to achieve equilibrium, the cells swell and may explode (*lyse*) and die.

In this chapter, we will explore not only how materials move into and out of the cell but also the basic properties of energy and how cells use metabolic pathways and enzymes to conduct the complex reactions needed to sustain life. This chapter provides a foundation for the more detailed examination of photosynthesis and cellular respiration in the following chapters.

BEFORE YOU BEGIN

Before beginning this chapter, take a few moments to review the following discussions.

Section 1.1 Define metabolism, and explain its relationship to energy.

Section 3.2 What are the four levels of structure associated with the three-dimensional shape of a protein?

Section 4.3 What are the roles of proteins in the plasma membrane of a cell?

5.1 What Is Energy?

Learning Outcomes

Upon completion of this section, you should be able to

1. Distinguish between potential and kinetic forms of energy.
2. Define the two laws of thermodynamics.
3. Summarize how the laws of thermodynamics and the concept of entropy relate to living organisms.

Energy is defined as the capacity to do work—to make things happen. Without a source of energy, we humans would not be here on Earth—nor would any other living thing. Our biosphere gets its energy from the sun, and thereafter one form of energy is changed to another form as life processes take place.

The two basic forms of energy are potential energy and kinetic energy. **Potential energy** is stored energy, and **kinetic energy** is the energy of motion. Potential energy is constantly being converted to kinetic energy, and vice versa. An example is shown in **Figure 5.1.** The food a cross-country skier has for breakfast contains chemical energy, which is a form of potential energy. When the skier hits the slopes, she may have to ascend a hill. During her climb, the potential energy of food is converted to the kinetic energy of motion, a type of mechanical energy. By the time she reaches the hilltop, kinetic energy has been converted to the potential energy of location. As she skis down the hill, this potential energy is converted to the kinetic energy of motion again. But with each conversion, some energy is lost as heat and other unusable forms.

Animation
Energy

Measuring Energy

Chemists use a unit of measurement called the joule to measure energy, but it is common to measure food energy in terms of calories. A **calorie** is the amount of heat required to raise the temperature of 1 gram of water by one degree Celsius. This isn't much energy, so the caloric value of food is listed in nutrition labels and diet charts in terms of **kilocalories** (1,000 calories). In this text, we will use *Calorie* to mean 1,000 calories.

Figure 5.1 Potential energy versus kinetic energy.

Food contains potential energy, which a skier can convert to kinetic energy in order to climb a hill. Height is potential energy due to location, which the skier converts to kinetic energy as she descends the hill. With every conversion to kinetic energy, some potential energy is lost as heat.

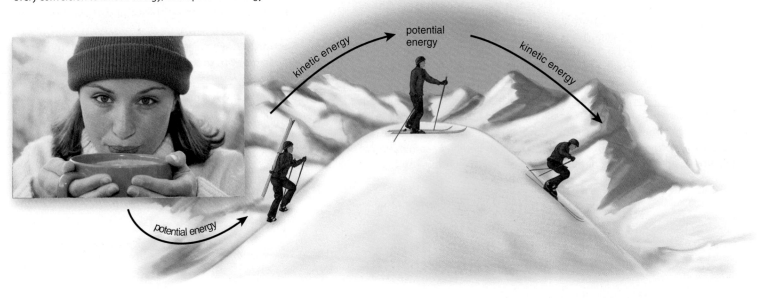

Two Energy Laws

Two **energy laws** govern energy flow and help us understand the principles of energy conversion. Collectively, these are called the laws of thermodynamics. The first law of thermodynamics, also called the law of conservation of energy, tells us *energy cannot be created or destroyed, but it can be changed from one form to another.* Relating the law to our previous example, we know that the skier had to acquire energy by eating food before she could climb the hill and that energy conversions occurred before she reached the bottom of the hill again.

The second law of thermodynamics tells us *energy cannot be changed from one form to another without a loss of usable energy.* Many forms of energy are usable, such as the energy of the sun, food, and ATP. **Heat** is diffuse energy and the least usable form. Every energy conversion results in a loss of usable energy in the form of heat. In our example, much of the potential energy stored in the food is lost as heat as the skier converts it into the kinetic energy of motion. The generation of this heat allows us to maintain a constant body temperature, but eventually the heat is lost to the environment.

MP3
Laws of
Thermodynamics

Entropy

The second law of thermodynamics can be stated another way: Every energy transformation leads to an increase in the amount of disorganization or disorder. The term **entropy** refers to the relative amount of disorganization. The only way to bring about or maintain order is to add more energy to a system. To take an example from your own experience, a tidy room is more organized and less stable than a messy room, which is disorganized and more stable **(Fig. 5.2a)**. In other words, your room is much more likely to stay messy than it is to stay tidy. Why? Unless you continually add energy to keep your room organized and neat, it will inevitably become less organized and messy.

Because our universe is a *closed system* (energy is not entering or leaving the system), all energy transformations, including those in cells, increase the total entropy of the universe. Figure 5.2b shows a process that occurs in cells because it proceeds from a more ordered state to a more disordered state. Just as a tidy room tends to become messy, hydrogen ions (H^+) that have accumulated on one side of a membrane tend to move to the other side unless they are prevented from doing so by the addition of energy. Why? Because when hydrogen ions are distributed equally on both sides of the membrane, no additional energy is needed to keep them that way, and the entropy, or disorder, of their arrangement has increased. The result is a more stable arrangement of H^+ ions in the cell.

What about reactions in cells that apparently proceed from disorder to order? For example, plant cells can make glucose out of carbon dioxide and water. How do they do it? In order to overcome the natural tendency toward disorder, energy input is required, just as it requires energy to organize a messy room. Likewise, energy provided by the sun allows plants to make glucose, a highly organized molecule, from the more disorganized water and carbon dioxide. Even this process, however, involves a loss of some potential energy.

Figure 5.2 Cells and entropy.

The second law of thermodynamics states that entropy (disorder) always increases. Therefore, (**a**) a tidy room tends to become messy and disorganized, and (**b**) hydrogen ions (H^+) on one side of a membrane tend to move to the other side, so that the ions are equally distributed. Both processes result in a loss of potential energy and an increase in entropy.

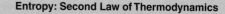
Entropy: Second Law of Thermodynamics

Tidy bedroom | **Messy bedroom**

a.

- more organized
- more potential energy
- less stable
- less entropy

- less organized
- less potential energy
- more stable
- more entropy

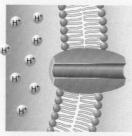 

Unequal distribution of hydrogen ions | **Equal distribution of hydrogen ions**

b.

Connecting the Concepts

For more information on the relationship between energy and life, refer to the following discussions.

Section 6.1 explores how photosynthesis converts the kinetic energy of the sun into the potential energy of carbohydrates.

Section 7.1 provides an overview of how living organisms convert potential energy to kinetic energy.

Section 31.2 examines how energy moves within an ecosystem.

When light energy is converted to chemical energy in plant cells, some of the sun's energy is always lost as heat. In other words, the organization of a cell has a constant energy cost that also results in an increase in the entropy of the universe.

Check Your Progress 5.1

1. Contrast potential energy with kinetic energy; give an example of potential energy moving to kinetic energy.
2. Describe the two laws of thermodynamics.
3. Hypothesize how cells avoid entropy (disorder) and maintain their organization.

5.2 ATP: Energy for Cells

Learning Outcomes

Upon completion of this section, you should be able to

1. Summarize the role of adenosine triphosphate (ATP) in a cell.
2. Describe the phases of the ATP cycle.
3. Describe the flow of energy between photosynthesis and cellular respiration.

ATP (adenosine triphosphate) is the energy currency of cells. Just as you use coins to purchase all sorts of products, a cell uses ATP to carry out nearly all of its activities, including synthesizing proteins, transporting ions across the plasma membranes, and causing organelles and cilia to move. Cells use the readily accessible energy supplied by ATP to provide energy wherever it is needed.

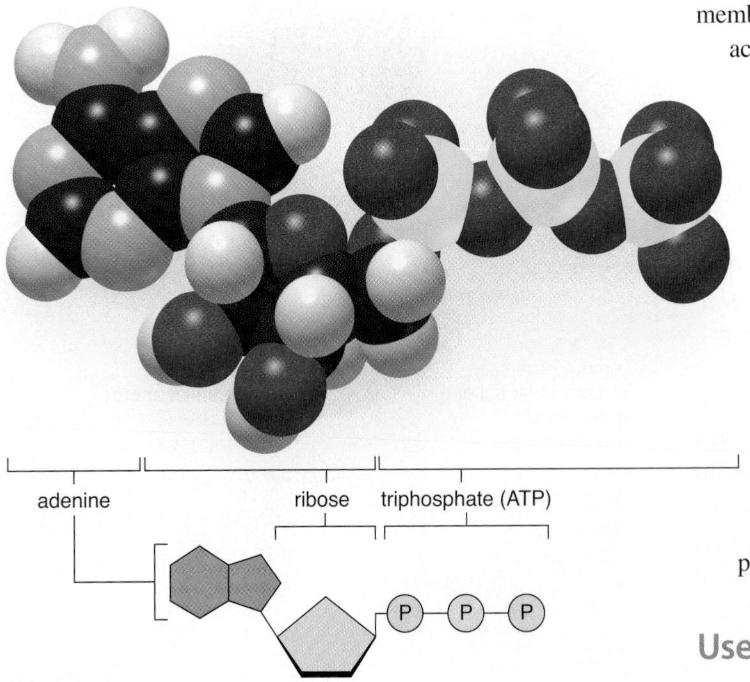

Structure of ATP

ATP is a nucleotide, the type of molecule that serves as a monomer for the construction of DNA and RNA. ATP's name, adenosine triphosphate, means that it contains the sugar ribose, the nitrogen-containing base adenine, and three phosphate groups (**Fig. 5.3**). The three phosphate groups are negatively charged and repel one another. It takes energy to overcome their repulsion; thus, these phosphate groups make the molecule unstable. ATP easily loses the last phosphate group because the breakdown products, ADP (adenosine diphosphate) and a separate phosphate group symbolized as P, are more stable than ATP. This reaction is written as ATP $\rightarrow$ ADP + P and energy is released as ATP breaks down. ADP can also lose a phosphate group to become AMP (adenosine monophosphate).

Use and Production of ATP

The continual breakdown and regeneration of ATP is known as the ATP cycle (**Fig. 5.4**). ATP holds energy for only a short period of time before it is used in a reaction that requires energy. Then ATP is rebuilt from ADP + P. Each ATP molecule undergoes about 10,000 cycles of synthesis and breakdown every day.

Figure 5.3 ATP.

ATP, the universal energy currency of cells, is composed of the nucleotide adenine, the sugar ribose, and three phosphate groups (called a triphosphate).

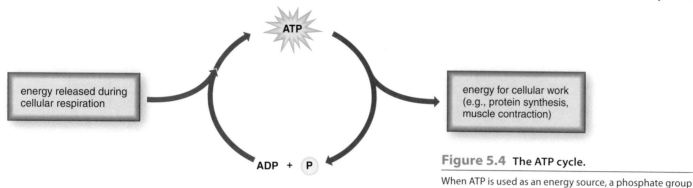

Figure 5.4 **The ATP cycle.**

When ATP is used as an energy source, a phosphate group is removed by hydrolysis. ATP is primarily regenerated in the mitochondria by cellular respiration.

Our body uses some 40 kg (about 88 lb) of ATP daily, and the amount on hand at any one moment is sufficiently high to meet only current metabolic needs.

ATP's instability, the very feature that makes it an effective energy carrier, keeps it from being an energy storage molecule. Instead, the many H—C bonds of carbohydrates and fats make them the energy storage molecules of choice. Their energy is extracted during cellular respiration and used to rebuild ATP, mostly within mitochondria. You will learn in Chapter 7 that the breakdown of 1 molecule of glucose permits the buildup of some 38 molecules of ATP. During cellular respiration, only 39% of the potential energy of glucose is converted to the potential energy of ATP; the rest is lost as heat.

The production of ATP is still worthwhile for the cell for the following reasons:

1. ATP releases energy quickly, which facilitates the speed of enzymatic reactions.
2. When ATP becomes ADP + (P), the amount of energy released is usually just enough for a biological purpose. Breaking down an entire carbohydrate or fat molecule would be wasteful, since it would release much more energy than is needed.
3. The structure of ATP allows its breakdown to be easily *coupled* to an energy-requiring reaction, as described next.

MP3
ATP

Coupled Reactions

Many metabolic reactions require energy. Energy can be supplied when a reaction that requires energy (e.g., building a protein) occurs in the vicinity of a reaction that gives up energy (e.g., ATP breakdown). A **coupled reaction** brings them together in such a way that the energy-releasing reaction can drive the energy-requiring reaction. Usually, the energy-releasing reaction is ATP breakdown, which generally releases more energy than the amount consumed by the energy-requiring reaction. This increases entropy, but both reactions will proceed. The simplest way to represent a coupled reaction is like this:

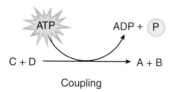

Coupling

This reaction tells you that coupling occurs, but it does not show how coupling is achieved. A cell has two main ways to couple ATP breakdown to an energy-requiring reaction: ATP may be used either to energize a reactant or to change its

Connections and Misconceptions

What is creatine and is it safe?

In humans, creatine is found in muscle cells as creatine phosphate (sometimes called phosphocreatine). Its function is to provide a brief, quick, recharge to ATP molecules in muscle cells. Over the past several years, creatine supplements have gained popularity with athletes for muscle-building as a performance enhancement. While there is some evidence that creatine supplements may increase performance for some individuals, there have been no long-term studies by the FDA on potential detrimental effects of creatine use on organs such as the kidneys, which are responsible for the secretion of excess creatine in the urine. In addition, creatine supplements can produce dangerous interactions when used with over-the-counter drugs—such as acetaminophen, and sometimes even caffeine. Additional studies are needed to determine if creatine supplements are safe.

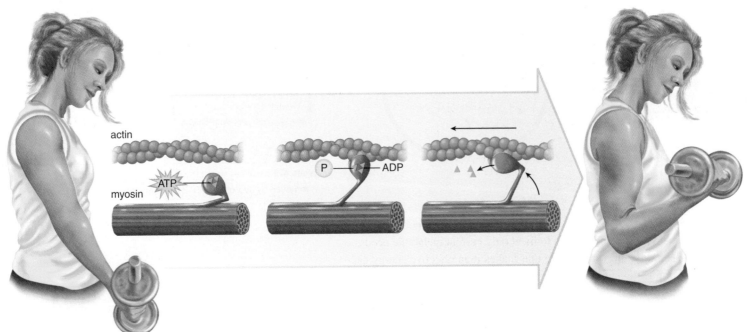

Figure 5.5 Coupled reaction.

Muscle contraction occurs only when it is coupled to ATP breakdown. Myosin combines with ATP prior to its breakdown. Release of ADP + Ⓟ causes myosin to change position and pull on an actin filament.

shape. Both are often achieved by transferring a phosphate group to the reactant. For example, when polypeptide synthesis occurs at a ribosome, an enzyme transfers a phosphate group from ATP to each amino acid in turn, and this transfer activates the amino acid, causing it to bond with another amino acid.

Figure 5.5 shows how ATP breakdown provides the energy necessary for muscular contraction. During muscle contraction, myosin filaments pull actin filaments to the center of the cell, and the muscle shortens. First, myosin combines with ATP, and only then does ATP break down to ADP + Ⓟ. The release of ADP + Ⓟ from the molecule causes myosin to change shape and pull on the actin filament.

Animation
Muscle Contraction

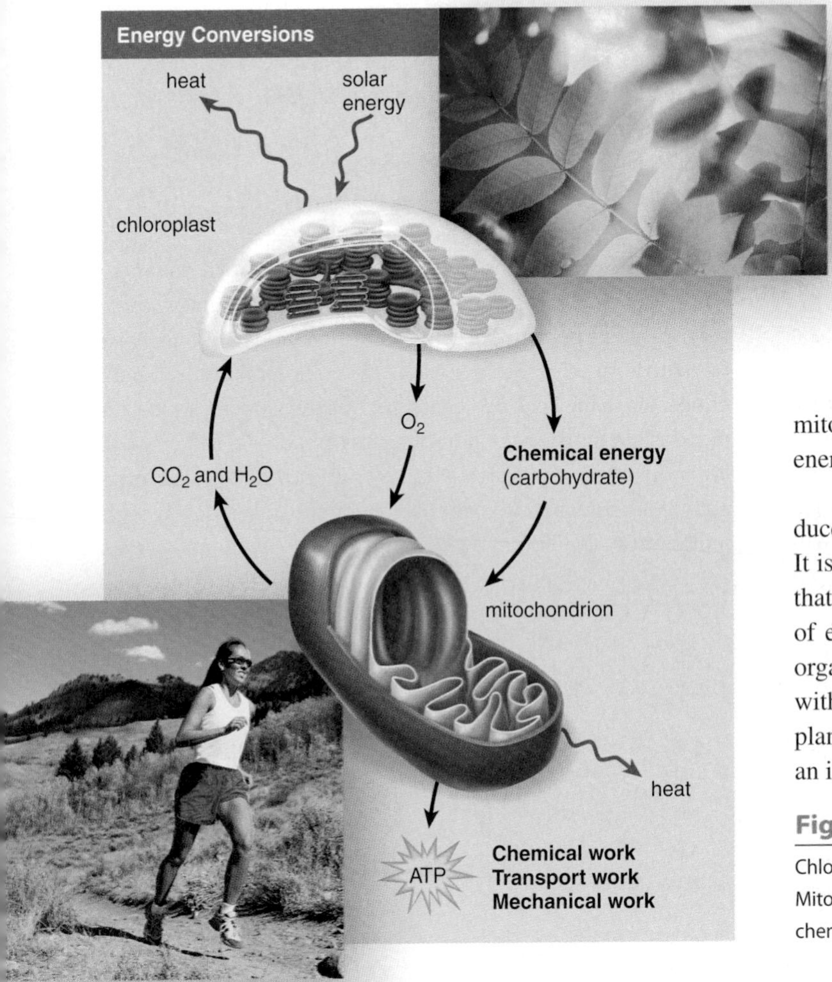

The Flow of Energy

In the biosphere, the activities of chloroplasts and mitochondria enable energy to flow from the sun through all living things. During photosynthesis, the chloroplasts in plants capture solar energy and use it to convert water and carbon dioxide to carbohydrates, which serve as food for themselves and for other organisms. During cellular respiration, mitochondria complete the breakdown of carbohydrates and use the released energy to build ATP molecules.

Notice in **Figure 5.6** that cellular respiration requires oxygen and produces carbon dioxide and water, the very molecules taken up by chloroplasts. It is actually the cycling of molecules between chloroplasts and mitochondria that allows a flow of energy from the sun through all living things. This flow of energy maintains the levels of biological organization from molecules to organisms to ecosystems. In keeping with the energy laws, useful energy is lost with each chemical transformation, and eventually the solar energy captured by plants is lost in the form of heat. In this way, living things are dependent upon an input of solar energy.

Energy Conversions

heat   solar energy

chloroplast

O_2

CO_2 and H_2O

Chemical energy
(carbohydrate)

mitochondrion

heat

ATP   **Chemical work**
Transport work
Mechanical work

Figure 5.6 Flow of energy.

Chloroplasts convert solar energy to the chemical energy stored in nutrient molecules. Mitochondria convert this chemical energy to ATP molecules, which cells use to perform chemical, transport, and mechanical work.

Human beings are also involved in the cycling of molecules between plants and animals and in the flow of energy from the sun. We inhale oxygen and eat plants and their stored carbohydrates, or we eat other animals that have eaten plants. Oxygen and nutrient molecules enter our mitochondria, which produce ATP and release carbon dioxide and water. Without a supply of energy-rich foods, we could not produce the ATP molecules needed to maintain our body and carry on activities.

Check Your Progress 5.2

1. Identify how ATP is produced and explain why ATP cannot be used as an energy storage molecule.
2. Illustrate a coupled reaction and explain the role of ATP in a coupled reaction.
3. Describe the flow of energy through the biosphere.

Connecting the Concepts

ATP plays an important role in many cellular and physiological processes. For examples, refer to the following discussions.

Section 6.3 explores how plants store ATP energy as glucose molecules.

Section 7.1 demonstrates the mitochondria's role in the generation of ATP.

Section 28.2 examines how ATP is used to power muscle contractions.

5.3 Metabolic Pathways and Enzymes

Learning Outcomes

Upon completion of this section, you should be able to

1. Illustrate how metabolic reactions are catalyzed by specific enzymes.
2. Identify the role that enzymes play in metabolic pathways.
3. Explain the induced fit model of enzymatic action.
4. Detail the processes that inhibit enzyme activity.
5. Relate the role of enzymes in lowering the energy of activation needed for a reaction.

Reactions do not occur haphazardly in cells. Just as a house is built in a series of ordered steps, chemical reactions in a cell usually occur in a particular order. They are part of a **metabolic pathway,** a series of linked reactions. Metabolic pathways begin with a particular reactant and terminate with an end product. For example, glucose is broken down by a series of consecutive reactions until the end products CO_2 and H_2O result. In the pathway, one reaction leads to the next reaction, which leads to the next reaction, and so forth in an organized, highly structured manner.

An example of a metabolic pathway is shown to the right. In this diagram, the letters A–F are reactants, and the letters B–G are products. In other words, the product from the previous reaction becomes the reactant of the next reaction. In the first reaction, A is the reactant and B is the product. Then B becomes the reactant in the next reaction of the pathway, and C is the product. This process continues until the final product (G) forms.

Animation
Biochemical
Pathways

In the preceding diagram, the letters e_1–e_6 represent **enzymes,** which are protein molecules that function as organic catalysts to speed chemical reactions. Enzymes can only speed reactions that are possible to begin with. In the cell, an enzyme is analogous to a mutual friend who causes two people to meet and interact, because an enzyme brings together particular molecules and causes them to react with one another.

$$\begin{array}{cccccc} e_1 & e_2 & e_3 & e_4 & e_5 & e_6 \\ A \rightarrow & B \rightarrow & C \rightarrow & D \rightarrow & E \rightarrow & F \rightarrow G \end{array}$$

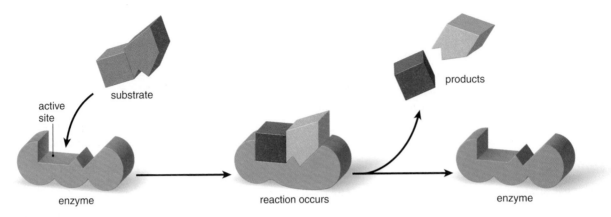

active site

substrate

enzyme

reaction occurs

products

enzyme

Figure 5.7 **Enzymatic action.**

An enzyme has an active site where the substrates and enzyme fit together in such a way that the substrates are oriented to react. Following the reaction, the products are released, and the enzyme is free to act again.

The molecules acted on by an enzyme are called its **substrates.** An enzyme converts substrates into products. The substrates and products of an enzymatic reaction vary greatly (**Fig. 5.7**). Many enzymes facilitate the breakdown of a substrate into multiple products. Or an enzyme may convert a single substrate into a single product. Still others may combine two or more substrates into a single product.

An Enzyme's Active Site

In most instances, only one small part of the enzyme, called the **active site,** accommodates the substrate(s). At the active site, the substrate fits into the enzyme seemingly as a key fits a lock; thus, most enzymes can fit only one substrate. However, the active site undergoes a slight change in shape in order to accommodate the substrate(s). This is called the **induced fit model** because the enzyme is induced to undergo a slight alteration to achieve optimal fit (see Fig. 5.7).

The change in the shape of the active site facilitates the reaction that occurs next. After the reaction has been completed, the product(s) is(are) released, and the active site returns to its original state, ready to bind to another substrate molecule. Only a very small amount of each enzyme is needed in a cell because enzymes are not used up by the reaction.

Animation
Enzyme-Controlled Reactions

Enzyme Inhibition

Enzyme inhibition occurs when an active enzyme is prevented from combining with its substrate. Enzyme inhibitors are often poisonous to certain organisms. Cyanide, for example, is an inhibitor of the enzyme cytochrome *c* oxidase, which performs a vital function in cells because it is involved in making ATP. Cyanide is a poison because it binds the enzyme, blocking its activity. But some enzyme inhibitors are useful drugs. For example, penicillin (**Fig. 5.8**) is a poison for bacteria, but not humans, because it blocks the active site of an enzyme unique to bacteria. Many other antibiotic drugs also act as enzyme inhibitors.

The activity of almost every enzyme in a cell is regulated by **feedback inhibition.** In the simplest case, when a product is in abundance, it competes with the substrate for the enzyme's active site. As the product is used up, inhibition is reduced, and then more product can be produced. In this way, the concentration of the product always stays within a certain range.

500mg
(800,000 units)
NDC 0029-8160-32

BEEPEN-VK®
PENICILLIN V POTASSIUM
TABLETS

500 Tablets

SB SmithKline Beecham

Figure 5.8 **Action of antibiotics.**

Some antibiotics, such as penicillin, work by inhibiting enzymes that are found only in bacteria.

Most metabolic pathways are regulated by more complex types of feed-back inhibition (**Fig. 5.9**). In these instances, when the end product is plentiful, it binds to a site other than the active site of the first enzyme in the pathway. This binding changes the shape of the active site, preventing the enzyme from binding to its substrate. Without the activity of the first enzyme, the entire pathway shuts down.

Animation
Feedback
Inhibition

Energy of Activation

Molecules frequently do not react with one another unless they are activated in some way. In the lab, activation is very often achieved by heating a mixture to increase the number of effective collisions between molecules. The added heat causes the molecules to move more quickly, increasing the likelihood of effective collisions between molecules. The energy needed to cause molecules to react with one another is called the **energy of activation (E_a). Figure 5.10** compares the E_a without an enzyme to the E_a with an enzyme, illustrating that the enzyme lowers the amount of energy required for a reaction to occur. Enzymes lower the energy of activation by bringing the substrates into contact and at times by participating in the reaction.

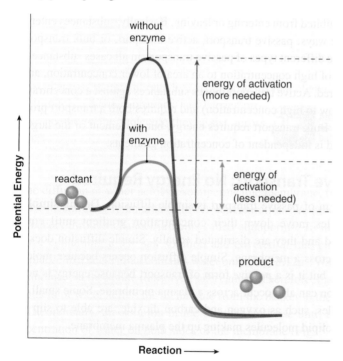

Figure 5.10 **Energy of activation (E_a).**

Enzymes speed the rate of reactions because they lower the amount of energy required for the reactants to react. Even reactions like this one, in which the energy of the product is less than the energy of the reactant, speed up when an enzyme is present.

Check Your Progress 5.3

1. Explain the benefit of metabolic pathways in cells.
2. Describe how the induced fit model explains the binding of a substrate to an enzyme's active site.
3. Summarize the benefit of using feedback inhibition to control metabolic pathways.

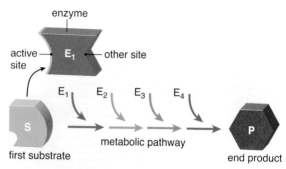

Active enzyme and active pathway

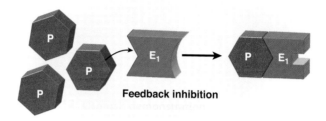

Feedback inhibition

Inactive enzyme and inactive pathway

Figure 5.9 **Feedback inhibition.**

This type of feedback inhibition occurs when the end product (P) of an active enzyme pathway is plentiful and binds to the first enzyme (E_1) of the pathway at a site other than the active site. This changes the shape of the active site, so that the substrate (S) can no longer bind to the enzyme. Now the entire pathway becomes inactive.

Connecting the Concepts

For more examples of how enzymes work in living organisms, refer to the following discussions.

Section 11.1 demonstrates the enzymes that are involved in the replication of DNA.

Section 24.3 examines some of the enzymes that are active in the human digestive system.

Section 27.1 investigates how enzymes are involved in the synapses between neurons.

Red blood cells

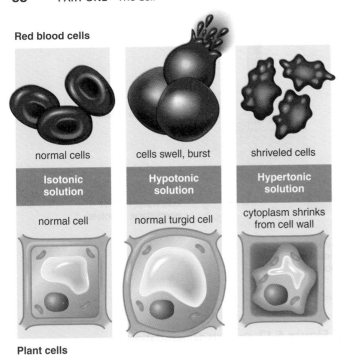

normal cells	cells swell, burst	shriveled cells
Isotonic solution	**Hypotonic solution**	**Hypertonic solution**
normal cell	normal turgid cell	cytoplasm shrinks from cell wall

Plant cells

Figure 5.13 Osmosis in animal and plant cells.

In an isotonic solution, cells neither gain nor lose water. In a hypotonic solution, cells gain water. Red blood cells swell to bursting, and plant cells become turgid. In a hypertonic solution, cells lose water. Red blood cells shrivel, and plant cell cytoplasm shrinks away from the cell wall.

Cells placed in a **hypotonic solution** (*hypo,* less than) gain water (**Fig. 5.13**). Outside the cell, the concentration of solute is less, and the concentration of water is greater, than inside the cell. Animal cells placed in a hypotonic solution expand and sometimes burst. The term *lysis* refers to disrupted cells; *hemolysis,* then, is disrupted red blood cells.

When a plant cell is placed in a hypotonic solution, the large central vacuole gains water, and the plasma membrane pushes against the rigid cell wall as the plant cell becomes *turgid.* The plant cell does not burst because the cell wall does not give way. Turgor pressure in plant cells is extremely important in maintaining their erect position.

Cells placed in a **hypertonic solution** (*hyper,* more than) lose water. Outside the cell, the concentration of solute is higher, and the concentration of water is lower, than inside the cell. Animal cells placed in a hypertonic solution shrink. For example, meats are sometimes preserved by being salted. Bacteria are killed not by the salt but by the lack of water in the meat.

When a plant cell is placed in a hypertonic solution, the plasma membrane pulls away from the cell wall as the large central vacuole loses water. This is an example of **plasmolysis,** shrinking of the cytoplasm due to osmosis. Cut flowers placed into salty water will wilt due to plasmolysis. The dead plants you may see along a roadside could have died due to exposure to a hypertonic solution during the winter, when salt was used on the road.

Active Transport: Energy Required

During **active transport,** molecules or ions move through the plasma membrane, accumulating on one side of the cell (**Fig. 5.14**). For example, iodine collects in the cells of the thyroid gland; glucose is completely absorbed from the digestive tract by the cells lining the digestive tract; and sodium can be almost completely withdrawn from urine by cells lining the kidney tubules. In these instances, molecules move against their concentration gradients, a situation that requires both a transport protein and much ATP. Therefore, cells involved in active transport, such as kidney cells, have many mitochondria near their plasma membranes to generate the ATP that is needed.

The passage of salt (NaCl) across a plasma membrane is of primary importance in cells because the salt causes water to move to that side of the plasma membrane. First, sodium ions are actively transported across a membrane, and then chloride ions simply diffuse through channels that allow their passage. Chloride ion channels malfunction in persons with cystic fibrosis, leading to the symptoms of this inherited disorder. Proteins engaged in active transport are often called *pumps.* The **sodium-potassium pump,** vital in nerve conduction, undergoes a change in shape that allows it to combine alternately with sodium ions and potassium ions.

Animation
Sodium-Potassium Pump

Low concentration

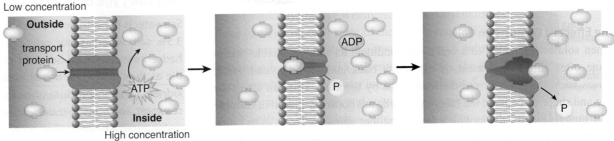

Outside

transport protein

ATP

Inside

ADP

P

P

High concentration

Figure 5.14 Active transport.

During active transport, a transport protein uses energy to move a solute across the plasma membranes toward a higher concentration. Note that the transport protein uses energy to change shape during the process.

Connections and Misconceptions

What causes cystic fibrosis?

In 1989, scientists determined that defects in a gene on chromosome 7 causes cystic fibrosis (CF). This gene, called *CFTR* (short for cystic fibrosis conductance regulator), codes for a protein that is responsible for the movement of chloride ions across the membranes of cells that produce mucus, sweat, and saliva. Defects in this gene cause an improper water/salt balance in the excretions of these cells, which in turn leads to the symptoms of CF. Currently, there are over 1,400 known mutations in the CF gene. This tremendous amount of variation accounts for the differences in the severity of the symptoms in CF patients. By knowing the precise gene that causes the disease, scientists have been able to develop new treatment options for people with CF. At one time, people with CF rarely saw their 20th birthday; now it is routine for people to live into their 30s and 40s. New treatments, such as gene therapy, are being explored for sufferers of CF.

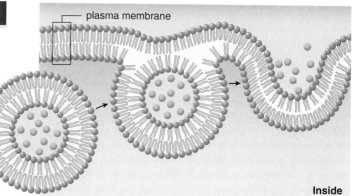

a. Exocytosis

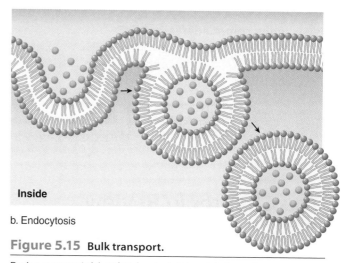

b. Endocytosis

Figure 5.15 Bulk transport.

During exocytosis **(a)** and endocytosis **(b)**, vesicle formation transports substances out of or into a cell, respectively.

Bulk Transport

Macromolecules, such as polypeptides, polysaccharides, and polynucleotides, are often too large to be moved by transport proteins. Instead, vesicle formation takes them into or out of a cell. For example, digestive enzymes and hormones use molecules transported out of the cell by **exocytosis** (**Fig. 5.15***a*). In cells that synthesize these products, secretory vesicles accumulate near the plasma membrane. These vesicles release their contents when the cell is stimulated to do so.

When cells take in substances by vesicle formation, the process is known as **endocytosis** (Fig. 5.15*b*). If the material taken in is large, such as a food particle or another cell, the process is called **phagocytosis.** Phagocytosis is common in unicellular organisms, such as amoebas. It also occurs in humans. Certain types of human white blood cells are amoeboid—that is, they are mobile, like an amoeba, and are able to engulf debris, such as worn-out red blood cells and bacteria. When an endocytic vesicle fuses with a lysosome, digestion occurs. In Chapter 26, we will see that this process is a necessary and preliminary step toward the development of immunity to bacterial diseases.

 **Animation** Endocytosis and Exocytosis

Pinocytosis occurs when vesicles form around a liquid or around very small particles. White blood cells, cells that line the kidney tubules and the intestinal wall, and plant root cells all use pinocytosis to ingest substances.

During **receptor-mediated endocytosis,** receptors for particular substances are found at one location in the plasma membrane. This location is called a coated pit because there is a layer of protein on its intracellular side (**Fig. 5.16**). Receptor-mediated endocytosis is selective and much more efficient than ordinary pinocytosis. It is involved when substances move from maternal blood into fetal blood at the placenta, for example.

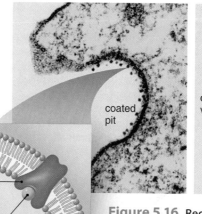
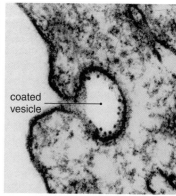

Figure 5.16 Receptor-mediated endocytosis.

During receptor-mediated endocytosis, molecules first bind to specific receptor proteins that are in a coated pit. The vesicle that forms contains the molecules and their receptors.

Connecting the Concepts

For more information on how living organisms use the transport mechanisms mentioned in this section, refer to the following discussions.

Section 20.7 describes how plants use osmotic pressure to move nutrients.

Section 24.2 examines how the sodium-potassium pump in the human kidney helps regulate fluid levels.

Section 27.1 investigates the role of the sodium-potassium pump in the movement of signals in the nervous system.

Media Study Tools

www.mhhe.com/maderessentials3

Enhance your study of this chapter with study tools and practice tests. Also ask your instructor about the resources available through ConnectPlus, including the media-rich eBook, interactive learning tools, and animations.

Enzyme-Controlled Reactions

The virtual lab "Enzyme-Controlled Reactions" provides an interactive investigation of how environmental conditions regulate enzyme activity.

The Chapter in Review

Summary

5.1 What Is Energy?

Potential energy can be converted to kinetic energy, and vice versa. Solar and chemical energy are forms of potential energy; mechanical energy is a form of kinetic energy. Two energy laws hold true universally:

- Energy cannot be created or destroyed, but it can be transferred or transformed.
- Energy can be converted from one form to another, but some is lost as unusable forms. Therefore, the entropy of the universe is increasing, and only a constant input of energy maintains the organization of living things.

5.2 ATP: Energy for Cells

Energy flows from the sun through chloroplasts and mitochondria, which produce ATP. Because ATP has three adjoining negative phosphate groups, it is a high-energy molecule that tends to break down to ADP + P, releasing energy. ATP breakdown is coupled to various energy-requiring cellular reactions, including protein synthesis, active transport, and muscle contraction. Cellular respiration provides the energy for the production of ATP. The following diagram summarizes the ATP cycle:

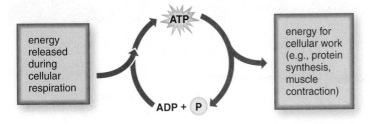

5.3 Metabolic Pathways and Enzymes

A metabolic pathway is a series of reactions that proceed in an orderly, step-by-step manner. Each reaction requires an enzyme that is specific to its substrate. Enzymes bring substrates together at an enzyme's active site and speed reactions by lowering the energy of activation. The activity of most enzymes and metabolic pathways is regulated by feedback inhibition.

5.4 Cell Transport

The plasma membrane is semipermeable; some substances can freely cross the membrane, and some must be assisted across if they are to enter the cell.

Passive transport requires no metabolic energy and moves substances from a higher to a lower concentration.

- In simple diffusion, molecules move from an area of higher concentration to the area of lower concentration until the concentration of molecules is the same at both sites. Some molecules, such as dissolved gases, cross plasma membranes by simple diffusion.
- In facilitated diffusion, molecules diffuse across a plasma membrane through a channel protein (aquaporins are channel proteins for water) or with the assistance of carrier proteins.
- Osmosis is the simple diffusion of water through aquaporins across a membrane toward the area of lower water concentration (higher solute concentration).

Active transport requires metabolic energy (ATP) and moves substances from a lower to a higher concentration across a membrane.

- A transport protein acts as a pump that causes a substance to move against its concentration gradient. For example, the sodium-potassium pump carries Na^+ to the outside of the cell and K^+ to the inside of the cell.

Bulk transport requires vesicle formation and metabolic energy. It occurs independently of concentration gradients.

- Exocytosis transports macromolecules out of a cell via vesicle formation and often results in secretion.
- Endocytosis transports macromolecules into a cell via vesicle formation.
- Receptor-mediated endocytosis makes use of receptor proteins in the plasma membrane.

The following diagram illustrates the types of passive and active transport:

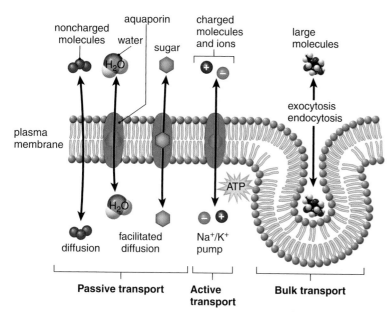

Passive transport   **Active transport**   **Bulk transport**

Key Terms

active site   84
active transport   88
aquaporin   86
calorie   78
coupled reaction   81
endocytosis   89
energy   78
energy laws   79
energy of activation (E$_a$)   85
entropy   79
enzyme   83
enzyme inhibition   84
exocytosis   89
facilitated diffusion   86
feedback inhibition   84
heat   79
hypertonic solution   88
hypotonic solution   88

induced fit model   84
isotonic solution   87
kilocalorie   78
kinetic energy   78
metabolic pathway   83
osmosis   87
phagocytosis   89
pinocytosis   89
plasmolysis   88
potential energy   78
receptor-mediated
  endocytosis   89
simple diffusion   86
sodium-potassium pump   88
solute   86
solution   86
solvent   86
substrate   84

Testing Yourself

Choose the best answer for each question.

1. Which of the following is not a fundamental law of energy?
   a. Energy cannot be created or destroyed.
   b. Energy can be changed from one form to another.
   c. Energy is lost when it is converted from one form to another.
   d. Potential energy cannot be converted to kinetic energy.

2. As a result of energy transformation,
   a. entropy increases.
   b. entropy decreases.
   c. potential energy increases.
   d. energy is lost in the form of heat.
   e. Both a and d are correct.

3. Identify the incorrect statement:
   a. When ATP becomes ADP + (P), the amount of energy released is enough for a biological purpose.
   b. The structure of ATP allows its breakdown to be easily coupled.
   c. ATP can be easily recycled.
   d. ATP consists of an adenine, a sugar, and two phosphates.

4. ATP is a good source of energy for a cell because
   a. it is versatile—able to be used in many types of reactions.
   b. its breakdown is easily coupled with energy-requiring reactions.
   c. it provides just the right amount of energy for cellular reactions.
   d. All of these are correct.

5. Compared with carbon dioxide and water, glucose
   a. is less organized.   c. is more stable.
   b. has less potential energy.   d. exhibits lower entropy.

For questions 6–9, match the items to the correct answers in the key. Each answer may include more than one item.

Key:
   a. oxygen
   b. carbon dioxide
   c. water
   d. carbohydrates

6. Consumed by cellular respiration.
7. Produced by cellular respiration.
8. Consumed by photosynthesis.
9. Produced by photosynthesis.

10. In the following figure, one curve illustrates the energy of activation required for a reaction "with an enzyme," while the other illustrates the energy of activation "without an enzyme." Label each curve.

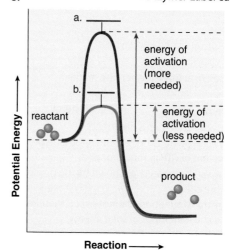

11. In the induced fit model of enzyme catalysis,
    a. the substrate exactly fits multiple enzymes.
    b. many substrates fit every enzyme.
    c. the enzyme changes shape slightly to accommodate the substrate.
    d. the enzyme doesn't fit the substrate until it is induced by another substrate.

For questions 12–18, match the items to the correct answers in the key. Each answer may include more than one item.

Key:

    a. simple diffusion
    b. facilitated diffusion
    c. osmosis
    d. active transport

12. Movement of molecules from high concentration to low concentration.
13. Requires a membrane.
14. Requires energy input.
15. Requires a protein pump.
16. Mechanism by which oxygen enters cells.
17. Mechanism by which glucose enters cells.
18. Movement of water across a membrane.
19. Cells involved in active transport have a large number of _____ near their plasma membrane.
    a. vacuoles
    b. mitochondria
    c. actin filaments
    d. aquaporins
20. The movement of water across the plasma membrane of a cell is facilitated by
    a. aquaporins.
    b. substrate-mediated endocytosis.
    c. receptor-mediated diffusion.
    d. channel proteins.
    e. Both a and d are correct.
21. A cell specializing in endocytosis would likely have many
    a. lysosomes.
    b. nuclei.
    c. mitochondria.
    d. aquaporins.
22. Seawater is _____ to blood because it contains a(n) _____ concentration of solutes.
    a. isotonic, higher
    b. hypotonic, higher
    c. isotonic, equal
    d. hypertonic, higher
23. A coated pit is associated with
    a. simple diffusion.
    b. osmosis.
    c. receptor-mediated endocytosis.
    d. pinocytosis.
24. Pumps used in active transport are made of
    a. sugars.
    b. proteins.
    c. lipids.
    d. cholesterol.
25. Which of these statements is correct?
    a. Energy can be transformed into another type, as shown by the production of ATP in mitochondria.
    b. Energy can be created, as shown by the production of ATP in mitochondria.
    c. Energy transformations cannot occur; therefore, ecosystems need a constant input of solar energy.
    d. Energy lost as heat is recaptured as decreased entropy.
26. Coupled reactions
    a. create energy through the hydrolysis of ATP.
    b. couple favorable reactions with unfavorable reactions.
    c. release energy trapped in enzymes.
    d. always result in decreased entropy.
27. *Entropy* is a term used to indicate the relative amount of
    a. organization.
    b. disorganization.
    c. enzyme action.
    d. None of these are correct.
28. Enzymes catalyze reactions by
    a. bringing the reactants together, so that they can bind.
    b. lowering the activation energy.
    c. Both a and b are correct.
    d. Neither a nor b is correct.
9. The active site of an enzyme
    a. is identical in structure for all enzymes.
    b. is the part of the enzyme where its substrate can fit.
    c. can be used only once.
    d. is not affected by environmental factors, such as pH and temperature.

Thinking Scientifically

1. Some coupled reactions in cells, including many involved in protein synthesis, use the nucleotide GTP as an energy source instead of ATP. What would be the advantage of using GTP instead of ATP as an energy source for these cellular reactions?

2. Cystic fibrosis is a genetic disorder caused by a defective membrane transport protein. The defective protein closes chloride channels in membranes, preventing chloride from being exported out of cells. This results in the development of a thick mucus on the outer surfaces of cells. This mucus clogs the ducts that carry digestive enzymes from the pancreas to the small intestine, clogs the airways in the lungs, and promotes lung infections. How do you think the defective protein results in a thick, sticky mucus outside the cells, instead of a loose, fluid covering?

3. Some prokaryotes, especially archaea, are capable of living in extreme environments, such as deep-sea heat vents, where temperatures can reach 80°C (176°F). Few organisms can survive at this temperature. What adaptations might archaea possess that allow them to survive in such extreme heat?

Bioethical Issue

Arsenic in Drinking Water

Arsenic is chemically similar to phosphorus, and it may form arsenate ions (AsO_4^{3-}) within the body that can substitute for phosphate (PO_4^{3-}) in many enzymatic reactions. Arsenate blocks some of the processes in the cell that are responsible for producing ATP, resulting in cell death. Metabolically active cells are particularly vulnerable to this toxin.

Arsenic occurs naturally in the water and soil; thus, arsenic levels in drinking water are a concern in many parts of the world. In 2000, the Clinton administration directed the Environmental Protection Agency (EPA) to reset the standard for maximal allowable levels in water to 10 parts per billion (ppb) from the previous level of 50 ppb set in 1942. These rules were suspended by the Bush administration in 2001, citing the high cost of compliance by utilities. Supporters asserted that the costs of compliance would be significant and that smaller utilities serving rural communities would be particularly hard hit. Other critics objected on the basis that the cost-to-benefit ratio of the new standards did not justify the new rules. But many health and environmental organizations and many members of Congress blasted the decision to rescind the new rules, citing studies demonstrating health problems associated with arsenic levels in water below 50 ppb. Proponents also pointed out that rural water customers would be the primary beneficiaries of the new rules. The Bush administration ultimately relented and reinstated the newer, lower standard.

Do you think the government did the right thing in reinstating the tougher standard? Why or why not? And should politics play a role in making scientific decisions, especially when health and environmental safety are at risk?

6

Energy for Life

Photosynthesis Is Essential to Life on Earth

Most people are aware that plants produce oxygen gas. Further, many think plants produce this oxygen as some sort of altruistic act to benefit humans and other animals! But the facts don't bear this out. During the process of photosynthesis, plants are actually making food for themselves. To do this, they reduce the carbon dioxide we exhale to carbohydrates after converting energy from the sun into chemical energy. This conversion process is complex, and it depends on the energizing of electrons removed from water. In the process, water splits and releases oxygen. This oxygen is the gas we and other animals depend upon. Thus, plants provide the oxygen we inhale and the glucose we use as a source of energy only because they are in the process of photosynthesizing to keep themselves alive. Plants are solar powered, and indirectly so are all other living things on Earth, including us.

Understanding the process of photosynthesis and its critical importance to nearly all living organisms will help you appreciate just how dependent we humans are on plants. Scientists are studying the complex process of photosynthesis to better understand it and how the process affects life, and even the environment. In fact, scientists are developing ways of using photosynthesis to produce electricity and reduce our dependence on fossil fuels! In this chapter, you will learn about basic plant cell structure and how plants perform photosynthesis.

Video Plants

BEFORE YOU BEGIN

Before beginning this chapter, take a few moments to review the following discussions.

Section 1.3 What is the role of the producers in an ecosystem?

Section 4.4 What is the structure of a chloroplast?

Section 5.2 How is ADP energized to form ATP?

6.1 Overview of Photosynthesis

Learning Outcomes

Upon completion of this section, you should be able to

1. Identify the cellular structures where photosynthesis occurs and list their functions.
2. State the overall chemical equation for photosynthesis.
3. Compare and contrast the light reactions of photosynthesis with the Calvin cycle reactions.

Photosynthesis transforms solar energy into the chemical energy of a carbohydrate. Photosynthetic organisms, including plants, algae, and cyanobacteria, produce an enormous amount of carbohydrate (**Fig. 6.1**). If the amount of carbohydrate were instantly converted to coal, and the coal loaded into standard railroad cars (each car holding about 50 tons), the photosynthesizers of the biosphere would fill more than 100 cars *per second* with coal.

Figure 6.1 Photosynthetic organisms.

Photosynthetic organisms include plants, such as flowers and mosses, that typically live on land; photosynthetic protists, such as *Euglena*, diatoms, and kelp, which typically live in water; and cyanobacteria, a type of bacterium that lives everywhere.

It is no wonder, then, that photosynthetic organisms are able to sustain themselves and, with a few exceptions,[1] all of the other living things on Earth. To appreciate this, consider that most food chains lead back to plants. In other words, producers, which have the ability to synthesize carbohydrates, feed not only themselves but also consumers, which must take in preformed organic molecules.

Our analogy about photosynthetic products becoming coal is apt because the bodies of plants formed the coal we burn today. This process occurred hundreds of thousands of years ago, and that is why coal is called a fossil fuel. The wood of trees is also commonly used as fuel. In addition, the fermentation of plant materials produces alcohol, which can be used to fuel automobiles directly or as a gasoline additive.

Plants as Photosynthesizers

The green portions of plants, particularly the leaves, carry on photosynthesis. Carbon dioxide in the air enters the many spaces of mesophyll tissue through small openings called **stomata** (sing., stoma). These spaces ensure that carbon dioxide, a raw material for photosynthesis, is efficiently delivered to mesophyll cells. The roots of a plant absorb water, which then moves in vascular tissue up the stem to the leaves, where it exits at leaf veins. Carbon dioxide and water diffuse into mesophyll cells and then into **chloroplasts,** the organelles that carry on photosynthesis (**Fig. 6.2**).

In a chloroplast, a double membrane surrounds a fluid-filled area called the **stroma.** A third membrane system within the stroma forms flattened sacs called **thylakoids,** which in some places are stacked to form **grana** (sing., granum), so called because early microscopists thought they looked like piles of seeds. The space within each thylakoid is believed to be connected to the space within every other thylakoid, thereby forming an inner compartment within chloroplasts called the *thylakoid space.*

Chlorophyll and other pigments reside within the membranes of the thylakoids. These pigments are capable of absorbing solar energy, the energy that drives photosynthesis. A thylakoid membrane also contains complexes that convert solar energy into a chemical form useable by the enzymes in the stroma. The stroma is an enzyme-rich region in which carbon dioxide is first attached to an organic compound and then reduced to a carbohydrate using the chemical energy provided by the thylakoid membranes.

This carbohydrate is the only source of energy for most organisms on Earth. When they metabolize the carbohydrate produced by chloroplasts, they release carbon dioxide into the air. This carbon dioxide enters leaves and is converted to a carbohydrate. The relationship between producers (e.g., plants) and consumers (e.g., animals) is a fundamental part of the intricate web of life.

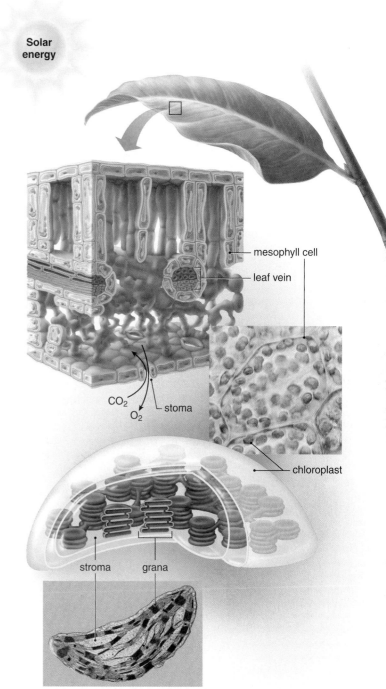

Solar energy

mesophyll cell

leaf vein

CO_2

O_2

stoma

chloroplast

stroma   grana

Figure 6.2 Leaves and photosynthesis.

The raw materials for photosynthesis are carbon dioxide and water. Water enters a leaf by way of leaf veins. Carbon dioxide enters a leaf by way of stomata. Chloroplasts have two major parts. The grana are made up of thylakoids, membranous disks that contain photosynthetic pigments, such as chlorophyll. These pigments absorb solar energy. The stroma is a fluid-filled space where carbon dioxide is enzymatically processed to form carbohydrates, such as glucose.

[1]A few types of bacteria are chemosynthetic organisms, which obtain the necessary energy to produce their own organic nutrients by oxidizing inorganic compounds.

Connections and Misconceptions

Where did the oxygen in the atmosphere come from?

The atmosphere of the early Earth was lacking in oxygen but rich in carbon dioxide and other gases. The first photosynthetic organisms are believed to have appeared around 3.5 billion years ago (BYA). Most scientists think that these were the cyanobacteria (see Fig. 6.1) or their close relatives. For almost a billion years, these organisms churned out oxygen as a by-product of photosynthesis. Then, beginning around 1.5 BYA, single-celled photosynthetic eukaryotes (such as algae, and the diatoms shown here) began to make a contribution. It was not until between 400 and 450 BYA that land plants participated in the production of oxygen.

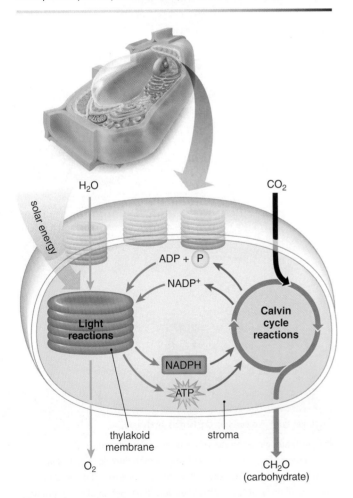

Figure 6.3 The photosynthetic process.

The process of photosynthesis consists of the light reactions and the Calvin cycle reactions. The light reactions, which produce ATP and NADPH, occur in a thylakoid membrane. These molecules are used by the Calvin cycle in the stroma to reduce carbon dioxide to a carbohydrate.

The Photosynthetic Process

The product of photosynthesis, carbohydrate (CH_2O), contains a lot more energy than the reactants carbon dioxide (CO_2) and water (H_2O). Why? Electrons are removed from H_2O and energized by solar energy before they are transferred to CO_2. These electrons are accompanied by hydrogen ions (H^+), and that's why hydrogen atoms are apparently transferred during photosynthesis:

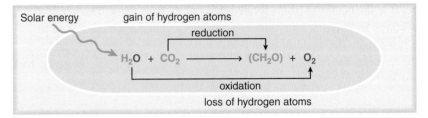

When a molecule gains hydrogen atoms ($e^- + H^+$), **reduction** occurs and the molecule is said to be reduced. Therefore, carbon dioxide is reduced when it becomes CH_2O. When a molecule gives up hydrogen atoms ($e^- + H^+$), **oxidation** occurs and the molecule is said to be oxidized. Therefore, water is oxidized when it splits and releases O_2. In metabolism, reduction goes along with oxidation and the reaction is called a **redox reaction.** Oxidation and reduction reactions are usually coupled together within a cell.

Notice that if you multiplied (CH_2O) by 6 you would get $C_6H_{12}O_6$, which is the formula for glucose. While glucose is often described as the final product of photosynthesis, in reality the reduction of CO_2 produces a molecule called G3P (glyceraldehyde 3-phosphate). G3P molecules are often combined to form glucose shortly after the photosynthetic reactions.

Two Sets of Reactions

An overall equation for photosynthesis tells us the beginning reactant and the end products of the pathway. But much goes on in between. The word *photosynthesis* suggests that the process requires two sets of reactions. *Photo,* which means light, refers to the reactions that capture solar energy, and *synthesis* refers to the reactions that produce a carbohydrate. The two sets of reactions are called the **light reactions** and the **Calvin cycle reactions (Fig. 6.3).**

Light Reactions The following events occur in the thylakoid membrane during the light reactions:

- Chlorophyll within the thylakoid membranes absorbs solar energy and energizes electrons.
- Water is oxidized, releasing electrons, hydrogen ions (H^+), and oxygen.
- ATP is produced from ADP + ⓟ with the help of an electron transport chain.
- $NADP^+$, an enzyme helper,[2] accepts electrons (is reduced) and becomes NADPH.

Calvin Cycle Reactions The following events occur in the stroma during the Calvin cycle reactions:

- CO_2 is taken up by one of the molecules in the cycle.
- ATP and NADPH from the light reactions reduce CO_2 to a carbohydrate (G3P).

In the following sections, we discuss the details of these two reactions.

[2]NADP = nicotinamide adenine dinucleotide phosphate.

Connecting the Concepts

For more information on photosynthesis, including additional examples of photosynthetic organisms, refer to the following discussions.

Section 17.3 provides additional information on the photosynthetic autotrophs called cyanobacteria.

Section 17.4 examines photosynthetic protists, such as the algae and diatoms.

Section 31.2 explores the role of photosynthetic organisms as producers in an ecosystem.

Check Your Progress 6.1

1 Identify the reasons that photosynthesis in plants is an important process for life on Earth.

2 Describe all the structures in plants that allow photosynthesis to occur.

3 Distinguish between a reduction and an oxidation reaction.

6.2 The Light Reactions—Harvesting Energy

Learning Outcomes

Upon completion of this section, you should be able to

1. List the types of photosynthetic pigments and briefly explain their functions.
2. Explain the electron pathway of the light reactions and list the order in which each complex participates in the pathway.
3. Explain how the energy harvested by the electron transport chain is utilized to make ATP and NADPH.

During the light reactions, the pigments within the thylakoid membranes absorb solar (radiant) energy. Solar energy can be described in terms of its wavelength and its energy content. **Figure 6.4** lists the types of radiant energy, from the shortest wavelength, gamma rays, to the longest, radio waves. *Visible light* is only a small portion of this spectrum.

Visible light contains various wavelengths of light, as can be proven by passing it through a prism; the different wavelengths appear to us as the colors of the rainbow, ranging from violet (the shortest wavelength) to blue, green, yellow, orange, and red (the longest wavelength). The energy content is highest for violet light and lowest for red light.

Only about 42% of the solar radiation that hits the Earth's atmosphere ever reaches its surface, and most of this radiation is within the visible-light range. Higher-energy wavelengths are screened out by the ozone layer in the atmosphere, and lower-energy wavelengths are screened out by water vapor and CO_2 before they reach the Earth's surface. Both the organic molecules within organisms and certain life processes, such as vision and photosynthesis, are adapted to utilize the radiation that is most prevalent in the environment.

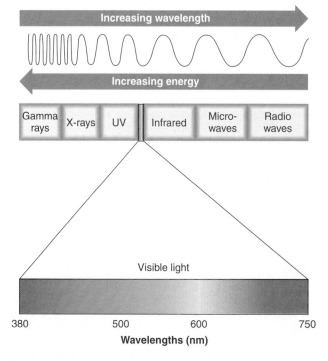

Figure 6.4   Radiant energy.

Radiant energy exists in a range of wavelengths extending from the very short wavelengths of gamma rays through the very long wavelengths of radio waves. Visible light, which drives photosynthesis, is expanded to show its component colors. The colors differ according to wavelength and energy content.

Connections and Misconceptions

How do the grow-lights in a greenhouse work?

The grow-lights in commercial greenhouses do not emit all of the wavelengths of visible light. Instead, these lamps have been designed to emit more strongly in the blue and red wavelengths of light. Since these  are the preferred wavelengths of the chlorophyll molecules, the growers can optimize plant growth. You will also notice that the color of your clothes may appear slightly different under a grow-light. This is because the light reaching the pigments in the cloth does not contain all of the wavelengths, so not all the colors are reflected back to your eyes.

Photosynthetic Pigments

The pigments found within most types of photosynthesizing cells are **chlorophylls** and **carotenoids.** These pigments are capable of absorbing portions of visible light. Both chlorophyll *a* and chlorophyll *b* absorb violet, blue, and red wavelengths better than those of other colors. Because green light is reflected and only minimally absorbed, leaves appear green to us. Accessory pigments, such as the carotenoids, appear yellow or orange because they are able to absorb light in the violet-blue-green range, but not the yellow-orange range. These pigments and others become noticeable in the fall when chlorophyll breaks down and the other pigments are uncovered (**Fig. 6.5**).

Animation
Absorption of Light

The Electron Pathway of the Light Reactions

Much as a solar energy panel captures the sun's energy and stores it in a battery, the light reactions capture the sun's energy and store it in the form of a hydrogen ion (H^+) gradient. This hydrogen ion gradient is used to produce ATP molecules. The light reactions also produce NADPH (**Fig. 6.6**).

The light reactions use two **photosystems,** called photosystem I (PS I) and photosystem II (PS II). The photosystems are named for the order in which they were discovered, not for the order in which they participate in the photosynthetic process. Here is a summary of how the pathway works:

- *Both photosystems receive photons.* PS II and PS I consist of a pigment complex (containing chlorophyll and carotenoid molecules) and an electron acceptor. The pigment complex serves as an "antenna" for gathering solar energy. The energy is passed from one pigment to the other until it is concentrated in a particular pair of chlorophyll *a* molecules, called the reaction center.

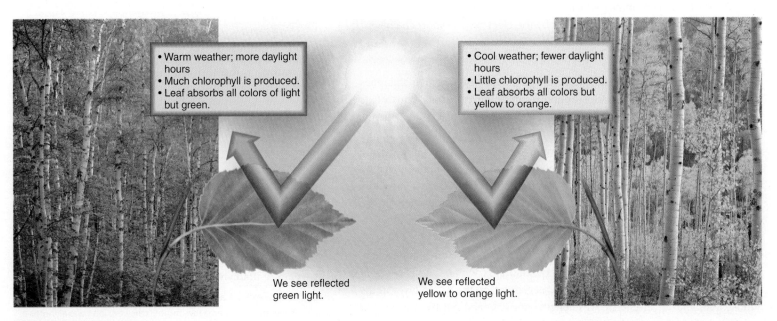

- Warm weather; more daylight hours
- Much chlorophyll is produced.
- Leaf absorbs all colors of light but green.

We see reflected green light.

- Cool weather; fewer daylight hours
- Little chlorophyll is produced.
- Leaf absorbs all colors but yellow to orange.

We see reflected yellow to orange light.

Figure 6.5   Leaf colors.

During the summer, leaves appear green because the chlorophylls absorb other portions of the visible spectrum better than they absorb green. During the fall, chlorophyll breaks down, and the remaining carotenoids cause leaves to appear yellow to orange because they do not absorb these colors.

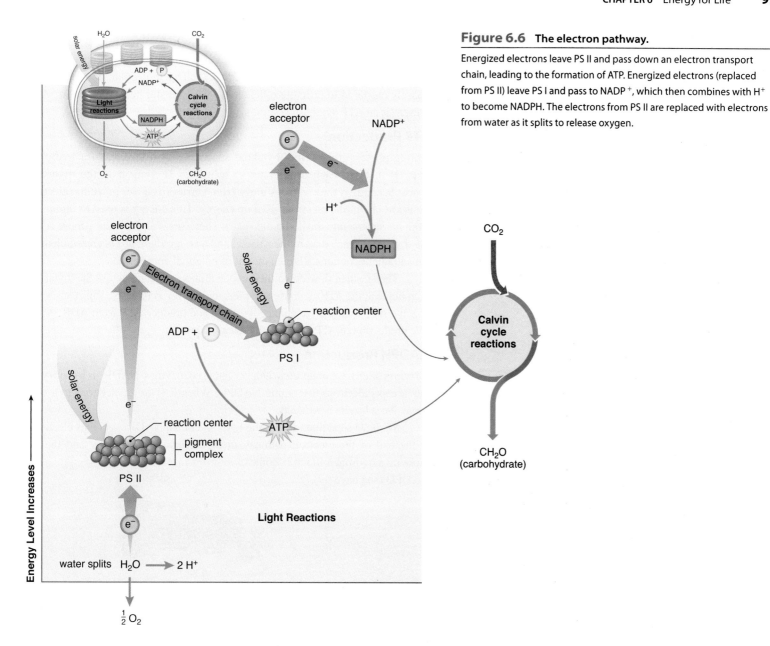

Figure 6.6 The electron pathway.

Energized electrons leave PS II and pass down an electron transport chain, leading to the formation of ATP. Energized electrons (replaced from PS II) leave PS I and pass to $NADP^+$, which then combines with H^+ to become NADPH. The electrons from PS II are replaced with electrons from water as it splits to release oxygen.

- *PS II splits water.* Due to the absorption of solar energy, electrons (e^-) in the reaction center of PS II become so energized that they escape and move to a nearby electron-acceptor molecule. Replacement electrons are removed from water, which splits, releasing oxygen and two hydrogen ions (H^+). The electron acceptor sends energized electrons, received from the reaction center, down an electron transport chain.

- *The electron transport chain establishes an energy gradient.* In an **electron transport chain**, a series of carriers pass electrons from one to the other, releasing energy stored in the form of a hydrogen ion (H^+) gradient. Later, ATP is produced (see page 100).

- *PS I produces NADPH.* When the PS I pigment complex absorbs solar energy, energized electrons leave its reaction center and are captured by a different electron acceptor. (Low-energy electrons from the electron transport chain adjacent to PS II replace those lost by PS I.) The electron acceptor in PS I passes its electrons to a $NADP^+$ molecule. $NADP^+$ accepts electrons and a hydrogen ion and is reduced, becoming NADPH.

Organization of the Thylakoid Membrane

PS I, PS II, and the electron transport chain are molecular complexes located within the thylakoid membrane (**Fig. 6.7**). The molecular complexes are densely clustered to promote efficient electron transfer between carriers. Also present is an ATP synthase complex.

ATP Production

During photosynthesis, the thylakoid space acts as a reservoir for hydrogen ions (H$^+$). First, each time water is split, two H$^+$ remain in the thylakoid space. Second, as the electrons move from carrier to carrier down the electron transport chain, the electrons give up energy. This energy is used to pump H$^+$ from the stroma into the thylakoid space. Therefore, there are many more H$^+$ in the thylakoid space than in the stroma, and a H$^+$ gradient has been established. This gradient contains a large amount of potential energy.

The H$^+$ then flow down their concentration gradient, across the thylakoid membrane at the ATP synthase complex, and energy is released. This causes the enzyme **ATP synthase** to change its shape and produce ATP from ADP + (P). The production of ATP captures the released energy.

NADPH Production

Enzymes often have nonprotein helpers called **coenzymes**. NADP$^+$ is a coenzyme that accepts electrons, becoming NADPH. When it gives up electrons to a substrate, the substrate is reduced.

We have seen that, during the light reactions, NADP$^+$ receives electrons at the end of the electron pathway in the thylakoid membrane, and then it picks up a hydrogen ion to become NADPH (see page 99).

Animation
Proton Pump

Video
Spinach Battery

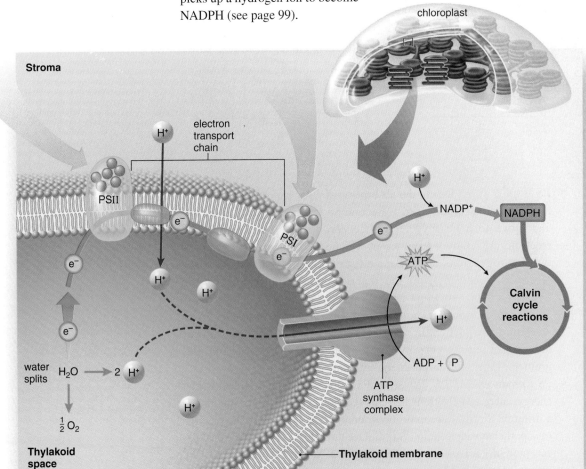

Figure 6.7 Organization of a thylakoid.

Molecular complexes of the electron transport chain within the thylakoid membrane pump hydrogen ions from the stroma into the thylakoid space. When hydrogen ions flow back out of the space into the stroma through the ATP synthase complex, ATP is produced from ADP + (P). NADP$^+$ accepts two electrons and joins with H$^+$ to become NADPH.

Check Your Progress 6.2

1. Identify the events that occur during the light reactions.
2. Illustrate the pathway electrons take during the light reactions.
3. Hypothesize how the structure of the thylakoid membrane allows the light reactions to function.

Connecting the Concepts

For more on the chemical principles decribed in this section, refer to the following discussions.

Figure 5.4 illustrates the ATP cycle.

Section 5.4 describes how proteins aid in the movement of ions across membranes.

6.3 The Calvin Cycle Reactions— Making Sugars

Learning Outcomes

Upon completion of this section, you should be able to

1. List the three stages of the Calvin cycle and describe the major event that occurs during each stage.
2. Describe how ATP and NADPH are utilized in the reduction of CO_2.
3. Summarize how the output of the Calvin cycle is used to make other carbohydrates.

The ATP and NADPH generated by the light reactions power the Calvin cycle reactions. The Calvin cycle reactions occur in the stroma of chloroplasts and consist of a series of reactions that produce carbohydrate before returning to the starting point (**Fig. 6.8**). The end product of the Calvin cycle is often considered to be glucose, as we shall see. The cycle is named for Melvin Calvin, who, with colleagues, used the radioactive isotope ^{14}C as a tracer to discover the reactions that make up the cycle.

This series of reactions utilizes carbon dioxide from the atmosphere and consists of these three steps: (1) carbon dioxide fixation, (2) carbon dioxide reduction, and (3) regeneration of the first substrate, RuBP (ribulose-1, 5-bisphosphate). The last step also results in a molecule of G3P (glyceraldehyde 3-phosphate), which plants use to produce glucose and other types of organic molecules. The Calvin cycle requires a lot of energy, which is supplied by ATP and NADPH from the light reactions.

Fixation of Carbon Dioxide

Carbon dioxide (CO_2) fixation is the first step of the Calvin cycle. During this reaction, CO_2 from the atmosphere is attached to RuBP, a 5-carbon molecule. The enzyme for this reaction is called **RuBP carboxylase (rubisco),** and the result is a 6-carbon molecule that splits into two 3-carbon molecules.

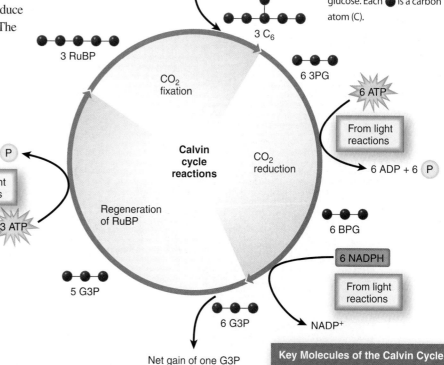

Figure 6.8   The Calvin cycle reactions.

The Calvin cycle is divided into three portions: CO_2 fixation, CO_2 reduction, and regeneration of RuBP. Because five G3P are needed to re-form three RuBP, it takes three turns of the cycle to achieve a net gain of one G3P. Two G3P molecules are needed to form glucose. Each ● is a carbon atom (C).

Key Molecules of the Calvin Cycle	
RuBP	ribulose 1,5-bisphosphate
3PG	3-phosphoglycerate
BPG	1,3-bisphosphoglycerate
G3P	glyceraldehyde-3-phosphate

Connections and Misconceptions

Can forests absorb enough carbon dioxide to offset climate change?

When speaking of carbon absorption by forests, you are actually talking about the process of carbon dioxide fixation in the Calvin cycle. The Environmental Protection Agency estimates that, in 2007, forests in the United States absorbed over 1,062.5 million metric tons of CO_2. However, during the same period, the combustion of fossil fuels released about 5,235.8 million metric tons of CO_2 into the atmosphere. While other factors are involved in the cycling of carbon (see Chapter 31), it is unlikely that in the United States that forests alone can absorb enough carbon to balance the equation.

Figure 6.9 The fate of G3P.

G3P is the first reactant in a number of plant cell metabolic pathways, such as those that lead to products other than carbohydrates (blue box) and those that are carbohydrates (green boxes). Two G3Ps are needed to form glucose phosphate; glucose is often considered the end product of photosynthesis. Sucrose is the transport sugar in plants; starch is the storage form of glucose; and cellulose is a major constituent of plant cell walls.

Reduction of Carbon Dioxide

Reduction of CO_2 is the sequence of reactions that uses NADPH and some of the ATP from the light reactions. Electrons are added, and carbon dioxide is reduced to a carbohydrate as R–CO_2 becomes R–CH_2O. NADPH and ATP supply the needed electrons and energy for CO_2 reduction, respectively.

Regeneration of RuBP

The product of the Calvin cycle is actually G3P, which is used to form glucose. Notice that the Calvin cycle reactions in Figure 6.8 are multiplied by three because it takes three turns of the Calvin cycle to allow one G3P to exit for making glucose and other carbohydrates. Why? For every three turns of the Calvin cycle, five molecules of G3P are used to re-form three molecules of RuBP, a 5-carbon molecule. These reactions also utilize some of the ATP produced by the light reactions.

▶ **Animation** The Calvin Cycle

The Fate of G3P

Compared with animal cells, algae and plants have enormous biochemical capabilities. From a G3P molecule, they can make all the molecules they need (**Fig. 6.9**). A plant can utilize the hydrocarbon skeleton of G3P to form fatty acids and glycerol, which are combined in plant oils. We are all familiar with corn oil, sunflower oil, and olive oil used in cooking. Also, when nitrogen is added to the hydrocarbon skeleton derived from G3P, amino acids are formed.

Notice also that glucose phosphate is among the organic molecules that result from G3P metabolism. Glucose is the molecule that plants and animals most often metabolize to produce ATP molecules to meet their energy needs. Glucose phosphate can be combined with fructose (and the phosphate removed) to form sucrose, the molecule that plants use to transport carbohydrates from one part of the plant to another. Glucose phosphate is also the starting point for the synthesis of starch and cellulose. Starch is the storage form of glucose. Some starch is stored in chloroplasts, but most starch is stored in amyloplasts in plant roots. Cellulose is a structural component of plant cell walls; it serves as fiber in our diet because we are unable to digest it.

Connecting the Concepts

For more information on how plants transport the outputs of photosynthesis, refer to the following discussions.

Section 20.7 examines the movement of carbohydrates and other nutrients in a plant.

Section 31.2 explores the role of forests in the carbon cycle.

Check Your Progress 6.3

1. Describe the steps of the Calvin cycle.
2. Contrast the Calvin cycle with the light reactions.
3. Hypothesize why plants have the ability to change G3P into so many different molecules.

6.4 Other Types of Photosynthesis

Learning Outcomes

Upon completion of this section, you should be able to

1. Define C_4 photosynthesis and explain why some plants must use this type of photosynthesis.
2. Describe both the advantages and the disadvantages of C_4 photosynthesis over C_3 photosynthesis.
3. Compare and contrast the leaf structure of a C_3 plant with that of a C_4 plant.
4. Explain CAM photosynthesis and describe the conditions under which plants can use it.

Figure 6.10   Carbon dioxide fixation in C_3 plants.

In C_3 plants, such as these wildflowers, CO_2 is taken up by the Calvin cycle directly in mesophyll cells, and the first detectable molecule is a C_3 molecule (red).

Plants are physically adapted to their environments. In the cold, windy climates of the north, evergreen trees have small, narrow leaves that look like needles. In the warm, wet climates of the south, some evergreen trees have large, flat leaves to catch the rays of the sun.

In the same way, plants are metabolically adapted to their environments. Where temperature and rainfall tend to be moderate, plants carry on C_3 photosynthesis, and are therefore called C_3 plants. In a **C_3 plant,** the first detectable molecule after CO_2 fixation is a C_3 molecule composed of three carbon atoms (**Fig. 6.10**). Look again at the Calvin cycle (see Fig. 6.8), and notice that the C_6 molecule that forms when RuBP combines with carbon dioxide immediately breaks down to two C_3 molecules. In a C_3 leaf, **mesophyll cells,** arranged in parallel rows, contain well-formed chloroplasts. The Calvin cycle, including CO_2 fixation, occurs in the chloroplasts of mesophyll cells. The majority of plant species carry out C_3 photosynthesis.

If the weather is hot and dry, stomata close, preventing the loss of water. (Water loss might cause the plant to wilt and die.) But this also prevents CO_2 from entering the leaf and traps O_2, a by-product of photosynthesis, within the leaf spaces, where it can diffuse back into mesophyll cells. In C_3 plants, this O_2 competes with CO_2 for the active site of *rubisco*, the first enzyme of the Calvin cycle, and less C_3 is produced. Such yield decreases are of concern because many food crops are C_3 plants.

Some plants have evolved an adaptation that allows them to be successful in hot, dry conditions. These plants carry out C_4 photosynthesis instead of C_3 photosynthesis. In a **C_4 plant,** the first detectable molecule following CO_2 fixation is a C_4 molecule composed of four carbon atoms. C_4 plants are able to avoid the uptake of O_2 by rubisco. Let's explore how C_4 plants do this.

C_4 Photosynthesis

The anatomy of a C_4 plant is different from that of a C_3 plant (**Fig. 6.11**). In a C_4 leaf, chloroplasts are located in the mesophyll cells, but they are also located in bundle sheath cells, which surround the leaf vein. Further, the mesophyll cells are arranged concentrically around the bundle sheath cells, shielding the bundle sheath cells from O_2 in the leaf spaces.

In C_4 plants, the Calvin cycle occurs only in the bundle sheath cells and not in the mesophyll cells. Because the bundle sheath cells are not accessible to leaf spaces, the CO_2 needed for the Calvin cycle is not directly taken from the air. Instead, CO_2 is fixed in mesophyll cells by a C_3 molecule, and a C_4 molecule forms. The C_4 molecule is modified and then pumped into bundle

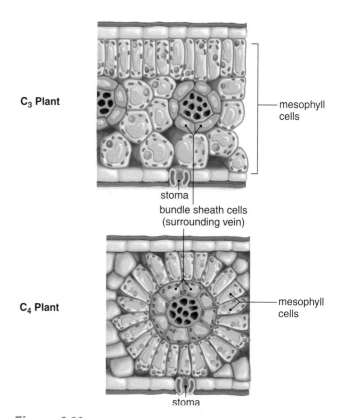

Figure 6.11   Comparison of C_3 and C_4 plant anatomy.

In C_3 plants, mesophyll cells are arranged in rows and contain chloroplasts. The Calvin cycle, including CO_2 fixation, occurs in mesophyll cells. In C_4 plants, mesophyll cells are arranged in rings around bundle sheath cells, and both contain chloroplasts. CO_2 fixation occurs within mesophyll cells, and the Calvin cycle occurs in the bundle sheath cells.

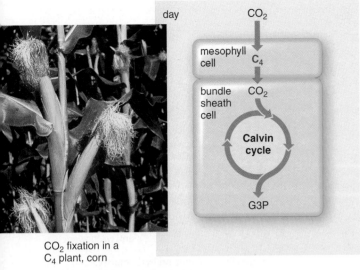

CO$_2$ fixation in a
C$_4$ plant, corn

Figure 6.12 **Carbon dioxide fixation in C$_4$ plants.**

In C$_4$ plants, such as corn, CO$_2$ fixation results in C$_4$ molecules (red) in mesophyll cells. The C$_4$ molecule releases CO$_2$ to the Calvin cycle in bundle sheath cells.

CO$_2$ fixation in a
CAM plant, pineapple

Figure 6.13 **Carbon dioxide fixation in a CAM plant.**

CAM plants, such as pineapple, fix CO$_2$ at night, forming a C$_4$ molecule (red). The C$_4$ molecule releases CO$_2$ to the Calvin cycle during the day.

Connecting the Concepts

For more information on the material presented in this section, refer to the following discussions.

Section 20.3 explores the structure of a plant leaf, and describes the location of the mesophyll cells.

Section 31.3 examines the relationship of temperature and rainfall to the types of organisms found in an ecosystem.

sheath cells (**Fig. 6.12**). Then CO$_2$ enters the Calvin cycle. This represents partitioning of CO$_2$ fixation and the Calvin cycle in space.

It takes energy to pump molecules, and you would think that the C$_4$ pathway would be disadvantageous. But in hot, dry climates, the net photosynthetic rate of C$_4$ plants, such as sugarcane, corn, and Bermuda grass, is two to three times that of C$_3$ plants, such as wheat, rice, and oats. Why do C$_4$ plants enjoy such an advantage? When the weather is hot and dry and stomata close, rubisco is not exposed to O$_2$, and yield is maintained.

When the weather is moderate, C$_3$ plants ordinarily have the advantage, but when the weather becomes hot and dry, C$_4$ plants have their chance to take over, and we can expect them to predominate. In the early summer, C$_3$ plants, such as Kentucky bluegrass and creeping bent grass, predominate in lawns in the cooler parts of the United States, but in midsummer, crabgrass, a C$_4$ plant, begins to take over.

CAM Photosynthesis

Another type of photosynthesis is called **CAM,** which stands for crassulacean-acid metabolism. It gets its name from the Crassulaceae, a family of flowering succulent (water-containing) plants that live in warm, arid regions. CAM was first discovered in these plants, but now it is known to be prevalent among most succulent plants that grow in desert environments, including cactuses.

Whereas a C$_4$ plant represents partitioning in space (that is, CO$_2$ fixation occurs in mesophyll cells, and the Calvin cycle occurs in bundle sheath cells), CAM is partitioning by the use of time. During the night, CAM plants use C$_3$ molecules to fix CO$_2$, forming C$_4$ molecules. These molecules are stored in large vacuoles in mesophyll cells. During the day, the C$_4$ molecules release CO$_2$ to the Calvin cycle when NADPH and ATP are available from the light reactions (**Fig. 6.13**).

The primary advantage of this partitioning again relates to the conservation of water. CAM plants open their stomata only at night; therefore, only at that time is atmospheric CO$_2$ available. During the day, the stomata close. This conserves water, but CO$_2$ cannot enter the plant.

Photosynthesis in a CAM plant is minimal because of the limited amount of CO$_2$ fixed at night, but it does allow CAM plants to live under stressful conditions.

Evolutionary Trends

C$_4$ plants most likely evolved in, and are adapted to, areas of high light intensities, high temperatures, and limited rainfall. However, C$_4$ plants are more sensitive to cold, and C$_3$ plants probably do better than C$_4$ plants below 25°C.

CAM plants, on the other hand, compete well with either type of plant when the environment is extremely arid. Surprisingly, CAM is quite widespread and has evolved in 30 families of flowering plants, including cactuses, stonecrops, orchids, and bromeliads. It is also found among nonflowering plants, such as some ferns and cone-bearing trees.

Check Your Progress 6.4

1. Identify the differences between C$_3$ and C$_4$ plants and give an example of each type.

2. Describe why C$_4$ plants need to fix CO$_2$ in their mesophyll cells.

3. Distinguish some advantages CAM plants have over C$_4$ plants.

Connections and Misconceptions

What is known about C_4 plant evolution?

Scientists are trying to understand the evolutionary triggers that gave rise to the C_4 plants. They do know that the first evidence of the C_4 plants dates back to around 25 MYA, which makes them a relatively young group of organisms. The exact location in which they first evolved is still a matter of debate, but most scientists agree ii was probably a hot, dry environment. Despite being newcomers on the evolutionary scene, C_4 plants are very successful. Although only 4% of plant species are C_4 plants, over 20% of total annual plant growth is conducted by these plants. Of the 18 most problematic weed plants on the planet, 14 are C_4 plants.

Media Study Tools

www.mhhe.com/maderessentials3

Enhance your study of this chapter with study tools and practice tests. Also ask your instructor about the resources available through ConnectPlus, including the media-rich eBook, interactive learning tools, and animations.

The Chapter in Review

Summary

6.1 Overview of Photosynthesis

Cyanobacteria, algae, and plants carry on photosynthesis, a process in which water is oxidized (split) and carbon dioxide is reduced using solar energy. The end products of photosynthesis include carbohydrate and oxygen:

$$\text{solar energy} + CO_2 + H_2O \longrightarrow (CH_2O) + O_2$$

In plants, photosynthesis takes place in chloroplasts. A chloroplast is bounded by a double membrane and contains two main components: (1) the liquid stroma and (2) the membranous grana made up of thylakoids. During photosynthesis, the light reactions take place in the thylakoid membrane, and the Calvin cycle reactions take place in the stroma:

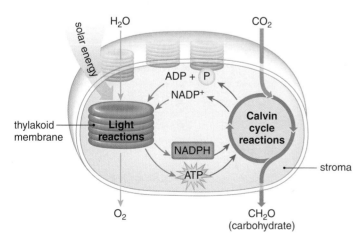

6.2 The Light Reactions—Harvesting Energy

The light reactions use solar energy in the visible-light range. Pigment complexes in two photosystems (PS II and PS I) absorb various wavelengths of light. The collected solar energy results in high-energy electrons at the reaction centers of the photosystems.

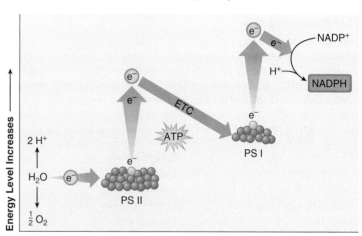

- Solar energy enters PS II, and energized electrons are picked up by an electron acceptor. The oxidation (splitting) of water replaces these electrons in the reaction center. Oxygen (O_2) is released to the atmosphere, and hydrogen ions (H^+) remain in the thylakoid space.
- As electrons pass from one acceptor to another in an electron transport chain, the release of energy allows the carriers to pump H^+ into the thylakoid space. The buildup of H^+ establishes an electrochemical gradient.
- When solar energy is absorbed by PS I, energized electrons leave and are ultimately received by $NADP^+$, which also combines with H^+ from the stroma to become NADPH. Electrons from PS II replace those lost by PS I.
- When H^+ flows down its concentration gradient through the channel present in ATP synthase complexes, ATP is synthesized from ADP and (P) by ATP synthase.

6.3 The Calvin Cycle Reactions—Making Sugars

The Calvin cycle consists of the following stages:

- **CO_2 fixation** The enzyme rubisco fixes CO_2 to RuBP, producing a 6-carbon molecule that immediately breaks down to two C_3 molecules.

- **CO₂ reduction** CO_2 (incorporated into an organic molecule) is reduced to carbohydrate (CH_2O). This step requires the NADPH and some of the ATP from the light reactions.
- **Regeneration of RuBP** For every three turns of the Calvin cycle, the net gain is one G3P molecule; the other five G3P molecules are used to re-form three molecules of RuBP. This step also requires the energy of ATP. G3P is then converted to all the organic molecules a plant needs. It takes two G3P molecules to make one glucose molecule.

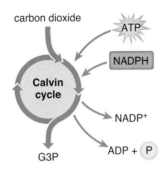

6.4 Other Types of Photosynthesis

C₄ Photosynthesis

When the weather is hot and dry, C_3 plants are at a disadvantage because stomata close and O_2 from photosynthesis competes with CO_2 for the active site of rubisco. C_4 plants avoid this drawback. In C_4 plants, CO_2 fixation occurs in mesophyll cells, and the Calvin cycle reactions occur in bundle sheath cells. In mesophyll cells, a C_3 molecule fixes CO_2, and the result is a C_4 molecule, for which this type of photosynthesis is named. A modified form of this molecule is pumped into bundle sheath cells, where CO_2 is released to the Calvin cycle. This represents partitioning of CO_2 fixation and the Calvin cycle in space.

CAM Photosynthesis

CAM plants live in hot, dry environments. At night when stomata remain open, a C_3 molecule fixes CO_2 to produce a C_4 molecule. The next day, CO_2 is released and enters the Calvin cycle within the same cells. This represents a partitioning of CO_2 fixation and the Calvin cycle in time: Carbon dioxide fixation occurs at night, and the Calvin cycle occurs during the day.

Key Terms

ATP synthase 100	light reactions 96
C_3 plant 103	mesophyll cells 103
C_4 plant 103	oxidation 96
Calvin cycle reactions 96	photosynthesis 94
CAM 104	photosystem 98
carbon dioxide (CO_2) fixation 101	redox reaction 96
carotenoid 98	reduction 96
chlorophyll 95, 98	RuBP carboxylase (rubisco) 101
chloroplast 95	stomata (sing., stoma) 95
coenzyme 100	stroma 95
electron transport chain 99	thylakoid 95
grana (sing., granum) 95	

Testing Yourself

Choose the best answer for each question.

1. The raw materials for photosynthesis are
   a. oxygen and water.
   b. oxygen and carbon dioxide.
   c. carbon dioxide and water.
   d. carbohydrates and water.
   e. carbohydrates and carbon dioxide.

2. During photosynthesis, carbon dioxide is _____ and water is _____.
   a. reduced, oxidized
   b. oxidized, reduced
   c. reduced, reduced
   d. oxidized, oxidized

3. When electrons in the reaction center of PS I are passed to an energy-acceptor molecule, they are replaced by electrons that have been given up by
   a. oxygen.
   c. carbon dioxide.
   b. glucose.
   d. water.

4. Which of the following is associated with the thylakoid membrane?
   a. PS II
   b. ATP production
   c. electron transport chain
   d. All of the above are correct.

5. During the light reactions of photosynthesis, ATP is produced when hydrogen ions move
   a. down a concentration gradient from the thylakoid space to the stroma.
   b. against a concentration gradient from the thylakoid space to the stroma.
   c. down a concentration gradient from the stroma to the thylakoid space.
   d. against a concentration gradient from the stroma to the thylakoid space.

6. Oxygen is generated by the
   a. light reactions.
   b. Calvin cycle.
   c. light reactions and the Calvin cycle reactions.
   d. closing of stomata.

7. Energy for the Calvin cycle is supplied by _____ from the light reactions.
   a. carbon dioxide and water
   d. ATP and water
   b. ATP and NADPH
   e. electrons and hydrogen ions
   c. carbon dioxide and ATP

8. The product(s) of the Calvin cycle and a reactant(s) in many plant metabolic pathways is/are
   a. ribulose-1,5-bisphosphate, RuBP.
   b. adenosine diphosphate, ADP.
   c. fructose and glucose.
   d. glyceraldehyde 3-phosphate, G3P.
   e. carbon dioxide, CO_2.

9. Use these terms to label the following diagram of the light reactions.
   ATP   thylakoid space
   thylakoid membrane   NADPH
   electron transport chain   ATP synthase complex
   stroma

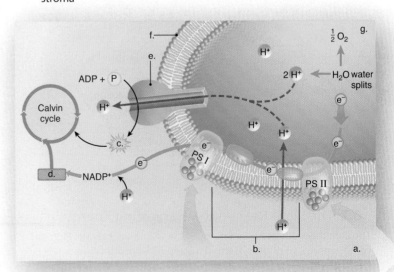

10. In C_4 plants, CO_2 fixation occurs
    a. by adding electrons and a hydrogen ion when splitting water in bundle sheath cells.
    b. by joining CO_2 to a C_3 molecule and pumping it into bundle sheath cells.
    c. by splitting water only in bundle sheath cells.
    d. by oxidizing a C_4 molecule with O_2 derived from bundle sheath cells.

11. A leaf of a C_4 plant differs from that of a C_3 plant because
    a. stomata cannot close.
    b. the mesophyll has no leaf spaces.
    c. bundle sheath cells are protected from leaf spaces by a ring of mesophyll cells.
    d. C_4 plant leaves are narrower to allow more O_2 to diffuse away.

For questions 12–15, match the items to those in the key. Answers can be used more than once, and each question can have more than one answer.

Key:

    a. C_3 plant
    b. C_4 plant
    c. CAM plant

12. CO_2 fixation occurs in mesophyll cells.
13. The Calvin cycle occurs in bundle sheath cells.
14. CO_2 fixation and the Calvin cycle are separated by space.
15. CO_2 fixation and the Calvin cycle are separated by time.
16. Carbon dioxide fixation occurs when CO_2 combines with
    a. ATP.
    c. G3P.
    b. NADHP.
    d. RuBP.
17. Photosynthesis occurs best at wavelengths that are
    a. blue.
    c. infrared.
    b. gamma.
    d. ultraviolet.
18. Which of these descriptions is NOT true of chlorophyll?
    a. absorbs solar energy
    b. located in the grana
    c. located in the thylakoid membranes
    d. passes electrons directly to NADP
    e. passes electrons to an acceptor molecule
19. The enzyme that produces ATP from ADP + (P) in the thylakoid is
    a. RuBP carboxylase.
    d. ATP synthase.
    b. G3P.
    e. coenzyme A.
    c. ATPase.
20. Which of the following statements is NOT true?
    a. H^+ concentration is higher in the stroma than in the thylakoid space.
    b. The electron transport chain pumps H^+ from the stroma into the thylakoid space.
    c. The ATP synthase complex is present in the thylakoid membrane.
    d. All of these are true.
21. Each of the following is a product of light reactions EXCEPT
    a. ATP.
    c. oxygen.
    b. NADHP.
    d. sugar.

Thinking Scientifically

1. In 1882, T. W. Engelmann carried out an ingenious experiment to demonstrate that chlorophyll absorbs light in the blue and red portions of the spectrum. He placed a single filament of a green alga in a drop of water on a microscope slide. Then he passed light through a prism and onto the string of algal cells. The slide also contained aerobic bacterial cells. After some time, he peered into the microscope and saw the bacteria clustered around the regions of the algal filament that were receiving blue light and red light, as shown in the following illustration. Why do you suppose the bacterial cells were clustered in this manner?

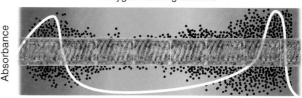

Oxygen-seeking bacteria

Absorbance

Filament of green alga

2. Photosynthesis makes use of light in the visual range, the same range that is suitable to our vision. What explanation can you give to account for this observation?

Bioethical Issue

Are Carbon Offset Credits Effective?

With the prospect of climate change looming, many leading corporations, public figures, and concerned citizens have made an effort to fight global climate change by reducing the impact of their activities on the environment. Many have turned to the planting of trees and other plants, which can remove excess carbon dioxide from the atmosphere by photosynthesis, as a way to offset lifestyle choices, such as driving a car or flying on a plane, that cause carbon dioxide emissions. In response, many new companies and organizations have begun selling "carbon offset credits" in return for planting trees to offset the purchaser's carbon dioxide–emitting activities. But are carbon offset credits effective?

Advocates of these programs contend that the selling of carbon offset credits is effective at removing excess carbon dioxide from the atmosphere. They point out that carbon offset credits discourage behaviors that result in carbon dioxide emissions, such as excessive travel or energy use, by attaching a monetary penalty to the activity. Indeed, such credits systems are encouraged by the 1997 Kyoto Protocol, and usage-based taxes are generally supported by politicians of all stripes. Furthermore, many environmentally conscious citizens insist that it is their duty to ensure that they do whatever is necessary to minimize their impact on the environment.

However, not everyone believes that carbon offset credits are an effective way to address the problem. Environmentalist George Monbiot compares the largely unregulated carbon offset credit system to the indulgences sold by the Catholic Church in the Middle Ages. Many skeptics view offsets as snake oil and ridicule purchasers as people trying to assuage their guilt while failing to address the real problem of their carbon dioxide–emitting activities. Many also point out that the current system has failed to reduce carbon dioxide emissions in countries where the credits are being sold. Furthermore, some environmentalists have become concerned that the planting of large stands of nonnative trees in some tropical areas by carbon offset credit suppliers may reduce biodiversity and disrupt native ecosystems.

Do you believe that carbon offset credits are an effective way to reduce excess carbon dioxide emissions and discourage excess energy consumption? Or are they, as some environmentalists contend, simply an excuse to pollute?

7

Energy for Cells

OUTLINE

BEFORE YOU BEGIN

Before beginning this chapter, take a few moments to review the following discussions.

Section 3.2 What are the roles of carbohydrates in living organisms?

Section 5.2 What is the ATP cycle?

Section 6.1 What do the terms *oxidation* and *reduction* mean?

Not All Diets Are Created Equal

An obesity epidemic rages unfettered in the United States. Not surprisingly, an entire industry has emerged, hawking weight-loss supplements, "miracle" fat burners, and fad diets. Many diets may help you initially lose weight, but have you ever considered the science behind how these diets claim to work? When you do, you will see that many of these claims simply can't be verified. Many fad diets rob the body of the essential nutrients it needs to stay healthy, yet millions of people spend a fortune, hoping for a miracle, and many of them suffer health problems as a result.

Your cells break down glucose in a process called cellular respiration to extract the energy from your food to create the ATP they need. To lose weight, it all comes down to regulating the input of energy while providing adequate levels of important vitamins and minerals for the cells to convert these nutrients to ATP. When cells have excess energy, or a deficiency in vitamins, the excess is converted to fat. In this chapter, you will learn how cells use glucose, fat, and protein to produce ATP. You will also learn how the body uses metabolites to build glucose, protein, and fat. Understanding these processes will give you the information you need to make informed decisions about weight loss, diets, and nutritional supplements.

7.1 Cellular Respiration

Learning Outcomes

Upon completion of this section, you should be able to

1. Recognize the overall reaction for glucose breakdown.
2. Explain how the breakdown of glucose to CO_2 and H_2O during cellular respiration drives the synthesis of ATP.
3. Identify the four phases of cellular respiration and the location of each within the cell.

Whether you go skiing, take an aerobics class, or just hang out, ATP molecules provide the energy needed for your muscles to contract. ATP molecules are produced during cellular respiration, a process that requires the participation of mitochondria. Cellular respiration is aptly named because, just as you take in oxygen (O_2) and give off carbon dioxide (CO_2) during breathing, so do the mitochondria in your cells (**Fig. 7.1**). In fact, cellular respiration, which occurs in all the cells of the body, is the reason you breathe.

Oxidation of substrates is a fundamental part of cellular respiration. In living things, oxidation doesn't occur by the addition of oxygen (O_2). Instead, oxidation is the removal of hydrogen atoms from a molecule. As cellular respiration occurs, hydrogen atoms are removed from glucose (and glucose products) and transferred to oxygen atoms, forming carbon dioxide (CO_2) and water (H_2O):

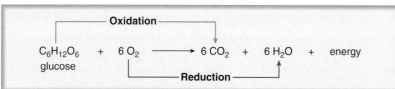

$$\underset{\text{glucose}}{C_6H_{12}O_6} + 6\,O_2 \longrightarrow 6\,CO_2 + 6\,H_2O + \text{energy}$$

with **Oxidation** from $C_6H_{12}O_6$ to $6\,CO_2$ and **Reduction** from $6\,O_2$ to $6\,H_2O$.

Note that respiration is essentially the reverse of photosynthesis. Carbohydrates are oxidized to release energy. Oxygen is consumed, and carbon dioxide and water are formed as by-products. In photosynthesis, carbon dioxide is reduced to make carbohydrate; in respiration, carbohydrate is oxidized and carbon dioxide is formed.

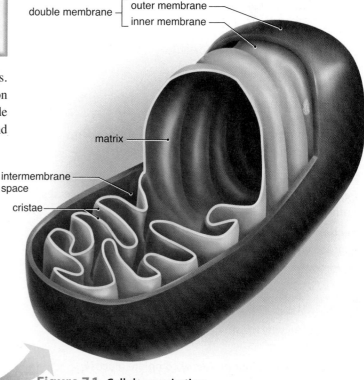

double membrane — outer membrane / inner membrane

matrix

intermembrane space

cristae

Figure 7.1 Cellular respiration.

Glucose from our food and the oxygen we breathe are requirements for cellular respiration. Carbon dioxide and water are released as by-products. The process begins in the cytoplasm but is completed in the mitochondria.

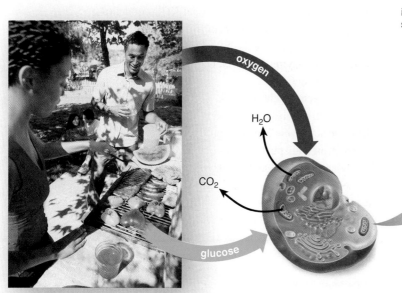

oxygen

H_2O

CO_2

glucose

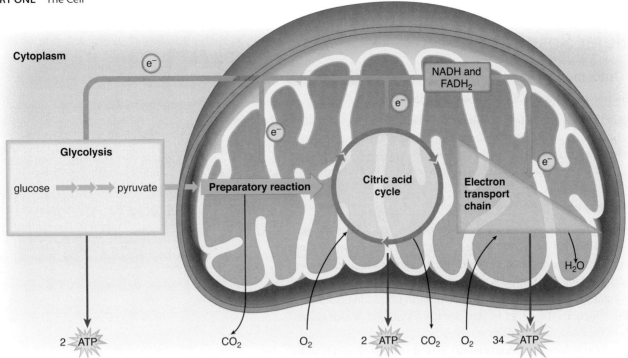

Cytoplasm

e⁻

NADH and FADH₂

e⁻

e⁻

Glycolysis

glucose → pyruvate

Preparatory reaction

Citric acid cycle

Electron transport chain

e⁻

H₂O

2 ATP   CO₂   O₂   2 ATP   CO₂   O₂   34 ATP

Figure 7.2 The four phases of complete glucose breakdown.

The enzymatic reactions of glycolysis take place in the cytoplasm. The preparatory reaction, the citric acid cycle, and the electron transport chain occur in mitochondria.

Connecting the Concepts

For more information on the use of nutrients for energy, refer to the following discussions.

Section 3.2 provides an introduction to carbohydrates.

Section 24.3 examines how the digestive system processes nutrients for use by the cells.

Section 25.2 reviews the classes of nutrients and how they relate to energy use in the body.

Check Your Progress 7.1

① Explain why breathing is necessary for cellular respiration.

② Describe the four phases of complete glucose breakdown.

③ Differentiate between the phases of glucose breakdown that form CO₂ and H₂O and those that do not.

Phases of Complete Glucose Breakdown

The breakdown of glucose releases a lot of energy. If you mistakenly burn sugar in a skillet, the energy escapes into the atmosphere as heat. A cell is more sophisticated than that. In a cell, glucose is broken down slowly—not all at once—and the energy given off isn't all lost as heat. Hydrogen atoms are removed bit by bit, and this allows energy to be captured and used to make ATP molecules.

The enzymes that carry out oxidation during cellular respiration are assisted by nonprotein helpers called **coenzymes.** As glucose is oxidized, the coenzymes NAD⁺ and FAD⁺ receive hydrogen atoms (H⁺ + e⁻) and become NADH and FADH₂, respectively (**Fig. 7.2**)[1].

Animation
How the NAD⁺ Works

During these phases, notice where CO₂ and H₂O are produced.

- **Glycolysis,** which occurs in the cytoplasm outside the mitochondria, is the breakdown of glucose to 2 molecules of pyruvate. Energy is invested to activate glucose, 2 ATP are gained, and oxidation results in NADH, which will be used later for additional ATP production.

- The **preparatory (prep) reaction** takes place in the matrix of mitochondria. Pyruvate is broken down to a 2-carbon acetyl group carried by **coenzyme A (CoA).** Oxidation of pyruvate results in not only NADH but also CO₂.

- The **citric acid cycle** also takes place in the matrix of mitochondria. As oxidation occurs, NADH and FADH₂ result and more CO₂ is released. The citric acid cycle is able to produce 2 ATP per glucose molecule.

- The **electron transport chain** is a series of electron carriers in the cristae of mitochondria. NADH and FADH₂ give up electrons to the chain. Energy is released and captured as the electrons move from a higher energy to a lower energy state. Later, this energy will be used for the production of ATP. Oxygen (O₂) is the final electron acceptor. It then combines with hydrogen ions (H⁺) to produce water (H₂O).

MP3
Cellular Respiration

---

[1]NAD = Nicotinamide adenine dinucleotide; FAD = Flavin adenine dinucleotide

7.2 Outside the Mitochondria: Glycolysis

Learning Outcomes

Upon completion of this section, you should be able to

1. List the inputs and outputs of glycolysis.
2. Describe the energy-investment and energy-harvesting steps of glycolysis.
3. Describe how the metabolic pathway of glycolysis partially breaks down glucose.

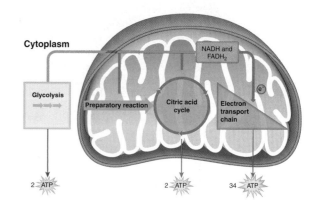

In eukaryotes, such as plants and animals, glycolysis takes place within the cytoplasm outside the mitochondria. During glycolysis, glucose, a C_6 molecule, is broken down to 2 molecules of pyruvate, a C_3 molecule. Glycolysis is divided into (1) the energy-investment step, when ATP is used; and (2) the energy-harvesting steps, when NADH and ATP are produced. **Figure 7.3** outlines the basic steps of glycolysis, including some of the more important intermediate molecules. As you examine this figure, notice how it includes an initial investment of ATP, followed by an ATP-harvesting phase.

Energy-Investment Step

During the energy-investment step, 2 ATP transfer phosphate groups to substrates, and $2 ADP + \text{(P)}$ result. In other words, ATP has been broken down, not built up. However, the phosphate groups activate the substrates, so that they can undergo reactions.

3PG	3-phosphoglycerate
BPG	1,3-bisphosphoglycerate
G3P	glyceraldehyde-3-phosphate

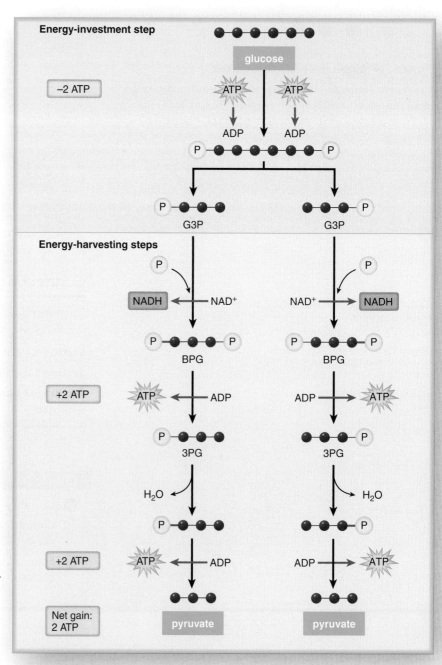

Figure 7.3 Glycolysis.

This metabolic pathway begins with glucose (C_6) and ends with 2 pyruvate (C_3) molecules. A net gain of 2 ATP can be calculated by subtracting those expended during the energy-investment step from those produced during the energy-harvesting steps. Each ● is a carbon atom (C).

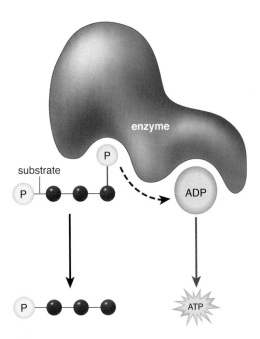

Figure 7.4 Substrate-level ATP synthesis.

At an enzyme's active site, ADP acquires an energized phosphate group from a substrate of a metabolic pathway, and ATP results. The net gain of 2 ATP from glycolysis is the result of substrate-level ATP synthesis.

Energy-Harvesting Steps

During the energy-harvesting steps, substrates are oxidized by the removal of hydrogen atoms, and 2 NADH result.

Oxidation produces substrates with energized phosphate groups, which are used to synthesize 4 ATP. As a phosphate group is transferred to ADP, ATP results. The process is called substrate-level ATP synthesis (**Fig. 7.4**).

What is the net gain of ATP from glycolysis? Notice in Figure 7.3 that 2 ATP are used to get started, and 4 ATP are produced. Therefore, there is a net gain of 2 ATP from glycolysis.

If oxygen is available, pyruvate, the end product of glycolysis, enters mitochondria, where it undergoes further breakdown. If oxygen is not available, pyruvate remains in the cytoplasm and undergoes reduction. In humans, if oxygen is not available, pyruvate is reduced to lactate, as discussed on page 118.

 Animation
Glycolysis

Glycolysis	
inputs	outputs
glucose	2 pyruvate
2 NAD⁺	2 NADH
2 ATP	2 ADP
4 ADP + 4 P	4 ATP
	2 ATP — net

Connecting the Concepts

For more information on the inputs and outputs of glycolysis, refer to the following discussions.

Section 3.1 examines the role of carbohydrates as energy molecules.

Section 5.2 reviews the ATP cycle.

Section 5.3 provides information on how enzymatic pathways are regulated.

Check Your Progress 7.2

1   Identify where in the cell glycolysis occurs.

2   Contrast the energy-investment step of glycolysis with the energy-harvesting steps.

3   Summarize the net yield of ATP per glucose molecule as a result of glycolysis.

7.3 Inside the Mitochondria

Learning Outcomes

Upon completion of this section, you should be able to

1. Identify the role of the preparatory reaction and the citric acid cycle in the breakdown of glucose.
2. Detail how the electron transport chain produces most of the ATP during cellular respiration.
3. Name the inputs and outputs of each pathway of aerobic cellular respiration.
4. Explain the importance of oxygen during cellular respiration.
5. Recognize how alternate metabolic pathways allow the utilization of protein and fats within cellular respiration.

Following glycolysis, the other three phases of cellular respiration occur inside the mitochondria (**Fig. 7.5**).

Preparatory Reaction

Occurring in the matrix, the preparatory (prep) reaction is so called because it produces a substrate that can enter the citric acid cycle. The preparatory reaction occurs twice per glucose molecule because glycolysis results in 2 pyruvate molecules (see Fig. 7.3). During a prep reaction,

- Pyruvate is oxidized, and a CO_2 molecule is given off. This is part of the CO_2 we breathe out!
- NAD^+ accepts a hydrogen atom, and NADH results.
- The product, a C_2 acetyl group, is attached to coenzyme A (CoA), forming **acetyl-CoA.**

Therefore,

- Per glucose molecule, the total is 2 CO_2, 2 NADH, and 2 acetyl-CoA.

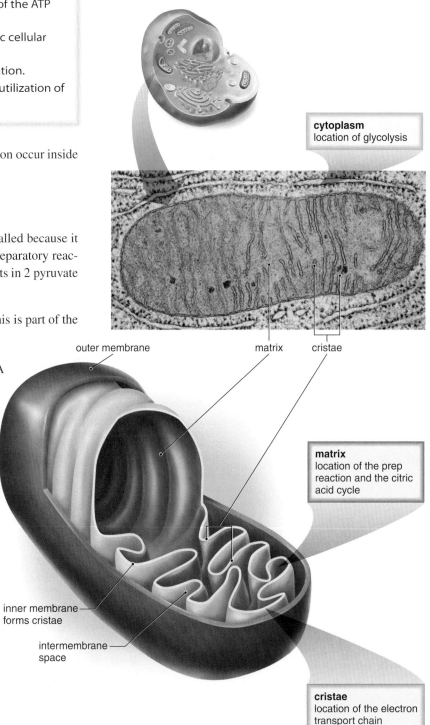

cytoplasm
location of glycolysis

outer membrane   matrix   cristae

matrix
location of the prep reaction and the citric acid cycle

inner membrane forms cristae

intermembrane space

cristae
location of the electron transport chain

Figure 7.5 Mitochondrion structure and function.

A mitochondrion is bounded by a double membrane. The inner membrane folds inward to form the shelf-like cristae. Glycolysis takes place in the cytoplasm outside the mitochondrion. The preparatory reaction and the citric acid cycle occur within the mitochondrial matrix. The electron transport chain is located on the cristae of a mitochondrion.

The Citric Acid Cycle

The citric acid cycle is a cyclical metabolic pathway located in the matrix of mitochondria (**Fig. 7.6**). It was originally called the *Krebs cycle* to honor the scientist who first studied it. At the start of the citric acid cycle, the C_2 acetyl group carried by CoA joins with a C_4 molecule, and a C_6 citrate molecule results. The CoA returns to the preparatory reaction to be used again.

Animation
How the Krebs Cycle Works

During the citric acid cycle,

- The acetyl group is oxidized, and the rest of the CO_2 we breathe out per glucose molecule is released.
- Both NAD^+ and FAD accept hydrogen atoms, resulting in NADH and $FADH_2$.
- Substrate-level ATP synthesis occurs (see Fig. 7.4), and an ATP results.

Because the citric acid cycle turns twice for each original glucose, the inputs and outputs of the citric acid cycle per glucose are as follows:

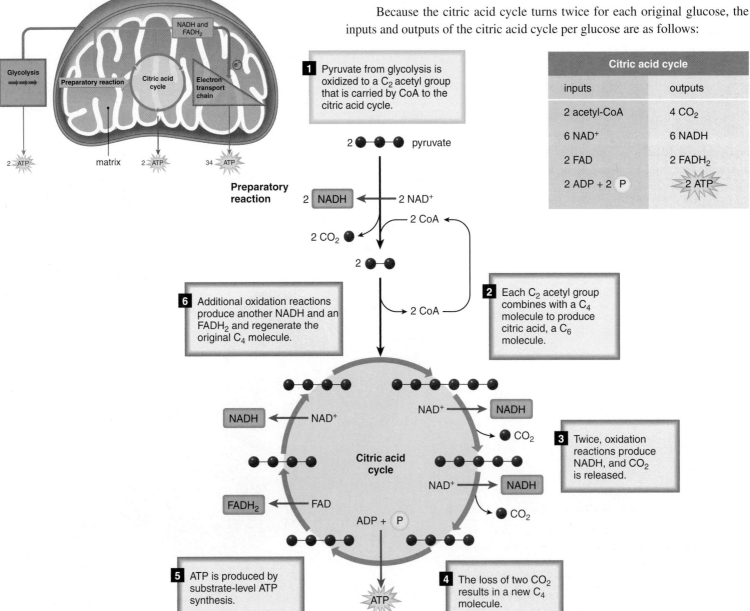

1 Pyruvate from glycolysis is oxidized to a C_2 acetyl group that is carried by CoA to the citric acid cycle.

2 Each C_2 acetyl group combines with a C_4 molecule to produce citric acid, a C_6 molecule.

3 Twice, oxidation reactions produce NADH, and CO_2 is released.

4 The loss of two CO_2 results in a new C_4 molecule.

5 ATP is produced by substrate-level ATP synthesis.

6 Additional oxidation reactions produce another NADH and an $FADH_2$ and regenerate the original C_4 molecule.

Citric acid cycle	
inputs	outputs
2 acetyl-CoA	4 CO_2
6 NAD^+	6 NADH
2 FAD	2 $FADH_2$
2 ADP + 2 P	2 ATP

Figure 7.6 The preparatory reaction and the citric acid cycle.

Each acetyl-CoA from the preparatory reaction enters the citric acid cycle. The net result of one turn of this cycle of reactions is the oxidation of the acetyl group to 2 CO_2 and the formation of 3 NADH and 1 $FADH_2$. Substrate-level ATP synthesis occurs, and the result is 1 ATP. For each glucose, the citric acid cycle turns twice, doubling each of these amounts.

The Electron Transport Chain

The electron transport chain located in the cristae of mitochondria is a series of carriers that pass electrons from one to the other. NADH and $FADH_2$ deliver electrons to the chain. Consider that the hydrogen atoms attached to NADH and $FADH_2$ consist of an electron (e^-) and a hydrogen ion (H^+). The members of the electron transport chain accept only e^- and not H^+.

In **Figure 7.7,** high-energy electrons enter the chain, and low-energy electrons leave the chain. When NADH gives up its electrons, the next carrier gains the electrons and is reduced. This oxidation-reduction reaction starts the process, and each of the carriers in turn becomes reduced and then oxidized as the electrons move down the system. As the pair of electrons is passed from carrier to carrier, energy is released and captured for ATP production. The final acceptor of electrons is oxygen (O_2). Your breathing provides the O_2 for cellular respiration. Overall, the role of oxygen in cellular respiration is to keep the electrons moving from the first to the last carrier. Why can oxygen play this role? Because oxygen attracts electrons to a greater degree than the carriers of the chain. Once oxygen accepts electrons, it combines with H^+ to form a water molecule, the other end product of cellular respiration (see the equation on page 109).

When each NADH delivers electrons to the first carrier of the electron transport chain, enough energy has been captured by the time the electrons are received by O_2 to permit the production of 3 ATP molecules. When each $FADH_2$ delivers electrons to the electron transport chain, only 2 ATP are produced.

Animation
Electron Transport System and ATP Synthesis

Once NADH has delivered electrons to the electron transport chain, NAD^+ is regenerated and can be used again. In the same manner, FAD is regenerated and can be used again. The recycling of coenzymes, and for that matter ADP, increases cellular efficiency, since it does away with the need to synthesize NAD^+, FAD, and ADP anew.

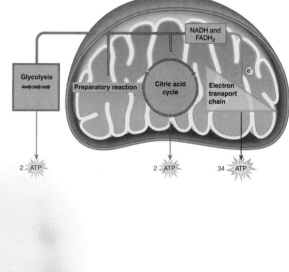

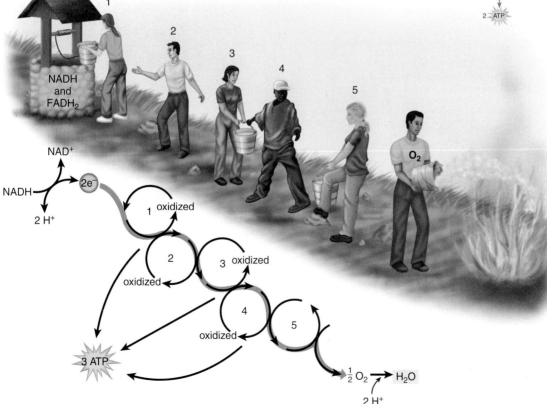

Figure 7.7   The electron transport chain.

An electron transport chain operates as a bucket brigade. Each electron carrier (steps 1–5) is alternatively reduced (orange flow of electrons) and oxidized, as if the electrons were a bucket being passed from person to person. As oxidation-reduction occurs, energy is released that will be used to make ATP. Oxygen is the final electron acceptor.

Connections and Misconceptions

Why is carbon monoxide so dangerous?

Carbon monoxide (CO) is a colorless, odorless gas that is produced by the incomplete burning
of organic material, such as fossil
fuels and wood. Carbon mon-
oxide binds more strongly than
oxygen to the hemoglobin in your red blood cells. When
this happens, the red blood cells cannot transport adequate
oxygen to the mitochondria of your cells, and ATP production shuts down. Over 400 people die annually in the
United States due to carbon monoxide poisoning. For a list
of ways to prevent CO poisoning, visit the CDC website at
www.cdc.gov/co/faqs.htm.

The Cristae of a Mitochondrion

The carriers of the electron transport chain are located in molecular complexes
within the inner mitochondrial membrane. ATP synthesis is carried out by ATP
synthase complexes also located in this membrane (**Fig. 7.8**).

The carriers of the electron transport chain accept electrons from NADH
or $FADH_2$ and then pass them from one to the other by way of two additional
mobile carriers (orange arrow). What happens to the hydrogen ions (H^+) carried
by NADH and $FADH_2$? The complexes use the energy released by oxidation-
reduction to pump H^+ from the mitochondrial matrix into the **intermembrane
space** located between the outer and inner membrane of a mitochondrion. The
pumping of H^+ into the intermembrane space establishes an unequal distribution of H^+; in other words, there are many H^+ in the intermembrane space but
few in the matrix of a mitochondrion.

Just as in photosynthesis, the H^+ gradient contains a large amount of
stored energy that can be used to drive forward ATP synthesis. The cristae of
mitochondria (like the thylakoid membrane of chloroplasts) contain an *ATP
synthase complex* that allows H^+ to return to the matrix. The flow of H^+ through
the ATP synthase complex brings about a conformational change, which causes
the enzyme ATP synthase to synthesize ATP from ADP + (P). ATP leaves the
matrix by way of a channel protein. This ATP remains in the cell and is used
for cellular work.

Energy Yield from Glucose Metabolism

Figure 7.9 calculates the ATP yield for the complete breakdown of glucose to
CO_2 and H_2O. Per glucose, there is a net gain of 2 ATP from glycolysis, which
takes place in the cytoplasm. The citric acid cycle, which occurs in the matrix of
mitochondria, accounts for 2 ATP per glucose. This means that a total of 4 ATP
form due to substrate-level ATP synthesis outside the electron transport chain.

Animation
Electron Transport
Chain

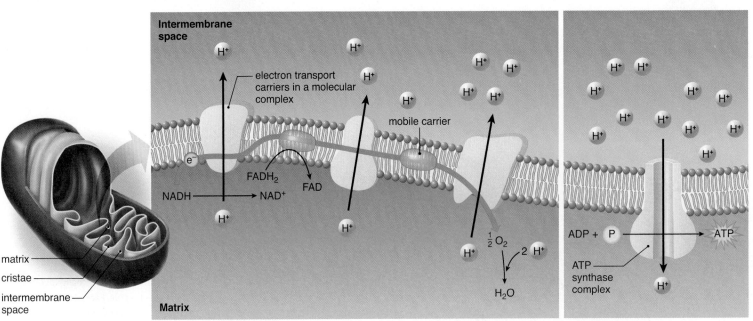

a. Electron transport chain

b. ATP synthesis

Figure 7.8 The organization of cristae.

Molecular complexes that contain the electron transport carriers are located in the cristae, as are ATP synthase complexes. **a.** As electrons move from one carrier to the other,
hydrogen ions (H^+) are pumped from the matrix into the intermembrane space. **b.** As hydrogen ions flow back down a concentration gradient through an ATP synthase
complex, ATP is synthesized by the enzyme ATP synthase.

Most of the ATP produced comes from the electron transport chain and the ATP synthase complex. Per glucose, 10 NADH and 2 $FADH_2$ take electrons from glycolysis and the citric acid cycle to the electron transport chain. The maximum number of ATP produced by the chain is therefore 34 ATP, and the maximum number produced by both the chain and substrate-level ATP synthesis is 38. However, for reasons beyond the scope of this book, the maximum number of ATP produced per glucose in some cells is only 36 ATP or lower. A yield of 36–38 ATP represents about 40% of the energy that was initially available in the glucose molecule. The rest of the energy is lost in the form of heat.

Alternative Metabolic Pathways

Let's say you are on a low-carbohydrate diet. Will you then run out of ATP? No, because your cells can also utilize other energy sources—the components of fats and oils—namely, glycerol and fatty acids—and amino acids, which are derived from proteins (**Fig. 7.10**).

Because glycerol is a carbohydrate, it enters the process of cellular respiration during glycolysis. Fatty acids can be metabolized to acetyl groups, which enter the citric acid cycle. A fatty acid with a chain of 18 carbons can make three times the number of acetyl groups that glucose does. For this reason, fats are an efficient form of stored energy. The complete breakdown of glycerol and fatty acids results in many more ATP per fat molecule than does the breakdown of glucose.

Only the hydrocarbon backbone of amino acids can be used by the cellular respiration pathways. The amino group becomes ammonia (NH_3), which becomes part of urea, the primary excretory product of humans. Just where the hydrocarbon backbone from an amino acid begins degradation to produce ATP depends on its length. Figure 7.10 shows that the hydrocarbon backbone from an amino acid can enter cellular respiration pathways at pyruvate, at acetyl-CoA, or during the citric acid cycle.

The smaller molecules in Figure 7.10 can also be used to synthesize larger molecules. In such instances, ATP is used instead of generated. For example, some substrates of the citric acid cycle can become amino acids through the addition of an amino group, and these amino acids can be employed to synthesize proteins. Similarly, substrates from glycolysis can become glycerol, and acetyl groups can be used to produce fatty acids. When glycerol and three fatty acids join, the result is a molecule of fat (triglyceride). This explains why you can gain weight from eating carbohydrate-rich foods.

Connections and Misconceptions

What is chromium?

Chromium is a trace mineral that is found naturally in many vegetables. Most commonly, people are familiar with chromium as a component of over-the-counter "fat-burning" medications. Chromium is known to have an influence on the activity of insulin, allowing the body to use carbohydrates and fats more efficiently as energy sources. However, most people are not deficient in chromium, and scientific studies have not shown a conclusive link between increased levels of chromium and weight loss. While chromium supplements are sometimes prescribed to treat some medical conditions, such as type II diabetes, it is important to note that the FDA does not regulate dietary supplements.

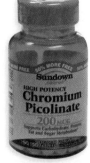

Phase	NADH	$FADH_2$	ATP Yield
Glycolysis	2	–	2
Prep reaction	2	–	–
Citric acid cycle	6	2	2
Electron transport chain	10 ⟶	2 ⟶	30 / 4
Total ATP			38

Figure 7.9 Calculating ATP energy yield per glucose molecule.

Substrate-level ATP synthesis during glycolysis and the citric acid cycle accounts for 4 ATP. The electron transport chain produces a maximum of 34 ATP, and the maximum total is 38 ATP. Some cells, however, produce only 36 ATP per glucose, or even less.

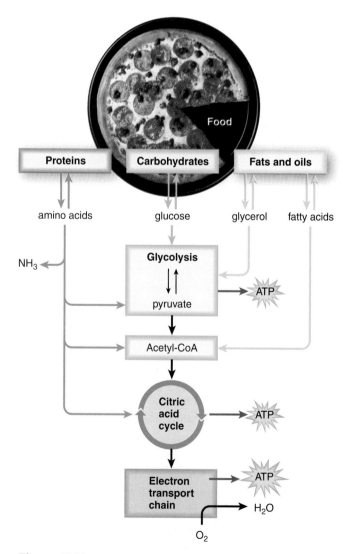

Figure 7.10 Alternative metabolic pathways.

All the types of food in a pizza can be used to generate ATP.

Connecting the Concepts

For more information on the aerobic aspects of cellular respiration, refer to the following discussions.

Section 4.4 explores the structure of the mitochondria.

Section 17.4 examines the evolutionary events that gave rise to the mitochondria of eukaryotic cells.

Section 24.1 explains how the respiratory system moves oxygen to the cells of the body.

Connections and Misconceptions

Does lactic acid cause soreness in your muscles after exercise?

While the accumulation of lactic acid in muscles does cause the "burn" you may feel after prolonged exercise, it does not usually cause long-term muscle soreness. Instead, the tenderness that occurs a day or two after intense exercise is probably being caused by the anti-inflammatory components of the immune system responding to damage to the muscle tissues.

Check Your Progress 7.3

1 List the products of the citric acid cycle.

2 Explain how the electron transport chain results in the synthesis of ATP.

3 Explain why fats and oils are more efficient energy storage molecules than carbohydrates.

7.4 Fermentation

Learning Outcomes

Upon completion of this section, you should be able to

1. Compare and contrast fermentation with glycolysis.
2. Explain why fermentation pathways are beneficial when oxygen is not available.
3. Give examples of products made by fermenting yeast and bacteria.

As you have learned, oxygen is required for the complete breakdown of glucose. But what happens when oxygen is not available and the complete breakdown of glucose is not possible? **Fermentation,** which is the anaerobic breakdown of glucose, resulting in the production of 2 ATP per glucose, ensures that ATP is available for cellular processes (**Fig. 7.11**).

During fermentation in animal cells, the pyruvate formed by glycolysis accepts 2 hydrogen atoms and is reduced to lactate. Notice in Figure 7.11 that 2 NADH pass hydrogen atoms to pyruvate, reducing it. Why is it beneficial for pyruvate to be reduced to lactate when oxygen is not available? The answer is that this reaction regenerates NAD^+, which can then pick up more electrons during the earlier reactions of glycolysis. This keeps glycolysis going, during which ATP is produced by substrate-level ATP synthesis.

The 2 ATP produced by fermentation represent only a small fraction of the potential energy stored in a glucose molecule. Following fermentation, most of this potential energy is still waiting to be released. Despite its low yield of only 2 ATP, fermentation is essential. It can provide a rapid burst of ATP, and muscle cells are more apt than other cells to carry on fermentation. When our muscles are working vigorously over a short period of time, as when we run, fermentation is a way to produce ATP even though oxygen is temporarily in limited supply.

When we stop running, our bodies are in **oxygen deficit,** as signified by the fact that we continue to breathe very heavily for a time. Recovery is complete when enough oxygen is present to completely break down glucose. Blood carries away the lactate formed in muscles and transports it to the liver, where it is reconverted to pyruvate. Some of the pyruvate is oxidized completely, and the rest is converted back to glucose.

Microorganisms and Fermentation

Bacteria use fermentation to produce an organic acid, such as lactate, or an alcohol and CO_2, depending on the type of bacterium.

Yeasts (a type of fungi) are good examples of microorganisms that generate ethyl alcohol and CO_2 when they carry out fermentation. When yeasts are used to leaven bread, the CO_2 makes the bread rise. When yeasts are used to ferment grapes for wine production or to ferment wort—derived from barley—for beer production, ethyl alcohol is the desired product. However, the yeasts are killed by the very product they produce.

The inputs and outputs of fermentation are as follows:

Fermentation	
inputs	outputs
glucose	2 lactate or 2 alcohol and 2 CO_2
2 ATP	2 ADP
4 ADP + 4 P	4 ATP
	2 ATP net

Connecting the Concepts

For more information on some of the organisms that conduct fermentation reactions, refer to the following discussions.

Section 17.3 provides additional information on prokaryotic organisms and some of the beneficial products they produce.

Section 18.3 examines the biology of yeasts and other fungi.

Check Your Progress 7.4

1. Compare the processes of cellular respiration and fermentation.

2. Identify the products of fermentation by humans and yeasts.

3. Discuss why humans primarily use cellular respiration instead of fermentation to produce ATP.

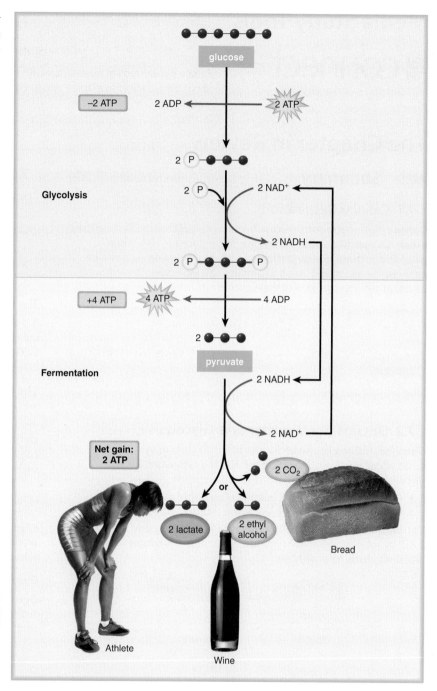

Figure 7.11 **Fermentation.**

Fermentation consists of glycolysis followed by a reduction of pyruvate by NADH. This regenerates NAD^+, which returns to the glycolytic pathway to pick up more hydrogen atoms.

Media Study Tools

The Chapter in Review

Summary

7.1 Cellular Respiration

During cellular respiration, glucose from food is oxidized to CO_2, which we exhale. Oxygen (O_2), which we breathe in, is reduced to H_2O. When glucose is oxidized, energy is released. Cellular respiration captures the energy of oxidation and uses it to produce ATP molecules. The following equation gives an overview of these events:

$$C_6H_{12}O_6 \; + \; 6\,O_2 \; \longrightarrow \; 6\,CO_2 \; + \; 6\,H_2O \; + \; \text{ATP}$$
$$\text{glucose}$$

7.2 Outside the Mitochondria: Glycolysis

Glycolysis, the breakdown of glucose to 2 molecules of pyruvate, is a series of enzymatic reactions that occur in the cytoplasm. During glycolysis,

- Glucose is oxidized by the removal of hydrogen atoms ($H^+ + e^-$).
- When NAD^+ accepts electrons, NADH results.

  Breakdown releases enough energy to immediately give a net gain of 2 ATP by substrate-level ATP synthesis. The inputs and outputs of glycolysis are summarized here:

Glycolysis	
inputs	outputs
glucose	2 pyruvate
2 NAD$^+$	2 NADH
2 ATP	2 ADP
4 ADP + 4 (P)	4 ATP
	2 ATP — net

When oxygen is available, pyruvate from glycolysis enters a mitochondrion.

7.3 Inside the Mitochondria

Preparatory Reaction

During the preparatory reaction in the matrix,

- Oxidation occurs as CO_2 is removed from pyruvate.
- NAD^+ accepts hydrogen atoms, and NADH results.
- An acetyl group, the end product, combines with CoA.
  This reaction takes place twice per glucose.

The Citric Acid Cycle

Acetyl groups enter the citric acid cycle, a series of reactions occurring in the mitochondrial matrix. During one turn of the cycle, oxidation results in 2 CO_2, 3 NADH, and 1 FADH$_2$. One turn also produces 1 ATP. The cycle turns twice and doubles these amounts per glucose.

The Electron Transport Chain

The final stage of cellular respiration involves the electron transport chain located in the cristae of the mitochondria. The chain is a series of electron carriers that accept high-energy electrons (e^-) from NADH and FADH$_2$ and pass electrons along until they are finally low-energy electrons received by oxygen, which combines with H^+ to produce water.

The carriers of the electron transport chain are located in molecular complexes on the cristae of mitochondria. These carriers capture energy from the passage of electrons and use it to pump H^+ into the intermembrane space of the mitochondrion. When H^+ flows down its gradient into the matrix through ATP synthase complexes, energy is released and used to form ATP molecules from ADP and (P).

Energy Yield from Glucose Metabolism

Of the maximum 38 ATP formed by complete glucose breakdown, 4 are the result of substrate-level ATP synthesis, and the rest are produced as a result of the electron transport chain and ATP synthase:

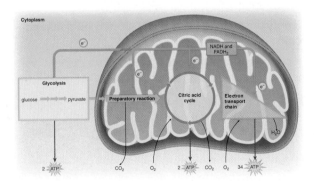

Alternative Metabolic Pathways

Besides carbohydrates, glycerol and fatty acids from fats and amino acids from proteins can undergo cellular respiration by entering glycolysis and/or the citric acid cycle. These metabolic pathways also provide substrates for the synthesis of fats and proteins.

7.4 Fermentation

Fermentation involves glycolysis followed by the reduction of pyruvate by NADH, either to lactate or to alcohol and CO_2. The reduction of pyruvate regenerates NAD^+, which can accept more hydrogen atoms during glycolysis.

- Although fermentation results in only 2 ATP, it still provides a quick burst of ATP energy for short-term, strenuous muscular activity.
- The accumulation of lactate puts an individual in oxygen deficit, which is the amount of oxygen needed when lactate is completely metabolized to CO_2 and H_2O.

Key Terms

acetyl-CoA   113
citric acid cycle   110
coenzyme   110
coenzyme A (CoA)   110
electron transport chain   110

fermentation   118
glycolysis   110
intermembrane space   116
oxygen deficit   118
preparatory (prep) reaction   110

Testing Yourself

Choose the best answer for each question.

1. During cellular respiration, _____ is oxidized and _____ is reduced.
   a. glucose, oxygen
   b. glucose, water
   c. oxygen, water
   d. water, oxygen
   e. oxygen, carbon dioxide

2. Cellular respiration requires _____, and _____ are produced.
   a. water and carbon dioxide, oxygen and glucose
   b. oxygen and carbon dioxide, glucose and water
   c. oxygen and glucose, carbon dioxide and water
   d. water and glucose, oxygen and carbon dioxide
   e. ATP and glucose, oxygen and water

3. During the energy-harvesting steps of glycolysis, which of the following are produced?
   a. ATP and NADH
   b. ADP and NADH
   c. ATP and NAD^+
   d. ADP and NAD^+

4. During the energy-investment step of glycolysis, which of the following is consumed?
   a. ATP
   b. NADH
   c. ADP + Ⓟ
   d. NAD^+

5. Which of the following is not true?
   a. Glycolysis can occur anaerobically.
   b. Glycolysis results in the production of pyruvate.
   c. Glycolysis can occur in the cytoplasm of the cell or the mitochondria.
   d. Glycolysis makes a net amount of 2 ATP.

6. The formation of acetyl-CoA is important because
   a. pyruvate cannot enter the citric acid cycle.
   b. it is the only step of cellular respiration that occurs anaerobically.
   c. it produces 4 molecules of ATP.
   d. it moves electrons directly into the electron transport chain.

7. The citric acid cycle results in the release of
   a. carbon dioxide.
   b. pyruvate.
   c. oxygen.
   d. water.

8. The energy needed for ATP synthesis by ATP synthase is obtained from which of the following?
   a. electrons
   b. NADH
   c. H^+ gradient in intermembrane space
   d. electrons from $FADH_2$

9. Which of the following reactions occur in the matrix of the mitochondria?
   a. glycolysis and the preparatory reaction
   b. the preparatory reaction and the citric acid cycle
   c. the citric acid cycle and the electron transport chain
   d. the electron transport chain and glycolysis

10. Match the following descriptions to the lettered events in the preparatory reaction and citric acid cycle:
    Pyruvate is broken down to an acetyl group.
    An acetyl group is taken up and a C_6 molecule results.
    Oxidation results in NADH and CO_2.
    ATP is produced by substrate-level ATP synthesis.
    Oxidation produces more NADH and $FADH_2$.

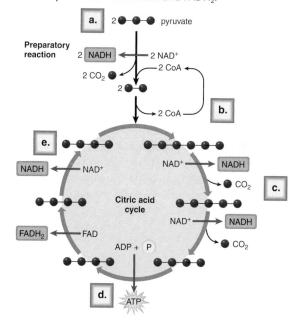

11. The strongest and final electron acceptor in the electron transport chain is
    a. NADH.
    b. $FADH_2$.
    c. oxygen.
    d. water.

For questions 12–18, match the items to those in the key. Answers can be used more than once, and each question can have more than one answer.

Key:

a. glycolysis
b. preparatory reaction
c. citric acid cycle
d. electron transport chain

12. produces ATP
13. uses ATP
14. produces NADH
15. uses NADH
16. produces carbon dioxide
17. occurs in cytoplasm
18. occurs in mitochondria

19. The carriers in the electron transport chain undergo
    a. oxidation only.
    b. reduction only.
    c. oxidation and reduction.
    d. the loss of hydrogen ions.
    e. the gain of hydrogen ions.

20. The final acceptor for hydrogen atoms during fermentation is
    a. O_2.
    b. acetyl-CoA.
    c. FAD.
    d. pyruvic acid.

21. Which of the following do not enter the cellular respiration pathways?
    a. fats
    b. amino acids
    c. nucleic acids
    d. carbohydrates

22. When animals carry out fermentation, they produce _____, while yeasts produce _____.
    a. lactate, NADH
    b. lactate, ethyl alcohol and CO_2
    c. NADH, ethyl alcohol and CO_2
    d. $FADH_2$, lactate
    e. ethyl alcohol and CO_2, lactate

23. Fermentation does not yield as much ATP as cellular respiration does because fermentation
    a. generates mostly heat.
    b. makes use of only a small amount of the potential energy in glucose.
    c. creates by-products that require large amounts of ATP to break down.
    d. creates ATP molecules that leak into the cytoplasm and are broken down.

24. Which type of human cell carries on the most fermentation?
    a. muscle
    b. fat
    c. nerve
    d. bone

25. The metabolic process that produces the most ATP molecules is
    a. glycolysis.
    b. the citric acid cycle.
    c. the electron transport chain.
    d. fermentation.

26. The greatest contributor of electrons to the electron transport chain is
    a. oxygen.
    b. glycolysis.
    c. the citric acid cycle.
    d. the preparatory reaction.
    e. fermentation.

27. Substrate-level ATP synthesis takes place in
    a. glycolysis and the citric acid cycle.
    b. the electron transport chain and the preparatory reaction.
    c. glycolysis and the electron transport chain.
    d. the citric acid cycle and the preparatory reaction.

28. Match the terms to their definitions. Only four of these terms are needed:

    anaerobic   oxygen deficit
    citric acid cycle   pyruvate
    fermentation   preparatory reaction

    a. occurs in mitochondria and produces CO_2, ATP, NADH, and $FADH_2$
    b. growing or metabolizing in the absence of oxygen
    c. end product of glycolysis
    d. anaerobic breakdown of glucose that results in a gain of 2 ATP and end products such as alcohol and lactate

29. Which of the following is not true of fermentation?
    a. It has a net gain of only 2 ATP.
    b. It occurs in the cytoplasm.
    c. It donates electrons to the electron transport chain.
    d. It is carried on by humans.
    e. It is carried on by yeast.

Thinking Scientifically

1. The compound malonate, a substrate of the citric acid cycle, can be poisonous at high concentrations because it can block cellular respiration. How might a substrate of the citric acid cycle block respiration when it is present in great excess?

2. Insulin resistance can lead to diabetes. Insulin normally causes cells to take in glucose from the blood, but insulin resistance causes cells to fail to respond properly to this signal. The pancreas then compensates by secreting even more insulin into the body. Researchers have found a connection among a high-fat diet, decreased mitochondrial function, and insulin resistance in some elderly patients. What might be the connection among these three events?

3. You observe that prokaryotic cells sometimes have invaginations of the plasma membrane where metabolic pathways take place. How could this observation be used to support your hypothesis that mitochondria evolved when a prokaryote cell was engulfed by a nucleated cell?

Bioethical Issue

Regulating Dietary Supplements

Americans spend millions of dollars every year on products and diets that claim to boost metabolic rates and assist in weight loss. These and other supplements are available in supermarkets, by mail order, and on the Internet. Drug makers are required to prove that a drug is safe and effective before it can be sold to the public. But the Food and Drug Administration (FDA) cannot legally supervise the supplement industry because supplements are not defined as drugs. Thus, regulators must prove that a supplement is dangerous before it can be removed from the market. Although some of the more dangerous supplements, such as ephedra, have been banned, many questionable and harmful products remain freely available to anyone who wants them. Since the FDA does not have the time or resources to perform the necessary research, most dietary supplements receive scant supervision.

Many scientists and other concerned individuals contend that FDA regulation of the supplement industry is needed to protect public safety. They argue that FDA regulation would protect the public from potentially dangerous products and ensure the efficacy of dietary supplements and diet products. However, some are concerned that such regulation would undermine the public interest by reducing choices in the marketplace. Furthermore, they argue that the FDA regulation of supplements would create the need for additional taxes and divert resources that would be better spent regulating prescription and over-the-counter drugs.

Do you believe that the government should increase its scrutiny of dietary and weight-loss supplements? Or do you believe that FDA regulation of these products would be a waste of time and money?

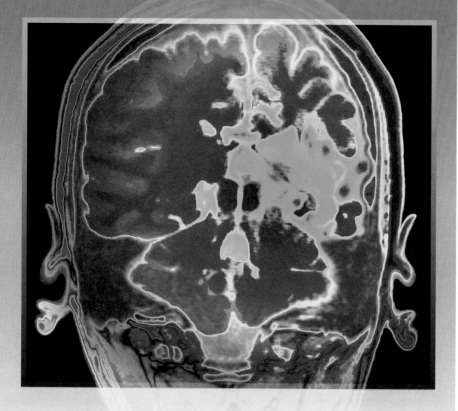

8

Cellular Reproduction

Regenerating Neurons

Many of our body tissues and organs can repair themselves if they become injured or damaged. This is because their cells are capable of undergoing mitosis, a process of nuclear division that enables cells and tissues to grow and be repaired. Exceptions are nerve cells, the cells that make up the nervous system. Injuries to the brain and spinal cord are very serious because nerve cells normally last a lifetime and seldom undergo mitosis. When nerve cells die, they are not likely to be replaced. The brain has approximately 35 billion nerve cells, and the loss of some of these is normal. However, in the case of brain injury, or a condition such as Alzheimer disease, nerve cells die at a rapid rate and brain function declines.

Scientists want to find a way to coerce differentiated nerve cells into undergoing mitosis. They also hypothesize that human stem cells, which do divide, can be coaxed into becoming new nerve cells. Although the use of embryonic stem cells is controversial and limited by legislation in the United States, adult stem cells are easily accessible and already show promise of becoming nerve cells under the proper circumstances. All cells in the body of an organism have the same genetic composition, which explains why stem cells can be used to form all cell types.

In this chapter, you will learn about the process of cell division called mitosis, how it is regulated, and what happens if it occurs either not at all or too frequently.

OUTLINE

BEFORE YOU BEGIN

Before beginning this chapter, take a few moments to review the following discussions.

Section 3.2 What is the structure of DNA?

Section 4.4 What are the roles of the microtubules and nucleus in a eukaryotic cell?

Section 4.5 What is a cell wall and what types of organisms is it found in?

a. Children grow

b. Tissues repair

c. Amoebas reproduce

d. Zygotes develop

Figure 8.1 Cellular reproduction.

Cellular reproduction occurs (**a**) when small children grow to be adults and (**b**) when mature organisms repair their tissues. Cellular reproduction also occurs (**c**) when unicellular organisms reproduce and (**d**) when zygotes, the product of sperm and egg fusion, develop into multicellular organisms capable of independent existence.

8.1 The Basics of Cellular Reproduction

Learning Outcomes

Upon completion of this section, you should be able to

1. Summarize the purpose of cellular reproduction.
2. Explain the role of histones and the nucleosome in the compaction of the chromatin.
3. Explain the role of the centromere in relation to the sister chromatids.

We humans, like other multicellular organisms, begin life as a single cell. However, in less than 10 months, we become trillions of cells because cellular reproduction has occurred over and over again. Even after we are born, cellular reproduction doesn't stop—it continues as we grow (**Fig. 8.1a**), and when we are adults, it replaces worn-out or damaged tissues (Fig. 8.1b). Right now, your body is producing thousands of new red blood cells, skin cells, and cells that line your respiratory and digestive tracts. If you suffer a cut, cellular reproduction helps repair the injury.

Cellular reproduction is also necessary for the reproduction of certain organisms. When an amoeba splits (Fig. 8.1c), two new individual amoebas are produced. **Binary fission** is a type of cell division in bacteria (Chapter 17) that produces two bacterial cells. These processes are forms of *asexual reproduction* because they produce new cells that are identical to the original. Sexual reproduction, which involves a sperm and an egg, introduces variation to the offspring. It is covered in Chapter 9.

Animation
Binary Fission

The cell theory (Chapter 4) emphasizes the importance of cellular reproduction when it states that "all cells come from preexisting cells." Therefore, you can't have a new cell without a preexisting cell, and you can't have a new organism without a preexisting organism (Fig. 8.1d). Cellular reproduction is necessary for the production of both new cells and new organisms.

Cellular reproduction involves two important processes: growth and cell division. During growth, a cell duplicates its contents, including the organelles and its DNA. Then, during cell division, the DNA and other cellular contents of the *parent cell* are distributed to the *daughter cells*. (These terms have nothing to do with gender; they are simply a way to designate the beginning cell and the resulting cells.) Both processes are heavily regulated to prevent runaway cellular reproduction, which can have serious consequences.

Chromosomes

An important event that occurs in preparation for cell division is **DNA replication,** when a cell copies its DNA. Then, a full copy of all the DNA is passed to both daughter cells. This step is critical because cells cannot continue to live without a copy of the genetic material. But a human cell contains an astonishing 5 centimeters (2 in.) of DNA, and moving this large quantity of DNA is no easy task. However, DNA and associated proteins are packaged into a set of **chromosomes,** which allow it to be distributed to the daughter cells. We shall see that proteins are involved in packaging the DNA and that these proteins are associated with still other proteins involved in DNA replication and function.

Chromatin to Chromosomes

When a eukaryotic cell is not undergoing cell division, the DNA and associated proteins have the appearance of thin threads called **chromatin.** Closer examination reveals that chromatin is periodically wound around a core of eight protein molecules, so that it looks like beads on a string. The protein molecules are **histones,** and each bead is called a **nucleosome (Fig. 8.2).** Chromatin normally adopts a zigzag structure, and then it is folded into loops for further compaction. This looped chromatin more easily fits within the nucleus.

Just before cell division occurs, the chromatin condenses multiple times into large loops, which produce highly compacted chromosomes. Each species has a characteristic number of chromosomes; a human cell has 46. We can easily see chromosomes with a light microscope because, just before division occurs, a chromosome is over 10,000 times more compact than is chromatin.

By the time we can clearly see the chromosomes, they are duplicated. A duplicated chromosome is composed of two identical halves, called **sister chromatids,** held together at a constricted region called a **centromere.** Each sister chromatid contains an identical DNA double helix.

Animation
Overview of Cell Division

Connecting the Concepts

For a closer examination of the concepts presented in this section, refer to the following discussions.

Section 4.2 examines two main classifications of cells and provides an overview of the cell theory.

Section 9.1 explores how the exchange of information between the sister chromatids provides genetic variation.

Section 11.1 provides a more detailed look at DNA replication.

Check Your Progress 8.1

1. Describe the ways cellular reproduction is necessary to the continued existence of a mature organism and the production of a new organism.

2. Detail the two processes that typically occur when a cell reproduces.

3. Contrast the structure of chromatin when the cell is not dividing with that of a chromosome just before cellular reproduction.

Connections and Misconceptions

Is the number of chromosomes related to the complexity of the organism?

In eukaryotes, like humans, the number of chromosomes varies considerably. A fruit fly has 8 chromosomes, and humans have 46. So at first glance, you may think that a relationship exists between chromosome number and complexity. However, yeasts (a one-celled eukaryote) have 32 chromosomes, and horses have 64. There is a species of fern with 1,252 chromosomes. Thus, there is no definite relationship between the number of chromosomes and the complexity of the organism.

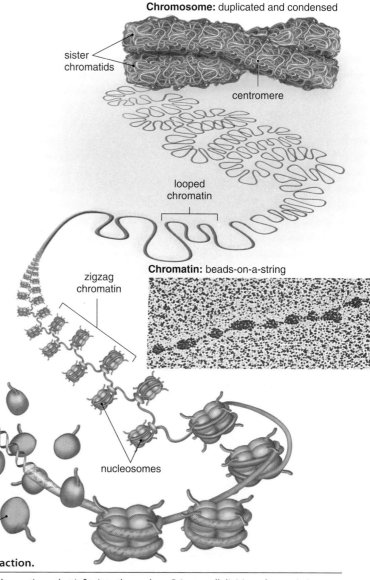

Figure 8.2   Chromosome compaction.

Histones are responsible for packaging chromatin so that it fits into the nucleus. Prior to cell division, chromatin is further condensed into chromosomes, so that the genetic material can be easily divided between the daughter cells.

8.2 The Cell Cycle

Learning Outcomes

Upon completion of this section, you should be able to

1. Summarize the activities that occur in the cell during each stage of the cell cycle.
2. Explain the significance of the G_0 phase.
3. Summarize the events that occur during the M phase of the cell cycle.

We have already seen that cellular reproduction involves duplication of cell contents followed by cell division. For cellular reproduction to be orderly, you would expect the first event to occur before the second event, and that's just what happens during the cell cycle. The **cell cycle** is an orderly sequence of stages that takes place between the time a new cell has arisen from the division of the parent cell to the point when it has given rise to two daughter cells.

Interphase

As **Figure 8.3** shows, most of the cell cycle is spent in **interphase.** This is the time when a cell performs its usual functions, depending on its location in the body. The amount of time the cell takes for interphase varies widely. Embryonic cells complete the entire cell cycle in just a few hours. A rapidly dividing mammalian cell, such as an adult stem cell, typically takes about 24 hours to complete the cell cycle and spends 22 hours in interphase.

DNA replication occurs in the middle of interphase and serves as a way to divide interphase into three phases: G_1, S, and G_2. G_1 is the phase before DNA replication, and G_2 is the phase following DNA synthesis. Originally, G stood for "gap," but now that we know how metabolically active the cell is, it is better to think of G as standing for "growth." Protein synthesis is very much a part of these growth stages.

During G_1, a cell doubles its organelles (such as mitochondria and ribosomes) and accumulates materials that will be used for DNA replication. At this point, the cell integrates internal and external signals and "decides" whether to continue with the cell cycle. Some cells, such as muscle cells, typically remain in interphase, and cell division is permanently arrested. These cells are said to have entered a G_0 phase. If an injury occurs, many cells in the G_0 phase can reenter the cell cycle and divide again to repair the damage. But a few cell types, such as nerve cells, almost never divide again once they have entered G_0.

Following G_1, the cell enters the S phase. The S stands for "synthesis," and certainly DNA synthesis is required for DNA replication. At the beginning of the S phase each chromosome has one **chromatid** consisting of a single DNA double helix. At the end of this stage, each chromosome is composed of two *sister chromatids,* each having one double helix. The two chromatids of each chromosome remain attached at the centromere. DNA replication results in duplicated chromosomes.

The G_2 phase extends from the completion of DNA replication to the onset of mitosis. During this stage, the cell synthesizes the proteins that will be needed for cell division, such as the protein found in microtubules. The role of microtubules in cell division is described in a later section.

M (Mitotic) Phase

Cell division occurs during the M phase, which encompasses both division of the nucleus and division of the cytoplasm. The type of nuclear division associated

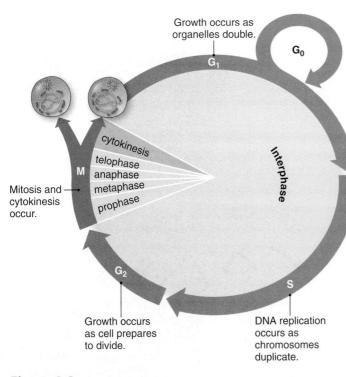

Figure 8.3 **The cell cycle.**

Cells go through a cycle consisting of four phases: G_1, S, G_2, and M. Interphase includes G_1, S, and G_2 phases. Some cells can exit G_1 and enter a G_0 phase.

with the cell cycle is called **mitosis,** which accounts for why this stage is called the M phase. As a result of mitosis, the daughter nuclei are identical to the parent cell and to each other—they all have the same number and kinds of chromosomes. Division of the cytoplasm, which starts even before mitosis is finished, is called **cytokinesis.**

Animation
The Cell Cycle

Check Your Progress 8.2

1. Identify the importance of the cell cycle.

2. Compare and contrast what is occurring during each phase of interphase: G_1, S, and G_2.

3. Explain what would occur if interphase did not happen; could mitosis take place? Why or why not?

8.3 Mitosis and Cytokinesis

Learning Outcomes

Upon completion of this section, you should be able to

1. Summarize the role of the spindle in cell division.
2. Describe the phases of mitosis and the process of cytokinesis.
3. Compare cytokinesis in a plant cell and an animal cell.

During mitosis, the duplicated nuclear contents of the parent cell are distributed equally to the daughter cells. Recall that, during the S phase, the DNA of each chromosome is replicated to produce a duplicated chromosome that contains two identical sister chromatids, which remain attached at the centromere (**Fig. 8.4**). Each chromatid is a single DNA double helix containing the same sequence of base pairs as the original chromosome. During mitosis, the sister chromatids of each chromosome separate and are now called *daughter chromosomes.* Because each original chromosome goes through the same process of DNA replication followed by separation of the sister chromatids, the daughter nuclei produced by mitosis are genetically identical to each other and to the parent nucleus. Thus, if the parent nucleus has four chromosomes, each daughter nucleus also has four chromosomes of exactly the same type. One way to keep track of the number of chromosomes in drawings is to count the number of centromeres, because every chromosome has a centromere.

Most eukaryotic cells have an even number of chromosomes because each parent has contributed half of the chromosomes to the new individual.

The Spindle

While it may seem easy to separate the chromatids of only 4 duplicated chromosomes, imagine the task when there are 46 chromosomes, as in humans, or 78, as in dogs. Certainly, it is helpful if chromosomes are highly condensed before the task begins, but some mechanism is needed to complete separation in an organized manner. Most eukaryotic cells rely on a **spindle,** a cytoskeletal structure, to pull the chromatids apart. A spindle has spindle fibers made of microtubules that are able to assemble and disassemble. First, the microtubules assemble to form the spindle that takes over the center of the cell and separates the chromatids. Later, they disassemble.

Connecting the Concepts

For more information on eukaryotic cells and DNA, refer to the following discussions.

Section 4.4 describes the structure and function of a typical eukaryotic cell.

Section 11.1 explores how the DNA is replicated during the S phase.

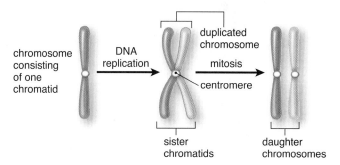

chromosome consisting of one chromatid — DNA replication → duplicated chromosome — mitosis → daughter chromosomes

centromere

sister chromatids

Figure 8.4 Overview of mitosis.

DNA replication forms the sister chromatids, which are separated to form daughter chromosomes by mitosis.

A **centrosome** is the primary microtubule organizing center of a cell. In an animal cell, each centrosome has two barrel-like structures, called **centrioles**, and an array of microtubules called an **aster.** Plant cells have centrosomes, but they are not clearly visible because they lack centrioles. Centrosome duplication occurs at the start of the S phase of the cell cycle and has been completed by G_2. During the first part of the M phase, the centrosomes separate and move to opposite sides of the nucleus, where they form the poles of the spindle. As the nuclear envelope breaks down, spindle fibers take over the center of the cell. Some overlap at the **spindle equator,** which is midway between the poles. Others attach to duplicated chromosomes in a way that ensures the separation of the sister chromatids and their proper distribution to the daughter cells. Whereas the chromosomes will be inside the newly formed daughter nuclei, a centrosome will be just outside.

Traditionally, mitosis is divided into a sequence of events, even though it is a continuous process. We will describe mitosis as having four phases: **prophase, metaphase, anaphase,** and **telophase.** These phases are labeled in **Figure 8.5** for a dividing plant cell nucleus. With the exception of the movement of the centrioles, the following descriptions of the phases of mitosis apply to both animal and plant cells.

Phases of Mitosis in Animal and Plant Cells

Although we have divided mitosis into four phases, it is a continuous process in the cell. **Figure 8.6** illustrates the phases of mitosis in an animal cell. Before studying the descriptions of these phases in Figure 8.6, recall that, before mitosis begins, DNA has been replicated. Each chromatid contains a double helix of DNA, and each chromosome consists of two sister chromatids attached at a centromere. Notice that, in Figure 8.6, some chromosomes are colored red and some are colored blue. The red chromosomes were inherited from one parent and the blue chromosomes from the other parent.

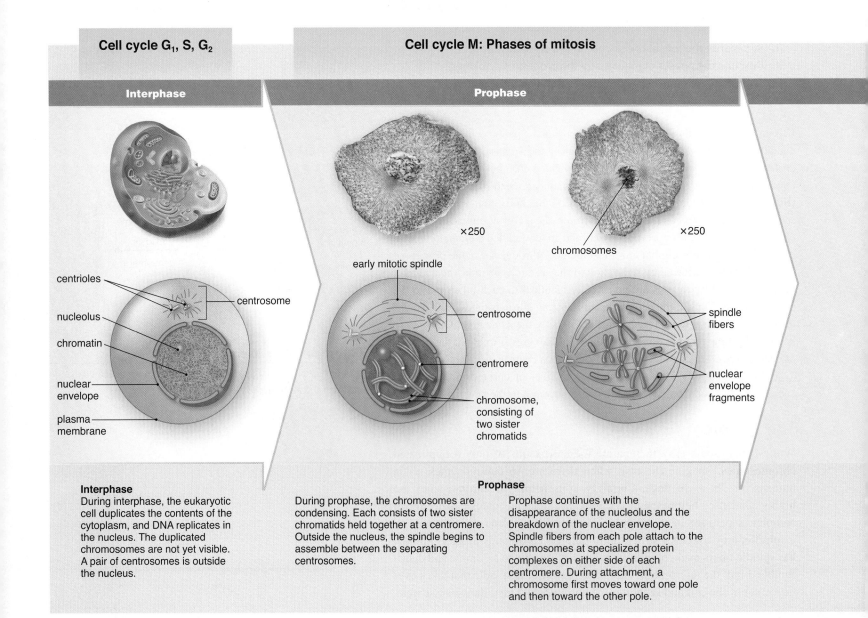

Cell cycle G₁, S, G₂

Cell cycle M: Phases of mitosis

Interphase

Prophase

×250

×250

chromosomes

early mitotic spindle

centrioles

nucleolus

chromatin

nuclear envelope

plasma membrane

centrosome

centrosome

centromere

chromosome, consisting of two sister chromatids

spindle fibers

nuclear envelope fragments

Interphase
During interphase, the eukaryotic cell duplicates the contents of the cytoplasm, and DNA replicates in the nucleus. The duplicated chromosomes are not yet visible. A pair of centrosomes is outside the nucleus.

Prophase
During prophase, the chromosomes are condensing. Each consists of two sister chromatids held together at a centromere. Outside the nucleus, the spindle begins to assemble between the separating centrosomes.

Prophase continues with the disappearance of the nucleolus and the breakdown of the nuclear envelope. Spindle fibers from each pole attach to the chromosomes at specialized protein complexes on either side of each centromere. During attachment, a chromosome first moves toward one pole and then toward the other pole.

Figure 8.5 Mitosis in a plant cell.

The chromosomes are stained blue, and microtubules of the spindle fibers are stained pink in these photos of a dividing African blood lily. For a description of these phases, see Figure 8.6.

chromosomes

condensing chromosomes

spindle equator

nucleus with chromatin

spindle fibers (pink)

Metaphase ×5

Telophase ×5

pole

Interphase ×5 **Prophase** ×5

Anaphase ×5

Figure 8.6 Phases of mitosis in animal cells.

The images on the top row are micrographs of an animal cell during mitosis. In the lower images, the red chromosomes were inherited from one parent and the blue chromosomes from the other parent.

Phases of mitosis

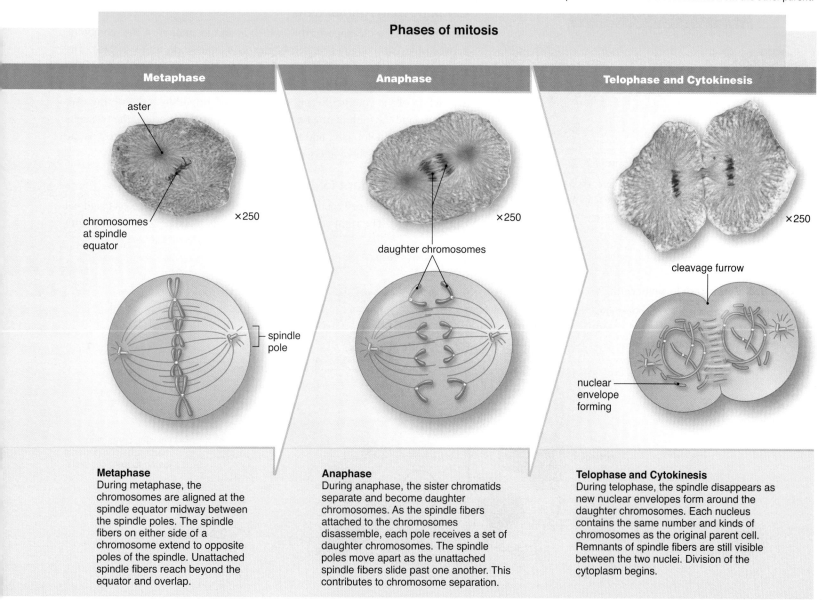

Metaphase	Anaphase	Telophase and Cytokinesis

aster

×250

×250

×250

chromosomes at spindle equator

daughter chromosomes

cleavage furrow

spindle pole

nuclear envelope forming

Metaphase
During metaphase, the chromosomes are aligned at the spindle equator midway between the spindle poles. The spindle fibers on either side of a chromosome extend to opposite poles of the spindle. Unattached spindle fibers reach beyond the equator and overlap.

Anaphase
During anaphase, the sister chromatids separate and become daughter chromosomes. As the spindle fibers attached to the chromosomes disassemble, each pole receives a set of daughter chromosomes. The spindle poles move apart as the unattached spindle fibers slide past one another. This contributes to chromosome separation.

Telophase and Cytokinesis
During telophase, the spindle disappears as new nuclear envelopes form around the daughter chromosomes. Each nucleus contains the same number and kinds of chromosomes as the original parent cell. Remnants of spindle fibers are still visible between the two nuclei. Division of the cytoplasm begins.

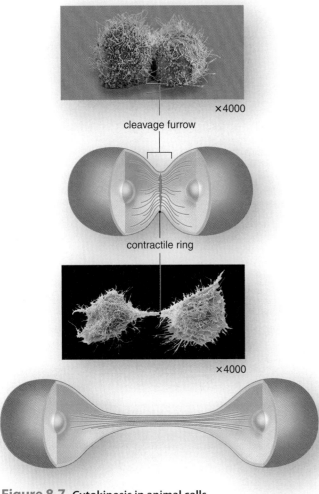

cleavage furrow

×4000

contractile ring

×4000

Figure 8.7 **Cytokinesis in animal cells.**

A single cell becomes two cells by a furrowing process. A cleavage furrow appears as early as anaphase, and a contractile ring, composed of actin filaments, gradually gets smaller until there are two cells.

During mitosis, a spindle arises that will separate the sister chromatids of each duplicated chromosome. Once separated, the sister chromatids become daughter chromosomes. In this way, the resulting daughter nuclei are identical to each other and to the parental nucleus. In Figure 8.6, the daughter nuclei not only have four chromosomes; they each have one red short, one blue short, one red long, and one blue long, the same as the parent cell had. In other words, each daughter nucleus is genetically the same as the original parent nucleus. Mitosis is usually followed by division of the cytoplasm, or cytokinesis. Cytokinesis begins during telophase and continues after the nuclei have formed in the daughter cells. The cell cycle is now complete, and the daughter cells enter interphase, during which the cell will grow and DNA will replicate once again.

Animation
Mitosis

Cytokinesis in Animal and Plant Cells

In most cells, cytokinesis follows mitosis. When mitosis occurs, but cytokinesis doesn't occur, the result is a multinucleated cell. For example, skeletal muscle cells in vertebrate animals and the embryo sac in flowering plants are multinucleated.

Cytokinesis in Animal Cells

In animal cells, a cleavage furrow, which is an indentation of the membrane between the two daughter nuclei, begins as anaphase draws to a close. The cleavage furrow deepens when a band of actin filaments, called the contractile ring, slowly forms a circular constriction between the two daughter cells. The action of the contractile ring can be likened to pulling a drawstring ever tighter around the middle of a balloon. A narrow bridge between the two cells is visible during telophase, and then the contractile ring continues to separate the cytoplasm until there are two independent daughter cells (**Fig. 8.7**).

Cytokinesis in Plant Cells

Cytokinesis in plant cells occurs by a process different from that seen in animal cells (**Fig. 8.8**). The rigid cell wall that surrounds plant cells does not permit

Figure 8.8 **Cytokinesis in plant cells.**

During cytokinesis in a plant cell, a cell plate forms midway between two daughter nuclei and extends to the original plasma membrane.

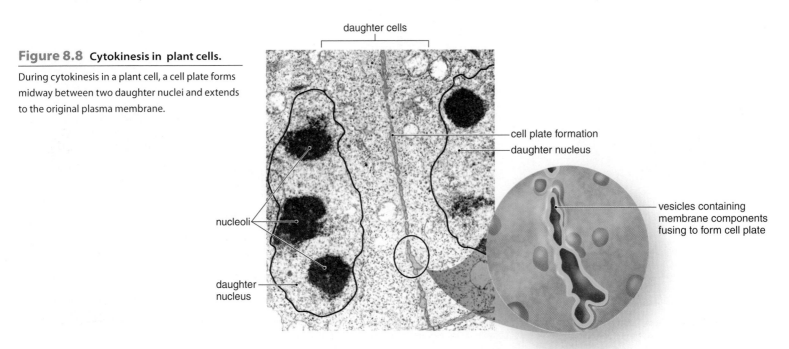

daughter cells

cell plate formation

daughter nucleus

vesicles containing membrane components fusing to form cell plate

nucleoli

daughter nucleus

cytokinesis by furrowing. Instead, cytokinesis in plant cells involves the building of new plasma membrane and cell walls between the daughter cells.

Cytokinesis is apparent when a small, flattened disk appears between the two daughter plant cells. In electron micrographs, it is possible to see that the disk is composed of vesicles. The Golgi apparatus produces these vesicles, which move along microtubules to the region of the disk. As more vesicles arrive and fuse, a cell plate can be seen. The **cell plate** is simply newly formed plasma membrane that expands outward until it reaches the old plasma membrane and fuses with it. The new membrane releases the molecules that form the new plant cell walls. These cell walls are later strengthened by the addition of cellulose fibrils.

 Animation Cytokinesis
 MP3 Mitosis

Check Your Progress 8.3

1. Identify the phases of mitosis and explain what happens to the chromosomes during each phase.
2. Describe the structure of the spindle and illustrate its importance to mitosis.
3. Predict what the daughter cells of a mitotic division would look like if cytokinesis did not occur properly. Would the daughter cells still be able to function correctly if this occurred?

Connecting the Concepts

For more information on the concepts presented in this section, refer to the following discussions.

Section 4.4 describes the structure of mictotubules and microfilaments.

Section 9.4 examines the consequences if the chromosomes do not separate during anaphase.

Section 28.2 explores the function of the multinucleated muscle cells in vertebrate animals.

8.4 The Cell Cycle Control System

Learning Outcomes

Upon completion of this section, you should be able to

1. Summarize the role of checkpoints in the cell cycle.
2. Explain how checkpoints are regulated by internal and external signals.
3. Describe the process of apoptosis.

In order for a cell to reproduce successfully, the cell cycle must be controlled. The importance of cell cycle control can be appreciated by comparing the cell cycle to the events that occur in a washing machine. The washer's control system starts to wash only when the tub is full of water, doesn't spin until the water has been emptied, and so forth. Similarly, the cell cycle's control system ensures that the G_1, S, G_2, and M phases occur in order and only when the previous phase has been successfully completed.

Cell Cycle Checkpoints

Just as a washing machine will not begin to agitate the load until the tub is full, the cell cycle has **checkpoints** that can delay the cell cycle until certain conditions are met. The cell cycle has many checkpoints, but we will consider only three: G_1, G_2, and the mitotic checkpoint (**Fig. 8.9**).

The G_1 *checkpoint* is especially significant because, if the cell cycle passes this checkpoint, the cell is committed to divide. If the cell does not pass

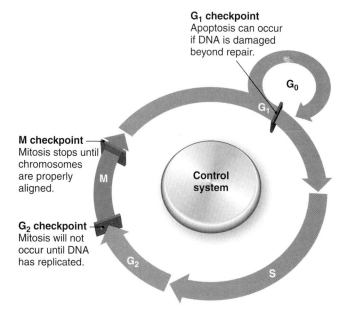

Figure 8.9 Cell cycle checkpoints.

Three important checkpoints are designated. Internal and external signals determine whether the cell is ready to proceed past cell cycle checkpoints.

this checkpoint, it can enter G_0, during which it performs its normal functions but does not divide. The proper growth signals, such as certain growth factors, must be present for a cell to pass the G_1 checkpoint. Additionally, the integrity of the cell's DNA is checked. If the DNA is damaged, the protein p53 can stop the cycle at this checkpoint and initiate DNA repair. If that is not possible, it can cause the cell to undergo programmed cell death, or **apoptosis.**

The cell cycle halts momentarily at the G_2 *checkpoint* until the cell verifies that DNA has replicated. This prevents the initiation of the M phase unless the chromosomes are duplicated. Also, if DNA is damaged, as from exposure to solar radiation or X-rays, arresting the cell cycle at this checkpoint allows time for the damage to be repaired, so that it is not passed on to daughter cells.

Another cell cycle checkpoint occurs during the mitotic stage. The cycle hesitates between metaphase and anaphase to make sure the chromosomes are properly attached to the spindle and will be distributed accurately to the daughter cells. The cell cycle does not continue until every chromosome is ready for the nuclear division process.

Animation
Cell Cycle Control

Connections and Misconceptions

Is apoptosis the only cause of cell death?

Apoptosis allows an organism to control cell death. It is an important aspect of embryological development. The opposite of apoptosis is necrosis, or unprogrammed cell death,

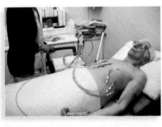

the death of cells due to disease or injury. For example, during a heart attack, the cells of the heart are deprived of oxygen and nutrients. The cells then burst and die, causing inflammation of the surrounding tissues. This can cause the death of the surrounding cells. In contrast, apoptosis does not cause inflammation and does not spread to surrounding cells.

Internal and External Signals

The checkpoints of the cell cycle are controlled by internal and external signals. A **signal** is a molecule that stimulates or inhibits an event. Signals outside the cell are called external signals; those inside the cell are called internal signals. Researchers have identified a series of internal signals called **cyclins.** The levels of these proteins increase and decrease as the cell cycle progresses. The appropriate cyclin has to be present at the correct levels for the cell to proceed from the G_1 phase to the S phase, and from the G_2 phase into the M phase. In addition, enzymes called **kinases** remove phosphate from ATP and add it to another molecule. The addition of the energized phosphate from ATP often acts as an off/on switch for cellular activities. Kinases are active in the removal of the nuclear membrane and the condensation of the chromosomes early in prophase.

Some external signals, such as growth factors and hormones, stimulate cells to go through the cell cycle. Growth factors also stimulate tissue repair. Even cells that are arrested in G_0 will finish the cell cycle if growth factors stimulate them to do so. For example, epidermal growth factor (EGF) stimulates skin in the vicinity of an injury to finish the cell cycle, thereby repairing the damage. Hormones act on tissues at a distance, and some signal cells to divide. For example, at a certain time in the menstrual cycle of females, the hormone estrogen stimulates cells lining the uterus to divide and prepares the lining for implantation of a fertilized egg.

The cell cycle can be inhibited by cells coming into close contact with other cells. In the laboratory, eukaryotic cells will divide until they line a container in a one-cell-thick sheet. Then they stop dividing, due to a phenomenon termed **contact inhibition.** Some researchers suggest that contact inhibition prevents cells from overgrowing within the body. When tissues are damaged, cells divide to repair the damage. But once the tissue has been repaired, contact inhibition prevents cell overgrowth by halting the cell cycle.

Some years ago, it was noted that mammalian cells in cell culture divide about 70 times, and then they die. Cells seem to "remember" the number of times they have already divided, and they stop when they have reached the usual number of cell divisions. It's as if *senescence,* the aging of cells, is dependent on an internal battery-operated clock that runs down and then stops. We now know that senescence is due to the shortening of telomeres. A **telomere** is a repeating DNA base sequence (TTAGGG) at the end of the chromosomes that can be as long as 15,000 base pairs. Telomeres have been likened to the protective caps on the ends of shoelaces because they ensure chromosomal stability. Each time a cell divides, a portion of a telomere is lost. When telomeres become too short, the cell is "old" and dies by the process of apoptosis.

Apoptosis

Apoptosis is often defined as programmed cell death because the cell progresses through a typical series of events that bring about its destruction (**Fig. 8.10**). The cell rounds up and loses contact with its neighbors. The nucleus fragments, and the plasma membrane develops blisters, called blebs. Finally, the cell breaks into fragments, which are engulfed by white blood cells and/or neighboring cells. Oddly enough, healthy cells routinely harbor the enzymes, called caspases, that bring about apoptosis. These enzymes are ordinarily held in check by inhibitors but are unleashed by either internal or external signals.

Cell division and apoptosis are two opposing processes that keep the number of cells in the body at an appropriate level. In other words, cell division increases, and apoptosis decreases, the number of **somatic cells** (body cells). Both the cell cycle and apoptosis are normal parts of growth and development. An organism begins as a single cell that repeatedly undergoes the cell cycle to produce many cells, but eventually some cells must die for the organism to take shape. For example, when a tadpole becomes a frog, the tail disappears as apoptosis occurs. In humans, the fingers and toes of an embryo are at first webbed, but later they are freed from one another as a result of apoptosis.

Apoptosis occurs all the time, particularly if an abnormal cell that could become cancerous appears. Death through apoptosis prevents a tumor from developing.

Connecting the Concepts

For more information on cell cycle controls, refer to the following discussions.

Section 12.2 examines how cancer cells become immortal by repairing their telomeres.

Section 26.3 describes how the immune system uses apoptosis to destroy infected cells.

Section 27.2 lists the roles of the major hormones of the human body.

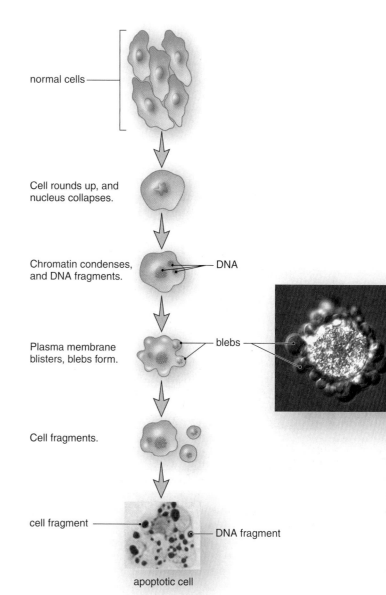

normal cells

Cell rounds up, and nucleus collapses.

Chromatin condenses, and DNA fragments. — DNA

Plasma membrane blisters, blebs form. — blebs

Cell fragments.

cell fragment — — DNA fragment

apoptotic cell

Figure 8.10 Apoptosis.

Apoptosis is a sequence of events that results in a fragmented cell. The fragments are phagocytized by white blood cells and neighboring tissue cells.

Check Your Progress 8.4

1. Explain the significance of checkpoints in the cell cycle.
2. Clarify how signals are used for cell cycle control, and describe the two types of signals used by cells.
3. Predict what might happen to an organism if apoptosis is not regulated.

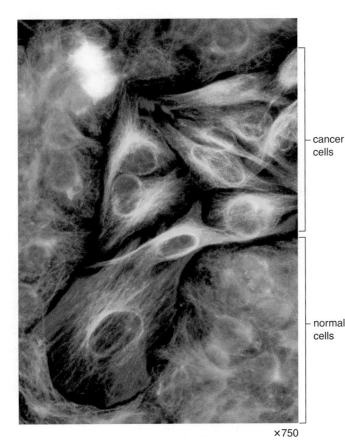

cancer cells

normal cells

×750

Figure 8.11 Cancer cells.

Compared with the other cells, the brightly colored cancer cells appear distinctly abnormal.

Connections and Misconceptions

What in cigarette smoke causes cancer?

There are over 4,000 known chemicals in tobacco smoke. Of these, 60 are known to be carcinogens, or cancer-causing agents. Included in this list are arsenic (a heavy metal), benzene (an additive in gasoline), and cadmium (a metal found in batteries). In addition to the carcinogens, cigarette smoke contains high levels of carbon monoxide and hydrogen cyanide, both of which reduce the body's ability to process oxygen. Exposure to these chemicals is estimated to reduce life expectancy by 13.2 years for males, 14.5 years for females.

8.5 The Cell Cycle and Cancer

Learning Outcomes

Upon completion of this section, you should be able to

1. Describe the characteristics of cancer cells.
2. Summarize the types of treatment for cancer.
3. Describe the factors that reduce the risk of cancer.

The cell cycle is regulated by signals that either inhibit or promote the cell cycle. When these signals become unbalanced, **cancer** may result. Cancer is a disease of the cell cycle in which cellular reproduction occurs repeatedly without end. Cancers are classified according to their location. *Carcinomas* are cancers of the epithelial tissue that lines organs; *sarcomas* are cancers arising in muscle or connective tissue (especially bone or cartilage); and *leukemias* are cancers of the blood. In this section, we'll consider some general characteristics of cancer cells. Chapter 12 explores specific genetic changes that lead to cancerous growth.

Characteristics of Cancer Cells

Carcinogenesis, the development of cancer, is gradual; it may be decades before a tumor is visible. Cancer cells then have the following characteristics.

Cancer cells lack differentiation. Cancer cells lose their specialization and do not contribute to the functioning of a body part. A cancer cell does not look like a differentiated epithelial, muscle, nervous, or connective tissue cell; instead, it looks distinctly abnormal (**Fig. 8.11**). As mentioned, normal cells can enter the cell cycle about 70 times, and then they die. Cancer cells can enter the cell cycle repeatedly; in this way, they are immortal.

Cancer cells have abnormal nuclei. The nuclei of cancer cells are enlarged and may contain an abnormal number of chromosomes. The chromosomes are also abnormal; some parts may be duplicated, or some may be deleted. In addition, gene amplification (extra copies of specific genes) is seen much more frequently than in normal cells.

Cancer cells do not undergo apoptosis. Ordinarily, cells with damaged DNA undergo apoptosis, preventing tumors from developing. Cancer cells, however, do not respond normally to the signals to initiate apoptosis, and thus they continue to divide.

Cancer cells form tumors. Normal cells anchor themselves to a substratum and/or adhere to their neighbors. Then, they exhibit contact inhibition and stop dividing. Cancer cells, on the other hand, have lost all restraint; they pile on top of one another and grow in multiple layers, forming a **tumor.** They have a reduced need for growth factors, and they no longer respond to inhibitory signals. As cancer develops, the most aggressive cell becomes the dominant cell of the tumor.

Cancer cells undergo metastasis and promote angiogenesis. A **benign tumor** is usually contained within a capsule and therefore cannot invade adjacent tissue. Additional changes may cause cancer cells to produce enzymes that allow tumor cells to escape the capsule and invade nearby tissues. The tumor has then become **malignant,** meaning that it is invasive and may spread. Cells from a malignant tumor may travel through the blood or lymph to start new tumors elsewhere within the body. This process is known as **metastasis (Fig. 8.12)**.

An actively growing tumor can grow only so large before it becomes unable to obtain sufficient nutrients to support further growth. Additional mutations in cancer cells allow them to secrete factors that promote **angiogenesis,** the formation of new blood vessels. Additional nutrients and oxygen reach the tumor, allowing it to grow larger. Some modes of cancer treatment are aimed at preventing angiogenesis. The patient's prognosis (probable outcome) depends on (1) whether the tumor has invaded surrounding tissues and (2) whether there are metastatic tumors in distant parts of the body.

Cancer Treatment

Cancer treatments either remove the tumor or interfere with the cancer cells' ability to reproduce. For many solid tumors, removal by surgery is often the first line of treatment. When the cancer is detected at an early stage, surgery may be sufficient to cure the patient by removing all cancerous cells.

Because cancer cells are rapidly dividing, they are susceptible to radiation therapy and chemotherapy. The goal of radiation is to kill cancer cells within a specific tumor by directing high-energy beams at the tumor. DNA is damaged to the point that replication can no longer occur, and the cell undergoes apoptosis. Chemotherapy is a way to kill cancer cells that have spread throughout the body. Like radiation, most chemotherapeutic drugs lead to the death of cells by damaging their DNA or interfering with DNA synthesis. Others interfere with the functioning of the mitotic spindle. The drug vinblastine, first obtained from a flowering plant called a periwinkle, prevents the spindle from forming. Taxol, extracted from the bark of the Pacific yew tree, prevents the spindle from functioning as it should. Unfortunately, radiation and chemotherapy often damage cells other than cancer cells, leading to side effects, such as nausea and hair loss.

Hormonal therapy also prevents cell division, but in a different way. These drugs are designed to prevent cancer cell growth by preventing the cells from receiving the signals they need for continued growth and division. For example, the drug tamoxifen modifies the receptor for estrogen, so that this hormone cannot bind to it and promote the cell cycle. Other drug therapies are also being investigated. One proposed therapy that is now well underway uses drugs designed to inhibit angiogenesis, limiting a tumor's ability to grow and spread.

Prevention of Cancer

Evidence is clear that the risk of certain types of cancer can be reduced by adopting protective behaviors and following recommended dietary guidelines.

Protective Behaviors

To lower the risk of developing certain cancers, people are advised to avoid smoking, sunbathing, and excessive alcohol consumption.

Cigarette smoking accounts for about 30% of all cancer deaths. Smoking is responsible for 90% of lung cancer cases among men and 79% among women—about 87% altogether. People who smoke two or more packs of cigarettes a day have lung cancer mortality rates 15 to 25 times greater than those of nonsmokers. Smokeless tobacco (chewing tobacco or snuff) increases the risk of cancers of the mouth, larynx, throat, and esophagus.

Many skin cancers are sun-related. Sun exposure is a major factor in the development of the most dangerous type of skin cancer, melanoma, and the incidence of this cancer increases in people living near the equator. Similarly, excessive radon gas[1] exposure in homes increases the risk of lung cancer, especially in cigarette smokers. It is best to test your home and, if necessary, take the proper remedial actions.

---

[1]Radon gas results from the natural radioactive breakdown of uranium in soil, rocks, and water.

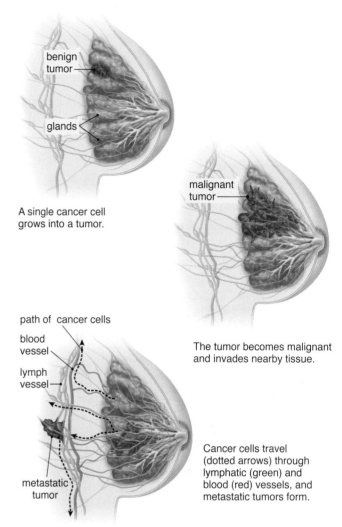

A single cancer cell grows into a tumor.

The tumor becomes malignant and invades nearby tissue.

path of cancer cells

blood vessel

lymph vessel

metastatic tumor

Cancer cells travel (dotted arrows) through lymphatic (green) and blood (red) vessels, and metastatic tumors form.

a. Development of breast cancer and metastatic tumors

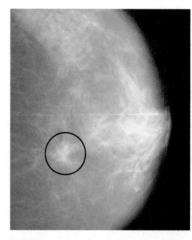

b. Mammogram showing tumor

Figure 8.12 Development of breast cancer.

a. Breast cancer begins as a single abnormal cell grows into a benign tumor. Additional changes allow the tumor to become invasive, or malignant, and invade nearby tissue. Metastasis occurs when the tumor invades blood and lymphatic vessels and new tumors develop some distance from the original tumor. **b.** A mammogram (X-ray image of the breast) can find a tumor (circled) too small to be felt.

Connecting the Concepts

For more information on cancer, refer to the following discussions.

Section 12.2 details the relationship between cancer and mutations in the cell cycle.

Section 25.2 describes the role of vitamins in the body, including the antioxidant vitamins A and C.

Check Your Progress 8.5

1. Describe how cancer relates to the cell cycle.

2. List the characteristic features of cancer cells, and list the major protective strategies that can be used to reduce the risk of cancer.

3. Contrast the modes of action of radiation and chemotherapy with the mode of action of hormonal therapy.

Video
Melanoma
Markers

Cancers of the mouth, throat, esophagus, larynx, and liver occur more frequently among heavy drinkers, especially when accompanied by tobacco use (cigarettes, cigars, or chewing tobacco).

Protective Diet

The risk of cancer (especially colon, breast, and uterine cancers) is 55% greater among obese women. The risk of cancer is 33% greater among obese men, compared with people of normal weight. Thus, weight loss in these groups can reduce cancer risk. In addition, the following dietary guidelines are recommended (**Fig. 8.13**):

- Increase your consumption of foods that are rich in vitamins A and C. Beta-carotene, a precursor of vitamin A, is found in dark green leafy vegetables, carrots, and various fruits. Vitamin C is present in citrus fruits. These vitamins are called antioxidants, because in cells they prevent the formation of free radicals (organic ions having an unpaired electron), which can damage DNA. Vitamin C also prevents the conversion of nitrates and nitrites into carcinogenic nitrosamines in the digestive tract.

- Avoid salt-cured or pickled foods because they may increase the risk of stomach and esophageal cancers. Smoked foods, such as ham and sausage, contain chemical carcinogens similar to those in tobacco smoke. Nitrites are sometimes added to processed meats (e.g., hot dogs and cold cuts) and other foods to protect them from spoilage; as mentioned, nitrites are sometimes converted to cancer-causing nitrosamines in the digestive tract.

- Include in the diet vegetables from the cabbage family, which includes cabbage, broccoli, brussels sprouts, kohlrabi, and cauliflower. Eating these vegetables may reduce the risk of gastrointestinal and respiratory tract cancers.

Figure 8.13 The right diet helps prevent cancer.

Foods that help protect against cancer include fruits and dark green leafy vegetables (such as swiss chard [right]), which are rich in vitamins A and C, and vegetables from the cabbage family, including broccoli.

Media Study Tools

www.mhhe.com/maderessentials3

Enhance your study of this chapter with study tools and practice tests. Also ask your instructor about the resources available through ConnectPlus, including the media-rich eBook, interactive learning tools, and animations.

The Cell Cycle and Cancer

The virtual lab "The Cell Cycle and Cancer" provides a more detailed look at how a failure in the control system of the cell cycle can lead to cancer.

The Chapter in Review

Summary

8.1 The Basics of Cellular Reproduction

Cellular reproduction occurs when growth and repair of tissues take place. Cellular reproduction is also necessary to both asexual and sexual reproduction. Following fertilization, a single cell becomes a multicellular organism by cellular reproduction.

Cellular reproduction involves two processes: growth, during which a cell grows and doubles its contents, and cell division, during which the parent cell's contents are split between two daughter cells. During cell division, one-half of the cytoplasm and a copy of the cell's DNA are passed on to each of two daughter cells.

When a cell is not dividing, the DNA is wound around histones, then arranged into a zigzag configuration to form chromatin. Just before cell division, duplicated chromatin condenses into large loops to form duplicated chromosomes consisting of two chromatids held together at a centromere.

8.2 The Cell Cycle

The cell cycle consists of interphase and the M (mitotic) phase.

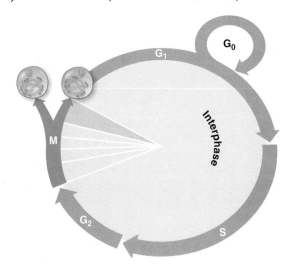

Interphase has the following phases:
- **G_1:** The cell doubles its organelles and accumulates materials that will be used for DNA replication.

- **S:** The cell replicates its DNA.
- **G_2:** The cell synthesizes the proteins that will be needed for cell division.

The M (mitotic) phase consists of mitosis and cytokinesis. As a result of mitosis, the daughter nuclei are genetically identical to the parent nucleus and to each other. If the parent nucleus has four chromosomes, the daughter nuclei each have four chromosomes.

8.3 Mitosis and Cytokinesis

Cell division consists of mitosis and cytokinesis.

Mitosis has four phases:
- During **prophase,** the chromosomes are condensing. Outside the nucleus, the spindle begins to assemble between the separating centrosomes. Prophase continues with the disappearance of the nucleolus and the breakdown of the nuclear envelope. Spindle microtubules from each pole attach to the chromosomes in the region of a centromere.
- During **metaphase,** the chromosomes are aligned at the spindle equator midway between the spindle poles.
- During **anaphase,** the sister chromatids separate and become daughter chromosomes. As the microtubules attached to the chromosomes disassemble, each pole receives a set of daughter chromosomes.
- During **telophase,** the spindle disappears as new nuclear envelopes form around the daughter chromosomes. Each nucleus contains the same number and kinds of chromosomes as the original parent nucleus. Division of the cytoplasm begins.

Cytokinesis differs for animal and plant cells:
- In animal cells, a furrowing process involving actin filaments divides the cytoplasm.
- In plant cells, a cell plate forms, from which the plasma membrane and cell wall develop.

8.4 The Cell Cycle Control System

Checkpoints are interacting signals that promote or inhibit the continuance of the cell cycle. Important checkpoints include these three:
- At G_1, the cell can enter G_0 or undergo apoptosis if DNA is damaged beyond repair. If the cell cycle passes this checkpoint, the cell is committed to complete the cell cycle.
- At G_2, the cell checks to make sure DNA has been replicated properly.

- At the mitotic (M) checkpoint, the cell makes sure the chromosomes are properly aligned and ready to be partitioned to the daughter cells.

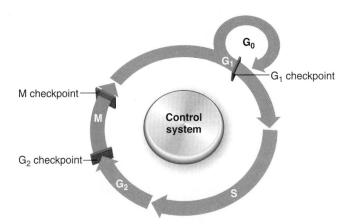

Both internal and external signals converge on checkpoints to control the cell cycle. Cyclin-kinase complexes are internal signals that promote either DNA replication or mitosis. Growth factors and hormones are external signals that promote the cell cycle in connection with growth. Inhibitory external signals are responsible for contact inhibition. Internally, when telomeres are too short, the cell cycle stops because the chromosomes become unstable.

When DNA cannot be repaired, apoptosis occurs as enzymes called caspases bring about destruction of the nucleus and the rest of the cell.

8.5 The Cell Cycle and Cancer

Cancer results when control of the cell cycle is lost and cells divide uncontrollably. Cancer cells are undifferentiated, have abnormal nuclei, do not undergo apoptosis, and form tumors. A tumor may stimulate angiogenesis, allowing it to obtain more nutrients and grow larger. Tumors may also undergo metastasis, which allows them to spread throughout the body. Certain behaviors, such as avoiding smoking and sunbathing and adopting a diet rich in fruits and vegetables, are protective against cancer.

Key Terms

anaphase 128	cytokinesis 127
angiogenesis 134	DNA replication 125
apoptosis 132	histone 125
aster 128	interphase 126
benign tumor 134	kinase 132
binary fission 124	malignant 134
cancer 134	metaphase 128
carcinogenesis 134	metastasis 134
cell cycle 126	mitosis 127
cell plate 131	nucleosome 125
centriole 128	prophase 128
centromere 125	signal 132
centrosome 128	sister chromatid 125
checkpoint 131	somatic cell 133
chromatid 126	spindle 127
chromatin 125	spindle equator 128
chromosome 124	telomere 133
contact inhibition 132	telophase 128
cyclin 132	tumor 134

Testing Yourself

Choose the best answer for each question.

1. _____ is nuclear division, whereas _____ is division of the cytoplasm.
   a. Cytokinesis, mitosis
   b. Apoptosis, mitosis
   c. Mitosis, apoptosis
   d. Mitosis, cytokinesis

2. A duplicated chromosome is composed of identical
   a. chromosome arms.   c. nucleosomes.
   b. chromatids.   d. homologues.

For questions 3–6, match the items to those in the key. Answers can be used more than once.

Key:
   a. G_1 phase of interphase
   b. G_2 phase of interphase
   c. S phase of interphase

3. This phase follows mitosis.
4. DNA is replicated during this phase.
5. Organelles are doubled in number during this phase.
6. During this phase, the cell produces the proteins that will be needed for cell division.
7. Label the phases of the cell cycle on the following diagram. Include anaphase, cytokinesis, G_1 phase, G_2 phase, metaphase, prophase, S phase, and telophase.

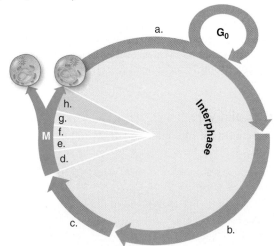

For questions 8–13, match the items to those in the key. Answers can be used more than once, and each question can have more than one answer.

Key:
   a. prophase
   b. metaphase
   c. anaphase
   d. telophase

8. The nucleolus disappears, and the nuclear membrane breaks down.
9. The spindle disappears, and the nuclear envelope forms.
10. Sister chromatids separate.
11. Spindle poles move apart as spindle fibers slide past each other.
12. Chromosomes are aligned on the spindle equator.
13. Each chromosome is composed of two sister chromatids.

14. Plants cannot use a cleavage furrow to undergo cytokinesis because they
  a. lack a plasma membrane.
  b. have a cell wall.
  c. have too many chromosomes.
  d. are too small.

15. Which cell cycle checkpoint allows damaged DNA to be repaired before it is passed on to daughter cells?
  a. G_1
  c. S
  b. G_2
  d. mitotic

16. Which of the following is not an event of apoptosis?
  a. loss of contact with neighboring cells
  b. blistering of the plasma membrane
  c. increase in the number of mitochondria
  d. fragmentation of the nucleus

17. Rats have 42 chromosomes. How many total chromatids does a rat cell have during the G_1 phase? Explain your answer.
  a. 21
  d. 84
  b. 42
  e. 126
  c. 63

18. Rats have 42 chromosomes. How many total chromatids does a rat cell have during the G_2 phase? Explain your answer.
  a. 21
  d. 84
  b. 42
  e. 126
  c. 63

19. Which of the following is not a feature of cancer cells?
  a. exhibit contact inhibition
  b. have enlarged nuclei
  c. stimulate the formation of new blood vessels
  d. are capable of traveling through blood and lymph

20. The spindle begins to assemble during
  a. prophase.
  c. anaphase.
  b. metaphase.
  d. interphase.

21. What is the checkpoint for the completion of DNA synthesis?
  a. G_1
  c. G_2
  b. S
  d. M

22. In human beings, mitosis is necessary for
  a. the growth and repair of tissues.
  b. the formation of gametes.
  c. maintenance of the chromosome number in all body cells.
  d. the death of unnecessary cells.
  e. Both a and c are correct.

23. In which phase of mitosis is evidence of cytokinesis present?
  a. prophase
  d. telophase
  b. metaphase
  e. Both c and d are correct.
  c. anaphase

24. Which of these is not a behavior that could help prevent cancer?
  a. maintaining a healthy weight
  b. eating more dark green leafy vegetables, carrots, and various fruits
  c. not smoking
  d. maintaining estrogen levels through hormone replacement therapy
  e. consuming alcohol only in moderation

25. Cytokinesis in plant cells differs from cytokinesis in animal cells because
  a. plant cells form a cleavage furrow that constricts the daughter cells.
  b. in plant cells, many small vesicles form a cell plate that eventually becomes a new cell wall.
  c. animal cells seldom undergo cytokinesis.
  d. Both b and c are correct.

Thinking Scientifically

1. The survivors of the atomic bombs that were dropped on Hiroshima and Nagasaki have been the subjects of long-term studies of the effects of ionizing radiation on cancer incidence. The frequencies of different types of cancer in these individuals varied across the decades. In the 1950s, high levels of leukemia and cancers of the lung and thyroid gland were observed. The 1960s and 1970s brought high levels of breast and salivary gland cancers. In the 1980s, rates of colon cancer were especially high. Why do you suppose the rates of different types of cancer varied across time?

2. During prophase of mitosis, the nuclear membrane breaks down. This is undoubtedly a complex, energy-consuming process that has to be carried out again, in reverse, at the end of mitosis. Why do you suppose the nuclear membrane must break down at the beginning of mitosis? What mechanism can you envision for the dismantling of the nuclear membrane?

3. Why is it advantageous that checkpoints evolved during the cell cycle?

Bioethical Issue

Paying for Cancer Treatment

As you have read, new drugs and treatment options are now available for cancers once thought incurable. However, many of these new cures are very expensive. A new form of treatment for prostate cancer is estimated to cost over $30,000 per dose, although it has not been shown to prolong life for more than a few months. Most of these costs are passed on to the public, largely through higher health insurance rates and higher taxes. Since many of the risk factors for cancer, such as smoking and obesity, are controllable, this has led to an obvious question—should some cancer patients be held responsible for the cost of their treatment?

Many individuals feel that people who develop cancer because of poor lifestyle choices should be held fully responsible for paying for treatment. By doing so, they argue, people have a direct financial incentive to avoid risky behaviors and are no longer a financial burden on society. But while an individual's poor lifestyle can often be implicated, cancer sometimes strikes otherwise healthy people without any known risk factors. And it can be difficult to determine whether a person's poor behavior is to blame, even when certain risk factors are present. In such cases, is it really fair to incriminate the patient for the disease when there may be other causes?

Many people also disagree on who should bear the high costs of developing new drugs and treatments. Many, including insurance companies, feel that cancer patients who benefit from new cancer treatments should bear at least some of those costs. Others, however, argue that we as a society owe it to our fellow citizens to fund the research that can often save many innocent lives. There are no easy answers for any of these questions, but as cancer continues to exact a high toll in both human life and financial resources, future generations may soon face some difficult choices.

9

Sexual Reproduction

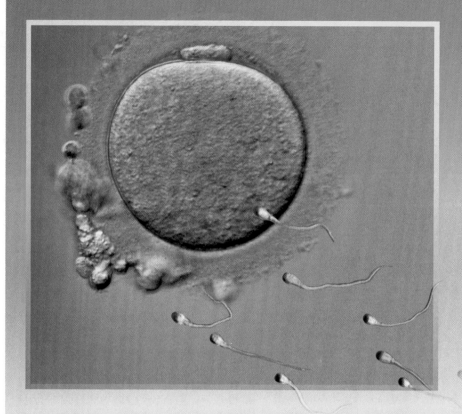

OUTLINE

BEFORE YOU BEGIN

Before beginning this chapter, take a few moments to review the following discussions.

Section 4.4 What is the role of the cytoskeleton in a eukaryotic cell?

Section 8.2 At what point of the cell cycle is the DNA replicated?

Section 8.3 What events occur during prophase, metaphase, telophase, and anaphase of mitosis?

Sexual Reproduction Involves Meiosis

Because of a process called meiosis, two individuals can create offspring that are genetically different from themselves and from each other. This process creates an enormous amount of diversity; in humans, more than 70 trillion different genetic combinations are possible from the mating of just two individuals!

Meiosis is the type of cell division that occurs during sexual reproduction. The process plus maturation result in four cells, which in animals are called gametes. In humans, the male gametes are sperm, and the female gametes are eggs. Meiosis that helps produce sperm is called spermatogenesis, and meiosis that helps produce eggs is called oogenesis.

Although spermatogenesis and oogenesis follow the same steps overall, they differ in some ways. One major difference pertains to the age at which the process begins and ends. In males, sperm production does not begin until puberty, but then continues throughout a male's lifetime. In females, the process of producing eggs has started before the female is even born and ends around the age of 50, a time called menopause.

Another difference concerns the number of gametes that can be produced. In males, sperm production is unlimited, whereas females produce only one egg a month. Many more eggs start the process than ever finish it. Several of the birth control methods available to females, such as the pill or "the patch," work by manipulating hormones, so that cells called oocytes do not complete meiosis and an egg is never released for fertilization.

In this chapter, you will see how meiosis is involved in the production of gametes. You will also learn how meiosis compares with mitosis, the cell division process discussed in Chapter 8.

9.1 The Basics of Meiosis

Learning Outcomes

Upon completion of this section, you should be able to

1. Distinguish between autosomes and sex chromosomes.
2. Illustrate the human life cycle and describe the role of meiosis in this process.
3. Differentiate between sister chromatids and nonsister chromatids of a homologous pair of chromosomes.
4. Describe the processes of synapsis and crossing-over, and explain when these processes occur during meiosis.

Human beings, like most other animals, practice sexual reproduction in which two parents pass chromosomes to their offspring. But why are the genetic and physical characteristics of siblings different from those of each other and from those of either parent? The answer is **meiosis,** a type of nuclear division that is important in sexual reproduction.

Meiosis serves two major functions: reducing the chromosome number and shuffling the chromosomes and genes to produce genetically different gametes, called sperm in males and eggs in females. With random **fertilization** of the egg by a single sperm, even more shuffling occurs. In the end, each child is unique and has a different combination of chromosomes and genes than either parent. But how does meiosis bring about the distribution of chromosomes to offspring in a way that ensures not only the correct number of chromosomes but also a unique combination of chromosomes and genes? For the answer, let's start by examining the chromosomes.

Homologous Chromosomes

Geneticists and genetic counselors can visualize chromosomes by looking at a picture of the chromosomes called a **karyotype** (**Fig. 9.1**). Notice how the chromosomes occur in pairs. Both males and females normally have 23 pairs of chromosomes; 22 pairs of these, called **autosomes,** are the same in both males and females. The remaining pair are called the **sex chromosomes** because they contain the genes that determine gender. The larger chromosome of this pair is the X chromosome, and the smaller is the Y chromosome. Females have two X chromosomes; males have a single X and Y.

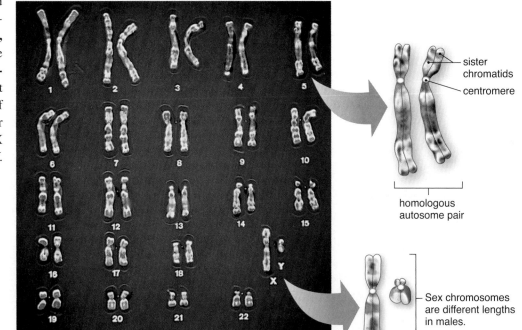

sister chromatids

centromere

homologous autosome pair

Sex chromosomes are different lengths in males.

Figure 9.1 **Homologous chromosomes.**

In body cells, the chromosomes occur in pairs called homologous chromosomes. In this karyotype of chromosomes from a human cell, the pairs have been numbered. These chromosomes are duplicated, and each one is composed of two sister chromatids.

The 46 chromosomes of a male

Twenty-three pairs of chromosomes, or 46 altogether, constitute the **diploid (2n) number** of chromosomes in humans. Half this number is the **haploid (n) number** of chromosomes.

The members of a chromosome pair are called **homologous chromosomes,** or **homologues,** because they have the same size, shape, and location of the centromere (Fig. 9.1). When chromosomes are stained and viewed under a microscope, homologous chromosomes have the same characteristic banding pattern. Homologous chromosomes contain copies of the same genes arranged in the same order. One homologue of each pair was contributed by each parent. Just as your mother may have long fingers and your father may have short fingers, each homologue may have different versions of a gene. Alternate versions of a gene for a particular trait (e.g., finger length) are called **alleles.** Alleles occur at the same location on each homologue. Your mother's allele for long fingers is on one homologue, and your father's allele for short fingers is on the other homologue.

The Human Life Cycle

The term **life cycle** in sexually reproducing organisms refers to all the reproductive events that occur from one generation to the next. The human life cycle involves two types of nuclear division: mitosis and meiosis (**Fig. 9.2**).

During development and after birth, mitosis is involved in the continued growth of the child and the repair of tissues at any time. As a result of mitosis, each somatic (body) cell has the diploid number of chromosomes.

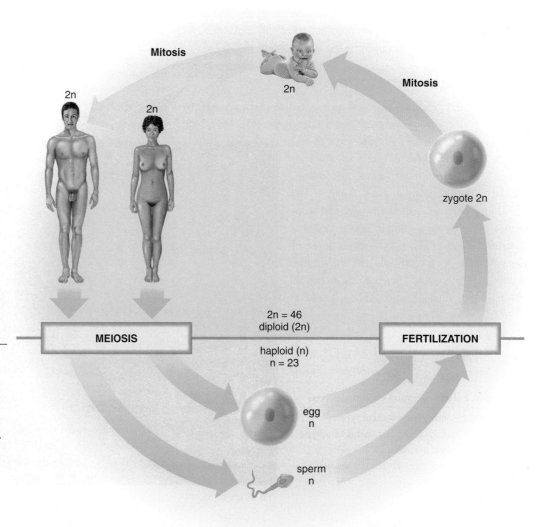

Figure 9.2 Life cycle of humans.

Meiosis in males is a part of sperm production, and meiosis in females is a part of egg production. When a haploid sperm fertilizes a haploid egg, the zygote is diploid. The zygote undergoes mitosis as it eventually develops into a newborn child. Mitosis continues throughout life during growth and repair.

During sexual reproduction, meiosis reduces the chromosome number from the diploid to the haploid number in such a way that the gametes (sperm and egg) have one chromosome derived from each homologous pair of chromosomes. In males, meiosis is a part of **spermatogenesis,** which occurs in the testes and produces sperm. In females, meiosis is a part of **oogenesis,** which occurs in the ovaries and produces eggs. After the sperm and egg join during fertilization, the resulting cell, called the **zygote,** again has homologous pairs of chromosomes. The zygote then undergoes mitosis and differentiation of cells to become a fetus, and eventually a new human being.

Meiosis is important because, if it did not halve the chromosome number, the gametes would contain the same number of chromosomes as the body cells and the number of chromosomes would double with each new generation. Within a few generations, the cells of sexually reproducing organisms would be nothing but chromosomes! But meiosis, followed by fertilization, keeps the chromosome number constant in each generation.

Connections and Misconceptions

Does meiosis always produce gametes?

While meiosis is the first step in the production of gametes in animals, the life cycle of other organisms, such as plants, is slightly different. In plants, meiosis still reduces the number of chromosomes in half, but instead of producing another reproductive structure called a spore. Spores are an important part of the plant life cycle (see Fig. 21.10). Although plants do use gametes, such as pollen, for sexual reproduction, they are produced by mitosis, not meiosis.

Overview of Meiosis

Meiosis results in four daughter cells because it consists of two divisions, called **meiosis I** and **meiosis II** (**Fig 9.3**). Before meiosis I begins, each chromosome has duplicated and is composed of two sister chromatids. During meiosis I, the homologous chromosomes of each pair come together and line up side by side

Figure 9.3 **Overview of meiosis.**

Following duplication of chromosomes, the parent cell undergoes two divisions, meiosis I and meiosis II. During meiosis I, homologous chromosomes separate, and during meiosis II, sister chromatids separate. The final daughter cells are haploid. (The blue chromosomes were originally inherited from one parent, and the red chromosomes were originally inherited from the other parent.)

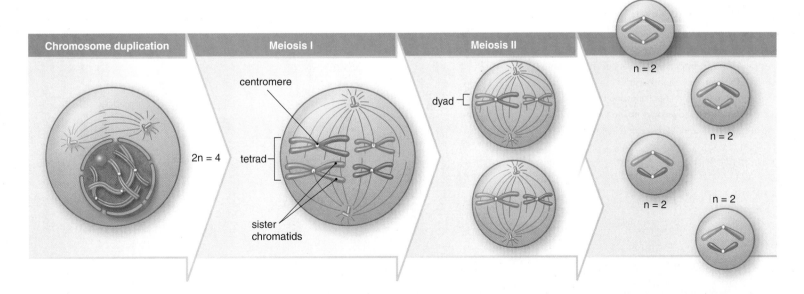

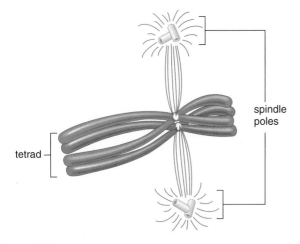

Figure 9.4 **Synapsis.**

During prophase I, homologous chromosomes line up side by side and form a tetrad, consisting of two chromosomes each, containing two chromatids, for a total of four chromatids.

in an event called **synapsis** (**Fig 9.4**). Despite a considerable amount of research, it is still unknown what draws the chromosomes together. Synapsis results in a **tetrad,** an association of four chromatids (two homologous chromosomes consisting of two chromatids each). The chromosomes of a tetrad stay in close proximity until they separate. Later in meiosis I, where the homologous chromosomes of each pair separate, one chromosome from each homologous pair goes to each daughter nucleus. No rules restrict which chromosome goes to which daughter nucleus. Therefore, all possible combinations of chromosomes may occur within the gametes.

No duplication of chromosomes is needed between meiosis I and meiosis II. Following meiosis I, the daughter nuclei each have half the number of chromosomes, but the chromosomes are still duplicated. The chromosomes are called **dyads** because each one is composed of two sister chromatids. During meiosis II, the sister chromatids of each dyad separate and become daughter chromosomes. The resulting four new daughter cells have the haploid number of chromosomes. If the parent cell has four chromosomes, then following meiosis each daughter cell has two chromosomes. (Remember that counting the centromeres tells you the number of chromosomes in a nucleus.)

Because of meiosis, the gametes have all possible combinations of chromosomes. Notice in Figure 9.3 (bottom) that the daughter cells to the left do not have the same combinations of chromosomes as those to the right. Why not? Because the homologous chromosomes of each pair separated during meiosis I. Other chromosome combinations are possible in addition to those depicted. It's possible for the daughter cells to have only red or only blue chromosomes. The daughter cells from meiosis will complete either spermatogenesis (and become sperm) or oogenesis (and become eggs).

Crossing-Over

Meiosis not only reduces the chromosome number but also shuffles the genetic information between the homologous chromosomes. When a tetrad forms during synapsis, the **nonsister chromatids** may exchange genetic material, an event called **crossing-over** (**Fig. 9.5**). Crossing-over occurs between nonsister

Figure 9.5 **Crossing-over.**

When homologous chromosomes are in synapsis, the nonsister chromatids exchange genetic material. This illustration shows only one crossover per chromosome pair, but the average is slightly more than two per homologous pair in humans. Following crossing-over, the sister chromatids of a dyad may no longer be identical and instead may have different combinations of alleles.

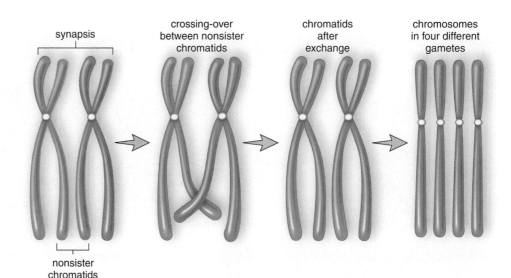

chromatids of homologous chromosomes in a tetrad. Notice in Figure 9.5 how one of the blue sister chromatids is exchanging information with the red non-sister chromatid of its tetrad. Crossing-over between the nonsister chromatids shuffles the alleles and serves as the way that meiosis brings about genetic recombination in the gametes.

Recall that the homologous chromosomes carry genes for traits, such as finger length, but that one allele might call for short fingers, for example, while the other allele calls for long fingers. Therefore, when the nonsister chromatids exchange genetic material, the sister chromatids then have a different combination of alleles, and the resulting gametes will be genetically different. Thus, in Figure 9.5, even though two of the gametes will have the same chromosomes, these chromosomes may not have the same combination of alleles as before because of crossing-over. Crossing-over increases the diversity of the gametes and, therefore, of the offspring.

Animation
Meiosis
Crossing-Over

The Importance of Meiosis

Meiosis is important for the following reasons:

- It helps keep the chromosome number constant by producing haploid daughter cells that become the gametes. When a haploid sperm fertilizes a haploid egg, the new individual has the diploid number of chromosomes.
- It introduces genetic variations because (1) crossing-over can result in different types of alleles on the sister chromatids of a homologue, and (2) every possible combination of chromosomes can occur in the daughter cells.

At fertilization, new combinations of chromosomes can occur, and a zygote can have any one of a vast number of combinations of chromosomes. In humans, $(2^{23})^2$, or 70,368,744,000,000 chromosomally different zygotes are possible, even assuming no crossing-over.

Animation
Genetic Diversity

Connecting the Concepts

For more on the life cycles of other eukaryotic organisms, refer to the following discussions.

Section 17.4 examines the life cycle of selected protistans.

Section 18.1 describes the general life cycle of a plant.

Section 18.3 provides an overview of some of the life cycles found in the fungi.

Check Your Progress 9.1

1. Describe homologous chromosomes and why they occur in pairs.
2. Detail some of the significant events in meiosis I and meiosis II.
3. Describe the two means by which meiosis results in genetically different gametes, and explain why this is important.

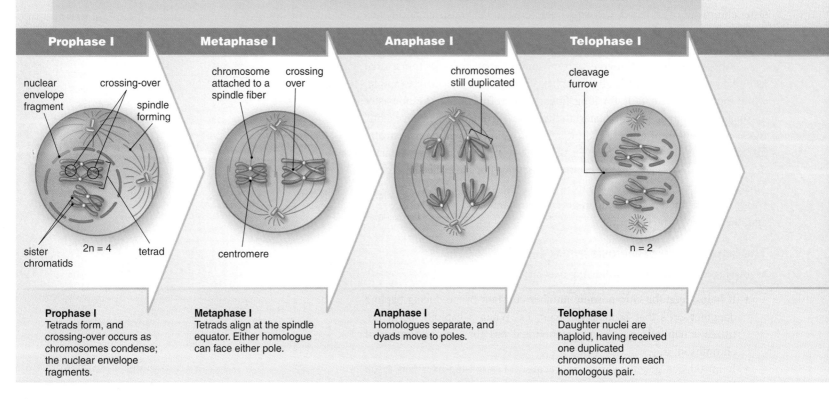

Meiosis I: Homologous chromosomes separate

Prophase I

nuclear envelope fragment

crossing-over

spindle forming

sister chromatids

2n = 4

tetrad

Metaphase I

chromosome attached to a spindle fiber

crossing over

centromere

Anaphase I

chromosomes still duplicated

Telophase I

cleavage furrow

n = 2

Prophase I
Tetrads form, and crossing-over occurs as chromosomes condense; the nuclear envelope fragments.

Metaphase I
Tetrads align at the spindle equator. Either homologue can face either pole.

Anaphase I
Homologues separate, and dyads move to poles.

Telophase I
Daughter nuclei are haploid, having received one duplicated chromosome from each homologous pair.

Figure 9.6 Meiosis I.

As a result of meiosis I, the daughter nuclei are haploid and are genetically dissimilar. They contain a different combination of chromosomes and may have a different combination of alleles on the sister chromatids. The number of centromeres in a cell equals the number of chromosomes.

9.2 The Phases of Meiosis

Learning Outcomes

Upon completion of this section, you should be able to

1. List the phases of meiosis and briefly explain what events occur during each phase.
2. Contrast the alignment of chromosomes during metaphase I and metaphase II.

The same four stages of mitosis—prophase, metaphase, anaphase, and telophase—occur during both meiosis I (**Fig. 9.6**) and meiosis II (**Fig. 9.7**). However, meiosis I and meiosis II differ from each other and from mitosis in the way chromosomes align during metaphase.

The First Division—Meiosis I

To help you recall the events of meiosis, keep in mind what meiosis accomplishes in humans—namely, the production of gametes that have a reduced chromosome number and are genetically different from each other and from the parent cell. During prophase I, the nuclear envelope fragments and the nucleolus disappears as the spindle appears. As the chromosomes condense, homologues undergo synapsis to produce tetrads. Crossing-over between nonsister chromatids occurs during synapsis, and this "shuffles" the alleles on chromosomes.

During metaphase I, the tetrads attach to the spindle and align at the spindle equator, with each homologue facing opposite spindle poles. It does not matter which homologous chromosome faces which pole; therefore, all possible combinations of chromosomes will occur in the gametes. In effect, metaphase of meiosis I shuffles the chromosomes into new combinations. The homologous

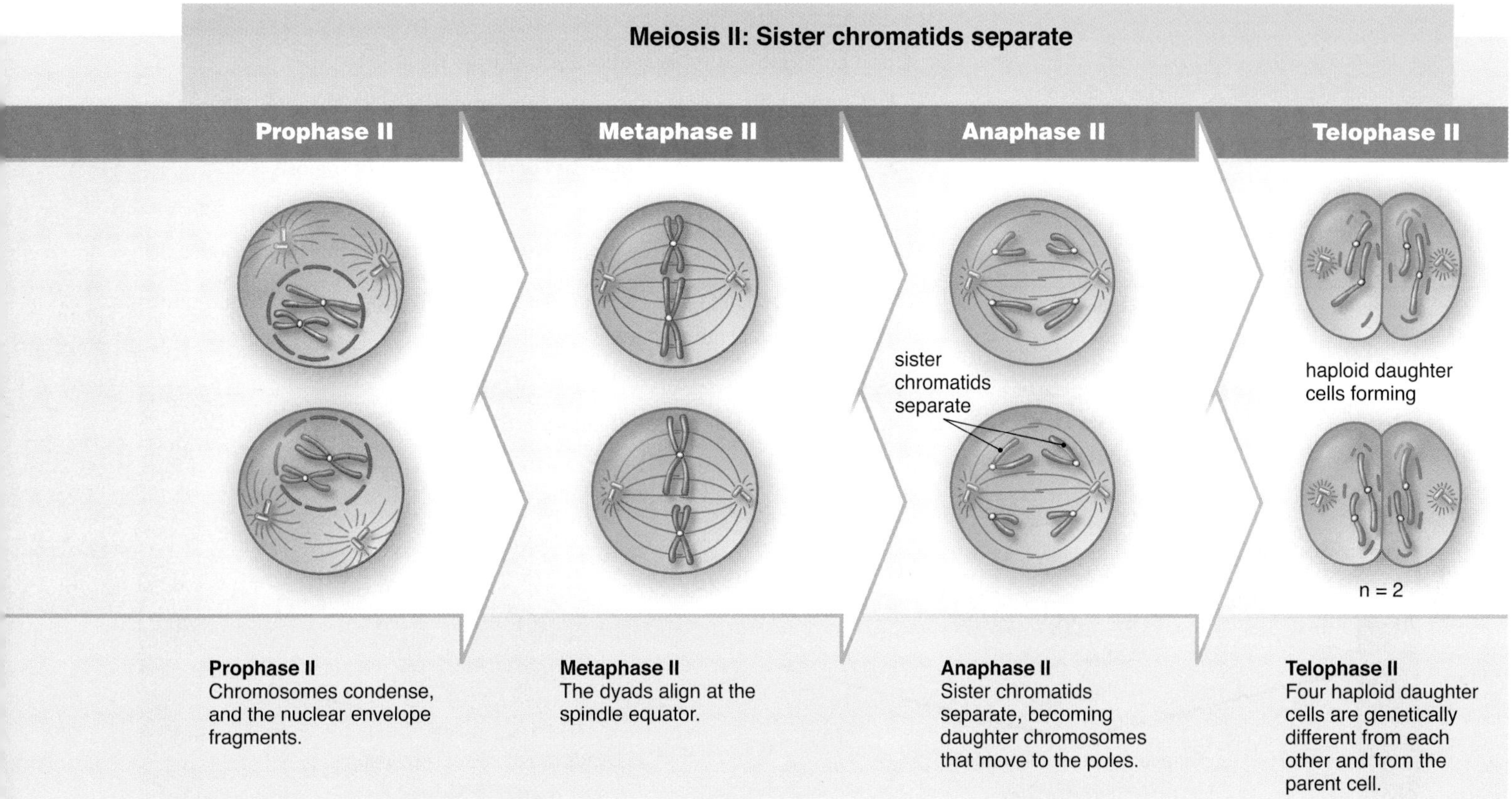

Meiosis II: Sister chromatids separate

Prophase II	Metaphase II	Anaphase II	Telophase II

sister
chromatids
separate

haploid daughter
cells forming

n = 2

Prophase II
Chromosomes condense,
and the nuclear envelope
fragments.

Metaphase II
The dyads align at the
spindle equator.

Anaphase II
Sister chromatids
separate, becoming
daughter chromosomes
that move to the poles.

Telophase II
Four haploid daughter
cells are genetically
different from each
other and from the
parent cell.

chromosomes then separate during anaphase I. Following re-formation of the nuclear envelopes during telophase and cytokinesis, the daughter nuclei are haploid: Each daughter cell contains only one chromosome from each homologous pair. The chromosomes are now dyads, and each still has two sister chromatids. No replication of DNA occurs between meiosis I and II, a period called **interkinesis.**

Animation
Meiosis I

Animation
Random Orientation
of Chromosomes
During Meiosis

The Second Division—Meiosis II

Essentially, the events of meiosis II are the same as those for mitosis, except that the cells are haploid. At the beginning of prophase II, a spindle appears while the nuclear envelope fragments and the nucleolus disappears. Dyads are present, and each attaches to the spindle. During metaphase II, the dyads are lined up at the spindle equator, with sister chromatids facing opposite spindle poles. During anaphase II, the sister chromatids of each dyad separate and move toward the poles. Both poles receive the same number and kinds of chromosomes. In telophase II, the spindle disappears as nuclear envelopes form.

During cytokinesis, the plasma membrane pinches off to form two complete cells, each of which has the haploid, or n, number of chromosomes. The gametes are genetically dissimilar because they can contain different combinations of chromosomes and because crossing-over changes which alleles are together on a chromosome. Because both cells from meiosis I undergo meiosis II, four daughter cells are produced from the original diploid parent cell. Each daughter cell contains a unique combination of genes.

Animation
Meiosis II

MP3
Meiosis

Figure 9.7 Meiosis II.

During anaphase II, daughter chromosomes consisting of one chromatid each move to the poles. Following meiosis II, there are four haploid daughter cells. Comparing the number of centromeres in the daughter cells with the number in the parent cell at the start of meiosis I verifies that the daughter cells are haploid. Notice that the daughter cells are genetically different from each other and from the original parent cell because they have a different combination of chromosomes and crossing-over occurred during meiosis I.

Connecting the Concepts

For more on the stages of meiosis, refer to the following discussions.

Section 10.1 links the process of meiosis to Gregor Mendel's studies on patterns of inheritance.

Section 14.1 introduces the importance of genetic variation to the process of natural selection.

Section 29.2 explores some of the differences in meiosis between males and females.

Check Your Progress 9.2

1. Describe the significance of tetrad formation during meiosis I.
2. Contrast the chromosome composition and number in a parent cell before meiosis begins with the chromosome composition and number in the daughter cells when meiosis is complete.
3. Discuss what the result of meiosis II might be if the chromosomes do not separate correctly during meiosis I.

9.3 Meiosis Compared with Mitosis

Figure 9.8 compares meiosis with mitosis. As you examine this figure, notice that

- Meiosis requires two consecutive nuclear divisions, but mitosis requires only one nuclear division.

Figure 9.8 Meiosis compared with mitosis.

Notice that meiosis produces four haploid cells, whereas mitosis produces two diploid cells.

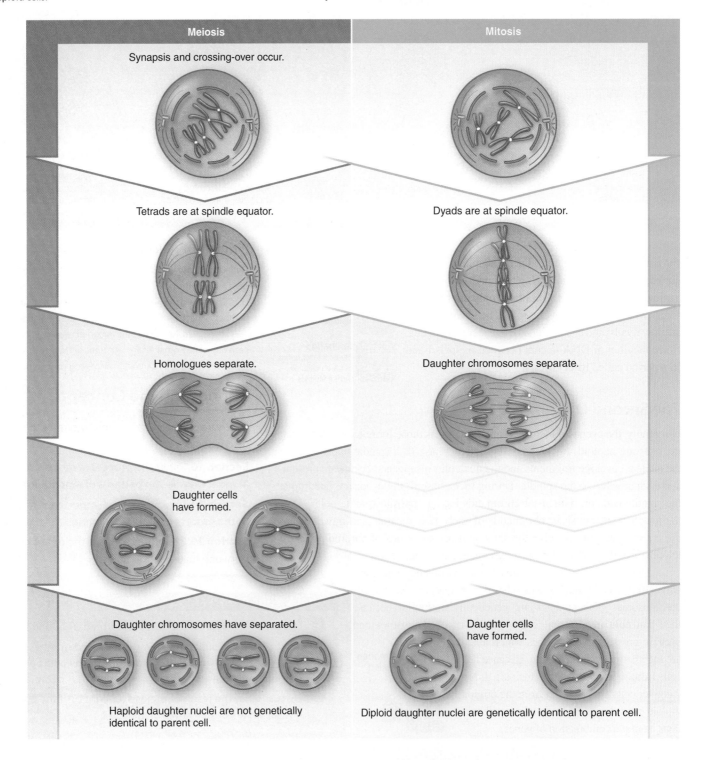

Meiosis

Synapsis and crossing-over occur.

Tetrads are at spindle equator.

Homologues separate.

Daughter cells have formed.

Daughter chromosomes have separated.

Haploid daughter nuclei are not genetically identical to parent cell.

Mitosis

Dyads are at spindle equator.

Daughter chromosomes separate.

Daughter cells have formed.

Diploid daughter nuclei are genetically identical to parent cell.

- Meiosis produces four daughter nuclei, and there are four daughter cells following cytokinesis. Mitosis followed by cytokinesis results in two daughter cells.
- Following meiosis, the four daughter cells are haploid and have half the chromosome number of the parent cell. Following mitosis, the daughter cells have the same chromosome number as the parent cell.
- Following meiosis, the daughter cells are genetically dissimilar to each other and to the parent cell. Following mitosis, the daughter cells are genetically identical to each other and to the parent cell.

The specific differences between these nuclear divisions can be categorized according to process and occurrence.

Process

To summarize the processes, **Table 9.1** compares meiosis I with mitosis, and **Table 9.2** compares meiosis II with mitosis.

Meiosis I Compared with Mitosis

The following events distinguish meiosis I from mitosis:

- During prophase I of meiosis, synapsis occurs. During synapsis, tetrads form and crossing-over occurs. These events do not occur during mitosis.
- During metaphase I of meiosis, tetrads align at the spindle equator, with homologous chromosomes facing opposite spindle poles. The paired chromosomes have a total of four chromatids each. During metaphase in mitosis, dyads align separately at the spindle equator.
- During anaphase I of meiosis, the homologous chromosomes of each tetrad separate, and dyads (with centromeres intact) move to opposite poles. Sister chromatids do not separate during anaphase I. During anaphase of mitosis, sister chromatids separate, becoming daughter chromosomes that move to opposite poles.

Meiosis II Compared with Mitosis

The events of meiosis II are like those of mitosis except that, in meiosis II, the cells have the haploid number of chromosomes.

Occurrence

Meiosis occurs only at certain times in the life cycle of sexually reproducing organisms, and only in specialized tissues. In humans, meiosis occurs only in the testes and ovaries, where it is involved in the production of gametes. The functions of meiosis are to provide gamete variation and to keep the chromosome number constant generation after generation. With fertilization, the full chromosome number is restored. Because unlike gametes fuse, fertilization introduces great genetic diversity into the offspring.

Mitosis is much more common because it occurs in all tissues during embryonic development and during growth and repair. The function of mitosis is to keep the chromosome number constant in all the cells of the body, so that every cell has the same genetic material. Consider that reproductive cloning results in an individual with the same genes as the parent because a single diploid nucleus gives genes to the new individual (see Fig. 12.2).

Table 9.1 Meiosis I Compared with Mitosis

Meiosis I	Mitosis
Prophase I	*Prophase*
Pairing of homologous chromosomes; crossing-over	No pairing of chromosomes; no crossing-over
Metaphase I	*Metaphase*
Tetrads at spindle equator	Dyads at spindle equator
Anaphase I	*Anaphase*
Homologues of each tetrad separate, and dyads move to poles	Sister chromatids separate, becoming daughter chromosomes that move to the poles
Telophase I	*Telophase*
Two haploid daughter cells, not identical to parent cell	Two diploid daughter cells, identical to the parent cell

Table 9.2 Meiosis II Compared with Mitosis

Meiosis II	Mitosis
Prophase II	*Prophase*
No pairing of chromosomes	No pairing of chromosomes
Metaphase II	*Metaphase*
Haploid number of dyads at spindle equator	Diploid number of dyads at spindle equator
Anaphase II	*Anaphase*
Sister chromatids separate, becoming daughter chromosomes that move to the poles	Sister chromatids separate, becoming daughter chromosomes that move to the poles
Telophase II	*Telophase*
Four haploid daughter cells, different from each other genetically and from the parent cell	Two diploid daughter cells, genetically identical to the parent cell

Connecting the Concepts

For more information on the processes of meiosis and mitosis in humans, refer to the following discussions.

Section 8.3 examines mitosis as a form of cell division.

Section 29.2 explores the human reproductive system and the location of meiosis in males and females.

Check Your Progress 9.3

1. Identify the differences between mitosis and meiosis.
2. Describe the purpose and location of mitosis and meiosis in humans.
3. Compare the attachment of chromosomes with the spindle during metaphase I and metaphase II (or mitosis).

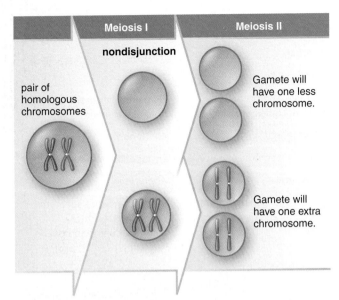

a. Nondisjunction during meiosis I

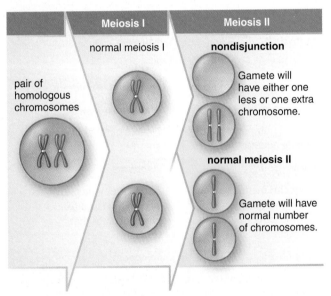

b. Nondisjunction during meiosis II

Figure 9.9 Nondisjunction during meiosis.

Because of nondisjunction, gametes either lack a chromosome or have an extra chromosome. Nondisjunction can occur **(a)** during meiosis I if homologous chromosomes fail to separate and **(b)** during meiosis II if the sister chromatids fail to separate completely.

9.4 Abnormal Chromosome Inheritance

Learning Outcomes

Upon completion of this section, you should be able to

1. Define nondisjunction and briefly explain how nondisjunction may bring about an abnormal chromosome number.
2. List the causes and symptoms of Down syndrome.
3. List two syndromes that may result from abnormal sex chromosome number, and briefly explain the cause of each.

The normal number of chromosomes in human cells is 46, but occasionally humans are born with an abnormal number of chromosomes because the chromosomes fail to separate correctly (called **nondisjunction**) during meiosis. Nondisjunction occurs during meiosis I when both members of a homologous pair go into the same daughter cell, or during meiosis II when the sister chromatids fail to separate and both daughter chromosomes go into the same gamete (**Fig. 9.9**). If an egg that ends up with 24 chromosomes instead of 23 is fertilized with a normal sperm, the result is a *trisomy*, so called because one type of chromosome is present in three copies. If an egg that has 22 chromosomes instead of 23 is fertilized by a normal sperm, the result is a *monosomy*, so called because one type of chromosome is present in a single copy.

Down Syndrome

Down syndrome is *trisomy 21*, in which an individual has three copies of chromosome 21. In most instances, the egg contained two copies of this chromosome instead of one. (In 23% of the cases studied, however, the sperm had the extra chromosome 21.)

Down syndrome is easily recognized by the following characteristics: short stature; an eyelid fold; stubby fingers; a wide gap between the first and second toes; a large, fissured tongue; a round head; a palm crease; and, unfortunately, mental disabilities, which can be severe (**Fig. 9.10**).

The chance of a woman having a Down syndrome child increases rapidly with age, starting at about age 40. The frequency of Down syndrome is 1 in

Connections and Misconceptions

Why is the age of a female a factor in Down syndrome?

There is a difference in the timing of meiosis between males and females. Following puberty, males produce sperm continuously throughout their lives. In contrast, meiosis for females begins about 5 months after being conceived. However, the process is paused at prophase I of meiosis. Only after puberty are a few of these  cells allowed to continue meiosis as part of the female menstrual cycle. Since long periods of time may occur between the start and completion of meiosis, there is a greater chance that nondisjunction will occur, and thus there is a greater chance of producing a child with Down syndrome as a female ages.

800 births for mothers under 40 years of age and 1 in 80 for mothers over 40. However, most Down syndrome babies are born to women younger than age 40, because this is the age group having the most babies.

Abnormal Sex Chromosome Number

Nondisjunction during oogenesis or spermatogenesis can result in gametes that have too few or too many X or Y chromosomes. Figure 9.9 can be used to illustrate nondisjunction of the sex chromosomes during oogenesis if we assume that the chromosomes shown represent X chromosomes.

Just as an extra copy of chromosome 21 causes Down syndrome, additional or missing X or Y chromosomes cause certain syndromes. Newborns with an abnormal sex chromosome number are more likely to survive than are those with an abnormal autosome number, and the explanation is quite surprising. Normal females, like normal males, have only one functioning X chromosome. The other X chromosome (or additional X chromosomes) becomes an inactive mass called a **Barr body** (after the person who discovered it).

In humans, the presence of a Y chromosome, not the number of X chromosomes, almost always determines maleness. The *SRY* (*s*ex-determining *r*egion *Y*) gene located on the short arm of the Y chromosome produces a hormone called testis-determining factor, which plays a critical role in the development of male genitals. No matter how many X chromosomes are involved, an individual with a Y chromosome is a male, assuming that a functional *SRY* is on the Y chromosome. Individuals lacking a functional *SRY* on their Y chromosome have Swyer syndrome, also known as an "XY female."

A person with **Turner syndrome** (45, XO) is a female. The number indicates the total number of chromosomes the individual has, and the O signifies the absence of a second sex chromosome. Turner syndrome females are short, with a broad chest and webbed neck. The ovaries, oviducts, and uterus are very small and underdeveloped. Turner females do not undergo puberty or menstruate, and their breasts do not develop (**Fig. 9.11***a*). However, some have given birth following in vitro fertilization using donor eggs. These women usually have normal intelligence and can lead fairly normal lives if they receive hormone supplements.

A person with **Klinefelter syndrome** (47, XXY) is a male. A Klinefelter male has two or more X chromosomes in addition to a Y chromosome. The extra X chromosomes become Barr bodies. In Klinefelter males, the testes and prostate gland are underdeveloped. There is no facial hair, but some breast development may occur (Fig. 9.11*b*). Affected individuals generally have large hands and feet and very long arms and legs. They are usually slow to learn but not mentally handicapped unless they inherit more than two X chromosomes. As with Turner syndrome, it is best for parents to know as soon as possible that their child has Klinefelter syndrome because much can be done to help the child lead a normal life.

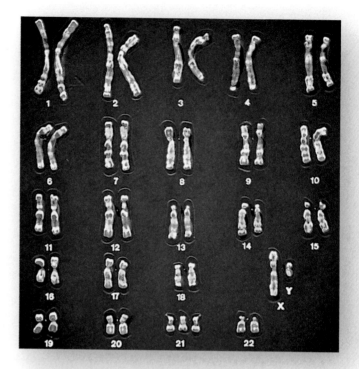

Figure 9.10   Down syndrome.

Down syndrome occurs when the egg or the sperm has an extra chromosome 21 due to nondisjunction in either meiosis I or meiosis II. Characteristics include a wide, rounded face and narrow, slanting eyelids. Mental disabilities are often present, but may vary greatly.

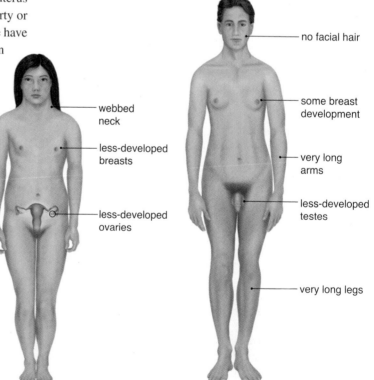

Figure 9.11   Abnormal sex chromosome number.

a. A female with Turner syndrome (XO) has a short, thick neck; short stature; and a lack of breast development. **b.** A male with Klinefelter syndrome (XXY) has immature sex organs and some development of the breasts.

a. A female with Turner (XO) syndrome   b. A male with Klinefelter (XXY) syndrome

Connecting the Concepts

For more about chromosomes, refer to the following discussions.

Section 10.4 explores how the location of genes can be determined on a chromosome.

Section 13.1 examines how karyotypes are made and how they can indicate chromosomal problems.

Check Your Progress 9.4

1. Describe how a zygote can receive an abnormal chromosome number.

2. Contrast monosomy with trisomy.

3. Distinguish between a chromosomal abnormality in the sex chromosomes from one in the autosomes.

Connections and Misconceptions

Are individuals who are XYY more aggressive?

While the presence of the Y chromosome is a determining factor in the development of male sexual characteristics, the presence of two Y chromosomes, called Jacobs syndrome (XYY), does not appear to make an individual have a more aggressive personality. Early studies of Jacobs syndrome in the 1960s suggested that XYY males may be more prone to violent crimes; however, the study was conducted using a very small sample size, and subsequent studies quickly dispelled the link. While XYY males are known to have great height, and sometimes learning problems, there is very little evidence suggesting an increase in aggressive behavior. It is estimated that 1 in 1,000 males is actually an XYY individual, and 96% of them are normal.

Media Study Tools

www.mhhe.com/maderessentials3

Enhance your study of this chapter with study tools and practice tests. Also ask your instructor about the resources available through ConnectPlus, including the media-rich eBook, interactive learning tools, and animations.

The Chapter in Review

Summary

9.1 The Basics of Meiosis

Meiosis is a type of nuclear division that reduces the chromosome number and shuffles the genes between chromosomes. It produces haploid daughter cells with a unique combination of chromosomes and alleles.

Animation
How Meiosis Works

Homologous Chromosomes

Homologous chromosomes are the same size with the same centromere position, and they contain the same genes. The alleles of a gene may differ between homologues. Diploid cells in humans have 22 homologous pairs of autosomes and 1 pair of sex chromosomes, for a total of 46 chromosomes. Males are XY and females are XX. Haploid cells have 22 autosomes and 1 sex chromosome, either an X or a Y, for a total of 23 chromosomes.

The Human Life Cycle

The human life cycle has two types of nuclear division:

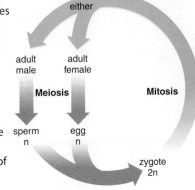

- **Mitosis** ensures that every body cell has 23 pairs of chromosomes. It occurs during growth and repair.
- **Meiosis** occurs during the formation of gametes. It ensures that the gametes are haploid and have 23 chromosomes, 1 from each of the pairs of chromosomes.

Overview of Meiosis

Meiosis has two nuclear divisions called meiosis I and meiosis II. Therefore, meiosis results in four daughter cells.

- **Meiosis I:** Synapsis and the formation of tetrads leads to separation of homologous chromosomes. Because no restrictions govern which member goes to which pole, the gametes will contain all possible combinations of chromosomes.
- **Meiosis II:** The chromosomes are still duplicated at the beginning of meiosis II. Sister chromatids later separate, forming daughter chromosomes. The daughter cells are haploid. If the parent cell has four chromosomes, each of the daughter cells has two chromosomes. All possible combinations of chromosomes occur among the daughter cells.

Crossing-Over Crossing-over between nonsister chromatids during meiosis I shuffles the alleles on the chromosomes, so that each sister chromatid of a homologue has a different mix of alleles.

The Importance of Meiosis Meiosis is important because the chromosome number stays constant between the generations of individuals and because the daughter cells, and therefore the gametes, are genetically different. When fertilization occurs, the offspring are genetically unique.

9.2 The Phases of Meiosis

The First Division—Meiosis I

- **Prophase I:** Tetrads form, and crossing-over occurs as chromosomes condense; the nuclear envelope fragments.
- **Metaphase I:** Tetrads align at the spindle equator. Either homologue can face either pole.
- **Anaphase I:** Homologues of each tetrad separate, and dyads move to the poles.

- **Telophase I:** Daughter nuclei are haploid, having received one duplicated chromosome from each homologous pair.

The Second Division—Meiosis II

- **Prophase II:** Chromosomes condense, and the nuclear envelope fragments.
- **Metaphase II:** The haploid number of dyads align at the spindle equator.
- **Anaphase II:** Sister chromatids separate, becoming daughter chromosomes that move to the poles.
- **Telophase II:** Four haploid daughter cells are genetically different from each other and from the parent cell.

9.3 Meiosis Compared with Mitosis

The following diagrams show the differences between meiosis and mitosis. Duplication of chromosomes occurs before each begins.

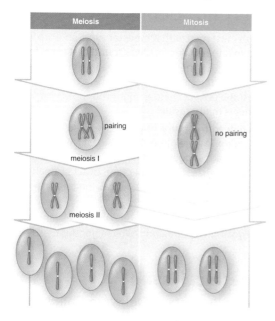

Meiosis

- In humans, meiosis occurs in the testes and ovaries, where it produces the gametes.
- Because meiosis has two nuclear divisions, it produces four daughter cells. The daughter cells are haploid and genetically different from each other and from the parent cell.

Mitosis

- Mitosis occurs in the body cells and accounts for growth and repair. Because mitosis has one nuclear division, it produces two daughter cells. The daughter cells are diploid and are genetically identical to each other and to the parent cell.

Meiosis I Compared with Mitosis

Meiosis I differs from mitosis in the following ways:

- During prophase I, tetrads form and crossing-over occurs. No such events occur in mitosis.
- During metaphase I, tetrads are at the spindle equator. During mitosis, dyads are at the spindle equator.
- During anaphase I, homologous chromosomes separate. During mitosis, sister chromatids separate.

Meiosis II Compared with Mitosis

- The events of meiosis II are the same as those of mitosis, except that the cells undergoing meiosis II are haploid, while those undergoing mitosis are diploid.

9.4 Abnormal Chromosome Inheritance

Nondisjunction accounts for the inheritance of an abnormal chromosome number. Nondisjunction can occur during meiosis I if homologous chromosomes fail to separate and if both go into one daughter cell and the other daughter cell receives neither. It can also occur during meiosis II if chromatids fail to separate and if both go into one daughter cell and the other daughter cell receives neither.

Down Syndrome

In Down syndrome, an example of an autosomal syndrome, the individual inherits three copies of chromosome 21.

Abnormal Sex Chromosome Number

Examples of syndromes caused by inheritance of abnormal sex chromosomes are Turner syndrome (XO) and Klinefelter syndrome (XXY).

- **Turner syndrome:** An XO individual inherits only one X chromosome. This individual survives because, even in an XX individual, one X becomes a Barr body.
- **Klinefelter syndrome:** An XXY individual inherits two (or more) X chromosomes and a Y chromosome. This individual survives because one (or more) X chromosome becomes a Barr body.

Key Terms

allele 142	life cycle 142
autosome 141	meiosis 141
Barr body 151	meiosis I 143
crossing-over 144	meiosis II 143
diploid (2n) number 142	nondisjunction 150
Down syndrome 150	nonsister chromatid 144
dyad 144	oogenesis 143
fertilization 141	sex chromosome 141
haploid (n) number 142	spermatogenesis 143
homologous chromosome 142	synapsis 144
homologue 142	tetrad 144
interkinesis 147	Turner syndrome 151
karyotype 141	zygote 143
Klinefelter syndrome 151	

Testing Yourself

Choose the best answer for each question.

1. A human cell contains _____ pair(s) of sex chromosomes.
   - a. 1
   - b. 2
   - c. 22
   - d. 23
2. Mitosis _____ chromosome number, whereas meiosis _____ the chromosome number of the daughter cells.
   - a. maintains, increases
   - b. increases, maintains
   - c. increases, decreases
   - d. maintains, decreases

For questions 3–6, match the items to those in the key. Answers can be used more than once.

Key:
   - a. haploid number
   - b. three times the haploid number
   - c. diploid number
   - d. two times the diploid number

3. How many chromatids are in a parent cell entering meiosis?
4. How many chromosomes are in each daughter cell following meiosis I?
5. How many chromatids are in each daughter cell following meiosis I?

6. How many chromosomes are in a parent cell entering meiosis?

7. Which of the following best describes the attachment of chromosomes to the spindle apparatus during meiosis I?
   a. The homologous pair lines up together, with homologues facing opposite spindle poles.
   b. The homologous pair lines up together, with homologues facing the same spindle pole.
   c. Homologous pairs do not line up together, but homologues face opposite spindle poles.
   d. Homologous pairs do not line up together, with homologues facing the same spindle pole.

For questions 8–15, match the items to those in the key.

Key:

   a. prophase I
   b. metaphase I
   c. anaphase I
   d. telophase I
   e. prophase II
   f. metaphase II
   g. anaphase II
   h. telophase II

8. A cleavage furrow forms, resulting in haploid nuclei. Each chromosome contains two chromatids.

9. Tetrads form, and crossing-over occurs.

10. Dyads align at the spindle equator.

11. Four haploid daughter cells are created.

12. Homologous chromosomes separate and move to opposite poles.

13. Sister chromatids separate.

14. Tetrads align on the spindle equator.

15. Chromosomes in haploid nuclei condense.

16. At each letter in the following diagram, indicate how mitosis differs from meiosis.

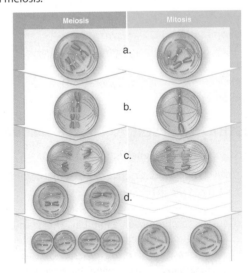

17. Nondisjunction can result in
    a. monosomy.
    b. isomy.
    c. trisomy.
    d. Both a and b are correct.

18. An individual with Turner syndrome is conceived when a normal gamete unites with
    a. an egg that was produced by nondisjunction during oogenesis.
    b. a sperm that was produced by nondisjunction during spermatogenesis.
    c. Either a or b is correct.
    d. Neither a nor b is correct.

19. An individual with triple X karyotype (47, XXX) has how many Barr bodies?
    a. 0
    b. 1
    c. 2
    d. 3

20. Crossing-over
    a. involves recombination between nonhomologous chromosomes.
    b. involves recombination between sister chromatids.
    c. occurs only between members of a tetrad.
    d. occurs between adjacent tetrads.

21. The separation of sister chromatids of a dyad in anaphase II results in
    a. dyads.
    b. daughter chromosomes.
    c. tetrads.
    d. sister chromatids.

Thinking Scientifically

1. Although most men with Klinefelter syndrome are infertile, some are able to father children. It was found that most fertile individuals with Klinefelter syndrome exhibit mosaicism, in which some cells are normal (46, XY) but others contain the extra chromosome (47, XXY). How might this mosaicism come about? What effects might result?

2. In the nineteenth century, physicians noticed that people with Down syndrome were often the youngest children in large families. Some physicians suggested that the disorder was due to "maternal reproductive exhaustion." What might be a reasonable explanation for the relationship between advanced maternal age and incidence of Down syndrome?

3. As you can reason, meiosis creates a very large amount of genetic variability within a population. Why might this be advantageous?

Bioethical Issue

The Risks and Benefits of Delaying Childbirth

Many women are delaying childbirth until their thirties and forties. Between 1991 and 2001, the birthrate among women aged 35 to 39 increased over 30%, while the birthrate among women aged 40 to 44 jumped by almost 70%. Fewer social stigmas, better prenatal care, and improved medical technologies have all enabled this trend. However, many feel that this should be discouraged due to the perceived costs to society.

The decision to delay childbirth carries risks. The risk of many disorders associated with nondisjunction, such as Down syndrome, increase greatly with age, rising from nearly 1 in 400 at age 35 to 1 in 35 by age 45. The risk of health complications to the mother also increase greatly with age, necessitating more thorough prenatal care. Many of these costs are handed down in the form of higher taxes and health insurance premiums.

Despite the risks, there are some benefits to delaying childbirth. Women over age 35 are usually more affluent, do not often divorce, and are less likely to give birth out of wedlock, lessening the need for social welfare programs. Furthermore, many women at more advanced stages of their careers are able to devote more time to their children than are younger women just embarking on their careers. Advocates point out that the children of older parents are usually more well cared for. Thus, any extra health-care costs are offset by savings on social programs.

Should we as a society discourage delayed childbirth because of the high costs incurred? Or do the benefits of advanced maternal age make up for the costs incurred?

The Environment Influences Our Genetic Traits

Just how much do genes control the development of particular genetic diseases? In cases such as sickle cell disease and cystic fibrosis, the genes control whether a disease develops. Therefore, the probable occurrence of the trait is predictable. In other cases, the influence of the genes alone does not forecast a particular trait. Rather, environmental factors influence the activity of genes and whether certain specific traits develop. Determining the pattern of inheritance for traits that are influenced by the environment is a difficult task.

Tests are widely becoming available for alleles associated with some genetic diseases. For example, a woman can be tested to see if she has the *BRCA-1* and *BRCA-2* alleles, both of which are associated with breast and ovarian cancer. However, the allele alone is not sufficient to cause the development of cancer—environmental factors are also involved. Therefore, even if a woman tests positive for these alleles, she may not develop cancer; it requires additional undetermined environmental factors for cancer to develop. Unfortunately, it is impossible to predict for sure whether cancer will occur. Still, some women have resorted to extraordinary methods, such as radical mastectomy to remove all breast tissue, to prevent the possibility of cancer.

In this chapter, you will learn about the basic principles of inheritance. We will focus on the rules of genetics that allow us to predict the chances of a trait being passed on from one generation to the next. However, we have to bear in mind that sometimes environmental influences affect the passage of traits.

10

Patterns of Inheritance

OUTLINE

BEFORE YOU BEGIN

Before beginning this chapter, take a moment to review the following discussions.

Section 3.2 What are the roles of proteins and nucleic acids in a cell?

Section 8.3 How are sister chromatids and homologous chromosomes related?

Section 9.2 How does meiosis introduce new genetic combinations?

10.1 Mendel's Laws

Learning Outcomes

Upon completion of this section, you should be able to

1. Explain Mendel's laws of inheritance.
2. Distinguish between genotype and phenotype.
3. Distinguish between dominant and recessive traits.
4. Apply Mendel's laws to solve and interpret monohybrid and dihybrid genetic crosses.
5. Recognize and explain the relationship between Mendel's laws and meiosis.

Today, most people know that DNA is the genetic material, and they may have heard that scientists have determined the DNA base sequence of human chromosomes. In contrast, they may not know about Gregor Mendel, an Austrian

monk who developed certain laws of heredity after doing crosses between garden pea plants in 1860 (**Fig. 10.1**). Gregor Mendel investigated genetics at the organismal level, and this is still the level that intrigues most of us on a daily basis. We observe, for example, that facial and other features run in families, and we would like some way of explaining this observation. And so, it is appropriate to begin our study of genetics at the organismal level and to learn Mendel's laws of heredity.

When Mendel began his work, most plant and animal breeders acknowledged that both sexes contribute equally to a new

a.

b.

Figure 10.1 Mendel working in his garden.

a. Mendel grew and tended the garden peas, *Pisum sativum,* he used for his experiments. **b.** Mendel selected these seven traits for study. Before he did any crosses, he made sure the parents bred true. All the offspring had either the dominant trait or the recessive trait.

individual. However, they were unable to account for the presence of definite variations (differences) among the members of a family, generation after generation. Mendel's model of heredity does account for such variations. Therefore, Mendel's model is compatible with the theory of evolution, which states that various combinations of traits are tested by the environment, and those combinations that lead to reproductive success are the ones that are passed on.

Mendel's Experimental Procedure

Mendel's parents were farmers, so he no doubt acquired the practical experience he needed to grow pea plants during childhood. Mendel was also a mathematician; most likely, his background in mathematics prompted him to use a statistical basis for his breeding experiments. He prepared for his experiments carefully and conducted preliminary studies with various animals and plants. He then chose to work with the garden pea, *Pisum sativum.*

The garden pea was a good choice. The plants were easy to cultivate and had a short generation time. And although peas normally self-pollinate (pollen goes only to the same flower), they could be cross-pollinated by hand. Many varieties of peas were available, and Mendel initially grew 22 of them. For his experiments, Mendel chose 7 varieties that produced easily identifiable differences (Fig. 10.1*b*). When these varieties self-pollinated, they were *true-breeding*—meaning that the offspring were like the parent plants and like each other. In contrast to his predecessors, Mendel studied the inheritance of relatively simple and easily detected traits, such as seed shape, seed color, and flower color, and he observed no intermediate characteristics among the offspring. **Figure 10.2** shows Mendel's procedure.

Figure 10.2 **Garden pea anatomy and traits.**

a. In the garden pea, pollen grains produced in the anther contain sperm, and ovules in the ovary contain eggs. **b.** When Mendel performed crosses, he brushed pollen from one type of pea plant onto the stigma developed into seeds (peas).

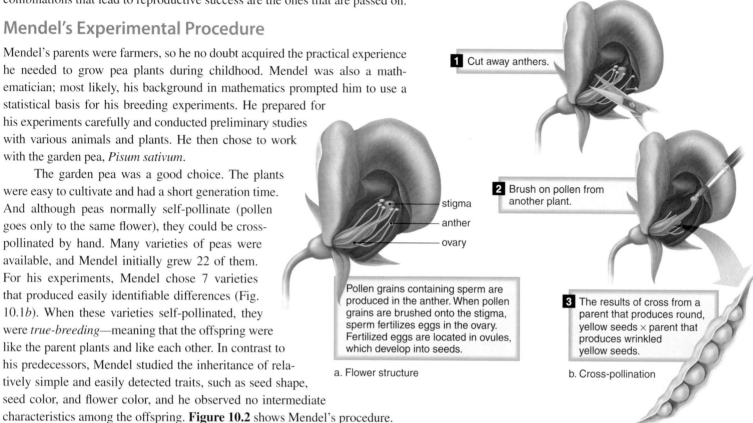

1 Cut away anthers.

2 Brush on pollen from another plant.

— stigma
— anther
— ovary

Pollen grains containing sperm are produced in the anther. When pollen grains are brushed onto the stigma, sperm fertilizes eggs in the ovary. Fertilized eggs are located in ovules, which develop into seeds.

a. Flower structure

3 The results of cross from a parent that produces round, yellow seeds × parent that produces wrinkled yellow seeds.

b. Cross-pollination

Trait				
Seed shape	Seed color	Flower position	Flower color	Pod color
Round	Yellow	Axial	Purple	Green
Wrinkled	Green	Terminal	White	Yellow

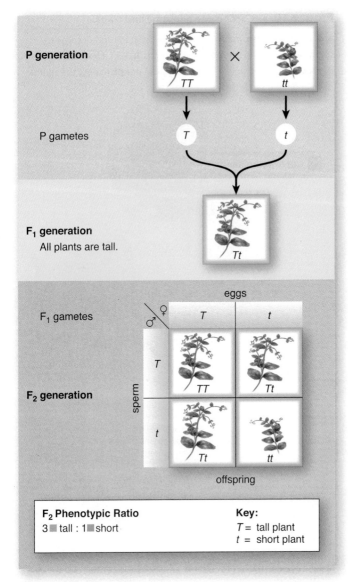

Figure 10.3 One-trait cross.

The P generation plants differ in one regard—length of the stem. All the F₁ generation plants are tall, but the factor for short has not disappeared because ¼ of the F₂ generation are short. The 3:1 ratio allowed Mendel to deduce that individuals have two discrete genetic factors for each trait.

Connections and Misconceptions

Why so many peas?

Mendel did not have an obsession with pea plants. Instead, as a mathematician, he recognized that, in order for his results to be statistically significant, he needed to examine a large sample size. Peas are ideal for this type of analysis, since each pea is an individual zygote. In his initial experiments, Mendel examined the shapes and colors of over 15,300 seeds and 1,700 pea pods! Only by doing so could Mendel be assured that the observed 3:1 ratios were not simply the result of randomness.

As Mendel followed the inheritance of individual traits, he kept careful records. He used his understanding of the mathematical laws of probability to interpret his results and to arrive at a theory that has since been supported by innumerable experiments. This theory is the *particulate theory of inheritance,* so called because it is based on the existence of minute particles we now call genes. Inheritance involves the reshuffling of the same genes from generation to generation.

One-Trait Inheritance

After ensuring that his pea plants were true-breeding—for example, that his tall plants always had tall offspring and his short plants always had short offspring—Mendel was ready to perform a cross between these two strains. Mendel called the original parents the *P generation*, the first-generation offspring the *F₁* (for filial) *generation*, and the second-generation offspring the *F₂ generation*. The diagram in **Figure 10.3** representing Mendel's F₁ cross is called a Punnett square (named after Reginald Punnett, an early twentieth-century geneticist). In a **Punnett square,** all possible types of sperm are lined up vertically, and all possible types of eggs are lined up horizontally, or vice versa, so that every possible combination of gametes the offspring may inherit occurs within the square.

As Figure 10.3 shows, when Mendel crossed tall pea plants with short pea plants, all the F₁ offspring resembled the tall parent. Did this mean that the other characteristic, shortness, had disappeared permanently? No, because when Mendel allowed the F₁ plants to self-pollinate, ¾ of the F₂ generation were tall and ¼ were short, a 3:1 ratio. Therefore, the F₁ plants had been able to pass on a factor for shortness—it didn't just disappear. Perhaps the F₁ plants were tall because tallness was dominant to shortness?

Mendel's mathematical approach led him to interpret his results differently than previous breeders had done. He reasoned that a 3:1 ratio among the F₂ offspring was possible only if (1) the F₁ parents contained two separate copies of each hereditary factor, one dominant and the other recessive; (2) the factors separated when the gametes were formed, and each gamete carried only one copy of each factor; and (3) random fusion of all possible gametes occurred upon fertilization. Only in this way would shortness reoccur in the F₂ generation.

One-Trait Testcross

Mendel's experimental use of simple dominant and recessive characteristics allowed him to test his interpretation of his crosses. To see if the F₁ carries a recessive factor, Mendel crossed his F₁ generation tall plants with true-breeding, short plants. He reasoned that half the offspring would be tall and half would be short (**Fig. 10.4 a**). And, indeed, those were the results he obtained; therefore, his hypothesis that factors segregate when gametes are formed was supported.

Today, a one-trait **testcross** is used to determine whether an individual with the dominant trait has two dominant factors for a particular trait. This is not possible to tell by observation because an individual can exhibit the dominant appearance while having only one dominant factor. Figure 10.4b shows that, if the individual has two dominant factors, all the offspring will be tall.

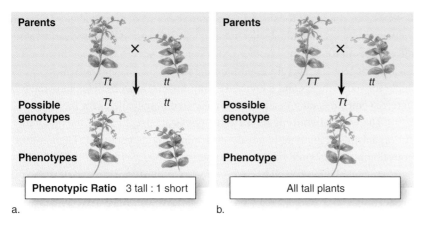

a.

b.

Figure 10.4 **One-trait testcross.**

Crossing an individual with a dominant appearance (phenotype) with a recessive individual indicates the genotype. **a.** If a parent with the dominant phenotype has only one dominant factor, the results among the offspring are 1:1. **b.** If a parent with the dominant phenotype has two dominant factors, all offspring have the dominant phenotype.

After doing one-trait crosses, Mendel arrived at his first law of inheritance—the **law of segregation,** which is a cornerstone of his particulate theory of inheritance. The law of segregation states the following:

- Each individual has two factors for each trait.
- The factors segregate (separate) during the formation of the gametes.
- Each gamete contains only one factor from each pair of factors.
- Fertilization gives each new individual two factors for each trait.

The Modern Genetics View

In the early twentieth century, scientists noted the parallel behavior of Mendel's particulate factors and chromosomes and proposed the *chromosomal theory of inheritance,* which states that chromosomes are carriers of genetic information. Today, we recognize that traits are controlled by alleles (alternate forms of a gene) that occur on the chromosomes at particular locations called the **gene locus.** The **dominant allele** is so named because of its ability to mask the expression of the other allele, called the **recessive allele.** The dominant allele is identified by an uppercase (capital) letter, the recessive allele by the same but lowercase (small) letter. In humans, for example, alleles for finger length might be *S* for short fingers and *s* for long fingers. While the alleles on one homologue can be the alternates of the alleles on the other homologue, the sister chromatids have the same types of alleles (**Fig. 10.5**). Excluding environmental influences, an individual's traits are determined by the alleles that he or she inherits.

As you learned in Chapter 9, meiosis is the type of cell division that reduces the chromosome number. During meiosis I, the homologues separate. Therefore, the process of meiosis explains Mendel's law of segregation and why only one allele for each trait is in a gamete (see Fig. 10.8). When fertilization occurs, the resulting offspring again have two alleles for each trait, one from each parent.

Genotype Versus Phenotype Different combinations of alleles for a trait can give an organism the same outward appearance. For instance, *SS* and *Ss* individuals both have short fingers. For this reason, it is necessary to distinguish between the alleles present in an organism and the appearance of that organism.

The word **genotype** refers to the alleles an individual receives at fertilization. Genotype may be indicated by letters or by short, descriptive phrases. When an organism has two identical alleles, it is termed **homozygous.** A person who is

Connections and Misconceptions

Why are some alleles dominant and others recessive?

Genes contain information for the production of a specific protein. In Mendel's peas, there is a gene for flower color that has two variations, or alleles. The dominant allele produces a purple flower color. However, the recessive allele codes for a variation of the protein that does not produce the correct pigmentation. In other words, it has lost its function. This results in a white color. In a flower with one dominant and one recessive allele (a heterozygote), the dominant purple color masks the recessive white color. As you will see in the next section, not all dominant alleles completely mask the recessive allele.

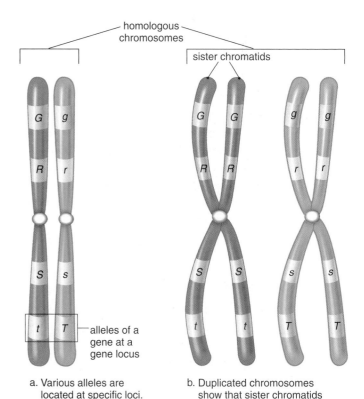

a. Various alleles are located at specific loci.

b. Duplicated chromosomes show that sister chromatids have identical alleles.

Figure 10.5 **Homologous chromosomes.**

a. The letters represent alleles—that is, alternate forms of a gene. Each allelic pair, such as *Gg* or *Tt,* is located on homologous chromosomes at a particular gene locus. **b.** Sister chromatids carry the same alleles in the same order.

Table 10.1 Genotype Versus Phenotype

Genotype	Genotype	Phenotype
SS	Homozygous dominant	Short fingers
Ss	Heterozygous	Short fingers
ss	Homozygous recessive	Long fingers

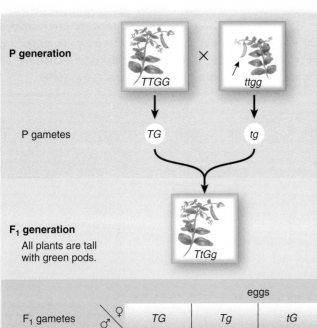

P generation

TTGG × ttgg

P gametes

TG   tg

F₁ generation

All plants are tall with green pods.

TtGg

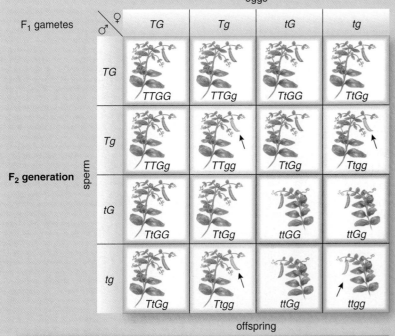

F₂ generation

eggs

F₁ gametes ♂\♀	TG	Tg	tG	tg
TG	TTGG	TTGg	TtGG	TtGg
Tg	TTGg	TTgg	TtGg	Ttgg
tG	TtGG	TtGg	ttGG	ttGg
tg	TtGg	Ttgg	ttGg	ttgg

sperm

offspring

F₂ Phenotypic Ratio

9 ■ tall plant, green pod
3 ■ tall plant, yellow pod
3 ■ short plant, green pod
1 ■ short plant, yellow pod

Key:

T = tall plant
t = short plant
G = green pod
g = yellow pod

homozygous dominant for short fingers possesses two dominant alleles (*SS*). All gametes from this individual will contain an allele for short fingers (*S*). Likewise, all gametes produced by a homozygous recessive parent (*ss*) contain an allele for long fingers (*s*). Therefore, all the offspring from this couple will have one allele for short fingers and another for long fingers, as in *Ss*. This individual is **heterozygous,** meaning that he or she has two different alleles for the trait.

The word **phenotype** refers to the physical appearance of the individual. An organism's phenotype is mostly determined by its genotype. The homozygous dominant (*SS*) individual and the heterozygous (*Ss*) individual both show the dominant phenotype and have short fingers, while the homozygous recessive individual shows the recessive phenotype and has long fingers (**Table 10.1**).

Two-Trait Inheritance

Mendel performed a second series of crosses in which he crossed true-breeding plants that differed in two traits. For example, he crossed tall plants having green pods (*TTGG*) with short plants having yellow pods (*ttgg*) (**Fig. 10.6**). The F₁ plants showed both dominant characteristics (tall with green pods). As before, Mendel then allowed the F₁ plants to self-pollinate. Two possible results could occur in the F₂ generation:

1. If the dominant factors (*TG*) always go together into the F₁ gametes, and the recessive factors (*tg*) always stay together, then two phenotypes result among the F₂ plants—tall plants with green pods and short plants with yellow pods.
2. If the four factors segregate into the F₁ gametes independently, then four phenotypes result among the F₂ plants—tall plants with green pods, tall plants with yellow pods, short plants with green pods, and short plants with yellow pods.

Figure 10.6 shows that Mendel observed four phenotypes among the F₂ plants, supporting the second hypothesis. Therefore, Mendel formulated his second law of heredity—the **law of independent assortment**—which states the following:

- Each pair of factors segregates (assorts) independently of the other pairs.
- All possible combinations of factors can occur in the gametes.

Note that when all possible sperm have an opportunity to fertilize all possible eggs, such as the example in Figure 10.6, *the expected phenotypic results of a two-trait cross are always 9:3:3:1 when both parents are heterozygous for the two traits.*

Two-Trait Testcross

The fruit fly *Drosophila melanogaster,* less than one-fifth the size of a housefly, is a favorite subject for genetic

Figure 10.6 Two-trait cross done by Mendel.

P generation plants differ in two regards—stem length and pod color. The F₁ generation shows only the dominant phenotypes, but all possible phenotypes appear among the F₂ generation. The 9:3:3:1 ratio allowed Mendel to deduce that factors segregate into gametes independently of other factors. (Arrow indicates yellow pods.)

research because it has several mutant characteristics that are easily determined. The **wild-type** fly—the type you are most likely to find in nature—has long wings and a gray body, while some mutant flies have short (vestigial) wings and black (ebony) bodies. The key for a cross involving these traits is *L* = long wings, *l* = short wings, *G* = gray body, and *g* = black body.

A two-trait testcross can be used to determine whether an individual is homozygous dominant or heterozygous for either of the two traits. Because it is not possible to determine the genotype of a long-winged, gray-bodied fly by inspection, the genotype may be represented as *L__ G__*.

In a two-trait testcross, an individual with the dominant phenotype for both traits is crossed with an individual with the recessive phenotype for both traits because this fly has a *known* genotype. For example, a long-winged, gray-bodied fly is crossed with a short-winged, black-bodied fly. The heterozygous parent fly (*LlGg*) forms four different types of gametes. The homozygous parent fly (*llgg*) forms only one type of gamete:

P: *LlGg*   ×   *llgg*
Gametes: *LG, Lg, lG, lg*   *lg*

As **Figure 10.7** shows, all possible combinations of phenotypes occur among the offspring. This 1:1:1:1 phenotypic ratio shows that the *L__G__* fly is heterozygous for both traits and has the genotype *LlGg*. Such an individual is called a **dihybrid.** A Punnett square can also be used to predict the chances of an offspring having a particular phenotype (Fig. 10.7). What are the chances of an offspring with long wings and a gray body? The chances are one in four, or 25%. What are the chances of an offspring with short wings and gray body? The chances are also one in four, or 25%.

Mendel's Laws and Probability

When we use a Punnett square to calculate the results of genetic crosses, we assume that each gamete contains one allele for each trait (law of segregation) and that collectively the gametes have all possible combinations of alleles (law of independent assortment). Further, we assume that the male and female gametes combine at random—that is, all possible sperm have an equal chance to fertilize all possible eggs. Under these circumstances, it is possible to use the rules of probability to calculate the expected phenotypic ratio. The **rule of multiplication** says that the chance of two (or more) independent events occurring together is the product of their chances of occurring separately. For example, the chance of getting tails when you toss a coin is ½. The chance of getting two tails when you toss two coins at once is ½ × ½ = ¼.

Let's use the rule of multiplication to calculate the expected results in Figure 10.7. Because each allele pair separates independently, we can treat the cross as two separate one-trait crosses:

Ll × *ll*:   Probability of *ll* offspring   =   ½
Gg × *gg*:   Probability of *gg* offspring   =   ½

The probability of the offspring's genotype being *llgg*:

½ *ll* × ½ *gg* = ¼ *llgg*

And the same results are obtained for the other possible genotypes among the offspring in Figure 10.7.

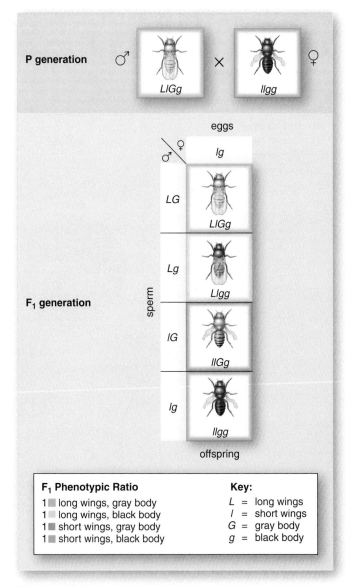

Figure 10.7 Two-trait testcross.

If a fly heterozygous for both traits is crossed with a fly that is recessive for both traits, the expected ratio of phenotypes is 1:1:1:1.

Dominant

Recessive

Unattached earlobes: *EE* or *Ee*

Attached earlobes: *ee*

Widow's peak: *WW* or *Ww*

Straight hairline: *ww*

Mendel's Laws and Meiosis

Today, we are aware that Mendel's laws relate to the process of meiosis. In **Figure 10.8,** a human cell has two pairs of homologous chromosomes, recognized by length—one pair of homologues is short and the other is long. (The color signifies that we inherit chromosomes from our parents; one homologue of each pair is the "paternal" chromosome, and the other is the "maternal" chromosome.) When the homologues separate (segregate), each gamete receives one member from each pair. The homologues separate (assort) independently; it does not matter which member of each pair goes into which gamete. In the simplest of terms, a gamete in Figure 10.8 can receive one short and one long chromosome of either color. Therefore, all possible combinations of chromosomes are in the gametes.

The alleles *E* for unattached earlobes and *e* for attached earlobes are on one pair of homologues, and the alleles *W* for widow's peak and *w* for straight hairline are on the other pair of homologues. Because there are no restrictions as to which homologue goes into which gamete, a gamete can receive either an *E* or an *e* and either a *W* or a *w* in any combination because the chromosomes they are located on assort independently. In the end, collectively, the gametes will have all possible combinations of alleles.

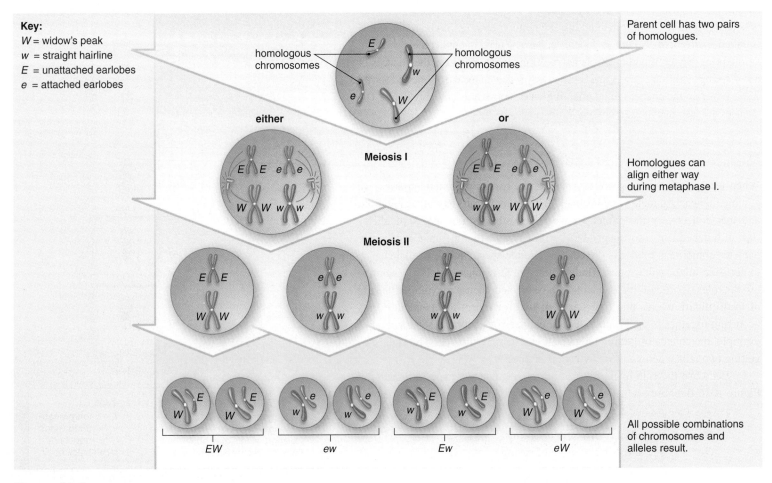

Key:
W = widow's peak
w = straight hairline
E = unattached earlobes
e = attached earlobes

Parent cell has two pairs of homologues.

homologous chromosomes

homologous chromosomes

either   **Meiosis I**   **or**

Homologues can align either way during metaphase I.

Meiosis II

All possible combinations of chromosomes and alleles result.

EW   *ew*   *Ew*   *eW*

Figure 10.8 **Mendel's laws and meiosis.**

A human cell has 23 pairs of homologous chromosomes (homologues), of which 2 pairs are represented here. The homologues, and the alleles they carry, segregate independently during gamete formation. Therefore, all possible combinations of chromosomes and alleles occur in the gametes.

Check Your Progress 10.1

1 Detail Mendel's experiments with peas, and list the terms and theories he discovered during these experiments.

2 Contrast genotype with phenotype.

3 Explain why you are unable to determine the genotype of a dominant phenotype in humans.

4 Compare the phenotypic and genotypic ratios of a cross between two heterozygotes.

5 Solve the following: A black mouse with a straight tail reproduces with a yellow mouse with a bent tail. If the phenotypic ratio among offspring is 1:1:1:1, what is the genotype of the black mouse?

6 Predict the result if the test individual in Figure 10.7 were (a) homozygous dominant for both traits or (b) homozygous dominant for one trait but heterozygous for the other.

Connecting the Concepts

Although Mendel didn't know it, the laws of inheritance that he studied were due to the actions of genes and chromosomes. For more information on these topics, refer to the following discussions.

Section 8.2 illustrates where in the cell cycle sister chromatids are formed.

Section 9.2 describes the independent assortment of chromosomes during meiosis.

Section 11.2 examines how the information in an allele is processed to form a protein.

10.2 Beyond Mendel's Laws

Learning Outcomes

Upon completion of this section, you should be able to

1. Solve and interpret genetic crosses that exhibit incomplete dominance and codominance.

2. Explain and provide examples of polygenic inheritance and pleiotropy.

3. Explain how the environment may influence the phenotype.

Since Mendel's time, variations in the dominant/recessive relationship he described have been discovered. Some alleles are neither dominant nor recessive, and some genes have more than two alleles. Furthermore, some traits are affected by more than one pair of genes and by the environment. Together, these variations make it clear that the concept of the genotype alone cannot account for all the observable traits of an organism.

Incomplete Dominance

Incomplete dominance is exhibited when the heterozygote has an intermediate phenotype between that of either homozygote. In the flowering plant known as the four-o'clock, a cross between red-flowered four-o'clocks and white-flowered four-o'clocks produces offspring with pink flowers (**Fig. 10.9**). But this is not an example of blending inheritance, because when the pink-flowered plants self-pollinate, ¼ of the offspring have red flowers, ¼ have white flowers, and the rest have pink flowers (a 1:2:1 ratio). The reappearance of the original phenotypes makes it clear that we are still dealing with particulate inheritance of the type Mendel described.

People with curly hair have the homozygous recessive genotype, while those with straight hair have the homozygous dominant condition. The heterozygote has wavy hair. We can sometimes explain incomplete dominance by assuming that only the dominant allele codes for a gene product and that the single dose of the product gives the intermediate result.

$C^R C^R$ $C^R C^W$ $C^W C^W$

Figure 10.9 Incomplete dominance in four-o'clocks.

In incomplete dominance, the heterozygote is intermediate between the homozygotes. In this case, the heterozygote is pink, whereas the homozygotes are red or white.

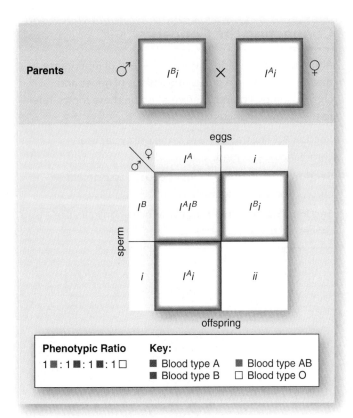

Figure 10.10   Inheritance of ABO blood type.

A mating between individuals with type A blood and type B blood can result in any one of the four blood types. Why? Because the parents are I^Bi and I^Ai. The *i* allele is recessive; both I^A and I^B are dominant.

Multiple-Allele Traits

In ABO blood group inheritance, three alleles determine the presence or absence of antigens on red blood cells and therefore blood type:

I^A = A antigen on red blood cells
I^B = B antigen on red blood cells
i = Neither A nor B antigen on red blood cells

Each person has only two of the three possible alleles, and both I^A and I^B are dominant over *i*. Therefore, there are two possible genotypes for type A blood (I^AI^A, I^Ai) and two possible genotypes for type B blood (I^BI^B, I^Bi). Type O blood can only result from one genotype (*ii*) because the *i* allele is recessive. But I^A and I^B are fully expressed in the presence of each other. Therefore, if a person inherits one of each of these alleles, that person will have a fourth blood type, AB. **Figure 10.10** shows that matings between certain genotypes can have surprising results in terms of blood type.

Notice that human blood type inheritance is also an example of **codominance,** another type of inheritance that differs from Mendel's findings because more than one allele is fully expressed. When an individual has blood type AB, both A and B antigens appear on the red blood cells. The two different capital letters signify that both alleles are coding for an antigen.

Polygenic Inheritance

Polygenic inheritance occurs when a trait is governed by two or more sets of alleles. The individual has a copy of all allelic pairs, possibly located on many different pairs of chromosomes. Each dominant allele has a quantitative effect on the phenotype, and these effects are additive. The result is a continuous variation of phenotypes, resulting in a distribution that resembles a bell-shaped curve. The more genes involved, the more continuous are the variations and distribution of the phenotypes. Also, environmental effects are involved; in the case of human height, differences in childhood nutrition may modify the phenotype to bring about a smooth, bell-shaped curve (**Fig. 10.11**).

Multifactorial traits are those controlled by polygenes subject to environmental influences. Many genetic disorders, such as cleft lip and/or palate, clubfoot, congenital dislocations of the hip, hypertension, diabetes, schizophrenia, and even allergies and cancers, are most likely multifactorial because they are likely due to the combined action of many genes plus environmental influences. In recent years, reports have surfaced that all sorts of behaviors, including alcoholism, phobias, and even suicide, can be associated with particular genes. No doubt, behavioral traits are somewhat controlled by genes, but again, it is impossible at this time to determine to what degree. And very few scientists would support the idea that these behavioral traits are predetermined by our genes.

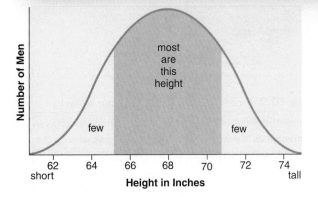

Figure 10.11   Height in humans, a polygenic trait.

When you record the heights of a large group of people chosen at random, the values follow a bell-shaped curve because multiple gene pairs control the trait. Environmental influences, such as nutrition, also affect the phenotype.

Environment and the Phenotype

The relative importance of genetic and environmental influences on the phenotype can vary, but in some instances the environment seems to have an extreme effect. In the water buttercup, the submerged part of the plant has a different appearance from the part above water. This difference is thought to be related to a difference in water intake by the cells.

Temperature can also have a dramatic effect on the phenotypes of plants and animals. Primroses have white flowers when grown above 32°C but red flowers when grown at 24°C. The coats of Siamese cats and Himalayan rabbits are darker in color at the ears, nose, paws, and tail. Himalayan rabbits are known to be homozygous for the allele *ch,* which is involved in the production of melanin. Experimental evidence suggests that the enzyme encoded by this gene is active only at a low temperature and that, therefore, black fur occurs only at the extremities where body heat is lost to the environment (**Fig. 10.12**). When the animal is placed in a warmer environment, new fur on these body parts is light in color.

These examples lend additional support to the observation that human traits controlled by polygenes are also subject to environmental influences. Therefore, many investigators are trying to determine what percentage of various traits is due to nature (inheritance) and what percentage is due to nurture (the environment). Some studies use twins separated from birth, because if identical twins in different environments share a trait, that trait is most likely inherited. Identical twins are more similar in their intellectual talents, personality traits, and levels of lifelong happiness than are fraternal twins separated from birth. Biologists conclude that all behavioral traits are partly determined by inheritance and that the genes for these traits act together in complex combinations to produce a phenotype that is modified by environmental influences.

Figure 10.12 Coat color in Himalayan rabbits.

Hair growing under an ice pack in these rabbits is black. The dark color on the ears, nose, and feet of these rabbits is believed to be due to a lower body temperature in these areas.

Connections and Misconceptions

Are all dominant alleles "normal"?

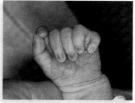

In most examples, the dominant allele represents the "normal" allele and the recessive allele represents the altered version. However, this is not always the case. For example, one form of polydactyly—the presense of extra fingers or toes—is due to the inheritance of a dominant allele. The "normal" allele is recessive, meaning that an individual only needs to inherit one copy of the polydactyly allele in order to display the phenotype. Polydactyly is also an example of a trait that has variable expressivity, meaning that for one person the presence of the allele may cause an extra finger but in another person it may be an extra toe.

Connections and Misconceptions

What is the genetic basis of skin color?

The color of human skin is both polygenic and multifactorial. Skin color is determined by the production of a pigment called melanin. Many genes and multiple alleles contribute to the amount of melanin in a person's skin. In addition, melanin production is increased by exposure to the ultravioloet radiation in sunlight, which explains changes in skin color during the summer months. Interestingly, scientists are discovering that skin color is not necessarily the best indication of a person's race, since people from very different regions of the globe can have similar levels of melanin in their skin.

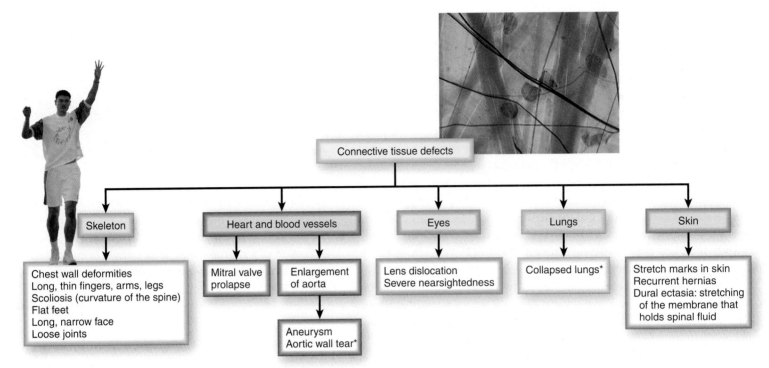

Figure 10.13 **Marfan syndrome, multiple effects of a single human gene.**

Individuals with Marfan syndrome exhibit defects in the connective tissue throughout the body. Important changes occur in the skeleton, heart, blood vessels, eyes, lungs, and skin. All these conditions are due to mutation of the gene *FBN1*, which codes for a constituent of connective tissue. *life-threatening.

Connecting the Concepts

For more on some of the conditions presented in this section, refer to the following discussions.

Section 15.2 examines how natural selection may favor heterozygotes, as is the case with sickle cell disease.

Section 22.1 describes the role of connective tissue in humans.

Section 23.3 provides more information on the role of blood in the body.

Check Your Progress 10.2

1. Define incomplete dominance and give an example.

2. Describe codominance and list all the possible genotypes for the ABO blood group in humans.

3. Compare and contrast polygenic inheritance, multifactorial traits, and pleiotropy.

Pleiotropy

Pleiotropy occurs when a single gene has more than one effect. Often, this leads to a *syndrome*, a group of symptoms that appear together and indicate the presence of a particular genetic mutation. For example, persons with Marfan syndrome have disproportionately long arms, legs, hands, and feet; a weakened aorta; and poor eyesight (**Fig. 10.13**). All of these characteristics are due to the production of abnormal connective tissue. Marfan syndrome has been linked to a mutated gene (*FBN1*) on chromosome 15 that ordinarily specifies a functional protein called fibrillin. This protein is essential for the formation of elastic fibers in connective tissue. Without the structural support of normal connective tissue, the aorta can burst, particularly if the person is engaged in a strenuous sport, such as volleyball or basketball. Flo Hyman may have been the best American woman volleyball player ever, but she fell to the floor and died when only 31 years old because her aorta gave way during a game. Now that coaches are aware of Marfan syndrome, they are on the lookout for it among very tall basketball players. Chris Weisheit, whose career was cut short after he was diagnosed with Marfan syndrome, said, "I don't want to die playing basketball."

Many other disorders, including sickle cell disease and porphyria, are also examples of pleiotropic traits. Porphyria is caused by a chemical insufficiency in the production of hemoglobin, the pigment that makes red blood cells red. The symptoms of porphyria are photosensitivity, strong abdominal pain, port-wine-colored urine, and paralysis in the arms and legs. Many members of the British royal family in the late 1700s and early 1800s suffered from this disorder, which can lead to epileptic convulsions, bizarre behavior, and coma. The vampire legends are most likely also based on individuals with a specific form of porphyria.

10.3 Sex-Linked Inheritance

> ### Learning Outcomes
>
> Upon completion of this section, you should be able to
>
> 1. Explain differences in the inheritance of sex-linked traits between male and female offspring.
> 2. Solve and interpret genetic crosses that exhibit sex-linked inheritance.

Geneticists of the early twentieth century were convinced that the genes are on the chromosomes because the genes and chromosomes behave similarly during gamete formation (see Fig. 10.8). They also knew that the chromosomes differ between the sexes. As you learned in Chapter 9, the sex chromosomes in females are XX, and those in males are XY (**Fig. 10.14**). Notice that males produce two different types of gametes during meiosis—those that contain an X and those that contain a Y. Therefore, the chromosomal contribution of the male determines the sex of the new individual.

The much shorter Y chromosome contains only about 80 genes, and most of these genes are concerned with sex differences between men and women. One of the genes on the Y chromosome, *SRY*, does not have a copy on the X chromosome. If the functional *SRY* gene is present, the individual becomes a male, and if it is absent, the individual becomes a female. Thus, female sex development is the "default setting."

In contrast, the X chromosome is quite large and contains nearly 2,000 genes, most of which have nothing to do with the gender of the individual. By tradition, the term **X-linked** refers to such genes carried on the X chromosome. Examples of X-linked traits in humans include hemophilia and red-green color-blindness. The Y chromosome does not carry these genes, and that makes for an interesting inheritance pattern.

Video
Why a Guy Is a Guy

X-Linked Alleles

We have already mentioned that the fruit fly is a favorite subject for genetic studies. Flies can be easily and inexpensively raised in simple laboratory glassware; females mate only once and then lay hundreds of eggs during their lifetime; and the generation time is short, taking only about 10 days when conditions are favorable. Early *Drosophila* geneticists noticed that, when a mutant male with white eyes was crossed with a red-eyed female, all the F_1 had red eyes:

	♀		♂
P	red-eyed	×	white-eyed
F_1	red-eyed		red-eyed

From these results, researchers knew that red eyes are the dominant characteristic and white eyes are the recessive characteristic. However, the F_2 generation gave an unusual result: The expected 3:1 ratio resulted, but all of the white-eyed flies were males:

	♀		♂
$F_1 \times F_1$	red-eyed	×	red-eyed
F_2	red-eyed		1 red-eyed : 1 white-eyed

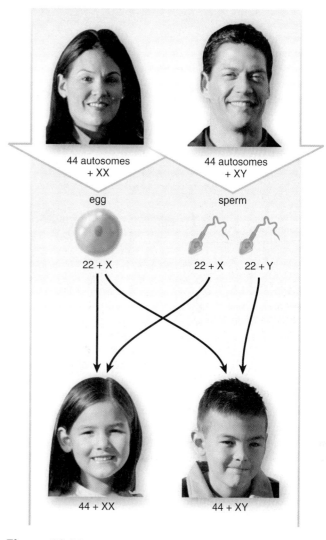

Figure 10.14 Sex determination in human beings.
Sperm determine the sex of an offspring because sperm can carry an X or a Y chromosome. Males inherit a Y chromosome. Genes on the Y chromosome also affect the gender of an individual.

Connections and Misconceptions

What is an example of a sex-linked trait in humans?

Perhaps the best-known example of an X-linked trait in humans is hemophilia A. This disease is due to a recessive mutation in one of the genes associated with blood clotting. Since the gene is located on the X chromosome, males who inherit the defective allele from their mothers have hemophilia. Females must inherit a defective gene from both their mother and their father. Hemophilia A has been a problem in the royal families of England, Spain, Russia, and Germany for several centuries.

Sex chromosomes contain genes for traits unrelated to gender.

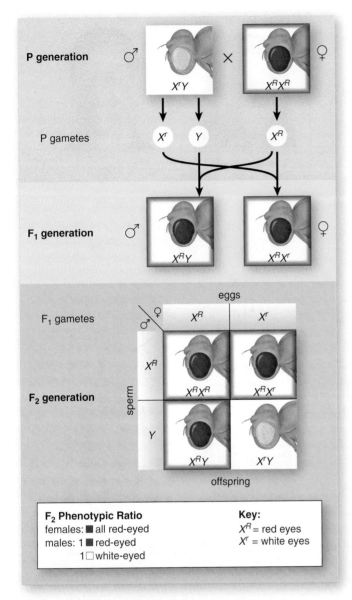

Figure 10.15   X-linked inheritance.

Once researchers deduced that the allele for red/white eye color is only on the X chromosome in *Drosophila*, they were able to explain their experimental results. Males with white eyes in the F_2 inherit the recessive allele only from the female parent; they receive a blank Y chromosome from the male parent.

Connecting the Concepts

For more information on the X and Y chromosomes, refer to the following discussions.

Section 9.1 distinguishes between the sex chromosomes and autosomes in humans.

Section 13.2 provides examples of sex-linked disorders in humans.

Obviously, a major difference between the male flies and the female flies was their sex chromosomes. Could it be possible that an allele for eye color was on the Y chromosome but not on the X? This idea was quickly discarded because females usually have red eyes, and they have no Y chromosome. Next, investigators hypothesized that perhaps an allele for eye color was on the X chromosome, but not on the Y chromosome. This explanation turned out to match the results obtained in the experiment (**Fig. 10.15**). These results also support the chromosome theory of inheritance by showing that the behavior of a specific allele corresponds exactly with that of a specific chromosome—the X chromosome in *Drosophila.*

Notice that X-linked alleles have a different pattern of inheritance than alleles on the autosomes. The Y chromosome is blank for these alleles, so the inheritance of a Y chromosome cannot offset the inheritance of an X-linked recessive allele. For the same reason, *males always receive an X-linked recessive mutant allele from their female parent;* they receive the Y chromosome, which does not have an allele for the trait, from their male parent.

An X-Linked Problem

When solving autosomal genetics problems involving fruit flies, the alleles and genotypes are represented as follows:

Key: L =   long wings   Genotypes: = LL, Ll, ll
      l =   short wings

As noted in Figure 10.15, however, an X-linked gene is represented by attaching the allele to an X:

Key: X^R =   red eyes
      X^r =   white eyes

The possible genotypes in both males and females are as follows:

Genotypes: $X^R X^R$ =   red-eyed female
           $X^R X^r$ =   red-eyed female
           $X^r X^r$ =   white-eyed female
           $X^R Y$ =   red-eyed male
           $X^r Y$ =   white-eyed male

Notice that three genotypes are possible for females, but only two are possible for males. Females can be heterozygous $X^R X^r$, in which case they are carriers. **Carriers** usually do not exhibit a recessive trait, but they are capable of passing on a recessive allele for a trait. Males cannot be carriers; if the dominant allele is on the single X chromosome, they show the dominant phenotype, and if the recessive allele is on the single X chromosome, they show the recessive phenotype.

Males have white eyes when they receive the mutant recessive allele from the female parent. Females can have white eyes only when they receive a recessive allele from both parents.

Check Your Progress 10.3

1. Clarify why males produce two different gametes with respect to sex chromosomes.

2. Explain why a son inherits X-linked recessive traits only from his mother and not his father.

3. Discuss why males cannot be carriers of X-linked disorders.

10.4 Inheritance of Linked Genes

Learning Outcomes

Upon completion of this section, you should be able to

1. Describe what is meant when two genes are said to be linked.
2. Compare and contrast the inheritance patterns of linked genes to those of genes that are not linked.
3. Explain how genes can be mapped on a chromosome.

Consider that an organism usually has far fewer chromosomes than genes. For example, *Drosophila* has 14,000 genes, but only 4 chromosomes. Therefore, many alleles must be on each chromosome. The alleles that occur on the same chromosome form a *linkage group* because these alleles tend to be inherited together. **Figure 10.16** considers a heterozygous individual in which the dominant alleles (*GR*) occur on one homologue and the recessive alleles (*gr*) occur on the other homologue. It shows you that, if the alleles remain linked during meiosis (Fig. 10.16*a*), the gametes have to contain either a daughter chromosome with the dominant alleles or a daughter chromosome with the recessive alleles. However, if crossing-over occurs between the two alleles (Fig. 10.16*b*), the daughter chromosomes contain new combinations of alleles. These chromosomes are said to be recombinant, since they were produced by crossing-over between the homologous chromosomes. **Recombinant gametes** contain recombined alleles, but they occur in reduced numbers because crossing-over is infrequent.

Figure 10.16   Linked alleles and crossing-over.

a. Alleles on the same chromosome form a linkage group, and linked alleles tend to be found together in the same daughter cells and, therefore, gametes. **b.** Crossing-over can cause linked alleles to separate and go into different daughter cells, and these become recombinant gametes.

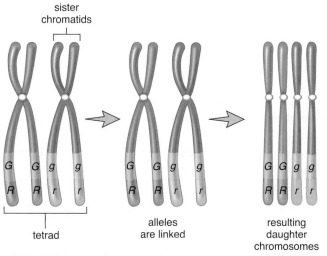

a. Linked alleles usually stay together

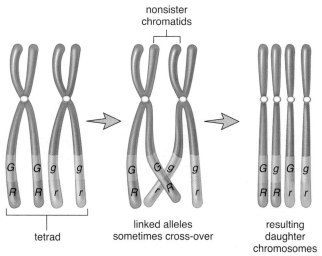

b. Crossing-over results in recombination of alleles

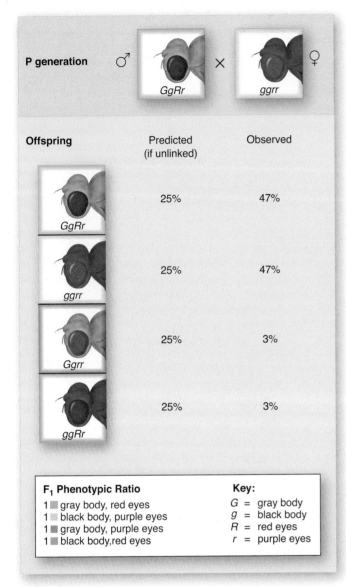

Offspring	Predicted (if unlinked)	Observed
GgRr	25%	47%
ggrr	25%	47%
Ggrr	25%	3%
ggRr	25%	3%

F₁ Phenotypic Ratio

1 ▢ gray body, red eyes
1 ▢ black body, purple eyes
1 ▢ gray body, purple eyes
1 ▢ black body, red eyes

Key:

G = gray body
g = black body
R = red eyes
r = purple eyes

Figure 10.17 Linked alleles do not assort independently.

When gametes are formed in the heterozygote shown in Figure 10.16, the two alleles that occur together on each chromosome segregate together.

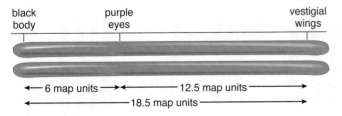

black body · purple eyes · vestigial wings

← 6 map units → ← 12.5 map units →
← 18.5 map units →

Figure 10.18 Mapping chromosomes.

Genes are arranged linearly on the chromosome at specific gene loci. Using the crossing-over data supplied by the text, these genes are sequences as shown.

Gene linkage was discovered by scientists performing *Drosophila* crosses. For example, when scientists performed a two-trait testcross between a gray-bodied, red-eyed fly heterozygous for both traits and a black-bodied, purple-eyed recessive fly, as in **Figure 10.17,** most of the flies resembled one or the other parent. This was unexpected because, if the two pairs of alleles are on different chromosomes, independent assortment always occurs, and the heterozygote produces four different types of gametes in equal numbers. In this instance, if the alleles for body color assorted independently of the alleles for eye color, then a 1:1:1:1 ratio would occur among the offspring and 25% of the offspring would be one of four possible genotypes. Instead, only 3% of the offspring showed the recombinant phenotypes. Why? Because the linked alleles usually remained together in the gametes made by the heterozygote, as shown in Figure 10.16*a,* and generally did not resort, as in Figure 10.16*b.* Recombination did occur sometimes, however, and that is why a total of 6% of the offspring had a recombinant phenotype. Their results allowed scientists to conclude that gene linkage was almost complete for these two sets of alleles.

Animation
Crossing-over in Meiosis

Distance Between Gene Loci

It stands to reason that, the closer together two genes are, the less likely they are to cross over. Numerous experiments have repeatedly shown that this is the case. Therefore, you can use the percentage of recombinant phenotypes to determine the distance between genes, because there is a direct relationship between the frequency of crossing-over (the percentage of recombinant phenotypes) and the distance between alleles. In the previous example (see Fig. 10.17), a total of 6% of the offspring are recombinants, and this means that the rate of crossing-over is 6%. For the sake of mapping the chromosomes, it is assumed that 1% of crossing-over equals one map unit. Therefore, the allele for black body and the allele for purple eyes are six map units apart.

Mapping the Chromosome

Suppose you want to determine the order of three gene loci on the chromosomes. To do so, you perform crosses that involve the three genes. The results will tell you the map distance between the three gene loci. Assume, for instance, the following:

1. The distance between the black-body and purple-eye alleles = 6 map units.
2. The distance between the purple-eye and vestigial-wing alleles = 12.5 map units.
3. The distance between the black-body and vestigial-wing alleles = 18.5 map units.

Therefore, the order of the alleles must be as shown in **Figure 10.18.** Black body must be 6 map units away from purple eyes, and purple eyes must be 12.5 map units away from vestigial wings.

Scientists used linkage data in *Drosophila* to construct a **chromosome map,** which shows the relative distance between the gene loci on a chromosome. However, the possibility of using linkage data to map human chromosomes is limited because we can work only with matings that occur by chance. This, coupled with the fact that humans tend not to have numerous offspring, means that additional methods are needed to sequence the genes on human chromosomes. Today, it is customary also to rely on biochemical methods to map the human chromosomes.

Check Your Progress 10.4

1. State how crossing-over produces recombinant gametes.
2. Explain which of Mendel's laws is not accurate for linked genes.
3. Predict how crossing-over can tell you the sequence of gene loci on a chromosome.

Connecting the Concepts

For more information on how DNA is studied, refer to the following discussions.

Section 11.3 describes how recombinant DNA technology and the polymerase chain reaction are used to study DNA.

Section 11.4 provides an overview of the science of genomics.

Media Study Tools

www.mhhe.com/maderessentials3

Enhance your study of this chapter with study tools and practice tests. Also ask your instructor about the resources available through ConnectPlus, including the media-rich eBook, interactive learning tools, and animations.

Punnett Squares and Sex-Linked Traits

Two virtual labs, Punnett Squares and Sex-Linked Traits, can be used to help you understand how to analyze the patterns of inheritance presented in this chapter.

The Chapter in Review

Summary

10.1 Mendel's Laws

In 1860, Gregor Mendel, an Austrian monk, developed two laws of heredity based on crosses utilizing the garden pea.

Law of Segregation

Mendel's law of segregation states the following:

- Each individual has two factors for each trait.
- The factors segregate (separate) during the formation of the gametes.
- Each gamete contains only one factor from each pair of factors.
- Fertilization gives each new individual two factors for each trait.

In the context of genetics today,

- Gene pairs are on the chromosomes, one allele on each homologue.
- Alternative forms of a gene are called alleles. An individual may have different alleles of a gene on each homologue.
- Alleles are assigned uppercase letters if they are dominant, lowercase letters if they are recessive. An individual's genotype may be homozygous dominant (*AA*), heterozygous (*Aa*), or homozygous recessive (*aa*).
- Homologues separate during meiosis, and the gametes have only one allele for each trait—either an *A* or an *a*.
- Fertilization gives each new individual two alleles for each trait.

Law of Independent Assortment

Mendel's law of independent assortment states the following:

- Each pair of factors segregates (assorts) independently of the other pairs.
- All possible combinations of factors can occur in the gametes.

In the context of genetics today,

- Each pair of homologues separates independently of the other pairs.
- All possible combinations of chromosomes and their alleles occur in the gametes.
- Mendel's laws are consistent with the manner in which homologues and their alleles separate during meiosis.

Common Autosomal Genetic Crosses

A Punnett square allows all types of sperm to fertilize all types of eggs and gives these results:

Aa $\times$ *Aa*	3:1 phenotypic ratio
Aa $\times$ *aa*	1:1 phenotypic ratio
AaBb $\times$ *AaBb*	9:3:3:1 phenotypic ratio
AaBb $\times$ *aabb*	1:1:1:1 phenotypic ratio

10.2 Beyond Mendel's Laws

In some patterns of inheritance, the alleles are not just dominant or recessive.

Incomplete Dominance

In incomplete dominance, the heterozygote is intermediate between the two homozygotes. For example, the offspring of red and white four-o'clocks produce pink flowers. The red and white phenotypes reappear when pink four-o'clocks are crossed.

Multiple-Allele Traits

The multiple-allele inheritance pattern is exemplified in humans by blood type inheritance. Every individual has two out of three possible alleles: I^A, I^B, *i*. Both I^A and I^B are expressed; therefore, this is also a case of codominance.

Polygenic Inheritance

In polygenic inheritance, a trait is controlled by more than one set of alleles. The dominant alleles have an additive effect on the phenotype.

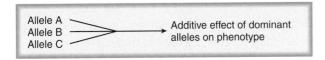

Environment and the Phenotype

Many phenotypes, especially those governed by polygenes, are modified by the environment.

Pleiotropy

In pleiotropy, one gene (consisting of two alleles) has multiple effects on the body. For example, all the disorders common to Marfan syndrome are due to a mutation that leads to a defect in the composition of connective tissue.

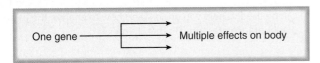

10.3 Sex-Linked Inheritance

Males produce two different types of gametes—those that contain an X and those that contain a Y. The contribution of the male determines the sex and gender of the new individual. XX = female, XY = male.

Certain alleles are carried on the X chromosome, but the Y has very few genes. Therefore, a male needs to inherit only one recessive allele on the X chromosome to have a recessive genetic disorder.

Common X-Linked Genetic Crosses

$X^B X^b \times X^B Y$ All daughters will be normal, even though they have a 50% chance of being carriers, but sons have a 50% chance of being color blind.

$X^B X^B \times X^b Y$ All children are normal (daughters will be carriers).

10.4 Inheritance of Linked Genes

Alleles on the same chromosome are linked, and they are generally inherited together, but crossing-over can shuffle them. Crossing-over data can be used to determine the distance between gene loci and to construct a chromosome map, which shows the sequence of gene loci along the chromosome.

Key Terms

Testing Yourself

Choose the best answer for each question.

1. The offspring ratio from a testcross ($F_1 \times$ homozygous recessive) should be
   a. all dominant.
   b. ¾ dominant, ¼ recessive.
   c. ½ dominant, ½ recessive.
   d. all recessive.

2. Which of the following is not a component of the law of segregation?
   a. Each gamete contains one factor from each pair of factors in the parent.
   b. Factors segregate during gamete formation.
   c. Following fertilization, the new individual carries two factors for each trait.
   d. Each individual has one factor for each trait.

3. When using a Punnett square to predict offspring ratios, we assume that
   a. each gamete contains one allele of each gene.
   b. the gametes have all possible combinations of alleles.
   c. male and female gametes combine at random.
   d. All of these are correct.

4. If you cross a black spaniel with a red spaniel and get a litter of eight black and one red, what is the genotype of the black parent?
   a. BB
   b. Bb
   c. bb
   d. The genotype is impossible to determine.

5. Cystic fibrosis (CF) is an autosomal recessive disorder in humans. If two unaffected individuals have a child with CF, what is the chance that their second child will have CF?
   a. 25%
   b. 50%
   c. 75%
   d. 100%
   e. It is impossible to determine.

6. When one physical trait is affected by two or more pairs of alleles, the condition is called
   a. incomplete dominance.
   b. codominance.
   c. homozygous dominant.
   d. multiple allele.
   e. polygenic inheritance.

7. Fill in the blank spaces in the following Punnett square.

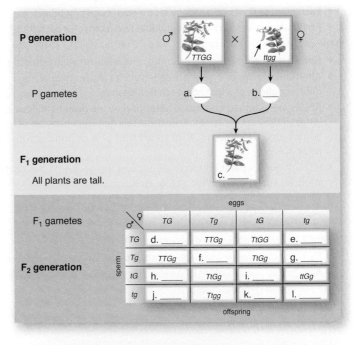

For questions 8–10, consider that coat color and spotting pattern in cocker spaniels depend on two genes. Black (*B*) is dominant to red (*b*), and solid color (*S*) is dominant to spotted (*s*).

8. What phenotypic ratio do you expect for a dihybrid cross?
   a. 1 black solid : 1 red solid : 1 black spotted : 1 red spotted
   b. 9 black solid : 3 red solid : 3 black spotted : 1 red spotted
   c. 1 black solid : 3 red solid : 3 black spotted : 9 red spotted
   d. all black solid

9. If you cross a black spotted dog with a black solid dog and get a ratio of 3 black solid : 3 black spotted : 1 red solid : 1 red spotted, what is the genotype of the black spotted parent?
   a. *BBss*   d. *bbSs*
   b. *BbSs*   e. *bbss*
   c. *Bbss*

10. What is the genotype of the black solid parent in question 9?
    a. *BBSS*   d. *Bbss*
    b. *BbSS*   e. *bbSs*
    c. *BbSs*

For questions 11–17, match the cross with the predicted phenotypic ratios in the key.

Key:

    a. 3:1   d. 1:1:1:1
    b. 9:3:3:1   e. none of these
    c. 1:1

11. *AaBb* × *aabb*
12. *Aa* × *Aa*
13. *AAbb* × *aaBB*
14. *Aa* × *aa*
15. *AaBb* × *aabb*
16. *AABB* × *AaBb*
17. *AaBb* × *AaBb*

18. When two monohybrid round squashes are crossed, the offspring ratio is ¼ flat: ½ oblong: ¼ round. Squash shape, therefore, is controlled by incomplete dominance. What offspring ratio would you expect from a cross between a plant with oblong fruit and one with round fruit?
    a. all oblong   d. ¾ round: ¼ oblong
    b. all round   e. ½ oblong: ½ round
    c. ¾ oblong: ¼ round

19. If a man of blood group AB marries a woman of blood group B whose father was type O, what phenotypes could their children be?
    a. A only   d. A, AB, and B
    b. A, AB, B, and O   e. O only
    c. AB only

For questions 20–23, list the progeny phenotypes from the following key that would result from each of the crosses in *Drosophila*. Red eye color is dominant over white. The gene for eye color is on the X chromosome. Answers can be used more than once.

Key:

    a. red-eyed female   c. white-eyed female
    b. red-eyed male   d. white-eyed male

20. homozygous red-eyed female × white-eyed male
21. heterozygous female × white-eyed male
22. white-eyed female × white-eyed male
23. heterozygous female × red-eyed male

24. Alice and Henry are at the opposite extremes for height, a polygenic trait. Their children will
    a. exhibit the middle height between their two parents.
    b. exhibit a 9:3:3:1 ratio of heights between that of their parents.
    c. exhibit a 3:1 ratio of tall to short heights.
    d. be the same height as Alice or Henry.

25. Mary's son is color-blind, but Mary is not color-blind. Which of these is a more likely genotype for a relative?
    a. mother X^cX^c c. brother X^cX^c
    b. father X^cY d. Both b and c are likely.

Thinking Scientifically

1. Multiple gene pairs may also interact to produce a single phenotype. In peas, genes *C* and *P* are required for pigment production in flowers. Gene *C* codes for an enzyme that converts a compound into a colorless intermediate product. Gene *P* codes for an enzyme that converts the colorless intermediate product into anthocyanin, a purple pigment. A flower, therefore, will be purple only if it contains at least one dominant allele for each of the two genes (*C__ P__*). What phenotypic ratio would you expect in the F_2 generation following a cross between two double heterozygotes (*CcPp*)?

2. Geneticists often look for unusual events to provide insight into genetic mechanisms. In one such instance, researchers studied XX men and XY women. They found that the XX men contained a chromosomal segment normally found in men but not in women, while the XY women were missing that region. What gene do you suppose is on that chromosome piece?

3. What would cause a particular polygenic trait to become more frequent in a population over time?

Bioethical Issue

Selecting the Sex of Your Child

As you know, the sex of a child depends upon whether an X-bearing sperm or a Y-bearing sperm enters the egg. But a new technology that can separate X-bearing and Y-bearing sperm offers prospective parents the opportunity to choose the sex of their child. First, the sperm are dosed with a DNA-staining chemical. Because the X chromosome has slightly more DNA than the Y chromosome, it takes up more dye. When a laser beam shines on the sperm, the X-bearing sperm shine more brightly than the Y-bearing sperm. A machine sorts the sperm into two groups on this basis. The results are not perfect. Following artificial insemination, there's about an 85% success rate for a girl and about a 65% success rate for a boy. But is it morally acceptable to select the sex of your child?

Those who find this practice unethical contend that using such technology is akin to "playing God." They are greatly concerned that this new ability may lead to a society with far more members of one sex than another, which could lead to serious problems. Furthermore, they contend that allowing parents to select the sex of their children could lead to other ethical concerns, such as selecting for specific traits in children.

Proponents of sex-selection technology argue that there are many instances in which the ability to choose the sex of the child may benefit society. For instance, if a mother is a carrier of an X-linked genetic disorder, such as hemophilia or Duchenne muscular dystrophy, this would be the simplest way to ensure a healthy child. Previously, a pregnant woman with these concerns had to wait for the results of an amniocentesis test and then decide whether to abort the fetus if it was a boy. In such cases, is it better for all involved to ensure that a child does not have a specific genetic disorder than to take the risk?

11

DNA Biology and Technology

BEFORE YOU BEGIN

Before beginning this chapter, take a few moments to review the following discussions.

Section 3.2 What is the general structure of a DNA and RNA molecule?

Sections 4.2 and 4.4 What are the differences between prokaryotic and eukaryotic cells?

Section 8.1 What is the end result of DNA replication?

Pig Organs for Transplantation

One recent advance in DNA technology is the production of transgenic animals—animals that contain genes from another species. For example, researchers have created a variety of miniature pigs for the purpose of providing organs for human transplants. A unique feature of these pigs is a yellow snout and hooves due to a gene introduced from jellyfish! This makes the pigs easy to identify.

An alternative source of transplant organs is needed because human organs are in limited supply, and many people die each year, waiting for transplants. If organs from a normal pig are transplanted into a human, two major problems occur: The organs from pigs are too large for humans and the organs are quickly rejected. However, transplants from miniature genetically altered pigs have the potential to avoid both of these complications.

First, the miniature size of the pigs makes their organs more appropriate for humans. Second, rejection is avoided because the genes coding for plasma membrane proteins that normally trigger rejection have been "knocked out"— that is, they are not expressed. Pig organs may eventually become a life-saving alternative for people who are unable to acquire human organ transplants.

In this chapter, you will learn about the structure and function of DNA and RNA. This knowledge will allow you to appreciate the incredible advances in the field of DNA technology.

11.1 DNA and RNA Structure and Function

Learning Outcomes

Upon completion of this section, you should be able to

1. Briefly describe the Watson and Crick structure of DNA and list the evidence used in proposing this structure.
2. List the steps involved in replicating DNA.
3. Compare and contrast the structure of RNA with that of DNA.
4. List the three major types of RNA and describe their functions.

Mendel knew nothing about DNA. It took many years for investigators to come to the conclusion that Mendel's factors, now called genes, are on the chromosomes. Then, researchers wanted to show that DNA, not proteins, is responsible for heredity. One experiment, by Alfred Hershey and Martha Chase, involved the use of a virus that attacks bacteria, such as *E. coli* (**Fig. 11.1**). A virus is composed of an outer capsid made of protein and an inner core of DNA. The use of radioactive tracers showed that DNA, but not protein, enters bacteria and directs the formation of new viruses. By the early 1950s, investigators had learned that genes are composed of DNA and that mutated genes result in errors of metabolism. Therefore, DNA in some way must control the cell.

Animation
Hershey and Chase Experiment

Even though DNA took its name—deoxyribonucleic acid—from the chemical components of its nucleotides, its detailed structure was still to be determined. Finding the structure of DNA was the first step toward understanding how DNA is able to do the following:

- Be variable in order to account for species differences
- Replicate so that every cell gets a copy during cell division
- Store information needed to control the cell
- Undergo mutations, accounting for evolution of new species

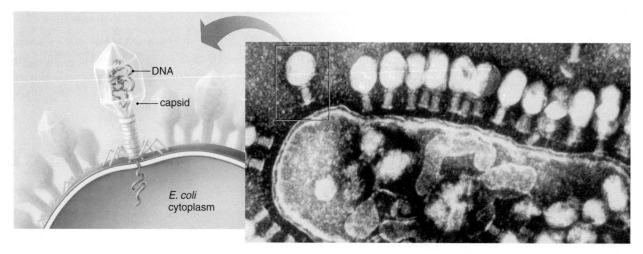

Figure 11.1 The genes are composed of DNA.

An early experiment by Hershey and Chase helped determine that DNA is the genetic material. Their experiment involved a virus that infects bacteria, such as *E. coli*. They wanted to know which part of the virus—the capsid made of protein or the DNA inside the capsid—enters the bacterium. Radioactive tracers showed that DNA, not protein, enters the bacterium and guides the formation of new viruses. Therefore, DNA must be the genetic material.

Structure of DNA

Once researchers knew that genes are composed of DNA, they were racing against time and each other to determine the structure of DNA. They believed that whoever discovered it first would get a Nobel Prize. How James Watson and Francis Crick determined the structure of DNA (and eventually received a Nobel Prize) resembles a mystery, in which each clue was added to the total picture until the breathtaking design of DNA—a double helix—was finally revealed. To achieve this success, Watson and Crick particularly relied on studies done by Erwin Chargaff and Rosalind Franklin.

Chargaff's Rules

Before Erwin Chargaff began his work, it was known that DNA contains four different types of nucleotides based on their nitrogen-containing bases (**Fig. 11.2** *d*). The bases **adenine (A)** and **guanine (G)** are purines with a double ring, and the bases **thymine (T)** and **cytosine (C)** are pyrimidines with a single ring. With the development of new chemical techniques in the 1940s, Chargaff decided to analyze in detail the base content of DNA in various species.

In contrast to accepted beliefs, Chargaff found that each species has its own percentages of each type of nucleotide. For example, in a human cell, 31% of bases are adenine; 31% are thymine; 19% are guanine; and 19% are cytosine. In all the species Chargaff studied, the amount of A always equaled the amount of T, and the amount of G always equaled the amount of C. These relationships are called *Chargaff's rules:*

1. The amount of A, T, G, and C in DNA varies from species to species.
2. In each species, the amount of A = T and the amount of G = C.

Chargaff's data suggest that DNA has a means to be stable, in that A can pair only with T, and G can pair only with C. His data also show that DNA can be variable as required for the genetic material. Today, we know that the paired bases may occur in any order, and the amount of variability in their sequences is overwhelming. For example, suppose a chromosome contains 140 million base pairs. Since any of the four possible nucleotide pairs can be present at each pair location, the total number of possible nucleotide pair sequences is $4^{140 \times 10^6}$ or $4^{140,000,000}$.

Franklin's X-Ray Diffraction Data

Rosalind Franklin was a researcher at King's College in London in the early 1950s (**Fig. 11.3***a*). She was studying the structure of DNA using X-ray crystallography. When

Figure 11.2 Nucleotide composition of DNA and RNA.

a. All nucleotides contain phosphate, a 5-carbon sugar, and a nitrogen-containing base, such as cytosine (C). **b.** In DNA, the nitrogen-containing bases are adenine, guanine, cytosine, and thymine; in RNA, the bases are adenine, guanine, cytosine, and uracil. **c.** Structure of phosphate. **d.** In DNA, the sugar is deoxyribose; in RNA, the sugar is ribose.

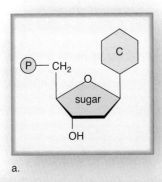

a.

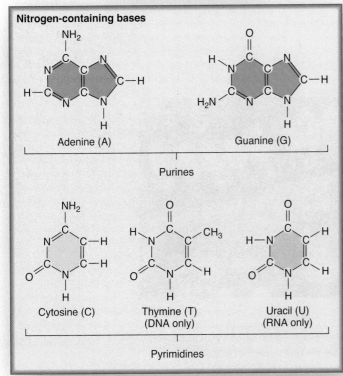

b.

c.

d.

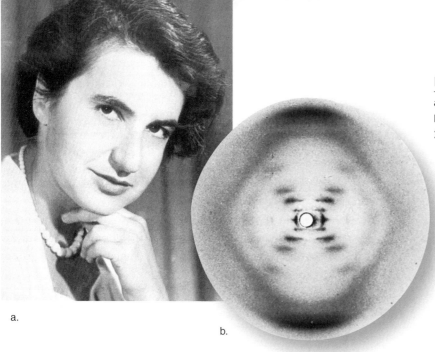

a.

b.

Figure 11.3 **X-ray diffraction pattern of DNA.**

a. Rosalind Franklin X-rayed DNA using crystallography techniques.
b. The resulting pattern indicated that DNA is a double helix (see X pattern in the center) and that part of the molecule is repeated over and over again (see the dark portions at top and bottom). Watson and Crick said that this repeating feature is the paired bases.

a crystal (a solid substance whose atoms are arranged in a definite manner) is X-rayed, the X-ray beam is diffracted (deflected), and the pattern that results shows how the atoms are arranged in the crystal.

First, Franklin made a concentrated, viscous solution of DNA and then saw that it could be separated into fibers. Under the right conditions, the fibers were enough like a crystal that, when they were X-rayed, a diffraction pattern resulted. The X-ray diffraction pattern of DNA shows that DNA is a double helix. The helical shape is indicated by the crossed (X) pattern in the center of the photograph in Figure 11.3*b*. The dark areas at the top and bottom of the photograph indicate that some portion of the helix is repeated many times.

Video
DNA Dark Lady

The Watson and Crick Model

In 1951, James Watson, having just earned a Ph.D., began an internship at the University of Cambridge, England. There, he met Francis Crick, a British physicist who was interested in molecular structures. Together, they set out to determine the structure of DNA and to build a model that would explain how DNA varies from species to species, replicates, stores information, and undergoes mutation.

Based on the available data, they knew the following:

1. DNA is a polymer of four types of nucleotides with the bases adenine (A), guanine (G), cytosine (C), and thymine (T).
2. Based on Chargaff's rules, the amount of A = T, and the amount of G = C.
3. Based on Franklin's X-ray diffraction photograph, DNA is a double helix with a repeating pattern.

Using these data, Watson and Crick built a model of DNA out of wire and tin (**Fig. 11.4**). The model showed that the deoxyribose sugar–phosphate molecules are bonded to one another to make up the sides of a twisted ladder. The nitrogenous bases make up the rungs of the ladder—they project into the middle and hydrogen-bond with bases on the other strand. Indeed, the pairing of A with T and G with C—now called **complementary base pairing**—results in rungs of a consistent width, as elucidated by the X-ray diffraction data.

Figure 11.4 **Watson and Crick model of DNA.**

James Watson (left) and Francis Crick with their model of DNA.

Figure 11.5 shows two ways to represent the structure of DNA. Notice that DNA is a double helix because it is double-stranded. The two strands are antiparallel and run in opposite directions, as best seen in Figure 11.5a, where the carbon atoms in deoxyribose are numbered. The 5′ carbon has an attached Ⓟ group, and the 3′ carbon has an attached —OH group, which is circled and colored pink for easy recognition. Further, in the double helix (Fig. 11.5b), each strand has a 5′ end where a free Ⓟ appears and a 3′ end where a free —OH group appears.

The double-helix model of DNA permits the base pairs to be in any order, a necessity for genetic variability between species. Also, the model suggests that complementary base pairing may play a role in the replication of DNA. As Watson and Crick pointed out in their original paper, published in *Nature* in 1953, "It has not escaped our notice that the specific pairing we have postulated immediately suggests a possible copying mechanism for the genetic material."

Animation
DNA Structure

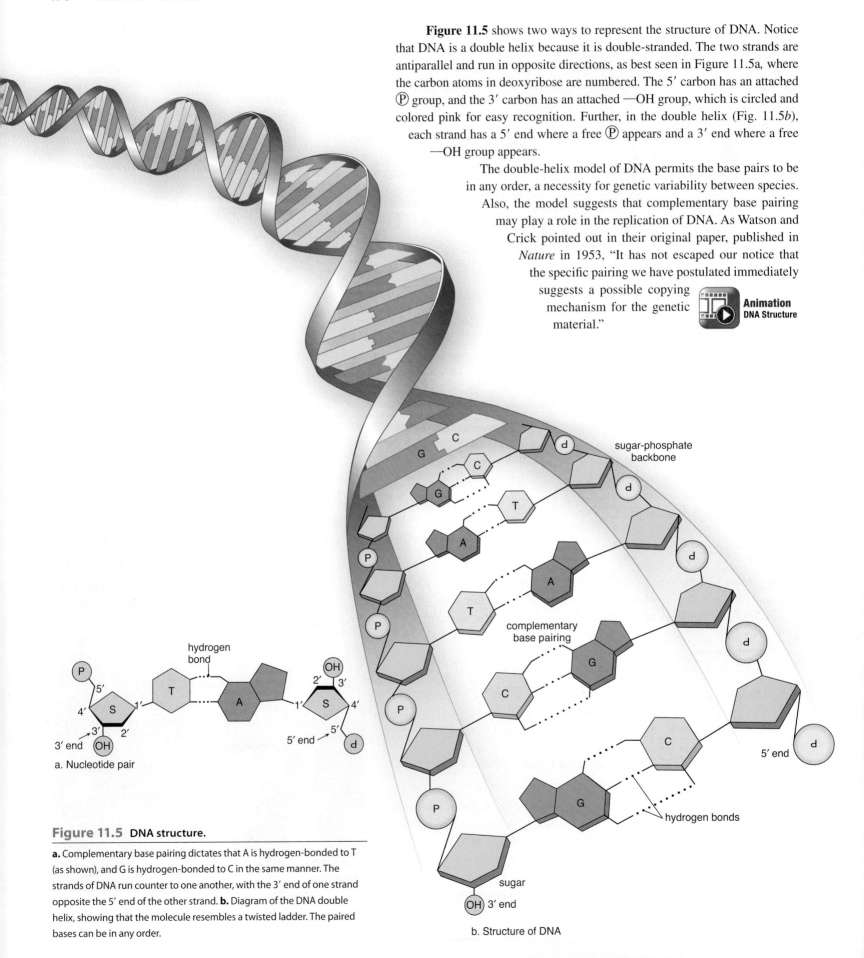

Figure 11.5 **DNA structure.**

a. Complementary base pairing dictates that A is hydrogen-bonded to T (as shown), and G is hydrogen-bonded to C in the same manner. The strands of DNA run counter to one another, with the 3′ end of one strand opposite the 5′ end of the other strand. **b.** Diagram of the DNA double helix, showing that the molecule resembles a twisted ladder. The paired bases can be in any order.

Replication of DNA

Cells need to make identical copies of themselves for the growth and repair of tissues. Before division occurs, each new cell requires an exact copy of the parent cell's DNA, or it will not be able to function as the original cell did. **DNA replication** refers to the process of making an identical copy of a DNA molecule.

During DNA replication, the two DNA strands, which are held together by hydrogen bonds, are separated and each old strand of the parent molecule serves as a **template** for a new strand in a daughter molecule (**Fig. 11.6**). This process is referred to as **semiconservative,** since one of the two old strands is conserved, or present, in each daughter molecule.

To begin replication, the DNA double helix must separate and unwind. This is accomplished by breaking the hydrogen bonds between the nucleotides, then unwinding the helix structure using an enzyme called **helicase**. At this point, new nucleotides are added to the parental template strand. Nucleotides, ever present in the nucleus, will complementary base pair onto the now single-stranded parental strand. The addition of the new strand is completed using an enzyme complex called **DNA polymerase**. The daughter strand is synthesized by DNA polymerase in a 5′–3′ direction as shown in Figure 11.6. Any breaks in the deoxyribose-phosphate backbone are sealed by the enzyme **DNA ligase**.

Animation
DNA Replication

In Figure 11.6, the backbones of the parent molecule (original double strand) are blue. Following replication, the daughter molecules each have a green backbone (new strand) and a blue backbone (old strand). A daughter DNA double helix has the same sequence of base pairs as the parent DNA double helix had. Complementary base pairing has allowed this sequence to be maintained.

In eukaryotes, DNA replication begins at numerous sites, called origins of replication, along the length of the chromosome. At each origin of replication, replication forks form, allowing replication to proceed in both directions. Around each replication fork, a "replication bubble" forms. Within the replication bubble, the process of DNA replication occurs. The replication bubbles spread in both directions until they meet (**Fig. 11.7**). Although eukaryotes replicate their DNA at a fairly slow rate—500 to 5,000 base pairs per minute—there are many individual origins of replication throughout the DNA molecule. Therefore, eukaryotic cells complete the replication of the diploid amount of DNA (in humans, over 3 billion base pairs) in a matter of hours!

Animation
Bidirectional DNA Replication

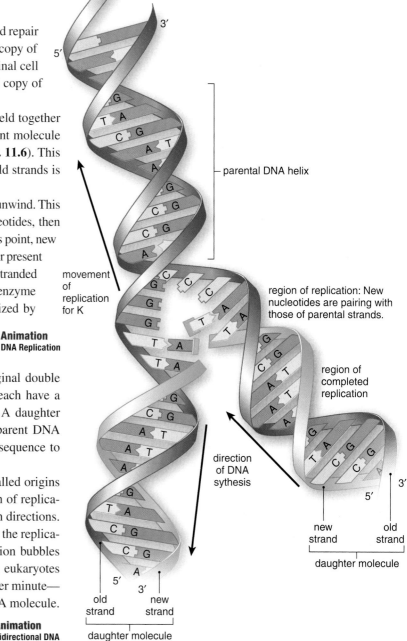

Figure 11.6 Semiconservative replication.

After the DNA molecule unwinds, each old strand serves as a template for the formation of a new strand. After replication is complete, there are two daughter DNA molecules, identical to each other and to the original double helix.

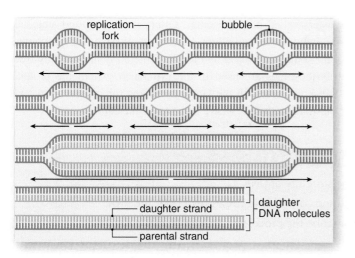

Figure 11.7 Eukaryotic replication.

In eukaryotes, replication occurs at numerous replication forks. The replication bubbles that are created spread out until they meet.

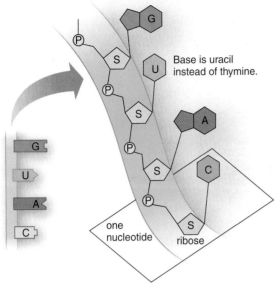

Figure 11.8 Structure of RNA.

Like DNA, RNA is a polymer of nucleotides. In an RNA nucleotide, the sugar ribose is attached to a phosphate molecule and to a base, either G, U, A, or C. Notice that, in RNA, the base uracil (U) replaces thymine as one of the bases. RNA is usually single-stranded, whereas DNA is double-stranded.

Table 11.1 Comparison of DNA and RNA

DNA-RNA SIMILARITIES	
Both are nucleic acids.	
Both are composed of nucleotides.	
Both have a sugar-phosphate backbone.	
Both have four different types of bases.	

DNA-RNA DIFFERENCES	
DNA	**RNA**
Found in nucleus	Found in nucleus and cytoplasm
Genetic material	Helper to DNA
Sugar is deoxyribose	Sugar is ribose
Bases are A, T, C, G	Bases are A, U, C, G
Double-stranded	Single-stranded
Is transcribed (to give mRNA, tRNA, and rRNA)	mRNA is translated (to give proteins)

Connecting the Concepts

For more information on DNA and RNA, refer to the following discussions.

Section 3.2 describes the structure of DNA and RNA molecules.

Section 8.1 explains how DNA is packaged into chromosomes.

RNA Structure and Function

RNA (ribonucleic acid) is made up of nucleotides containing the sugar ribose, thus accounting for its name. The four nucleotides that make up an RNA molecule have the following bases: adenine (A), **uracil (U)**, cytosine (C), and guanine (G). Notice that, in RNA, uracil replaces the thymine in DNA (**Fig. 11.8**).

RNA, unlike DNA, is single-stranded, but the single RNA strand sometimes doubles back on itself, allowing complementary base pairing to occur. The similarities and differences between these two nucleic acid molecules are listed in **Table 11.1.**

In general, RNA is a helper to DNA, allowing protein synthesis to occur according to the genetic information that DNA provides. There are three major types of RNA, each with a specific function in protein synthesis.

Messenger RNA

Messenger RNA (mRNA) is produced in the nucleus of eukaryotes, and in the nucleoid of prokaryotes. DNA serves as a template for the formation of mRNA during a process called transcription. Which DNA genes are transcribed into mRNA is highly regulated in each type of cell and accounts for the specific functions of all cell types. Once formed, mRNA carries genetic information from DNA in the nucleus to the ribosomes in the cytoplasm, where protein synthesis occurs through a process called translation.

Transfer RNA

Transfer RNA (tRNA) is also produced in the nucleus of eukaryotes, and a portion of DNA serves as a template for its production. Appropriate to its name, tRNA transfers amino acids present in the cytoplasm to the ribosomes, where the amino acids are joined to form a protein: a process called translation. Twenty different amino acids make up proteins. Each amino acid has its own tRNA molecule, used to transfer the amino acid to the ribosome complex to form a protein.

Ribosomal RNA

In eukaryotic cells, **ribosomal RNA (rRNA)** is produced in the nucleolus of a nucleus, where a portion of DNA serves as a template for its formation. Ribosomal RNA joins with proteins made in the cytoplasm to form the subunits of **ribosomes,** one large and one small. Each subunit has its own mix of proteins and rRNA. The subunits leave the nucleus and come together in the cytoplasm when protein synthesis is about to begin.

Proteins are synthesized at the ribosomes, which look like small granules in low-power electron micrographs. Ribosomes in the cytoplasm may be free floating or in clusters called polyribosomes. Often, they are found attached to the edge of the endoplasmic reticulum. Proteins synthesized by ribosomes attached to the ER normally are used by the ER. Proteins synthesized by free ribosomes or polyribosomes are used in the cytoplasm, where the protein is carried in a transport vesicle to the Golgi apparatus for modification and transport to the plasma membrane, where it can leave the cell.

Check Your Progress 11.1

1. Detail the structure of DNA.
2. Compare and contrast the structure of DNA and RNA.
3. Briefly describe the process of DNA replication.

11.2 Gene Expression

In the early 1900s, the English physician Sir Archibald Garrod observed that family members often have the same metabolic disorder, and he said most likely they all lacked the same functioning enzyme in a metabolic pathway. He introduced the phrase "inborn error of metabolism" to describe this relationship. Garrod's findings were generally overlooked until George Beadle and Edward Tatum devised a way in 1940 to confirm his hypothesis. These investigators performed a series of experiments utilizing red bread mold and found that each of their mutant molds was indeed unable to produce a particular enzyme. This led them to propose that one gene directs the synthesis of one enzyme. Today we know that genes are also responsible for specifying any type of protein in a cell, not just enzymes, so their finding has been modified to "one gene–one polypeptide."

From DNA to RNA to Protein

It's one thing to know that genes specify proteins and another to explain how they do it. Molecular genetics, which largely began when Watson and Crick discovered the structure of DNA in the 1950s, is able to explain exactly how genes control the building of a specific type of protein. Consider that, in eukaryotes, DNA resides in the nucleus but RNA is found both in the nucleus and in the cytoplasm where protein synthesis occurs. This means that DNA must pass its genetic information to mRNA, which then actively participates in protein synthesis. The *central dogma of molecular biology* states that genetic information flows from DNA to RNA to protein (**Fig. 11.9**).

Gene expression has occurred when a gene's product—the protein it specifies—is functioning in a cell. Specifically, gene expression requires two processes, called transcription and translation. In eukaryotes, transcription takes place in the nucleus, and translation takes place in the cytoplasm. During **transcription,** a portion of DNA serves as a template for mRNA formation. During **translation,** the sequence of mRNA bases (which are complementary to those in template DNA) determines the sequence of amino acids in a polypeptide. So, in effect, molecular genetics tells us that genetic information lies in the sequence of the bases in DNA, which through mRNA determines the sequence of amino acids in a protein. tRNA assists mRNA during protein synthesis by bringing amino acids to the ribosomes.

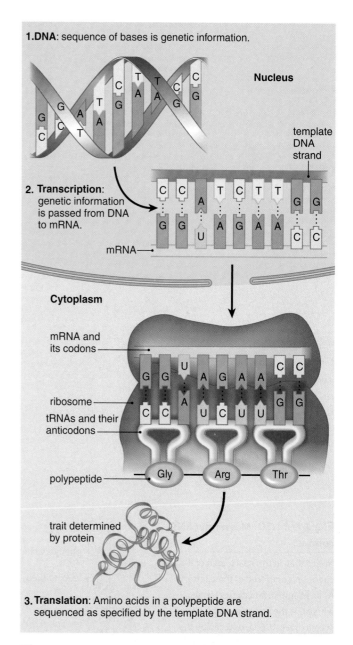

Figure 11.9   Flow of genetic information.

1. Genetic information (genes) consists of a particular sequence of bases. **2.** The process of transcription in the nucleus passes this genetic information to an mRNA molecule. The mRNA moves to a ribosome in the cytoplasm. **3.** During the process of translation at a ribosome in the cytoplasm, the genetic information results in a particular sequence of amino acids in a protein. The activity or inactivity of this protein contributes to a trait and often whether the phenotype is normal or not normal.

Proteins differ from one another by the sequence of their amino acids, and proteins determine the structure and function of cells and indeed the phenotype of the organism.

MP3
Protein Synthesis

The Genetic Code

The information contained in DNA and RNA is written in a chemical language different from that in the protein specified by the DNA and RNA. The cell needs a way to translate one language into the other, and it uses the genetic code.

But how can the four bases of RNA (A, C, G, U) provide enough combinations to code for the 20 amino acids found in proteins? If the code were a singlet code (one base stands for an amino acid), only four amino acids could be encoded. If the code were a doublet (any two bases stand for one amino acid), it still would not be possible to code for 20 amino acids. But if the code were a **triplet code,** the four bases could supply 64 different triplets, far more than needed to code for 20 different amino acids. Each three-letter (nucleotide) unit of an mRNA molecule is called a **codon,** which codes for a single amino acid (**Fig. 11.10**). Sixty-one triplets correspond to a particular amino acid; the remaining three are stop codons that signal the end of a polypeptide. The codon that stands for the amino acid methionine is also used as a start codon that signals the initiation of translation. Most amino acids have more than one codon, which offers some protection against possibly harmful mutations that could otherwise change the sequence of the bases.

Cracking the genetic code was no simple matter. Researchers performed a series of experiments in which they added artificial mRNA to a medium containing bacterial ribosomes and a mixture of amino acids. By comparing the bases in the mRNA with the resulting polypeptide, they were able to learn the code. For example, an mRNA with a sequence of repeating guanines (GGG′GGG′…) would encode a string of glycine amino acids, so they concluded that the mRNA codon GGG specifies the amino acid glycine in a protein.

The genetic code is almost universal in all living things. This suggests an evolutionary aspect to the genetic code: that the code dates back to the very first organisms on Earth and that all living things are related.

Transcription

During transcription of DNA, a strand of RNA forms that is complementary to a portion of DNA. While all three classes of RNA are formed by transcription, we will focus on transcription to create mRNA.

Figure 11.10 Messenger RNA codons.

Notice that, in this chart, each of the codons is composed of three letters. As an example, find the rectangle where C is the first base and A is the second base. U, C, A, or G can be the third base. CAU and CAC are codons for histidine; CAA and CAG are codons for glutamine.

Second base

First base		U	C	A	G	Third base
U		UUU ⎤ phenylalanine (Phe) UUC ⎦ UUA ⎤ leucine (Leu) UUG ⎦	UCU ⎤ UCC ⎥ serine (Ser) UCA ⎥ UCG ⎦	UAU ⎤ tyrosine (Tyr) UAC ⎦ **UAA** stop **UAG** stop	UGU ⎤ cysteine (Cys) UGC ⎦ **UGA** stop UGG tryptophan (Trp)	U C A G
C		CUU ⎤ CUC ⎥ leucine (Leu) CUA ⎥ CUG ⎦	CCU ⎤ CCC ⎥ proline (Pro) CCA ⎥ CCG ⎦	CAU ⎤ histidine (His) CAC ⎦ CAA ⎤ glutamine (Gln) CAG ⎦	CGU ⎤ CGC ⎥ arginine (Arg) CGA ⎥ CGG ⎦	U C A G
A		AUU ⎤ AUC ⎥ isoleucine (Ile) AUA ⎦ **AUG** methionine (Met) (**start**)	ACU ⎤ ACC ⎥ threonine (Thr) ACA ⎥ ACG ⎦	AAU ⎤ asparagine (Asn) AAC ⎦ AAA ⎤ lysine (Lys) AAG ⎦	AGU ⎤ serine (Ser) AGC ⎦ AGA ⎤ arginine (Arg) AGG ⎦	U C A G
G		GUU ⎤ GUC ⎥ valine (Val) GUA ⎥ GUG ⎦	GCU ⎤ GCC ⎥ alanine (Ala) GCA ⎥ GCG ⎦	GAU ⎤ aspartic acid (Asp) GAC ⎦ GAA ⎤ glutamic acid (Glu) GAG ⎦	GGU ⎤ GGC ⎥ glycine (Gly) GGA ⎥ GGG ⎦	U C A G

mRNA Is Formed Transcription begins when the enzyme **RNA polymerase** binds tightly to a **promoter,** a region of DNA with a special nucleotide sequence that marks the beginning of a gene. RNA polymerase opens up the DNA helix just in front of it, so that complementary base pairing can occur. Then the enzyme adds new RNA nucleotides that are complementary to the template DNA strand, and an mRNA molecule results (**Fig. 11.11**). The resulting **mRNA transcript** is a complementary copy of the sequence of bases in the template DNA strand. Once transcription is completed, the mRNA is ready to be processed before it leaves the nucleus for the cytoplasm.

mRNA Is Processed The newly synthesized *primary-mRNA* must be processed in order for it to be used properly. Processing occurs in the nucleus of eukaryotic cells. Three steps are required: capping, the addition of a poly-A tail, and splicing (**Fig. 11.12**). After processing, the mRNA is called a *mature mRNA* molecule.

The first nucleotide of the primary-mRNA is modified by the addition of a cap that is composed of an altered guanine nucleotide. On the 3′ end, enzymes add a poly-A tail, a series of adenosine nucleotides. These modifications provide stability to the mRNA; only those that have a cap and tail remain active in the cell.

Most genes in humans are interrupted by segments of DNA that do not code for protein. These portions are called *introns* because they are intervening segments. The other portions of the gene, called *exons,* contain the protein-coding portion of the gene. In *primary-mRNA splicing,* the introns are removed and the exons joined together. The result is a mature mRNA molecule consisting of continuous exons. A surprising finding of late has been that introns may play a regulatory function in gene expression.

Ordinarily, processing brings together all the exons of a gene. In some instances, however, cells use only certain exons rather than all of them to form the mature RNA transcript. The result is a different protein product in each cell. In other words, this so-called *alternative mRNA splicing* can increase the number of protein products that can be made from a single gene.

After the mRNA strand is processed, it passes from the cell nucleus into the cytoplasm for translation.

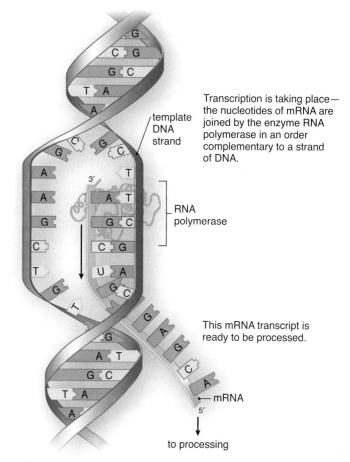

Transcription is taking place—the nucleotides of mRNA are joined by the enzyme RNA polymerase in an order complementary to a strand of DNA.

template DNA strand

RNA polymerase

This mRNA transcript is ready to be processed.

mRNA

to processing

Figure 11.11 Transcription to form mRNA.

During transcription, complementary RNA is made from a DNA template. A portion of DNA unwinds and unzips at the point of attachment of RNA polymerase. A strand of RNA, such as mRNA, is produced when complementary bases join in the order dictated by the sequence of bases in the template DNA strand.

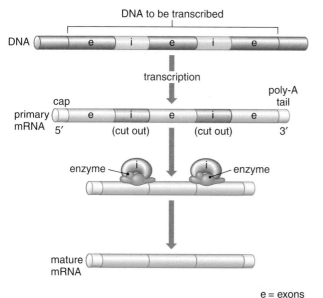

DNA to be transcribed

DNA

transcription

cap
primary mRNA
5′ (cut out) (cut out) 3′
poly-A tail

enzyme — enzyme

mature mRNA

e = exons
i = introns

Figure 11.12 mRNA processing.

Primary mRNA results when both exons and introns are transcribed from DNA. A "cap" and a poly-A "tail" are attached to the ends of the primary RNA transcript, and the introns are removed, so that only the exons remain. This mature mRNA molecule moves into the cytoplasm of the cell, where translation occurs.

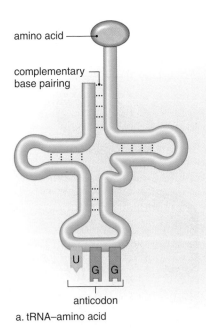

amino acid

complementary base pairing

U G G

anticodon

a. tRNA–amino acid

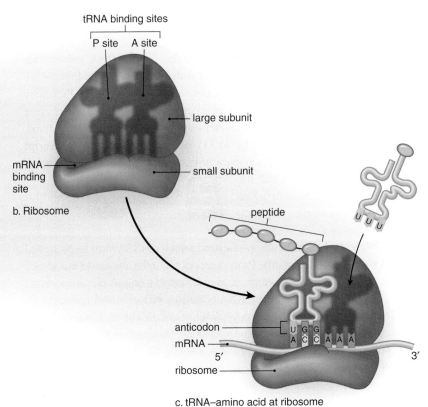

tRNA binding sites

P site A site

large subunit

mRNA binding site

small subunit

b. Ribosome

peptide

anticodon

mRNA

5′   3′

ribosome

U G G
A C C A A A

c. tRNA–amino acid at ribosome

Figure 11.13 tRNA structure and function.

a. A tRNA, which is single-stranded but folded, has an amino acid attached to one end and an anticodon at the other end. The anticodon is complementary to a codon. **b.** A ribosome has two binding sites for tRNA, called the P site and the A site. A polypeptide attached to tRNA at the P site will be passed to an amino acid attached to a tRNA as soon as it arrives at the A site. **c.** The pairing between codon and anticodon at a ribosome ensures that the sequence of amino acids in a polypeptide is the same sequence directed originally by DNA.

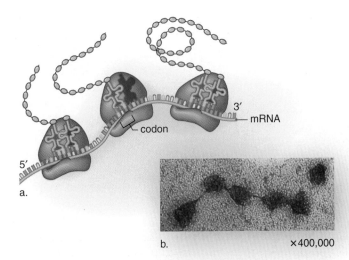

3′
mRNA

codon

5′

a.

b.   ×400,000

Figure 11.14 Polyribosome structure and function.

a. Several ribosomes, collectively called a polyribosome, move along the same mRNA. Therefore, several proteins can be made at the same time. **b.** Electron micrograph of a polyribosome.

Translation: An Overview

Translation is the second step by which gene expression leads to protein (poly-peptide) synthesis. Translation requires several enzymes, mRNA, and the other two types of RNA: transfer RNA and ribosomal RNA.

Transfer RNA Brings Amino Acids to the Ribosomes   Each tRNA is a single-stranded nucleic acid that doubles back on itself such that complementary base pairing results in the cloverleaf-like shape shown in **Figure 11.13**. There is at least one tRNA molecule for each of the 20 amino acids found in proteins. The amino acid binds to one end of the molecule. The opposite end of the molecule contains an **anticodon,** a group of three bases that is complementary to a specific codon of mRNA.

During translation, the order of codons in mRNA determines the order in which tRNAs bind at the ribosomes. When a tRNA–amino acid complex comes to the ribosome, its anticodon pairs with an mRNA codon. For example, if the codon is ACC, what is the anticodon, and what amino acid will be attached to the tRNA molecule? From Figure 11.10, we can determine this:

Codon (mRNA)	Anticodon (tRNA)	Amino Acid (protein)
ACC	UGG	Threonine

After translation is complete, a protein contains the sequence of amino acids originally specified by DNA. This is the genetic information that DNA stores and passes on to each cell during the cell cycle, then to the next genera-tion of individuals. DNA's sequence of bases determines the proteins in a cell, and the proteins determine the function of each cell.

Ribosomal RNA Is in Ribosomes   Ribosomes are the small structural bodies where translation occurs. Ribosomes are composed of many proteins and several ribosomal RNAs (rRNAs). In eukaryotic cells, rRNA is produced in a nucleolus within the nucleus. There, it joins with proteins manufactured in the cytoplasm to form two ribosomal subunits, one large and one small. The subunits leave the nucleus and join together in the cytoplasm to form a ribosome just as protein synthesis begins.

A ribosome has a binding site for mRNA as well as binding sites for two tRNA molecules at a time (Fig. 11.13). These binding sites facilitate complementary base pairing between tRNA anticodons and mRNA codons. The P binding site is for a tRNA attached to a *p*eptide, and the A binding site is for a newly arrived tRNA attached to an *a*mino acid.

As soon as the initial portion of mRNA has been translated by one ribosome and the ribosome has begun to move down the mRNA, another ribosome attaches to the same mRNA. Therefore, several ribosomes are often attached to and translating the same mRNA, allowing the cell to produce many copies of the same protein at a time. The entire complex is called a **polyribosome** (**Fig. 11.14**).

Translation Has Three Phases

Polypeptide synthesis has three phases: initiation, an elongation cycle, and termination. Enzymes are required for each of the steps to occur, and energy is needed for the first two steps.

1. During *initiation*, mRNA binds to the smaller of the two ribosomal subunits; then the larger subunit associates with the smaller one.
2. During an *elongation cycle*, a peptide lengthens one amino acid at a time. The growing peptide is transferred from the outgoing tRNA to the incoming tRNA–amino acid complex, and then the outgoing tRNA leaves. The ribosome then translates the next mRNA codon as it receives a new incoming tRNA–amino acid complex.
3. *Termination* occurs at any one of three special codons that mean "stop." The ribosomal subunits and mRNA dissociate, and the completed polypeptide is released.

Initiation   During initiation, a small ribosomal subunit, the mRNA, an *initiator tRNA* bound to the amino acid methionine, and a large ribosomal subunit all come together (**Fig. 11.15**):

- The small ribosomal subunit attaches to the mRNA in the vicinity of the start codon (AUG).
- The anticodon of the initiator tRNA–methionine complex pairs with this codon.
- The large ribosomal subunit joins to the small subunit.

Elongation Cycle   As discussed, a ribosome has two binding sites for tRNA where the tRNA's anticodon binds to a codon of mRNA. During the elongation cycle (**Fig. 11.16**),

- tRNA at the P site contains the growing peptide chain. (See **1**.)
- This tRNA passes its peptide to tRNA–amino acid at the A site. The tRNA at the P site leaves. (See **2** and **3**.)
- During **translocation,** the tRNA-peptide moves to the P site, and the codon at the A site is ready for the next tRNA–amino acid. (See **4**.)

The complete cycle—complementary base pairing of new tRNA, transfer of the growing peptide chain, and translocation—is repeated at a rapid rate. The outgoing tRNA is recycled and can pick up another amino acid in the cytoplasm to take to the ribosome.

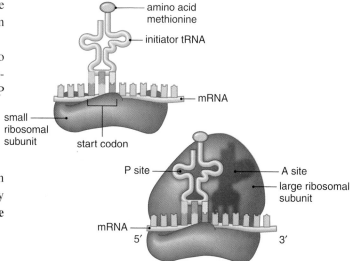

Figure 11.15 Initiation.

A small ribosomal subunit binds to mRNA; an initiator tRNA's anticodon binds to its codon, and the large ribosomal subunit completes the ribosome.

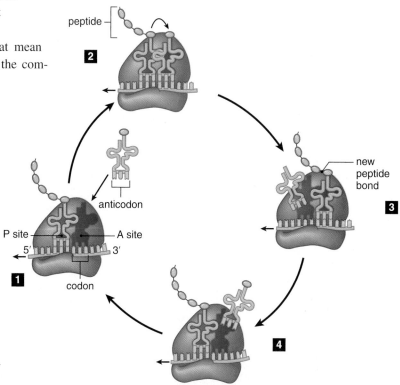

Figure 11.16 Elongation cycle.

1 Two tRNAs can be at a ribosome at one time. **2** The tRNA at the P site passes its peptide to the tRNA at the A site. **3** The tRNA at the P site leaves. **4** The ribosome moves forward (translocation), and the tRNA-peptide complex is now at the P site. A new tRNA–amino acid complex comes to the A site.

Connections and Misconceptions

How long does it take to copy the DNA in one human cell?

The enzyme DNA polymerase in humans can copy approximately 50 bases per second. If only one DNA polymerase were used to copy human DNA, it would take almost three weeks. However, multiple DNA polymerases copy the human genome by starting at many different places. All 3 billion base pairs can be copied in eight hours in a rapidly dividing cell.

Termination   When a stop codon (see Fig. 11.10) appears at the A site of a ribosome, termination occurs. A protein called a **release factor** binds to the stop codon and cleaves the polypeptide from the last tRNA. Then, the polypeptide and the assembled components that carried out protein synthesis are separated from one another. The mRNA, ribosomes, and tRNA molecules can then be used for another round of translation.

Review of Gene Expression

Genes are segments of DNA that code for proteins. A gene is expressed when its protein product has been made. **Figure 11.17** reviews transcription and mRNA processing in the nucleus, as well as translation during protein synthesis in the cytoplasm of a eukaryotic cell.

Some ribosomes remain free in the cytoplasm, and others become attached to rough ER. In the latter case, the polypeptide enters the lumen of the ER by way of a channel, where it can be further processed by the addition of sugars. Transport vesicles carry the protein to other locations in the cell, including the Golgi apparatus, which may modify it further and package it in a vesicle for transport out of the cell or cause it to become embedded in the plasma membrane. Proteins have innumerable functions in cells, from enzymatic to structural. Proteins also have functions outside the cell. Together, they account for the structure and function of cells, tissues, organs, and the organism.

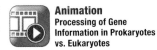
Animation
Processing of Gene Information in Prokaryotes vs. Eukaryotes

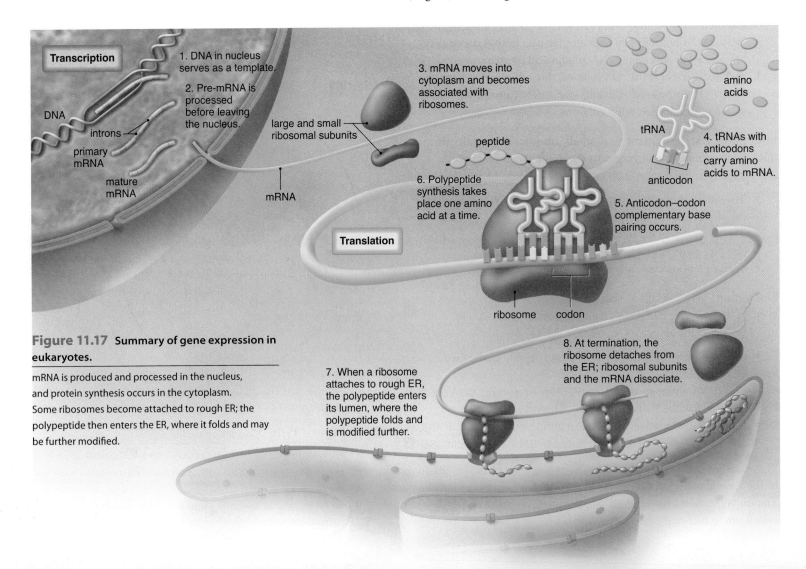

Figure 11.17 **Summary of gene expression in eukaryotes.**

mRNA is produced and processed in the nucleus, and protein synthesis occurs in the cytoplasm. Some ribosomes become attached to rough ER; the polypeptide then enters the ER, where it folds and may be further modified.

Genes and Gene Mutations

A gene is a sequence of DNA bases that codes for a product, most often a protein. A **gene mutation** is a change in the sequence of those bases. A mutation can increase the diversity of organisms by creating an entirely new product with a positive function to the organism; however, the results of a mutation can be negative.

Causes of Gene Mutations

A gene mutation can be caused by an error in replication, a transposon, or an environmental mutagen. Mutations due to DNA replication errors are rare: a frequency of 1 in 100 million per cell division on average in most eukaryotes. This average is low due to DNA polymerase, the enzyme that carries out replication, proofreading the new strand against the old strand, detecting and correcting any mismatched pairs. **Transposons (Fig. 11.18)** are specific DNA sequences that have the remarkable ability to move within and between chromosomes. Their movement often disrupts genes, rendering them nonfunctional. These so-called jumping genes have now been discovered in bacteria, plants, fruit flies, and humans, and it is likely all organisms have such elements.

Mutagens are environmental influences that cause mutations. Different forms of radiation, such as radioactivity, X-rays, and ultraviolet (UV) light, and chemical mutagens, such as pesticides and compounds in cigarette smoke, may cause breaks or chemical changes in DNA. If mutagens bring about a mutation in the gametes, the offspring of the individual may be affected. If the mutation occurs in the body cells, cancer may result. The overall rate of mutation is low, however, because *DNA repair enzymes* constantly monitor and repair any irregularities.

Animation
Transposons:
Shifting Segments
of the Genome

Types and Effects of Mutations

The effects of mutations vary greatly. The severity of a mutation usually depends on whether it affects one or more codons or a gene. In general, we know a mutation has occurred when the organism has a malfunctioning protein that leads to a genetic disorder or to the development of cancer. But many mutations go undetected because they have no observable effect or have no detectable effect on the protein's function. These are called silent mutations.

Point mutations involve a change in a single DNA nucleotide, and the severity of the results depends on the particular base change that occurs. A single base change can result in a change in the amino acid at that location of the gene. For example, if valine (coded for by a GAA codon) instead of glutamate (coded for by a GUA codon) occurs due to a point mutation in the β chain of hemoglobin, sickle cell disease results. The abnormal hemoglobin stacks up inside cells, causing them to become sickle-shaped and preventing adequate circulation through the body.

Animation
Addition and
Deletion
Mutations

A **frameshift mutation** is caused by extra or missing nucleotides in a DNA sequence. They are usually much more severe than point mutations because codons are read from a specific starting point. Therefore, all downstream codons are affected. For instance, if the letter C is deleted from the sentence THE CAT ATE THE RAT, the "reading frame" is shifted. The sentence now reads THE ATA TET HER AT—something that doesn't make sense. Likewise, a frameshift mutation in a gene often renders the protein nonfunctional because it no longer makes sense. The movement of a transposon can cause a frameshift mutation.

a.

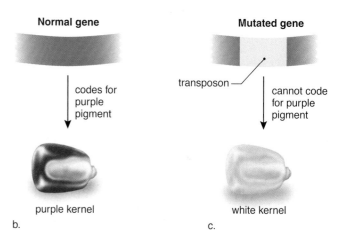

Normal gene	Mutated gene
codes for purple pigment	transposon — cannot code for purple pigment
purple kernel	white kernel

b.　　　　　　　　　　c.

Figure 11.18 Transposon.

a. Barbara McClintock, shown here instructing a student, won the 1983 Nobel Prize in Physiology or Medicine for being the first to discover transposons. She worked with maize (corn). **b.** In corn, a purple-coding gene ordinarily codes for a purple pigment. **c.** A transposon "jumps" into the purple-coding gene. This mutated gene is unable to code for purple pigment, and a white kernel results.

Connecting the Concepts

For more on gene expression, refer to the following discussions.

Section 12.1 explores how gene expression can be controlled.

Section 13.3 examines the developing technology of gene therapy as a mechanism for controlling gene expression.

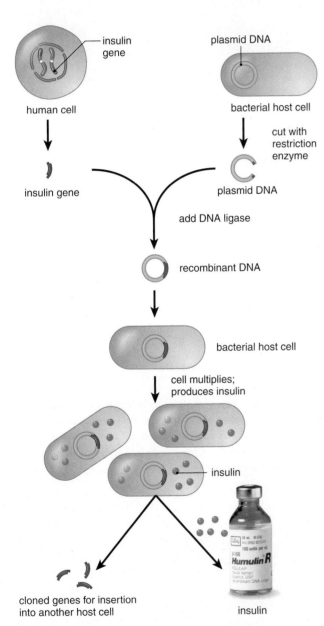

Figure 11.19 Recombinant DNA technology.

The production of insulin is one example of how recombinant DNA technology can benefit humans. Human DNA and plasmid DNA are spliced together. Gene cloning is achieved when a host cell takes up the recombinant plasmid; as the cell reproduces, the plasmid is replicated. Multiple copies of the gene are now available. If the insulin gene functions normally as expected, insulin may also be retrieved.

Check Your Progress 11.2

1 Briefly describe the process of transcription.

2 Detail how all three types of RNA function in translation.

3 Describe a few types of mutations, detail how they occur, and predict what effects they might have on DNA.

11.3 DNA Technology

Learning Outcomes

Upon completion of this section, you should be able to

1. Describe the steps involved in making a recombinant DNA molecule.
2. Explain how a transgenic organism is made and give some applications of transgenic organisms.
3. Describe the applications of polymerase chain reaction (PCR) and DNA fingerprinting.

An understanding of how DNA functions in cells allowed scientists to manipulate the genes of organisms. Genes can be cloned (identical copies made) and then used for various purposes. During **genetic engineering,** a cloned gene can be inserted into the genome of an organism, which is then called a **transgenic organism.** Because the genetic code is almost universal, it's possible to transfer cloned genes from virtually any organism into bacteria, plants, and animals.

Recombinant DNA Technology

Recombinant DNA (rDNA) contains DNA from two or more different sources (**Fig. 11.19**). To make rDNA, a researcher needs a **vector,** a piece of DNA that acts as a carrier for the foreign DNA. One common vector is a *plasmid,* which is a small accessory ring of DNA found in bacteria.

Two enzymes are needed to introduce foreign DNA into plasmid DNA: (1) **restriction enzymes** that can cleave, or cut, DNA at specific places (for example, the restriction enzyme *EcoRI* always cuts DNA at the base sequence GAATTC) and (2) DNA ligase, which can seal the foreign DNA into an opening in a cut plasmid.

Animation Restriction Endonucleases

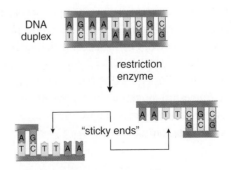

If a plasmid is cut with *EcoRI,* a gap now exists into which a piece of foreign DNA can be placed if it ends in bases complementary to those exposed by the restriction enzyme. To ensure this, it is necessary to cleave the foreign DNA with the same restriction enzyme. The overhanging bases at the ends of the two DNA molecules are called "sticky ends," because they can bind a piece of foreign DNA by complementary base pairing. Sticky ends facilitate the insertion of foreign DNA into vector DNA, very similar to the way puzzle pieces fit together.

DNA ligase, the enzyme that functions in DNA replication to seal breaks in a double-stranded helix, seals the foreign piece of DNA into the plasmid. Molecular biologists often give the rDNA to bacterial cells, which readily take up recombinant plasmids when the cells are treated to make them more permeable. Thereafter, as the bacteria replicate the plasmid, the gene is cloned. Cloned genes have many uses. A scientist may allow the **genetically modified** bacterial cells to express the cloned gene and retrieve the protein. Or copies of the cloned gene may be removed from the bacterial cells and then introduced into another organism, such as a corn plant, to produce a transgenic organism.

Transgenic Organisms

The term **biotechnology** refers to the use of natural biological systems to create a product or achieve some other end desired by human beings. Today, bacteria, plants, and animals can be genetically engineered to make biotechnology products. Various uses of transgenic organisms are shown in **Figure 11.20.**

Transgenic Bacteria

Recombinant DNA technology is used to produce transgenic bacteria, which are grown in huge vats called bioreactors. The gene product is usually collected from the growth medium. Products now on the market that are produced by bacteria include insulin, human growth hormone, t-PA (tissue plasminogen activator, used to dissolve blood clots), and hepatitis B vaccine. Bacteria can be selected for their ability to degrade a particular substance, such as oil, and this ability can then be enhanced by genetic engineering. Similarly, some mining companies are testing genetically engineered organisms that have improved bioleaching capabilities. Bacteria can also be engineered to produce organic chemicals used in industry.

Transgenic Plants and Animals

Foreign genes have been transferred to cotton, corn, and potato strains to make these plants resistant to pests by causing their cells to produce an insect toxin. Similarly, soybeans have been made resistant to a common herbicide. Some corn and cotton plants are both pest- and herbicide-resistant. These and other genetically engineered crops have increased yields. Genetically engineered plants are revolutionizing both agriculture and medicine—there are now plants that can produce human proteins, such as hormones, clotting factors, and antibodies!

Video
Pesticide Plants

Figure 11.20 **Uses of transgenic organisms.**

a. Transgenic bacteria, grown industrially, are used to produce medicines. Transgenic animals and plants increase the food supply. **b.** These salmon (right) are much heavier and reproduce much sooner because of the growth hormone genes they received as embryos. **c.** Pests didn't consume unblemished peas because the plants that produced them received genes for pest inhibitors. **d.** Transgenic bacteria can also be used for environmental cleanup. These bacteria are able to break down oil. **e.** Transgenic corn can resist tough herbicides used to control weeds, greatly increasing crop yields.

Foreign genes have also been inserted into the eggs of animals, often to give them the gene for bovine growth hormone (bGH). The procedure has resulted in larger fishes, cows, pigs, rabbits, and sheep. Gene "pharming," the use of transgenic farm animals to produce pharmaceuticals, is being pursued by a number of firms. Genes that code for therapeutic and diagnostic proteins are incorporated into an animal's DNA, and the proteins appear in the animal's milk.

To achieve a sufficient number of animals that can produce the product, transgenic animals are sometimes cloned. For many years, it was believed that adult vertebrate animals could not be cloned. Although each cell contains a copy of all the genes, certain genes are turned off in mature, specialized cells, and cloning an adult vertebrate would require that all the genes of an adult cell be turned on again, which had long been thought impossible. In 1997, however, Scottish scientists produced a cloned sheep, which they named Dolly. Since then, calves, goats, and many other animals, including some endangered species, have also been cloned. After enucleated eggs (egg cells that have the nucleus removed) from a donor are microinjected with 2n nuclei from a transgenic animal, they are coaxed to begin development in vitro. Development continues in host females until the clones are born. The offspring are clones because all have the identical genotype of the adult that donated the 2n nuclei. Now that scientists have a way to clone animals, this procedure will undoubtedly be used routinely to procure biotechnology products.

Polymerase Chain Reaction

The **polymerase chain reaction (PCR)** can create millions of copies of a segment of DNA very quickly in a test tube without the use of a vector or a host cell. PCR is very sensitive—it *amplifies* (makes copies of) a targeted DNA sequence that can be less than one part in a million of the total DNA sample!

PCR requires the following: DNA polymerase, the enzyme that carries out DNA replication; a set of *primers* that flank both sides of the segment of DNA to be amplified; and a supply of nucleotides for the new DNA strands. PCR is automated, thanks to the discovery of a heat-tolerant DNA polymerase from the bacterium *Thermus aquaticus,* which lives in hot springs. The process is shown in **Figure 11.21.** During a PCR cycle, the mixture of DNA and primers is heated to 95°C to separate the two strands of the double helix. Next, the temperature is lowered, so that the primers can bind to the single strands of DNA and DNA polymerase can copy the DNA. Then, the procedure is repeated 20 to 30 times. The amount of DNA doubles with each replication cycle; after one cycle there are two copies of the targeted DNA sequence, after two cycles there are four copies, and so forth. Twenty to 30 PCR cycles can yield billions of copies of the specified DNA segment.

DNA amplified by PCR has many uses. PCR is often utilized to copy genes for creating recombinant DNA or increasing the amount of sample DNA needed for forensic testing. PCR has also been used to amplify mitochondrial DNA sequences in order to decipher the evolutionary history of human populations. Because PCR is so efficient, it has even been possible to sequence DNA taken from a 76,000-year-old mummified human brain.

Animation
Polymerase
Chain Reaction

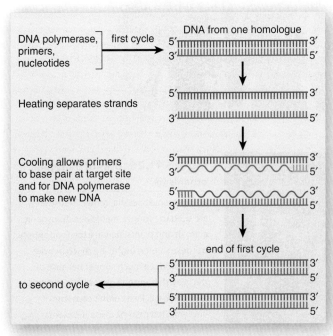

Figure 11.21 **Polymerase chain reaction.**

At the start of PCR, double-stranded DNA is mixed with DNA polymerase, two primers (gold and pink) that will flank the target DNA sequence, and a supply of the four nucleotides in DNA. At the end of the first cycle, there are two copies of the target DNA sequence. After 20–30 cycles, millions of copies of the target DNA sequence are present.

DNA Fingerprinting

DNA fingerprinting makes use of noncoding sections of DNA that consist of two to five bases repeated over and over again, as in ATCATCATCATC. People differ by how many times such a sequence is repeated, but how can this difference be detected? PCR is used to amplify the entire repeated region. The greater the number of repeats, the longer the segment of DNA that will result after PCR amplification is done. People can be heterozygous for the number of repeats, as in **Figure 11.22,** so the DNA from both homologues must be amplified separately. Following PCR, the DNA is subjected to gel electrophoresis, a process that separates DNA molecules according to their size with an electric current through a jelly-like material. The shorter the segment of DNA, the farther it migrates through the gel material in a banding pattern.

DNA fingerprinting has many uses. Medically, it can identify the presence of a viral infection or a mutated gene that could predispose someone to cancer. In forensics, DNA fingerprinting from a single sperm is enough to identify a suspected rapist because the DNA is amplified by PCR. It can also be used to identify the parents of a child or identify the remains of someone who died, such as a victim of the September 11, 2001, terrorist attacks. In the future, we will undoubtedly see more applications of recombinant DNA technology that will greatly enrich our lives and improve our health.

Video World Trade Center   **Animation** DNA Fingerprinting

Connections and Misconceptions

How is a paternity test done, and why are the results always a percentage instead of a certainty?

In a paternity test, a sample from the mother, child, and possible father are taken and amplified by PCR; then the samples are electrophoresed and the banding patterns compared. In a typical paternity test, 13–16 human identity markers (HID) are examined. These markers, or genes, are found in all humans, but in differing lengths of repeats that are inherited from parent to child.

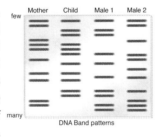

When a certain amount of these markers match in the child and possible father, the result is a percentage of certainty that the man is the father (normally, 99.0% or higher). Only by comparing entire genomes, which contain hundreds of millions of base pairs and approximately 20,000–25,000 genes, can a paternity test result in 100% certainty.

Connecting the Concepts

The DNA technologies discussed in this section are used extensively in the study of biology. For additional examples of DNA technologies, refer to the following discussions.

Section 12.1 examines how DNA cloning can be used in both agriculture and medicine.

Section 14.2 explores how molecular evidence is used to study evolutionary patterns.

Section 16.3 investigates the science of systematics, which uses molecular data to establish evolutionary relationships among species.

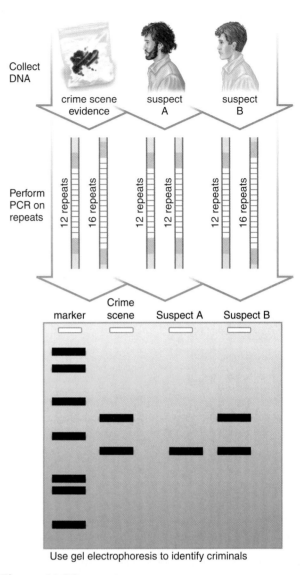

Use gel electrophoresis to identify criminals

Figure 11.22   DNA fingerprinting.

Following PCR, gel electrophoresis reveals the comparative pattern of repeats in the DNA by separating different sizes of DNA segments, resulting in a banding pattern on the gel. Here, DNA fingerprinting is being used to identify a criminal by comparing the DNA found at the crime scene with the DNA of two suspects.

Check Your Progress 11.3

1. State how a restriction enzyme can be used in recombinant DNA technology.
2. Explain why transgenic organisms can be useful.
3. Relate how PCR can be a useful tool in biotechnology.

11.4 Genomics and Proteomics

In the previous century, researchers discovered the structure of DNA, how DNA replicates, and how protein synthesis occurs. Genetics in the twenty-first century concerns **genomics,** the study of genomes—our genes and **intergenic DNA** sequences, as well as those of other organisms. Researchers now know the sequence of all the base pairs along the length of the human chromosomes. The enormity of the task can be appreciated by knowing that, at the very least, our DNA contains 3.4 billion base pairs and approximately 20,500 genes. Many organisms have even larger genomes. Genes make up only 3–5% of the human genome, so genomics may help reveal the function and evolutionary history of many intergenic DNA sequences.

Sequencing the Bases of the Human Genome

We now have a working draft of the base-pair sequence in the DNA of all our chromosomes. This feat was accomplished by the Human Genome Project, a 13-year effort that involved both university and private laboratories around the world. How did they do it? First, investigators developed a laboratory procedure that would allow them to decipher a short sequence of base pairs, and then they devised an instrument that would carry out this procedure automatically. Over the project's 13-year span, DNA sequencers were constantly improved to the point where modern automated sequencers can analyze and sequence up to 2 million base pairs of DNA in a 24-hour period.

Variations in Base Sequence

Scientists studying the human genome found that many small regions of DNA vary among individuals. For instance, there may be a difference in a single base within a gene (**Fig. 11.23**a, b) or within an uncharacterized sequence between genes (Fig. 11.23c).

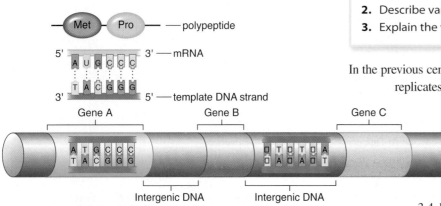

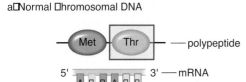

a Normal Chromosomal DNA

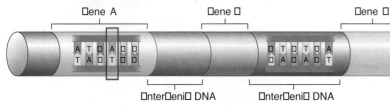

b Variation in the order of the bases within a gene due to a mutation

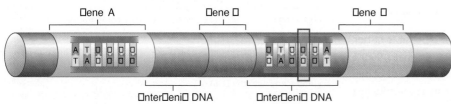

c Variation in the order of the bases within an intergenic sequence due to a mutation

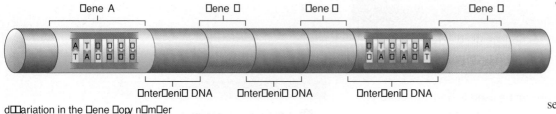

d Variation in the gene copy number

Figure 11.23 **Variations in DNA sequence.**
a. A section of DNA containing three genes (A, B, and C) and two intergenic sequences may vary between individuals. **b.** Variation in the sequence of bases within a gene due to a mutation may alter the sequence of amino acids of the protein specified by the gene. **c.** An individual may possess a variation in the order of bases within an intergenic sequence due to a mutation. This mutation would not change the proteins encoded by gene A, B, or C. **d.** In addition to changes in base sequence, DNA is also subject to changes in gene copy number—that is, the number of copies of a particular gene. Note that now there are two copies of gene B.

One surprising finding is that some individuals even have additional copies of some genes (Fig. 11.23*d*)! Many of these differences have no ill effects, but some of them may increase or decrease an individual's susceptibility to disease.

Genome Comparisons

Researchers are comparing the human genome with the genomes of other species for clues to our evolutionary origins. One surprising discovery is that the genomes of all vertebrates are similar. Researchers were not surprised to find that the genes of chimpanzees and humans are 98% alike, but they did not expect to find that our sequence is also 85% like that of a mouse. Scientists also discovered that we share a number of genes with much simpler organisms, including bacteria! As the genomes of more organisms are sequenced, genome comparisons will likely reveal evolutionary relationships between organisms never previously considered.

One study compared the human genome with that of chromosome 22 in chimpanzees. Among the many genes that differed in sequence were several of particular interest: a gene for proper speech development, several for hearing, and several for smell. The gene necessary for proper speech development is thought to have played an important role in human evolution. You can suppose that changes in hearing may have facilitated the use of language for communication between people. Changes in smell genes are a little more problematic. The researchers speculated that the olfaction genes may have affected dietary changes or sexual selection. Or they may have been involved in traits other than smell (**Fig. 11.24**). It was a surprise to find that many of the other genes they located and studied are known to cause human diseases if abnormal. Perhaps comparing genomes is a way of finding genes associated with human diseases.

Proteomics and Bioinformatics

The known sequence of bases in the human genome predicts that about 20,500 genes are translated into approximately 100,000 different proteins due to alternative mRNA splicing; collectively, all of these proteins are referred to as the human **proteome.** The field of **proteomics** explores the structure and function of these cellular proteins and examines how they interact to contribute to the production of traits. Because drugs tend to be proteins or small molecules that affect the behavior of proteins, proteomics is crucial in the development of new drugs for the treatment of disease.

Computer modeling of the three-dimensional shape of these proteins is an important part of proteomics, since structure relates to function. Researchers may then use the models to predict which molecules will be effective drugs. For example, a compound that fits the active site of an enzyme without being converted into a product may be an effective drug to inhibit that enzyme.

Bioinformatics is the application of computer technologies to the study of the genome and proteome. Genomics and proteomics produce raw data, and these fields depend on computer analysis to find significant patterns in the data. As a result of bioinformatics, scientists hope to find cause-and-effect relationships between an individual's overall genetic makeup and resulting genetic disorders.

More than half of the human genome consists of uncharacterized sequences that contain no genes and have no known function. Bioinformatics might find that these regions have functions by correlating any sequence changes with resulting phenotypes, or that some sequences are an evolutionary relic that once coded for a protein that we no longer need. New computational tools will most likely be needed in order to accomplish these goals.

a.

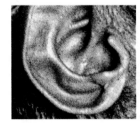
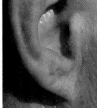

b.

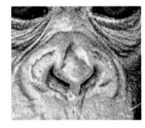

c.

Figure 11.24 **Studying genomic differences between chimpanzees and humans.**

By comparing the human genome with that of chimpanzees, researchers were able to discover genes that cause human disease when they are abnormal. They also concluded that the genes for **(a)** speech, **(b)** hearing, and **(c)** smell may have influenced the evolution of humans.

Connecting the Concepts

For more information on how genomics and proteomics contribute to our understanding of evolution, refer to the following discussions.

Section 14.2 examines how molecular evidence supports the evolutionary theory.

Section 16.3 describes the science of systematics, which characterizes the evolutionary relationships among organisms.

Check Your Progress 11.4

1. Describe the goals of genomics research.

2. Explain why it is important to compare genomes of different organisms.

3. Discuss why proteomics can be beneficial to the future of health care.

Media Study Tools

Enhance your study of this chapter with study tools and practice tests. Also ask your instructor about the resources available through ConnectPlus, including the media-rich eBook, interactive learning tools, and animations.

DNA and Genes

The virtual lab "DNA and Genes" provides an interactive look at how mutations influence the amino acid sequence of a protein.

The Chapter in Review

Summary

11.1 DNA and RNA Structure and Function

Structure of DNA

DNA, the genetic material, has the following structure:

- DNA is a polymer in which four nucleotides differ by their bases. The bases are symbolized by the letters A, G, C, and T.
- A is always paired with T, and G is always paired with C, a pattern called complementary base pairing:

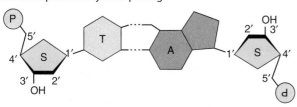

- DNA is a double helix. The deoxyribose sugar–phosphate molecules make up the sides of a twisted ladder and run counter to one another. The paired bases are the rungs of the ladder. Base pairs can be in any order in each type of species.

Replication of DNA

DNA replicates and every new cell gets a copy. During DNA replication, each old strand gives rise to a new strand. Therefore, the process is called semiconservative. Because of complementary base pairing, all these double-helical molecules are identical:

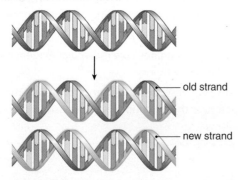

— old strand

— new strand

RNA Structure and Function

RNA has the following characteristics:

- It is found in the nucleus and cytoplasm of eukaryotic cells.
- It contains the sugar ribose and the bases A, U, C, and G.
- It is single-stranded.
- The three forms are mRNA, which carries the DNA message to the ribosomes; tRNA, which transfers amino acids to the ribosomes, where protein synthesis occurs; and rRNA, which is found in the ribosomes.

11.2 Gene Expression

DNA specifies the sequence of amino acids in a protein; therefore, gene expression occurs once the protein product is present. Gene expression requires two steps: transcription and translation.

From DNA to RNA to Protein

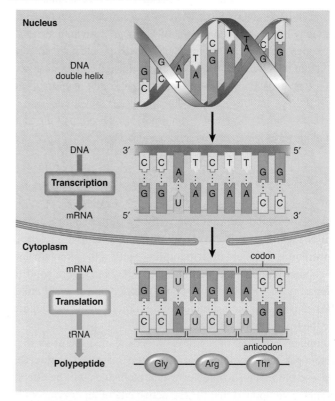

Transcription: During transcription, a primary-mRNA forms and is then processed. Processing involves (1) the addition of a cap to one end, (2) the addition of a poly-A tail to the other end, and (3) the removal of introns, so that only exons remain. The mature mRNA now contains instructions for assembling a protein.

Translation: Translation requires three types of RNA:

mRNA contains codons in which every three bases code for a particular amino acid except when the codon means stop.

tRNA brings amino acids to the ribosomes. One end of the molecule binds to the amino acid, and the other end is the anticodon, three bases that pair with a codon. Each tRNA binds with only one of the 20 types of amino acids in a protein.

rRNA is located in the ribosomes. The P site of a ribosome contains a tRNA attached to a peptide, and the A site contains a tRNA attached to an amino acid.

Translation Has Three Phases

The three phases of translation are initiation, the elongation cycle, and termination.

Initiation: During chain initiation at the start codon, the ribosomal subunits, the mRNA, and the tRNA-methionine complex come together.

Elongation cycle: The chain elongation cycle consists of these events:
A tRNA at the P site passes its peptide to tRNA–amino acid at the A site. The tRNA at the P site leaves.
The ribosome moves forward one codon (called translocation), and the codon at the A site is ready for the next tRNA–amino acid.

Termination: During chain termination, a stop codon is reached on the mRNA. A release factor binds, the ribosome dissociates, the mRNA departs, and the polypeptide is released.

Genes and Gene Mutations

A gene mutation is a change in the sequence of bases. Mutations can be due to errors in replication, transposons, or environmental mutagens. The results of mutations vary from no effect to a nonfunctional protein.

11.3 DNA Technology

Recombinant DNA Technology

Recombinant DNA technology uses restriction enzymes to cleave DNA, so that a foreign gene can be inserted into a vector, such as a plasmid:

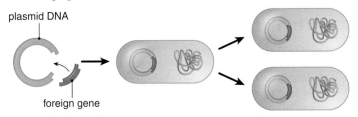

plasmid DNA

foreign gene

Transgenic Organisms

Biotechnology uses natural biological systems to create a product or to achieve an end desired by human beings. Foreign genes have been transferred to bacteria, crops, and farm animals to improve their characteristics and produce commercial products.

Polymerase Chain Reaction

The polymerase chain reaction occurs in a test tube and is often used in conjunction with recombinant DNA technology. Heat separates double-stranded DNA. Primers flank the target DNA, and heat-insensitive DNA polymerase copies the target DNA. The cycle is repeated until millions of copies are produced.

DNA Fingerprinting

DNA fingerprinting helps identify relatives, remains, and criminals. Following PCR if needed, restriction enzymes are used to fragment samples of DNA. When separated by gel electrophoresis, a pattern results that is unique to the individual.

11.4 Genomics and Proteomics

Sequencing the Bases of the Human Genome

Scientists have sequenced all 3.4 billion bases of the human genome, and they are now employing genomics to study the information it contains. Differences in base sequence between individuals, including single base changes within genes, single base changes in unknown sequences, and differences in copy number of certain genes, together create variation within the population and affect disease susceptibility.

Genome Comparisons The genomes of many organisms, including humans, have been compared to study the evolutionary relationships between them.

Proteomics and Bioinformatics

Proteomics, the study of all the proteins active in a cell, is being used to predict the structures of proteins and to design drugs that may affect them. Bioinformatics applies computer technology to the study of the genome and proteome. With these tools, scientists hope to sift through the massive amounts of data generated through genomics and proteomics to discover underlying patterns.

Key Terms

adenine (A) 176	polyribosome 185
anticodon 184	promoter 183
bioinformatics 193	proteome 193
biotechnology 189	proteomics 193
codon 182	recombinant DNA
complementary base pairing 177	(rDNA) 188
cytosine (C) 176	release factor 186
DNA fingerprinting 191	restriction enzyme 188
DNA ligase 179	ribosomal RNA (rRNA) 180
DNA polymerase 179	ribosome 180
DNA replication 179	RNA (ribonucleic acid) 180
frameshift mutation 187	RNA polymerase 183
gene mutation 187	semiconservative 179
genetically modified 189	template 179
genetic engineering 188	thymine (T) 176
genomics 192	transfer RNA (tRNA) 180
guanine (G) 176	transgenic organism 188
helicase 179	translocation 185
intergenic DNA 192	transposon 187
messenger RNA (mRNA) 180	triplet code 182
mRNA transcript 183	uracil (U) 180
mutagen 187	vector 188
point mutation 187	
polymerase chain reaction (PCR) 190	

Testing Yourself

1. In the following diagram, label all parts of the DNA molecule.

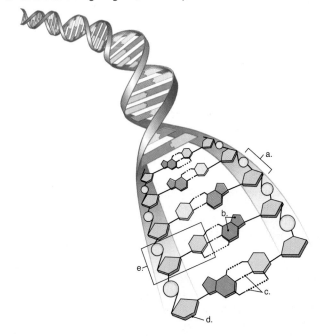

Choose the best answer for each question.

2. Chargaff's rules state that the amount of A, T, G, and C in DNA
   a. varies from species to species, and the amount of A = T and G = C.
   b. varies from species to species, and the amount of A = G and T = C.
   c. is the same from species to species, and the amount of A = T and G = C.
   d. is the same from species to species, and the amount of A = G and T = C.

3. DNA replication is said to be semiconservative because
   a. each new DNA molecule contains one new strand and one old one.
   b. every other new DNA molecule contains two old strands.
   c. each new DNA strand contains two new strands.
   d. every other new DNA molecule contains one new strand and one old one.

4. Which of the following is **not** a feature of eukaryotic DNA replication?
   a. Replication bubbles spread in two directions.
   b. A new strand is synthesized using an old one as a template.
   c. Complementary base pairing determines which nucleotides should be added to the new strand.
   d. Each chromosome has one origin of replication.

For questions 5–8, match the items to those in the key. Answers can be used more than once.

Key:
   a. messenger RNA
   b. transfer RNA
   c. ribosomal RNA

5. contains an anticodon
6. carries amino acids to the ribosome
7. carries genetic information from DNA to ribosomes
8. produced in the nucleolus

9. Which of the following statements about the genetic code is not true?
   a. The genetic code is almost universal.
   b. The genetic code is a doublet code.
   c. Multiple codons may encode the same amino acid.
   d. Some special codons mean "stop."

10. Transcription produces _____, while translation produces _____.
    a. DNA, RNA
    b. RNA, polypeptides
    c. polypeptides, RNA
    d. RNA, DNA

11. The application of computer techniques to analyze the genome is called
    a. proteomics.
    b. genomics.
    c. bioinformatics.
    d. genetic code.

12. If a DNA sequence were AATCGGTAT, what would be the (1) replicated DNA strand produced, (2) mRNA strand produced from the original sequence, and (3) sequence of the amino acids from the mRNA strand?

For questions 13–17, match the items to the best answer from the key.

Key:
   a. not harmful
   b. sometimes harmful
   c. usually harmful

13. point mutation that changes one codon of a gene
14. frameshift mutation that changes all following codons
15. point mutation that changes a codon into a "stop" codon
16. movement of a transposon into the middle of a gene
17. point mutation that does not change the amino acid specified by the codon

18. Which of the following is not useful in creating recombinant DNA?
    a. restriction endonuclease
    b. DNA ligase
    c. RNA polymerase
    d. plasmid

19. Genome comparisons may yield information about which of the following?
    a. evolutionary relationships between organisms
    b. changes in genes over evolutionary history
    c. links between genes and disease
    d. All of these are correct.

Thinking Scientifically

1. Experiments with bacteria in the 1930s showed that exposing a nonvirulent (noninfectious) strain of bacteria to a heat-killed virulent strain can convert the nonvirulent strain into a virulent one. What might explain this phenomenon?

2. As you have just read, humans and other mammals have a large amount of intergenic DNA. It requires much energy to maintain this "extra" DNA; therefore, scientists reason it must have a function. Explain.

Bioethical Issue

Mandatory DNA Fingerprinting

Arthur Lee Whitfield had served 22 years of a 63-year sentence for rape when he was notified that he had gained the right to use DNA fingerprinting to establish his innocence. Luckily, biological evidence had been saved from his jury trial, and DNA fingerprinting showed that his DNA did not match that of the rapist. However, this DNA did match that of another inmate, who was serving a life sentence for an unrelated rape conviction.

In response to Whitfield's and similar cases, both law enforcement agencies and criminal defense attorneys are advocating that all individuals who are arrested submit samples for DNA fingerprinting. Some are even pushing to require DNA fingerprinting of all citizens, citing the potential to resolve many cases and exonerate people wrongfully convicted of crimes. However, many civil libertarians insist that this constitutes illegal search without reasonable cause. Furthermore, others insist that the government cannot be trusted with such a wealth of information, because it might be accessed by hackers or used for purposes other than that for which it was intended.

Should the government require all citizens, or those who are arrested for any reason, to submit samples for DNA fingerprinting? Or do you believe that it would be a violation of the U.S. Constitution?

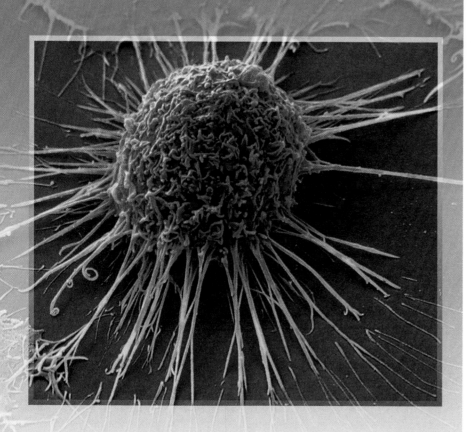

12

Gene Regulation and Cancer

Faulty Gene Regulation and Cancer

The term *cancer* describes a family of diseases having the common characteristic of uncontrolled cell division. You are probably aware that lifestyle choices such as smoking or tanning might increase your odds of developing certain types of cancer. However, have you ever considered that some viral infections can also increase your cancer risk? A group of viruses, termed *oncogenic viruses,* can contribute to that risk. These viruses are characterized by their ability to insert viral DNA into human chromosomes. Humans have several genes, called proto-oncogenes, that help regulate cell division. When oncogenic viral DNA inserts a mutated proto-oncogene, called an oncogene, into a human chromosome, the cell cycle may occur repeatedly, creating a tumor.

Human papillomavirus (HPV) is a well-known oncogenic virus. It takes many forms and often causes benign warts. However, several strains are known to be oncogenic and are the primary cause of cervical cancer. Women infected with oncogenic strains of this HPV typically have no symptoms, but an annual Pap smear can detect changes in cervical cells that are consistent with HPV infection. When detected early, infected cells can be removed.

Although many individuals equate genetic mutations with cancer, many more of these changes are usually required for cancer to develop. The changes cause a breakdown in gene regulation, which can lead to a loss of control of the cell cycle. In this chapter, you will learn about how genes are regulated and some of the genetic changes that contribute to the development of cancer.

BEFORE YOU BEGIN

Before beginning this chapter, take a few moments to review the following discussions.

Section 3.2 How does the shape of a protein relate to its function?

Section 8.4 How is the cell cycle controlled?

Section 11.2 What are the stages in the expression of a gene?

12.1 Stem Cells and Cloning

Learning Outcomes

Upon completion of this section, you should be able to

1. Distinguish between reproductive and therapeutic cloning.
2. Differentiate between embryonic and adult stem cells.

The events of the cell cycle and DNA replication ensure that every cell of the body receives a copy of all the genes. This means that every one of your cells has the potential to become a complete organism. While some cells, such as stem cells, retain their ability to form any other type of cell, most other cells differentiate, or specialize, to become specific types of cells, such as muscle cells. This specialization is based on the expression of certain groups of genes at specific times in development. One of the best ways to understand how specialization influences the fate of a cell is to take a look at the processes of reproductive and therapeutic cloning.

Reproductive and Therapeutic Cloning

In **reproductive cloning,** the desired end is an individual that is exactly like the original individual. The cloning of plants has been routine for some time (**Fig. 12.1**). The cloning of some animals, such as amphibians, has been underway since the 1950s. However, at one time it was thought that the cloning of adult mammals would be impossible because investigators found it difficult to have the nucleus of an adult cell "start over," even when it was placed in another egg that has had its own nucleus removed (an enucleated egg cell).

In March 1997, Scottish investigators announced they had cloned a Dorset sheep, which they named Dolly. How was their procedure different from all the others that had been attempted? Again, an adult nucleus was placed in an enucleated egg cell; however, the donor cells had been starved. Starving the donor cells caused them to stop

1 Tiny disks are obtained from carrot root.

2 Each disk produces an undifferentiated tissue mass.

3 Many like carrot plantlets are cloned from each tissue mass.

Figure 12.1   Cloning carrots.

Carrot cells from a root will produce cloned copies of carrot plants in cell culture. This demonstrates that all the nuclei in a carrot root contain all the genes of a carrot plant.

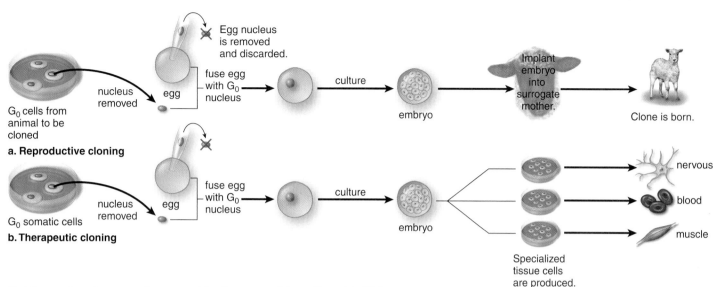

a. Reproductive cloning

b. Therapeutic cloning

dividing and go into a resting stage (the G_0 stage of the cell cycle). This was the change needed, because G_0 stage nuclei are open to cytoplasmic signals for the initiation of development (**Fig. 12.2**). Now it is common practice to clone farm animals that have desirable genetic traits (**Fig. 12.3**), and even to clone rare animals that might otherwise become extinct.

Currently in the United States, no federal funds can be used for experiments to clone human beings. And at this point, even cloning farm animals is wasteful—in the case of Dolly, out of 29 attempted clones, only 1 was successful. Also, there are concerns that cloned animals may not have the same life expectancy as noncloned animals. Some, but not all, cloned animals have demonstrated symptoms of abnormal aging. For example, Dolly was put down by lethal injection in 2003 because she was suffering from lung cancer and crippling arthritis. She had lived only half the normal life span for a Dorset sheep.

Video
Cloning
Endangered
Species

Figure 12.2 Two types of cloning.

a. The purpose of reproductive cloning is to produce an individual that is genetically identical to the one that donated a diploid (2n) nucleus. The 2n nucleus is placed in an enucleated egg; after several divisions, the embryo comes to term in a surrogate mother. **b.** The purpose of therapeutic cloning is to produce specialized tissue cells. A 2n nucleus is placed in an enucleated egg; after several divisions, the embryonic cells (called embryonic stem cells) are separated and treated to become specialized cells.

Connections and Misconceptions

Are food products from cloned animals nutritious?

A comprehensive study by the Food and Drug Administration (FDA) in 2008 indicated that meat and milk from cloned animals, such as cows, goats, and pigs, are just as nutritious as products from noncloned animals. Many people confuse cloned with "transgenic" organisms— those that contain genes from multiple species. While scientific studies are still underway to assess the health aspects of transgenics, a cloned animal is no different from its parent. The meat and milk from cloned animals have the same nutritional content as those of noncloned animals.

In **therapeutic cloning,** the desired end is not an individual organism but various types of mature cells. The purposes of therapeutic cloning are (1) to learn more about how cell specialization occurs and (2) to provide cells and tissues that could be used to treat human illnesses, such as diabetes, spinal cord injuries, and Parkinson disease.

Figure 12.3 Cloned farm animals.

These pigs were cloned using diploid (2n) nuclei from an adult swine. All nuclei contain all the genes of an organism.

Connections and Misconceptions

Have stem cells been used to cure human disease?

Stem cells have been used since the 1960s as a treatment for leukemia, a form of cancer in which the stem cells in bone marrow produce a large number of nonfunctional blood cells. In a bone marrow transplant, the patient's stem cells are destroyed using radiation or chemotherapy. Then, healthy stem cells from a donor are introduced into the patient's bone marrow. Stem cell therapy is available for a variety of human diseases, including type 2 diabetes, autoimmune diseases, and heart conditions.

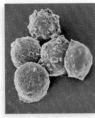

Connecting the Concepts

For more information on transgenic organisms, refer to the following discussion.

Section 11.3 examines how scientists generate and study transgenic organisms.

Section 32.1 explores how transgenic organisms may be used to increase the food resources that are needed for human population growth.

Therapeutic cloning can be carried out in several ways. The most common method is to isolate **embryonic stem cells** and subject them to treatments that cause them to become particular cell types, such as red blood cells, muscle cells, or nerve cells (see Fig. 12.2b). Since embryonic stem cells have the potential to become any other type of cell in an organism, they are said to be **totipotent.** Eventually, it may be possible to produce entire tissues and organs from totipotent stem cells. Ethical concerns exist about this type of therapeutic cloning because, if the embryo had been allowed to continue development, it would have become an individual.

Another way to carry out therapeutic cloning is to use **adult stem cells,** found in many organs of an adult's body. Adult stem cells are said to be **multipotent,** since they have already started to specialize and are not able to produce every type of cell in the organism. For example, the skin has stem cells that constantly divide and produce new skin cells, while the bone marrow has stem cells that produce new blood cells. Currently, adult stem cells are limited in the number of cell types that they may become. However, scientists have been able to coax adult stem cells from skin into becoming more like embryonic stem cells by adding only four genes. Researchers are investigating ways of controlling gene expression in adult stem cells, so that they can be used in place of the more controversial embryonic stem cells in fighting disease in humans.

 **Video** Heart Stem Cells   **Video** Stem Cells

Check Your Progress 12.1

❶ Describe the differences between reproductive and therapeutic cloning.

❷ Briefly describe the process of reproductive cloning.

❸ Briefly describe the process of therapeutic cloning.

12.2 Control of Gene Expression

Learning Outcomes

Upon completion of this section, you should be able to

1. Summarize the operation of the *lac* operon in prokaryotes.
2. List the levels of control of gene expression in eukaryotes.
3. Distinguish between euchromatin and heterochromatin.
4. Explain the role of transcription factors in eukaryotic gene regulation.

Levels of Gene Expression Control

The human body consists of many types of cells that differ in structure and function. Each cell type must contain its own mix of proteins that makes it different from all other cell types. Therefore, only certain genes are active in cells that perform specialized functions, such as nerve, muscle, gland, and blood cells.

Some of these active genes are called housekeeping genes because they govern functions that are common to many types of cells, such as glucose metabolism. But the activity of some genes accounts for the specialization of cells. In other words, gene expression is controlled in a cell, and this control accounts for its specialization (**Fig. 12.4**). Let's begin by examining a simpler system—the control of transcription in prokaryotes.

Gene Expression in Prokaryotes

The bacterium *Escherichia coli* lives in the human large intestine and can quickly adjust its production of enzymes according to what we eat. If we drink a glass of milk, *E. coli* immediately begins to make three enzymes needed to metabolize lactose. The transcription of all three enzymes is under the control of one **promoter,** a short DNA sequence where RNA polymerase attaches. French microbiologists François Jacob and Jacques Monod called such a cluster of bacterial genes, along with the DNA sequences that control their transcription, an **operon.** They received a Nobel Prize in 1961 for their investigations because they were the first to show how gene expression is controlled—specifically, how the *lac* operon is controlled in lactose metabolism (**Fig. 12.5**). Jacob and Monod proposed that a **regulatory gene** located outside the operon codes for a **repressor**—a protein that, in the *lac* operon, normally binds to the *operator,* which lies next to the promoter. When the repressor is attached to the operator, transcription of the lactose metabolizing genes does not take place because RNA polymerase is unable to bind to the promoter. The *lac* operon is normally turned off in this way because lactose is usually absent (Fig. 12.5*a*).

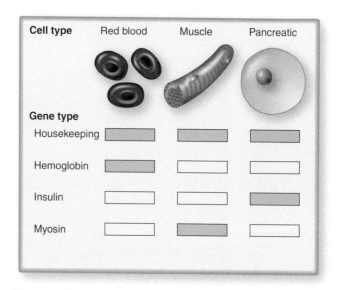

Figure 12.4   Gene expression in specialized cells.

All the cells in your body contain all the genes, but gene expression is controlled, and only certain genes are expressed in each type of cell. Housekeeping genes are expressed in all cells; in addition, cells express those genes that account for their specialization.

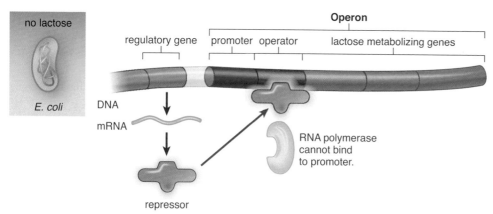

a. **Lactose is absent—operon is turned off.**
   Enzymes needed to metabolize lactose are not produced.

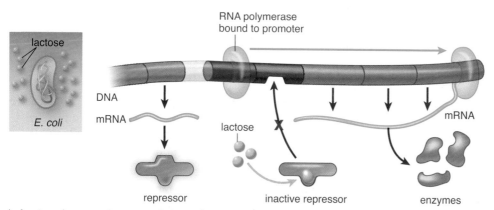

b. **Lactose is present—operon is turned on.**
   Enzymes needed to metabolize lactose are produced.

Figure 12.5   The *lac* operon.

a. The regulatory gene codes for a repressor. The repressor binds to the operator and prevents RNA polymerase from attaching to the promoter. Therefore, transcription of the lactose metabolizing genes does not occur. **b.** When lactose is present, it binds to the repressor, changing its shape, so that it becomes inactive and can no longer bind to the operator. Now RNA polymerase binds to the promoter, and the lactose metabolizing genes are expressed.

What turns the operon on when lactose is present? Lactose binds to the repressor and it changes shape. Now the repressor is unable to bind to the operator and RNA polymerase is able to bind to the promoter (Fig. 12.5*b*). Transcription of the genes needed for lactose metabolism occurs. When lactose is present and glucose, the preferred sugar, is absent, a protein called CAP (not shown in Fig. 12.5) assists in the binding of RNA polymerase to the promoter. This further ensures that the lactose metabolizing enzymes are transcribed when they are needed.

Animation *lac* Operon Regulation

Other bacterial operons, such as those that control amino acid synthesis, are usually turned on. For example, in the *trp* operon, the regulatory gene codes for a repressor that ordinarily is unable to attach to the operator. Therefore, the genes needed to make the amino acid tryptophan are ordinarily expressed. When tryptophan is present, it binds to the repressor. A change in shape activates the repressor and allows it to bind to the operator. Now the operon is turned off.

Animation Tryptophan Repressor

Gene Expression in Eukaryotes

In bacteria, a single promoter serves several genes that make up a transcription unit, while in eukaryotes, each gene has its own promoter where RNA polymerase binds. Bacteria rely mostly on transcriptional control, but eukaryotes employ a variety of mechanisms to regulate gene expression. These mechanisms affect whether the gene is expressed, the speed with which it is expressed, and how long it is expressed.

Animation Control of Gene Expression in Eukaryotes

Some mechanisms of gene expression occur in the nucleus; others occur in the cytoplasm (**Fig. 12.6**). In the nucleus, chromatin condensation, DNA transcription, and mRNA processing all play a role in determining which genes are expressed in a particular cell type.

In the cytoplasm, mRNA translation into a polypeptide at the ribosomes can occur right away or be delayed. The mRNA can last a long time or be destroyed immediately, and the same holds true for a protein. These mechanisms control the quantity of gene product and/or how long it is active.

Chromatin Condensation Eukaryotes utilize chromatin condensation as a way to keep genes turned on or off. The more tightly chromatin is compacted, the less often genes within it are expressed. Darkly staining portions of chromatin, called **heterochromatin,** represent tightly compacted, inactive chromatin. A dramatic example of this is the *Barr body* in mammalian females. Females have a small, darkly staining mass of condensed chromatin adhering to the inner edge of the nuclear envelope. This structure is an inactive X chromosome.

How do we know that Barr bodies are inactive X chromosomes that are not producing gene product? Suppose 50% of the cells in a female have one X chromosome active, and 50% have the other X chromosome active. Wouldn't the body of a heterozygous female be a mosaic, with "patches" of genetically different cells? This is exactly what happens. For example, human females who are heterozygous for an X-linked recessive form of ocular albinism have patches of pigmented and nonpigmented cells at the back of the eye. And women who are heterozygous for the hereditary absence of sweat glands have patches of skin lacking sweat glands. The female calico cat also provides dramatic support for a difference in X-inactivation in its cells (**Fig. 12.7**).

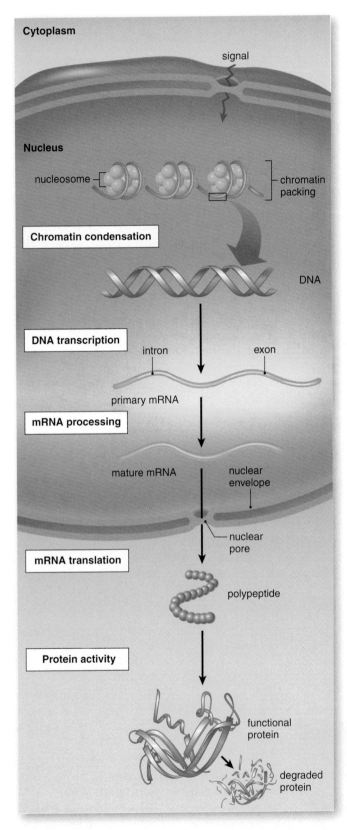

Figure 12.6 **Control of gene expression in eukaryotic cells.**

Gene expression is controlled at various levels in eukaryotic cells. Also, external signals (red arrow), such as growth factors, may alter gene expression.

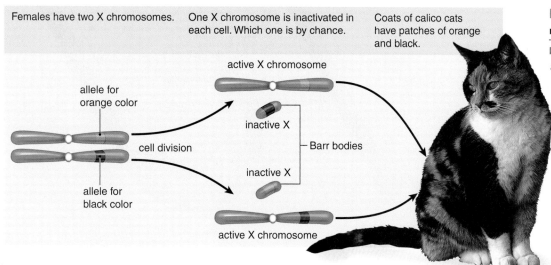

Females have two X chromosomes. One X chromosome is inactivated in each cell. Which one is by chance. Coats of calico cats have patches of orange and black.

active X chromosome

allele for orange color

cell division

inactive X

inactive X — Barr bodies

allele for black color

active X chromosome

Figure 12.7 X-inactivation in mammalian females.

In cats, the alleles for black or orange coat color are carried on the X chromosome. Random X-inactivation occurs in females. Therefore, in heterozygous females, 50% of the cells have an allele for black coat color, and 50% of cells have an allele for orange coat color. The result is calico or tortoiseshell cats that have coats with patches of both black and orange. A calico cat, but not a tortoiseshell cat, has patches of white due to the activity of another gene.

Connections and Misconceptions

Why does X-inactivation occur?

While the X chromosome contains over 1,600 genes, the majority of these are not involved in anything related to sex. Therefore, in humans and other mammals, a female with two X chromosomes normally produces twice as much gene product (proteins) as a male. Too much protein can negatively influence many metabolic pathways, so in order to regulate the amount of protein produced, females inactivate all but one of their X chromsomes. Most females possess a single Barr body, but the cells of XXX females have two Barr bodies, and males with Klinefelter syndrome (XXY) have a single Barr body in their cells, even though they are males.

Animation
X-inactivation

When heterochromatin undergoes unpacking, it becomes **euchromatin,** a more loosely packed form of chromatin that contains active genes. You learned in Chapter 8 that, in eukaryotes, a *nucleosome* is a portion of DNA wrapped around a group of histone molecules. When DNA is transcribed, a chromatin remodeling complex pushes aside the histone portions of nucleosomes, so that transcription can begin (**Fig. 12.8**). In other words, even euchromatin needs further modification before transcription can begin. The presence of histones limits access to DNA, and euchromatin becomes genetically active when histones no longer bar access to DNA. Only then is it possible for a gene to be turned on and expressed in a eukaryotic cell.

DNA Transcription Transcription in eukaryotes follows the same principles as in bacteria, except that many more regulatory proteins per gene are involved. The occurrence of so many regulatory proteins allows for not only greater control but also a greater chance of malfunction.

In eukaryotes, **transcription factors** are DNA-binding proteins that help RNA polymerase bind to a promoter. Several transcription factors are needed in each case; if one is missing, transcription cannot take place. All the transcription factors form a *complex* that also helps pull double-stranded DNA apart and even acts to position RNA polymerase so that transcription can begin. The same transcription factors, in different combinations, are used over again at other promoters, so it is easy to imagine that, if one malfunctions, the result

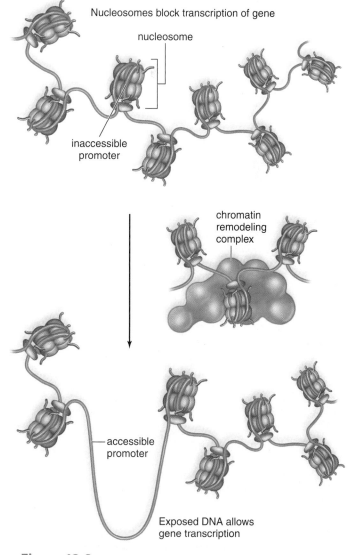

Nucleosomes block transcription of gene

nucleosome

inaccessible promoter

chromatin remodeling complex

accessible promoter

Exposed DNA allows gene transcription

Figure 12.8 Histone displacement.

In euchromatin, a chromatin remodeling complex pushes aside the histone portions of nucleosomes, so that transcription can begin.

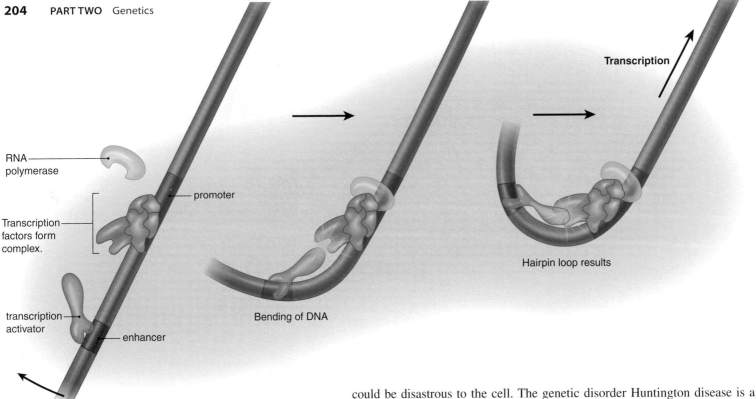

Figure 12.9 **Transcription factors and transcription activators.**

Transcription factors form a complex at a promoter, and a transcription activator binds to an enhancer. RNA polymerase binds to the promoter, but transcription does not begin until a hairpin DNA loop brings all regulatory proteins together.

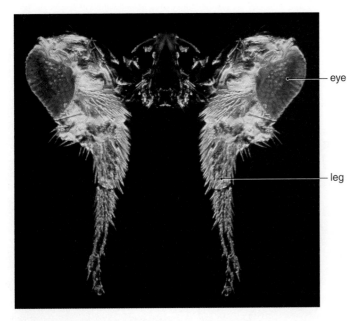

Figure 12.10 *Ey* gene.

In fruit flies, the expression of the *Ey* gene, in the precursor cells of the leg, triggers the development of an eye on the leg.

could be disastrous to the cell. The genetic disorder Huntington disease is a devastating psychomotor ailment caused by a defect in a transcription factor.

In eukaryotes, **transcription activators** are DNA-binding proteins that speed transcription dramatically. They bind to a DNA region, called an **enhancer,** that can be quite a distance from the promoter. A hairpin loop in the DNA can bring the transcription activators attached to enhancers into contact with the transcription factor complex (**Fig. 12.9**).

Animation
Transcription
Factors

A single transcription activator can have a dramatic effect on gene expression. For example, investigators have found one DNA-binding protein, MyoD, that alone can activate the genes necessary for fibroblasts to become muscle cells in various vertebrates. Another DNA-binding protein, called Ey, can bring about the formation of not just a single cell type but a complete eye in flies (**Fig. 12.10**).

mRNA Processing After transcription, the removal of introns and splicing of exons occur before mature mRNA leaves the nucleus and passes into the cytoplasm. **Alternative mRNA processing** is a mechanism by which the same primary-mRNA can produce different protein products according to which exons are spliced together to form mature mRNAs (**Fig. 12.11**).

The fruit fly gene *DScam* offers a dramatic example of alternative mRNA processing. As many as 38,000 different proteins can be produced according to which of the gene's exons are combined in mature mRNAs. Evidence suggests that these proteins provide each nerve cell with a unique identity as it communicates with other nerve cells in the brain.

Animation
Exon
Shuffling

mRNA Translation The cytoplasm contains proteins that can control whether translation of mRNA takes place. For example, an initiation factor, known as IF-2, inhibits the start of protein synthesis when it is phosphorylated by a specific protein kinase. Environmental conditions can also delay translation.

Red blood cells do not produce hemoglobin unless heme, an iron-containing group, is available.

The longer an mRNA remains in the cytoplasm before it is broken down, the more gene product is produced. During maturation, mammalian red blood cells eject their nuclei, yet they continue to synthesize hemoglobin for several months. The necessary mRNAs must be able to persist the entire time. Differences in the length of the poly-A tail can determine how long a particular transcript remains active. Hormones may also affect the stability of certain mRNA transcripts. An mRNA called vitellin persists for three weeks, instead of 15 hours, if it is exposed to estrogen.

Protein Activity Some proteins are not active immediately after synthesis. For example, insulin is a single, long polypeptide that folds into a three-dimensional structure. Only then is a sequence of about 30 amino acids enzymatically removed from the middle of the molecule. This leaves two polypeptide chains bonded together by disulfide (S—S) bonds, and an active insulin molecule results (**Fig. 12.12**). This mechanism allows a protein's activity to be delayed until it is needed.

Many proteins are short-lived in cells because they are degraded or destroyed. Cyclins, which are proteins involved in regulating the cell cycle, are destroyed by giant enzyme complexes called *proteasomes* when they are no longer needed. Proteasomes break down other proteins as well.

Signaling Between Cells in Eukaryotes

In multicellular organisms, cells are constantly sending out chemical signals that influence the behavior of other cells. During animal development, these signals determine the specialized role a cell will play in the organism. Later, the signals help coordinate growth and day-to-day functions. Plant cells also signal each other, so that their responses to environmental stimuli, such as direct sunlight, are coordinated.

Typically, **cell-signaling** occurs because a chemical signal binds to a receptor protein in a target cell's plasma membrane. The signal causes the receptor protein to initiate a series of reactions within a **signal transduction pathway.** The end product of the pathway (not the signal) directly affects the metabolism of the cell. For example, growth is possible only if certain genes have been turned on by regulatory proteins.

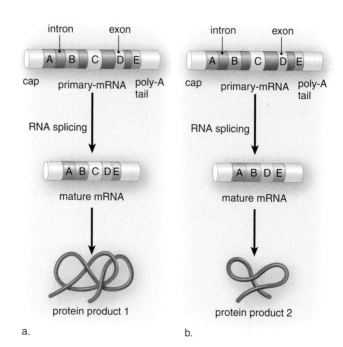

Figure 12.11 Processing of mRNA transcripts.

Because the primary-mRNAs are processed differently in these two cells, distinct proteins (a and b) result.

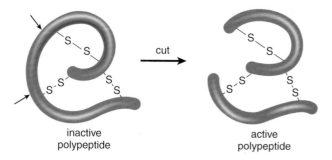

Figure 12.12 Protein activity.

The protein insulin is not active until an enzyme removes a portion of the initial polypeptide. The resulting two polypeptide chains are held together by two disulfide (S—S) bonds between like amino acids.

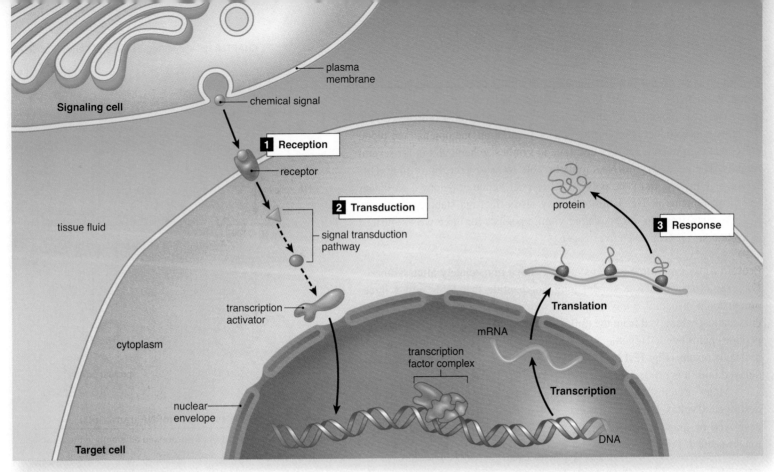

Figure 12.13 Cell-signaling.

1 Reception: A chemical signal is received at a specific receptor protein. **2** Transduction: A signal transduction pathway terminates when a transcription activator is stimulated. It enhances transcription of a specific gene. **3** Response: Translation of the mRNA transcript follows and a protein product results. The protein product is the response to the signal.

In **Figure 12.13,** a signaling cell secretes a chemical signal that binds to a specific receptor located in the receiving cell's plasma membrane. The binding activates a series of reactions within a signal transduction pathway. The last reaction activates a transcription activator that enhances the transcription of a specific gene. Transcription leads to the translation of mRNA and a protein product that, in this case, stimulates the cell cycle, so that growth occurs.

A protein called Ras functions in signal transduction pathways that lead to the transcription of many genes, several of which promote the cell cycle. Ras is normally inactive, but the reception of a growth factor leads to its activation. If Ras is continually activated, cancer will develop because cell division will occur continuously.

Connecting the Concepts

For more information on gene expression, refer to the following discussion.

Section 11.2 describes how the information in DNA is processed to form a functional protein.

Check Your Progress 12.2

1 Compare and contrast gene expression in prokaryotes and eukaryotes.

2 Detail the levels at which eukaryotes regulate gene expression.

3 Describe the relationship between signal transduction and gene expression.

12.3 Cancer: A Failure of Genetic Control

Learning Outcomes

Upon completion of this section, you should be able to

1. Distinguish between proto-oncogenes and tumor suppressor genes in regard to cancer.
2. Explain the role of telomerase in stem cells and cancer cells.
3. Summarize how chromosomal rearrangements may cause some forms of cancer.
4. Identify the relationship between certain genes and cancer.

As described in Chapter 8, cancer is a genetic disease caused by a lack of control in the cell cycle (**Fig. 12.14**). The development of cancer requires several mutations, each propelling cells toward the development of a tumor. These mutations disrupt the many redundant regulatory pathways that prevent normal cells from taking on the characteristics of cancer cells (**Table 12.1**). Because of this sequence, it takes several years for cancer to develop, but the likelihood of cancer increases as we age.

 **Animation**
Control of the Cell Cycle

Cells that are already highly specialized, such as nerve cells and cardiac muscle cells, seldom become cancer cells because they rarely divide. Carcinogenesis (the development of cancer) is more likely to begin in cells that have the capacity to enter the cell cycle. Fibroblasts and cells lining the cavities of the lungs, liver, uterus, and kidneys are able to divide when stimulated to do so. Adult stem cells continue to divide throughout life. These include blood-forming cells in the bone marrow and basal cells of the skin and digestive tract. Continuous division of these cells is required because blood cells live only a short while, and the cells that line the intestines and the cells that form the outer layer of the skin are continually sloughed off.

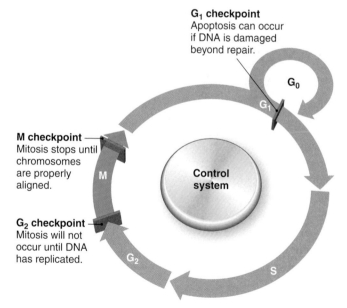

Figure 12.14   Cancer is a loss of cell cycle control.

The checkpoints of the cell cycle regulate how frequently a cell may replicate. A loss of these control mechanisms may lead to cancer.

Table 12.1   Normal Cells Compared with Cancer Cells

Normal Cells	Cancer Cells
Contact inhibition	No contact inhibition
Controlled growth	Uncontrolled growth (tumor)
Specialized cells	Nonspecialized cells
Normal chromosomes	Abnormal chromosomes
Undergo apoptosis	No apoptosis

Cell (red) acquires a mutation for repeated cell division.

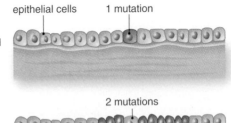

epithelial cells   1 mutation

New mutations arise, and one cell (teal) has the ability to start a tumor.

2 mutations

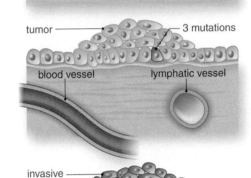

tumor ———————— 3 mutations

blood vessel   lymphatic vessel

Cancer in situ. The tumor is at its place of origin. One cell (purple) mutates further.

invasive tumor

Cells have gained the ability to invade underlying tissues by producing a proteinase enzyme.

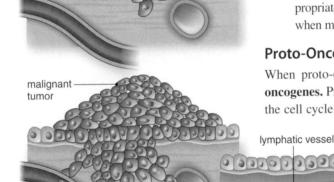

malignant tumor

Cancer cells now have the ability to invade lymphatic and blood vessels.

lymphatic vessel   distant tumor

New metastatic tumors are found some distance from the original tumor.

Figure 12.15 Development of cancer.

The development of cancer requires several mutations, each contributing to the development of a tumor. Normally, cells exhibit contact inhibition and controlled growth. They stop dividing when they are bordered by other cells. Tumor cells lack contact inhibition and pile up, forming a tumor. Cancer cells do not fulfill the functions of a specialized tissue, and they have abnormal chromosomes. As more mutations accumulate, the cancer cells may spread and form tumors in other parts of the body.

The development of colon cancer has been studied in detail, and **Figure 12.15** is based on those studies. This figure shows that a single abnormal cell begins the process toward the development of a tumor. Along the way, the most aggressive cell becomes the dominant cell of the tumor. As additional mutations occur, the tumor cells release a growth factor, which causes neighboring blood vessels to branch into the cancerous tissue, a process called *angiogenesis*. Additional mutations allow cancer cells to produce enzymes that degrade the basement membrane and invade underlying tissues. Cancer cells are motile—able to travel through the blood or lymphatic vessels to other parts of the body—where they start distant tumors, a process called *metastasis*.

Proto-Oncogenes and Tumor Suppressor Genes

Recall that the cell cycle (see Fig. 12.14) consists of interphase followed by mitosis. Special proteins help regulate the cell cycle at checkpoints. When cancer develops, the cell cycle occurs repeatedly, due in large part to mutations in two types of genes. As shown in **Figure 12.16,**

1. **Proto-oncogenes** code for proteins that promote the cell cycle and inhibit apoptosis. They are often likened to the gas pedal of a car because they promote the cell cycle.
2. **Tumor suppressor genes** code for proteins that inhibit the cell cycle and promote apoptosis. They are often likened to the brakes of a car because they inhibit acceleration of the cell cycle and stop cells from dividing inappropriately. They are called tumor suppressors because tumors may occur when mutations cause these genes to become nonfunctional.

Proto-Oncogenes Become Oncogenes

When proto-oncogenes mutate, they become cancer-causing genes called **oncogenes.** Proto-oncogenes promote the cell cycle, and oncogenes accelerate the cell cycle. Therefore, a mutation that causes a proto-oncogene to become an oncogene is a "gain of function" mutation.

A **growth factor** is a signal that activates a cell-signaling pathway by bringing about phosphorylation of a signaling protein (Fig. 12.16*a*). The cell-signaling pathway then activates numerous proteins, many of which promote the cell cycle (Fig. 12.16*b*). Some proto-oncogenes code for a growth factor or for a receptor protein that receives a growth factor. When these proto-oncogenes become oncogenes, receptor proteins are easy to activate and may even be stimulated by a growth factor produced by the receiving cell. For example, the *Ras* proto-oncogenes promote mitosis when a growth factor binds to a receptor. When *Ras* proto-oncogenes become oncogenes, they promote mitosis even when growth factors are not present. *Ras* oncogenes are found in 20–30% of human cancers.

Tumor Suppressor Genes Become Inactive

When tumor suppressor genes mutate, their products no longer inhibit the cell cycle or promote apoptosis. Therefore, these mutations can be called "loss of function" mutations (see Fig. 12.16).

The retinoblastoma protein (RB) controls the activity of a transcription activator called E2F. In the absence of growth factors, RB binds E2F and inhibits entry into the S stage of the cell cycle, but mutations in RB may cause it to release E2F and promote the cell cycle when it is not appropriate.

Cancer develops gradually; multiple mutations usually occur before a cell becomes cancerous. As you can imagine, the conversion of proto-oncogenes into oncogenes, coupled with the inactivation of tumor suppressor genes, causes the cell cycle to continue unabated, just as a car with a sticking gas pedal and faulty brakes soon careens out of control.

 Animation
How Tumor Suppressor Genes Block Cell Division

Other Genetic Changes and Cancer

Other genetic influences can also contribute to cancer development.

Absence of Telomere Shortening

A **telomere** is a repeating DNA sequence (TTAGGG) at the end of the chromosomes that can be as long as 15,000 base pairs. Just as the caps at the ends of shoelaces protect them from unraveling, so telomeres promote chromosomal stability, so that replication can occur. Each time a cell divides, some portion of a telomere is lost; when telomeres become too short, the chromosomes cannot be replicated properly and the cell cycle is stopped.

Embryonic cells and certain adult cells, such as stem cells and germ cells, have an enzyme, called *telomerase,* that can rebuild telomeres. The gene that codes for telomerase is turned on in cancer cells. Now telomeres do not shorten, and cells divide over and over again. Telomerase is believed to become active only after a cell has already started proliferating wildly.

Animation
Telomerase Function

Connections and Misconceptions

What is a colonoscopy?

A colonoscopy is a medical procedure in which a special camera, called a colonoscope, is used to examine the large intestine (colon) and rectum for signs of cancer and other disease. Colon cancer is frequently caused by cancer cells (such as the one shown here) that form small growths called polyps. During a colonoscopy, the doctor may remove the polyps for additional study (called a biopsy). A colonoscopy is recommended for everyone over the age of 50, or if there is a family history of colon cancer or inflammatory bowel disease.

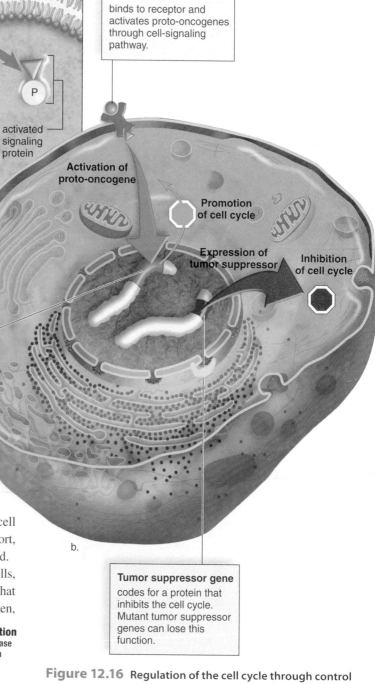

growth factor
receptor

P
P
P
activated signaling protein
signaling protein   phosphate

a.

Growth factor
binds to receptor and activates proto-oncogenes through cell-signaling pathway.

Activation of proto-oncogene

Promotion of cell cycle

Expression of tumor suppressor

Inhibition of cell cycle

Proto-oncogene
codes for a protein that promotes the cell cycle. If a proto-oncogene mutates, the resulting oncogenes may lead to uncontrolled cell division.

b.

Tumor suppressor gene
codes for a protein that inhibits the cell cycle. Mutant tumor suppressor genes can lose this function.

Figure 12.16 Regulation of the cell cycle through control of gene expression.

a. A growth factor binds to a receptor, stimulating a signal transduction pathway (green arrow). **b.** The signal transduction pathway activates the expression of proto-oncogenes that promote the cell cycle. Tumor suppressor genes limit cell division by inhibiting the cell cycle (red arrow).

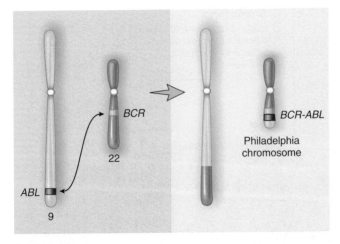

Figure 12.17 **A translocation can create the Philadelphia chromosome.**

The Philadelphia chromosome results when a portion of chromosome 9 is swapped for a portion of chromosome 22. This creates the *BCR-ABL* oncogene on chromosome 22. The Philadelphia chromosome is responsible for 95% of cases of chronic myelogenous leukemia.

SEM of breast cancer cell

Figure 12.18 **Breast cancer can run in families.**

These three sisters have all had breast cancer. Genetic tests can identify women at risk for breast cancer, so that they can choose to have frequent examinations to allow for early detection.

Chromosomal Rearrangements

When the chromosomes of cancer cells become unstable, portions of the DNA double helix may be lost, duplicated, or scrambled. For example, a portion of a chromosome may break off and reattach to another chromosome. Such a **translocation** may lead to cancer, especially if it disrupts genes that regulate the cell cycle.

The so-called *Philadelphia chromosome* is the result of a translocation between chromosomes 9 and 22 (**Fig. 12.17**). The *BCR* (*b*reakpoint *c*luster *r*egion) and *ABL* genes are fused, creating an oncogene, called *BCR-ABL*, that promotes cell division. This translocation causes nearly 95% of cases of chronic myelogenous leukemia (CML), a cancer of the bone marrow. Recently, researchers have used a drug called Gleevec to successfully treat CML. Gleevec inhibits the activity of the protein coded for by *BCR-ABL*. Encouraged by this success, scientists are now seeking to develop similar drugs that can inhibit the products of other oncogenes. Although cancer is usually a somatic disease, meaning that it develops only in body cells, in some cases, individuals may inherit a predisposition for developing some forms of cancer.

Connections and Misconceptions

Can transposons cause cancer?

Transposons are small, mobile sequences of DNA that have the ability to move throughout the genome. Also known as "jumping genes," transposons are known to cause mutation as they move (see Section 11.2). While the chances of a transposon disrupting the activity of a proto-oncogene or a tumor supressor gene in a specific cell are very small (our genome has over 3.4 billion nucleotides), there have been cases when a transposon has caused a loss of cell cycle control and been a factor in the development of cancer. Transposon activity has also been associated with the development of other diseases, such as some forms of hemophilia and muscular dystrophy.

Inheritance of Cancer-Causing Alleles

BRCA1* and *BRCA2 In 1990, DNA linkage studies on large families in which females tended to develop breast cancer identified the first gene allele associated with that disease. Scientists named that gene *b*reast *ca*ncer 1, or *BRCA1* (pronounced brak-uh). Later, they found that breast cancer in other families was due to a faulty allele of another gene, which they called *BRCA2*. Both alleles are mutant tumor suppressor genes that are inherited in an autosomal recessive manner. If one mutated allele is inherited from either parent, a mutation in the other allele is required before the predisposition to cancer is increased. Because the first mutated gene is inherited, it is present in all cells of the body, and then cancer is more likely wherever the second mutation occurs. If the second mutation occurs in the breast, breast cancer may develop (**Fig. 12.18**). If the second mutation is in the ovary, ovarian cancer may develop if additional cancer-causing mutations occur.

***RB* Gene** The *RB* gene is also a tumor suppressor gene. It takes its name from its association with an eye tumor called a *r*etino*b*lastoma, which first appears as a white mass in the retina. A tumor in one eye is most common because it takes mutations in both alleles before cancer can develop (**Fig. 12.19**). Children who inherit a mutated allele are more likely to have tumors in both eyes.

***RET* Gene**   An abnormal allele of the *RET* gene, which predisposes a person to thyroid cancer, can be passed from parent to child. *RET* is a proto-oncogene known to be inherited in an autosomal dominant manner—only one mutated allele is needed to increase a predisposition to cancer. The remaining mutations necessary for thyroid cancer to develop are acquired (not inherited).

Testing for These and Other Genes

Genetic tests can detect the presence of the *BRCA*, *RET*, and *RB* genes. Persons who have inherited these genes may decide to have surgery, by choice rather than necessity, or be examined frequently for signs of cancer.

Genetic tests are also available for other types of mutated genes that help a physician diagnose cancer. For example, a *ras* oncogene can be detected in stool and urine samples. If the test for *ras* oncogene in the stool is positive, the physician suspects colon cancer, and if the test for *ras* oncogene in the urine is positive, the physician suspects bladder cancer. Tests are even being developed that can screen for many cancers using only saliva.

 Video Melanoma Marker   **Video** New Cancer Clue

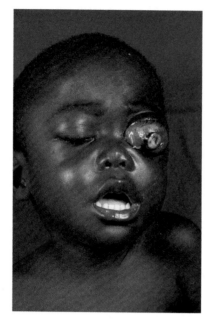

Figure 12.19 Inherited retinoblastoma.

A child is at risk for an eye tumor when a mutated *RB* allele is inherited, even though a second mutation in the normal allele is required before the tumor develops.

Connecting the Concepts

For more information on this material, refer to the following discussions.

Section 8.2 examines how the cell controls the cell cycle.

Section 8.5 lists the characteristics of a cancer cell.

Section 9.4 investigates how changes in chromosome structure are related to disease.

Check Your Progress 12.3

1. Explain why cancer incidence increases with age.
2. Describe the different influences that can contribute to cancer development and give an example of each.
3. Contrast the activity of proto-oncogenes with that of tumor suppressor genes.

Media Study Tools

www.mhhe.com/maderessentials3

Enhance your study of this chapter with study tools and practice tests. Also ask your instructor about the resources available through ConnectPlus, including the media-rich eBook, interactive learning tools, and animations.

The Cell Cycle and Cancer

The virtual lab "The Cell Cycle and Cancer" provides a more detailed look at how a failure in the control system of the cell cycle can lead to cancer.

The Chapter in Review

Summary

12.1 Stem Cells and Cloning

The study of stem cells has provided insight into how gene expression results in the formation of the tissues of an organism. Cloning, or the production of genetically identical cells and organisms, is possible with an understanding of gene expression.

Reproductive and Therapeutic Cloning

In reproductive cloning, the desired end is an individual genetically identical to the original individual. To achieve this type of cloning, the early embryo is placed in a surrogate mother until the individual comes to term.

In therapeutic cloning, the cells of the embryo are separated and treated, so that they develop into specialized tissues that can be used to treat human disorders. Alternately, adult stem cells can be treated to become specialized tissues.

12.2 Control of Gene Expression

Levels of Gene Expression Control

The specialization of cells is not due to the presence or absence of genes; it is due to the activity or inactivity of genes.

Gene Expression in Prokaryotes In prokaryotes, gene expression is controlled at the level of transcription.

- In the *lac* operon, the repressor usually binds to the operator. Then RNA polymerase cannot bind, and the operon is turned off. When lactose is present, it binds to the repressor, and now RNA polymerase binds to the promoter, and the genes for lactose metabolism are transcribed.
- In the *trp* operon, the genes are usually transcribed because the repressor is inactive. When tryptophan is present, the repressor becomes active, and RNA polymerase cannot bind to the promoter.

Gene Expression in Eukaryotes Eukaryotes regulate gene expression at various levels:

- Chromatin condensation—conversion of heterochromatin to euchromatin, so that genes can be expressed
- DNA transcription—transcription factors present at the promoter, and transcription activators (bound to enhancers) are brought into contact and promote transcription
- mRNA processing—alternative mRNA processing leads to a sequence of different exons in mature RNA and, therefore, different proteins
- mRNA translation—various factors affect how long an mRNA stays active and how many times it is translated
- Protein activity—activation of a protein may require further steps, and degradation of the protein can occur at once or after a while

Signaling Between Cells in Eukaryotes Eukaryotic cells constantly communicate with their neighbors by secreting signals.

- Signals are received at receptor proteins located in the plasma membrane.
- A signal transduction pathway (enzymatic reactions) leads to a change in the behavior of the cell. For example, when stimulated via a transduction pathway, a transcription activator stimulates transcription of a particular gene.

12.3 Cancer: A Failure of Genetic Control

Evidence is strong that cancer is a genetic disease. A series of mutations, particularly in cells that can divide, lead to cancer cells with these characteristics:

- Have no contact inhibition
- Have uncontrolled growth
- Are nonspecialized
- Have abnormal chromosomes
- Do not undergo apoptosis

Proto-Oncogenes and Tumor Suppressor Genes

- Proto-oncogenes ordinarily promote the cell cycle and inhibit apoptosis. Mutation results in a gain of function for these activities.
- Tumor suppressor genes ordinarily suppress the cell cycle and promote apoptosis. Mutation results in loss of function for these activities.

The development of cancer is a gradual process. As control of gene expression is further compromised, cancer cells develop additional characteristics that lead to angiogenesis and metastasis.

Other Genetic Changes and Cancer

Other genetic influences contribute to cancer development:

- When telomerase is present, telomeres don't shorten and cells can continue to divide.
- Chromosomal rearrangements, such as translocations, may disrupt genes that regulate the cell cycle.
- Certain inherited genes are known to cause cancer: *BRCA1* and *BRCA2* are mutant genes associated with breast cancer; a mutant *RB* gene is associated with the development of eye tumors; a mutant *RET* gene predisposes a person to thyroid cancer.

Genetic tests are available to help a physician diagnose cancer. These genes are turned off in normal cells but are active in cancer cells. For example, if a test for the presence of telomerase is positive, the cell is cancerous.

Key Terms

adult stem cell 200	proto-oncogene 209
alternative mRNA processing 204	regulatory gene 201
cell-signaling 205	repressor 201
embryonic stem cell 200	reproductive cloning 198
enhancer 204	signal transduction pathway 205
euchromatin 203	telomere 209
growth factor 208	therapeutic cloning 199
heterochromatin 202	totipotent 200
multipotent 200	transcription factor 203
oncogene 208	translocation 210
operon 201	tumor suppressor gene 209
promoter 201	

Testing Yourself

Choose the best answer for each question.

1. Reproductive cloning differs from therapeutic cloning in that the goal of reproductive cloning is
   a. the production of specific cell types for medical purposes.
   b. an individual that is genetically identical to the original individual.
   c. an individual that is genetically different from the original individual.
   d. the production of specific tissues and organs for medical transplantation.

2. The major challenge to therapeutic cloning using adult stem cells is
   a. finding appropriate cell types.
   b. obtaining enough tissue.
   c. controlling gene expression.
   d. keeping cells alive in culture.

3. Cancer cells exhibit abnormal characteristics because
   a. they do not produce telomerase.
   b. their tumor suppressor genes stimulate apoptosis.
   c. they have an unlimited food supply.
   d. control of gene expression has been lost.

4. Label this operon with these terms: active repressor, lactose metabolizing genes, operator, RNA polymerase, promoter, regulatory gene, mRNA.

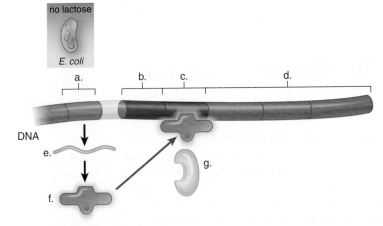

5. Which of the following is not a gene expression control mechanism seen in eukaryotes?
   a. alternative mRNA processing
   b. rate of ribosome synthesis
   c. longevity of mRNA
   d. chromatin condensation

For questions 6–10, match the examples to the gene expression control mechanisms in the key. Each answer can be used more than once.

Key:

  a. chromatin condensation
  b. alternative mRNA processing
  c. translation of mRNA
  d. protein activity

6. Insulin does not become active until 30 amino acids are cleaved from the middle of the molecule.

7. The mRNA for vitellin is longer-lived if it is exposed to estrogen.

8. Genes in Barr bodies are inactivated.

9. Each neuron combines the exons of the *DScam* primary-mRNA in different ways to produce a unique combination in mature mRNA.

10. A translation initiation factor inhibits translation when it is phosphorylated by kinase.

11. A cell may be cancerous if it tests positive for a(n)
  a. elongation factor.
  b. transcription factor.
  c. enhancer.
  d. chromosomal translocation.

12. Which of the following statements is true?
  a. Euchromatin contains inactive genes.
  b. Heterochromatin contains both active and inactive genes.
  c. The rearrangement of histones in euchromatin is necessary for transcription to occur.
  d. Barr bodies are composed mainly of genetically active euchromatin.
  e. All of these are correct.

13. How is transcription controlled in eukaryotic cells?
  a. through the use of proto-oncogenes
  b. by means of alternative mRNA processing
  c. through the use of transcription factors and activators
  d. when chromatin is packed to keep genes turned on
  e. None of these are correct.

14. An operon is a short sequence of DNA containing genes regulated together
  a. that prevent RNA polymerase from binding to the promoter.
  b. that prevent transcription from occurring.
  c. and the sequences that control its transcription.
  d. that code for the repressor protein.
  e. that prevent the repressor from binding to the operator.

15. The regulatory gene for the *lac* operon
  a. promotes transcription of the operon.
  b. is a sequence of DNA that codes for the repressor.
  c. prevents the repressor from binding to the operator.
  d. keeps the operon off until lactose is present.
  e. Both b and d are correct.

16. What is the purpose of cell-signaling in eukaryotes?
  a. to change the behavior of the receiving cell
  b. to initiate a set of reactions through the receptor protein
  c. to turn on a gene
  d. to transmit a nerve impulse, for example
  e. All of these are correct.

17. A transposon may cause cancer if
  a. it stops the cell from producing telomerase.
  b. it no longer stimulates the production of enzymes that degrade the basement membrane.
  c. cell division stops.
  d. a gene needed for cell cycle control is disrupted.

18. Which of the following statements is true?
  a. The gene product of a proto-oncogene inhibits the cell cycle.
  b. Proto-oncogenes are mutant versions of oncogenes that cause cancer.
  c. When mutated tumor suppressor genes lose their function, cancer may result.
  d. When tumor suppressor genes mutate, they become oncogenes.
  e. All of these are true.

19. Which of the following is not an example of why cancer cells are a failure of genetic control?
  a. They exhibit contact inhibition and do not grow in multiple layers.
  b. They lack specialization and do not contribute to the function of a body part.
  c. They have enlarged nuclei and may have an abnormal number of chromosomes.
  d. They fail to undergo apoptosis, even though they are abnormal cells.
  e. They release a growth factor that causes neighboring blood vessels to branch into the tumor and therefore feed the cancerous tissue.

Thinking Scientifically

1. Cyclin D is a proto-oncogene that is necessary for entry into the S stage of the cell cycle. When bound to the correct kinase, *RB* is modified, and cell division is promoted. Then, cyclin D is destroyed. How might a mutation in cyclin D lead to uncontrolled cell growth?

2. Researchers have observed more alternative mRNA splicing in complex animals, such as humans, than in less complex animals, such as fruit flies. Why might alternative mRNA splicing be advantageous in complex animals?

Bioethical Issue

Should BPA Be Banned?

Environmental estrogens are hormone-like chemicals that can compete with sex hormones for receptors, altering gene expression and promoting cell division. Recently, a link has been established between the estrogen-like chemical bisphenol A (BPA) and an increased risk of certain cancers, leading to calls from many to regulate or ban the compound.

BPA is used in the manufacture of polycarbonate plastics, found in many consumer products, such as food containers and baby bottles. Studies demonstrate that cleaning polycarbonate with alkali or bleach may cause leaching of BPA, potentially exposing consumers to the chemical. Studies found BPA responsible for enlarged female reproductive organs in mice kept in polycarbonate cages; follow-up studies indicated that mice exposed to BPA had a three- to fourfold elevated risk of breast cancer. Additional studies suggest that exposure to BPA during fetal development may have similar effects. Citing these and other government-funded studies, many have called for government regulation of BPA, with many more demanding a ban on the use of polycarbonate baby bottles. In 2008, Health Canada weighed in on the issue, stating that BPA "is toxic to human health."

However, not all scientists agree. Industry scientists and other critics point to industry-funded studies that show little to no risk to humans, especially to adults. Furthermore, they insist that diligence by consumers when cleaning polycarbonate plastics can greatly decrease whatever risk is posed by the plastic, since avoiding bleach or alkali almost totally eliminates BPA leaching.

Do you believe that polycarbonate plastic food containers should be banned? If not, would a ban on polycarbonate baby bottles be reasonable? Or, as some contend, does polycarbonate pose no more risk to humans than other common household materials?

13

Genetic Counseling

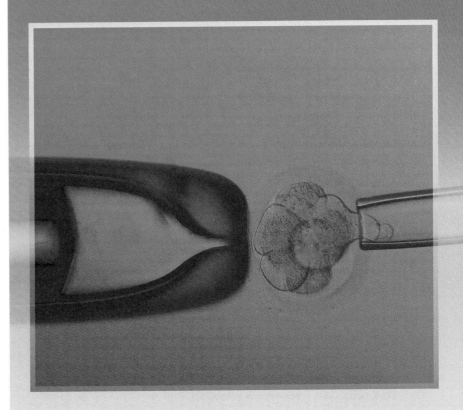

BEFORE YOU BEGIN

Before beginning this chapter, take a few moments to review the following discussions.

Section 9.4 How does nondisjunction cause a change in chromosome number?

Section 10.1 How is probability used in analyzing patterns of inheritance?

Section 11.3 How are restriction enzymes and the polymerase chain reaction used in the study of DNA?

Ensuring a Healthy Baby

Today it is possible to screen embryos for genetic abnormalities before they are implanted in the uterus. This process is called preimplantation genetic diagnosis (PGD). Couples who are at high risk of having a child with a genetic disorder can use PGD to increase their chances of having a normal child. The couple donates a number of egg cells and sperm cells, and the embryos are created in a lab by a process called in vitro fertilization. Once the embryos are large enough, one cell from each embryo can be removed and tested for genetic mutations. If an embryo is found to have a mutation for a genetic disorder of concern, it can be discarded. Embryos that appear free of genetic flaws can be implanted into the mother. If all goes well, a healthy baby will be born.

Some people have ethical concerns about using PGD to select genetically healthy embryos. Even more controversial, however, is the use of this technology purely to choose the sex of an embryo. Some couples so desperately want a boy or a girl that they use PGD to guarantee the gender of their child, and some physicians justify this use of the technology for the achievement of "gender-balance" within a family. This use of PGD is legal in the United States, but it has been banned in many countries.

In this chapter, you will learn about some of the techniques currently being used to counsel couples about their chances of passing on a genetic disorder, as well as technologies that may be used to test and treat parents and children for genetic disorders in the future.

13.1 Counseling for Chromosomal Disorders

> ## Learning Outcomes
>
> Upon completion of this section, you should be able to
>
> 1. Explain what a karyotype is and describe its applications in genetic counseling.
> 2. Compare and contrast amniocentesis with chorionic villus sampling (CVS) and describe the benefits and drawbacks of each procedure.
> 3. Define chromosomal deletions and duplications, explain why either may cause a syndrome, and list some known syndromes associated with them.
> 4. Explain the consequences of a translocation or an inversion and give an example of a condition or syndrome caused by each.

Potential parents are becoming aware that many illnesses are caused by abnormal chromosomal inheritance or by gene mutations. Therefore, more couples are seeking **genetic counseling,** which is available at many major hospitals as a means to determine the risk of inherited disorders in a family. For example, a couple might be prompted to seek counseling after several miscarriages, when several relatives have a particular medical condition, or if they already have a child with a genetic defect. The counselor helps the couple understand the mode of inheritance, the medical consequences of a particular genetic disorder, and the decisions they might wish to make (**Fig. 13.1**).

Figure 13.1 Genetic counseling.

A genetic counselor uses genetic information about the family to predict the chances a couple will have a child affected by a genetic disorder. If the woman is pregnant, tests can determine whether the child will be born free of genetic disorders.

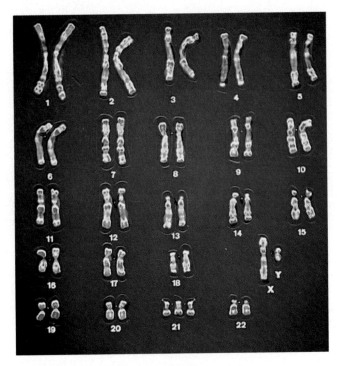

Figure 13.2 Karyotype analysis.

A karyotype can reveal chromosomal mutations. In this case, the karyotype shows that the newborn will have Down syndrome (trisomy 21).

Various human disorders may result from abnormal chromosome number or structure. When a pregnant woman is concerned that her unborn child might have a chromosomal defect, the counselor may recommend karyotyping the fetus's chromosomes.

Karyotyping

A **karyotype** is a visual display of the chromosomes arranged by size, shape, and banding pattern. Any cell in the body except red blood cells, which lack a nucleus, can be a source of chromosomes for karyotyping. In adults, it is easiest to use white blood cells separated from a blood sample for this purpose. In fetuses, whose chromosomes are often examined to detect a syndrome, cells can be obtained by either amniocentesis or chorionic villus sampling. As described in Chapter 9, the karyotype of a person who has Down syndrome usually has three copies of chromosome number 21 instead of two (**Fig. 13.2**).

Amniocentesis is a procedure for obtaining a sample of amniotic fluid from the uterus of a pregnant woman. Blood tests and the age of the mother are considered when determining whether the procedure should be done. The risk of spontaneous abortion increases by about 0.3% due to amniocentesis, and doctors use the procedure only if it is medically warranted. Amniocentesis is not usually performed until the fourteenth to the seventeenth week of pregnancy.

A long needle is passed through the abdominal and uterine walls to withdraw a small amount of fluid, which also contains fetal cells (**Fig. 13.3a**). Tests are done on the amniotic fluid, and the cells are cultured for karyotyping. Karyotyping the chromosomes may be delayed as long as four weeks, so that the cells can be cultured to increase their number.

Chorionic villus sampling (CVS) is a procedure for obtaining chorionic cells in the region where the placenta will develop. This procedure can be done as early as the fifth week of pregnancy. A long, thin suction tube is inserted through the vagina into the uterus (Fig. 13.3b). Ultrasound, which gives a picture of the uterine contents, is used to place the tube between the uterine lining and the chorionic villi. Then, a sampling of chorionic cells is obtained by suction. The cells do not have to be cultured, and karyotyping can be done immediately. But testing amniotic fluid is not possible because no amniotic fluid is collected. Also, CVS carries a greater risk of miscarriage than amniocentesis—0.8% compared with 0.3%. The advantage of CVS is getting the results of karyotyping at an earlier date.

Figure 13.3 Testing for chromosomal mutations.

To test the fetus for an alteration in the chromosome number or structure, fetal cells can be acquired by (**a**) amniocentesis or (**b**) chorionic villus sampling.

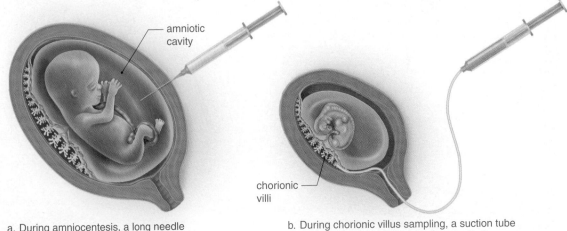

a. During amniocentesis, a long needle is used to withdraw amniotic fluid containing fetal cells.

b. During chorionic villus sampling, a suction tube is used to remove cells from the chorion, where the placenta will develop.

After a cell sample has been obtained, the cells are stimulated to divide in a culture medium. A chemical is used to stop mitosis during metaphase when chromosomes are the most highly compacted and condensed. The cells are then killed, spread on a microscope slide, and dried. In a traditional karyotype, stains are applied to the slides, and the cells are photographed. Staining causes the chromosome to have dark and light crossbands of varying widths, and these can be used, in addition to size and shape, to help pair up the chromosomes. Today, technicians use fluorescent dyes and computers to arrange the chromosomes in pairs.

Chromosomal Mutations

Most chromosomal mutations can be detected microscopically. A karyotype reveals changes in chromosome number, and a skilled technician is able to detect differences in chromosome structure based on a change in the normal banding patterns of the chromosomes (see Fig. 13.2).

In humans, only a few variations in chromosome number, such as Down syndrome, Turner syndrome, and Klinefelter syndrome, are typically seen. Changes in chromosome structure, however, are much more common in the population. Syndromes that result from changes in chromosome structure are due to the breakage of chromosomes and their failure to reunite properly. Various environmental agents—radiation, certain organic chemicals, and even viruses—can cause chromosomes to break apart. Ordinarily, when breaks occur in chromosomes, the segments reunite to give the same sequence of genes. But their failure to do so results in one of several types of mutations: deletion, duplication, translocation, or inversion. Chromosomal mutations can occur during meiosis, and if the offspring inherits the abnormal chromosome, a syndrome may result.

Deletions and Duplications

A **deletion** occurs when a single break causes a chromosome to lose an end piece, or when two simultaneous breaks lead to the loss of an internal chromosome segment. An individual who inherits a normal chromosome from one parent and a chromosome with a deletion from the other parent no longer has a pair of alleles for each trait, and a syndrome can result.

Williams syndrome occurs when chromosome 7 loses a tiny end piece (**Fig. 13.4**). Children who have this syndrome look like pixies, with turned-up noses, wide mouths, a small chin, and large ears. Although their academic skills are poor, they exhibit excellent verbal and musical abilities. The gene that governs the production of the protein elastin is missing, and this affects the health of the cardiovascular system and causes their skin to age prematurely. Such individuals are very friendly but need an ordered life, perhaps because of the loss of a gene for a protein that is normally active in the brain.

Connections and Misconceptions

What do the colored bands on a karyotype represent?

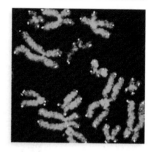

The colored bands (see Fig. 13.2) are not the natural color of the chromosomes. Instead, using a procedure called *fluorescent immunohistochemistry in situ hybridization* (FISH for short), short pieces of DNA are labeled with a fluorescent marker. These pieces of DNA bind to their complementary sequences on the chromosomes. When exposed to specific wavelengths of light, the markers emit different colors of light. This helps counselors more easily identify the chromosomes. A tagged piece of DNA can also be developed for a specific gene (see the accompanying figure), which helps researchers identify the chromosomal location of a gene of interest.

b.

Figure 13.4 **Deletion.**

a. When chromosome 7 loses an end piece, the result is Williams syndrome. **b.** These children, although unrelated, have the same appearance, health, and behavioral problems that are characteristic of Williams syndrome.

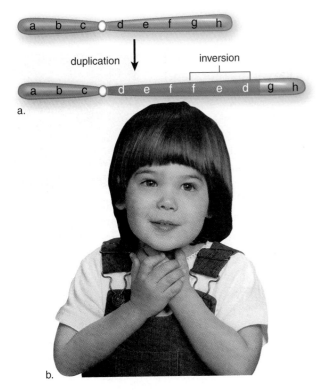

Figure 13.5 Duplication.

a. When a piece of chromosome 15 is duplicated and inverted, inv dup 15 syndrome results. **b.** Children with this syndrome have poor muscle tone and autistic characteristics.

Cri du chat (cat's cry) syndrome occurs when chromosome 5 is missing an end piece. The affected individual has a small head, is mentally disabled, and has facial abnormalities. Abnormal development of the glottis and larynx results in the most characteristic symptom—the infant's cry resembles that of a cat.

In a **duplication,** a chromosome segment is repeated, so that the individual has more than two alleles for certain traits. An inverted duplication is known to occur in chromosome 15. (*Inversion* means that a segment joins in the direction opposite from normal.) Children with this syndrome, called *inv dup 15 syndrome,* have poor muscle tone, mental disabilities, seizures, a curved spine, and autistic characteristics that include poor speech, hand flapping, and lack of eye contact (**Fig. 13.5**).

Animation
Changes in
Chromosome
Structure

Translocation

A **translocation** is the exchange of chromosome segments between two nonhomologous chromosomes. A person who has both of the involved chromosomes has the normal amount of genetic material and a normal phenotype, unless the chromosome exchange breaks an allele into two pieces or fuses two genes together. The person who inherits only one of the translocated chromosomes will no doubt have only one copy of certain alleles and three copies of other alleles. A genetic counselor begins to suspect a translocation has occurred when spontaneous abortions are commonplace and family members suffer from various syndromes. A special microscopic technique allows a technician to determine that a translocation has occurred.

In 5% of Down syndrome cases, a translocation that occurred in a previous generation between chromosomes 21 and 14 is the cause. As long as the two chromosomes are inherited together, the individual is normal. But in future generations a person may inherit two normal copies of chromosome 21 and the abnormal chromosome 14 that contains a segment of chromosome 21. In these cases, Down syndrome is not related to parental age but instead tends to run in the family of either the father or the mother.

Figure 13.6 shows a father and daughter who have a translocation between chromosomes 2 and 20. Although they have the normal amount of genetic material, they have the distinctive face, abnormalities of the eyes and internal organs, and severe itching characteristic of Alagille syndrome. People with this syndrome ordinarily have a deletion on chromosome 20; therefore, it can be deduced that the translocation disrupted an allele on chromosome 20 in the father. The symptoms of Alagille syndrome range from mild to severe, so some people may not be aware they have the syndrome. This father did not realize it until he had a child with the syndrome.

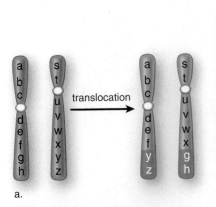

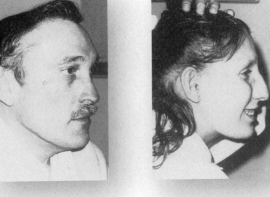

Figure 13.6 Translocation.

a. When chromosomes 2 and 20 exchange segments, Alagille syndrome results because the translocation disrupts as allele on chromosome 20.
b. Distinctive facial features are one of the characteristics of Alagille syndrome.

Inversion

An **inversion** occurs when a segment of a chromosome is turned 180°. You might think this is not a problem because the same genes are present, but the reverse sequence of alleles can lead to altered gene activity if it disrupts control of gene expression.

Inversions usually do not cause problems, but may lead to an increased occurrence of abnormal chromosomes during sexual reproduction. Crossing-over between an inverted chromosome and the noninverted homologue can lead to recombinant chromosomes that have both duplicated and deleted segments. This happens because alignment between the two homologues is only possible when the inverted chromosome forms a loop (**Fig. 13.7**).

Animation
The Consequence of Inversion

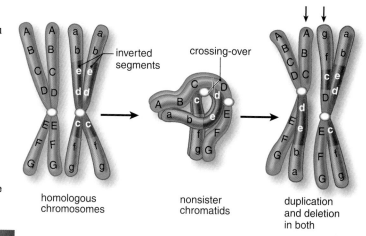

Figure 13.7 **Inversion.**

(Left) A segment is inverted in the sister chromatids of one homologue. Notice that, in the red chromosome, *edc* occurs instead of *cde*. (Middle) The nonsister chromatids can align only when the inverted sequence forms an internal loop. After crossing-over, a duplication and a deletion can occur. (Right) The nonsister chromatid at the left arrow has *AB* and *ab* alleles and neither *fg* nor *FG* alleles. The nonsister chromatid at the right arrow has *gf* and *GF* alleles and neither *AB* nor *ab* alleles.

Connecting the Concepts

For more information on the contents of this section, refer to the following discussions.

Section 9.4 describes how nondisjunction causes Down, Turner, and Klinefeltner syndromes.

Section 29.4 explores the stages of human development.

Check Your Progress 13.1

1. Describe the relationship between amniocentesis and karyotyping.
2. List an advantage and a disadvantage of chorionic villus sampling compared with amniocentesis.
3. Explain why karyotypes utilize metaphase chromosomes of mitosis.

13.2 Counseling for Genetic Disorders

Learning Outcomes

Upon completion of this section, you should be able to

1. Interpret a pedigree to determine if the pattern of inheritance is autosomal dominant or recessive.
2. List the characteristics of autosomal dominant and autosomal recessive pedigrees.
3. Distinguish between a pedigree of autosomal and X-linked disorders.
4. List some common genetic disorders, state the symptoms of each, and describe the inheritance pattern that each one exhibits.

Connections and Misconceptions

Do we know what genes cause Down syndrome?

Since individuals who inherit Down syndrome by a translocation event receive only a portion of chromosome 21, it has been possible to determine which genes on chromosome 21 are contributing to the symptoms of Down syndrome.

One of the more significant genes appears to be *GART*, a gene involved in the processing of a type of nucleotide called purines. Extra copies of *GART* are believed to be a major contributing factor in the symptoms of mental retardation shown in Down syndrome patients. Another gene that has been identified encodes for the protein collagen (*COL6A1*), a component of connective tissue. Too many copies of this gene cause the heart defects common in Down syndrome.

Even if no chromosomal abnormality is likely, amniocentesis still might be done because it is now possible to perform biochemical tests on amniotic fluid to detect over 400 different disorders caused by specific genes in a fetus. The genetic counselor determines ahead of time what tests might be warranted. To do this, the counselor needs to know the medical history of the family in order to construct a pedigree.

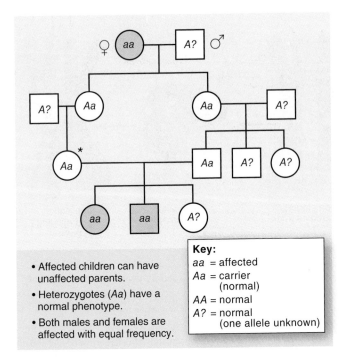

Figure 13.8 lists other ways that a counselor may recognize an autosomal recessive pattern of inheritance. Notice that, in this pedigree, cousins are the parents of three children, two of whom have the disorder. Aside from illustrating that reproduction between related individuals is more likely to bring out recessive traits, this pedigree also shows that "chance has no memory"; therefore, each child born to heterozygous parents has a 25% chance of having the disorder. In other words, if a heterozygous couple has four children, each child might have the condition.

Figure 13.9 shows an *autosomal dominant* pattern of inheritance. In this pattern, a child can be unaffected even when the parents are heterozygous and therefore affected. Figure 13.9 lists other ways to recognize an autosomal dominant pattern of inheritance. This pedigree illustrates that, when both parents are unaffected, all their children are unaffected. Why? Because neither parent has a dominant gene that causes the condition to be passed on.

Pedigrees for Sex-Linked Disorders

Figure 13.10 gives a pedigree for an *X-linked recessive disorder*. Recall that sons inherit X-linked recessive traits from their mothers because their only X chromosome comes from their mothers. More males than females have the disorder because recessive alleles on the X chromosome are always expressed in males—the Y chromosome lacks an allele. Females who have the condition inherited the allele from both their mother and their father, and all the sons of such a female will have the condition.

If a male has an X-linked recessive condition, his daughters are all carriers even if his partner is normal. Therefore, X-linked recessive conditions often appear to pass from grandfather to grandson. Figure 13.10 lists other ways to recognize an X-linked recessive disorder.

Family Pedigrees

A **pedigree** is a chart of a family's history with regard to a particular genetic trait. Some traits are carried on the autosomes (nonsex chromosomes) and are called **autosomal traits;** other traits are carried on the sex chromosomes and are, therefore, called **sex-linked traits.** In a pedigree, males are designated by squares and females by circles. Shaded circles and squares represent individuals expressing the genetic disorder. A line between a square and a circle represents a union. A vertical line going downward leads directly to a single child; if there are more children, they are placed off a horizontal line.

From the counselor's knowledge of genetic disorders, he or she might already know the *pattern of inheritance* of a trait—that is, whether it is autosomal dominant, autosomal recessive, or X-linked recessive. The counselor can then determine the chances that any child born to the couple will have the abnormal phenotype.

Pedigrees for Autosomal Disorders

A family pedigree for an *autosomal recessive disorder* is shown in **Figure 13.8.** In this pattern, a child can be affected when neither parent is affected. Such heterozygous parents are *carriers* because, although they are unaffected, they are capable of having a child with the genetic disorder. If the family pedigree suggests that the parents are carriers for an autosomal recessive disorder, the counselor might suggest confirming this by doing the appropriate genetic test. Then, if the parents so desire, it would be possible to do prenatal testing of the fetus for the genetic disorder.

Figure 13.8 Autosomal recessive pedigree.

The list gives ways to recognize an autosomal recessive disorder. How would you know that the individual at the * is heterozygous?

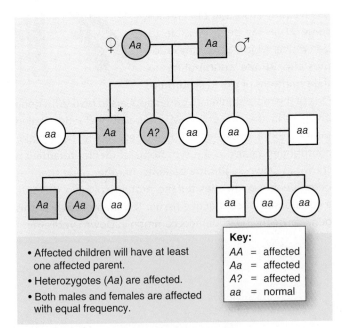

Figure 13.9 Autosomal dominant pedigree.

The list gives ways to recognize an autosomal dominant disorder. How would you know that the individual at the * is heterozygous?

*The designated individual in Figure 13.8 does not have the disorder but she has children with the disorder. The designated individual in Figure 13.9 does have the disorder but he has children without the disorder.

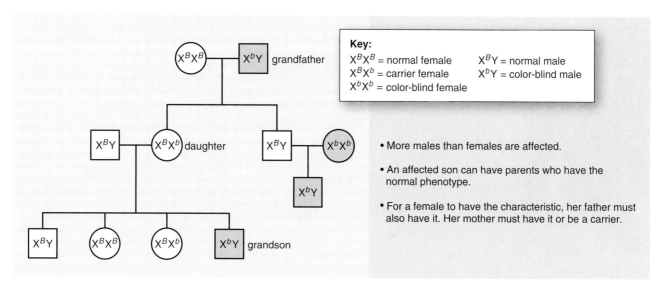

Key:
X^BX^B = normal female X^BY = normal male
X^BX^b = carrier female X^bY = color-blind male
X^bX^b = color-blind female

- More males than females are affected.

- An affected son can have parents who have the normal phenotype.

- For a female to have the characteristic, her father must also have it. Her mother must have it or be a carrier.

Figure 13.10 **X-linked recessive pedigree.**

The list gives ways of recognizing an X-linked recessive disorder—in this case, color blindness.

Still fewer traits are known to be *X-linked dominant.* If a condition is X-linked dominant, daughters of affected males have a 100% chance of having the condition. Females can pass an X-linked dominant allele to both sons and daughters. If a female is heterozygous and her partner is normal, each child has a 50% chance of escaping an X-linked dominant disorder, depending on which of the mother's X chromosomes is inherited.

Still fewer genetic disorders involve genes carried on the Y chromosome. For example, the *SRY* gene (sex determining region Y) is involved in determining gender during development. A counselor would recognize a *Y-linked pattern of inheritance* because Y-linked disorders are present only in males and are passed directly from father to *all* sons but not to daughters.

Genetic Disorders of Interest

Although many conditions are caused by interactions of genes and the environment (see Chapter 10), historically medical genetics has focused on disorders caused by single gene mutations. A few examples are provided in this section.

Autosomal Disorders

Autosomal disorders are caused by mutated alleles on the autosomal chromosomes (all the chromosomes except the sex chromosomes). Some of these disorders are recessive, and therefore an individual must inherit two affected alleles before having the disorder. Others are dominant, meaning that it takes only one affected allele to cause the disorder. Dominant and recessive inheritance was discussed in Chapter 10.

Methemoglobinemia

Methemoglobinemia is a relatively harmless disorder that results from an accumulation of methemoglobin, an alternative form of hemoglobin, in the blood. Since methemoglobin is blue instead of red, the skin of people with the disorder appears bluish-purple in color (**Fig. 13.11**).

Connections and Misconceptions

What is an example of a Y-linked disorder?

Perhaps the most widely recognized Y-linked disorder is azoospermia, a condition in which males do not release any sperm during ejaculation. Azoospermia may be caused by either an obstruction of the pathway that releases sperm from the testis or a problem with the maturation of sperm cells. In some males, a small deletion in the *DAZ* (deleted in *a*zoospermia) gene on the Y chromosome causes azoospermia, although the exact function of this gene is still unknown.

Video Human Sperm

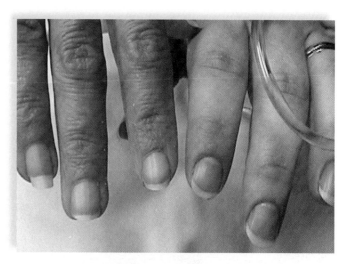

Figure 13.11 **Methemoglobinemia.**

The hand of the woman on the right appears blue due to chemically induced methemoglobinemia. Individuals with the disorder lack the enzyme diaphorase and are unable to clear methemoglobin from the blood.

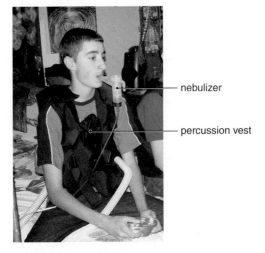

Figure 13.12 **Cystic fibrosis.**

This person is undergoing antibiotic and percussion therapy for cystic fibrosis. Antibiotic therapy is used to control lung infections of cystic fibrosis patients. The antibiotic can be aerosolized and administered using a nebulizer. A percussion vest loosens mucus in the lungs.

Figure 13.13
Alkaptonuria.

Alkaptonuria is caused by the inability to metabolize a compound called homogentisic acid. Excess homogentisic acid accumulates in the urine, and its presence causes the urine to turn black when the urine is exposed to air.

Normal   Alkaptonuria

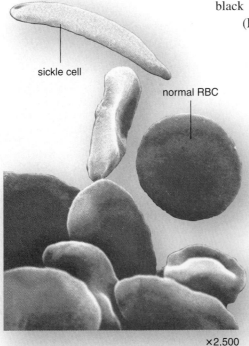

sickle cell

normal RBC

Figure 13.14 **Sickle cell disease.**

Persons with sickle cell disease have sickle-shaped red blood cells because of an abnormal hemoglobin molecule.

×2,500

A persistent and determined physician finally solved the age-old mystery of what causes methemoglobinemia through blood tests and pedigree analysis of a family with the disorder, the Fugates of Troublesome Creek in Kentucky. On a hunch, the physician tested the Fugates for the enzyme diaphorase, which normally converts methemoglobin back to hemoglobin, and found that they indeed lacked the enzyme. Next, he treated a patient with the disorder in a simple but rather unconventional manner—by injecting a dye called methylene blue! The dye is a strong reducing agent capable of donating electrons to methemoglobin, converting it back into hemoglobin. The results were striking but immediate—the patient's skin quickly turned pink again.

Cystic Fibrosis

Cystic fibrosis is an autosomal recessive disorder that occurs among all ethnic groups, but it is the most common lethal genetic disorder among Caucasians in the United States. Research has demonstrated that chloride ions (Cl^-) fail to pass through a plasma membrane channel protein in the cells of these patients. Ordinarily, after chloride ions have passed through the membrane, sodium ions (Na^+) and water follow. It is believed that the lack of water then causes abnormally thick mucus in the bronchial tubes and pancreatic ducts, thus interfering with the function of the lungs and pancreas. To ease breathing in affected children, the thick mucus in the lungs must be loosened periodically, but still the lungs become infected frequently (**Fig. 13.12**). Clogged pancreatic ducts prevent digestive enzymes from reaching the small intestine, and to improve digestion, patients take digestive enzymes mixed with applesauce before every meal.

Alkaptonuria

Black urine disease, or **alkaptonuria,** is a rare genetic disorder that follows an autosomal recessive inheritance pattern. People with alkaptonuria lack a functional copy of the homogentisate oxygenase (*HGD*) gene found on chromosome 3. The HGD enzyme normally breaks down a compound called homogentisic acid. When the enzyme is missing, homogentisic acid accumulates in the blood and is passed into the urine. The compound turns black on exposure to air, giving the urine a characteristic color and odor (**Fig. 13.13**). Homogentisic acid also accumulates in joint spaces and connective tissues, leading to darkening of the tissues and eventual arthritis by adulthood.

Sickle Cell Disease

Sickle cell disease is an autosomal recessive disorder in which the red blood cells are not biconcave disks like normal red blood cells, but are irregular in shape (**Fig. 13.14**). In fact, many are sickle-shaped. A single base change in the globin gene causes hemoglobin to differ from normal hemoglobin by a single amino acid. Now, the abnormal hemoglobin molecules stack up and form insoluble rods, and the red blood cells become sickle-shaped.

Because sickle-shaped cells can't pass along narrow capillary passageways as well as disk-shaped cells can, they clog the vessels and break down. This is why persons with sickle cell disease suffer from poor circulation, anemia, and low resistance to infection. Internal hemorrhaging leads to further complications, such as jaundice, episodic pain in the abdomen and joints, and damage to internal organs.

Sickle cell heterozygotes have normal blood cells but carry the sickle cell trait. Most experts believe that persons that are heterozygous for the sickle cell trait are generally healthy and do not need to restrict their physical activity. However, there occasionally may be problems if they experience dehydration or mild oxygen deprivation.

Marfan Syndrome

Marfan syndrome (see Chapter 10, page 166), an autosomal dominant disorder, is caused by a defect in an elastic connective tissue protein called fibrillin. This protein is normally abundant in the lens of the eye; the bones of limbs, fingers, and ribs; and the wall of the aorta. Thus, the affected person often has a dislocated lens, tall stature, long limbs and fingers, and a caved-in chest. The aorta wall is weak and can burst without warning. A tissue graft can strengthen the aorta, but affected individuals should not overexert themselves.

Huntington Disease

Huntington disease is a dominant neurological disorder that leads to progressive degeneration of neurons in the brain (**Fig. 13.15**). The disease is caused by a single mutated copy of the gene for a protein called huntingtin. Most patients appear normal until they are middle-aged and have already had children, who may also have the inherited disorder. There is no effective treatment, and death usually occurs 10 to 15 years after the onset of symptoms.

Several years ago, researchers found that the gene for Huntington disease is located on chromosome 4. A test was developed for the presence of the gene, but few people want to know if they have inherited the gene because there is no cure. But now we know that the disease stems from an unusual mutation. Extra codons cause the huntingtin protein to have a series of extra glutamines. Whereas the normal version of huntingtin has stretches of between 10 and 25 glutamines, mutant huntingtin may contain 36 or more. Because of the extra glutamines, the huntingtin protein changes shape and forms large clumps inside neurons. Even worse, it attracts and causes other proteins to clump with it. One of these proteins, called CBP, helps nerve cells survive. Researchers hope to combat the disease by boosting CBP levels.

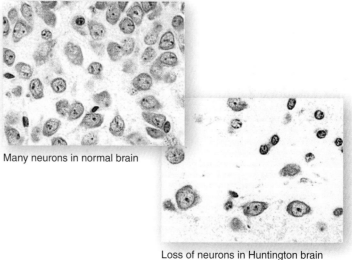

Many neurons in normal brain

Loss of neurons in Huntington brain

Figure 13.15 Huntington disease.

Huntington disease is characterized by increasingly serious psychomotor and mental disturbances because of a loss of neurons in the brain.

X-Linked Recessive Disorders

As you know, X-linked recessive disorders are caused by mutated alleles on the X chromosome. A son inherits an X-linked recessive condition from his mother.

Color Blindness   **Color blindness** is a common X-linked recessive disorder. About 8% of Caucasian men have red-green color blindness. Most of them see brighter greens as tans, olive greens as browns, and reds as reddish-browns. A few cannot tell reds from greens at all; they see only yellows, blues, blacks, whites, and grays.

Duchenne Muscular Dystrophy   **Duchenne muscular dystrophy** is an X-linked recessive disorder characterized by wasting away of the muscles. The absence of a protein, now called dystrophin, is the cause of the disorder. Much investigative work determined that dystrophin is involved in the release of calcium from the sarcoplasmic reticulum in muscle fibers. The lack of dystrophin causes calcium to leak into the cell, which promotes the action of an enzyme that dissolves muscle fibers. When the body attempts to repair the tissue, fibrous tissue forms (**Fig. 13.16**), and this cuts off the blood supply, so that more and more cells die.

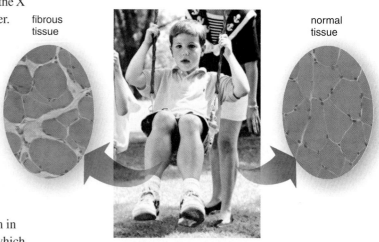

fibrous tissue

normal tissue

Figure 13.16 Muscular dystrophy.

In muscular dystrophy, the calves enlarge because fibrous tissue develops as muscles waste away due to a lack of the protein dystrophin.

Symptoms such as waddling gait, toe walking, frequent falls, and difficulty in rising may appear as soon as the child starts to walk. Muscle weakness intensifies until the individual is confined to a wheelchair. Death usually occurs by age 20; therefore, affected males are rarely fathers. The recessive allele remains in the population through passage from carrier mother to carrier daughter.

As therapy, immature muscle cells can be injected into muscles, and for every 100,000 cells injected, dystrophin production occurs in 30–40% of muscle fibers.

Connecting the Concepts

For more on the physiology of the diseases presented in this section, refer to the following discussions.

Section 23.3 describes the function of blood and how it contributes to homeostasis in the body.

Section 27.1 examines how the neurons of the nervous system process information.

Section 28.1 explores the operation of muscle tissue.

Check Your Progress 13.2

1 Contrast a duplication chromosome with a deletion chromosome and contrast a translocation chromosome with an inversion chromosome.

2 Explain why an individual with a translocation might appear normal.

3 Summarize the differences between the inheritance of an autosomal disorder and a sex-linked disorder. Give a few examples of each.

13.3 Testing for Genetic Disorders

Learning Outcomes

Upon completion of this section, you should be able to

1. Summarize how genetic markers and DNA microarrays may be used to diagnose a genetic disorder.
2. Distinguish among the procedures used to test DNA, the fetus, and the embryo for specific genetic disorders.

Following genetic testing, a genetic counselor can explain to prospective parents the chances a child of theirs will have a disorder that runs in their family. If a woman is already pregnant, the parents may want to know whether the unborn child has the disorder. If the woman is not pregnant, the parents may opt for testing of the embryo or egg before she does become pregnant, as described shortly.

Testing depends on the genetic disorder of interest. In some instances, it is appropriate to test for a particular protein, and in others to test for the mutated gene.

Testing for a Protein

Some genetic mutations lead to disorders caused by a lack of enzyme activity. For example, in the case of methemoglobinemia, it is possible to test for the quantity of the enzyme diaphorase in a blood sample and, from that, determine whether the individual is likely homozygous normal, is a carrier, or has methemoglobinemia. If the parents are carriers, each child has a 25% chance of having methemoglobinemia. This knowledge may lead prospective parents to opt for testing of the embryo or egg, as described in Figures 13.20 and 13.21.

Testing the DNA

Two types of DNA testing are possible: testing for a genetic marker and using a DNA probe.

Genetic Markers

Testing for a genetic marker relies on a difference in the DNA due to the presence of the abnormal allele. As an example, consider that individuals with sickle cell trait or Huntington disease have an abnormality in a gene's base sequence. This abnormality in sequence is a **genetic marker.** As you know, restriction enzymes cleave DNA at particular base sequences (see Chapter 11). Therefore, the fragments that result from the use of a restriction enzyme may be different for people who are normal than for those who are heterozygous or homozygous for a mutation (**Fig. 13.17**).

Animation
Restriction
Endonucleases

Genetic Profiling

New technologies have made DNA testing easy and inexpensive. For example, it is now possible to place thousands of known disease-associated mutant alleles onto a **DNA microarray**—a small silicon chip containing many DNA samples—in this case, the mutant alleles (**Fig. 13.18**). Genomic DNA from the subject to be tested is labeled with a fluorescent dye, then added to the microarray. The spots on the microarray fluoresce if the DNA binds to the mutant alleles on the chip, indicating that the subject may have a particular disorder or is at risk of developing it later in life. An individual's complete genotype, including all the various mutations, is called the genetic profile.

Animation
Using a DNA
Microarray

Video
Melanoma
Marker

With the help of a genetic counselor, individuals can be educated about their genetic profile. It's possible that a person has or will have a genetic disorder caused by a single pair of alleles. However, polygenic traits are more common, and in these instances, the genetic profile can indicate an increased or decreased risk for a disorder. Risk information can be used to design a program of medical surveillance and to foster a lifestyle aimed at reducing the risk. For example, suppose an individual has mutations common to people with colon cancer. It would be helpful for him or her to have an annual colonoscopy, so that any abnormal growths can be detected and removed before they become invasive.

Connections and Misconceptions

Are over-the-counter (OTC) genetic tests accurate?

With advances in DNA technology that have provided the ability for DNA microarrays to be assembled inexpensively, a number of companies are now offering OTC genetic tests. Currently, the Food and Drug Administration (FDA) does not regulate OTC genetic tests, and there are a number of concerns regarding the validity of the information obtained from these tests. In addition, some people worry that the test results may be used to discriminate against persons applying for insurance or jobs. It is recommended that genetic tests be performed by licensed labs, and only after all of the patient's rights have been discussed.

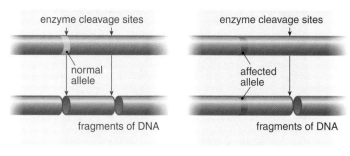

a. Normal fragmentation pattern

b. Genetic disorder fragmentation pattern

Figure 13.17 Use of a genetic marker to test for a genetic mutation.

a. In this example, DNA from a normal individual has certain restriction enzyme cleavage sites. **b.** DNA from another individual lacks one of the cleavage sites, and this loss indicates that the person has a mutated gene. In heterozygotes, half of their DNA would have the cleavage site and half would not have it. (In other instances, the gain in a cleavage site could be an indication of a mutation.)

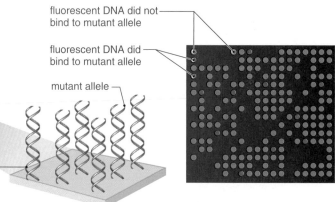

Figure 13.18 Use of a DNA microarray to test for mutated genes.

This DNA microarray contains many disease-associated mutant alleles. Fluorescently labeled genomic DNA from an individual has been added to the microarray. Any fluorescent spots indicate that binding has occurred and that the individual may have the genetic disorder or is at risk for developing it later in life.

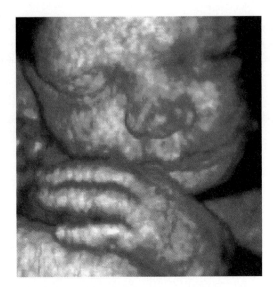

Figure 13.19 Ultrasound.

An ultrasound exam can produce three-dimensional images, such as the normal one shown here. Ultrasound exams can allow physicians to detect serious abnormalities, such as neural tube abnormalities.

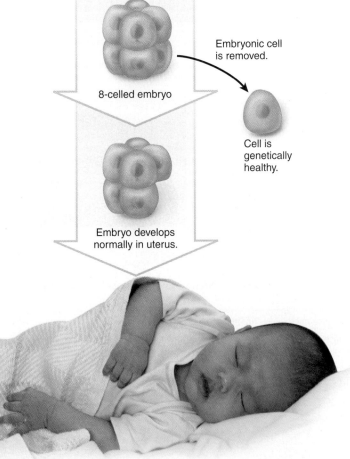

Embryonic cell is removed.

8-celled embryo

Cell is genetically healthy.

Embryo develops normally in uterus.

Figure 13.20 Testing the embryo.

Genetic diagnosis is performed on one cell removed from an 8-celled embryo. If this cell is found to be free of the genetic defect of concern, and the 7-celled embryo is implanted in the uterus, it develops into a newborn with a normal phenotype.

Testing the Fetus

If a woman is already pregnant, ultrasound can detect serious fetal abnormalities, and it is also possible to obtain and test the DNA of fetal cells for genetic defects.

Ultrasound

Ultrasound images help doctors evaluate fetal anatomy. An ultrasound probe scans the mother's abdomen, and a transducer transmits high-frequency sound waves, which are transformed into a picture on a video screen. This picture shows the fetus inside the uterus (**Fig. 13.19**). Ultrasound can be used to determine the baby's age and size, as well as whether there is more than one baby. Also, some chromosomal abnormalities, such as Down syndrome, Edwards syndrome (three copies of chromosome 18), and Patua syndrome (three copies of chromosome 13), cause anatomical abnormalities during fetal development that may be detected by ultrasound by the twentieth week of pregnancy. For this reason, a routine ultrasound at this time is considered essential to good prenatal care.

Many other conditions, such as spina bifida, can be diagnosed by an ultrasound. Spina bifida results when the spine fails to close properly during the first month of pregnancy. Surgery to close a newborn's spine in such a case is generally performed within 24 hours after birth.

Testing Fetal Cells

Fetal cells can be tested for various genetic disorders. If the fetus has an incurable disorder, the parents may wish to consider an abortion.

For testing purposes, fetal cells may be acquired through amniocentesis or chorionic villus sampling, as described earlier in this chapter. In addition, fetal cells may be collected from the mother's blood. As early as nine weeks into the pregnancy, a small number of fetal cells can be isolated from the mother's blood using a cell sorter. While mature red blood cells lack a nucleus, immature red blood cells do have a nucleus, and they have a shorter life span than mature red blood cells. Therefore, if nucleated fetal red blood cells are collected from the mother's blood, they are known to be from this pregnancy.

Video Safer Fetal Test

Only about 1/70,000 blood cells in a mother's blood are fetal cells, and therefore the polymerase chain reaction (PCR) is used to amplify the DNA from the few cells collected. The procedure poses no risk to the fetus.

Testing the Embryo and Egg

As discussed in Chapter 29, in vitro fertilization (IVF) is carried out in laboratory glassware. The physician obtains eggs from the prospective mother and sperm from the prospective father and places them in the same receptacle, where fertilization occurs. Following IVF, now a routine procedure, it is possible to test the embryo. Prior to IVF, it is possible to test the egg for any genetic defect. In any case, only normal embryos are transferred to the uterus for further development.

Video In Vitro Fertilization

Testing the Embryo

If prospective parents are carriers for one of the genetic disorders discussed earlier, they may want assurance that their offspring will be free of the disorder. Genetic diagnosis of the embryo will provide this assurance.

Following IVF, the zygote (fertilized egg) divides. When the embryo has six to eight cells, one of these cells can be removed for diagnosis, with no effect on normal development (**Fig. 13.20**). Only embryos that test negative for the genetic disorders of interest are placed in the uterus to continue developing.

So far, over 1,000 children worldwide have been born free of alleles for genetic disorders that run in their families following embryo testing. In the future, embryos that test positive for a disorder could be treated by gene therapy, so that those embryos, too, would be allowed to continue to term.

Testing the Egg

Unlike males, which produce four sperm cells following meiosis, meiosis in females results in the formation of a single egg and at least two nonfunctional cells called polar bodies. Polar bodies, which later disintegrate, receive very little cytoplasm, but they do receive a haploid number of chromosomes, and thus can be useful for genetic diagnosis. When a woman is heterozygous for a recessive genetic disorder, about half the polar bodies receive the mutated allele, and in these instances the egg receives the normal allele. Therefore, if a polar body tests positive for a mutated allele, the egg probably received the normal allele (**Fig. 13.21**). Only normal eggs are then used for IVF. Even if the sperm should happen to carry the mutation, the zygote will, at worst, be heterozygous. But the phenotype will appear normal.

If gene therapy becomes routine in the future, it's possible that an egg will be given genes that control traits desired by the parents, such as musical or athletic ability, prior to IVF. Such genetic manipulation, called **eugenics,** carries many ethical concerns.

Connecting the Concepts

For more information on this material, refer to the following discussions.

Section 11.3 examines how scientists use restriction enzymes and the polymerase chain reaction to study DNA.

Section 29.3 explores assisted reproductive technologies, such as IVF.

Figure 29.7 explains how polar bodies are generated during meiosis in females.

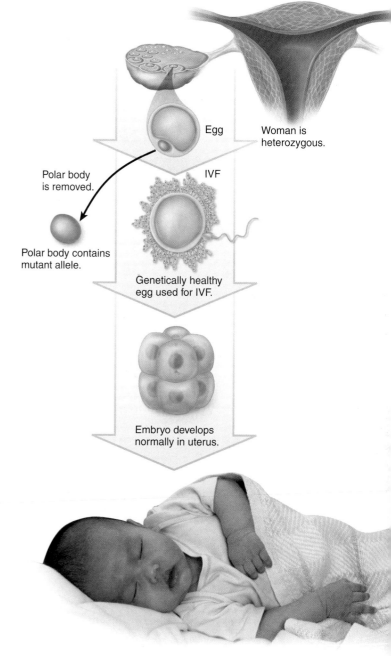

Figure 13.21 **Testing the egg.**

Genetic diagnosis is performed on a polar body removed from an egg. If the egg is free of a genetic defect, it is used for IVF, and the embryo is transferred to the uterus for further development.

Check Your Progress 13.3

1. Explain the differences in testing for a protein and testing the DNA for genetic disorders.

2. Describe the differences between an ultrasound, amniocentesis, and chorionic villus sampling. List the limitations of each.

3. Summarize the advantages of testing the egg for genetic disorders and testing the fetus. Describe any disadvantages for each as well.

4. Describe how fetal blood cells can be separated from the mother's cells.

5. Predict how changes in base sequence can be used to test for genetic disorders.

6. Discuss the rationale for testing a polar body to determine if the egg carries a genetic disorder.

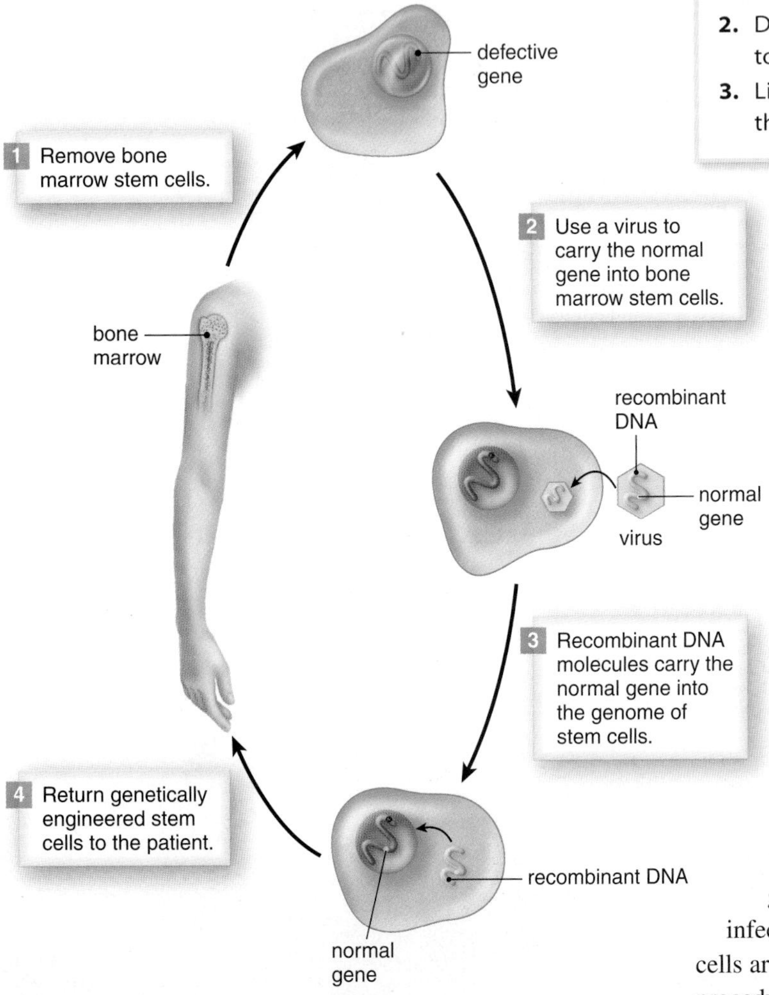

13.4 Gene Therapy

1 Remove bone marrow stem cells.

defective gene

2 Use a virus to carry the normal gene into bone marrow stem cells.

bone marrow

recombinant DNA

normal gene

virus

3 Recombinant DNA molecules carry the normal gene into the genome of stem cells.

4 Return genetically engineered stem cells to the patient.

recombinant DNA

normal gene

Figure 13.22 Ex vivo gene therapy in humans.

Bone marrow stem cells are withdrawn from the body. A virus is used to insert a normal gene into the host genome, and then the cells are returned to the body.

Gene therapy is the insertion of genetic material into human cells for the treatment of a disorder. It includes procedures that give a patient healthy genes to make up for faulty genes, as well as the use of genes to treat various other human illnesses, such as cardiovascular disease and cancer. Gene therapy includes both ex vivo (outside the body) and in vivo (inside the body) methods. Viruses genetically modified to be safe can be used to ferry a normal gene into cells (**Fig. 13.22**), and so can liposomes, which are microscopic globules of lipids specially prepared to enclose the normal gene. On the other hand, sometimes the gene is injected directly into a particular region of the body. This section discusses examples of **ex vivo gene therapy** (the gene is inserted into cells that have been removed and then returned to the body) and **in vivo gene therapy** (the gene is delivered directly into the body).

Ex Vivo Gene Therapy

Children who have severe combined immunodeficiency (SCID) lack the enzyme adenosine deaminase (ADA), which is involved in the maturation of cells that produce antibodies. In order to carry out gene therapy, bone marrow stem cells are removed from the blood and infected with a virus that carries a normal gene for the enzyme. Then the cells are returned to the patient. Bone marrow stem cells are preferred for this procedure because they divide to produce more cells with the same genes. Patients who have undergone this procedure show significantly improved immune function associated with a sustained rise in the level of ADA enzyme activity in the blood (**Fig. 13.23***a*).

Ex vivo gene therapy is also used in the treatment of familial hypercholesterolemia, a genetic

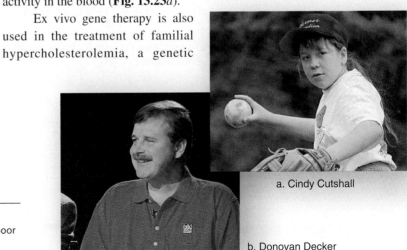

a. Cindy Cutshall

b. Donovan Decker

Figure 13.23 Gene therapy patients.

a. This patient was treated with the *ADA* gene to cure severe combined immunodeficiency. **b.** This patient was treated with the *VEGF* gene to alleviate poor coronary circulation.

disorder in which high levels of plasma cholesterol make the patient subject to fatal heart attacks at a young age. A small portion of the liver is surgically excised and then infected with a retrovirus containing a normal gene for a cholesterol receptor before the tissue is returned to the patient. Several patients have experienced lowered plasma cholesterol levels following this procedure.

Some cancers are being treated with ex vivo gene therapy procedures. In one procedure, immune system cells are removed from a cancer patient and genetically engineered to display tumor antigens. After these cells are returned to the patient, they stimulate the immune system to kill tumor cells.

In Vivo Gene Therapy

Most cystic fibrosis patients lack a gene that codes for a regulator of a transmembrane carrier for the chloride ion. In gene therapy trials, the gene needed to cure cystic fibrosis is sprayed into the nose or delivered to the lower respiratory tract by adenoviruses or by the use of liposomes (microscopic vesicles that spontaneously form when lipoproteins are put into a solution). Investigators are trying to improve uptake of the genes and are hypothesizing that a combination of all three vectors might be more successful.

Genes are also being used to treat medical conditions such as poor coronary circulation. It has been known for some time that vascular endothelial growth factor (VEGF) can cause the growth of new blood vessels. The gene that codes for this growth factor can be injected alone, or within a virus, into the heart to stimulate branching of coronary blood vessels. Patients who have received this treatment report that they have less chest pain and can run longer on a treadmill (Fig. 13.23b).

Rheumatoid arthritis, a crippling disorder in which the immune system turns against a person's own body and destroys joint tissue, has recently been treated with in vivo gene therapy methods. Clinicians inject adenoviruses that contain an anti-inflammatory gene into the affected joint. The added gene reduces inflammation within the joint space and lessens the patient's pain and suffering. Clinical trials have been promising, and animal studies have even shown that gene therapy may stave off arthritis in at-risk individuals.

Figure 13.24 summarizes how gene therapy is being used to treat various illnesses.

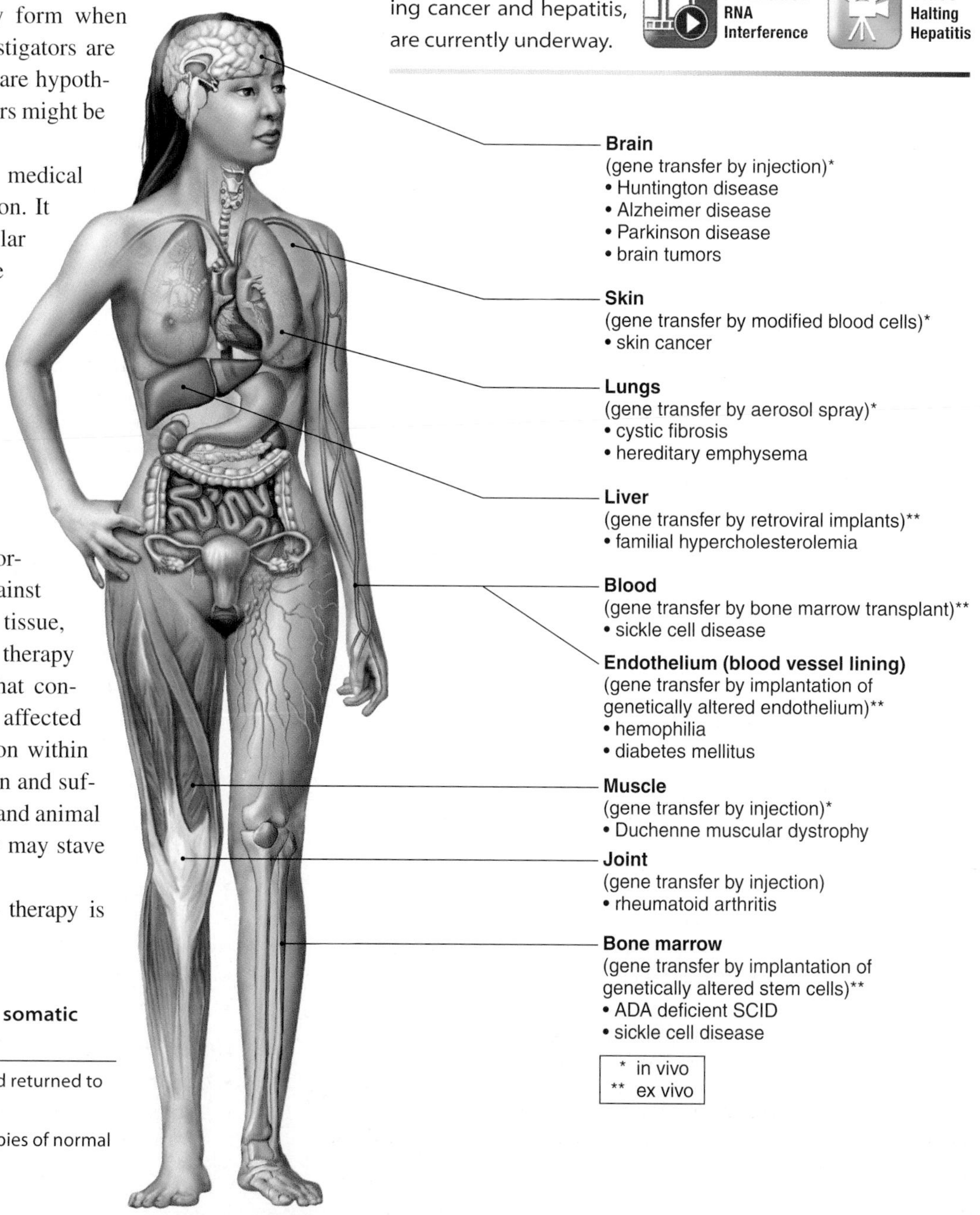

Figure 13.24 Sites of ex vivo and in vivo somatic gene therapy.

Ex vivo: Tissue is removed, genetically modified, and returned to patient.

In vivo: Targeted areas (e.g., brain, lungs) receive copies of normal genes by various methods of gene transfer.

Connections and Misconceptions

What is RNA interference?

RNA interference, or RNAi, is an experimental procedure in which small pieces of RNA are used to "silence" the expression of specific alleles. These RNA sequences are designed to be complementary to the mRNA transcribed by a gene of interest. Once the complementary RNA sequences enter the cell, they bind with the target RNA, producing double-stranded RNA molecules. These double-stranded RNA molecules are then broken down by a series of enzymes within the cell. First discovered in worms, RNAi is believed to have evolved in eukaryotic organisms as a protection against certain types of viruses. Research into developing RNAi treatments for a number of human diseases, including cancer and hepatitis, are currently underway.

 Animation RNA Interference

 Video Halting Hepatitis

Brain
(gene transfer by injection)*
• Huntington disease
• Alzheimer disease
• Parkinson disease
• brain tumors

Skin
(gene transfer by modified blood cells)*
• skin cancer

Lungs
(gene transfer by aerosol spray)*
• cystic fibrosis
• hereditary emphysema

Liver
(gene transfer by retroviral implants)**
• familial hypercholesterolemia

Blood
(gene transfer by bone marrow transplant)**
• sickle cell disease

Endothelium (blood vessel lining)
(gene transfer by implantation of genetically altered endothelium)**
• hemophilia
• diabetes mellitus

Muscle
(gene transfer by injection)*
• Duchenne muscular dystrophy

Joint
(gene transfer by injection)
• rheumatoid arthritis

Bone marrow
(gene transfer by implantation of genetically altered stem cells)**
• ADA deficient SCID
• sickle cell disease

* in vivo
** ex vivo

Connecting the Concepts

For more information on these topics, refer to the following discussions.

Section 12.1 provides more information on stem cells.

Section 17.1 describes the structure of a virus and how it invades a host cell.

Sections 26.2 and **26.3** examine how the immune system defends the body against viruses.

Media Study Tools

www.mhhe.com/maderessentials3

Enhance your study of this chapter with study tools and practice tests. Also ask your instructor about the resources available through ConnectPlus, including the media-rich eBook, interactive learning tools, and animations.

The Chapter in Review

Summary

13.1 Counseling for Chromosomal Disorders

A counselor can detect chromosomal mutations by studying a karyotype of the individual.

Karyotyping

A karyotype is a display of the chromosomes arranged by pairs; the autosomes are numbered from 1 to 22. The sex chromosomes are not numbered.

Chromosomal Mutations

Chromosomal mutations involve changes in chromosome number or structure. Abnormal chromosome number results from nondisjunction during meiosis. Changes in chromosome structure include deletions, duplications, translocations, and inversions:

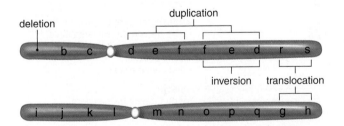

In Williams syndrome, one copy of chromosome 7 has a deletion; in cri du chat syndrome, one copy of chromosome 5 has a deletion; and in inv dup 15 syndrome, chromosome 15 has an inverted duplication. Down syndrome can be due to a translocation between chromosomes 14 and 21 in a previous generation.

An inversion can lead to chromosomes that have a deletion and a duplication. During synapsis, the homologue with the inverted sequence must loop back to align with the normal homologue. Following crossing-over, one nonsister chromatid has a deletion and the other has a duplication.

13.2 Counseling for Genetic Disorders

To assist a couple in determining whether they have or their prospective children will have a genetic disorder, a genetic counselor can (1) construct family pedigrees for genetic disorders of interest, (2) order genetic testing of prospective parents, the fetus, the embryo, or the egg, and (3) suggest gene therapy.

Family Pedigrees

A family pedigree is a visual representation of the history of a genetic disorder in a family. Constructing pedigrees helps a genetic counselor decide whether a genetic disorder that runs in a family is autosomal recessive (see Fig. 13.8) or dominant (see Fig. 13.9); X-linked (see Fig. 13.10); or some other pattern of inheritance.

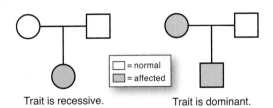

Trait is recessive.   Trait is dominant.

Genetic Disorders of Interest

Autosomal Disorders

- Methemoglobinemia (inability to clear methemoglobin from the blood)
- Cystic fibrosis (faulty regulator of chloride channel)
- Alkaptonuria (inability to metabolize homogentisic acid)
- Sickle cell disease (sickle-shaped red blood cells)
- Marfan syndrome (defective elastic connective tissue)
- Huntington disease (abnormal huntingtin protein)

X-Linked Disorders

- Color blindness (inability to detect colors normally)
- Duchenne muscular dystrophy (absence of dystrophin leading to muscle weakness)

13.3 Testing for Genetic Disorders

A counselor can order the appropriate test to detect a disorder.

Testing for a Protein Blood or tissue samples can be tested for enzyme activity. Lack of enzyme activity can sometimes indicate that a genetic disorder exists.

Testing the DNA Cut DNA with restriction enzymes and then compare the fragment pattern to the normal pattern. Label DNA with fluorescent tags, and see if fragments bind to DNA probes on a DNA chip that contain the mutation.

Testing the Fetus An ultrasound can detect disorders due to certain chromosomal abnormalities and other conditions, such as spina bifida, that are due to the inheritance of mutant alleles. Fetal cells can be obtained by amniocentesis or chorionic villus sampling, or by sorting out fetal cells from the mother's blood.

Testing the Embryo Following in vitro fertilization (IVF), it is possible to test the embryo. A cell is removed from an eight-celled embryo, and if it is found to be genetically healthy, the embryo is implanted in the uterus, where it develops to term.

Testing the Egg Before IVF, a polar body can be tested. If the woman is heterozygous, and the polar body has the genetic defect, the egg does not have it. Following fertilization, the embryo is implanted in the uterus.

13.4 Gene Therapy

During gene therapy, a genetic defect is treated by giving the patient a foreign gene.

- **Ex vivo therapy:** Cells are removed from the patient, treated, and returned to the patient.
- **In vivo therapy:** A foreign gene is given directly to the patient via nasal spray, liposomes, or adenovirus.

▉ Key Terms

▉ Testing Yourself

Choose the best answer for each question.

1. The major advantage of chorionic villus sampling over amniocentesis is that it
   a. allows karyotyping to be done earlier.
   b. produces karyotypes with better images of chromosomes.
   c. carries a lower risk of spontaneous abortion.
   d. produces a sample that is not contaminated by cells from the mother.

2. Which of the following disorders is not caused by a change in chromosome structure?
   a. cri du chat syndrome   c. Klinefelter syndrome
   b. inv dup 15 syndrome   d. Alagille syndrome

3. Which of the following chromosomal mutations would not result in an individual possessing three alleles of a gene?
   a. an extra chromosome   c. a duplication
   b. a translocation   d. a deletion

4. Fill in the genotypes a.–i. of the family members in the following pedigree for an autosomal recessive trait. Shaded individuals are affected.

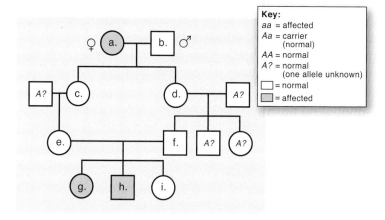

Key:
aa = affected
Aa = carrier (normal)
AA = normal
A? = normal (one allele unknown)
☐ = normal
▨ = affected

5. Fill in the genotypes a.–h. of the family members in the following pedigree for an autosomal dominant trait. Shaded individuals are affected.

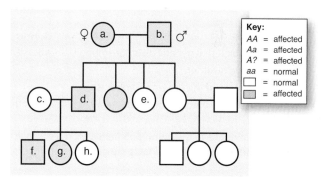

Key:
AA = affected
Aa = affected
A? = affected
aa = normal
☐ = normal
▨ = affected

6. Fill in the genotypes a.–e. of the family members in the following pedigree for an X-linked recessive trait. Shaded individuals are affected.

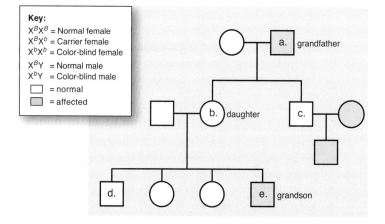

Key:
X^BX^B = Normal female
X^BX^b = Carrier female
X^bX^b = Color-blind female
X^BY = Normal male
X^bY = Color-blind male
☐ = normal
▨ = affected

For questions 7–12, match the pedigree characteristics to those in the key. Answers can be used more than once. Some questions may have more than one answer.

Key:

a. X-linked recessive
b. X-linked dominant
c. autosomal recessive
d. autosomal dominant
e. Y-linked

7. All daughters and no sons from affected males express the trait.
8. Males express the trait more frequently than females.
9. Affected children always have at least one affected parent.
10. Affected children can have unaffected parents.
11. Two affected parents may have unaffected children.
12. The trait is passed directly from father to son, but never from mother to son.

For questions 13–20, match the descriptions to the conditions in the key. Answers can be used more than once. Some questions may have more than one answer.

Key:

a. methemoglobinemia
b. cystic fibrosis
c. alkaptonuria
d. sickle cell disease
e. Marfan syndrome
f. Huntington disease
g. color blindness
h. Duchenne muscular dystrophy

13. autosomal dominant disorder
14. autosomal recessive disorder
15. neurological disorder that leads to progressive degeneration of brain cells
16. X-linked recessive disorder in which muscle tissue wastes away
17. the most common lethal genetic disorder among U.S. Caucasians
18. X-linked recessive disorder
19. results from the lack of the enzyme diaphorase
20. caused by abnormal hemoglobin, resulting in red blood cells that are not round
21. Parents who do not have methemoglobinemia produce a child who has methemoglobinemia. What is the genotype of all persons involved?
22. A 25-year-old man has Huntington disease. Is it possible for him to father a normal son? Explain.
23. A girl has color blindness. What is her genotype? What are the possible genotypes of the parents?
24. The advantage of using a DNA microarray over testing for genetic markers is that a DNA microarray
    a. tests for additional or missing restriction enzyme cleavage sites.
    b. can test for many different mutations at once.
    c. is the only method that can distinguish between a heterozygous and homozygous individual.
    d. does not require the use of PCR to amplify DNA.
25. An ultrasound can be used to
    a. determine if a fetus is heterozygous or homozygous for a trait.
    b. identify anatomical abnormalities that may suggest a genetic defect.
    c. amplify fetal blood cells for DNA extraction.
    d. perform a karyotype.

26. Testing a polar body may be done if
    a. both parents are homozygous for a genetic disorder.
    b. the male parent has a translocation.
    c. the male parent is a carrier for a genetic disorder.
    d. the female parent is a carrier for a genetic disorder.
27. Ex vivo gene therapy involves
    a. delivery of genes to cells within the body.
    b. only the use of liposomes to deliver genes.
    c. removal of cells from the body before treatment.
    d. All of the above are correct.

Thinking Scientifically

1. As you learned in Chapter 10, some individuals possess extra copies of certain genes. Recently, a few genetic diseases were linked to having extra copies of specific genes. How might extra copies of a gene be harmful? Can you think of any instances in which extra genes would not be harmful?
2. The use of adenoviruses to deliver genes used in gene therapy has sometimes proven problematic. Recently, adenoviruses were used in gene therapy trials performed on 10 infants with X-linked severe combined immunodeficiency syndrome (XSCID), also known as "bubble boy disease." Although the trial was considered a success because the gene was expressed and the children's immune systems were restored, researchers were shocked and disappointed when two children developed leukemia. How might you explain the development of leukemia in these gene therapy patients?
3. Genetic counselors note that certain debilitating diseases are maintained in a population even when they are disadvantageous. Why wouldn't they be removed by the evolutionary process? Can you think of any advantageous reason they remain in the gene pool?

Bioethical Issue

Gene Therapy Trials

Despite past successes and lofty promises, some gene therapy clinical trials have had disappointing or even tragic results. Eighteen-year-old Jesse Gelsinger died in a 1999 gene therapy trial because of a massive immune response to the adenovirus used to deliver the gene. In 2007, 36-year-old Jolee Mohr died from a massive fungal infection while undergoing gene therapy treatment for rheumatoid arthritis.

These highly publicized cases raised concern among scientists, government regulators, and patient advocates. Many of them insist that increased scrutiny of gene therapy clinical trials is needed. Indeed, a subsequent review of gene therapy clinical trials revealed that 37 of 970 adverse effects were reported to the National Institutes of Health (NIH), constituting serious legal and ethical violations. However, many researchers contend that these and other cases were isolated events, and they insist that current safeguards are sufficient to protect test subjects. For example, they point out that none of the other 125 patients participating in Mohr's clinical trial suffered adverse effects from the treatment.

Do you believe that the NIH should increase supervision of gene therapy trials and increase penalties for violations, even if it limits gene therapy trials and hampers research? Or are current safeguards sufficient? Are the small number of adverse side effects, even though they have led to a few deaths, worth the advances in gene therapy overall?

Akiapolaau

14

Darwin and Evolution

Evolution Accounts for Diversity

What do the many breeds of dogs, the honeycreepers of Hawaii, and a child's antibiotic-resistant ear infection have in common? Evolution! Without **evolution**—change in a line of descent over time—we wouldn't see such a great variety of living things about us. But aside from its many benefits, evolution also sometimes causes problems for humans.

Some bacteria have evolved to the point that they are resistant to the antibiotics once successfully used to cure the diseases they cause. For example, antibiotics originally cured bacterial ear infections within a few days. Unseen, however, were the one or two bacteria with just the right mutation to resist a particular drug. All the descendants of these bacteria were also resistant, causing the antibiotic to be useless as a cure for this type of ear infection. The antibiotic is considered the *selective agent* because it allowed the resistant bacteria to flourish while killing their relatives.

What was the selective agent for the many breeds of dogs available as pets today? Humans were, of course. Over the years, humans selected which dogs to mate to produce today's varieties. This process is called *artificial selection*.

In nature, the selective agent isn't usually a human or some foreign substance, such as an antibiotic. It is an aspect of the natural environment, and the process is called *natural selection*. An environment may offer a variety of food sources to any given species. Different individuals may utilize some resources but not others. Over time, specializations to acquire certain foods can occur. At least 50 species of birds called honeycreepers were once found on the Hawaiian Islands and nowhere else on Earth. Of the few species that remain today, some feed on insects, others on flower nectar, and still others on seeds. All honeycreepers have been traced back to one unspecialized ancestor arriving on the islands an estimated 4–7 million years ago.

This chapter describes how the work of Charles Darwin contributes to our current knowledge of the evolutionary process and how other studies give evidence that all life has a common source.

OUTLINE

BEFORE YOU BEGIN

Before beginning this chapter, take a few moments to review the following discussions.

Section 1.2 Why is evolution a core concept of biology?

Section 9.1 What is an allele?

Section 9.2 How does meiosis increase variation?

14.1 Darwin's Theory of Evolution

Learning Outcomes

Upon completion of this section, you should be able to

1. Summarize the contributions of Cuvier and Lamarck to the study of evolutionary change.
2. Explain how Darwin's study of fossils and biogeography contributed to the development of the theory of natural selection.
3. Explain the steps in the theory of natural selection.
4. Distinguish between natural and artificial selection.
5. State Wallace's contributions to the theory of natural selection.

Charles Darwin was only 22 in 1831 when he accepted the position of naturalist aboard the HMS *Beagle*, a British naval ship about to sail around the world (**Fig. 14.1**). Darwin had a suitable background for this position. He was a dedicated student of nature and had long been a collector of insects. His sensitive nature had prevented him from studying medicine, and he went to divinity school at Cambridge instead. Even so, he attended many lectures in both biology and geology, and he was tutored in these subjects by the Reverend John Henslow. Darwin spent the summer of 1831 doing fieldwork with Cambridge geologist Adam Sedgwick, before Henslow recommended him to the captain of the *Beagle* as the ship's naturalist. The voyage was to take two years, but it ended up taking five, traversing the Southern Hemisphere, where

Figure 14.1 **Voyage of the HMS *Beagle*.**

The map shows the journey of the HMS *Beagle* around the world. As Darwin traveled along the east coast of South America, he noted that a bird called a rhea looked like the African ostrich. On the Galápagos Islands, marine iguanas, found no other place on Earth, use their large claws to cling to rocks and their blunt snouts for eating marine algae.

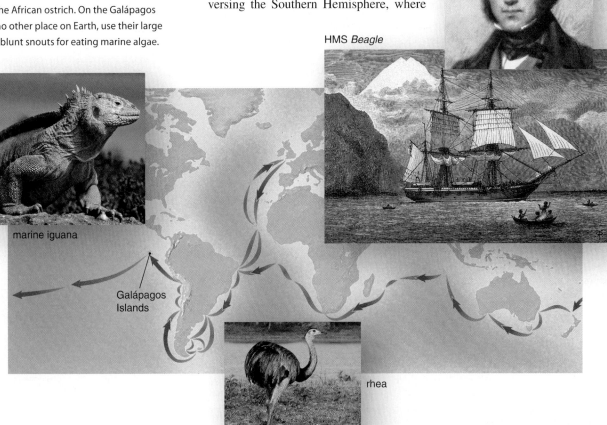

Charles Darwin

HMS *Beagle*

marine iguana

Galápagos Islands

rhea

we now know that life is most abundant and varied. Along the way, Darwin encountered forms of life very different from those in his native England.

Darwin's major mission was to expand the navy's knowledge of natural resources (e.g., water and food) in foreign lands. By the time the trip was over, Darwin had made observations that led him to conclude that biological evolution occurs, contrary to most beliefs at the time.

Before Darwin

Prior to Darwin, most people had an entirely different way of viewing the world. They believed that the Earth was only a few thousand years old and that, since the time of creation, species had remained exactly the same. Even so, studying the anatomy of organisms and then classifying them interested many investigators who wished to show that species were created to be suitable to their environment. Explorers and collectors traveled the world and brought back currently existing species and fossils to be classified. **Fossils** are the remains of once-living species often found in strata. **Strata** are layers of rock formed from sedimentary material (**Fig. 14.2**a). A newfound fossil shows that snakes had hip bones and hindlimbs some 90 million years ago (MYA) and that they evolved on land (**Fig. 14.2**b).

A noted zoologist of the early nineteenth century, Georges Cuvier, founded the science of *paleontology*, the study of fossils. Cuvier was faced with a problem. He believed in the fixity of species, yet Earth's strata clearly showed a succession of different life-forms over time. To explain these observations, he hypothesized that a local catastrophe had caused a mass extinction whenever a new stratum of that region showed a new mix of fossils. After each catastrophe, the region was repopulated by species from surrounding areas, which accounted for the appearance of new fossils in the new stratum. The result of all these catastrophes was change appearing over time. Some of Cuvier's followers, who came to be called *catastrophists*, even suggested that worldwide catastrophes had occurred and that, after each of these events, new sets of species had been created.

In contrast to Cuvier, Jean-Baptiste de Lamarck, another biologist, hypothesized that evolution occurs and that adaptation to the environment is the cause of diversity. Therefore, after studying the succession of life-forms in strata, Lamarck concluded that more complex organisms are descended from less complex organisms. To explain the process of adaptation to the environment, Lamarck proposed the idea of *inheritance of acquired characteristics*, in which the use and disuse of a structure can bring about inherited change. One example Lamarck gave—and the one for which he is most famous—is that the long neck of giraffes developed over time because giraffes stretched their necks to reach food high in trees and then passed on a long neck to their offspring (see Fig. 14.8). However, this hypothesis of the inheritance of acquired

a. Visible strata

b. A fossil snake with hip bones

ribs

hip bone

Figure 14.2 Fossils and strata.

a. Due to erosion, it is often possible to see a number of strata, layers of rock or sedimentary material that contain fossils. The oldest fossils are in the lowest stratum. **b.** A fossil snake dated 90 MYA.

Reprinted by permission from Macmillan Publishers Ltd: Nature (A Cretaceous Terrestrial Snake With Robust Hindlimbs and a Sacrum), copyright 2006.

characteristics has never been substantiated by experimentation. For example, if acquired characteristics were inherited, people who become blind by accident would have blind children, and circumcised males would have boys that lack a foreskin from birth. Modern genetics explains why the idea of acquired characteristics cannot be substantiated. Phenotypic changes acquired during an organism's lifetime do not result in genetic changes that can be passed to subsequent generations.

Darwin's ideas were close to those of Lamarck. For example, Darwin said that living things share common characteristics because they have a common ancestry. One of the most unfortunate misinterpretations of this statement was that humans are descendants of apes. For Darwin, however, humans and apes share a common ancestor, just as, say, you and your cousins can trace your ancestry back to the same grandparents. In contrast to Lamarck, Darwin's observations led him to conclude that species are suited to the environment through no will of their own but by natural selection. He saw the process of natural selection as the means by which different species come about (see Fig. 14.8).

Darwin's Conclusions

Darwin's conclusions that organisms are related through common descent and that adaptation to various environments results in diversity were based on several types of data, including his study of geology, fossils, and biogeography. **Biogeography** is the study of the distribution of life-forms on Earth.

Darwin's Study of Geology and Fossils

Darwin took Charles Lyell's book *Principles of Geology* on the *Beagle* voyage. In contrast to former beliefs, this book gave evidence that Earth is subject to slow but continuous cycles of erosion and uplift. Weathering causes erosion; thereafter, dirt and rock debris are washed into the rivers and transported to oceans. When these loose sediments are deposited, strata result (**Fig. 14.3***a*). Then the strata, which often contain fossils, are uplifted over long periods of time from below sea level to form land. Given enough time, slow natural processes can account for extreme geological changes. Lyell went on to propose the theory of **uniformitarianism,** which stated that these slow changes occurred at a uniform rate. Even though uniformitarianism has been rejected, modern geology certainly substantiates a hypothesis of slow and continual geological change. Darwin, too, was convinced that Earth's massive geological changes are the result of slow processes and, therefore, Earth must be very old.

Figure 14.3   Formation of strata.

a. This diagram shows how water takes sediments into the sea; the sediments then become compacted to form a stratum (singular). Fossils are often trapped in strata, and as a result of a later geological upheaval, the strata may be located on land. **b.** Fossil remains of freshwater snails, *Turritella*, in a stratum.

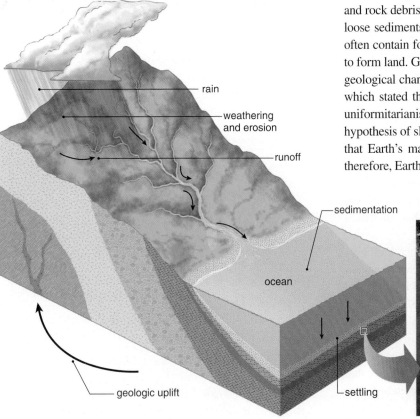

rain

weathering and erosion

runoff

sedimentation

ocean

geologic uplift

settling

a.

b.

On his trip, Darwin observed massive geological changes firsthand. When he explored what is now Argentina, he saw raised beaches for great distances along the coast. In Chile, he witnessed the effects of an earthquake that caused the land to rise several feet and left marine shells inland, well above sea level. When Darwin also found marine shells high in the cliffs of the impressive Andes Mountains, he became even more convinced that Earth is subject to slow geological changes. While Darwin was making geological observations, he also collected fossil specimens that differed somewhat from modern species **(Fig. 14.4)**. Once Darwin accepted the supposition that Earth must be very old, he began to think that there would have been enough time for descent with modification—that is, evolution—to occur. Living forms could be descended from extinct forms known only from the fossil record. It would seem that species are not fixed; instead, they change over time.

Darwin's Study of Biogeography

Darwin could not help but compare the animals of South America with those he observed in England. For example, instead of rabbits, he found Patagonian hares in the grasslands of South America. The Patagonian hare, a rodent also native to South America **(Fig. 14.5)**, has long legs and ears but the face of a guinea pig. Did the Patagonian hare resemble a rabbit because the two types of animals were adapted to the same type of environment—an outcome called convergent evolution today? Did the two distantly related species converge on the same overall body form because they live in similar habitats and have similar behaviors? Did the Patagonian hare have the face of a guinea pig because of common ancestry with guinea pigs?

As Darwin sailed southward along the eastern coast of South America, he saw how similar species replaced one another. For example, the greater rhea (an ostrich-like bird) found in the north was replaced by the lesser rhea in the south. Therefore, Darwin reasoned that related species can be modified according to environmental differences caused by change in latitude. When he reached the

a.

b.

Figure 14.4 A glyptodont compared with an armadillo.

a. The giant armadillo-like glyptodont is known only by the study of its fossil remains. Darwin found such fossils and came to the conclusion that this extinct animal must be related to living armadillos. The glyptodont weighed 2,000 kilograms. **b.** A modern armadillo weighs about 4.5 kilograms.

a. Patagonian hare

b. European rabbit

Figure 14.5 The Patagonian hare and European rabbit.

a. The Patagonian hare, native to South America, has long legs and other adaptations similar to those of a rabbit but has the face of a guinea pig. **b.** The characteristics of the Patagonian hare resemble those of the European rabbit, which does not occur naturally in South America.

a. Ground-dwelling finch

b. Woodpecker finch

c. Cactus finch

Figure 14.6 **Galápagos finches.**

Each of the present-day 13 species of finches has a beak adapted to a particular way of life. **a.** The heavy beak of the large ground-dwelling finch is suited to a diet of seeds. **b.** The beak of the woodpecker finch is suited for using tools to probe for insects in holes or crevices. **c.** The long, somewhat decurved beak and split tongue of the cactus finch are suited to probing cactus flowers for nectar.

Connections and Misconceptions

What has happened to Darwin's finches since the nineteenth century?

The finches of the Galápagos Islands have continued to provide a wealth of information on evolutionary processes. Starting in 1973, a team of researchers led by Drs. Peter and Rosemary Grant began a 30-year study of a species of groundfinch on the island of Daphne Major in the Galápagos. Through detailed measurements and observations of over 19,000 birds, the Grants were able to document the evolutionary change of a species in response to changes in the environment in action. The details of their study were reported in the Pulitzer Prize–winning book *The Beak of the Finches: A Story of Evolution in Our Time* by Jonathan Weiner.

Video
Finches Natural Selection

Galápagos Islands, he found further evidence of this. The Galápagos Islands are a small group of volcanic islands located 965 kilometers off the western coast of South America. These islands are too far from the mainland for most terrestrial animals and plants to colonize, yet life was present. The types of plants and animals found there were slightly different from species Darwin had observed on the mainland; even more important, they also varied from island to island according to their own unique environment. Where did the species inhabiting these islands come from, and what caused the islands to have different species?

Finches Although some of the finches on the Galápagos Islands seemed like mainland finches, others were quite different (**Fig. 14.6**). Today, there are ground-dwelling finches with beak sizes dependent on the sizes of the seeds they eat. Tree-dwelling finches have beak sizes and shapes dependent on the sizes of their insect prey. A cactus-eating finch possesses a more pointed beak, enabling access to nectar within cactus flowers. The most unusual of the finches is a woodpecker-type finch. A woodpecker normally has a sharp beak to chisel through tree bark and a long tongue to probe for insects. The Galápagos woodpecker-type finch has the sharp beak but lacks the long tongue. To make up for this, the bird carries a twig or cactus spine in its beak and uses it to poke into crevices. Once an insect emerges, the finch drops this tool and seizes the insect with its beak.

Later, Darwin speculated as to whether all the different species of finches he had seen could have descended from a mainland finch. In other words, he wondered if a mainland finch was the common ancestor of all the types on the Galápagos Islands. Had **speciation,** the formation of a new species, occurred because the islands allowed isolated populations of birds to evolve independently? Could the present-day species have resulted from accumulated changes occurring within each of these isolated populations?

Tortoises Each of the Galápagos Islands also seemed to have its own type of tortoise, and Darwin began to wonder if this could be correlated with a difference in vegetation among the islands. Long-necked tortoises seemed to inhabit only dry areas, where food was scarce, and most likely the longer neck was helpful in reaching tall-growing cactuses. In moist regions with relatively abundant ground foliage, short-necked tortoises were found. Had an ancestral tortoise from the mainland of South America given rise to these different types, each adapted to a different environment? An **adaptation** is any characteristic that makes an organism more suited to its environment. It often takes many generations for an adaptation to develop.

Natural Selection and Adaptation

Once Darwin realized that adaptations develop over time, he began to think about a mechanism by which adaptations might arise. Eventually, he proposed **natural selection** as the mechanism. Natural selection is a process that results in the evolution of organisms well adapted to their environment. Natural selection requires the following steps:

1. The members of a population have heritable variations (**Fig. 14.7**).
2. The population produces more offspring than the resources of an environment can support.
3. The individuals that have favorable traits survive and reproduce to a greater extent than those that lack these traits.
4. Across generations, a larger proportion of the population possesses the favorable traits, and the population becomes adapted to the environment.

Notice that, because natural selection utilizes only variations that happen to be provided by genetic changes, it lacks any directedness or anticipation of future needs. Natural selection is an ongoing process because the environment of living things is constantly changing. Extinction (loss of a species) can occur when previous adaptations are no longer suitable to a changed environment.

Organisms Have Variations

Darwin emphasized that the members of a population vary in their functional, physical, and behavioral characteristics. Prior to Darwin, variations were considered imperfections that should be ignored, since they were not important to the description of a species. Darwin, on the other hand, realized that variations are essential to the natural selection process. Darwin suspected—but did not have the evidence we have today—that the occurrence of variations is completely random; they arise by accident and for no particular purpose. Also, new variations are more likely to be harmful than beneficial to an organism.

The variations that make adaptation to the environment possible are those that are passed on from generation to generation. The science of genetics was not yet well established, so Darwin was never able to determine the cause of variations or how they are passed on. Today, we realize that genes, together with the environment, determine the phenotypes of an organism. Mutations, along with chromosomal rearrangements and assortment of chromosomes during meiosis and fertilization, can cause new variations to arise.

Organisms Struggle to Exist

In Darwin's time, a socioeconomist named Thomas Malthus stressed the reproductive potential of human beings. He proposed that death and famine are inevitable because the human population tends to increase faster than the supply of food. Darwin applied this concept to all organisms and saw that the available resources were not sufficient to allow all members of a population to survive. He calculated the reproductive potential of elephants and concluded that, after only 750 years, the descendants of a single pair of elephants would number about 19 million! Obviously, no environment has the resources to support an elephant population of this magnitude. Because each generation has the same reproductive potential as the previous generation, there is a constant struggle for existence, and only certain members of a population survive and reproduce each generation.

Organisms Differ in Fitness

Fitness is the reproductive success of an individual relative to other members of the population. Fitness is determined by comparing the number of surviving fertile offspring that are produced with each member of the population. The most fit individuals are the ones that capture a disproportionate amount of resources and convert these resources into a larger number of viable offspring. Because organisms vary anatomically and physiologically, and because the

Figure 14.7 Variations in shells of a marine snail, *Liguus fascitus*.

For Darwin, variations such as these in a species of snails were highly significant and were required in order for natural selection to result in adaptation to the environment.

Connections and Misconceptions

Are there examples of artificial selection in animals?

Almost all animals that are currently used in modern agriculture are the result of thousands of years of artificial selection by humans. But perhaps the greatest example of artifical selection in animals is the modern dog. Analysis of canine DNA indicates that the dog (*Canis familiaris*) is a direct descendent of the grey wolf (*Canis lupus*). This domestication and subsequent selection for desirable traits appears to have begun over 130,000 years ago. Artificial selection of dogs continues to this day, with over 150 variations (or breeds) currently known.

Due to artificial selection, you can choose from more than 150 breeds of dogs.

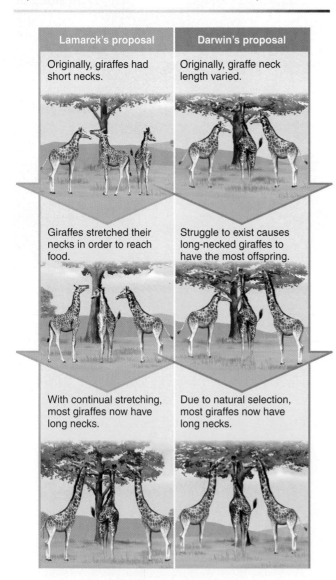

Lamarck's proposal	Darwin's proposal
Originally, giraffes had short necks.	Originally, giraffe neck length varied.
Giraffes stretched their necks in order to reach food.	Struggle to exist causes long-necked giraffes to have the most offspring.
With continual stretching, most giraffes now have long necks.	Due to natural selection, most giraffes now have long necks.

Figure 14.8   Mechanism of evolution.

This diagram contrasts Jean-Baptiste Lamarck's proposal of acquired characteristics with Charles Darwin's proposal of natural selection. Only natural selection is supported by data.

challenges of local environments vary, what determines fitness varies for different populations. For example, among western diamondback rattlesnakes living on lava flows, the most fit are those that are black. Among those living on desert soil, the most fit are typically light-colored with brown blotching. Background matching helps an animal both capture prey and avoid being captured; therefore, it is expected to lead to survival and increased reproduction.

Darwin noted that, when humans help carry out **artificial selection,** they breed selected animals with particular traits to reproduce. For example, prehistoric humans probably noted desirable variations among wolves and selected particular individuals for breeding. Therefore, the desired traits increased in frequency in the next generation. The same process was repeated many times, resulting in today's numerous varieties of dogs, all descended from the wolf. In a similar way, several varieties of vegetables can be traced to a single ancestor. Chinese cabbage, brussels sprouts, and kohlrabi are all derived from a single species, *Brassica oleracea.*

In nature, interactions with the environment determine which members of a population reproduce to a greater degree than other members. In contrast to artificial selection, the result of natural selection is not predesired. Natural selection occurs because certain members of a population happen to have a variation that allows them to survive and reproduce to a greater extent than other members. For example, any variation that increases the speed of a hoofed animal helps it escape predators and live longer; a variation that reduces water loss is beneficial to a desert plant; and one that increases the sense of smell helps a wild dog find its prey. Therefore, we expect the organisms with these characteristics to have increased fitness. **Figure 14.8** contrasts Lamarck's ideas with those of Darwin.

Organisms Become Adapted

An adaptation (see page 238) may take many generations to evolve. We can especially recognize an adaptation when unrelated organisms living in a particular environment display similar characteristics. For example, manatees, penguins, and sea turtles all have flippers, which help them move through the water—also an example of convergent evolution. Adaptations also account for why organisms are able to escape their predators **(Fig. 14.9)** and why they are suited to their way of life **(Fig. 14.10).** Natural selection causes adaptive traits to be increasingly represented in each succeeding generation. There are other processes of evolution aside from natural selection (see Chapter 15), but natural selection is the only process that results in adaptation to the environment.

Darwin and Wallace

After the HMS *Beagle* returned to England in 1836, Darwin waited more than 20 years to publish his book *On the Origin of Species.* During the intervening years, he used the scientific process to support his hypothesis that today's diverse life-forms arose by descent from a common ancestor and that natural selection is a mechanism by which species can change and new species can arise. Darwin was prompted to publish his book after reading a similar hypothesis formulated by Alfred Russel Wallace.

Wallace was an English naturalist who, like Darwin, was a collector at home and abroad. He went on collecting trips, each of which lasted several years, to the Amazon and the Malay Archipelago. After studying the animals of every island within the Malay Archipelago, he divided the islands into a western group, with organisms like those found in Asia, and an eastern group,

false head

eye

false
eyespot

b.

a.

Figure 14.9 Adaptations of the alligator bug.

The alligator bug of the Brazilian rain forest has antipredator adaptations. **a.** The insect blends into its background, but if discovered, the false head, which resembles a miniature alligator, may frighten a predator. **b.** If the predator is not frightened, the insect suddenly reveals huge false eyespots on its hindwings.

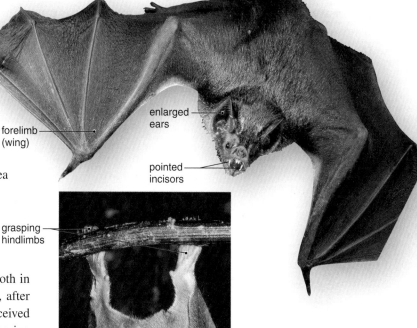

enlarged
ears

forelimb
(wing)

pointed
incisors

grasping
hindlimbs

Figure 14.10 Adaptations of the vampire bat.

Vampire bats of the rain forests of Central and South America have the adaptations of a nocturnal, winged predator. The bat uses its enlarged ears and echolocation to locate prey in the dark. It bites its prey with two pointed incisors. Saliva containing an anticoagulant called draculin runs into the bite; the bat then licks the flowing blood. The vampire bat's forelimbs are modified to form wings, and it roosts by using its grasping hindlimbs.

with organisms like those of Australia. The sharp line dividing these two island groups within the archipelago is now known as Wallace's Line (**Fig. 14.11**). A narrow but deep strait occurs at Wallace's Line. At times during the past 50 million years, this strait persisted even when sea levels were low and land bridges appeared between the other islands. Therefore, the strait would have always served as a barrier to the dispersal of organisms between the two groups of islands.

While traveling, Wallace also wrote an essay called "On the Law Which Has Regulated the Introduction of New Species." In this essay, he said that "every species has come into existence coincident both in time and space with a preexisting closely allied species." A year later, after reading Malthus's treatise on human population increase, Wallace conceived the idea of "survival of the fittest." He quickly completed an essay proposing natural selection as an agent for evolutionary change and sent it to Darwin for comment. Darwin was stunned. Here was the hypothesis he had formulated but had never dared to publish. He told his friend and colleague, Charles Lyell, that Wallace's ideas were so similar to his own that even Wallace's "terms now stand as heads of my chapters."

Darwin suggested that Wallace's paper be published immediately, even though Darwin himself as yet had nothing in print. However, Lyell and others who knew of Darwin's detailed work substantiating the process of natural

Figure 14.11 The Wallace Line

Located in the southern Pacific Ocean, the Wallace Line was one of the earliest attempts to explain the biogeographical distribution of species.

selection suggested that a joint paper be read to the renowned Linnean Society. On July 1, 1858, Darwin presented an abstract of *On the Origin of Species,* which was published in 1859. The title of Wallace's section was "On the Tendency of Varieties to Depart Indefinitely from the Original Type." This professional presentation served as the announcement to the world that species share a common descent and have diverged through natural selection.

Connecting the Concepts

Evolution by natural selection plays an important role in the study of the biological sciences. For more information on this topic, refer to the following discussions.

Section 9.2 describes how sexual reproduction and meiosis increase the variation of a species.

Section 15.2 examines how populations evolve over time.

Section 16.3 explores how the science of systematics identifies the evolutionary relationships between species.

Check Your Progress 14.1

1. Describe the pre-Darwinian view of the world.
2. Paraphrase how catastrophists explain change within the fossil record.
3. Summarize Cuvier's contribution to evolutionary theory.
4. Describe Lamarck's idea of inheritance of acquired characteristics.
5. Explain how natural selection can lead to adaptation.
6. Predict what might happen to a population of individuals over time if there is no diversity and variation in the gene pool.

14.2 Evidence for Evolution

Learning Outcome

Upon completion of this section, you should be able to

1. Explain how the fossil record, biogeography, comparative anatomy, development, and biochemistry support the hypothesis of common descent.

The theory of evolution states that all living things have a common ancestor, but each is adapted to a particular way of life. Many lines of evidence consistently support this hypothesis that organisms are related through common descent. A hypothesis becomes a scientific theory only when a variety of evidence from independent investigators supports the hypothesis. The theory of evolution is a unifying principle in biology because it can explain so many different observations in various fields of biology. The theory of evolution has the same status in biology that the germ theory of disease has in medicine.

Darwin cited much of the evidence for evolution we will discuss, except that he had no knowledge of the biochemical data that became available after his time.

Fossil Evidence

The fossils trapped in rock strata are the **fossil record** that tell us about the history of life. One of the most striking patterns in the fossil record is a succession of life-forms from the simple to the more complex. Occasionally, this pattern is reversed, showing that evolution is not unidirectional. Particularly interesting are the fossils that serve as transitional links between groups. Even in Darwin's day, scientists knew of the *Archaeopteryx* fossils, which show that birds have reptilian features, including jaws with teeth, and long, jointed tails. *Archaeopteryx* also had feathers and wings **(Fig. 14.12)**.

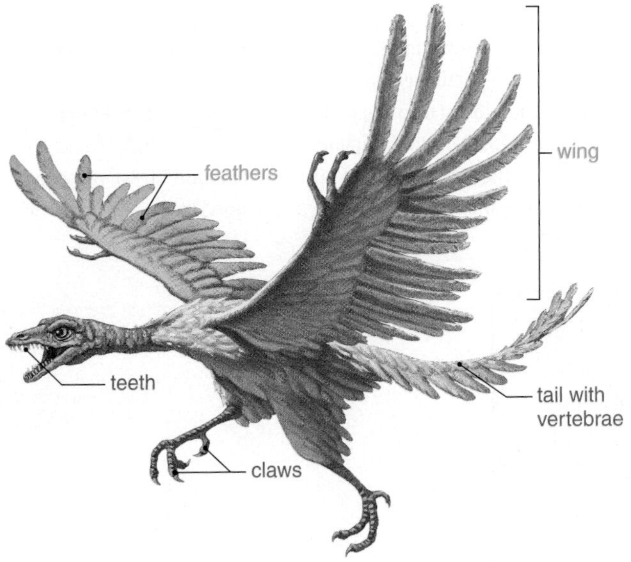

Other transitional links between fossil vertebrates have more recently been found, including the lobe-finned fish *Tiktaalik roseae,* the reptile-like

Figure 14.12 Re-creation of *Archaeopteryx.*

The fossil record suggests that *Archaeopteryx* had a feather-covered, reptilian-type tail, which shows up well in this artist's representation. (Red labels = reptilian characteristics; green labels = bird characteristics.)

feathers
wing
teeth
tail with vertebrae
claws

amphibian *Seymouria*, and the mammal-like synapsids. These fossils allow us to deduce that fishes evolved before amphibians, which evolved before reptiles, including birds. The synapsids show that reptiles and mammals share an amniote ancestry.

Recently, four-legged aquatic mammals, the mesonychids, were discovered. They provide important insights into the evolution of whales from land-living, hoofed ancestors **(Fig. 14.13)**. The fossilized whale *Ambulocetus* may have been amphibious, walking on land and swimming in the sea. *Rodhocetus* swam with an up-and-down tail motion, as modern whales do; its reduced hindlimbs could not have helped in swimming.

In 2006, a snake fossil dated 90 MYA was discovered with hip bones and hind limbs—a trait absent in all living snakes (see Fig. 14.2*b*). Some snakes, such as pythons, have vestigial hindlimbs, but these snakes lack the hip bones present in this fossil. Since all lizards have hip bones and most have limbs, this fossil is now serving as a transitional fossil between lizards and snakes.

Biogeographical Evidence

Biogeography is the study of the distribution of organisms throughout the world. Such distributions are consistent with the hypothesis that, when forms are related, they evolved in one locale and then spread to accessible regions. Therefore, you would expect a different mix of plants and animals whenever geography separated continents, islands, or seas. As previously mentioned, Darwin noted that South America lacked rabbits, even though the environment

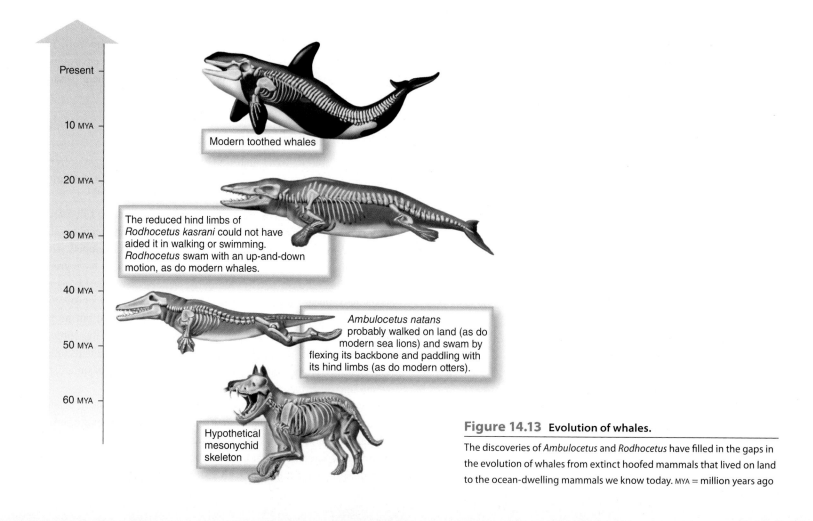

Present

10 MYA

Modern toothed whales

20 MYA

The reduced hind limbs of *Rodhocetus kasrani* could not have aided it in walking or swimming. *Rodhocetus* swam with an up-and-down motion, as do modern whales.

30 MYA

40 MYA

Ambulocetus natans probably walked on land (as do modern sea lions) and swam by flexing its backbone and paddling with its hind limbs (as do modern otters).

50 MYA

60 MYA

Hypothetical mesonychid skeleton

Figure 14.13 Evolution of whales.

The discoveries of *Ambulocetus* and *Rodhocetus* have filled in the gaps in the evolution of whales from extinct hoofed mammals that lived on land to the ocean-dwelling mammals we know today. MYA = million years ago

Sugar glider, a tree dweller

Kangaroo, a herbivore of plains and forests

Koala, a tree dweller

Coarse-haired wombat, nocturnal and living in burrows

Figure 14.14 Marsupials of Australia.

Each type of marsupial in Australia is adapted to a different way of life. All of them presumably evolved from a common ancestor that entered Australia some 60 MYA.

was quite suitable for them. He concluded that rabbits evolved elsewhere and had no means of reaching South America.

To take another example, both cactuses and spurges (*Euphorbia*) are plants adapted to a hot, dry environment—both are succulent, spiny, flowering plants. Why do cactuses grow in American deserts and most *Euphorbia* grow in African deserts, when each would do well on the other continent? It seems obvious that they evolved similar adaptations because their respective continents had similar environments.

The islands of the world are home to many unique species of animals and plants found nowhere else, even when the soil and climate are the same. Why do so many species of finches live on the Galápagos Islands and so many species of honeycreepers, a type of finch, live in the Hawaiian Islands when these species are not on the mainland? The reasonable explanation is that an ancestral finch migrated to all the different islands. Then geographic isolation allowed the ancestral finch to evolve into a different species on each island.

Video
Finches Adaptive Radiation

Also, long ago, South America, Antarctica, and Australia were connected (see Fig. 16.13*a*). Marsupials (pouched mammals) and placental mammals arose at this time, but today marsupials are plentiful only in Australia, and placental mammals are plentiful in South America. Why are marsupials plentiful only in Australia (**Fig. 14.14**)? After marsupials arose, Australia separated and drifted away, and marsupials were free to evolve into many different forms because they had no competition from placental mammals. In the Americas, the placental mammals competed successfully against the marsupials, and the opossum is the only marsupial in the Americas. In some cases, marsupial and placental mammals physically resemble one another—for example, the marsupial wombat and the marmot, and the marsupial Tasmanian wolf and the wolf. This supports the hypothesis that evolution is influenced by the environment and by the mix of plants and animals in a particular continent—that is, by biogeography, not by design.

Anatomical Evidence

Darwin was able to show that a common descent hypothesis offers a plausible explanation for vestigial structures and anatomical similarities among organisms.

Vestigial structures are anatomical features that are fully developed in one group of organisms but reduced and nonfunctional in other, similar groups. Most birds, for example, have well-developed wings used for flight. However, some bird species (e.g., ostrich) have greatly reduced wings and do not fly. Similarly, whales (see Fig. 14.13) and snakes have no use for hindlimbs, yet extinct whales and snakes have remnants of hip bones and legs. Humans have a tailbone but no tail. The presence of vestigial structures can be explained by the common descent hypothesis. Vestigial structures occur because organisms inherit their anatomy from their ancestors; they are traces of an organism's evolutionary history.

Vertebrate forelimbs are used for flight (birds and bats), orientation during swimming (whales and seals), running (horses), climbing (arboreal lizards), and swinging from tree branches (monkeys). However, all vertebrate forelimbs contain the same sets of bones organized in similar ways, despite their dissimilar functions (**Fig. 14.15**). The most plausible explanation

Connections and Misconceptions

What are some other vestigial organs in humans?

The human body is littered with vestigial organs from our evolutionary past—for example, the tiny muscles (called piloerectors) that surround each hair follicle. During times of stress, these muscles causes the hair to stand straight up—a useful defense mechanism for small mammals trying to escape predators but one that has little fuction in humans. Wisdom teeth are also considered to be vestigial organs, since most people now retain their teeth for the majority of their lives.

for this unity is that the basic forelimb plan belonged to a common ancestor, and then the plan became modified in the succeeding groups as each continued along its own evolutionary pathway. Anatomically similar structures explainable by inheritance from a common ancestor are called **homologous structures.** In contrast, **analogous structures** serve the same function but are not constructed similarly, and therefore could not have a common ancestry. The wings of birds and insects are analogous structures.

The homology shared by vertebrates extends to their embryological development (**Fig. 14.16**). At some point during development, all vertebrates have a postanal tail and exhibit paired pharyngeal pouches supported by cartilaginous bars. In fishes and amphibian larvae, these pouches develop into functioning gills. In humans, the first pair of pouches becomes the cavity of the middle ear and the auditory tube. The second pair becomes the tonsils; the third and fourth pairs become the thymus and parathyroid glands. Why should pharyngeal pouches, which have lost their original function, develop and then become modified in terrestrial vertebrates? The most likely explanation is that new structures (or structures with novel functions) originate by "modifying" the preexisting structures of one's ancestors.

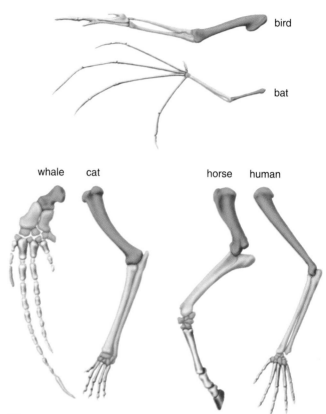

Figure 14.15   Significance of structural similarities.

Although the specific details of vertebrate forelimbs are different, the same basic bone stucture and position are present (color-coded here). This unity of anatomy is evidence of a common ancestor.

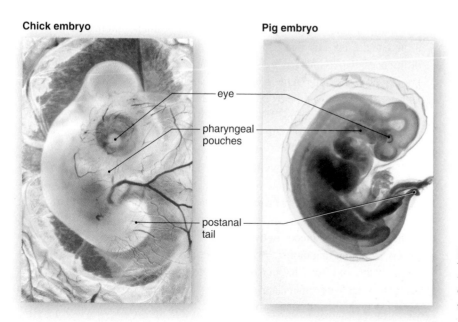

Figure 14.16   Significance of developmental similarities.

At this comparable developmental stage, a chick embryo and a pig embryo have many features in common, which suggests they evolved from a common ancestor.

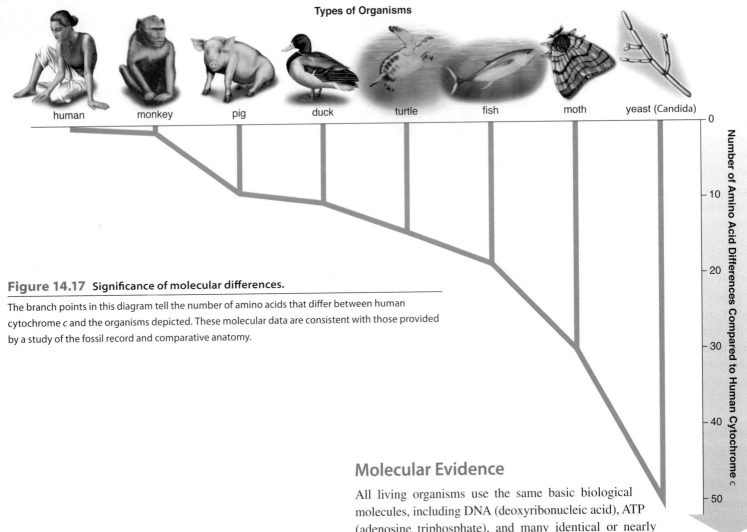

Types of Organisms

human   monkey   pig   duck   turtle   fish   moth   yeast (Candida)

Number of Amino Acid Differences Compared to Human Cytochrome c

Figure 14.17 **Significance of molecular differences.**

The branch points in this diagram tell the number of amino acids that differ between human cytochrome *c* and the organisms depicted. These molecular data are consistent with those provided by a study of the fossil record and comparative anatomy.

Molecular Evidence

All living organisms use the same basic biological molecules, including DNA (deoxyribonucleic acid), ATP (adenosine triphosphate), and many identical or nearly identical enzymes. Further, organisms utilize the same DNA triplet code and the same 20 amino acids in their proteins. Now that we know the sequence of DNA bases in the genomes of many organisms, it has become clear that humans share a large number of genes with much simpler organisms.

Also of interest, evolutionary developmental biologists have found that many developmental genes are shared in animals ranging from worms to humans. It appears that life's vast diversity has come about by only slight differences in the same genes. The result has been widely divergent body plans. For example, a similar gene in arthropods and vertebrates determines the back to front axis. But, although the base sequences are similar, the genes have opposite effects. In arthropods, such as fruit flies and crayfish, the nerve cord is toward the front; in vertebrates, such as chickens and humans, the nerve cord is toward the back.

When the degree of similarity in DNA base sequences of genes or in amino acid sequences of proteins is examined, the data are as expected, assuming common descent. Cytochrome *c* is a molecule used in the electron transport chain of all the organisms shown in **Figure 14.17**. Data regarding differences in the amino acid sequence of cytochrome *c* show that the sequence in a human differs from that in a monkey by only 1 amino acid, from that in a duck by 11 amino acids, and from that in *Candida*, a yeast, by 51 amino acids. These data are consistent with other data regarding the anatomical similarities of these organisms, and therefore their relatedness.

Connecting the Concepts

For examples of how scientists have applied these concepts to the study of how organisms have evolved over time, refer to the following discussions.

Section 17.3 examines how molecular studies have been used to develop two distinct prokaryotic domains of life.

Section 18.2 provides an overview of the evolution of plants.

Section 19.6 explores the fossil evidence supporting human evolution.

Check Your Progress 14.2

1. Explain how biogeographical information about Galápagos finches supports the theory of evolution.

2. Describe how vestigial structures support the theory of evolution.

3. Contrast homologous structures with analogous structures.

Media Study Tools

www.mhhe.com/maderessentials3

Enhance your study of this chapter with study tools and practice tests. Also ask your instructor about the resources available through ConnectPlus, including the media-rich eBook, interactive learning tools, and animations.

The Chapter in Review

Summary

14.1 Darwin's Theory of Evolution

Charles Darwin took a position as a naturalist aboard the HMS *Beagle* and made a trip around the world, largely in the Southern Hemisphere.

Before Darwin

Before Darwin, people believed that Earth was young, species did not change, variations were imperfections, and observations could substantiate these views.

- Cuvier was an early paleontologist who believed that species do not change. He observed species come and go in the fossil record, and he said these changes were due to catastrophies.
- Lamarck was a zoologist who hypothesized that evolution and adaptation to the environment do occur. He suggested that acquired characteristics could be inherited. For example, he said giraffes stretched their necks to reach food in trees, and then this change was inherited by the next generation.

Darwin's Conclusions

Darwin's conclusions based on geology and fossils are

- The Earth is very old, giving time for evolution to occur.
- Living things are descended from extinct forms known only from the fossil record.

Darwin's conclusions based on biogeography are

- Living things evolve where they are. This explains, for example, why South America has the Patagonian hare, whereas England has the rabbit.
- Living things are adapted to local environments. This explains why there are many types of finches and tortoises in the Galápagos Islands.

Natural Selection and Adaptation

According to Darwin, the result of natural selection is a population better adapted to its local environment than previous generations:

	Observation	Result	Conclusion
1	a. Organisms have variations.	b. New adaptations to the environment arise.	Organisms become more adapted with each generation.
2	a. Organisms struggle to exist.	b. More organisms are present than can survive.	
3	a. Organisms differ in fitness.	b. Organisms best suited to the environment survive and reproduce.	

Darwin and Wallace

Alfred Russel Wallace was a naturalist who, like Darwin, traveled to other continents in the Southern Hemisphere. He also collected evidence of common descent, and his reading of Malthus caused him to develop the same mechanism for adaptation (natural selection) as Darwin. Darwin's work was more thorough, as evidenced by his book *On the Origin of Species*.

14.2 Evidence for Evolution

A theory in science is a concept supported by much evidence, and the theory of evolution is supported by several types of evidence:

- The fossil record indicates the history of life in general and allows us to trace the descent of a particular group.
- Biogeography shows that the distribution of organisms on Earth is explainable by assuming organisms evolved in one locale.
- Anatomy and development of organisms reveals homologies explainable only by common ancestry.
- Biochemical molecules of the same type occur in all organisms. Differences indicate the degree of relatedness.

Key Terms

adaptation 238	fossils 235
analogous structure 245	homologous structure 245
artificial selection 240	natural selection 238
biogeography 236	speciation 238
evolution 233	strata 235
fitness 239	uniformitarianism 236
fossil record 242	vestigial structure 244

Testing Yourself

Choose the best answer for each question.

1. _____ developed the idea that acquired characteristics can be inherited.
   - a. Darwin
   - b. Lamarck
   - c. Wallace
   - d. Sedgwick
   - e. Cuvier

2. Which of the following is not an example of natural selection?
   - a. Insect populations exposed to pesticides become resistant to the chemicals.
   - b. Plant species that produce fragrances to attract pollinators produce more offspring.
   - c. Rabbits that sprint quickly are more likely to escape predation.
   - d. On a tree, leaves that grow in the shade are larger than those that grow in the sun.

3. The variations necessary for natural selection
   - a. occur randomly.
   - b. are influenced by the environment.
   - c. can be caused by mutation.
   - d. can be caused by recombination during meiosis.
   - e. All of these are correct.

4. Which is most likely to be favored during natural selection, but not artificial selection?
   a. fast seed germination rate
   b. short generation time
   c. efficient seed dispersal
   d. lean pork meat production

5. Natural selection is the only process that results in
   a. genetic variation.
   b. adaptation to the environment.
   c. phenotypic change.
   d. competition among individuals in a population.

6. Why was it helpful to Darwin to learn that Lyell concluded the Earth is very old?
   a. An old Earth has more fossils than a new Earth.
   b. It meant there was enough time for evolution to have occurred slowly.
   c. There was enough time for the same species to spread out into all continents.
   d. Darwin said artificial selection occurs slowly.
   e. All of these are correct.

7. In the following diagram, contrast Lamarck's proposal with Darwin's proposal by matching the phrases in the key to the letters in the diagram.

Key:

Originally, giraffe neck length varied.

Giraffes stretched their necks in order to reach food.

Struggle to exist caused long-necked giraffes to have the most offspring.

Originally, giraffes had short necks.

Today most giraffes have long necks. (used twice)

8. All the finches on the Galápagos Islands
   a. are unrelated but descended from a common ancestor.
   b. are descended from a common ancestor and therefore are related.
   c. rarely compete for the same food source.
   d. Both a and c are correct.
   e. Both b and c are correct.

9. Evolution is considered a
   a. hypothesis because it is supported by data from the fossil record.
   b. hypothesis because it is supported by multiple types of data.
   c. theory because it is supported by data from the fossil record.
   d. theory because it is supported by multiple types of data.

10. Catastrophists were not able to explain
    a. multiple extinctions.
    b. the replacement of one group of organisms by another.
    c. successive changes that link groups of organisms in the fossil record.
    d. More than one of these are correct.

For questions 11–14, match the description with the type of evidence for evolution it supports, as listed in the key. Answers can be used more than once.

    a. biogeographical
    b. anatomical
    c. biochemical

11. The genetic code is the same for all organisms.

12. The human knee bone and spine were derived from ancestral structures that supported four-legged animals.

13. The South American continent lacks rabbits, even though the environment is quite suitable.

14. The amino acid sequence of human hemoglobin is more similar to that of rhesus monkeys than to that of mice.

15. Fossils that serve as transitional links allow scientists to
    a. determine how prehistoric animals interacted with each other.
    b. deduce the order in which various groups of animals arose.
    c. relate climate change to evolutionary trends.
    d. determine why evolutionary changes occur.

16. Among vertebrates, the flipper of a dolphin and the fin of a tuna are
    a. homologous structures.
    b. homogeneous structures.
    c. analogous structures.
    d. reciprocal structures.

17. Which of these pairs is mismatched?
    a. Charles Darwin—natural selection
    b. Cuvier—series of catastrophes explains the fossil record
    c. Lamarck—uniformitarianism
    d. All of these are matched correctly.

18. According to the inheritance of acquired characteristics hypothesis,
    a. if a man loses his hand, then his children will also be missing a hand.
    b. changes in phenotype are passed on by way of the genotype to the next generation.
    c. organisms are able to bring about a change in their phenotype.
    d. evolution is striving toward particular traits.
    e. All of these are correct.

19. Organisms
    a. compete with other members of their species.
    b. differ in fitness.
    c. are adapted to their environment.
    d. are related by descent from common ancestors.
    e. All of these are correct.

20. DNA nucleotide differences between organisms
    a. indicate how closely related organisms are.
    b. indicate that evolution occurs.
    c. explain why there are phenotypic differences.
    d. are to be expected.
    e. All of these are correct.

21. The fossil record offers direct evidence for common descent because you can
    a. see that the types of fossils change over time.
    b. sometimes find common ancestors.
    c. trace the ancestry of a particular group.
    d. trace the biological history of living things.
    e. All of these are correct.

22. Molecular evidence is increasingly used today to establish relationships because
    a. genes are composed of DNA.
    b. the more similar the genes, the more likely two organisms are related.
    c. this evidence is not subject to opinion.
    d. All of these are correct.

23. For there to be homologous structures,
    a. a common ancestor had to have existed.
    b. analogous structures also have to exist.
    c. the bones have to be used similarly.
    d. All of these are correct.

Thinking Scientifically

1. The human appendix, a vestigial extension off the large intestine, is homologous to a structure called a caecum in other mammals. A caecum, generally larger than our appendix, houses bacteria that aid in digesting cellulose, the main component of plants. How might the presence of the appendix be used to show our common ancestry with other mammals, and what might it tell us about the dietary history of humans?

2. Geneticists compare DNA base sequences among organisms and from these data determine a gene's rate of evolution. Different genes have been found to evolve at different rates. Explain why some genes have faster rates of evolution than other genes as populations adapt to their environments.

3. Both Darwin and Wallace concluded that natural selection is the mechanism for biological evolution while observing life on islands. The Hawaiian and nearby islands once had at least 50 species of honeycreepers, and they lived nowhere else on Earth. Natural selection occurs everywhere and in all species. What characteristics of islands allow the outcome of natural selection to be so obvious?

15

Evolution on a Small Scale

OUTLINE

BEFORE YOU BEGIN

Before beginning this chapter, take a few moments to review the following discussions.

Section 1.2 Why is evolution considered to be the core concept of biology?

Section 9.3 How does sexual reproduction and meiosis increase variation in a population?

Section 14.1 How does natural selection act as the mechanism of evolutionary change?

Sexual Selection

You might think that fitness means keeping in shape, but to an evolutionary biologist it means having more fertile offspring than other individuals. Think about it—only if an animal reproduces can that animal's genes be passed on and become prevalent in the next generation. Adaptation to the environment increases the chance of reproducing, but so does *sexual selection*, which occurs because of an advantage that helps an animal acquire a mate.

Such advantages as increased size, brilliant feathers, evolution of horns, and enlarged canine teeth help males fight for and attract females. Females, in turn, must choose carefully. Perhaps to a female, the showier male is healthier or more appealing. If so, the same characteristics will be advantageous for her sons! Sexual selection increases the chances of reproducing, but it can shorten the life span of males possessing such advantageous traits. A large, showy male that is consistently fighting most likely doesn't live as long as a small, inconspicuous male.

Do male competition and female choice occur among humans? Some think so. They point out that human males tend to be larger and more aggressive than females, and that statistically males have a shorter life span. Also, wealthy and successful males are more apt to be attractive to women, and sometimes older men marry younger women who are still fertile, thereby increasing their own fitness.

Sexual selection is one of the factors that influences evolution on a small scale (microevolution), which is the subject of this chapter.

15.1 Natural Selection

Learning Outcomes

Upon completion of this chapter, you should be able to

1. Describe the three types of natural selection—directional, stabilizing, and disruptive.
2. Explain how heterozygotes maintain variation in a population and summarize the concept of a heterozygote advantage.

Natural selection is the process that results in adaptation of a population to the biotic (living) and abiotic (nonliving) environments. In the biotic environment, organisms acquire resources through competition, predation, and parasitism. The abiotic environment includes weather conditions, dependent chiefly on temperatures and precipitation. Charles Darwin became convinced that species evolve with time and suggested natural selection as the mechanism for adaptation to the environment (see Chapter 14). In **Table 15.1,** Darwin's hypothesis of natural selection is stated in a way that is consistent with modern genetics.

As a result of natural selection, the most *fit* individuals become more prevalent in a population, and in this way, a population changes over time. The most fit individuals are those that reproduce more than others. The more fit individuals are likely those that are better adapted to the environment.

Types of Selection

Most of the traits on which natural selection acts are **polygenic** and controlled by more than one pair of alleles located at different gene loci. Such traits have a range of phenotypes, the frequency distribution of which usually resembles a bell-shaped curve (see Figs. 15.1–15.3).

Three types of natural selection are possible for any trait: directional selection, stabilizing selection, and disruptive selection.

Directional Selection

Directional selection occurs when an extreme phenotype is favored and the distribution curve shifts in that direction. Such a shift can occur when a population is adapting to a changing environment.

Resistance to antibiotics and insecticides are classic examples of directional selection. As you may know, the widespread use of antibiotics and pesticides results in populations of bacteria and insects that are resistant to these chemicals. When an antibiotic is administered, some bacteria may survive because they are genetically resistant to the antibiotic. These are the bacteria that are likely to pass on their genes to the next generation. As a result, the number of resistant bacteria keeps increasing. Drug-resistant strains of bacteria that cause tuberculosis have now become a serious threat to the health of people worldwide.

Another example of directional selection is the human struggle against malaria, a disease caused by an infection of the liver and the red blood cells. The *Anopheles* mosquito transmits the disease-causing protozoan *Plasmodium* from person to person. In the early 1960s, international health authorities thought malaria would soon be eradicated. A new drug, chloroquine, seemed effective against *Plasmodium,* and spraying of DDT (an insecticide) had reduced the

Table 15.1   Natural Selection

Evolution by natural selection requires

1. Variation. The members of a population differ from one another.

2. Inheritance. Many of these differences are heritable genetic differences.

3. Degrees of successful reproduction. Individuals that are better adapted to their environment are more likely to reproduce, and their fertile offspring will make up a greater proportion of the next generation.

mosquito population. But by the mid-1960s, *Plasmodium* was showing signs of chloroquine resistance, and worse yet, mosquitoes were becoming resistant to DDT. A few drug-resistant parasites and a few DDT-resistant mosquitoes had survived and multiplied, shifting the distribution curve toward the resistant type of parasite.

The gradual increase in the size of the modern horse, *Equus*, is an example of directional selection that can also be correlated with a change in the environment—in this case, from forest conditions to grassland conditions (**Fig. 15.1**). Even so, as discussed previously, the evolution of the horse should not be viewed as a straight line of descent because we know of many side branches that became extinct.

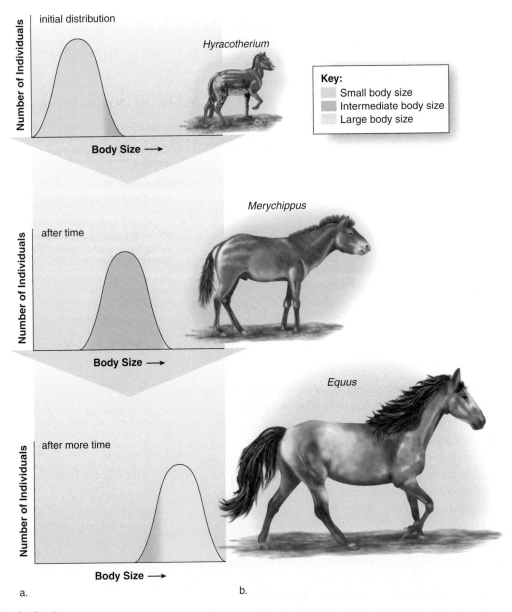

Key:
- Small body size
- Intermediate body size
- Large body size

a.   b.

Figure 15.1 Directional selection.

a. Natural selection favors one extreme phenotype, and the distribution curve shifts. **b.** *Equus*, the modern-day horse, evolved from *Hyracotherium*, which was about the size of a dog. This small animal could have hidden among trees, and it had low-crowned teeth for browsing. When grasslands began to replace forests, the ancestors of *Equus* may have been subject to selection pressure for the development of strength, intelligence, speed, and durable grinding teeth. Larger animals were stronger and more successful in combat, those with larger skulls and brains had better sensory processing, those with longer legs and better developed hooves could escape enemies, and animals with durable teeth could feed more efficiently on grass. Animals with these characteristics tended to produce more offspring.

Stabilizing Selection

Stabilizing selection occurs when an intermediate phenotype is favored. With stabilizing selection, extreme phenotypes are selected against, and individuals near the average are favored. This is the most common form of selection because the average individual is well adapted to its environment.

As an example, consider that when Swiss starlings (*Sturnus vulgaris*) lay four or five eggs, more young survive than when the female lays more or less than this number (**Fig. 15.2**). Genes determining physiological characteristics, such as the production of yolk, and behavioral characteristics, such as how long the female will mate, are involved in determining clutch size.

Disruptive Selection

In **disruptive selection,** two or more extreme phenotypes are favored over any intermediate phenotype. Therefore, disruptive selection favors *polymorphism,* the occurrence of different forms in a population of the same species. For example, British land snails (*Cepaea nemoralis*) are found in low-vegetation areas (grass fields and hedgerows) and in forests. In low-vegetation areas, thrushes feed mainly on snails with dark shells that lack light bands; in forest areas, they feed mainly on snails with light-banded shells. Therefore, these two distinctly different phenotypes, each adapted to its own environment, are found in this population (**Fig. 15.3**).

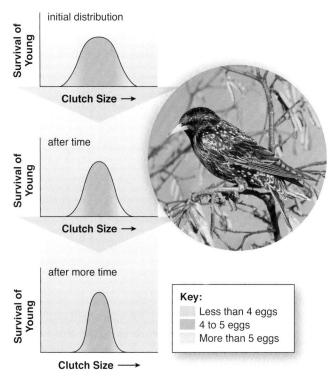

Figure 15.2 Stabilizing selection.

Stabilizing selection occurs when natural selection favors the intermediate phenotype over the extremes. For example, Swiss starling birds that lay four or five eggs (usual clutch size) have more young survive than those that lay fewer than four eggs or more than five eggs.

Figure 15.3 Disruptive selection.

a. Disruptive selection favors two extreme phenotypes, no banding and banding. **b.** Today, British land snails mainly comprise these two different phenotypes, each adapted to a different habitat.

Connections and Misconceptions

Are there examples of stabilizing selection in humans?

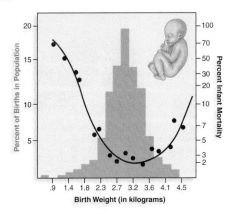

Perhaps the best example of stabilizing selection in humans is in regard to birth weight. Studies in England and the United States in the mid-twentieth century indicated that infants with birth weights beween 6 and 8 pounds had a higher rate of survival. Interestingly, advances in medical care for premature babies with low birth weights, and the increased use of cesarean sections for delivery of high-birth-weight babies, have relaxed the effects of stabilizing selection in some parts of the world.

Sexual Selection

As indicated in the chapter opener, *sexual selection* refers to adaptive changes in males and females that lead to an increased ability to secure a mate. Each sex has a different strategy with regard to sexual selection. Since females produce few eggs, the choice of a mate is a serious consideration. However, males can father many offspring because they continuously produce sperm in great quantity. Therefore, males often compete in order to inseminate as many females as possible. Because of this, sexual selection in males usually results in an increased ability to compete with other males for a mate. On the other hand, sexual selection in females favors the choice of a single male with the best **fitness,** or the ability to produce surviving offspring. Males often demonstrate their fitness by coloration or elaborate mating rituals (**Fig.15.4**). By choosing a male with optimal fitness, the female increases the chances that her traits will be passed on to the next generation. Because of this, many consider sexual selection a form of natural selection.

Video
Cichlid
Territoriality

Video
Romance Genes

Adaptations Are Not Perfect

Natural selection doesn't always produce organisms that are perfectly adapted to their environment. Why not? First, it is important to realize that evolution doesn't start from scratch. Just as you can only bake a cake with the ingredients available to you, evolution is constrained by the available variations. Light-weight titanium bones might benefit birds, but their bones contain calcium and other minerals, the same as those of other reptiles. Similarly, the processes of development prevent the emergence of novel features; therefore, the wing of a bird has the same bones as those of other vertebrate forelimbs.

Imperfections are common because of necessary compromises. The success of humans is attributable to their dexterous hands, but the spine is subject to injury because the vertebrate spine did not originally evolve to stand erect. A feature that evolves has a benefit that is worth the cost. For example, the benefit of freeing the hands must have been worth the increased cost of spinal injuries from assuming an erect posture.

Maintenance of Variations

A population always shows some genotypic variation. The maintenance of variation is beneficial because populations with limited variation may not be able to adapt to new conditions, should the environment change, and may become extinct. How can variation be maintained in spite of selection constantly working to reduce it?

Figure 15.4. Sexual selection.

The elaborate coloration in the males of some species is a form of sexual selection that is intended to demonstrate an increased level of fitness in the male.

First, we must remember that the forces that promote variation are still at work: Mutation still generates new alleles, recombination and independent assortment still shuffle the alleles during gametogenesis, and fertilization still creates new combinations of alleles from those present in the gene pool. Second, gene flow might be occurring between two populations (see Section 15.2). If the receiving population is small and is mostly homozygous, gene flow can be a significant source of new alleles. Finally, natural selection favors certain phenotypes, but the other types may still remain in reduced frequency. Disruptive selection even promotes polymorphism in a population. For diploid species, heterozygotes may also help maintain variation because they conserve recessive alleles in the population.

The Heterozygote Advantage

Only alleles that are expressed (cause a phenotypic difference) are subject to natural selection. In diploid organisms, this fact makes the heterozygote a potential protector of recessive alleles that might otherwise be weeded out of the gene pool. Because the heterozygote remains in a population, so does the possibility of the recessive phenotype, which might have greater fitness in a changing environment. When natural selection favors the ratio of two or more phenotypes in generation after generation, it is called *balanced polymorphism*. Sickle cell disease offers an example of balanced polymorphism.

Sickle Cell Disease   Individuals with sickle cell disease have the genotype Hb^SHb^S (Hb = hemoglobin, the oxygen-carrying protein in red blood cells; s = sickle cell) and tend to die at an early age due to hemorrhaging and organ destruction. Those who are heterozygous and have sickle cell trait (Hb^AHb^S; A = normal) are better off because their red blood cells usually become sickle-shaped only when the oxygen content of the environment is low. Ordinarily, those with a normal genotype (Hb^AHb^A) are the most fit.

Geneticists studying the distribution of sickle cell disease in Africa have found that the recessive allele (Hb^S) has a higher frequency (0.2 to as high as 0.4 in a few areas) in regions where malaria is also prevalent (**Fig. 15.5**). What is the connection between higher frequency of the recessive allele and malaria? Malaria is caused by a parasite that lives in and destroys the red blood cells of the normal homozygote (Hb^AHb^A). However, the parasite is unable to live in the red blood cells of the heterozygote (Hb^AHb^S) because the infection causes the red blood cells to become sickle-shaped. Sickle-shaped red blood cells lose potassium, and this causes the parasite to die. In an environment where malaria is prevalent, the heterozygote is favored. Each of the homozygotes is selected against but is maintained because the heterozygote is favored in parts of Africa subject to malaria. **Table 15.2** summarizes the effects of the three possible genotypes.

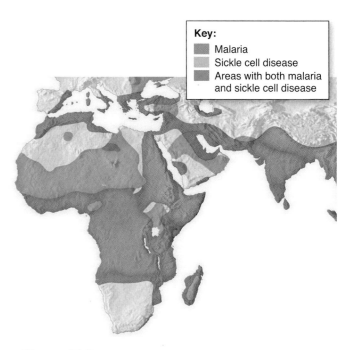

Figure 15.5 **Sickle cell disease.**

Red shows the areas where malaria was prevalent in Africa, the Middle East, southern Europe, and southern Asia in 1920, before eradication programs began; shown in blue are the areas where sickle cell disease most often occurred. The overlap of these two distributions (purple) suggested a causal connection.

Table 15.2 **Sickle Cell Disease**

Genotype	Phenotype	Result
Hb^AHb^A	Normal	Dies due to malarial infection
Hb^AHb^S	Sickle cell trait	Lives due to protection from both
Hb^SHb^S	Sickle cell disease	Dies due to sickle cell disease

Connecting the Concepts

For more on natural selection, refer to the following discussions.

Section 1.2 describes the importance of evolution and natural selection in the study of biology.

Section 14.1 examines how Charles Darwin and Alfred Wallace developed the theory of natural selection.

Section 16.1 explains how natural selection plays a role in evolution over long periods of time.

Check Your Progress 15.1

1. Contrast directional selection with stabilizing selection.
2. Describe the effect of disruptive selection.
3. List the forces that help maintain genetic variability in a population.
4. Explain the higher incidence of sickle cell disease in populations subject to malaria.

Connections and Misconceptions

Why don't individuals evolve?

Evolution represents the genetic change of a population over periods of time. While individuals, such as humans, may develop new skills and abilities (such as learning a new language or playing a guitar), their genetic material remains unchanged. These new abilities are not passed on to the next generation and do not change the genetic composition of the population.

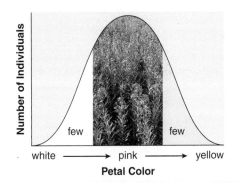

Figure 15.6 Population of perennial flowers.

Among the members of a flowering plant population, petal color can be a continuous variation, so that there is a range of phenotypes. Few individuals have the extreme phenotypes, white petals and yellow petals, and most individuals have the intermediate phenotype, pink petals. A graph of the phenotypes results in the bell-shaped curve shown above a photo of the population.

15.2 Microevolution

Learning Outcomes

Upon completion of this section, you should be able to

1. Explain how microevolution occurs when allele frequencies change from one generation to the next.
2. Demonstrate that the Hardy-Weinberg principle describes a nonevolving population in terms of allele frequencies.
3. Describe how mutations, gene flow, nonrandom mating, genetic drift, and natural selection can cause allele frequency changes in a population.

The idea that individuals evolve is a common misconception. Instead, as evolution occurs, genetic and therefore phenotypic changes occur within a population. A **population** is all the members of a single species occupying a particular area at the same time and reproducing with one another. **Microevolution** pertains to small, measurable evolutionary changes within a population from generation to generation.

Darwin stressed that the members of a population vary, as in **Figure 15.6,** but he did not know how variations are generated or how they are passed down from parents to offspring. Today, DNA sequencing in a number of plants and animals shows that each gene in sexually reproducing organisms has many alleles. The reshuffling of alleles during sexual reproduction can result in a range of phenotypes among the members of a population (see Chapter 10). Even though sexual reproduction can cause genotypes and phenotypes to vary, it cannot in and of itself bring about microevolution. Microevolution is influenced by several other circumstantial factors.

Evolution in a Genetic Context

It was not until the 1930s that biologists were able to apply the principles of genetics to populations and thereafter to develop a way to recognize when evolution has occurred and measure how much population has changed.

In **population genetics,** the various alleles at all the gene loci in all individuals make up the **gene pool** of the population. It is customary to describe the gene pool of a population in terms of genotype and allele frequencies. The genotype frequency is the percent of a specific genotype—for example, homozygous dominant individuals—in a population. The allele frequency represents how much a specific allele is represented in the gene pool of the population. Let's take an example based on peppered moths, which can be light-colored or dark-colored (**Fig. 15.7**). Suppose you research the literature and find that the color of peppered moths is controlled by a single set of alleles and you decide to use the following key:

$$D = \text{dark color} \qquad d = \text{light color}$$

Furthermore, you find that, in one Great Britain population before pollution fully darkened the trees (Fig.15.7a), only 4% (0.04) of the moths were homozygous dominant (*DD*); 32% (0.32) were heterozygous (*Dd*), and 64% (0.64) were homozygous recessive (*dd*). From these genotype frequencies, you can calculate the allele frequencies in the population:

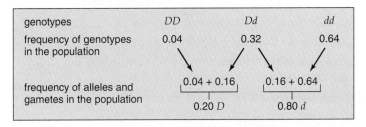

genotypes	*DD*	*Dd*	*dd*
frequency of genotypes in the population	0.04	0.32	0.64
frequency of alleles and gametes in the population	0.04 + 0.16	0.16 + 0.64	
	0.20 *D*	0.80 *d*	

a.

b.

Figure 15.7 **Industrial melanism and microevolution.**

Coloration in the peppered moth (*Biston betularia*) is due to two alleles in the gene pool. **a.** Before widespread pollution in England, the light-colored phenotype was more frequent in the population because the birds were unable to see the light-colored moths on the light tree trunks. **b.** After pollution darkened the trunks of the trees, the dark-colored phenotype became more frequent in the population. Microevolution has occurred with changes in gene pool frequencies—in this case, due to natural selection .

In this population, the frequency of the *D* allele (dark) in the gene pool is 20% (0.20) and the frequency of the *d* allele (light) is 80% (0.80). Therefore, the gametes (sperm and egg) produced by this population will have a 20% chance of carrying the *D* allele and an 80% chance of carrying the *d* allele. Assuming random mating (all possible gametes have an equal chance to combine with any other), we can use these frequencies to calculate the ratio of genotypes in the next generation by using a Punnett square (**Fig. 15.8**). For example, to produce a homozygous dominant (*DD*) moth, both parents must contribute the *D* allele. Since this allele is present in only 20% of the gene pool, the chances that the male will contribute a sperm cell with the *D* allele is 20% (0.20) and that the female will contribute an egg with the *D* allele is 20% (0.20). The chances that both of these events will occur is 0.20 times 0.20, or 0.04 (4%). Therefore, if the moths are randomly mating, 4% of the next generation should be homozygous dominant (*DD*).

There is an important difference between a Punnett square that represents a cross between individuals (as was the case with one-trait and two-trait inheritance; see Chapter 10) and the one shown in Figure 15.8. In Figure 15.8, we are using the gamete frequencies in the *population* to determine the genotype frequencies in the next generation. As you can see, the results show that the genotype frequencies (and therefore the allele frequencies) in the next generation are the same as they were in the previous generation. In other words, the homozygous dominant moths (*DD*) are still 0.04; the heterozygous moths (*Dd*) are still 0.32, and the homozygous recessive moths (*dd*) are still 0.64 of the population. This remarkable finding tells us that *sexual reproduction alone cannot bring about a change in genotype and allele frequencies.* Also, the dominant allele need not increase from one generation to the next. Dominance does not cause an allele to become a common allele.

The fact that the allele frequencies of the gene pool appear to remain at equilibrium from one generation to the next was independently recognized in 1908 by G. H. Hardy, an English mathematician, and W. Weinberg, a German physician. They developed a binomial

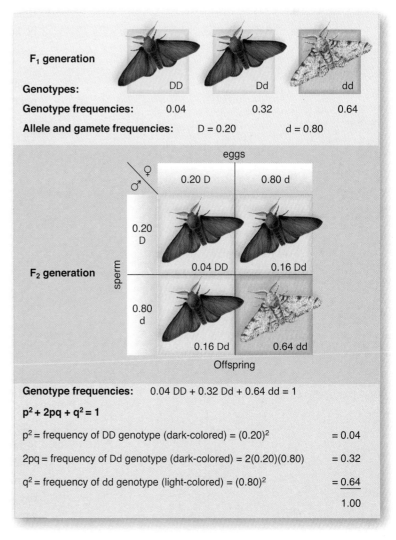

Figure 15.8 **Hardy-Weinberg equilibrium.**

Using the gamete frequencies in a population, it is possible to employ a Punnett square to calculate the genotype frequencies of the next generation. When this is done, it can be shown that sexual reproduction alone does not alter a Hardy-Weinberg equilibrium; the genotype and therefore allele frequencies remain the same.

equation to calculate the genotype and allele frequencies of a population (Fig. 15.8). In this equation,

p = frequency of the dominant allele (in the case of the moths, the D allele)

q = frequency of the recessive allele (the d allele for the moths)

The *Hardy-Weinberg principle* states that an equilibrium of genotype frequencies exists in a gene pool and may be represented by the expression $p^2 + 2pq + q^2 = 1$.

Let's take a look at this equation in relation to our example with the peppered moths (Fig. 15.8). In our example, the dark allele (D) was the dominant allele, and it was present in 20% (0.20) of the population. Therefore, p = 0.20, and the probability that both parents would contribute the allele would be $p \times p$ (p^2), or 0.04 (4%). In order for an individual to be homozygous recessive (dd), he or she must inherit a recessive allele from both parents. Since the recessive allele (d) has a frequency of 0.80 in the gene pool (represented by q), then the probability is 0.8×0.8 (q^2), or 0.64 (64%). Notice from the Punnett square in Figure 15.8 that there are two possible ways that an individual may be heterozygous, which in our equation is represented by $2pq$. Therefore, the probability of being heterozygous is $2 \times 0.2 \times 0.8$, or .32 (32%).

The mathematical relationships of the Hardy-Weinberg principle will remain in effect in each succeeding generation of a sexually reproducing population as long as five conditions are met:

1. No mutations: Allelic changes do not occur, or changes in one direction are balanced by changes in the opposite direction.
2. No gene flow: Migration of alleles into or out of the population does not occur.
3. Random mating: Individuals pair by chance, not according to their genotypes or phenotypes.
4. No genetic drift: The population is very large, and changes in allele frequencies due to chance alone are insignificant.
5. No selection: No selective agent favors one genotype over another.

These conditions are rarely, if ever, met, and genotype and allele frequencies in the gene pool of a population do change from one generation to the next. Therefore, microevolution does occur, and the extent of change can be measured. The significance of the Hardy-Weinberg principle is that it tells us what factors cause evolution—those that violate the conditions listed. Microevolution can be detected and measured by noting the amount of deviation from a Hardy-Weinberg equilibrium of genotype frequencies in the gene pool of a population.

For genotype frequencies to be subject to natural selection, they must result in a change of phenotype frequencies. **Industrial melanism,** an increase in the frequency of a dark phenotype due to pollution, provides us with an example. We supposed that only 36% of our moth population was dark-colored (homozygous dominant plus heterozygous). Why might that be? Before the rise of industry, dark-colored moths rested on light tree trunks, where they were seen and eaten by birds. However, with the advent of industry, the trunks of trees darkened, and the light-colored moths became visible and were eaten (see Fig. 15.7*b*). Predatory birds acted as a selective agent, and microevolution occurred—in the mid-1950s, the number of dark-colored moths in some Great Britain populations exceeded 80%. Aside from showing that natural selection can occur within a short period of time, our example illustrates that a change in gene pool frequencies does occur as microevolution occurs.

Causes of Microevolution

Any conditions that deviate from the list of conditions for allelic equilibrium cause evolutionary change. Thus, these five conditions can cause a divergence from the Hardy-Weinberg equilibrium: genetic mutation, gene flow, nonrandom mating, genetic drift, and natural selection.

Genetic Mutation

Mutations, which are permanent genetic changes, are the raw material for evolutionary change. Without mutations, there could be no new variations among members of a population on which natural selection can act. However, the rate of mutations is generally very low—on the order of 1 mutation per 100,000 cell divisions. In addition, many mutations are neutral (**Fig. 15.9**), meaning that they are not selected for or against by natural selection. Prokaryotes do not reproduce sexually and therefore are more dependent on mutations to introduce variations. All mutations that occur and result in phenotypic differences can be tested by the environment. However, in sexually reproducing organisms, mutations, if recessive, do not immediately affect the phenotype.

In a changing environment, even a seemingly harmful mutation that results in a phenotypic difference can be the source of an adaptive variation. For example, the water flea *Daphnia* ordinarily thrives at temperatures around 20°C, but a mutation exists that requires *Daphnia* to live at temperatures between 25°C and 30°C. The adaptive value of this mutation is entirely dependent on environmental conditions.

Gene Flow

Gene flow, also called *gene migration,* is the movement of alleles among populations by migration of breeding individuals. Gene flow can increase the variation within a population by introducing novel alleles that were produced by mutation in another population. Continued gene flow due to migration of individuals makes gene pools similar and reduces the possibility of allele frequency differences among populations now and in the future. Indeed, gene flow among populations can prevent speciation from occurring. Due to gene flow, the snake populations featured in **Figure 15.10** are

Figure 15.9   Freckles.

A dominant allele causes freckles, so why doesn't everyone have freckles? The Hardy-Weinberg principle, which states that sexual reproduction in and of itself doesn't change allele frequencies, explains why dominant alleles don't become more prevalent with each generation.

P.o. obsoleta

P.o. quadrivittata

P.o. lindheimeri

P.o. spiloides

P.o. rossaleni

Figure 15.10   Gene flow.

Each rat snake represents a separate population of snakes. Because the populations are adjacent to one another, interbreeding occurs, and so does gene flow between the populations. This keeps their gene pools somewhat similar, and each of these populations is considered a subspecies of the species *Pantherophis obsoleta* (as indicated by the three-part name).

subspecies—different populations within the same species. Despite somewhat distinctive characteristics, there is enough genetic similarity between the populations that these subspecies of *Pantherophis obsoleta* can readily interbreed when they come in contact with one another.

Nonrandom Mating

Random mating occurs when individuals select mates and pair by chance and not according to their genotypes or phenotypes. Inbreeding, or mating between relatives, is an example of **nonrandom mating.** Inbreeding does not change allele frequencies, but it does gradually increase the proportion of homozygotes, because the homozygotes that result must produce only homozygotes.

Assortative mating occurs when individuals tend to mate with those that have the same phenotype with respect to a certain characteristic. For example, in humans, tall people tend to mate with each other. Assortative mating causes the population to subdivide into two phenotypic classes, between which gene exchange is reduced. Homozygotes for the gene loci that control the trait in question increase in frequency, and heterozygotes for these loci decrease in frequency.

Sexual selection favors characteristics that increase the likelihood of obtaining mates, and in this way it promotes nonrandom mating. In most species, males that compete best for access to females and/or have a phenotype that attracts females are more apt to mate and have increased fitness (see Section 15.1 and the introduction to this chapter).

Genetic Drift

Genetic drift refers to changes in the allele frequencies of a gene pool due to chance. This pattern of evolution is called genetic drift because allele frequencies "drift" over time. They can increase or decrease due to which members of a population die, survive, or reproduce with one another. Although genetic drift occurs in both large and small populations, a larger population is expected to suffer less of a sampling error than a smaller population. Suppose you had a large bag containing 1,000 green balls and 1,000 blue balls, and you randomly drew 10%, or 200, of the balls. Because there is a large number of balls of each color in the bag, you can reasonably expect to draw 100 green balls and 100 blue balls, or at least a ratio close to this. It is extremely unlikely that you would draw 200 green or 200 blue balls. But suppose you had a bag containing only 10 green balls and 10 blue balls and you drew 10%, or only 2 balls. You could easily draw two green balls or two blue balls, or one of each color.

When a population is small, random events may reduce the ability of one genotype in the production of the next generation. Suppose that, in a small population of frogs, certain frogs by chance do not pass on their traits. Certainly, the next generation will have a change in allele frequencies (**Fig. 15.11**). When genetic drift leads to a loss of one or more alleles, other alleles over time become *fixed* in the population.

In an experiment involving brown eye color, each of 107 *Drosophila* populations was kept in its own culture bottle. Every bottle contained eight heterozygous flies of each sex. There were no homozygous recessive or homozygous dominant flies. From the many offspring, the experimenter chose at random eight males and eight females. This action, which represented genetic drift, continued for 19 generations. By the nineteenth generation, 25% of the populations contained only homozygous recessive flies, and 25% contained only homozygous dominant flies having the allele for brown eye color.

Genetic drift is a random process, and therefore it is not likely to produce the same results in several populations. In California, there are a number of

10% of population

natural disaster kills five green frogs

20% of population

Figure 15.11 Genetic drift.

Genetic drift occurs when a random event changes the frequency of alleles in a population. The allele frequencies of the next generation's gene pool may be markedly different from those of the previous generation.

cypress groves, each a separate population. The phenotypes within each grove are more similar to one another than they are to the phenotypes in the other groves. Some groves have longitudinally shaped trees, and others have pyramidally shaped trees. The bark is rough in some colonies and smooth in others. The leaves are gray to bright green or bluish, and the cones are small or large. Because the environmental conditions are similar for all the groves, and no correlation has been found between phenotype and environment across groves, scientists hypothesize that these variations among populations are due to genetic drift.

Bottleneck Effect Sometimes a species is subjected to near extinction because of a natural disaster (e.g., earthquake or fire) or because of over-harvesting and habitat loss. It is as though most of the population has stayed behind and only a few survivors have passed through the neck of a bottle (**Fig. 15.12**). This so-called **bottleneck effect** prevents the majority of genotypes from participating in the production of the next generation.

The extreme genetic similarity found in cheetahs is believed to be due to a bottleneck. In a study of 47 different enzymes, each of which can occur in several different forms in other types of cats, all the cheetahs studied had exactly the same form. This demonstrates that genetic drift can cause certain alleles to be lost from a population. Exactly what caused the cheetah bottleneck is not known. It is speculated that cheetahs were slaughtered by nineteenth-century cattle farmers protecting their herds, were captured by Egyptians as pets 4,000 years ago, or were decimated by a mass extinction tens of thousands of years ago. Today, cheetahs suffer from relative infertility because of the intense inbreeding that occurred after the bottleneck.

Founder Effect The **founder effect** is an example of genetic drift in which rare alleles, or combinations of alleles, occur at a higher frequency in a population isolated from the general population. After all, founding individuals contain only a fraction of the total genetic diversity of the original gene pool. The alleles carried by their founder or founders are dictated by chance alone. The Amish of Lancaster County, Pennsylvania, are an isolated group that was founded by German settlers. Today, as many as 1 in 14 individuals carries a recessive allele that causes an unusual form of dwarfism (affecting only the lower arms and legs) and polydactylism (extra fingers; **Fig. 15.13**). In most populations, only 1 in 1,000 individuals has this allele.

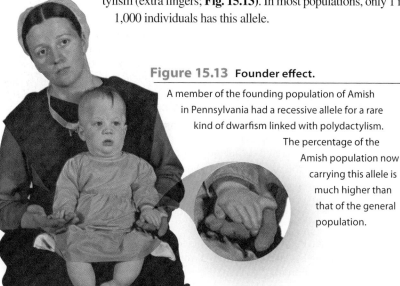

Figure 15.13 Founder effect.

A member of the founding population of Amish in Pennsylvania had a recessive allele for a rare kind of dwarfism linked with polydactylism. The percentage of the Amish population now carrying this allele is much higher than that of the general population.

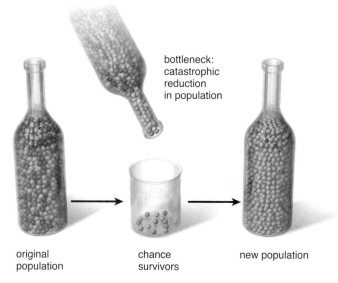

bottleneck: catastrophic reduction in population

original population   chance survivors   new population

Figure 15.12 Bottleneck effect.

A bottleneck effect may occur when a catastrophic reduction in a population occurs, such as after a major epidemic or a hurricane. For example, a parent population contains roughly equal numbers of genotypes, represented by blue, yellow, purple, green, and red marbles. The chance survivors of the catastrophe have genotypes represented mostly by green and yellow, resulting in a new population with altered gene pool frequencies.

Connecting the Concepts

For more on evolutionary changes in populations, refer to the following discussions.

Section 14.1 examines how Charles Darwin and Alfred Wallace recognized that natural selection is the mechanism of evolutionary change.

Section 16.1 describes how microevolution relates to long-term changes in populations, or macroevolution.

Section 16.3 explores the science of systematics, which interprets evolutionary change in populations over time.

Check Your Progress 15.2

1. Define the term *gene pool* and explain how it relates to allele frequencies in a population.

2. Explain the Hardy-Weinberg equilibrium. What happens to this equilibrium when microevolution occurs?

3. List the five conditions that can cause microevolution to occur.

4. Describe the significance of mutations in terms of evolution.

5. Explain how gene flow and nonrandom mating cause microevolution.

6. Explain the consequences of genetic drift and why it is more likely to happen in a small population.

Media Study Tools

www.mhhe.com/maderessentials3

Enhance your study of this chapter with study tools and practice tests. Also ask your instructor about the resources available through ConnectPlus, including the media-rich eBook, interactive learning tools, and animations.

The Chapter in Review

Summary

15.1 Natural Selection

Adaptation occurs when the more fit individuals reproduce more than others. These individuals usually possess traits better suited for survival in the environment, and over generations the frequency of this adaptive trait increases within the population.

Types of Selection

Most of the traits of evolutionary significance are polygenic, and a range of phenotypes in a population results in a bell-shaped curve. Three types of selection occur:

- **Directional selection:** The curve shifts in one direction, as when dark-colored peppered moths become prevalent in polluted areas.
- **Stabilizing selection:** The peak of the curve increases, as when most human babies have an intermediate birth weight. Babies that are very small or very large are less fit than those of intermediate weight.
- **Disruptive selection:** The curve has two peaks, as when British land snails vary because a wide geographic range causes selection to vary.

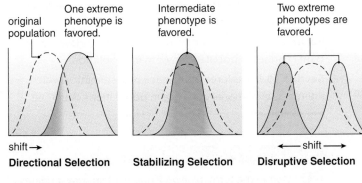

Directional Selection   **Stabilizing Selection**   **Disruptive Selection**

Sexual Selection

Sexual selection is different between males and females:

- Males compete to mate with as many females as possible.
- Females select males that exhibit the greatest fitness.

Adaptations Are Not Perfect

Adaptations are not perfect because evolution builds on what variation exists. Only certain types of variations are available, and developmental procedures tend toward the same types of results. The result is often a compromise between a benefit and a cost. Sexual selection has a reproductive benefit that may not pertain to adaptation.

Maintenance of Variations

Despite constant natural selection, variation is maintained because

- Mutations and recombination still occur; gene flow among populations can introduce new alleles; and natural selection may not eliminate less favored phenotypes.
- In sexually reproducing diploid organisms, the heterozygote acts as a repository for recessive alleles whose frequency is low. In sickle cell disease, the heterozygote is more fit in areas where malaria occurs; therefore, both homozygotes are maintained in the population.

15.2 Microevolution

Evolution in a Genetic Context

Microevolution involves several elements:

- All the various genes of a population make up its gene pool.
- The Hardy-Weinberg equilibrium is present when gene pool allele frequencies remain the same from generation to generation. Certain conditions have to be met to achieve an equilibrium:
- The conditions are (1) no mutations, (2) no gene flow, (3) random mating, (4) no genetic drift, and (5) no selection. Since these conditions are rarely met, a change in gene pool frequencies is likely.
- When gene pool frequencies change, microevolution has occurred. Deviations from a Hardy-Weinberg equilibrium allow us to determine when evolution has taken place and to measure the extent of change.

Causes of Microevolution

Microevolution is caused by five conditions:

- Mutations are the ultimate source of variation. Certain genotypic variations may be of evolutionary significance, only if the environment changes. Genetic diversity is promoted when there are several alleles for each locus and when single nucleotide polymorphisms (SNPs) exist.
- Gene flow occurs when a breeding individual (in animals) migrates to another population or when gametes and seeds (in plants) are carried into another population. Constant gene flow between two populations causes their gene pools to become similar.
- Nonrandom mating occurs when relatives mate (inbreeding) or when assortative mating takes place. Both of these cause an increase in homozygotes. Sexual selection, which occurs when a characteristic that increases the chances of mating is favored, favors random mating.
- Genetic drift occurs when allele frequencies are altered by chance— that is, by random sampling errors. Genetic drift is particularly evident after a bottleneck, when severe inbreeding occurs, or when founders start a new population.
- Natural selection (see Section 15.1).

Key Terms

Testing Yourself

Choose the best answer for each question.

1. A population consists of 48 *AA*, 54 *Aa*, and 22 *aa* individuals. What is the frequency of the *A* allele?
   a. 0.60
   b. 0.40
   c. 0.62
   d. 0.42
   e. 0.58

2. Which of the following is the binomial equation for the Hardy-Weinberg principle?
   a. $2p^2 + 2pq + 2q^2$
   b. $p^2 + pq + q^2$
   c. $2p^2 + pq + 2q^2$
   d. $p^2 + 2pq + q^2$

For questions 3 and 4, consider that about 70% of white North Americans can taste the chemical phenylthiocarbamide. The ability to taste is due to the dominant allele *T*. Nontasters are *tt*. Assume this population is in Hardy-Weinberg equilibrium.

3. What is the frequency of *t*?
   a. 0.30
   b. 0.70
   c. 0.55
   d. 0.09
   e. 0.60

4. What is the frequency of heterozygous tasters?
   a. 0.495
   b. 0.21
   c. 0.42
   d. 0.2475
   e. 0.45

5. Typically, mutations are immediately expressed and tested by the environment in
   a. prokaryotes.
   b. eukaryotes.
   c. prokaryotes and eukaryotes.
   d. neither prokaryotes nor eukaryotes.

6. The offspring of better adapted individuals are expected to make up a larger proportion of the next generation. The most likely explanation is
   a. mutations and nonrandom mating.
   b. gene flow and genetic drift.
   c. mutations and natural selection.
   d. mutations and genetic drift.

7. The northern elephant seal went through a severe population decline as a result of hunting in the late 1800s. The population has rebounded but is now homozygous for nearly every gene studied. This is an example of
   a. negative assortative mating.
   b. migration.
   c. mutation.
   d. a bottleneck.
   e. disruptive selection.

8. Which of the following generally results in a net gain in genetic variability?
   a. genetic drift
   b. mutation
   c. directional selection
   d. bottleneck
   e. stabilizing selection

For questions 9–15, indicate the effect of each of the conditions of the Hardy-Weinberg principle on genotype and allele frequencies. Each answer may be used more than once.

Key:
   a. alters genotype and allele frequencies
   b. alters genotype frequency only
   c. alters allele frequency only
   d. does not alter genotype or allele frequency

9. mutation
10. gene flow
11. inbreeding
12. assortative mating
13. genetic drift
14. bottleneck
15. natural selection

16. A small, reproductively isolated religious sect called the Dunkers was established by 27 families that came to the United States from Germany 200 years ago. The frequencies for blood group alleles in this population differ significantly from those in the general U.S. population. This is an example of
    a. negative assortative mating.
    b. natural selection.
    c. the founder effect.
    d. the bottleneck effect.
    e. gene flow.

17. Assuming a population is in Hardy-Weinberg equilibrium, 21% of the population is homozygous dominant, 50% is heterozygous, and 29% is homozygous recessive. What percentage of the next generation is predicted to be homozygous recessive?
    a. 21%
    b. 50%
    c. 29%
    d. 42%
    e. 58%

18. When a population is small, there is a greater chance of
    a. gene flow.
    b. genetic drift.
    c. natural selection.
    d. mutations.
    e. sexual selection.

19. Which of the following is not expected to help maintain genetic variability?
    a. gene flow
    b. mutation
    c. recombination
    d. disruptive selection
    e. genetic drift

20. The sickle cell allele is maintained in regions where malaria is prevalent because
    a. the allele confers resistance to the parasite.
    b. gene flow is high in those regions.
    c. disruptive selection is occurring.
    d. genetic drift randomly selects for the allele.

21. Complete the following by drawing a curve in the other two graphs to show the effect of directional selection.

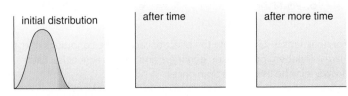

22. Complete the following by drawing a curve in the other two graphs to show the effect of stabilizing selection.

23. Complete the following by drawing curves in the other two graphs to show the effect of disruptive selection.

Thinking Scientifically

1. A behaviorist studying mountain bluebird reproductive behavior found that male aggression toward another male was highest during nest construction, and then it lessened after the first egg was laid, and again after hatching occurred. **a.** Knowing that natural selection favors behaviors that save energy, explain the reproductive behavior of the male mountain bluebirds. **b.** What might make you rule out genetic drift as an explanation?

Bioethical Issue

Who Should Reproduce?

The highly regarded population geneticist Sir Ronald Aylmer Fisher published a book in 1930 entitled *The Genetical Theory of Natural Selection*. In the book, he claims that civilizations fail because individuals with the highest level of fitness (those at the top of the societal ladder) do not reproduce as often as less affluent individuals. He suggests that high-income couples be paid to have children in order to improve the population. In other words, Fisher suggests that we carry out selection within the human population.

Can you envision any scenario in which our society should encourage reproduction by the most fit and discourage it by the least fit? If so, how should it be done? What characteristics would be appropriate to select for or against?

Iguanas as Examples of Evolutionary Change

Iguanas are a type of lizard known for a very long tail, which can be up to three times their body length. The green iguana is common throughout the South American continent. Its claws enable it to climb trees, where it feeds on succulent leaves and soft fruits along riverbanks. It is an excellent swimmer and uses rivers to travel to new feeding areas.

Green iguanas do not inhabit islands to the west and northeast of South America. Instead, the Galápagos Islands to the west have one type of land iguana and one type of marine iguana. Marine iguanas, black in color, are unique to the Galápagos Islands—they occur no place else on Earth. These air-breathing reptiles can dive to a depth exceeding 10 meters (33 feet) and remain submerged for more than 30 minutes. They use their claws to cling to the bottom rocks while feeding on algae.

Hispaniola, an island to the northeast of South America, is inhabited by the rhinoceros iguana, which has three horny bumps on its snout and is dark brown to black in color. This iguana lives mainly in dry forests with rocky, limestone habitats and feeds on a wide variety of plants.

The theory of evolution explains the occurrence of the unique iguanas on the Galápagos Islands and Hispaniola. It is hypothesized and supported by various data that ancestral iguanas swam or hitchhiked on floating driftwood from South America to the Galápagos to the west and to Hispaniola to the northeast. After arrival, these iguana populations were cut off from other iguana populations, and this allowed them to evolve into these new and different species. The origin of species and how they are classified and studied are the topics of this chapter.

16

Evolution on a Large Scale

BEFORE YOU BEGIN

Before beginning this chapter, take a few moments to review the following discussions.

Section 14.2 What roles do the fossil records and the study of comparative anatomy have in understanding evolutionary change?

Section 15.1 How does natural selection act as the mechanism of evolutionary change?

Section 15.2 What is microevolution?

16.1 Speciation and Macroevolution

Learning Outcomes

Upon completion of this section, you should be able to

1. Explain how scientists define a biological species and describe the limitation of this definition.
2. List five types of prezygotic isolating mechanisms and explain how they prevent members of two different species from reproducing.
3. List three types of postzygotic isolating mechanisms and explain how they prevent members of two different species from reproducing.
4. Contrast allopatric speciation with sympatric speciation and explain how each method may result in the creation of a new species.
5. Explain how new species may arise by adaptive radiation.

Chapter 15 considered evolution on a small scale—that is, microevolution, small changes over a short period of time. In this chapter, we turn our attention to evolution on a large scale—that is, **macroevolution,** large changes over a very long period of time. The history of life on Earth is a part of macroevolution. Macroevolution requires **speciation,** the splitting of one species into two or more new species. Speciation involves the gene pool changes that we studied in Chapter 15.

Species originate, evolve adaptations to their environments, and then may become extinct (**Fig. 16.1**). Without the origin and extinction of species, life on Earth would not have the ever-changing history that we find in the fossil record.

Defining Species

Before we consider the origin of species, we first need to define a species. Appearance is not always a good criterion. The members of different species can look quite similar, while the members of a single species can be diverse in appearance. Although many definitions have been proposed, the

Figure 16.1 Dinosaurs.

Artwork of a Mesozoic landscape, Cretaceous period, including *saltasaurus, brachiosaurus, mononykus, protoceratops, nyctosaurus, styracosaurus, tarbosaurus, maiasaura, stegoceras, evoplocephalus, deltatheridium,* and *ichthyornis* (145 to 65 MYA). All of these species are now extinct.

biological species concept offers a testable way to define a species that does not depend on appearance: The members of a species interbreed and have a shared gene pool, and each species is reproductively isolated from every other species. For example, the flycatchers in **Figure 16.2** are members of separate species because they do not interbreed in nature.

According to the biological species concept, gene flow occurs between the populations of a species, but not between populations of different species. The red maple and the sugar maple are found over a wide geographic range in the eastern half of the United States, and each species is made up of many populations. However, the members of each species' population rarely hybridize in nature. Therefore, these two types of plants are separate species. In contrast, the human species has many populations, which certainly differ in physical appearance (**Fig. 16.3**). We know, however, that all humans belong to one species because the members of these populations can produce fertile offspring.

The biological species concept is useful, as we shall see, but even so, it has its limitations. For example, it applies only to sexually reproducing organisms and cannot apply to asexually reproducing organisms. Then, too, sexually reproducing organisms are not always as reproductively isolated as we would expect. Some North American orioles live in the western half of the continent, some in the eastern half, yet even the two most genetically distant oriole species, as recognized by analyzing their mitochondrial DNA, will hybridize where they meet in the middle of the continent.

There are other definitions of species aside from the biological definition. Later in this chapter, we will define species as a category of classification below the rank of genus. Species in the same genus share a recent common ancestor.

Acadian flycatcher,
Empidonax virescens

Traill's flycatcher,
Empidonax trailli

Least flycatcher,
Empidonax minimus

Figure 16.2 Three species of flycatchers.

Although these flycatcher species are nearly identical in appearance, we know they are separate species because they are reproductively isolated—the members of each species reproduce only with one another. Each species has a characteristic song and its own habitat during the mating season as well.

a.   b.

Figure 16.3 Human populations.

The Maassai of East Africa (**a**) and the Kuna Indians from the San Blas Islands of Panama (**b**) are both members of the species *Homo sapiens* because the Maassai and Kuna can reproduce and produce fertile offspring.

Connections and Misconceptions

How can we determine if an organism that does not reproduce sexually is a distinct species?

Many organisms either do not or very rarely reproduce sexually. For example, there are species of mosses that reproduce sexually only every 200 to 300 years! To determine if two populations of asexual organisms are distinct species, scientists rely on DNA analysis, morphological studies, and a close examination of the organisms' ecology to determine whether the two populations would reproduce naturally. Often, scientists have to revisit the classification of a species as research unveils new information.

A **common ancestor** is a single ancestor for two or more different groups. For example, your father's mother is the common ancestor for you, your siblings, and your paternal cousins. Similarly, there is a common ancestor for all species of roses.

Reproductive Barriers

As mentioned, for two species to be separate, they must be reproductively isolated—that is, gene flow must not occur between them. Reproductive barriers are isolating mechanisms that prevent successful reproduction (**Fig. 16.4**). In evolution, reproduction is successful only when it produces fertile offspring.

Prezygotic (before the formation of a zygote) **isolating mechanisms** prevent reproductive attempts and make it unlikely that fertilization will be successful if mating is attempted. Habitat isolation, temporal isolation, behavioral isolation, mechanical isolation, and gamete isolation make it highly unlikely that particular genotypes will contribute to the gene pool of a population.

Habitat isolation When two species occupy different habitats, even within the same geographic range, they are less likely to meet and attempt to reproduce. This is one of the reasons that the flycatchers in Figure 16.2 do not mate and the red maple and sugar maple do not exchange pollen. In tropical rain forests, many animal species are restricted to a particular level of the forest canopy; in this way, they are isolated from similar species.

Temporal isolation Two species can live in the same locale, but if they reproduce at different times of year, they do not attempt to mate. For example, *Reticulitermes hageni* and *R. virginicus* are two species of termites. The former has mating flights in March through May, whereas the latter mates in the fall and winter months.

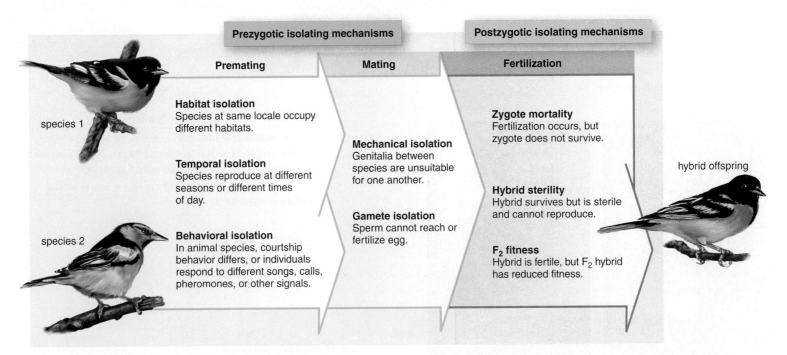

Figure 16.4   Reproductive barriers.

Prezygotic isolating mechanisms prevent mating attempts or a successful outcome, should mating take place—for example, between two species of orioles. No zygote is ever formed if these mechanisms are successful. Postzygotic isolating mechanisms prevent offspring from reproducing—that is, if a hybrid oriole should result, it would be unable to breed successfully.

Behavioral isolation Many animal species have courtship patterns that allow males and females to recognize one another (**Fig. 16.5**). Female fireflies recognize males of their species by the pattern of the males' flashings; similarly, female crickets recognize males of their species by the males' chirping. Many males recognize females of their species by sensing chemical signals, called pheromones. For example, female gypsy moths secrete chemicals from abdominal glands. These chemicals are detected downwind by receptors on the antennae of males.

Video Flirting Flies

Mechanical isolation When animal genitalia or plant floral structures are incompatible, reproduction cannot occur. Inaccessibility of pollen to certain pollinators can prevent cross-fertilization in plants, and the sexes of many insect species have genitalia that do not match, or other characteristics that make mating impossible. For example, male dragonflies have claspers that are suitable for holding only the females of their own species.

Gamete isolation Even if the gametes of two different species meet, they may not fuse to become a zygote. In animals, the sperm of one species may not be able to survive in the reproductive tract of another species, or the egg may have receptors only for sperm of its species. Also, in each type of flower, only certain pollen grains can germinate, so that sperm successfully reach the egg.

Postzygotic (after the formation of a zygote) **isolating mechanisms** prevent hybrid offspring (reproductive product of two different species) from developing or breeding, even if reproduction attempts have been successful.

Zygote mortality The hybrid zygote may not be viable, so it dies. A zygote with two different chromosome sets may fail to go through mitosis properly, or the developing embryo may receive incompatible instructions from the maternal and paternal genes, so that it cannot continue to exist.

Hybrid sterility The hybrid zygote may develop into a sterile adult. As is well known, a cross between a horse and a donkey produces a mule, which is usually sterile—it cannot reproduce (**Fig. 16.6**). Sterility of hybrids generally results from complications in meiosis, which lead to an inability to produce viable gametes. A cross between a cabbage and a radish produces offspring that cannot form gametes, even though the diploid number is 18, an even number, most likely because the cabbage chromosomes and the radish chromosomes cannot align during meiosis.

F_2 fitness If hybrids can reproduce, their offspring are unable to reproduce. In some cases, mules are fertile, but their offspring (the F_2 generation) are not fertile.

Models of Speciation

The introduction to this chapter suggests that iguanas of South America may be the common ancestor for both the marine iguana on the Galápagos Islands (to the west of South America) and the rhinoceros iguana on Hispaniola (the Caribbean island containing the countries of Haiti and the Dominican Republic). If so, how could it have happened? Green iguanas are strong swimmers, so by chance a few could have migrated to these islands, where they formed populations separate from each other and from the parent population in South America. Each population continued on its own evolutionary path as new mutations, genetic drift, and natural selection occurred. Eventually, reproductive isolation developed, and there

Figure 16.5 Prezygotic isolating mechanism.

An elaborate courtship display allows the blue-footed boobies of the Galápagos Islands to select a mate. The male lifts his feet in a ritualized manner that shows off their bright blue color.

horse ♀   ♂ donkey

mating

fertilization

mule (F_1 hybrid)

Usually mules cannot reproduce. If an F_2 offspring does result, it cannot reproduce.

Figure 16.6 Postzygotic isolating mechanism.

Mules are horse-donkey hybrids. Mules are infertile due to a difference in the chromosomes inherited from their parents.

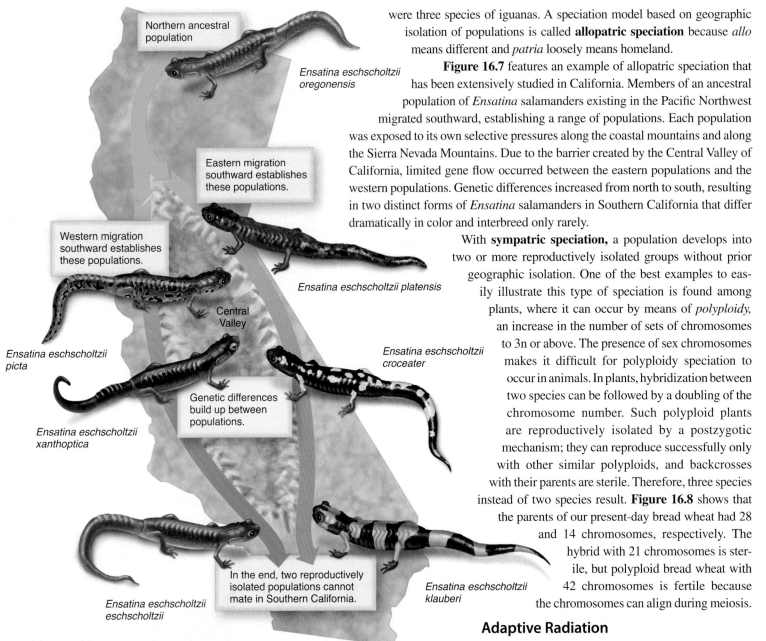

Figure 16.7 Allopatric speciation.

In this example of allopatric speciation, the Central Valley of California is separating a range of populations descended from the same northern ancestral species. The limited contact between the populations on the west and those on the east allow genetic changes to build up to such an extent that members of the two southern populations rarely reproduce with each other and are designated as subspecies.

were three species of iguanas. A speciation model based on geographic isolation of populations is called **allopatric speciation** because *allo* means different and *patria* loosely means homeland.

Figure 16.7 features an example of allopatric speciation that has been extensively studied in California. Members of an ancestral population of *Ensatina* salamanders existing in the Pacific Northwest migrated southward, establishing a range of populations. Each population was exposed to its own selective pressures along the coastal mountains and along the Sierra Nevada Mountains. Due to the barrier created by the Central Valley of California, limited gene flow occurred between the eastern populations and the western populations. Genetic differences increased from north to south, resulting in two distinct forms of *Ensatina* salamanders in Southern California that differ dramatically in color and interbreed only rarely.

With **sympatric speciation,** a population develops into two or more reproductively isolated groups without prior geographic isolation. One of the best examples to easily illustrate this type of speciation is found among plants, where it can occur by means of *polyploidy,* an increase in the number of sets of chromosomes to 3n or above. The presence of sex chromosomes makes it difficult for polyploidy speciation to occur in animals. In plants, hybridization between two species can be followed by a doubling of the chromosome number. Such polyploid plants are reproductively isolated by a postzygotic mechanism; they can reproduce successfully only with other similar polyploids, and backcrosses with their parents are sterile. Therefore, three species instead of two species result. **Figure 16.8** shows that the parents of our present-day bread wheat had 28 and 14 chromosomes, respectively. The hybrid with 21 chromosomes is sterile, but polyploid bread wheat with 42 chromosomes is fertile because the chromosomes can align during meiosis.

Adaptive Radiation

A clear example of speciation through adaptive radiation is provided by the finches on the Galápagos Islands, which are often called Darwin's

Figure 16.8 Sympatric speciation.

In this example of sympatric speciation, two populations of wild wheat hybridized many years ago. The hybrid is sterile, but chromosome doubling allowed some plants to reproduce. These plants became today's bread wheat.

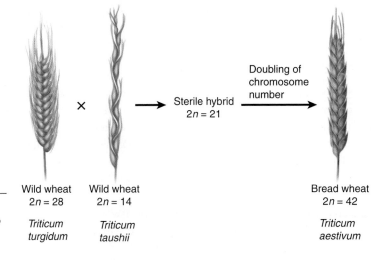

Wild wheat
2n = 28

Triticum turgidum

Wild wheat
2n = 14

Triticum taushii

Sterile hybrid
2n = 21

Doubling of chromosome number

Bread wheat
2n = 42

Triticum aestivum

finches because Darwin first realized their significance as an example of how evolution works. During **adaptive radiation,** many new species evolve from a single ancestral species. The many species of finches that live on the Galápagos Islands are hypothesized to be descendants of a single type of ancestral finch from the mainland (**Fig. 16.9**). The populations on the various islands were subjected to the founder effect involving genetic drift, genetic mutations, and the process of natural selection. Because of natural selection, each population became adapted to a particular habitat on its island. In time, the various populations became so genotypically different that now, when by chance they reside on the same island, they do not interbreed and are therefore separate species. There is evidence that the finches use beak shape to recognize members of the same species during courtship. Rejection of suitors with the wrong type of beak is a behavioral prezygotic isolating mechanism.

Video
Finches Adaptive Radiation

Similarly, inhabiting the Hawaiian Islands is a wide variety of honeycreepers, all descended from a common goldfinch-like ancestor that arrived from Asia or North America about 5 MYA. Today, honeycreepers have a range of beak sizes and shapes (see Fig. 1.11) for feeding on various food sources, including seeds, fruits, flowers, and insects.

Check Your Progress 16.1

1. Define species according to the biological species concept.
2. Describe the limitations of the biological species concept.
3. Categorize the different types of reproductive barriers as being either a prezygotic or postzygotic barrier, and give an example of each.
4. Compare and contrast allopatric speciation with sympatric speciation. Give an example of each.
5. Relate adaptive radiation to Darwin's finches.

Connections and Misconceptions

Are there examples of polyploid species in animals?

In general, polyploidy is rarer in animals than in plants. However, there are examples of polyploid insects and fish, and polyploidy also appears to occur frequently in the amphibians, specifically in salamanders. In 1999, scientists reported a polyploid rat species (*Typanoctomys barrerae*) in Argentina, but later genetic analysis refuted this claim. Most geneticists believe that polyploidy in mammals is unlikely due to the well-defined role of mammalian sex chromosomes and the balance between the number of autosomes and sex chromosomes.

Connecting the Concepts

For more information on the concepts presented in this section, refer to the following discussions.

Section 9.4 provides background information on how nondisjunction causes changes in chromosome number.

Section 14.1 examines the role that the Galápagos finches played in establishing natural selection as the mechanism for evolutionary change.

Figure 16.9 Darwin's finches.

Each of Darwin's finches is adapted to gathering and eating a different type of food. Tree finches have beaks largely adapted to eating insects and, at times, plants. Ground finches have beaks adapted to eating the flesh of the prickly-pear cactus or different-sized seeds.

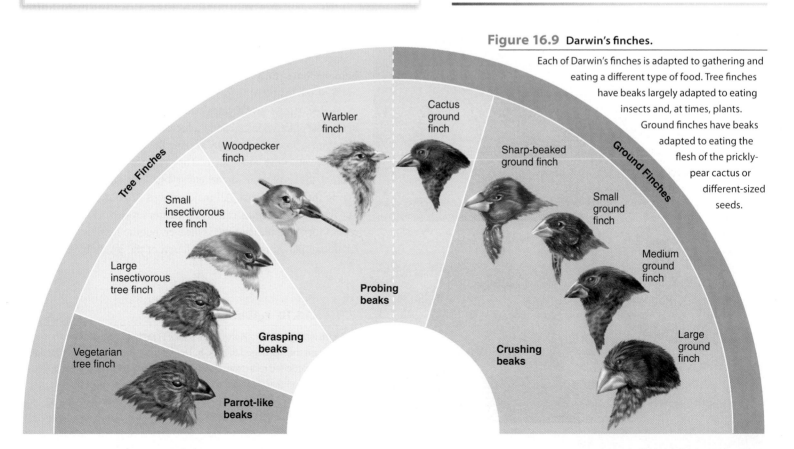

16.2 The Fossil Record

Learning Outcomes

Upon completion of this section, you should be able to

1. Describe how the geological timescale is divided into eras, periods, and epochs.
2. Contrast the gradualistic model of evolution with the punctuated equilibrium model of evolution.
3. Explain some of the ways in which mass extinctions of organisms may have occurred.

The history of the origin and extinction of species on Earth is best discovered by studying the fossil record (**Fig. 16.10**). **Fossils** are the traces and remains of past life or any other direct evidence of past life. **Paleontology** is the science of discovering and studying the fossil record and, from it, making decisions about the history of species.

a.

b.

c.

d.

The Geological Timescale

Because life-forms have evolved over time, the strata (layers of sedimentary rock, see Fig. 14.2*a*) of the Earth's crust contain different fossils. By studying the strata and the fossils they contain, geologists have been able to construct a geological timescale (**Table 16.1**). It divides the history of life on Earth into eras, then periods, and then epochs and describes the types of fossils common to each of these divisions of time. Notice that, in the geological timescale, only the periods of the Cenozoic era are divided into epochs, meaning that more attention is given to the evolution of primates and flowering plants than to the earlier evolving organisms. Despite an epoch assigned to modern civilization, humans have only been around about .04% of the history of life.

Animation
Geological History of the Earth

In contrast, prokaryotes existed alone for some 2 billion years before the eukaryotic cell and multicellularity arose during Precambrian time. Some prokaryotes became the first photosynthesizers to add oxygen to the atmosphere. The presence of oxygen may have spurred the evolution of the eukaryotic cell and multicellularity during the Precambrian. All major groups of animals evolved during what is sometimes called the Cambrian explosion. The fossil record for Precambrian time is meager, but the fossil record for the Cambrian period is rich. The evolution of the invertebrate external

Figure 16.10 Fossils.

a. A fern leaf from 245 million years ago (MYA) remains because it was buried in sediment that hardened to rock. **b.** This midge (40 MYA) became embedded in amber (hardened resin from a tree). **c.** Most fossils, such as this early insectivore mammal (47 MYA) are remains of hard parts because they do not decay as the soft parts do. **d.** This dinosaur footprint (135 MYA) is a sign of ancient life. **e.** This tree trunk (190 MYA) is petrified because minerals have replaced the original tissues.

e.

Table 16.1   The Geological Timescale: Major Divisions of Geological Time and Some of the Major Evolutionary Events That Occurred

Era	Period	Epoch	MYA	Plant and Animal Life
Cenozoic	Quaternary	Holocene	0–0.01	AGE OF HUMAN CIVILIZATION; Destruction of ecosystems accelerates extinctions.
			SIGNIFICANT MAMMALIAN EXTINCTION	
		Pleistocene	0.01–2	Modern humans appear; modern plants spread and diversify.
		Pliocene	2–6	First hominids appear; modern angiosperms flourish.
		Miocene	6–24	Ape-like mammals, grazing mammals, and insects flourish; grasslands spread; and forests contract.
	Tertiary	Oligocene	24–37	Monkey-like primates appear; modern angiosperms appear.
		Eocene	37–58	All modern orders of mammals are present; subtropical forests flourish.
		Paleocene	58–65	Primates, herbivores, carnivores, insectivores are present; angiosperms diversify.
	MASS EXTINCTION: 50% OF ALL SPECIES, DINOSAURS, AND MOST REPTILES			
Mesozoic	Cretaceous		65–144	Placental mammals and modern insects appear; angiosperms spread and conifers persist.
	Jurassic		144–208	Dinosaurs flourish; birds and angiosperms appear.
	MASS EXTINCTION: 48% OF ALL SPECIES, INCLUDING CORALS AND FERNS			
	Triassic		208–250	First mammals and dinosaurs appear; forests of conifers and cycads dominate land; corals and molluscs dominate seas.
	MASS EXTINCTION (THE "GREAT DYING"): 83% OF ALL SPECIES ON LAND AND SEA			
Paleozoic	Permian		250–286	Reptiles diversify; amphibians decline; and gymnosperms diversify.
	Carboniferous		286–360	Amphibians diversify; reptiles appear; and insects diversify. Age of great coal-forming forests.
	MASS EXTINCTION: OVER 50% OF COASTAL MARINE SPECIES, CORALS			
	Devonian		360–408	Jawed fishes diversify; insects and amphibians appear; seedless vascular plants diversify and seed plants appear.
	Silurian		408–438	First jawed fishes and seedless vascular plants appear.
	MASS EXTINCTION: OVER 57% OF MARINE SPECIES			
	Ordovician		438–510	Invertebrates spread and diversify; jawless fishes appear; nonvascular plants appear on land.
	Cambrian		510–543	Marine invertebrates with skeletons are dominant and invade land, and marine algae flourish.
Precambrian time			600	Oldest soft-bodied invertebrate fossils.
			1,400–700	Protists evolve and diversify.
			2,000	Oldest eukaryotic fossils.
			2,500	O_2 accumulates in atmosphere.
			3,500	Oldest known fossils (prokaryotes).
			4,500	Earth forms.

Connections and Misconceptions

What is the Burgess Shale?

The Burgess Shale is the name for a rock formation in the Canadian Rocky Mountains near the Burgess Pass. Around 525 MYA, this region was located along the coast. It is believed that an  earthquake caused a landslide that almost instantly buried much of the marine life living in the shallow coastal waters. Unlike many fossil beds, the Burgess Shale contains the remains of soft-shelled organisms, such as worms and sea cucumbers, as well as other organisms from the Cambrian explosion—a period of rapid diversification in marine life around 545 MYA. Over 60,000 unique types of fossils have been found in the Burgess Shale (including the trilobite fossil shown here), making this fossil bed one of our most valuable assets for studying the early evolution of life in the oceans.

skeleton accounts for this increase in the number of fossils. Perhaps this skeleton, which impedes the uptake of oxygen, couldn't evolve until oxygen was plentiful. Or perhaps the external skeleton was merely a defense against predation.

The origin of life on land is another interesting topic. During the Paleozoic era, plants were present on land before animals. Nonvascular plants preceded vascular plants, and among these, cone-bearing plants (gymnosperms) preceded flowering plants (angiosperms). Among vertebrates, the fishes were aquatic, and the amphibians invaded land. The reptiles, including dinosaurs and birds, shared an amniote ancestor with the mammals. The number of species in the world has continued to increase until the present time, despite the occurrence of five mass extinctions, including one significant mammalian extinction during the history of life on Earth.

The Pace of Speciation

Darwin theorized that evolutionary changes occur gradually. In other words, he supported a *gradualistic model* to explain the pace of evolution. Speciation probably occurs after populations become isolated, with each group continuing slowly on its own evolutionary pathway. The gradualism model often shows the evolutionary history of groups of organisms by drawing the type of **evolutionary tree** shown in **Figure 16.11***a*. In this diagram, note that an ancestral species has given rise to two separate species, represented by a slow change in plumage color. The gradualistic model suggests that it is difficult to indicate when speciation has occurred because there would be so many transitional links. In some cases, it has been possible to trace the evolution of a group of organisms by finding transitional links.

More often, however, species appear quite suddenly in the fossil record, and then they remain essentially unchanged phenotypically until they undergo extinction. Some paleontologists have therefore developed a *punctuated equilibrium model* to explain the pace of evolution. The model says that a period of equilib-

Figure 16.11 Pace of evolution.

a. According to the gradualistic model, new species evolve slowly from an ancestral species.
b. According to the punctuated equilibrium model, new species appear suddenly and then remain largely unchanged until they become extinct.

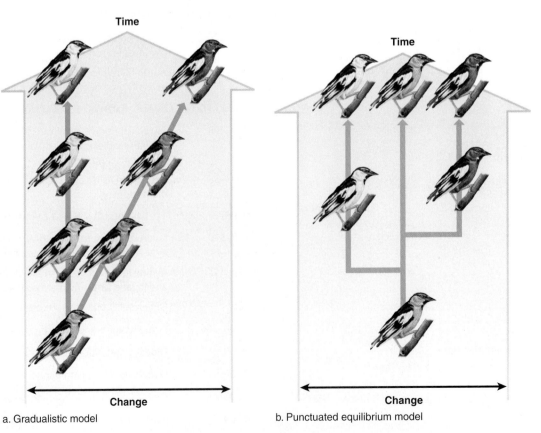

rium (no change) is punctuated (interrupted) by speciation. Figure 16.11b shows the type of diagram paleontologists prefer to use when representing the history of evolution over time. This model suggests that transitional links are less likely to become fossils and less likely to be found. Speciation most likely involves only an isolated subpopulation at one locale. Only when this new subpopulation expands and replaces existing species is it apt to show up in the fossil record.

The differences between these two models are subtle, especially when we consider that the "sudden" appearance of a new species in the fossil record could represent many thousands of years because geological time is measured in millions of years.

Mass Extinctions of Species

As researchers have noted, most species exist for only a limited amount of time (measured in millions of years), and then they die out (become extinct). **Mass extinctions** are the disappearance of a large number of species or a higher taxonomic group within a relatively short period of time. The geological timescale in Table 16.1 shows the occurrence of five mass extinctions: at the ends of the Ordovician, Devonian, Permian, Triassic, and Cretaceous periods. Also, there was a significant mammalian extinction at the end of the Pleistocene epoch. While many factors contribute to mass extinctions, two possible types of events (continental drift and meteorite impacts) have particular significance.

Continental drift—the movement of continents—has contributed to several extinctions. You may have noticed that the coastlines of several continents are mirror images of each other. For example, the outline of the west coast of Africa matches that of the east coast of South America. Also, the same geological structures are found in many of the areas where the continents touched at one time. A single mountain range runs through South America, Antarctica, and Australia, for example. But the mountain range is no longer continuous because the continents have drifted apart. The reason the continents drift is explained by a branch of geology known as *plate tectonics,* which is based on the fact that the Earth's crust is fragmented into slab-like plates that float on a lower, hot mantle layer (**Fig. 16.12**). The continents and the ocean basins are a part of these rigid plates, which move like conveyor belts.

Figure 16.12 Plate tectonics.

The Earth's surface is divided into several solid tectonic plates floating on the fluid magma beneath them. At rifts in the ocean floor, two plates gradually separate as fresh magma wells up and cools, enlarging the plates. Mountains, including volcanoes, are raised where one plate pushes beneath another at subduction zones. Where two plates slowly grind past each other at a fault line, tension builds up, which is released occasionally in the form of eathquakes.

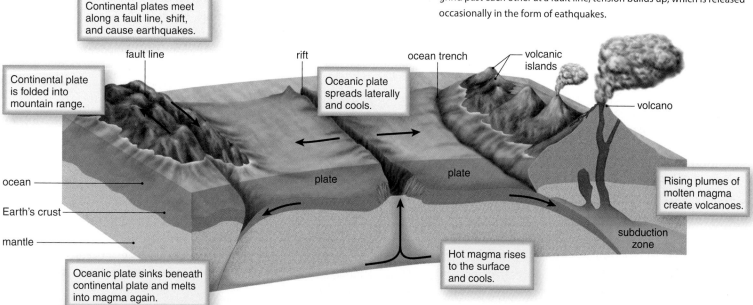

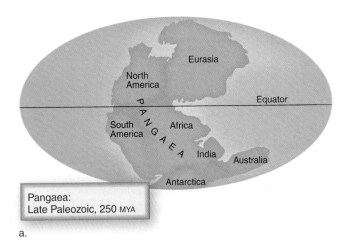

Pangaea:
Late Paleozoic, 250 MYA

a.

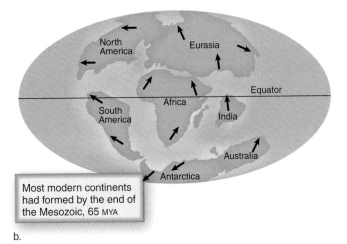

Most modern continents
had formed by the end of
the Mesozoic, 65 MYA

b.

Figure 16.13   Continental drift.

a. About 250 MYA, all the continents were joined into a supercontinent called Pangaea. **b.** By 65 MYA, all the continents had begun to separate. This process is continuing today. North America and Europe are separating at a rate of about 2 centimeters per year.

The loss of habitat is a significant cause of extinctions, and continental drift can lead to massive habitat changes. We know that 250 million years ago (MYA), at the time of the Permian mass extinction, all the Earth's continents came together to form one supercontinent called Pangaea (**Fig. 16.13a**). The result was dramatic environmental change through the shifting of wind patterns, ocean currents, and most importantly the amount of available shallow marine habitat. Marine organisms suffered as the oceans merged, and the amount of shoreline, where many marine organisms lived, was drastically reduced. Species diversity did not recover until some continents drifted away from the poles, shorelines increased, and warmth returned (Fig. 16.13b). Terrestrial organisms were affected as well because the amount of interior land, which tends to have a drier and more erratic climate, increased. Immense glaciers developing at the poles withdrew water from the oceans and chilled even once tropical land.

The other event that is known to have contributed to mass extinctions is the impact of a meteorite as it crashed into the Earth. The result of a large meteorite striking Earth could have been similar to that of a worldwide atomic bomb explosion: A cloud of dust would have mushroomed into the atmosphere, blocking out the sun and causing plants to freeze and die. This type of event has been proposed as a primary cause of the Cretaceous extinction that saw the demise of the dinosaurs. Cretaceous clay contains an abnormally high level of iridium, an element that is rare in the Earth's crust but more common in meteorites. A layer of soot has been identified in the strata alongside the iridium, and a huge crater that could have been caused by a meteorite was found in the Caribbean–Gulf of Mexico region on the Yucatán peninsula.

Connections and Misconceptions

What is the "Sixth Mass Extinction Event"?

Many ecologists now support the concept that we are currently involved in the Earth's sixth mass extinction event. However, unlike the first five major events, this one is caused not by geological or astronomical events but by human actions. Pollution, land use, invasive species, and global  climate change associated with the burning of fossil fuels are all recognized as contributing factors. The exact rate of species loss can be difficult to determine, but international agencies report that the current loss of species is between 100 and 1,000 times faster than the pre-human rates recorded by the fossil record.

Connecting the Concepts

For more information on the material in this section, refer to the following discussions.

Section 14.2 describes how the fossil record is used as evidence of evolutionary change.

Sections 18.1 and **19.1** provide an overview of the evolution of the plants and animals, respectively.

Section 32.1 examines some of the influences that humans are having on the diversity of life.

Check Your Progress 16.2

1 Summarize how the fossil record is the best evidence for macroevolution.

2 Explain the phrase "punctuated equilibrium model."

3 Relate the mass extinctions of species with the types of events that can cause them.

4 Discuss some of the major evolutionary events that occurred during different periods of geological time to form the geological timescale (Table 16.1).

16.3 Systematics

Learning Outcomes

Upon completion of this section, you should be able to

1. List the hierarchical levels of Linnaean classification from the most inclusive to the least inclusive and explain how this type of classification is useful to biologists.
2. Explain what information can be learned from a phylogenetic tree and list some of the types of information that is used in constructing them.
3. Contrast a homologous structure with an analogous structure.
4. Define cladistics and explain how this method may be used to study the evolutionary relationships between groups of organisms.
5. List the three domains of living organisms, and describe the general characteristics of organisms contained within each domain.

All fields of biology, but especially **systematics,** are dedicated to understanding the evolutionary history of life on Earth. Systematics is very analytical and relies on a combination of data from the fossil record and comparative anatomy and development, with an emphasis today on molecular data, to determine **phylogeny,** the evolutionary history of a group of organisms. Classification is a part of systematics because ideally organisms are classified according to our present understanding of evolutionary relationships.

Linnaean Classification

Taxonomy is the branch of biology concerned with identifying, naming, and classifying organisms. A **taxon** (pl., taxa) is an organism or a group of organisms at a particular level in a classification system. The binomial system of nomenclature assigns a two-part name to each type of organism. For example, the plant in **Figure 16.14** has been named *Cypripedium acaule.* This name means that the plant is in the genus *Cypripedium* and that the specific epithet is *acaule.* Notice that the scientific name is in italics and only the genus is capitalized. The genus can be abbreviated to a single letter if the full name has been given previously and if it is used with a specific epithet. For example, *C. acaule* is an acceptable way to designate this plant. The name of an organism usually tells you something about the organism. In this instance, the genus name, *Cypripedium,* refers to the slipper shape of the flower, and the specific epithet, *acaule,* says that the flower has no independent stem.

Why do organisms need scientific names? And why do scientists use Latin, rather than common names, to describe organisms? There are several reasons. First, a common name will vary from country to country because different countries use different languages. Second, even people who speak the same language sometimes use different common names to describe the same organism—for example, bowfin,

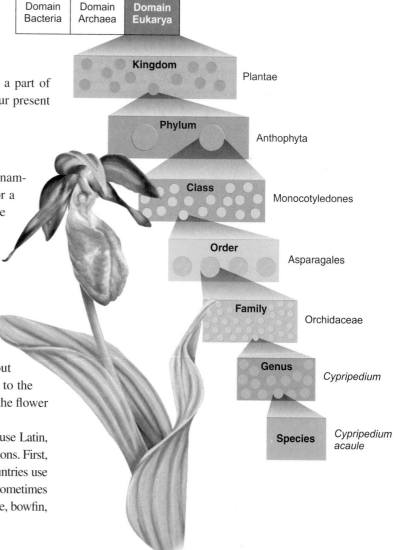

Figure 16.14 Taxonomy hierarchy.

A domain is the most inclusive of the classification categories. The plant kingdom is in the domain Eukarya. In the plant kingdom are several phyla, each represented here by lavender circles. The phylum Anthophyta has only two classes (the monocots and eudicots). The class Monocotyledones encompasses many orders. In the order Orchidales are many families; in the family Orchidaceae are many genera; and in the genus *Cypripedium* are many species—for example, *Cypripedium acaule.* (This illustration is diagrammatic and doesn't necessarily show the correct number of subcategories.)

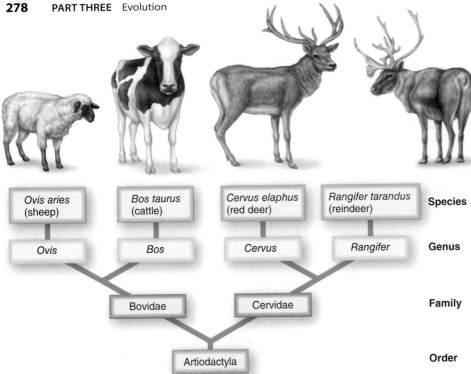

				Species
Ovis aries (sheep)	Bos taurus (cattle)	Cervus elaphus (red deer)	Rangifer tarandus (reindeer)	
Ovis	Bos	Cervus	Rangifer	Genus
Bovidae		Cervidae		Family
Artiodactyla				Order

Figure 16.15 Classification and phylogeny.

The classification and phylogenetic tree for a group of organisms is ideally constructed to reflect their phylogenetic history. A species is most closely related to other species in the same genus, more distantly related to species in other genera of the same family, and so forth on through order, class, phylum, and kingdom.

grindle, choupique, and cypress trout are all the same common fish, *Amia calva.* Furthermore, between countries the same common name is sometimes given to different organisms. A robin in England is very different from a robin in the United States, for example. Latin, on the other hand, is a universal language that not too long ago was well known by most scholars, many of whom were physicians or clerics. When scientists throughout the world use a scientific binomial name, they know they are speaking of the same organism.

Today, taxonomists use several categories of classification created by Swedish biologist Carl Linnaeus in the 1700s to show varying levels of similarity: **species, genus, family, order, class, phylum,** and **kingdom.** More recently, a higher taxonomic category, the **domain,** has been added to this list. There can be several species within a genus, several genera within a family, and so forth. In this hierarchy, the higher the category, the more inclusive it is (Fig. 16.14). Therefore species in the same domain have general traits in common, while those in the same genus have quite specific traits in common.

In most cases, each category of classification can be subdivided into three additional categories, as in superorder, order, suborder, and infraorder. Considering these, there are more than 30 categories of classification.

Phylogenetic Trees

Figure 16.15 shows how the classification of groups of organisms allows us to construct a **phylogenetic tree,** a diagram that indicates common ancestors and lines of descent (lineages). The common ancestor at the base of the tree has traits that are shared by all the other groups in the tree. For example, the Artiodactyla are characterized by having hoofs with an even number of toes. On the other hand, notice that the Cervidae have antlers but the Bovidae have no antlers. Finally, among the Cervidae, the antlers are highly branched in the red deer but palmate (having the shape of a hand) in reindeer. As the lineage moves from common ancestor to common ancestor, the traits become more specific to just particular groups of animals. It is this progression in specificity that allows classification categories to serve as a basis for depicting a phylogenetic tree.

Animation Phylogenetic Trees

Tracing Phylogeny

While Figure 16.15 makes use of morphological data, systematists today use a multitude of data in order to discover the evolutionary relationships between species. They rely heavily on a combination of fossil record data, morphological data, and molecular data to determine the correct sequence of common ancestors in any group of organisms.

Morphological data include homologies, which are similarities among organisms that stem from having a common ancestor. Comparative anatomy, including embryological evidence and fossil data, provides information regarding homology. **Homologous structures** are related to each other through common descent. The forelimbs of vertebrates are homologous because they contain the same bones organized in the same general way as in a common ancestor (see Fig. 14.15). This is the case even though a horse has but a single digit and toe (the hoof), while a bat has four lengthened digits that support its membranous wings.

Deciphering homology is sometimes difficult because of convergent evolution. **Convergent evolution** is the acquisition of the same or similar traits in distantly related lines of descent. Similarity due to convergence is termed **analogy.** The wings of an insect and the wings of a bat are analogous. You may recall from Chapter 14 that **analogous structures** have the same function in different groups but organisms with these structures do not have a recent common ancestor. *Analogous structures arise because of adaptations to the same type of environment.* Both cactuses and spurges are adapted similarly to a hot, dry environment, and both are succulent, spiny, flowering plants. However, the details of their flower structure indicate that these plants are not closely related.

Speciation occurs when mutations bring about changes in the base-pair sequences of DNA. Systematists, therefore, assume that, the more closely species are related, the fewer changes will be found in DNA base-pair sequences. Because molecular data are straightforward and numerical, they can sometimes sort out relationships obscured by inconsequential anatomical variations or convergence. Computer software breakthroughs have made it possible to analyze nucleotide sequences quickly and accurately. Also, these analyses are available to anyone doing comparative studies through the Internet, so each investigator doesn't have to start from scratch. The combination of accuracy and availability of vast amounts of data, even entire genomes, has made molecular systematics a standard way to study the relatedness of organisms.

All cells have ribosomes essential for protein synthesis, and the genes that code for ribosomal RNA (rRNA) have changed very slowly during evolution because drastic changes lead to malfunctioning cells. Therefore, comparative rRNA sequencing provides a reliable indicator of the similarity between organisms. Ribosomal RNA sequencing helped investigators conclude that all living things can be divided into the three domains.

One study involving DNA differences produced the data shown in **Figure 16.16.** Notice the close relationship between chimpanzees and humans.

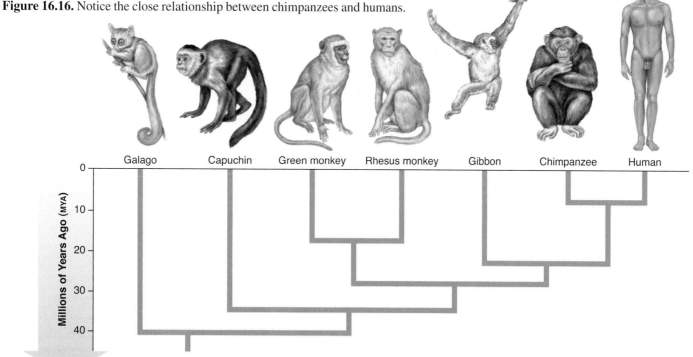

Figure 16.16   Molecular data.

The relationships of certain primate species are based on a study of their genomes. The length of the branches indicates the relative number of DNA base-pair differences between the groups. These data, along with knowledge of the fossil record for one divergence, make it possible to suggest a date for the other divergences in the tree.

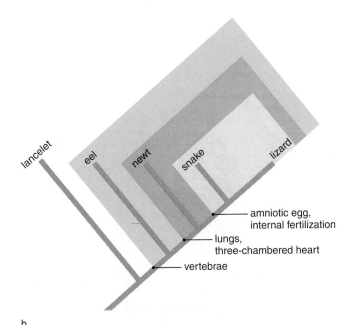

	lancelet	eel	newt	snake	lizard
Notochord in embryo	■	■	■	■	■
Vertebrae		■	■	■	■
Lungs			■	■	■
Three-chambered heart			■	■	■
Internal fertilization				■	■
Amniotic membrane in egg				■	■
Four bony limbs			■		■
Long cylindrical body		■		■	

a.

lancelet
eel
newt
snake
lizard

amniotic egg,
internal fertilization

lungs,
three-chambered heart

vertebrae

b.

Figure 16.17 Constructing a cladogram.

a. First, a table is drawn up, listing characters for all the taxa. An examination of the table shows which characters are ancestral (notochord, aqua) and which are derived (lavender, orange, and yellow). The shared derived characters distinguish the taxa. **b.** In a cladogram, the shared derived characters are sequenced in the order they evolved and are used to define clades. A clade contains a common ancestor and all the species that share the same derived characters (homologies). Four bony limbs and a long, cylindrical body were not used in constructing the cladogram because they are in scattered taxa.

This relationship has recently been recognized by the designation of a new subfamily, Homininae, that includes not only chimpanzees and humans but also gorillas. Molecular data indicate that gorillas and chimpanzees are more closely related to humans than they are to orangutans. Below the taxon subfamily, humans and chimpanzees are placed together in their own tribe, a rarely used classification category.

Cladistics and Cladograms

Cladistics is a way to trace the evolutionary history of a group by using shared traits derived from a common ancestor to determine relationships. These traits are then used to construct phylogenetic trees called cladograms. A **cladogram** depicts the evolutionary history of a group based on the available data.

The first step in constructing a cladogram is to draw up a table that summarizes the traits of the species being compared. At least one but preferably several species are considered an *outgroup*. The outgroup is not part of the study group, also called the *ingroup*. In **Figure 16.17a**, lancelets are the outgroup because, unlike the species in the ingroup, they are not vertebrates. Any trait, such as a notochord, found in both the outgroup and the ingroup is a shared *ancestral trait,* presumed to have been present in a common ancestor to both the outgroup and ingroup. Ancestral traits are not shared derived traits and therefore are not used to construct a cladogram. They merely help us determine which traits will be used to construct the cladogram.

A rule that many cladists follow is the principle of *parsimony,* which states that the least number of assumptions is the most probable. Thus, they construct a cladogram that minimizes the number of assumed evolutionary changes or that leaves the fewest number of derived traits unexplained. Therefore, any trait in the table found in scattered species (in this case, four bony limbs and a long, cylindrical body) is not used to construct the cladogram because we would have to assume that these traits evolved more than once among the species of the study group. The other differences are designated as *shared derived traits*—that is, they are homologies shared by only certain species of the study group. Combining the data regarding shared derived traits will tell us how the members of the ingroup are related to one another.

The Cladogram

A cladogram contains several clades; each **clade** includes a common ancestor (at the circles) and all of its descendant species. The cladogram in Figure 16.17b has three clades (1–3), which differ in size because the first includes the other two and so forth. All the species in the study group belong to a clade that has vertebrae; only newts, snakes, and lizards are in a clade that has lungs and a three-chambered heart; and only snakes and lizards are in a clade that has an amniotic egg and internal fertilization. (An amniotic egg has a sac that surrounds and protects the embryo—fish and amphibian eggs do not have this sac.) Following the principle of parsimony, this is the sequence in which these traits must have evolved during the evolutionary history of vertebrates. Any other arrangement of species would produce a less parsimonious evolutionary sequence—that is, a tree that would be more complicated.

A cladogram is objective—it lists the traits used to construct the cladogram. Cladists typically use much morphological, fossil, and molecular data to construct a cladogram. Still, cladists regard a cladogram as a hypothesis. Whether our tree is consistent with the one, true evolutionary history of life can be tested, and modifications can be made on the basis of additional data.

Linnaean Classification Versus Cladistics

Figure 16.18 illustrates the types of problems that arise when trying to reconcile Linnaean classification with the principles of cladistics. Figure 16.18, which is based on cladistics, shows that birds are in a clade with crocodiles, with which they share a recent common ancestor. This ancestor had a gizzard. An examination of the skulls of crocodiles and birds would show other derived traits that they share. Birds have scaly skin and share this ancestor with other reptiles as well. However, Linnaean classification places birds in their own group, separate from crocodiles and from reptiles in general. In many other instances, Linnaean classification is not consistent with new understandings about phylogenetic relationships. Therefore, some cladists have proposed a different system of classification, called the International Code of Phylogenetic Nomenclature, or PhyloCode, which sets forth rules for the naming of clades. Other biologists are hoping to modify Linnaean classification to be consistent with the principles of cladistics.

Two major problems may be unsolvable: (1) Clades are hierarchical, as are Linnaean categories. However, there may be more clades than Linnaean taxonomic categories, and it is therefore difficult to equate clades with taxons. (2) The taxons are not necessarily equivalent in the Linnaean system. For example, the family taxon within Kingdom Plantae may not be equivalent to the family taxon in Kingdom Animalia. Because of such problems, some cladists recommend abandoning Linnaeus altogether.

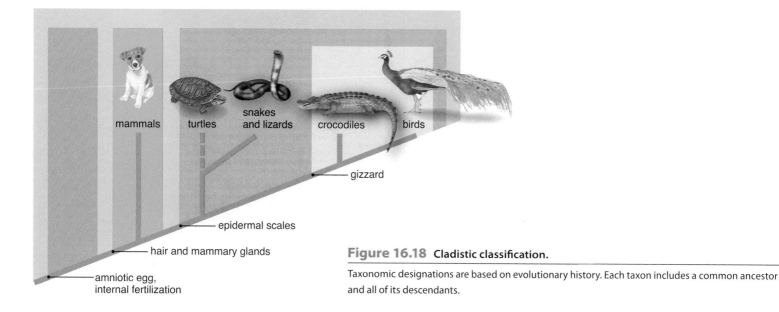

mammals   turtles   snakes and lizards   crocodiles   birds

gizzard

epidermal scales

hair and mammary glands

amniotic egg, internal fertilization

Figure 16.18 Cladistic classification.

Taxonomic designations are based on evolutionary history. Each taxon includes a common ancestor and all of its descendants.

The Three-Domain System

Classification systems change over time. Historically, most biologists utilized the **five-kingdom system** of classification, which contains kingdoms for the plants, animals, fungi, protists, and monerans. Organisms were placed into these kingdoms based on type of cell (prokaryotic or eukaryotic), level of organization (unicellular or multicellular), and type of nutrition. In the five-kingdom system, the monerans were distinguished by their structure—they were prokaryotic (lack a membrane-bounded nucleus)—whereas the organisms in the other kingdoms were eukaryotic (have a membrane-bounded nucleus). The kingdom Monera contained all prokaryotes, which evolved first, according to the fossil record.

Figure 16.19 **Three-domain system.**

In this system, the prokaryotes are in the domains Bacteria and Archaea. The eukaryotes are in the domain Eukarya, which contains four kingdoms for the protists, animals, fungi, and plants.

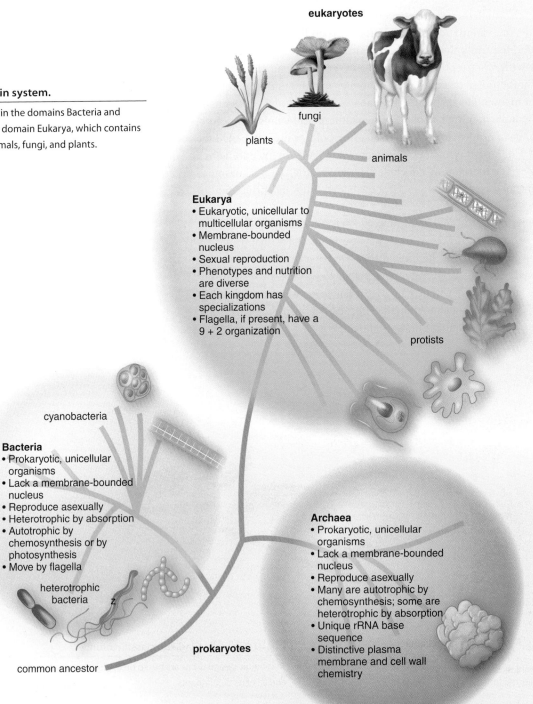

eukaryotes

fungi

plants

animals

Eukarya
• Eukaryotic, unicellular to multicellular organisms
• Membrane-bounded nucleus
• Sexual reproduction
• Phenotypes and nutrition are diverse
• Each kingdom has specializations
• Flagella, if present, have a 9 + 2 organization

protists

cyanobacteria

Bacteria
• Prokaryotic, unicellular organisms
• Lack a membrane-bounded nucleus
• Reproduce asexually
• Heterotrophic by absorption
• Autotrophic by chemosynthesis or by photosynthesis
• Move by flagella

Archaea
• Prokaryotic, unicellular organisms
• Lack a membrane-bounded nucleus
• Reproduce asexually
• Many are autotrophic by chemosynthesis; some are heterotrophic by absorption
• Unique rRNA base sequence
• Distinctive plasma membrane and cell wall chemistry

heterotrophic bacteria

prokaryotes

common ancestor

Sequencing the genes for rRNA has provided new information that calls into question the five-kingdom system of classification. Aside from molecular data, cellular data also suggest that there are two groups of prokaryotes, named the Bacteria and the Archaea. These two groups are so fundamentally different from each other they have been assigned to separate domains, a category of classification that is higher than the kingdom category. The Bacteria arose first, followed by the Archaea and then the Eukarya (**Fig. 16.19**). The Archaea and Eukarya are more closely related to each other than either is to the Bacteria. Systematists, using the **three-domain system** of classification, are in the process of sorting out what kingdoms belong within **domain Bacteria** and **domain Archaea. Domain Eukarya** contains kingdoms for protists, animals, fungi, and plants. Later in this text, we will study the individual kingdoms that occur within the domain Eukarya. The protists do not share one common ancestor, and some suggest that the kingdom should be divided into many different kingdoms. The number of kingdoms is still being determined among systematists, illustrating that classification can be changed as new data become available.

Animation
Three Domains

Connecting the Concepts

For more information on the topics presented in this section, refer to the following discussions.

Section 14.2 provides an overview of how comparative anatomy is used to study evolution.

Section 17.3 examines the evolution of the prokaryotic domains of life.

Figure 19.4 illustrates the phylogenetic tree of animal evolution.

Check Your Progress 16.3

1. List the categories of classification in order from smallest to largest.

2. Name the types of data used to determine evolutionary relationships.

3. Contrast homologous structure with analogous structure.

4. Explain why the sequencing of ribosomal RNA (rRNA) is done for evolutionary studies.

5. Explain why the Linnaean classification is difficult to reconcile with the principles of cladistics.

6. Explain the following: If B and D are fishes, what other animal (designated by a letter) in this cladogram is also a fish? Explain.

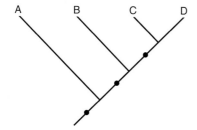

Media Study Tools

www.mhhe.com/maderessentials3

Enhance your study of this chapter with study tools and practice tests. Also ask your instructor about the resources available through ConnectPlus, including the media-rich eBook, interactive learning tools, and animations.

The Chapter in Review

Summary

16.1 Speciation and Macroevolution

Macroevolution is evolution on a large scale because it considers the history of life on Earth. Macroevolution begins with speciation, the origin of new species. Without speciation and the extinction of species, life on Earth would not have a history.

Defining Species

The biological concept of species

- recognizes a species by its inability to produce viable fertile offspring with members of another group.
- is useful because species can look similar, and members of the same species can have different appearances.
- has its limitations. Hybridization does occur between some species, and it is only relevant to extant (living or not extinct) sexually reproducing organisms.

Reproductive Barriers

Prezygotic and postzygotic barriers keep species from reproducing with one another.

- Prezygotic isolating mechanisms involve habitat isolation, temporal isolation, and behavioral isolation.
- Postzygotic isolating mechanisms prevent hybrid offspring from developing or breeding if reproduction has been successful.

Models of Speciation

Allopatric speciation and sympatric speciation are two models of speciation.

- The allopatric speciation model proposes that a geographic barrier keeps groups of populations apart. Meanwhile, prezygotic and postzygotic isolating mechanisms develop, and these prevent successful reproduction if these two groups come in future contact.
- The sympatric speciation model proposes that a geographic barrier is not required for speciation to occur.

16.2 The Fossil Record

The fossil record, as outlined by the geological timescale, traces the history of life in broad terms. It has been possible to absolutely date fossils by using radioactive dating techniques.

The fossil record

- can be used to support a gradualistic model: Two groups of organisms arise from an ancestral species and gradually become two different species.

- also supports the punctuated equilibrium model: A period of equilibrium (no change) is interrupted by speciation within a relatively short period of time.
- shows that at least five mass extinctions, including one significant mammalian extinction, have occurred during the history of life on Earth. Two major contributors to mass extinctions are the loss of habitat due to continental drift and the disastrous results from a meteorite impacting Earth.

16.3 Systematics

- Systematics is the study of the evolutionary relationships among all organisms, past and present. Systematics includes classification. In the Linnaean system of classification, every organism is assigned a scientific name, which indicates its genus and specific epithet. Species are also assigned to a family, order, class, phylum, kingdom, and domain according to their molecular and structural similarities as well as evolutionary relationships to other species.
- Phylogeny depicts the evolutionary history of a group of organisms. Systematics relies on the fossil record, homology, and molecular data to determine relationships among organisms.
- Cladists use shared derived characters to construct cladograms. In a cladogram, a clade consisting of a common ancestor and all the species derived from that ancestor, the species have shared derived characteristics.
- Linnaean classification has come under severe criticism because it does not always follow the principles of cladistics in the grouping of organisms.

Classification Systems

- The *three-domain system* uses molecular data to designate three evolutionary domains: Bacteria, Archaea, and Eukarya:

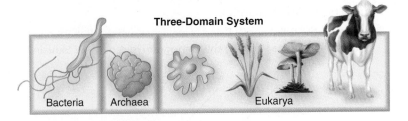

Three-Domain System

Bacteria   Archaea   Eukarya

- Domains Bacteria and Archaea contain prokaryotes.
- Domain Eukarya contains kingdoms for the protists, animals, fungi, and plants.

Key Terms

adaptive radiation   271
allopatric speciation   270
analogous structure   279
analogy   279
clade   280
cladistics   280
cladogram   280
class   278
common ancestor   268
convergent evolution   279
domain   278
domain Archaea   283
domain Bacteria   283
domain Eukarya   283
evolutionary tree   274
family   278
five-kingdom system   282
fossil   272
genus   278
homologous structure   278

kingdom   278
macroevolution   266
mass extinction   275
order   278
paleontology   272
phylogenetic tree   278
phylogeny   277
phylum   278
postzygotic isolating
   mechanism   269
prezygotic isolating
   mechanism   268
speciation   266
species   278
sympatric speciation   270
systematics   277
taxon (pl., taxa)   277
taxonomy   277
three-domain system   283

Testing Yourself

Choose the best answer for each question.

1. A biological species
   a. always looks different from other species.
   b. always has a different chromosome number from that of other species.
   c. is reproductively isolated from other species.
   d. never occupies the same niche as other species.

For questions 2–9, indicate the type of isolating mechanism described in each scenario.

Key:

   a. habitat isolation
   b. temporal isolation
   c. behavioral isolation
   d. mechanical isolation
   e. gamete isolation
   f. zygote mortality
   g. hybrid sterility
   h. low F_2 fitness

2. Females of one species do not recognize the courtship behaviors of males of another species.
3. One species reproduces at a different time of year than another species.
4. A cross between two species produces a zygote that always dies.
5. Two species do not interbreed because they occupy different areas.
6. A hybrid between two species produces gametes that are not viable.
7. Two species of plants do not hybridize because they are visited by different pollinators.
8. The sperm of one species cannot survive in the reproductive tract of another species.
9. The offspring of two hybrid individuals exhibit poor vigor.

10. Complete the following diagram illustrating allopatric speciation by using these phrases: genetic changes (used twice), geographic barrier, species 1, species 2, species 3.

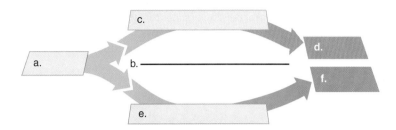

11. Transitional links are least likely to be found if evolution proceeds according to the
    a. gradualistic model.
    b. punctuated equilibrium model.
12. Which of the following is the scientific name of an organism?
    a. *Rosa rugosa*
    b. *Rosa*
    c. *rugosa*
    d. *rugosa rugosa*
    e. Both a and d are correct.
13. Which of these statements best pertains to taxonomy?
    a. Species always have three-part names, such as *Homo sapiens sapiens*.
    b. Species are always reproductively isolated from other species.
    c. Species share ancestral traits but may have their own unique derived traits.
    d. Species always look exactly alike.
    e. Both c and d are correct.
14. Which of the following groups are domains? Choose more than one answer if correct.
    a. bacteria
    b. archaea
    c. eukarya
    d. animals
    e. plants
15. Which pair is mismatched?
    a. homology—character similarity due to a common ancestor
    b. molecular data—DNA strands match
    c. fossil record—bones and teeth
    d. homology—functions always differ
    e. molecular data—DNA and RNA data
16. One benefit of the fossil record is
    a. that hard parts are more likely to fossilize.
    b. that fossils can be dated.
    c. its completeness.
    d. that fossils congregate in one place.
    e. All of these are correct.
17. The discovery of common ancestors in the fossil record, the presence of homologies, and molecular data similarities help scientists determine
    a. how to classify organisms.
    b. the proper cladogram.
    c. how to construct phylogenetic trees.
    d. how evolution occurred.
    e. All of these are correct.

18. In cladistics,
    a. a clade must contain the common ancestor plus all its descendants.
    b. shared derived characters help construct cladograms.
    c. the principle of parsimony states that the simpler hypothesis is preferred.
    d. the species in a clade share homologous structures.
    e. All of these are correct.

19. Answer these questions about the following cladogram.

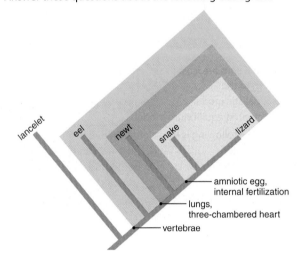

    a. This cladogram contains how many clades? How are they designated in the diagram?
    b. What character is shared by all animals in the study group? What characters are shared by only snakes and lizards?
    c. Which animals share the most recent common ancestor? How do you know?

20. Allopatric, but not sympatric, speciation requires
    a. reproductive isolation.
    b. geographic isolation.
    c. spontaneous differences in males and females.
    d. prior hybridization.
    e. rapid rate of mutation.

21. Which kingdom is mismatched?
    a. Protista—domain Bacteria
    b. Protista—single-celled algae
    c. Plantae—flowers and mosses
    d. Animalia—arthropods and humans
    e. Fungi—molds and mushrooms

22. Many new species evolving in various environments from a common ancestor is called
    a. cladistics.
    b. the gradualistic model of evolution.
    c. adaptive radiation.
    d. convergent evolution.
    e. the PhyloCode.

Thinking Scientifically

1. Using as many terms as necessary (from both X and Y axes), fill in the proposed phylogenetic tree for vascular plants.

	Ferns	Conifers	Ginkgos	Monocots	Eudicots
vascular tissue	X	X	X	X	X
produce seeds		X	X	X	X
naked seeds		X	X		
needle-like leaves		X			
fan-shaped leaves			X		
enclosed seeds				X	X
one embryonic leaf				X	
two embryonic leaves					X

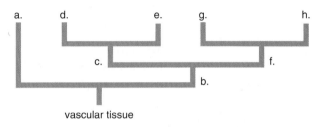

vascular tissue

2. The Hawaiian Islands are located thousands of kilometers from any mainland. Each island arose from the sea bottom and was colonized by plants and animals that drifted in on ocean currents or winds. Each island has a unique environment in which its inhabitants have evolved. Consequently, most of the plant and animal species on the islands do not exist anywhere else in the world.

   In contrast, on the islands of the Florida Keys, there are no unique or indigenous species. All of the species on those islands also exist on the mainland. Suggest an explanation for these two different patterns of speciation.

Bioethical Issue

Funding for Phylogenies

Reconstructing evolutionary relationships can have its benefits. For example, an emerging virus is apt to jump to a related species rather than to an unrelated species. We now know, for example, that the HIV virus jumped from the chimpanzee—with which we share 95% of our DNA sequence—to humans. Chimpanzees don't become ill from the virus; therefore, studying their immune system might help us develop strategies to combat AIDS in humans.

Cladists want to acquire more information about how genetic and environmental factors influence bone development. Cladograms based on bone shape and arrangement do not match up well with those based on DNA base-pair sequencing. If we knew more about bone development under different environmental conditions, we might be able to discover the reason they don't match and develop evolutionary (phylogenetic) trees in which all have confidence. To acquire the necessary data could cost millions of dollars in research funding.

Should the public be willing to fund all types of research or only research that has immediate medical benefits, as was described for HIV research? Would you be willing to fund research that helps us understand our evolutionary past, even if the medical benefit is not immediately known? Do you think public education in cladistic research and its benefits would be beneficial?

17

The Microorganisms: Viruses, Bacteria, and Protists

Disease-Causing Microbes

AIDS, polio, tuberculosis, gonorrhea, SARS, H1N1, malaria: These diseases are caused by *microbes* (Greek; *mikros,* small, and *bios,* life). Disease-causing microbes, discussed in this chapter, include tiny viruses, some bacteria, and certain protists. Although microscopic size unifies the microbes, their structures differ. Viruses are noncellular, bacteria are prokaryotes, and protists are eukaryotes. Viruses always cause an infection, but only some bacteria and a few protists do. The parasitic way of life allows the disease-causing bacteria and protists to get their food and reproduce. Viruses don't need food—after entering a cell, they hijack the cellular machinery and cause the cell to make hundreds of copies of the virus at a time.

We didn't always know that microbes cause disease. The germ theory of disease didn't take hold until the twentieth century. After that, rapid progress was made in identifying which microbes cause which diseases and how they are transmitted. For example, we now know that AIDS and gonorrhea are sexually transmitted, tuberculosis and SARS are contracted by way of the respiratory tract, and the *Anopheles* mosquito spreads malaria. Each microbe lives in a different part of the body: HIV replicates in just certain types of white blood cells, tuberculosis usually centers in the lungs, and the protist that causes malaria reproduces in red blood cells.

This chapter provides basic information about microbes and the various ways they reproduce. Such knowledge enables us not only to fight the diseases microbes sometimes cause but also to take advantage of the many benefits they provide.

OUTLINE

BEFORE YOU BEGIN

Before beginning this chapter, take a few moments to review the following discussions.

Figure 4.2 How much smaller is a virus than a bacterial cell? An animal or plant cell?

Section 4.2 What are the differences between the structure of a prokaryotic and eukaryotic cell?

Figure 16.19 What two domains of life contain prokaryotic organisms?

17.1 The Viruses

Learning Outcomes

Upon completion of this section, you should be able to

1. Describe the structure of a virus.
2. Explain the basis of viral host specificity.
3. Describe the process of viral reproduction.

The study of **viruses** has contributed much to our understanding of disease, genetics, and even the characteristics of living things. Their contribution is surprising because viruses are not included in the classification of organisms. They are noncellular, while all organisms are cellular. Viruses can be amazingly small; at 0.2 micron, a virus is only about one-fifth the size of a bacterium.

Each type of virus always has at least two parts: an outer **capsid,** composed of protein subunits, and an inner core of nucleic acid—either DNA or RNA. The adenovirus shown in **Figure 17.1** also has spikes (formed from a glycoprotein), which are involved in attaching the virus to the host cell. In some animal viruses, the capsid is surrounded by an outer membranous envelope with glycoprotein spikes. The envelope is actually a piece of the host's plasma membrane that also contains proteins produced by the virus. The interior of a virus can contain various enzymes, such as the polymerases, which are needed to produce viral DNA and/or RNA. The viral genome has at most several hundred genes; by contrast, a human cell contains tens of thousands of genes.

Should viruses be considered living? Both scientific and philosophical debates have raged with regard to this question. Viruses are *obligate intracellular parasites* because they can reproduce only inside a living cell (*obligate* means restricted to a specific form). Outside a living cell, viruses can be stored independently of living cells or even synthesized in the laboratory from chemicals. Still, viruses do have a genome that mutates and functions to direct their reproduction when inside a cell.

Viral Reproduction

Viruses are specific to a particular host cell because a spike or some portion of the capsid adheres in a lock-and-key manner to a specific molecule (called a receptor) on the host cell's outer surface. A virus cannot infect a host cell to which it is unable to attach. For example, the tobacco mosaic virus (discussed on page 290) cannot infect an exposed human because its capsid cannot attach to the receptors on the surfaces of human cells. Once inside a host cell, the viral genome *takes over the metabolic machinery of the host cell.* In large measure, the virus uses the host's enzymes, ribosomes, transfer RNA (tRNA), and ATP for its reproduction.

Video
How Viruses Attack

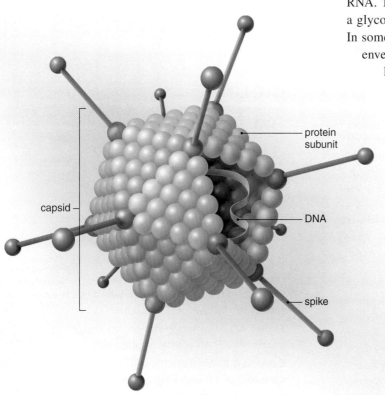

protein subunit

capsid

DNA

spike

|←————— 60–90 nm —————→|

Figure 17.1 Adenovirus anatomy.

Typical of viruses, adenoviruses have a nucleic acid core and a coat of protein, the so-called capsid. Note the projections called spikes—they help a virus enter a cell.

Reproduction of Bacteriophages

A **bacteriophage,** or simply *phage,* is a virus that reproduces in a bacterium. Bacteriophages are named in different ways; the one shown in **Figure 17.2** is called lambda (λ). When phage λ reproduces, it can undergo the lytic cycle or the lysogenic cycle. The **lytic cycle** (**Fig. 17.3***a*) may be divided into five stages: attachment, penetration, biosynthesis, maturation, and release. **1** During *attachment,* the capsid combines with a receptor in the bacterial cell wall. **2** During *penetration,* a viral enzyme digests away part of the cell wall, and viral DNA is injected into the bacterial cell. **3** *Biosynthesis* of viral components begins after the virus inactivates host genes not necessary to viral replication. The machinery of the host cell then carries out viral DNA replication and production of multiple copies of the capsid protein subunits. **4** During *maturation,* viral DNA and capsids assemble to produce several hundred viral particles. Lysozyme, an enzyme coded for by a viral gene, disrupts the cell wall, and **5** *release* of phage particles occurs. The bacterial cell dies as a result.

Video
Virus Lytic Cycle

In the **lysogenic cycle** (Fig. 17.3*b*), the infected bacterium does not immediately produce phage, but it may do so in the future. In the meantime, the phage is *latent*—not actively reproducing. Following attachment and penetration, *integration* occurs: Viral DNA becomes incorporated into bacterial DNA with no destruction of host DNA. While latent, the viral DNA is called a *prophage.* The prophage is replicated along with the host DNA, and all subsequent cells, called lysogenic cells, carry a copy of the prophage. Certain environmental factors, such as ultraviolet radiation, can induce the prophage to enter the lytic stage of biosynthesis, followed by maturation and release.

Animation
Lambda Phage Replication

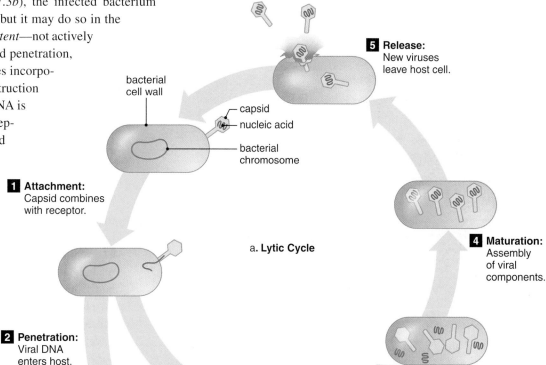

Figure 17.2 **Bacteriophage lambda (λ).**

The micrograph shows many viral particles attached to a bacterium, and the blow-up shows how DNA from a virus enters the bacterium.

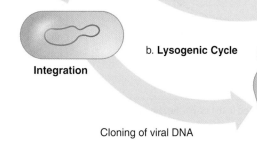

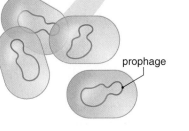

a. **Lytic Cycle**

1 **Attachment:** Capsid combines with receptor.

2 **Penetration:** Viral DNA enters host.

3 **Biosynthesis:** Viral components are synthesized.

4 **Maturation:** Assembly of viral components.

5 **Release:** New viruses leave host cell.

b. **Lysogenic Cycle**

Integration

Cloning of viral DNA

prophage

Figure 17.3 Lytic and lysogenic cycles of a virus called lambda.

a. In the lytic cycle, viral particles escape when the cell is lysed (broken open). **b.** In the lysogenic cycle, viral DNA is integrated into host DNA. At a future time, the lysogenic cycle can be followed by the last three steps of the lytic cycle.

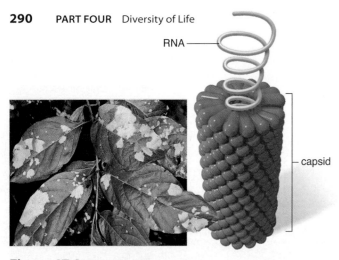

RNA

capsid

Figure 17.4 **Infected tobacco plant.**

Mottling and a distorted leaf shape are typical of a tobacco mosaic virus infection.

Plant Viruses

Crops and garden plants are also subject to viral infections. Plant viruses tend to enter through damaged tissues and then move about the plant through *plasmodesmata,* cytoplasmic strands that extend between plant cell walls. The best studied plant virus is tobacco mosaic virus, a long, rod-shaped virus with only one type of protein subunit in its capsid (**Fig. 17.4**). Not all viruses are deadly, but over time, they often debilitate a plant.

Viruses are often passed from one plant to another by insects and gardening tools, which move sap from one plant to another. Viral particles are also transmitted by way of seeds and pollen. Unfortunately, no chemical can control viral diseases. Until recently, the only way to deal with viral diseases was to destroy symptomatic plants and to control the insect vector, if there is one. Now that bioengineering is routine, it is possible to transfer genes conferring disease resistance between plants. One of the most successful examples is the creation of papaya plants resistant to papaya ring spot virus (PRSV) in Hawaii. One transgenic line is now completely resistant to PRSV.

Animal Viruses

Viruses that cause disease in animals, including humans, reproduce in a manner similar to that of bacteriophages (**Fig. 17.5**). However, there are modifications. In particular, some, but not all, animal viruses have an outer membranous envelope beyond their capsid. After attachment to a receptor in the plasma membrane, viruses with an envelope either fuse with the plasma membrane or enter by endocytosis. After an enveloped virus enters, uncoating follows—that is, the capsid is removed by enzymes within the host cell (see **1** to **3**). Once the viral genome, either DNA or RNA, is free of its covering, biosynthesis plus the other steps then proceed (see **7** to **9**). While a naked animal virus exits the host cell in the same manner as does a bacteriophage, those with an envelope bud from the cell. During budding, the virus picks up its envelope, consisting mainly of lipids and proteins, from the host plasma membrane. Spikes, those portions of the envelope that allow the virus to enter a host cell, are coded for by viral genes.

The herpesviruses, which cause cold sores, genital herpes, and chickenpox in humans, are examples of infections that remain latent much of the time. Herpesviruses linger in spinal ganglia until stress, excessive sunlight, or some other stimulus causes them to undergo the lytic cycle. The human immunodeficiency virus (HIV), the cause of AIDS, remains relatively latent in lymphocytes, slowly releasing new viruses.

1 Attachment

2 Fusion and entry

9 Release

HIV   envelope

spike

receptor

3 Uncoating

viral RNA

4 Reverse transcription

reverse transcriptase

single-stranded DNA

8 Maturation

5 Replication

7 Biosynthesis

double-stranded DNA

viral RNA

ribosome

6 Integration

host DNA

provirus DNA

viral RNA

Nucleus

Figure 17.5 **Reproduction of HIV.**

HIV, the virus that causes AIDS, goes through the steps noted in the boxes. Because HIV is a retrovirus with an RNA genome, the enzyme reverse transcriptase is utilized to produce a single-stranded DNA copy of the genome. After synthesis of a complementary strand, the double-stranded viral DNA integrates into the host DNA.

Animation
Entry of a Virus Into a Host Cell

Video
Halting Hepatitis

Retroviruses

Retroviruses are RNA animal viruses that have a DNA stage. Figure 17.5 illustrates the reproduction of the retrovirus HIV. A retrovirus contains a special enzyme called reverse transcriptase, which carries out transcription of RNA to DNA. The enzyme synthesizes one strand of DNA using the viral RNA as a template, and another DNA strand that is complementary to the first one. Using host enzymes, the resulting double-stranded DNA is integrated into the host genome. The viral DNA (or *provirus*) remains in the host genome and is replicated when host DNA is replicated (see **4** to **6**). When and if this DNA is transcribed, new viruses are produced by the steps we have already cited: biosynthesis, maturation, and release. Being an animal virus with an envelope, HIV buds from the cell.

Animation
Replication Cycle
of a Retrovirus

Emerging Viruses

HIV is an **emerging virus,** the causative agent of a disease that only recently has infected large numbers of people. Other examples of emerging viruses are West Nile virus, SARS virus, hantavirus, Ebola virus, and avian influenza (H5N1 or bird flu) virus (**Fig. 17.6**).

Infectious diseases emerge in several different ways. In some cases, the virus is simply transported from one location to another. The West Nile virus is making headlines because it changed its range: It was transported into the United States and is taking hold in bird and mosquito populations. Severe acute respiratory syndrome (SARS) was clearly transported from Asia to Toronto, Canada. A world in which you can begin your day in Bangkok and end it in Los Angeles is a world in which disease can spread at an unprecedented rate.

Video
Killer Flu

Other factors can also cause infectious viruses to emerge. Viruses are well known for their high mutation rates. Some of these mutations affect the structure of the spikes or capsids, so a virus that previously could infect only a particular animal species can now also infect the human species. For example, the diseases AIDS and ebola fever are caused by viruses that at one time infected only monkeys and apes. SARS is another mutant virus that most likely jumped species. A related virus was isolated from the palm civet, a cat-like carnivore sold for food in China. Wild ducks are resistant to avian influenza viruses that can spread from them to chickens, which increases the likelihood the disease, often called bird flu, will spread to humans. Another way a virus could emerge is by a change in the mode of transmission. What would happen if a little-known virus were suddenly able to be transmitted like the common cold?

Video
Virus Crisis

Connections and Misconceptions

Is H1N1 an emerging virus?

The strain of influenza virus called 2009 H1N1 is a classic example of an emerging virus. Although sometimes called "swine flu," H1N1 is actually a form of the human influenza A virus. However, genetic analysis has indicated that it is a combination of influenza viruses from pigs, birds, and humans. H1N1 viruses have created problems for humans for a considerable amount of time (the 1918 pandemic that killed 50 million people worldwide was an H1N1 virus). The 2009 H1N1 virus is of special concern to the medical community because it is so new, and therefore there is a relatively low level of immunity to this form of virus in the population.

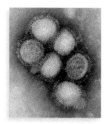

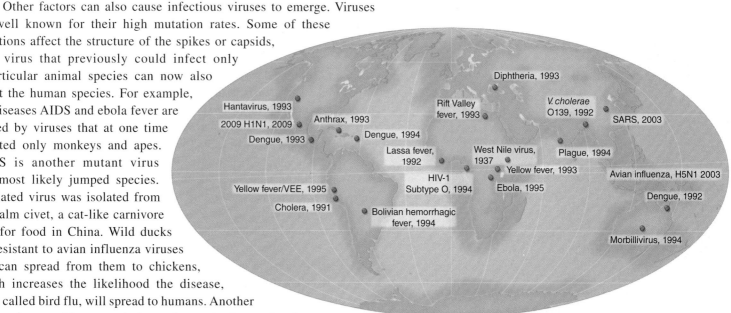

Figure 17.6 Emerging diseases.

Emerging diseases, such as those noted here according to their country of origin, are new or demonstrate increased prevalence. These disease-causing agents may have acquired new virulence factors, or environmental factors may have encouraged their spread to an increased number of hosts.

Drug Control of Human Viral Diseases

Because viruses reproduce using the metabolic machinery of the cell, it has been difficult to develop antiviral drugs. If a patient has not been vaccinated against the agent, he or she is usually told to simply let the virus run its course. However, some antiviral drugs are available. Most antiviral compounds, such as ribavirin and acyclovir, are structurally similar to nucleotides; therefore, they interfere with viral genome synthesis. Compounds related to acyclovir are commonly used to suppress herpes outbreaks. HIV is treated with antiviral compounds specific to a retrovirus. The well-publicized drug AZT and others block reverse transcriptase. And HIV protease inhibitors block the enzymes required for the maturation of viral proteins.

 Video Flu Jail Cell
 **Animation** Antiviral Agents

Connecting the Concepts

For more on viruses, refer to the following discussions.

Section 13.3 explores how viruses may be used in gene therapy.

Section 26.5 provides additional information on the HIV virus and the disease AIDS.

Section 29.3 examines some of the sexually transmitted diseases (STDs) that are caused by viruses.

Check Your Progress 17.1

1. Describe the structure of a virus.
2. Categorize the types of viral reproduction.
3. Summarize the ways viruses can emerge as an infectious disease.

17.2 Viroids and Prions

Learning Outcomes

Upon completion of this section, you should be able to

1. Describe the composition of viroids and prions.
2. Compare and constrast the structure of a viroid with that of a virus.

About a dozen crop diseases have been attributed not to viruses but to **viroids,** which are naked strands of RNA (not covered by a capsid). Like viruses, though, viroids direct the cell to produce more viroids.

Some diseases in humans have been attributed to **prions,** a term coined for *proteinaceous infectious* particles. The discovery of prions began when it was observed that members of a primitive tribe in the highlands of Papua New Guinea died from a disease called kuru (meaning trembling with fear) after participating in the cannibalistic practice of eating a deceased person's brain (**Fig. 17.7**). The causative agent was smaller than a virus—it was a misshapen protein. It appears that a normal protein changes shape, so that its polypeptide chain is in a different configuration. The result is a fatal prion infection and a neurodegenerative disorder.

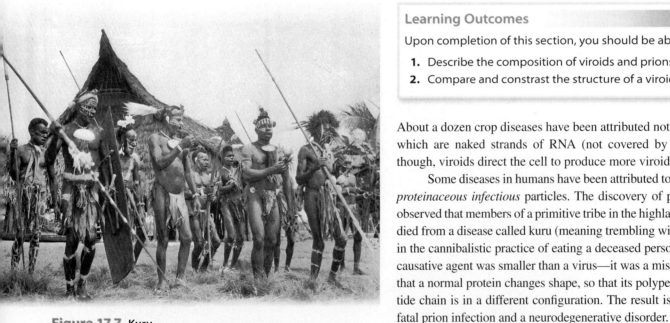

Figure 17.7 Kuru.

Kuru is a fatal prion disease that infected the Fore (pronounced foray), a tribe of the remote highlands of Papua New Guinea, prior to the 1950s. Following their death, family members were ritualistically cooked and eaten, with the closest female relatives and children usually consuming the brain, the organ most likely to pass on kuru. A misshapen protein is the causative agent of all prion diseases, whether they occur in animals (e.g., mad cow disease) or in humans (e.g., kuru).

Animation How Prions Arise

It is believed that a prion can interact with normal proteins to "turn them over to the dark side," but the mechanism is unclear. The process has been best studied in a disease called scrapie, which attacks sheep. Other prion diseases include the widely publicized mad cow disease; human maladies, such as Creutzfeldt-Jakob syndrome (CJD); and a variety of chronic wasting syndromes in several other animal species.

Animation Prion Diseases

Connecting the Concepts

For more on viroids and prions, refer to the following discussions.

Figure 3.19 illustrates the four levels of protein structure.

Section 27.1 examines the structures of the brain influenced by prion-related diseases, such as CJD and Alzheimer disease.

Check Your Progress 17.2

1 Contrast a viroid with a prion.

2 Describe a few prion disorders.

3 Explain the following: Since prions cause disease, why, then, can proteins that become prions be found in healthy brains?

Connections and Misconceptions

Do prions cause Alzheimer disease?

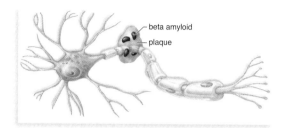

One of the symptoms of Alzheimer disease is an accumulation of a protein called beta amyloid. This accumulation forms tangled structures in the brain called plaques. In normal brain cells, a protein prevents plaques of beta amyloid from forming. In some individuals, a mutation in a gene (*PrP*) causes this protein to malfunction and, in effect, become a prion. For people who have this mutation, beta amyloid accumulates and causes the death of nerve cells associated with Alzheimer disease.

17.3 The Prokaryotes

Learning Outcomes

Upon completion of this section, you should be able to

1. Describe the abiotic emergence of cellular life.
2. Describe the structure and reproductive process of a prokaryotic cell.
3. List and discuss the roles of bacteria in ecosystems, bioremediation, food science, biotechnology, and human health.
4. Distinguish between bacteria and archaea.

What was the first cellular life on Earth like? What features in modern organisms are the most primitive? The answers to these questions will probably come from a study of prokaryotes. These cells harbor no nucleus to contain their genome. They have no sealed compartments and no membrane-bounded organelles to perform specific functions, yet these structurally simple, unicellular organisms are endowed with far greater metabolic capabilities than more structurally complex organisms; some of them need neither air nor organic matter to survive. The two types of **prokaryotes** are the bacteria and the archaea. First, we will discuss the origin of cells, then examine the bacteria, the best-known prokaryotes, followed by the archaea.

Until the nineteenth century, many thought that prokaryotes could arise spontaneously. But in 1850, Louis Pasteur showed that previously sterilized broth cannot become cloudy with growth unless it is exposed directly to the air, where bacteria are abundant. The cell theory formulated about this time states that all organisms are composed of cells, and cells only come from preexisting cells. How, then, did the first cells arise? Much work has gone into studying this question, which pertains to the origin of life, because all living organisms are composed of cells.

The Origin of Cells

The first living cells were prokaryotes, possessing DNA but lacking a nucleus. Fossilized prokaryotes have been found in rocks that are 3.5 billion years old; scientists think that prokaryotes are likely to have existed for millions of years prior to these most ancient fossils. The first cells were preceded by biological macromolecules, such as proteins and nucleic acids. Today, amino acids, nucleotides, and other building blocks for biological macromolecules are routinely produced by living cells; this is known as *biotic synthesis*. Prior to cellular life, macromolecules must have formed by *abiotic synthesis.*

Conditions on the early Earth were very different than they are today. Initially, temperatures were very high. Although there was little free oxygen (O_2) in the atmosphere, there would have been an abundance of other gases, such as water vapor (H_2O), carbon dioxide (CO_2), and nitrogen (N_2), along with smaller amounts of hydrogen (H_2), methane (CH_4), ammonia (NH_3), hydrogen sulfide (H_2S), and carbon monoxide (CO). As the early Earth cooled, water vapor condensed to liquid water, and rain began to fall, producing the oceans. The abiotic synthesis of organic monomers under these special conditions may have occurred with the input of energy from a variety of possible sources, including lightning, sunlight, meteorite impacts, volcanic activity, or radioactive decay (**Fig. 17.8**). The monomers polymerized to form macromolecules, perhaps enabled by inorganic catalysts, such as iron and zinc.

Protocells, cell-like structures complete with an outer membrane, may have resulted from the self-assembly of macromolecules and eventually given rise to cellular life. In fact, researchers have studied cell-like structures that arise from collections of biological macromolecules under laboratory conditions. Some are water-filled spheres with an outer layer similar to that of a cell. If enzymes are trapped inside a sphere, they can catalyze chemical reactions, such as those of cellular metabolism.

An important development in the origin of life would have been a molecule capable of passing information about metabolism and structure from one generation to the next, the role fulfilled by DNA in today's cells. Some researchers think the first hereditary molecule may have been RNA. It is hypothesized that, over time, the more stable DNA replaced RNA as a long-term repository for genetic information, leading to the self-replicating system seen in all living cells today.

Bacteria

Bacteria are the most diverse and prevalent organisms on Earth. Billions of bacteria exist in nearly every square meter of soil, water, and air. They also make a home on your skin and in your intestines. Although tens of thousands of different bacteria have been identified, this is likely only a very small fraction of living bacteria. Less than 1% of bacteria in the soil can be grown in the laboratory. Molecular genetic techniques are being used to discover the extent of bacterial diversity.

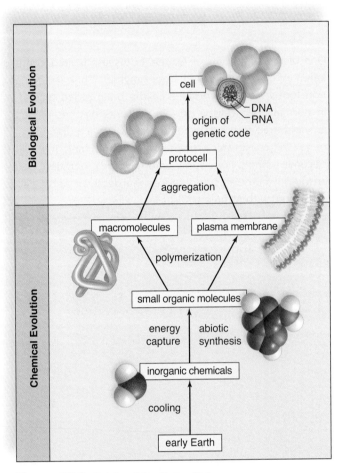

Figure 17.8 Origin of the first cell(s).

There was an increase in the complexity of macromolecules, leading to a self-replicating system (DNA → RNA → protein) enclosed by a plasma membrane. The protocell underwent biological evolution, becoming a true cell.

General Biology of Bacteria

Bacterial Structure   Bacteria have a variety of shapes. However, most bacteria are spheres (called cocci), rods (called bacilli), or spirals (called spirilla; sing., spirillum) if they are rigid; they are called spirochetes if they are flexible (**Fig. 17.9**). Intermediate forms between a sphere and a rod are coccobacilli; a slightly curved rod is called a vibrio. Many bacteria grow as single cells, but some form doublets (the diplococci and diplobacilli). Others form chains as do the streptococci, the cause of strep throat. A third growth habit resembles a bunch of grapes, as in the staphylococci, a cause of food poisoning.

Animation
Prokaryotic Cell Shape

Figure **17.10***a* shows the structure of a bacterium. A bacterium, being a prokaryote, has no nucleus. A single, closed circle of double-stranded DNA constitutes the chromosome, which occurs in a limited area of the cell called the **nucleoid.** In some cases, extrachromosomal DNA molecules called **plasmids** are also found in bacteria.

Bacteria have ribosomes but not membrane-bounded organelles, such as mitochondria and chloroplasts. Those that are photosynthetic have thylakoid membranes, but these are not enclosed by another membrane. Motile bacteria generally use **flagella** for locomotion, but never cilia. The bacterial flagellum is not structured like a eukaryotic flagellum (Fig. 17.10*b*). It has a filament composed of three strands of the protein flagellin wound in a helix. The filament is inserted into a hook that is anchored by a basal body. The fully reversible 360° rotation causes the bacterium to spin as it moves forward and backward.

Animation
Bacterial Locomotion

Bacteria have an outer cell wall strengthened not by cellulose but by **peptidoglycan,** a complex of polysaccharides linked by amino acids. The cell wall prevents bacteria from bursting or collapsing due to osmotic changes. Parasitic bacteria are further protected from host defenses by a polysaccharide capsule that surrounds the cell wall.

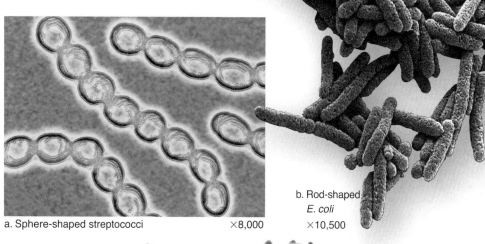

a. Sphere-shaped streptococci   ×8,000

b. Rod-shaped *E. coli*   ×10,500

c. Spirochete, *T. pallidum*

Figure 17.9 **Shapes of bacteria.**

a. Streptococci, which exist as chains of cocci, cause a number of illnesses, including strep throat.
b. *Escherichia coli*, which lives in your intestines, is a rod-shaped bacillus.
c. *Treponema pallidum*, the cause of syphilis , is a spirochete.

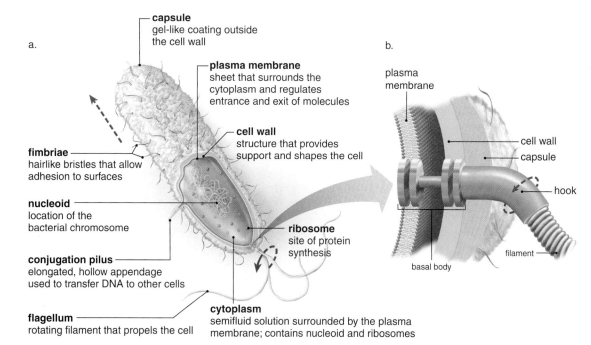

a.

capsule
gel-like coating outside the cell wall

plasma membrane
sheet that surrounds the cytoplasm and regulates entrance and exit of molecules

cell wall
structure that provides support and shapes the cell

fimbriae
hairlike bristles that allow adhesion to surfaces

nucleoid
location of the bacterial chromosome

ribosome
site of protein synthesis

conjugation pilus
elongated, hollow appendage used to transfer DNA to other cells

flagellum
rotating filament that propels the cell

cytoplasm
semifluid solution surrounded by the plasma membrane; contains nucleoid and ribosomes

b.

plasma membrane

cell wall

capsule

hook

filament

basal body

Figure 17.10 **Flagella.**

a. The structure of a prokaryotic cell.
b. Each flagellum of a bacterium contains a basal body, a hook, and a filament. The red-dashed arrows indicate that the hook and filament rotate 360°.

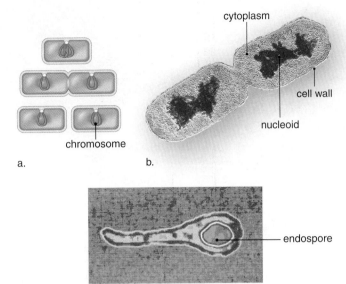

cytoplasm

cell wall

nucleoid

chromosome

a.

b.

endospore

c.   ×5,000

Figure 17.11 **Bacterial reproduction and survival.**

a, b. When conditions are favorable to growth, prokaryotes divide to multiply. **c.** Formation of endospores allows bacteria to survive unfavorable environmental conditions.

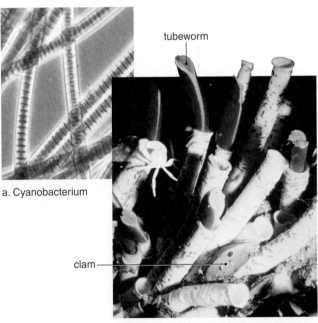

tubeworm

a. Cyanobacterium

clam

b. Hosts for chemoautotrophic bacteria

Figure 17.12 **Bacterial autotrophs.**

a. Cyanobacteria are photoautotrophs that photosynthesize in the same manner as plants—they split water and release oxygen. **b.** Certain chemoautotrophic bacteria live inside tubeworms, where they produce organic compounds without the need of sunlight. In this way, they help support ecosystems at hydrothermal vents deep in the ocean.

Bacterial Reproduction   Bacteria (and archaea) reproduce asexually by means of **binary fission.** The single, circular chromosome replicates, and then the two copies separate as the cell enlarges. The newly formed plasma membrane and cell wall partition the two new cells, with a chromosome in each one (**Fig. 17.11***a, b*). Mitosis, which requires the formation of a spindle apparatus, does not occur in prokaryotes. Binary fission turns one cell into two cells, two cells into four cells, four cells into eight cells, and so on until billions of cells have been produced.

Animation Binary Fission

In eukaryotes, genetic recombination occurs as a result of sexual reproduction. Sexual reproduction does not occur among prokaryotes, but three means of genetic recombination have been observed in bacteria. **Conjugation** occurs when a donor cell passes DNA directly to a recipient cell. During conjugation, donor and recipient are temporarily linked together, often by means of a *conjugation pilus.* While they are linked, the donor cell passes DNA to the recipient cell. **Transformation** occurs when a bacterium picks up (from the surroundings) free pieces of DNA secreted by live prokaryotes or released by dead prokaryotes. During **transduction,** bacteriophages carry portions of bacterial DNA from one cell to another. Plasmids, which sometimes carry genes for resistance to antibiotics, can be transferred between infectious bacteria by any of these means.

Animation Bacterial Conjugation

Animation Bacterial Transformation

When faced with unfavorable environmental conditions, some bacteria form **endospores** (Fig. 17.11*c*). A portion of the cytoplasm and a copy of the chromosome dehydrate and are then encased by three heavy, protective spore coats. The rest of the bacterial cell deteriorates, and the endospore is released. Spores survive in the harshest of environments—desert heat and desiccation, boiling temperatures, polar ice, and extreme ultraviolet radiation. They also survive for very long periods. When anthrax spores 1,300 years old germinate, they can still cause a severe infection (usually seen in cattle and sheep). Humans also fear a deadly but uncommon type of food poisoning called botulism, which is caused by the germination of endospores inside cans of food. Spore formation is not a means of reproduction, but it does allow the survival and dispersal of bacteria to new places.

Animation Bacterial Spore Formation

Bacterial Nutrition   Prokaryotes are more metabolically diverse than eukaryotes. For example, plants are **photoautotrophs** (many times called autotrophs or photosynthesizers) that can only perform oxygenic photosynthesis: They depend on solar energy to split water and energize electrons for the reduction of carbon dioxide. Among bacteria, the **cyanobacteria** are also photoautotrophs and do the same. The cyanobacteria may well represent the oldest lineage of oxygenic organisms (**Fig. 17.12***a*). Some fossil cyanobacteria have been dated at 3.5 billion years old. The evolution of cyanobacteria drastically altered the atmosphere of the early Earth by adding vast amounts of oxygen. Many cyanobacteria are capable of fixing atmospheric nitrogen and reducing it to an organic form. Therefore, they need only minerals, air, sunlight, and water for growth.

Other bacterial photosynthesizers don't release oxygen because they take electrons from a source other than water; some split hydrogen sulfide (H_2S) and release sulfur (S) in marshes where they live anaerobically.

The **chemoautotrophs** (many times called chemosynthesizers) don't use solar energy at all. They reduce carbon dioxide using energetic electrons derived from inorganic molecules, such as ammonia or hydrogen gas. Electrons can also be extracted from certain minerals, such as iron. Some chemoautotrophs oxidize sulfur compounds spewing from deep-sea vents 2.5 kilometers below sea level. The organic compounds they produce support the growth of communities of organisms found at vents, where only darkness prevails (Fig. 17.12*b*).

Most bacteria are **chemoheterotrophs** (often referred to as simply heterotrophs) and, like animals, take in organic nutrients, which they use as a source of energy and building blocks to synthesize macromolecules. The first cells were most likely chemoheterotrophs that fed on the abundant organic molecules in their environment. (Autotrophs would have appeared later as the nutrient supply was depleted and the ability to make one's own food became advantageous.) Unlike animals, chemoheterotrophic bacteria are **saprotrophs** that send enzymes into the environment and decompose almost any large organic molecule to smaller ones that are absorbable. There is probably no natural organic molecule that cannot be digested by at least one bacterial species. Bacteria play a critical role in recycling matter and making inorganic molecules available to photosynthesizers.

Heterotrophic bacteria may be either free-living or **symbiotic,** meaning that they form relationships that are (1) mutualistic (both partners benefit), (2) commensalistic (one partner benefits, the other is not harmed), or (3) parasitic (one partner benefits, the other is harmed). Mutualistic bacteria that live in human intestines release vitamins K and B_{12}, which we can use to help produce blood components. In the stomachs of cows and goats, mutualistic prokaryotes digest cellulose, enabling these animals to feed on grass.

Commensalism often occurs when one population modifies the environment in such a way that a second population benefits. Obligate anaerobes can live in our intestines only because the bacterium *Escherichia coli* uses up the available oxygen. The parasitic bacteria cause diseases, as discussed on page 298.

Environmental and Medical Importance of Bacteria

Bacteria in the Environment   For an ecosystem to sustain its populations, the chemical elements available to living things must eventually be recycled. A fixed and limited amount of elements are available to living things—the rest are either buried too deep in the Earth's crust or present in forms that are not usable. All living organisms, including producers, consumers, and decomposers, are involved in the important process of cycling elements to sustain life (see Fig. 1.12). Bacteria are *decomposers* that digest dead organic remains and return inorganic nutrients to producers. Without the work of decomposers, life would soon come to a halt.

Video Decomposers

While decomposing, bacteria perform the reactions needed for biogeochemical cycling, such as for the carbon and nitrogen cycles. Let's examine how bacteria participate in the nitrogen cycle. Plants are unable to fix atmospheric nitrogen (N_2), but they need a source of ammonia or nitrate in order to produce proteins. Bacteria in the soil can fix atmospheric nitrogen and/or change nitrogen compounds into forms that plants can use. In addition, mutualistic bacteria live in the root nodules of soybean, clover, and alfalfa plants, where they reduce atmospheric nitrogen to ammonia, which is used by plants (**Fig. 17.13**). Without the work of bacteria, nitrogen would not be available for plants to produce proteins or available to animals that feed on plants or other animals.

Bioremediation is the biological cleanup of an environment that contains harmful chemicals called *pollutants*. To help correct the situation, the vast ability of bacteria to break down almost any substance, including sewage, is

Figure 17.13 **Nodules of a legume.**

Although some free-living bacteria carry on nitrogen fixation, those of the genus *Rhizobium* invade the roots of legumes, with the resultant formation of nodules. Here the bacteria convert atmospheric nitrogen to an organic nitrogen that the plant can use. These are nodules on the roots of a soybean plant.

a.

Figure 17.14
Bioremediation.

a. Bacteria have been used for many years in sewage treatment plants to break down human wastes.
b. Increasingly, the ability of bacteria to break down pollutants, such as oil spills, is being researched and enhanced.

b.

Figure 17.15 Bacteria in food processing.

Bacteria are used to help produce food products. Bacterial fermentation results in acids that give some types of cheeses their characteristic taste.

being exploited (**Fig. 17.14***a*). People have added thousands of tons of slowly degradable pesticides and herbicides, nonbiodegradable detergents, and plastics to the environment. Strains of bacteria are being developed specifically for cleaning up these types of pollutants. Some strains have been used to remove agent orange, a potent herbicide, from soil samples, and dual cultures of two types of bacteria have been shown to degrade PCBs, chemicals formerly used as coolants and industrial lubricants. The occasional oil spill spoils beaches and kills wildlife. The ability of bacteria to degrade petroleum has been improved by biotechnology and the addition of a growth-promoting fertilizer (Fig. 17.14*b*). Without the fertilizer, the lack of nitrogen and phosphate in seawater limits their growth.

Bacteria in Food Science and Biotechnology   A wide variety of food products are created through the action of bacteria (**Fig. 17.15**). Under anaerobic conditions, bacteria carry out fermentation, which results in a variety of acids. One of these acids is lactate, a product that pickles cucumbers, curdles milk into cheese, and gives these foods their characteristic tangy flavor. Other bacterial fermentations can produce flavor compounds, such as the propionic acid in swiss cheese. Bacterial fermentation is also useful in the manufacture of such products as vitamins and antibiotics—in fact, most antibiotics known today were discovered in soil bacteria.

As you know, biotechnology can be used to alter the genome and the products generated by bacterial cultures. Bacteria can be genetically engineered to produce medically important products, such as insulin, human growth hormone, and vaccines against a number of human diseases. The natural ability of bacteria to perform all manner of reactions is also enhanced through biotechnology.

Bacterial Diseases in Humans   Microbes that can cause disease are called **pathogens.** Pathogens are able to (1) produce a toxin, (2) adhere to surfaces, and sometimes (3) invade organs or cells.

Toxins are small organic molecules, or small pieces of protein or parts of the bacterial cell wall that are released when bacteria die. Toxins are poisonous, and bacteria that produce a toxin usually cause serious diseases. In almost all cases, the growth of the microbes themselves does not cause disease; the toxins they release cause disease. When someone steps on a rusty nail, bacteria may be introduced deep into damaged tissue. The damaged area does not have good blood flow and can become anaerobic. *Clostridium tetani,* the cause of tetanus (lockjaw), proliferates under these conditions. The bacteria never leave the site of the wound, but the tetanus toxin they produce does move throughout the body. This toxin prevents the relaxation of muscles. In time, the body contorts because all the muscles have contracted. Eventually, the person suffocates.

Adhesion factors allow a pathogen to bind to certain cells, and this determines which organs or cells of the body will be its host. Like many bacteria that cause dysentery (severe diarrhea), *Shigella dysenteriae* is able to stick to the intestinal wall. In addition, *S. dysenteriae* produces a toxin called Shiga toxin, which makes it even more life-threatening. Also, invasive mechanisms that give a pathogen the ability to move through tissues and into the bloodstream result in a more medically significant disease than if it were localized. Usually, a person can recover from food poisoning caused by *Salmonella*. But some strains of *Salmonella* have virulence factors—including a needle-shaped apparatus that injects toxin into body cells—that allow the bacteria to penetrate the lining of the colon and move beyond this organ. Typhoid fever, a life-threatening disease, can then result.

Because bacteria are cells in their own right, a number of antibiotic compounds are active against bacteria and are widely prescribed. One problem with antibiotic therapy has been increasing bacterial resistance to antibiotics (see Bioethical Focus, page 308).

Archaea

As discussed in Chapter 16, the tree of life now contains three domains: Archaea, Bacteria, and Eukarya. Because many **archaea** and some bacteria are found in extreme environments (hot springs, thermal vents, salt basins), they may have diverged from a common ancestor relatively soon after life began. Later, the eukarya are believed to have split off from the archaeal line of descent. Archaea and eukarya share some of the same ribosomal proteins (not found in bacteria), initiate transcription in the same manner, and have similar types of tRNA.

 **Animation** Three Domains

Structure

The plasma membranes of archaea contain unusual lipids that allow many of them to function at high temperatures. The archaea have also evolved diverse cell wall types, which facilitate their survival under extreme conditions. The cell walls of archaea do not contain peptidoglycan, as do the cell walls of bacteria. In some archaea, the cell wall is largely composed of polysaccharides; in others, the wall is pure protein. A few have no cell wall.

Types of Archaea

Archaea are often discussed in terms of their unique habitats. The **methanogens** (methane makers) are found in anaerobic environments, such as in swamps, marshes, and the intestinal tracts of animals. Those found in animal intestines exist as mutualists or commensals, not as parasites—that is, archaea are not known to cause infectious diseases. Methanogens are chemoautotrophs that couple the production of methane (CH_4) from hydrogen gas (H_2) and carbon dioxide to the formation of ATP (**Fig. 17.16**). This methane, which is also called *biogas*, is released into the atmosphere, where it contributes to the greenhouse effect and global warming. About 65% of the methane in our atmosphere is produced by methanogenic archaea.

The **halophiles** require high-salt concentrations for growth (usually 12–15%; by contrast, the ocean is about 3.5% salt). Halophiles have been isolated from highly saline environments, such as the Great Salt Lake in Utah, the Dead Sea, solar salt ponds, and hypersaline soils (**Fig. 17.17**). These archaea have evolved a number of mechanisms to survive in high-salt environments. They depend on a pigment related to the rhodopsin in our eyes to absorb light energy for pumping chloride, and another, similar pigment for synthesizing ATP.

A third major type of archaea are the **thermoacidophiles** (**Fig. 17.18**). These archaea are isolated from extremely hot, acidic environments, such as hot springs, geysers, submarine thermal vents, and the area around volcanoes. They reduce sulfur to sulfides and survive best at temperatures above 80°C; some can even grow at 105°C (remember that water boils at 100°C). The metabolism of sulfides results in acidic sulfates, and these bacteria grow best at pH 1 to 2.

a.

Figure 17.16 Methanogen habitat and structure.

a. A swamp where methanogens live. **b.** Micrograph of *Methanosarcina mazei*, a methanogen.

a.

Figure 17.17 Halophile habitat and structure.

a. Great Salt Lake, Utah, where halophiles live. **b.** Micrograph of *Halobacterium salinarium*, a halophile.

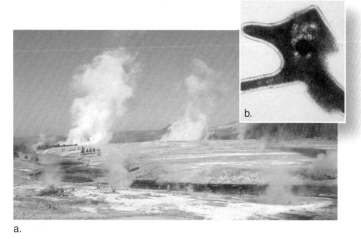

a.

Figure 17.18 Thermoacidophile habitat and structure.

a. Boiling springs and geysers in Yellowstone National Park, where thermoacidophiles live. **b.** Micrograph of *Sulfolobus acidocaldarius*, a thermoacidophile.

Connecting the Concepts

For more on the beneficial uses of bacteria, refer to the following discussions.

Section 11.3 describes how bacteria are used in biotechnology.

Section 24.3 examines the role of bacteria in the digestive system.

Section 31.2 includes information on how bacteria are involved in nutrient cycling—specifically, the nitrogen cycle.

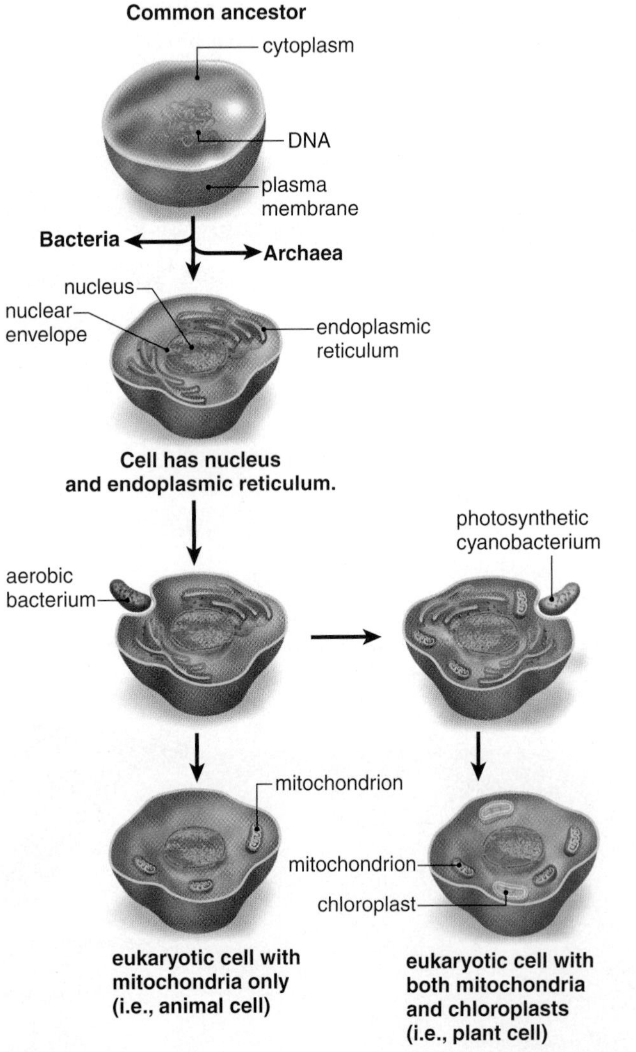

Figure 17.19 Evolution of the eukaryotic cell.

Invagination of the plasma membrane accounts for the formation of the nucleus and certain other organelles. The endosymbiotic theory states that mitochondria and chloroplasts are derived from prokaryotes that were taken up by a much larger eukaryotic cell.

Check Your Progress 17.3

❶ Explain the following: Since the first cells arose in an environment without oxygen, what can you conclude about their oxygen-using capabilities?

❷ List and describe the three common shapes a bacteria cell can take.

❸ Differentiate the three means of genetic recombination in bacteria cells.

❹ Describe the chromosome of a bacterium. Contrast how they are different from eukaryotic chromosomes.

❺ Describe a protocell and hypothesize if a protocell was first formed by biotic or abiotic synthesis.

❻ Explain how some archaea are able to survive in extreme environments. Why is this beneficial?

17.4 The Protists

Learning Outcomes

Upon completion of this section, you should be able to

1. Define endosymbiosis and explain its role in the origin of the eukaryotic cell.
2. Discuss the diversity of protist structures and life cycles.
3. List the groups of algae based on their pigments.
4. List the groups of protozoans based on locomotion.
5. Relate the impact of selected protozoans on human health.

Like ancient creatures from another planet, the **protists** inhabit the oceans and other watery environments of the world. Their morphological diversity is their most outstanding feature—unicellular diatoms are encrusted in silica "petri dishes"; dinoflagellates have plates of armor; and ciliates shaped like slippers have complex structures.

Many protists are unicellular, but all are eukaryotes with a nucleus and a wide range of organelles. We now have ample evidence that the organelles of eukaryotic cells arose from close symbiotic associations between bacteria and primitive eukaryotes (**Fig. 17.19**). This so-called **endosymbiotic theory** is supported by the presence of double membranes around mitochondria and chloroplasts. Also, these organelles have their own genomes, although incomplete, and their ribosomal genes point to bacterial origins. The mitochondria appear closely related to certain bacteria, and the chloroplasts are most closely related to cyanobacteria.

To explain the diversity of protists, we can well imagine that, once the eukaryotic cell arose, it provided the opportunity for many different lineages to begin. Some unicellular protists have organelles not seen in other eukaryotes. For example, food is digested in food vacuoles, and excess water is expelled when contractile vacuoles discharge their contents.

Protists also possibly give us insight into the evolution of a multicellular organism with differentiated tissues. Some protists are a colony of single cells, with certain cells specialized to produce eggs and sperm, and others are multicellular, with tissues specialized for various purposes. Perhaps the first type of organization preceded the second in a progression toward multicellular organisms.

Animation
Endosymbiosis

General Biology of Protists

The complexity and diversity of protists make it difficult to classify them. The variety of protists is so great that it's been suggested they could be split into more than a dozen kingdoms. Due to limited space, this text groups the protists according to modes of nutrition.

Traditionally, the term **algae** (sing., alga) means aquatic photosynthesizer. At one time, botanists classified algae as plants because they contain chlorophyll *a* and carry on photosynthesis. In aquatic environments, algae are a part of the phytoplankton, photosynthesizers that lie suspended in the water. They are producers, which serve as a source of food for other organisms and pour oxygen into the environment. In terrestrial systems, algae are found in soils, on rocks, and in trees. One type of alga is a symbiote of animals called corals, which depend on them for food as they build the coral reefs of the world. Others partner with fungi in lichens capable of living in harsh terrestrial environments.

The definition of a **protozoan** as a unicellular chemoheterotroph explains why protozoans were originally classified with the animals. Often, a protozoan has some form of locomotion, by either flagella, pseudopods, or cilia. In aquatic environments, protozoans are a part of the zooplankton, suspended microscopic heterotrophs that serve as a food source for animals. While most are free-living, some protozoans are human pathogens, often causing diseases of the blood or intestines. In many cases, their complex life cycles inhibit the development of suitable treatments.

There is still one other group of protists: the slime molds and water molds. These microorganisms are chemoheterotrophs, but the slime molds ingest their food in the same manner as the protozoan called an amoeba, while the water molds are saprotrophic, like fungi.

Algae

The green alga *Chlamydomonas* serves as our model for algal structure (**Fig. 17.20**). The most conspicuous organelle in the algal cell is the chloroplast. Algal chloroplasts share many features with those of plants, and the two groups likely share a common origin; for example, the photosynthetic pigments are housed in thylakoid membranes. Not surprisingly, then, algae perform photosynthesis in the same manner as plants. Pyrenoids are organelles found in algae that are active in starch storage and metabolism. Vacuoles are seen in algae, along with mitochondria. Algae generally have a cell wall, and many produce a slime layer that can be harvested and used for food processing. Some algae are nonmotile, while others possess flagella.

Algae can reproduce asexually or sexually in most cases. Asexual reproduction can occur by binary fission, as in bacteria. Some proliferate by forming flagellated spores called zoospores, while others simply fragment, with each

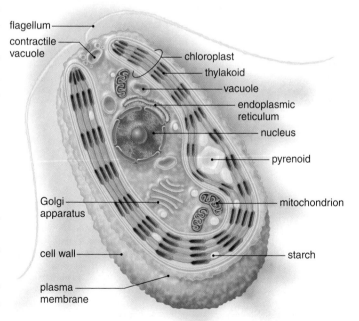

Figure 17.20 *Chlamydomonas.*

Chlamydomonas, a green alga, has the organelles and other structures typical of a motile algal cell.

fragment becoming a progeny alga. Sexual reproduction generally requires the formation of gametes that combine to form a zygote.

One traditional way to classify algae is based on the color of the pigments in their chloroplasts: **green algae, red algae, golden-brown algae,** and **brown algae** (**Fig. 17.21**). The green algae are most closely related to plants, and they are commonly represented by three species: *Chlamydomonas*; colonial *Volvox*, a large, hollow sphere with dozens to hundreds of cells; and *Spirogyra*, a filamentous alga in which the chloroplasts form a green spiral ribbon. The coralline red algae deposit calcium carbonate in their cell walls and contribute to the formation of coral reefs. The golden-brown algae are represented by the diatoms, which have a petri dish structure with each half an elaborate shell composed of silica. The green algae, red algae, and brown algae include the multicellular seaweeds.

The Protozoans

Protozoans are unicellular, but their cells are very complex. Many of the functions we normally associate with the organs of multicellular organisms are performed by organelles in protozoans. Some protozoans have more than one nucleus. In some cases, the two nuclei are identical in size and function. Other protozoans, such

Figure 17.21 **Algal diversity.**

(a) *Acetabularia*, a unicellular green alga; **(b)** *Bossiella*, a coralline red alga; **(c)** *Fucus*, a brown alga; **(d)** *Licmorpha*, a stalked diatom; **(e)** *Volvox*, a colonial green alga with daughter colonies inside; **(f)** *Spirogyra*, a filamentous green alga undergoing conjugation to form zygotes; **(g)** *Ceratium*, an armored dinoflagellate; **(h)** *Macrocystis*, a brown alga; **(i)** *Sargassum*, a brown alga.

zygote

as *Paramecium*, have a large macronucleus and a small micronucleus (**Fig. 17.22**). The macronucleus produces mRNA and directs metabolic functions. The micronucleus is important for reproduction. Protozoans usually reproduce asexually by binary fission.

Protozoans are heterotrophic, and many feed by engulfing food particles. Phagocytic vacuoles act as their "stomachs," into which digestive enzymes and acid are added. Secretory vacuoles release enzymes that may enhance any pathogenicity. Contractile vacuoles permit osmoregulation, particularly in fresh water.

The various protozoans shown in **Figure 17.23** illustrate that they are usually motile. The **ciliates,** so named because they move by cilia, are represented by *Paramecium*, and the **amoeboids** by amoebas, which move by cytoplasmic extensions called **pseudopods.** The **radiolarians** and **foraminiferans** are two types of marine amoeboids with calcium carbonate skeletons important in limestone formations—including the White Cliffs of Dover in Britain. The amoeba *Entamoeba histolytica* causes amoebic dysentery. The **zooflagellates,** which move by flagella, also cause diseases. In this group, a **trypanosome** is the cause of African sleeping sickness, a blood disease that cuts off circulation to the brain. Apicomplexans, commonly called **sporozoans** because they produce spores, are unlike other

Video Amoeba Locomotion

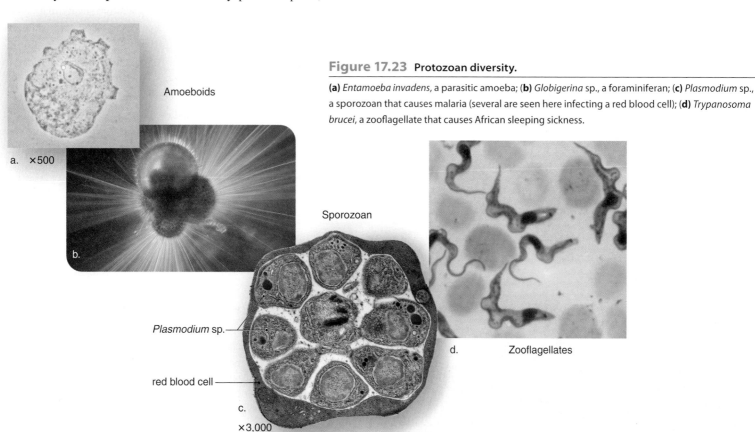

food vacuole

contractile vacuole

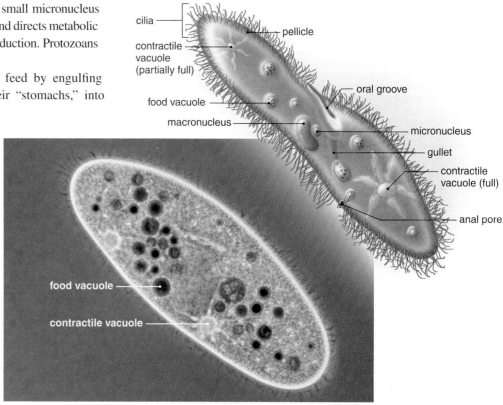

cilia — pellicle
contractile vacuole (partially full)
food vacuole — oral groove
macronucleus — micronucleus
— gullet
— contractile vacuole (full)
— anal pore

Figure 17.22 A paramecium.

A paramecium is a ciliate, a type of complex protozoan that moves by cilia.

Figure 17.23 Protozoan diversity.

(a) *Entamoeba invadens*, a parasitic amoeba; **(b)** *Globigerina* sp., a foraminiferan; **(c)** *Plasmodium* sp., a sporozoan that causes malaria (several are seen here infecting a red blood cell); **(d)** *Trypanosoma brucei*, a zooflagellate that causes African sleeping sickness.

Amoeboids

a. ×500

b.

Sporozoan

Plasmodium sp.—

red blood cell—

c.

×3,000

d. Zooflagellates

protozoan groups because they are not motile. One genus, *Plasmodium*, causes malaria, the most widespread and dangerous protozoan disease. According to the Centers for Disease Control and Prevention (CDC), there are approximately 350–500 million cases of malaria worldwide every year; more than a million people die of the infection. Malaria is transmitted by a mosquito. Toxoplasmosis, another protozoan disease, is commonly transmitted by cats.

Connections and Misconceptions

What is giardiasis?

The intestinal disorder known as giardiasis is caused by a zooflagellate called *Giardia lamblia*. *Giardia* lives in the intestines of a variety of animals, including humans, cattle, deer, dogs, and cats. It is commonly spread by drinking water that has been contaminated by the feces of an infected animal, although it is possible to contract *Giardia* from the soil or food. Symptoms of giardiasis include diarrhea, stomach and/or intestinal cramping, and excess gas (flatulence). For most people, a *Giardia* infection lasts several weeks, but the parasite has been known to cause complications in the elderly, young children, and individuals with compromised immune systems. Since *Giardia* is a eukaryote, antibiotics are ineffective, although other medications are available.

Slime Molds and Water Molds

In forests and woodlands, **slime molds** feed on, and therefore help dispose of, dead plant material. They also feed on bacteria, keeping their population sizes under control. Usually, plasmodial slime molds exist as a plasmodium—a diploid, multinucleated, cytoplasmic mass enveloped by a slime sheath that creeps along, phagocytizing decaying plant material in a forest or an agricultural field. At times unfavorable to growth, such as during a drought, the plasmodium develops many sporangia. A sporangium is a reproductive structure that produces spores resistant to dry conditions. When favorable moist conditions return, the spores germinate, releasing a flagellated cell or an amoeboid cell. Eventually, two of them fuse to form a zygote that feeds and grows, producing a multinucleated plasmodium once again. **Figure 17.24** shows the life cycle of a plasmodial slime mold.

Video
Decomposers

The **water molds** decompose remains but are also significant parasites of plants and animals in ecosystems. The potato blight that

Plasmodium phagocytizes food

Sporangia produce spores

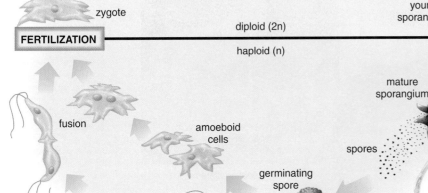

mature plasmodium

young plasmodium

sporangia formation begins

zygote

young sporangium

diploid (2n)

FERTILIZATION

MEIOSIS

haploid (n)

mature sporangium

fusion

amoeboid cells

spores

germinating spore

flagellated cells

Figure 17.24 Plasmodial slime molds.

As long as conditions are favorable, a plasmodial slime mold exists as a multinucleated diploid plasmodium that creeps along the forest floor, phagocytizing organic remains. When conditions become unfavorable, sporangia form where meiosis produces haploid spores, structures that can survive unfavorable times. Spores germinate to release independent haploid cells. Fusion of these cells produces a zygote that becomes a mature plasmodium once again.

brought about famine in nineteenth-century Ireland was caused by a water mold. Slime molds and water molds were once classified as fungi, but unlike fungi, all have flagellated cells at some time during their life cycles. Only the water molds have a cell wall, but it contains cellulose, not the chitin of fungal cell walls. Both slime molds and water molds form spores, each a small, single-celled reproductive body capable of becoming a new organism. The spores of water molds are flagellated, but those of slime molds are windblown.

Although grouped together here, slime molds and water molds may not be very closely related to one another. Some experts suggest that water molds may be near relatives of diatoms and brown algae, while slime molds are associated with the amoeboids. Indeed, the vegetative state of the slime molds is mobile and amoeboid, and like the amoeboids, they ingest their food by phagocytosis.

Check Your Progress 17.4

1 List the features of mitochondria and chloroplasts that support the endosymbiotic theory.

2 Compare and contrast algae and protozoans.

3 Contrast the structure of slime and water molds.

Connecting the Concepts

For more on the protists, refer to the following discussions.

Section 4.4 examines the general structure of eukaryotic cells.

Section 15.1 examines the relationship between the worldwide distribution of malaria and sickle cell disease.

Section 31.2 explores the causes of algal blooms in water supplies.

Media Study Tools

www.mhhe.com/maderessentials3

Enhance your study of this chapter with study tools and practice tests. Also ask your instructor about the resources available through ConnectPlus, including the media-rich eBook, interactive learning tools, and animations.

The Chapter in Review

Summary

17.1 The Viruses

Viruses are noncellular particles.

```
                    ┌ Capsid ──┬─ Protein subunits
                    │          └─ Envelope (in some)
  Virus particle ──┤
                    │             ┌─ Nucleic acids (DNA or RNA)
                    └ Inner core ─┤
                                  └─ Various proteins, especially enzymes
```

Structure

- Viruses have at least two parts: an outer capsid composed of protein subunits and an inner core of nucleic acid, either DNA or RNA.
- Animal viruses either are naked (no outer envelope) or have an outer membranous envelope.

Reproduction

Viruses are obligate intracellular parasites that can reproduce only inside living cells. Bacteriophages can have a lytic or lysogenic life cycle.

The *lytic cycle* consists of these steps:

- **1** Attachment
- **2** Penetration
- **3** Biosynthesis
- **4** Maturation
- **5** Release

In the *lysogenic cycle,* viral DNA is integrated into bacterial DNA for an indefinite period of time, but it can undergo the last three steps of the lytic cycle at any time.

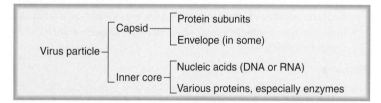

Plant Viruses

Crops and garden plants are subject to viral infections.
Not all viruses are deadly, but over time they debilitate a plant.

Animal Viruses

The reproductive cycle in animal viruses has the same steps as in a bacteriophage, with modifications if the virus has an envelope, in which case:

- Fusion or endocytosis brings the virus into the cell.
- Uncoating is needed to free the genome from the capsid.
- Budding releases the viral particles from the cell.

  HIV, the AIDS virus, is an RNA retrovirus. These viruses have an enzyme, reverse transcriptase, which carries out reverse transcription. This produces single-stranded DNA, which replicates, forming a double helix that becomes integrated into host DNA.

Drug Control of Human Viral Diseases

Antiviral drugs are structurally similar to a nucleotide and interfere with viral genome synthesis. HIV protease inhibitors block the enzymes required for the maturation of viral proteins.

17.2 Viroids and Prions

Viroids are naked (not covered by a capsid) strands of RNA that can cause disease.

Prions are protein molecules that have a misshapen tertiary structure. Prions cause such diseases as CJD in humans and mad cow disease in cattle when they cause other proteins of their own kind to also become misshapen.

17.3 The Prokaryotes

The bacteria and archaea are prokaryotes. Prokaryotes lack a nucleus and most of the other cytoplasmic organelles found in eukaryotic cells.

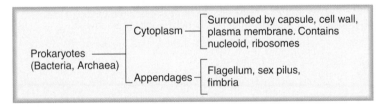

The Origin of Cells

The first cells were prokaryotes. Before the first cells appeared, biological macromolecules formed spontaneously in the unique conditions of the early Earth's atmosphere. Scientists have shown that collections of biological macromolecules assemble to form nonliving, cell-like structures under laboratory conditions.

General Biology of Bacteria
Bacterial Structure

- Structures include rods (bacilli), spheres (cocci), and spirals (spirilla or spirochetes).
- Single, closed circle of double-stranded DNA (chromosome) is in a nucleoid.
- Flagellum is unique and rotates, causing the organism to spin.
- Cell wall contains peptidoglycan.

Bacterial Reproduction and Survival

- production is asexual by binary fission.
- Genetic recombination occurs by means of conjugation, transformation, and transduction.

- Endospore formation allows bacteria to survive an unfavorable environment.
- Endospores are extremely resistant to destruction; the genetic material can thereby survive unfavorable conditions.

Table 17.1–Comparison of Viruses and Prokaryotes

Characteristic of Life	Viruses	Prokaryotes
Consist of cell	No	Yes
Metabolize	No	Yes
Respond to stimuli	No	Yes
Multiply	Yes (always inside living cell)	Yes (usually independently)
Evolve	Yes	Yes

Bacterial Nutrition

- Bacteria can be autotrophic. Cyanobacteria are photoautotrophs—they photosynthesize, as do plants. Chemoautotrophs oxidize inorganic compounds, such as hydrogen gas, hydrogen sulfide, and ammonia, to acquire energy to make their own food. Chemoautotrophs are chemosynthesizers that support communities at deep-sea vents.
- Like animals, most bacteria are chemoheterotrophs (heterotrophs), but they are saprotrophic decomposers. Many heterotrophic prokaryotes are symbiotic. The mutualistic nitrogen-fixing bacteria live in nodules on the roots of legumes.

Environmental and Medical Importance of Bacteria
Bacteria in the Environment

- As decomposers, bacteria keep inorganic nutrients cycling in ecosystems.
- The reactions they perform keep the nitrogen cycle going.
- Bacteria play an important role in bioremediation.

Bacteria in Food Science and Biotechnology

- Bacterial fermentations are important in the production of foods.
- Genetic engineering allows bacteria to produce medically important products.

Bacterial Diseases in Humans

- Bacterial pathogens that can cause diseases are able to (1) produce a toxin, (2) adhere to surfaces, and (3) sometimes invade organs or cells.
- Indiscriminate antibiotic therapy has led to bacterial resistance to some antibiotics.

Archaea

The archaea (domain Archaea) are a second type of prokaryote. The following are some characteristics of archaea:

- They appear to be more closely related to the eukarya than to the bacteria.
- They do not have peptidoglycan in their cell walls, as do the bacteria, and they share more biochemical characteristics with the eukarya than do bacteria.
- Some are well known for living under harsh conditions, such as anaerobic marshes (methanogens), salty lakes (halophiles), and hot sulfur springs (thermoacidophiles).

17.4 The Protists

General Biology of Protists

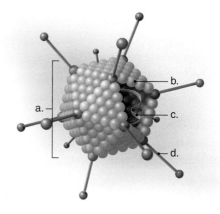

- Protists are eukaryotes. Endosymbiotic events account for the presence of mitochondria and chloroplasts in eukaryotic cells. Some are multicellular with differentiated tissues.
- Protists have great ecological importance because in largely aquatic environments they are the producers (algae) or sources (algae and protozoans) of food for other organisms.

Algae

- Algae possess chlorophylls. They store reserve food as starch and have cell walls, as do plants.
- Algae are divided according to their pigments, being either green, brown, golden-brown, or red. Many are unicellular, but the green, red, and brown algae include multicellular seaweeds.

Protozoans

- Protozoans are heterotrophic and usually motile by means of cilia (paramecia), pseudopods (amoebas), or flagella (zooflagellates).
- Sporozoans (apicomplexans) cause diseases, including malaria, the most serious protozoan disease.
- A zooflagellate (trypanosome) causes African sleeping sickness; *Giardia* causes severe diarrhea.

Slime Molds and Water Molds

- Slime molds are decomposers in forests and woodlands.
- Water molds have cell walls of cellulose, which distinguish them from fungi.

Key Terms

algae (sing., alga) 301
amoeboid 303
archaea 299
bacteria 294
bacteriophage 289
binary fission 296
bioremediation 297
brown algae 302
capsid 288
chemoautotroph 296
chemoheterotroph 297
ciliate 303
conjugation 296
cyanobacteria 296
emerging virus 291
endospore 296
endosymbiotic theory 300
flagella 295
foraminiferan 303
golden-brown algae 302
green algae 302
halophile 299
lysogenic cycle 289
lytic cycle 289
methanogen 299
nucleoid 295

pathogen 298
peptidoglycan 295
photoautotroph 296
plasmid 295
prion 292
prokaryote 293
protist 300
protocell 294
protozoan 301
pseudopod 303
radiolarian 303
red algae 302
retrovirus 291
saprotroph 297
slime mold 304
sporozoan 303
symbiotic 297
thermoacidophile 299
transduction 296
transformation 296
trypanosome 303
viroid 292
virus 288
water mold 304
zooflagellate 303

Testing Yourself

Choose the best answer for each question.

1. A virus contains which of the following?
   a. a cell wall
   b. a plasma membrane
   c. nucleic acid
   d. cytoplasm

2. Label the parts of a virus in the following illustration.

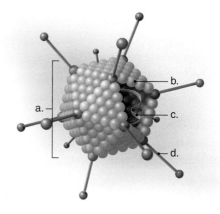

3. The five stages of the bacteriophage lytic cycle occur in this order:
   a. penetration, attachment, release, maturation, biosynthesis
   b. attachment, penetration, release, biosynthesis, maturation
   c. biosynthesis, attachment, penetration, maturation, release
   d. attachment, penetration, biosynthesis, maturation, release
   e. penetration, biosynthesis, attachment, maturation, release

4. Animal viruses
   a. contain both DNA and RNA.
   b. sometimes have an envelope.
   c. sometimes infect bacteria.
   d. do not reproduce inside cells.

5. The enzyme that is unique to retroviruses is
   a. reverse transcriptase.   c. DNA gyrase.
   b. DNA polymerase.   d. RNA polymerase.

6. The _____ cycle is a phage life cycle in which the infected bacterium does not immediately produce phage.
   a. lytic   c. pathogenic
   b. lysogenic   d. transformation

7. Which of these is mismatched?
   a. amoeboids—pseudopods   c. algae—variously colored
   b. sporozoans—disease agents   d. slime molds—trypanosomes

8. Which of the following statements about viroids is false?
   a. They are composed of naked RNA.
   b. They cause plant diseases.
   c. They die once they reproduce.
   d. They cause infected cells to produce more viroids.

9. Prion proteins cause disease when they
   a. enlarge in size.
   b. break into small pieces.
   c. cause normal proteins to change shape.
   d. interact with DNA.

10.   A bacterium contains all of the following, except
   a.   ribosomes.
   b.   DNA.
   c.   mitochondria.
   d.   cytoplasm.
   e.   a plasma membrane.

11.   The primary producers at deep-sea vents are
   a.   heterotrophs.
   b.   symbionts.
   c.   chemoautotrophs.
   d.   photoautotrophs.

For questions 12–17, determine which type of organism is being described. Each answer in the key may be used more than once.

Key:

   a.   bacteria
   b.   archaea
   c.   both bacteria and archaea
   d.   neither bacteria nor archaea

12.   peptidoglycan in cell wall
13.   methanogens
14.   sometimes parasitic
15.   contain a nucleus
16.   plasma membrane contains lipids
17.   reproduction by binary fission

18.   Unlike plants, algae contain
   a.   chloroplasts.
   b.   vacuoles.
   c.   mitochondria.
   d.   pyrenoids.
   e.   cell walls.

19.   Which of the following is incorrect about protists?
   a.   Some protists are prokaryotic.
   b.   Protozoans can be unicellular.
   c.   Algae can reproduce asexually or sexually.
   d.   Slime molds and water molds are decomposers.
   e.   Protozoans are heterotrophic.

20.   Which of the following are unicellular, golden-brown algae with a silica shell?
   a.   diatoms
   b.   radiolarians
   c.   trypanosomes
   d.   sporozoans
   e.   foraminiferans

21.   Label the parts of the paramecium in the following illustration.

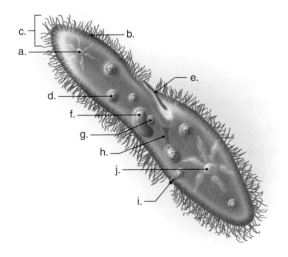

Thinking Scientifically

1.   Based on the accompanying tree, are the protists more closely related to bacteria or to archaea? What evidence supports this relationship? Why are viruses not shown on the tree?

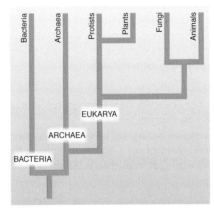

2.   In the former Soviet Union and Eastern Europe, bacteriophage therapy has long been used as an alternative to antibiotic drugs for treating bacterial infections. How could introducing bacteriophages into a patient's body help fight a bacterial infection? What are some potential benefits and shortcomings of bacteriophage therapy?

Bioethical Issue

Misuse of Antibiotics

The Food and Drug Administration (FDA) estimates that physicians annually write 50 million unnecessary prescriptions for antibiotics to treat viral infections. Of course, these antibiotics are ineffective against viruses. Furthermore, the frequent exposure of bacteria to these drugs has resulted in the development of numerous strains with antibiotic resistance. Doctors may prescribe antibiotics because patients demand them. In addition, if the cause of the illness is not known, it may be safer to prescribe an antibiotic that turns out to be ineffective for the symptoms than to withhold the antibiotic when it would have helped. The FDA has, therefore, initiated a new policy requiring manufacturers to label antibiotics with precautions about their misuse.

Do you think this type of labeling information will educate consumers enough to significantly reduce the inappropriate use of antibiotics? Physicians already know about the dangers of antibiotic resistance but may prescribe them, anyway. This is probably due, in large part, to the fact that they are liable for erring on the side of withholding an antibiotic but not for inappropriately prescribing one. Should physicians be held more accountable for prescribing antibiotics? If so, how?

18

Land Environment: Plants and Fungi

Thanking Plants and Fungi Too

Have you thanked a green plant today? Without plants we and our civilized way of life could not exist. Land plants provide us with our food, either directly or indirectly, and with the oxygen we breathe. Oxygen also rises into the stratosphere and becomes the ozone shield that absorbs ultraviolet rays and makes life on land possible.

The bulk of the human diet comes from just 12 plants. Wheat is associated with Europe, corn with the Americas, and rice with the Far East. Other significant food crops include white and sweet potatoes, cassava, soybeans, common beans, sugarcane, sugar beets, coconuts, and bananas.

Between 354 and 290 million years ago (MYA), the bodies of land plants began to transform slowly into the coal we now burn to produce much of our electricity. The wood of trees is also commonly used as fuel. Then, too, fermentation of plant materials produces alcohol, which can be used directly to fuel automobiles or as a gasoline additive.

Much of our clothing comes from land plants. People still like the feel of cotton next to their skin, while linen from flax makes a stronger cloth. Cotton is used in all sorts of other cloth products, such as towels, sheets, and upholstered furniture. Even the manufacture of rayon depends on cellulose from plant cell walls.

Thousands of other household products come from land plants. Wood is used to make furniture, as well as the houses that contain the furniture. Jute is used for rope, carpet, insulation, and burlap bags. Coconut oil is found in soaps, shampoos, and suntan lotions. Even toothpastes contain such plant flavorings as peppermint and spearmint. Suffice it to say, it would be impossible to mention all the products from plants that we depend on for our daily needs.

Fungi also have great economic importance, from the mushrooms topping your pizza to the yeast that makes the dough rise. More importantly, the roles of fungi in ecosystems are indispensable to life on Earth. The majority of fungi are decomposers that help break down dead organisms and recycle their nutrients. Others have symbiotic relationships that enhance the growth of land plants.

In this chapter, you will explore the evolution and diversity of two kingdoms: Plantae and Fungi.

OUTLINE

BEFORE YOU BEGIN

Before beginning this chapter, take a few moments to review the following discussions.

Section 6.2 What process produces oxygen during photosynthesis?

Section 9.3 What are the differences between mitosis and meiosis with regard to chromosome number?

Section 16.3 What domain of life do the plants and fungi belong to?

18.1 Overview of the Land Plants

Learning Outcomes

Upon completion of this section, you should be able to

1. Describe the evolutionary relationship between green algae and land plants.
2. List the five significant events in the evolution of land plants.
3. Describe the alternation of generations life cycle in plants.

Plants (kingdom Plantae) are multicellular, photosynthetic eukaryotes that have become adapted to a land existence. Although a land environment does offer advantages, such as plentiful light, it also has challenges, such as the constant threat of desiccation (drying out). Most importantly, all stages of reproduction—gametes, zygote, and embryo—must be protected from the drying effects of air. To keep the internal environment of cells moist, a land plant must acquire water and transport it to all parts of the body, while keeping the body in an erect position. We will see how plants have adapted to these problems by evolving an internal vascular system.

The evolutionary history of plants begins in the water. Most likely, land plants evolved from a form of freshwater green algae some 500 MYA (million years ago): Both green algae and plants (1) contain chlorophylls *a* and *b* and various accessory pigments, (2) store excess carbohydrates as starch, and (3) have cellulose in their cell walls. A comparison of DNA and RNA base sequences suggests that land plants are most closely related to a group of freshwater green algae known as **charophytes.** Although *Spirogyra* (see Fig. 17.21) is a charophyte, molecular scientists tell us that the ancestors of land plants were more closely related to the charophytes shown in **Figure 18.1.** Although the common ancestor of modern charophytes and land plants no longer exists, if it did, it would have features resembling those of the *Chara* and *Coleochaete.*

Let's take a look at these filamentous green algae. *Chara* are commonly known as stoneworts because they are encrusted with calcium carbonate deposits. The body consists of a single file of very long cells anchored in mud by thin filaments. Whorls of branches occur at regions called nodes, located between the cells of the main axis. Male and female reproductive structures grow at the nodes. A *Coleochaete* looks flat, like a pancake, but the body is actually composed of long, branched filaments of cells. Most important to the evolution of plants, charophytes protect the zygote. Land plants not only protect the zygote but also protect and nourish the resulting embryo—this may be the first derived feature that separates land plants from green algae.

To illustrate the evolution of land plants, we will discuss five groups. It is possible to associate each group with one of five evolutionary events, each one representing a major adaptation to existence on land.

Mosses are low-lying plants that lack vascular tissue and therefore have no means of transporting water, but they do protect the body of the plant from drying out and protect the embryo within a special structure.

The lycophytes, which evolved around 420 MYA, are among the first plants to have a vascular system that transports water and solutes from the roots to the leaves of land plants. Plants with vascular tissue have true roots, stems, and

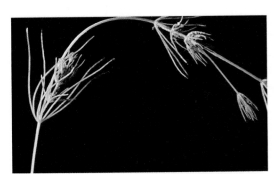
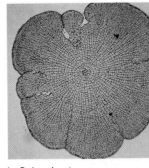

a. *Chara*   b. *Coleochaete*

Figure 18.1 **Close algal relatives of plants.**

The closest living relatives of land plants are the filamentous green algae known as charophytes. **a.** *Chara*, commonly called stoneworts, favor an environment rich in the mineral calcium carbonate, which becomes deposited on the surface of the alga. **b.** The body of this *Coleochaete* is also composed of filaments of cells; even so, it is only the size of a pinhead and one cell layer thick.

leaves. The leaves of lycophytes, called **microphylls,** are very narrow. Ferns are well-known plants with large leaves called **megaphylls.** The evolution of branching and megaphylls allows a plant to increase the amount of photosynthesis and carbohydrate produced. Without an adequate production of food, a plant can't increase in size. The next evolutionary event was the evolution of **seeds.** A seed contains an embryo and stored organic nutrients within a protective coat (see Fig. 18.10). Seeds are highly resistant structures well suited to protect a plant embryo from drying out until conditions are favorable for germination. The gymnosperms were the first seed plants to appear, about 360 MYA. The final evolutionary event of interest to us is the evolution of the **flower,** a reproductive structure found in angiosperms. Flowers attract pollinators, such as insects, and they give rise to fruits that cover seeds. Plants with flowers evolved between 120 and 140 MYA. **Figure 18.2** traces the evolutionary history of land plants and will serve as a backdrop as we discuss the major groups of plants in the pages that follow.

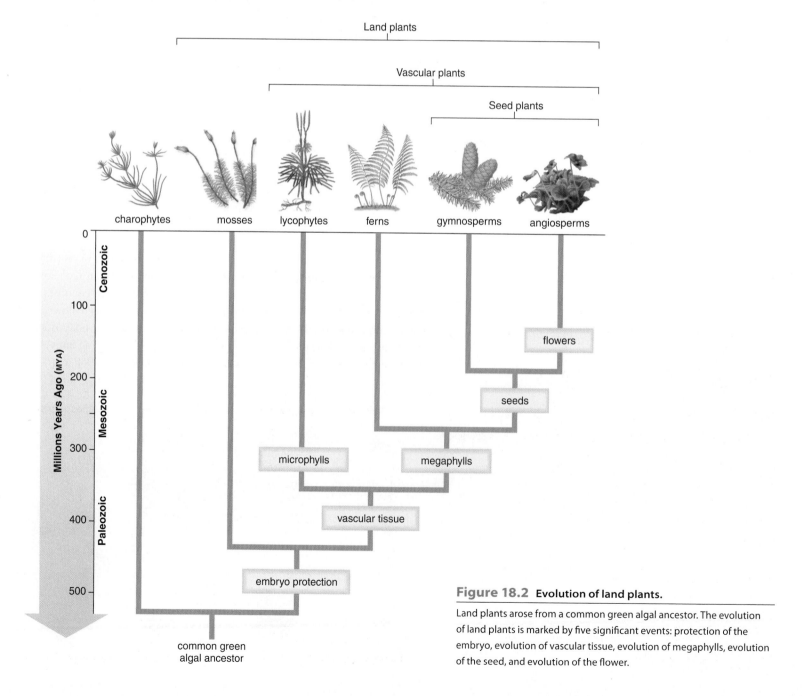

Figure 18.2   Evolution of land plants.

Land plants arose from a common green algal ancestor. The evolution of land plants is marked by five significant events: protection of the embryo, evolution of vascular tissue, evolution of megaphylls, evolution of the seed, and evolution of the flower.

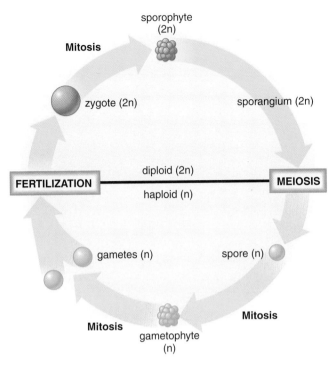

Figure 18.3 Alternation of generations life cycle.

Plants alternate between a sporophyte (2n) stage and a gametophyte (n) stage. The dominant stage is photosynthetic.

Alternation of Generations

The life cycle of land plants is quite different from that of animals. All land plants undergo an **alternation of generations,** which means that each type of plant exists in two forms (**Fig. 18.3**). One form is the diploid (2n) sporophyte, and the other is the haploid (n) gametophyte. The **sporophyte** is named for its production of haploid spores by meiosis. A **spore** is a reproductive cell that develops into a new organism without the need to fuse with another reproductive cell. In the plant life cycle, a spore undergoes mitosis and becomes a gametophyte. The **gametophyte** is named for its production of gametes. A sperm and an egg fuse, forming a zygote that undergoes mitosis and becomes the sporophyte.

You'll want to make two observations in Figure 18.3. First, note that, in plants, meiosis produces haploid spores. This is consistent with the realization that the sporophyte is the diploid generation, and spores are haploid. Second, note that mitosis occurs as a spore becomes a gametophyte, and it occurs as a zygote becomes a sporophyte. Mitosis must happen in both places in order to have two generations.

The Dominant Generation

When you bring a land plant to mind, you're probably thinking of the dominant generation. The dominant generation is the generation that conducts photosynthesis. In other words, the dominant generation is responsible for the growth and nourishment of the plant tissues. In nonvascular plants, the gametophyte is the dominant generation, and in vascular plants, the sporophyte is the dominant generation. In the history of plants, *only the sporophyte evolves vascular tissue;* therefore, the trend toward sporophyte dominance is an adaptation to life on land. As the sporophyte gains in dominance, the gametophyte becomes microscopic. It also becomes dependent on the sporophyte, the generation best adapted to a dry land environment.

Note the appearance of the generations in **Figure 18.4.** In mosses (**bryophytes**), the gametophyte is much larger than the sporophyte. In lycophytes and ferns, the gametophyte is a small, independent structure. In contrast, the female gametophyte of cone-bearing plants (**gymnosperms**) and flowering plants (**angiosperms**) is microscopic—it is retained within the body of the sporophyte plant. This protects the female gametophyte from drying out. Also, the male gametophyte of seed plants lies within a pollen grain. Pollen grains

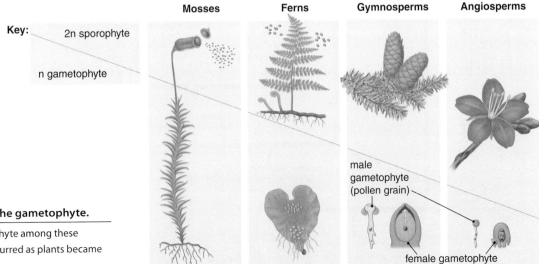

Figure 18.4 Reduction in the size of the gametophyte.

Notice the reduction in the size of the gametophyte among these representatives of today's plants. This trend occurred as plants became adapted for life on land.

have strong protective walls and are transported by wind, insects, and birds to reach the egg. In the life cycle of seed plants, the spores, the gametes, and the zygote are protected from drying out in the land environment.

Check Your Progress 18.1

1 List the similarities between charophytes and land plants.

2 Describe the five evolutionary events that allowed plants to successfully inhabit land.

3 Summarize alternation of generations, identifying the stages that are haploid versus diploid.

18.2 Diversity of Land Plants

Learning Outcomes

Upon completion of this section, you should be able to

1. Characterize and give examples of the various groups of land plants.
2. Describe the life cycles and reproductive strategies used by each group of land plants.
3. Summarize the economic and ecological significance of plants.

In the evolution of the land plants, the nonvascular plants evolved first. They are represented today by the mosses. The vascular plants, which include plants such as the ferns, pines, and flowering plants, evolved several important adaptations that allowed them to succeed in the land environment.

Nonvascular Plants

The **nonvascular plants** include the liverworts, hornworts, and mosses (**Fig. 18.5**). Collectively, they are often called the *bryophytes*. Bryophytes, in general, do not have true roots, stems, and leaves—all of which, by definition, must contain well-developed vascular tissue. In bryophytes, the gametophyte is the dominant generation.

The most familiar bryophytes are the liverworts and mosses, which are low-lying plants. In bryophytes, the gametophyte consists of leafy shoots, which produce the gametes. The gametophyte stage of a bryophyte is completely dependent on water for reproduction. Flagellated sperm swim in a film of water to reach an egg. After a sperm fertilizes an egg, the resulting zygote becomes an embryo that develops into a sporophyte. The sporophyte is attached to, and derives its nourishment from, the photosynthetic gametophyte. The sporophyte produces spores in a structure called a **sporangium.** The spores are released into the air, where they can be dispersed by the wind, an adaptation to life on land. The spores will germinate if they land in moist surroundings. Upon germination, male and female gametophytes develop.

The common name of several organisms implies that they are mosses, when they are not. Irish moss is an edible red alga that grows in leathery tufts along northern seacoasts. Reindeer moss, a lichen, is the dietary mainstay of reindeer and caribou in northern lands. Club mosses, discussed later in this chapter, are vascular plants, and Spanish moss, which hangs in grayish clusters from trees in the southeastern United States, is a flowering plant of the pineapple family.

Connecting the Concepts

For more on the evolution of the land plants, refer to the following discussion.

Table 16.1 outlines the major events in the evolution of life, including the plants.

Hornwort

Liverwort gametophyte

Figure 18.5 **Mosses.**

In bryophytes, exemplified by mosses, the gametophyte is the dominant generation. Sperm swim from male shoots to female shoots, and the zygote develops into an attached sporophyte. Within structures called sporangia, the sporophyte produces haploid spores that develop into gametophyte shoots.

Moss gametophyte

Adaptations and Uses of Bryophytes

The lack of well-developed vascular tissue and the presence of swimming sperm largely account for the short height of bryophytes, such as mosses. Still, mosses can be found from the Antarctic through the tropics to parts of the Arctic. Although most mosses prefer damp, shaded locations in the temperate zone, some survive in deserts, and others inhabit bogs and streams. In forests, they frequently form a mat that covers the ground and rotting logs. In dry environments, they may become shriveled, turn brown, and look dead. As soon as it rains, the plant becomes green and resumes metabolic activity. Mosses are much better than flowering plants at living on stone walls, on fences, and in shady cracks of hot, exposed rocks. When bryophytes colonize bare rock, the rock degrades to soil that can be used for their own growth and the growth of other organisms.

In areas such as bogs, where the ground is wet and acidic, dead mosses, especially *Sphagnum,* do not decay. The accumulated moss, called peat or bog moss, can be used as fuel. Peat moss is also commercially important in another way. Because it has special nonliving cells that can absorb moisture, peat moss is often used in gardens to improve the water-holding capacity of the soil. The so-called copper mosses live only in the vicinity of copper and can serve as an indicator of copper deposits.

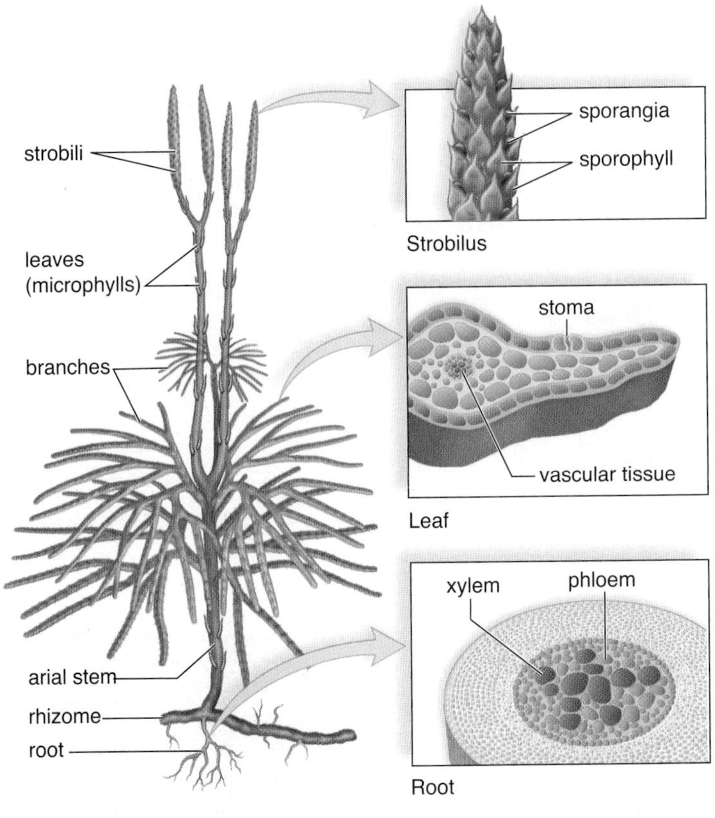

Vascular Plants

All the other land plants we will study are **vascular plants.** The vascular plants usually have *true* roots, stems, and leaves. The roots absorb water from the soil, and the stem conducts water to the leaves. The leaves are fully covered by a waxy cuticle, except where it is interrupted by stomata, little pores for gas exchange, the opening and closing of which can be regulated to control water loss.

Vascular tissue consists of **xylem,** which conducts water and minerals up from the roots, and **phloem,** which conducts organic nutrients from one part of a plant to another. The walls of conducting cells in xylem are strengthened by **lignin,** an organic compound that makes them stronger, more waterproof, and resistant to attack by parasites and predators. Only because of strong cell walls can plants reach great heights.

Animation
Vascular System of Plants

Seedless Vascular Plants

Certain vascular plants (e.g., lycophytes and ferns) are seedless; the other two groups of vascular plants (gymnosperms and angiosperms) are seed plants. In seedless vascular plants, the dominant sporophyte produces windblown spores, and the independent gametophyte produces flagellated sperm that require outside moisture to swim to an egg (see Fig. 18.8).

Lycophytes

Lycophytes, also called **club mosses,** were among the first land plants to have vascular tissue. Unlike true mosses (bryophytes), the lycophytes have well-developed vascular tissue in roots, stems, and leaves (**Fig. 18.6**). Typically, a fleshy underground and horizontal stem, called a rhizome, sends up upright aerial stems. Tightly packed, scale-like leaves cover the stems and branches, giving the plant a mossy look. The small leaves,

Figure 18.6 Ground pine, *Lycopodium.*

Lycophytes, such as *Lycopodium,* have vascular tissue and thus true roots, stems, and leaves. The *Lycopodium* sporophyte develops an underground rhizome system. A rhizome is an underground stem; this rhizome produces roots along its length.

termed *microphylls,* each have a single vein composed of xylem and phloem. The sporangia are borne on terminal clusters of leaves, called strobili, which are club-shaped. The spores are sometimes harvested and sold as lycopodium powder, or vegetable sulfur, for use in pharmaceuticals and in fireworks because it is highly flammable. The *Lycopodium* featured in Figure 18.6 are common in moist woodlands in temperate climates, where they are called ground pines; they are also abundant in the tropics and subtropics.

Ferns

Ferns are a widespread group of plants that are well known for their attractiveness. Unlike lycophytes, ferns have megaphylls, or large leaves with branched veins. Megaphylls provide a large surface area for capturing the sunlight needed for photosynthesis, and the veins conduct water and minerals throughout the leaf tissue. Ferns and other land plants with megaphylls are better able to produce food and thus can grow and reproduce more efficiently than plants with microphylls. Fern megaphylls are called **fronds.** The maidenhair fern has fronds that are broad, with subdivided leaflets; those of the royal fern stand about 1.8 meters tall; and those of the hart's tongue fern are strap-like and leathery (**Fig. 18.7**). Sporangia are often located in clusters, called **sori** (sing., sorus), on the undersides of the fronds, where they may be shielded by thin, protective structures called *indusia* (**Fig. 18.8**).

maidenhair fern

royal fern

hart's tongue fern

Figure 18.7
Diversity of ferns.

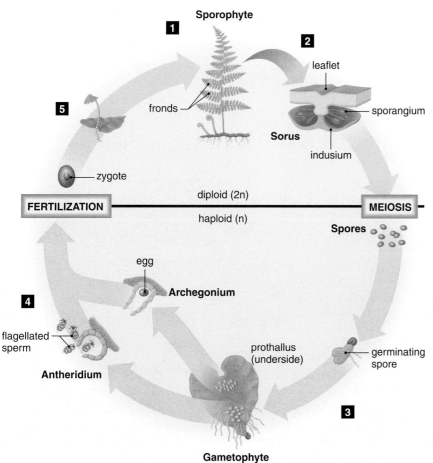

Figure 18.8 Fern life cycle.

1 The sporophyte is dominant in ferns. **2** In the fern shown here, sori are on the underside of the fern. Each sorus, protected by an indusium, contains sporangia, in which meiosis occurs and spores are produced and then released. **3** A spore germinates into a prothallus (the gametophyte), which has sperm-bearing and egg-bearing structures on its underside. **4** Fertilization takes place when moisture is present, because the flagellated sperm must swim in a film of water to the egg. **5** The resulting zygote begins its development inside an archegonium, and eventually a young sporophyte becomes visible. The young sporophyte develops, and fronds appear.

Connections and Misconceptions

What do horsetail dietary supplements do?

Horsetail belongs to a genus of plants called *Equisetum*, which are close relatives of the ferns. The use of horsetail supplements to treat a variety of illnesses, including tuberculosis and ulcers, goes back to the time of the ancient Romans. While some people have suggested that horsetail supplements may be used to prevent osteoporosis (since horsetails contain the mineral silicon), there have been very few studies on the long-term effects of horsetail use. For unknown reasons, horsetail causes a reduction in some B vitamins in the body, and it may interact with other medications and supplements. As always, you should consult with a physician before you start taking any new dietary supplements.

Figure 18.9 **The Carboniferous period.**

Growing in the swamp forests of the Carboniferous period were tree-like club mosses (left), tree-like horsetails (right), and lower, fern-like foliage (left). When the trees fell, they were covered by water and did not decompose completely. Sediment built up and turned to rock, the pressure of which caused the organic material to become coal, a fossil fuel that helps run our industrialized society.

Adaptations and Uses of Ferns   Ferns are most often found in moist environments because the small, water-dependent gametophyte, which lacks vascular tissue, is separate from the sporophyte. Also, flagellated sperm require an outside source of moisture in which to swim to the eggs (see Fig. 18.8). Once established, some ferns, such as the bracken fern, can spread into drier areas because their rhizomes, which grow horizontally in the soil, produce new plants.

At first it may seem that ferns do not have much economic value, but they are frequently used by florists in decorative bouquets and as ornamental plants in the home and garden. Wood from tropical tree ferns is often used as a building material because it resists decay, particularly by termites. Ferns, especially the ostrich fern, are used as food—in the northeastern United States, many restaurants feature fiddleheads (that season's first growth) as a special treat. Ferns also have medicinal value; many Native Americans use ferns as an astringent during childbirth to stop bleeding, and the maidenhair fern is the source of a cold medicine.

Coal Age Plants

Ferns and the other seedless vascular plants we have been discussing were as large as trees and more abundant during the Carboniferous period, when a great swamp forest encompassed what is now northern Europe, the Ukraine, and the Appalachian Mountains in the United States (**Fig. 18.9**). A large number of these plants died but did not decompose completely. Instead, they were compressed to form the coal that we still mine and burn today; therefore, seedless vascular plants are sometimes called the *Coal Age* plants. (Oil has a similar origin, but it most likely formed in marine sedimentary rocks and includes animal remains.)

Seed Plants

Seed plants are the most plentiful land plants in the biosphere. Most trees are seed plants, and so are almost all garden plants. The major parts of a seed are shown in **Figure 18.10.** The seed coat and stored food protect the sporophyte embryo and allow it to survive harsh conditions during a period of dormancy (arrested state), until environmental conditions become favorable for growth. Seeds can even remain dormant for hundreds of years. When a seed germinates, the stored food is a source of nutrients for the growing seedling. The survival value of seeds largely accounts for the dominance of seed plants today.

Seed plants have two types of spores and produce two kinds of gametophytes—male and female. The gametophytes are microscopic and consist of just a few cells. **Pollen grains** are drought-resistant male gametophytes. **Pollination** occurs when a pollen grain is brought to the vicinity of the female gametophyte by wind or a pollinator. Later, sperm move toward the female gametophyte through a growing pollen tube, and fertilization occurs. This represents a major adaptation to the land environment, since seed plants do not need external water for fertilization to occur. The whole male gametophyte, rather than just the sperm (as in seedless plants), moves to the female gametophyte.

A female gametophyte develops within an **ovule,** which eventually becomes a seed (**Fig. 18.11**). The embryo in a seed will be the sporophyte plant. In gymnosperms, the ovules are not completely enclosed by sporophyte tissue at the time of pollination. In angiosperms, the ovules are completely enclosed within diploid sporophyte tissue (an ovary), which becomes a fruit. We will see that the flower has many advantages, one of which is the production of fruit.

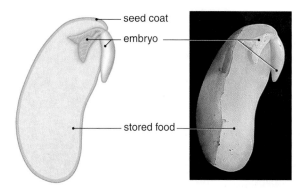

Figure 18.10 Seed anatomy.

A split bean seed showing the seed coat, sporophyte embryo, and stored food.

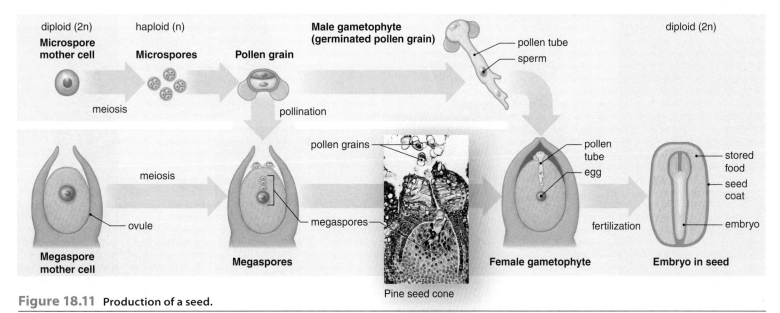

Figure 18.11 Production of a seed.

Pine seed cone

Development of the male gametophyte (above) begins when a microspore mother cell undergoes meiosis to produce microspores, each of which becomes a pollen grain. Development of the female gametophyte (below) begins in an ovule, where a megaspore mother cell undergoes meiosis to produce megaspores, only one of which will undergo mitosis to become the female gametophyte. During pollination, the pollen grain is carried to the vicinity of the ovule. The pollen grain germinates, and a nonflagellated sperm travels in a pollen tube to the egg produced by the female gametophyte. Following fertilization, the zygote becomes the sporophyte embryo; tissue within the ovule becomes the stored food, and the ovule wall becomes the seed coat.

Male cycad with a pollen-bearing cone.

Female cycad with a seed-bearing cone.

Figure 18.12 **Cycads.**

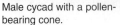

Cycads are an ancient group of gymnosperms that are threatened today because they grow slowly. Herbivorous dinosaurs of the Mesozoic era most likely fed on cycad seeds.

Gymnosperms

The term *gymnosperm* means naked seed. In gymnosperms, ovules and seeds are exposed on the surface of a cone scale (modified leaf). Ancient gymnosperms, including cycads, were present in the swamp forests of the Carboniferous period (**Fig. 18.12**). Of these, the conifers have become a dominant plant group.

Conifers

Pines, spruces, firs, cedars, hemlocks, redwoods, and cypresses are all **conifers** (**Fig. 18.13**). The name *conifer* signifies plants that bear **cones** containing the reproductive structures of the plant, but other types of gymnosperms are also cone-bearing. The coastal redwood, a conifer native to northwestern California and southwestern Oregon, is the tallest living vascular plant; it can attain nearly 100 meters in height. Another conifer, the bristlecone pine of the White Mountains of California, is the oldest living tree. One living specimen is over 4,500 years old, and there is evidence that some bristlecone pines have lived as many as 4,900 years.

Adaptations and Uses of Pine Trees Pine trees are well adapted for dry conditions. For instance, vast areas of northern temperate regions are covered in evergreen coniferous forests. The tough, needle-like leaves of pines conserve water because they have a thick cuticle and recessed stomata. This type of leaf helps them live in areas where frozen topsoil makes it difficult for the roots to obtain plentiful water.

pollen cones

seed cones

a.

fleshy seed cones (juniper berries)

c.

Figure 18.13 **Conifers.**

b.

a. Pine trees are the most common of the conifers. The pollen cones (male) are smaller than the seed cones (female) and produce plentiful pollen. A cluster of pollen cones may produce more than 1 million pollen grains. Other conifers include (**b**) the spruces, which make beautiful Christmas trees, and (**c**) the junipers, which possess fleshy seed cones.

A substance called resin protects leaves and other parts of the tree from insect and fungal attacks. The resin of certain pines is harvested; the liquid portion, called turpentine, is a paint thinner, while the solid portion is used on stringed instruments. The wood of pines is used extensively in construction, and vast forests of pines are planted for this purpose. The wood consists primarily of xylem tissue that lacks some of the more rigid cell types found in flowering trees. Therefore, it is considered a "soft" rather than a "hard" wood.

Angiosperms

The angiosperms (meaning covered seeds) are an exceptionally large and successful group of land plants, with 240,000 known species—six times the number of species of all the other plant groups combined. Angiosperms, also called the flowering plants, live in all sorts of habitats, from freshwater to desert, and from the frigid north to the torrid tropics. They range in size from the tiny, almost microscopic duckweed to *Eucalyptus* trees over 35 meters tall. Most garden plants produce flowers and therefore are angiosperms. In northern climates, the trees that lose their leaves are flowering plants. In subtropical and tropical climates, flowering trees as well as gymnosperms tend to keep their leaves all year.

Although the first fossils of angiosperms are no older than about 135 million years, the angiosperms probably arose much earlier. Indirect evidence suggests that the possible ancestors of angiosperms may have originated as long ago as 160 MYA. To help solve the mystery of their origin, botanists have turned to DNA comparisons to find a living plant that is most closely related to the first angiosperms. Their data point to *Amborella trichopoda* as having the oldest lineage among today's angiosperms (**Fig. 18.14**). This small shrub, which has small, cream-colored flowers, lives only on the island of New Caledonia in the South Pacific.

Figure 18.14 *Amborella trichopoda.*

Molecular data suggest that this plant is most closely related to the first flowering plants.

The Flower

Most flowers have certain parts in common, despite their dissimilar appearances. The flower parts, called **sepals, petals, stamens,** and **carpels,** occur in whorls (circles) (**Fig. 18.15**). The sepals, collectively called the **calyx,** protect the flower bud before it opens. The sepals may drop off or may be colored like the petals. Usually, however, sepals are green and remain in place. The petals, collectively called the **corolla,** are quite diverse in size, shape, and color. The petals often attract a particular pollinator. Each stamen consists of a stalk, called a **filament,** and an **anther,** where pollen is produced in pollen sacs. In most flowers, the anther is positioned where the pollen can be carried away by wind or a pollinator. One or more carpels are at the center of a flower. A carpel has three major regions: ovary, style, and stigma. The swollen base is the **ovary,** which contains from one to hundreds of ovules. The **style** elevates the **stigma,** which is sticky or otherwise adapted for the reception of pollen grains. Glands located in the region of the ovary produce nectar, a nutrient that is gathered by pollinators as they go from flower to flower.

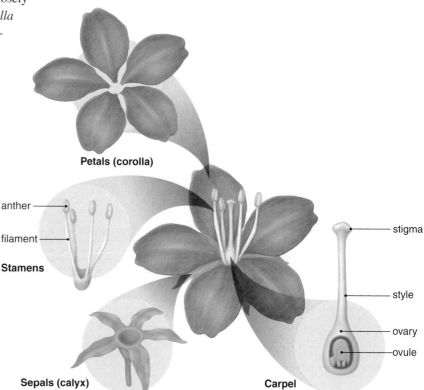

Figure 18.15 Generalized flower.

A flower has four main parts: sepals, petals, stamens, and carpels. A stamen has a filament and an anther. A carpel has an ovary, a style, and a stigma. The ovary contains ovules.

Flowering Plant Life Cycle

In angiosperms, the flower produces seeds enclosed by fruit. The ovary of a carpel contains several ovules, and each of these eventually holds an egg-bearing female gametophyte called the embryo sac. During pollination, a pollen grain is transported by various means from the anther of a stamen to the stigma of a carpel, where it germinates. The **pollen tube** carries the two sperm into a small opening of an ovule. During **double fertilization,** one sperm unites with an egg nucleus, forming a diploid zygote, and the other sperm unites with two other nuclei, forming a triploid (3n) **endosperm** (**Fig. 18.16**). In angiosperms, the endosperm is the stored food.

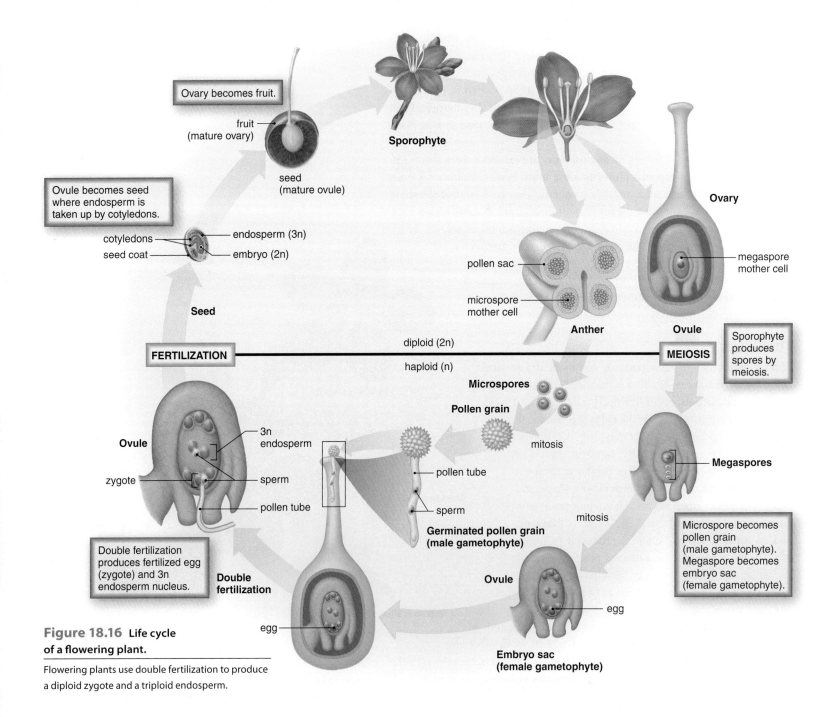

Ovary becomes fruit.

fruit
(mature ovary)

seed
(mature ovule)

Sporophyte

Ovary

Ovule becomes seed where endosperm is taken up by cotyledons.

cotyledons
seed coat

endosperm (3n)
embryo (2n)

pollen sac

microspore
mother cell

megaspore
mother cell

Seed

Anther

Ovule

Sporophyte produces spores by meiosis.

FERTILIZATION

diploid (2n)

MEIOSIS

haploid (n)

Microspores

Pollen grain

mitosis

Megaspores

Ovule

3n
endosperm

zygote

sperm

pollen tube

pollen tube

sperm

**Germinated pollen grain
(male gametophyte)**

mitosis

Microspore becomes pollen grain (male gametophyte). Megaspore becomes embryo sac (female gametophyte).

Double fertilization produces fertilized egg (zygote) and 3n endosperm nucleus.

**Double
fertilization**

Ovule

egg

egg

**Embryo sac
(female gametophyte)**

**Figure 18.16 Life cycle
of a flowering plant.**

Flowering plants use double fertilization to produce a diploid zygote and a triploid endosperm.

Ultimately, the ovule becomes a seed that contains a sporophyte embryo. In some seeds, the endosperm is absorbed by the seed leaves, called **cotyledons,** whereas in other seeds, endosperm is digested as the seed germinates. When you open a peanut, the two halves are the cotyledons. If you look closely, you will see the embryo between the cotyledons. A **fruit** is derived from an ovary and possibly accessory parts of the flower. Some fruits (e.g., apple) provide a fleshy covering for their seeds, and other fruits provide a dry covering (e.g., pea pod, peanut shell).

Adaptations and Uses of Angiosperms

Successful completion of sexual reproduction in angiosperms requires the effective dispersal of pollen and then seeds. Adaptations have resulted in various means of dispersal of pollen and seeds. Wind-pollinated flowers are usually are not showy, whereas many insect- and bird-pollinated flowers are colorful (**Fig. 18.17***a–c*). Night-blooming flowers attract nocturnal mammals and insects; these flowers are usually aromatic and white or cream-colored (Fig. 18.17*d*). Although some flowers disperse their pollen by wind, many are adapted to attract specific pollinators, such as bees, wasps, flies, butterflies, moths, and even bats, which carry only particular pollen from flower to flower. For example, bee-pollinated flowers are usually blue or yellow and have ultraviolet shadings that lead the pollinator to the location of nectar at the base of the flower. In turn, the mouthparts of bees are fused into a long tube that is able to obtain nectar from this location. Today, there are some 240,000 species of flowering plants and over 900,000 species of insects. This diversity suggests that the success of angiosperms has contributed to the success of insects and vice versa.

> **Video** Pollinators

The fruits of flowers protect and aid in the dispersal of seeds. Dispersal occurs when seeds are transported by wind, gravity, water, and animals to another location. Fleshy fruits may be eaten by animals, which transport the seeds to a new location and then deposit them when they defecate. Because animals live in particular habitats and/or have particular migration patterns, they are apt to deliver the fruit-enclosed seeds to a suitable location for seed germination (initiation of growth) and development of the plant.

 Video Dung Seed Dispersal

 Video Fruit Bat Seed Dispersal

Figure 18.17 Pollinators.

a. A bee-pollinated flower is typically a color other than red (bees cannot see red). **b.** Butterfly-pollinated flowers are wide, allowing the butterfly to land. **c.** Hummingbird-pollinated flowers are curved back, allowing the bird's beak to reach the nectar. **d.** Bat-pollinated flowers are large and sturdy and able to withstand rough treatment.

Connections and Misconceptions

Do carnivorous plants capture insects for food?

All plants, including carnivorous plants, such as the sundew plant, are autotrophic organisms that get their energy from photosynthesis. However, some plants that live in mineral-poor soils have evolved adaptations to allow them to capture insects and sometimes small amphibians. The goal of these plants is not to extract energy from their prey. Instead the plant is after a mineral or nutrient (often nitrogen) that is lacking in the soil of their environment. There are over 600 known species of carnivorous plants.

> **Video** Carnivorous Plants

Ovules become the seeds.

a.

b.   c.

Ovary becomes
the fruit.

d.

Figure 18.18   Pea flower and the development of a pea pod.

a. Pea flower. **b.** Pea flower still has its petals soon after fertilization. **c.** Petals fall and the pea pod is quite noticeable. **d.** Pea pod is clearly visible developing from an ovary.

Economic Benefits of Plants

One of the primary economic benefits of land plants is the use of their fruits as food. Botanists use the term *fruit* in a much broader way than do laypeople. Among the foods mentioned in the introduction, you would have no trouble recognizing a banana as a fruit. A coconut is also a fruit, as are grains (corn, wheat, rice) and pods that contain beans or peas (**Fig. 18.18**). Cotton is derived from the cotton boll, a fruit containing seeds with seed hairs that become textile fibers used to make cloth.

Video
Warming Hurts
Rice

Other economic benefits of land plants include foods and commercial products made from roots, stems, and leaves. Cassava and sweet potatoes are edible roots; white potatoes are the tubers of underground stems. Most furniture and paper are made from the wood of a tree trunk (**Fig. 18.19**). Also, the many chemicals produced by plants make up 50% of all pharmaceuticals and various other types of products we can use. The cancer drug taxol originally came from the bark of the Pacific yew tree. Today, plants are even bioengineered to produce certain substances of interest.

Indirectly, the economic benefits of land plants are often dependent on pollinators. Only if pollination occurs can these plants produce a fruit and propagate themselves. In recent years, the populations of honeybees and other pollinators have been declining worldwide, principally due to a parasitic mite but partly because of the widespread use of pesticides. Consequently, some plants are endangered because they have lost their normal pollinators. Because of our dependence on flowering plants, we should protect pollinators!

Figure 18.19   Humans use wood and wood products in daily life.

Ecological Benefits of Plants

The ecological benefits of flowering plants are so important that we could not exist without them. Land plants produce food for themselves and directly or indirectly for all other organisms in the biosphere. And all organisms that carry out cellular respiration use the oxygen that land plants produce through photosynthesis.

Forests are an important part of the water cycle and the carbon cycle. In particular, the roots of trees hold soil in place and absorb water, which returns to the atmosphere. Without these functions of trees and other plants, rainwater runs off and contributes to flooding. Plants' absorption of carbon dioxide lessens the amount in the atmosphere. CO_2 in the atmosphere contributes to global warming because it and other gases trap heat near the surface of the Earth. The burning of tropical rain forests is double trouble for global warming because it adds CO_2 to the atmosphere and removes trees that otherwise would absorb CO_2. Some plants can also be used to clean up toxic messes. For example, poplar, mustard, and mulberry species take up lead, uranium, and other pollutants from the soil.

In addition to all the other uses of land plants, we should not forget their aesthetic value. Almost everyone prefers to vacation in a natural setting and enjoys the sight of trees and flowers.

Video
Plants

18.3 The Fungi

Learning Outcomes

Upon completion of this section, you should be able to

1. Describe the general biology of a fungus.
2. Compare and contrast fungi with animals and plants; explain what makes chytrids unique among fungi.
3. Explain the life cycles of black bread molds and mushrooms.
4. Summarize the economic and ecological significance of fungi.
5. Provide examples of fungal diseases.

Asked whether **fungi** are more closely related to animals or plants, most would choose plants. But this would be wrong, because fungi do not have chloroplasts, and they can't photosynthesize. Then, too, fungi are not animals, even though they are chemoheterotrophs, like animals. Animals ingest their food, but fungi must grow into their food. Fungi release digestive enzymes into their immediate environment and then absorb the products of digestion. Also, animals are motile, but most fungi are nonmotile and do not have flagella at any stage in their life cycle. The fungal life cycle differs from that of both animals and plants because fungi produce windblown spores during both an asexual and a sexual life cycle.

Table 18.1 contrasts fungi with land plants and animals. The many unique features of fungi indicate that, although fungi are multicellular eukaryotes (except for the unicellular yeasts and chytrids), they are not closely related to any other group of organisms. DNA sequence data suggest that fungi are distantly related to animals rather than plants and may be descendants of a flagellated protist.

Connecting the Concepts

For more on the ecology of plants, refer to the following discussions.

Section 31.1 provides an overview of the role of plants in an ecological community.

Section 31.2 examines the role of plants in chemical and energy cycling in an ecosystem.

Section 32.2 describes the direct and indirect benefits of preserving plant biodiversity.

Check Your Progress 18.2

1. List the differences between nonvascular and vascular plant structure.
2. List the ecological benefits of plants.
3. Describe the production of a seed.
4. Detail the life cycle of a flowering plant.
5. Predict what might occur if nonvascular plants are not near a water source.
6. Discuss the potential economic hardships that may occur if either gymnosperms or angiosperms went extinct.

Table 18.1 How Fungi Differ from Land Plants and Animals

Feature	Fungi	Land Plants	Animals
Nutrition	Chemoheterotrophic by absorption	Photosynthetic	Chemoheterotrophic by ingestion
Movement	Most nonmotile	Nonmotile	Motile
Body	Mycelium of hyphae	Specialized tissues/organs	Specialized tissues/organs
Adult chromosome number	Haploid	Haploid/diploid	Diploid
Cell wall	Composed of chitin	Composed of cellulose	No cell wall
Reproduction	Most have spores/mating hyphae	Spores/gametes	Gametes

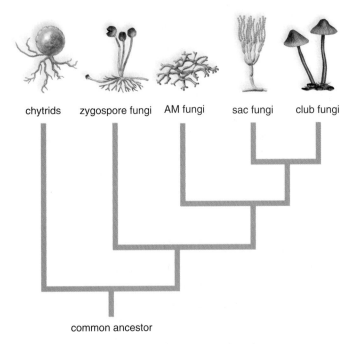

chytrids   zygospore fungi   AM fungi   sac fungi   club fungi

common ancestor

Figure 18.20   Evolutionary relationships of the fungi.

The evolutionary relationships of the major groups of fungi are shown here. The AM fungi consist of those species that form mycorrhizal relationships with plants.

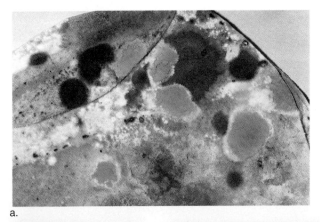

a.

b.   c.

General Biology of a Fungus

The evolutionary relationships of the major groups of fungi are illustrated in **Figure 18.20.** Representative members of kingdom Fungi are shown in **Figure 18.21.** Our description of fungal structure applies best to the zygospore fungi, sac fungi, and club fungi. All parts of a typical fungus are composed of **hyphae** (sing., hypha), which are thin filaments of cells. The hyphae are packed closely together to form a complex structure, such as a mushroom. However, the main body of a fungus is not the mushroom or the puffball or the morel; these are just temporary reproductive structures. Mold shows best that the main body of a fungus is a mass of hyphae called a **mycelium** (**Fig. 18.22**). The mycelium penetrates the soil, the wood, or the bread in which the fungus is growing. Mycelia in the soil can become large enough to cover acres, making them the largest organisms on Earth and earning them the label "the humongous fungus among us."

Fungal cells typically have thick cell walls, but unlike plants, the fungal cell walls do not contain cellulose. They are made of another polysaccharide in which the glucose monomers contain amino groups (amino sugars) and form a polymer called chitin. This polymer is also the major structural component of the exoskeleton of insects and arthropods, such as lobsters and crabs. Walls, or *septa* (sing., septum), divide the cells of a hypha in many types of fungi. Septa have pores that allow the cytoplasm to pass from one cell to the other along the length of the hypha. The hyphae give the mycelium quite a large surface area, which facilitates the ability of the mycelium to absorb nutrients. Hyphae extend toward a food source by growing at their tips, and the hyphae of a mycelium absorb and then pass nutrients on to the growing tips.

Zygospore Fungi—Black Bread Mold

Multicellular organisms are characterized by specialized cells. Black bread mold, a type of zygospore fungi, demonstrates that the hyphae of a fungus may be specialized for various purposes (**Fig. 18.23**). In this fungus, horizontal hyphae exist on the surface of the bread; other hyphae grow into the bread, anchoring the mycelium and carrying out digestion; and some form stalks that bear sporangia.

The mycelia of two different mating types are featured in the center and bottom of Figure 18.23. During asexual reproduction, each mycelium produces sporangia, where spore formation occurs. Spores are resistant to environmental damage, and they are often made in large numbers. Fungal spores are windblown, a distinct advantage for a nonmotile organism living on land. When spores encounter a moist environment, they germinate into new mycelia without going through any developmental stages, another feature that distinguishes fungi from animals.

Sexual reproduction in fungi involves the conjugation of hyphae from different mating types (usually designated + and −). In black bread mold, the tips of + and − hyphae join, the nuclei fuse, and a thick-walled zygospore results. The zygospore undergoes a period of

Figure 18.21   Diversity of fungi.

a. A zygospore fungi, the common bread mold. **b.** A morel, an edible sac fungus. **c.** A club fungi, the white button mushroom.

dormancy before it germinates, producing sporangia. Meiosis occurs within the sporangia, producing spores of both mating types. The spores, which are dispersed by air currents, give rise to new mycelia. In fungi, only the zygote is diploid, and all other stages of the life cycle are haploid, a distinct difference from animals.

Club and Sac Fungi—the "Mushrooms"

The common term *mushroom* is often used to describe the part of the fungi called the *fruiting body*. Most mushrooms belong to the club fungi (see Fig. 18.20), although a few, such as the morel (see Fig. 18.21*b*) are actually sac fungi. The function of the fruiting body is to produce spores. In mushrooms, when the tips of + and – hyphae fuse, the haploid nuclei do not fuse immediately. Instead, so-called dikaryotic (two nuclei) hyphae form a mushroom consisting of a stalk and a cap. Club-shaped structures called basidia (sing., basidium) project from the gills located on the underside of

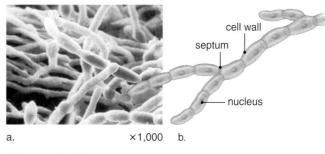

a.   ×1,000   b.

Figure 18.22 Body of a fungus.

a. The body of a fungus is called a mycelium. **b.** A mycelium contains many individual chains of cells, and each chain is called a hypha.

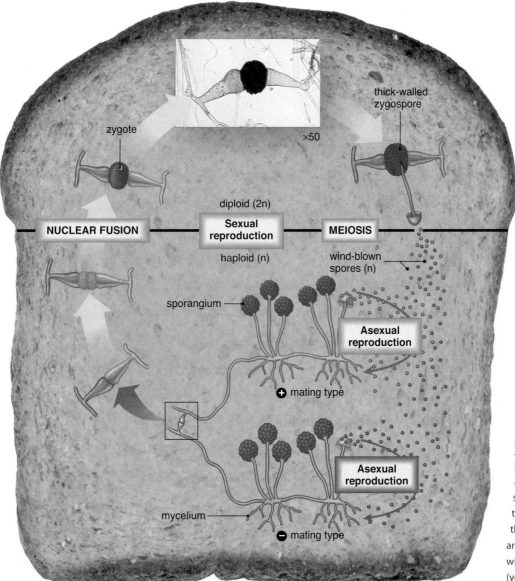

Figure 18.23 Life cycle of black bread mold.

During asexual reproduction, sporangia produce asexual spores (purple and blue arrows). During sexual reproduction, two hyphae tips fuse, and then two nuclei fuse, forming a zygote that develops a thick, resistant wall (zygospore). When conditions are favorable, the zygospore germinates, and meiosis within a sporangium produces windblown spores (yellow arrows).

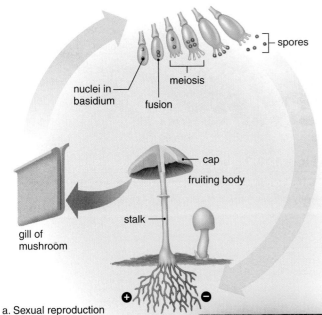

a. Sexual reproduction

Figure 18.24 Sexual reproduction in club fungi produces mushrooms.

a. For club fungi, fusion of + and − hyphae tips results in hyphae that form the mushroom (a fruiting body). The nuclei fuse in club-like structures (basidia) attached to the gills of a mushroom, and meiosis produces spores. **b.** A circle of mushrooms is called a fairy ring.

b.

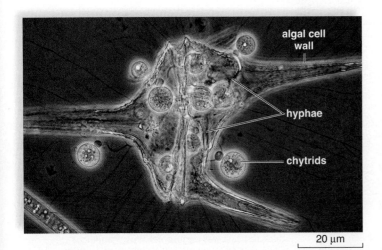

20 μm

Figure 18.25 Chytrids.

Chytriomyces hyalinus, a chytrid, attacking a dinoflagellate, a unicellular alga.

the cap (**Fig. 18.24**a). Fusion of nuclei inside these structures is followed by meiosis and production of windblown spores. Each cap of a mushroom produces tens of thousands of spores.

In the soil, a mycelium is absorbing nutrients and may form a ring. Therefore, when the weather turns warm and rain is plentiful, the mycelium may put forth a circle of mushrooms on your lawn. At one time, people believed this ring indicated where fairies had joined hands and danced the night before (Fig. 18.24b).

Chytrids

Chytrids are a unique group of fungi characterized by their motility (**Fig. 18.25**). Both spores and gametes of chytrids have flagella, a feature which was lost at some point in the evolution of the other fungi. Some of the chytrids are unicellular, while others form hyphae without septa. Another oddity distinguishes the chytrids: Some have an alternation of generations life cycle, much like that of green plants and certain algae, but very uncommon among fungi. Most inhabit water and soil, although some live as parasites of plants and animals.

Ecological Benefits of Fungi

Most fungi are **saprotrophs** that decompose the remains of plants, animals, and microbes in the soil. Fungal enzymes can degrade cellulose and even lignin in the woody parts of plants. That is why fungi so often grow on dead trees. This also means that fungi can be used to remove excess lignin from paper pulp. Ordinarily, lignin is difficult to extract from pulp and ends up being a pollutant once it is removed.

Along with the bacteria, which are also decomposers, fungi play an indispensable role in the environment by returning inorganic nutrients to photosynthesizers. Many people take advantage of the activities of bacteria and fungi by composting their food scraps or yard waste. When a gardener makes a compost pile and provides good conditions for decomposition to occur, the result is a dark, crumbly material that serves as an excellent fertilizer. And while the material may smell bad as decomposition is occurring, the finished compost looks and smells like rich, moist earth.

Some fungi eat animals that they encounter as they feed on their usual meals of dead organic remains. For example, the oyster fungus secretes a substance that anesthetizes any nematodes (roundworms) feeding on it (**Fig. 18.26**). After the worms become inactive, the fungal hyphae penetrate and digest their bodies, absorbing the nutrients. Other fungi snare, trap, or fire projectiles into nematodes and other small animals before digesting them. The animals serve as a source of nitrogen for the fungus.

Mutualistic Relationships

In a mutualistic relationship, two different species live together and help each other. **Lichens** are a mutualistic association between a particular fungus and cyanobacteria or green algae (**Fig. 18.27**). The fungal partner is efficient at acquiring nutrients and moisture, and therefore lichens can survive in poor soils, as well as on rocks with no soil. The organic acids given off by fungi release from rocks the minerals that can be used by the photosynthetic partner. Lichens are ecologically important because they produce organic matter and create new soil, allowing plants to invade the area.

Figure 18.26 Carnivorous fungus.

Fungi like this oyster fungus, a type of bracket fungus, grow on trees because they can digest cellulose and lignin. If this fungus meets a roundworm in the process, it immobilizes the worm and digests it also. The worm is a source of nitrogen for the fungus.

Lichens occur in three varieties: compact crustose lichens, often seen on bare rocks or tree bark; shrub-like fruticose lichens; and leaf-like foliose lichens. Regardless, the body of a lichen has three layers. The fungal hyphae form a thin, tough upper layer and a loosely packed lower layer. These layers shield the photosynthetic cells in the middle layer. Specialized fungal hyphae that penetrate or envelop the photosynthetic cells transfer organic nutrients to the rest of the mycelium. The fungus not only provides minerals and water to the photosynthesizer but also offers protection from predation and desiccation. Lichens can reproduce asexually by releasing fragments that contain hyphae and an algal cell. At first, the relationship between the fungi and algae was likely a parasite-and-host interaction. Over evolutionary time, the relationship apparently became more mutually beneficial, although how to test this hypothesis is a matter of debate at the present time.

Mycorrhizal fungi, also called the *AM fungi* (see Fig. 18.20), form mutualistic relationships with the roots of most plants, helping the plants grow more successfully in dry or poor soils, particularly those deficient in inorganic nutrients (**Fig. 18.28**). The fungal hyphae greatly increase the surface area from which the plant can absorb water and nutrients. It has been found beneficial to encourage the growth of mycorrhizal fungi when restoring lands damaged by strip mining or chemical pollution.

Mycorrhizal fungi may live on the outside of roots, enter between root cells, or penetrate root cells. The fungus and plant cells exchange nutrients, with the fungus bringing water and minerals to the plant and the plant providing organic carbon to the fungus. Early plant fossils indicate that the relationship between fungi and plant roots is an ancient one, and therefore it may have helped plants adapt to life on dry land. The general public is not familiar with mycorrhizal fungi, but a few people relish truffles, the fruiting bodies of a mycorrhizal fungus that grows in oak and beech forests. Truffles are considered a gourmet delicacy.

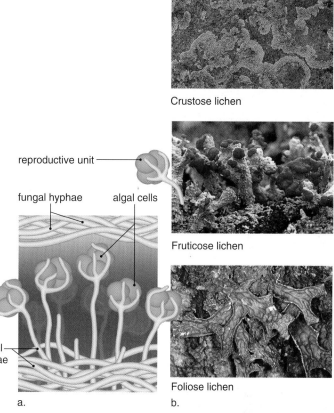

Crustose lichen

Fruticose lichen

Foliose lichen

reproductive unit

fungal hyphae   algal cells

fungal hyphae

a.

b.

Figure 18.27 Lichens.

a. Morphology of a lichen. **b.** Examples of lichens.

a.   b.   c.

Figure 18.28 Plant growth experiment.

A soybean plant without mycorrhizal fungi (**a**) grows poorly compared with two others (**b, c**) infected with different strains of mycorrhizal fungi.

Figure 18.29 Commercial importance of fungi.

Delicacies such as various types of mushrooms are fungi, and some of our other favorite foods require the participation of fungi to produce them. Some antibiotics, such as amoxicillin, are derived from fungi.

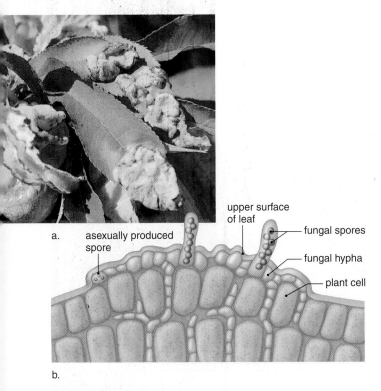

a. asexually produced spore

upper surface of leaf

fungal spores

fungal hypha

plant cell

b.

Figure 18.30 Plant fungal disease.

Peach leaf curl invades the leaf, causing lesions, as shown in (**a**) the photo and (**b**) the diagram.

Economic Benefits of Fungi

Fungi help us produce medicines and many types of foods (**Fig. 18.29**). The mold *Penicillium* was the original source of penicillin, a breakthrough antibiotic that led to an important class of cillin antibiotics. Cillin antibiotics have saved millions of lives.

Yeast fermentation is utilized to make bread, beer, wine, and distilled spirits. During fermentation, yeast cells produce the carbon dioxide that causes dough to rise when gas pockets are preserved as the bread bakes. Ethanol is also a product of yeast fermentation and is used to make alcoholic beverages. Other types of fungal fermentation contribute to the manufacture of various cheeses and soy sauce from soybeans. Another commercial application of interest is the use of fungi to soften the centers of certain candies.

Fungi as Food

In the United States, the consumption of mushrooms has been steadily increasing. In 2001, consumption of all mushrooms totaled 1.13 billion pounds—21% greater than in 1991. In addition to adding taste and texture to soups, salads, and omelets, and being used in stir-fries, mushrooms are an excellent low-calorie meat substitute with great nutritional value and lots of vitamins. Although there are thousands of mushroom varieties in the world, the white button mushroom, *Agaricus bisporus*, dominates the U.S. market. However, in recent years, sales of brown-colored variants have surged in popularity and have been one of the fastest-growing segments of the mushroom industry. Portabella is a marketing name used by the mushroom industry for the more flavorful brown strains of *A. bisporus*. The mushroom is brown because it is allowed to open, exposing the mature gills with their brown spores; crimini is the same brown strain, but it is not allowed to open before it is harvested. Non-*Agaricus* varieties, especially shiitake and oyster, have slowly gained in popularity over the past decade. Shiitake is touted for lowering cholesterol levels and having antitumor and antiviral properties.

Fungi as Disease-Causing Organisms

Fungi cause diseases in both plants and animals, including humans.

Fungi and Plant Diseases

Fungal pathogens, which usually gain access to plants by way of the stomata or a wound, are a major concern for farmers. Serious crop losses occur each year due to fungal disease. As much as a third of the world's rice crop is destroyed each year by rice blast disease. Corn smut is a major problem in the midwestern United States. Various rusts attack grains, and leaf curl is a disease of fruit trees (**Fig. 18.30**).

The life cycle of rusts may be particularly complex, since it requires two different host species to complete the cycle. Black stem rust of wheat uses barberry bushes as an alternate host. Eradication of barberry bushes in areas where wheat is grown helps control this rust. Fungicides are regularly applied to crops to limit the negative effects of fungal pathogens. Wheat rust can also be controlled by producing new and resistant strains of wheat.

Video
Christmas Tree Threat

Fungi and Animal Diseases

As is well known, certain mushrooms are poisonous. The ergot fungus that grows on grain can result in ergotism when a person eats contaminated bread. Ergotism is characterized by hysteria, convulsions, and sometimes death.

Mycoses are diseases caused by fungi. Mycoses have three possible levels of invasion: Cutaneous mycoses affect only the epidermis; subcutaneous mycoses affect deeper skin layers; and systemic mycoses spread their effects throughout the body by traveling in the bloodstream. Fungal diseases that can be contracted from the environment include ringworm from soil, rose gardener's disease from thorns, Chicago disease from old buildings, and basketweaver's disease from grass cuttings. Opportunistic fungal infections now seen in AIDS patients stem from fungi that are always present in the body but take the opportunity to cause disease when the immune system becomes weakened.

Candida albicans causes the widest variety of fungal infections. Disease occurs when antibacterial treatments kill off the microflora community, allowing *Candida* to proliferate. Vaginal *Candida* infections are commonly called "yeast infections" in women. Oral thrush is a *Candida* infection of the mouth common in newborns and AIDS patients (**Fig. 18.31**a). In individuals with inadequate immune systems, *Candida* can move throughout the body, causing a systemic infection that can damage the heart, the brain, and other organs.

Connections and Misconceptions

Are all fungi edible?

No! In fact, very few species of fungi are edible. Each year, many people are poisoned and some die from eating poisonous mushrooms. One of the most common inedible mushrooms is the death cap, or *Amanita phalloides*. The poison in the death cap causes liver and kidney failure, and most people who ingest these fungi die within a few weeks. At the cellular level, the death cap's poison is a polypeptide that shuts down the process of transcription, causing cell death. The similar color and shape of the death cap to many edible mushrooms is believed to be the main reason that people mistakenly believe that it is edible.

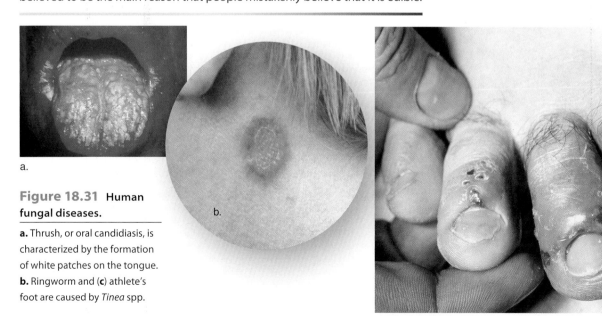

Figure 18.31 Human fungal diseases.

a. Thrush, or oral candidiasis, is characterized by the formation of white patches on the tongue. **b.** Ringworm and (**c**) athlete's foot are caused by *Tinea* spp.

Ringworm is a group of related diseases caused, for the most part, by fungi in the genus *Tinea*. Ringworm is a cutaneous infection that does not penetrate the skin. The fungal colony grows outward, forming a ring of inflammation. The center of the lesion begins to heal, giving the lesion its characteristic appearance, a red ring surrounding an area of healed skin (Fig. 18.31*b*). Athlete's foot is a form of tinea that affects the foot, mainly causing itching and peeling of the skin between the toes (Fig. 18.31*c*).

Batrachochytrium dendrobatidis is a parasitic chytrid that causes a cutaneous infection called chytridiomycosis in frogs around the world. The disease is thought to have originated in South Africa. It began to spread in the 1930s after African clawed frogs were captured and sold as pets and for use in medicine and research. Chytridiomycosis has recently decimated frog populations in Australia and Central and South America.

The vast majority of people living in the midwestern United States have been infected with *Histoplasma capsulatum*. This common soil fungus, often associated with bird droppings, leads in most cases to a mild "fungal flu." However, about 3,000 of these cases develop into a severe disease. About 50 persons die each year from histoplasmosis when the fungus grows within cells of the immune system. Lesions are formed in the lungs, leaving calcifications that are visible in X-ray images and resemble those of tuberculosis.

Because fungi are eukaryotes and more closely related to animals than to bacteria, it is hard to design an antibiotic against fungi that does not also harm animals. Thus, researchers exploit any biochemical differences they can discover between animals and fungi.

Connecting the Concepts

For more information on the fungi, refer to the following discussions.

Section 16.3 describes how fungi are related to plants and animals in domain Eukarya.

Section 29.3 describes how some species of fungi can cause sexually transmitted diseases (STDs).

Section 31.2 examines how fungi function as decomposers in ecosystems.

Check Your Progress 18.3

1. Contrast sexual reproduction in black bread mold with sexual reproduction in a mushroom.
2. Summarize how fungi contribute to ecological cycling.
3. List the organisms that make up a lichen, and relate how the relationship between the organisms is mutualistic.

Media Study Tools

www.mhhe.com/maderessentials3

Enhance your study of this chapter with study tools and practice tests. Also ask your instructor about the resources available through ConnectPlus, including the media-rich eBook, interactive learning tools, and animations.

The Chapter in Review

▓▓ Summary

18.1 Overview of the Land Plants

The Ancestry of Plants

Plants (kingdom Plantae) probably evolved from a multicellular, freshwater green alga about 500 MYA. The freshwater green algae known as charophytes appear to be the closest living relatives of land plants. Whereas algae are adapted to life in the water, plants are adapted to living on land. The charophytes have some characteristics that could be helpful on land. During the evolution of plants, five significant events are associated with adaptation to a land existence: evolution of (1) embryo protection, (2) vascular tissue, (3) megaphylls, (4) seeds, and (5) the flower.

Alternation of Generations

Land plants have an alternation of generations life cycle, in which each type of plant exists in two forms, the sporophyte (2n) and the gametophyte (n):

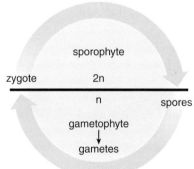

18.2 Diversity of Land Plants

Nonvascular Plants

Bryophytes, represented by the mosses, are plants with the following characteristics:

- There is no well-developed vascular tissue.
- The gametophyte is dominant, and flagellated sperm swim in external moisture to the egg.
- The sporophyte is dependent on the gametophyte.
- Windblown spores disperse the gametophyte.

Vascular Plants

In vascular plants, the dominant sporophyte has two kinds of well-defined conducting tissues. Xylem is specialized to conduct water and dissolved minerals, and phloem is specialized to conduct organic nutrients and

hormones. Certain vascular plants are seedless (lycophytes and ferns). They were large and abundant during the Carboniferous period.

Lycophytes have true roots, stems, and leaves because they contain vascular tissue. Vascular tissue allows lycophytes and other vascular plants to attain greater size due to more efficient transport of water, minerals, nutrients, and hormones. Lycophytes have narrow leaves called microphylls.

Ferns have large leaves with branching veins called megaphylls. Megaphylls enhance food production by photosynthesis. Ferns also have the following life cycle, which is typical of seedless vascular plants:

- The sporophyte generation is dominant and produces windblown spores.

sporophyte

- The gametophyte is separate and independent. Flagellated sperm swim in external moisture to the egg.

Seed plants have reproductive structures that are protected from drying out. Seed plants have male and female gametophytes. Gametophytes are reduced in size. The female gametophyte is retained within an ovule, and the male gametophyte is the mature pollen grain. The fertilized ovule becomes the seed, which contains a sporophyte embryo, food, and a seed coat.

- **Gymnosperms** are cone-bearing plants, represented by the pine tree. They have "naked seeds" because the seeds are not enclosed by fruit, as are those of flowering plants.
- **Angiosperms** are the flowering plants. Angiosperm reproductive organs are in the flower. Pollen is produced in pollen sacs inside an anther. Pollen is transported by wind or from flower to flower by birds, insects, or bats. Fertilized ovules in the ovary become seeds, and the ovary becomes the fruit. Thus, angiosperms have "covered seeds."

18.3 The Fungi

Kingdom Fungi includes unicellular yeasts, multicellular mushrooms and molds, and flagellated chytrids.

General Biology of a Fungus

Fungi are neither animals nor plants.

- The body of a typical fungus is composed of thin filaments of cells, called hyphae, that form a mass called a mycelium.
- The cell wall contains chitin.
- Most fungi produce windblown spores during both asexual and sexual reproduction.

Ecological Benefits of Fungi

- Fungi are saprotrophs that carry on external digestion. As decomposers, fungi keep ecological cycles going in the biosphere.
- Lichens (fungi plus cyanobacteria or green algae) are primary colonizers in poor soils or on rocks.
- Mycorrhizal fungi grow on or in plant roots and help the plant absorb minerals and water.

Economic Benefits of Fungi

- Fungi help produce foods and medicines, as well as serving as a source of food themselves.

Fungi as Disease-Causing Organisms

- Fungal pathogens of plants include blasts, smuts, and rusts that attack crops of great economic importance, such as rice and wheat.
- Animal diseases caused by fungi include thrush, ringworm, chytridiomycosis, and histoplasmosis.

Key Terms

alternation of generations 312	megaphylls 311
angiosperm 312	microphylls 311
anther 319	mycelium 324
bryophyte 312	mycorrhizal fungi 327
calyx 319	nonvascular plant 313
carpel 319	ovary 319
charophytes 310	ovule 317
club mosses 314	petal 319
cone 318	phloem 314
conifer 318	pollen grain 317
corolla 319	pollen tube 320
cotyledon 321	pollination 317
double fertilization 320	saprotroph 326
endosperm 320	seed 311
fern 315	sepal 319
filament 319	sori 315
flower 311	sporangium 313
frond 315	spore 312
fruit 321	sporophyte 312
fungi 323	stamen 319
gametophyte 312	stigma 319
gymnosperm 312	style 319
hyphae 324	vascular plant 314
lichen 326	vascular tissue 314
lignin 314	xylem 314
lycophyte 314	

Testing Yourself

Choose the best answer for each question.

1. Which of the following is not a plant adaptation to land?
   a. recirculation of water
   b. protection of embryo in maternal tissue
   c. development of flowers
   d. presence of vascular tissue
   e. seed production

2. Plant spores are
   a. haploid (n).
   b. produced by gametophytes.
   c. produced by sporophytes.
   d. diploid (2n).
   e. Both a and c are correct.

3. Charophytes
   a. are freshwater green algae.
   b. lack vascular tissue.
   c. are the closest living relatives of land plants.
   d. enclose their zygotes within protective structures.
   e. All of these are correct.

4. Which of the following is a true statement?
   a. People don't eat mushrooms because they might be poisonous.
   b. Penicillin is derived from a fungus.
   c. The alcohol from yeast fermentation makes bread rise.
   d. Fungi are prokaryotes like bacteria.
   e. Fungi ingest their food in the same way animals do.

5. Label the parts of the generalized plant life cycle in the following illustration.

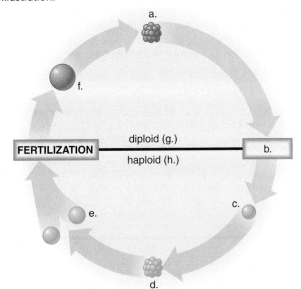

For questions 6–13, identify the group(s) to which each feature belongs. Each answer in the key may be used more than once. Each question may have more than one answer.

Key:

   a. mosses
   b. lycophytes
   c. ferns
   d. gymnosperms
   e. angiosperms

6. have megaphylls
7. exhibit alternation of generations
8. produce seeds
9. lack true roots, stems, and leaves
10. produce ovules that are not completely surrounded by sporophyte tissue
11. produce swimming sperm
12. produce flowers
13. protect the embryo as well as the zygote

14. Label the parts of the flower in the following illustration.

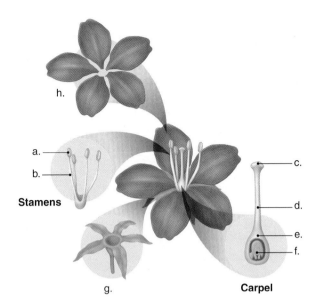

Stamens

Carpel

15. Lichens
   a. can live on bare rocks.
   b. can only survive in nutrient-rich soils.
   c. are parasitic on trees.
   d. can reproduce asexually.
   e. Both a and d are correct.

16. A fruit is derived from
   a. the corolla.   **c.** an ovule.
   b. an ovary.   **d.** the calyx.

17. Which of the following statements about fungi is false?
   a. Most fungi are multicellular.
   b. Fungal cell walls are composed of cellulose.
   c. Most fungi are nonmotile.
   d. Fungi digest their food before ingesting it.

18. Chytrids are
   a. a group of fungi with flagellated spores and gametes.
   b. responsible for a serious infection in frogs.
   c. a group of flowering plants that produce "naked seeds."
   d. the immature leaves of ferns.
   e. Both a and b are correct.

19. A fungal spore
   a. contains an embryonic organism.
   b. germinates directly into an organism.
   c. is often windblown.
   d. is most often diploid.
   e. Both b and c are correct.

20. Mycorrhizal fungi
   a. are a type of lichen.
   b. help plants gather solar energy.
   c. help plants gather inorganic nutrients.
   d. All of these are correct.

21. The gametophyte is the dominant generation in
   a. ferns.   **d.** angiosperms.
   b. mosses.   **e.** More than one of these are correct.
   c. gymnosperms.

22. A seed is a mature
   a. embryo.   **c.** ovary.
   b. ovule.   **d.** pollen grain.

23. Xylem is
   a. a polysaccharide found in the cell walls of all fungi.
   b. a plant vascular tissue that transports water and minerals.
   c. a plant vascular tissue that transports organic compounds.
   d. a polysaccharide found in the cell walls of all plants.
   e. the part of a flower that produces pollen.

24. Which of the following is a saprotroph?
   a. photosynthetic organism   **d.** decomposer
   b. pine tree   **e.** fern leaf
   c. nonvascular plant

25. A mycelium is
   a. a mass of fungal filaments.
   b. a type of fungus with flagellated spores and gametes.
   c. the main body of a typical fungus.
   d. a mutualistic association between a fungus and a green alga or cyanobacterium.
   e. Both a and c are correct.

Thinking Scientifically

1. Bare-root pine tree seedlings transplanted into open fields often grow very slowly. However, pine seedlings grow much more vigorously if they are dug from their native environment and then transplanted into a field, as long as some of the original soil is retained on the seedlings. Why is it so important to retain some native soil on the seedlings?

2. Evolutionary trees, such as this one, indicate that members of kingdoms Plantae and Fungi both had protist ancestors. All three groups of organisms belong to which domain? What characteristics would distinguish the protist ancestors of plants from those of fungi?

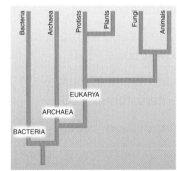

Bioethical Issue

Scientific Limitations

Bristlecone pines are among the oldest trees on Earth. In 1964, a graduate student working in the southwestern United States took core samples from several trees to determine their age. One tree was found to be over 4,000 years old. When the student's coring tool broke, the U.S. Forest Service gave him permission to cut down the tree in order to accurately determine its age. The tree was found to be 4,862 years old—the oldest known living creature on Earth.

   Did the student do anything wrong, scientifically or morally, considering that he was given permission to cut down the tree? Who should be responsible for protecting unique trees like the bristlecone pine?

19

Both Water and Land: Animals

BEFORE YOU BEGIN

Before beginning this chapter, take a few moments to review the following discussions.

Section 14.2 What is the difference between a homologous and an analogous structure?

Table 16.1 During what geological time frame did the first animals appear? The first land vertebrates?

Section 16.3 To what domain of life do the animals belong?

If It Looks Like a Duck and Lays Eggs Like a Duck, It May Be a Mammal

Finding relationships isn't all that easy. A duckbill platypus has webbed feet and a paddle-like tail for swimming. It uses its bill to find and collect food. Millions of tiny, jelly-filled pits lining the toothless bill are electroreceptors that detect tiny electrical charges given off by the platypus's prey. A platypus eats worms, crustaceans, fish, tadpoles, and adult frogs, which it grinds with two horn-like plates at the back of its bill. To reproduce, the duckbill platypus lays hard-shelled eggs.

Somewhat surprisingly, the duckbill platypus is a mammal, the same as a dog or cat. The two characteristics of a mammal are hair and mammary glands, and the duckbill platypus has both. Thick hair covers its body, except for its bill and feet. Female platypuses produce milk from mammary glands, but because they have no nipples, the young lick the milk from their fur.

It's clear to us that gulls are birds. After all, they fly in the air. But penguins don't fly. When on land, they walk around on webbed hind feet with a flipper hanging at each side. Penguins are excellent divers and spend most of their time at sea, feeding on fish, squid, and small crustaceans called krill. The Emperor penguin has been known to dive to a depth of more than 1,500 feet and stay down for over 18 minutes. Penguins' solid bones—not hollow, like those of other birds—help them stay under water.

So what makes a penguin a bird? Like the gull, the attractive coat of a penguin is made of feathers, and its flipper, built like a bird's wing, allows it to "fly" through the water. Penguins lay hard-shelled eggs and raise their chicks on land. A penguin might look like a mammal from a distance, but it is clearly a bird!

Throughout this chapter we will be exploring the evolutionary relationships of the major groups of animals.

19.1 Evolution of Animals

Learning Outcomes

Upon completion of this section, you should be able to

1. Explain how animals are distinguished from other groups of organisms.
2. Identify the key events in the evolution of the animals.
3. List the characteristics that distinguish protostomes from deuterostomes.

Animals, along with protists, plants, and fungi, are members of the domain Eukarya. Like the plants and fungi, animals are multicellular eukaryotes, but unlike plants, which make their food through photosynthesis, they are chemoheterotrophs and must acquire nutrients from an external source. So must fungi, but fungi digest their food externally and absorb the breakdown products. Animals ingest (eat) their food and digest it internally.

Animation
Three Domains

Animals usually carry on sexual reproduction and typically begin life as a fertilized diploid egg. From this starting point, they undergo a series of developmental stages to produce an organism that has specialized tissues, usually within organs that carry on specific functions. Two types of tissues in particular—muscles and nerves—characterize animals. The presence of these tissues allows an animal to exhibit motility and a variety of flexible movements. The evolution of these tissues enables many types of animals to search actively for their food and to prey on other organisms. Coordinated movements also allow animals to seek mates, shelter, and a suitable climate—behaviors that have resulted in the vast diversity of animals. The more than 30 animal phyla we recognize today have all evolved from a single ancestor.

Figure 19.1 illustrates the development of an animal using the frog as an example. A frog goes through a number of embryonic stages to become a larval

Figure 19.1 Animals.

Most animals begin life as a fertilized egg. The egg undergoes development to produce a multicellular organism that has specialized tissues. Animals depend on a source of external food to carry on life's processes. This series of images shows the development and metamorphosis of the frog, a complex animal.

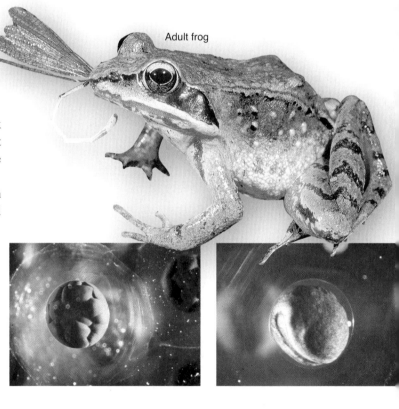

Adult frog

Stages in development, from zygote to embryo.

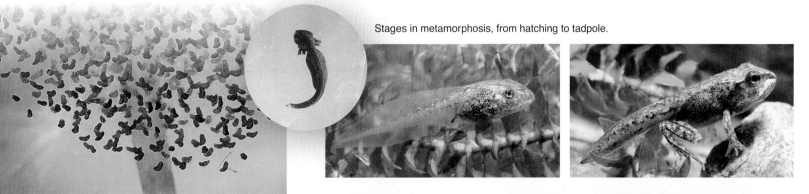

Stages in metamorphosis, from hatching to tadpole.

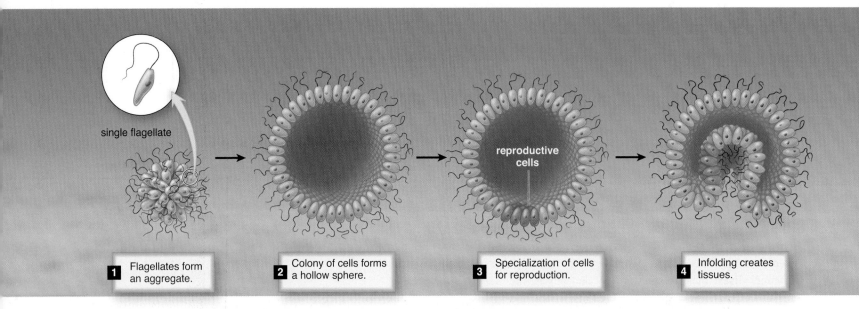

1. Flagellates form an aggregate.
2. Colony of cells forms a hollow sphere.
3. Specialization of cells for reproduction.
4. Infolding creates tissues.

single flagellate

reproductive cells

Figure 19.2 **The colonial flagellate hypothesis.**

The hypothesis explains how a colony of protistans may have formed some of the specialized structures characteristic of the first animals.

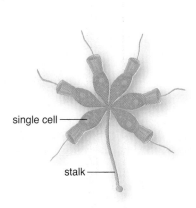

single cell

stalk

Figure 19.3 **Choanoflagellates.**

The choanoflagellates are the living protozoans most closely related to animals and may resemble their last unicellular ancestor. Some choanoflagellates live in colonies like these, in which a group of cells is attached to a surface by means of a stalk.

form (the tadpole) with specialized organs, including muscular and nervous systems that enable it to swim. A larva is an immature stage that typically lives in a different habitat and feeds on different foods than the adult. By means of a change in body form called *metamorphosis*, the larva, which only swims, turns into a sexually mature adult frog that swims and hops. The aquatic tadpole lives on plankton, and the terrestrial adult typically feeds on insects and worms. A large African bullfrog will try to eat just about anything, including other frogs, as well as small fish, reptiles, and mammals.

Video
Tadpole
Development

Ancestry of Animals

In Chapter 18, we discussed evidence that plants most likely share a green algal ancestor with the charophytes. Most scientists agree that animals also evolved from a protist, most likely a protozoan. The *colonial flagellate hypothesis* states that animals are descended from an ancestor that resembled a hollow, spherical colony of flagellated cells. **Figure 19.2** shows how the process would have begun with an aggregate of a few flagellated cells. From there, a larger number of cells could have formed a hollow sphere. Individual cells within the colony would have become specialized for particular functions, such as reproduction. Two tissue layers could have arisen by an infolding of certain cells into a hollow sphere. Tissue layers do arise in this manner during the development of animals today. The colonial flagellate hypothesis is also attractive because it implies that **radial symmetry** preceded **bilateral symmetry,** as the evolutionary tree of animals predicts (see Fig. 19.4).

Among the protists, choanoflagellates (collared flagellates) most likely resemble the last unicellular ancestor of animals, and molecular data tell us that they are the closest living protist relative of animals. A choanoflagellate is a single cell, 3–10 μm in diameter, with a flagellum surrounded by a collar of 30–40 microvilli (**Fig. 19.3**).

The Evolutionary Tree of Animals

Because so many types of animals arose during the Cambrian period in such a short time (see Table 16.1), historically it was almost impossible to trace the evolutionary history of animals with absolute certainty. However, systemicists were able to group animals based on major evolutionary trends (**Fig. 19.4**). Within the past decade, molecular analyses and studies of the developmental stages of the

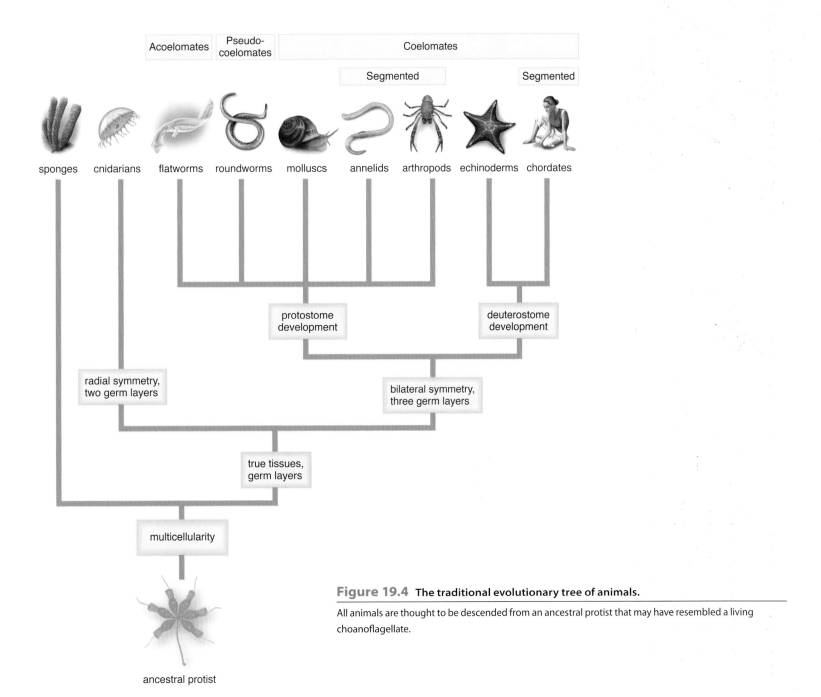

Figure 19.4   The traditional evolutionary tree of animals.

All animals are thought to be descended from an ancestral protist that may have resembled a living choanoflagellate.

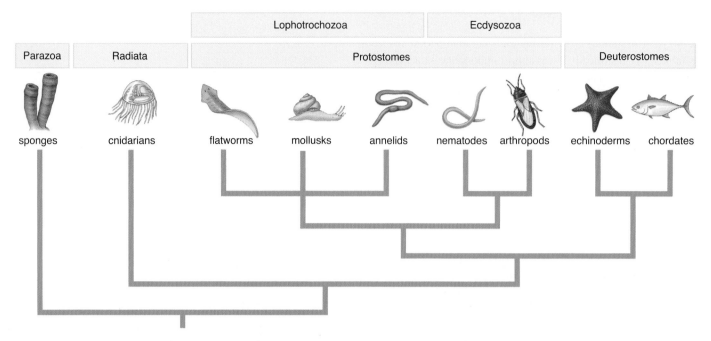

Figure 19.5 A modern look at the animal evolutionary tree.

The relationships in this diagram are based on research into the developmental biology of each groups, as well as molecular studies of DNA, RNA, and protein similarity.

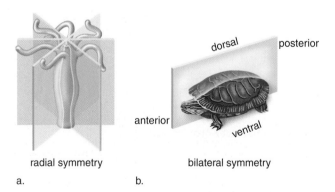

Figure 19.6 Radial versus bilateral symmetry.

a. With radial symmetry, two mirror images are obtained no matter how the animal is sliced longitudinally. Radially symmetrical animals tend to stay in one place and reach out in all directions to get their food. **b.** With bilateral symmetry, mirror images are obtained only if the animal is sliced down the middle. Bilaterally symmetrical animals tend to actively go after their food.

animals have challenged the traditional relationships. Based on this information, scientists have begun to rethink the evolutionary relationships between the major groups of animals (**Fig. 19.5**). We will use the relationships illustrated in Figure 19.5 as the basis of our exploration of the animals.

Evolutionary Trends

Animals differ in biological organization. Sponges, like all animals, are *multicellular*, but they have no true tissues and therefore have the cellular level of organization. Because of the characteristics, sponges are classifed as parazoans, which means "beside the animals," which refers to their position on the evolutionary tree alongside the "true animals" or eumetazoans. Cnidarians, such as *Hydra*, are eumetazoans. They have true tissues, which are formed from two germ layers when they are embryos. All other eumetazoans have three germ layers as embryos. These are called the ectoderm, endoderm, and mesoderm. **Germ layers** are so called because they give rise to all other tissues and organs in an animal's body.

Animals differ in symmetry. Many sponges have no particular symmetry and are therefore asymmetrical. *Radial symmetry*, as seen in cnidarians, means that the animal is organized circularly, similar to a wheel. No matter where the animal is sliced longitudinally, two mirror images are obtained (**Fig. 19.6***a*). *Bilateral symmetry*, as seen in flatworms, means that the animal has definite right and left halves; only a longitudinal cut down the center of the animal will produce mirror images (Fig. 19.6*b*). During the evolution of animals, the trend toward bilateral symmetry is accompanied by **cephalization,** localization of a brain and specialized sensory organs at the anterior end of an animal. Therefore, bilateral symmetry was an important precursor for the evolutionary trend of increasing complexity of the nervous system and sensory organs. The appearance of bilateral symmetry marked the evolution of animals that were able to exploit environmental resources in new ways through their more active lifestyles.

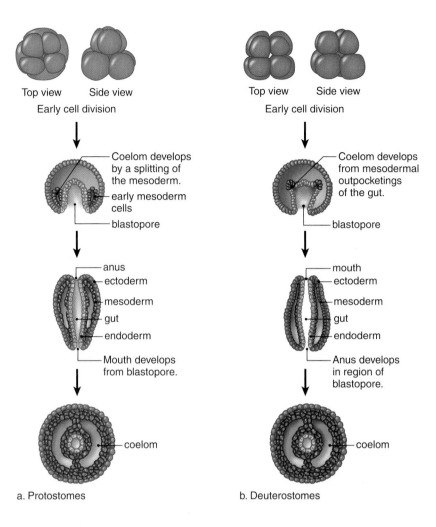

Top view   Side view
Early cell division

Top view   Side view
Early cell division

Coelom develops
by a splitting of
the mesoderm.
early mesoderm
cells
blastopore

Coelom develops
from mesodermal
outpocketings
of the gut.
blastopore

anus
ectoderm
mesoderm
gut
endoderm
Mouth develops
from blastopore.

mouth
ectoderm
mesoderm
gut
endoderm
Anus develops
in region of
blastopore.

coelom

coelom

a. Protostomes

b. Deuterostomes

Figure 19.7 Development in protostomes and deuterostomes.

a. Protostomes are characterized by spiral cell division when the embryo first forms and a blastopore that becomes the mouth. Further, if a coelom is present, it forms by a splitting of the mesoderm. **b.** Deuterostomes are characterized by radial cell division when the embryo first forms and a blastopore that becomes the anus. Also, the primitive gut outpockets to form the coelom.

When using molecular data, it is assumed that, the more closely related two organisms are, the more nucleotide sequences they will have in common. Molecular data and developmental data suggest that the groups designated as **protostomes** are more closely related to each other than they are to the **deuterostomes.** Protostomes and deuterostomes can be distinguished on the basis of embryological development. Notice that cell division to form an embryo is different in protostomes (flatworms, roundworms, molluscs, annelids, and arthropods) than in deuterostomes (echinoderms and chordates). In protostomes and in deuterostomes, the first embryonic opening is called the blastopore (**Fig. 19.7**). In protostomes, the blastopore becomes the mouth, and in deuterostomes, it becomes the anus. Actually, this is the feature that explains their names; the mouth is the first opening in protostomes (*proto*, before) and the second opening in deuterostomes (*deutero*, second). The protostomes are further divided into two subgroups, the lophotrochozoa and the ecdyozoa (see Fig. 19.5). The difference between the two is based on how they grow. Lophotrochozoans (flatworms, molluscs, and annelids) grow by adding additional mass to their existing body, while the ecdysozoans (nematodes and arthropods) grow by molting. The protostome group also differs with regard to the presence of a **coelom,** or body cavity. Some protostomes have no body cavity and are **acoelomates,** as are flatworms (**Fig. 19.8**a). Acoelomates are packed solid with mesoderm. In contrast, a body cavity provides a space for the various internal organs. Roundworms are **pseudocoelomates,** and their body cavity is incompletely lined by mesoderm—that is, a layer of mesoderm exists beneath the body wall but not around the gut (Fig. 19.8b). The other protostomes

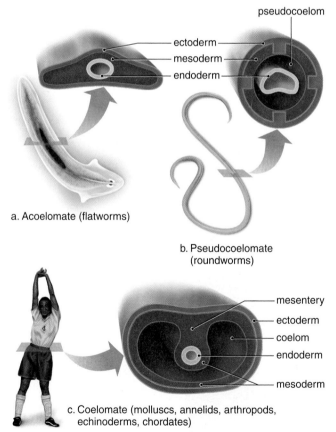

pseudocoelom
ectoderm
mesoderm
endoderm

a. Acoelomate (flatworms)

b. Pseudocoelomate
(roundworms)

mesentery
ectoderm
coelom
endoderm
mesoderm

c. Coelomate (molluscs, annelids, arthropods,
echinoderms, chordates)

Figure 19.8 Type of body cavity.

a. Flatworms don't have a body cavity; they are acoelomates, and mesoderm fills the space between ectoderm and endoderm. **b.** Roundworms are pseudocoelomates; they have a body cavity, and mesoderm lies next to the ectoderm but not the endoderm. **c.** Other animals are coelomates, and mesoderm lines the entire body cavity.

Connecting the Concepts

For additional information on animal evolution, refer to the following discussions.

Section 1.2 compares the general characteristics of animals with that of the other groups of living oganisms.

Table 16.1 places the major events in the evolution of the animals in the context of the geological time scale.

Section 29.4 details the events that occur during human embryonic development.

Check Your Progress 19.1

1 Summarize the key events in the evolution of the animals.

2 Compare and contrast the differences in cell divisions to form an embryo in protostomes and deuterostomes.

3 Predict what might have happened to the evolutionary tree of animals if bilateral symmetry did not exist.

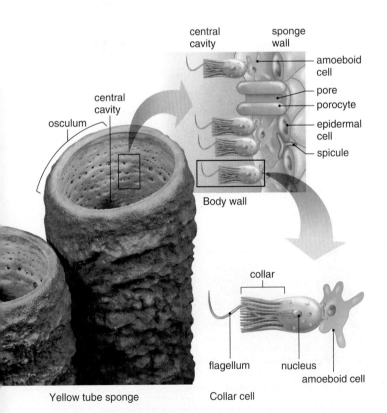

Figure 19.9 Sponge anatomy.

Water enters a sponge through pores and circulates past collar cells before exiting at the mouth, or osculum. Collar cells digest small particles that become trapped by their appendages, flagella in the microvilli of their collar. Amoeboid cells transport nutrients from cell to cell.

and the deuterostomes are coelomates in which the body cavity is completely lined with mesoderm (Fig. 19.8c). In protostomes, the coelom develops by a splitting of the mesoderm, while in dueterostomes the coelum develops as outpockets of the primitive gut (see Fig. 19.7). In animals with a coelom, mesentery, which is composed of strings of mesoderm, supports the internal organs. In coelomate animals, such as earthworms, lobsters, and humans, the mesoderm can interact not only with the ectoderm but also with the endoderm. Therefore, body movements are freer because the outer wall can move independently of the organs, and the organs have the space to become more complex. In animals without a skeleton, a coelom even acts as a so-called hydrostatic skeleton.

Animals can be nonsegmented or *segmented*. **Segmentation** is the repetition of body parts along the length of the body. Annelids, such as earthworms; arthropods, such as lobsters; and chordates, such as ourselves, are segmented. To illustrate your segmentation, run your hand along your backbone, which is composed of a series of vertebrae. Segmentation leads to specialization of parts because the various segments can become differentiated for specific purposes. In the case of your backbone, the two small vertebrae just beneath your skull are specialized to permit you to move your head up and down and side to side. The vertebrae of your lower back are much larger and sturdier, as they support the weight of your upper body.

19.2 Sponges and Cnidarians: The Early Animals

Learning Outcomes

Upon completion of this section, you should be able to

1. Describe the evolutionary trends among sponges and cnidarians.
2. Distinguish between a parazoan and an eumetazoan and give an example of each.

This section describes the first two groups of **invertebrate** animals—the sponges and cnidarians. The term *invertebrate* refers to any group of animals that lack a backbone. While you may be more familiar with the **vertebrate** animals, those with a backbone, almost 95% of all species on the planet are invertebrates.

Sponges: Multicellularity

Sponges (phylum Porifera) have sac-like bodies perforated by many pores (**Fig. 19.9**). Sponges are aquatic, largely marine animals that vary greatly in size, shape, and color. Sponges are multicellular but lack organized tissues. Therefore, *sponges have the cellular level of organization*. Molecular data place sponges as Parazoans, which are located at the base of the evolutionary tree of animals.

The body wall of a sponge is lined internally with flagellated cells called *collar cells*, or choanocytes (Fig. 19.9). The beating of the flagella produces water currents that flow through the pores into the central cavity and out through the osculum, the upper opening of the body. Even a simple sponge only 10 centimeters tall is estimated to filter as much as 100 liters of water each day. It takes this much water to supply the needs of the sponge. A sponge is a sedentary **filter feeder,** an organism that filters its food from the water by means of a straining device—in this case, the pores of the walls and the microvilli making up the collar of collar cells. Microscopic food particles that pass between the microvilli are engulfed by the collar cells and digested by them in food vacuoles.

Sponges can reproduce both asexually and sexually. They reproduce asexually by fragmentation or by *budding*. During budding, a small protuberance appears and gradually increases in size until a complete organism forms. Budding produces colonies of sponges that can become quite large. During sexual reproduction, eggs and sperm are released into the central cavity, and the zygote develops into a flagellated larva that may swim to a new location. If the cells of a sponge are mechanically separated, they will reassemble into a complete and functioning organism! Like many less specialized organisms, sponges are also capable of regeneration, or growth of a whole from a small part.

Some sponges have an endoskeleton composed of *spicules*, small, needle-shaped structures with one to six rays. Most sponges have fibers of spongin, a modified form of collagen; a bath sponge is the dried spongin skeleton from which all living tissue has been removed. Today, however, commercial "sponges" are usually synthetic.

Cnidarians: True Tissues

Cnidarians (phylum Cnidaria) are an ancient group of invertebrates with a rich fossil record. Cnidarians are radially symmetrical and capture their prey with a ring of tentacles that bear specialized stinging cells, called cnidocytes (**Fig. 19.10**). Each cnidocyte has a capsule called a **nematocyst,** containing a long, spirally coiled, hollow thread. When the trigger of the cnidocyte is touched, the nematocyst is discharged. Some nematocysts merely trap a prey or predator; others have spines that penetrate and inject paralyzing toxins before the prey is captured and drawn into the gastrovascular cavity. Most cnidarians live in the sea, though there are a few freshwater species.

 Video Portuguese Man-of-War

During development, cnidarians have two germ layers (ectoderm and endoderm), and as adults *cnidarians have the tissue level of organization.* Therefore, cnidarians are classified as the first of the eumetazoans. Two basic body forms are seen among cnidarians—the polyp and the medusa. The mouth of a polyp is directed upward from the substrate, while the mouth of the medusa is directed downward. A medusa has much jelly-like packing material and is commonly called a "jellyfish." Polyps are tubular and generally attached to a rock (**Fig. 19.11**).

Cnidarians, as well as other marine animals, have been the source of medicines, particularly drugs that counter inflammation.

Connecting the Concepts

For more information on sponges and cnidarians, refer to the following discussions.

Table 16.1 describes the geological time frame in which these animals first appeared.

Section 31.2 examines the importance of the coral reefs.

Table 16.1 describes the geological time frame in which these animals first appeared.

Section 31.2 examines the importance of the coral reefs.

Check Your Progress 19.2

1. List the key evolutionary trends seen in sponges and cnidarians.
2. Compare and contrast the body style and development of sponges and cnidarians.
3. Discuss why sponges are capable of regeneration but cnidarians are not.

a. Hydra

Figure 19.10 **Cnidarians.**

Hydras (**a**) and sea anemones (**b**) are solitary polyps that use tentacles laden with stinging cells to capture their food.

b. Sea anemone

a. Portuguese man-of-war   b. Cup coral

Figure 19.11 **More cnidarians.**

a. The Portuguese man-of-war is a colony of polyp and medusa types of individuals. One polyp becomes a gas-filled float, and the other polyps are specialized for feeding. **b.** The calcium carbonate skeletons of corals form the coral reefs.

Connections and Misconceptions

What is causing the loss of the coral reefs?

Coral reefs are widely recognized as biodiversity hotspots—areas where large numbers of species can be found. However, over the past few decades, scientists estimate that over 25% of the world's coral reefs have been lost, and another 33% are in danger. What is causing this loss? Destructive fishing practices, pollution and sediment from poor coastal land management, and the removal of coastal mangrove forests are all contributing factors. In addition, the elevation of ocean temperatures associated with global climate change are causing coral reefs to die, a phenomenon called "coral bleaching."

Video Coral Reef Ecosystems

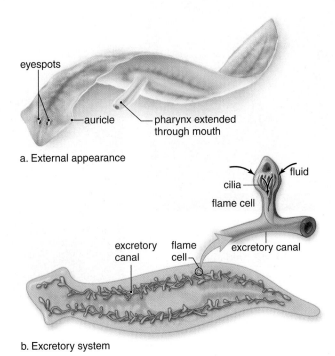

a. External appearance

b. Excretory system

c. Nervous system

d. Reproductive and digestive systems

Figure 19.12 **Anatomy of a planarian.**

a. This drawing shows that flatworms are bilaterally symmetrical and have a head region with eyespots. **b.** The excretory system with flame cells is shown in detail. **c.** The nervous system has a ladder-like appearance. **d.** The reproductive system (shown in brown) has both male and female organs, and the digestive system (shown in pink) has a single opening. When the pharynx is extended, as shown in (a), a planarian sucks food up into a gastrovascular cavity, which branches throughout its body.

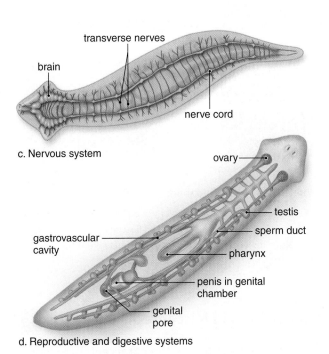

19.3 Flatworms, Molluscs, and Annelids: The Lophotrochozoans

Learning Outcomes

Upon completion of this section, you should be able to

1. List the distinguishing characteristics of a lophotrochozoan.
2. Distinguish among the flatworms, molluscs, and annelids based on body plan.

Flatworms, molluscs, and annelids all belong to the group of protostome animals called lophotrochozoans. The *lopho* portion of their name is derived from a tentacle-like feeding structure called a lophophore. The *trocho* portion of the name refers to a larval stage, called a trochophore, that is characterized by a distinct band of cilia. While not all members of this group have both a lophophore and a trochophore larval stage, molecular analyses have supported the hypothesis that all members share a common evolutionary ancestor. Lophotrochozoans are also distinct from other protostomes, such as the arthropods and nematodes, in that they increase their body mass gradually without molting.

Flatworms: Bilateral Symmetry

Flatworms (phylum Platyhelminthes) have *bilateral symmetry*. Like all the other animal phyla we will study, they also have three germ layers. However, the flatworms have no coelom—they are acoelomates. The presence of mesoderm in addition to ectoderm and endoderm gives bulk to the animal and leads to greater complexity.

Free-living flatworms, called **planarians,** have several body systems (**Fig. 19.12**), including a ladder-like nervous system. A small anterior brain and two lateral nerve cords are joined by cross-branches called transverse nerves. Planarians exhibit cephalization; aside from a brain, the "head" end has light-sensitive organs (the eyespots) and chemosensitive organs located on the auricles. Their three muscle layers—an outer circular layer, an inner longitudinal layer, and a diagonal layer—allow for varied movement. Their ciliated lower epidermis allows planarians to glide along a film of mucus.

The animal captures food by wrapping itself around the prey, entangling it in slime, and pinning it down. Then the planarian extends a muscular pharynx and, by a sucking motion, tears up and swallows its food. The pharynx leads into a three-branched gastrovascular cavity, where digestion occurs. The digestive tract is incomplete because it has only one opening.

Planarians are **hermaphrodites,** meaning that they possess both male and female sex organs. The worms practice cross-fertilization: The penis of one is inserted into the genital pore of the other, and a reciprocal transfer of sperm takes place. The fertilized eggs hatch in two to three weeks as tiny worms.

The parasitic flatworms belong to two classes: the tapeworms and the flukes. As adults, tapeworms are endoparasites (internal parasites) of various vertebrates, including humans (**Fig. 19.13***a*). They vary in size from a few millimeters to nearly 20 meters. Tapeworms have a well-developed anterior region called the scolex, which bears hooks and suckers for attachment to the intestinal wall of the host. Behind the scolex, a series of reproductive units called proglottids contains a full set of female and male sex organs. After fertilization, the organs within a proglottid disintegrate, and it becomes filled with mature eggs. The eggs, or the mature proglottids, are eliminated in the feces of the host.

Flukes are all endoparasites of various vertebrates. The anterior end of these animals has an oral sucker and at least one other sucker used for attachment to the host. Flukes are usually named for the organ they inhabit; for example, there are blood, liver, and lung flukes. Blood flukes (*Schistosoma* spp.) occur predominantly in the Middle East, Asia, and Africa. Nearly 800,000 persons die each year from blood fluke infection, called schistosomiasis. Adults are small (approximately 2.5 centimeters long) and may live for years in their human hosts (Fig. 19.13*b*).

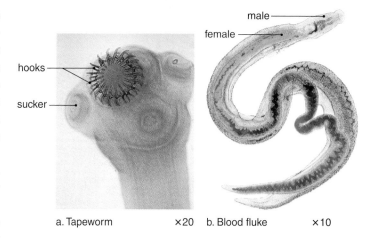

a. Tapeworm   ×20   b. Blood fluke   ×10

Figure 19.13 **Parasitic flatworms.**

a. The anterior end of a tapeworm has hooks and suckers for attachment to the intestinal wall. **b.** Sexes are separate in blood flukes, which cause schistosomiasis.

Molluscs

Molluscs (phylum Mollusca) are coelomate organisms with a complete digestive tract. Despite being a very large and diversified group, all molluscs have a body composed of at least three distinct parts: (1) The *visceral mass* is the soft-bodied portion that contains internal organs; (2) the *foot* is the strong, muscular portion used for locomotion; and (3) the *mantle* is a membranous or sometimes muscular covering that envelops, but does not completely enclose, the visceral mass. In addition, the *mantle cavity* is the space between the two folds of the mantle. The mantle may secrete an exoskeleton called a *shell*. Another feature often present is a rasping, tongue-like *radula*, an organ that bears many rows of teeth and is used to obtain food (**Fig. 19.14**).

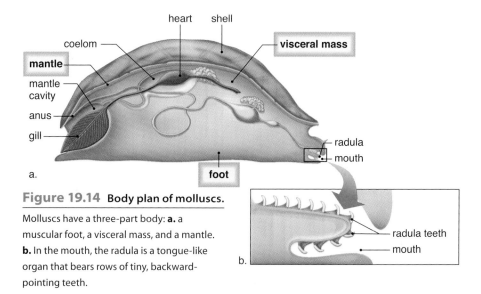

Figure 19.14 **Body plan of molluscs.**

Molluscs have a three-part body: **a.** a muscular foot, a visceral mass, and a mantle. **b.** In the mouth, the radula is a tongue-like organ that bears rows of tiny, backward-pointing teeth.

Land snail

Nudibranch

Octopus

Nautilus

Scallop

Mussels

Figure 19.15 Molluscan diversity.

Top: Snails (left) and nudibranchs (right) are gastropods. Middle: Octopuses (left) and nautiluses (right) are cephalopods. Bottom: Scallops (left) and mussels (right) are bivalves.

In **gastropods** (meaning stomach-footed), including nudibranchs, conchs, and snails, the foot is ventrally flattened, and the animal moves by muscle contractions that pass along the foot (**Fig. 19.15**). Many gastropods are herbivores that use their radulas to scrape food from surfaces. Others are carnivores, using their radulas to bore through surfaces, such as bivalve shells, to obtain food. In snails that are terrestrial, the mantle is richly supplied with blood vessels and functions as a lung.

In **cephalopods** (meaning head-footed), including octopuses, squids, and nautiluses, the foot has evolved into tentacles about the head (Fig. 19.15). The tentacles seize prey, and then a powerful beak and a radula tear it apart. Cephalopods possess well-developed nervous systems and complex sensory organs. The large brain is formed from a fusion of ganglia, and nerves leaving the brain supply various parts of the body. An especially large pair of nerves controls the rapid contraction of the mantle, allowing these animals to move quickly by jet propulsion of water. Rapid movement and the secretion of a brown or black pigment from an ink gland help cephalopods escape their enemies. In the squid and octopus, a well-developed eye resembles that of vertebrates, having a cornea, a lens, a retina, and an iris. Octopuses have no shell, and squids have only a remnant of one concealed beneath the skin. Scientific experiments reveal that octopuses, in particular, are highly intelligent.

Clams, oysters, scallops, and mussels are called **bivalves** because of the two parts to their shells (Fig. 19.15). They have a muscular foot that projects ventrally from the shell. In a clam, such as the freshwater clam, the calcium carbonate shell has an inner layer of mother-of-pearl. The clam is a filter feeder. Food particles and water enter the mantle cavity by way of the incurrent siphon, a posterior opening between the two valves. Mucous secretions cause smaller particles to adhere to the gills, and ciliary action sweeps them toward the mouth.

Molluscs have some economic importance as a source of food and pearls. If a foreign body is placed between the mantle and the shell of a clam, concentric layers of shell are deposited about the particle to form a pearl.

Annelids: Segmented Worms

Annelids (phylum Annelida) are *segmented*, as can be seen externally by the rings that encircle the body of an earthworm. Partitions called septa divide the well-developed, fluid-filled coelom, which is used as a hydrostatic skeleton to facilitate movement. In annelids, the complete digestive tract body plan has led to the specialization of parts (**Fig. 19.16**). For example, the digestive system may include pharynx, esophagus, crop, gizzard, intestine, and accessory glands. Annelids have an extensive closed circulatory system with blood vessels that run the length of the body and branch to every segment. The nervous system consists of a brain connected to a ventral nerve cord, with ganglia in each segment. The excretory system consists of nephridia in most segments. A **nephridium** is a tubule that collects waste material and excretes it through an opening in the body wall.

Most annelids are polychaetes (having many setae per segment) that live in marine environments. Setae are bristles that anchor the worm or help it move. A clam worm is a predator. It preys on crustaceans and other small animals, capturing them with a pair of strong, chitinous jaws that extend with a part of the pharynx (**Fig. 19.17***a*). In support of its predatory way of life, a clam worm has a well-defined head region, with eyes and other sense organs. Other polychaetes are sedentary (sessile) tube worms, with tentacles that form a funnel-shaped fan. Water currents created by the action of cilia trap food particles that are directed toward the mouth (Fig. 19.17*b*).

The oligochaetes (few setae per segment) include the earthworms (see Fig. 19.16). Earthworms do not have a well-developed head, and they reside in soil where there is adequate moisture to keep the body wall moist for gas exchange. They are scavengers that feed on leaves and any other organic matter, living or dead, that can conveniently be taken into the mouth along with dirt.

Leeches have no setae but have the same body plan as other annelids. Most are found in fresh water, but some are marine or even terrestrial. The medicinal leech can be as long as 20 centimeters, but most leeches are much shorter. A leech has two *suckers*, a small one around the mouth and a large posterior one. While some leeches are free-living, most are fluid feeders that penetrate the surface of an animal by using a proboscis or their jaws, and then suck in fluids with their powerful pharynx. Leeches are able to keep blood flowing and prevent clotting by means of a substance in their saliva known as hirudin, a powerful anticoagulant. This secretion has added to their potential usefulness in the field of medicine today (Fig. 19.17*c*).

Connecting the Concepts

For more information on the material presented in this section, refer to the following discussions.

Table 16.1 describes the geological time frame in which the annelids, molluscs, and flatworms first appeared.

Section 32.2 outlines the direct and indirect benefits that invertebrate biodiversity has on humans.

Check Your Progress 19.3

1. Explain why flatworms, molluscs, and annelids are all classified as lophotrochozoans.
2. Contrast the body styles of a typical flatworm and annelid.
3. Summarize the differences between gastropods, cephalopods, and bivalves.

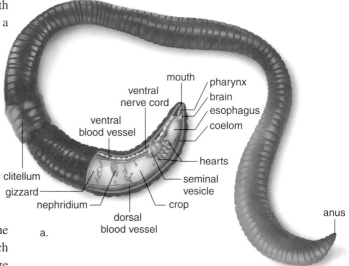

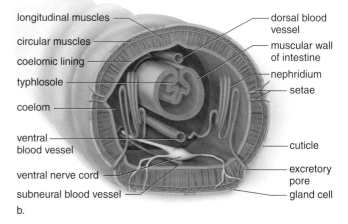

Figure 19.16   Earthworm anatomy.

a. Internal anatomy of the anterior part of an earthworm. **b.** Cross section of an earthworm.

Figure 19.17   Other annelids.

A polychaete can be predacious, like this marine clam worm (**a**), or live in a tube, like this fan worm called the Christmas tree worm (**b**). **c.** The medicinal leech, also an annelid, is sometimes used to remove blood from tissues after surgery.

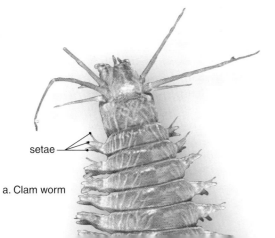

a. Clam worm

b. Christmas tree worm

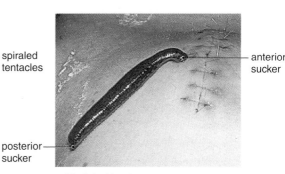

c. Medicinal leech

19.4 Roundworms and Arthropods: The Ecdysozoans

Learning Outcomes

Upon completion of this section, you should be able to

1. List the characteristics that classify an animal as a ecdysozoan.
2. Distinguish between the roundworms and arthropods with regard to their body plan.
3. Explain the success of the arthropods.

Like the lophotrochozoans in the previous section, the ecdysozoans are invertebrate protostomes. The name ecdysozoan is derived from the fact that these organisms all secrete a nonliving exoskeleton (or cuticle) that must be shed in order for the organism to grow. This process is called molting, or *ecdysis*. The ecdysozoans consist of the roundworms, or Nematodes, and the highly-successful Arthropods.

Roundworms: Pseudocoelomates

The **roundworms** (phylum Nematoda) are nonsegmented, meaning that they have a smooth outside body wall. They possess a pseudocoelom that is incompletely lined with mesoderm (see Fig. 19.8). The fluid-filled pseudocoelom supports muscle contraction and enhances flexibility. The digestive tract is complete because it has both a mouth and an anus (**Fig. 19.18**). Roundworms are generally colorless and less than 5 centimeters long, and they occur almost everywhere—in the sea, in fresh water, and in the soil—in such numbers that thousands of them can be found in a small area. Many are free-living and feed on algae, fungi, microscopic animals, dead organisms, and plant juices, causing great agricultural damage. Parasitic roundworms live anaerobically in every type of animal and many plants. Several parasitic roundworms infect humans.

A female *Ascaris lumbricoides,* a human parasite, is very prolific, producing over 200,000 eggs daily. The eggs are eliminated with host feces, and under the right conditions they can develop into a worm within two weeks. The eggs enter the body via uncooked vegetables, soiled fingers, or ingested fecal material and hatch in the intestines. The juveniles make their way into the cardiovascular

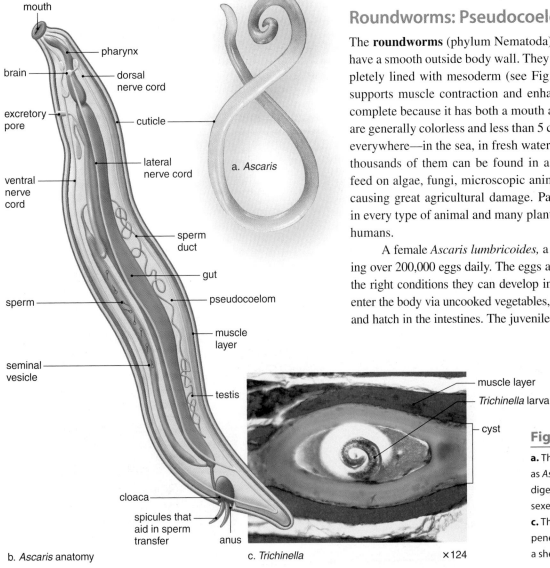

Figure 19.18 Roundworm anatomy.

a. The roundworm *Ascaris.* **b.** Roundworms, such as *Ascaris,* have a pseudocoelom and a complete digestive tract with a mouth and an anus. The sexes are separate; this is a male roundworm. **c.** The larvae of the roundworm *Trichinella* penetrate striated muscle fibers, where they coil in a sheath formed from the muscle fiber.

a. *Ascaris*

b. *Ascaris* anatomy

c. *Trichinella* ×124

system and are carried to the heart and lungs. From the lungs, the larvae travel up the trachea, where they are swallowed and eventually reach the intestines. There the larvae mature and begin feeding on intestinal contents.

Trichinosis is a fairly serious human infection rarely seen in the United States. The female trichina worm burrows into the wall of the host's small intestine; there she deposits live larvae, which are carried by the bloodstream to the skeletal muscles (Fig. 19.18*b*), where they encyst. Once the adults are in the small intestine, digestive disorders, fatigue, and fever occur. After the larvae encyst, the symptoms include aching joints, muscle pain, and itchy skin. Humans catch the disease when they eat infected meat.

Elephantiasis is caused by a roundworm called a filarial worm, which utilizes mosquitoes as a secondary host. Because the adult worms reside in lymphatic vessels, fluid return is impeded and the limbs of an infected human can swell to an enormous size, even resembling those of an elephant. When a mosquito bites an infected person, it can transport larvae to a new host.

Other roundworm infections are more common in the United States. Children frequently acquire pinworm infection, and hookworm is seen in the southern states, as well as worldwide. A hookworm infection can be very debilitating because the worms attach to the intestinal wall and feed on blood. Good hygiene, proper disposal of sewage, thorough cooking of meat, and regular deworming of pets usually protect people from parasitic roundworms.

Arthropods: Jointed Appendages

Arthropods (phylum Arthropoda) are extremely diverse. Over 1 million species have been discovered and described, but some experts suggest that as many as 30 million arthropod species may exist—most of them insects. The success of arthropods can be attributed to the following six characteristics:

1. *Jointed appendages.* Basically hollow tubes moved by muscles, jointed appendages have become adapted to different means of locomotion, food gathering, and reproduction (**Fig. 19.19**). These modifications account for much of the diversity of arthropods.

2. *Exoskeleton.* A rigid but jointed exoskeleton is composed primarily of **chitin,** a strong, flexible, nitrogenous polysaccharide. The exoskeleton serves many functions, including protection, prevention of desiccation, attachment for muscles, and locomotion. Because an exoskeleton is hard and nonexpandable, arthropods must undergo **molting,** or shedding of the exoskeleton, as they grow larger.

3. *Segmentation.* In many species, the repeating units of the body are called segments. Each has a pair of jointed appendages. In others, the segments are fused into a head, a thorax, and an abdomen.

4. *Well-developed nervous system.* Arthropods have a brain and a ventral nerve cord. The head bears various types of sense organs, including compound and simple eyes. Many arthropods also have well-developed touch, smell, taste, balance, and hearing capabilities. Arthropods display many complex behaviors and communication skills.

5. *Variety of respiratory organs.* Marine forms utilize gills; terrestrial forms have book lungs (e.g., spiders) or air tubes called tracheae. Tracheae serve as a rapid way to transport oxygen directly to the cells. The circulatory system is open, with the dorsal heart pumping blood into various sinuses throughout the body.

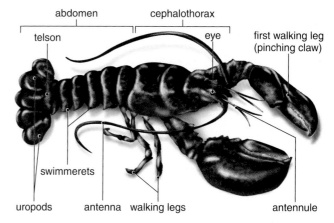

Figure 19.19 Exoskeleton and jointed appendages.

Arthropods, such as a lobster, have various appendages attached to the head region of a cephalothorax and five pairs of walking legs attached to the thorax region. Appendages called swimmerets, used in reproduction and swimming, are attached to the abdomen. The uropods and telson make up a fan-shaped tail.

Figure 19.20 **Monarch butterfly metamorphosis.**

a. A caterpillar (larva) eats and grows. **b.** After the larva goes through several molts, it builds a chrysalis around itself and becomes a pupa. **c.** Inside the pupa, the larva undergoes changes in organ structure to become an adult. **d, e.** The adult butterfly emerges from the chrysalis and reproduces, and the cycle begins again.

6. *Metamorphosis.* Many arthropods undergo a change in form and physiology as a larva becomes an adult. Metamorphosis allows the larva to have a different lifestyle than the adult (**Fig. 19.20**). For example, larval crabs live among and feed on plankton, while adult crabs are bottom dwellers that catch live prey or scavenge dead organic matter. Among insects such as butterflies, the caterpillar feeds on leafy vegetation, while the adult feeds on nectar.

Crustaceans, a name derived from their hard, crusty exoskeleton, are a group of largely marine arthropods that include barnacles, shrimps, lobsters, and crabs (**Fig. 19.21**). There are also some freshwater crustaceans, including the crayfish, and some terrestrial ones, including the sowbug, or pillbug. Although crustacean anatomy is extremely diverse, the head usually bears a pair of compound eyes and five pairs of appendages. The first two pairs of appendages, called antennae and antennules, respectively, lie in front of the mouth and have sensory functions. The other three pairs are mouthparts used in feeding. In a lobster, the thorax bears five pairs of walking legs. The first walking leg is a pinching claw. The *gills* are situated above the walking legs. The head and thorax are fused into a cephalothorax, which is covered on the top and sides by a nonsegmented carapace. The abdominal segments, which are largely muscular, are equipped with swimmerets—small, paddle-like structures. The last two segments bear the uropods and the telson, which make up a fan-shaped tail to propel the lobster backward (see Fig. 19.19).

Crustaceans play a vital role in the food chain. Tiny crustaceans known as krill are a major source of food for baleen whales, sea birds, and seals. Countries such as Japan are harvesting krill for human use. Copepods and other small crustaceans are primary consumers in marine and aquatic ecosystems. Many species of lobsters, crabs, and shrimp are important in the seafood industry. Some barnacles are destructive to wharfs, piers, and boats.

Among arthropods, the **arachnids** include spiders, scorpions, ticks, mites, and harvestmen ("daddy longlegs"; **Fig. 19.22***a*). Spiders have a narrow waist that separates the cephalothorax from the abdomen. Most spiders inject venom into their prey and digest their food externally before sucking it into the stomach. Spiders use silk threads for all sorts of purposes, from lining their nests to catching prey. The internal organs of spiders also show how they are adapted to a terrestrial way of life. Malpighian tubules work in conjunction with rectal glands to reabsorb ions and water before a relatively dry nitrogenous waste (uric acid) is excreted. Invaginations of the inner body wall form lamellae ("pages") of spiders' so-called book lungs.

Scorpions are among the oldest terrestrial arthropods (Fig. 19.22*b*) and may be the direct descendants of the first arthropods to leave the aquatic environments. Today, they occur in the tropics, subtropics, and temperate regions worldwide. They are nocturnal and spend most of the day hidden under a log or rock. Ticks and mites are parasites. Ticks suck the blood of vertebrates and sometimes transmit diseases, such as Rocky Mountain spotted fever or Lyme disease. Chiggers, the larvae of certain mites, feed on the skin of vertebrates.

Video Lobster Larvae

Video Voice of the Lobster

Video Barnacle-Free Boats

Video Lyme Disease

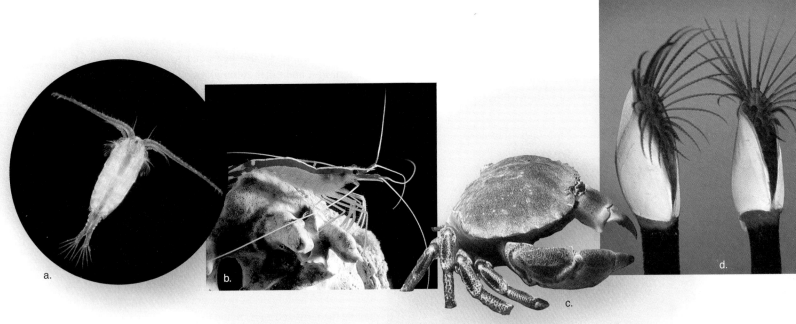

Figure 19.21 **Crustacean diversity.**

a. A copepod uses its long antennae for floating and its feathery maxillae for filter feeding. Shrimp (**b**) and crabs (**c**) are decapods—they have five pairs of walking legs. Shrimp resemble crayfish more closely than they do crabs, which have a reduced abdomen. Marine shrimp feed on copepods. **d.** The gooseneck barnacle is attached to an object by a long stalk. Barnacles have no abdomen and a reduced head; the thoracic legs project through a shell to filter feed. Barnacles often live on human-made objects, such as ships, buoys, and cables.

The horseshoe crab is grouped with the arachnids because the first pair of appendages are pincer-like structures used for feeding and defense. Horseshoe crabs have pedipalps, which they use as feeding and sensory structures, and four pairs of walking legs (Fig. 19.22c). Horseshoe crabs are of great interest to medical science. An extract from the horseshoe crab's blood cells is used to ensure that vaccines are free of bacterial toxins. Other compounds from the horseshoe crab are being investigated for antibiotic, antiviral, and anticancer properties.

While millipedes (Fig. 19.22d), with two pairs of legs on most segments, are herbivorous, centipedes (Fig. 19.22e), with a pair of appendages on every segment, are carnivorous. The head appendages of these animals are similar to those of insects, which are the largest group of arthropods, or indeed animals.

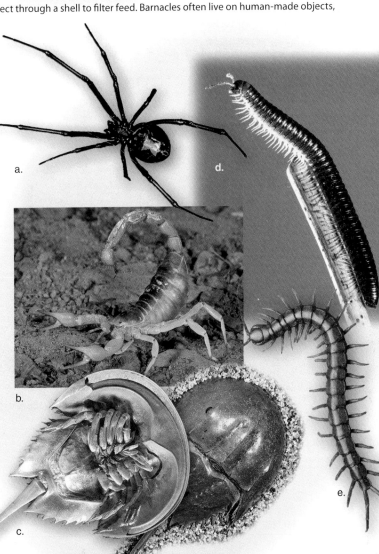

Figure 19.22 **More arthropods.**

a. The black widow spider is a venomous spider that spins a web. **b.** Scorpions have large pincers in front and a long abdomen, which ends with a stinger containing venom. **c.** Horseshoe crabs are common along the North American east coast. **d.** A millipede has two pairs of legs on most segments. **e.** A centipede has a pair of appendages on almost every segment.

Connecting the Concepts

For more information on insects, refer to the following discussions.

Section 24.1 examines the respiration system of insects.

Section 24.2 compares the excretory system of insects with that of humans.

Section 28.1 compares the vision of insects with that of other animals.

Insects are so numerous (well over 1 million species) and so diverse that the study of this one group is a major specialty in biology called entomology (**Fig. 19.23**). Some insects show remarkable behavior adaptations, as exemplified by the social systems of bees, ants, termites, and other colonial insects.

Video
Thorn Tree Ants

Insects are adapted to an active life on land, although some have secondarily invaded aquatic habitats. The body is divided into a head, a thorax, and an abdomen. The head usually bears a pair of sensory antennae, a pair of compound eyes, and several simple eyes. The mouthparts are adapted to each species' particular way of life: A grasshopper has mouthparts that chew (**Fig. 19.24**), and a butterfly has a long tube for siphoning the nectar from flowers.

The abdomen contains most of the internal organs; the thorax bears three pairs of legs and the wings—either one or two pairs, or none. Wings enhance an insect's ability to survive by providing a way of escaping enemies, finding food, facilitating mating, and dispersing the offspring. The exoskeleton of an insect is lighter and contains less chitin than that of many other arthropods. The male has a penis, which passes sperm to the female. The female, as in the grasshopper, may have an ovipositor for laying the fertilized eggs. Some insects, such as butterflies, undergo complete metamorphosis, involving a drastic change in form (see Fig. 19.20).

Check Your Progress 19.4

❶ Describe the two anatomical features seen in roundworms not seen in previously discussed organisms.

❷ List the six major characteristics of arthropods.

❸ Discuss why insects are the most numerous and diverse group of organisms on Earth.

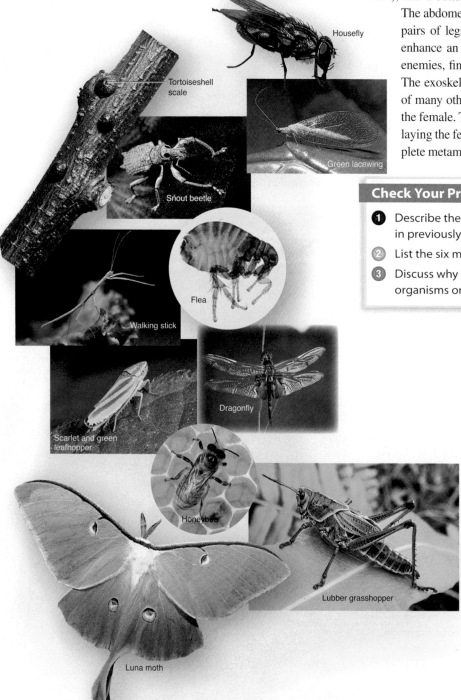

Figure 19.23 **Insect diversity.**

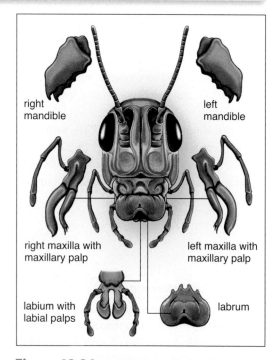

Figure 19.24 **Grasshopper mouthparts.**

The mouthparts of a grasshopper are specialized for chewing. The mouthparts of other insect species are adapted for their food source.

19.5 Echinoderms and Chordates: The Deuterostomes

As noted in the traditional (see Fig. 19.4) and modern (see Fig. 19.5) evolutionary tree of animals, the deuterostomes (see Fig. 19.7) include the echinoderms and the chordates.

Echinoderms

It may seem surprising that **echinoderms** (phylum Echinodermata) lack features associated with vertebrates, such as humans, yet of all the invertebrates they are most closely related to chordates (**Fig. 19.25**). However, there are some definite differences. For example, the echinoderms are often radially, not bilaterally, symmetrical. Their larva is a free-swimming filter feeder with bilateral symmetry, but it metamorphoses into a radially symmetrical adult. Also, adult echinoderms do not have a head, a brain, or segmentation. The nervous system consists of nerves in a ring around the mouth extending outward radially. Nevertheless, they are deuterostomes, as are the chordates.

a. Sea lily (above), feather star (right)

b. Sea cucumber

c. Brittle star

d. Sea urchins (left), sand dollar (right)

Figure 19.25 Echinoderm diversity.

a. Sea lilies are immobile, but feather stars can move about. They usually cling to coral or sponges, where they feed on plankton. **b.** Sea cucumbers have a long, leathery body that resembles a cucumber, except for the feeding tentacles about the mouth. **c.** Brittle stars have a central disk from which long, flexible arms radiate. **d.** Sea urchins and sand dollars have spines for locomotion, defense, and burrowing.

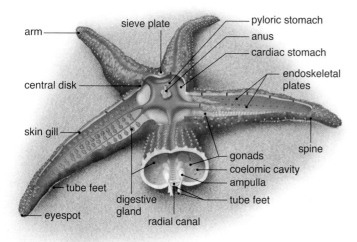

Figure 19.26 **Anatomy of a sea star.**

Many echinoderms, including this sea star, exhibit radial symmetry as adults.

Echinoderm locomotion depends on a water vascular system (**Fig. 19.26**). In the sea star, water enters this system through a sieve plate, or madreporite. Eventually, it is pumped into many tube feet, expanding them. When the foot touches a surface, the center withdraws, producing suction that causes the foot to adhere to the surface. By alternating the expansion and contraction of its many tube feet, a sea star moves slowly along.

Echinoderms don't have a complex respiratory, excretory, or circulatory system. Fluids within the coelomic cavity and a water vascular system carry out many of these functions. For example, gas exchange occurs across the skin gills and the tube feet. Nitrogenous wastes diffuse through the coelomic fluid and the body wall.

Most echinoderms feed variously on organic matter in the sea or substratum, but sea stars prey upon crustaceans, molluscs, and other invertebrates. From the human perspective, sea stars cause extensive economic loss because they consume oysters and clams before they can be harvested. However, they are important in many ways. In many ecosystems, fishes and sea otters eat echinoderms, and scientists favor echinoderms for embryological research.

Video
Sea Urchin
Reproduction

Chordates

At sometime in their life history, a **chordate** (phylum Chordata) has the four characteristics depicted in **Figure 19.27**:

1. A *dorsal supporting rod,* called a **notochord.** The notochord is located just below the nerve cord toward the back (i.e., dorsal). Vertebrates have an endoskeleton of cartilage or bone, including a vertebral column that has replaced the notochord during development.
2. A *dorsal tubular nerve cord. Tubular* means that the cord contains a canal filled with fluid. In vertebrates, the nerve cord is protected by the vertebrae. Therefore, it is called the spinal cord because the vertebrae form the spine.
3. *Pharyngeal pouches.* These structures are seen only during embryonic development in most vertebrates. In the invertebrate chordates, the fishes, and some amphibian larvae, the pharyngeal pouches become functioning gills. Water passing into the mouth and the pharynx goes through the gill slits, which are supported by gill arches. In terrestrial vertebrates that breathe with lungs, the pouches are modified for various purposes. In humans, the first pair of pouches become the auditory tubes. The second pair become the tonsils, while the third and fourth pairs become the thymus gland and the parathyroids.
4. A *tail.* Because the tail extends beyond the anus, it is called a *postanal tail.*

The Invertebrate Chordates

There are a few invertebrate chordates in which the notochord is never replaced by the vertebral column. **Tunicates** live on the ocean floor and take their name from a tunic that makes the adults look like thick-walled, squat sacs. They are also called sea squirts because they squirt water from one of their siphons when disturbed (**Fig. 19.28**a). The tunicate larva is bilaterally symmetrical and has the four chordate characteristics. Metamorphosis produces the sessile adult in which numerous cilia move water into the pharynx and out numerous gill slits, the only chordate characteristic that remains in the adult.

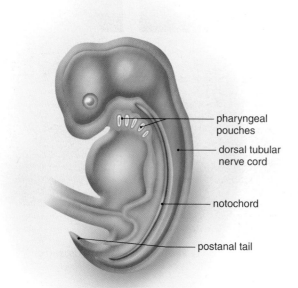

Figure 19.27 **The four chordate characteristics.**

Lancelets are marine chordates only a few centimeters long. They have the appearance of a lancet—a small, two-edged surgical knife (Fig. 19.28b). Lancelets are found in the shallow water along most coasts, where they usually lie partly buried in sandy or muddy substrates with only their anterior mouth and gill apparatus exposed. They feed on microscopic particles filtered out of the constant stream of water that enters the mouth and exits through the gill slits. Lancelets retain the four chordate characteristics as adults. In addition, segmentation is present, as witnessed by the fact that the muscles are segmentally arranged and the dorsal tubular nerve cord has periodic branches.

Evolutionary Trends Among the Chordates

Figure 19.29 depicts the evolutionary tree of the chordates and previews the animal groups we will discuss in the remainder of this chapter. The figure also lists at least one main evolutionary trend that distinguishes each group of animals from the preceding ones. The tunicates and lancelets are invertebrate chordates; they don't have vertebrae. The vertebrates are the fishes, amphibians, reptiles (including birds), and mammals.

As embryos, vertebrates have the four chordate characteristics. During embryonic development, the notochord is generally replaced by a vertebral column composed of individual vertebrae. Remnants of the notochord are

Figure 19.28 **Invertebrate chordates.**

a. Tunicates (sea squirts) have numerous gill slits, the only chordate characteristic that remains in the adult.
b. Lancelets have all four chordate characteristics as adults.

a.

b.

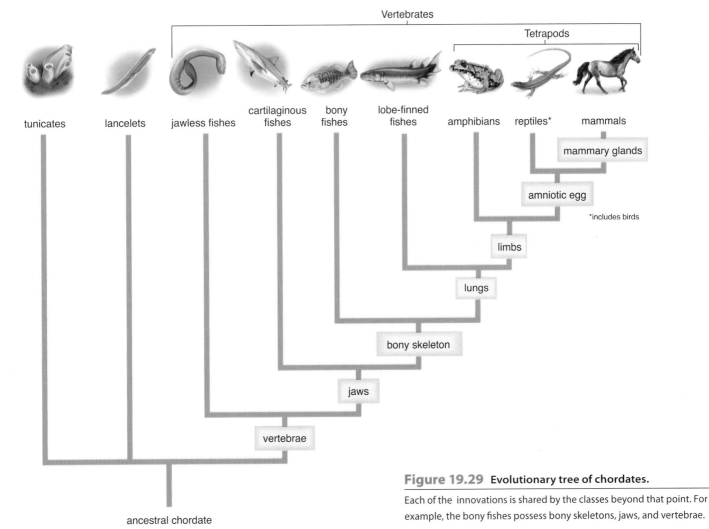

Figure 19.29 **Evolutionary tree of chordates.**

Each of the innovations is shared by the classes beyond that point. For example, the bony fishes possess bony skeletons, jaws, and vertebrae.

Figure 19.30

Evolution of jaws.

Jaws evolved from the anterior gill arches of ancient jawless fishes.

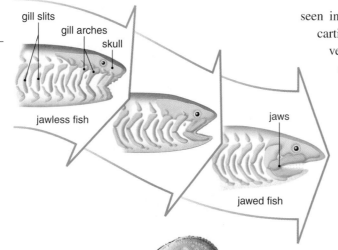

gill slits
gill arches
skull
jawless fish
jaws
jawed fish

Figure 19.31

Diversity of fishes.

a. The lamprey is a jawless fish. Note the toothed oral disk. **b.** The shark is a cartilaginous fish. **c.** The internal anatomy of a soldierfish, a bony fish.

toothed oral disk
gill slits (seven pairs)

a. Lamprey, a jawless fish

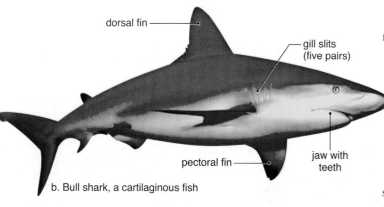

dorsal fin
gill slits (five pairs)
pectoral fin
jaw with teeth

b. Bull shark, a cartilaginous fish

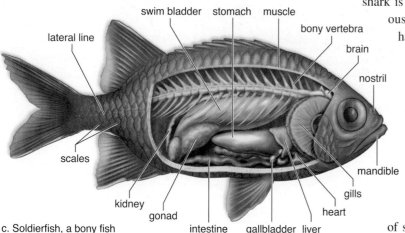

lateral line
swim bladder   stomach   muscle
bony vertebra
brain
nostril
scales
mandible
kidney
gonad
intestine   gallbladder   liver
gills
heart

c. Soldierfish, a bony fish

seen in the intervertebral discs, which are compressible cartilaginous pads found between the vertebrae. The vertebral column, which is a part of the flexible but strong endoskeleton, gives evidence that vertebrates are segmented.

Fishes with an endoskeleton of cartilage and some bone in their scales were the first to have jaws. Early bony fishes had lungs. Amphibians were the first group to clearly have jointed appendages and to invade land. The typical life cycle of amphibians includes a larval stage that lives in the water, as described in Figure 19.1. However, amphibians exploit a wide range of habitats and many reproduce on land. Reptiles, birds, and mammals have a means of reproduction more suited to land. During development, an amnion and other extra-embryonic membranes are present. These membranes carry out all the functions needed to support the embryo as it develops into a young offspring, capable of feeding on its own.

Fishes: First Jaws and Lungs

The first vertebrates were jawless fishes, which wiggled through the water and sucked up food from the ocean floor. Today, there are three living classes of fishes: jawless fishes, cartilaginous fishes, and bony fishes. The two last groups have jaws, tooth-bearing bones of the head. Jaws are believed to have evolved from the first pair of gill arches, structures that ordinarily support gills (**Fig. 19.30**). The presence of jaws permits a predatory way of life.

Living representatives of the **jawless fishes** are cylindrical and up to a meter long. They have smooth, scaleless skin and no jaws or paired fins. The two groups of living jawless fishes are *hagfishes* and *lampreys* (**Fig. 19.31***a*). The hagfishes are scavengers, feeding mainly on dead fish, while some lampreys are parasitic. When parasitic, the round mouth of the lamprey serves as a sucker. The lamprey attaches itself to another fish and taps into its circulatory system. Water cannot move in through the lamprey's mouth and out over the gills, as is common in all other fishes. Instead, water moves in and out through the gill openings.

Cartilaginous fishes are the sharks (Fig. 19.31*b*), the rays, and the skates, which have skeletons of cartilage instead of bone. The small dogfish shark is often dissected in biology laboratories. One of the most dangerous sharks inhabiting both tropical and temperate waters is the hammerhead shark. The largest sharks, the whale sharks, feed on small fishes and marine invertebrates and do not attack humans. Skates and rays are rather flat fishes that live partly buried in the sand and feed on mussels and clams.

Three well-developed senses enable sharks and rays to detect their prey: (1) They have the ability to sense electric currents in water—even those generated by the muscle movements of animals; (2) they, and all other types of fishes, have a lateral line system, a series of cells that lie within canals along both sides of the body and can sense pressure caused by a fish or another animal swimming nearby; and (3) they have a keen sense of smell—the part of the brain associated with this sense is twice as large as the other parts. Sharks can detect about one drop of blood in 115 liters (25 gallons) of water.

Bony fishes are by far the most numerous and diverse of all the vertebrates. Most of the bony fishes we eat, such as perch, trout, salmon, and haddock, are **ray-finned fishes** (Fig. 19.31*c*). Their fins, which are used in balancing and propelling the body, are thin and supported by bony spikes. Ray-finned fishes have various ways of life. Some, such as herring, are filter feeders; others, such as trout, are opportunists; and still others, such as piranhas and barracudas, are predaceous carnivores.

Video
Cichlid Specialization

Ray-finned fishes have a swim bladder, which usually serves as a buoyancy organ. By secreting gases into the bladder or absorbing gases from it, these fishes can change their density, and thus go up or down in the water. The streamlined shape, fins, and muscle action of ray-finned fishes are all suited to locomotion in the water. Their skin is covered by bony scales that protect the body but do not prevent water loss. When ray-finned fishes respire, the gills are kept continuously moist by the passage of water through the mouth and out the gill slits. As the water passes over the gills, oxygen is absorbed by the blood, and carbon dioxide is given off. Ray-finned fishes have a single-circuit circulatory system. The heart is a simple pump, and the blood flows through the chambers, including a nondivided atrium and ventricle, to the gills. Oxygenated blood leaves the gills and goes to the body proper, eventually returning to the heart for recirculation.

Another type of bony fish are called the **lobe-finned fishes.** Ancient lobe-finned fishes gave rise to the amphibians (**Fig. 19.32***a*). Not only did these fishes have fleshy appendages that could be adapted to land locomotion, but most also had a **lung,** which was used for respiration.

Animation
Bony Fish

Amphibians: Jointed Vertebrate Limbs

The first chordates to make their way to the land environment were the amphibians. **Amphibians,** whose name means living both on land and in the water, are represented today by frogs, toads, newts, and salamanders. Amphibians are variously colored. Some brightly colored ones have skin toxins that make predators sick or even die. Usually, the pattern of color is protective because it allows them to remain unnoticed.

Video
Amphibian Origins

Aside from *jointed limbs* (Fig. 19.32*b*), amphibians have other features not seen in bony fishes: eyelids for keeping their eyes moist, a sound-producing larynx, and ears adapted to picking up sound waves. The brain is larger than that of a fish. Adult amphibians usually have small lungs. Air enters the mouth by way of nostrils, and when the floor of the mouth is raised, air is forced into the relatively small lungs. Respiration is supplemented by gas exchange through the smooth, moist, and glandular skin. The amphibian heart has only three chambers, compared with the four of mammals. Mixed blood is sent to all parts of the body; some is sent to the skin, where it is further oxygenated.

Most members of this group lead an amphibious life—that is, the larval stage lives in the water, and the adult stage lives on the land (see Fig. 19.1). While metamorphosis is a distinctive characteristic of amphibians, some do not practice it. Some salamanders and even some frogs are direct developers; the adults do not

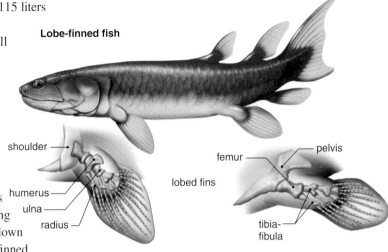

Lobe-finned fish

shoulder
humerus
ulna
radius
femur
pelvis
lobed fins
tibia-fibula

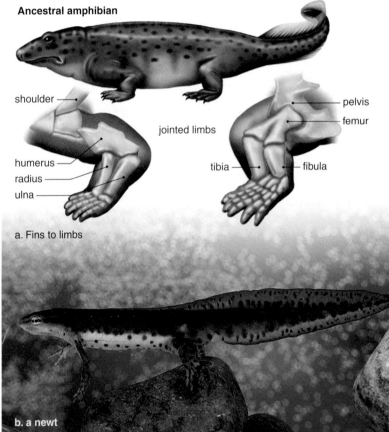

Ancestral amphibian

shoulder
humerus
radius
ulna
jointed limbs
pelvis
femur
tibia
fibula

a. Fins to limbs

b. a newt

Figure 19.32 Evolution of amphibians.

a. A lobe-finned fish compared with an amphibian. A shift in the position of the bones in the forelimbs and hindlimbs lifted and supported the body. **b.** Newts (shown here) and frogs (see Fig. 19.1) are types of living amphibians.

Figure 19.33 Keratinized skin of reptiles.

The keratin protein found in the scales of reptiles, and in the skin and hair of mammals, helps prevent water loss in the land environment.

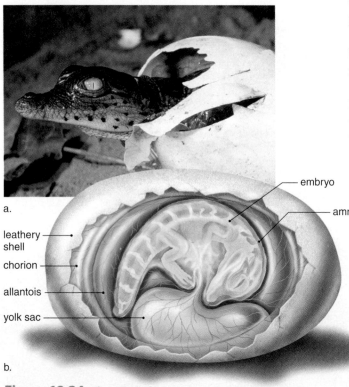

a.

- leathery shell
- chorion
- allantois
- yolk sac

embryo

amnion

b.

Figure 19.34 The amniotic egg.

a. A baby American crocodile hatching out of its shell. Note that the shell is leathery and flexible, not brittle, like a bird's egg. **b.** Inside the egg, the embryo is surrounded by extraembryonic membranes. The chorion aids gas exchange, the yolk sac provides nutrients, the allantois stores waste, and the amnion encloses a fluid that prevents drying out and provides protection.

return to the water to reproduce, and there is no tadpole stage. A great variety of reproductive strategies are seen among amphibians. In the gastric-brooding frogs, the female ingests as many as 20 fertilized eggs, which undergo development in her stomach, and then she vomits them up as tadpoles or froglets, depending on the species. The great variety of life histories observed among amphibians made them successful colonizers of the land environment. However, water pollution and human-made chemicals, such as pesticides, have caused a drastic reduction in amphibian populations worldwide. Many amphibian species are now in danger of extinction.

Video Frog Reproduction

Reptiles: Amniotic Egg

The **reptiles** diversified and were most abundant between 245 and 66 MYA. These animals included the dinosaurs, which are remembered for their great size. *Brachiosaurus,* an herbivore, was about 23 meters (75 feet) long and about 17 meters (56 feet) tall. *Tyrannosaurus rex,* a carnivore, was 5 meters (16 feet) tall when standing on its hind legs. The bipedal stance of some reptiles was preadaptive for the evolution of wings in birds.

The reptiles living today include turtles, crocodiles, snakes, lizards, and birds (discussed next). The typical reptilian body is covered with hard, keratinized scales, which protect the animal from desiccation and predators (**Fig. 19.33**). Reptiles have well-developed lungs enclosed by a protective rib cage. Most reptiles have a three-chambered heart because a septum that divides a third chamber is incomplete. This allows some mixing of O_2-rich and O_2-poor blood in this chamber.

Perhaps the most outstanding adaptation of the reptiles is that they have a means of reproduction suitable to a land existence. The penis of the male passes sperm directly to the female. Fertilization is internal, and the female lays leathery, flexible, shelled eggs. The *amniote egg* (**Fig. 19.34**) made development on land possible and eliminated the need for a water environment during development. The amniote egg provides the developing embryo with atmospheric oxygen, food, and water; removes nitrogenous wastes; and protects the embryo from drying out and from mechanical injury. This is accomplished by the presence of extraembryonic membranes, such as the chorion.

Fishes, amphibians, and reptiles (other than birds) are **ectothermic,** meaning that their body temperature matches the temperature of the external environment. If it is cold externally, their internal body temperature drops; if it is hot externally, their internal body temperature rises. Most reptiles regulate their body temperature by exposing themselves to the sun if they need warmth and hiding in the shadows if they need cooling off.

 Video Snake Eating

 Video Basilisk Lizard

Birds

Birds share a common ancestor with crocodiles and have traits such as a tail with vertebrae, clawed feet, and the presence of scales that show they are, in fact, reptiles. Perhaps you have noticed the scales on the legs of a chicken; in addition, a bird's *feathers* are actually modified reptilian scales. However, birds lay a hard-shelled amniote egg rather than the leathery egg of other reptiles. The exact history of birds is still in dispute, but gathering evidence indicates that birds are closely related to bipedal dinosaurs and that they should be classified as such.

Nearly every anatomical feature of a bird can be related to its ability to fly (**Fig. 19.35**). The forelimbs are modified as wings. The hollow, very light bones are

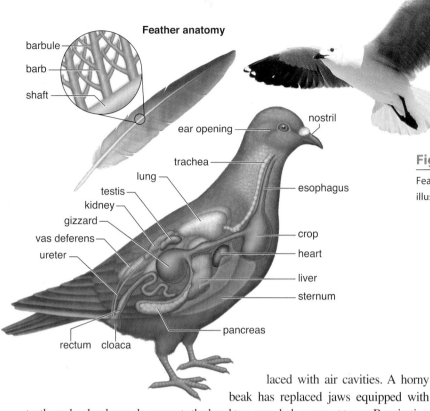

Feather anatomy

barbule
barb
shaft

ear opening — nostril
trachea
lung — esophagus
testis
kidney
gizzard — crop
vas deferens — heart
ureter — liver
— sternum
— pancreas
rectum cloaca

Figure 19.35 Birds are well adapted for flight.

Feathers and hollow bones are important adaptations for birds. This figure illustrates the internal anatomy of a typical bird.

laced with air cavities. A horny beak has replaced jaws equipped with teeth, and a slender neck connects the head to a rounded, compact torso. Respiration is efficient, since the lobular lungs form anterior and posterior air sacs. The presence of these sacs means that the air moves one way through the lungs, and gases are continuously exchanged across respiratory tissues. Another benefit of air sacs is that they lighten the body and aid flying.

Video Harris Hawk

Birds have a four-chambered heart that completely separates O_2-rich blood from O_2-poor blood. Birds are **endotherms** and generate internal heat. Many endotherms can use metabolic heat to maintain a constant internal temperature. This may be associated with their efficient nervous, respiratory, and circulatory systems. Also, their feathers provide insulation. Birds have no bladder and excrete uric acid in a semidry state.

Birds have particularly acute vision and well-developed brains. Their muscle reflexes are excellent. These adaptations are suited to flight. An enlarged portion of the brain seems to be the area responsible for instinctive behavior. A ritualized courtship often precedes mating. Many newly hatched birds require parental care before they are able to fly away and seek food for themselves. A remarkable aspect of bird behavior is the seasonal migration of many species over very long distances. Birds navigate by day and night, whether it's sunny or cloudy, by using the sun and stars and even the Earth's magnetic field to guide them.

Video Bird Radar

Traditionally, the classification of birds was particularly based on beak and foot types (**Fig. 19.36**), and to some extent on habitat and behavior. The various orders include birds of prey with notched beaks and sharp talons; shorebirds with long, slender, probing bills and long, stilt-like legs; woodpeckers with sharp, chisel-like bills and grasping feet; waterfowl with webbed toes and broad bills; penguins with wings modified as flippers; and songbirds with perching feet. Now genetics is used to determine relationships among birds.

Figure 19.36 Bird beaks.

a. A cardinal's beak allows it to crack tough seeds. **b.** A flamingo's beak strains food from the water with bristles that fringe the mandibles. **c.** A bald eagle's beak allows it to tear prey apart.

Mammals: Hair and Mammary Glands

Mammals (class Mammalia) are amniotes, and as such, they share a common ancestor with reptiles. However, mammals represent a separate evolutionary lineage from the reptile lineage that led to the birds. The first mammals evolved during the Triassic (208–250 MYA). They were small, about the size of mice. Due to many factors, including the dominance of the dinosaurs, mammals changed little during the Triassic and Jurassic (144–208 MYA) periods. Some of the earliest mammalian groups are still represented today by the monotremes and marsupials., However, neither of these groups is as abundant as the placental mammals, which can be found in most environments on the planet.

The two chief characteristics of mammals are hair and milk-producing mammary glands. Almost all mammals are endotherms and generate internal heat. Many of the adaptations of mammals are related to temperature control. Hair, for example, provides insulation against heat loss and allows mammals to be active, even in cold weather.

Mammary glands enable females to feed (nurse) their young without leaving them to find food. Nursing also creates a bond between mother and offspring that helps ensure parental care while the young are helpless. In most mammals, the young are born alive after a period of development in the uterus, a part of the female reproductive system. Internal development shelters the young and allows the female to move actively about while the young are maturing.

Monotremes are mammals that, like birds, have a *cloaca,* a terminal region of the digestive tract serving as a common chamber for feces, excretory wastes, and sex cells. They also lay hard-shelled amniote eggs. They are represented by the spiny anteater and the duckbill platypus, both of which live in Australia. The female duckbill platypus lays her eggs in a burrow in the ground. She incubates the eggs, and after hatching, the young lick up milk that seeps from mammary glands on her abdomen. The spiny anteater has a pouch on the belly side formed by swollen mammary glands and longitudinal muscle (**Fig. 19.37***a*). Hatching takes place in this pouch, and the young remain there for about 53 days. Then they stay in a burrow, where the mother periodically visits and nurses them.

The young of **marsupials** begin their development inside the female's body, but they are born in a very immature condition. Newborns crawl up into a pouch on their mother's abdomen. Inside the pouch, they attach to the nipples of

a.

Figure 19.37 Monotremes and marsupials.

a. The spiny anteater is a monotreme that lives in Australia.
b. The opossum is the only marsupial in the United States. The Virginia opossum is found in a variety of habitats. **c.** The koala is an Australian marsupial that lives in trees.

b.

c.

mammary glands and continue to develop. Frequently, more are born than can be accommodated by the number of nipples, and it's "first come, first served."

The Virginia opossum is the only marsupial that occurs north of Mexico (Fig. 19.37*b*). In Australia, however, marsupials underwent adaptive radiation for several million years without competition. Thus, marsupial mammals are now found mainly in Australia, with some in Central and South America as well. Among the herbivorous marsupials, koalas are tree-climbing browsers (Fig. 19.37*c*), and kangaroos are grazers. The Tasmanian wolf or tiger, thought to be extinct, was a carnivorous marsupial about the size of a collie dog.

The vast majority of living mammals are **placental mammals (Fig. 19.38)**. In these mammals, the extraembryonic membranes of the amniote egg (see Fig. 19.34) have been modified for internal development within the uterus of the female. The chorion contributes to the fetal portion of the placenta, while a part of the uterine wall contributes to the maternal portion. Here, nutrients, oxygen, and waste are exchanged between fetal and maternal blood.

Mammals are adapted to life on land and have limbs that allow them to move rapidly. In fact, an evaluation of mammalian features leads us to the obvious conclusion that they lead active lives. The brain is well developed; the lungs are expanded not only by the action of the rib cage but also by the contraction of the diaphragm, a horizontal muscle that divides the thoracic cavity from the abdominal cavity; and the heart has four chambers. The internal temperature is constant, and hair, when abundant, helps insulate the body.

The mammalian brain is enlarged due to the expansion of the cerebral hemispheres that control the rest of the brain. The brain is not fully developed until after birth, and young learn to take care of themselves during a period of dependency on their parents.

Mammals can be distinguished by their methods of obtaining food and their mode of locomotion. For example, bats have membranous wings supported by digits; horses have long, hoofed legs; and whales have paddle-like forelimbs. The specific shape and size of the teeth may be associated with whether the mammal is an herbivore (eats vegetation), a carnivore (eats meat), or an omnivore (eats both meat and vegetation). For example, mice have continuously growing incisors; horses have large, grinding molars; and dogs have long canine teeth.

 Video
Bat Echolocation

 Video
Meerkat Warning Calls

Connecting the Concepts

For more information on the physiology of the vertebrates, refer to the following discussions.

Figure 23.3 compares the path of blood flow in fish, amphibians, reptiles, and mammals.

Section 24.1 compares the process of respiration in birds, mammals, and fish.

Section 24.3 examines the differences in the digestive system of an herbivore and a carnivore.

Check Your Progress 19.5

1. List the unique features of the echinoderms.
2. Detail the evolutionary lineage of mammals.
3. Contrast the structure of a cartilaginous and bony fish.
4. Summarize the innovations shared by classes on the evolutionary tree of chordates.
5. Relate the special characteristics of amphibians to their adaptation to land.
6. Discuss the reason for the diversity of mammals.

Figure 19.38 Placental mammals.

Placental mammals have adapted to various ways of life. **a.** Deer are herbivores that live in forests. **b.** Lions are carnivores on the African plain. **c.** Monkeys typically inhabit tropical forests. **d.** Whales are sea-dwelling placental mammals.

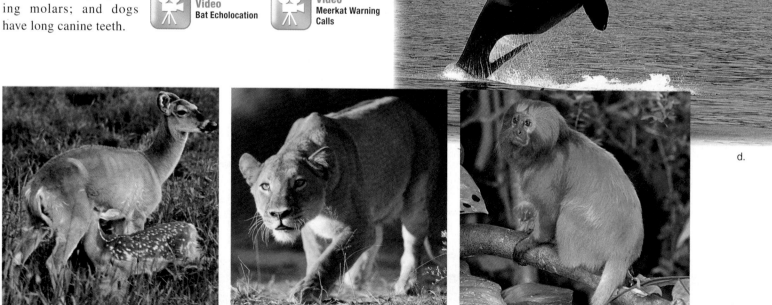

a.

b.

c.

d.

19.6 Human Evolution

Learning Outcomes

Upon completion of this section, you should be able to

1. Explain how the evolutionary tree of primates is constructed.
2. Describe evolutionary trends among the hominins.
3. Summarize the two major hypotheses for the evolution of modern humans.

The evolutionary tree in **Figure 19.39** shows that all primates share one common ancestor and that the other types of primates diverged from the human line of descent (called a **lineage**) over time. The **prosimians** include the lemurs, tarsiers, and lorises. The **anthropoids** include the monkeys, apes, and humans. The designation **hominid** includes the apes (African and Asian), chimpanzees, humans, and the closest extinct relatives of humans.

Primates are adapted to an arboreal life—life in trees. Primate limbs are mobile, and the hands and feet have five digits each. Many primates have both an opposable big toe and a thumb—that is, the big toe or thumb can touch each

Figure 19.39 Evolutionary tree of primates.

Primates are descended from an ancestor that may have resembled a tree shrew. The descendants of this ancestor adapted to the new way of life and developed traits such as a shortened snout and nails instead of claws. The time when each type of primate diverged from the main line of descent is known from the fossil record. A common ancestor was living at each point of divergence. For example, there was a common ancestor for hominids (humans and the African apes) about 7 MYA; and one for anthropoids about 45 MYA.

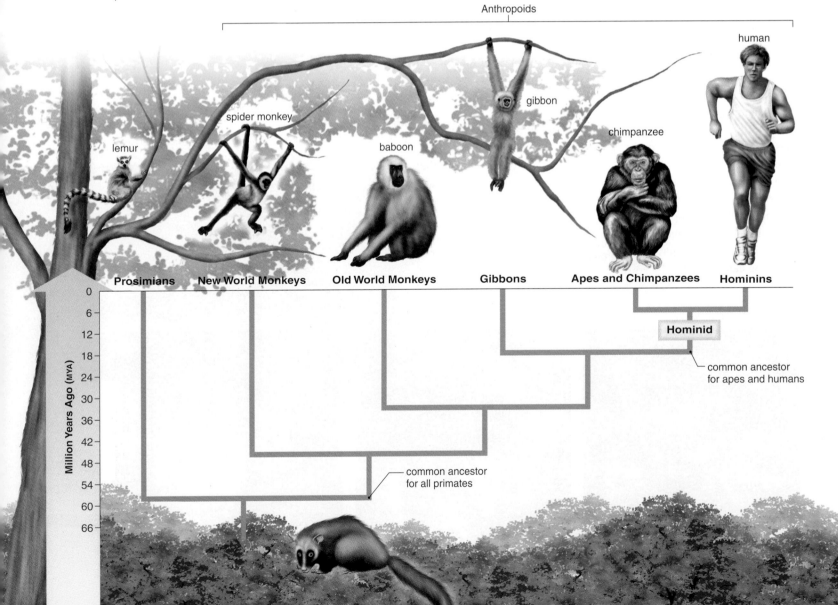

of the other toes or fingers. (Humans don't have an opposable big toe, but the thumb is opposable, resulting in a grip that is both powerful and precise.) The opposable thumb allows a primate to easily reach out and bring food, such as fruit, to the mouth. When locomoting, primates grasp and release tree limbs freely because nails have replaced claws.

The evolutionary trend among primates is toward a larger and more complex brain—the brain size is smallest in prosimians and largest in modern humans. In humans, the cerebral cortex has many association areas and expands so much that it becomes extensively folded. The portion of the brain devoted to smell gets smaller, and the portions devoted to sight increase in size and complexity during primate evolution. Also, more and more of the brain is involved in controlling and processing information received from the hands and the thumb. The result is good hand-eye coordination in humans.

Notice that prosimians were the first type of primate to diverge from the human line of descent, and African apes were the last group to diverge from our line of descent. The evolutionary tree also indicates that humans are most closely related to African apes. One of the most unfortunate misconceptions concerning human evolution is that Darwin and others suggested that humans evolved from apes. On the contrary, humans and apes share a common ape-like ancestor. Today's apes are our distant cousins, and we couldn't have evolved from our cousins because we are contemporaries—living on Earth at the same time. Presently, researchers hypothesize that the last common ancestor for African apes and humans lived about 7 MYA.

Evolution of Human-like Hominins

All of the fossils shown in **Figure 19.40** are all human-like **hominins.** This illustration tells when these various hominins lived and emphasizes their type of

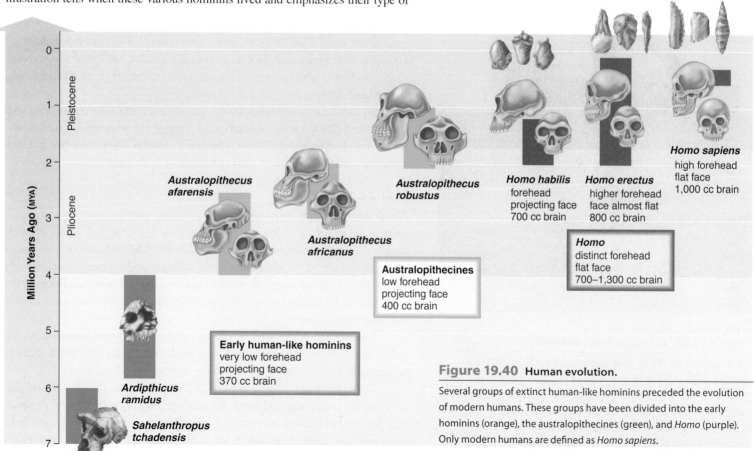

Figure 19.40 Human evolution.

Several groups of extinct human-like hominins preceded the evolution of modern humans. These groups have been divided into the early hominins (orange), the australopithecines (green), and *Homo* (purple). Only modern humans are defined as *Homo sapiens.*

face and brain size. However, to be a human-like hominin, it is only necessary that the fossil have an anatomy suitable for standing erect and walking on two feet, a characteristic called **bipedalism.** The bars in Figure 19.40 extend from the date of a species' appearance in the fossil record to the date it became extinct.

Early Human-like Hominins

In Figure 19.40, early hominins are represented by orange-colored bars. Scientists have now found several fossils dated around the time the ape lineage and the human lineage are believed to have split, and one of these is *Sahelanthropus tchadensis.* Only the braincase has been found and dated at 7 MYA. Although the braincase is very ape-like, the location of the opening for the spine at the back of the skull suggests *bipedalism.* Also, the canines are smaller and the tooth enamel is thicker than those of an ape.

Another early hominin, *Ardipithecus ramidus,* is representative of the ardipithecines of 4.5 MYA. So far, only skull fragments of *A. ramidus* have been described. Indirect evidence suggests that the species was bipedal and that some individuals may have been 122 centimeters tall. The teeth seem intermediate between those of earlier apes and later hominins. Recently, fossils dated 4 MYA show a direct link between *A. ramidus* and the australopithecines, discussed next.

Australopithecines

It's possible that one of the **australopithecines,** a group of hominins that evolved and diversified in Africa about 4 MYA, is a direct ancestor of humans. More than 20 years ago, a team led by Donald Johanson unearthed nearly 250 fossils of a hominid called *Australopithecus afarensis.* A now famous female skeleton dated at 3.18 MYA is known worldwide by its field name, Lucy. (The name derives from the Beatles song "Lucy in the Sky with Diamonds.") Although her brain was quite small (400 cc), the shapes and relative proportions of Lucy's limbs indicate that she stood upright and walked bipedally (**Fig. 19.41***a*). Even better evidence of bipedal locomotion comes from a trail of footprints dated about 3.7 MYA. The larger prints are double, as though a smaller individual was stepping in the footfalls of another—and there are additional small prints off to the side, within hand-holding distance (Fig. 19.41*b*).

Since the australopithecines were ape-like above the waist (small brain) and human-like below the waist (walked erect), it seems that human characteristics did not evolve all at one time. The term **mosaic evolution** is applied when different body parts change at different rates and therefore at different times.

Homo habilis

Homo habilis, dated between 2.0 and 1.9 MYA, may be ancestral to modern humans. Some of these fossils have a brain size as large as 775 cc, which is about 45% larger than the brain of *A. afarensis.* The cheek teeth are smaller than those of any of the australopithecines. Therefore, it is likely that this early *Homo* was omnivorous and ate meat in addition to plant material. Bones at the campsites of *H. habilis* bear cut marks, indicating that they used tools to strip the meat from bones.

The stone tools made by *H. habilis,* whose name means "handyman," are rather crude. These hominins may have used sharp flakes of broken rocks to scrape away hide, cut tendons, and easily remove meat from bones.

The skulls of *H. habilis* suggest that the portions of the brain associated with speech areas were enlarged. We can speculate that the ability to speak may have led to hunting cooperatively. Some members of the group may have remained plant gatherers, and if so, both hunters and gatherers most likely

b.

Figure 19.41 *Australopithecus afarensis.*

a. A reconstruction of Lucy on display at the St. Louis Zoo. **b.** These fossilized footprints occur in ash from a volcanic eruption some 3.7 MYA. The larger footprints are double, and a third, smaller individual was walking to the side. (A female holding the hand of a youngster may have been walking in the footprints of a male.) The footprints suggest that *A. afarensis* walked bipedally.

ate together and shared their food. In this way, society and culture could have begun. **Culture,** which encompasses human behavior and products (such as technology and the arts), is dependent on the capacity to speak and transmit knowledge. We can further speculate that the advantages of a culture to *H. habilis* may have hastened the extinction of the australopithecines.

Homo erectus

Homo erectus and like fossils are found in Africa, Asia, and Europe and dated between 1.9 and 0.3 MYA. Although all fossils assigned the name *H. erectus* are similar in appearance, there is enough discrepancy to suggest that several different species have been included in this group. Compared with *H. habilis, H. erectus* had a larger brain (about 1,000 cc) and a flatter face. The recovery of an almost complete skeleton of a 10-year-old boy indicates that *H. erectus* was much taller than the hominins discussed thus far. Males were 1.8 meters tall (about 6 feet), and females were 1.55 meters (approaching 5 feet). Indeed, these hominins were erect and most likely had a striding gait, like that of modern humans. The robust and most likely heavily muscled skeleton still retained some australopithecine features. Even so, the size of the birth canal indicates that infants were born in an immature state that required an extended period of care.

It is believed that *H. erectus* first appeared in Africa and then migrated into Asia and Europe. At one time, the migration was thought to have occurred about 1 MYA, but recently *H. erectus* fossil remains in Java and the Republic of Georgia have been dated at 1.9 and 1.6 MYA, respectively. These remains push the evolution of *H. erectus* in Africa to an earlier date than has yet been determined. In any case, such an extensive population movement is a first in the history of humankind and a tribute to the intellectual and physical skills of the species.

H. erectus was the first hominin to use fire, and it also fashioned more advanced tools than earlier *Homo* species. These hominins used heavy, teardrop-shaped axes and cleavers, as well as flakes that were probably used for cutting and scraping. Some investigators believe *H. erectus* was a systematic hunter that brought kills to the same site over and over again. In one location, researchers have found over 40,000 bones and 2,647 stones. These sites could have been "home bases," where social interaction occurred and a prolonged childhood allowed time for learning. Perhaps a language evolved and a culture more like our own developed.

Evolution of Modern Humans

Most researchers believe that **Homo sapiens** (modern humans) evolved from *H. erectus,* but they differ as to the details. The hypothesis that *H. sapiens* evolved separately from *H. erectus* in Asia, Africa, and Europe is called the **multiregional continuity hypothesis (Fig. 19.42***a***).** This hypothesis proposes that evolution to modern humans was essentially similar in several different places. If so, each region should show a continuity of its own anatomical characteristics from the time when *H. erectus* first arrived.

Opponents argue that it seems highly unlikely that evolution would have produced essentially the same result in these different places. They suggest, instead, the **replacement model,** or **out-of-Africa hypothesis,** which proposes that *H. sapiens* evolved from *H. erectus* only in Africa, and thereafter *H. sapiens* migrated to Europe and Asia about 100,000 years BP (before present) (Fig. 19.42*b*). If so, fossils dated 200,000 BP and 100,000 BP are expected to be markedly different from each other.

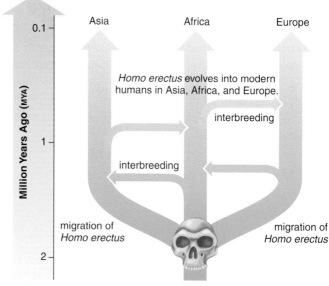

a. Multiregional continuity

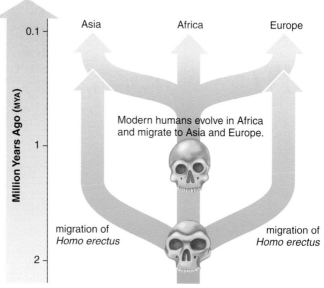

b. Out of Africa

Figure 19.42 Evolution of modern humans.

a. The multiregional continuity hypothesis proposes that *Homo sapiens* evolved separately in at least three places: Asia, Africa, and Europe. Therefore, continuity of genotypes and phenotypes is expected in each region, but not between regions. **b.** The replacement model (out-of-Africa hypothesis) proposes that *Homo sapiens* evolved only in Africa; then this species migrated and supplanted populations of *Homo* in Asia and Europe about 100,000 years ago.

Connections and Misconceptions

What are the "hobbits"?

In 2004, a new piece of our human heritage, *Homo floresiensis*, was discovered on the Indonesian island of Flores. Sometimes called "hobbits" due to their very small size (less than a meter in height), *H. floresiensis* also had a very small brain, but it is believed that they  used stone tools. Most scientists believe that *H. floresiensis* is more closely related to *H. erectus* than *Homo sapiens*. Interestingly, some dating techniques suggest that *H. floresiensis* may have occupied Flores as little as 15,000 years ago. If so, then it most likely would have interacted with *H. sapiens* in the not too distant past.

According to which hypothesis would modern humans be most genetically alike? The multiregional continuity hypothesis states that human populations have been evolving separately for a long time; therefore, genetic differences would be expected between groups. According to the replacement model, we are all descended from a few individuals from about 100,000 years BP. Therefore, the replacement model suggests that we are more genetically similar. The replacement model has been supported by analysis of mitochondrial DNA sequences. Unlike nuclear DNA, the DNA in the mitochondria does not undergo recombination and therefore only changes by the slow accumulation of mutations. In addition, mitochondrial DNA is only passed from the mother to her offspring, so the DNA in each of our mitochondria represents a genetic history that can be traced back through our mothers. The analysis of mitocondrial DNA has supported the hypothesis that humans are more genetically similar than what would be predicted by the multiregional continuity hypothesis.

Neandertals

Neandertals (*H. neandertalensis*) take their name from Germany's Neander Valley, where one of the first Neandertal skeletons, dated some 200,000 years ago, was discovered. It's possible that the Neandertals were already present in Eurasia when modern humans represented by Cro-Magnons (discussed next) arrived on the scene. In that case, competition with the Cro-Magnons may have made the Neandertals become extinct. The Neandertals had massive brow ridges, and their nose, jaws, and teeth protruded far forward. The forehead was low and sloping, and the lower jaw lacked a chin. The Neandertals were heavily muscled, especially in the shoulders and neck. The bones of their limbs were shorter and thicker than those of modern humans. The Neandertals lived in Europe and Asia during the last Ice Age, and their sturdy build could have helped conserve heat.

The Neandertals give evidence of being culturally advanced. Most lived in caves, but those living in the open may have built houses. They manufactured a variety of stone tools, including spear points, which could have been used for hunting, and scrapers and knives, which would have helped with food preparation. They most likely successfully hunted bears, woolly mammoths, rhinoceroses, reindeer, and other contemporary animals. They buried their dead with flowers and tools and may have had a religion.

Cro-Magnons

Cro-Magnons are the oldest fossils to be designated *H. sapiens*. Cro-Magnons, who are named after a fossil location in France, had a thoroughly modern appearance. They made advanced tools, including compound tools, as when stone flakes were fitted to a wooden handle. They may have been the first to make knife-like blades and to throw spears, enabling them to kill animals from a distance. They were such accomplished hunters that some researchers believe they were responsible for the extinction of many larger mammals, such as the giant sloth, the mammoth, the saber-toothed tiger, and the giant ox, during the late Pleistocene epoch.

Cro-Magnons hunted cooperatively and were perhaps the first to have a language. They are believed to have lived in small groups, with the men hunting by day while the women remained at home with the children, gathering and processing food items. Probably, the women also were engaged in maintenance tasks. The Cro-Magnon culture included art. They sculpted small figurines out of reindeer bones and antlers. They also painted beautiful drawings of animals, some of which have survived on cave walls in Spain and France (**Fig. 19.43**).

Figure 19.43 Cro-Magnons.

Cro-Magnon people are the first to be designated *Homo sapiens*. Their tool-making ability and other cultural attributes, including their artistic talents, are legendary.

Connecting the Concepts

For additional backgroud information on human evolution, refer to the following discussions.

Section 4.4 explores the role of the mitochondria in eukaryotic cells.

Section 14.2 examines how the fossil record is used to support evolutionary change over time.

Section 16.3 outlines how the science of systematics is involved in the study of evolutionary relationships.

Check Your Progress 19.6

1. List a few of the facts known about human evolution from fossils.
2. Briefly describe the two theories of evolution of modern humans.
3. Hypothesize where additional "missing links" might fit into the human evolutionary tree, and briefly describe what characteristics they might possess.

Connections and Misconceptions

How closely is *Homo sapiens* related to Neandertals?

In 2010, geneticists completed their first sequence analysis of the Neandertal genome. The results of this study revealed some very interesting facts regarding Neandertals and *Homo sapiens*. First, these two species were more genetically alike than previous thought. The initial analysis suggested that as few as 100 genes in *Homo sapiens* may show evidence of evolution since the Neanderthal-*sapiens* split, making the Neandertals one of our closest cousins. Second, some studies suggest that there is evidence that Neandertals and *H. sapiens* might have interbred. Although this is still being investigated, we know that these two species occupied overlapping territories in the Middle East and Europe for almost 14,000 years. Since Neandertals and *H. sapiens* were genetically similar, many scientists believe that interbreeding was not only possible but probable.

Media Study Tools

www.mhhe.com/maderessentials3

Enhance your study of this chapter with study tools and practice tests. Also ask your instructor about the resources available through ConnectPlus, including the media-rich eBook, interactive learning tools, and animations.

Classifying Arthropods
The virtual lab "Classifying Arthropods" provides an investigative look at how entomologists study one of the most successful groups of animals on the planet.

The Chapter in Review

Summary

19.1 Evolution of Animals

Animals are motile, multicellular heterotrophs that ingest their food.

Ancestry of Animals

The choanoflagellates are protists that most likely resemble the last unicellular ancestor of animals. A colony of flagellated cells could have led to a multicellular animal that formed tissues by invagination.

The Evolutionary Tree of Animals

The new evolutionary tree of animals is chiefly based on molecular and developmental data.

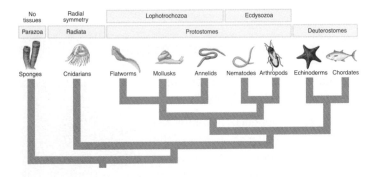

Evolutionary Trends

During the evolution of animals, multicellularity preceded true tissues and bilateral symmetry. Bilaterally symmetrical animals undergo either protostome development or deuterostome development. In protostomes, the first embryonic opening becomes the mouth. In deuterostomes, the second embryonic opening becomes the mouth. Protostomes are further divided based on whether the animal molts during development. Certain of the protostomes and all deuterostomes have an internal cavity called a coelom. Some of the coelomate animals are segmented.

19.2 Sponges and Cnidarians: The Early Animals

Sponges are multicellular (lack tissues) and have various symmetries.
Cnidarians have two tissue layers, are radially symmetrical, have a sac-like digestive cavity, and possess stinging cells and nematocysts.

19.3 Flatworms, Molluscs, and Annelids: The Lophotrochozoans

Lophotrochozoans are characterized by feeding structure and larval development. They increase their body mass gradually without molting.

Flatworms

Flatworms have ectoderm, endoderm, and mesoderm, but no coelom. They are bilaterally symmetrical and have a sac-like digestive cavity.

Molluscs

The body of a mollusc typically contains a visceral mass, a mantle, and a foot.

Gastropods Snails, representatives of this group, have a flat foot, a one-part shell, and a mantle cavity that carries on gas exchange.
Cephalopods Octopuses and squids display marked cephalization, move rapidly by jet propulsion, and have a closed circulatory system.
Bivalves Bivalves, such as clams, have a hatchet foot and a two-part shell and are filter feeders.

Annelids: Segmented Worms

Annelids are segmented worms; segmentation is seen both externally and internally.

Marine worms Polychaetes are worms that have many setae. A clam worm is a predaceous marine worm with a defined head region.
Earthworms Earthworms are oligochaetes that scavenge for food in the soil and do not have a well-defined head region.
Leeches Leeches are annelids that feed by sucking blood.

19.4 Roundworms and Arthropods: The Ecdysozoans

The ecdysozoans are invertebrate protostomes that increase their body mass by molting their exoskeleton or cuticle.

Roundworms

Roundworms have a pseudocoelom and a complete digestive tract.

Arthropods: Jointed Appendages

Arthropods are the most varied and numerous of animals. Their success is largely attributable to a flexible exoskeleton and specialized body regions.

Crustaceans have a head that bears compound eyes, antennae, and mouthparts. Five pairs of walking legs are present.

Arachnids include spiders, scorpions, ticks, mites, harvestmen, and horseshoe crabs. Spiders live on land and spin silk, which they use to capture prey as well as for other purposes.

Insects have three pairs of legs attached to the thorax. Insects have adaptations to a terrestrial life, such as wings for flying.

19.5 Echinoderms and Chordates: The Deuterostomes

Echinoderms

Echinoderms have radial symmetry as adults (not as larvae) and endoskeletal spines. Typical echinoderms have tiny skin gills, a central nerve ring with branches, and a water vascular system for locomotion, as exemplified by the sea star.

Chordates

Chordates (tunicates, lancelets, and vertebrates) have a notochord, a dorsal tubular nerve cord, pharyngeal pouches, and a postanal tail at some time in their life history.

Invertebrate chordates Adult tunicates lack chordate characteristics except gill slits, but adult lancelets have the four chordate characteristics and show obvious segmentation.

Vertebrate chordates Vertebrate chordates include the fishes, amphibians, reptiles (and birds), and mammals.

Fishes

- The first vertebrates, represented by hagfishes and lampreys, lacked jaws and fins.
- Cartilaginous fishes, represented by sharks and rays, have jaws and a skeleton made of cartilage.
- Bony fishes have jaws and fins supported by bony spikes; the bony fishes include those that are ray-finned and a few that are lobe-finned. Some of the lobe-finned fishes have lungs.

Amphibians

Amphibians, such as frogs, toads, newts, and salamanders, evolved from ancient lobe-finned fishes and have two pairs of jointed vertebrate limbs. Frogs usually return to the water to reproduce and metamorphose into terrestrial adults.

Reptiles

Reptiles, such as snakes, lizards, turtles, crocodiles, and birds, lay a shelled egg, which contains extraembryonic membranes, including an amnion that allows them to reproduce on land.

Birds are feathered reptiles, which helps them maintain a constant body temperature. They are adapted for flight; their bones are hollow with air cavities; lungs form air sacs that allow one-way ventilation; and they have well-developed sense organs.

Mammals

Mammals are amniotes that have hair and mammary glands. The former helps them maintain a constant body temperature, and the latter allows them to nurse their young.

- Monotremes lay eggs.
- Marsupials have a pouch in which the newborn matures.
- Placental mammals, which are far more varied and numerous, retain offspring inside the uterus until birth.

19.6 Human Evolution

Arboreal primates are mammals adapted to living in trees. During the evolution of primates, various groups diverged in a particular sequence. Prosimians (tarsiers and lemurs) diverged first, followed by the monkeys, then the apes, and then humans. Molecular biologists tell us we are most closely related to the African apes, with which we share a common ancestor dated at about 7 MYA.

Evolution of Human-like Hominins

Human evolution occurred in Africa, where it has been possible to find the remains of several early hominins that date back to 7 MYA. The evolution of hominins includes these innovations:

| Bipedal | • The most famous australopithecine is Lucy (3.18 MYA), whose brain was small but who walked bipedally. |

| Tool Use | • *Homo habilis,* present about 2 MYA, is certain to have made tools. |

| Brain Size Increased | • *Homo erectus,* with a brain capacity of 1,000 cc and a striding gait, was the first to migrate out of Africa. |

Evolution of Modern Humans

Two contradicting hypotheses have been suggested about the origin of modern humans:

Multiregional continuity hypothesis Modern humans originated separately in Asia, in Europe, and in Africa.

Replacement model (out-of-Africa hypothesis) Modern humans originated in Africa and, after migrating into Europe and Asia, replaced the other *Homo* species (including the Neandertals) found there. Cro-Magnon was the first hominid to be considered *Homo sapiens*. Mitochondrial DNA analysis supports the replacement model.

Key Terms

acoelomate   339
amphibian   355
annelid   344
anthropoid   360
arachnid   348
arthropod   347
australopithecine   362
bilateral symmetry   336
bipedalism   362
bird   356
bivalve   344
bony fishes   355
cartilaginous fishes   354
cephalization   338
cephalopod   344
chitin   347
chordate   352
cnidarian   341
coelom   339
Cro-Magnon   364
crustacean   348
culture   363
deuterostome   339

echinoderm   351
ectothermic   356
endotherm   357
filter feeder   340
flatworm   342
gastropod   344
germ layers   338
hermaphrodite   343
hominid   360
hominin   361
Homo erectus   363
Homo habilis   362
Homo sapiens   363
insect   350
invertebrate   340
jawless fishes   354
lancelet   353
leech   345
lineage   360
lobe-finned fishes   355
lung   355
mammal   358
marsupial   358

Testing Yourself

Choose the best answer for each question.

1. Which of these protists is hypothesized to be ancestral to animals?
   a. a green algal protist
   b. a choanoflagellate
   c. an amoeboid protist
   d. a slime mold

2. In the following diagram, label the major regions of the bodies of acoelomates, pseudocoelomates, and coelomates.

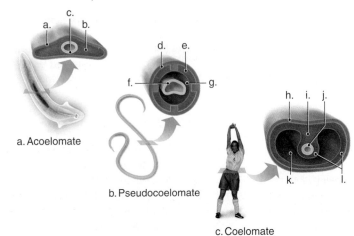

a. Acoelomate

b. Pseudocoelomate

c. Coelomate

3. Sponges are ancestors of
   a. cnidarians.   c. deuterostomes.
   b. protostomes.   d. None of these are correct.

For questions 4–10, identify the group(s) to which each feature belongs. Each answer may be used more than once. Each question may have more than one answer.

Key:
   a. flatworms
   b. roundworms
   c. molluscs
   d. annelids
   e. arthropods
   f. echinoderms
   g. chordates

4. most have an endoskeleton

5. typically bilaterally symmetrical as adults

6. nonsegmented

7. lack any kind of coelom

8. chitin exoskeleton

9. contain a mantle

10. move by pumping water

11. Insects have wings and three pairs of legs attached to the
    a. abdomen.   c. head.
    b. thorax.   d. midsection.

12. Cartilaginous fishes detect their prey by sensing
    a. electrical currents.
    b. odors.
    c. pressure changes.
    d. More than one of these are correct.
    e. All of these are correct.

13. Which of the following is not a feature of mammals?
    a. hair
    b. milk-producing glands
    c. ectothermic
    d. four-chambered heart
    e. diaphragm to help expand lungs

14. Which of these is matched correctly?
    a. *Australopithecus afarensis*—bipedal but small brain
    b. *Homo habilis*—made tools and migrated often
    c. *Homo erectus*—had fire and migrated out of Europe to Africa
    d. *Homo sapiens*—projecting face and had culture
    e. Both a and c are correct.

15. Cnidarians are considered to be organized at the tissue level because they contain
    a. ectoderm and endoderm.
    b. ectoderm.
    c. ectoderm and mesoderm.
    d. endoderm and mesoderm.
    e. mesoderm.

16. Which of the following is not a feature of a coelomate?
    a. radial symmetry
    b. three germ layers
    c. complete digestive tract body plan
    d. organ level of organization

17. The type of mollusc that has tentacles is a
    a. gastropod.   c. univalve.
    b. bivalve.   d. cephalopod.

18. A feature of annelids is
    a. a segmented body.   c. a sac body plan.
    b. acoelomate.   d. radial symmetry.

19. Which of the following is not a feature of an insect?
    a. compound eyes   d. an exoskeleton
    b. eight legs   e. jointed legs
    c. antennae

20. Which of the following is not an arachnid?
    a. spider   d. scorpion
    b. tick   e. beetle
    c. mite

21. Unlike bony fishes, amphibians have
    a. ears.   c. a circulatory system.
    b. jaws.   d. a heart.

22. Which of the following is not an adaptation for flight in birds?
    a. air sacs
    b. modified forelimbs
    c. bones with air cavities
    d. acute vision
    e. well-developed bladder

23. Examples of monotremes include the
    a. spiny anteater and duckbill platypus.
    b. opossum and koala.
    c. badger and skunk.
    d. porcupine and armadillo.

24. Mammals are distinguished based on
    a. size and hair type.
    b. mode of reproduction.
    c. number of limbs and method of caring for young.
    d. number of mammary glands and number of offspring.

25. The first human-like feature to evolve in the hominins was
    a. a large brain.
    b. massive jaws.
    c. a slender body.
    d. bipedal locomotion.

26. In *H. habilis*, enlargement of the portions of the brain associated with speech probably led to
    a. cooperative hunting.
    b. the sharing of food.
    c. the development of culture.
    d. All of these are correct.

27. Mitochondrial DNA data support which hypothesis for the evolution of humans?
    a. multiregional continuity
    b. replacement model (out-of-Africa)

28. Which of the following is an anthropoid, but not a hominid?
    a. human
    b. gibbon
    c. orangutan
    d. chimpanzee
    e. gorilla

29. Which of the following does not have cephalization?
    a. sponge
    b. planarian
    c. frog
    d. snail
    e. scorpion

30. Which of the following is a bilaterally symmetrical, hermaphroditic acoelomate?
    a. planarian
    b. sponge
    c. jellyfish
    d. earthworm
    e. sea star

Thinking Scientifically

1. Recently, three fossil skulls of *Homo sapiens,* dating to about 160,000 years BP, were discovered in eastern Africa. These skulls fill a gap between the 100,000-year-old *H. sapiens* skulls found in Africa and Israel and 500,000-year-old skulls of archaic *H. sapiens* found in Ethiopia. Supporters of the replacement model (out-of-Africa hypothesis) argue that this discovery strengthens their position by documenting the succession of human ancestors from 6 MYA through this most recent group. Does this finding prove that the replacement model is correct? If not, what fossil evidence might yet be found to support the multiregional continuity hypothesis?

2. Think of the animals in this chapter that are radially symmetrical (cnidarians, many adult echinoderms). How is their lifestyle different from that of bilaterally symmetrical animals? How does their body plan complement their lifestyle?

Bioethical Issue

Use of Animals in Research

People who approve of laboratory research involving animals point out that even today it would be difficult to develop new vaccines and medicines against infectious diseases, new surgical techniques for saving human lives, or new treatments for spinal cord injuries without the use of animals. Even so, most scientists favor what are now called the "three Rs": (1) replacement of animals by in vitro, or test-tube, methods whenever possible; (2) reduction of the number of animals used in experiments; and (3) refinement of experiments to cause less suffering to animals.

F. Barbara Orlans of the Kennedy Institute of Ethics at Georgetown University says, "It is possible to be both pro research and pro reform." She feels that animal activists need to accept that sometimes animal research is beneficial to humans, and all scientists need to consider the ethical dilemmas that arise when animals are used for laboratory research. Do you approve of this compromise?

20

Plant Anatomy and Growth

OUTLINE

BEFORE YOU BEGIN

Before beginning this chapter, take a few moments to review the following discussions.

Section 4.4 What are the differences between a plant cell and an animal cell?

Figure 6.1 What are the inputs and outputs of photosynthesis?

Section 18.1 What are the major adapations of plants to the land environment?

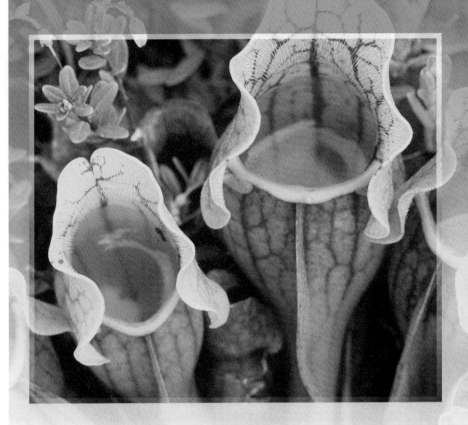

Carnivorous Plants Are Adapted to Harsh Conditions

Pitcher plants are unusual in that they like to grow in areas that lack nitrogen and phosphorus, which are required nutrients for plants. What's their secret? Pitcher plants feed on animals, particularly insects, to get the nutrients they need. That's right—like a few other types of plants, pitcher plants are carnivorous.

A pitcher plant is named for its leaves, which are shaped like a container we call a pitcher. The pitcher attracts flying insects because it has a scent, is brightly colored, and provides nectar that insects can eat. When an insect lands on the lip of the pitcher, a slippery substance and downward-pointing hairs encourage it to slide into a pit. There, juices secreted by the leaf begin digesting the insect. The plant then absorbs these nutrients into its tissues. Although the pitcher is designed to attract insects, animals as large as rats have been found in pitcher plants! The unique adaptations of pitcher plants allow them to thrive in hostile environments where few plants can survive.

Some animals are able to turn the tables on the pitcher plant by taking advantage of the liquid they provide. Larvae of mosquitoes and flies have been known to develop safely inside a pitcher plant.

Carnivorous plants, such as the pitcher plant, are fairly unusual in the plant world, but so are many others, such as the coastal redwoods—the tallest trees on planet Earth. In this chapter, you will learn about the organs and structures of flowering plants, the nutrition they need, and how they transport nutrients within plant tissues.

Video Carnivorous Plants

20.1 Plant Organs

Learning Outcomes

Upon completion of this section, you should be able to

1. List the three vegetative organs of a plant and briefly describe their functions.
2. Define primary growth and list the plant organ from which primary growth occurs.
3. Explain the function of leaves, stems, and roots in a plant and list some modifications that may occur to leaves.
4. Compare and contrast the seeds, roots, stems, leaves, and flowers of a monocot and a eudicot plant.

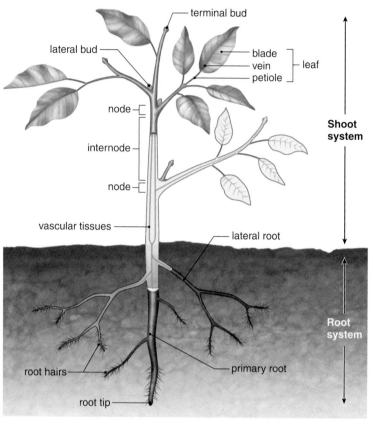

Figure 20.1 **The vegetative body of a plant.**

A plant has a root system, which extends below ground, and a shoot system, which is composed of the stem and leaves. Activity at the terminal bud and the root tip allows a plant to increase in length, a process called primary growth.

A flowering plant (or *Angiosperm*), whether a cactus, a daisy, or an apple tree, has a shoot system and a root system. The **shoot system** consists of the stem, the leaves, and the flowers, which carry out sexual reproduction, while the **root system** simply consists of the roots. The three vegetative organs—the leaf, the stem, and the root—perform functions that allow a plant to live and grow (**Fig. 20.1**). The structure and function of the flower are discussed in Chapter 21.

Activity at the **terminal bud,** the bud at the tip of a main shoot or a branch shoot, causes the shoot to increase in length, while activity at the tip of the root causes the root system to increase in length. This is called *primary growth,* and it would be equivalent to your increasing in height by growing from your head and your feet!

 Video Seedling Growth

Leaves

Leaves are usually the chief organs of photosynthesis, and as such they require a supply of solar energy, carbon dioxide, and water. Broad and thin foliage leaves have a maximum surface area for the collection of solar energy and the absorption of carbon dioxide. Leaves receive water from the root system by way of vascular tissue that terminates in the leaves. **Deciduous** plants lose their leaves, often due to a yearly dry season or the onset of winter. Other trees, called **evergreens,** retain their leaves for the entire year.

Leaves may have several other functions aside from photosynthesis (**Fig. 20.2**). Leaves may be modified as tendrils that allow the plant to attach to objects (Fig. 20.2*a*) or as traps for catching insects (Fig. 20.2*b*). The leaves of a cactus are spines that reduce water loss and protect the plant from browsing animals (Fig. 20.2*c*).

Stems

A **stem** supports the leaves, transports materials between the roots and leaves, and produces new tissue. At the end of a stem, a terminal bud, or apical bud, produces new leaves and other tissues during primary growth. Lateral (side) branches grow from a **lateral bud,** also called an axillary bud, located at the angle where a leaf joins the stem. A **node** occurs where a leaf or leaves are attached to the stem, and an **internode** is the region between the nodes (see Fig. 20.1). Vascular tissue transports water and minerals from the roots through the stem to the leaves and transports the products of photosynthesis, usually in the opposite direction.

Connections and Misconceptions

What is the world's tallest tree?

The record for the tallest living tree is currently held by a tree named Hyperion, a coastal redwood (*Sequoia sempervirens*) located in a remote region of the Redwood National and State Parks in northern California. Hyperion is 115.55 meters (379.1 feet) tall and is believed to be almost 800 years old. In comparison, the Statue of Liberty is only 93.12 meters (305.5 feet) tall!

tendril

leaves

stem

tuber

a. Cucumber

b. Venus's flytrap

c. Cactus

d. Potato

Figure 20.2 Leaf and stem diversity.

a. The tendrils of a cucumber are leaves modified to attach the plant to a support. **b.** Venus's flytrap leaves serve as the trap for unsuspecting insects. After digesting this fly, the plant will absorb the nutrients released. **c.** The spines of a cactus protect the fleshy stem from herbivores. The cactus uses its fleshy stem for both food production and water storage. **d.** The familiar potato is actually an underground stem, or tuber, that is used for food storage.

Stems may have functions other than those mentioned. In a cactus, the stem is the primary photosynthetic organ and serves as a water reservoir (see Fig. 20.2c), and the tuber of a potato plant is a food storage portion of an underground stem (Fig. 20.2d). **Perennial plants** are able to regrow each season from varied underground structures, such as tubers, bulbs, and rhizomes, all of which bear nodes that can produce a new shoot system.

Roots

A plant's root system supports the plant by anchoring it in the soil, as well as absorbing water and minerals from the soil for the entire plant. As a rule of thumb, the root system is at least equivalent in size and extent to its shoot system. Therefore, an apple tree has a much larger root system than a corn plant. Also, the extent of a root system depends on the environment. A single corn plant may have roots as deep as 2.5 meters (m), while a mesquite tree in the desert may have roots that penetrate to a depth of 20 m.

Roots have a cylindrical shape and a slimy surface, which allow them to penetrate the soil as they grow and permit water to be absorbed from all sides. In a special zone near the root tip, there are many **root hairs** that greatly increase the absorptive capacity of the root. Root hairs are so numerous that they increase the absorptive surface of a root tremendously. It has been estimated that a single rye plant has about 14 billion root-hair cells; if placed end to end, the root hairs would stretch over 10,600 kilometers. Root-hair cells are constantly being replaced, so the same rye plant forms about 100 million new root-hair cells every day. You probably know that a plant yanked out of the soil will not fare well when transplanted. This is because small lateral roots and root hairs are torn off. Transplantation is more apt to be successful if you take a part of the surrounding soil along with the plant, leaving as many of the lateral roots and root hairs intact as possible.

Just as stems and leaves are diverse, so are roots. A carrot plant has one main *taproot*, which stores the products of photosynthesis (**Fig. 20.3a**). Grass has fibrous roots that cling to the soil (Fig. 20.3b), and corn plants have *prop roots* that grow from the stem for better support (Fig. 20.3c).

a. Taproot

b. Fibrous root system

c. Prop roots

Figure 20.3 Root diversity.

a. A taproot has one main root. **b.** A fibrous root has many slender roots with no main root. **c.** Prop roots are organized for support.

Monocot Versus Eudicot Plants

Figure 20.4 shows how flowering plants can be divided into two major groups. One of the differences between the two groups concerns the **cotyledons,** which are embryonic leaves present in seeds. (The cotyledons wither after the first true leaves appear.) Plants whose embryo has one cotyledon are known as monocotyledons, or **monocots.** Other embryos have two cotyledons, and these plants are known as eudicotyledons, or **eudicots** (true dicots). The cotyledons of eudicots supply nutrients for seedlings, but the cotyledons of monocots store some nutrients and act as a transfer tissue for nutrients stored elsewhere.

Although the distinction between monocots and eudicots may seem minimal at first glance, there are many differences in their structures. For example, the location and arrangement of vascular tissue differ between monocots and eudicots. Vascular plants contain two main types of transport tissue: the **xylem** for water and minerals and the **phloem** for organic nutrients. In a sense, xylem and phloem are to plants what veins and arteries are to animals. In the monocot root, vascular tissue occurs in a ring around the center. But in the eudicot, vascular tissue is located in the center. The xylem forms a star shape, and phloem is located between the points of the star. In a stem, vascular tissue occurs in the **vascular bundles.** The vascular bundles are scattered in the monocot stem, and they occur in a ring in the eudicot stems.

In a leaf, vascular tissue forms **leaf veins.** In monocots, the veins are parallel, while in eudicots, the leaf veins form a net pattern. Monocots and eudicots also have different numbers of flower parts, and this difference will be discussed further in Chapter 21.

The eudicots are the larger group and include some of our most familiar flowering plants—from dandelions to oak trees. The monocots include grasses, lilies, orchids, and palm trees, among others. Some of our most significant food sources are monocots, including rice, wheat, and corn.

Connecting the Concepts

For more information on the material presented in this section, refer to the following discussions.

Figure 6.2 illustrates the internal structure of a typical leaf.

Section 18.1 outlines the adapations that have made the flowering plants successful on land.

Section 21.3 examines some of the differences between the flowers of monocot and eudicot plants.

Check Your Progress 20.1

1. List the three vegetative organs in a plant.
2. Illustrate some possible leaf modifications and describe the functions they serve.
3. Detail the major differences in the organization of vascular tissue between monocots and eudicots.

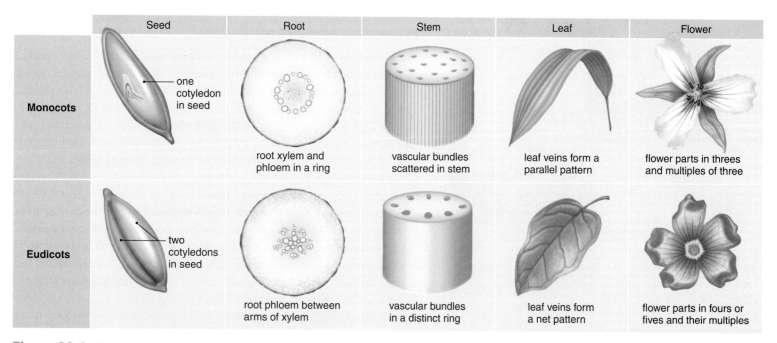

	Seed	Root	Stem	Leaf	Flower
Monocots	one cotyledon in seed	root xylem and phloem in a ring	vascular bundles scattered in stem	leaf veins form a parallel pattern	flower parts in threes and multiples of three
Eudicots	two cotyledons in seed	root phloem between arms of xylem	vascular bundles in a distinct ring	leaf veins form a net pattern	flower parts in fours or fives and their multiples

Figure 20.4   Flowering plants are either monocots or eudicots.

Five features are typically used to distinguish monocots from eudicots: number of cotyledons; arrangement of vascular tissue in roots, stems, and leaves; and number of flower parts.

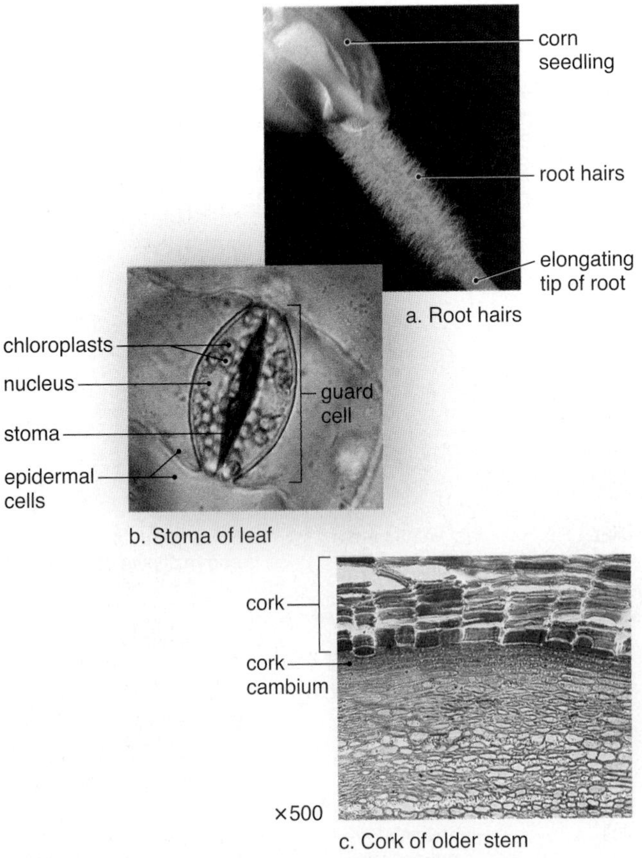

20.2 Plant Tissues and Cells

Learning Outcomes

Upon completion of this section, you should be able to

1. List the three types of specialized tissues of a plant and describe their functions.
2. Describe the modifications that may occur to epidermal tissue in a plant.
3. List the three types of ground tissue, and compare and contrast their structures and functions.
4. Explain how the vascular tissues of a plant are organized to move water and nutrients within a plant.

Unlike humans, flowering plants grow their entire lives because they have **meristem** (embryonic) tissue, composed of undifferentiated cells that divide. **Apical meristem** is located in the terminal bud of the shoot system and in the root tip. When apical meristem cells divide, one of the daughter cells remains a meristem cell, and the other differentiates into one of the three types of specialized tissues of a plant:

1. **Epidermal tissue** forms the outer protective covering of a plant.
2. **Ground tissue** fills the interior of a plant and helps carry out the functions of a particular organ.
3. **Vascular tissue** transports water and nutrients in a plant and provides support.

Another type of meristem tissue is associated with vascular tissue and is called **vascular cambium.** When the meristem cells of vascular cambium divide, they give rise to new vascular tissue, which causes a plant to increase in girth (called *secondary growth*).

Animation
Woody Dicot

Epidermis and Ground Tissue

The entire body of a plant is covered by an **epidermis,** a layer of closely packed cells that act as a barrier. The walls of epidermal cells that are exposed to air are covered with a waxy **cuticle** to minimize water loss. The cuticle also protects against bacteria and other organisms that might cause disease. As mentioned in Section 20.1, roots have root hairs, long, slender projections of epidermal cells that increase the surface area of the root for absorption of water and minerals (**Fig. 20.5***a*). In leaves, the epidermis often contains **stomata** (sing., stoma). A stoma is a small opening surrounded by two guard cells (Fig. 20.5*b*). When the stomata are open, gas exchange and water loss occur.

In the trunk of a tree, the epidermis is replaced by cork, which is a part of bark (Fig. 20.5*c*). New cork cells are made by a meristem called *cork cambium*. As the new cork cells mature, they increase slightly in volume, and their walls become encrusted with *suberin*, a lipid material, so that they are waterproof and chemically inert. These nonliving cells protect the plant and make it resistant to attack by fungi, bacteria, and animals.

Ground tissue forms the bulk of leaves, stems, and roots. Ground tissue contains three types of cells (**Fig. 20.6**). **Parenchyma** cells are the least specialized of the cell types and are found in all the organs of a plant. They may contain chloroplasts and carry on photosynthesis, or they may contain colorless plastids that store the products of photosynthesis. **Collenchyma** cells are like parenchyma

corn seedling

root hairs

elongating tip of root

a. Root hairs

chloroplasts

nucleus

stoma

epidermal cells

guard cell

b. Stoma of leaf

cork

cork cambium

×500

c. Cork of older stem

Figure 20.5   Modifications of epidermal tissue.

a. Root epidermis has root hairs, which help absorb water. **b.** Leaf epidermis contains stomata (sing., stoma) for gas exchange. **c.** Cork, which is a part of bark, replaces epidermis in older, woody stems.

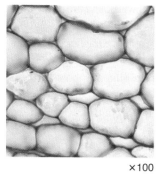
×100

a. Parenchyma cells

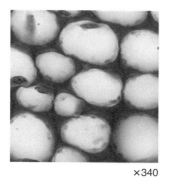
×340

b. Collenchyma cells

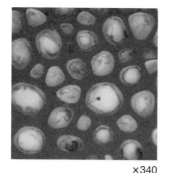
×340

c. Sclerenchyma cells

Figure 20.6 Ground tissue cells.

a. Parenchyma cells are the least specialized of the plant cells. **b.** Collenchyma cells have thicker and more irregular cell walls than parenchyma cells. **c.** Sclerenchyma cells have very thick walls and are nonliving—their only function is to give strong support.

cells except that they have irregularly shaped corners and thickened primary cell walls. Collenchyma cells often form bundles just beneath the epidermis and give flexible support to immature regions of a plant body. The familiar strands in celery stalks are composed mostly of collenchyma cells. **Sclerenchyma** cells have thick secondary cell walls containing *lignin,* which makes plant cell walls tough and hard. If we compare a cell wall to reinforced concrete, cellulose fibrils would play the role of steel rods, and lignin would be analogous to the cement. Most sclerenchyma cells are nonliving; their primary function is to support the mature regions of a plant. The hard outer shells of nuts are made of sclerenchyma cells. The long fibers in plants, composed of strings of sclerenchyma, make them useful for a number of purposes. For example, cotton and flax fibers can be woven into cloth, and hemp fibers can make strong rope.

Connections and Misconceptions

How is paper made?

Most paper is made from the cellulose fibers of trees. Recall from page 40 that cellulose is a component of the cell wall of plants. After the tree is harvested, it is debarked and cut into small chips to increase the surface area. However, before the cellulose can be extracted from the cells, the lignin and resins must  be removed. This is done either mechanically or by using a combination of steam and sulfur-based chemicals to separate the lignin and resins from the cellulose fibers. In the next step, the lignins and resins are removed by bleaching with chlorine gas (or chlorine dioxide), which also makes the fibers white. In the final steps, fibers are dried and treated in a machine that forms rolls of paper for commercial use.

Vascular Tissue

Vascular tissue extends from the root through the stem to the leaves and vice versa (see Fig. 20.1). In the root, the vascular tissue is located in a central cylinder; in the stem, vascular tissue is in vascular bundles; and in the leaves, it is found in leaf veins. Although both types of vascular tissue are usually found together, they have different functions. Xylem transports water and minerals from the roots to the leaves. Phloem transports sugar, in the form of sucrose, and other organic compounds, such as hormones, often from the leaves to the roots.

Xylem contains two types of conducting cells: vessel elements and tracheids (**Fig. 20.7***a*). Both types of conducting cells are hollow and nonliving,

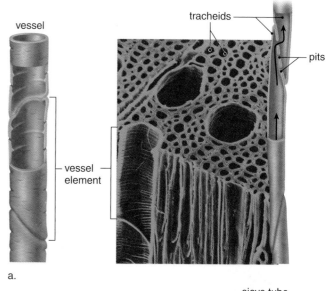
vessel
tracheids
pits
vessel element
a.

sieve tube
sieve-tube member
nucleus
companion cell
sieve plate
b.

Figure 20.7 Vascular tissue.

a. Photomicrograph of xylem vascular tissue, with (left) drawing of a vessel (composed of vessel elements) and (right) drawing of tracheids.
b. Photomicrograph of phloem vascular tissue, with drawing of a sieve tube. Each sieve-tube member has a companion cell.

but the **vessel elements** are larger, have perforated end walls, and are arranged to form a continuous pipeline for water and mineral transport. The end walls and side walls of **tracheids** have pits that allow water to move from one tracheid to another.

The conducting cells of phloem are **sieve-tube members,** so called because they contain a cluster of pores in their end walls. The pores are collectively known as a *sieve plate*. The sieve-tube members are arranged to form a continuous sieve tube (Fig. 20.7*b*). Sieve-tube members contain cytoplasm but no nuclei. Each sieve-tube member has a companion cell, which does have a nucleus. The companion cells may very well be involved in the transport function of phloem.

Animation
Vascular System of Plants

Connecting the Concepts

For more on the material covered in this section, refer to the following discussions.

Figure 3.10 illustrates the structure of cellulose fibers.

Section 18.2 examines the importance of vascular tissue in the evolution of the plants.

Section 20.7 explores how the xylem and phloem move water and nutrients in a plant.

Check Your Progress 20.2

1 List the three types of tissue in a plant, the cells that make up these tissues, and the functions of each.

2 Contrast the function of xylem with that of phloem.

3 Discuss what might occur to a plant if it did not include ground tissue. What if it did not include an epidermis?

20.3 Organization of Leaves

Learning Outcomes

Upon completion of this section, you should be able to

1. Compare and contrast the structure of a simple leaf with that of a compound leaf.
2. List the structures found in a typical eudicot leaf.
3. Describe the movement of water, oxygen, and carbon dioxide within a leaf.

Figure 20.8 shows the general structure of a leaf. The wide portion of a foliage leaf is called the **blade**. The **petiole** is a stalk that attaches the blade to the stem. The blade of a leaf is often undivided, or *simple,* as in a cottonwood

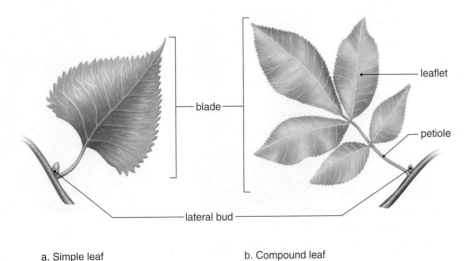

a. Simple leaf   b. Compound leaf

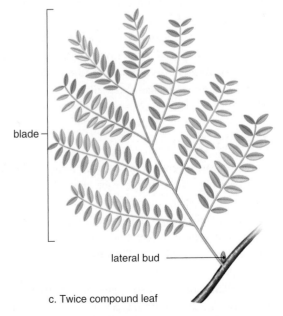

c. Twice compound leaf

Figure 20.8 Simple versus compound leaves.

a. The cottonwood tree has a simple leaf. **b.** The shagbark hickory has a compound leaf. **c.** The honey locust has a twice compound leaf. A single lateral bud appears where the leaf attaches to the stem.

(Fig. 20.8*a*). But in some leaves, the blade is divided, or *compound*, as in a shagbark hickory or honey locust (Fig. 20.8*b,c*). Notice that it is possible to tell from the placement of the lateral bud whether you are looking at several individual leaves or one compound leaf.

Figure 20.9 shows a typical eudicot leaf of a temperate-zone plant in cross section. The outer layer of the leaf, or epidermis, has an outer, waxy cuticle that prevents loss of water; it also prevents gas exchange because it is not gas permeable. The lower epidermis of eudicot leaves and both surfaces of monocot leaves contain stomata. At the stomata, carbon dioxide for photosynthesis enters a leaf, and oxygen, the by-product of photosynthesis, exits. The epidermis may also bear protective hairs and/or glands that secrete irritating substances. These adaptations discourage insects from eating leaves.

The interior of a leaf is composed of **mesophyll,** the tissue that carries on photosynthesis. Notice how leaf veins terminate at the mesophyll. Leaf veins transport water and minerals to a leaf and transport the product of photosynthesis, a sugar, away from the leaf. Mesophyll has two distinct regions: **palisade mesophyll,** containing elongated cells, and **spongy mesophyll,** containing irregular cells bounded by air spaces. The loosely packed arrangement of the cells in the spongy layer increases the amount of surface area for gas exchange and water loss. Water brought into a leaf by leaf veins evaporates from spongy mesophyll and exits at the stomata.

Connecting the Concepts

For more information on plant leaves, refer to the following discussions.

Figures 4.7 and **4.8** compare the structure of plant and animal cells.

Section 18.2 describes the evolutionary significance of microphyll and megaphyll leaves.

Section 21.3 examines how modified leaves form flowers in the angiosperms.

Check Your Progress 20.3

1. List the features of the epidermis of a leaf.
2. Detail the structural differences between a simple, a compound, and a twice compound leaf.
3. Correlate how the structure of a leaf aids in photosynthesis.

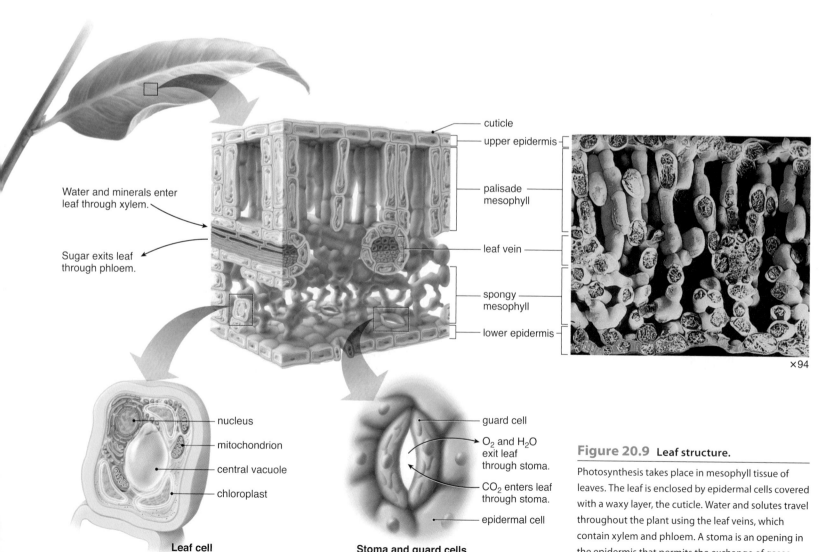

Water and minerals enter leaf through xylem.

Sugar exits leaf through phloem.

cuticle
upper epidermis
palisade mesophyll
leaf vein
spongy mesophyll
lower epidermis

×94

nucleus
mitochondrion
central vacuole
chloroplast

Leaf cell

guard cell
O₂ and H₂O exit leaf through stoma.
CO₂ enters leaf through stoma.
epidermal cell

Stoma and guard cells

Figure 20.9 **Leaf structure.**

Photosynthesis takes place in mesophyll tissue of leaves. The leaf is enclosed by epidermal cells covered with a waxy layer, the cuticle. Water and solutes travel throughout the plant using the leaf veins, which contain xylem and phloem. A stoma is an opening in the epidermis that permits the exchange of gases.

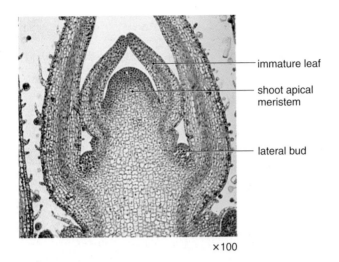

×100

Figure 20.10 Shoot tip.

The terminal bud contains the apical meristem that produces leaves and cells that become the tissues of stems.

immature leaf

shoot apical meristem

lateral bud

20.4 Organization of Stems

Learning Outcomes

Upon completion of this section, you should be able to

1. Contrast the organization of vascular bundles of a nonwoody (herbaceous) monocot with the organization of a eudicot stem.
2. Describe the organization of a woody eudicot stem.
3. Describe the formation of bark in a woody eudicot stem.
4. Explain how secondary growth of a woody eudicot stem occurs and contrast spring wood with summer wood.

During the primary growth of stems, the shoot apical meristem in the terminal bud produces new cells that elongate and then differentiate to become the tissues and organs of the shoot system—namely, the leaves and the stem. The shoot apical meristem is protected by the immature leaves that envelop it. Also, note the lateral buds from which lateral (side) branches (including leaves) develop (**Fig. 20.10**).

Nonwoody Stems

A stem that experiences only primary growth is nonwoody. Plants that have nonwoody stems, such as zinnias and daisies, are termed **herbaceous** plants. As with leaves, the outermost tissue of a herbaceous stem is the epidermis, which is covered by a waxy cuticle to prevent water loss. Beneath the epidermis of eudicot stems is the **cortex,** a narrow band of parenchyma cells. The cortex is sometimes

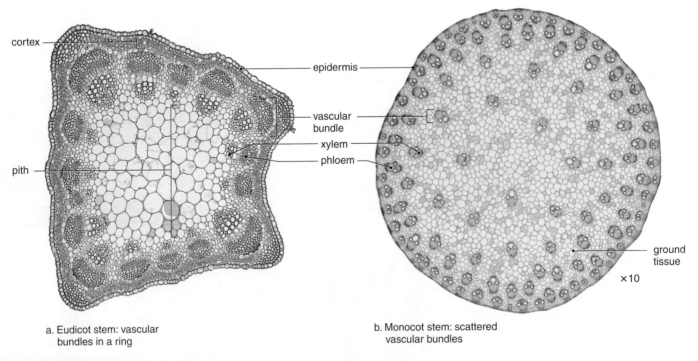

cortex

pith

epidermis

vascular bundle

xylem

phloem

ground tissue

×10

a. Eudicot stem: vascular bundles in a ring

b. Monocot stem: scattered vascular bundles

Figure 20.11 Nonwoody (herbaceous) stems.

a. In eudicot stems, the vascular bundles are arranged in a ring around well-defined ground tissue called pith. **b.** In monocot stems, the vascular bundles are scattered within the ground tissue.

green and carries on photosynthesis. The so-called **pith** consists of ground tissue and is found in the center. A monocot stem lacks an organized cortex or pith.

Herbaceous stems have distinctive vascular bundles, where xylem and phloem are found. In each bundle, xylem is typically found toward the inside of the stem, and phloem is found toward the outside. In the herbaceous eudicot stem, the vascular bundles are arranged in a distinct ring that separates the cortex from the pith. In a herbaceous monocot stem, the vascular bundles are scattered throughout the stem. **Figure 20.11** contrasts herbaceous eudicot and monocot stems.

The vascular tissue of a stem supports the growth of the shoot system. The sclerenchyma cells of vascular tissue and the strong walls of vessel members and tracheids help support the shoot system as it increases in length. Also, the vascular bundles take water and minerals to the leaf veins, which distribute them to the mesophyll cells of leaves, the primary organs of photosynthesis in most plants. The vascular bundles also distribute the products of photosynthesis to the root system and to any immature leaves that are not yet able to carry on photosynthesis.

Woody Stems

Plants with woody stems, such as trees and shrubs, experience both primary and secondary growth. *Secondary growth* increases the girth of stems, branches, and roots. It occurs because of a difference in the location and activity of vascular cambium, which, as you may recall from Section 20.2, is a type of meristem tissue. In herbaceous eudicot plants, vascular cambium is usually present between the xylem and phloem of each vascular bundle. In woody plants, the vascular cambium forms a ring of meristem that divides parallel to the surface of the plant and produces new xylem and phloem each year. Eventually, a woody eudicot stem has three distinct areas: the bark, the wood, and the pith. Vascular cambium occurs between the bark and the wood (**Fig. 20.12**).

Animation
Woody Dicot

Bark

The **bark** of a tree contains cork, cork cambium, cortex, and phloem. It is very harmful to remove the bark of a tree, because without phloem, organic nutrients cannot be transported. Although new phloem tissue is produced each year by vascular cambium, it does not build up in the same manner as xylem (Fig. 20.12).

Cork cambium is located beneath the epidermis. When cork cambium first begins to divide, it produces tissue that disrupts the epidermis and replaces it with cork cells. **Cork cells** are impregnated with suberin, a waxy layer that makes them waterproof but also causes them to die. In a woody stem, gas exchange is impeded, except at *lenticels,* which are pockets of loosely arranged cork cells not impregnated with suberin.

Wood

Wood is *secondary xylem* that builds up year after year, thereby increasing the girth of trees. In trees that have a growing season, vascular cambium is dormant during the winter. In the spring, when moisture is plentiful and leaves require much water for growth, vascular cambium produces secondary xylem tissue that contains wide vessels with thin walls. In this so-called *spring wood,* wide vessels transport sufficient water to the growing leaves. Later in the season, moisture is

Connections and Misconceptions

What trees are used to produce cork bottle stoppers?

Cork stoppers, such as those traditionally found in wine bottles, are manufactured almost exclusively in Spain and Portugal from the cork oak tree (*Quercus suber*). Cork from these trees can be harvested once the tree reaches an age of about 20 years. Then, every 9 years, the outer 1–2 inches of cork may be removed from the tree without harming it. This yields about 16 kilograms (35 pounds) of cork per tree, and cork trees can continue to produce cork for around 150 years. Cork is considered to be an environmentally friendly forest product because the trees are not harmed during its production.

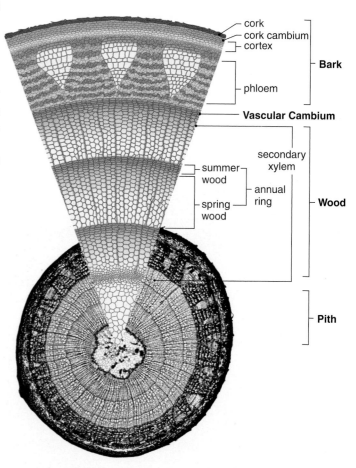

Figure 20.12 Organization of a woody stem.

In a woody stem, vascular cambium produces new secondary phloem and secondary xylem each year. Secondary xylem builds up to form wood consisting of annual rings. Counting the annual rings tells you that this woody stem is three years old.

scarce and the wood at this time, called *summer wood,* has a lower proportion of vessels. Strength is required because the tree is growing larger, and summer wood contains numerous thick-walled tracheids. At the end of the growing season, just before the cambium becomes dormant again, only heavy fibers with especially thick secondary walls may develop. When the trunk of a tree has spring wood followed by summer wood, the two together make up one year's growth, or an **annual ring** (Fig. 20.12). You can tell the age of a tree by counting the annual rings.

Wood is one of the most useful and versatile materials known. It is used for building structures and for making furniture; in addition to its functionality, its beauty gives us pleasure. Much wood is also used for heating and cooking, as well as for producing paper, chemicals, and pharmaceuticals. The resins derived from wood also have commercial importance.

Connecting the Concepts

For more information on the importance of forests and wood, refer to the following discussions.

Section 18.2 details the economic importance of the wood obtained from trees.

Section 32.2 examines some of the direct and indirect benefits of forest conservation.

Check Your Progress 20.4

1. Contrast a nonwoody stem with a woody stem.
2. List the components of bark.
3. Explain what causes the annual rings in wood and summarize the difference between summer wood and spring wood.

20.5 Organization of Roots

Learning Outcomes

Upon completion of this section, you should be able to

1. List the three zones of a eudicot root tip.
2. Contrast the organization of vascular tissue in a monocot and eudicot root.
3. List the five tissues of a root and briefly describe their functions.

Both monocot and eudicot roots contain the same growth zones. **Figure 20.13,** a longitudinal section of a eudicot root, reveals these zones where cells are in various stages of differentiation as primary growth occurs. The apical meristem of a root contains actively dividing cells and is protected by a **root cap** (Fig. 20.13). As the cells of the apical meristem divide, it provides cells to the zone of elongation. In the zone of elongation, the cells lengthen as they become specialized. In the zone of maturation, which contains fully differentiated cells, many of its epidermal cells bear root hairs.

Tissues of a Root

The structure and function of tissues in a root are examined in the following sections.

Vascular Tissue

Both monocot and eudicot roots have a **vascular cylinder** that contains xylem and phloem, but they are arranged differently. In a eudicot root (Fig. 20.13) the xylem is star-shaped because several xylem arms radiate from a common center. Phloem is found in separate regions between the points of the star. In a

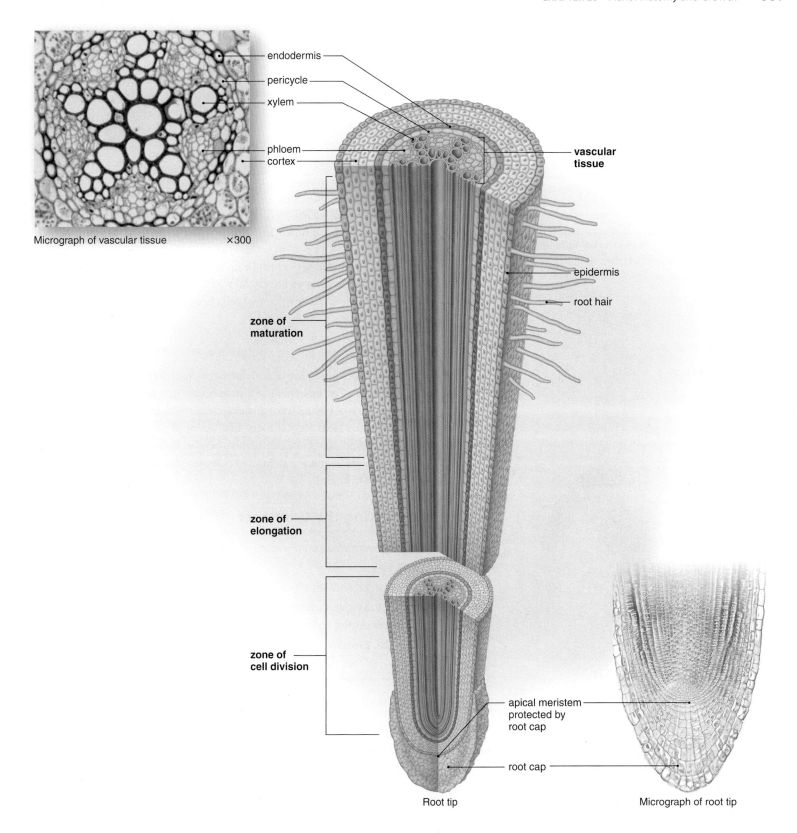

Micrograph of vascular tissue   ×300

endodermis
pericycle
xylem
phloem
cortex

vascular tissue

epidermis
root hair

zone of maturation

zone of elongation

zone of cell division

apical meristem protected by root cap

root cap

Root tip

Micrograph of root tip

Figure 20.13 Eudicot root tip.

The root tip is divided into three zones, best seen in a longitudinal section, such as this. In the eudicot, xylem is typically star-shaped, and phloem lies between the points of the star. Water and minerals must eventually pass through the cytoplasm of endodermal cells in order to enter xylem; endodermal cells regulate the passage of minerals into xylem.

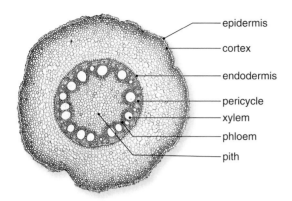

Figure 20.14 **Monocot root.**

In a monocot root, the same zones and tissues are present, but xylem and phloem form a ring surrounding the pith in the center.

monocot root (**Fig. 20.14**), the vascular cylinder consists of alternating xylem and phloem bundles that surround a pith. Pith can function as a storage site.

Pericycle The **pericycle** is the first layer of cells inside the endodermis. It has retained its capacity to divide and can start the development of branch, or lateral, roots (**Fig. 20.15**).

Endodermis The **endodermis** is a single layer of rectangular cells that fit snugly together. Further, a layer of impermeable material on all but two sides forces water and minerals to pass through endodermal cells. In this way, the endodermis regulates the entrance of minerals into the vascular tissue of the root.

Cortex Large, thin-walled parenchyma cells make up the cortex of the root. The cells contain starch granules, and the cortex may function in food storage.

Epidermis The epidermis, which forms the outer layer of the root, consists of only a single layer of largely thin-walled, rectangular cells. In the zone of maturation, many epidermal cells have root hairs.

Growth of Roots

Both roots and stems have apical meristem regions that account for their primary growth. The branching of a stem occurs at a lateral bud, whereas the branching of a root occurs some distance from apical meristem and is initiated by the pericycle. Most eudicot roots, like woody stems, experience secondary growth. Vascular cambium located between the arms of the xylem and the phloem produces secondary xylem and phloem. As secondary xylem builds up, a root increases in girth.

Connecting the Concepts

For more information on the roots of plants, refer to the following discussions.

Section 18.2 examines the evolution of roots in the plant kingdom.

Section 18.3 explores how the mutualistic relationship between plants and fungi increases the absorption of water and nutrients.

Section 31.2 explains how root nodules are involved in the cycling of nitrogen.

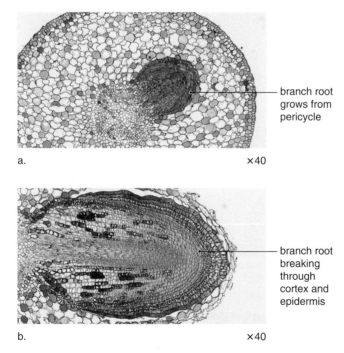

a. ×40

branch root grows from pericycle

branch root breaking through cortex and epidermis

b. ×40

Figure 20.15 **Branching of eudicot root.**

These cross sections of a willow show the origination (**a**) and growth (**b**) of a branch root from the pericycle.

Check Your Progress 20.5

1. Describe the relationship between the root apical meristem and the root cap.
2. List the functions of the tissues of a eudicot root.
3. Compare and contrast roots with stems.

20.6 Plant Nutrition

Learning Outcomes

Upon completion of this section, you should be able to

1. Differentiate between macronutrients and micronutrients and list the macronutrients and micronutrients that plants require.
2. Describe some adaptations of roots for obtaining minerals from the soil.

Plant nutrition is remarkable to us because plants require only inorganic nutrients, and from these they make all the organic compounds that compose their bodies. Of course, they require carbon, hydrogen, and oxygen, which they can acquire from carbon dioxide and water, but they also need other elements, or minerals, which they obtain from the environment. An element is termed an essential nutrient if a plant cannot live without it. The essential nutrients are divided into *macronutrients* and *micronutrients,* according to their relative concentrations in plant tissue. The following diagram indicates which are the macronutrients and which are the micronutrients:

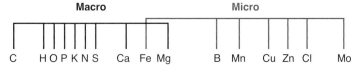

Notice that iron (Fe) is considered to be a micronutrient for some plants and a macronutrient for others. All of these elements play vital roles in plant cells. For example, nitrogen and phosphorus are required to make nucleic acids, and nitrogen and sulfur are important parts of amino acids. **Figure 20.16** shows how an insufficient supply of the macronutrient nitrogen can affect a plant. The micronutrients are often cofactors for enzymes in various metabolic pathways.

It's interesting to observe that humans make use of a plant's ability to take minerals from the soil. For example, we depend on plants for supplies of calcium to build our bones and teeth, as well as for iron to help carry oxygen to our cells. Minerals such as copper and zinc, which we acquire by eating plants, are also cofactors for our own enzymes.

Connections and Misconceptions

What do the numbers on a bag of fertilizer mean?

When you buy a bag of fertilizer, you will notice three numbers on the outside of the bag (for example, 18-12-14). These are called the NPK ratio, and the numbers refer to the amount of nitrogen (N), phosphorus (P), and potassium (K) in the fertilizer. All three of these are macronutrients that are important for plant health and growth. Nitrogen promotes the vegetative growth of the plant, while phosphorus is needed by the plant to maintain a healthy root system. Potassium is also involved in the general health of a plant and is needed for the formation of the chlorophyll molecules involved in photosynthesis. The optimal amount of each nutrient is dependent on the type of plant; thus, larger numbers do not always signify a better fertilizer.

Figure 20.16 Nitrogen deficiency.

This experiment shows that sunflower plants respond poorly if their growth medium lacks nitrogen.

Adaptations of Roots for Mineral Uptake

Minerals enter a plant at its root system, and two mutualistic relationships assist roots in fulfilling this function. The air all about us contains about 78% nitrogen (N_2), but plants can't make use of it. Most plants depend on bacteria in the soil to *fix* nitrogen—that is, the bacteria change atmospheric nitrogen (N_2) to nitrate (NO_3^-) or ammonium (NH_4^+), both of which plants can take up and use. Some plants, such as legumes, have roots colonized by bacteria that are able to take up atmospheric nitrogen and reduce it to a form suitable for incorporation into organic compounds (**Fig. 20.17***a*). The bacteria live in **root nodules,** and the plant supplies the bacteria with carbohydrates; the bacteria in turn furnish the plant with nitrogen compounds.

Animation
Root Nodule
Formation

Another mutualistic relationship involves fungi and almost all plant roots (Fig. 20.17*b*). This association is known as a **mycorrhizal association,** which literally means fungal roots. The hyphae of the fungus increase the surface area available for water uptake and break down organic matter, releasing inorganic nutrients that the plant can use. In return, the root furnishes the fungus with sugars and amino acids. Plants are extremely dependent on mycorrhizal fungi. For example, orchid seeds, which are quite small and contain limited nutrients, do not germinate until a mycorrhizal fungus has invaded their cells. The visual appearance of plants that do not have adequate nitrogen convinces us how important these associations are (see Fig. 20.16).

Connecting the Concepts

For more information on the material presented in this section, refer to the following discussions.

Figure 2.3 provides a periodic table of the elements.

Section 17.3 examines how soil bacteria benefit plants.

Section 18.3 explores how the mutualistic relationship between plants and fungi increases the absorption of nutrients.

Check Your Progress 20.6

1. List the macronutrients needed for plant growth.
2. Explain how plants obtain nitrogen from the air.
3. Discuss what might happen to a plant if there are no mutualistic relationships with fungi.

---

root —

root nodule —

Figure 20.17 **Adaptations of roots for mineral uptake.**

a. Nitrogen-fixing bacteria live in nodules on the roots of plants, particularly legumes. **b.** Rough lemon plants with mycorrhizal fungi (right) grow much better than plants without mycorrhizal fungi (left).

a.   b.

20.7 Transport of Nutrients

Water and Mineral Transport in Xylem

Water and minerals are taken up by root hairs at the same time, and they are both transported within xylem. The vessel elements, assisted by the tracheids, constitute a continuous system for water and mineral transport in a plant (see Fig. 20.7). In other words, a pipeline for water transport exists from the roots through the stem to the leaves. People have long wondered how plants, especially very tall trees, lift water to the leaves against gravity. Water enters root cells by osmosis, and sure enough, this does create a positive pressure called root pressure. But this pressure is not nearly enough to account for the movement of water and minerals all the way to the leaves.

The mechanism by which water and minerals travel up the xylem is called the **cohesion-tension model.** To understand how it works, we have to turn our attention to the anatomy of leaves and the properties of water. A leaf, as you know, has small openings called stomata, and the stomata open to air spaces in the leaf. Dry air passing across the leaves causes water to evaporate from the surface of spongy mesophyll cells. Water vapor in the air spaces of a leaf exits a leaf at the stomata, a phenomenon called **transpiration.** The evaporation of water creates a *tension* that is sufficient to pull a water column up from the roots to the leaves (**Fig. 20.18**). Why? Because water is both cohesive and adhesive. Water molecules are cohesive—they cling tightly to each other because of hydrogen bonding. As a result, the water molecules tend to stay together in a column as the molecules are pulled upward by transpiration. Also, water molecules are adhesive, because they are polar and are attracted to polar surfaces, such as the sides of the xylem vessels. The adhesion of water molecules to the vessel walls prevents them from slipping back.

The total amount of water a plant loses through transpiration over a long period of time is surprisingly large. At least 90% of the water taken up by roots is eventually lost at the leaves. A single corn plant loses between 135 and 200 liters of water through transpiration during a growing season. As transpiration occurs, the water column is pulled upward—first within the leaf, then from the stem, and finally from the roots.

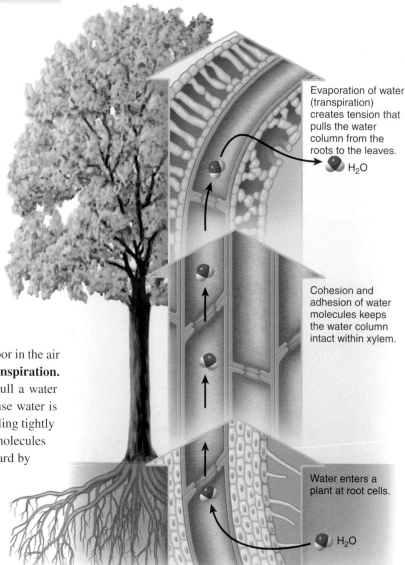

Evaporation of water (transpiration) creates tension that pulls the water column from the roots to the leaves.

H_2O

Cohesion and adhesion of water molecules keeps the water column intact within xylem.

Water enters a plant at root cells.

H_2O

Figure 20.18 **Cohesion-tension model of xylem transport.**

How does water rise to the top of tall trees? Xylem vessels are water-filled pipelines from the roots to the leaves. When water evaporates from spongy mesophyll into the air spaces of leaves, this water column is pulled upward due to the cohesion of water molecules with one another and the adhesion of water molecules to the sides of the vessels.

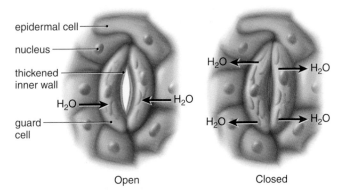

epidermal cell

nucleus

thickened inner wall

guard cell

H_2O

H_2O H_2O

H_2O H_2O

H_2O H_2O

Open   Closed

Figure 20.19 Opening and closing of stomata.

Stomata open when water enters guard cells and turgor pressure increases. Stomata close when water exits guard cells and turgor pressure is lost.

Opening and Closing of Stomata

Stomata open and close due to turgor pressure changes within guard cells (**Fig. 20.19**). When water enters the guard cells, turgor pressure is created; when water leaves the guard cells, turgor pressure is lost. For transpiration to occur, the stomata must stay open. But when a plant is under stress and about to wilt from lack of water, the stomata close. Now the plant is unable to take up carbon dioxide from the air, and photosynthesis ceases.

In plants kept in the dark, stomata open and close on a 24-hour basis, just as if they were responding to the presence of sunlight in the daytime and the absence of sunlight at night. Such circadian rhythms (behaviors that occur every 24 hours) are areas of intense investigation.

Organic Nutrient Transport in Phloem

Phloem transports organic nutrients from stored supplies in roots, or from mature leaves that are photosynthesizing, to all other parts of a plant, including young leaves that have not yet reached their full photosynthetic potential and flowers in the process of making seeds and fruits. Nutrients are transported to any part of a plant, including roots, when they lack a ready supply.

Just as xylem is continuous throughout a plant, so is phloem. In phloem, the sieve-tube members align end to end, and plasmodesmata extend through sieve plates from one sieve-tube member to the other, allowing direct cytoplasmic exchange of organic materials. Sieve tubes, therefore, form a continuous pathway for the transport of organic nutrients, primarily the sugar sucrose.

The mechanism by which nutrients are transported through phloem is called the **pressure-flow model** (**Fig. 20.20**). During the growing season, photosynthesizing leaves are making sugar and are therefore a *source* of sugar. This sugar is actively transported into sieve tubes. As the concentration of sugar rises inside the sieve tube cells, water follows by osmosis, since the concentration of water outside phloem is now lower than inside phloem. The buildup of water within sieve tubes creates a positive pressure that starts a flow of phloem contents. The other parts of a plant, such as roots that are not photosynthesizing, are a *sink* for sugar, because they require a supply of sugar. Here, sugar is actively transported out of phloem, and water again follows by osmosis. In this way, phloem contents continue to flow from a source (e.g., leaves) to a sink (e.g., root).

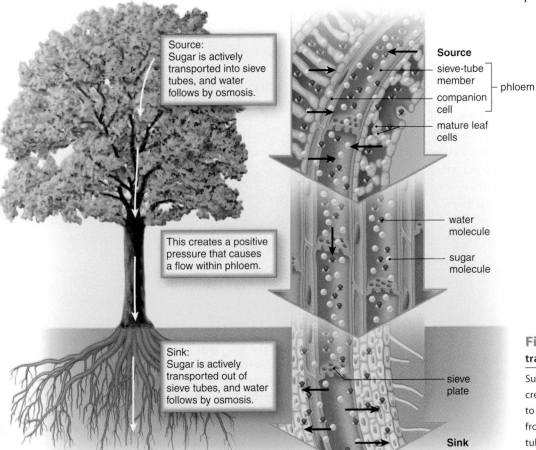

Source:
Sugar is actively transported into sieve tubes, and water follows by osmosis.

This creates a positive pressure that causes a flow within phloem.

Sink:
Sugar is actively transported out of sieve tubes, and water follows by osmosis.

Source

sieve-tube member

companion cell

phloem

mature leaf cells

water molecule

sugar molecule

sieve plate

Sink

Figure 20.20 Pressure-flow model of phloem transport.

Sugar and water enter sieve-tube members at a source. This creates a positive pressure, which causes phloem contents to flow. Sieve-tube members form a continuous pipeline from a source to a sink, where sugar and water exit sieve-tube members.

The experiment described in **Figure 20.21** supports the pressure-flow model of phloem transport. Two bulbs connected by a glass tube are placed in a split container of water. Each bulb is bounded by a membrane that is permeable to water, but not to sugar. The first bulb, the source, contains a concentrated sugar solution, while the second bulb, the sink, contains a dilute sugar solution. Analogous to phloem, water flows into the first bulb by osmosis because the first bulb contains a lower water concentration (higher solute concentration) than does the container. The entrance of water creates a *pressure* that causes water to *flow* toward the second bulb. Not only does this pressure drive the solution to the second bulb, but it is strong enough to overcome the force of osmosis. Water moves out through the membrane of the second bulb, even though the second bulb contains a lower water concentration than does the container.

Thus, the pressure-flow model can account for the flow of sugar in either direction, depending on which is the source and which is the sink. For example, in the spring, recently formed leaves can be a sink, and roots that have stored sugar over the winter can be a source. Thus, organic nutrients will flow from the roots up to the leaves, instead of the other way around.

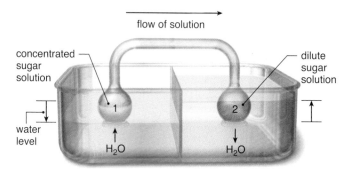

Figure 20.21   Experiment supporting the pressure-flow model.

When a concentrated sugar solution is placed in the first bulb, water enters and causes a flow of both water and sugar to the second bulb. The pressure is sufficient to cause water to exit the second bulb, even though the surrounding water contains no sugar. As water enters the first bulb, water level in the container decreases (on left); as water exits the second bulb, water level increases (on right).

Connecting the Concepts

For more information on the properties of water, refer to the following discussions.

Section 2.2 explains how hydrogen bonding between water molecules results in cohesion.

Section 5.4 examines how water and solute concentration contribute to osmotic pressure.

Check Your Progress 20.7

❶ Explain the role of transpiration in xylem transport.

❷ Detail how the cohesive and adhesive properties of water assist water transport in xylem.

❸ Describe how the pressure-flow model experiment explains the movement of sugar in the phloem.

Media Study Tools

www.mhhe.com/maderessentials3

Enhance your study of this chapter with study tools and practice tests. Also ask your instructor about the resources available through ConnectPlus, including the media-rich eBook, interactive learning tools, and animations.

Plant Transpiration

The virtual lab "Plant Transpiration" provides a more detailed look at the environmental conditions associated with the transpiration of water from the leaves of a plant.

The Chapter in Review

Summary

20.1 Plant Organs

A flowering plant has a shoot system and a root system. Both grow from meristems at their tips.

Leaves carry on photosynthesis but may be modified for other purposes.

Stems support leaves, conduct materials to and from roots and leaves, and produce new tissues.

Roots anchor a plant, absorb water and minerals, and store the products of photosynthesis.

Monocot Versus Eudicot Plants

Flowering plants are divided into monocots and eudicots according to the following

- Number of cotyledons in the seed.
- Arrangement of vascular tissue in leaves, stems, and roots.
- Number of flower parts.

monocot flower   eudicot flower

20.2 Plant Tissues and Cells

Plants have three types of specialized tissue:

Epidermal tissue is composed of only epidermal cells.

Ground tissue contains the following:
- Parenchyma cells, which are thin-walled and capable of photosynthesis when they contain chloroplasts
- Collenchyma cells, which have thicker walls for flexible support
- Sclerenchyma cells, which are hollow, nonliving support cells with secondary walls

Vascular tissue consists of the following:
- Xylem, which contains vessels composed of vessel elements and tracheids. Xylem transports water and minerals.
- Phloem, which contains sieve tubes composed of sieve-tube members, each of which has a companion cell. Phloem transports organic nutrients.

xylem   phloem

20.3 Organization of Leaves

- A leaf has a cuticle-covered epidermis, with stomata mostly in the lower layer.
- Stomata allow water vapor and oxygen to escape and carbon dioxide to enter the leaf.
- Mesophyll (palisade and spongy) forms the body of a leaf and carries on photosynthesis.

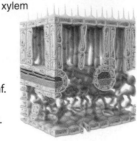

20.4 Organization of Stems

Primary growth of a stem is due to the activity of the shoot apical meristem, which is protected within a terminal bud.

Nonwoody Stems

- Nonwoody (herbaceous) eudicots have epidermis, cortex tissue, vascular bundles in a ring, and an inner pith.
- Monocot stems have scattered vascular bundles and lack a distinct cortex or pith.

eudicot stem   monocot stem

Woody Stems

- Secondary growth of a woody stem is due to vascular cambium, which produces new xylem and phloem every year.
- Cork replaces epidermis in woody plants.
- A woody stem has bark (contains cork, cork cambium, cortex, and phloem). Wood contains annual rings of xylem.

20.5 Organization of Roots

A monocot and a eudicot root tip has three zones:
- Zone of cell division (contains root apical meristem) protected by the root cap
- Zone of elongation, where cells elongate and differentiate
- Zone of maturation (has root hairs)

Tissues of a Root

Within the zone of maturation, a monocot and a eudicot root contains the following tissues:

- **Vascular tissue:** In eudicot roots, xylem is arranged in a star shape, with phloem between the arms of the xylem. In monocot roots, a central pith is surrounded by a ring of vascular tissue containing alternating bundles of xylem and phloem.
- **Pericycle:** Cells that can divide and form lateral roots
- **Endodermis:** Regulates entrance of minerals into vascular tissue
- **Cortex:** Parenchyma cells that function in food storage
- **Epidermis:** A single layer of cells that forms the outer layer and may possess root hairs

Growth of Roots

Like stems, roots undergo primary growth from an apical meristem. Stems branch from a lateral bud, but roots branch some distance from the apical meristem. Root branching is initiated by the pericycle.

20.6 Plant Nutrition

Plants need only inorganic nutrients to make all the organic compounds that make up their bodies. Some nutrients are essential, being either macronutrients or micronutrients.

Adaptations of Roots for Mineral Uptake

Some roots have nodules where bacteria fix nitrogen and produce forms that plants can use. Most roots have mycorrhizal fungi.

20.7 Transport of Nutrients

Water and Mineral Transport in Xylem

The cohesion-tension model of xylem transport states that transpiration (evaporation of water from spongy mesophyll at stomata) creates tension, which pulls water upward in xylem. This method works only because water molecules are cohesive and adhesive and form a water column in xylem.

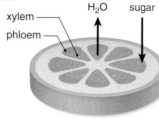

Organic Nutrient Transport in Phloem

The pressure-flow model of phloem transport states that sugar is actively transported into phloem at a source, and water follows by osmosis. The resulting increase in pressure creates a flow, which moves water and sugar to a sink.

Key Terms

annual ring 380	palisade mesophyll 377
apical meristem 374	parenchyma 374
bark 379	perennial plant 372
blade 376	pericycle 382
cohesion-tension model 385	petiole 376
collenchyma 374	phloem 373
cork cambium 379	pith 379
cork cell 379	pressure-flow model 386
cortex 378	root cap 380
cotyledon 373	root hair 372
cuticle 374	root nodule 384
deciduous 371	root system 371
endodermis 382	sclerenchyma 375
epidermal tissue 374	shoot system 371
epidermis 374	sieve-tube member 376
eudicot 373	spongy mesophyll 377
evergreen 371	stem 371
ground tissue 374	stomata (sing., stoma) 374
herbaceous 378	terminal bud 371
internode 371	tracheid 376
lateral bud 371	transpiration 385
leaf 371	vascular bundle 373
leaf vein 373	vascular cambium 374
meristem 374	vascular cylinder 380
mesophyll 377	vascular tissue 374
monocot 373	vessel element 376
mycorrhizal association 384	wood 379
node 371	xylem 373

Testing Yourself

Choose the best answer for each question.

1. It is possible to distinguish between stems and roots by looking for
   a. lateral branching on stems.   c. nodes on stems.
   b. petioles on roots.   d. nodes on roots.

2. A monocot stem differs from a eudicot stem because in the monocot stem,
   a. xylem and phloem are in a ring surrounding the pith.
   b. xylem and phloem are in bundles scattered throughout the stem.
   c. there is an organized cortex and pith.
   d. xylem forms a star in the center, with phloem between the points of the star.
   e. the epidermis does not have a waxy cuticle surrounding it.

3. Because of transpiration, water
   a. evaporates directly from the leaf surface.
   b. flows from leaf veins through the stem, toward the roots.
   c. exits the leaf through stomata, pulling water into the leaf from leaf veins.
   d. exits xylem in leaf veins and enters phloem by osmosis.

4. Hard materials in plants, such as the husks of nuts, are composed primarily of
   a. vascular tissue.   d. parenchyma cells.
   b. collenchyma cells.   e. epidermal tissue.
   c. sclerenchyma cells.

5. Label the parts of a plant in the following diagram.

For questions 6–10, identify the plant organ that contains each feature. Each answer may be used more than once. Each question may have more than one answer.

Key:
   a. nonwoody stem   c. root (nonwoody)
   b. woody stem   d. leaf

6. cork cambium
7. vascular bundles
8. ground tissue
9. vascular cambium
10. apical meristem

11. During secondary growth, a tree adds more xylem and phloem through the activity of the
    a. apical meristems.
    b. vascular cambium.
    c. pericycle.
    d. cork cambium.

12. Root nodules are important because they
    a. encourage the growth of mycorrhizal fungi.
    b. represent areas of extensive branch root growth.
    c. contain nitrogen-fixing bacteria.
    d. provide extra oxygen to the plant root system.

13. The closing of stomata and the presence of a waxy cuticle act similarly to
    a. minimize water loss.
    b. enhance gas exchange.
    c. enhance water absorption.
    d. allow nitrogen fixation.

14. According to the pressure-flow model, sugar is actively transported into phloem and
    a. enters xylem, where it is moved toward the leaves due to transpiration.
    b. creates pressure to move water toward the roots.
    c. is transported out of the leaves through stomata.
    d. water follows by osmosis, providing a pressure that moves sugar along.

15. Label the parts of a leaf in the following illustration.

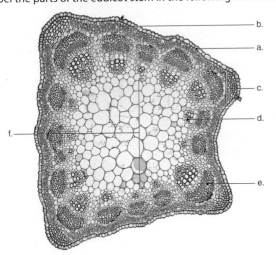

16. Label the parts of the eudicot stem in the following illustration.

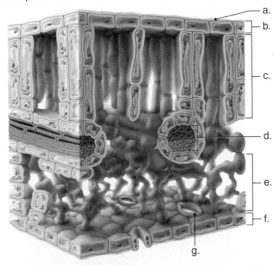

17. Label the parts of the root in the following illustration.

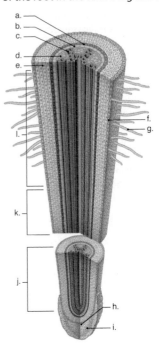

Thinking Scientifically

1. Scientists observe that the roots of legumes grow around nitrogen-fixing bacteria capable of forming nodules. Discuss how the roots might recognize these bacteria (see Fig. 4.6).

2. Plants of the genus *Welwitschia* live in the deserts in Africa. Annual rainfall averages only 2.5 centimeters per year, but every night a fog rolls in. Why might these plants be adapted to opening their stomata at night? What about the fog allows the plant to survive?

Bioethical Issue

Does Forest Thinning Prevent Wildfires?

In 2007, forest fires devastated large areas in California and revived the debate over forest fire prevention. Many blamed the numerous wildfires on poor management of national forests, proposing that limited logging of old-growth forests on federal lands occur. This logging, they contend, would clear flammable underbrush and thin out forests, so that fires could not spread as easily.

Many environmentalists, however, are skeptical. They point out that thinning the forests not only would disturb critical habitat for many endangered species but also could make forests even more combustible by allowing more sunlight to penetrate the canopy and dry out organic matter on the forest floor. Many feel that the proposed logging would merely enrich the timber industry without accomplishing much.

In light of the loss of life and property due to these devastating wildfires, should the federal government allow the thinning of national forests? Or would this only exacerbate the problem?

21

Plant Responses and Reproduction

Patents for Genetically Engineered Plants

Issuing patents for plants is not a new practice. The first plant to be patented was a peach tree in 1932, and since then numerous plants have been patented. In fact, the United States Patent and Trademark Office has a special patent class just for asexually reproducing plants.

Why might you patent a plant? Plant breeders develop plants that are more beautiful, more useful, or more resistant to disease. A patent gives the plant breeder exclusive rights to sell a particular plant for 20 years. One of the more controversial plant patents is the one for Roundup Ready® plants. These plants have been genetically modified to be resistant to Roundup, a herbicide widely used to kill weeds that compete with plants for nutrients and water. The benefit of Roundup Ready wheat, cotton, or soybeans is that, after fields are sprayed with Roundup, the crop survives but the weeds are destroyed. This method of fighting weeds saves time and labor compared with traditional means of controlling weeds. Although Roundup Ready crops have these benefits, what about possible risks? The biggest drawback to Roundup Ready crops is that their pollen could transmit the genes for herbicide resistance to other plants—weeds, in particular—and then these weeds would not be controlled by Roundup. A catastrophe could be in the making.

In this chapter, you will learn about other ways plants of commercial importance have been genetically engineered. But first, the chapter introduces the plant hormones involved in plant responses and flowering. It also expands on the content in Chapter 18 by providing more information on the processes of sexual and asexual reproduction in flowering plants.

OUTLINE

BEFORE YOU BEGIN

Before beginning this chapter, take a few moments to review the following discussions.

Section 4.5 What is the role of the cell wall in a plant cell?

Figure 18.16 How does the sporophyte stage differ from the gametophyte stage in a flowering plant?

Section 20.2 What are the four types of tissue found in a plant?

21.1 Plant Hormones

Learning Outcomes

Upon completion of this section, you should be able to

1. List the five classes of plant hormones and briefly describe their effects on plant growth.
2. Define apical dominance and explain the effect of auxin on it.
3. Explain how gibberellins cause a plant to break dormancy.
4. Define senescence and explain how cytokinins prevent it.
5. Explain the role of abscisic acid in dormancy and in the closure of stomata.
6. Describe the effect of ethylene on fruit and explain how it brings about abscission in a plant.

Plants usually respond to environmental stimuli, such as light, gravity, and seasonal changes, by altering their pattern of growth in some way. Most of these responses occur at the cellular level and are mediated by hormones. Plant hormones are small, organic molecules produced by the plant that serve as chemical signals between cells and tissues. Currently, the five commonly recognized groups of plant hormones are auxins, gibberellins, cytokinins, abscisic acid, and ethylene.

Video
No Mow Grass

Auxins

The most common naturally occurring **auxin** is indoleacetic acid (IAA). It is produced in the shoot apical meristem and is found in young leaves and in

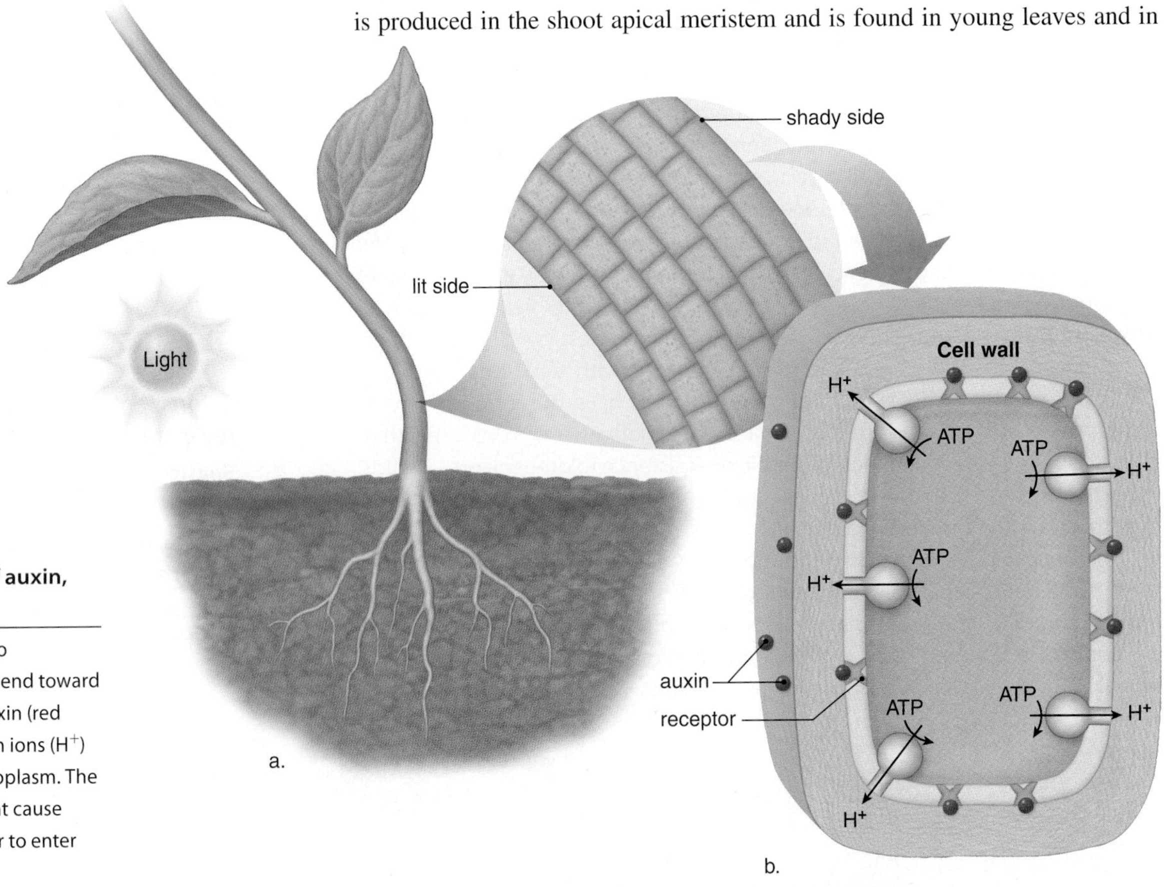

Figure 21.1 Mode of action of auxin, a plant hormone.

a. Plant cells on the shady side undergo elongation, and this causes a stem to bend toward the light. **b.** Elongation occurs after auxin (red balls) binds to a receptor and hydrogen ions (H$^+$) are actively transported out of the cytoplasm. The resulting acidity activates enzymes that cause the cell wall to weaken and allow water to enter the cell. The cell then elongates.

flowers and fruits. Therefore, you would expect auxin to affect many aspects of plant growth and development, and it does.

Auxin is involved in phototropism, observable when stems bend toward a light source (**Fig. 21.1**). When a plant is exposed to unidirectional light, auxin moves to the shady side, where it binds to receptors and activates an ATP-driven pump that transports hydrogen ions (H^+) out of the cell into the cell wall. The acidic environment weakens cellulose fibrils, and activated enzymes further degrade the cell wall. Water then enters the cell, and the resulting increase in turgor pressure causes the cells on the shady side to elongate and the stem to bend toward the light.

No doubt you have noticed that plants seldom produce branches at or near the shoot tip. This phenomenon is due to **apical dominance** caused by the action of auxin (**Fig. 21.2**). Auxin produced in the apical meristem of the terminal bud is transported downward, inhibiting the growth of lateral buds. Release from apical dominance occurs when pruning removes the shoot tip. Then, the lateral buds grow and the plant takes on a bushier appearance. When auxin is applied to the stump, apical dominance is restored.

Auxin also stimulates root formation. If you apply a paste that contains auxin to a stem cutting, roots begin to grow. Similarly, auxin production by seeds promotes the growth of fruit. And as long as auxin is concentrated in leaves or in fruits, they do not fall off the plant. Therefore, trees are often sprayed with auxin to keep mature fruit from falling to the ground.

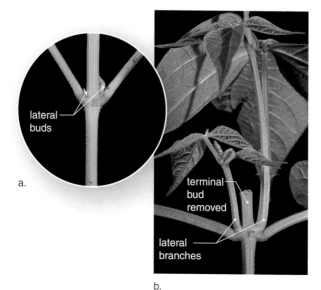

a.

b.

Figure 21.2 Apical dominance.

a. Lateral bud growth is inhibited when a plant retains its terminal bud. **b.** When the terminal bud is removed, lateral branches develop and the plant is bushier.

Gibberellins

Gibberellins were discovered in 1926 when a Japanese scientist was investigating a fungal disease of rice plants called "foolish seedling disease." Rapid stem elongation weakened the plants and caused them to collapse. The fungus infecting the plants produced an excess of a chemical called gibberellin, named after the fungus *Gibberella fujikuroi*. It wasn't until 1956 that a form of gibberellin now known as gibberellic acid was isolated from a flowering plant rather than from a fungus. We now know of about 70 gibberellins. The most common of these is gibberellic acid, GA_3 (the subscript designation distinguishes it from other gibberellins). Sources of gibberellin in flowering plant parts are young leaves, roots, embryos, seeds, and fruits.

Gibberellins are growth-promoting hormones that bring about elongation of the cells. When gibberellins are applied externally to plants, the most obvious effect is stem elongation between the nodes (**Fig. 21.3**). Gibberellins can cause dwarf plants to grow, cabbage plants to become as much as 2 meters tall, and bush beans to become pole beans.

Dormancy is a period during which a plant or seed does not grow, even though conditions may be favorable for growth. The dormancy of seeds and buds can be broken by applying gibberellins. Research with barley seeds has shown how GA_3 is involved in the germination of seeds. Endosperm is the tissue that serves as food for the embryo and seedling as they undergo development. Barley seeds have a large, starchy endosperm that must be broken down into sugars to provide energy for the embryo to grow. After GA_3 attaches to a receptor in the plasma membrane, a cell-signaling pathway (see Fig. 12.13) activates the gene that codes for amylase. Amylase then acts on starch to release sugars as a source of energy for embryonic and seedling growth.

Figure 21.3 Effect of gibberellins.

The plant on the right was treated with gibberellins; the plant on the left was not treated. Gibberellins are often used to promote stem elongation in economically important plants, but the exact mode of action remains unclear.

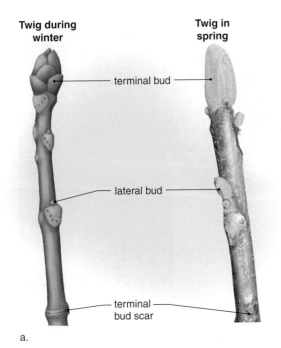

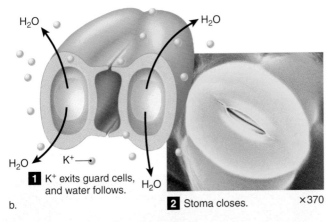

Figure 21.4 **Effects of abscisic acid.**

a. Abscisic acid encourages the formation of winter buds (left), and a reduction in the amount of abscisic acid breaks bud dormancy (right). **b.** Abscisic acid also brings about the closing of a stoma by influencing the movement of potassium ions (K^+) out of guard cells.

Cytokinins

Cytokinins were discovered as a result of attempts to grow plant tissues and organs in culture vessels in the 1940s. It was found that cell division occurs when coconut milk (a liquid endosperm) and yeast extract are added to the culture medium. Although the effective agent or agents could not be isolated at the time, they were collectively called cytokinins because, as you may recall, cytokinesis means division of the cytoplasm. A naturally occurring cytokinin was not isolated until 1967. Because it came from the kernels of maize (*Zea*), it was called zeatin.

Cytokinins influence plant growth by promoting cell division. Cytokinins are found in plant meristems, young leaves, root tips, and seeds and fruits. Whenever a plant grows, cytokinins are involved.

Other Effects of Cytokinins

Cytokinins also prevent **senescence,** or aging, of plant organs. As cytokinin levels drop within a plant organ, such as a leaf, growth slows or even stops. Then, the leaf loses its natural color as large molecules are broken down and transported to other parts of the plant. Senescence is a necessary part of a plant's growth. For example, as some plants grow taller, they naturally lose their lower leaves.

Plant organ formation is influenced by cytokinins. For example, researchers are well aware that the ratio of auxin to cytokinin and the acidity of the culture medium determine whether a plant tissue forms an undifferentiated mass, called a *callus,* or differentiates to form roots, vegetative shoots, leaves, or floral shoots. Some reports suggest that chemicals called oligosaccharins (chemical fragments released from the cell wall) are effective in directing differentiation. Perhaps the reception of auxin and cytokinins, which leads to the activation of enzymes, releases these fragments from the cell wall.

Abscisic Acid

Abscisic acid is sometimes called the stress hormone because it initiates and maintains seed and bud dormancy and brings about the closure of stomata. Dormancy has begun when a plant stops growing and prepares for adverse conditions (even though conditions at the time may be favorable for growth). For example, it is believed that abscisic acid moves from leaves to vegetative buds in the fall, and thereafter these buds are converted to winter buds. A winter bud is covered by thick, hardened scales (**Fig. 21.4***a*). A reduction in the level of abscisic acid and an increase in the level of gibberellins are believed to break seed and bud dormancy. Then seeds germinate, and buds send forth leaves.

Abscisic acid brings about the closing of stomata when a plant is under water stress (Fig. 21.4*b*). In some unknown way, abscisic acid causes potassium ions (K^+) to leave guard cells. Thereafter, the guard cells lose water, and a stoma closes.

It was once believed that abscisic acid functioned in **abscission,** the dropping of leaves, fruits, and flowers from a plant. Although the external application of abscisic acid promotes abscission, this hormone is no longer believed to function naturally in this process. Instead, the hormone ethylene, discussed next, is thought to bring about abscission.

Ethylene

Ethylene is a gas that can move freely in the air. Ethylene works with other hormones to bring about certain effects. For example, low levels of auxin and perhaps gibberellin, as compared with the levels in the stem, probably initiate abscission. But once the process of abscission has begun, ethylene stimulates certain enzymes, such as cellulase, which cause leaf, fruit, or flower drop (**Fig. 21.5***a,b*). Cellulase hydrolyzes cellulose in plant cell walls.

In the early 1900s, it was common practice to prepare citrus fruits for market by placing them in a room with a kerosene stove. Only later did researchers realize that ethylene, an incomplete combustion product of kerosene, was ripening the fruit. Because it is a gas, ethylene can act from a distance. A barrel of ripening apples can induce ripening in a bunch of bananas, even if they are in different containers. If a plant is wounded due to physical damage or infection, ethylene is released at the wound site. This is why one rotten apple spoils the whole barrel.

Table 21.1 summarizes the effects of the five groups of plant hormones.

a. No abscission   b. Abscission

c.

Figure 21.5 **Functions of ethylene.**

a. Normally, there is no abscission when a holly twig is placed under a glass jar for a week. **b.** When an ethylene-producing ripe apple is also under the jar, abscission of the holly leaves occurs. **c.** Ethylene causes fruits to ripen, making them luscious to eat.

Connecting the Concepts

For additional information on the topics presented in this section, refer to the following discussions.

Section 4.4 describes the structure of a general plant cell.

Section 20.3 provides additional information on the action of the stomata in the leaves.

Sections 20.4 and **20.5** describe the structure of plant stems and roots, respectively.

Check Your Progress 21.1

1. List the five commonly recognized groups of plant hormones.
2. Describe the main action of each of the five main plant hormones.
3. Discuss why, if all of the plant hormones have some effect on plant growth, they can be found in different areas of the plant.

Table 21.1 Functions of the Major Plant Hormones

Hormone	Functions	Where Produced or Found in Plant
Auxins	Maintain apical dominance; involved in phototropism and gravitropism; promote growth of roots in tissue culture; prevent leaf and fruit drop.	Apical meristems, immature leaves
Gibberellins	Promote stem elongation between nodes; break seed and bud dormancy and influence germination of seeds.	Apical meristem, immature leaves, seeds
Cytokinins	Promote cell division; prevent senescence; along with auxin, promote differentiation leading to roots, shoots, leaves, or floral shoots.	Root apical meristem
Abscisic acid	Initiate and maintain seed and bud dormancy; promote formation of winter buds; promote closure of stomata.	Endosperm, roots, tissues containing chloroplasts
Ethylene	Promote abscission (leaf, fruit, or flower drop); promote ripening of fruit.	Aging parts of plant and ripening fruit; apical meristem and nodes of stem

Figure 21.6 **Positive phototropism.**

The stem of a plant curves toward the light. This response is due to the accumulation of auxin on the shady side of the stem.

Figure 21.7 **Negative gravitropism.**

The stem of a plant curves away from the direction of gravity 24 hours after the plant is placed on its side. This response is due to the accumulation of auxin on the lower side of the stem.

21.2 Plant Responses

Plant responses are strongly influenced by such environmental stimuli as light, day length, gravity, and touch. The ability of a plant to respond to environmental signals fosters the survival of the plant and the species in that environment.

Plant responses to environmental signals can be rapid, as when stomata open in the presence of light, or they can take some time, as when a plant flowers in season. Despite their variety, most plant responses to environmental signals are due to growth and sometimes differentiation, brought about at least in part by certain hormones.

Video
Seedling Growth

Tropisms

Plant growth toward or away from a directional stimulus is called a **tropism.** Differential growth causes one side of an organ to elongate faster than the other, and the result is a curving toward or away from the stimulus. The following two well-known tropisms were each named for the stimulus that causes the response:

> Phototropism: a movement in response to a light stimulus
> Gravitropism: a movement in response to gravity

Growth toward a stimulus is called a positive tropism, and growth away from a stimulus is called a negative tropism. In **Figure 21.6**, a positive **phototropism** is illustrated as when a stem curves toward the light. **Figure 21.7** illustrates negative **gravitropism**—stems curve away from the direction of gravity. Roots, of course, exhibit positive gravitropism.

The role of auxin in the positive phototropism of stems has been studied for quite some time. Because blue light in particular causes phototropism to occur, it is believed that a yellow pigment related to the vitamin riboflavin acts as a photoreceptor for light. Following reception, auxin migrates from the bright side to the shady side of a stem. The cells on that side elongate faster than those on the bright side, causing the stem to curve toward the light (see Fig. 21.1). Negative gravitropism of stems occurs because auxin moves to the lower part of a stem when a plant is placed on its side.

Photoperiodism

Flowering in angiosperms is a striking response to environmental seasonal changes. In some plants, flowering occurs according to the **photoperiod,** which is the ratio of the length of day to the length of night over a 24-hour period. Plants can be divided into three groups:

1. **Short-day plants/long-night plants** flower when the day length is shorter and the night is longer than a definite length of time called the critical length. (Examples are cocklebur, poinsettia, and chrysanthemum.)

2. **Long-day plants/short-night plants** flower when the day is longer and the night is shorter than a critical length. (Examples are wheat, barley, clover, and spinach.)

3. **Day-neutral plants** do not depend on day/night length for flowering; instead, they rely on other environmental stimuli. (Examples are tomato and cucumber.)

Both long-day plants and short-day plants can have the same critical length. **Figure 21.8** illustrates that the cocklebur and the clover have the same critical length. The cocklebur flowers when the day is shorter (night is longer) than 8.5 hours, and clover flowers when the day is longer (night is shorter) than 8.5 hours.

Experiments have shown that the *length of continuous darkness, not light, controls flowering in many plants.* For example, the cocklebur will not flower if a suitable length of darkness is interrupted by a flash of light. On the other hand, clover will flower when an unsuitable length of darkness is interrupted by a flash of light. (Interrupting the light period with darkness has no effect on flowering.) Nurseries use these kinds of data to make all kinds of flowers available throughout the year (**Fig. 21.9**).

Phytochrome and Plant Flowering

If flowering is dependent on night length, plants must have some way to detect these periods. In some plants, this appears to be the role of a blue-green leaf pigment called **phytochrome.** The proportion of red light to far-red light determines the particular form of phytochrome:

P_r (phytochrome red) absorbs red light and is converted into P_{fr}.
P_{fr} (phytochrome far-red) absorbs far-red light and is converted into P_r.

During the day, sunlight contains more red light than far-red light, and P_r is converted into P_{fr}. But at dusk, more far-red light is available, and P_{fr} is converted to P_r. There is also a slow metabolic replacement of P_{fr} by P_r during the night. Is this the timing device that tells the plant the length of darkness? Researchers have been looking for a flowering hormone for many years, but perhaps phytochrome itself triggers flowering through a signaling pathway.

Animation
Phytochrome Signaling

Other Functions of Phytochrome

Apparently, the presence of P_{fr} indicates to some seeds that sunlight is present and conditions are favorable for germination. Such seeds must be only partly covered with soil when planted. Phytochrome may also affect leaf expansion and stem branching. In the absence of P_{fr}, stems elongate as a way to reach sunlight. Accordingly, seedlings grown in the dark have lengthened stems with smaller than normal leaves. But once the seedling is exposed to sunlight and P_{fr} is present, the seedling begins to grow normally—the leaves expand and the stem branches.

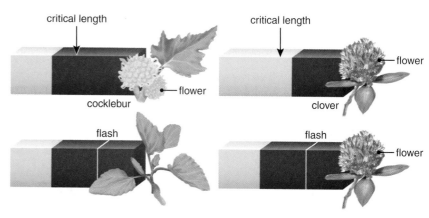

a. Short-day (long-night) plants   b. Long-day (short-night) plant

Figure 21.8 Length of darkness controls flowering.

a. The cocklebur flowers when days are short and nights are long (top). If a long night is interrupted by a flash of light, the cocklebur will not flower (bottom). **b.** Clover flowers when days are long and nights are short (top). If a long night is interrupted by a flash of light, clover will still flower (bottom). Therefore, we can conclude that the length of continuous darkness controls flowering.

Figure 21.9 Flowering.

Nurseries know how to regulate the photoperiod so that many types of flowers are available year-round.

Connecting the Concepts

For more information on the topics in this section, refer to the following discussions.

Figure 18.15 illustrates the structures of a flower.

Section 20.1 describes primary growth in plants.

Check Your Progress 21.2

❶ Compare and contrast positive and negative tropisms.

❷ Contrast short-day, long-day, and day-neutral plants.

❸ Predict what might occur without phytochromes in flowering plants.

21.3 Sexual Reproduction in Flowering Plants

Learning Outcomes

Upon completion of this section, you should be able to

1. Explain the alternation of generations between sporophyte and gametophyte in flowering plants.
2. Identify the parts of a flower and briefly define their functions.
3. Contrast a monoecious plant with a dioecious plant.
4. Describe the processes of pollen grain germination and double fertilization.
5. List some methods of seed dispersal.
6. Compare and contrast seed germination in a eudicot versus in a monocot.

In Chapter 18, we noted that plants have two multicellular stages in their life cycle, and therefore their life cycle is called an **alternation of generations.** In this life cycle, a diploid (2n) sporophyte alternates with a haploid (n) gametophyte:

The **sporophyte** (2n) produces haploid spores by meiosis. The spores develop into gametophytes.

The **gametophytes** (n) produce gametes. Upon fertilization, the cycle returns to the 2n sporophyte.

Overview of the Plant Life Cycle

Flowering plants have an alternation of generations life cycle, but with the modifications shown in **Figure 21.10.** First, we will give an overview of the flowering plant life cycle, and then we will discuss the life cycle in more depth. In flowering plants, the sporophyte is dominant, and it is the generation that bears flowers. The flower is the reproductive structure of angiosperms. The flower of the sporophyte produces two types of spores: microspores and megaspores. A **microspore** develops into a male gametophyte, which is a pollen grain. A **megaspore** develops into a female gametophyte, the embryo sac, which is microscopic and retained within the flower.

A pollen grain is either windblown or carried by an animal to the vicinity of the embryo sac. At maturity, a pollen grain contains two nonflagellated sperm. The embryo sac contains an egg.

A pollen grain develops a pollen tube, and the sperm move down the pollen tube to the embryo sac. After a sperm fertilizes an egg, the zygote becomes

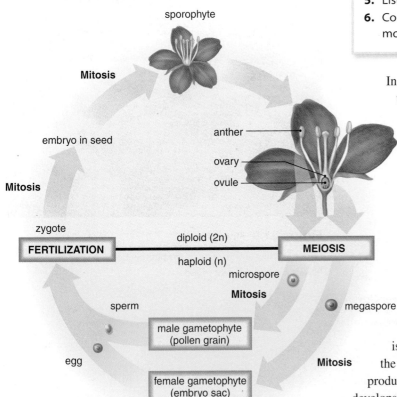

Figure 21.10 Alternation of generations in flowering plants.

In flowering plants, there are two types of spores and two gametophytes, male and female. Flowering plants are adapted to a land existence: The spores, the gametophytes, and the zygote are protected from drying out, in large part by the sporophyte.

an embryo, still within the flower. The structure that houses the embryo develops into a seed. The seed also contains stored food and is surrounded by a seed coat. The seeds are often enclosed by a fruit, which aids in dispersing the seeds. When a seed germinates, a new sporophyte emerges and develops.

As you learned in Chapter 18, the life cycle of flowering plants is well adapted to a land existence. No external water is needed to transport the pollen grain to the embryo sac, or to enable the sperm to reach the egg. All stages of the life cycle are protected from drying out.

Flowers

The **flower** is a unique reproductive structure found only in angiosperms (**Fig. 21.11**). Flowers produce the spores and protect the gametophytes. They often attract pollinators, which help transport pollen from plant to plant. Flowers also produce the fruits that enclose the seeds. The success of angiosperms, with over 240,000 species, is largely attributable to the evolution of the flower.

In monocots, flower parts occur in threes and multiples of three; in eudicots, flower parts are in fours or fives and multiples of four or five (**Fig. 21.12**).

A typical flower has four whorls of modified leaves attached to a **receptacle** at the end of a flower stalk.

1. The **sepals** are the most leaf-like of all the flower parts. They are usually green but some resemble petals (Fig. 21.12*a*). Sepals protect the bud as the flower develops.
2. An open flower also has a whorl of **petals,** whose color accounts for the attractiveness of many flowers. The size, the shape, and the color of petals are attractive to specific pollinators. Wind-pollinated flowers may have no petals at all.
3. **Stamens** are the "male" portion of the flower. Each stamen has two parts: the **anther,** a sac-like container, and the **filament,** a slender stalk. Pollen grains develop from the microspores produced in the anther.
4. At the very center of a flower is the **carpel,** a vase-like structure that represents the "female" portion of the flower. A carpel usually has three parts: the **stigma,** an enlarged sticky knob; the **style,** a slender stalk; and the **ovary,** an enlarged base that encloses one or more **ovules.** The ovule becomes the seed, and the ovary becomes the fruit.

A flower can have a single carpel or multiple carpels. Sometimes several carpels are fused into a single structure, in which case the ovary has several

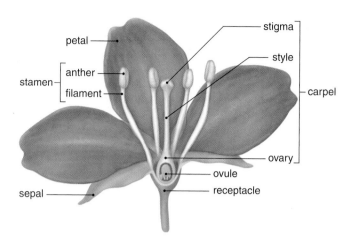

Figure 21.11   Anatomy of a flower.

A complete flower has all flower parts: sepals, petals, stamens, and at least one carpel.

a. Daylily, a monocot

b. Cranesbill geranium, a eudicot

Figure 21.12   Monocot versus eudicot flowers.

a. Monocots, such as daylilies, have flower parts in threes. In particular, note the three petals. **b.** Geraniums are eudicots. They have flower parts in fours or fives; note the five petals of this flower.

a. Male flowers   b. Female flowers

Figure 21.13 Corn plants are monoecious.

A single corn plant has clusters of male flowers (**a**) and female flowers (**b**). Male flowers possess stamens and produce the pollen that is carried by wind to the carpels of the female flowers, where an ear of corn develops.

chambers, each of which contains ovules. A carpel usually contains many ovules, which increases the number of seeds the plant may produce.

Not all flowers have sepals, petals, stamens, and a carpel. Those that do are said to be *complete,* and those that do not are said to be *incomplete*. Flowers that have both stamens and carpels are called bisexual flowers; those with only stamens are male flowers. Those with only carpels are female flowers. If both male and female flowers are on one plant, the plant is called *monoecious* (**Fig. 21.13**). But if male and female flowers occur on separate plants, the plant is called *dioecious*. Holly trees are dioecious, and if red berries are a priority, it is necessary to acquire a plant with male flowers and another plant with female flowers.

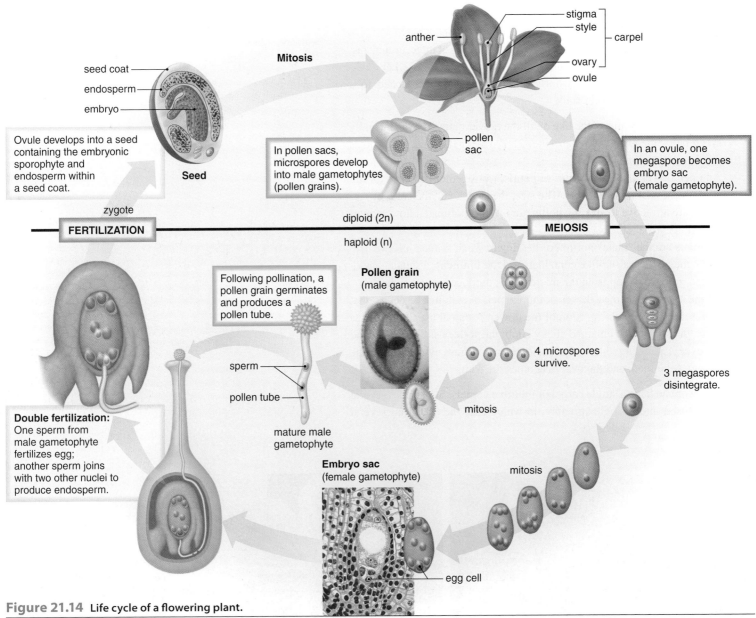

Figure 21.14 Life cycle of a flowering plant.

Follow the development of a microspore into a germinated pollen grain that contains sperm. Also follow the development of a megaspore into the female gametophyte that contains an egg. Note the occurrence of double fertilization, in which one sperm from the pollen tube fertilizes the egg, producing a zygote, and the other joins with two other female gametophyte cells to become endosperm. Now the ovule becomes a seed.

From Spores to Fertilization

Now that we have some acquaintance with the flowering plant life cycle, we will examine it in more detail. As you may recall, the sporophyte of seed plants produces two types of spores: microspores, which become **male gametophytes** (mature pollen grains); and megaspores, which become the **female gameto-phyte,** or **embryo sac.** Just exactly where does this happen, and how do these events contribute to the life cycle of flowering plants?

Microspores develop into pollen grains in the anthers of stamens (**Fig. 21.14**). A **pollen grain** at first consists of two cells. The larger cell will eventually produce a pollen tube. The smaller cell divides, either now or later, to become two sperm. This is why the stamen is called the "male" portion of the flower.

Pollen grains are distinctive to the particular plant (**Fig. 21.15**), and **pollination** is simply the transfer of pollen from the anther to the stigma of a carpel. Plants often have adaptations that favor *cross-pollination,* which occurs when the pollen landing on the stigma is from a different plant of the same species. For example, the carpels may mature only after the anthers have released their pollen. Cross-pollination may also be brought about with the assistance of an animal pollinator. If a pollinator, such as a bee, goes from flower to flower of only one type of plant, cross-pollination is more likely to occur in an efficient manner. The secretion of nectar is one way that insects are attracted to plants, and over time, certain pollinators have become adapted to reach the nectar of only one type of flower. In the process, pollen is inadvertently picked up and taken to another plant of the same type. Plants attract particular pollinators in still other ways. For example, through the evolutionary process, orchids of the genus *Ophrys* have flowers that look like female wasps. Males of that species pick up pollen when they attempt to mate with these flowers!

 Video Pollinators

Figure 21.14 also allows you to follow the development of the megaspore. In an ovule, within the ovary of a carpel, meiosis produces four megaspores. One of the cells develops into the female gametophyte, or so-called embryo sac, which is a seven-celled structure containing a single egg cell. Fertilization of the egg occurs after a pollen grain lands on the stigma of a carpel and develops a pollen tube. A pollen grain that has germinated and produced a pollen tube is the mature male gametophyte (Fig. 21.14, *middle*). A pollen tube contains two sperm. Once it reaches the ovule, **double fertilization** occurs. One sperm unites with the egg, forming a 2n zygote. The other sperm unites with two nuclei centrally placed in the embryo sac, forming a 3n endosperm cell.

Development of the Seed in a Eudicot

It is now possible to account for the three parts of a seed. The ovule wall will become a protective covering called the *seed coat.* Double fertilization, as discussed earlier, has resulted in an endosperm nucleus and a zygote. Cell division produces a multicellular *embryo* and a multicellular endosperm, which is the *stored food* of a seed.

a.

b.

c.   ×533

Figure 21.15 Pollen.

a. Pollen grains are so distinctive that a paleontologist can use fossilized pollen to date the appearance of a plant in a particular area. Pollen grains can become fossils because their strong walls are resistant to chemical and mechanical damage. **b.** Pollen grains of Canadian goldenrod. **c.** Pollen grains of a pussy willow.

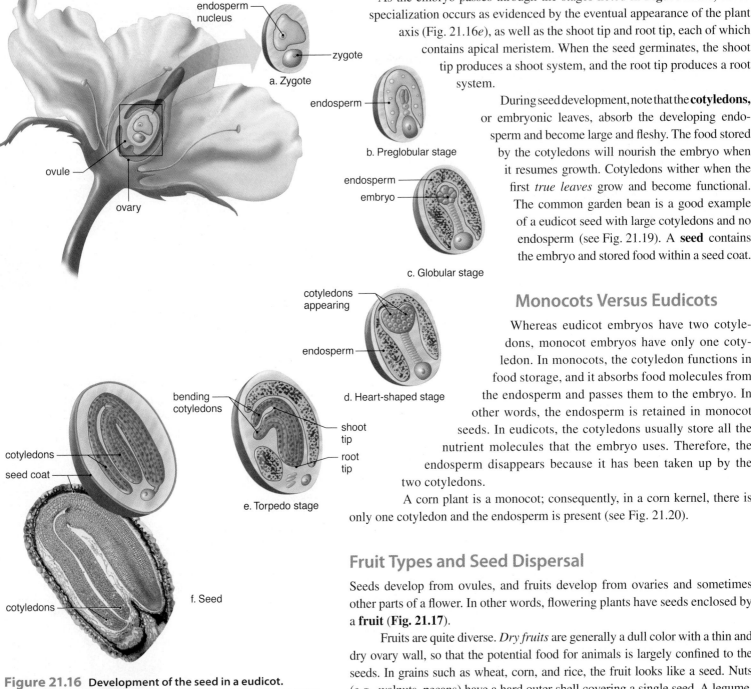

As the embryo passes through the stages noted in **Figure 21.16,** tissue specialization occurs as evidenced by the eventual appearance of the plant axis (Fig. 21.16*e*), as well as the shoot tip and root tip, each of which contains apical meristem. When the seed germinates, the shoot tip produces a shoot system, and the root tip produces a root system.

During seed development, note that the **cotyledons,** or embryonic leaves, absorb the developing endosperm and become large and fleshy. The food stored by the cotyledons will nourish the embryo when it resumes growth. Cotyledons wither when the first *true leaves* grow and become functional. The common garden bean is a good example of a eudicot seed with large cotyledons and no endosperm (see Fig. 21.19). A **seed** contains the embryo and stored food within a seed coat.

Monocots Versus Eudicots

Whereas eudicot embryos have two cotyledons, monocot embryos have only one cotyledon. In monocots, the cotyledon functions in food storage, and it absorbs food molecules from the endosperm and passes them to the embryo. In other words, the endosperm is retained in monocot seeds. In eudicots, the cotyledons usually store all the nutrient molecules that the embryo uses. Therefore, the endosperm disappears because it has been taken up by the two cotyledons.

A corn plant is a monocot; consequently, in a corn kernel, there is only one cotyledon and the endosperm is present (see Fig. 21.20).

Fruit Types and Seed Dispersal

Seeds develop from ovules, and fruits develop from ovaries and sometimes other parts of a flower. In other words, flowering plants have seeds enclosed by a **fruit** (**Fig. 21.17**).

Fruits are quite diverse. *Dry fruits* are generally a dull color with a thin and dry ovary wall, so that the potential food for animals is largely confined to the seeds. In grains such as wheat, corn, and rice, the fruit looks like a seed. Nuts (e.g., walnuts, pecans) have a hard outer shell covering a single seed. A legume, such as a pea, has a several-seeded fruit that splits open to release the seeds. A legume illustrates that what we call a vegetable may actually be a fruit.

In contrast to dry fruits, *fleshy fruits* have an usually juicy portion that is brightly colored to attract animals. A drupe (e.g., peach, cherry, olive) is a "stone fruit"—the outer part of the ovary wall is fleshy, but there is an inner stony layer. Inside the stony layer is the seed. A berry, such as a tomato, contains many seeds. An apple is a *pome,* in which a dry ovary covers the seeds, and the fleshy part is derived from the receptacle of the flower. A strawberry is an interesting fruit because the flesh is derived from the receptacle, and what appear to be the seeds are actually dry fruits!

Figure 21.16 Development of the seed in a eudicot.

Development begins with (**a**) the zygote and ends with (**f**) the seed. As the embryo develops, it progresses through the stages noted from (**b**) to (**e**).

Figure 21.17 Examples of fruits.

Some fruits are dry, as in walnuts and peas. Some fruits are fleshy, as in peaches and apples. To a botanist, any plant product derived from an ovary plus perhaps other flower parts is a fruit.

Connections and Misconceptions

Is a tomato a fruit or a vegetable?

From a strict botanical perspective, a tomato is a fruit, since it produces seeds and develops from the ovaries of the flower. In fact, all vegetables are fruits based on this definition. But the story of tomatoes is a little more complicated. The U.S. Supreme Court in 1893 declared that the tomato is a vegetable (at least for taxation purposes), and Louisiana has made it the state vegetable. However, Ohio and Tennessee have declared the tomato a state fruit. Arkansas has solved the problem by declaring it both the state fruit and vegetable.

Dispersal of Seeds

For plants to be widely distributed, their seeds have to be dispersed—that is, distributed preferably long distances from the parent plant. Various means of dispersal are well known. Birds and mammals sometimes eat fruits, including the seeds, which then pass out of the digestive tract with the feces some distance from the parent plant (**Fig. 21.18***a*). Squirrels and other animals gather seeds and fruits, which they bury some distance away. Some plants have evolved unusual ways to ensure dispersal. The hooks and spines of clover, cocklebur, and burdock fruits attach to the fur of animals and the clothing of humans, which carry them far away from the parent plant (Fig. 21.18*b*). Other seeds are dispersed by wind. Woolly hairs, plumes, and wings are all adaptations for this type of dispersal. The dandelion fruit is attached to several hairs that function as a parachute and aid dispersal (Fig. 21.18*c*). The winged fruit of a maple tree, which contains two seeds, has been known to travel up to 10 kilometers from its parent (Fig. 21.18*d*). Different still, a touch-me-not plant has seed pods that swell as they mature. A passing animal may cause the swollen pods to burst, hurling the ripe seeds a great distance away from the plant.

Video
Fruit Bat Seed Dispersal

a. Cherry   b. Burdock

c. Dandelion   d. Maple

Figure 21.18 Methods of seed dispersal.

Many plants have adaptations to ensure that their seeds are spread some distance from the parent plant. **a.** The waxwing will carry the seed of a cherry some distance away. **b.** The spines of burdock fruit stick to a passerby. **c.** Dandelion and (**d**) maple fruits have adaptations that allow them to be carried long distances by wind.

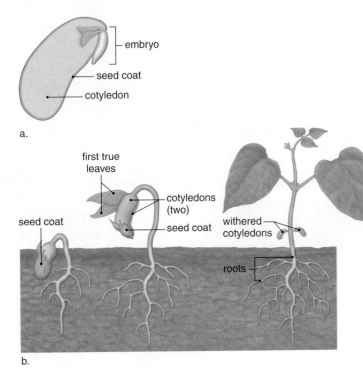

Figure 21.19 Common garden bean, a eudicot.

a. Seed structure. **b.** Germination and development of the seedling. Notice that there are two cotyledons and that the leaves are net-veined.

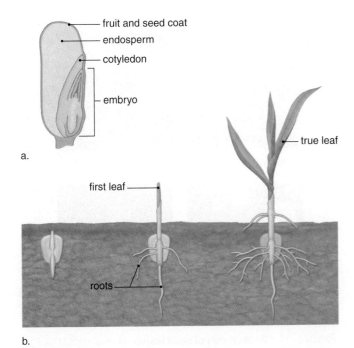

Figure 21.20 Corn, a monocot.

a. Corn kernel structure. **b.** Germination and development of the seedling. Notice that there is one cotyledon and that the leaves are parallel-veined.

Germination of Seeds

Following dispersal, if all goes well, the seeds will **germinate.** As growth occurs, a seedling appears. Germination doesn't usually take place until there is sufficient water, warmth, and oxygen to sustain growth. In deserts, germination does not occur until there is adequate moisture. These requirements help ensure that seeds do not germinate until the most favorable growing season has arrived. Some seeds do not germinate until they have been dormant for a period of time.

For seeds, dormancy is the time during which no growth occurs, even though conditions may be favorable for growth. In the temperate zone, seeds often have to be exposed to a period of cold weather before dormancy is broken. Fleshy fruits (e.g., apples, pears, oranges, and tomatoes) contain inhibitors, so that germination does not occur until the seeds are removed and washed. Aside from water, bacterial action and even fire can act on the seed coat, allowing it to become permeable to water. The uptake of water causes the seed coat to burst and germination to occur.

Eudicot Versus Monocot Germination

If the two cotyledons of a bean seed are parted, you can see the cotyledons and a rudimentary plant with immature leaves. As the eudicot seedling emerges from the soil, the shoot is hook-shaped to protect the immature leaves as they start to grow. The cotyledons shrivel up as the true leaves of the plant begin photosynthesizing (**Fig. 21.19**). A corn kernel is actually a fruit, and therefore the outer covering is the fruit and seed coat combined (**Fig. 21.20**). Inside is the single cotyledon. Also, both the immature leaves and the root are covered by sheaths. The sheaths are discarded when the seedling begins growing, and the immature leaves become the first true leaves of the corn plant.

Connecting the Concepts

For more information on the topics presented in this section, refer to the following discussions.

Section 18.2 discusses the evolution of the angiosperms.

Figure 20.4 illustrates some additional differences between monocot and eudicot plants.

Check Your Progress 21.3

1 Describe the anatomy of a flower.

2 Detail the life cycle of a flowering plant.

3 Discuss the importance of seeds and seed dispersal.

21.4 Asexual Reproduction in Flowering Plants

Learning Outcomes

Upon completion of this section, you should be able to

1. List some natural methods of asexual reproduction in flowering plants and describe the structures involved.
2. Describe the propagation of plants in tissue culture and list the benefits and drawbacks to this method.
3. Explain how meristem culture and anther culture may be used to clone plants and list the benefits and drawbacks of each method.
4. List the possible benefits and concerns of using transgenic plants.

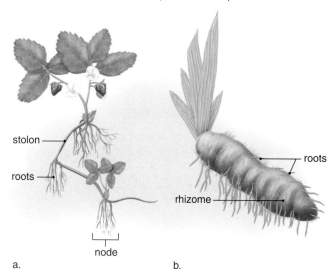

a.   b.

Figure 21.21   Asexual reproduction in plants.

a. A strawberry plant has an aboveground horizontal stem called a stolon. Every other node produces a new plant. **b.** An iris has an underground horizontal stem called a rhizome, whose nodes produce new plants.

Many plants can also reproduce asexually because they contain undifferentiated cells in the meristem. In asexual reproduction, there is only one parent, instead of two, as in sexual reproduction. For example, complete strawberry plants can grow from the nodes of stolons, and iris plants can grow from the nodes of rhizomes (**Fig. 21.21**). White, or Irish, potatoes are actually portions of underground stems, and each eye is a bud that will produce a new potato plant if it is planted with a portion of the swollen tuber. Sweet potatoes are modified roots; they can be propagated by planting sections of the root. You may have noticed that the roots of some fruit trees, such as cherry and apple trees, produce "suckers," small plants that can be used to grow new trees.

There are several ways that seedless fruits can arise, but usually some sort of stimulation causes the plant to produce fruit, even though fertilization never occurred. In the seedless watermelon, which is triploid and cannot go through meiosis to form sperm and eggs, pollination is used as a stimulus to cause the plants to form fruit. Often, hormones, such as auxins, gibberellins, or cytokinins, are used to stimulate seedless fruit formation. Seedless grapes actually contain seeds, but the hard seed casing fails to develop and this causes the embryo to abort. Then, the grapes are sprayed with gibberellin to increase their size.

Propagation of Plants in Tissue Culture

Tissue culture is the growth of a tissue in an artificial liquid or solid culture medium. Plant cells are **totipotent,** which means that each plant cell can become an entire plant. This has led to the commercial production of embryos from the somatic cells of plants (**Fig. 21.22**). Adult cells are first treated to remove their cell walls. Then, thousands or even millions of embryos can be produced by following the procedure shown in Figure 21.22. The embryos can be encapsulated in artificial seeds and shipped anywhere. Many vegetable plants found in nurseries, such as tomato, celery, and

Figure 21.22   Tissue culture.

a. When plant cell walls are removed by digestive enzyme action, cells without cell walls, called protoplasts, result. **b.** Cell walls regenerate, and cell division begins. **c.** Cell division produces aggregates of cells. **d.** An undifferentiated mass, called a callus, develops. **e.** From the callus, somatic embryos such as this one appear. **f.** The embryos develop into plantlets that can be transferred to soil for growth into adult plants.

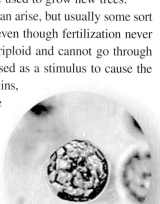

a. Protoplasts (naked cells)

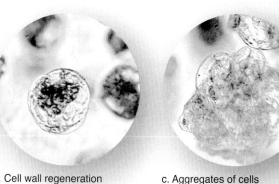

b. Cell wall regeneration   c. Aggregates of cells

d. Callus (undifferentiated mass)

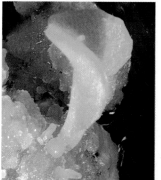

e. Somatic embryo

f. Plantlet

Figure 21.23 **Whole plants from meristem tissue.**

Each of these flasks contains meristem tissue and a growth medium. In the presence of light and with slow rotation of the flasks, a new plant will appear in a few weeks.

asparagus, are produced in this manner. Occasionally, mutations arise during the production process, and this is another way that plant breeders can produce new plants with desirable traits.

Instead of using mature plant tissues as the starting material, you can use meristem tissue as a source of plant cells. In this case, the end result is a population of *clonal plants* that have the same traits as the original plant (**Fig. 21.23**). If the correct proportions of hormones are added to the liquid medium, many new shoots will develop from a single shoot tip. When these are removed, more shoots form. An advantage to using meristem culture is that meristem tissue, unlike other portions of a plant, is virus-free. Therefore, the plants produced will also be virus-free. (The presence of plant viruses weakens plants and makes them less productive.)

Anther culture is a technique in which mature anthers are cultured in a medium containing vitamins and growth factors. The haploid cells within the pollen grains divide, producing *proembryos* consisting of as many as 20 to 40 cells. Finally, the pollen grains rupture, releasing haploid embryos. The experimenter can then generate a haploid plant, or chemical agents can be added that encourage chromosomal doubling. After chromosomal doubling, the resulting plants are diploid, and the homologous chromosomes carry the same genes. Anther culture is another way to produce plants that are certain to have the same traits as the parent plant.

When the desired product is not an entire plant, but merely a substance the plant produces, scientists may use a technique called *cell suspension culture*. Rapidly growing calluses are cut into small pieces and shaken in a liquid nutrient medium, so that single cells or small clumps of cells break off and form a suspension. These cells often produce the same chemicals as the entire plant. For example, cell suspension cultures of cells from the quinine tree produce quinine, and those of the wooly foxglove produce digitoxin, both of which are useful drugs for humans.

Genetic Engineering of Plants

Traditionally, **hybridization,** the crossing of different varieties of plants or even species, was used to produce plants with desirable traits. Hybridization, followed by vegetative propagation of the mature plants, generated a large number of identical plants with these traits. Today, it is possible to directly alter the genes of organisms and in that way produce new varieties with desirable traits.

Tissue Culture and Genetic Engineering

Genetic engineering can be done utilizing plant cells lacking cell walls in tissue culture (see Fig. 21.22). A foreign piece of DNA isolated from any type of organism—plant, animal, or bacteria—is placed in the tissue culture medium. High-voltage electrical pulses can then be used to create pores in the plasma membrane, so that the foreign gene enters the cells. In one such procedure, a gene for the production of the jellyfish green fluorescent protein was inserted into tobacco, and the adult plants glowed when exposed to ultraviolet light. Likewise, scientists have even attempted to produce vaccines by inserting nonlethal viral genes into crop plants!

Video
Potato Vaccine

Today, it is possible to use a gene gun to bombard a callus with microscopic DNA-coated metal particles. Then, genetically altered embryos develop into genetically altered adult plants. Many plants, including corn and wheat varieties, have been genetically engineered by this method. Such plants are called **genetically modified plants (GMPs),** or **transgenic plants,** because they carry a foreign gene and have new and different traits. **Figure 21.24** shows two types of transgenic plants.

Agricultural Plants with Improved Traits

Corn and cotton plants, in addition to soybean and potato plants, have been genetically engineered to be resistant to either herbicides or insect pests. Some corn and cotton plants have been developed that are resistant to both insects and herbicides. In 2001, transgenic crops were planted on more than 74 million acres in the United States, and the acreage has been increasing annually. If crops are resistant to a broad-spectrum herbicide and weeds are not, then the herbicide can be used to kill the weeds. When herbicide-resistant plants were planted, weeds were easily controlled, less tillage was needed, and soil erosion was minimized. However, wildlife and native plants may decline when increased amounts of herbicides are used.

Some citizens are concerned about GMPs because of possible effects on their health. This concern has been prompted by limited laboratory data suggesting that some people may be allergic to GMPs (**Fig. 21.25**). Some feel that drastic consequences might occur in the future, even though no life-threatening effects of GMPs have so far been recorded. In 2000, such fears even prompted the recall of several brands of taco shells from stores after it was discovered that the shells contained traces of a genetically modified variety of corn not approved for humans. Ecologists also worry that GMPs may harm the environment. Beneficial insects feeding on the GMPs might be harmed by them, or GMPs might pass their herbicide resistance genes to certain weeds.

 **Video Pesticide Plants**    **Video GM Food Safety**

In the meantime, researchers are trying to produce more types of GMPs (**Fig. 21.26**). A salt-tolerant tomato will soon be field-tested. First, scientists identified a gene coding for a channel protein that transports Na^+ into a vacuole, preventing it from interfering with plant metabolism. Then the scientists used the gene to engineer tomato plants that maximally produce the channel protein. The GMPs thrived when watered with a salty solution. This technology may turn once unusable land into arable farmland. Salt-tolerant and drought- and cold-tolerant cereals, rice, and sugarcane might help provide enough food for the growing world population.

Potato blight is the most serious potato disease in the world. About 150 years ago, it was responsible for the Irish potato famine, which caused the deaths of millions of people. By placing a gene from a naturally blight-resistant wild potato into a farmed variety, researchers have created potato plants that are no longer vulnerable to a range of blight strains. A similar technique is being employed in the fight against the chestnut blight, which nearly wiped out the American chestnut in the early twentieth century.

a. Herbicide-resistant soybean plant

b. Nonresistant potato plant

c. Pest-resistant potato plant

Figure 21.24 Transgenic plants.

a. These soybean plants have been genetically modified to be resistant to herbicides. Nonresistant potato plants (**b**) can be genetically modified to be resistant to pests (**c**).

Figure 21.25 Regulation of genetically modified plants.

Activists protest the present regulation of genetically modified plants, which they believe to be too permissive.

Transgenic Crops of the Future

Improved Agricultural Traits

Herbicide resistant	Wheat, rice, sugar beets, canola
Salt tolerant	Cereals, rice, sugarcane, canola
Drought tolerant	Cereals, rice, sugarcane
Cold tolerant	Cereals, rice, sugarcane
Improved yield	Cereals, rice, corn, cotton
Disease protected	Wheat, corn, potatoes

Improved Food-Quality Traits

Fatty acid/oil content	Corn, soybeans
Protein/starch content	Cereals, potatoes, soybeans, rice, corn
Amino acid content	Corn, soybeans

Medicinal Benefits

Vaccine production	Corn, soybeans, tomatoes

a. Desirable traits

Salt intolerant   c. Transgenic corn for antibody production

b. Salt tolerant

Figure 21.26 Genetically modified crops of the future.
a. Genetically modified crops of the future include those with improved agricultural traits, better food quality, or medicinal benefits. **b.** A salt-tolerant tomato plant has been engineered. The plant to the right does poorly when watered with a salty solution, but the engineered plant to the left is tolerant of the solution. The development of salt-tolerant crops could increase food production in the future. **c.** Corn plants have been engineered to produce vaccines for many diseases.

Some progress has also been made in increasing the food quality of crops. Soybeans have been developed that produce mainly monounsaturated fatty acids, a change that may improve human health. These altered plants also produce acids that can be used as hardeners in paints and plastics. The necessary genes were taken from other plants and transferred into the soybean DNA.

Other types of genetically engineered plants are expected to increase productivity. Stomata might be altered to take in more carbon dioxide or to lose less water. A team of Japanese scientists is working on introducing the C_4 photosynthetic capability into rice. Unlike C_3 plants, C_4 plants do well in hot, dry weather (see Chapter 6). These modifications would require a more complete reengineering of plant cells than the single-gene transfers that have been done so far.

Video
Warming Hurts Rice

Commercial Products

Single-gene transfers also have allowed plants to produce various products, including human hormones, clotting factors, and antibodies. One type of antibody made by corn can deliver radioisotopes to tumor cells, and another antibody may help prevent HIV infection (Fig. 21.26c). The tobacco mosaic virus has been used as a vector to introduce a human gene into adult tobacco plants in the field. (Note that this technology bypasses the need for tissue culture.) Tens of grams of α-galactosidase, an enzyme needed for the treatment of a human lysosome storage disease, were harvested per acre of tobacco plants. And it took only 30 days to get tobacco plants to produce antibodies to treat non-Hodgkin lymphoma after being sprayed with a genetically engineered virus.

Connections and Misconceptions

Are transgenic crops organic?

According to the U.S. Department of Agriculture (USDA), an organic crop is one that is produced without the use of pesticides, irradiation, hormones, antibiotics, or bioengineering. Therefore, a transgenic crop may not be marketed as an organic crop, since it is a product of artificial technology.

Connecting the Concepts

For additional information on the topics in this section, refer to the following discussions.

Section 11.3 examines the technologies that are used to produce transgenic plants.

Figure 17.4 illustrates the structure of the tobacco mosaic virus.

Section 20.2 describes the role of meristem tissue in a plant.

Check Your Progress 21.4

1. Explain how plants with seedless fruits can be propagated.
2. Compare and contrast tissue culture plant propagation with meristem culture.
3. Summarize the pros and cons of genetically modified plants (GMPs).

Media Study Tools

 plus+

|BIOLOGY

www.mhhe.com/maderessentials3

Enhance your study of this chapter with study tools and practice tests. Also ask your instructor about the resources available through ConnectPlus, including the media-rich eBook, interactive learning tools, and animations.

The Chapter in Review

Summary

21.1 Plant Hormones

Plant hormones lead to physiological changes within the cell. The five commonly recognized groups of plant hormones are

- **Auxins:** Affect growth patterns; cause apical dominance and phototropism
- **Gibberellins:** Promote stem elongation, break seed dormancy
- **Cytokinins:** Promote cell division, prevent senescence of leaves, influence differentiation of plant tissues
- **Abscisic acid:** Initiates and maintains seed and bud dormancy, closing of stomata
- **Ethylene:** Causes abscission of leaves, fruits, and flowers; ripens fruits

21.2 Plant Responses

Environmental signals play a significant role in plant growth and development.

Tropisms

Tropisms are growth responses toward (positive) or away from (negative) unidirectional stimuli.

> Phototropism: Response to a light
> Gravitropism: Response to gravity

- Auxin is responsible for the negative gravitropism exhibited by stems. Stems grow upward opposite the direction of gravity.

Photoperiodism

Flowering is a response to seasonal changes—namely, length of the night.

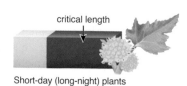

critical length

Short-day (long-night) plants

- Short-day plants flower when nights are longer than a critical length.
- Long-day plants flower when nights are shorter than a critical length.
- Some plants are day/night-neutral.

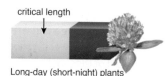

critical length

Long-day (short-night) plants

Phytochrome and Plant Flowering   Phytochrome is a plant pigment that responds to daylight. Phytochrome in plant cells brings about flowering and encourages seed germination. In addition, phytochrome influences leaf expansion and affects stem branching.

21.3 Sexual Reproduction in Flowering Plants

The life cycle of flowering plants is adapted to a land existence.

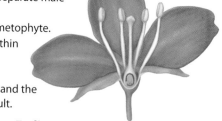

- Flowering plants have an alternation of generations life cycle with separate male and female gametophytes.
- Pollen grain is the male gametophyte.
- The embryo sac, located within the ovule of a flower, is the female gametophyte.
- Double fertilization occurs, and the zygote and endosperm result.

Development of the Seed in a Eudicot

The zygote undergoes a series of stages to become an embryo. In eudicots, the embryo has two cotyledons, which absorb the endosperm. In monocots, the embryo has a single cotyledon and endosperm. Aside from the embryo and stored food, a seed has a seed coat.

eudicot seed   monocot seed

Fruit Types and Seed Dispersal

A fruit is a mature, ripened ovary and may be composed of other flower parts. Some fruits are dry (e.g., nuts, legumes), and some are fleshy (e.g., apples, peaches). In general, fruits aid the dispersal of seeds. Following dispersal, a seed germinates.

21.4 Asexual Reproduction in Flowering Plants

Many flowering plants reproduce asexually.

- Nodes located on stems (either aboveground or underground) give rise to entire plants.
- Roots produce new shoots.

Propagation of Plants in Tissue Culture

The production of clonal plants utilizing tissue culture is now a commercial venture. Plant cells in tissue culture can produce chemicals of medical and commercial importance.

Genetic Engineering of Plants

The practice of plant tissue culture facilitates genetic engineering to produce plants that have improved agricultural or food-quality traits. Plants can also be engineered to produce chemicals of use to humans.

Key Terms

Testing Yourself

Choose the best answer for each question. For questions 1–8, identify the plant hormone in the key that is associated with each phenomenon. Each answer may be used more than once.

Key:

   a. auxin

   b. gibberellin

   c. cytokinin

   d. abscisic acid

   e. ethylene

1. initiates and maintains seed and bud dormancy
2. stimulates root development
3. capable of moving from plant to plant through the air
4. causes phototropism in stems
5. responsible for apical dominance
6. stimulates leaf, fruit, and flower drop
7. needed to break seed and bud dormancy
8. prevents senescence

9. A tropism is defined as
   a. a growth response to a directional stimulus.
   b. a plant response to hormones.
   c. a hormonal response to overcrowding by other plants.
   d. abscission of leaves due to plant hormones.

10. Stigma is to carpel as anther is to
    a. sepal.
    b. stamen.
    c. ovary.
    d. style.

11. Label the parts of the flowering plant life cycle in the following illustration.

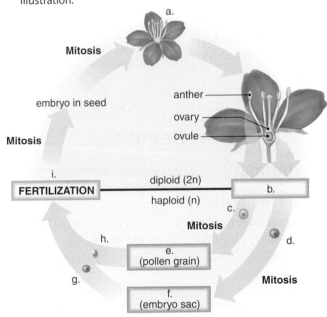

12. Plant tissue culture takes advantage of
    a. phototropism.
    b. gravitropism.
    c. asexual reproduction from stolons.
    d. totipotency.

13. Double fertilization is the formation of a _____ and a(n) _____.
    a. zygote, zygote
    b. zygote, pollen grain
    c. zygote, megaspore
    d. zygote, endosperm

14. Which is the correct order of the following events?:
    (1) megaspore becomes embryo sac, (2) embryo formed,
    (3) double fertilization, (4) meiosis?
    a. 1, 2, 3, 4
    b. 4, 1, 3, 2
    c. 4, 3, 2, 1
    d. 2, 3, 4, 1

15. Phytochrome plays a role in
    a. flowering.
    b. stem growth.
    c. leaf growth.
    d. All of these are correct.

16. A plant requiring a dark period of at least 14 hours will
    a. flower if a 14-hour night is interrupted by a flash of light.
    b. not flower if a 14-hour night is interrupted by a flash of light.
    c. flower if the days are 14 hours long.
    d. flower when the nights drop below 14 hours long.

17. Short-day plants
    a. are apt to flower in the fall.
    b. are apt to flower in the summer.
    c. do not have a critical photoperiod.
    d. will not flower if a short day is interrupted by bright light.
    e. All of these are correct.

18. Which of the following is a natural method of asexual reproduction in plants?
    a. meristem culture
    b. vegetative propagation from stolons or rhizomes
    c. anther culture
    d. cell suspension culture

19. Which of the following types of modified leaves is not a part of a flower?
    a. tendrils
    c. sepals
    b. anthers
    d. carpels

20. The megaspore is similar to the microspore in that both
    a. have the diploid number of chromosomes.
    b. become an embryo sac.
    c. become a gametophyte that produces a gamete.
    d. are necessary for seed production.
    e. Both c and d are correct.

21. Label the parts of the flower in the following diagram.

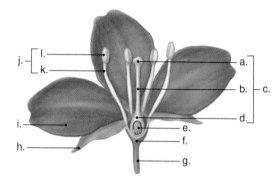

22. The embryo sac is derived from
    a. one of four megaspores.
    b. the second sperm nucleus of the pollen grain.
    c. the endosperm.
    d. the cotyledons.

23. In contrast to eudicots, monocot
    a. embryos have two cotyledons.
    b. seeds contain endosperm.
    c. cotyledons store food.
    d. embryos undergo a heart stage of development.

24. Foreign genes may be introduced into plants by using
    a. microinjected seeds.
    b. genetically modified fungi.
    c. electrical pulses.
    d. DNA replication.

25. Plant biotechnology can lead to
    a. increased crop production.
    b. disease-resistant plants.
    c. treatment of human disease.
    d. All of these are correct.

26. Plants may be propagated asexually in the laboratory through
    a. tissue culture.
    b. anther culture.
    c. root or stem cutting.
    d. cell suspension culture.
    e. All of these are correct.

27. Plants propagated by tissue culture rather than anther culture differ because plants produced by tissue culture
    a. are subject to mutations and may not be identical.
    b. are haploid rather than diploid.
    c. have exactly the same characteristics as the parent plant.
    d. are produced from calluses.

28. Transgenic plants
    a. are produced by anther culture or tissue culture.
    b. are genetically identical.
    c. are a new plant species.
    d. may possess improved traits.

29. Label the parts of the eudicot seed and seedling in the following diagram.

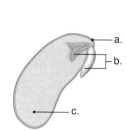
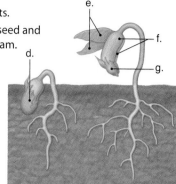

Thinking Scientifically

1. In late November every year, florists ship truckloads of poinsettia plants to stores around the country. Typically, the plants are individually wrapped in plastic sleeves. If the plants remain in the sleeves for too long during shipping and storage, their leaves begin to curl under and eventually fall off. What plant hormone do you think causes this response? How do you suppose the plastic sleeves affect the response?

2. Snow buttercups (*Ranunculus adoneus*) live in alpine regions and produce sun-tracking flowers. The flowers face east in the morning to absorb the sun's warmth in order to attract pollinators and speed the growth of fertilized ovules. The flowers track the sun all day, continually bending to face the sun. How might you determine whether the flowers or stems are responsible for sun tracking? Assuming you have determined that the stem is responsible, how would you determine which region of the stem follows the sun?

3. Some plants, such as Hepatica, bloom in the late winter or very early spring when weather conditions are harsh and there are relatively few insects available to pollinate their flowers. What advantage might this give a plant? And what types of modifications would you expect the flowers of these plants to possess?

Bioethical Issue

Unintended Consequences of Genetic Engineering

Witchweed is a parasitic weed that destroys 40% of Africa's cereal crop annually. Because it is intimately associated with its host (the cereal crop), witchweed is difficult to selectively destroy with herbicides. One control strategy is to create genetically modified cereals with herbicide resistance. Then, herbicides will kill the weed without harming the crop. Herbicide-resistant sorghum was created to help solve the witchweed problem. However, scientists discovered that the herbicide resistance gene could be carried via pollen into Johnson grass, a relative of sorghum and a serious problem in the United States. Farmers would have a new challenge if Johnson grass could no longer be effectively controlled with herbicides. Consequently, efforts to create herbicide-resistant sorghum were put on hold.

Do you think farmers in Africa should be able to grow herbicide-resistant sorghum, allowing them to control witchweed? Or should the production of herbicide-resistant sorghum be banned worldwide in order to avoid the risk of introducing the herbicide resistance gene into Johnson grass?

22

Being Organized and Steady

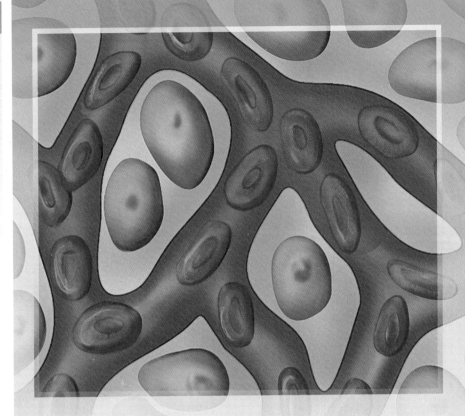

OUTLINE

BEFORE YOU BEGIN

Before beginning this chapter, take a few moments to review the following discussions.

Section1.1 How do tissues and organs fit into the levels of biological organization?

Section 4.3 What are the functions of proteins in the plasma membrane?

Section 4.5 What types of junctions link cells together to form tissues?

How Homeostasis Works

To keep you healthy, your body is always maintaining homeostasis, the relative constancy, or equilibrium, of the internal environment. Your cells are surrounded by tissue fluid called interstitial fluid, which continually exchanges substances with the cytoplasm of the cell and with the blood. In order for homeostasis to be maintained, substances must be able to move between these fluids.

How might homeostasis be maintained? Let's take an example. Despite extreme fluctuations in our diet, the blood glucose (sugar) concentration usually stays within an acceptable range, day in and day out. This is primarily due to the work of two hormones, called insulin and glucagon. Just after you have eaten, the blood glucose level usually rises. In response, your pancreas releases the hormone insulin, which causes the body's cells, including those of the liver, to take up glucose from the blood. Now the blood glucose level lowers back into the normal range. The liver ordinarily stores glucose as the branched molecule glycogen, so if your blood runs out of glucose before you eat again, the other pancreatic hormone, glucagon, causes the liver to release glucose into the blood.

If by chance the liver runs out of stored glucose, it will convert other molecules, such as amino acids from protein, to glucose in order to maintain the blood glucose level. Liver glucose stores might become extremely low if you have diabetes mellitus and your pancreas does not produce enough insulin or if you are on a low-carbohydrate diet. The inability of the pancreas to produce insulin and the inability of the liver to maintain the blood glucose concentration are serious conditions that require good medical care. If homeostasis cannot be maintained, death can occur due to either an extremely high or an extremely low blood glucose level. This occurs mainly because, of all the organs in the body, it is particularly important that the brain receive a constant supply of glucose.

Understanding the structure and function of the various types of tissues in the body is critical to understanding the functions of the organ systems. In this chapter, you will learn about basic tissue types and how they come together to form organs and organ systems. In addition, you will learn how all these components of the body maintain homeostasis.

22.1 The Body's Organization

Learning Outcomes

Upon completion of this section, you should be able to

1. List in order of increasing complexity the levels of organization of an animal body.
2. Describe epithelial tissue and explain its functions.
3. Describe the primary characteristics of connective tissue.
4. Compare and contrast the three types of muscle tissue.
5. Describe the function of nervous tissue.

In Part One of this text, you studied the general structure and function of a plant cell and an animal cell. Here we will take that knowledge of animal cell structure and see how millions of individual cells of different types come together to make an organism. Cells of the same structural and functional type occur within a tissue. Simply stated, a **tissue** is a group of similar cells performing a similar function. An **organ** contains different types of tissues, each performing a function to aid in the overall action of the organ. In other words, the structure and function of an organ are dependent on the tissues it contains. That is why it is sometimes said that tissues, not organs, are the structural and functional units of the body. An organ system contains multiple organs that work together to perform a specific physiological function within the organism.

Let's take an example (**Fig. 22.1**). In humans, one purpose of the urinary system is to filter wastes out of the blood, produce urine from those waste products, and permanently remove the urine from the body. The urinary system is composed of several indvidual organs, each with a particular function in the process; the kidneys, ureters, bladder, and urethra. The kidneys are composed of several different tissue types; one type in particular, the epithelial tissue, contains cells that function in filtration. These cells can filter blood and remove the waste products (forming urine) into another organ, the ureters. The tissues in the ureters form a tube-like structure that allows urine to pass from the kidneys into the bladder. The bladder, also composed of many different types of tissues, has one tissue made of cells that allow the entire organ to distend, or expand, when full of urine. And finally the urethra, like the ureters, is composed of tissues that form a cylindrical structure, allowing urine to flow from the bladder out of the body. The overall functions of the urinary system—to produce, store, and rid the body of metabolic wastes—are dependent on the cells that make up the tissues, the tissues that make up the organs, and the organs that make up the organ system.

The structure of the cells, the tissues, and the organs they compose also directly aid function. A common

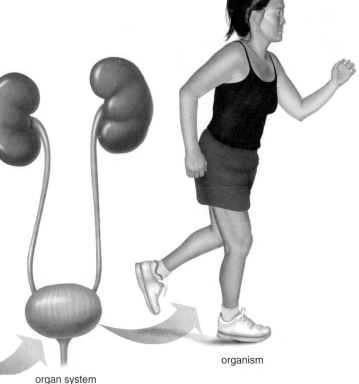
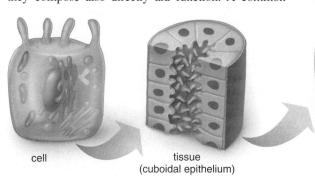

organism

organ system
(urinary system)

organ
(kidney)

cell

tissue
(cuboidal epithelium)

Figure 22.1 Levels of biological organization.

A tissue is composed of specialized cells, all having the same structure and performing the same functions. An organ is composed of those types of tissues that help it perform particular functions. An organ system contains several organs and has functions necessary to the continued existence of an organism.

saying in biology is "structure equals function," meaning that the structure of an organ (and hence the tissues and cells that compose it) dictates its function. For example, the small intestine functions in the absorption of nutrients from the digestive tract. The larger the surface area of each cell, the more absorption can occur. Some intestinal cells have areas covered in microvilli (small, finger-like projections that are extensions of the plasma membrane) that increase the surface area of the cell, thus increasing areas where absorption can occur, without increasing the overall size of the cell itself. A skeletal muscle cell has an internal arrangement of contractile fibers that can slide past each other when the cell contracts, instead of having an arrangement of fibers that would curl or twist in order for a contraction to occur. This strucure enables the entire muscle tissue to contract, without wear and tear on the fibers that might increase the possibility of damage.

From the many different types of animal cells, biologists have been able to categorize tissues into just four major types:

MP3
Overview of Tissues

- *Epithelial tissue* (epithelium) covers body surfaces and lines body cavities.
- *Connective tissue* binds and supports body parts.
- *Muscular tissue* moves the body and its parts.
- *Nervous tissue* receives stimuli and conducts nerve impulses.

Except for nervous tissue, each type of tissue is subdivided into even more types (**Fig. 22.2**). This chapter looks at the particular structure and function of each of these tissue types, as well as the organs and organ systems where they are used.

Figure 22.2 Types of vertebrate tissues.

The four classes of vertebrate tissues are epithelial (pink), connective (blue), muscular (tan), and nervous (yellow).

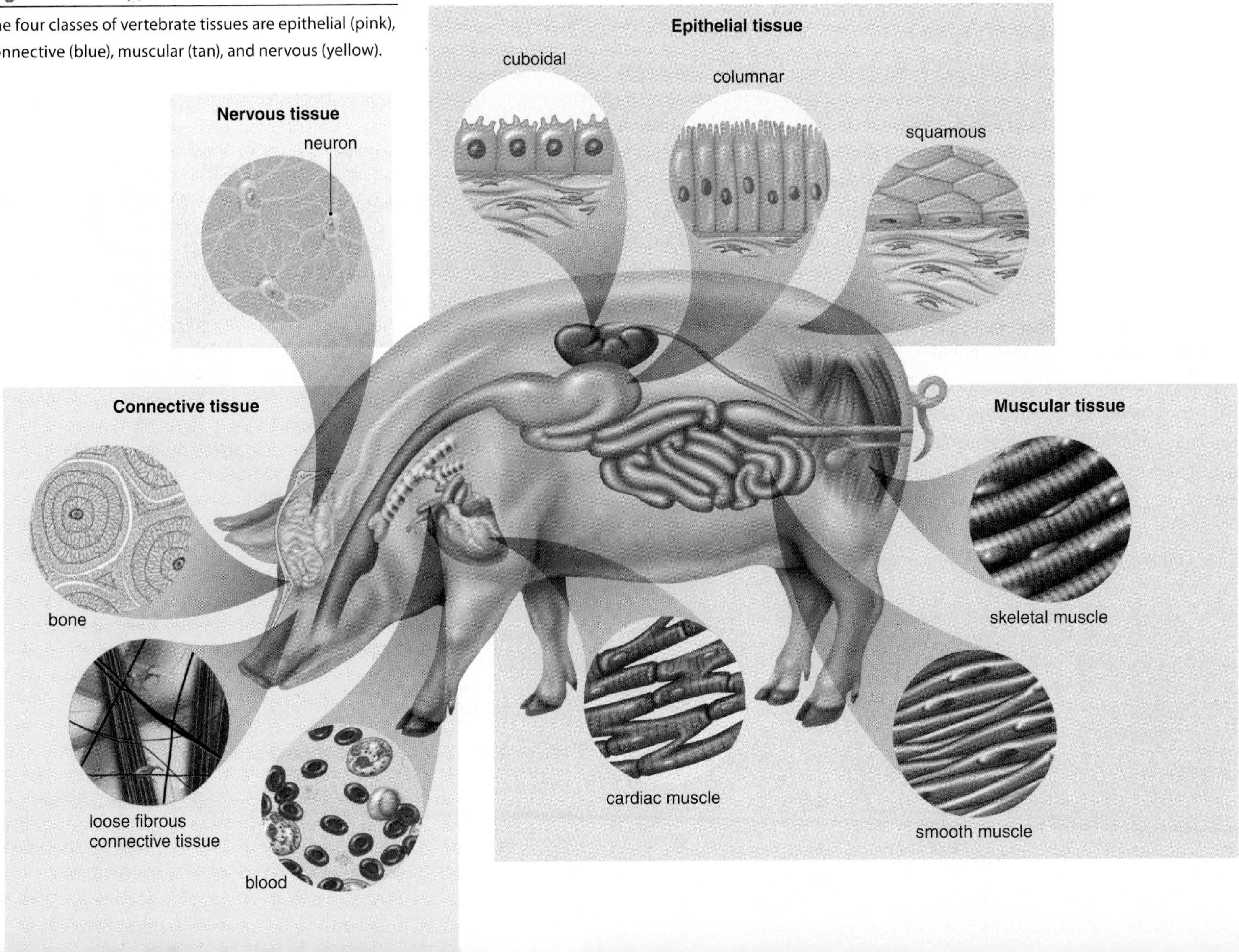

Nervous tissue
neuron

Epithelial tissue
cuboidal
columnar
squamous

Connective tissue
bone
loose fibrous connective tissue
blood

Muscular tissue
skeletal muscle
cardiac muscle
smooth muscle

Epithelial Tissue Protects

Epithelial tissue, also called *epithelium,* forms the external coverings and internal linings of many organs and covers the entire surface of the body. Therefore, for a substance to enter or exit the body—for example, in the digestive tract, the lungs, or the genital tract—it must cross an epithelial tissue. Epithelial cells adhere to one another, but an epithelium is generally only one cell layer thick. This enables an epithelium to serve a protective function, as substances have to pass through epithelial cells in order to reach a tissue beneath them.

Epithelial cells differ in shape (**Fig. 22.3**). **Squamous epithelium,** such as that lining blood vessels and areas of gas exchange in the lungs, is composed of thin, flattened cells. **Cuboidal epithelium,** which lines the **lumen** (cavity) of a portion of the kidney, contains cube-shaped cells that are roughly the same height as width. **Columnar epithelium** has cells resembling rectangular pillars

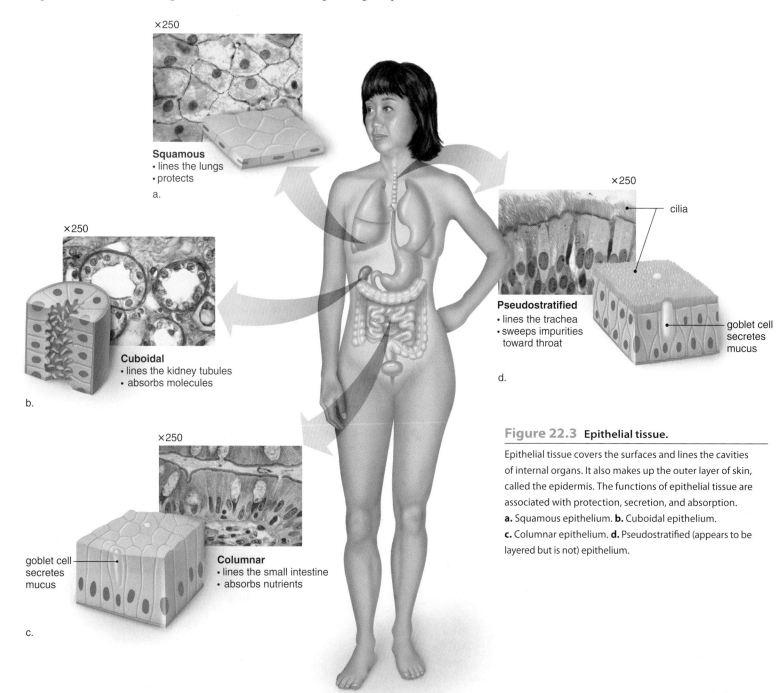

×250

Squamous
• lines the lungs
• protects

a.

×250

Cuboidal
• lines the kidney tubules
• absorbs molecules

b.

×250

goblet cell secretes mucus

Columnar
• lines the small intestine
• absorbs nutrients

c.

×250

cilia

Pseudostratified
• lines the trachea
• sweeps impurities toward throat

goblet cell secretes mucus

d.

Figure 22.3 Epithelial tissue.

Epithelial tissue covers the surfaces and lines the cavities of internal organs. It also makes up the outer layer of skin, called the epidermis. The functions of epithelial tissue are associated with protection, secretion, and absorption.
a. Squamous epithelium. **b.** Cuboidal epithelium.
c. Columnar epithelium. **d.** Pseudostratified (appears to be layered but is not) epithelium.

×250

Stratified
• epidermis of skin
• protects

Figure 22.4 Skin.

The outer portion of skin, called the epidermis, is a stratified epithelium. The many layers of tightly packed cells reinforced by the protein keratin make the skin protective against water loss and pathogen invasion. New epidermal cells arise in the innermost layer and are shed at the outer layer. In reptiles, such as this gila monster, the epidermis forms scales, which are simply projections hardened with keratin. When the epidermis is shed, the scales have to be replaced. Pigmented epidermal cells account for coloration of the skin.

Connections and Misconceptions

How quickly does epithelial tissue renew?

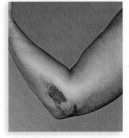

On average, if there is no injury and the body is simply replacing worn-out cells, a new epithelial cell, such as the kind in your skin, can renew and move to the top of the five layers of skin (in thick skin) in about a month. When there is an injury or damage, various hormones, such as epidermal growth factor, will speed up the process during wound healing, and it can take only a week or two.

or columns, with nuclei usually located near the bottom of each cell. Columnar epithelium lines portions of the lumen of the digestive tract. In addition to the shape difference, epithelial cells can be classified by the number of layers these cells make in tissues. A layer that is only one cell thick is referred to as simple. Multiple layers of cells is called stratified. Pseudostratified epithelium is a special classification in which the tissue appears to be layered but actually has only one layer of cells. This is normally found in columnar cells, where the nuclei of the cells, instead of all being near the bottom of each cell, are in various locations in each cell, giving the appearance of multiple layers (Fig. 22.3).

The outer region of the skin, called the epidermis, is stratified squamous epithelium in which the cells have been reinforced by keratin, a protein that provides strength and waterproofing (**Fig. 22.4**). A stratified epithelium allows skin to protect the body from injury, drying out, and possible pathogen (virus and bacterium) invasion.

The lining of the urinary bladder consists of transitional epithelium, whose structure suits its function. When the walls of the bladder are relaxed, the transitional epithelium consists of several layers of cuboidal cells. When the bladder is distended with urine, the epithelium stretches, and the outer cells take on a squamous appearance. The cells are able to slide in relation to each other while forming a barrier that prevents urine from diffusing into other regions of the body.

One or more types of epithelial cells are the primary components of glands, which produce and secrete products (mainly hormones). For example, each mucus-secreting goblet cell in the lining of the digestive tract is a single-celled gland that produces mucus that protects the digestive tract from acidic gastric juices (see Fig. 22.3c). Pseudostratified epithelium lines the trachea (windpipe) (see Fig. 22.3d), where mucus traps foreign particles and the upward motion of cilia on the cells carries the mucus to the back of the throat, where it may be either swallowed or expelled. Smoking can cause a change in mucus production and secretion and inhibit ciliary action, resulting in an inflammatory condition called chronic bronchitis.

Epithelial tissue cells can go through mitosis frequently and quickly, which is why epithelial tissue is found in places that get a lot of wear and tear. This feature is particularly useful along the digestive tract, where rough food particles and enzymes can damage the lining. Swallowing a potato chip and having a sharp edge scrape down the esophagus and burning the roof of your mouth on a hot bite of food are typical injuries that can heal quickly due to the epithelial cell lining of the digestive tract. The liver, which is composed of cells of epithelial origin, can regenerate whole portions that have been removed due to injury or surgery. But there is a price to pay for the ability of epithelial tissue to divide constantly—it is more likely than other tissue types to become cancerous. Cancers of epithelial tissue within the digestive tract, lungs, and breast are called carcinomas.

Connective Tissue Connects and Supports

The many types of **connective tissue** are all involved in binding organs together and providing support and protection. As a rule, connective tissue cells are widely separated by a **matrix,** a noncellular material that varies from solid to semifluid to fluid. The matrix usually has fibers—notably, collagen fibers. Collagen, used mainly for structural support, is the most common protein in the human body, which gives you some idea of the prevalence of connective tissue.

Loose Fibrous and Related Connective Tissues

Let's consider **loose fibrous connective tissue** first and then compare the other types with it (see Fig. 22.2). This tissue occurs beneath an epithelium and connects it to the other tissues within an organ. It also forms a protective covering for many internal organs, such as muscles, blood vessels, and nerves. Its cells are called **fibroblasts** because they produce a matrix that contains fibers, including collagen fibers and elastic fibers, which stretch under tension and return to their original shape when released. The presence of loose fibrous connective tissue in the walls of lungs and arteries gives these organs resilience, the ability to expand and then return to their original shape without damage.

Adipose tissue is a type of loose connective tissue in which the fibroblasts enlarge and store fat, and there is limited matrix (**Fig. 22.5**). Adipose tissue is located beneath the skin and around organs, such as the heart and kidneys, where it cushions and protects the organs and serves as long-term energy storage.

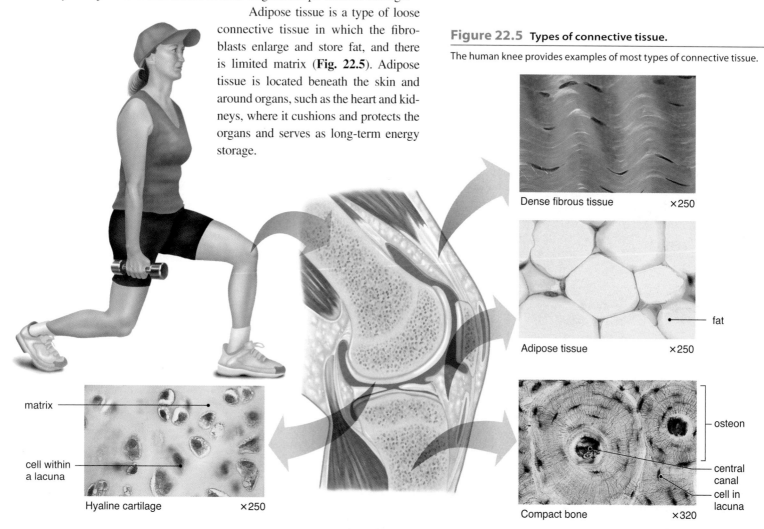

Figure 22.5 Types of connective tissue.

The human knee provides examples of most types of connective tissue.

Dense fibrous tissue   ×250

Adipose tissue   ×250

fat

osteon

central canal

cell in lacuna

Compact bone   ×320

matrix

cell within a lacuna

Hyaline cartilage   ×250

Compared with loose fibrous connective tissue, **dense fibrous connective tissue** contains more collagen fibers, which are packed closely together (Fig. 22.5). This type of tissue has more specific functions than does loose fibrous connective tissue. For example, dense fibrous connective tissue is found in **tendons,** which connect muscles to bones, and in **ligaments,** which connect bones to other bones at joints.

In **cartilage,** the cells lie in small, open cavities called **lacunae,** separated by a matrix that is semisolid yet flexible. **Hyaline cartilage,** the most common type of cartilage, contains only very fine collagen fibers (Fig. 22.5). The matrix has a white, translucent appearance when unstained. Hyaline cartilage is found in the nose and at the ends of the long bones and the ribs, and it forms rings in the walls of respiratory passages. The human fetal skeleton is also made of this type of cartilage, which is later replaced by bone. Cartilaginous fishes, such as sharks, have a cartilaginous skeleton throughout their lives.

Bone is the most rigid connective tissue (Fig. 22.5). It consists of an extremely hard matrix of inorganic salts—notably, calcium salts—deposited around collagen fibers. The inorganic salts give bone rigidity, and the protein fibers provide elasticity and strength, much as steel rods do in reinforced concrete. The inorganic salts found in bone also act as storage for calcium and phosphate ions for the entire body. **Compact bone,** the most common type of bone in humans, consists of cylindrical structural units called osteons. The central canal of each osteon is surrounded by rings of hard matrix. Bone cells are located in lacunae between the rings of matrix. Blood vessels in the central canal carry nutrients that allow bone to renew itself.

Blood

Blood is composed of several types of cells suspended in a liquid matrix called plasma. Blood is unlike other types of connective tissue in that the matrix (i.e., plasma) is not made by the cells of the connective tissue (**Fig. 22.6**). Even though the blood is liquid, it is a connective tissue by definition; it consists of cells within a matrix. Blood has many functions for overall homeostasis of the body. It transports nutrients and oxygen to cells and removes their wastes. It helps distribute heat and plays a role in fluid, ion, and pH balance. Also, various components of blood help protect us from disease, and blood's ability to clot prevents fluid loss.

Blood contains three formed elements; red blood cells, white blood cells, and platelets. **Red blood cells** are small, biconcave, disk-shaped cells without nuclei. The presence of the red pigment hemoglobin makes the cells red, and blood as a whole red. Hemoglobin binds oxygen and allows the red blood cells to transport oxygen to the cells of the body.

White blood cells can be distinguished from red blood cells by the fact that they are usually larger, have a nucleus, and would appear translucent without staining. White blood cells fight infection in two primary ways: (1) Some are phagocytic and engulf infectious pathogens; (2) others produce antibodies, molecules that combine with foreign substances to inactivate them.

Platelets are another component of blood, but they are not complete cells; rather, they are fragments of giant cells present only in bone marrow. When a blood vessel is damaged, platelets form a plug that seals the vessel, and along with injured tissues, platelets release molecules that help the clotting process.

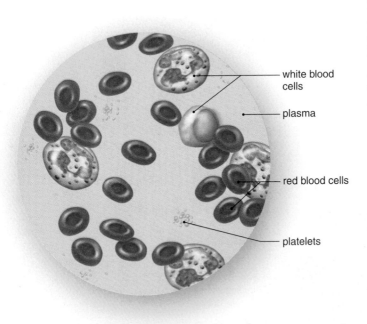

white blood cells

plasma

red blood cells

platelets

Figure 22.6 Blood, a liquid tissue.

Blood is classified as connective tissue because the cells are separated by a matrix—plasma, the liquid portion of blood. The cellular components of blood are red blood cells, white blood cells, and platelets (which are actually fragments of larger cells).

Muscular Tissue Moves the Body

Muscular tissue and nervous tissue work together to enable animals to move. **Muscular tissue** contains contractile protein filaments, called actin and myosin filaments, that interact to produce movement. The three types of vertebrate muscles are skeletal, cardiac, and smooth.

Skeletal muscle, which works under voluntary movement, is attached by tendons to the bones of the skeleton, and when it contracts, bones move. Contraction of skeletal muscle, being under voluntary control, occurs faster than in the other two muscle types. The cells of skeletal muscle, called fibers, are cylindrical and quite long—sometimes they run the entire length of the muscle (**Fig. 22.7a**). They arise during development when several cells fuse, resulting in one fiber with multiple nuclei. The nuclei are located at the edge of the cell, just inside the plasma membrane. The fibers have alternating light and dark bands running across the cell that gives them a **striated** appearance. These bands are due to the arrangement of actin filaments and myosin filaments in the cell.

Cardiac muscle is found only in the walls of the heart, and its contraction pumps blood and accounts for the heartbeat. Like skeletal muscle, cardiac muscle has striations, but the contraction of the heart is autorhythmic (occurring at a set pace) and involuntary (Fig. 22.7b). Cardiac muscle cells also differ from skeletal muscle cells in that they have a single, centrally placed nucleus. The cells are branched and seemingly fused one with the others. The heart appears to be composed of one large, interconnecting mass of muscular cells. Actually, cardiac muscle cells are separate and individual, but they are bound end to end at **intercalated disks,** areas where folded plasma membranes allow the contraction impulse to spread from cell to cell.

Smooth muscle is so named because the cells lack striations. The spindle-shaped cells form layers in which the thick middle portion of one cell is opposite the thin ends of adjacent cells. Consequently, the nuclei form an irregular pattern in the tissue (Fig. 22.7c). Like cardiac muscle, smooth muscle is involuntary. Smooth muscle is also sometimes called *visceral* muscle because it is found in the walls of the viscera (intestines, stomach, and other internal organs) and blood vessels. Smooth muscle contracts more slowly than skeletal muscle but can remain contracted for a longer time. When the smooth muscle of the intestines contracts, food moves along its lumen. When the smooth muscle of the blood vessels contracts, blood vessels constrict, helping raise blood pressure.

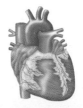
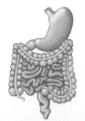

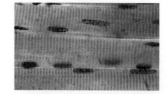

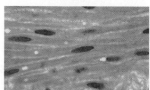

×250

intercalated disk   ×100

×100

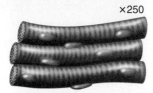

a. **Skeletal muscle**
- has striated, tubular, multinucleated fibers.
- is usually attached to skeleton.
- is voluntary.

b. **Cardiac muscle**
- has striated, branched, uninucleated fibers.
- occurs in walls of heart.
- is involuntary.

c. **Smooth muscle**
- has spindle-shaped, nonstriated, uninucleated fibers.
- occurs in walls of internal organs.
- is involuntary.

Figure 22.7 Types of muscular tissue.

Figure 22.8 **Nervous tissue.**

Neurons are surrounded by neuroglia, such as Schwann cells, which envelop axons and form the myelin sheath. Only axons conduct nerve impulses.

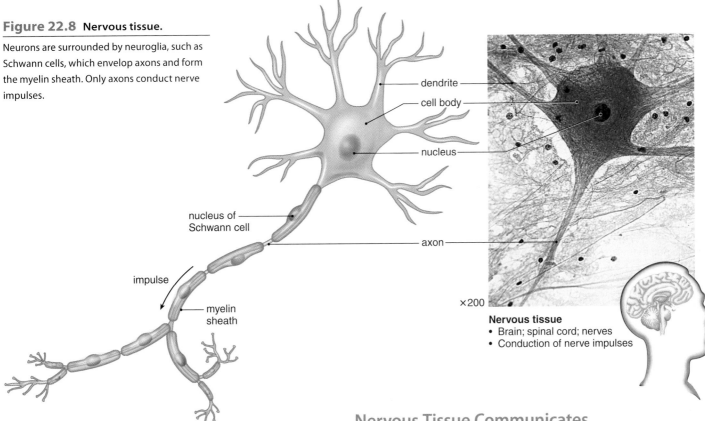

dendrite

cell body

nucleus

nucleus of Schwann cell

impulse

axon

myelin sheath

×200

Nervous tissue
- Brain; spinal cord; nerves
- Conduction of nerve impulses

Connections and Misconceptions

How big are neurons and how quickly do they communicate?

The average adult human brain weighs about 3 pounds and contains over 100 billion neurons. Some of the neurons—specifically, the axons of the neurons—are less than a millimeter in length; others, such as the axons from the spinal cord to a muscle in the foot, can be 3 feet long or more. The speed at which transmission occurs can vary as well. It can be as slow as 0.5 meter/second or as fast as 120 meters/second—268 mph!

Nervous Tissue Communicates

Nervous tissue coordinates the functions of body parts and allows an animal to respond to external and internal environments. The nervous system depends on (1) sensory input, (2) integration of data, and (3) motor output to carry out its functions. Nerves conduct impulses from sensory receptors, such as pain receptors in the skin, to the spinal cord and brain, where integration occurs. The phenomenon called sensation occurs only in the brain, however. Nerves then conduct nerve impulses away from the spinal cord and brain to the muscles and glands, causing them to contract or secrete in response. In this way, a coordinated response to both internal and external stimuli is achieved.

A nerve cell is called a **neuron.** Every neuron has three parts: dendrites, a cell body, and an axon (**Fig. 22.8**). A *dendrite* is an extension of the neuron cell body that conducts signals toward the cell body. The *cell body* contains the major concentration of the cytoplasm and the nucleus of the neuron. An *axon* is an extension that conducts nerve impulses away from the neuron cell body to other cells. The brain and spinal cord contain many neurons, whereas **nerves** contain only bundles of axons of neurons. The dendrites and cell bodies of these neurons are located in the spinal cord or brain, depending on whether it is a spinal nerve or a cranial nerve.

In addition to neurons, nervous tissue contains **neuroglia,** cells that support and nourish neurons. They outnumber neurons nine to one and take up more than half the volume of the brain. Although their primary function is support, research is currently being conducted to determine how much neuroglia directly contribute to brain function. Schwann cells are the type of neuroglia that encircle long nerve fibers within nerves, forming a protective coating on the axons called a myelin sheath. The presence of myelin sheaths allows nerve impulses to travel much more quickly down the axon, similar to insulation on the outer walls of a house.

Connecting the Concepts

For examples of each of these tissue types, refer to the following discussions.

Section 23.2 examines the function of blood, an example of a connective tissue.

Section 24.3 examines the role of epithelial tissue in nutrient absorption.

Section 27.1 explores the structure and function of neurons.

Section 28.2 examines how the muscular tissue is involved in movement.

Check Your Progress 22.1

1. List the four basic tissue types, and describe the functions of each.
2. Recall the basic shapes of cells and the terms for the layers of cells in tissues.
3. Explain why epithelial tissue is found where it is in the human body.
4. Compare and contrast epithelial and connective tissue.
5. Analyze why there is a difference in densities in the matrixes of connective tissues. Does the matrix relate to the overall function of each tissue?
6. Relate how the nervous system functions in conjunction with all three types of muscular tissue.

22.2 Organs and Organ Systems

Learning Outcomes

Upon completion of this section, you should be able to

1. Classify each organ system according to its involvement in transport, body maintenance, control, sensory input and motor output, or reproduction.
2. List the general functions of each organ system.

Organs are composed of a number of tissues, and the structure and function of an organ are dependent on those tissues. Since an organ has many types of tissues, it can perform a function that none of the tissues can do alone. For example, the function of the bladder is to store urine and to expel it when convenient. But this function is dependent on the individual tissues making up the bladder. The epithelium lining the bladder helps the organ store urine by stretching while preventing urine from leaking into the internal body cavity. The muscles of the bladder propel the urine forward into the urethra, so that it can be removed from the body. Similarly, the functions of an organ system are dependent on its organs. The function of the urinary system is to produce urine, store it, and then transport it out of the body. The kidneys produce urine, the bladder stores it, and various tubes transport it from the kidneys to the bladder (the ureters) and out of the body (through the urethra).

This text divides the systems of the body into those involved in (1) the transport of fluids throughout the body, (2) maintenance of the body, (3) control of the body's systems, (4) sensory input and motor output, and (5) reproduction. All these systems have functions that contribute to homeostasis, the relative constancy of the internal environment (see Section 22.3).

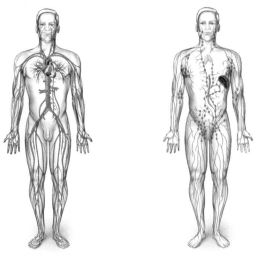

a. Cardiovascular system   b. Lymphatic and immune systems

Figure 22.9 **The body's transport systems.**

Both the cardiovascular and lymphatic systems are involved in transport. The lymphatic systems is also involved in defense against disease.

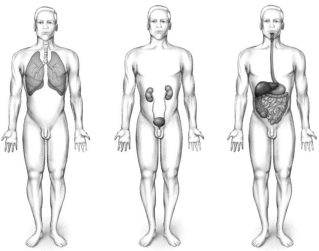

a. Respiratory system   b. Urinary system   c. Digestive system

Figure 22.10 **The body's maintenance systems.**

The respiratory, urinary, and digestive systems keep the body's internal environment constant.

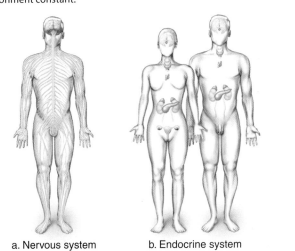

a. Nervous system   b. Endocrine system

Figure 22.11 **The control systems.**

The nervous and endocrine systems coordinate the other systems of the body.

Transport

The **cardiovascular system** (**Fig. 22.9**a) consists of blood, the heart, and the blood vessels that carry blood throughout the body. The body's cells are surrounded by a liquid called interstitial fluid, or **tissue fluid;** blood transports nutrients and oxygen to tissue fluid for the cells and removes waste molecules excreted by cells from the tissue fluid. Recall that the internal environment of the body consists of the blood within the blood vessels and the tissue fluid that surrounds the cells.

The **lymphatic system** (Fig 22.9b) consists of lymphatic vessels, lymph, lymph nodes, and other lymphatic organs. Lymphatic vessels absorb fat from the digestive system and collect excess tissue fluid, which is returned to the blood in the cardiovascular system.

The cardiovascular and lymphatic systems are also involved in defense against disease. Along with the thymus and spleen, certain blood cells in the lymph and blood are part of the **immune system,** which specifically protects the body from disease.

Maintenance of the Body

Three systems (respiratory, urinary, and digestive) add substances to and/or remove substances from the blood. If the composition of the blood remains constant, so does that of the tissue fluid.

The **respiratory system** (**Fig. 22.10**a) consists of the lungs, the trachea, and other tubes that take air to and from the lungs. The respiratory system brings oxygen into the body and takes carbon dioxide out through the lungs. It also exchanges gases with the blood.

The **urinary system** (Fig. 22.10b) contains the kidneys and the urinary bladder, along with tubes that transport urine. This system rids blood of wastes and helps regulate the fluid level and chemical content of the blood.

The **digestive system** (Fig. 22.10c) consists of the organs along the digestive tract, together with the associated organs, such as the teeth, salivary glands, liver, and pancreas. This system receives food and digests it into nutrient molecules, which then enter the blood.

Control

The **nervous system** (**Fig. 22.11**a) consists of the brain, the spinal cord, and associated nerves. The nerves conduct nerve impulses from receptors to the brain and spinal cord. They also conduct nerve impulses from the brain and spinal cord to the muscles and glands, allowing us to respond to both external and internal stimuli. Sensory receptors and sense organs are sometimes considered a part of the nervous system.

The **endocrine system** (Fig. 22.11b) consists of the hormonal glands, such as the thyroid and adrenal glands, which secrete hormones, chemicals that serve as messengers between body parts. Both the nervous and endocrine systems coordinate and regulate the functions of the body's other systems. The nervous system tends to cause quick responses in the body, while the body's responses to hormones released by the endocrine system tend to last much longer. The endocrine system also helps maintain the proper functioning of the male and female reproductive organs.

Sensory Input and Motor Output

The **integumentary system** (**Fig. 22.12***a*) consists of the skin and its accessory structures. The sensory receptors in the skin, and in organs such as the eyes and ears, respond to specific external stimuli and communicate with the brain and spinal cord by way of nerve fibers. These messages may cause the brain to respond to a stimulus.

The **skeletal system** (Fig. 22.12*b*) and the **muscular system** (Fig. 22.12*c*) enable the body and its parts to move as a result of motor output. The skeleton, as a whole, serves as a place of attachment for the skeletal muscles. Contraction of the muscles in the muscular system accounts for the movement of body parts.

These three systems protect and support the body. The skeletal system, consisting of the bones of the skeleton, protects body parts. For example, the skull forms a protective encasement for the brain, as does the rib cage for the heart and lungs. The skin serves as a barrier between the outside world and the body's tissues.

Reproduction

The **reproductive system** (**Fig. 22.13**) involves different organs in the male and female. The male reproductive system consists of the testes, other glands, and various ducts, such as the ductus deferens, that conduct semen to and through the penis. The testes produce sex cells called sperm. The female reproductive system consists of the ovaries, oviducts, uterus, vagina, and external genitals. The ovaries produce sex cells called eggs. When a sperm fertilizes an egg, an offspring begins development.

Connecting the Concepts

For more on these organ systems, refer to the following discussions.

Chapter 23 explores the circulatory and lymphatic systems.

Chapter 24 examines the structure and function of the respiratory, urinary, and digestive systems.

Chapter 26 provides an overview of how the body defends itself against disease.

Chapter 27 explores the organization and operation of the nervous and endocrine systems.

Chapter 28 investigates how the senses work, as well as the operation of the muscular and skeletal systems.

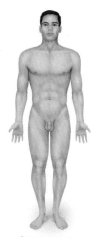 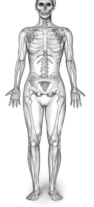 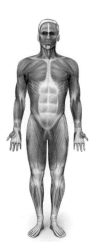

a. Integumentary system   b. Skeletal system   c. Muscular system

Figure 22.12   Sensory input and motor output.
Sensory receptors in the skin (integumentary system) and the sense organs send input to the skeletal and muscular systems, the control systems that cause a response to stimuli.

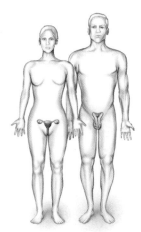

Reproductive system

Figure 22.13   The reproductive system.
The reproductive organs in the female and male differ. The reproductive system ensures the survival of the species.

Check Your Progress 22.2

1. Compare and contrast the function of the cardiovascular system with that of the lymphatic system.
2. Describe the common function of the nervous and endocrine systems.
3. Distinguish all 11 body systems by the overall functions they perform, such as transport and sensory input.

22.3 Homeostasis

Learning Outcomes

Upon completion of this section, you should be able to

1. Identify the role of each of the organ systems in maintaining homeostasis.
2. Use the concept of negative feedback to explain body temperature control in humans.

When discussing homeostasis, there are two types of environments to consider: The external environment, which includes everything outside our body, and the internal environment, which includes our cells, tissues, fluids, and organs. Recall that cells live in a liquid environment called tissue fluid. Tissue fluid is constantly renewed with nutrients and gases by exchanges with the blood (**Fig. 22.14**). Therefore, blood and tissue fluid constitute part of the body's internal environment of the body. The volume and composition of tissue fluid remains relatively constant only as long as blood composition remains near normal levels. *Relatively constant* means that the composition of both tissue fluid and blood usually falls within a certain range of normality. Maintenance of the relatively constant condition of the internal environment is **homeostasis.**

Because of homeostasis, even though external conditions may change dramatically, internal conditions stay within a narrow range. For example, the temperature of the body is maintained near 37°C (97°F to 99°F), even if the surrounding temperature is lower or higher. If you eat acidic foods, the pH of your blood still stays about 7.4, and even if you eat a candy bar, the amount of sugar in your blood remains at just about 0.1%.

All the systems of the body contribute to maintaining homeostasis. The cardiovascular system conducts blood to and away from capillaries, where exchange occurs. The heart pumps the blood, thereby keeping it moving toward the capillaries. Red blood cells transport oxygen and participate in the transport of carbon dioxide. White blood cells fight infection, and platelets participate in the clotting process. The lymphatic system is accessory to the cardiovascular system. Lymphatic capillaries collect excess tissue fluid and return it via lymphatic vessels to the cardiovascular system. Lymph nodes help purify lymph and keep it free of pathogens.

The digestive system takes in and digests food, providing nutrient molecules that enter the blood to replace those that are constantly being used by the body's cells. The respiratory system removes carbon dioxide from and adds oxygen to the blood. The chief regulators of blood composition are the kidneys and the liver. Urine formation by the kidneys is extremely critical to the body, not only because it rids the body of metabolic wastes but also because the kidneys carefully regulate blood volume, salt balance, and pH. The liver, among other functions, regulates the glucose concentration of the blood. Immediately after glucose enters the blood, the liver removes the excess for storage as glycogen. Later, the glycogen is broken down to replace the glucose that was used by body cells. In this way, the glucose composition of the blood remains relatively constant. The liver makes urea, a nitrogenous end product of protein metabolism.

The nervous system and endocrine system regulate the other systems of the body. They work together to control body systems, so that homeostasis is maintained. In mechanisms involving the nervous system, sensory receptors send nerve impulses to control centers in the brain, which then direct effectors (muscles or glands) to become active. Muscles bring about an immediate change. Endocrine glands secrete hormones that bring about a slower, more lasting change that keeps the internal environment relatively stable.

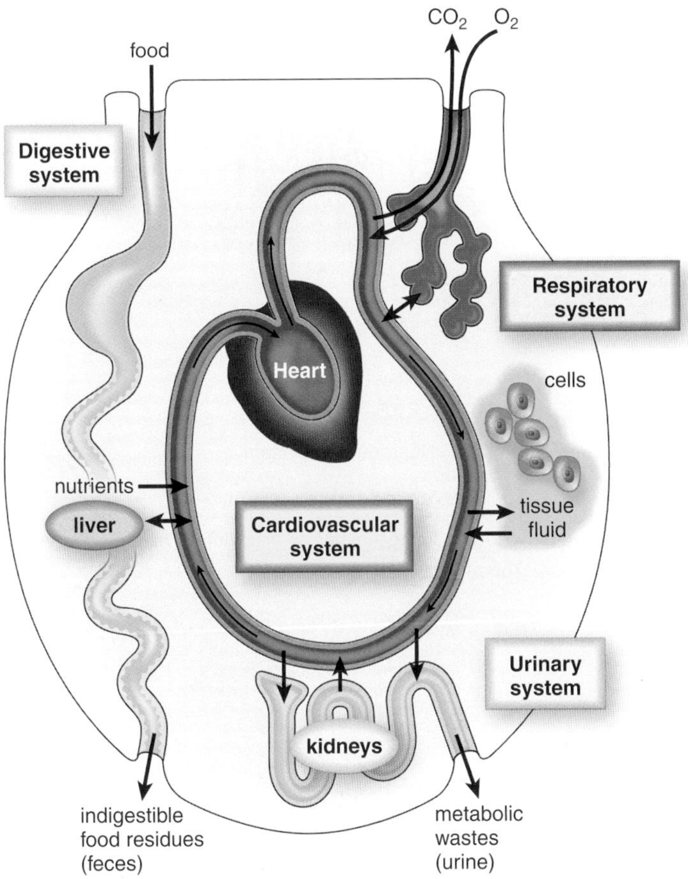

Figure 22.14 **Constancy of the internal environment (blood and tissue fluid).**

The respiratory system exchanges gases with the external environment and with the blood. The digestive system takes in food and adds nutrients to the blood, and the urinary system removes metabolic wastes from the blood and excretes them. When blood exchanges nutrients and oxygen for carbon dioxide and other wastes with tissue fluid, the composition of the tissue fluid stays within normal limits.

Negative Feedback

Negative feedback is the primary homeostatic mechanism that allows the body to keep the internal environment constant. A negative feedback mechanism has at least two components: a sensor and a control center (**Fig. 22.15**). The sensor detects a change in the internal environment (a stimulus); the control center initiates an effect that brings conditions back to normal. At that point, the sensor is no longer activated. In other words, a negative feedback mechanism is present when the output of the system dampens the original stimulus.

Consider a simple example. When the pancreas detects that the blood glucose level is too high, it secretes insulin, a hormone that causes cells to take up glucose. Then the blood sugar level returns to normal, and the pancreas is no longer stimulated to secrete insulin.

When conditions exceed their limits and feedback mechanisms cannot compensate, illness results. For example, if the pancreas is unable to produce insulin, as in diabetes mellitus, the blood sugar level becomes dangerously high and the individual can become seriously ill. The study of homeostatic mechanisms is therefore medically important.

Animation
Positive and
Negative Feedback

Mechanical Example

A home heating system is often used to illustrate how a more complicated negative feedback mechanism works (**Fig. 22.16**). Suppose you set the thermostat at 20°C (68°F). This is the *set point*. The thermostat contains a thermometer, a sensor that detects when the room temperature is above or below the set point. The thermostat also contains a control center; it turns the furnace off when the room is warm and turns it on when the room is cool. When the furnace is off, the room cools a bit, and when the furnace is on, the room warms a bit. In other words, a negative feedback system results in fluctuation above and below the set point.

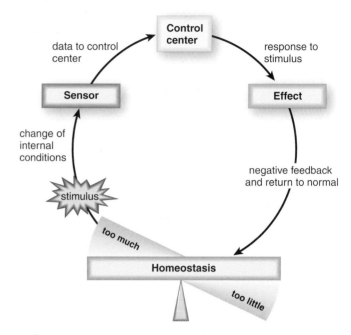

Figure 22.15 Negative feedback mechanism.

This diagram shows how the basic elements of a negative feedback mechanism work.

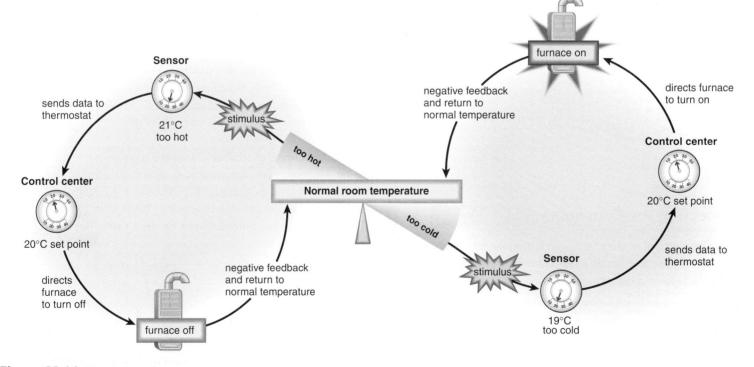

Figure 22.16 Regulation of room temperature.

This diagram shows how room temperature is returned to normal when the room becomes too hot. A contrary cycle in which the furnace turns on and gives off heat returns the room temperature to normal when the room becomes too cold.

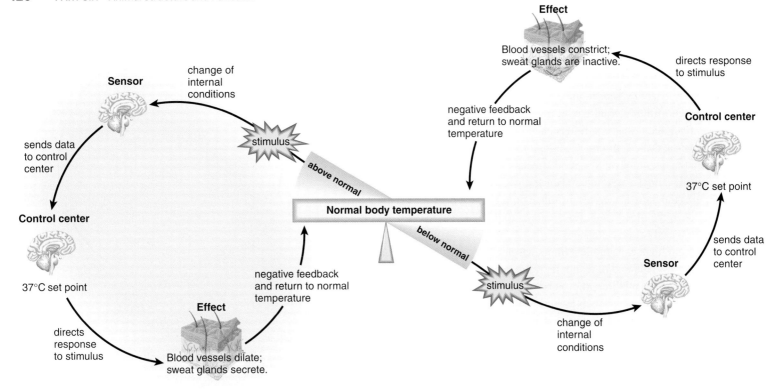

Figure 22.17   Regulation of body temperature.

This diagram shows how normal body temperature is maintained by a negative feedback system.

Connecting the Concepts

The maintenance of homeostasis is an important function of all organ systems. For more examples of homeostasis, refer to the following discussions.

Section 23.3 describes how the cardiovascular system helps maintain blood homeostasis.

Section 24.2 explores the role of the urinary system in maintaining homeostasis.

Section 27.2 describes the role of the hypothalamus in regulating the internal conditions of the body.

Human Example: Regulation of Body Temperature

The thermostat for body temperature is located in a part of the brain called the hypothalamus. When the core body temperature falls below normal, the control center directs (via nerve impulses) the blood vessels of the skin to constrict (**Fig. 22.17**). This action conserves heat. If the core body temperature falls even lower, the control center sends nerve impulses to the skeletal muscles, and shivering occurs. Shivering generates heat, and gradually body temperature rises toward 37°C (98.6°F). When the temperature rises to normal, the control center is inactivated.

When the body temperature is higher than normal, the control center directs the blood vessels of the skin to dilate. More blood is then able to flow near the surface of the body, where heat can be lost to the environment. In addition, the nervous system activates the sweat glands, and the evaporation of sweat helps lower body temperature. Gradually, body temperature decreases to 37°C (98.6°F).

Notice that a negative feedback mechanism prevents continued change in the same direction; body temperature does not get warmer and warmer, because warmth stimulates changes that decrease body temperature. Also, body temperature does not get colder and colder, because a body temperature below normal causes changes that bring body temperature up.

MP3
Homeostasis

Check Your Progress 22.3

1. Explain the differences between the external and internal environments as they relate to homeostasis.

2. Summarize the components of a negative feedback mechanism.

3. Describe how the control center in the brain responds to a drop as well as a rise in internal temperature.

Media Study Tools

www.mhhe.com/maderessentials3

Enhance your study of this chapter with study tools and practice tests. Also ask your instructor about the resources available through ConnectPlus, including the media-rich eBook, interactive learning tools, and animations.

The Chapter in Review

Summary

22.1 The Body's Organization

Levels of animal organization:

Four groups of vertebrate tissues:

Epithelial Tissue Protects

Epithelial tissue covers the body and lines its cavities.

- Types of epithelial tissue are squamous, cuboidal, and columnar.
- Epithelial cells sometimes form glands that secrete either into ducts or into the blood.

Connective Tissue Connects and Supports

In connective tissue, cells are separated by a matrix that contains fibers (e.g., collagen fibers). The four types are

- Loose fibrous connective tissue, including adipose tissue
- Dense fibrous connective tissue (tendons and ligaments)
- Cartilage and bone (matrix for cartilage is more flexible than that for bone)
- Blood (matrix is a liquid called plasma)

Muscular Tissue Moves the Body

Muscular tissue is of three types: skeletal, cardiac, and smooth.

- Both skeletal muscle (attached to bones) and cardiac muscle (forms the heart wall) are striated.

- Both cardiac muscle (wall of heart) and smooth muscle (wall of internal organs) are involuntary.

Nervous Tissue Communicates

- Nervous tissue is composed of neurons and several types of neuroglia.
- Each neuron has dendrites, a cell body, and an axon. Axons conduct nerve impulses.

22.2 Organs and Organ Systems

Organs make up organ systems, which are summarized in the accompanying table.

22.3 Homeostasis

Homeostasis is the relative constancy of the internal environment. The internal environment consists of blood and tissue fluid. Due to exchange of nutrients and wastes with the blood, tissue fluid stays constant:

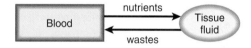

All organ systems contribute to the relative constancy of tissue fluid and blood.

- The cardiovascular system transports nutrients to cells and wastes from cells.
- The lymphatic system absorbs excess tissue fluid and functions in immunity.
- The digestive system takes in food and adds nutrients to the blood.
- The respiratory system carries out gas exchange with the external environment and the blood.
- The urinary system (i.e., the kidneys) removes metabolic wastes and regulates the pH and salt content of the blood.
- The nervous system and endocrine system regulate the other systems.

ORGAN SYSTEMS

Transport	Sensory
Cardiovascular (heart and blood vessels) Lymphatic (lymphatic vessels)	Integumentary (skin)
Maintenance	*Motor*
Digestive (e.g., stomach, intestines) Respiratory (tubes and lungs) Urinary (tubes, urinary bladder, and kidneys)	Skeletal (bones and cartilage) Muscular (muscles)
Control	*Reproduction*
Nervous (brain, spinal cord, and nerves) Endocrine (glands—e.g., thyroid and adrenal glands)	Reproductive (tubes and testes in males; tubes, uterus, and ovaries in females)

Negative Feedback

Negative feedback mechanisms keep the internal environment relatively stable. When a sensor detects a change above or below a set point, a control center brings about an effect that reverses the change, brings conditions back to normal, and shuts off the response. Examples include the following:

- Regulation of blood glucose level by insulin
- Regulation of room temperature by a thermostat and furnace
- Regulation of body temperature by the brain and sweat glands.

Key Terms

blood 418
bone 418
cardiac muscle 419
cardiovascular system 422
cartilage 418
columnar epithelium 415
compact bone 418
connective tissue 417
cuboidal epithelium 415
dense fibrous connective tissue 418
digestive system 422
endocrine system 422
epithelial tissue 415
fibroblast 417
homeostasis 424
hyaline cartilage 418
immune system 422
integumentary system 423
intercalated disks 419
lacuna 418
ligament 418
loose fibrous connective tissue 417
lumen 415
lymphatic system 422

matrix 417
muscular system 423
muscular tissue 418
negative feedback 425
nerve 420
nervous system 422
nervous tissue 420
neuroglia 420
neuron 420
organ 413
platelet 418
red blood cell 418
reproductive system 423
respiratory system 422
skeletal muscle 419
skeletal system 423
smooth muscle 419
squamous epithelium 415
striated 419
tendon 418
tissue 413
tissue fluid 422
urinary system 422
white blood cell 418

Testing Yourself

Choose the best answer for each question.

1. Label the levels of biological organization in the following illustration.

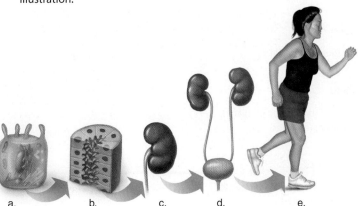

2. Which of the following is not a feature of epidermis?
   a. contains keratin for strength
   b. stratified
   c. contains a matrix
   d. composed of epithelial cells

For questions 3–9, identify the tissue in the key that matches the description. Each answer may be used more than once. Each question may have more than one answer.

Key:
   a. epithelial tissue   c. muscular tissue
   b. connective tissue   d. nervous tissue

3. contains cells that are separated by a matrix
4. has a protective role
5. covers the body surface and forms the linings of organs
6. composed of a group of cells with a similar function
7. best capable of replacing its cells
8. allows animals to respond to their environment
9. contains actin and myosin filaments

10. Label the blood smear in the following illustration.

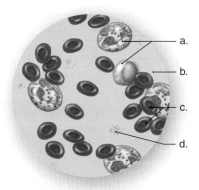

11. Which of the following systems does not add substances to or remove substances from the blood?
   a. digestive   c. urinary
   b. cardiovascular   d. respiratory

12. A major function of tissue fluid is to
   a. provide nutrients and oxygen to cells and remove wastes.
   b. keep cells warm.
   c. prevent cells from touching each other.
   d. provide flexibility by allowing cells to slide over each other.

13. Label the parts of a neuron in the following illustration.

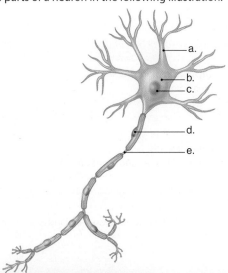

14. Which of the following is a function of skin?
    a. temperature regulation
    b. protection against water loss
    c. collection of sensory input
    d. protection from invading pathogens
    e. All of these are correct.

15. When a human being is cold, the blood vessels
    a. dilate, and the sweat glands are inactive.
    b. dilate, and the sweat glands are active.
    c. constrict, and the sweat glands are inactive.
    d. constrict, and the sweat glands are active.
    e. contract so that shivering occurs.

16. Blood is a(n) _____ tissue because it has a _____ .
    a. connective, gap junction
    b. muscle, matrix
    c. epithelial, gap junction
    d. connective, matrix

17. The skeletal system functions in
    a. blood cell production.
    b. mineral storage.
    c. movement.
    d. All of these are correct.

18. Which of the following is an example of negative feedback?
    a. Air conditioning goes off when room temperature lowers.
    b. Insulin decreases blood sugar levels after a meal.
    c. Heart rate increases when blood pressure drops.
    d. All of these are examples of negative feedback.

19. Which of these is not a type of epithelial tissue?
    a. cuboidal and columnar
    b. bone and cartilage
    c. squamous
    d. pseudostratified
    e. All of these are epithelial tissue.

20. Which tissue is more apt to line a lumen?
    a. epithelial tissue
    b. connective tissue
    c. nervous tissue
    d. muscular tissue
    e. only smooth muscle

21. Tendons and ligaments
    a. are connective tissue.
    b. are associated with the bones.
    c. are found in vertebrates.
    d. contain collagen.
    e. All of these are correct.

22. Which of these components of blood fight infection?
    a. red blood cells
    b. white blood cells
    c. platelets
    d. hydrogen ions
    e. All of these are correct.

23. Which of these body systems contribute to homeostasis?
    a. digestive and urinary systems
    b. respiratory and nervous systems
    c. nervous and endocrine systems
    d. immune and cardiovascular systems
    e. All of these are correct.

24. Which of these is a loose connective tissue?
    a. tendons
    b. ligaments
    c. adipose tissue
    d. All of these are correct.

25. Cells working together form
    a. organs.
    b. tissues.
    c. systems.
    d. organisms.

26. Which of these is a connective tissue?
    a. bone
    b. cartilage
    c. blood
    d. All of these are correct.

27. Which of these is an example of homeostasis?
    a. Muscular tissue is specialized to contract.
    b. Normal body temperature is always about 37°C.
    c. There are more red blood cells than white blood cells.
    d. All of these are correct.

28. What type of epithelial cells are found in the epidermis?
    a. squamous
    b. columnar
    c. cuboidal
    d. Both b and c are correct.

29. Which of these types of muscles helps maintain posture?
    a. skeletal
    b. smooth
    c. cardiac
    d. involuntary

30. In a negative feedback control system,
    a. homeostasis is impossible.
    b. there is a constancy of the internal environment.
    c. there is a fluctuation about a mean.
    d. Both a and b are correct.

31. A muscular tissue that is both striated and involuntary is
    a. smooth.
    b. skeletal.
    c. cardiac.
    d. All of these are correct.

Thinking Scientifically

1. Patients with muscular dystrophy often develop heart disease. However, heart disease in these patients may not be detected until it has progressed past the point at which treatment options are effective. New research indicates that the link between muscular dystrophy and heart disease is so strong that muscular dystrophy patients should be screened for heart disease early, so that treatment can be started. What do you suppose is the link between muscular dystrophy and heart disease?

2. If homeostasis of our internal environment is so important to keep us alive, why do we interact with the outside environment at the risk of an imbalance to the internal environment? Would it be possible to survive with no interactions between the internal and external environments? Give a few examples of what could occur if no interactions were possible.

3. Not all animals have organ systems or even organs. What are the advantages to having specific tissues, organs, and organ systems that perform specific functions in the body? In other words, what might have been some of the selective pressures that led to the evolution of more complex organisms?

Bioethical Issue

Organ Insurance

Despite widespread efforts to convince people of the need for organ donors, supply continues to lag far behind demand. One proposed strategy to bring supply and demand into better balance is to develop an "insurance" program for organs. In this program, participants would pay the "premium" by promising to donate their organs at death. They, in turn, would receive priority for transplants as their "benefit." To avoid the problem of too many high-risk people applying for this type of insurance, a medical exam would be required, so that only people with a normal risk of requiring a transplant would be accepted. Do you suppose this system would be an improvement over the current system, in which organ donation is voluntary, with no tangible benefit? Is this proposal ethical? What if the premiums are so high that only the wealthy can afford it? Can you think of any other strategy that would be more effective at increasing organ donations?

23

The Transport Systems

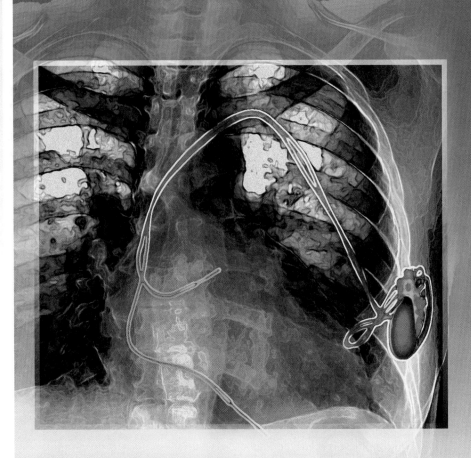

BEFORE YOU BEGIN

Before beginning this chapter, take a few moments to review the following discussions.

Section 22.1 Why is blood considered to be a connective tissue?

Section 22.2 What is the overall role of the circulatory system in the body?

Figure 22.14 How does the cardiovascular system play a central role in maintaining homeostasis?

Our Heart's Pacemaker

The beat of the heart is under involuntary control, so we rarely consider the complex series of events that occurs when the heart moves blood to the lungs and the body tissues. Ordinarily, contraction of the heart is controlled by its natural pacemaker (referred to as the SA [sinoatrial] node), which automatically triggers the heart's rhythmic beating. At rest, the heart contracts about 70 times per minute, and each contraction moves the entire blood volume of approximately 4–6 liters through the system once every minute. The nervous system gets involved when the body requires an increased amount of oxygen, as when you exercise vigorously. A particular nerve releases an adrenaline-like chemical in order to increase the activity of the SA node and therefore the heart rate. This action of the nervous system is mimicked whenever adrenaline is injected directly into the heart muscle, as is sometimes shown in the movies.

When damage occurs to the SA node, physicians often implant an artificial pacemaker. The heart responds to electrical signals generated by the SA node, and the artificial pacemaker emits electrical signals at the proper rate in order to duplicate the activity of a normal SA node. It is amazing to think that a simple electrical device can keep the entire cardiovascular system running.

The cardiovascular system is one of the body's transport systems. It must move blood to the tissues at a rate appropriate to meet the oxygen and nutrient demands of the cells. In this chapter, you will learn how the cardiovascular system works and how its functioning can be adjusted to meet the body's needs. Some of the consequences of a malfunctioning system are also discussed.

23.1 Open and Closed Circulatory Systems

Learning Outcomes

Upon completion of this section, you should be able to

1. Describe how organisms without a circulatory system exchange materials with the external environment.
2. Distinguish between an open and a closed circulatory system and give an example of each.
3. Compare the circulatory systems of fish, amphibians, birds, and mammals.
4. Summarize the differences between the pulmonary and systemic circulatory systems.

Most animals have a circulatory system to deliver oxygen and nutrients to cells and to take carbon dioxide and wastes away from cells. In some animals, the body plan makes a circulatory system unnecessary (**Fig. 23.1**). In a hydra, cells either are part of a single layer of external cells or line the gastrovascular cavity. In either case, each cell is exposed to water and can independently exchange gases and rid itself of wastes. The cells that line the gastrovascular cavity are specialized to carry out digestion. They pass nutrient molecules to other cells by diffusion. In a planarian, a trilobed gastrovascular cavity branches throughout the small, flattened body. No cell is very far from one of the digestive branches, so nutrient molecules can diffuse from cell to cell. Similarly, diffusion meets the respiratory and excretory needs of the cells.

Some other animals, such as nematodes and echinoderms, rely on movement of fluid within a coelom to circulate gases, nutrients, and wastes.

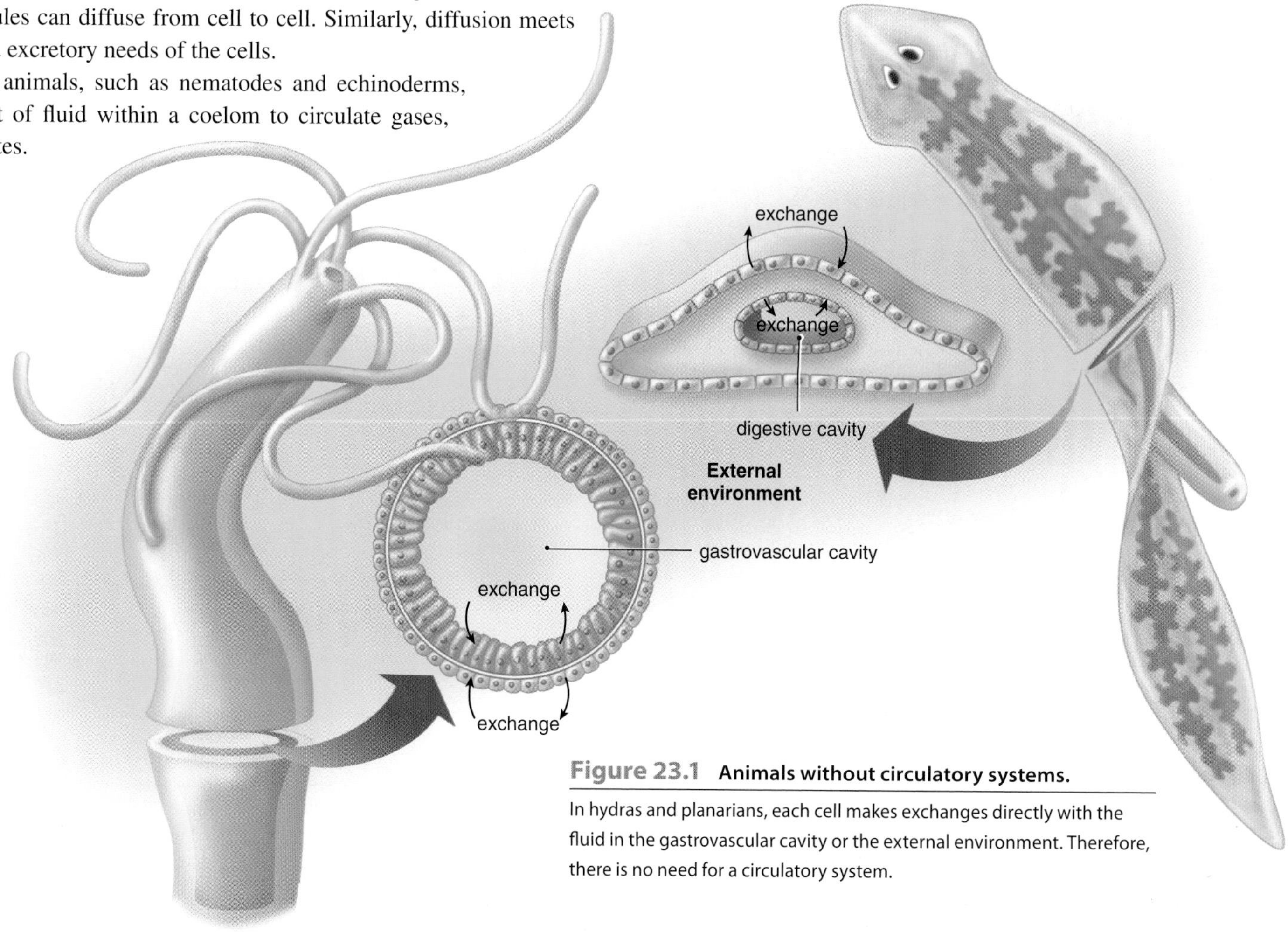

Figure 23.1   Animals without circulatory systems.

In hydras and planarians, each cell makes exchanges directly with the fluid in the gastrovascular cavity or the external environment. Therefore, there is no need for a circulatory system.

Open Circulatory Systems

More complex animals have a circulatory system in which a pumping **heart** moves fluid into blood vessels. The grasshopper is an example of an invertebrate animal that has an **open circulatory system** (**Fig. 23.2a**), meaning that the blood is not always confined to the blood vessels. A tubular heart pumps a fluid called hemolymph through a network of channels and cavities in the body. Collectively known as the **hemocoel,** these cavities, or **sinuses,** contain the animal's organs. Eventually, hemolymph (a combination of blood and tissue fluid) drains back to the heart. When the heart contracts, openings called ostia (sing., **ostium**) are closed; when the heart relaxes, the hemolymph is sucked back into the heart by way of the ostia. The hemolymph of a grasshopper is colorless because it does not contain hemoglobin or any other respiratory pigment that combines with and carries oxygen. Oxygen is taken to cells, and carbon dioxide is removed from them by way of air tubes, called tracheae, which are found throughout the body. The tracheae provide efficient transport and delivery of respiratory gases while restricting water loss.

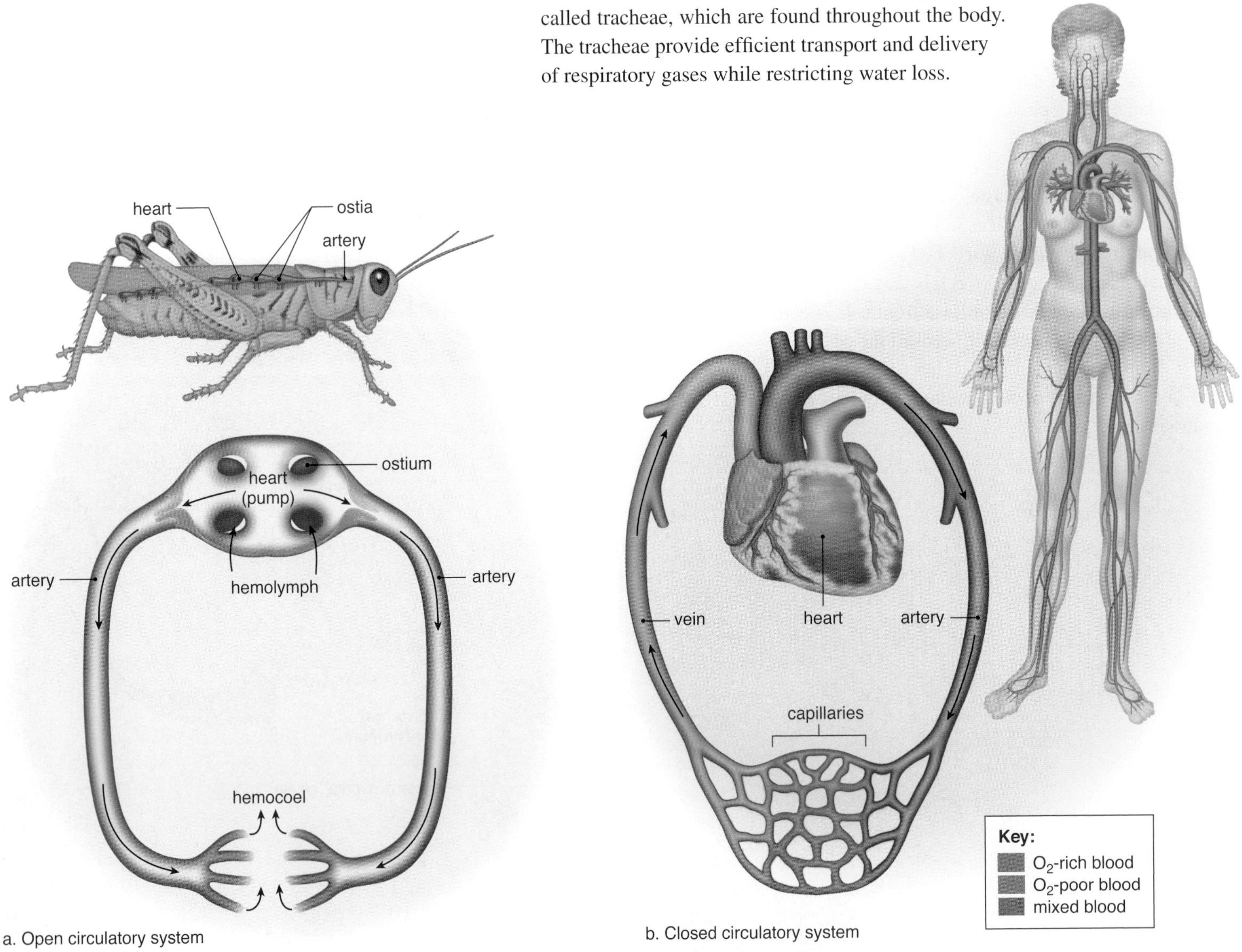

a. Open circulatory system

b. Closed circulatory system

Key:
- O$_2$-rich blood
- O$_2$-poor blood
- mixed blood

Figure 23.2 Open versus closed circulatory systems.

a. The grasshopper has an open circulatory system. Hemolymph freely bathes the internal organs. The heart, a pump, keeps the hemolymph moving, but an open system probably could not rapidly supply oxygen to wing muscles. These muscles receive oxygen directly from tracheae (air tubes). **b.** Vertebrates and some invertebrates have a closed circulatory system. The heart pumps blood into the arteries, which take blood away from the heart to the capillaries, where exchange takes place. Veins then return blood to the heart.

Closed Circulatory Systems

All vertebrates and some invertebrates have a **closed circulatory system** (Fig. 23.2b), which is more properly called a **cardiovascular system** because it consists of a strong, muscular heart and blood vessels. In a closed circulatory system, the blood remains within the blood vessels at all times. In humans, the heart has two receiving chambers, called atria (sing., atrium), and two pumping chambers, called ventricles. There are three kinds of vessels: arteries, which carry blood away from the heart; capillaries, which exchange materials with tissue fluid; and veins, which return blood to the heart. Blood is always contained within these vessels and never runs freely into the body unless an injury occurs.

As blood passes through capillaries, the pressure of blood forces some water out of the blood and into the tissue fluid. Some of this fluid returns directly to a capillary, and some is picked up by lymphatic capillaries in the vicinity. The fluid, now called lymph, is returned to the cardiovascular system by lymphatic vessels. The function of the lymphatic system is discussed in Section 23.2.

Comparison of Vertebrate Circulatory Pathways

Two types of circulatory pathways are seen among vertebrate animals. In fishes, blood follows a one-circuit (single-loop) pathway through the body. The heart has a single atrium and a single ventricle (**Fig. 23.3***a*). The pumping action of the ventricle sends blood under pressure to the gills, where gas exchange occurs. After passing through the gills, blood is under reduced pressure and flow. However, this single circulatory loop has advantages in that the gill capillaries receive O₂-poor blood and the systemic capillaries receive O₂-rich blood.

As a result of evolutionary adaptations to life on land, the other vertebrates have a two-circuit (double-loop) circulatory pathway. The heart pumps blood to the tissues through the **systemic circuit,** and also pumps blood to the lungs through the **pulmonary circuit.** This double pumping action is seen in terrestrial animals that utilize lungs to breathe air.

In amphibians, the heart has two atria, but only a single ventricle (Fig. 23.3*b*), and some mixing of O₂-rich and O₂-poor blood does occur. The same holds true for most reptiles, except that the ventricle has a partial septum, so this mixing is reduced. The hearts of some reptiles (e.g., crocodilians and birds) and mammals are divided into right and left halves (Fig. 23.3*c*). The right ventricle pumps blood to the lungs, and the left ventricle, which is larger than the right ventricle, pumps blood to the rest of the body. This arrangement provides adequate blood pressure for both the pulmonary and systemic circuits.

Connecting the Concepts

For more information on the material presented in this section, refer to the following discussions.

Section 19.1 describes the major trends in the evolution of the animals.

Section 19.5 examines the evolution of the vertebrate animals.

Check Your Progress 23.1

1. Compare and contrast an open circulatory system with a closed circulatory system.

2. Contrast a one-circuit circulatory pathway with a two-circuit pathway.

3. Compare and contrast the amphibian heart with the mammalian heart.

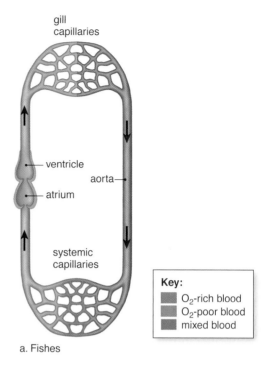

Key:
- O₂-rich blood
- O₂-poor blood
- mixed blood

a. Fishes

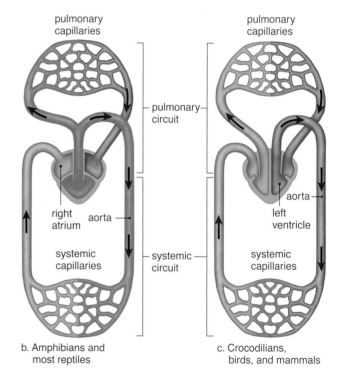

b. Amphibians and most reptiles

c. Crocodilians, birds, and mammals

Figure 23.3 Comparison of circulatory circuits in vertebrates.

a. In a fish, the blood moves in a single circuit. The heart has a single atrium and ventricle, and it pumps the blood into the gill region, where gas exchange takes place. Blood pressure created by the pumping of the heart is dissipated after the blood passes through the gill capillaries. **b.** Amphibians and most reptiles have a two-circuit system in which the heart pumps blood to both the lungs and the body itself. There is a single ventricle, and some mixing of O₂-rich and O₂-poor blood takes place. **c.** The pulmonary and systemic circuits are completely separate in crocodilians, birds, and mammals. The right side pumps blood to the lungs, and the left side pumps blood to the body proper.

23.2 Transport in Humans

Learning Outcomes

Upon completion of this section, you should be able to

1. Illustrate the anatomy of the human heart.
2. Trace the flow of blood through the heart and pulmonary circulation.
3. Detail the cardiac cycle and describe the electrical activity associated with it.
4. Identify and describe the types of blood vessels.
5. Compare and contrast the pulmonary circuit and the systemic circuit.
6. Explain the role of the lymphatic system in circulation within the human body.
7. Identify the forces that cause the movement of substances into and out of the capillaries.

In the human cardiovascular system, like that of other vertebrates, the heart pumps blood into blood vessels, which take it to capillaries, where exchanges take place. In the lungs, carbon dioxide is exchanged for oxygen, and in the tissues, nutrients and oxygen are exchanged for carbon dioxide and other wastes. These exchanges in the lungs and tissues are so important that, if the heart stops pumping, death results.

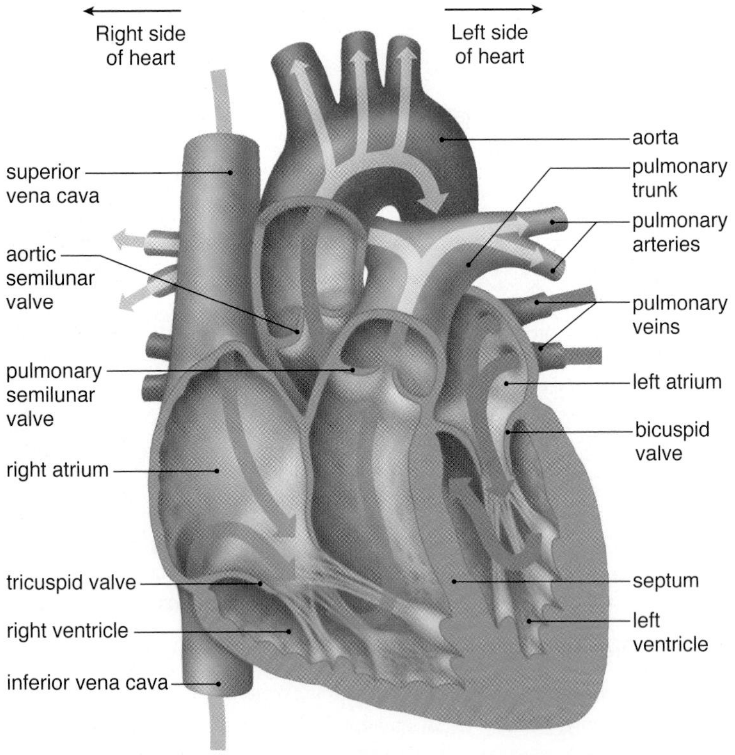

The Human Heart

In humans, the heart is a double pump: The right side of the heart pumps O_2-poor blood to the lungs, and the left side of the heart pumps O_2-rich blood to the tissues (**Fig. 23.4**). The heart can be a double pump because a **septum** separates the right from the left side. Further, the septum is complete and prevents O_2-poor blood from mixing with O_2-rich blood.

Each side of the heart has two chambers. The upper, thin-walled chambers are called atria (sing., **atrium**), and they receive blood. The lower chambers are the thick-walled **ventricles,** which pump the blood away from the heart.

Valves occur between the atria and the ventricles, and between the ventricles and attached vessels. Because these valves close after the blood moves through, they keep the blood moving in the correct direction. The valves between the atria and ventricles are called the **atrioventricular valves,** and the valves between the ventricles and their attached vessels are called **semilunar valves** because their cusps look like half-moons.

This sequence traces the path of blood through the heart. The right atrium receives blood from attached veins (the **venae cavae**) that are returning O_2-poor blood to the heart from the tissues. After the blood passes through an atrioventricular valve (also called the *tricuspid valve* because of its three flaps), the right ventricle pumps it through the *pulmonary semilunar valve* into the **pulmonary trunk** and **pulmonary arteries,** which take it to the lungs. The **pulmonary veins** bring O_2-rich blood to the left atrium. After this blood passes through an atrioventricular valve (also called the *bicuspid valve*), the left ventricle pumps it through the *aortic semilunar valve* into the **aorta,** which takes it to the tissues.

O_2-poor blood is often associated with all veins and O_2-rich blood with all arteries, but this idea is incorrect: Pulmonary arteries and pulmonary veins are just the reverse. That is why the pulmonary arteries are colored blue and

Figure 23.4 **Heart anatomy and path of blood through the heart.**

The right side of the heart receives and pumps O_2-poor blood to the lungs, and therefore is colored blue. The left side of the heart receives and pumps O_2-rich blood to tissues. The directions "right side" and "left side" of the heart refer to how the heart is positioned in your body, not to the left and right sides of the illustration.

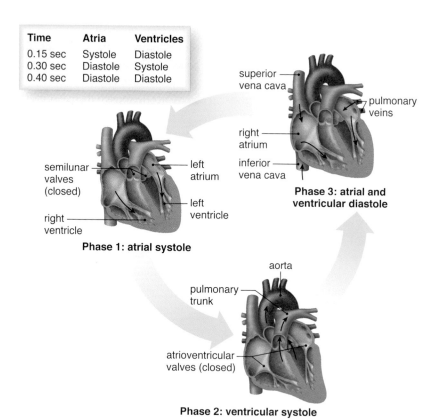

Time	Atria	Ventricles
0.15 sec	Systole	Diastole
0.30 sec	Diastole	Systole
0.40 sec	Diastole	Diastole

Phase 1: atrial systole

Phase 2: ventricular systole

Phase 3: atrial and ventricular diastole

Figure 23.5 The heartbeat.

A heartbeat is a cycle of events. **Phase 1:** The atria contract and pass blood to the ventricles. **Phase 2:** The ventricles contract and blood moves into the attached arteries. **Phase 3:** Both the atria and ventricles relax while the heart fills with blood.

the pulmonary veins are colored red in Figure 23.4. The correct definitions are that an **artery** is a vessel that takes blood away from the heart, and a **vein** is a vessel that takes blood to the heart.

The Heartbeat

The heart's pumping action, known as the heartbeat or **cardiac cycle,** consists of a series of events: First the atria contract, then the ventricles contract, and then they both rest. **Figure 23.5** lists and depicts the events of a heartbeat, using the term **systole** to mean contraction and **diastole** to mean relaxation. When the heart beats, the familiar *lub-dub* sound is caused by the closing of the heart valves. The longer and lower-pitched *lub* occurs when the atrioventricular valves close, and the shorter and sharper *dub* is heard when the semilunar valves close. The **pulse** is a wave effect that passes down the walls of the arterial blood vessels when the aorta expands and then recoils following the ventricular systole. Because there is one pulse per ventricular systole, the pulse rate can be used to determine the heart rate. The heart beats about 70 times per minute, although a normal adult heart rate can vary from 60 to 100 beats per minute.

 Animation Cardiac Cycle **MP3** Cardiac Cycle

The beat of the heart is regular because it has an intrinsic pacemaker, called the **SA (sinoatrial) node.** The nodal tissue of the heart, located in two regions of the atrial wall, is a unique type of cardiac muscle tissue. Every 0.85 second, the SA node automatically sends out an excitation impulse that causes the atria to contract (**Fig. 23.6***a*). When this impulse is picked up by the

Connections and Misconceptions

What is a heart murmur?

Like mechanical valves, the heart valves are sometimes leaky; they may not close properly, and there is a backflow of blood. A *heart murmur* is often due to leaky atrioventricular valves, which allow blood to pass back into the atria after the valves have closed. These

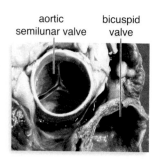

may be caused by a number of factors, high blood pressure (hypertension), heart disease, or rheumatic fever. Rheumatic fever is a bacterial infection that begins in the throat and spreads throughout the body. The bacteria attack various organs, including the heart valves. When damage is severe, the valve can be replaced with a synthetic valve or one taken from a pig's heart.

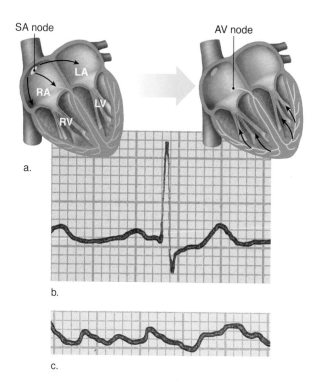

Figure 23.6 Control of the heartbeat.

a. The heart beats regularly because the SA node (called the pacemaker) automatically sends out an impulse that causes the atria (RA, LA) to contract and is picked up by the AV node. Thereafter, the ventricles (RV, LV) contract. **b.** An electrocardiogram records the electrical changes that occur as the heart beats. The large spike is associated with ventricular activation. **c.** Ventricular fibrillation produces an irregular ECG.

AV (atrioventricular) node, it passes to the *Purkinje fibers,* which cause the ventricles to contract. If the SA node fails to work properly, the ventricles still beat, due to impulses generated by the AV node, but the beat is slower (40–60 beats per minute). To correct this condition, it is possible to implant an artificial pacemaker, which automatically gives an electrical stimulus to the heart every 0.85 second. Pacemakers are available today that can adjust to the person's level of physical activity. Activity sensors in these self-adjusting pacemakers generate variable electrical signals, depending on the level of activity of the person. The pacemakers then change their output, depending on these signals.

Although the beat of the heart is intrinsic, it is regulated by the nervous system and various hormones. Activities such as yoga and meditation lead to activation of the vagus nerve, which slows the heart rate. Exercise or anxiety leads to release of the hormones norepinephrine and epinephrine by the adrenal glands, which cause the heart rate to speed up.

An **electrocardiogram (ECG)** is a recording of the electrical changes that occur in the wall of the heart during a cardiac cycle. Body fluids contain ions that conduct electrical currents, and therefore the electrical changes in heart muscle can be detected on the skin's surface. When an electrocardiogram is being taken, electrodes placed on the skin are connected by wires to an instrument that detects these electrical changes (Fig. 23.6b). Various types of abnormalities can be detected by an electrocardiogram. One of them, called ventricular fibrillation, is due to uncoordinated contraction of the ventricles (Fig. 23.6c). Ventricular fibrillation is of special interest because it can be caused by an injury or a drug overdose. It is the most common cause of sudden cardiac death in a seemingly healthy person. To stop fibrillation, a defibrillator can be used to apply a strong electrical current for a short period. Then, the SA node may be able to reestablish a coordinated beat. Easy-to-use defibrillators are becoming increasingly available in public places, such as airports and college campuses.

Blood Vessels

Arteries transport blood away from the heart. When the heart contracts, blood is sent under pressure into the arteries; thus, **blood pressure** accounts for the flow of blood in the arteries. Arteries have a much thicker wall than veins because of a well-developed middle layer composed of smooth muscle and elastic fibers. The elastic fibers allow arteries to expand and accommodate the sudden increase in blood volume that results after each heartbeat. The smooth muscle strengthens the wall and prevents overexpansion (**Fig. 23.7a**).

Arteries branch into arterioles. **Arterioles** are small arteries just visible to the naked eye, and their diameter can be regulated by the nervous system, depending on the needs of the body. When arterioles are dilated, more blood flows through them, and when they are constricted, less blood flows. The constriction of arterioles can also raise blood pressure. Arterioles branch into **capillaries,** which are extremely narrow, microscopic tubes with a wall composed of only epithelium, often called endothelium (Fig. 23.7b). Capillaries, which are usually so narrow that red blood cells pass through in single file, allow the exchange of nutrient and waste molecules across their thin walls. Capillary beds (many interconnected capillaries) are so widespread that, in humans, all cells are less than a millimeter from a capillary. Because the number of capillaries is so extensive, blood pressure drops and blood flows slowly along. The slow movement of blood

Figure 23.7 Blood vessels.

a. Arteries have well-developed walls with a thick middle layer of elastic fibers and smooth muscle. **b.** Capillary walls are composed of an epithelium only one cell thick. **c.** Veins have flabby walls, particularly because the middle layer is not as thick as in arteries.

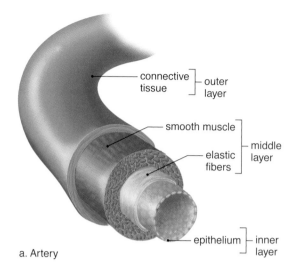

connective tissue — outer layer

smooth muscle
elastic fibers — middle layer

epithelium — inner layer

a. Artery

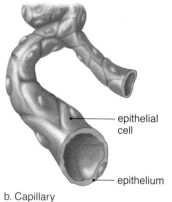

epithelial cell

epithelium

b. Capillary

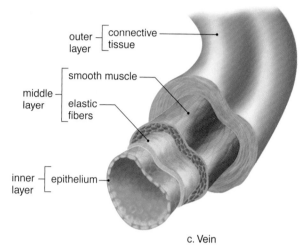

outer layer — connective tissue

middle layer — smooth muscle / elastic fibers

inner layer — epithelium

c. Vein

through the capillaries also facilitates efficient exchange of substances between the blood and tissue fluid. The entrance to a capillary bed is controlled by bands of muscle called precapillary sphincters. During muscular exercise, sphincters relax, and the capillary beds of the muscles are open. Also, after an animal has eaten, the capillary beds in the digestive tract are open. Otherwise, blood moves through a shunt that takes the blood from arteriole to venule (see Fig. 23.9*b*).

Venules collect blood from capillary beds and join as they deliver blood to veins. Veins carry blood back to the heart. Blood pressure is much reduced by the time blood reaches the veins. The walls of veins are much thinner and their diameter wider than those of arteries (Fig. 23.7*c*). This allows skeletal muscle contraction to push on the veins, forcing the blood past a valve (**Fig. 23.8**). Valves within the veins point, or open, toward the heart, preventing a backflow of blood when they close. When inhalation occurs and the chest expands, the thoracic pressure falls and abdominal pressure rises. This action also aids the flow of venous blood back to the heart because blood flows in the direction of reduced pressure.

 MP3 **Blood Flow & Blood Pressure**

Connections and Misconceptions

What are varicose veins?

Veins are thin-walled tubes, divided into many separate chambers by vein valves. Excessive stretching occurs if veins are overfilled with blood. For example, if a person stands in one place for a long time, leg veins can't drain properly and blood pools in them. As the vein expands, vein valves become distended and fail to function. These two mechanisms cause the veins to bulge and be visible on the skin's surface. Obesity, a sedentary lifestyle, female gender, genetic predisposition, and increasing age are risk factors for varicose veins. Hemorrhoids are varicose veins in the rectum.

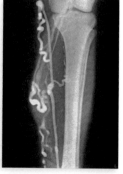

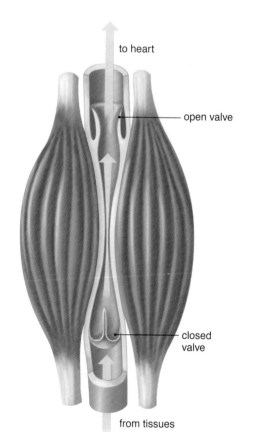

to heart

open valve

closed valve

from tissues

Figure 23.8 Movement of blood in a vein.

Veins have valves, which point toward the heart. Pressure on the walls of a vein, exerted by skeletal muscles, increases blood pressure within the vein and forces a valve open. When the muscles relax, gravity pulls the blood downward, closing the valves and preventing the blood from flowing in the opposite direction.

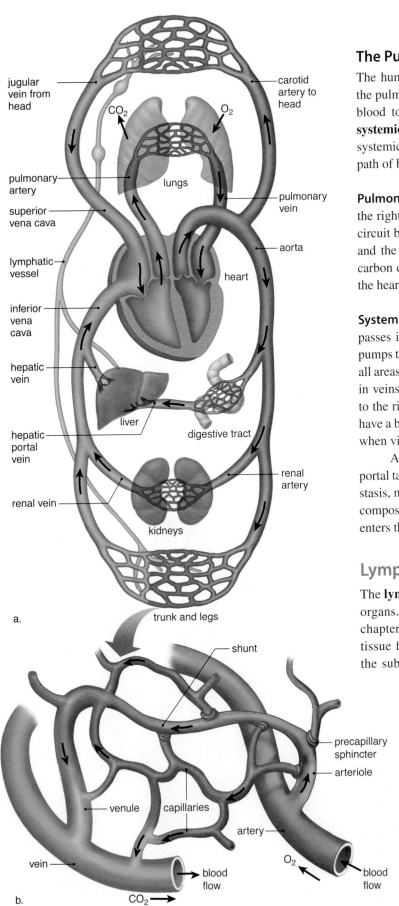

jugular vein from head

carotid artery to head

CO_2

O_2

pulmonary artery

lungs

superior vena cava

pulmonary vein

lymphatic vessel

aorta

heart

inferior vena cava

hepatic vein

liver

digestive tract

hepatic portal vein

renal artery

renal vein

kidneys

a.

trunk and legs

shunt

precapillary sphincter

arteriole

venule   capillaries

artery

vein

O_2

blood flow

blood flow

b.

CO_2

The Pulmonary and Systemic Circuits

The human cardiovascular system includes two major circulatory pathways, the pulmonary circuit and the systemic circuit. The **pulmonary circuit** moves blood to and from the lungs, where O_2-poor becomes O_2-rich blood. The **systemic circuit** moves blood to and from the body proper. The function of the systemic circuit is to serve the needs of the body's cells. **Figure 23.9** traces the path of blood in both circuits.

Pulmonary Circuit O_2-poor blood from all regions of the body collects in the right atrium and then passes into the right ventricle (RV). The pulmonary circuit begins when the RV pumps blood to the lungs via the pulmonary trunk and the pulmonary arteries. As blood passes through pulmonary capillaries, carbon dioxide is given off and oxygen is picked up. O_2-rich blood returns to the heart via the pulmonary veins. Pulmonary veins enter the left atrium.

Systemic Circuit O_2-rich blood enters the left atrium from the lungs and passes into the left ventricle (LV). The systemic circuit begins when the LV pumps the blood into the aorta. Arteries branching from the aorta carry blood to all areas and organs of the body, where it passes through capillaries and collects in veins. Veins converge on the venae cavae, which return the O_2-poor blood to the right atrium. In the systemic circuit, arteries contain O_2-rich blood and have a bright red color, and veins contain O_2-poor blood and appear dull red or, when viewed through the skin, blue.

A **portal system** begins and ends in capillaries. For example, the hepatic portal takes blood from the intestines to the liver. The liver, an organ of homeostasis, modifies substances absorbed by the intestines and monitors the normal composition of the blood. The hepatic vein (see Fig. 24.27) leaves the liver and enters the inferior vena cava.

Lymphatic System

The **lymphatic system** consists of lymphatic vessels and various lymphatic organs. The lymphatic system serves many functions in the body. In this chapter, we are interested in the lymphatic vessels that take up excess tissue fluid and return it to cardiovascular veins in the shoulders, namely the subclavian veins (**Fig. 23.10**). The lymphatic vessels also take up fat

Figure 23.9 Path of blood.

a. Overview of cardiovascular system. When tracing blood from the right to the left side of the heart in the pulmonary circuit, you must include the pulmonary vessels. When tracing blood from the digestive tract to the right atrium in the systemic circuit, you must include the hepatic portal vein, the hepatic vein, and the inferior vena cava. **b.** To move from an artery to a vein, blood must move through a capillary bed, where exchange occurs between blood and tissue fluid. When precapillary sphincters shut down a capillary bed, blood moves through a shunt from arteriole to venule.

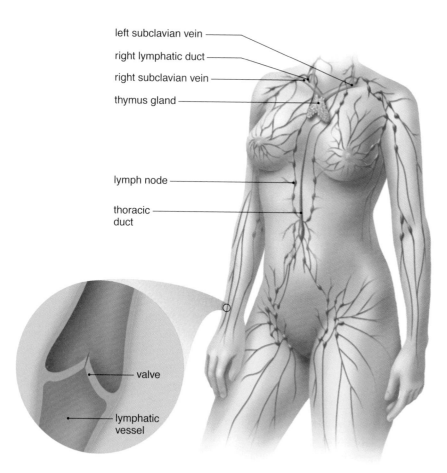

left subclavian vein
right lymphatic duct
right subclavian vein
thymus gland
lymph node
thoracic duct
valve
lymphatic vessel

Figure 23.10 Lymphatic vessels.

Lymphatic vessels drain excess fluid from the tissues and return it to the cardiovascular system. The enlargement shows that lymphatic vessels, like cardiovascular veins, have valves to prevent backward flow. Lymph nodes filter lymph and remove impurities.

in the form of lipoproteins at the digestive tract and transport it to these veins. Refer to **Figure 23.11** to see how the lymphatic vessels fit into the organization of the cardiovascular system. As you will see in Chapter 26, the lymphatic system works with the immune system to help defend the body against disease.

Lymphatic vessels are quite extensive; most regions of the body are richly supplied with lymphatic capillaries. The construction of the larger lymphatic vessels is similar to that of cardiovascular veins, including the presence of valves. Also, the movement of lymph within these vessels is dependent on skeletal muscle contraction. When the muscles contract, the lymph is squeezed past a valve that closes, preventing the lymph from flowing backward.

The lymphatic system is a one-way system that begins at the lymphatic capillaries. These capillaries take up fluid that has diffused from and not been reabsorbed by the blood capillaries. Once tissue fluid enters the lymphatic vessels, it is called **lymph.** The lymphatic capillaries join to form larger lymphatic vessels that merge before entering one of two ducts: the thoracic duct or the right lymphatic duct. The thoracic duct is much larger than the right lymphatic duct. It serves the lower limbs, the abdomen, the left arm, and the left side of both the head and the neck. The right lymphatic duct serves the right arm, the right side of both the head and the neck, and the right thoracic area. The lymphatic ducts enter the subclavian veins.

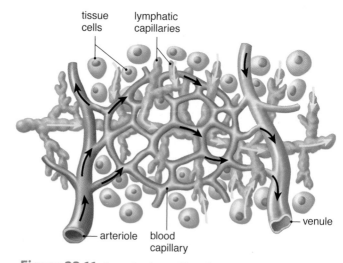

tissue cells
lymphatic capillaries
arteriole
blood capillary
venule

Figure 23.11 Lymphatic capillary bed.

A lymphatic capillary bed lies near a blood capillary bed. The heavy black arrows show the flow of blood. The yellow arrows show that lymph is formed when lymphatic capillaries take up excess tissue fluid.

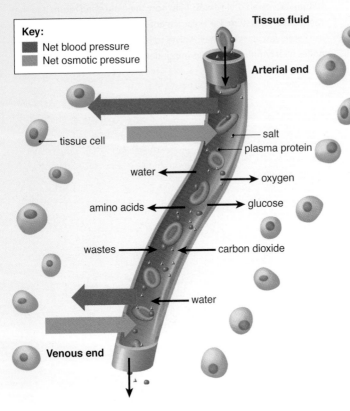

Figure 23.12 **Capillary exchange.**

A capillary, illustrating the exchanges that take place and the forces that aid the process. At the arterial end of a capillary (*top*), the blood pressure is higher than the osmotic pressure; therefore, water tends to leave the bloodstream. In the midsection, molecules, including oxygen and carbon dioxide, follow their concentration gradients. At the venous end of a capillary (*bottom*), the osmotic pressure is higher than the blood pressure; therefore, water tends to enter the bloodstream. Notice that the red blood cells and the plasma proteins are too large to exit a capillary.

Capillary Exchange in the Tissues

In Figure 23.1, we noted that, in some animals, exchange is carried out by each cell individually because there is no cardiovascular system. When an animal does have a cardiovascular system, tissue fluid (the fluid between the cells) makes exchanges with blood within a capillary. Notice in **Figure 23.12** that amino acids, oxygen, and glucose exit a capillary and enter tissue fluid to be used by cells. On the other hand, carbon dioxide and wastes exit tissue fluid and enter a capillary to be taken away and excreted from the body. A chief purpose of the cardiovascular system is to take blood to the capillaries, where exchange occurs. Without this exchange, homeostasis is not maintained, and the cells of the body perish.

Figure 23.12 illustrates certain mechanics of capillary exchange. Blood pressure and osmotic pressure are two opposing forces at work along the length of a capillary. Blood pressure is caused by the beating of the heart, while osmotic pressure is due to the salt and protein content of the blood. Blood pressure holds sway at the arterial end of a capillary and water exits. Blood pressure is reduced by the time blood reaches the venous end of a capillary, and osmotic pressure now causes water to enter. Midway between the arterial and venous ends of a capillary, blood pressure pretty much equals osmotic pressure, and passive diffusion alone causes nutrients to exit and wastes to enter. This works because tissue fluid always has the lesser amount of nutrients and the greater amount of wastes. After all, cells use nutrients and thereby create wastes.

The exchange of water at a capillary is not exact, and the result is always excess tissue fluid. Excess tissue fluid is collected by lymphatic capillaries, and in this way it becomes lymph (see Fig. 23.11). Lymph contains all the components of plasma except much smaller amounts of protein. Lymph is returned to the cardiovascular system when the major lymphatic vessels enter the subclavian veins in the shoulder region (see Fig. 23.10).

Animation
Fluid Exchange
Across the Walls
of Capillaries

MP3
Capillary Exchange
and Bulk Flow

In addition to nutrients and wastes, blood distributes heat to body parts. When you are warm, many capillaries that serve the skin are open, and your face is flushed. This helps rid the body of excess heat. When you are cold, skin capillaries close, conserving heat.

Connecting the Concepts

For more information on the movement of blood and lymph, refer to the following discussions.

Section 24.2 explores how waste material is filtered from the blood by the nephrons of the kidneys.

Figure 24.27 illustrates the hepatic portal system that moves blood and nutrients from the small intestine to the liver.

Section 29.4 examines the development of the circulatory system in the fetus.

Check Your Progress 23.2

1. Describe the events that occur in the generation of a heartbeat.
2. Describe the structures of the different types of blood vessels and detail why capillaries are used for gas exchange.
3. Predict what might happen to the homeostasis of the blood without the action of the lymphatic system.

23.3 Blood: A Transport Medium

The **blood** of mammals helps maintain their homeostasis. Blood's numerous functions include the following:

1. Transports substances to and from the capillaries, where exchange with tissue fluid takes place
2. Helps defend the body against invasion by pathogens (e.g., disease-causing viruses and bacteria)
3. Helps regulate body temperature
4. Forms clots, preventing a potentially life-threatening loss of blood

In humans, blood has two main portions: the liquid portion, called plasma, and the formed elements, consisting of various cells and platelets (**Fig. 23.13**).

Plasma

Plasma is composed mostly of water (90–92%) and proteins (7–8%), but it also contains smaller quantities of many types of molecules, including nutrients, wastes, and salts. The salts and proteins are involved in buffering the blood, effectively keeping the pH near 7.4, slightly basic. They also maintain the blood's osmotic pressure, so that water has an automatic tendency to enter blood capillaries. Several plasma proteins are involved in blood clotting, and others transport large organic molecules in the blood. Albumin, the most plentiful of the plasma proteins, transports bilirubin, a breakdown product of hemoglobin. Lipoproteins transport cholesterol.

Formed Elements

The **formed elements** are red blood cells, white blood cells, and platelets. Among the formed elements, **red blood cells,** also called erythrocytes, transport oxygen. Red blood cells are small, biconcave disks that at maturity lack a nucleus and contain the respiratory pigment hemoglobin. There are 6 million red blood cells in a small drop of whole blood, and each one of these cells contains about 250 million hemoglobin molecules. **Hemoglobin** contains iron, which combines loosely with oxygen; in this way, red blood cells transport oxygen. If the number of red blood cells is insufficient, or if the cells do not have enough hemoglobin, the individual suffers from anemia and has a tired, run-down feeling. At high altitudes, where oxygen levels are lower, red blood cell formation is stimulated. If an individual lives and works in a high altitude long enough, he or she will have a greater number of red blood cells.

Red blood cells are manufactured continuously within certain bones, namely the skull, the ribs, the vertebrae, and the ends of the long bones. The hormone *erythropoietin (EPO)* stimulates the production of red blood cells.

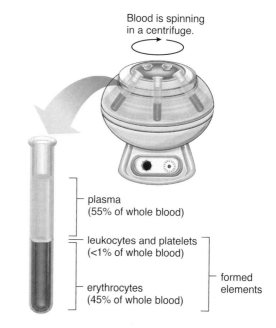

Blood is spinning in a centrifuge.

plasma (55% of whole blood)

leukocytes and platelets (<1% of whole blood)

erythrocytes (45% of whole blood)

formed elements

Figure 23.13   Components of blood.

After blood is centrifuged and settling occurs, it is apparent that blood is composed of plasma and formed elements.

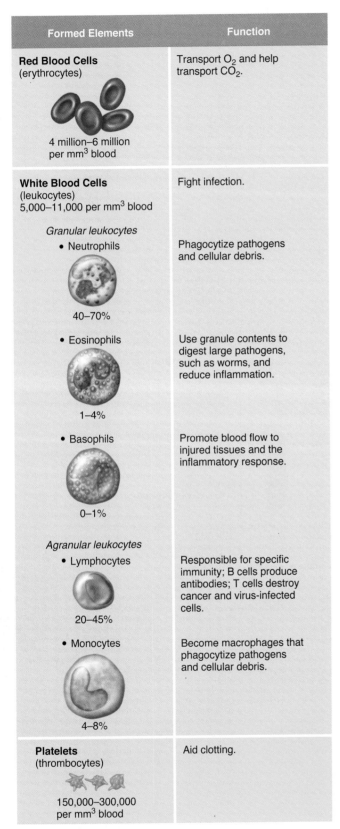

Formed Elements	Function
Red Blood Cells (erythrocytes) 4 million–6 million per mm³ blood	Transport O_2 and help transport CO_2.
White Blood Cells (leukocytes) 5,000–11,000 per mm³ blood	Fight infection.
Granular leukocytes • Neutrophils 40–70%	Phagocytize pathogens and cellular debris.
• Eosinophils 1–4%	Use granule contents to digest large pathogens, such as worms, and reduce inflammation.
• Basophils 0–1%	Promote blood flow to injured tissues and the inflammatory response.
Agranular leukocytes • Lymphocytes 20–45%	Responsible for specific immunity; B cells produce antibodies; T cells destroy cancer and virus-infected cells.
• Monocytes 4–8%	Become macrophages that phagocytize pathogens and cellular debris.
Platelets (thrombocytes) 150,000–300,000 per mm³ blood	Aid clotting.

• Appearance with Wright's stain.

Figure 23.14 Formed elements.

White blood cells are quite varied and have different functions, which are all associated with defense of the body against infections.

The kidneys produce erythropoietin when they act on a precursor made by the liver. Now available as a drug, erythropoietin is helpful to persons with anemia, but it has also been abused by athletes to enhance their performance.

Before they are released from the bone marrow into the blood, red blood cells synthesize hemoglobin and lose their nuclei. After living about 120 days, they are destroyed, chiefly in the liver and spleen, where they are engulfed by large, phagocytic cells. When red blood cells are destroyed, hemoglobin is released. The iron is recovered and returned to the red bone marrow for reuse. Another portion of the molecules (i.e., heme) undergoes chemical degradation and is excreted by the liver as bile pigments in the bile. The bile pigments are primarily responsible for the color of feces.

Connections and Misconceptions

What is blood doping?

Blood doping is any method of increasing the normal supply of RBCs for the purpose of delivering oxygen more efficiently, reducing fatigue, and giving athletes a competitive edge. To accomplish blood doping, athletes can inject themselves with EPO some months before the competition. These injections will increase the

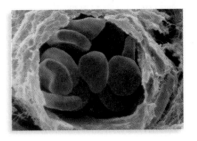

number of RBCs in their blood. Several weeks later, units of their blood are removed and centrifuged to concentrate the RBCs. The concentrated RBCs are reinfused shortly before the athletic event. Blood doping is a dangerous, illegal practice. Several cyclists died in the 1990s from heart failure, probably due to blood that was too thick with cells for the heart to pump.

White blood cells, also called leukocytes, help fight infections. White blood cells differ from red blood cells in that they are usually larger and have a nucleus, they lack hemoglobin, and without staining they appear translucent. **Figure 23.14** shows the appearance of the various types of white blood cells. With staining, white blood cells appear light blue unless they have granules that bind with certain stains. The white blood cells that have granules also have a lobed nucleus. The agranular leukocytes have no granules and a spherical or indented nucleus. There are approximately 5,000–11,000 white blood cells in a small drop of whole blood. Growth factors are available to increase the production of all white blood cells, and these are helpful to people with low immunity, such as AIDS patients.

Red blood cells are confined to the blood, but white blood cells are able to squeeze between the cells of a capillary wall. Therefore, they are found in tissue fluid and lymph and in lymphatic organs. When an infection is present, white blood cells greatly increase in number. Many white blood cells live only a few days—they probably die while engaging pathogens. Others live months or even years.

When microorganisms enter the body due to an injury, the body's response is called an *inflammatory response* because swelling, reddening, heat, and pain occur at the injured site. Damaged tissue releases kinins, which dilate capillaries, and histamines, which increase capillary permeability. White blood cells called **neutrophils,** which are amoeboid, squeeze through the capillary wall and enter the tissue fluid, where they phagocytize foreign material. White blood cells called **monocytes** appear and are transformed into

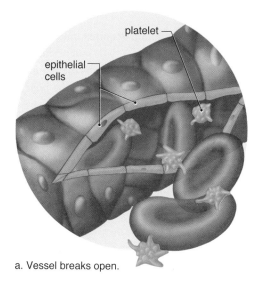

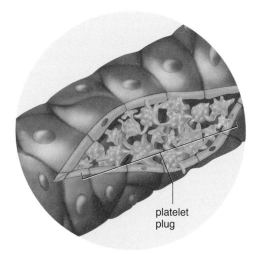

a. Vessel breaks open.

b. Platelet plug forms.

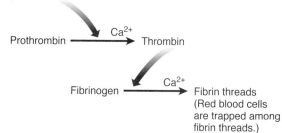

c. Clotting occurs.

macrophages—large, phagocytizing cells that release white blood cell growth factors. Soon the number of white blood cells increases explosively. A thick, yellowish fluid called pus contains a large proportion of dead white blood cells that have fought the infection.

Lymphocytes, another type of white blood cell, also play an important role in fighting infection. Certain lymphocytes called T cells attack infected cells that contain viruses. Other lymphocytes, called B cells, produce antibodies. Each B cell produces just one type of antibody, which is specific for one type of antigen. An **antigen,** which is most often a protein but sometimes a polysaccharide, causes the body to produce an **antibody** because the antigen does not belong to the body. Antigens are present in the outer covering of parasites or in their toxins. When antibodies combine with antigens, the complex is often phagocytized by a macrophage. An individual is actively immune when a large number of B cells are all producing the antibody needed for a particular infection.

Platelets and Blood Clotting

Platelets (also called thrombocytes) result from fragmentation of certain large cells, called megakaryocytes, in the red bone marrow. Platelets are produced at a rate of 200 billion a day, and a small drop of whole blood contains 150,000–300,000. These formed elements are involved in blood clotting, or coagulation.

Blood contains at least 12 clotting factors that participate in the formation of a blood clot. *Hemophilia* is an inherited clotting disorder in which the liver is unable to produce one of the clotting factors. The slightest bump can cause the affected person to bleed into the joints, and this leads to degeneration of the joints. Bleeding into muscles can lead to nerve damage and muscular atrophy. The most frequent cause of death due to hemophilia is bleeding into the brain.

Prothrombin and fibrinogen, two proteins involved in blood clotting, are manufactured and deposited in blood by the liver. Vitamin K, found in green vegetables and formed by intestinal bacteria, is necessary for the production of prothrombin, and if this vitamin is missing from the diet, hemorrhagic disorders can develop.

A series of reactions leads to the formation of a blood clot (**Fig. 23.15**a–c). When a blood vessel in the body is damaged, platelets clump at the site of the puncture and form a plug, which temporarily seals the leak. Platelets and the injured tissues release a clotting factor, called prothrombin activator,

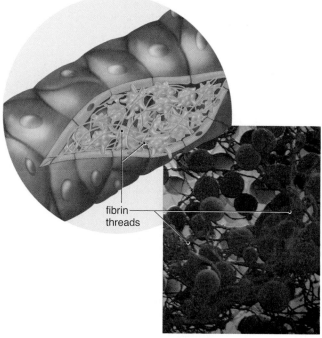

d. Blood clot is present.   ×1430

Figure 23.15 Blood clotting.

a. When a capillary is injured, blood begins leaking out. **b.** Platelets congregate to form a platelet plug, and this temporarily seals the leak. **c.** Platelets and damaged tissue cells release an activator, which sets in motion a series of reactions, ending with (**d**) a blood clot.

that converts prothrombin to thrombin. This reaction requires calcium ions (Ca^{2+}). **Thrombin,** in turn, acts as an enzyme that severs two short amino acid chains from each fibrinogen molecule. These activated fragments then join end to end, forming long threads of fibrin. Fibrin threads wind around the platelet plug in the damaged area of the blood vessel and provide the framework for the clot. Red blood cells also are trapped within the fibrin threads; these cells make a clot appear red (Fig. 23.15*d*). Clot retraction follows, during which the clot gets smaller as platelets contract. A fluid called serum is squeezed from the clot. A fibrin clot is present only temporarily. As soon as blood vessel repair is initiated, an enzyme called plasmin destroys the fibrin network and restores the fluidity of plasma.

Cardiovascular Disorders

Cardiovascular disease is the leading cause of untimely death in Western countries. In the United States, it is estimated that about 20% of the population suffers from **hypertension,** which is high blood pressure. Normal blood pressure is 120/80 mm Hg. The top number is called *systolic* blood pressure because it is due to the contraction of the ventricles, and the bottom number is called *diastolic* blood pressure because it is measured during the relaxation of the ventricles. Hypertension occurs when blood pressure readings are higher than these numbers—say, 160/100. Hypertension is sometimes called a silent killer because it may not be detected until a stroke or heart attack occurs.

Video
Cardiac Repair

Inheritance and lifestyle contribute to hypertension. For example, hypertension is often seen in individuals who have atherosclerosis (formerly called arteriosclerosis), an accumulation of soft masses of fatty materials, particularly cholesterol, beneath the inner linings of arteries (**Fig. 23.16**). Such deposits, called **plaque**, tend to protrude into the lumen of the vessel, interfering with the flow of blood. Atherosclerosis begins in early adulthood and develops progressively through middle age, but symptoms may not appear until an individual is 50 or older. To prevent its onset and development, the American Heart Association and other organizations recommend a diet low in saturated fat and cholesterol and rich in fruits and vegetables. Smoking, alcohol or other drug abuse, obesity, and lack of exercise contribute to the risk of atherosclerosis.

Plaque can cause a clot to form on the irregular arterial wall. As long as the clot remains stationary, it is called a *thrombus,* but when and if it dislodges and moves along with the blood, it is called an *embolus.* If *thromboembolism* is not treated, serious health problems can result. A cardiovascular accident, also called a **stroke,** often occurs when a small cranial arteriole bursts or is blocked by an embolus. Lack of oxygen causes a portion of the brain to die, and paralysis or death can result. A person is sometimes forewarned of a stroke by a feeling of numbness in the hands or face, difficulty in speaking, or temporary blindness in one eye. If a coronary artery becomes completely blocked due to thromboembolism, a heart attack can result.

Video
Heart Stem Cells

The coronary arteries bring O_2-rich blood from the aorta to capillaries in the wall of the heart, and the cardiac veins return O_2-poor blood from the capillaries to the right ventricle. If the coronary arteries narrow due to cardiovascular disease, the individual may first suffer from angina pectoris, chest pain that is often accompanied by a radiating pain in the left arm. When a coronary artery

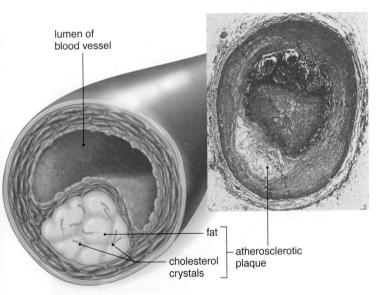

lumen of
blood vessel

fat

cholesterol
crystals

atherosclerotic
plaque

Figure 23.16 **Plaque.**

Plaque is an irregular accumulation of cholesterol and fat in the wall of an artery that closes off blood vessels. Of special concern is the closure of coronary arteries.

is completely blocked, a portion of the heart muscle dies due to a lack of oxygen. This is known as a **heart attack.** Two surgical procedures are frequently performed to correct a blockage or facilitate blood flow. In a coronary bypass operation, a portion of a blood vessel from another part of the body is sutured from the aorta to the coronary artery, past the point of obstruction (**Fig. 23.17***a*). Now blood flows normally again from the aorta to the wall of the heart. In balloon angioplasty, a plastic tube is threaded through an artery to the blockage, and a balloon attached to the end of the tube is inflated to break through the blockage. A stent is often used to keep the vessel open (Fig. 23.17*b*).

Connecting the Concepts

For more information on the topics presented in this section, refer to the following discussions.

Section 22.1 explains how blood acts as a type of connective tissue in the body.

Section 26.2 examines how the components of the blood are involved in the inflammatory response.

Section 26.3 explores the role of the lymphocytes in protecting the body against infection.

Check Your Progress 23.3

1. Compare and contrast all the types of white blood cells.

2. Discuss a few cardiovascular disorders and detail any available treatments.

3. Hypothesize what would happen if an injury occurred and the cardiovascular system was deficient in platelets.

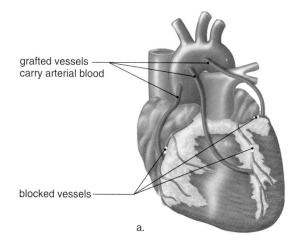
grafted vessels
carry arterial blood

blocked vessels

a.

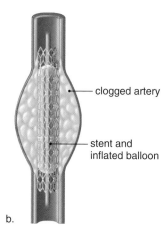
clogged artery

stent and
inflated balloon

b.

Figure 23.17   Treatment for clogged coronary arteries.

Many heart procedures, such as coronary bypass and stent insertion, can be performed using robotic surgery techniques. **a.** During a coronary bypass operation, blood vessels (usually veins from the leg) are stitched to the heart, taking blood past the region of obstruction. **b.** During stenting, a cylinder of expandable metal mesh is positioned inside the coronary artery by using a catheter. Then, a balloon is inflated, so that the stent expands and opens the artery.

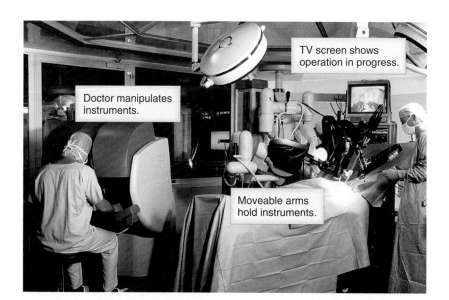
TV screen shows
operation in progress.

Doctor manipulates
instruments.

Moveable arms
hold instruments.

Media Study Tools

www.mhhe.com/maderessentials3

Enhance your study of this chapter with study tools and practice tests. Also ask your instructor about the resources available through ConnectPlus, including the media-rich eBook, interactive learning tools, and animations.

Blood Pressure

The Virtual Lab "Blood Pressure" provides an interactive look at how factors such as age and gender influence the risk of hypertension.

The Chapter in Review

Summary

23.1 Open and Closed Circulatory Systems

- Some invertebrates do not have a transport system because their body plan allows each cell to exchange molecules with the external environment.
- Other invertebrates do have a transport system. Some have an open circulatory system, and some have a closed circulatory system.
- All vertebrates have a closed circulatory system in which arteries carry blood away from the heart to capillaries, where exchange takes place, and veins carry blood to the heart.

Comparison of Vertebrate Circulatory Pathways

- Fishes have a one-circuit pathway because the heart, with a single atrium and ventricle, pumps blood only to the gills.
- Other vertebrates have both pulmonary and systemic circuits. Amphibians have two atria but a single ventricle. Crocodilians, birds, and mammals, including humans, have a heart with two atria and two ventricles, in which O_2-rich blood is always separate from O_2-poor blood.

23.2 Transport in Humans

The cardiovascular system consists of the heart and the blood vessels.

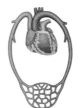

The Human Heart

The heart has a right and a left side. Each side has an atrium and a ventricle. Valves keep the blood moving in the correct direction.

Right side of the heart: The atrium receives O_2-poor blood from the tissues, and the ventricle pumps it to the lungs.

Left side of the heart: The atrium receives O_2-rich blood from the lungs, and the ventricle pumps it to the tissues.

Heartbeat: During a heartbeat, first the atria contract and then the ventricles contract. The heart sounds, *lub-dub*, are caused by the closing of valves.

SA node: The SA node (pacemaker) causes the two atria to contract. The SA node also stimulates the AV node.

AV node: The AV node causes the two ventricles to contract.

Blood Vessels

Arteries, with thick walls, take blood away from the heart to arterioles, which take blood to capillaries that have walls composed only of epithelial cells. Venules take blood from capillaries and merge to form veins, which have thinner walls than arteries have.

- Blood pressure, created by the beat of the heart, accounts for the flow of blood in the arteries.
- Skeletal muscle contraction is largely responsible for the flow of blood in the veins, which have valves preventing backward flow.

Blood Has Two Circuits

- In the pulmonary circuit, blood travels to and from the lungs.
- In the systemic circuit, the aorta divides into blood vessels that serve the body's cells. Venae cavae return O_2-poor blood to the heart.

Lymphatic System

The lymphatic system is a one-way system that consists of lymphatic vessels and lymphatic organs. The lymphatic vessels receive fat at the digestive tract and excess tissue fluid at blood capillaries; the lymphatic vessels carry these to the subclavian veins (cardiovascular veins in the shoulders).

Capillary Exchange in the Tissues

Capillary exchange in the tissues helps keep the internal environment constant. When blood reaches a capillary,

- Water moves out at the arterial end due to blood pressure.
- Water moves in at the venous end due to osmotic pressure.
- Nutrients diffuse out of and wastes diffuse into the capillary between the arterial end and the venous end.
- Lymphatic capillaries in the vicinity pick up excess tissue fluid and return it to cardiovascular veins.

23.3 Blood: A Transport Medium

Blood has two main parts: plasma and formed elements.

Plasma

Plasma contains mostly water (90–92%) and proteins (7–8%) but also nutrients, wastes, and salts.

- The proteins and salts buffer the blood and maintain its osmotic pressure.
- The proteins also have specific functions, such as participating in blood clotting and transporting molecules.

Formed Elements

The red blood cells contain hemoglobin and function in oxygen transport. Defense against disease depends on the various types of white blood cells:

Red blood cells

- Neutrophils and monocytes are phagocytic and are especially responsible for the inflammatory response.
- Lymphocytes are involved in the development of specific immunity to disease.

White blood cells

The platelets and two plasma proteins, prothrombin and fibrinogen, function in blood clotting, an enzymatic process that results in fibrin threads.

Cardiovascular Disorders

Platelets

Hypertension and atherosclerosis are two conditions that lead to heart attack and stroke. Following a heart-healthy diet, getting regular exercise, maintaining a proper weight, and not smoking cigarettes are protective against these conditions.

 ## Key Terms

antibody   443
antigen   443
aorta   434
arteriole   436
artery   435, 436
atrioventricular valve   434
atrium (pl., atria)   434
AV (atrioventricular) node   436
blood   441
blood pressure   436
capillary   436
cardiac cycle   435
cardiovascular system   433
closed circulatory system   433
diastole   435
electrocardiogram (ECG)   436
formed elements   441
heart   432
heart attack   445
hemocoel   432
hemoglobin   441
hypertension   444
lymph   439
lymphatic system   438
lymphatic vessel   439
lymphocyte   443
macrophage   443
monocyte   442

neutrophil   442
open circulatory system   432
ostium (pl., ostia)   432
plaque   444
plasma   441
platelet   443
portal system   438
pulmonary artery   434
pulmonary circuit   433, 438
pulmonary trunk   434
pulmonary vein   434
pulse   435
red blood cell   441
SA (sinoatrial) node   435
semilunar valve   434
septum   434
sinus   432
stroke   444
systemic circuit   433, 438
systole   435
thrombin   444
vein   435
vena cava (pl., venae cavae)   434
ventricle   434
venule   437
white blood cell   442

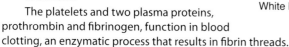 ## Testing Yourself

Choose the best answer for each question.

1. In insects with an open circulatory system, oxygen is taken to cells by
   - **a.** blood.
   - **b.** hemolymph.
   - **c.** tracheae.
   - **d.** capillaries.

2. Label the components of the cardiovascular system in the following diagram.

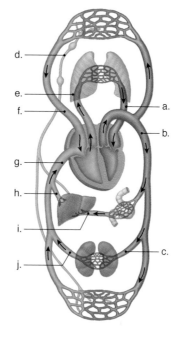

3. Which of the following statements is true?
   - **a.** Arteries carry blood away from the heart, and veins carry blood to the heart.
   - **b.** Arteries carry blood to the heart, and veins carry blood away from the heart.
   - **c.** All arteries carry O_2-rich blood, and veins carry O_2-poor blood.
   - **d.** Arteries usually carry O_2-poor blood, and veins carry O_2-rich blood.

4. Label the following diagram of the heart.

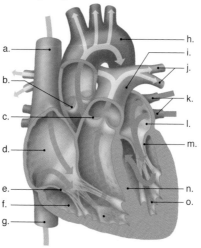

5. The _____ circuit takes blood to and from the tissues.
   a. systemic
   b. pulmonary
   c. coronary
   d. atrial

6. Atria are
   a. the main pumps of the heart.
   b. the lower chambers of the heart.
   c. the holes in the heart of an invertebrate.
   d. the upper chambers of the heart.

7. An artificial pacemaker replaces the effect of the
   a. vagus nerve.
   b. SA (sinoatrial) node.
   c. AV (atrioventricular) node.
   d. ventricular systole.

8. Which of the following lists the events of the cardiac cycle in the correct order?
   a. contraction of atria, rest, contraction of ventricles
   b. contraction of ventricles, rest, contraction of atria
   c. contraction of atria, contraction of ventricles, rest
   d. contraction of ventricles, contraction of atria, rest

9. The *lub,* the first heart sound, is produced by the closing of
   a. the aortic semilunar valve.
   b. the pulmonary semilunar valve.
   c. the atrioventricular (tricuspid) valve.
   d. the atrioventricular (bicuspid) valve.
   e. both atrioventricular valves.

10. A _____ is a condition generally caused by leaky heart valves.
    a. heart murmur
    b. hypertension
    c. fibrillation
    d. plaque

11. An electrocardiogram measures
    a. chemical signals in the brain and heart.
    b. electrical activity in the brain and heart.
    c. chemical signals in the heart.
    d. electrical changes in the wall of the heart.

12. The _____ is the heart's intrinsic natural pacemaker.
    a. atrioventricular node
    b. sinoatrial node
    c. vagus nerve
    d. superior vena cava

13. The lymphatic system
    a. returns excess tissue fluid to cardiovascular veins.
    b. is found in limited regions of the body.
    c. relies most extensively on diffusion to move lymph.
    d. is a two-way system.
    e. More than one of these are correct.

For questions 14–17, identify the cardiovascular disorder in the key that matches the description. Each answer may be used more than once. Each question may have more than one answer.

Key:
   a. hypertension
   b. atherosclerosis
   c. stroke
   d. heart attack

14. often not detected until after a stroke or heart attack
15. sometimes called a silent killer
16. results from the accumulation of plaque
17. may be caused by an embolism

18. Exchange of the gases oxygen and carbon dioxide occurs across the _____ of the _____ .
    a. veins, lungs
    b. capillaries, tissues
    c. arteries, tissues
    d. All of these are correct.

19. The _____ and the _____ are the main artery and veins in the systemic circuit.
    a. aorta, venae cavae
    b. pulmonary trunk, pulmonary veins
    c. vena cava, aortas
    d. atrium, ventricles

20. The average heart rate in humans is about _____ beats per minute.
    a. 100
    b. 45
    c. 60
    d. 70

21. If the SA node fails and the AV node takes over, the result will be a heart rate that is
    a. slower.
    b. faster.
    c. the same.
    d. None of these are correct.

22. Which association is incorrect?
    a. white blood cells—infection fighting
    b. red blood cells—blood clotting
    c. plasma—water, nutrients, and wastes
    d. red blood cells—hemoglobin
    e. platelets—blood clotting

23. Plasma
    a. is composed mostly of proteins.
    b. has a pH of 12.
    c. is the liquid portion of blood.
    d. is approximately 10% water.

24. In the tissues, nutrients and _____ are exchanged for _____ and other wastes.
    a. blood, oxygen
    b. oxygen, carbon dioxide
    c. hemoglobin, tissue fluid
    d. None of these are correct.

25. Lymph is formed from
    a. urine.
    b. fats.
    c. excess tissue fluid.
    d. All of these are correct.

26. A decrease in lymphocytes would result in problems associated with
    a. clotting.
    b. immunity.
    c. oxygen transportation.
    d. All of these are correct.

27. Blood pressure is lowest in which of the following?
    a. veins
    b. arteries
    c. capillaries

28. Which of the following assists in the return of venous blood to the heart?
    a. valves
    b. skeletal muscle contractions
    c. respiratory movements
    d. blood flow in the direction of reduced pressure
    e. All of these are correct.

29. Water enters the venous end of capillaries because
    a. osmotic pressure is higher than blood pressure.
    b. of an osmotic pressure gradient.
    c. of higher blood pressure on the venous side.
    d. of higher blood pressure on the arterial side.
    e. of higher red blood cell concentration on the venous side.

30. One-way conduction in lymphatic vessels is aided by
    a. valves.
    b. skeletal muscle contractions.
    c. Both a and b are correct.
    d. None of these are correct.

31. Red blood cells
    a. reproduce themselves by mitosis.
    b. live for several years.
    c. continually synthesize hemoglobin.
    d. are destroyed in the liver and spleen.
    e. More than one of these are correct.

32. The last step in blood clotting
    a. is the only step that requires calcium ions.
    b. occurs outside the bloodstream.
    c. is the same as the first step.
    d. converts prothrombin to thrombin.
    e. converts fibrinogen to fibrin.

33. Hemorrhagic disorders can result when which vitamin is missing from the diet?
    a. A   d. D
    b. C   e. thiamine
    c. K

34. In the following diagram, label arrows a–d as either blood pressure or osmotic pressure.

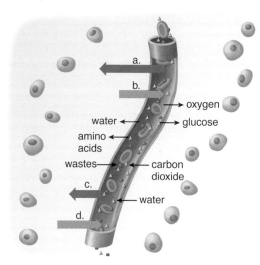

Thinking Scientifically

1. Explain why the evolution of the four-chambered heart was critical for the development of an endothermic lifestyle—the generation of internal heat—in birds and mammals.

2. Provide a physiological explanation for the benefit gained by athletes who train at high altitudes.

Bioethical Issue

Blood Doping

In the highly competitive world of professional athletics, individuals are constantly striving to gain an edge over the competition. With seven-figure salaries, international recognition, and multimillion-dollar endorsements as their incentives, some athletes resort to life-threatening measures to enhance their performance.

In recent years, various means of increasing the red blood cell count in athletes have been employed. By increasing the number of red blood cells in the body, more oxygen can be delivered to the muscles. This enhances athletic ability, especially in events requiring a high level of endurance.

In the past, blood doping was commonly used. Red blood cells (RBCs) were removed from the athlete's body and stored for several weeks. During this time, the body would release increased levels of the hormone erythropoietin to stimulate RBC production. When the RBC count returned to normal, the blood cells previously removed were injected back into the body. This was done a few days before the competition to get maximum effect. In more recent times, injections of artificially produced erythropoietin are being used to stimulate RBC production far above normal levels.

The artificial form of erythropoietin is known as epoetin alfa and is structurally identical to the naturally occurring form. Therefore, it is very difficult to determine when it is being used.

However, many athletes have paid a high price for the competitive advantage offered by epoetin alfa. The use of this hormone causes the blood to become thicker than normal. Hypertension and an increased risk of heart attack and stroke result. In recent decades, the suspected use of epoetin alfa has been linked to the deaths of numerous competitive cyclists.

What measures, if any, do you think should be taken to prevent athletes' use of epoetin alfa? If you were a professional athlete and knew you would benefit from using epoetin alfa, how would you weigh the benefits against the risks?

24

The Maintenance Systems

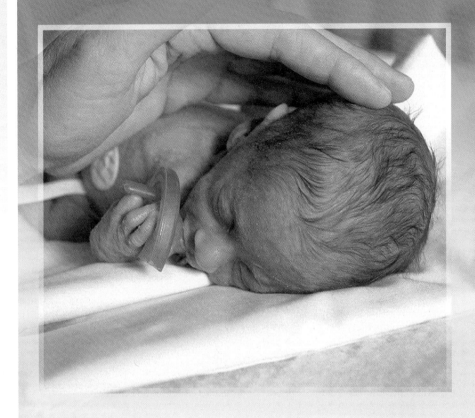

OUTLINE

BEFORE YOU BEGIN

Before beginning this chapter, take a few moments to review the following discussions.

Section 5.4 What influences the diffusion of a gas across a membrane?

Section 22.1 What are the four types of tissue, and what is the general function of each?

Section 22.3 How does negative feedback assist in the maintenance of homeostasis?

Respiratory Distress Syndrome

During human development, organ systems begin to function on an "as needed" basis. Within the first month of embryonic development, many organ systems (such as the cardiovascular and nervous systems) are present and functioning. However, some organ systems (such as the respiratory and digestive systems) that appear may not be refined until much later because they are not critical to early development. When babies are born prematurely (before the 37th week of gestation), numerous health problems can occur, particularly with regard to these systems. One condition that is common in very premature babies (those born at 32 weeks or earlier) is respiratory distress syndrome (RDS). Because the lungs of a fetus do not normally function until full term (40 weeks), lung development is not complete when a baby is born prematurely. Usually, a natural surfactant that helps keep the air sacs of the lungs inflated so that gas exchange can occur is not being produced. Without treatment, RDS is often fatal because the blood is not being oxygenated. Now, an artificial surfactant is available, and premature babies are given the surfactant they lack. This treatment greatly increases the likelihood that premature babies will survive. Other medical breakthroughs are also now available to encourage the functioning of a premature baby's organ systems.

The maintenance systems have vital roles in the overall functioning of the body, and a malfunction in one of these systems can lead to problems in other systems. In this chapter, you will learn about the structure and function of the respiratory, urinary, and digestive systems and how they support the body's other organ systems.

24.1 Respiratory System

Learning Outcomes

Upon completion of this section, you should be able to

1. List the three primary steps of respiration in terrestrial vertebrates.
2. List the components of the upper and lower respiratory tracts.
3. Compare inspiration and exhalation.
4. Compare and contrast the process of respiration in humans, insects, and fish.
5. Explain how oxygen and carbon dioxide are transported in the blood.

The cells of your body are bathed in a fluid, called tissue fluid. The cells acquire oxygen and nutrients and get rid of carbon dioxide and other wastes through exchanges with the tissue fluid. In turn, tissue fluid makes the same exchanges with the blood. Blood is refreshed because the respiratory, urinary, and digestive systems make exchanges with the external environment. Only in this way is blood cleansed of waste molecules and supplied with the oxygen and nutrients the cells require. Notice in **Figure 24.1** that, when blood enters the lungs of the respiratory system, it gives up carbon dioxide and picks up oxygen. Carbon dioxide exits the body through exhalation, and the oxygen, obtained through inhalation, is delivered to the body's cells. In this way, respiration, also referred to as ventilation, contributes to homeostasis. **Respiration** in terrestrial vertebrates (including humans) requires these steps:

1. Breathing: inspiration (entrance of air into the lungs) and expiration (exit of air from the lungs)
2. External exchange of gases between the air and the blood within the lungs
3. Internal exchange of gases between blood and tissue fluid; the body's cells exchange gases with tissue fluid

Regardless of the particular gas-exchange surfaces of animals and the manner in which gases are delivered to the cells, in the end oxygen enters mitochondria, where aerobic cellular respiration takes place. Without the delivery of oxygen to the body's cells, ATP is not produced and life ceases. Carbon dioxide (CO_2), a waste molecule given off by cells, is a by-product of cellular respiration.

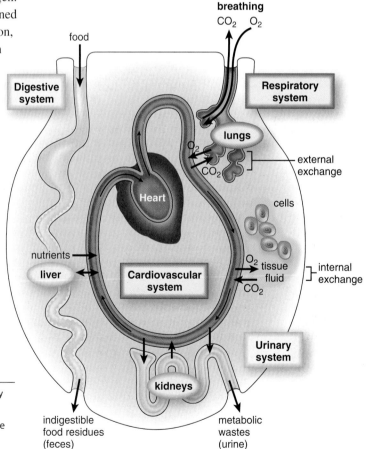

Figure 24.1   Keeping the internal environment steady.

The respiratory system functions in gas exchange. CO_2-laden blood enters the pulmonary capillaries, and then CO_2 diffuses into the lungs and exits the body by way of respiratory passages. O_2-laden air enters the respiratory passages and lungs. Then O_2 diffuses into the blood at the pulmonary capillaries.

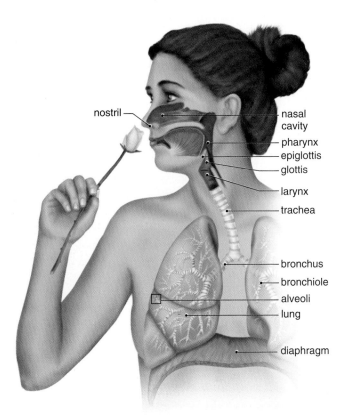

nostril

nasal cavity

pharynx

epiglottis

glottis

larynx

trachea

bronchus

bronchiole

alveoli

lung

diaphragm

Figure 24.2 **The human respiratory tract.**

To reach the lungs, air moves from the nasal cavities through the pharynx, larynx, trachea, bronchi, and bronchioles, which end in the lungs.

cilia

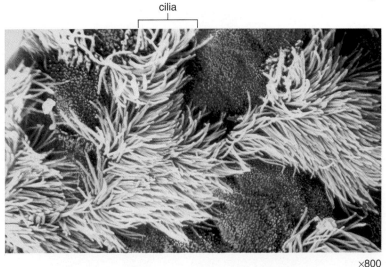

×800

Figure 24.3 **Ciliated cells of the respiratory passages.**

These cilia sweep impurities up, away from the lungs toward the throat, where they can be swallowed. Smoking first inactivates and then destroys the cilia.

The Human Respiratory Tract

The human respiratory system includes all of the structures that conduct air in a continuous pathway to and from the **lungs** (**Fig. 24.2**). As air moves through the respiratory tract, it is filtered, so that it is free of debris, warmed, and humidified. By the time the air reaches the lungs, it is at body temperature and saturated with water. In the nose, hairs and cilia act as filtering devices. In the respiratory passages, cilia beat upward, carrying mucus, dust, and occasional bits of food that "went down the wrong way" into the throat, where the accumulation may be swallowed or spit out (**Fig. 24.3**). Smoking cigarettes and cigars inactivates and eventually destroys these cilia, so that the lungs become laden with soot and debris. This is the first step toward various lung disorders.

Conversely, as air moves out of the tract, it cools and loses its moisture. As air cools, it deposits its moisture on the lining of the tract, and the nose may even drip as a result of this condensation. However, the air still retains so much moisture that, on a cold day, it forms a small cloud when we breathe out.

Upper Respiratory Tract

The upper respiratory tract consists of the nasal cavities, pharynx, and larynx (Fig. 24.2). The nose, a prominent feature of the face, is the only external portion of the respiratory system. The nose contains the **nasal cavities,** narrow canals separated from one another by a septum composed of bone and cartilage. Tears from the eyes drain into the nasal cavities by way of tear ducts. For this reason, crying produces a runny nose. The nasal cavities communicate with the **sinuses,** air-filled spaces that reduce the weight of the skull and act as resonating chambers for the voice. If the ducts leading from the sinuses become inflamed, fluid may accumulate, causing a sinus headache. The nasal cavities are separated from the mouth by a partition called the palate. The palate has two portions. Anteriorly, the hard palate is supported by bone; posteriorly, the soft palate is made solely of soft tissue and muscle.

The **pharynx** is a funnel-shaped passageway that connects the nasal cavity and mouth to the **larynx,** or voice box. The tonsils form a protective ring of lymphatic tissue at the junction of the mouth and the pharynx. *Tonsillitis* occurs when the tonsils become inflamed and enlarged. If tonsillitis occurs frequently and enlargement makes breathing difficult, the tonsils can be removed surgically, a procedure called a tonsillectomy. In the pharynx, the air passage and food passage cross because the larynx, which receives air, is anterior to the esophagus, which receives food. This arrangement may seem inefficient, since there is danger of choking if food accidentally enters the trachea, but it does have the advantage of letting you breathe through your mouth if your nose is plugged. In addition, it permits greater intake of air during heavy exercise, when greater gas exchange is required. When swallowing occurs, the epiglottis, an elastic flap of cartilage, covers the glottis, the opening into the larynx, and this helps prevent choking.

Air passes from the pharynx through the **glottis.** The larynx is always open because it is formed by a complex of cartilages, among them the one that forms the "Adam's apple." At the edges of the glottis, embedded in mucous membrane, are the **vocal cords.** These flexible bands of connective tissue vibrate and produce sound when air is expelled past them through the glottis from the larynx. *Laryngitis* is an infection of the larynx with accompanying hoarseness, leading to the inability to speak audibly.

Lower Respiratory Tract

The lower respiratory tract contains the respiratory tree, consisting of the trachea, bronchi, and bronchioles (see Fig. 24.2). The **trachea,** commonly called the windpipe, is a tube connecting the larynx to the bronchi. The trachea is held open by a series of C-shaped, cartilaginous rings that do not completely meet in the rear. This arrangement allows food to pass down through the esophagus, which lies right behind the trachea in the neck, without the rings of cartilage damaging the outer tissue of the esophagus. The trachea divides into two primary **bronchi,** which enter the right and left lungs. *Bronchitis* is an infection of the bronchi. As bronchitis develops, a nonproductive cough becomes a deep cough that produces mucus and perhaps pus. The deep cough of smokers indicates that they have bronchitis and the respiratory tract is irritated. When a person stops smoking, this progression reverses and the airways become healthy again. Chronic bronchitis is the second step toward emphysema and lung cancer caused by smoking cigarettes. Lung cancer often begins in the bronchi, and from there it spreads to the lungs.

The bronchi continue to branch until there are a great number of smaller passages called **bronchioles.** The two bronchi resemble the trachea in structure, but as the passages divide and subdivide, their walls become thinner, the rings of cartilage disappear, and the amount of smooth muscle increases. During an attack of *asthma,* the smooth muscle of the bronchioles contracts, causing constriction of the bronchioles and characteristic wheezing. Each bronchiole terminates in an elongated space enclosed by a multitude of little air pockets, or sacs, called **alveoli** (sing., alveolus), which make up the lungs (see Fig. 24.7).

Animation
Asthma

Video
911 Pollution

Respiration in Insects   While humans have one trachea, insects have many tracheae, little air tubes supported by rings of chitin that branch into every part of the body (**Fig. 24.4**). The tracheal system begins at spiracles, openings that perforate the insect's body wall, and ends in very fine, fluid-filled tubules, which may actually indent the plasma membranes of cells to come close to mitochondria. Ventilation is assisted by the presence of air sacs that draw in the air. Since no cell is very far from the site of gas exchange, the bloodstream does not transport oxygen, and no oxygen-carrying pigment is required.

Connections and Misconceptions

Why does your nose run when you are cold?

One of the functions of the mucosal lining in the nose is to use the mucus to warm the air entering the lungs. The tissue lining the nose is highly vascularized (lots of blood vessels), and combining the circulation of the blood with the mucus will warm the cold air in the nostrils, so that it will be closer to body temperature when it gets inside the lungs. This is done so the internal body temperature does not fluctuate too greatly and cause adverse effects on homeostasis.

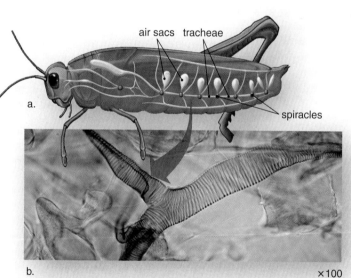

Figure 24.4   Tracheae of insects.

A system of air tubes extending throughout the body of an insect carries oxygen to the cells. Air enters the tracheae at openings called spiracles. From there, the air moves to smaller tubes, which take it to the cells where gas exchange takes place. **a.** Diagram of tracheae. **b.** The photomicrograph shows how the walls of the tracheae are stiffened with bands of chitin.

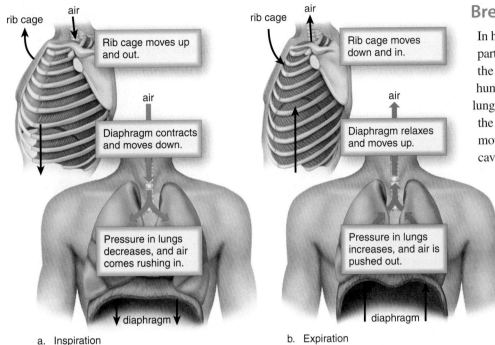

a. Inspiration

b. Expiration

Figure 24.5 Inspiration versus expiration.

a. During inspiration, the thoracic cavity and lungs expand, so that air is drawn in. **b.** During expiration, the thoracic cavity and lungs resume their original positions and pressures, forcing air out.

Figure 24.6 Breathing in birds.

Two types of air sacs are attached to the lungs of birds. When a bird inhales, most of the air enters the abdominal air sacs; when a bird exhales, air moves through the lungs to the thoracic air sacs before exiting the trachea. This one-way flow of air through the lungs allows more fresh air to be present in the lungs with each breath, leading to greater uptake of oxygen from one breath of air.

Breathing

In humans, the **diaphragm** is a muscular, membranous partition that divides the **thoracic** (upper) **cavity** from the **abdominal** (lower) **cavity** of the body. When humans breathe, the volume of the thoracic cavity and lungs is increased by muscle contractions that both lower the diaphragm and raise the ribs (**Fig. 24.5**). These movements create a negative pressure in the thoracic cavity and lungs, and air then flows into the lungs, a process called *inspiration*. It is important to realize that air comes in because the lungs have already opened up; air does not force the lungs open. When the rib and diaphragm muscles relax, the lungs recoil and air moves out as a result of increased pressure in the lungs, a process called *expiration*.

In mammals and most reptiles, air moves in and out by the same route; therefore, some residual air, low in oxygen, is always left in the lungs of humans. Birds, on the other hand, use a one-way ventilation mechanism (**Fig. 24.6**). Incoming air is carried past the lungs by a trachea, which takes it to a set of abdominal air sacs. Then air passes forward through the lungs into a set of thoracic air sacs. Fresh air never mixes with used air in the lungs of birds, and thereby gas-exchange efficiency is greatly improved.

Increased concentrations of hydrogen ions (H^+) and carbon dioxide (CO_2) in the blood are the primary stimuli that increase the breathing rate in humans. The chemical content of the blood is monitored by chemoreceptors called **aortic bodies** and **carotid bodies**—specialized structures in the walls of arteries, the aorta, and common carotid arteries. These receptors are very sensitive to changes in H^+ and CO_2 concentrations, but they are only minimally sensitive to lower oxygen (O_2) concentrations. The need to breathe comes from a buildup of CO_2 in the bloodstream, not necessarily from a lack of oxygen. Information from the chemoreceptors goes to the *breathing center* in the brain, which increases the breathing rate when concentrations of hydrogen ions and carbon dioxide rise (see Fig. 24.10). The breathing center is also directly sensitive to the chemical content of the blood, including its oxygen content.

Lungs and External Exchange of Gases

The lungs of humans and other mammals are more elaborately subdivided than those of amphibians and reptiles. Frogs and salamanders have a moist skin that allows them to use the surface of their body for gas exchange in addition to the lungs. It has been estimated that human lungs have a total surface area at least 50 times the skin's surface area because of the presence of alveoli (**Fig. 24.7**).

An alveolus, like the capillary that surrounds it, is bounded by squamous epithelium. Diffusion alone accounts for gas exchange between the alveolus and the capillary. Carbon dioxide, being more plentiful in the pulmonary venule, diffuses from a pulmonary capillary to enter an alveolus, while oxygen, being more plentiful in the lungs, diffuses from an alveolus into a pulmonary capillary.

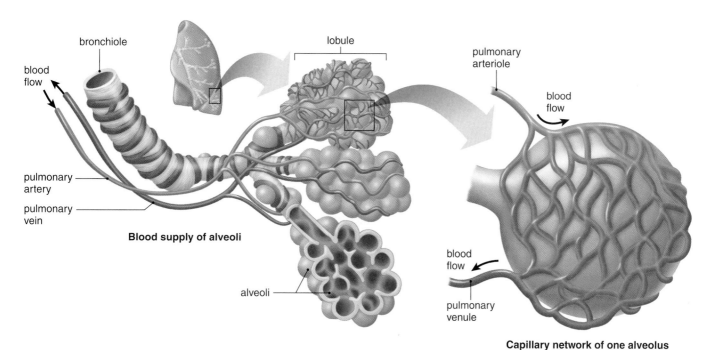

Figure 24.7 **Gas exchange in the lungs.**

Bronchioles lead to the alveoli, each of which is surrounded by an extensive capillary network. The pulmonary artery and arteriole carry O_2-poor blood (colored blue) and the pulmonary venule and vein carry O_2-rich blood (colored red).

The process of diffusion requires a gas-exchange region to be not only large but also moist and thin. The alveoli are lined with surfactant, a film of lipoprotein that lowers the surface tension of water, thereby preventing the alveoli from collapsing. Some newborns, especially if premature, lack this film. Surfactant replacement therapy is now available to treat this condition (see chapter opener).

The blood within pulmonary capillaries is indeed spread thin, and the red blood cells are pressed up against their narrow walls. The alveolar epithelium and the capillary epithelium are so close that together they are called the *respiratory membrane*. Hemoglobin in the red blood cells quickly picks up oxygen molecules as they diffuse into the blood.

Animation
Gas Exchange
During Respiration

Emphysema is a serious lung condition in which the walls of many alveoli have been destroyed. The lungs have less recoil, and there is less surface area for gas exchange to occur.

Gills of Fish   In contrast to the lungs of terrestrial vertebrates, fish and other aquatic animals use gills as their respiratory organ (**Fig. 24.8**). In fish, water is drawn into the mouth and out from the pharynx across the gills. The flow of blood in gill capillaries is opposite the flow of water across the gills; therefore, the blood is always exposed to water having a higher oxygen content. In the end, about 80–90% of the dissolved oxygen in water is absorbed.

Transport and Internal Exchange of Gases

Recall from Chapter 23 that hemoglobin molecules within red blood cells carry oxygen to the body's tissues. If hemoglobin did not transport oxygen, it would take about three years for an oxygen molecule to move from your lungs to your toes. Each hemoglobin molecule contains four polypeptide chains: two

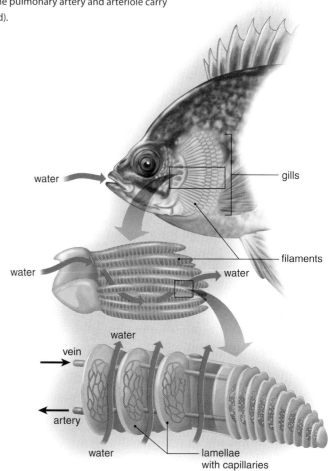

Figure 24.8 **Gills of bony fishes.**

Gills are finely divided into filaments, and each filament has many thin, plate-like lamellae. Gases are exchanged between the capillaries inside the lamellae and the water that flows between the lamellae.

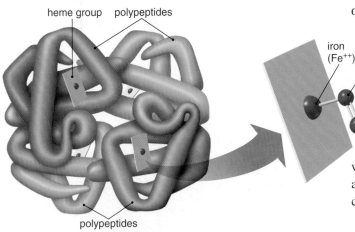

Figure 24.9 **Hemoglobin.**

Hemoglobin has two each of two different polypeptides. Each one is associated with a heme group. Oxygen bonds with the central iron atom of a heme group.

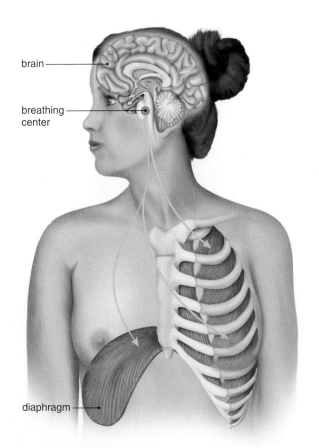

Figure 24.10 **Neural control of breathing rate.**

The brain regulates breathing rate by controlling contraction of the rib cage muscles and the diaphragm. When the breathing rate increases, H⁺ lowers as CO₂ is removed from the blood, and the pH of the blood returns to normal.

α and two β chains. Each polypeptide is folded around an iron-containing group called **heme.** It is actually the iron that bonds with oxygen and carries it to the tissues (**Fig. 24.9**). Since there are about 250 million hemoglobin molecules in each red blood cell, each cell is capable of carrying at least 1 billion molecules of oxygen. Hemoglobin gives up its oxygen in the tissues during internal exchange primarily because tissue fluid always has a lower oxygen concentration than blood does. This difference occurs because cells take up and utilize oxygen when they carry on cellular respiration. Another reason hemoglobin gives up oxygen is the warmer temperature and lower pH in the tissues, environmental conditions that are also caused by cellular respiration. When cells respire, they give off heat and carbon dioxide as by-products.

Carbon dioxide enters the blood during internal exchange because the tissue fluid always has a higher concentration of carbon dioxide. Most of the carbon dioxide is transported in the form of the **bicarbonate ion** (HCO_3^-). First, carbon dioxide combines with water, forming carbonic acid, and then this acid dissociates to a hydrogen ion (H^+) and HCO_3^-:

$$CO_2 \;+\; H_2O \longrightarrow H_2CO_3 \longrightarrow H^+ \;+\; HCO_3^-$$

| carbon dioxide | water | carbonic acid | hydrogen ion | bicarbonate ion |

The H^+ does cause the pH to lower, but only slightly because much of the H^+ is absorbed by the globin portions of hemoglobin. The HCO_3^- is carried in the plasma.

What happens to the preceding equation in the lungs? As blood enters the pulmonary capillaries, carbon dioxide diffuses out of the blood into the alveoli. Hemoglobin gives up the H^+ it has been carrying as this reaction occurs:

$$H^+ + HCO_2^- \longrightarrow H_2CO_3 \longrightarrow H_2O + CO_2$$

Now, much of the carbon dioxide diffuses out of the blood into the alveoli of the lungs. Should the blood level of H^+ rise, the breathing center in the brain increases the breathing rate, and as more CO_2 leaves the blood, the pH of blood is corrected (**Fig. 24.10**).

MP3
Control of Respiration

Connecting the Concepts

For more information on topics associated with respiration, refer to the following discussions.

Section 5.4 explains the factors that determine the rates of diffusion across a membrane.

Section 7.1 provides an overview of cellular respiration.

Section 23.3 provides more information on the role of red blood cells.

Check Your Progress 24.1

1. List the three steps of respiration in a terrestrial vertebrate.

2. Contrast the anatomical differences between the upper and lower respiratory tracts.

3. Detail what causes oxygen to diffuse into the blood from the lungs and what happens when carbon dioxide enters the blood.

24.2 Urinary System and Excretion

Learning Outcomes

Upon completion of this section, you should be able to

1. Connect the three functions of the urinary system with homeostasis in the human body.
2. Describe the anatomy of the kidney and nephron.
3. Detail the three processes involved in urine formation.
4. Compare and contrast excretion in insects with excretion in mammals.
5. Describe the causes of kidney disease and its treatment.

In complex animals, **kidneys** excrete nitrogenous wastes and are involved in regulating the water-salt balance of the body (**Fig. 24.11**). In addition, the mammalian kidney helps regulate the pH of blood. To summarize, these three functions are associated with the kidney:

1. Excretion of nitrogenous wastes, such as urea and uric acid
2. Maintenance of the water-salt balance of the blood
3. Maintenance of the acid-base balance of the blood

In humans, the kidneys are bean-shaped, reddish-brown organs, each about the size of a fist. They are located on each side of the vertebral column, just below the diaphragm, where they are partially protected by the lower rib cage. **Urine** made by the kidneys is conducted from the body by the other organs in the urinary system. Each kidney is connected to a **ureter,** a tube that takes urine from the kidney to the **urinary bladder,** where it is stored until it is voided from the body through the single **urethra** (**Fig. 24.12**a). In

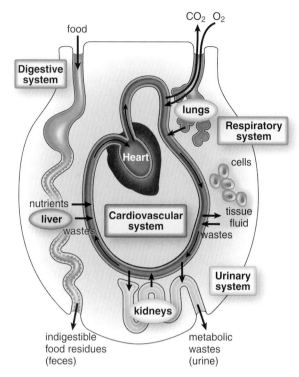

Figure 24.11 **Keeping the internal environment steady.**

Excretion carried out by the kidneys serves three functions: excretion of urea (produced by the liver) and other nitrogenous wastes, maintenance of water-salt balance, and maintenance of pH.

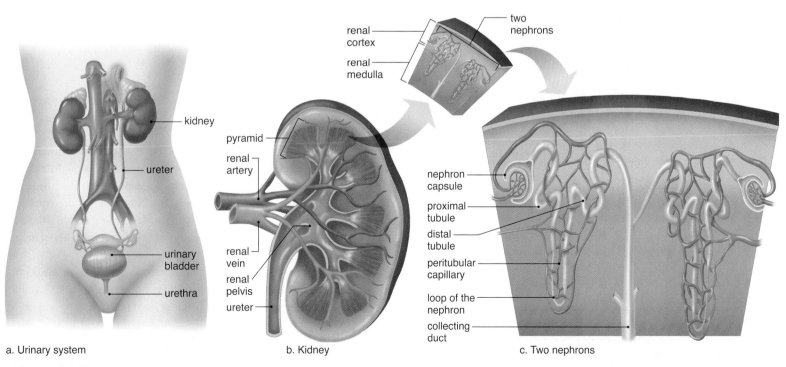

a. Urinary system

b. Kidney

c. Two nephrons

Figure 24.12 **The human urinary system.**

a. The human urinary system produces, stores, and expels urine from the body. **b.** The human kidney has three sections, which can be correlated with (**c**) the presence of about 1 million nephrons. Nephrons do the work of the kidney.

males, the urethra passes through the penis; in females, it opens in front of the opening of the vagina. In females, there is no connection between the genital (reproductive) and urinary systems, but there is a connection in males—that is, the urethra also carries sperm during ejaculation.

In amphibians, birds, reptiles, and some fishes, the bladder empties into the cloaca, a common chamber and outlet for the digestive, urinary, and genital tracts.

Kidneys

If a kidney is sectioned longitudinally, three major parts can be distinguished (Fig. 24.12b). The **renal cortex,** the outer region of a kidney, has a somewhat granular appearance. The **renal medulla** consists of the cone-shaped renal pyramids, which lie inside the renal cortex. The innermost part of the kidney is a hollow chamber called the **renal pelvis.** Urine collects in the renal pelvis and then is carried to the bladder by a ureter. Microscopically, each kidney is composed of about 1 million tiny tubules called **nephrons** (Fig. 24.12c). The nephrons of a kidney produce urine.

Nephrons

Each nephron is made up of several parts (**Fig. 24.13**). The blind, or closed, end of a nephron is pushed in on itself to form the **nephron capsule.** Filtration of the blood occurs at this portion of the nephron. The inner layer of the capsule is composed of specialized cells that allow the easy passage of molecules. Leading from the capsule is a portion of the nephron called the **proximal tubule,** which is lined by cells with many mitochondria and tightly packed microvilli. Then comes the **nephron loop,** with a descending limb and an ascending limb. This is followed by the **distal tubule.** Several distal tubules enter one **collecting duct.** A collecting duct delivers urine to the renal pelvis. The nephron loop and the collecting duct give the pyramids of the renal medulla their striped appearance (see Fig. 24.12b).

Each nephron has its own blood supply, and various exchanges occur between parts of the nephron and a blood capillary as urine forms.

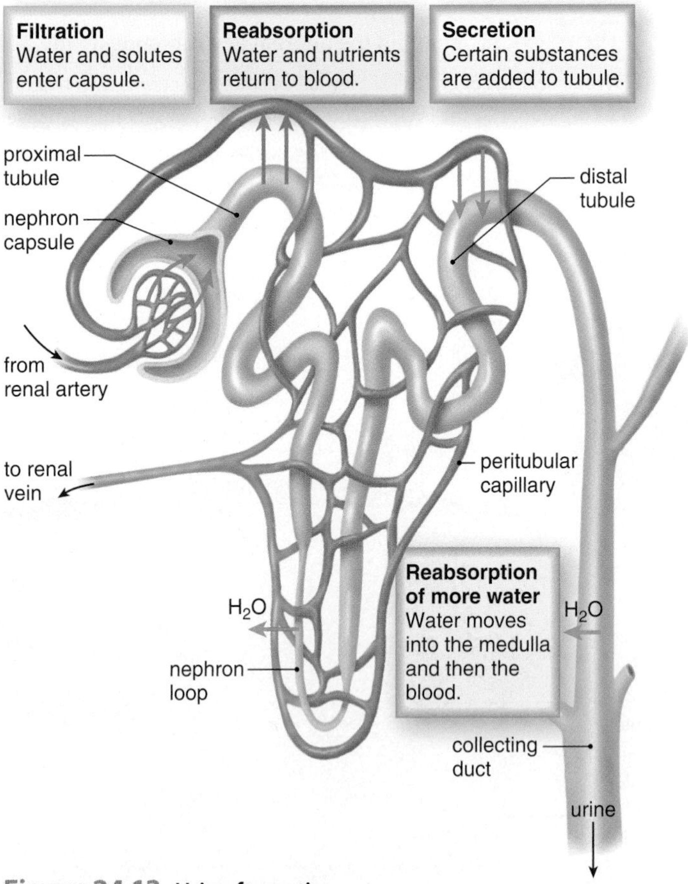

Urine Formation

Urine formation requires three steps: filtration, reabsorption, and secretion (Fig. 24.13).

Filtration Filtration occurs whenever small substances pass through a filter and large substances are left behind. During urine formation, **filtration** is the movement of small molecules from a blood capillary to the inside of the capsule as a result of adequate blood pressure. Small molecules, such as water, nutrients, salts, and urea, move to the inside of the capsule. Plasma proteins and blood cells are too large to be part of this filtrate, so they remain in the blood.

If the composition of the filtrate were not altered in other parts of the nephron, death from loss of nutrients (starvation) and loss of water (dehydration) would quickly follow. The next step, reabsorption, helps prevent this.

Reabsorption of Solutes **Reabsorption** takes place when substances from the proximal tubule move into the blood. Nutrients such as glucose and amino acids also return to the blood. This process is selective because some

Figure 24.13 **Urine formation.**

Urine formation requires three steps: filtration, reabsorption, and secretion. Production of a hypertonic urine occurs when water is reabsorbed into the blood along the length of the nephron, and especially at the nephron loop and collecting duct.

molecules, such as glucose, are both passively and actively reabsorbed. The cells of the proximal tubule have numerous microvilli, which increase the surface area, and numerous mitochondria, which supply the energy needed for active transport. However, if there is more glucose in the filtrate than there are carriers to handle it, glucose will appear in the urine. Glucose in the urine is a sign of diabetes mellitus, sometimes caused by a lack of the hormone insulin.

Sodium ions (Na^+) are also actively pumped into the peritubular capillary, and then chloride ions (Cl^-) follow passively. Now water moves by osmosis from the tubule into the blood. About 60–70% of salt and water are reabsorbed at the proximal tubule.

Some of the urea, the primary nitrogenous waste product of human metabolism, and other types of nitrogenous wastes excreted by humans are passively reabsorbed, but most remain in the filtrate.

Secretion **Secretion** is the transport of substances into the nephron by means other than filtration. For our purposes, secretion may be particularly associated with the distal tubule. Substances such as uric acid, hydrogen ions, ammonia, and penicillin are eliminated by secretion. The process of secretion helps rid the body of potentially harmful compounds that were not filtered into the capsule.

 **Animation** Kidney Function   **MP3** Overview of Urine Formation

Regulation of Water-Salt Balance and pH   Typically, animals have some means of regulating the osmolarity of the internal environment, so that the water-salt balance stays within normal limits. Insects have a unique excretory system consisting of long, thin tubules called **Malpighian tubules** attached to the gut (**Fig. 24.14**). Uric acid, their primary nitrogenous waste product, simply diffuses from the surrounding hemolymph into these tubules, and water follows a salt gradient established by active transport of potassium (K^+). Water and other useful substances are reabsorbed at the rectum, but the uric acid leaves the body at the anus. Insects that live in water or eat large quantities of moist food reabsorb little water. But insects in dry environments reabsorb most of the water and excrete a dry, semisolid mass of uric acid. Most animals can regulate the blood levels of both water and salt. For example, freshwater fishes take up salt in the digestive tract and gills and produce large amounts of dilute urine. Saltwater fishes take in salt water by drinking, but then they pump ions out at the gills and produce only small amounts of concentrated urine.

In mammals, the long nephron loop allows secretion of a hypertonic urine (see Fig. 24.13). The ascending limb of the nephron loop pumps salt and urea into the renal medulla, and water follows by osmosis both at the descending limb of the nephron loop and at the collecting duct. As you will see in Chapter 27, at least three hormones are involved in regulating water-salt reabsorption by the kidneys. Drinking coffee interferes with one of these hormones, and that is why coffee is a *diuretic*, a substance that causes the production of more urine.

 **MP3** Water Conservation

Most mammals can also regulate the pH of the blood. Both the bicarbonate (HCO_3^-) buffer system of the blood and the regulation of breathing rate (see Fig. 24.10) helps rid the body of CO_2 and contribute to maintaining blood pH. As helpful as these mechanisms might be, only the kidneys can excrete a wide range of acidic and basic substances. The kidneys are slower-acting than the buffer/breathing mechanism, but they have a more powerful effect on pH.

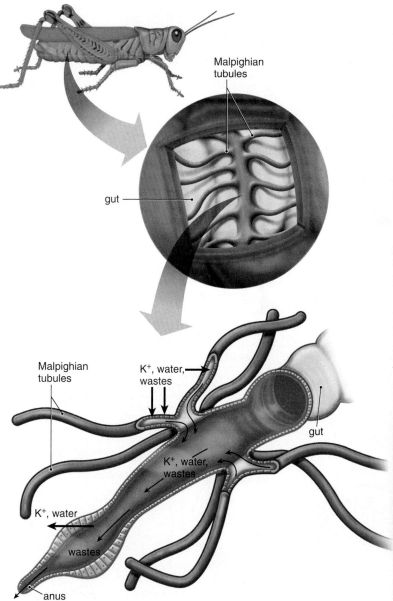

Figure 24.14   Malpighian tubules.

The Malpighian tubules of insects are attached to the gut and surrounded by the hemolymph of the open circulatory system. Wastes diffuse into tubules. K^+ is secreted into these tubules, drawing in water by osmosis. Much of the water and K^+ is reabsorbed across the wall of the rectum.

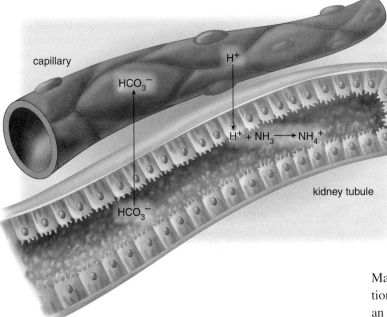

capillary

HCO_3^-

H^+

$H^+ + NH_3 \longrightarrow NH_4^+$

HCO_3^-

kidney tubule

Figure 24.15 **Acid-base balance.**

In the kidneys, bicarbonate ions (HCO_3^-) are reabsorbed and hydrogen ions (H^+) are excreted as needed to maintain the pH of the blood. Excess hydrogen ions are buffered—for example, by ammonia (NH_3), which becomes ammonium (NH_4^+). Ammonia is produced in tubule cells by the breakdown of amino acids.

For the sake of simplicity, we can think of the kidneys as reabsorbing bicarbonate ions and excreting hydrogen ions as needed to maintain the normal pH of the blood (**Fig. 24.15**). If the blood is acidic, hydrogen ions are excreted and bicarbonate ions are reabsorbed. If the blood is basic, hydrogen ions are not excreted and bicarbonate ions are not reabsorbed. The fact that urine is most often acidic shows that usually an excess of hydrogen ions is produced by the body and is excreted. Ammonia (NH_3) provides a means for buffering these hydrogen ions in urine: ($NH_3 + H^+ \longrightarrow NH_4^+$). Ammonia (the presence of which is quite obvious in the diaper pail or kitty litter box) is produced in tubule cells by the breakdown of amino acids. Phosphate provides another means of buffering hydrogen ions in urine.

Problems with Kidney Function

Many types of illnesses, especially diabetes, hypertension, and inherited conditions, cause progressive renal disease and renal failure. Urinary tract infections, an enlarged prostate gland, pH imbalances, or simply an intake of too much calcium can lead to kidney stones. Kidney stones form in the renal pelvis and usually pass unnoticed in the urine flow. If they grow to several centimeters and block the renal pelvis or ureter, back pressure builds up and destroys nephrons. One of the first signs of nephron damage is the presence of albumin, white blood cells, or even red blood cells in the urine, as detected by a urinalysis. If damage is so extensive that more than two-thirds of the nephrons are inoperative, urea and other waste substances accumulate in the blood. Although nitrogenous waste in the blood is a threat to homeostasis, the retention of water and salts is of even greater concern. **Edema,** fluid accumulation in the body tissues, may occur. Imbalance in the ionic composition of body fluids can lead to loss of consciousness and even heart failure.

Video
Cranberries vs Bacteria

Hemodialysis and Kidney Transplants

Patients with renal failure can undergo **hemodialysis,** utilizing either an artificial kidney machine or **continuous ambulatory peritoneal dialysis (CAPD).** *Dialysis* is defined as the diffusion of dissolved molecules through a semipermeable membrane with pore sizes that allow only small molecules to pass through. In an artificial kidney machine, the patient's blood is passed through a membranous tube, which is in contact with a dialysis solution, or **dialysate** (**Fig. 24.16**). Substances more concentrated in the blood diffuse into the dialysate, and substances more concentrated in the dialysate diffuse into the blood. The dialysate is continuously replaced to maintain favorable concentration gradients. In this way, the artificial kidney can be utilized either to extract substances from the blood, including waste products or toxic chemicals and drugs, or to add substances to the blood—for example, bicarbonate ions (HCO_3^-) if the blood is acidic. In the course of a three- to six-hour hemodialysis procedure, from 50 to 250 grams of urea can be removed from a patient, which greatly exceeds the amount excreted by our kidneys within the same time frame. Therefore, a patient needs to undergo treatment only about twice a week.

CAPD is so named because the peritoneum, the epithelium that lines the abdominal cavity, is the dialysis membrane. A fresh amount of dialysate is introduced directly into the abdominal cavity from a bag that is temporarily

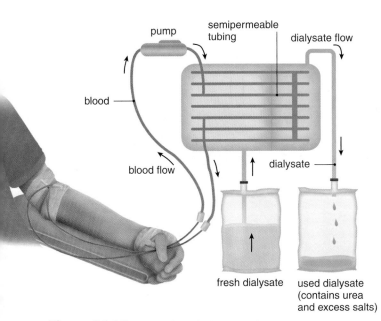

pump

semipermeable tubing

dialysate flow

blood

blood flow

dialysate

fresh dialysate

used dialysate (contains urea and excess salts)

Figure 24.16 **An artificial kidney machine.**

As the patient's blood is pumped through dialysis tubing, the tubing is exposed to a dialysate (dialysis solution). Wastes exit from blood into the solution because of a preestablished concentration gradient. In this way, not only is blood cleansed but its water-salt and acid-base balances can be adjusted as well.

attached to a permanently implanted plastic tube. The dialysate flows into the peritoneal cavity by gravity. Waste and salt molecules pass from the blood vessels in the abdominal wall into the dialysate before the fluid is collected four or eight hours later. The solution is drained into a bag from the abdominal cavity by gravity, and then it is discarded. One advantage of CAPD over an artificial kidney machine is that the individual can go about his or her normal activities during CAPD.

Patients with renal failure may undergo a transplant operation in which a functioning kidney from a donor is received. A person needs only one functioning kidney. The possibility of organ rejection exists, as it does with all organ transplants. Receiving a kidney from a close relative has the highest chance of success. The current one-year survival rate is 97% if the kidney comes from a relative, 90% if it is from a nonrelative. In the future, it may be possible to use kidneys from pigs or kidneys created in the laboratory for transplant operations.

Connecting the Concepts

For more information on homeostasis and regulation of the kidneys, refer to the following discussions.

Section 22.3 provides a summary of how the body maintains homeostasis.

Section 27.1 describes the role of the hypothalamus and medulla oblongata as regulatory centers in the central nervous system.

Section 27.2 describes the hormones produced by the endocrine system that regulate kidney function.

Check Your Progress 24.2

1. List the three functions of the kidney.
2. Describe the parts of a nephron and the function of each part.
3. Summarize the importance of water-salt balance and pH to homeostasis.

Connections and Misconceptions

During a kidney transplant, is the failing kidney removed?

Normally no. When a patient needs a kidney transplant, the donor kidney is normally placed in the lower abdomen, below one of the two original kidneys, with the blood vessels from the donor kidney connected to the recipient's arteries and veins and the ureter from the donor kidney connected to the recipient's bladder. The patient's failing kidney is kept in most cases because, even if the nephrons are not functioning in cleansing the blood and producing urine, the kidney can 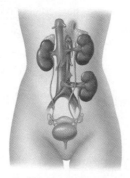 still produce other chemicals and substances the body needs to function. The failing kidney will be removed in cases of severe infection, cancer, or polycystic kidney disease (PKD).

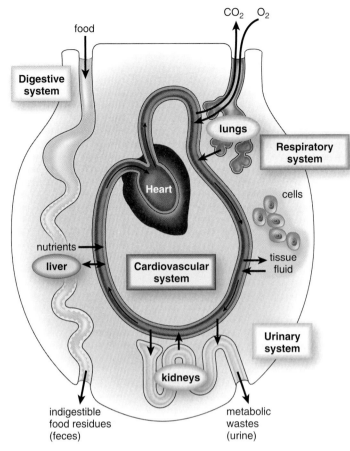

Figure 24.17 **Keeping the internal environment steady.**

The digestive system takes in food and digests it to nutrient molecules that enter the blood. The blood transports nutrients to the tissues, where exchange occurs with tissue fluid.

24.3 Digestive System

Learning Outcomes

Upon completion of this section, you should be able to

1. Compare the digestive tracts of herbivores, carnivores, and omnivores.
2. Identify the organs of the human digestive system and provide a function for each.
3. List the accessory organs of the digestive system and explain their functions.
4. Detail the digestion and absorption of specific nutrients (e.g., carbohydrates) in the digestive tract.

Digestion contributes to homeostasis by providing the body's cells with the nutrients they need to continue functioning. A digestive tract performs the following functions (**Fig. 24.17**):

1. Ingests food
2. Breaks food down into small molecules that can cross plasma membranes
3. Absorbs nutrient molecules
4. Eliminates indigestible remains

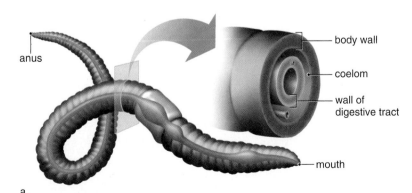

a.

Figure 24.18 **Complete digestive system.**

The earthworm (**a**) and humans (**b**) have complete digestive systems. A complete digestive system leads to specialization of organs along the digestive tract.

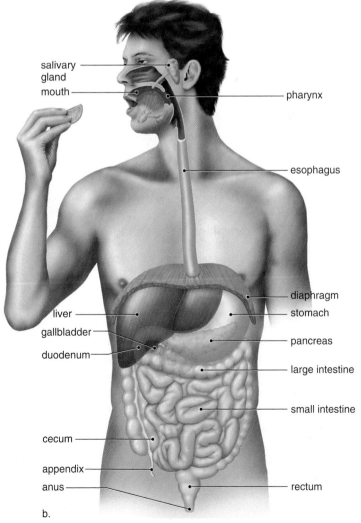

b.

Tube-Within-a-Tube Body Plan

Hydras and planarians (see Chapter 19) have a sac body plan—that is, the mouth, like the top of a sack, serves as both an entrance and an exit. Most other animals, such as the earthworm, have a tube-within-a-tube plan, so called because the inner tube has both an entrance and an exit—the mouth and the anus. Therefore, it is called a complete digestive tract (**Fig. 24.18***a*). Notice that the inner tube (the digestive tract, sometimes called the alimentary canal) is separated from the outer tube (the body wall) by the **coelom.** The digestive tract of humans and other vertebrates consists of so many specialized organs that it might be hard to realize that the basic plan of vertebrates is the same as that of the earthworm (Fig. 24.18*b*). The tube-within-a-tube plan and a complete tract result in this specialization of parts.

Digestion of food in earthworms and humans is an extracellular process. Digestive enzymes, produced by glands in the wall of the tract or by accessory glands that lie nearby, are released into the tract. Food is never found within these accessory glands, only within the tract itself.

Mouth

In humans, the digestive system begins with the **mouth,** which chews food into pieces, the beginning of mechanical digestion. Many vertebrates have teeth, an exception being birds, which lack teeth and depend on the churning of small pebbles within a gizzard to break up their food. The teeth (dentition) of mammals reflect their diet (**Fig. 24.19**). **Carnivores** eat meat, which is easily digestible because the cells do not have a cellulose wall. **Herbivores** eat plant material, which needs a lot of chewing and other processing to break up the cellulose walls. Humans are **omnivores;** they eat both meat and plant material. The four front teeth (top and bottom) of humans are sharp, chisel-shaped incisors used for biting. On each side of the incisors are the pointed canines used for tearing food. The premolars and molars grind and crush food. It is as though humans are carnivores in the front of their mouths and herbivores in the back.

Food contains cells composed of molecules of carbohydrates, proteins, nucleic acids, and lipids. Digestive enzymes break down these large molecules to smaller molecules. In the mouth, three pairs of **salivary glands** send saliva by way of ducts to the mouth, to begin the process of chemical digestion. One of these digestive enzymes is **salivary amylase,** which breaks down starch, a carbohydrate, to maltose, a disaccharide. While in the mouth, food is manipulated by a muscular tongue (mechanical digestion), mixing the chewed food with saliva (chemical digestion) and then forming this mixture into a mass called a **bolus,** which is swallowed.

Figure 24.19 Dentition among mammals.

Key:
- incisors
- canines
- premolars
- molars

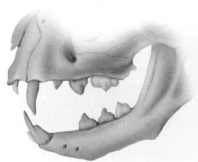

a. Carnivores, such as cats, have enlarged canines for killing their prey, short incisors for scraping bones, and jagged molars for tearing apart flesh.

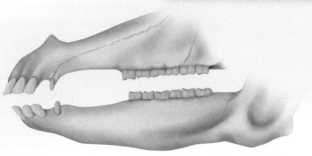

b. Herbivores, such as a horse, have sharp incisors for clipping grasses and large, flat premolars and molars for grinding.

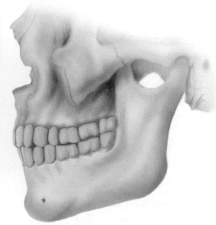

c. The front teeth of humans resemble those of carnivores, and the back teeth are like those of herbivores.

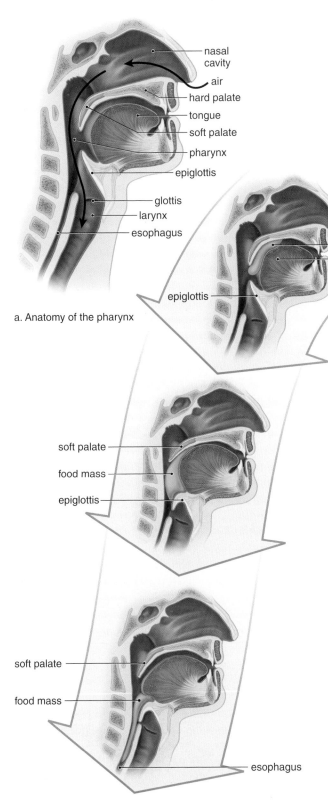

a. Anatomy of the pharynx

b. Swallowing

Figure 24.20 The human mouth, pharynx, esophagus, and larynx.

a. The palates (both the hard and the soft) separate the mouth from the nasal cavities. The pharynx leads to the esophagus and the larynx.
b. When food is swallowed, the soft palate closes off the nasal cavities and the epiglottis closes off the larynx.

Swallowing

The human digestive and respiratory passages come together in the pharynx and then separate (**Fig. 24.20***a*). When food is swallowed, the soft palate, the rear portion of the mouth's roof, moves up to close off the nasal cavities (Fig. 24.20*b*). A flap of tissue called the **epiglottis** covers the glottis, an opening into the larynx. Ordinarily, the bolus must move through the pharynx and into the esophagus because the air passages are blocked. Unfortunately, we have all had the unpleasant experience of having food "go the wrong way." The wrong way may be either into the nasal cavities or into the trachea. If it is the latter, coughing will most likely force the food up out of the trachea and into the pharynx again.

The **esophagus** is a muscular tube that takes food to the stomach, which lies below the diaphragm. When food enters the esophagus, peristalsis begins. **Peristalsis** is a rhythmic contraction of smooth muscles that moves the contents along in tubular organs—in this case, those of the digestive tract.

Stomach

The human **stomach** is a thick-walled, J-shaped organ (**Fig. 24.21**) on the left side of the abdominal cavity below the liver. The stomach is continuous with the esophagus above and the duodenum of the small intestine below. The stomach is about 25 cm (10 in.) long, regardless of the amount of food it holds, but the diameter varies, depending on how full it is. The stomach receives food from the esophagus, stores food, starts the digestion of proteins, and moves food into the small intestine.

The wall of the stomach has deep folds, which disappear as the stomach fills to an approximate capacity of 1 liter. Therefore, humans can periodically eat relatively large meals, freeing the rest of their time for other activities. But the stomach is much more than a mere storage organ. The muscular walls of the stomach contract vigorously and mix food with juices that are secreted whenever food enters the stomach. The epithelial lining of the stomach, called a mucosa, has millions of gastric glands. These gastric glands produce gastric juice containing so much hydrochloric acid that the stomach routinely has a pH of about 2. Such strong acidity is usually sufficient to kill any microbes that might be in food. This low pH also promotes the activity of **pepsin,** a hydrolytic enzyme that acts on protein to produce peptides. In addition, strong acidity causes *heartburn* and *gastric reflux disease* when gastric juice backs up into the esophagus.

As with the rest of the digestive tract, a thick layer of mucus protects the wall of the stomach from enzymatic action. Still, an ulcer, which is an open sore in the wall caused by the gradual destruction of tissues, may occur in some individuals. Ulcers are due to an infection by an acid-resistant bacterium, *Helicobacter pylori,* which is able to attach to the epithelial lining. Wherever the bacterium attaches, the lining stops producing mucus and the area becomes exposed to digestive action, allowing an ulcer to develop.

The stomach acts chemically but also acts physically on food. The wall of the stomach contains three muscle layers: one is longitudinal, another is circular, and the third is obliquely arranged. The muscular wall not only moves food along but also churns, mixing the food with gastric juice and breaking it down into small pieces.

Alcohol and other liquids are absorbed in the stomach, but food particles are not. Peristalsis pushes food along in the stomach, as it does in other digestive organs (Fig. 24.21*b*). At the base of the stomach is a narrow

opening controlled by a *sphincter,* a muscle that surrounds a tube and closes or opens it by contracting and relaxing. When food leaves the stomach, it is a thick, soupy liquid called chyme. Whenever the sphincter relaxes, a small quantity of chyme squirts through the opening into the small intestine.

**Animation
Three Phases
of Gastric
Secretion**

Ruminants   Ruminants, a type of mammal that includes cattle, sheep, goats, deer, and buffalo, are named for a part of their stomach, the rumen (**Fig. 24.22**). The rumen contains symbiotic bacteria and protozoans that, unlike in mammals, produce enzymes that can digest cellulose. After these herbivores feed on grass, it goes to the rumen, where it is broken down by the enzymes released by the microbes, and then it becomes small balls of cud. The cud returns to the mouth, where the animal "chews the cud." The cud may return to the rumen for a second go-round before passing through the other chambers of the stomach. The rumen is an adaptation to a diet rich in fiber that may have been promoted by competition among the many types of animals that feed on grass. The last chamber in ruminants is analogous to the human stomach, being the place where protein is digested to peptides.

Small Intestine

Processing food in humans is more complicated than one might think. So far, the food has been chewed in the mouth and worked on by the enzyme salivary amylase, which digests starch to maltose. In addition, the digestion of protein has begun in the stomach as pepsin digests proteins to peptides. At this point, the contents of the digestive tract are called **chyme.** Chyme passes from the stomach to the **small intestine,** a long, coiled tube that has two functions: (1) digestion of all the molecules in chyme, including polymers of carbohydrate, protein, nucleic acid, and fat, and (2) absorption of the nutrient molecules into the body.

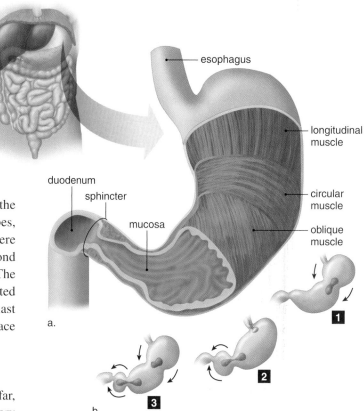

Figure 24.21   Anatomy of the human stomach.

a. The stomach has a thick wall that expands as it fills with food. The wall contains three layers of muscle, and their presence allows the stomach to churn and mix food with gastric juices. The mucosa of the stomach wall secretes mucus and contains gastric glands, which secrete gastric juices active in the digestion of protein. **b.** Peristalsis, a rhythmic contraction, occurs along the length of the digestive tract.

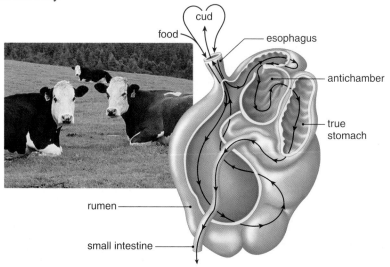

Figure 24.22   A ruminant's stomach.

Ruminants eat grass, which is made of cells with strong cellulose walls. The first chamber of a ruminant's stomach, called the rumen, contains symbiotic bacteria and protozoans, which release enzymes that digest cellulose. After a first pass through the rumen, the "cud" returns to the mouth, where it is leisurely chewed. Then, it may return to the rumen for a second go-round of digestion before passing through to the true stomach.

Connections and Misconceptions

Why does your stomach "growl" when you are hungry?

Borborygmi is the medical term for the "growl" sound in your stomach. It is actually the stomach walls squeezing together in an attempt to mix digestive juices and gases for digestion. When your stomach is empty, the result is the sound of these juices bouncing off the walls of the hollow stomach. The "hunger center" of the brain will send a message to your stomach to begin the process of digestion, sometimes initiating borborygmi and signaling the need to eat.

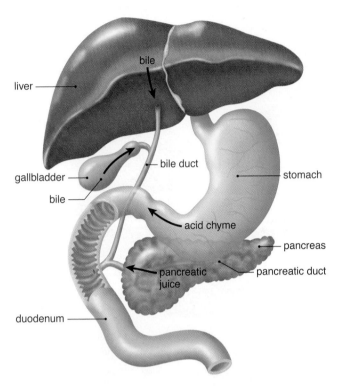

Figure 24.23   The pancreatic and bile ducts empty into the duodenum.

Bile, made by the liver and stored in the gallbladder, and pancreatic juice, which contains enzymes, enter the duodenum by way of ducts.

The first part of the small intestine is called the duodenum. Two important accessory glands—the **pancreas,** located behind the stomach, and the **liver**—send secretions to the duodenum by way of ducts (**Fig. 24.23**). The liver produces **bile,** which is stored in the **gallbladder.** Bile looks green because it contains pigments that are the products of hemoglobin breakdown. This green color is familiar to anyone who has observed how bruised tissue changes color. Hemoglobin within the bruise is breaking down into the same types of pigments found in bile. Bile also contains bile salts, which break up fat into fat droplets by a process called **emulsification.** Fat droplets mix with water and have more surface area for digestion by enzymes.

The pancreas produces pancreatic juice, which contains sodium bicarbonate ($NaHCO_3$) and digestive enzymes. $NaHCO_3$ neutralizes chyme and makes the pH of the small intestine slightly basic. The higher pH helps prevent autodigestion of the intestinal lining by pepsin and optimizes the pH for pancreatic enzymes. **Pancreatic amylase** digests starch to maltose; **trypsin** digests proteins to peptides; **lipase** digests fat droplets to glycerol and fatty acids; and **nuclease** digests nucleic acids to nucleotides.

Still more digestive enzymes are present in the small intestine. The wall of the small intestine contains finger-like projections called **villi** (**Fig. 24.24***b*). The epithelial cells of the villi produce **intestinal enzymes,** which remain attached to them. These enzymes complete the digestion of peptides and sugars. Peptides, which result from the first step in protein digestion, are digested by peptidase to amino acids. Maltose, which results from the first step in starch digestion, is digested by maltase to glucose. Other disaccharides, each of which is acted upon by a specific enzyme, are digested to simple sugars as well.

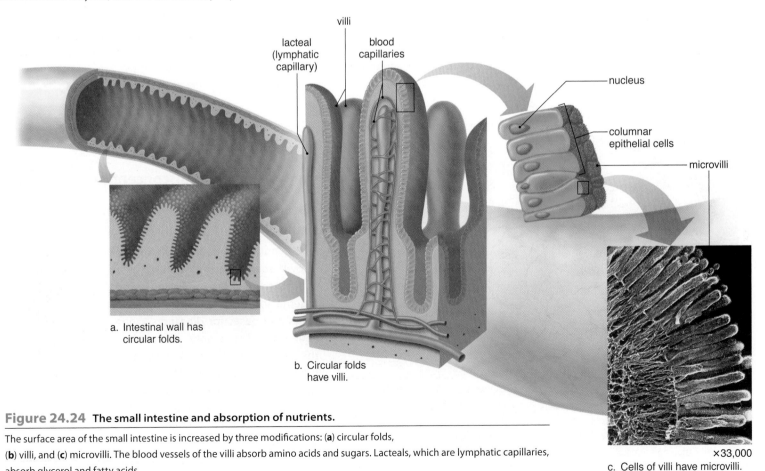

Figure 24.24   The small intestine and absorption of nutrients.

The surface area of the small intestine is increased by three modifications: (**a**) circular folds, (**b**) villi, and (**c**) microvilli. The blood vessels of the villi absorb amino acids and sugars. Lacteals, which are lymphatic capillaries, absorb glycerol and fatty acids.

a. Intestinal wall has circular folds.

b. Circular folds have villi.

×33,000
c. Cells of villi have microvilli.

Finally, these small nutrient molecules can be absorbed into the body through the bloodstream. Our cells use these molecules as a source of energy and as building blocks to make their own macromolecules.

Absorption by Villi The wall of the small intestine is adapted to absorbing nutrient molecules because it has an extensive surface area—approximately that of a tennis court! First, the mucous membrane layer of the small intestine has circular folds that give it an almost corrugated appearance (Fig. 24.24*a*). Second, on the surface of these circular folds are the villi. Finally, the cells on the surface of the villi have minute projections called **microvilli** (Fig. 24.24*c*). If the human small intestine were simply a smooth tube, it would have to be 500 to 600 meters long to have a comparable surface area for absorption. Carnivores have a much shorter digestive tract than herbivores because meat is easier to process than plant material (**Fig. 24.25**).

The villi of the small intestine absorb small nutrient molecules into the body. Each villus contains an extensive network of blood capillaries and a lymphatic capillary called a **lacteal**. As discussed in Chapter 23, the lymphatic system is an adjunct to the cardiovascular system—its vessels carry fluid, called lymph, to the cardiovascular veins. Sugars and amino acids enter the blood capillaries of a villus and are carried to the liver by way of the hepatic portal system. In contrast, glycerol and fatty acids (digested from fats) enter the epithelial cells of the villi and, within them, are joined and packaged as lipoprotein droplets, which enter a lacteal. Absorption continues until almost all nutrient molecules have been absorbed. Absorption occurs by diffusion, as well as by active transport, which requires an expenditure of cellular energy. Lymphatic vessels transport lymph to cardiovascular veins. Eventually, the bloodstream carries the nutrients absorbed by the digestive system to all the cells of the body.

 MP3 Absorption of Nutrients and Water

 Animation Enzymatic Action and the Hydrolysis of Sucrose

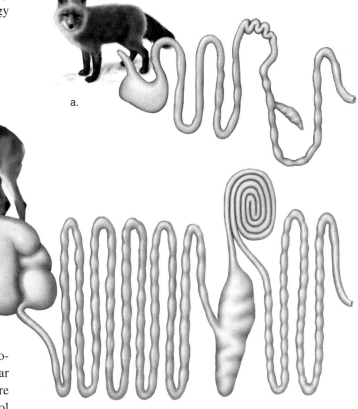

Figure 24.25 Digestive tract of a carnivore compared to that of a ruminant herbivore.

The digestive tract of a carnivore (**a**) is much shorter than that of a ruminant herbivore (**b**) because proteins can be more easily digested than plant matter.

Large Intestine

The word *bowel* technically means the part of the digestive tract between the stomach and the anus, but it is sometimes used to mean only the large intestine. The **large intestine** absorbs water, salts, and some vitamins. It also stores indigestible material until it is eliminated at the anus. The large intestine takes its name from its diameter rather than its length, which is shorter than that of the small intestine. The large intestine has a blind pouch, the cecum, below the entry of the small intestine with a small projection containing lymphatic tissue called the **appendix.** In humans, the appendix may play a role in fighting infections by acting as a reservoir of beneficial bacteria. In the condition called *appendicitis,* the appendix becomes infected and so filled with fluid that it may burst. If an infected appendix bursts before it can be removed, a serious, generalized infection of the abdominal lining, called peritonitis, may result.

The large intestine has a large population of bacteria, notably *Escherichia coli*. The bacteria break down indigestible material, and they produce some vitamins, including vitamin K. Vitamin K is necessary for blood clotting. Digestive wastes (feces) eventually leave the body through the **anus,** the opening of the anal canal. Feces are about 75% water and 25% solid matter. Almost one-third of this

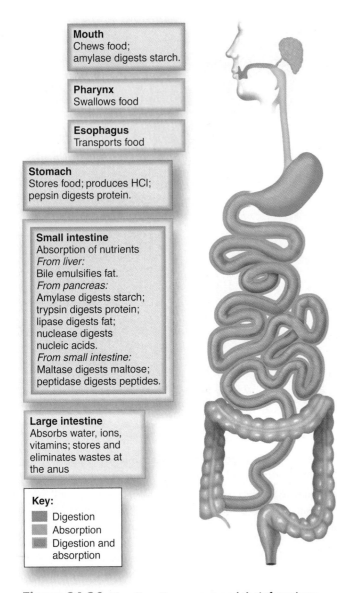

Mouth
Chews food;
amylase digests starch.

Pharynx
Swallows food

Esophagus
Transports food

Stomach
Stores food; produces HCl;
pepsin digests protein.

Small intestine
Absorption of nutrients
From liver:
Bile emulsifies fat.
From pancreas:
Amylase digests starch;
trypsin digests protein;
lipase digests fat;
nuclease digests
nucleic acids.
From small intestine:
Maltase digests maltose;
peptidase digests peptides.

Large intestine
Absorbs water, ions,
vitamins; stores and
eliminates wastes at
the anus

Key:
■ Digestion
■ Absorption
■ Digestion and
absorption

Figure 24.26 The digestive organs and their functions.

A review of the processing of food to nutrient molecules and their absorption into the body.

solid matter is made up of intestinal bacteria. The remainder is indigestible plant material (also called fiber), fats, waste products (such as bile pigments), inorganic material, mucus, and dead cells from the intestinal lining. A diet that includes fiber adds bulk to the feces, improves regularity of elimination, and prevents *constipation*.

About 1.5 liters of water enter the digestive tract daily as a result of eating and drinking. An additional 8.5 liters also enter the digestive tract each day, carrying the various substances secreted by the digestive glands. About 95% of this water is absorbed by the small intestine, and much of the remaining portion is absorbed by the large intestine. If this water is not reabsorbed, **diarrhea** can occur, leading to serious dehydration and ion loss, especially in children.

The large intestine (also called the colon) is subject to the development of **polyps,** which are small growths arising from the mucosa. Polyps, whether benign or cancerous, can be removed surgically.

Figure 24.26 reviews the digestive process.

 **Animation**
Organs of Digestion

 **MP3**
Overview of the
Digestive System

Accessory Organs

The pancreas and the liver are accessory organs of digestion, along with the teeth, salivary glands, and gallbladder.

Pancreas

The pancreas (see Fig. 24.23) functions as both an endocrine gland and an exocrine gland. **Endocrine glands** are ductless and secrete their products into the blood. The pancreas is an endocrine gland when it produces and secretes the hormones **insulin** and **glucagon** into the bloodstream. **Exocrine glands** secrete into ducts. The pancreas is an exocrine gland when it produces and secretes pancreatic juice into the **duodenum** of the small intestine through the common bile duct.

Liver

The liver has numerous functions, including the following:

1. Detoxifies the blood by removing and metabolizing poisonous substances
2. Produces the plasma proteins, such as albumin and fibrinogen
3. Destroys old red blood cells and converts hemoglobin to the breakdown products in bile (bilirubin and biliverdin)
4. Produces bile, which is stored in the gallbladder before entering the small intestine, where it emulsifies fats
5. Stores glucose as glycogen and breaks down glycogen to glucose between meals to maintain a constant glucose concentration in the blood
6. Produces urea from amino acids and ammonia

Blood vessels from the large and small intestines, as well as others, merge to form the hepatic portal vein, which leads to the liver (**Fig. 24.27**). The liver helps maintain the glucose concentration in blood at about 0.1% by removing excess glucose from the hepatic portal vein and storing it as glycogen. Between meals, glycogen is broken down to glucose, and glucose enters the hepatic veins. Like plant-made starch, glycogen is made up of glucose molecules; thus, it is sometimes called animal starch. If the supply of glycogen and glucose runs short, the liver converts amino acids to glucose molecules.

Amino acids contain nitrogen in the form of amino groups, whereas glucose contains only carbon, oxygen, and hydrogen. Therefore, before amino acids can be converted to glucose molecules, deamination, the removal of amino groups from amino acids, must occur. By a complex metabolic pathway, the liver converts the amino groups to urea, the most common nitrogenous (nitrogen-containing) waste product of humans. After urea is formed in the liver, it is transported by the bloodstream to the kidneys, where it is excreted.

Liver Disorders When a person is jaundiced, the skin has a yellowish tint due to an abnormally large quantity of bile pigments in the blood. In hemolytic jaundice, red blood cells are broken down in abnormally large amounts; in obstructive jaundice, the bile duct is obstructed or the liver cells are damaged. Obstructive jaundice often occurs when crystals of cholesterol precipitate out of bile and form gallstones.

Jaundice can also result from viral infection of the liver, called *hepatitis*. Hepatitis A is most often caused by eating contaminated food. Hepatitis B and C are commonly spread by blood transfusions, kidney dialysis, and injection with unsterilized needles. These three types of hepatitis can also be spread by sexual contact.

Cirrhosis is a chronic liver disease in which the organ first becomes fatty, and then the liver tissue is replaced by inactive fibrous scar tissue. Alcoholics often get cirrhosis, most likely due at least in part to the excessive amounts of alcohol the liver is forced to break down.

Digestive Juices

Does your mouth water when you smell food cooking? Even the thought of food can sometimes cause the nervous system to order the secretion of digestive juices. The secretion of these juices is also under the influence of several peptide hormones, so called because each is a small sequence of amino acids. When you eat a meal rich in protein, the stomach wall produces a peptide hormone that enters the bloodstream and doubles back to cause the stomach to produce more gastric juices. When protein and fat are present in the small intestine, another peptide hormone made in the intestinal wall stimulates the secretion of bile and pancreatic juices. In this way, the organs of digestion regulate their own needs.

The various digestive enzymes present in the digestive juices help break down carbohydrates, proteins, nucleic acids, and fats, the major components of food. Starch is a carbohydrate, and its digestion begins in the mouth. Saliva from the salivary glands has a neutral pH and contains salivary amylase, the first enzyme to act on starch.

$$\text{starch} + H_2O \xrightarrow{\text{salivary amylase}} \text{maltose}$$

Maltose, a disaccharide, cannot be absorbed by the intestine; additional digestive action in the small intestine converts maltose to glucose, which can be absorbed.

Protein digestion begins in the stomach. Gastric juice secreted by gastric glands has a very low pH—about 2—because it contains hydrochloric acid (HCl). Pepsin, which is also present in gastric juice, acts on protein to produce peptides.

$$\text{protein} + H_2O \xrightarrow{\text{pepsin}} \text{peptides}$$

Peptides are usually too large to be absorbed by the intestinal lining, but later they are broken down to amino acids in the small intestine.

Starch, proteins, fats, and nucleic acids are all enzymatically broken down in the small intestine (**Fig. 24.28**). Pancreatic juice, which enters the duodenum, has a basic pH because it contains sodium bicarbonate ($NaHCO_3$). One pancreatic enzyme, pancreatic amylase, digests starch (Fig. 24.28*a*).

$$\text{starch} + H_2O \xrightarrow{\text{pancreatic amylase}} \text{maltose}$$

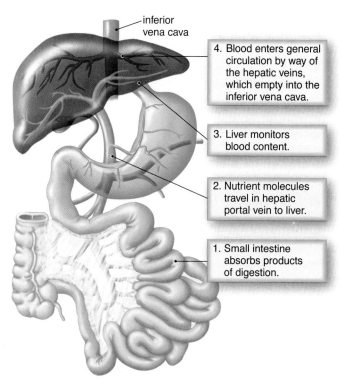

inferior vena cava

4. Blood enters general circulation by way of the hepatic veins, which empty into the inferior vena cava.

3. Liver monitors blood content.

2. Nutrient molecules travel in hepatic portal vein to liver.

1. Small intestine absorbs products of digestion.

Figure 24.27 Hepatic portal system.

The hepatic portal vein takes the products of digestion from the digestive system to the liver, where they are processed before entering hepatic veins.

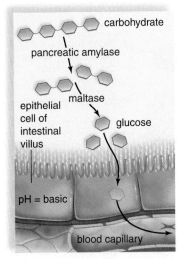

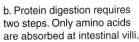

a. Carbohydrate (starch) digestion requires two steps. Only glucose molecules are absorbed at intestinal villi.

b. Protein digestion requires two steps. Only amino acids are absorbed at intestinal villi.

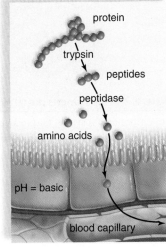

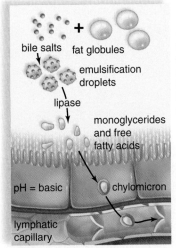

c. Fat digestion requires emulsification before digestion occurs. Then monoglycerides (glycerol bonded to one fatty acid) and free fatty acids are absorbed at intestinal villi.

Figure 24.28 **Digestion and absorption of nutrients in the small intestine.**

Another pancreatic enzyme, trypsin, digests protein (Fig. 24.28b).

$$\text{protein} + H_2O \xrightarrow{\text{trypsin}} \text{peptides}$$

Maltase and **peptidases,** enzymes produced by the small intestine, complete the digestion of starch to glucose and protein to amino acids, respectively. Glucose and amino acids are small molecules that cross into the cells of the villi and enter the blood (Fig. 24.28a, b).

Maltose, a disaccharide that results from the first step in starch digestion, is digested to glucose by maltase.

$$\text{maltose} + H_2O \xrightarrow{\text{maltase}} \text{glucose} + \text{glucose}$$

Other disaccharides have their own enzyme and are digested in the small intestine. The absence of any one of these enzymes can cause illness.

Peptides, which result from the first step in protein digestion, are digested to amino acids by peptidases.

$$\text{peptides} + H_2O \xrightarrow{\text{peptidases}} \text{amino acids}$$

Lipase, a third pancreatic enzyme, digests fat molecules in fat droplets after they have been emulsified by bile salts.

$$\text{fat droplets} + H_2O \xrightarrow{\text{lipase}} \text{glycerol} + 3 \text{ fatty acids}$$

Specifically, the end products of lipase digestion are monoglycerides (glycerol + one fatty acid) and fatty acids. These enter the cells of the villi, and within these cells they are rejoined and packaged as lipoprotein droplets, called chylomicrons. Chylomicrons enter the lacteals (Fig. 24.28c).

Connecting the Concepts

For more topics related to the content of this section, refer to the following discussions.

Section 12.2 outlines some of the more common causes of cancer.

Section 25.2 examines the dietary requirements for each class of nutrients.

Section 26.2 examines how the bacteria in the large intestine act as innate defense against disease.

Check Your Progress 24.3

1. List the functions of a digestive tract.
2. Describe peristalsis and explain why it is a needed function in digestion.
3. Describe the specializations of the stomach of ruminants.
4. Contrast the anatomy and functions of the small and large intestine.
5. Describe the relationship among the duodenum, the liver, and the pancreas.
6. Detail the activity of the following digestive enzymes: salivary amylase, pepsin, trypsin, peptidase, and lipase.

Media Study Tools

Mc Graw Hill **connect** (plus+)
|BIOLOGY

www.mhhe.com/maderessentials3

Enhance your study of this chapter with study tools and practice tests. Also ask your instructor about the resources available through ConnectPlus, including the media-rich eBook, interactive learning tools, and animations.

The Chapter in Review

Summary

24.1 Respiratory System

A respiratory system has these functions: (1) breathing, (2) external exchange of gases, and (3) internal exchange of gases.

The respiratory tract consists of several parts:

- **Nasal cavities:** In the nasal cavities of the nose, air is moistened and hairs trap debris.
- **Pharynx:** Air crosses from front to back.
- **Larynx:** The larynx contains the vocal cords.
- **Respiratory tree:** The respiratory tree consists of the trachea, bronchi, and bronchioles, which terminate in the alveoli of the lungs.

Insects have many tracheae. A tracheal system has no need for a respiratory pigment because air is delivered directly to cells.

- **Lungs:** Contain many alveoli, air sacs surrounded by a capillary network

Breathing

- **Inspiration:** The diaphragm lowers and the rib cage moves up and out; the lungs expand and air rushes in.
- **Expiration:** The diaphragm relaxes and moves up; the rib cage moves down and in; pressure in the lungs increases; air is pushed out of the lungs.

Lungs and External Exchange of Gases

- Alveoli are surrounded by pulmonary capillaries.
- CO_2 diffuses from the blood into the alveoli, and O_2 diffuses into the blood from the alveoli, because of their respective concentration gradients.
- Hemoglobin takes up oxygen.

Transport and Internal Exchange of Gases

Oxygen for cellular respiration follows this path:

- Transported by the iron portion of hemoglobin
- Exits blood at tissues by diffusion
- Given up by hemoglobin in capillaries because cellular respiration makes tissues warmer and more acidic

Carbon dioxide, from cellular respiration, follows this path:

- Enters blood by diffusion
- Taken up by red blood cells or joins with water to form carbonic acid
- Carbonic acid breaks down to H^+ and bicarbonate ion (HCO_3^-):

$$CO_2 + H_2O \longrightarrow H_2CO_3 \longrightarrow H^+ + HCO_3^-$$
carbon dioxide   water   carbonic acid   hydrogen ion   bicarbonate ion

Bicarbonate ions are carried in plasma. H^+ combines with globin of hemoglobin.

- In the lungs, H^+ joins with bicarbonate ion to form carbonic acid, which breaks down to water and carbon dioxide:

$$H^+ + HCO_3^- \longrightarrow H_2CO_3 \longrightarrow H_2O + CO_2$$

24.2 Urinary System and Excretion

The kidneys perform the following functions:

- Excrete nitrogenous wastes, such as urea and uric acid
- Maintain the normal water-salt balance of blood
- Maintain the acid-base balance of blood

The urinary system consists of these parts:

- **Kidneys:** Produce urine
- **Ureters:** Take urine to the bladder
- **Urinary bladder:** Stores urine
- **Urethra:** Releases urine to the outside

Kidneys

Macroscopically, the kidneys have three parts: renal cortex, renal medulla, and renal pelvis. Microscopically, they contain the nephrons. A nephron has a nephron capsule, proximal tubule, nephron loop, and distal tubule.

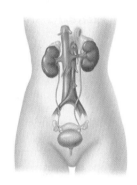

Urine formation requires three steps:

- **Filtration:** Water, nutrients, and wastes move from the blood to the inside of the nephron capsule.
- **Reabsorption:** Primarily salts, water, and nutrients are reabsorbed at the proximal tubule.
- **Secretion:** Certain substances (e.g., hydrogen ions) are transported into the distal tubule from blood.

Regulation of Water-Salt Balance and pH Animals regulate their osmolarity; examples are the Malpighian tubules of insects and the specialized organs of saltwater versus freshwater fishes. The reabsorption of water and the production of a hypertonic urine involve establishment of a solute gradient that pulls water from the descending limb of the nephron loop and from the collecting duct. The kidneys keep blood pH at about 7.4 by reabsorbing HCO_3^- and excreting H^+ as needed. Ammonia buffers H^+ in the urine.

Problems with Kidney Function

Various medical conditions, including diabetes, kidney stones, and kidney infections, can lead to renal failure. Renal failure can be treated by hemodialysis using a kidney machine or CAPD or by a kidney transplant.

24.3 Digestive System

The digestive system, along with the respiratory and urinary systems, makes exchanges with the external environment and blood, thereby supplying the blood with nutrients and oxygen and cleansing it of waste molecules. Then the blood makes exchanges with the tissue fluid, and in this way cells acquire nutrients and oxygen and rid themselves of wastes.

Tube-Within-a-Tube Body Plan

Like the earthworm, humans have a complete digestive system that (1) ingests food, (2) breaks food down to small molecules that can cross plasma membranes, (3) absorbs these nutrient molecules, and (4) eliminates indigestible remains.

The digestive tract consists of several specialized parts:

- **Mouth:** Teeth chew the food, saliva contains salivary amylase for digesting starch, and the tongue forms a bolus for swallowing.
- **Pharynx:** The air and food passages cross in the pharynx. During swallowing, the air passage is blocked off by the soft palate and epiglottis; peristalsis begins.
- **Stomach:** The stomach expands and stores food and it churns, mixing food with the acidic gastric juices. This juice contains pepsin, an enzyme that digests protein. The stomach of ruminants has a chamber, the rumen, where symbiotic bacteria and protozoans release enzymes that digest grass.
- **Small intestine:** The duodenum of the small intestine receives bile from the liver and pancreatic juice from the pancreas. Pancreatic juice contains trypsin (digests protein), lipase (digests fat), and pancreatic amylase (digests starch). The small intestine produces enzymes that finish digestion, breaking food down to small molecules that cross the villi. Amino acids and glucose enter blood capillaries. Glycerol and fatty acids are joined and packaged as lipoproteins before entering lymphatic vessels called lacteals.
- **Large intestine:** The large intestine stores the remains of digestion until they can be eliminated. It also absorbs water, salts, and some vitamins. Reduced water absorption results in diarrhea. The intake of water and fiber helps prevent constipation.

Accessory Organs

The pancreas is both an exocrine gland that produces pancreatic juice and an endocrine gland that produces the hormones insulin and glucagon. The liver produces bile, which is stored in the gallbladder. Pancreatic juice and bile enter the small intestine. The nervous system

and the peptide hormones regulate the secretion of digestive juices and bile.

The hepatic portal vein carries absorbed molecules from the small intestine to the liver, an organ that performs many important functions.

Key Terms

abdominal cavity   454
alveolus (pl., alveoli)   453
anus   467
aortic body   454
appendix   467
bicarbonate ion   456
bile   466
bolus   463
bronchiole   453
bronchus (pl., bronchi)   453
carnivore   463
carotid body   454
chyme   465
coelom   463
collecting duct   458
continuous ambulatory
    peritoneal dialysis (CAPD)   460
dialysate   460
diaphragm   454
diarrhea   468
distal tubule   458
duodenum   468
edema   460
emulsification   466
endocrine gland   468
epiglottis   464
esophagus   464
exocrine gland   468
filtration   458
gallbladder   466
glottis   453
glucagon   468
heme   456
hemodialysis   460
herbivore   463
insulin   468
intestinal enzyme   466
kidneys   457
lacteal   467
large intestine   467
larynx   452

lipase   466
liver   466
lungs   452
Malpighian tubule   459
maltase   470
microvillus (pl., microvilli)   467
mouth   463
nasal cavity   452
nephron   458
nephron capsule   458
nephron loop   458
nuclease   466
omnivore   463
pancreas   466
pancreatic amylase   466
pepsin   464
peptidase   470
peristalsis   464
pharynx   452
polyp   468
proximal tubule   458
reabsorption   458
renal cortex   458
renal medulla   458
renal pelvis   458
respiration   451
salivary amylase   463
salivary gland   463
secretion   459
sinus   452
small intestine   465
stomach   464
thoracic cavity   454
trachea (pl., tracheae)   453
trypsin   466
ureter   457
urethra   457
urinary bladder   457
urine   457
villus (pl., villi)   466
vocal cord   453

Testing Yourself

Choose the best answer for each question.

1. Label the components of the human respiratory system in the following illustration.

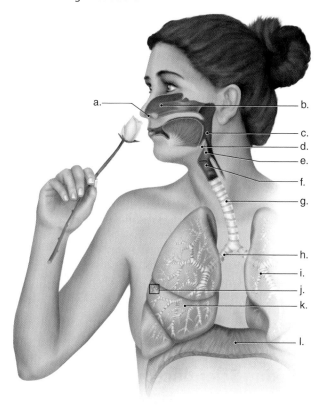

2. In insects, _____ are small air tubes that carry air into every part of the body.
   a. bronchi
   c. tracheae
   b. spiracles
   d. alveoli

3. Birds have a higher gas-exchange capacity than other vertebrates because they
   a. have more alveoli in their lungs.
   b. have a stronger diaphragm.
   c. have a shorter trachea.
   d. never mix fresh air with used air.
   e. breathe at a more rapid rate.

4. Without this structure, the larynx cannot connect to the bronchi.
   a. glottis
   c. pharynx
   b. trachea
   d. epiglottis

5. Gas exchange occurs in the _____, small pockets in the bronchioles.
   a. alveoli
   c. tonsils
   b. tracheae
   d. nephrons

6. The larynx is
   a. the site of gas exchange in the human respiratory system.
   b. located between the trachea and the bronchi.
   c. a flap of cartilage that keeps food out of the respiratory tract during swallowing.
   d. the voice box.

7. Heme
   a. is the protein portion of a hemoglobin molecule.
   b. is the lead-containing group in a hemoglobin molecule.
   c. is the portion of a hemoglobin molecule that directly combines with oxygen molecules.
   d. floats free in blood plasma.

8. Label the components of the human urinary system in the following illustration.

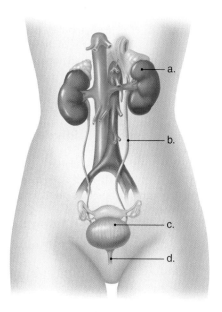

For questions 9–12, identify the kidney component in the key that matches the description.

Key:
   a. nephron capsule
   b. proximal tubule
   c. nephron loop
   d. distal tubule

9. This pumps salt into the renal medulla, and water follows by osmosis.

10. Uric acid is eliminated by secretion here.

11. Microvilli reabsorb molecules, such as glucose, here.

12. Filtration occurs here.

13. The presence of albumin in the urine is indicative of
   a. edema.
   b. exposure to diuretics.
   c. the inability to adjust blood pH.
   d. hypertension.

14. _____ are the excretory organs of insects.
   a. Malpighian tubules
   c. Kidneys
   b. Tracheae
   d. Nephrons

15. Cleansing inhaled air involves
   a. hairs.
   c. mucus.
   b. cilia.
   d. All of these are correct.

16. Both food and air travel through the
   a. lungs.
   c. larynx.
   b. pharynx.
   d. trachea.

In questions 17–20, match the functions to the structures in the key.

Key:

    a. glottis   **c.** bronchi

    b. larynx   **d.** lungs

17. passage of air into larynx
18. passage of air into lungs
19. sound production
20. gas exchange

21. The use of an artificial kidney is known as
    - **a.** filtration.
    - **b.** excretion.
    - **c.** hemodialysis.
    - **d.** None of these are correct.

22. Which of the following materials would not be filtered from the blood at the nephron capsule?
    - **a.** water
    - **b.** urea
    - **c.** protein
    - **d.** nutrients
    - **e.** salts

23. The renal medulla has a striped appearance due to the presence of which structure(s)?
    - **a.** nephron loop
    - **b.** collecting duct
    - **c.** capillaries
    - **d.** Both a and b are correct.

24. Label the components of the human digestive system in the following illustration.

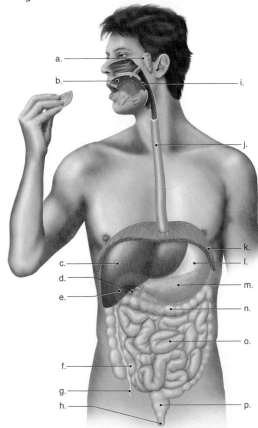

25. In the pharynx,
    - **a.** the digestive and respiratory passages come together.
    - **b.** the glottis covers the epiglottis during swallowing.
    - **c.** the digestive and urinary passages come together.
    - **d.** chyme passes into the esophagus.

26. If plants did not have cellulose in their cell walls,
    - **a.** an important nutrient would be missing in the diet of omnivores.
    - **b.** the teeth of herbivores would more closely resemble those of carnivores.
    - **c.** more birds would be carnivores.
    - **d.** herbivores would have a longer digestive tract.

27. The stomach
    - **a.** is lined with a thick layer of mucus.
    - **b.** contains sphincter glands.
    - **c.** has a pH of about 6.
    - **d.** digests pepsin.
    - **e.** More than one of these are correct.

28. A rumen in cows
    - **a.** provides them with water.
    - **b.** provides a home for symbiotic bacteria.
    - **c.** helps them digest plant material.
    - **d.** Both b and c are correct.

29. Pepsin
    - **a.** breaks down protein in the small intestine.
    - **b.** breaks down protein in the stomach.
    - **c.** is found in saliva.
    - **d.** breaks down fats in the stomach.

30. The contents of the digestive tract are in the form of _____ when they pass from the stomach to the small intestine.
    - **a.** chyme
    - **b.** a bolus
    - **c.** feces
    - **d.** a small, round mass

31. Which of the following is not a function of the liver?
    - **a.** removal of poisonous substances from the blood
    - **b.** secretion of digestive juices
    - **c.** production of albumin
    - **d.** storage of glucose
    - **e.** production of bile

Thinking Scientifically

1. Currently, the only way people with diabetes mellitus can take their insulin is as an injection; no insulin pill is available. Taking into account what happens in the stomach and small intestine in terms of digestion and absorption, and that insulin is a protein hormone, explain why insulin cannot be taken in pill form.

2. Individuals who have had their gallbladders removed are told to lower the fat intake in their diet after surgery. What effect will removal of the gallbladder have on their ability to digest and absorb fats?

3. Grasshoppers, like most other insects, are highly adapted to life on land. What adaptations to life on land are present in their respiratory and urinary systems?

Bioethical Issue

When to Terminate Life

As a result of serious injury or illness, a person may enter a persistent vegetative state. In this condition, the person does not recognize anyone and feels no pain. Advances in medical care now allow us to use ventilators and feeding tubes to keep vegetative patients alive for years. A living will allows patients to declare that they do not wish to be maintained in this way, giving doctors the legal authority to discontinue life-sustaining treatments. But for a patient without a living will, family members are left with a difficult decision and sometimes a court battle over what they all think is best. One recent case involved a woman who was kept in a vegetative state since suffering a heart attack in 1990. If her feeding tube were removed, she would die within two weeks. Her husband asked to have the tube removed, saying that would have been his wife's wish. Her parents, on the other hand, wanted her to be kept alive. They believed she was trying to communicate with noises and facial expressions, although her doctors said those activities were simply reflexes. What would you do if you were a judge and had to determine whether the husband's or the parents' wishes should be carried out?

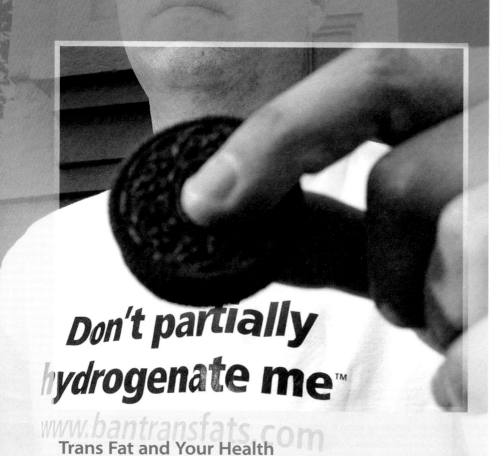

Trans Fat and Your Health

Did you know that an extremely high-protein diet may eventually cause liver and kidney disease? That some vitamins can harm the body if taken in extreme excess? Or that trans fats are the worst possible kind of fat you can eat?

For some time now, we have known that cardiovascular health problems can result from eating high-fat and high-cholesterol foods. Further, saturated fats had to be listed on the Nutrition Facts panel of all food packages, and many consumers became smart enough to check nutrition labels to avoid eating saturated fats. Enter trans fats. *Trans fats* are formed from liquid unsaturated fats when hydrogen has been added to them. This process, called hydrogenation, makes a liquid fat resemble a solid saturated fat. Why did manufacturers go through the trouble of creating trans fats when they could have just used a natural saturated fat? Well, we know that, until recently, the trans fat content of food did not have to be listed on nutrition labels. Therefore, a food product could appear healthier than it actually was if trans fats, instead of saturated fats, were used in processed foods.

But it turns out that trans fats are worse offenders than saturated fats when it comes to cardiovascular health. In 2006, the Food and Drug Administration (FDA) decided that trans fat content *must* be disclosed on Nutrition Facts panels. Now, many manufacturers of products high in trans fats have lowered the trans fat content of their foods, but consumers should still check the nutrition labels for trans fat content.

In this chapter, you will learn about the components of a healthy, balanced diet, as well as some of the consequences of an unhealthy one.

25

Human Nutrition

OUTLINE

BEFORE YOU BEGIN

Before beginning this chapter, take a few moments to review the following discussions.

Section 2.2 What are the properties of water that make it an important nutrient for all life?

Section 3.2 What are the roles of carbohydrates, lipids, and proteins in a cell?

Section 7.1 How does aerobic respiration release the energy found in organic molecules, such as glucose?

Figure 25.1 A healthy diet.

Making the right food choices leads to better health.

25.1 Nutrition

The vigilance of your immune system, the strength of your muscles and bones, the ease with which your blood circulates—all aspects of your body's functioning—depend on proper **nutrition (Fig. 25.1)**. A **nutrient** is a component of food that performs a physiological function in the body. Nutrients provide us with energy, promote growth and development by supplying the material for cellular structures, and regulate cellular metabolism. They are also involved in homeostasis. For example, nutrients help maintain the fluid balance and proper pH of blood. Your body can make up for a nutrient deficiency to a degree, but eventually signs and symptoms of a **deficiency disorder** will appear. As an example, vitamin C is needed to synthesize and maintain *collagen,* the protein that holds tissues together. When the body lacks vitamin C, collagen weakens and capillaries break easily. Gums may bleed, especially when the teeth are brushed, or tiny bruises may form under the skin when it is gently pressed. In other words, early signs of vitamin C deficiency are gums that bleed and skin that bruises easily **(Fig. 25.2)**.

By learning about nutrition, you can improve your diet and increase the likelihood of enjoying a longer, more active, and productive life. Conversely, poor diet and lack of physical activity are responsible for 400,000 deaths in the United States annually. Such lifestyle factors may soon overtake smoking as the major cause of preventable death. We all can benefit from learning what constitutes a poor diet versus a healthy diet, so that we can choose foods that supply all the nutrients in proper balance.

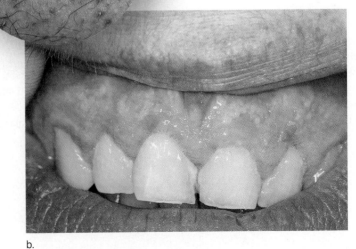

a.

b.

Figure 25.2 Vitamin C deficiency.

a. Pinpoint hemorrhages (tiny bruises that appear as red spots in the skin) are an early indication of vitamin C deficiency. **b.** Bleeding gums are another early sign of vitamin C deficiency.

Introducing the Nutrients

A **diet** is composed of a person's typical food choices. Several factors, including cultural and ethnic backgrounds, financial situations, environmental conditions, and psychological states, influence what we eat. A *balanced diet* supplies all the nutrients in the proper proportions necessary for a healthy, functioning body.

There are six classes of nutrients: carbohydrates, lipids, proteins, minerals, vitamins, and water. An **essential nutrient** must be supplied by the diet because the body is not able to produce it, or at least not in sufficient quantity to meet the body's needs. For example, amino acids are needed for protein synthesis and the body is unable to produce certain of them. Therefore, there are essential amino acids needed in the diet.

Carbohydrates, lipids, and proteins are called **macronutrients** because the body requires relatively large quantities of them. **Micronutrients,** such as vitamins and minerals, are needed in small quantities only. Macronutrients, not micronutrients, supply our energy needs. Although advertisements often imply that people can boost their energy levels by taking vitamin or mineral supplements, the body does not metabolize these nutrients for energy. Water does not provide energy, either. Therefore, foods with high water content, such as fruits and vegetables, are usually lower in energy content than foods with less water and more macronutrient content.

Nearly every food is a mixture of nutrients. A slice of bread, for example, is about 50% carbohydrates, 35% water, 10% proteins, and 4% lipids. Vitamins and minerals make up less than 1% of the bread's nutrient content (**Fig. 25.3**). No single, naturally occurring food contains enough essential nutrients to meet all of our nutrient needs.

Contrary to popular belief, so-called bad foods or junk foods do have nutritional value. If a food contains water, sugar, or fat, it has nutritional value. However, such foods as sugar-sweetened soft drinks, cookies, and pastries have high amounts of fat and/or sugar in relation to their vitamin and mineral content. Therefore, these foods are more appropriately called *empty-calorie foods,* rather than junk foods. Diets that contain too many empty-calorie foods will assuredly lack enough vitamins and minerals.

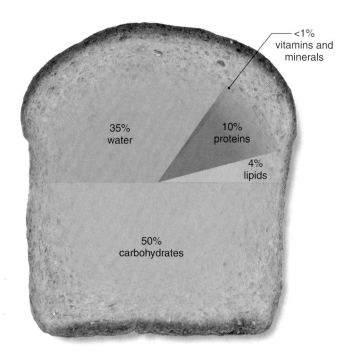

Figure 25.3 **Nutrient composition of a slice of bread.**

Most foods are mixtures of nutrients.

Connecting the Concepts

For more information on the nutrients described in this section, refer to the following discussions.

Section 3.2 describes the structure and function of carbohydrates, lipids, and proteins in a cell.

Section 7.1 provides an overview of how aerobic respiration releases the energy of macronutrients, such as carbohydrates.

Check Your Progress 25.1

1 List the classes of nutrients that do not provide energy.

2 Explain why the term *junk food* may be inappropriate.

3 Describe an empty-calorie food, and explain why it fits this description.

Figure 25.4 Major food sources.

Carbohydrates, lipids, and proteins are represented.

25.2 The Classes of Nutrients

Learning Outcomes

Upon completion of this section, you should be able to

1. List the dietary sources of carbohydrates.
2. Explain the importance of fiber in the diet.
3. List the types of lipids and their dietary sources.
4. List the dietary sources of protein.
5. Give the dietary sources of the major and trace minerals.
6. Define the term *vitamin* and list the essential vitamins.
7. Explain why the intake of water is important.

Each type of nutrient usually has more than one function in the body and can be supplied by several food sources (**Fig. 25.4** and **Table 25.1**).

Carbohydrates

Carbohydrates are present in food as sugars, starch, and fiber. Fruits, vegetables, milk, and honey are natural sources of sugars. Glucose and fructose are monosaccharide sugars, and lactose (milk sugar) and sucrose (table sugar) are disaccharides. After absorption into the body, all sugars are converted to glucose for transport in the blood. Glucose is the preferred direct energy source in cells.

Plants store glucose as starch, and animals store glucose as glycogen. High-starch foods are beans, peas, cereal grains, and potatoes. Starch is digested to glucose in the digestive tract, and any excess glucose is stored as glycogen. Although other animals likewise store glucose as glycogen in liver or muscle tissue (meat), it is gone by the time an animal is eaten for food. Except for honey and milk, which contain sugars, animal foods do not contain carbohydrates.

Table 25.1 Summarizing the Classes of Nutrients

Class of Nutrient	Major Food Sources	Primary Physiological Roles	Macronutrient or Micronutrient
Carbohydrates	Sugars and starches in fruits, vegetables, cereals and other grains	Glucose is metabolized for energy; fiber adds to fecal bulk, preventing constipation.	Macronutrient
Lipids	Oil, margarine, salad dressings, meat, fish, poultry, nuts, fried foods, dairy products made with whole milk	Triglycerides are metabolized for energy and stored for insulation and protection of organs. Cholesterol is used to make certain hormones, bile, and vitamin D.	Macronutrient
Proteins	Meat, fish, poultry, eggs, nuts, dried beans and peas, soybeans, cereal products, milk, cheese, yogurt	Proteins are needed for building, repairing, and maintaining tissues; synthesizing enzymes and certain hormones; and producing antibodies.	Macronutrient
Minerals	Widespread in vegetables and other foods	Minerals regulate energy metabolism, maintain fluid balance, and produce certain structures, enzymes, and hormones.	Micronutrient
Vitamins	Widespread in foods; plant foods are good sources of antioxidants	Vitamins regulate metabolism and physiological development.	Micronutrient
Water	Widespread in foods, beverages	Water participates in many chemical reactions; it is needed for maintenance of body fluids, temperature regulation, joint lubrication, and transportation of material in cells and the body.	(Not applicable)

Fiber

Fiber includes various nondigestible carbohydrates derived from plants. Food sources rich in fiber include beans, peas, nuts, fruits, and vegetables. Whole-grain products are also a good source of fiber and are therefore more nutritious than food products made from refined grains. During *refinement,* fiber and vitamins and minerals are removed from grains, so that primarily starch remains. A slice of bread made from whole-wheat flour, for example, contains 3 grams (g) of fiber; a slice of bread made from refined wheat flour contains less than a gram of fiber.

Technically, fiber is not a nutrient for humans because it cannot be digested to small molecules that enter the bloodstream. Insoluble fiber, however, adds bulk to fecal material, which stimulates movements of the large intestine, preventing constipation. Soluble fiber combines with bile acids and cholesterol in the small intestine and prevents them from being absorbed. In this way, high-fiber diets may protect against heart disease. The typical American consumes only about 15 g of fiber each day; the recommended daily intake is 25 g for women and 38 g for men. To increase your fiber intake, eat whole-grain foods, snack on fresh fruits and raw vegetables, and include nuts and beans in your diet (**Fig. 25.5**).

Can Carbohydrates Be Harmful?

If you or someone you know has lost weight by following the Atkins or South Beach diet, you may think "carbs" are unhealthy and should be avoided. According to nutritionists, however, carbohydrates should supply a large portion of your energy needs. Evidence suggests that Americans are not eating the right kind of carbohydrates. In some countries, the traditional diet is 60–70% high-fiber carbohydrates, and these people have a low incidence of the diseases that plague Americans.

Obesity is associated with *type 2 diabetes* and *cardiovascular disease,* as discussed in Section 25.3. Some nutritionists hypothesize that the high intake of foods, such as those rich in refined carbohydrates and fructose sweeteners processed from cornstarch, may be responsible for the prevalence of obesity in the United States. These are empty-calorie foods that provide sugars but no vitamins or minerals. **Table 25.2** tells you how to reduce dietary sugars in your diet. Other nutritionists point out that consuming too much energy from any source contributes to body fat, which increases a person's risk of obesity and associated illnesses. Because many foods, such as doughnuts, cakes, pies, and cookies, are high in both refined carbohydrates and fat, it is difficult to determine which dietary component is responsible for the current epidemic of obesity among Americans.

Many people mistakenly believe children become hyperactive after eating sugar. There is no scientific basis for this belief because sucrose is broken down into glucose and fructose in the small intestine, and these sugars are absorbed into the bloodstream. The spike caused by their absorption does not last long. Excess glucose and fructose enters the liver, and fructose is converted to glucose. As you know, the liver stores glucose as glycogen, and glycogen is broken down to maintain the proper glucose level.

Lipids

Like carbohydrates, **triglycerides** (fats and oils) supply energy for cells, but **fat** is stored for the long term in the body. Fat deposits under the skin, called *subcutaneous fat,* insulate the body from cold temperatures; deeper fat deposits in the trunk protect organs against injury.

Figure 25.5 Fiber-rich foods.

Plants are a good source of carbohydrates. They also provide fiber when they are not processed (refined).

Table 25.2 Reducing Dietary Sugars

TO REDUCE DIETARY SUGAR:
1. Eat fewer sweets, such as candy, soft drinks, ice cream, and pastry.
2. Eat fresh fruits or fruits canned without heavy syrup.
3. Use less sugar—white, brown, or raw—and less honey and syrups.
4. Avoid sweetened breakfast cereals.
5. Eat less jelly, jam, and preserves.
6. Drink pure fruit juices, not imitations.
7. When cooking, use spices such as cinnamon, instead of sugar, to flavor foods.
8. Do not put sugar in tea or coffee.
9. Avoid potatoes and processed foods made from refined carbohydrates, such as white bread, rice, and pasta. These foods are immediately broken down to sugar during digestion.

Table 25.3 Reducing Certain Lipids

TO REDUCE DIETARY SATURATED FAT:

1. Choose poultry, fish, or dry beans and peas as a protein source.

2. Remove skin from poultry and trim fat from red meats before cooking; place on a rack, so that fat drains off.

3. Broil, boil, or bake rather than frying.

4. Limit your intake of butter, cream, trans fats, shortening, and tropical oils (coconut and palm oils).*

5. To season vegetables, use herbs and spices instead of butter, margarine, or sauces. Use lemon juice instead of salad dressing.

6. Drink skim milk instead of whole milk, and use skim milk in cooking and baking.

TO REDUCE DIETARY CHOLESTEROL:

1. Eat white fish and poultry in preference to cheese, egg yolks, liver, and certain shellfish (shrimp and lobster).

2. Substitute egg whites for egg yolks in both cooking and eating.

3. Include soluble fiber in the diet. Oat bran, oatmeal, beans, corn, and fruits, such as apples, citrus fruits, and cranberries, are high in soluble fiber.

*Although coconut and palm oils are from plant sources, they are mostly saturated fat.

Connections and Misconceptions

Does "0% trans fat" on a food label really mean that the item is free of trans fat?

When the Food and Drug Administration required the addition of trans fat information to food labels in 2006, many food companies created labels touting their products as "trans fat free." A check of the label details would list 0 grams of trans fat, in the area where fat grams are listed. But a more thorough check of the list of ingredients might reveal a bit of trans fat lurking in the food. If you see "partially hydrogenated oil" listed with the ingredients, there are some trans fats in that food. Trans fats have to be listed with the breakdown of fat grams only when there is 0.5 gram (g) or more per serving. Limiting trans fats to 1% of daily calories is recommended by the American Heart Association. Unfortunately, eating more than one serving of a food with "hidden" trans fats might push some people over the recommended daily intake of trans fats.

Nutrition Facts

Serving Size: 1 cup (228g)
Servings Per Container about 2

Amount Per Serving	
Calories 260	**Calories from Fat** 120

	% Daily Value*
Total Fat 13g	20%
Saturated Fat 5g	25%
Trans Fat 0g	0%
Cholesterol 30mg	10%
Sodium 660mg	28%
Total Carbohydrate 31g	10%
Dietary Fiber 0g	0%
Sugars 5g	
Protein 5g	

Nutritionists generally recommend that unsaturated rather than saturated fats be included in the diet. Two unsaturated fatty acids (alpha-linolenic and linoleic acids) are *essential dietary fatty acids.* Delayed growth and skin problems can develop when the diet lacks these essential fatty acids, which can be supplied by eating fatty fish and by including plant oils, such as canola and soybean oils, in the diet.

Animal foods such as butter, meat, whole milk, and cheeses contain saturated fats. Plant oils contain unsaturated fats. Each type of oil has a particular percentage of monounsaturated and polyunsaturated fatty acids (see Fig. 3.13).

Cholesterol, a lipid, can be synthesized by the body. Cells use cholesterol to make various compounds, including bile, steroid hormones, and vitamin D. Plant foods do not contain cholesterol; only animal foods, such as cheese, egg yolks, liver, and certain shellfish (shrimp and lobster), are rich in cholesterol.

Can Lipids Be Harmful?

Elevated blood cholesterol levels are associated with an increased risk of cardiovascular disease, the number one killer of Americans. A diet rich in cholesterol and saturated fats increases the risk of cardiovascular disease (see also Section 25.3). Statistical studies suggest that trans fats are even more harmful than saturated fats. Trans fats arise when unsaturated oils are hydrogenated to produce a solid fat found largely in processed foods. Trans fatty acids may reduce the function of the plasma membrane receptors that clear cholesterol from the bloodstream. Trans fatty acids are found in commercially packaged foods, such as cookies and crackers; in commercially fried foods, such as french fries; in packaged snacks, such as microwave popcorn; and in vegetable shortening and some margarines. **Table 25.3** tells you how to reduce harmful lipids in the diet.

Proteins

Dietary **proteins** are digested to amino acids, which cells use to synthesize hundreds of cellular proteins. Of the 20 different amino acids, 9 are *essential amino acids* that must be present in the diet. Children will not grow if their diets lack the essential amino acids. Eggs, milk products, meat, poultry, and most other foods derived from animals contain all 9 essential amino acids and are "complete," or "high-quality," protein sources.

Foods derived from plants generally do not have as much protein per serving as those derived from animals, and each type of plant food generally lacks one or more of the essential amino acids. Therefore, most plant foods are "incomplete," or "low-quality," protein sources. Vegetarians, however, do not have to rely on animal sources of protein. To meet their protein needs, total vegetarians (vegans) can eat grains, beans, and nuts in various combinations (**Fig. 25.6**). Also, tofu, soymilk, and other foods made from processed soybeans are complete protein sources. A balanced vegetarian diet is quite possible with a little planning.

Can Proteins Be Harmful?

According to nutritionists, protein should not supply the bulk of dietary calories. The average American eats about twice as much protein as he or she needs, and some people may be on a diet that encourages the intake of proteins instead of carbohydrates as an energy source. Also, body builders, weight lifters, and other athletes may include amino acid or protein supplements in the diet because they think these

Animation
Protein
and Muscle
Production

supplements will increase muscle mass. However, excess amino acids are not always converted into muscle tissue. When they are used as an energy source, the liver removes the nitrogen portion (*deamination*) and uses it to form urea, which is excreted in urine. The water needed for the excretion of urea can cause dehydration when a person is exercising and losing water by sweating. High-protein diets can also increase calcium loss in the urine and encourage the formation of kidney stones. Furthermore, many high-protein foods contain a high amount of fat.

Minerals

The body needs about 20 elements called **minerals** for numerous physiological roles, including regulation of biochemical reactions, maintenance of fluid balance, and incorporation into certain structures and compounds. **Major minerals** are needed in the body at a higher level than **trace minerals.** More than 100 milligrams (mg) per day of each major mineral is required in the diet, and less than 100 mg per day of each trace mineral is required in the diet. **Table 25.4** lists the major minerals and a sample of the trace minerals. Their

Figure 25.6 **Beans as food.**

Beans are a good source of protein, but they do not supply all nine of the essential amino acids. To ensure a complete source of protein in the diet, they should be eaten in combination with a grain, such as rice.

Table 25.4 Minerals

Mineral	Function(s)	Food Source(s)	Conditions Caused By:	
			Too Little	**Too Much**
Major Minerals				
Calcium (Ca^{2+})	Strong bones and teeth, nerve conduction, muscle contraction	Dairy products, leafy green vegetables	Stunted growth in children, low bone density in adults	Kidney stones; interferes with iron and zinc absorption
Phosphorus (PO_4^{3-})	Bone and soft tissue growth; part of phospholipids, ATP, and nucleic acids	Meat, dairy products, sunflower seeds, food additives	Weakness, confusion, pain in bones and joints	Low blood and bone calcium levels
Potassium (K^+)	Nerve conduction, muscle contraction	Many fruits and vegetables, bran	Paralysis, irregular heartbeat, eventual death	Vomiting, heart attack, death
Sodium (Na^+)	Nerve conduction, pH and water balance	Table salt	Lethargy, muscle cramps, loss of appetite	High blood pressure, calcium loss
Chloride (Cl^-)	Water balance	Table salt	Not likely	Vomiting, dehydration
Magnesium (Mg^{2+})	Part of various enzymes for nerve and muscle contraction, protein synthesis	Whole grains, leafy green vegetables	Muscle spasm, irregular heartbeat, convulsions, confusion, personality changes	Diarrhea
Sulfur (S^{2-})	Part of some amino acids and some vitamins	Meats, legumes, whole grains	Not likely	Not likely
Trace Minerals				
Zinc (Zn^{2+})	Protein synthesis, wound healing, fetal development and growth, immune function	Meats, legumes, whole grains	Delayed wound healing, night blindness, diarrhea, mental lethargy	Anemia, diarrhea, vomiting, renal failure, abnormal cholesterol levels
Iron (Fe^{2+})	Hemoglobin synthesis	Whole grains, meats, prune juice	Anemia, physical and mental sluggishness	Iron toxicity disease, organ failure, eventual death
Copper (Cu^{2+})	Iron metabolism; part of various antioxidant enzymes	Meat, nuts, legumes	Anemia, stunted growth in children	Damage to internal organs if not excreted
Iodine (I^-)	Thyroid hormone synthesis	Iodized table salt, seafood	Thyroid deficiency	Depressed thyroid function, anxiety
Selenium (SeO_4^{2-})	Part of antioxidant enzyme	Seafood, meats, eggs	Vascular collapse, possible cancer development	Hair and fingernail loss, discolored skin

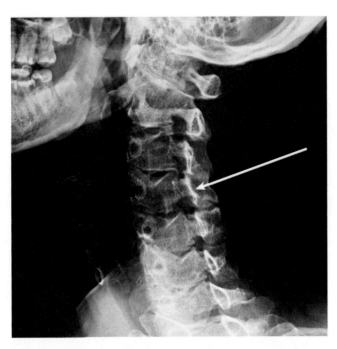

Figure 25.7 **An X-ray of an individual with osteoporosis.**

The fractures in this X-ray are caused by osteoporosis, which reduces the thickness and strength of bones.

Connections and Misconceptions

Where does most of the sodium in the diet come from?

Contrary to popular belief, the majority of the sodium in your diet does not come from the salt you put on your food when you are eating. Instead, most of our sodium (over three-quarters!) comes from processed foods and condiments. Sodium is used in these items both to preserve the food or condiment and to make it taste better. But sometimes the amount of sodium is phenomenal. A single teaspoon of soy sauce contains almost 1,000 mg of sodium, and a half-cup of prepared tomato sauce typically has over 400 mg of sodium. Websites such as nutritiondata.com can help you track your daily sodium intake.

6%   5%

12%

77%

Key:
From processed foods
In foods naturally
Added at eating
Added at cooking

functions and food sources are given, as well as the health effects of too little or too much intake.

Some individuals (especially women) do not get enough iron, calcium, magnesium, or zinc in their diets. Adult females need more iron in the diet than males (18 mg compared to 10 mg) if they are menstruating each month. *Anemia,* characterized by a run-down feeling due to insufficient red blood cells, results when the diet lacks iron. Also, many people take calcium supplements as directed by a physician to counteract *osteoporosis,* a degenerative bone disease (**Fig. 25.7**) that affects an estimated one-quarter of older men and one-half of older women in the United States.

Animation
Osteoporosis

One mineral that people consume too much of is sodium. The recommended amount of sodium intake per day is 2,300 mg (1,500 mg if you have high blood pressure), although the average American takes in 3,400–4,000 mg each day. The American Heart Association recommends that you aim to decrease your sodium intake to less than 1,500 mg per day. **Table 25.5** gives recommendations for reducing the amount of sodium in the diet.

Vitamins

Vitamins are organic compounds (other than carbohydrates, fats, and proteins, including amino acids) that regulate various metabolic activities. **Tables 25.6** and **25.7** list the major vitamins and some of their functions and food sources.

Although many people think vitamins can enhance health dramatically, prevent aging, and cure diseases such as arthritis and cancer, there is no scientific evidence that vitamins are "wonder drugs." However, vitamins C, E, and A have been shown to defend the body against free radicals, and therefore they are termed **antioxidants.** These vitamins are especially abundant in fruits and vegetables, so it is suggested that we eat about 4½ cups of fruits and vegetables per day. To achieve this goal, we should consume salad greens, raw or cooked vegetables, dried fruit, and fruit juice in addition to traditional apples and oranges and other fresh foods.

Video
Apples for
Alzheimers

Vitamin deficiencies can lead to disorders, and even death, in humans. Although many foods in the United States are now enriched, or fortified, with vitamins, some individuals, especially the elderly, young children, alcoholics, and low-income people, are still at risk for vitamin deficiencies, generally as a result of poor food choices. For example, skin cells normally contain a precursor cholesterol molecule that is converted to vitamin D after UV exposure. But a vitamin D deficiency leads to a condition called rickets, in which defective mineralization of the skeleton causes bowing of the legs. Most milk today is

Table 25.5 **Reducing Dietary Sodium**

TO REDUCE DIETARY SODIUM:

1. Use spices instead of salt to flavor foods.

2. Add little or no salt to foods at the table, and add only small amounts of salt when you cook.

3. Eat unsalted crackers, pretzels, potato chips, nuts, and popcorn.

4. Avoid hot dogs, ham, bacon, luncheon meats, smoked salmon, sardines, and anchovies.

5. Avoid processed cheese and canned or dehydrated soups.

6. Avoid brine-soaked foods, such as pickles and olives.

7. Read labels to avoid high-salt products.

fortified with vitamin D, which helps prevent the occurrence of rickets. Another example is vitamin C deficiency, the effects of which are illustrated in Figure 25.2. If a diet involves high alcohol consumption, then vitamin deficiencies may occur even if the intake of vitamins is adequate. Deficiencies occur because alcohol interferes with the absorption of certain vitamins, such as vitamin B_{12}, folacin, and vitamin A, and it increases the excretion of other vitamins, such as vitamin C.

Animation
B Vitamins

Table 25.6 Water-Soluble Vitamins

Vitamin	Functions	Food Sources	Conditions Caused By:	
			Too Little	Too Much
Vitamin C	Antioxidant; needed for forming collagen; helps maintain capillaries, bones, and teeth	Citrus fruits, leafy green vegetables, tomatoes, potatoes, cabbage	Scurvy, delayed wound healing, infections	Gout, kidney stones, diarrhea, decreased copper
Thiamine (vitamin B_1)	Part of coenzyme needed for cellular respiration; promotes activity of the nervous system	Whole-grain cereals, dried beans and peas, sunflower seeds, nuts	Beriberi, muscular weakness, enlarged heart	Can interfere with absorption of other vitamins
Riboflavin (vitamin B_2)	Part of coenzymes, such as FAD; aids cellular respiration, including oxidation of protein and fat	Nuts, dairy products, whole-grain cereals, poultry, leafy green vegetables	Dermatitis, blurred vision, growth failure	Unknown
Niacin (nicotinic acid)	Part of coenzymes NAD and NADP; needed for cellular respiration, including oxidation of protein and fat	Peanuts, poultry, whole-grain cereals, leafy green vegetables, beans	Pellagra, diarrhea, mental disorders	High blood sugar and uric acid, vasodilation
Folacin (folic acid)	Coenzyme needed for production of hemoglobin and formation of DNA	Leafy dark green vegetables, nuts, beans, whole-grain cereals	Megaloblastic anemia, spina bifida	May mask B_{12} deficiency
Vitamin B_6	Coenzyme needed for synthesis of hormones and hemoglobin; CNS control	Whole-grain cereals, bananas, beans, poultry, nuts, leafy green vegetables	Rarely, convulsions, vomiting, seborrhea, muscular weakness	Insomnia, neuropathy
Pantothenic acid	Part of coenzyme A needed for oxidation of carbohydrates and fats; aids in the formation of hormones and certain neurotransmitters	Nuts, beans, dark green vegetables, poultry, fruits, milk	Rarely, loss of appetite, mental depression, numbness	Unknown
Vitamin B_{12}	Complex, cobalt-containing compound; part of the coenzyme needed for synthesis of nucleic acids and myelin	Dairy products, fish, poultry, eggs, fortified cereals	Pernicious anemia	Unknown
Biotin	Coenzyme needed for metabolism of amino acids and fatty acids	Generally in foods, especially eggs	Skin rash, nausea, fatigue	Unknown

Table 25.7 Fat-Soluble Vitamins

Vitamin	Functions	Food Sources	Conditions Caused By:	
			Too Little	Too Much
Vitamin A	Antioxidant needed for normal vision, immune system function, and cellular growth	Yellow-orange and dark green fruits and vegetables, milk and cereals	Night blindness, poor immune system functioning	Nausea, headache, bone fractures, hair loss
Vitamin D	A group of steroids needed for development and maintenance of bones and teeth	Milk fortified with vitamin D, fish liver oil	Rickets, bone decalcification and weakening	Calcification of soft tissues, diarrhea, possible renal damage
Vitamin E	Antioxidant that prevents oxidation of vitamin A and polyunsaturated fatty acids	Leafy green vegetables, fruits, vegetable oils, nuts, whole-grain breads and cereals	Unknown	Diarrhea, nausea, headache, fatigue, muscle weakness
Vitamin K	Synthesis of substances active in clotting of blood	Leafy green vegetables, cabbage, cauliflower	Easy bruising and bleeding	Can interfere with anticoagulant medication

Water

Water constitutes about 60% of an adult's body. Water participates in many chemical reactions; in addition, watery fluids lubricate joints, transport other nutrients, and help maintain body temperature. Beverages, soups, fruits, and vegetables are sources of water, and most solid foods contain some water. The amount of *total* water (water from beverages and foods) that you need to consume depends on your physical activity level, your diet, and environmental conditions. On average, men should consume about 125 ounces (oz.) and women should consume about 90 oz of total water each day. Thirst is a healthy person's best guide for meeting water needs and avoiding dehydration.

Connecting the Concepts

For more on the diseases and conditions mentioned in this section, refer to the following discussions.

Section 23.2 provides additional information on the causes of cardiovascular disease.

Section 24.3 examines the role of fiber in the large intestine.

Section 28.2 illustrates the structure of human bones and describes how osteoporosis influences bone strength.

Check Your Progress 25.2

1. List the classes of nutrients, and give an example of each.
2. Describe the potential health consequences of overindulging in each of the following: carbohydrates, lipids, and proteins.
3. Predict what might happen to homeostasis without the proper amount of vitamins and minerals.

25.3 Nutrition and Health

Learning Outcomes

Upon completion of this section, you should be able to

1. Calculate a body mass index and determine if this value reflects a healthy weight.
2. Discuss the balance between energy intake and energy output.
3. Describe diseases associated with obesity.
4. List and describe eating disorders.

Many serious disorders in Americans are linked to a diet that results in excess body fat. In the United States, the number of people who are overweight or obese has reached epidemic proportions. Nearly two-thirds of adult Americans have too much body fat.

Excess body fat increases the risk of type 2 diabetes, cardiovascular disease, and possibly certain cancers. These conditions are among the leading causes of disability and death in the United States. Therefore, it is important for us all to stay within the recommended weight for our height.

Body Mass Index

Medical researchers use the **body mass index (BMI)** to determine if a person is overweight or obese. On the whole, the height to which we grow is determined genetically, while our weight is influenced by other factors as well, particularly diet and lifestyle. BMI is the relationship between a person's weight and height. Here's how to calculate your BMI:

$$\frac{\text{weight (pounds)} \times 703.1}{\text{height}^2 \text{ (inches)}}$$

Example: A woman whose height is 63 inches and weighs 133 pounds has a BMI of 23.56.

$$\frac{133 \times 703.1}{63^2} = \frac{93512.3}{3969} = 23.56 \text{ (BMI)}$$

Underweight BMI	<	18.5
Healthy BMI	=	18.5 to 24.9
Overweight BMI	=	25.0 to 29.9
Obese BMI	=	30.0 to 39.9
Morbidly obese BMI	=	40.0 or more

Energy Intake Versus Energy Output

While genetics is a factor in being overweight, a person cannot become fat without taking in more food energy (calories) than are expended.

Energy Intake   Scientists use a bomb calorimeter to measure a food's caloric value (**Fig. 25.8**). A portion of food is placed inside the chamber and then ignited. As the food burns, it raises the temperature of the water that surrounds the chamber. Scientists measure the change in water temperature and calculate the number of calories in the food. The energy value of food is often reported in kilocalories (kcal).[1] A kilocalorie is the amount of heat that raises the temperature of a liter of water by 1°C.

Animation
Bomb Calorimeter

For practical purposes, you can estimate a food's caloric value if you know how many grams of carbohydrate, fat, protein, and alcohol it contains. Each gram of carbohydrate or protein supplies 4 kcal, and each gram of fat supplies 9 kcal. Although alcohol is not a nutrient, it is considered a food, and each gram (g) supplies 7 kcal. Therefore, if a serving of food contains 30 g of carbohydrate, 9 g of fat, and 5 g of protein, it supplies 221 kcal:

carbohydrate	30 g	× 4 kcal	=	120 kcal
fat	9 g	× 9 kcal	=	81 kcal
protein	5 g	× 4 kcal	=	20 kcal
Total				221 kcal

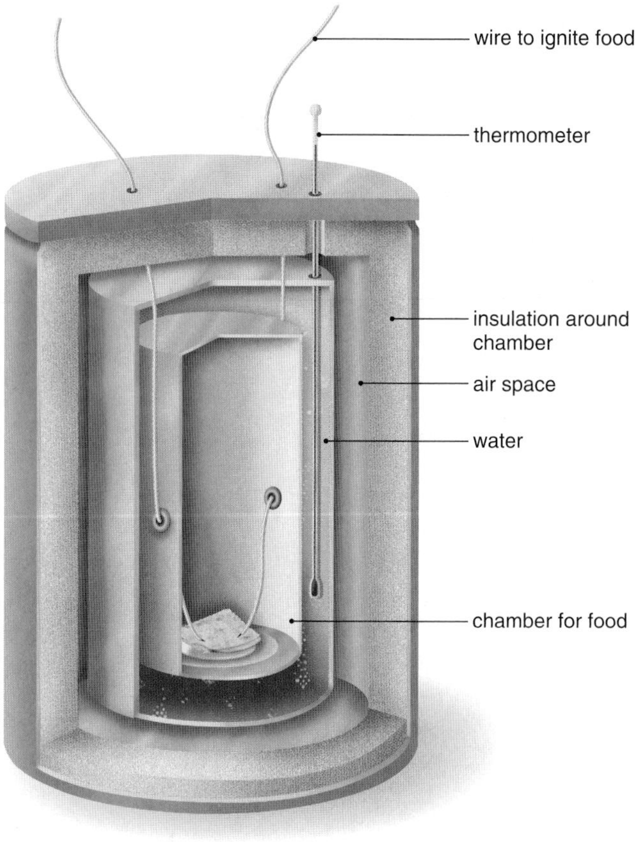

- wire to ignite food
- thermometer
- insulation around chamber
- air space
- water
- chamber for food

Figure 25.8 **Measuring the energy content of food.**

Scientists use a bomb calorimeter to determine the amount of calories supplied by a particular food.

---

[1]Nutritionists use the term *Calorie*, while scientists prefer *kcal*.

a.

b.

c.

Figure 25.9 **Measuring energy needed for a physical activity.**

a. A sedentary job requires much less energy than (**b**) a physical job. **c.** Scientists measure oxygen intake and carbon dioxide output to determine the kcal required for a particular physical activity.

Energy Output The body expends energy primarily for (1) metabolic functions; (2) physical activity; and (3) digestion, absorption, and processing of nutrients from food. Scientists can access a person's energy expenditure for a particular physical activity by measuring oxygen intake and carbon dioxide output during performance of the activity (**Fig. 25.9**).

For practical purposes, here's how to estimate your daily energy needs:

1. *Kcal needed daily for metabolic functions:* Multiply your weight in kilograms (weight in pounds divided by 2.2) times 1.0 if you are a man, and times 0.9 if you are a woman. Then multiply that number by 24. Example: Meghan, a woman who weighs 130 pounds (about 59 kg), would use the following steps to calculate her daily caloric use:

   $$0.9 \text{ kcal} \times 59 \text{ kg} \times 24 \text{ hours} = \text{approximately } 1{,}274 \text{ kcal/day}$$

2. *Kcal needed daily for physical activity:* Choose a multiplication factor from one of the follow categories:

   Sedentary (little or no physical activity) = .20 to .40
   Light (walk daily) = .55 to .65
   Moderate (daily vigorous exercise) = .70 to .75
   Heavy (physical labor/endurance training) = .80 to 1.20

   Multiply this factor times the kcal value you obtained in step 1. Example: Meghan performs light physical activity daily. She multiplies .55 × 1,274 kcal to determine her physical activity use, which is 701 kcal.

3. *Kcal needed for digestion, absorption, and processing of nutrients:* Example: Meghan adds 1,274 kcal and 701 kcal to calculate her total energy needs so far. The answer is 1,975 kcal. Multiply this value by 0.1, and add it to the total kcal from steps 1 and 2 to get your total daily energy needs. Example: Meghan multiplies 1,975 kcal by 0.1 and adds that value (197.5) to 1,975 to obtain her total daily energy needs of 2,172 kcal. Therefore, Meghan will maintain her weight of 130 pounds if she continues to consume about 2,170 kcal a day and to perform light physical activities.

Maintaining a Healthy Weight

To maintain weight at an appropriate level, the kcal intake (eating) should not exceed the daily kcal output (metabolism + physical activity + processing food). For many Americans, this ratio is out of sync; they take in more calories than they need. The extra energy is converted to fat stored in *adipose tissue,* and they become overweight. To lose weight, an overweight person needs to lower the kcal intake and increase the kcal output in the form of physical activity. Only then does the body metabolize its stored fat for energy needs, allowing the person to lose weight. **Figure 25.10** illustrates how body weight changes in relation to kcal intake and kcal output.

Dieting Fad weight-reduction diets—high-protein, low-carb, high-fiber, and even cabbage soup diets—come and go. During the first few weeks of a fad diet, overweight people often lose weight rapidly, because they consume fewer calories than usual, and excess body fat is metabolized for energy needs. In most cases, however, dieters become bored with eating the same foods and avoiding their favorite foods, which may be high in fat and sugars. When most dieters go off their diets, they regain the weight they lost, and they often feel frustrated and angry at themselves for failing to maintain the weight loss.

There are no quick and easy solutions for losing weight. The typical fad diet is nutritionally unbalanced and difficult to follow over the long term. Weight loss and weight maintenance require permanent lifestyle changes, such as increasing the level of physical activity and reducing portion sizes. Behavior modification allows an overweight person to lose weight safely, generally at a reasonable rate of about ½ to 2 pounds per week. Once body weight is under control, it needs to be maintained by continuing to eat sensibly.

Connections and Misconceptions

What is the 10,000 step program?

Research suggests that a minimum of 10,000 (10K) steps per day is necessary for weight maintenance and good health. That number of steps per day is roughly equivalent to the recommended 30 minutes of daily exercise. To get started, you'll need a pedometer. You'll probably find you need to increase the amount of walking you do to reach the goal of 10K steps a day. There are a number of easy ways to add steps to your routine. Park a little farther away from your office or the store. Take the stairs instead of the elevator, or go for a walk after a meal. If your goal is to lose weight, 12–15K steps a day have been shown to promote weight loss.

Intake	Output	Weight Change

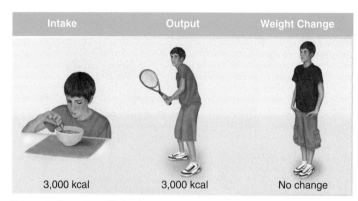

3,000 kcal   3,000 kcal   No change

Energy balance (equilibrium)

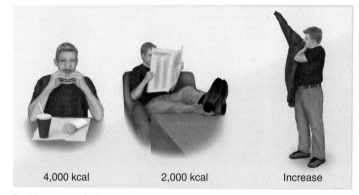

4,000 kcal   2,000 kcal   Increase

Positive energy balance

2,000 kcal   3,000 kcal   Decrease

Negative energy balance

Figure 25.10 How body weight changes.

These illustrations show the relationships among food intake, energy use, and weight change.

Figure 25.11   Exercising for good health.

Regular exercise helps prevent and control type 2 diabetes.

Figure 25.12   French fries.

Most people enjoy eating french fries, but fast-food fries are apt to contain saturated fats and trans fats.

Disorders Associated with Obesity

Type 2 diabetes and cardiovascular disease are often seen in people who are obese.

Type 2 Diabetes

As discussed in Chapter 27, diabetes comes in two forms, type 1 and type 2. When a person has type 1 diabetes, the pancreas does not produce insulin, and the patient has to have daily insulin injections.

In contrast to type 1 diabetes, children and more often adults with type 2 diabetes are usually obese and display impaired insulin production and insulin resistance. Normally, the presence of insulin causes the cells of the body to take up and metabolize glucose. In a person with insulin resistance, the body's cells fail to take up glucose even when insulin is present. Therefore, the blood glucose value exceeds the normal level, and glucose appears in the urine.

Type 2 diabetes is increasing rapidly in most industrialized countries of the world. Because type 2 diabetes is most often seen in people who are obese, dietary factors are generally believed to contribute to the development of type 2 diabetes. Further, a healthy diet, increased physical activity, and weight loss have been seen to improve the ability of insulin to function properly in type 2 diabetics (**Fig. 25.11**). How might diet contribute to the occurrence of type 2 diabetes? Simple sugars in foods, such as candy and ice cream, immediately enter the bloodstream, as do sugars from the digestion of starch within white bread and potatoes. When the blood glucose level rises rapidly, the pancreas produces an overload of insulin to bring the level under control. Chronically high insulin levels apparently lead to insulin resistance, increased fat deposition, and a high blood fatty acid level. Over the years, the body's cells become insulin resistant, and thus type 2 diabetes can occur. In addition, high fatty acid levels can lead to increased risk of cardiovascular disease.

It is well worth the effort to control type 2 diabetes because all diabetics, whether type 1 or type 2, are at risk for blindness, kidney disease, and cardiovascular disease.

Cardiovascular Disease

In the United States, cardiovascular disease, which includes hypertension, heart attack, and stroke, is among the leading causes of death. Cardiovascular disease is often due to arteries blocked by plaque, which contains saturated fats and cholesterol. Cholesterol is carried in the blood by two types of lipoproteins: low-density lipoprotein (LDL) and high-density lipoprotein (HDL). LDL is thought of as "bad" because it carries cholesterol from the liver to the cells, while HDL is thought of as "good" because it carries cholesterol from the cells to the liver, which takes it up and converts it to bile salts.

Saturated fats, including trans fats, tend to raise LDL cholesterol levels, while unsaturated fats lower LDL cholesterol levels. Beef, dairy foods, and coconut oil are rich sources of saturated fat. Foods containing partially hydrogenated oils (e.g., vegetable shortening and stick margarine) are sources of trans fats (**Fig. 25.12**). Unsaturated fatty acids in olive and canola oils, most nuts, and coldwater fish tend to lower LDL cholesterol levels. Furthermore, coldwater fish (e.g., herring, sardines, tuna, and salmon) contain polyunsaturated fatty acids, and especially *omega-3 unsaturated fatty acids,* which can reduce the risk of cardiovascular disease. Taking fish oil supplements to obtain omega-3s is not recommended without a physician's approval, because too much of these fatty acids can interfere with normal blood clotting.

The American Heart Association recommends limiting total cholesterol intake to 300 mg per day. Eggs are very nutritious, but yolks contain about 210 mg of cholesterol. Most healthy people can eat a couple of whole eggs each week without experiencing an increase in their blood cholesterol levels. Overall, dietary saturated fats and trans fats raise LDL cholesterol levels more than dietary cholesterol.

A physician can determine if blood lipid levels are normal. If a person's cholesterol and triglyceride levels are elevated, modifying the fat content of the diet, losing excess body fat, and exercising regularly can reduce them. If lifestyle changes do not lower blood lipid levels enough to reduce the risk of cardiovascular disease, a physician may prescribe medication.

 Video A Burger a Day

Eating Disorders

People with eating disorders are dissatisfied with their body image. Social, cultural, emotional, and biological factors all contribute to the development of an eating disorder. These serious conditions can lead to malnutrition, disability, and death. Regardless of the eating disorder, early recognition and treatment are crucial. Treatment usually includes psychological counseling and antidepressant medications.

Anorexia nervosa is a severe psychological disorder characterized by an irrational fear of getting fat, causing a refusal to eat enough food to maintain a healthy body weight (**Fig. 25.13***a*). A self-imposed starvation diet is often accompanied by occasional binge eating, followed by purging and extreme physical activity to avoid weight gain. Binges usually include large amounts of high-calorie foods, and purging episodes involve self-induced vomiting and laxative abuse. About 90% of people suffering from anorexia nervosa are young women; an estimated 1 in 200 teenage girls is affected.

A person with **bulimia nervosa** binge-eats, then purges to avoid gaining weight (Fig. 25.13*b*). The binge-purge cyclic behavior can occur several times a day. People with bulimia nervosa can be difficult to identify because their body weight is often normal, and they tend to conceal their binging and purging practices. Women are more likely than men to develop bulimia; an estimated 4% of young women suffer from this condition.

Other abnormal eating practices include binge-eating disorder and muscle dysmorphia. Many obese people suffer from **binge-eating disorder,** a condition characterized by episodes of overeating that are not followed by purging. Stress, anxiety, anger, and depression can trigger food binges. A person suffering from **muscle dysmorphia** thinks his or her body is underdeveloped. Body-building activities and a preoccupation with diet and body form accompany this condition. Each day, the person may spend hours in the gym, working out on muscle-strengthening equipment. Unlike anorexia nervosa and bulimia, muscle dysmorphia affects more men than women.

Connecting the Concepts

For more on the diseases and conditions mentioned in this section, refer to the following discussions.

Section 12.2 explores the basis of cancer in the human body.

Section 23.2 provides additional information on the causes of cardiovascular disease.

Section 27.2 examines the role of hormones in regulating blood sugar and the disease diabetes mellitus.

a.

b.

Figure 25.13 Eating disorders.

a. People with anorexia nervosa have a mistaken body image and think they are fat even though they are thin. **b.** Those with bulimia nervosa overeat and then purge their bodies of the food they have eaten.

Check Your Progress 25.3

1. Detail the three types of requirements that determine daily caloric needs.

2. Describe a few disorders associated with obesity.

3. Explain why it is important to understand your individual daily energy input and output.

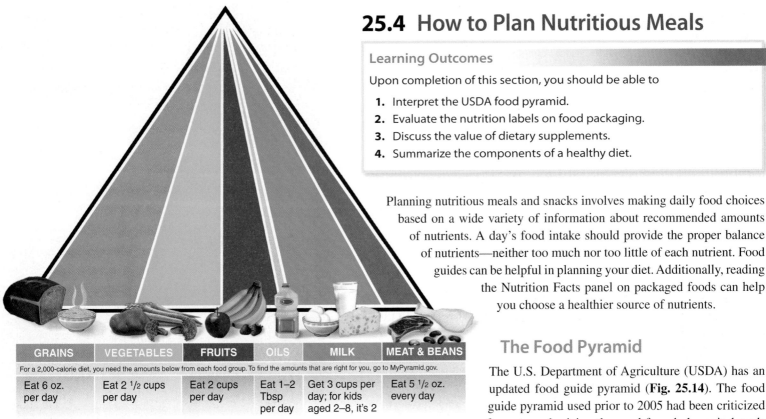

GRAINS | **VEGETABLES** | **FRUITS** | **OILS** | **MILK** | **MEAT & BEANS**

For a 2,000-calorie diet, you need the amounts below from each food group. To find the amounts that are right for you, go to MyPyramid.gov.

GRAINS	VEGETABLES	FRUITS	OILS	MILK	MEAT & BEANS
Eat 6 oz. per day	Eat 2 ½ cups per day	Eat 2 cups per day	Eat 1–2 Tbsp per day	Get 3 cups per day; for kids aged 2–8, it's 2	Eat 5 ½ oz. every day

Figure 25.14 **The 2008 USDA food guide pyramid.**

In 2005, the U.S. Department of Agriculture (USDA) first presented the current food guide pyramid. This food guide pyramid can be helpful when planning nutritious meals and snacks.

Source: Data from the U.S. Department of Agriculture.

25.4 How to Plan Nutritious Meals

Learning Outcomes

Upon completion of this section, you should be able to

1. Interpret the USDA food pyramid.
2. Evaluate the nutrition labels on food packaging.
3. Discuss the value of dietary supplements.
4. Summarize the components of a healthy diet.

Planning nutritious meals and snacks involves making daily food choices based on a wide variety of information about recommended amounts of nutrients. A day's food intake should provide the proper balance of nutrients—neither too much nor too little of each nutrient. Food guides can be helpful in planning your diet. Additionally, reading the Nutrition Facts panel on packaged foods can help you choose a healthier source of nutrients.

The Food Pyramid

The U.S. Department of Agriculture (USDA) has an updated food guide pyramid (**Fig. 25.14**). The food guide pyramid used prior to 2005 had been criticized for not emphasizing the need for whole-grain breads and cereals, low-fat dairy products, low-sugar fruits and vegetables, and unsaturated fats (such as olive and canola oil) in the diet. It also did not place an emphasis on exercise as a means to control weight gain.

The new pyramid, like the old one, groups foods with similar nutritional content and ranks food groups according to dietary emphasis, so that the diet will be balanced. The new pyramid emphasizes foods that should be eaten often and omits foods that should not be eaten on a regular basis. Additionally, the USDA provides recommendations concerning the minimum quantity of foods in each group that should be eaten daily. The sample menu in **Table 25.8** shows how the food guide pyramid can be used to plan a day's meals and snacks that supply about 2,200 kcal.

Table 25.8 **Preparing Meals and Snacks Using the Food Guide Pyramid**

Breakfast	Snack	Lunch	Snack	Dinner	Dessert
½ cup oat flakes cereal	Whole-wheat muffin	Toasted American cheese sandwich (whole-grain bread)	Apple	3 oz hamburger patty	½ cup frozen vanilla yogurt topped with sliced strawberries
8 oz nonfat milk	2 Tbsp peanut butter mixed with 1 tsp sunflower seeds	1 cup vegetable salad	4 graham crackers	Whole-wheat bun	
½ cup blueberries	8 oz nonfat milk	1 Tbsp olive oil and vinegar salad dressing	½ cup baked beans	1 cup steamed broccoli spears	
1 slice whole-grain toast		4 oz tomato juice		1 tsp soft margarine	
1 tsp soft margarine					
4 oz orange juice					

Making Sense of Nutrition Labels

A Nutrition Facts panel, such as the one shown in **Figure 25.15** provides specific dietary information about the product and general information about the nutrients the product contains.

Serving Size   The serving size is based on the typical serving size for the product and not necessarily what the food guide pyramid would consider a serving size. If you are comparing Calorie (kcal)[2] and other data about products of the same type, you want to be sure the serving size is the same for each product.

Calories   The total number of kcal is based on the serving size. Obviously, if you eat twice the serving size, you have taken in twice the number of calories.

The bottom of the panel lists the Calories (kcal) per gram (g) of fat, carbohydrate, and protein. You can use this information to calculate the calories per serving for each type of nutrient. The total for all nutrients should agree with the figure given for Calories (kcal) at the top of the panel.

% Daily Value   The % daily value (the % of the total amount needed in a 2,000 kcal diet) is calculated by comparing the specific information about this product with the information given at the bottom of the panel. For example, the product in Figure 25.15 has a fat content of 13 g, and the total daily recommended amount is less than 65 g, so 13/65 = 20%. *These % daily values are not applicable for people who require more or less than 2,000 Calories (kcal) per day.*

A % daily value for protein is generally not given because determining % daily value would require expensive testing of the protein quality of the product by the manufacturer. Also, notice there is a % daily value for carbohydrates but not sugars because there is no daily value for sugar.

How to Use the Panel   If the serving sizes are the same, you can use Nutrition Facts panels to compare two products of the same type. For example, if you wanted to reduce your caloric intake and increase your fiber and vitamin C intakes, comparing the panels from two different food products would allow you to see which one is lowest in calories and highest in fiber and vitamin C.

Animation
Anatomy
of a Food Label

Dietary Supplements

Dietary supplements are nutrients and plant products (such as herbal teas) that are used to enhance health. The U.S. government does not require dietary supplements to undergo the same safety and effectiveness testing that new prescription drugs must complete before they are approved. Therefore, many herbal products have not been tested scientifically to determine their benefits. Although people often think herbal products are safe because they are "natural," many plants, including lobelia, comfrey, and kava kava, can be poisonous.

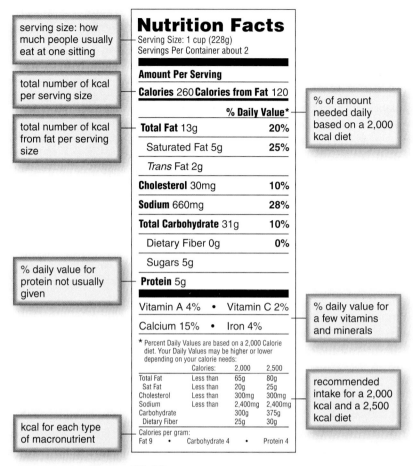

Figure 25.15 **Nutrition Facts panel.**

The Nutrition Facts panel provides information about the amounts of certain nutrients and other substances in a serving of food.

---

[2]Nutritionists use the word *Calorie*, while scientists prefer *kcal*.

Connections and Misconceptions

What health benefits are associated with drinking green tea?

All tea is derived from a plant native to China and India called *Camellia sinensis*. Green tea is made by steaming the leaves of this plant. Typically, green tea has less caffeine than black teas, which are made by fermenting the tea leaves. Research studies on the health benefits of green tea have shown that, overall, it has a very positive effect on the body. Antioxidants in green tea (called flavonoids) inhibit the growth of certain forms of cancer, prevent plaque buildup in the blood vessels, and may help improve blood cholesterol levels. However, most nutritionists still warn against taking green tea supplements. Instead, substitute a cup of tea for soda in your daily diet.

Dietary supplements that contain nutrients can also cause harm. Most fat-soluble vitamins are stored in the body and can accumulate to toxic levels, particularly vitamins A and D. Although excesses of water-soluble vitamins can be excreted, cases involving toxic amounts of vitamins B_6, thiamine, and vitamin C have been reported. Minerals can be harmful, even deadly, when ingested in amounts that exceed the body's needs.

Healthy people can take a daily supplement that contains recommended amounts of vitamins and minerals. Some people have metabolic diseases or physical conditions that interfere with their ability to absorb or metabolize certain nutrients. These individuals may need to add certain nutrient supplements to their diet. However, people should not take high doses of dietary supplements without checking with their physician.

The Bottom Line

While food guide pyramids differ in various ways, most nutritionists can agree that a healthy diet does the following:

- Has a moderate total fat intake and is low in saturated fat, trans fats, and cholesterol (see Table 25.3 for help in achieving this goal)
- Is rich in whole-grain products, vegetables, and legumes (e.g., beans and peas) as sources of complex carbohydrates and fiber
- Is low in refined carbohydrates, such as starches and sugars (see Table 25.2 for help in achieving this goal)
- Is low in salt and sodium intake (see Table 25.5 for help in achieving this goal)
- Contains only adequate amounts of protein, largely from poultry, fish, and plants
- Includes only moderate amounts of alcohol
- Contains adequate amounts of minerals and vitamins but avoids questionable food additives and supplements

Animation
Calories
in a Sandwich

Connecting the Concepts

For more information on these topics, refer to the following discussions.

Section 3.2 describes the differences between saturated and unsaturated fats.

Section 5.1 defines the terms *calorie* and *kilocalorie*.

Check Your Progress 25.4

1. Describe the importance of a balanced meal.
2. Summarize the categories of the food pyramid and the importance of each.
3. Explain the importance of a nutrition label and list the information it provides.

Media Study Tools

www.mhhe.com/maderessentials3

Enhance your study of this chapter with study tools and practice tests. Also ask your instructor about the resources available through ConnectPlus, including the media-rich eBook, interactive learning tools, and animations.

Nutrition

The virtual lab "Nutrition" provides an interactive investigation of how your food choices relate to your daily intake of select nutrients.

The Chapter in Review

Summary

25.1 Nutrition

Nutrients perform varied physiological functions in the body. A healthy diet can lead to a longer and more active life.

Introducing the Nutrients

Nearly every food is a mixture of nutrients. Balanced diets supply nutrients in proportions necessary for health. Essential nutrients must be supplied by the diet, or else deficiency disorders result. Macronutrients (carbohydrates, lipids, and proteins) supply energy. Micronutrients (vitamins and minerals) and water do not supply energy.

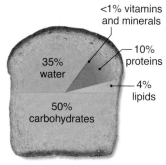

<1% vitamins and minerals

35% water

10% proteins

4% lipids

50% carbohydrates

25.2 The Classes of Nutrients

The six classes of nutrients are carbohydrates, lipids, proteins, minerals, vitamins, and water. Their functions and food sources follow.

Carbohydrates (Macronutrient)

- Carbohydrates in the form of sugars and starch provide energy for cells. Glucose is a simple carbohydrate the body uses directly for energy.
- Sources of carbohydrates are fruits, vegetables, cereals, and other grains.
- Fiber includes nondigestible carbohydrates from plants. Fiber in the diet can prevent constipation and may protect against cardiovascular disease, diabetes, and colon cancer. Refined carbohydrates, from which fiber, vitamins, and minerals have been removed, are a poor dietary choice.
- Nutritionists debate whether dietary fat or refined carbohydrates are responsible for the prevalence of obesity today.

Lipids (Macronutrient)

- Triglycerides (fats and oils) supply energy and are stored for insulation and protection of organs. Alpha-linolenic and linoleic acids are essential fatty acids.
- Cholesterol is used to make bile, steroid hormones, and vitamin D. The only food sources of cholesterol are foods derived from animals. Elevated blood cholesterol levels can lead to cardiovascular disease.
- Sources of lipids include oils, fats, whole-milk dairy products, meat, fish, poultry, and nuts.
- High intake of saturated fats, trans fats, and cholesterol is harmful to health.

Proteins (Macronutrient)

- The body uses the 20 different amino acids to synthesize hundreds of proteins. Nine amino acids are essential to the diet. Most animal foods are complete sources of protein because they contain all 9 essential amino acids.
- Sources of proteins include meat, fish, poultry, eggs, nuts, soybeans, and cheese.
- Healthy vegetarian diets rely on various sources of plant proteins.
- Consumption of excess protein can be harmful because the excretion of excess urea taxes the kidneys and can lead to kidney stones.

Minerals (Micronutrient)

- Minerals regulate metabolism and are incorporated into structures and compounds.
- About 20 minerals, obtained from most foods, are needed by the body.
- Lack of the mineral calcium can lead to osteoporosis.
- Most people consume too much sodium.

Vitamins (Micronutrient)

- Vitamins, obtained from most foods, regulate metabolism and physiological development. Vitamins C, E, and A also serve as antioxidants.
- Lack of any of the vitamins can lead to certain disorders.

Water

- Water participates in chemical reactions, lubricates joints, transports other nutrients, and helps maintain body temperature.

25.3 Nutrition and Health

Excess body fat increases the risk of type 2 diabetes, cardiovascular disease, and certain cancers.

Body Mass Index

Medical researchers use the body mass index (BMI) to determine if a person has a healthy weight, is overweight, or is obese.

Energy Intake Versus Energy Output

- Energy intake: The number of daily calories (kcal) consumed is based on grams of carbohydrate, fat, and protein in the foods eaten.
- Energy output: The number of daily calories (kcal) used is based on the amounts needed for metabolic functions, physical activity, and digestion, absorption, and processing of nutrients.

Maintaining a Healthy Weight

- To maintain a healthy weight, kcal intake should not exceed kcal output. To lose weight, decrease caloric intake and increase physical activity.
- Fad diets, in general, are nutritionally unbalanced and difficult to follow over the long term.

PART SIX Animal Structure and Function

Disorders Associated with Obesity

- Type 2 diabetes: A healthy diet, increased physical activity, and weight loss improve insulin function in type 2 diabetics.
- Cardiovascular disease: Saturated fats and trans fats are associated with high blood LDL cholesterol levels. To reduce risk of cardiovascular disease, cholesterol intake should be limited, and the diet should include sources of unsaturated, polyunsaturated, and omega-3 fatty acids to reduce cholesterol levels.

Eating Disorders

People suffering from eating disorders are dissatisfied with their body image. Anorexia nervosa and bulimia nervosa are serious psychological disturbances that can lead to malnutrition and death.

25.4 How to Plan Nutritious Meals

Planning nutritious meals and snacks in informed food choices.

The Food Pyramid

Food guide pyramids group foods according to similar nutrient content. Such guides can be helpful when planning nutritious meals and snacks. The USDA food guide pyramid emphasizes the need for exercise to prevent weight gain.

Making Sense of Nutrition Labels

The information in the Nutrition Facts panel on packaged foods can be useful when comparing foods for nutrient content, especially serving size, total calories, and % daily value.

Dietary Supplements

Dietary supplements are nutrients and plant products that are taken to enhance health. A multiple vitamin and mineral supplement that provides recommended amounts of nutrients can be taken daily, but herbal and nutritional supplements can be harmful if misused.

Key Terms

anorexia nervosa   489
antioxidant   482
binge-eating disorder   489
body mass index (BMI)   485
bulimia nervosa   489
carbohydrate   478
cholesterol   480
deficiency disorder   476
diet   477
dietary supplement   491
essential nutrient   477
fat   479
fiber   479

macronutrient   477
major mineral   481
micronutrient   477
mineral   481
muscle dysmorphia   489
nutrient   476
nutrition   476
obesity   479
protein   480
trace mineral   481
triglyceride   479
vitamin   482

Testing Yourself

Choose the best answer for each question.

1. Vitamins are considered
   a. micronutrients because they are small in size.
   b. micronutrients because they are needed in small quantities.
   c. macronutrients because they are large in size.
   d. macronutrients because they are needed in large quantities.

For questions 2–9, choose the class of nutrient from the key that matches the description. Each answer may be used more than once.

Key:

   a. carbohydrates
   b. lipids
   c. proteins
   d. minerals
   e. vitamins
   f. water

2. preferred source of direct energy for cells
3. constitutes the majority of human body mass
4. not present in most animal foods
5. include the essential fatty acids alpha-linolenic and linoleic
6. include antioxidants
7. generally found in higher levels in animal than in plant sources
8. an example is cholesterol
9. includes calcium, phosphorus, and potassium

10. The refining of whole-grain foods removes
    a. fiber.
    b. vitamins.
    c. minerals.
    d. More than one of these are correct.
    e. All of these are correct.

11. Cholesterol is
    a. a lipid that has only negative effects on the body.
    b. a carbohydrate.
    c. an essential vitamin.
    d. a lipid necessary for the synthesis of bile, steroid hormones, and vitamin D.

12. A serving of graham crackers contains 24 grams of carbohydrate, 3 grams of fat, and 2 grams of protein. How many calories are in a serving?
    a. 131
    b. 236
    c. 116
    d. 261
    e. 141

13. With reference to the food pyramid, fill in each component and the recommended daily dietary amount based on a 2,000 Calorie (kcal) diet.

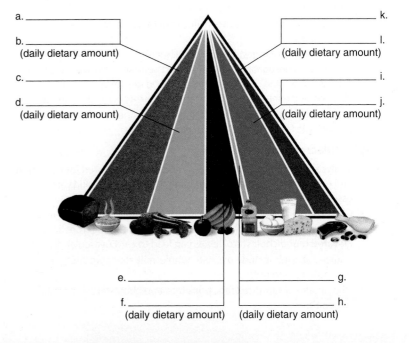

a. _____   k. _____
b. _____   l. _____
   (daily dietary amount)   (daily dietary amount)
c. _____   i. _____
d. _____   j. _____
   (daily dietary amount)   (daily dietary amount)

e. _____   g. _____
f. _____   h. _____
   (daily dietary amount)   (daily dietary amount)

14. Vitamins
    a. are inorganic compounds necessary in the diet.
    b. are found in high levels in the body and used as building blocks for body tissues.
    c. are organic compounds, found in low levels in the body, that regulate metabolic activities.
    d. can all be synthesized by the human body.

15. _____ are compounds that defend the body against free radicals.
    a. Macronutrients
    b. Proteins
    c. Minerals
    d. Antioxidants

16. The body mass index (BMI) is
    a. a measure of height relative to weight to determine whether or not a person is overweight.
    b. a measure of height relative to age to determine whether or not a person is of normal height.
    c. relatively low if a person is overweight.
    d. relatively high if a person is underweight.

17. You are comparing two breakfast cereals, and one has fewer calories than the other per serving. What should you check before deciding which one will lead to weight loss?
    a. the percentage of vitamins provided by each
    b. the grams of protein provided by each
    c. the grams of sugar provided by each
    d. the suggested serving size
    e. the size of the box

18. A % daily value for sugar is not included in a nutrition label because
    a. sugar is not a nutrient.
    b. it is too difficult to determine the caloric value of sugar.
    c. there is a daily value given for carbohydrates, but not for sugars.
    d. sugar quality varies from product to product.

19. A _____ is a component of food that provides a physiological function.
    a. macromolecule
    b. nutrient
    c. calorie
    d. chemical

20. The amino acids that must be consumed in the diet are called essential. Nonessential amino acids
    a. can be produced by the body.
    b. are needed only occasionally.
    c. are stored in the body until needed.
    d. can be taken in by supplements.

21. _____ is a nondigestible carbohydrate derived from plants.
    a. Fiber   b. Starch   c. Glycogen   d. Vitamin A

22. Bulimia nervosa is characterized by
    a. a restrictive diet but often includes excessive exercise.
    b. binge-eating followed by purging.
    c. an obsession about body shape and weight.
    d. a distorted body image, so that the person feels fat even when emaciated.

23. Which of these is a true statement?
    a. Technically speaking, junk foods have nutritional value.
    b. Water is not a nutrient because it contains no kcal.
    c. Even though fiber can be digested, it is not a nutrient.
    d. Cholesterol should be avoided because it plays no role in the body.
    e. A vegetarian diet does not provide all the essential amino acids.

24. Fats and oils are examples of
    a. sugars.   b. proteins.   c. triglycerides.   d. low-calorie foods.

Thinking Scientifically

1. One reason obesity is such a pervasive problem in the United States is that serving sizes have increased dramatically in the past two decades. For example, 20 years ago, a typical soda was 6.5 ounces, while today a 20-ounce (oz.) soda is considered average. A serving of french fries was 2.6 oz in the past but is at least 6 oz. today. Bagels have doubled in size. Complete the following table to determine the caloric values of today's servings, and then compare them.

	Soda Then 6.5 oz.	Soda Now 20 oz.	Calorie Increase
Fat (g) 0	0	_____	
Carbohydrate (g)	21	_____	
Protein (g)	0	_____	
Total calories	_____	_____	_____

	Fries Then 2.6 oz.	Fries Now 6 oz.	Calorie Increase
Fat (g) 0	11	_____	
Carbohydrate (g)	28	_____	
Protein (g)	3	_____	
Total calories	_____	_____	_____

	Bagels Then 3 oz.	Bagels Now 6 oz.	Calorie Increase
Fat (g) 0	0.5	_____	
Carbohydrate (g)	40	_____	
Protein (g)	6	_____	
Total calories	_____	_____	_____

2. Did humans evolve as herbivores, carnivores, or omnivores? Our dentition has both carnivore and herbivore characteristics. What about the data regarding the nine essential amino acids in our diet (see page 480)? Does that help you conclude what type diet we were originally adapted to? Why or why not?

Bioethical Issue

Misleading Food Advertisements

Some people believe that food and drink manufacturers are at least partly to blame for the obesity epidemic in the United States because of misleading advertising. For example, their advertisements show portion sizes that encourage excess consumption and make people believe that certain unhealthy substitutes are the equivalent of a well-balanced meal. Advertisements also lead children to desire sugar-coated cereals and fructose-loaded drinks. Do you think the government should regulate such advertising? If so, how extensive should the regulations be? If not, how might parents counter the effects of such advertising?

26

Defenses Against Disease

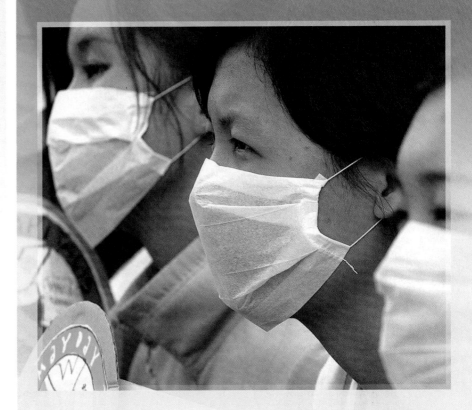

OUTLINE

BEFORE YOU BEGIN

Before beginning this chapter, take a few moments to review the following discussions.

Section 17.1 What is the general structure of a virus?

Section 17.3 What is the role of the peptidoglycan layer in the bacterial cell wall?

Section 23.2 What is the role of the lymphatic system with regard to circulation?

Are outbreaks of diseases more common today than in the past?

While it may seem that outbreaks of diseases, such as SARS, avian influenza, and the H1N1 virus, are unique to modern times, in fact, infectious disease outbreaks are described in the earliest written documents. Smallpox outbreaks were described 10,000 years ago in the Nile Valley. Malaria was described 4,000 years ago in China. The Chinese treated it with the Qinghao plant—the source of the drug currently used to treat malaria. Outbreaks of bubonic plague—the so-called Black Death—in Europe decimated the continent's population. The spread of the disease, as well as the public response to it, literally determined European history.

In more modern times, the battle between infectious agents and humans includes the development of vaccinations. The Chinese first developed vaccines around 200 BCE. In 1796, the smallpox vaccine was introduced to the Western world, and death rates from this virus dropped dramatically. The disease was eradicated from the human population in 1977. A rabies vaccine was developed in 1885, and vaccines for diphtheria, typhus, tetanus, and polio soon followed. Child mortality rates plummeted and remain low in countries where vaccination is routine.

The death rates from infectious diseases dropped still further in the 1940s with the development of antibiotics. With vaccines and antibiotics, scientists hoped they could eradicate other pathogens or, at least, significantly limit the suffering caused by infectious diseases. However, infectious diseases still kill large numbers of people. According to the World Health Organization (WHO), infectious diseases kill over 13 million people annually. In developing countries, infectious diseases account for 54% of deaths. Lower respiratory infections, HIV/AIDS, diarrheal diseases, malaria, and tuberculosis are currently the leading causes of infectious disease deaths.

Video
Virus Crisis

In this chapter, you will learn about the incredible array of defenses that your immune system utilizes to keep you free from infection. We will also examine how vaccinations and antibiotics assist the immune system in protecting you against viruses, bacteria, and other pathogens.

26.1 Organs, Tissues, and Cells of the Immune System

The **immune system** plays an important role in keeping us healthy because it fights infections and cancer. First and foremost, the immune system contains the **lymphatic organs:** red bone marrow, the thymus gland, the lymph nodes, and the spleen (**Fig. 26.1**). The **tonsils,** which are located in the pharynx, and the **appendix,** which is attached to a portion of the large intestine, are areas of lymphatic tissue that also belong to the immune system. The immune system not only contains a network of lymphatic organs and lymphatic tissues but also includes cells. In particular, we will discuss the cell types mentioned in **Table 26.1**. You will want to refer to this table as you read the chapter.

An important property of the immune system is its ability to distinguish between the cells of our body, or "self," and the antigens, or "nonself" presented by pathogens. **Immunity** is the body's ability to repel foreign substances, pathogens, and cancer cells. Nonspecific immunity indiscriminately repels pathogens, while specific immunity requires that a certain antigen be present. As previously discussed on page 443, an **antigen** is any molecule, usually a protein or carbohydrate, that stimulates the immune system.

Animation
Lymphatic System

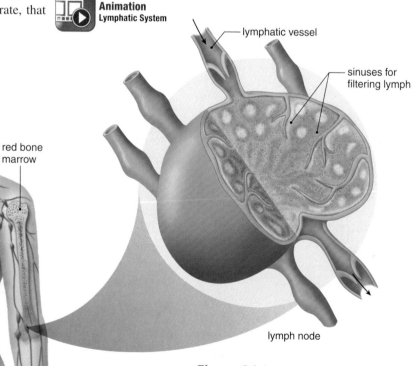

lymphatic vessel

sinuses for filtering lymph

red bone marrow

thymus gland

lymph nodes

spleen

lymphatic vessels

lymph node

Figure 26.1 Lymphatic organs.

The red bone marrow, thymus gland, lymph nodes, and spleen are lymphatic organs essential to immunity. The cells of the immune system, including lymphocytes, are found in these organs and in the lymph of lymphatic vessels.

Table 26.1 Immunocell Type and Function(s)

Cells	Function(s)
Macrophages	Phagocytize pathogens; inflammatory response and specific immunity
Mast cells	Release histamine, which promotes blood flow to injured tissues; inflammatory response
Neutrophils	Phagocytize pathogens; inflammatory response
Natural killer cells	Kill virus-infected and tumor cells by cell-to-cell contact
Lymphocytes	Are responsible for specific immunity
B cells	Produce plasma cells and memory cells
Plasma cells	Produce specific antibodies
Memory cells	Long-lived cells that may produce new B and T cells in the future
T cells	Regulate immune response; produce cytotoxic T cells and helper T cells
Cytotoxic T cells	Kill virus-infected and cancer cells
Helper T cells	Regulate immunity

Connections and Misconceptions

Why do tests for cancer often take biopsies of the lymph nodes?

In a biopsy, a physician uses a small needle to take a sample of a tissue for additional examination. For individuals with cancer, the doctor often takes a biopsy of the lymph nodes surrounding the original tumor. The reason for this is to see if the cancer has begun to spread, or metastasize, to other tissues. Since the lymphatic system filters the fluids returning from the tissues, metastasizing cancer cells can often be first detected in the lymph nodes closest to the tumor.

Lymphatic Organs

Each of the lymphatic organs has a particular function in immunity, and each is rich in lymphocytes, one of the types of white blood cells.

Red Bone Marrow

In a child, most bones have red bone marrow, and in an adult it is still present in the bones of the skull, the sternum (breastbone), the ribs, the clavicle, the pelvic bones, the vertebral column, and the ends of the humerus and femur nearest their attachment to the body. **Red bone marrow** produces all types of blood cells, but in this chapter we are interested in the cells listed in Table 26.1.

Lymphocytes differentiate into either **B lymphocytes (B cells)** or **T lymphocytes (T cells),** which are discussed at length later in the chapter. Bone marrow is not only the source of B lymphocytes but also the place where B lymphocytes mature. T lymphocytes mature in the thymus gland. You will learn that B cells produce antibodies and that T cells kill antigen-bearing cells outright.

Thymus Gland

The soft, bilobed **thymus gland** varies in size, but it is larger in children and shrinks as we get older. The thymus gland plays a role in the maturity of T lymphocytes. Immature T lymphocytes migrate from the bone marrow through the bloodstream to the thymus, where they develop, or "mature," into functioning T lymphocytes. Only about 5% of T lymphocytes ever leave the thymus. These T lymphocytes have survived a critical test: If any show the ability to react with "self" cells, they die. If they have the potential to attack a foreign cell, they leave the thymus and enter lymphatic vessels and organs.

The thymus gland also produces thymic hormones thought to aid in the maturation of T lymphocytes.

Lymph Nodes

Lymph nodes are small, ovoid structures along lymphatic vessels (Fig. 26.1). Lymph nodes filter lymph and keep it free of pathogens and antigens. Lymph is filtered as it flows through a lymph node because its many sinuses (open spaces) are lined by **macrophages**—large, phagocytic cells that engulf and then devour as many as a hundred pathogens and still survive (see Fig. 26.4a). Lymph nodes are also instrumental in fighting infections and cancer because they contain many lymphocytes.

Some lymph nodes are located near the surface of the body and are named for their location. For example, inguinal nodes are in the groin, and axillary nodes are in the armpits. Physicians often feel for the presence of swollen, tender lymph nodes in the neck as evidence that the body is fighting an infection. This method is a noninvasive, preliminary way to help them make a diagnosis.

Spleen

The **spleen,** which is about the size of a fist, is in the upper left abdominal cavity. The spleen's unique function is to filter the blood. This soft, spongy organ contains so-called red pulp and white pulp. Blood passing through the many sinuses in the red pulp is filtered of pathogens and debris, including worn-out red blood cells, because the sinuses are lined by macrophages. The white pulp contains lymphocytes, which are actively engaged in fighting infections and cancer.

The spleen's outer capsule is relatively thin, and an infection or a severe blow can cause the spleen to burst. The spleen's functions can be fulfilled by other organs, but individuals without a spleen are often slightly more susceptible to infections. As a result, they will receive certain vaccinations and may have to receive antibiotic therapy indefinitely.

Connecting the Concepts

For more information on the organs listed in this section, refer to the following discussions.

Section 23.2 describes the interaction of the lymphatic system with the circulatory system.

Figure 28.15 illustrates the structure of bone and shows the location of the red bone marrow.

Check Your Progress 26.1

1. Explain the relationship between antigens and immunity.
2. List the lymphatic organs, and identify their functions.
3. Summarize the roles of each class of immune system cells.

26.2 Nonspecific Defenses and Innate Immunity

Learning Outcomes

Upon completion of this section, you should be able to

1. Describe the barriers to pathogen entry into the body.
2. Explain the inflammatory response.
3. Describe the role of the complement system and natural killer cells in nonspecific immunity.

The body has an **innate immunity** composed of the various types of nonspecific defenses—our first line of defense against most types of infections. The nonspecific defenses are the barriers to entry, the inflammatory response, the complement system, and natural killer cells.

Barriers to Entry

Skin and the mucous membranes lining the respiratory, digestive, reproductive, and urinary tracts serve as mechanical barriers to entry by pathogens. Oil gland secretions contain chemicals that weaken or kill certain bacteria on the skin (**Fig. 26.2**). The upper respiratory tract is lined by ciliated cells that sweep mucus and trapped particles up into the throat, where they can be swallowed or expectorated (spit out). The stomach has an acidic pH, which inhibits growth or kills many types of bacteria. The various bacteria that normally reside in the large intestine and other areas, such as the vagina, prevent pathogens from taking up residence.

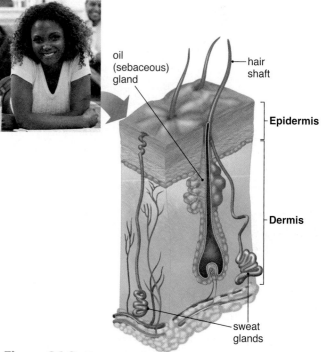

Figure 26.2   Skin.

The cells of the epidermis harden and die as they progress to the outer layer of skin. These outer dead cells form a protective barrier against invasion by pathogens. The sweat and oil glands are acidic enough to inhibit invasion by bacteria.

Connections and Misconceptions

How do antihistamines work?

Once histamine is released from mast cells, it binds to receptors on other body cells. This signals the nearby capillaries to become more "leaky," allowing fluid to leave the capillaries and enter the tissue. This excess fluid is the cause of the familiar symptoms of a runny nose and watery eyes. Antihistamines work by blocking the receptors on the cells, so that histamine can no longer bind. For allergy relief, antihistamines are most effective when taken before exposure to the allergen.

Figure 26.3 The inflammatory response.

If you cut your skin, the inflammatory response occurs immediately. As blood flow increases, the area gets red and warm. As mast cells, a type of white blood cell, release chemicals (such as histamine), the capillary becomes more permeable, localized swelling occurs, and pain receptors are stimulated. Neutrophils and macrophages begin phagocytizing the bacteria.

The Inflammatory Response

Whenever tissue is damaged, a series of events occurs that is known as the **inflammatory response** (**Fig. 26.3**). The inflamed area has four outward signs: redness, warmth, swelling, and pain. The inflammatory response involves the first three cell types listed in Table 26.1.

When an injury occurs, damaged tissue cells and **mast cells** release chemical mediators, such as **histamine**, which cause the capillaries to dilate and become more permeable. Excess blood flow due to enlarged capillaries causes the skin to redden and become warm. Increased permeability allows proteins and fluids to escape into the tissues, resulting in swelling. The swollen area stimulates free nerve endings, causing the sensation of pain.

Neutrophils are phagocytic white blood cells that migrate to the site of injury. They are amoeboid and can change shape to squeeze through capillary walls and enter tissue fluid. When neutrophils phagocytize pathogens, they are enclosed within a vesicle. The pathogens are later destroyed by hydrolytic enzymes when this vesicle combines with a lysosome, one of the cellular organelles. Neutrophils also have the ability to secrete a toxic array of chemicals, including hydrogen peroxide and hypochlorite, the active ingredient found in bleach. While this chemical attack is highly effective in killing bacteria, it also results in the death of the neutrophil.

Also present are *macrophages* (**Fig. 26.4***a*). As stated, lymph nodes contain macrophages, but so do some tissues, particularly connective tissues, which have resident macrophages that routinely act as scavengers, devouring old blood cells, bits of dead tissue, and other debris. During the inflammatory response, macrophages can also bring about an explosive increase in the number of leukocytes by liberating growth factors, which pass by way of blood to the red bone marrow. There, they stimulate the production and release of white blood cells, primarily neutrophils.

As the infection is being overcome, some neutrophils may die. These—along with dead cells, dead bacteria, and living white blood cells—form pus, a whitish material. The presence of pus indicates that the body is trying to overcome an infection.

The macrophages that have been fighting the infection move through the tissue fluid and lymph to the lymph nodes. Now lymphocytes are activated to react to the threat of an infection. Sometimes inflammation persists, and the result is chronic inflammation, which is often treated with anti-inflammatory agents, such as aspirin, ibuprofen, or cortisone. These medications act against the chemical mediators, such as histamine, released by the white blood cells in the area.

The Complement System

The **complement system**, often called simply complement, is composed of a number of blood plasma proteins designated by the letter C and a subscript. A limited amount of activated complement protein is needed because a domino effect occurs: As soon as a protein is activated, it stimulates many more.

The complement proteins are activated when pathogens enter the body. The proteins "complement" certain immune responses, which accounts for their name. For example, they are involved in and amplify the inflammatory response because complement proteins attract phagocytes to the scene. Some complement proteins bind to the surfaces of pathogens already coated with antibodies, which ensures that the pathogens will be phagocytized by a neutrophil or macrophage.

Certain other complement proteins join to form a membrane attack complex, which produces holes in the surfaces of microbes (Fig 26.4b). Fluids and salts then enter the pathogen to the point that it bursts.

Animation
Activation of
Complement

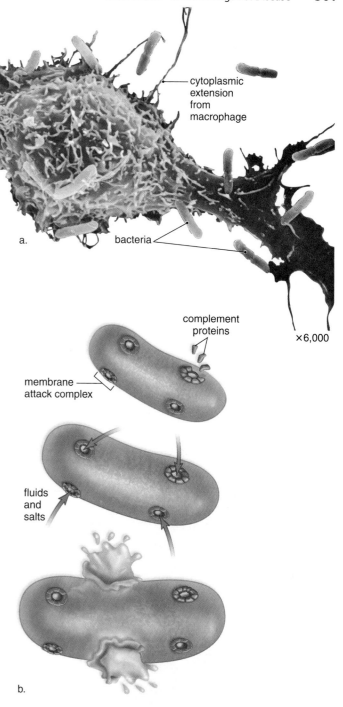

a.

cytoplasmic extension from macrophage

bacteria

complement proteins

×6,000

membrane attack complex

fluids and salts

b.

Natural Killer Cells

Natural killer (NK) cells are large, granular lymphocytes that kill virus-infected cells and tumor (cancer) cells by cell-to-cell contact (see Table 26.1).

What makes an NK cell attack and kill a cell? The cells of your body ordinarily have "self" proteins on their surface that bind to receptors on NK cells. Sometimes virus-infected cells and cancer cells undergo alterations and lose their ability to produce self proteins. When NK cells can find no self proteins to bind to, they kill the cell by the same method as T lymphocytes (see Fig. 26.8).

NK cells are not specific—their numbers do not increase when exposed to a particular antigen, and they have no means of "remembering" the antigen from previous contact.

Connections and Misconceptions

Is a fever always bad?

At the first sign of a fever, most people reach for over-the-counter (OTC) medicines to bring it under control. However, in many cases, using OTC medicines does more harm than good. Medical professionals now widely regard a low fever as a beneficial aspect of the immune system. When your body runs a fever, the elevated temperature increases your metabolic rate and slows down bacterial and viral reproduction. A low  fever also promotes the release of chemicals, called interferons, that prevent the infection from spreading. Of course, a high fever, or one that lasts for several days, should immediately be brought to the attention of your physician.

Figure 26.4 **Ways to get rid of pathogens.**

a. Macrophages, the body's scavengers, engulf pathogens and chop them up inside lysosomes. **b.** Complement proteins come together and form a membrane attack complex in the surface of a pathogen. Water and salts enter, and the pathogen bursts.

Connecting the Concepts

For more information on the material in this section, refer to the following discussions.

Section 22.1 describes the types of epithelial and connective tissues that make up human skin.

Section 23.3 explores how the components of blood participate in the inflammatory response.

Section 24.3 explains how bacteria in the appendix and large intestine protect the body against infection.

Check Your Progress 26.2

1. Describe the steps in the inflammatory response.
2. Describe the function of the complement system.
3. Explain how natural killer cells "know" what to attack.

26.3 Specific Defenses and Acquired Immunity

When nonspecific defenses have failed to prevent an infection, specific defenses, or **acquired immunity,** come into play. Specific defenses respond to antigens. Pathogens have antigens, and antigens can be part of a foreign cell or cancer cell. When our body learns to destroy a particular antigen, we have become immune to it. Because our immune system does not ordinarily respond to the proteins on the surface of our own cells (as if they were antigens), the immune system is said to be able to distinguish "self" from "nonself." Only in this way can the immune system aid, rather than disrupt, homeostasis.

Lymphocytes are capable of recognizing an antigen because they have plasma membrane receptor proteins shaped to allow them to combine with a specific antigen. Because we encounter millions of different antigens during our lifetime, we need a great diversity of lymphocytes to protect us against them. Remarkably, diversification occurs to such an extent during the maturation process that potentially a lymphocyte type exists for any possible antigen.

Immunity usually lasts for some time. For example, once we recover from the measles, we usually do not get the illness a second time. Immunity is primarily the result of the action of the B lymphocytes and T lymphocytes. B lymphocytes mature in the bone marrow, and T lymphocytes mature in the thymus gland. B lymphocytes, also called B cells, give rise to plasma cells, which produce antibodies (**Table 26.2**). These antibodies are secreted into the blood, lymph, and other body fluids. In contrast, T lymphocytes, also called T cells, do not produce antibodies. Some T cells regulate the immune response, and other T cells directly attack cells that bear antigens (see Table 26.1).

B Cells and the Antibody Response

Each B cell can bind only to a specific antigen—the antigen that fits the binding site of its receptor. The receptor is called a **B-cell receptor (BCR).** Some B cells never have anything to do because an antigen that fits the binding site of their type of receptor never shows up. But if an antigen does bind to a BCR, that B cell is activated, and it divides, producing many plasma cells and memory B cells (**Fig. 26.5**). This is called the **clonal selection model**, since the antigen selects which B cell is activated and this cell then divides to produce many clones of itself. A B cell is stimulated to divide and produce plasma cells by **cytokines,** which are signaling proteins that stimulate various immune cells to perform their functions. Plasma cells are larger than regular B cells because they have extensive rough endoplasmic reticulum for the mass production and secretion of antibodies specific to the antigen. The antibodies produced by plasma cells and secreted into the blood and lymph are identical to the BCR of the activated B cell.

Table 26.2 Characteristics of B Cells

- They provide an antibody response to a pathogen.
- They are produced and become mature in bone marrow.
- They reside in lymph nodes and the spleen; they circulate in blood and lymph.
- They directly recognize antigens and then undergo cell division.
- Cell division produces antibody-secreting plasma cells, as well as memory B cells.

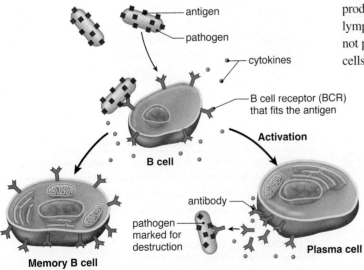

Figure 26.5 B cells and the antibody response.

When an antigen combines with a BCR, the B cell divides to produce plasma cells and memory B cells. Plasma cells produce antibodies but eventually disappear. Memory B cells remain in the body, ready to produce the same antibody in the future.

Memory B cells are the means by which long-term immunity is possible. If the same antigen enters the system again, memory B cells quickly divide and give rise to more plasma cells capable of producing the correct antibodies.

Once the threat of an infection has passed, the development of new plasma cells ceases, and those present undergo apoptosis. Apoptosis is a process of programmed cell death (PCD) involving a cascade of cellular events, leading to the death and destruction of the cell (see Fig. 8.10).

The Function of Antibodies

Antibodies are immunoglobulin proteins that are capable of combining with a specific antigen (**Fig. 26.6**). The antigen-antibody reaction can take several forms, but quite often the reaction produces complexes of antigens combined with antibodies. Such antigen-antibody complexes, sometimes called immune complexes, mark the antigens for destruction (see Fig. 26.5). An antigen-antibody complex may be engulfed by neutrophils or macrophages, or it may activate complement. Complement makes pathogens more susceptible to phagocytosis, as discussed previously.

Animation
Antibodies

ABO Blood Type Most likely, you know your ABO blood type—A, B, AB, or O. These letters stand for antigens on your red blood cells. If you have type O blood, you do not have either antigen A or antigen B on your red cells. Some blood types have antibodies in the plasma (**Table 26.3**). As an example, type O blood has both anti-A and anti-B antibodies in the plasma. You cannot give a person with type O blood a transfusion from an individual with type A blood. If you do, the antibodies in the recipient's plasma will react to type A red blood cells and agglutination will occur. **Agglutination,** the clumping of red blood cells, causes blood to stop circulating and red blood cells to burst. On the other hand, you can give type O blood to a person with any blood type because type O red blood cells bear neither A nor B antigens.

Animation
Agglutination and Precipitation

It is possible to determine who can give blood to whom based on the ABO system. However, other red blood cell antigens, in addition to A and B, are used in typing blood. Therefore, it is best to put the donor's blood on a slide with the recipient's blood to observe whether the two types match (no agglutination occurs) to determine whether blood can be safely transfused from one person to another.

T Cells and the Cellular Response

When T cells leave the thymus gland, they have unique **T-cell receptors (TCRs),** just as B cells do. Unlike B cells, however, T cells are unable to recognize an antigen without help. The antigen must be presented to them by an **antigen-presenting cell (APC),** such as a macrophage. After ingesting and destroying a pathogen, APCs travel to a lymph node or the spleen, where T cells also congregate. When a macrophage phagocytizes a virus, it is digested in a lysosome. An antigen from the virus is combined with a **major histocompatability complex (MHC)** protein. MHC proteins play a major role in identifying self cells. After the antigen binds to the MHC protein, the complex appears on the cell surface. Then the combined MHC + antigen complex is presented to a T cell. The importance of self proteins in plasma membranes was first recognized when it was discovered that they contribute to the specificity of tissues and make it difficult to transplant tissue from one human to another.

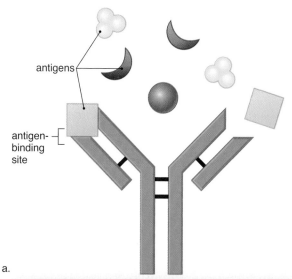
a.

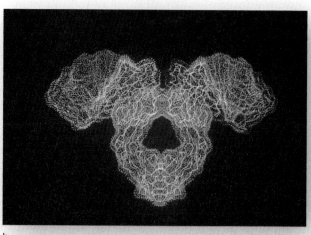
b.

Figure 26.6 Why an antibody is specific.

a. During a lifetime, a person will encounter a million different antigens that vary in shape. An antibody has two variable regions that end in an antigen-binding site. The variable regions differ so much that the binding site of each antibody has a shape that will fit only one specific antigen. **b.** Computer model of an antibody molecule. The antigen combines with the two side branches.

Table 26.3 Blood Types

Blood Type	Antigens on Red Blood Cells	Antibodies in Plasma	Compatible Donor(s)
A	A	Anti-B	A, O
B	B	Anti-A	B, O
AB	A, B	Neither anti-A nor anti-B	A, B, AB, O
O	—	Both anti-A and anti-B	O

Table 26.4 Characteristics of T Cells

- They provide a cellular response to virus-infected cells and cancer cells.
- They are produced in bone marrow and mature in the thymus gland.
- Antigen must be presented in groove of an MHC[1] molecule.
- Cytotoxic T cells destroy nonself protein-bearing cell.
- Helper T cells secrete cytokines that control the immune response.

[1]MHC = major histocompatibility complex.

The two main types of T cells are **helper T cells (T_H cells)** and **cytotoxic T cells (T_C cells)** (**Table 26.4**). Each of these types has a TCR that can recognize an antigen fragment in combination with an MHC molecule. However, the major difference in antigen recognition by these two types of cells is that the T_H cells recognize only antigens presented by APCs with MHC class II molecules on their surface, while T_C cells recognize only antigens presented by APCs with MHC class I molecules on their surface. In **Figure 26.7**, the antigen is represented by a red triangle, and the helper T cell that binds to the antigen has the specific TCR that can combine with this particular MHC II + antigen. Now the helper T cell is activated and divides to produce more helper T cells.

As the illness disappears, the immune reaction wanes and the activated T cells become susceptible to apoptosis. Apoptosis contributes to homeostasis by regulating the number of cells present in an organ, or in this case, the immune system. When apoptosis does not occur as it should, T-cell cancers (e.g., lymphomas and leukemias) can result. Also, in the thymus gland, any T cell that has the potential to destroy the body's own cells undergoes apoptosis.

Functions of Cytotoxic T Cells and Helper T Cells

Cytotoxic T cells specialize in cell-to-cell combat. They have storage vacuoles containing perforins or enzymes called granzymes. After a cytotoxic T cell binds to a virus-infected or cancer cell presenting the same antigen it has learned to recognize, it releases perforin molecules, which perforate the plasma membrane, forming a pore. Cytotoxic T cells then deliver a supply of granzymes into the pore, and these cause the cell to undergo apoptosis. Once cytotoxic T cells have released the perforins and granzymes, they move on to the next target cell. Cytotoxic T cells are responsible for a cellular response to virus-infected and cancer cells (**Fig. 26.8**).

 **Animation** Cytotoxic T Cell Activity Against Target Cells

 Video T Lymphocyte

Helper T cells specialize in regulating immunity by secreting cytokines that, in particular, stimulate B cells and cytotoxic T cells. Similar to B cells, cloned T cells include memory T cells that live for many years and can jump start an immune response to an antigen dealt with before. Because HIV, the virus that causes AIDS, infects helper T cells and other cells of the immune system, it inactivates the immune response and makes HIV-infected individuals susceptible to opportunistic infections. Infected macrophages serve as reservoirs for the HIV virus. AIDS is discussed in Section 26.5.

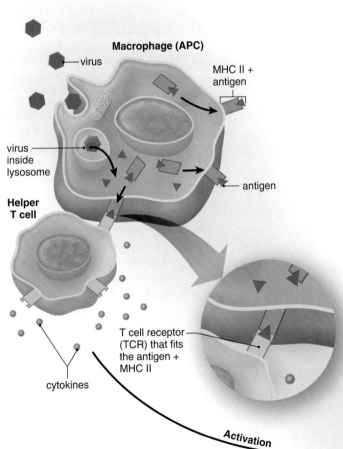

Figure 26.7 Activation of a T cell.

For a T cell to be activated, the antigen must be presented along with an MHC protein to the T cell by an APC, often a macrophage. Each type of T cell bears a specific receptor, and if its TCR fits the MHC II + antigen complex, it is activated to divide and produce more T cells—in this case, helper T cells, which also have this type of TCR. The helper T cells then produce and release cytokines.

Tissue Rejection

Certain organs, such as the skin, the heart, and the kidneys, could be transplanted easily from one person to another if the body did not attempt to reject them. Rejection occurs because cytotoxic T cells and antibodies bring about the destruction of foreign tissues in the body. When rejection occurs, the immune system is correctly distinguishing between self and nonself.

Organ rejection can be controlled by carefully selecting the organ to be transplanted and administering immunosuppressive drugs. It is best if the transplanted organ has the same type of MHC proteins as those of the recipient; otherwise, the transplanted organ would be antigenic to T cells. Two well-known immunosuppressive drugs, cyclosporine and tacrolimus, act by inhibiting the response of T cells to cytokines. Without cytokines, all types of immune responses are weak.

Connecting the Concepts

For more information on the material presented in this section, refer to the following discussions.

Section 8.4 describes the process of apoptosis, or programmed cell death.

Figure 10.10 illustrates the patterns of inheritance associated with human blood types.

Figure 17.5 illustrates the stages in HIV virus reproduction.

Check Your Progress 26.3

1. Explain the role of antibodies in acquired immunity.
2. Compare and contrast B cells with T cells.
3. Distinguish between the antibody and cellular responses, and list the types of cells involved in each.
4. Explain how MHC markers are associated with "self" and their role in tissue rejection.

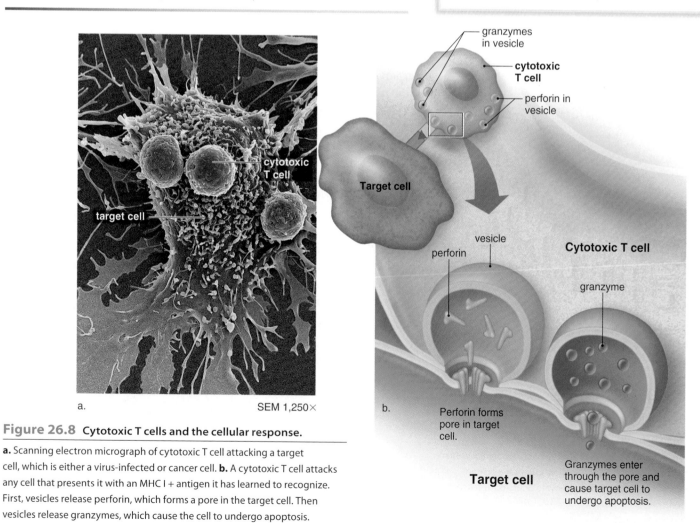

a. SEM 1,250×

b.

Figure 26.8 Cytotoxic T cells and the cellular response.

a. Scanning electron micrograph of cytotoxic T cell attacking a target cell, which is either a virus-infected or cancer cell. **b.** A cytotoxic T cell attacks any cell that presents it with an MHC I + antigen it has learned to recognize. First, vesicles release perforin, which forms a pore in the target cell. Then vesicles release granzymes, which cause the cell to undergo apoptosis.

26.4 Immunizations

Learning Outcomes

Upon completion of this section, you should be able to

1. Summarize the role of vaccines in providing immunity against disease.
2. Distinguish between active and passive immunity.

After you have had an infection, you may be immune to getting it again. Good examples of diseases against which you can acquire immunity are the childhood diseases measles and mumps. Unfortunately, few sexually transmitted diseases stimulate lasting immunity; for example, a person can get gonorrhea over and over again. If lasting immunity is possible, a vaccine for the disease likely exists or can be developed. **Vaccines** are substances that usually do not cause illness, but even so, the immune system responds to them. Traditionally, vaccines are the pathogens themselves, or their products, that have been treated so that they are no longer virulent (able to cause disease). Today, it is possible to genetically engineer bacteria to mass-produce a protein from pathogens, and this protein can be used as a vaccine. This method has now produced a vaccine against hepatitis B, a virus-induced disease, and it is being used to prepare a vaccine against malaria, a protozoan-induced disease.

 Animation Constructing Vaccines **Video** Potato Vaccine

Immunization promotes **active immunity.** After a vaccine is given, it is possible to follow an active immune response by determining the amount of antibody present in a sample of plasma; this is called the antibody titer. After the first exposure to a vaccine, a primary response occurs. For a period of several days, no antibodies are present; then their concentration rises slowly, levels off, and gradually declines as the antibodies bind to the antigen or simply break down (**Fig. 26.9**). After a second exposure to the vaccine, a secondary response is expected. The concentration now rises rapidly to a level much greater than before; then it slowly declines. The second exposure is called a "booster" because it boosts the antibody concentration to a high level. The high antibody concentration is expected to help prevent disease symptoms if the individual is exposed to the disease-causing antigen.

Active immunity is dependent on the presence of memory B cells and possibly memory T cells that are capable of responding to lower doses of antigen. Active immunity is usually long-lasting, but a booster may be required after a certain number of years.

Figure 26.9 Active immunity due to immunization.

Immunization often requires more than one injection. A minimal primary response occurs after the first vaccine injection, but after a second injection, the secondary response usually shows a dramatic rise in the amount of antibody present in plasma.

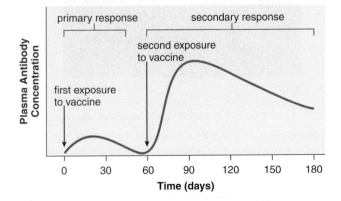

Connections and Misconceptions

Does the MMR vaccine cause autism?

Several large research studies have failed to find any connection among the measles, mumps, and rubella (MMR) vaccine, thimerosal (a preservative that was used for vaccines), and an increased risk of autism in children. While the cause of autism has not yet been identified, the evidence does not suggest that the vaccine or thimerosal is causing autism. The problem is that the first MMR vaccine is usually administered between 12 and 15 months of age, which is typically the age that an autistic child first presents symptoms. Most autism researchers believe that the factors causing autism are in place before this time frame, and the timing with the MMR vaccine is coincidental. Research is still ongoing, but most pediatricians agree that the threat of measles and rubella far outweighs the unsubstantiated risk of autism.

Although the body usually makes its own antibodies, it is sometimes possible to give an individual prepared antibodies (immunoglobulins) to combat a disease. Because these antibodies are not produced by the individual's plasma cells, this so-called **passive immunity** is temporary. For example, newborns are passively immune to some diseases because antibodies have crossed the placenta from the mother's blood. These antibodies soon disappear, however, so that within a few months infants become more susceptible to infections. Breast-feeding prolongs the natural passive immunity an infant receives from the mother because antibodies are present in the mother's milk.

Even though passive immunity does not last, it is sometimes used to prevent illness in a patient who has been unexpectedly exposed to an infectious disease. Usually, the patient receives an injection of gamma globulin, a portion of blood that contains antibodies, preferably taken from an individual who has recovered from the illness. In the past, horses were immunized, and gamma globulin was taken from them to provide the needed antibodies against such diseases as diphtheria, botulism, and tetanus. Unfortunately, patients who received these antibodies became ill about 50% of the time, because the serum contained proteins that the individual's immune system recognized as foreign. This condition was called serum sickness.

Connecting the Concepts

For more information on this topic, refer to the following discussion.

Section 29.3 explores how vaccines are being developed as contraceptives.

Check Your Progress 26.4

1 Explain why you cannot be vaccinated for most sexually transmitted diseases.

2 Explain why two doses of a vaccine are sometimes needed for immunity.

3 Distinguish between active and passive immunity, and explain how each occurs.

26.5 Immune System Problems

Learning Outcomes

Upon completion of this section, you should be able to

1. Distinguish between an immediate and a delayed allergic response.
2. List common autoimmune diseases and provide the causes of each.
3. Outline the effects of an HIV infection and summarize the treatments available.

The immune system usually protects us from disease because it can distinguish self from nonself. Sometimes, however, it responds in a manner that harms the body, as when individuals develop allergies or have an autoimmune disease.

Allergies

SEM of pollen

Figure 26.10 **Allergies.**

When people are allergic to pollen, they develop symptoms that include watery eyes, sinus headache, increased mucus production, labored breathing, and sneezing.

Allergies are hypersensitivities to substances in the environment, such as pollen, food, or animal hair, that ordinarily would not cause an immune reaction. The response to these antigens, called **allergens**, usually includes some unpleasant symptoms (**Fig. 26.10**). An allergic response is regulated by cytokines secreted by both T cells and macrophages.

Immediate allergic responses are caused by receptors attached to the plasma membrane of mast cells in the tissues. When an allergen attaches to receptors on mast cells, they release histamine and other substances that bring about the symptoms.

An immediate allergic response can occur within seconds of contact with the antigen. The symptoms can vary, but a dramatic example, anaphylactic shock, is a severe reaction characterized by a sudden and life-threatening drop in blood pressure.

Allergy shots, injections of the allergen in question, sometimes prevent the onset of an allergic response. It has been suggested that injections of the allergen may cause the body to build up large quantities of antibodies released by plasma cells, and these combine with allergens received from the environment before they have a chance to reach the receptors located in the membranes of mast cells.

Delayed allergic responses are probably initiated by memory T cells at the site of allergen contact in the body. A classic example of a delayed allergic response is the skin test for tuberculosis (TB). When the test result is positive, the tissue where the antigen was injected becomes red and hardened. This shows that the person has been previously exposed to tubercle bacillus, the cause of TB. Contact dermatitis, which occurs when a person is allergic to poison ivy, jewelry, cosmetics, and so forth, is also an example of a delayed allergic response.

For questions 13–17, identify the type of blood cell in the key that matches the description. Some answers may be used more than once.

Key:

   a. macrophage
   b. B lymphocyte
   c. cytotoxic T cell
   d. helper T cell
   e. mast cell

13. large white blood cell that phagocytizes pathogens and presents antigens to T cells
14. white blood cell produced in the bone marrow that gives rise to plasma cells
15. secretes granzymes into infected cells to cause apoptosis
16. responsible for production of histamine
17. releases cytokines and activates several components of the immune system

18. During agglutination,
   a. red blood cells clump.
   b. white blood cells attach pathogens.
   c. it is unlikely that a blood clot will form.
   d. platelets form a plug within a blood vessel.

19. Complement
   a. is a general defense mechanism.
   b. is involved in the inflammatory response.
   c. is a series of proteins present in the plasma.
   d. plays a role in destroying bacteria.
   e. All of these are correct.

20. Antibodies
   a. cause specific illnesses.
   b. result in immunity for a specific pathogen.
   c. are produced by helper T cells.
   d. can bond with many different types of pathogens.

21. Which applies to T lymphocytes?
   a. mature in bone marrow
   b. mature in thymus gland
   c. Both a and b are correct.
   d. None of these are correct.

22. Allergens
   a. cause an immune response in most people.
   b. are antigens to those people who are allergic to them.
   c. reduce the production of histamine by mast cells.
   d. are molecules that are usually rare in the environment.

23. AIDS is caused by which of the following viruses?
   a. SIV   **d.** AIDS
   b. HSV   **e.** HIV
   c. HPV

24. _____ is a condition that results when cytotoxic T cells attack the body's own cells.
   a. An allergic response   **c.** Passive immunity
   b. Autoimmune disease   **d.** Active immunity

25. Vaccines are associated with
   a. active immunity.   **c.** passive immunity.
   b. long-lasting immunity.   **d.** Both a and b are correct.

26. MHC proteins play a role in
   a. acquired immunity.   **c.** tissue transplantation.
   b. histamine release.   **d.** All of these are correct.

27. Type O blood has _____ antibodies in the plasma.
   a. anti-A   **c.** Neither a nor b is correct.
   b. anti-B   **d.** Both a and b are correct.

28. In the following diagram of skin, label the structures that contribute to the skin's protection of the body.

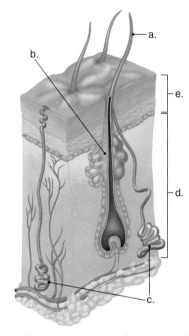

29. Which of the following is a nonspecific defense of the body?
   a. immunoglobulin   **d.** vaccine
   b. B cell   **e.** inflammatory response
   c. T cell

30. Antibodies combine with antigens
   a. at specific regions of the antibodies.
   b. at multiple regions of the antibodies.
   c. only if macrophages are present.
   d. Both a and c are correct.

31. An antigen-presenting cell (APC)
   a. presents antigens to T cells.
   b. secretes antibodies.
   c. marks each human cell as belonging to that person.
   d. secretes cytokines only.

32. Which of the following cells participate in the specific defense against an infection?
   a. mast cells   **d.** natural killer cells
   b. macrophages   **e.** B lymphocytes
   c. neutrophils

33. Which of the following is not an autoimmune disease?
   a. multiple sclerosis
   b. myasthenia gravis
   c. contact dermatitis
   d. systemic lupus erythematosus
   e. rheumatoid arthritis

34. Natural killer cells attack cells that
    a. have been determined to contain a virus.
    b. have lost their self proteins.
    c. are malformed.
    d. are dividing rapidly.

35. A person with type A blood can receive a transfusion from someone with type
    a. A blood only.
    b. A or B blood.
    c. A or AB blood.
    d. A or O blood.
    e. A, B, AB, or O blood.

36. Unlike B cells, T cells
    a. require help to recognize an antigen.
    b. are components of a specific defense response.
    c. are types of lymphocytes.
    d. contribute to homeostasis.

37. Passive immunity
    a. is permanent.
    b. may result from immunoglobulin injections.
    c. requires memory B cells.
    d. may be induced by vaccines.

38. Anaphylactic shock is an example of
    a. an immediate allergic response.
    b. an autoimmune disease.
    c. a type of serum shock.
    d. a delayed allergic response.

39. The HIV virus lives in and destroys
    a. B cells.
    b. cytotoxic T cells.
    c. helper T cells.
    d. All of these are correct.

40. AIDS patients generally die due to which of the following causes?
    a. rare diseases and infections
    b. HIV virus
    c. strokes
    d. heart attacks
    e. cancer

Thinking Scientifically

1. At one time, tonsillectomies (removal of the tonsils) were more commonplace than they are today. Why are the tonsils seemingly so susceptible to infection? Why should they not be removed at the first sign of infection?

2. The transplantation of organs from one person to another was impossible until the discovery of immunosuppressive drugs. Now, with the use of drugs such as cyclosporine, organs can be transplanted without rejection. Transplant patients must take immunosuppressive drugs for the remainder of their lives. What are the risks associated with long-term use of immunosuppressive drugs?

3. What role does the human immune system play in the evolution of viruses and bacteria? How does the evolution of pathogens affect our ability to achieve and maintain immunity against them?

Bioethical Issue

Access to AIDS Medications

Over 33 million people worldwide are living with AIDS. This disease is deadly without proper medical care, but it can be simply chronic if treated. Drug companies typically charge a high price for AIDS medications because Americans and their insurance companies can afford them. However, these drugs are out of reach in many countries, such as those in Africa, where AIDS is a widespread problem. Some people argue that drug companies should use the profits from other drugs (such as those for heart disease, depression, and impotence) to make AIDS drugs affordable to those who need them. That has not happened yet. In some countries, governments have allowed companies to infringe on foreign patents held by major drug companies, so that they can produce affordable AIDS drugs. Do drug companies have a moral obligation to provide low-cost AIDS drugs, even if they have to do so at a loss of revenue? Is it right for governments to ignore patent laws in order to provide their citizens with affordable drugs?

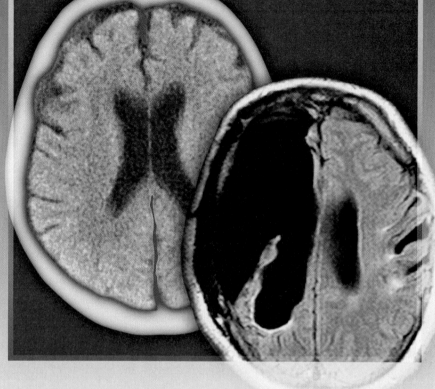

Life with Half a Brain?

Obviously, the brain's role in the human body is critical to survival. The brain controls nearly all the activities in the body, including those of the other organ systems. When particular areas of the brain malfunction, a multitude of problems, or even death, can occur. Certain conditions, such as epilepsy, inherited disease, or infection, can cause unusual electrical activity in areas of the brain, leading to severe and debilitating seizures. When the condition is serious enough and cannot be controlled by medication, physicians may consider performing a surgery termed a hemispherectomy. In this procedure, a portion of the brain (from a lobe to a complete half) is removed. Unbelievably, many of the functions linked to the side of the brain that was removed can be "re-mapped" to the remaining side. In fact, most children who have undergone this procedure have experienced minimal lasting side effects, other than some minor paralysis on one side of the body. Learning ability does not seem to be significantly damaged by the procedure. Such drastic surgeries have taught us that the brain is a highly adaptable organ that can recover from significant trauma.

Understanding the structure and function of the nervous system is critical to understanding the functioning of other organs and organ systems in the body. In this chapter, you will learn how the central and peripheral nervous systems operate, as well as how the central nervous system communicates with the endocrine system, which in turn regulates the body's internal environment.

27

The Control Systems

OUTLINE

BEFORE YOU BEGIN

Before beginning this chapter, take a few moments to review the following discussions.

Section 5.4 How does active transport move molecules and ions against their concentration gradient?

Section 19.1 What are the major trends in the evolution of the animals?

Section 22.3 What is the role of negative feedback in the maintenance of homeostasis?

27.1 Nervous Systems

Figure 27.1 Modes of action of the nervous and endocrine systems.

a. Nerve impulses passing along an axon cause the release of a neurotransmitter. The neurotransmitter, a chemical signal, causes the wall of an arteriole to constrict. **b.** The hormone insulin, a chemical signal, travels in the cardiovascular system from the pancreas to the liver, where it causes liver cells to store glucose as glycogen.

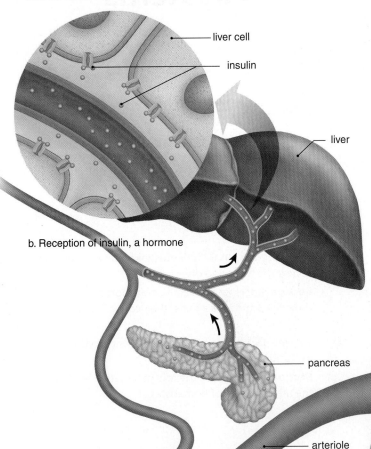

liver cell

insulin

liver

b. Reception of insulin, a hormone

The nervous system and the endocrine system work together to regulate the activities of the other systems. Both control systems use chemical signals when they respond to changes that might threaten homeostasis, but they have different means of delivering these signals (**Fig. 27.1**). The nervous system quickly sends a message along a nerve fiber directly to a target organ, such as skeletal or smooth muscle. Once a chemical signal is released, the muscle brings about an appropriate response.

The endocrine system uses the blood vessels of the cardiovascular system to send a hormone as a messenger to a target organ, such as the liver. The endocrine system is slower-acting because it takes time for the hormone to move through the bloodstream to a target organ. Also, hormones change the metabolism of cells, and this takes time; however, the response is longer-lasting. Cellular metabolism tends to remain the same for at least a limited period of time.

Let's first turn our attention to the human nervous system. As we examine it, we will also compare it with the nervous systems of other animals.

The Human Nervous System

The nervous system is always involved in an animal's ability to move around. In fact, about the only function of the nervous system in some animals, such as planarians, is movement, particularly to feed. Some planarians are predators, and

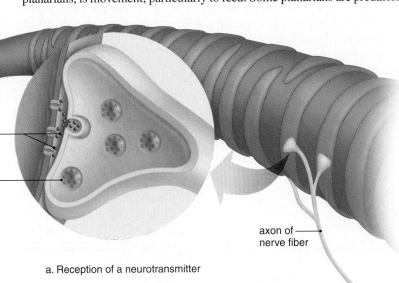

pancreas

arteriole

neurotransmitters

vesicle

axon of nerve fiber

a. Reception of a neurotransmitter

the worm wraps itself around its prey, entangling it in slime and pinning it down. Then a muscular pharynx extends out the mouth, and by a sucking motion, the prey is torn up and swallowed. The bilateral symmetry of planarians is reflected in the organization of their nervous system. They have two lateral nerve cords (bundles of nerves) joined together by transverse nerves. The arrangement is called a *ladder-like* nervous system. A "brain" receives sensory information from the eyespots and sensory cells in the auricles. The two lateral nerve cords allow a rapid transfer of information from the cerebral ganglia to the posterior end, and the transverse nerves between the nerve cords keep the movement of the two sides coordinated. The nervous organization in planarians is a foreshadowing of the central and peripheral nervous systems that are found in more complex invertebrates, such as earthworms, and in vertebrates, including humans (**Fig. 27.2**).

In humans, the nervous system controls the muscular system and works along with the endocrine system to maintain homeostasis. The **central nervous system (CNS)** includes the brain and spinal cord, which have a central location—they lie in the midline of the body. The **peripheral nervous system (PNS)** consists of nerves that lie outside the central nervous system. The brain gives off paired cranial nerves (one on each side of the body), and the spinal cord gives off paired spinal nerves. The division between the central nervous system and the peripheral nervous system is arbitrary; the two systems work together and are connected to one another.

What makes the human nervous system more complex than the planarian system? Five trends during the evolution of the vertebrate nervous system can be identified:

- A CNS developed that is able to summarize incoming messages before ordering outgoing messages.
- Nerve cells (neurons) became specialized to send messages to the CNS, between neurons in the CNS, or away from the CNS.
- A brain evolved that has special centers for receiving input from various regions of the body and for directing their activity.
- The CNS became connected to all parts of the body by peripheral nerves. Therefore, the central nervous system can respond to both external and internal stimuli.
- Complex sense organs, such as the human eye and ear, arose that can detect changes in the external environment.

This chapter discusses the first four aspects of the human nervous system. Sensory input and muscle response are the topics of Chapter 28.

Neurons

The shape of a nerve cell, or **neuron,** is suitable to its function. The **cell body** contains the nucleus and other organelles that allow a cell to function. The neuron's many short **dendrite** fibers fan out to receive signals from sensory receptors or other neurons. These signals can result in nerve impulses carried by an axon. The **axon,** a nerve fiber that is typically longer than a dendrite, is the portion of a neuron that conducts nerve impulses. An axon can reach all the way from the end of your spinal cord to the tip of your big toe or from your big toe to your spinal cord, depending on which way messages are being conducted. Long axons are covered by a white **myelin sheath** formed from the membranes of tightly spiraled cells that leave gaps called **nodes of Ranvier.** Myelin sheaths account for our impression that nerves are white and glistening.

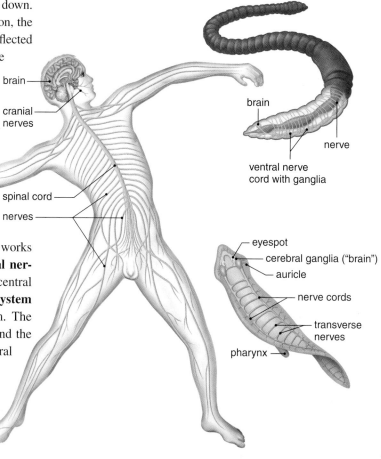

Figure 27.2 Comparison of nervous systems.

Invertebrates, such as a planarian and an earthworm, as well as vertebrates, such as humans, have a central nervous system (e.g., brain) and a peripheral nervous system (nerves).

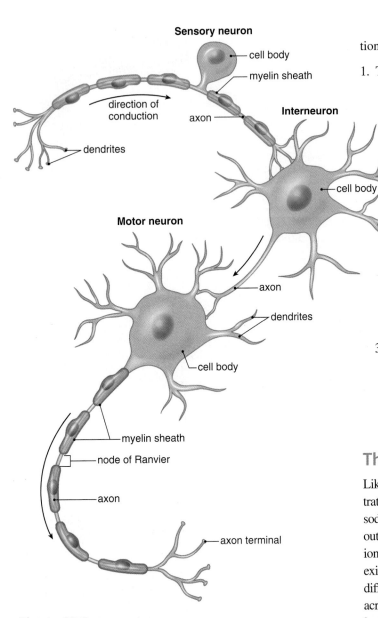

Figure 27.3 Types of neurons.

A sensory neuron, an interneuron, and a motor neuron are drawn here to show their arrangement in the body. Only axons conduct nerve impulses. In a sensory neuron, a process that extends from the cell body divides into an axon that takes nerve impulses all the way from the dendrites to the CNS. In a motor neuron and an interneuron, the axon extends directly from the cell body. The axon of sensory and motor neurons is covered by a myelin sheath. All long axons have a myelin sheath.

The nervous system has three types of neurons specific to its three functions (**Fig. 27.3**):

1. The nervous system receives sensory input. **Sensory neurons** perform this function. They take nerve impulses from sensory receptors to the CNS. The sensory receptor, which is the distal end of the axon of a sensory neuron, may be as simple as a naked nerve ending (a pain receptor), or it may be built into a highly complex organ, such as the eye or ear. In any case, the axon of a sensory neuron can be quite long if the sensory receptor is far from the CNS.

2. The nervous system performs integration—in other words, the CNS sums up the input it receives from all over the body. **Interneurons** occur entirely within the CNS and take nerve impulses between various parts of the CNS. Some interneurons lie between sensory neurons and motor neurons, and some take messages from one side of the spinal cord to the other or from the brain to the spinal cord, and vice versa. They also form complex pathways in the brain, where processes that account for thinking, memory, and language occur.

3. The nervous system generates motor output. **Motor neurons** take nerve impulses from the CNS to muscles or glands. Motor neurons cause muscle fibers to contract or glands to secrete, and therefore they are said to *innervate* these structures.

Video
Making Brain Cells

The Nerve Impulse

Like some other cellular processes, the nerve impulse is also dependent on concentration gradients. In neurons, these concentration gradients are maintained by the sodium-potassium pump. This pump actively transports sodium ions (Na^+) to the outside of the axon and actively transports potassium ions (K^+) inside. Aside from ion concentration differences across the axon's membrane, a charge difference also exists. *The inside of an axon is negative* compared with the outside. This charge difference is due in part to an unequal distribution of ions across the membrane, but is also due to the presence of large, negatively charged proteins in the axon cytoplasm.

Animation
How the Sodium-Potassium Pump Works

The charge difference across the axon's membrane offers a potential for change, or an **action potential,** as the nerve impulse is also called. The nerve impulse is a rapid, short-lived, self-propagating reversal in the charge difference across the axon's membrane. **Figure 27.4** shows how it works. A nerve impulse involves two types of gated channel proteins in the axon's membrane. In contrast to ungated channel proteins, which constantly allow ions across the membrane, gated channel proteins open and close in response to a stimulus, such as a signal from another neuron. One type of gated channel protein allows sodium (Na^+) to pass through the membrane, and the other allows potassium (K^+) to pass through the membrane. As an axon is conducting a nerve impulse, the Na^+ gates open at a particular location, and the inside of the axon becomes positive as Na^+ moves from outside the axon to the inside. The Na^+ gates close, and then the K^+ gates open. Now K^+ moves from inside the axon to outside the axon, and the charge reverses.

Animation
Nerve Impulse

In Figure 27.4, the axon is unmyelinated, and the action potential at one locale stimulates an adjacent part of the axon's membrane to produce an action potential. In myelinated axons, an action potential at one node of Ranvier causes an action potential at the next node (**Fig. 27.5**). This type of conduction, called **saltatory conduction,** is much faster than otherwise. Imagine running down a long hallway as quickly as you can; then picture yourself able to get to the end of the same hall in just a few leaping bounds. Leaping would enable you to travel the same distance in a much shorter time; likewise, saltatory conduction greatly speeds the conduction of nerve impulses. In thin, unmyelinated axons, the nerve impulse travels about 1.0 meter/second, but in thick, myelinated axons, the rate is more than 100 meters/second due to saltatory conduction. In any case, action potentials are self-propagating; each action potential generates another along the length of an axon.

Animation
Action Potential
Propagation

The conduction of a nerve impulse (action potential) is an all-or-none event—that is, either an axon conducts a nerve impulse or it does not. The intensity of a message is determined by how many nerve impulses are generated within a given time span. An axon can conduct a volley of nerve impulses because only a small number of ions are exchanged with each impulse. As soon as an impulse has passed by each successive portion of an axon, it undergoes a short refractory period, during which it is unable to conduct an impulse. During a refractory period, the sodium gates cannot yet open. This period ensures that nerve impulses travel in only one direction and do not reverse.

The Synapse

Each axon has many axon terminals. In the CNS, a terminal of one neuron, known as the *presynaptic cell,* lies very close to the dendrite (cell body) of another neuron, the *postsynaptic cell.* This region of close proximity is called a **synapse.** In the PNS, when the postsynaptic cell is a muscle cell, the region is called a neuromuscular junction. A small gap exists at a synapse, and this gap is called the **synaptic cleft.** How is it possible to excite the next neuron or a muscle cell across this gap? Nerve impulses cannot themselves pass across the synaptic cleft. Instead, transmission across a synaptic cleft is carried out by chemical signals

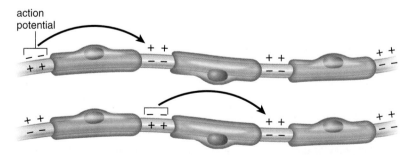

Figure 27.5 **Conduction of a nerve impulse in a myelinated axon.**

Action potentials can occur only at gaps in the myelin sheath, called nodes of Ranvier. This makes conduction much faster than in unmyelinated axons. In humans, all long axons are myelinated.

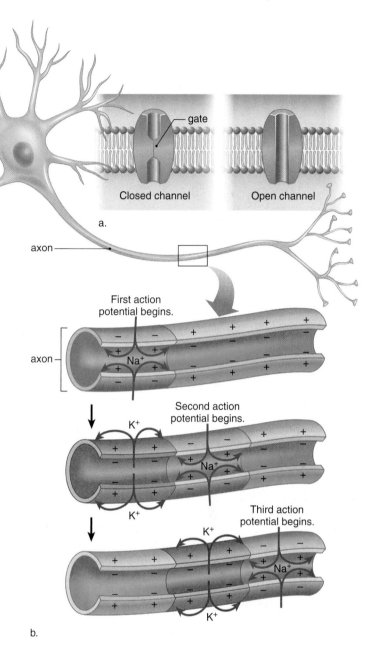

Figure 27.4 **Conduction of action potentials in an unmyelinated axon.**

a. Na$^+$ and K$^+$ each have their own gated channel protein by which they cross the axon's membrane. **b.** During an action potential, Na$^+$ enters the axon, and the charge difference between inside and outside reverses (blue); then K$^+$ exits, and the charge difference is restored (red). The action potential moves from section to section in an unmyelinated axon. Only a few ions are exchanged at a time.

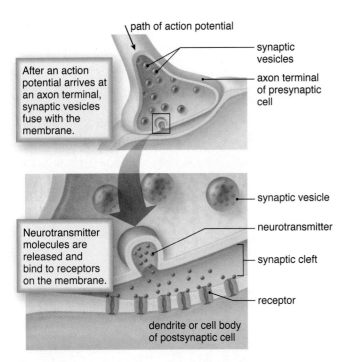

After an action potential arrives at an axon terminal, synaptic vesicles fuse with the membrane.

path of action potential

synaptic vesicles

axon terminal of presynaptic cell

Neurotransmitter molecules are released and bind to receptors on the membrane.

synaptic vesicle

neurotransmitter

synaptic cleft

receptor

dendrite or cell body of postsynaptic cell

Figure 27.6 Synapse structure and function.

Transmission across a synapse from one neuron (the presynaptic cell) to another occurs when a neurotransmitter is released, diffuses across a synaptic cleft, and binds to a receptor in the plasma membrane of the next neuron (the postsynaptic cell). Each axon releases only one type of neurotransmitter, symbolized here by a red ball.

Connections and Misconceptions

How do drugs that regulate depression and anxiety work?

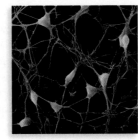

In general, pharmaceutical drugs that regulate behavior work by regulating the amount of neurotransmitter in the synapse. For example, drugs such as Xanax and Valium increase the levels of gamma-aminobutyric acid. These medications are used for panic attacks and anxiety. Reduced levels of norepinephrine and serotonin are linked to depression. Drugs such as Prozac, Paxil, and Cymbalta allow norepinephrine and/or serotonin to accumulate at the synapse, usually by blocking their reabsorption. By increasing the levels of these chemicals, the postsynaptic cell receives a more constant chemical message, which explains their effectiveness as antidepressants.

called **neurotransmitters,** which are stored in synaptic vesicles. When nerve impulses traveling along an axon reach an axon terminal, synaptic vesicles release a neurotransmitter into the synaptic cleft. Neurotransmitter molecules diffuse across the cleft and bind to a specific receptor protein on the postsynaptic cell (**Fig. 27.6**).

Depending on the type of neurotransmitter and/or the type of receptor, the response of the postsynaptic cell can be toward excitation or toward inhibition. At least 25 different neurotransmitters have been identified, but several well-known neurotransmitters are **acetylcholine (ACh), norepinephrine (NE), serotonin,** and gamma-aminobutyric acid (GABA)**.**

Once a neurotransmitter has been released into a synaptic cleft and has initiated a response, the neurotransmitter is removed from the cleft. In some synapses, the postsynaptic cell produces enzymes that rapidly inactivate the neurotransmitter. For example, the enzyme **acetylcholinesterase (AChE)** breaks down acetylcholine. In other synapses, the presynaptic cell rapidly reabsorbs the neurotransmitter. For example, norephinephrine is reabsorbed by the axon terminal. The short existence of neurotransmitters at a synapse prevents continuous stimulation (or inhibition) of the postsynaptic cell.

A single neuron has many dendrites plus the cell body, and both can have synapses with many other neurons; 1,000 to 10,000 synapses per single neuron are not uncommon. Therefore, a neuron is on the receiving end of many signals. An excitatory neurotransmitter produces a potential change that drives the neuron closer to an action potential, and an inhibitory neurotransmitter produces a potential change that drives the neuron farther from an action potential. Neurons integrate these incoming signals. **Integration** is the summing up of both excitatory and inhibitory signals. If a neuron receives many excitatory signals (either at different synapses or at a rapid rate from one synapse), chances are the axon will transmit a nerve impulse. On the other hand, if a neuron receives both inhibitory and excitatory signals, the integration of these signals may prohibit the axon from firing.

 Animation
Chemical Synapse

 MP3
Synapses

Drug Abuse

Many drugs that affect the nervous system act by interfering with or promoting the action of neurotransmitters. A drug can either enhance or block the release of a neurotransmitter, mimic the action of a neurotransmitter or block the receptor for it, or interfere with the removal of a neurotransmitter from a synaptic cleft. *Stimulants* are drugs that increase activity of the CNS, and *depressants* decrease the activity of the CNS. Increasingly, researchers are coming to believe that dopamine, a neurotransmitter in the brain, is responsible for mood. Many of the new medications developed to counter drug dependence and mental illness affect the release, reception, or breakdown of dopamine.

 Video
Caffeine Withdrawal

Drug abuse is apparent when a person takes a drug at a dose level and under circumstances that increase the potential for a harmful effect. A drug abuser often takes more of the drug than was intended. Drug abusers are apt to display a psychological and/or physical dependence on the drug. With physical dependence, formerly called "addiction," more of the drug is needed to get the same effect, and withdrawal symptoms occur when the user stops taking the drug.

Cocaine

Cocaine is an alkaloid derived from the shrub *Erythroxylon coca*. It is sold in powder form and as crack, a more potent extract. Because cocaine prevents the synaptic uptake of dopamine, the neurotransmitter remains in the synapse for a prolonged period of time and continues to stimulate the postsynaptic cell. As a result, the user experiences a "rush" sensation. The epinephrine-like effects of dopamine account for the state of arousal that lasts for several minutes after the rush experience.

A cocaine binge can go on for days, after which the individual suffers a crash. During the binge period, the user is hyperactive and has little desire for food or sleep but has an increased sex drive. During the crash period, the user is fatigued, depressed, and irritable; has memory and concentration problems; and displays no interest in sex.

Cocaine causes extreme physical dependence. With continued cocaine use, the postsynaptic cells become increasingly desensitized to dopamine. The user, therefore, experiences physical dependence withdrawal symptoms and an intense craving for cocaine. These are indications that the person is highly dependent on the drug.

Overdosing on cocaine can cause seizures and cardiac and respiratory arrest. It is possible that long-term cocaine abuse causes brain damage (**Fig. 27.7**). Babies born to addicts suffer withdrawal symptoms and may have neurological and developmental problems.

Methamphetamine

Methamphetamine is a synthetic drug made from amphetamine by the addition of a methyl group. Over 9 million U.S. residents have used methamphetamine at least once in their lifetime; teenagers and young adults represent approximately one-fourth of these. The addition of the methyl group is fairly simple, so methamphetamine is often produced from amphetamine in makeshift home laboratories. It is available as a powder (speed) or as crystals ("crystal meth," "glass," or "ice"). The crystals are smoked, and the effects are almost instantaneous and nearly as quick as when methamphetamine is snorted. The effects last four to eight hours when smoked. Methamphetamine has a structure similar to that of dopamine, and its stimulatory effect mimics cocaine. It reverses the effects of fatigue, maintains wakefulness, and temporarily elevates the mood of the user. After the initial rush, there is typically a state of high agitation that, in some individuals, leads to violent behavior. Chronic use can lead to what is called an amphetamine psychosis, resulting in paranoia; auditory and visual hallucinations; self-absorption; irritability; and aggressive, erratic behavior. Drug tolerance, dependence, and addiction are common. Hyperthermia, convulsions, and death can occur.

Video Meth and the Brain

Marijuana

The dried flowering tops, leaves, and stems of the Indian hemp plant, *Cannabis sativa,* contain and are covered by a resin that is rich in 9-tetrahydrocannabinol (THC). The names *cannabis* and *marijuana* apply to either the plant or THC. Usually, marijuana is smoked in a cigarette form called a "joint."

Recently, researchers have found that marijuana binds to a receptor for anandamide, a neurotransmitter that seems to create a feeling of peaceful contentment. The occasional marijuana user experiences a mild euphoria, along with alterations in vision and judgment, which result in distortions of space and time. Motor incoordination, including the inability to speak coherently, takes place. Heavy use can result in hallucinations, anxiety, depression, rapid flow of ideas, body image distortions, paranoid reactions, and similar psychotic symptoms. Craving and difficulty in stopping usage can occur as a result of regular use.

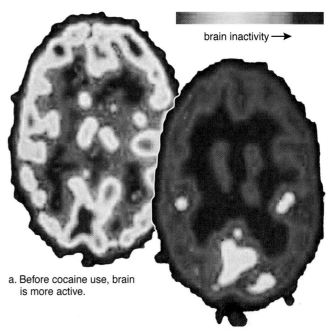

brain inactivity ⟶

a. Before cocaine use, brain is more active.

b. After cocaine use, brain is less active.

Figure 27.7   Cocaine's effect on the brain.

In a cocaine user, PET scans show that (**a**) the usual activity of the brain is (**b**) reduced. The color red indicates that brain tissue is active.

Connections and Misconceptions

What is medical marijuana used for?

Researchers continue to examine the potential medical benefits of THC, the active compound in marijuana. Initial studies have indicated that THC acts as an analgesic, or pain reducer. Medical marijuana is sometimes prescribed for patients in extreme  pain, such as those in the later stages of AIDS or those who have spinal cord injuries. However, it is important to note that it is the active compound in marijuana that has potential benefits. Smoking marijuana presents more health problems than the use of cigarettes. Marijuana smoke contains 50–70% more carcinogens than cigarette smoke In addition, in the hour following marijuana use, there is almost a fivefold increase in the risk of a heart attack.

The Central Nervous System

The organization of the brains in certain vertebrates—namely, reptiles (alligator and goose) and mammals (horse and human)—is compared in **Figure 27.8***a*. If we divide the brain into a hindbrain, midbrain, and forebrain, we can see that the forebrain is most prominent in humans. The forebrain of mammals also has an altered function in that it becomes the last depository for sensory information. This change accounts for why the forebrain carries on much of the integration for the entire nervous system before it sends out motor instructions to glands and muscles. In humans, the **spinal cord** provides a means of communication between the brain and the spinal nerves, which are a part of the PNS. Spinal nerves leave the spinal cord and take messages to and from the skin, glands, and muscles in all areas of the body, except the head and face. Long, myelinated fibers of interneurons in the spinal cord run together in bundles called **tracts.** These tracts connect the spinal cord to the brain. Because certain of the tracts cross over at one point, the left side of the brain controls the right side of the body, and vice versa. Also, as discussed on page 525, the spinal cord is involved in reflex actions, which are programmed, built-in circuits that allow for protection and survival. They are present at birth and require no conscious thought to take place.

The Brain

Our discussion will center on these parts of the brain: the cerebrum, the diencephalon, the cerebellum, and the brain stem (Fig. 27.8*b*).

Video
Brain Bank

Cerebrum   The cerebrum communicates with and coordinates the activities of the other parts of the brain. The cerebrum has two halves, and each half has a number of lobes. Most of the cerebrum is white matter where the long axons of interneurons are taking nerve impulses to and from the cerebrum. The highly convoluted outer layer of gray matter that covers the cerebrum is called the **cerebral cortex.** The cerebral cortex contains over 1 billion cell bodies, and it is the region of the cerebrum that interprets and initiates sensation, voluntary movement, and higher thought processes.

MP3
The Cerebrum

cerebellum

alligator

cerebrum

optic lobe

goose

medulla

olfactory lobe

horse

a.

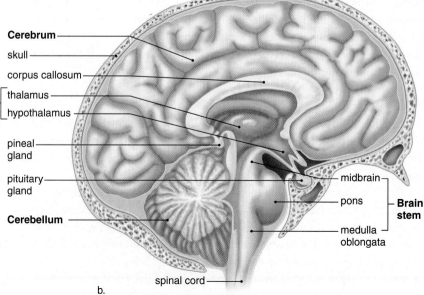

Cerebrum

skull

corpus callosum

Diencephalon — [thalamus

hypothalamus]

pineal gland

pituitary gland

Cerebellum

midbrain

pons

Brain stem

medulla oblongata

spinal cord

b.

Figure 27.8   Vertebrate brains.

a. A comparison of reptile and mammalian brains shows that the forebrain increased in size and complexity among these animals. **b.** The human brain.

Investigators have found that each part of the cerebrum has specific functions (**Fig. 27.9**). To take an example, the **primary sensory area** located in the parietal lobe receives information from the skin, skeletal muscles, and joints. Each part of the body has its own receiving area. The **primary motor area,** on the other hand, is in the frontal lobe just before the small cleft that divides the frontal lobe from the parietal lobe. Voluntary commands to skeletal muscles involve the primary motor area, and the muscles in each part of the body are controlled by a certain section of the primary motor area.

The lobes of the cerebral cortex have a number of specialized centers to receive information from the sensory receptors for sight, hearing, and smell. The lobes also have *association areas,* where integration occurs. The **prefrontal area,** an association area in the frontal lobe, receives information from the other association areas and uses this information to reason and plan our actions. Integration in this area accounts for our most cherished human abilities: to think critically and to formulate appropriate behaviors.

Video Brain Surgery

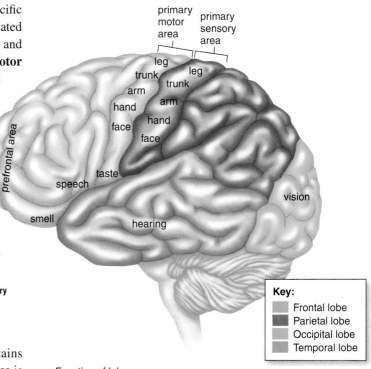

Key:
- Frontal lobe
- Parietal lobe
- Occipital lobe
- Temporal lobe

Function of lobes
- Frontal lobe—reasoning, planning, speech, movement, emotions, and problem solving
- Parietal lobe—integration of sensory input from skin and skeletal muscles, understanding speech
- Occipital lobe—seeing, perception of visual stimuli
- Temporal lobe—hearing, perception of auditory stimuli

Figure 27.9 **Functional regions of the cerebral cortex.**

Specific areas of the cerebral cortex receive sensory input from particular sensory receptors, integrate various types of information, or send out motor commands to particular areas of the body.

Diencephalon   Beneath the cerebrum is the diencephalon, which contains the hypothalamus and the thalamus (see Fig. 27.8*b*). The **hypothalamus** is an integrating center that helps maintain homeostasis by regulating hunger, sleep, thirst, body temperature, and water balance. The hypothalamus controls the pituitary gland, thereby serving as a link between the nervous and endocrine systems. The **thalamus** is on the receiving end for all sensory input except smell. Information from the eyes, ears, and skin arrives at the thalamus via the cranial nerves and tracts from the spinal cord. The thalamus integrates this information and sends it on to the appropriate portions of the cerebrum. The thalamus is involved in arousal of the cerebrum; it participates in motor functions and higher mental processes, such as memory and emotions.

The pineal gland, which secretes the hormone melatonin, is located in the diencephalon. Presently, there is much popular interest in melatonin because it is released at night when we are sleeping. Some researchers hypothesize that it can be used to help prevent jet lag or insomnia.

Video Winter Mood

Cerebellum   The **cerebellum** has two portions joined by a narrow median strip. Each portion is primarily composed of white matter, which in longitudinal section has a tree-like pattern (see Fig. 27.8*b*). Overlying the white matter is a thin layer of gray matter, which forms a series of complex folds.

The cerebellum receives sensory input from the eyes, ears, joints, and skeletal muscles about the present position of body parts, and it receives motor output from the cerebral cortex that specifies where these parts should be located. After integrating this information, the cerebellum sends motor impulses by way of the brain stem to the skeletal muscles. In this way, the cerebellum maintains posture and balance. It also ensures that all of the muscles work together to produce smooth, coordinated voluntary movements. The cerebellum helps us learn new motor skills, such as playing the piano or hitting a baseball.

Brain Stem   The **brain stem,** which contains the midbrain, the pons, and the medulla oblongata, connects the rest of the brain to the spinal cord (see Fig. 27.8*b*). It contains tracts that ascend or descend between the spinal cord and higher brain centers. The **midbrain** contains important visual and auditory reflex centers, and it coordinates responses such as the startle reflex. This occurs when you automatically turn your head in response to a sudden, loud noise, trying to see its source. The **medulla oblongata** contains a number of reflex centers for regulating heartbeat, breathing, and vasoconstriction (blood pressure). It also contains the reflex centers for vomiting, coughing, sneezing, hiccuping, and swallowing. In addition, the medulla oblongata helps control various internal organs. The **pons** links the medulla with the midbrain, and it is vital for the control of breathing.

The Limbic System

The **limbic system** is a complex network that includes the diencephalon and areas of the cerebrum (**Fig. 27.10**). The limbic system blends higher mental functions and primitive emotions into a united whole. It accounts for why activities such as sexual behavior and eating seem pleasurable and why, for instance, mental stress can cause high blood pressure.

Two significant structures within the limbic system are the hippocampus and the amygdala, which are essential for learning and memory. The hippocampus, a seahorse-shaped structure that lies deep in the temporal lobe, is well situated in the brain to make the prefrontal area aware of past experiences stored in sensory association areas. The amygdala, in particular, can cause these experiences to have emotional overtones. A connection between the frontal lobe and the limbic system means that reason can keep us from acting out strong feelings.

Learning and Memory   **Memory** is the ability to hold a thought in mind or recall events from the past, ranging from a word we learned only yesterday to an early emotional experience that has shaped our lives. Learning takes place when we retain and utilize past memories.

The prefrontal area in the frontal lobe is active during short-term memory, as when we temporarily recall a telephone number. Some telephone numbers go into long-term memory. Think of a telephone number you know by heart, and see if you can bring it to mind without also thinking about the place or person associated with that number. Most likely, you cannot, because typically long-term memory is a mixture of what is called *semantic memory* (numbers, words, and so on) and *episodic memory* (persons, events, and other associations). *Skill memory* is a type of memory that can exist independently of episodic memory. Skill memory is what allows us to perform motor activities, such as riding a bike or playing ice hockey. A person who has **Alzheimer disease (AD)** experiences a progressive loss of memory, particularly for recent events. Gradually, the person loses the ability to perform any type of daily activity and becomes bedridden. In AD patients, clusters of abnormal tissue develop among degenerating neurons, especially in the hippocampus and amygdala. Major research efforts are devoted to seeking a cure for AD.

What parts of the brain are functioning when we remember something from long ago? Our long-term memories are stored in bits and pieces throughout the sensory association areas of the cerebral cortex. The hippocampus gathers this information for use by the prefrontal area of the frontal lobe when we remember, for example, Uncle George or our summer holiday. Why

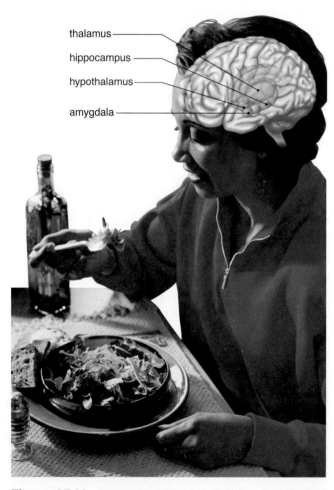

thalamus

hippocampus

hypothalamus

amygdala

Figure 27.10 The limbic system.

The limbic system includes the diencephalon and parts of the cerebrum. It joins higher mental functions, such as reasoning, with more instinctive feelings, such as fear and pleasure. Eating is a pleasurable activity for most people.

are some memories so emotionally charged? The amygdala is responsible for conditioning fear and for associating danger with sensory information received from the thalamus and the cortical sensory areas.

The Peripheral Nervous System

The peripheral nervous system (PNS) lies outside the central nervous system and contains **nerves,** which are bundles of axons (**Fig. 27.11**). The cell bodies of neurons are found in the CNS—that is, the brain and spinal cord—or in ganglia. Ganglia (sing., **ganglion**) are collections of cell bodies within the PNS.

Humans have 12 pairs of **cranial nerves** attached to the brain. Cranial nerves are largely concerned with the head, neck, and facial regions of the body. However, the vagus nerve is a cranial nerve that has branches not only to the pharynx and larynx but also to most of the internal organs.

Humans have 31 pairs of **spinal nerves,** and each contains many sensory and motor axons. The dorsal root of a spinal nerve contains the axons of sensory neurons, which conduct impulses to the spinal cord from sensory receptors. The cell body of a sensory neuron is in the **dorsal root ganglion.** The ventral root contains the axons of motor neurons, which conduct impulses away from the cord, largely to skeletal muscles. Each spinal nerve serves the region of the body in which it is located.

The Somatic System

The **somatic system** of the PNS includes the nerves that take sensory information from external sensory receptors to the CNS and motor commands away from the CNS to skeletal muscles. Voluntary control of skeletal muscles always originates in the brain. Involuntary responses to stimuli, called **reflexes,** can involve either the brain or just the spinal cord. Flying objects cause our eyes to blink, and sharp pins cause our hands to jerk away even without our having to think about it.

Figure 27.12 illustrates the path of a reflex that involves only the spinal cord. If your hand touches a sharp pin, sensory receptors in the skin generate nerve impulses that move along sensory axons toward the spinal cord (**Fig. 27.12**).

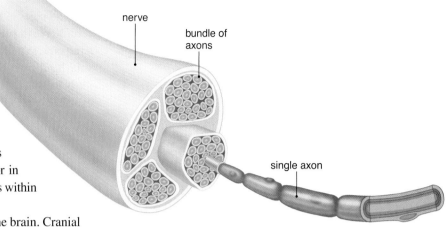

Figure 27.11   **A nerve.**

A nerve contains the axons of many neurons.

Figure 27.12   **A reflex arc showing the path of a spinal reflex.**

A stimulus (e.g., a pinprick) causes sensory receptors in the skin to generate nerve impulses, which travel in sensory axons to the spinal cord. Interneurons integrate data from sensory neurons and then relay signals to motor neurons. Motor axons convey nerve impulses from the spinal cord to a skeletal muscle, which contracts. Movement of the hand away from the pin is the response to the stimulus.

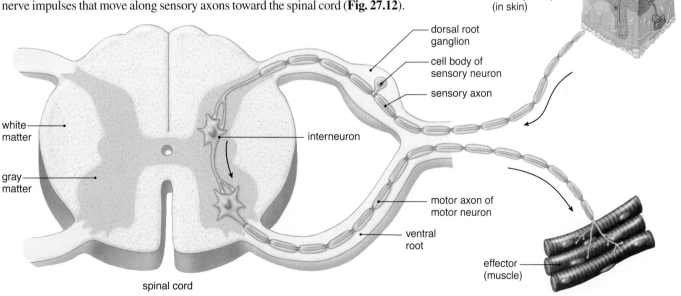

Sensory neurons that enter the spinal cord pass signals on to many interneurons. Some of these interneurons synapse with motor neurons. The short dendrites and the cell bodies of motor neurons are in the spinal cord, but their axons leave the cord. Nerve impulses travel along motor axons to an effector, which brings about a response to the stimulus. In this case, a muscle contracts, so that you withdraw your hand from the pin. Various other reactions are possible—you will most likely look at the pin, wince, and cry out in pain. This whole series of responses is explained by the fact that some of the interneurons involved carry nerve impulses to the brain. Also, sense organs send messages to the brain that make us aware of our actions. Your brain makes you aware of the stimulus and directs these other reactions to it.

The Autonomic System

The **autonomic system** of the PNS automatically and involuntarily regulates the activity of glands and cardiac and smooth muscle. The system is divided into the parasympathetic and sympathetic divisions (**Fig. 27.13**). Reflex actions, such as those that regulate blood pressure and breathing rate, are especially important to the maintenance of homeostasis. These reflexes begin when the sensory neurons in contact with internal organs send information to the CNS. They are completed by motor neurons within the autonomic system.

The **parasympathetic division** includes a few cranial nerves (e.g., the vagus nerve) and axons that arise from the last portion of the spinal cord. The parasympathetic division, sometimes called the "housekeeper division," pro-

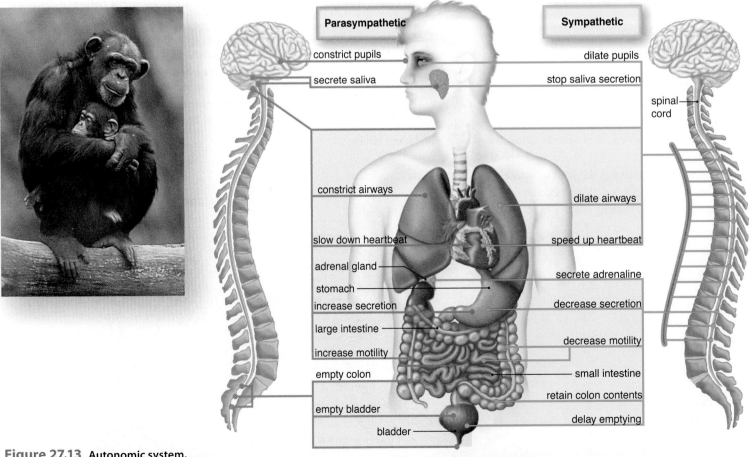

Figure 27.13 Autonomic system.

The parasympathetic and sympathetic motor axons go to the same organs, but they have opposite effects. The parasympathetic division is active when we, as mammals, feel warm and cozy in the arms of someone who loves us (see photograph). The sympathetic division is active when we are stressed and feel threatened.

motes all the internal responses we associate with a relaxed state (Fig. 27.13, photograph). For example, it causes the pupil of the eye to constrict, promotes the digestion of food, and retards the heartbeat. The neurotransmitter used by the parasympathetic division is acetylcholine (ACh).

Axons of the **sympathetic division** arise from portions of the spinal cord. The sympathetic division is especially important during emergency situations and is associated with "fight or flight." If you need to fend off a foe or flee from danger, active muscles require a ready supply of glucose and oxygen. On one hand, the sympathetic division accelerates the heartbeat and dilates the airways. On the other hand, the sympathetic division inhibits the digestive tract, since digestion is not an immediate necessity if you are under attack. The sympathetic nervous system uses the neurotransmitter norepinephrine, which has a structure like that of epinephrine (adrenaline) released by the adrenal medulla.

MP3
Organization of the Nervous System

Check Your Progress 27.1

1. List the three major types of neurons and their functions.
2. Identify the structural components of a neuron, and state the function of each.
3. List the steps in the generation of an action potential.
4. Explain how a signal is carried across the synaptic cleft.
5. Identify the structures of the central nervous system, and provide a function for each.
6. Explain the following: If a reflex action does not require the brain, why are we sometimes aware that a reflex has occurred?
7. Explain why the parasympathetic division of the autonomic system is called the "housekeeper division."

27.2 Endocrine System

Learning Outcomes

Upon completion of this section, you should be able to

1. Identify the major endocrine glands of the body, the hormones they secrete, and their effects in the body.
2. Compare and contrast the actions of steroid and peptide hormones.
3. Give examples of endocrine disorders.

The endocrine system consists of glands and tissues that secrete chemical signals called hormones (**Fig. 27.14**). **Endocrine glands** do not have ducts; they secrete their hormones directly into the bloodstream for distribution throughout the body. They can be contrasted with exocrine glands, which have ducts and secrete their products into these ducts for transport to body cavities. For example, the salivary glands are exocrine glands because they send saliva into the mouth by way of the salivary ducts.

The endocrine system and the nervous system are intimately involved in homeostasis, the relative stability of the internal environment. Hormones directly affect blood composition and pressure, body growth, and many more life processes. Certain hormones are involved in the maturation and function of the reproductive organs, and these are discussed in Chapter 29.

Connecting the Concepts

For more information on the interaction of the nervous system with other organ systems, refer to the following discussions.

Section 22.3 explores how the hypothalamus helps maintain temperature homeostasis.

Section 24.1 explains how the breathing center in the brain regulates respiration.

Section 28.1 examines the sensory systems that provide input to the central nervous system.

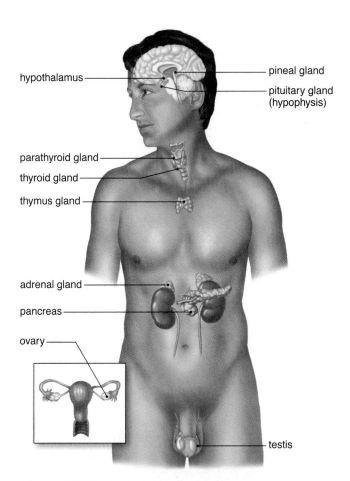

Figure 27.14 The endocrine system.

Location of major endocrine glands in the body.

Figure 27.15 How hormones work.

a. A steroid hormone (S) is a chemical signal that is able to enter the cell. Reception of this messenger causes the cell to synthesize a product by way of the cellular machinery for protein synthesis. **b.** A peptide hormone (P) is a "first messenger," which is received by a cell at the plasma membrane. Reception of the first messenger and a signal transduction pathway lead to a "second messenger," which changes the metabolism of the cell.

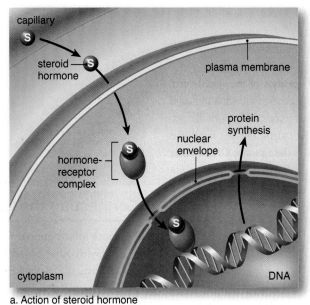

a. Action of steroid hormone

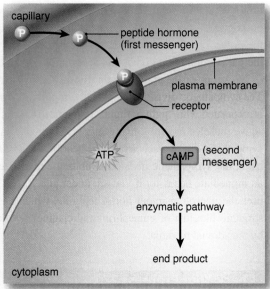

b. Action of peptide hormone

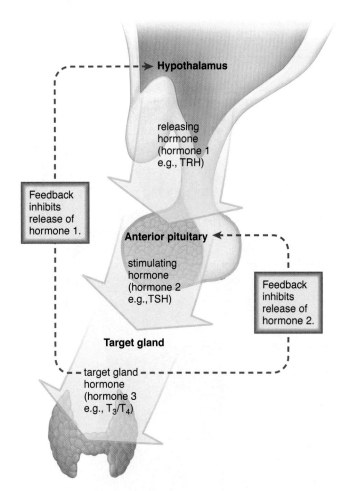

Figure 27.16 Negative feedback inhibition.

The hormones secreted by the thyroid (as well as the adrenal cortex and gonads) feed back to inhibit the anterior pituitary and hypothalamic-releasing hormones, so that their blood levels stay relatively constant.

The Action of Hormones

The cells that can respond to a hormone have receptor proteins that bind to the hormone. Hormones cause these cells to undergo a metabolic change. The type of change is dependent on the chemical structure of the hormone. **Steroid hormones** are lipids, and they can pass through the plasma membrane. The hormone-receptor complex then binds to DNA, and gene expression follows—for example, a protein (such as an enzyme) is made by the cell. The enzyme goes on to produce a change in the target cell's function (**Fig. 27.15**a).

Animation
Mechanism of Steroid Hormone Action

Peptide hormones comprise peptides, proteins, glycoproteins, and modified amino acids. Peptide hormones can't pass through the plasma membrane, so they bind to a receptor protein in the plasma membrane (Fig. 27.15b). The peptide hormone is called the "first messenger" because a signal transduction pathway leads to a second molecule—that is, the "second messenger"—that changes the metabolism of the cell. The second messenger sets in motion an enzyme pathway, which is sometimes called an enzyme cascade because each enzyme in turn activates another. Because enzymes work over and over, every step in an enzyme cascade leads to more reactions—the binding of a single peptide hormone molecule can result in as much as a thousandfold response.

Animation
Peptide Hormone Action

Animation
Second Messengers

Hypothalamus and Pituitary Gland

The hypothalamus, a part of the brain (see Fig. 27.8), helps regulate the internal environment. For example, it is on the receiving end of information about the heartbeat and body temperature. And to correct any abnormalities, the hypothalamus communicates with the medulla oblongata, where the brain centers that control the autonomic system are located. The hypothalamus is also a part of the endocrine system, containing specialized hormone-secreting neurons. It controls the glandular secretions of the **pituitary gland,** a small gland connected to the brain by a stalk-like structure. The pituitary has two portions, the anterior pituitary and the posterior pituitary, which are distinct from each other.

Anterior Pituitary

The hypothalamus controls the anterior pituitary by producing **hypothalamic-releasing hormones,** most of which are stimulatory (**Fig. 27.16**). The **anterior pituitary** in turn stimulates other glands (**Fig. 27.17**):

1. **Thyroid-stimulating hormone (TSH)** stimulates the thyroid to produce triiodothyronine (T_3) and thyroxine (T_4).
2. **Adrenocorticotropic hormone (ACTH)** stimulates the adrenal cortex to produce the glucocorticoids.
3. **Gonadotropic hormones (FSH** and **LH)** stimulate the gonads—the testes in males and the ovaries in females—to produce gametes and sex hormones.

In these instances, a three-tiered control system develops (see Fig. 27.16). For example, the secretion of thyroid-releasing hormone (TRH) by the hypothalamus stimulates the thyroid to produce the thyroid-stimulating hormone (TSH), and the thyroid produces its hormones (T_3 and T_4), which feed back to inhibit the release of the first two hormones mentioned.

Figure 27.17   The hypothalamus and the pituitary.

(Left) The hypothalamus controls the secretions of the anterior pituitary, and the anterior pituitary controls the secretions of the thyroid gland, adrenal cortex, and gonads, which are also endocrine glands. Growth hormone and prolactin are also produced by the anterior pituitary. (Right) The hypothalamus produces two hormones, ADH and oxytocin, which are stored and secreted by the posterior pituitary.

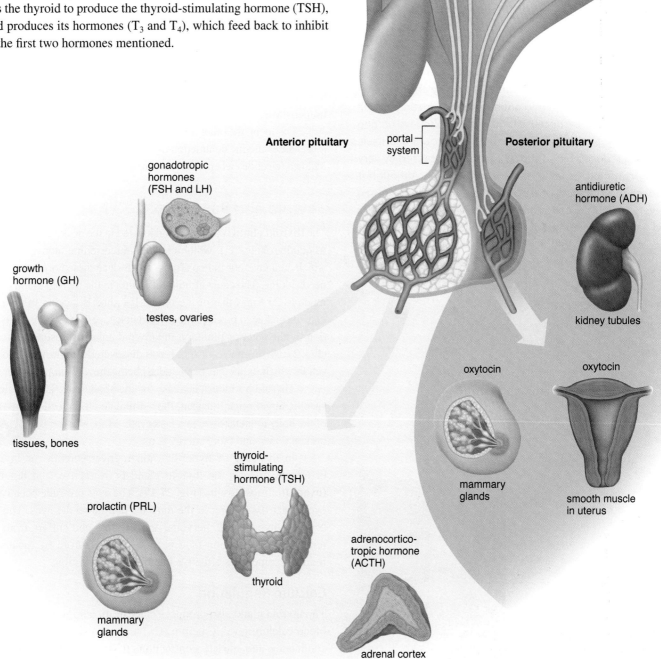

Connections and Misconceptions

How is labor induced if a woman's pregnancy extends past her due date?

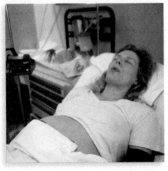

After the woman is given medication to prepare the birth canal for delivery, pitocin (a synthetic version of oxytocin) is used to induce labor. During labor, it may also be given to increase the strength of contractions. Stronger contractions speed the labor process, if necessary (for example, if the woman's uterus is contracting poorly or if the health of the mother or child is at risk during delivery). Pitocin is routinely used following delivery to minimize postpartum bleeding by ensuring that strong uterine contractions continue. Use of pitocin must be monitored carefully, because it may cause excessive uterine contractions. Should this occur, the uterus could tear itself. Further, a reduced blood supply to the fetus, caused by very strong contractions, may be fatal to the baby. Although it reduces the duration of labor, induction with pitocin can be very painful for the mother. Whenever possible, gentler and more natural methods should be used to induce labor and/or strengthen contractions.

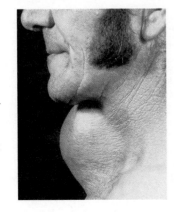

Figure 27.18
Simple goiter.

An enlarged thyroid gland can result from too little iodine in the diet. The thyroid is under constant stimulation to produce more of its hormones and so it enlarges, resulting in a simple goiter.

Two other hormones produced by the anterior pituitary do not affect other endocrine glands. **Prolactin (PRL)** is produced in quantity during pregnancy and after childbirth. It causes the mammary glands in the breasts to develop and produce milk. It also plays a role in carbohydrate and fat metabolism. **Growth hormone (GH)** promotes skeletal and muscular growth. It stimulates the rate at which amino acids enter cells and protein synthesis occurs. Underproduction of growth hormone leads to pituitary dwarfism, and overproduction can lead to pituitary gigantism.

Posterior Pituitary

The hypothalamus produces two hormones, **antidiuretic hormone (ADH)** and **oxytocin** (Fig. 27.17). The axons of hypothalamic secretory neurons extend into the **posterior pituitary,** where hormones are stored in the axon terminals. When the hypothalamus determines that the blood is too concentrated, ADH is released from the posterior pituitary. On reaching the kidneys, ADH causes water to be reabsorbed. As the blood becomes dilute, ADH is no longer released. This is also an example of control by **negative feedback** because the effect of the hormone (to dilute blood) is to shut down the hormone's release. Negative feedback, as discussed in Chapter 22, maintains homeostasis.

Oxytocin, the other hormone made in the hypothalamus, causes uterine contraction during childbirth and milk letdown (ejection) when a baby is nursing.

Animation
Hormonal Communication

Thyroid and Parathyroid Glands

The **thyroid gland** is a large gland located in the neck; it produces two hormones: triiodothyronine (T_3), which contains three iodine atoms, and **thyroxine (T_4)**, which contains four iodine atoms. To produce these hormones, the thyroid gland needs iodine. The concentration of iodine in the thyroid gland can increase to as much as 25 times that of the blood. If iodine is lacking in the diet, the thyroid gland is unable to produce the thyroid hormones. In response to constant stimulation by the anterior pituitary, the thyroid enlarges, resulting in a **simple goiter** (**Fig. 27.18**). Some years ago, it was discovered that the use of iodized salt (table salt to which iodine has been added) helps prevent simple goiter.

Thyroid hormones increase the metabolic rate. They do not have a single, specific target organ; instead, they stimulate all the cells of the body to metabolize at a faster rate. More glucose is broken down, and more energy is used.

Animation
Mechanism of Thyroxine Action

In the case of hyperthyroidism (oversecretion of thyroid hormone), or Graves disease, the thyroid gland is overactive, and bulging of the eyes (**exophthalmos**) results (**Fig. 27.19**). The eyes protrude because of swelling in eye socket tissues and in the muscles that move the eyes. The patient usually becomes hyperactive, nervous, and irritable and suffers from insomnia. The removal or destruction of a portion of the thyroid by means of radioactive iodine is sometimes effective in curing the condition.

Calcium Regulation

The thyroid gland also produces **calcitonin,** a hormone that helps regulate the blood calcium level. Calcium (Ca^{2+}) plays a significant role in both nervous conduction and muscle contraction. It is also necessary for blood clotting. Calcitonin temporarily reduces the activity and number of osteoclasts, cells that break down bone. Therefore, more calcium is deposited in bone. When the blood calcium level lowers to normal, the thyroid's release of calcitonin is inhibited by negative feedback, but a low level stimulates the parathyroid glands' release of **parathyroid hormone (PTH)**. The parathyroid glands are embedded in the posterior surface of the thyroid gland. Many years ago, the four **parathyroid glands** were sometimes mistakenly removed during thyroid surgery because of their size and location.

Parathyroid hormone promotes the activity of osteoclasts and the release of calcium from the bones. PTH also promotes the reabsorption of calcium by the kidneys, where it activates vitamin D. Vitamin D, in turn, stimulates the absorption of calcium from the small intestine. These effects bring the blood calcium level back to the normal range, so that the parathyroid glands no longer secrete PTH.

When insufficient parathyroid hormone production leads to a dramatic drop in the blood calcium level, tetany results. In **tetany,** the body shakes from continuous muscle contraction. This effect is brought about by increased excitability of the nerves, which, in turn, initiates nerve impulses spontaneously and without rest.

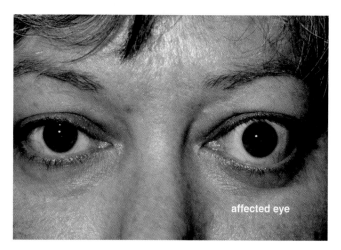

Figure 27.19 Exophthalmos.

Enlargement of the thyroid gland can cause the eyes to protrude, a condition called exophthalmos. In this individual, only the left eye is affected. The overactive thyroid produces too much of its hormones and the individual is hyperactive and nervous.

Adrenal Glands

Two **adrenal glands** sit atop the kidneys. Each adrenal gland consists of an inner portion called the **adrenal medulla** and an outer portion called the **adrenal cortex.** These portions, like the anterior pituitary and the posterior pituitary, have no functional connection with one another.

The hypothalamus exerts control over the activity of both portions of the adrenal glands. It initiates nerve impulses that travel by way of the brain stem, spinal cord, and sympathetic nerve fibers to the adrenal medulla, which then secretes its hormones. The hypothalamus, by means of ACTH-releasing hormone, controls the anterior pituitary's secretion of ACTH, which in turn stimulates the adrenal cortex. Stress of all types—including emotional and physical trauma, and even vigorous exercise—prompts the hypothalamus to stimulate both the adrenal medulla and the adrenal cortex.

Adrenal Medulla

Epinephrine (adrenaline) and norepinephrine (noradrenaline) produced by the adrenal medulla rapidly bring about all the body changes that occur when an individual reacts to an emergency situation. In so doing, these two hormones complement the actions of the sympathetic autonomic system. The effects of these hormones are short-term. In contrast, the hormones produced by the adrenal cortex provide a long-term response to stress.

Adrenal Cortex

The two major types of hormones produced by the adrenal cortex are the **mineralocorticoids,** such as aldosterone, and the **glucocorticoids,** such as cortisol. Aldosterone acts on the kidneys and thereby regulates salt and water balance, leading to increases in blood volume and blood pressure. Cortisol regulates carbohydrate, protein, and fat metabolism, leading to an increase in the blood glucose level. It is also an anti-inflammatory agent. The adrenal cortex also secretes a small amount of both male and female sex hormones in both sexes.

Animation
Glucocorticoid
Hormones

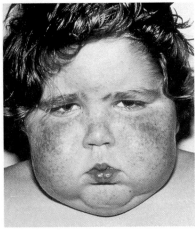

a. Before treatment

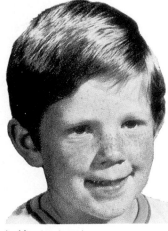

b. After treatment

Figure 27.20 Cushing syndrome.

Cushing syndrome results from hypersecretion of hormones by the adrenal cortex, possibly due to a tumor. **a.** Patient first diagnosed with Cushing syndrome. **b.** Four months later, after treatment.

When the level of adrenal cortex hormones is low due to hyposecretion, a person develops **Addison disease.** ACTH may build up as more is secreted to attempt to stimulate the adrenal cortex. The excess can cause bronzing of the skin, because ACTH in excess stimulates melanin production. Without cortisol, glucose cannot be replenished when a stressful situation arises. Even a mild infection can lead to death. The lack of aldosterone results in the loss of sodium and water by the kidneys, low blood pressure, and possibly severe dehydration. Left untreated, Addison disease can be fatal.

When the level of adrenal cortex hormones is high due to hypersecretion, a person develops **Cushing syndrome** (**Fig. 27.20**). The excess cortisol results in a tendency toward diabetes mellitus as muscle protein is metabolized and subcutaneous fat is deposited in the midsection. The trunk is obese, while the arms and legs remain a normal size. An excess of aldosterone and reabsorption of sodium and water by the kidneys lead to a basic blood pH and hypertension. The face swells and takes on a moon shape. Masculinization may occur in women because of excess adrenal male sex hormones.

Pancreas

The **pancreas** is composed of two types of tissue. Exocrine tissue produces and secretes digestive juices that pass through ducts to the small intestine. Endocrine tissue, called the **pancreatic islets** (islets of Langerhans), produces and secretes the hormones **insulin** and **glucagon** directly into the blood (**Fig. 27.21**).

Insulin is secreted when there is a high blood glucose level, which usually occurs just after eating. Insulin stimulates the uptake of glucose by cells, especially liver cells, muscle cells, and adipose tissue cells. In liver and muscle cells, glucose is then stored as glycogen. In muscle cells, glucose supplies energy for ATP production, leading to protein metabolism and muscle contraction. In fat cells, the breakdown of glucose supplies glycerol and acetyl groups for the formation of fat. In these ways, insulin lowers the blood glucose level.

Glucagon is usually secreted between meals, when the blood glucose level is low. The major target tissues of glucagon are the liver and adipose tissue. Glucagon stimulates the liver to break down glycogen to glucose and to use fat and protein in preference to glucose as energy sources. The latter spares glucose and makes more available to enter the blood. In these ways, glucagon raises the blood glucose level.

Diabetes Mellitus

Diabetes mellitus is a fairly common hormonal disease in which liver cells, and indeed all body cells, do not take up and/or metabolize glucose. Therefore, the cells are in need of glucose, even though there is plenty in the blood. As the blood glucose level rises, water and glucose are excreted in the urine. The loss of water in this way causes the diabetic person to be extremely thirsty.

Two types of diabetes mellitus have been distinguished. In *type 1 diabetes* (insulin-dependent diabetes), the pancreas is not producing insulin. The condition is believed to be brought on by exposure to an environmental agent, most likely a virus, whose presence causes cytotoxic T cells to destroy the pancreatic islets. The cells turn to the breakdown of protein and fat for energy. The metabolism of fat leads to acidosis (acid blood), which can eventually cause coma and death. As a result, the individual must have daily insulin injections. These injections control the diabetic symptoms but can still cause inconveniences, since either taking too much insulin or failing to eat regularly can bring on the symptoms of hypoglycemia (low blood sugar). These symptoms include perspiration, pale skin, shallow

Connections and Misconceptions

What is gestational diabetes, and what causes it?

Women who were not diabetic prior to pregnancy but have high blood glucose levels during pregnancy have gestational diabetes. Gestational diabetes affects a small percentage of pregnant women. This form of diabetes is caused by insulin resistance—body insulin concentration is normal, but the cells fail to respond normally. Gestational diabetes and insulin resistance generally develop later in the pregnancy. Carefully planned meals and exercise often control this form of diabetes, but insulin injections may be necessary.  If the woman is not treated, additional glucose crosses the placenta, causing high blood glucose in the fetus. The extra energy in the fetus is stored as fat, resulting in macrosomia, or a "fat" baby. Delivery of a very large baby can be dangerous for both the infant and the mother. Cesarean section is often necessary. Complications after birth are common for these babies. Further, there is a greater risk that the child will become obese and develop type 2 diabetes mellitus later in life. Usually, gestational diabetes goes away after the birth of the child. However, once a woman has experienced gestational diabetes, she has a greater chance of developing it again during future pregnancies. These women also tend to develop type 2 diabetes later in life.

breathing, and anxiety. Because the brain requires a constant supply of glucose, unconsciousness can result. The treatment is quite simple: Immediate ingestion of a sugar cube or fruit juice can very quickly counteract hypoglycemia.

Of the 16 million people who now have diabetes in the United States, most have *type 2 diabetes* (non-insulin-dependent diabetes). This type of diabetes mellitus usually occurs in people of any age who are obese and inactive, as discussed in Chapter 25. The pancreas produces insulin, but the liver and muscle cells do not respond to it in the usual manner. They are said to be insulin resistant. If type 2 diabetes is untreated, the results can be as serious as those of type 1. People with diabetes of either type are prone to blindness, kidney disease, and circulatory disorders. It is usually possible to prevent or at least control type 2 diabetes by adhering to a low-fat and low-sugar diet and exercising regularly.

Animation
Blood Sugar
Regulation in
Diabetics

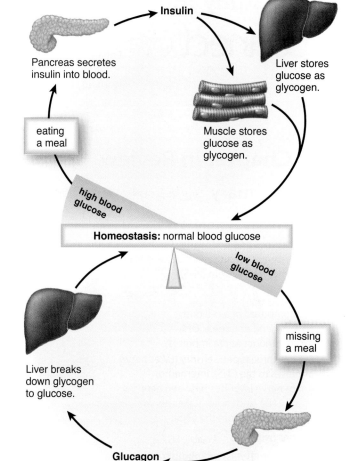

Figure 27.21   Regulation of blood glucose level.

Typical of hormones, insulin is regulated by negative feedback—once the blood glucose level is low, insulin is no longer secreted. The effect of insulin is countered by glucagon, which raises the blood glucose level. The two hormones work together to keep the blood glucose level relatively constant.

Check Your Progress 27.2

1 Explain how steroid and peptide hormones induce metabolic changes in cells.

2 Explain how the hypothalamus and pituitary gland work to control the endocrine system.

3 Compare and contrast calcitonin with parathyroid hormone (PTH).

4 Contrast the function of insulin with that of glucagon.

5 Compare and contrast type 1 diabetes with type 2 diabetes.

6 Explain how negative feedback mechanisms assist the endocrine system in the maintenance of homeostasis.

Connecting the Concepts

For more information on the topics discussed in this section, refer to the following discussions.

Section 22.2 explains how negative feedback mechanisms help maintain homeostasis.

Section 25.3 describes the link between diabetes and obesity.

Section 29.2 examines the hormones produced by the sex organs of males and females.

Media Study Tools

www.mhhe.com/maderessentials3

Enhance your study of this chapter with study tools and practice tests. Also ask your instructor about the resources available through ConnectPlus, including the media-rich eBook, interactive learning tools, and animations.

The Chapter in Review

Summary

27.1 Nervous System

Both the nervous and endocrine systems use chemical signals. The nervous system is organized into a central nervous system (CNS) and a peripheral nervous system (PNS).

Neurons

- Are composed of a cell body, an axon, and dendrites. Only axons conduct nerve impulses.
- Exist as three types: sensory (takes nerve impulses to the CNS), interneuron (takes nerve impulses between neurons of the CNS), and motor (takes nerve impulses away from the CNS).

The Nerve Impulse

- The sodium-potassium pump transports Na^+ out of the axon and K^+ into the axon. The inside of the axon has a negative charge; the outside has a positive charge.
- The nerve impulse is an action potential—there is a reversal of charge as Na^+ flows in, and then a return to the previous charge difference as K^+ flows out of an axon.
- The nerve impulse is much faster in myelinated axons because the impulse jumps from node to node.

The Synapse

- The synapse is a region of close proximity between an axon terminal and the next neuron (CNS), or between an axon terminal and a muscle cell (PNS).
- A nerve impulse causes the release of a neurotransmitter (excitatory or inhibitory) into the synaptic cleft.
- Neurotransmitters are ordinarily removed quickly from the synaptic cleft. AChE breaks down acetylcholine, for example.
- Drugs affect the action of neurotransmitters.

The Central Nervous System

The Brain

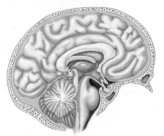

- The cerebrum functions in sensation, reasoning, learning and memory, language, and speech. The cerebral cortex has a primary sensory area in the parietal lobe, which receives sensory information from each part of the body, and a primary motor area in the frontal lobe,

which sends out motor commands to skeletal muscles. Association areas carry on integration.
- In the diencephalon, the hypothalamus helps control homeostasis; the thalamus specializes in sending sensory input to the cerebrum.
- The cerebellum primarily coordinates skeletal muscle contractions.
- In the brain stem, the medulla oblongata has centers for vital functions, such as breathing and the heartbeat, and helps control the internal organs.

The Limbic System

The limbic system blends higher functions into a united whole. The hippocampus and amygdala have roles in learning and memory and appear to be affected in Alzheimer disease.

The Peripheral Nervous System

- Somatic system: Reflexes (automatic responses) involve a sensory receptor, a sensory neuron, interneurons in the spinal cord, and a motor neuron.
- Autonomic system: The sympathetic division is active during times of stress, and the parasympathetic division is active during times of relaxation. Both divisions control the same internal organs.

27.2 Endocrine System

Endocrine glands secrete hormones into the bloodstream for distribution to target organs or tissues. Hormones are either steroids or peptides.

Hypothalamus and Pituitary Gland	
Hypothalamus secretes	
Releasing and inhibiting hormones	Control anterior pituitary
Antidiuretic hormone (ADH)	Released by posterior pituitary; causes water uptake by kidneys
Oxytocin	Released by posterior pituitary; causes uterine contractions
Anterior Pituitary secretes	
Gonadotropic hormones	Stimulate gonads (FSH and LH)
Thyroid-stimulating hormone (TSH)	Stimulates thyroid
Adrenocorticotropic hormone (ACTH)	Stimulates adrenal cortex
Prolactin	Causes milk production
Growth hormone (GH)	Causes cell division, protein synthesis, bone growth; too little = *pituitary dwarfism*; too much = *pituitary gigantism*

Thyroid and Parathyroid Glands

Thyroid Gland secretes

Thyroxine (T_4) and triiodothyronine (T_3)	Increase metabolic rate; *simple goiter* when iodine is lacking, *exophthalmos* when overactive
Calcitonin	Lowers blood calcium level

Parathyroid Glands secrete

Parathyroid hormone (PTH)	Raises blood calcium level

Adrenal Glands

Adrenal Medulla secretes

Epinephrine and norepinephrine	Response to emergency situations

Adrenal Cortex secretes

Mineralocorticoids (aldosterone)	Cause kidneys to reabsorb Na^+; too little, as in *Addison disease* = low blood pressure; too much, as in *Cushing syndrome* = high blood pressure
Glucocorticoids (cortisol)	Raises blood glucose level; too little, as in *Addison disease* = body can't respond to stress; too much, as in *Cushing syndrome* = diabetes
Sex hormones	Male/female differences

Pancreas secretes

Insulin	Causes cells to take up and liver to store glucose as glycogen; too little = *diabetes mellitus*
Glucagon	Causes liver to break down glycogen

Key Terms

acetylcholine (ACh)   520
acetylcholinesterase (AChE)   520
action potential   518
Addison disease   532
adrenal cortex   531
adrenal gland   531
adrenal medulla   531
adrenocorticotropic hormone (ACTH)   529
Alzheimer disease (AD)   524
anterior pituitary   529
antidiuretic hormone (ADH)   530
autonomic system   527
axon   517
brain stem   524
calcitonin   530
cell body   517
central nervous system (CNS)   517

cerebellum   523
cerebral cortex   522
cranial nerve   525
Cushing syndrome   532
dendrite   517
diabetes mellitus   532
dorsal root ganglion   525
endocrine gland   527
epinephrine   531
exophthalmos   530
ganglion   525
glucagon   532
glucocorticoid   531
gonadotropic hormones (FSH and LH)   529
growth hormone (GH)   530
hypothalamic-releasing hormone   529

hypothalamus   523
insulin   532
integration   520
interneuron   518
limbic system   524
medulla oblongata   524
memory   524
midbrain   524
mineralocorticoid   531
motor neuron   518
myelin sheath   517
negative feedback   530
nerve   525
neuron   517
neurotransmitter   520
nodes of Ranvier   518
norepinephrine (NE)   520
oxytocin   530
pancreas   532
pancreatic islets   532
parasympathetic division   527
parathyroid gland   531
parathyroid hormone (PTH)   531
peptide hormone   528
peripheral nervous system (PNS)   517

pituitary gland   528
pons   524
posterior pituitary   530
prefrontal area   523
primary motor area   523
primary sensory area   523
prolactin (PRL)   530
reflex   525
saltatory conduction   519
sensory neuron   518
serotonin   520
simple goiter   530
somatic system   525
spinal cord   522
spinal nerve   525
steroid hormone   528
sympathetic division   527
synapse   519
synaptic cleft   519
tetany   531
thalamus   523
thyroid gland   530
thyroid-stimulating hormone (TSH)   529
thyroxine (T_4)   530
tract   522

Testing Yourself

Choose the best answer for each question.

1. Unlike the nervous system, the endocrine system
   a. uses chemical signals as a means of communication.
   b. helps maintain equilibrium.
   c. sends messages to target organs.
   d. changes the metabolism of cells.
2. Pain receptors are at the distal ends of
   a. interneurons.   c. sensory neurons.
   b. intraneurons.   d. motor neurons.
3. Glucagon causes
   a. the use of fat for energy.
   b. glycogen to be converted to glucose.
   c. the use of amino acids to form fats.
   d. Both a and b are correct.
   e. None of these are correct.
4. When comparing the interior of an axon at rest with the exterior, there is
   a. a charge difference.
   b. an ion concentration difference.
   c. both a charge and an ion concentration difference.
   d. neither a charge nor an ion concentration difference.
5. Long-term use of cocaine causes an intense craving for the drug because the body has begun to make
   a. less dopamine.   c. less anandamide.
   b. more dopamine.   d. more anandamide.

6. Myelin is located where in a neuron?
  a. the dendrites   **c.** the axons
  b. the nucleus   **d.** within the synapse

7. Label the parts of the brain in the following illustration.

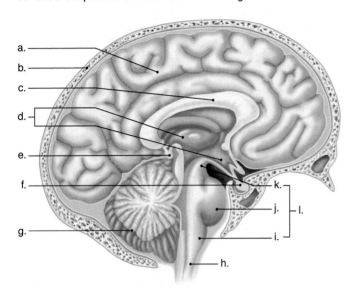

For questions 8–12, identify the part of the brain in the key that matches the description. Some answers may be used more than once. Some questions may have more than one answer.

Key:

  a. cerebrum
  b. diencephalon
  c. cerebellum
  d. brain stem

8. regulates hunger, thirst, and sleep
9. receives sensory information from eyes and ears
10. composed of two parts; mostly white matter
11. regulates heartbeat, breathing, and blood pressure
12. responsible for sensation, voluntary movement, and higher thought processes

13. In contrast to exocrine glands, endocrine glands
  a. secrete products.
  b. induce a response by the body.
  c. use ducts.
  d. produce hormones.

14. Both the adrenal medulla and the adrenal cortex are
  a. endocrine glands.
  b. in the same organ.
  c. involved in our response to stress.
  d. All of these are correct.

15. Type 1 diabetes is thought to result from a virus that
  a. interferes with gene expression in pancreatic cells.
  b. causes T cells to destroy the pancreatic islets.
  c. breaks down insulin.
  d. prevents the secretion of insulin by the pancreatic islets.

16. Label the parts of the endocrine system in the following illustration.

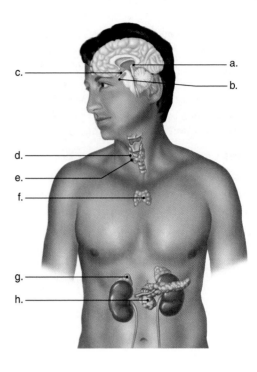

17. The cerebellum
  a. coordinates skeletal muscle movements.
  b. receives sensory input from the joints and muscles.
  c. receives motor input from the cerebral cortex.
  d. All of these are correct.

18. An interneuron can relay information
  a. from a sensory neuron to a motor neuron.
  b. only from a motor neuron to another motor neuron.
  c. only from a sensory neuron to another sensory neuron.
  d. None of these are correct.

19. The sympathetic division of the autonomic nervous system will
  a. increase heart rate and digestive activity.
  b. decrease heart rate and digestive activity.
  c. cause pupils to constrict.
  d. None of these are correct.

20. When the action potential begins, sodium gates open, allowing Na^+ to cross the membrane. Now the polarity changes to
  a. negative outside and positive inside.
  b. positive outside and negative inside.
  c. neutral outside and positive inside.
  d. There is no difference in charge between outside and inside.

21. Transmission of the nerve impulse across a synapse is accomplished by the
  a. movement of Na^+ and K^+.
  b. release of a neurotransmitter by a dendrite.
  c. release of a neurotransmitter by an axon.
  d. release of a neurotransmitter by a cell body.
  e. All of these are correct.

22. The autonomic system has two divisions, called the
  a. CNS and PNS.
  b. somatic and skeletal systems.
  c. efferent and afferent systems.
  d. sympathetic and parasympathetic divisions.

23. The limbic system
  a. involves portions of the cerebral lobes and the diencephalon.
  b. is responsible for our deepest emotions, including pleasure, rage, and fear.
  c. is a system necessary to memory storage.
  d. is not directly involved in language and speech.
  e. All of these are correct.

24. Which of these correctly describes the distribution of ions on either side of an axon when it is not conducting a nerve impulse?
  a. more sodium ions (Na^+) outside and more potassium ions (K^+) inside
  b. more K^+ outside and less Na^+ inside
  c. charged protein outside and Na^+ and K^+ inside
  d. Na^+ and K^+ outside and water only inside
  e. chloride ions (Cl^-) outside and K^+ and Na^+ inside

25. Growth hormone is produced by the
  a. posterior adrenal gland.
  b. posterior pituitary.
  c. anterior pituitary.
  d. kidneys.
  e. None of these are correct.

26. Which of the following is not true of the nerve impulse?
  a. It is subject to a short refractory period before it can occur again.
  b. It is slower in thick, myelinated fibers.
  c. It is also called an action potential.
  d. It moves from node to node in myelinated fibers.

27. PTH causes the blood level of calcium to _____, and calcitonin causes it to _____.
  a. increase, not change
  b. increase, decrease
  c. decrease, also decrease
  d. decrease, increase
  e. not change, increase

28. A bundle of axons in the PNS is called a
  a. ganglion.   **d.** node of Ranvier.
  b. tract.   **e.** synapse.
  c. nerve.

Thinking Scientifically

1. Recent research indicates that Parkinson disease damages the sympathetic division of the peripheral nervous system. One test for sympathetic division function, called the Valsalva maneuver, requires the patient to blow against resistance. A functional nervous system will compensate for the decrease in blood output from the heart by constricting blood vessels. How do you suppose Parkinson patients respond to the Valsalva maneuver? How does this relate to a common condition in Parkinson patients, called orthostatic hypotension, in which blood pressure falls suddenly when the person stands up, leading to dizziness and fainting?

2. Researchers have been trying to determine the reason for a dramatic rise in type 2 diabetes in recent decades. A sedentary lifestyle and poor eating habits certainly contribute to the risk of developing the disease. In addition, however, some researchers have observed a connection between childhood vaccination and type 1 diabetes. Epidemiological data from countries that have recently initiated mass immunization programs indicate that the incidence of type 1 diabetes has increased there as well. What might be the connection between vaccination and diabetes?

3. Scientists studying orangutans on the Indonesian islands of Borneo and Sumatra recently found that the subspecies inhabiting the food-scarce region of Borneo have smaller brains than those on Sumatra, where food is abundant. From this observation, what can we infer about the relationship between brain size and diet in orangutan evolution?

Bioethical Issue

Human Growth Hormone

Recently, the Food and Drug Administration (FDA) approved the use of human growth hormone (hGH) to treat "short stature." This decision implies that short stature is a medical condition with a legitimate need to be treated. In this case, medical care is not treating a disease but changing a feature of an otherwise healthy person. Proponents of the FDA decision say that administering hGH to short people will help them avoid discrimination and live a more normal life in a world designed for taller people. Opponents say that this decision may lead to slippery-slope scenarios in which medical treatments that make us smarter or faster are assumed to make us better. Do you think hGH should be used to "cure shortness"? Typically, parents will need to make decisions about hGH treatments for their children, since the treatments are generally administered to preschoolers. Should parents be allowed to make decisions about treatments that will determine their children's height?

At the other end of the life span, hGH has been widely advertised as a cure for aging. Levels of hGH naturally decline with age, and there is evidence that treating older adults with hGH boosts muscle mass and reduces body fat. However, the only effective way to administer hGH is by injection; the hGH pills hyped on numerous websites do not work. Injections of hGH are by prescription only and very expensive, around $1,000 a month. Furthermore, hGH therapy can have undesirable side effects, such as elevated blood sugar, fluid retention, and joint pain. Nevertheless, there is keen interest in hGH therapy, not only from those who wish to delay or avoid the effects of aging but also by athletes who wish to improve their performance. Should the physical decline of aging be accepted, or should we try to maintain a youthful body using hGH?

28

Sensory Input and Motor Output

BEFORE YOU BEGIN

Before beginning this chapter, take a few moments to review the following discussions.

Figure 5.4 What are the steps in the ADP-ATP cycle?

Figure 6.4 How does color relate to the wavelength of light?

Section 27.1 What areas of the brain are responsible for integrating sensory input and generating motor output?

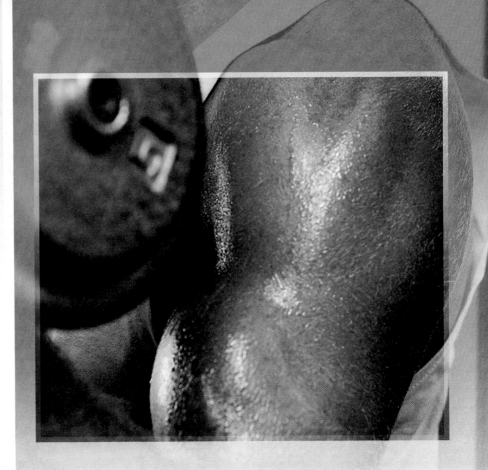

Your Body Needs Calcium

When you think of the importance of calcium in nutrition, you most likely think about bone. Bone serves as a depository for calcium, so that there is always a reserve in the body, and as an added benefit, storing calcium makes bone tissue strong. However, calcium plays other important roles in the body as well.

One of the most essential functions of calcium involves muscle contraction. No muscle in the body (including the heart and the diaphragm) can contract without the presence of calcium. If you were dependent solely on obtaining calcium from your daily diet and you ran out of calcium, your heart and diaphragm wouldn't contract. Then, the inability to circulate your blood and to breathe would cause you to die. Luckily, our bones stockpile calcium, so when we don't get enough in our diet, we get it from the reserve supply in our bones. The downside is that, in order to tap into the reserve of calcium in our bones, bone tissue must be degraded. This is why obtaining enough calcium in the diet is critical. When there is an imbalance between the amount of calcium taken from bones and the amount of calcium deposited in bones, conditions such as osteoporosis often result. In osteoporosis, the bones are fragile, and a broken hip, spine, or wrist often occurs. This is why we hear so much about including plenty of calcium in the diet.

The functions of bones and muscle are closely related to each other. Both are involved with executing motor output commands from the central nervous system. In this chapter, in addition to learning about these motor output systems, you will learn about how the senses provide input to the central nervous system.

28.1 The Senses

Learning Outcomes

Upon completion of this section, you should be able to

1. Summarize how the human senses of taste and smell work.
2. Describe the structures of the human ear and their functions.
3. Differentiate between rotational and gravitational equilibrium and the organs responsible for each.
4. Describe how statocysts and lateral line systems work.
5. Compare and contrast camera-type eyes and compound eyes.
6. Describe the structures of the human eye and their functions.
7. Explain what cutaneous receptors and proprioceptors are, where they are located in the human body, and how they work.

All living things respond to stimuli. Stimuli are environmental signals that tell us about the external environment or the internal environment. In Chapter 21, you learned that plants often respond to external stimuli, such as light, by changing their growth pattern. An animal's response often results in motion. Complex animals rely on sensory receptors to provide information to the central nervous system (brain and spinal cord), which integrates sensory input before directing a motor response (**Fig. 28.1**).

Sense organs, as a rule, are specialized to receive one kind of stimulus. The eyes ordinarily respond to light, ears to sound waves, pressure receptors to pressure, and chemoreceptors to chemical molecules. Sensory receptors transform the stimulus into nerve impulses that reach a particular section of the cerebral cortex. Those from the eye reach the visual areas, and those from the ears reach the auditory areas. The brain, not the sensory receptor, is responsible for sensation and perception, and each part of the brain interprets impulses in only one way. For example, if by accident the photoreceptors of the eye are stimulated by pressure and not light, the brain causes us to see "stars" or other visual patterns.

 MP3 Sensations and Receptors

Chemical Senses

The fundamental functions of sensory receptors include helping animals stay safe, find food, and find mates. *Chemoreceptors* give us the ability to detect chemicals in the environment, which is believed to be our most primitive sense. Chemoreceptors occur almost universally in animals. For example, they are present all over the body of planarians (flatworms) but concentrated on the auricles at the sides of the head. Male moths have receptors for a sex attractant on their antennae. The receptors on the antennae of the male silkworm moth are so sensitive that only 40 out of 40,000 receptor proteins need to be activated in order for the male to respond to a chemical released by the

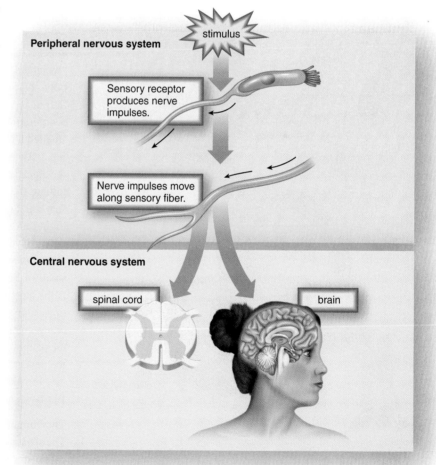

Figure 28.1   Sensory input.

After detecting a stimulus, sensory receptors initiate nerve impulses within the peripheral nervous system (PNS). These impulses give the central nervous system (CNS) information about the external and internal environments. The CNS integrates all incoming information and then initiates a motor response to the stimulus.

female (**Fig. 28.2***a*). Other insects, such as the housefly, have chemoreceptors largely on their feet—a fly tastes with its feet instead of its mouth. In mammals, the receptors for taste are located in the mouth, and the receptors for smell are in the nose.

Video
Sex and the Senses

Taste and Smell

In humans, **taste buds,** located primarily on the tongue, contain taste receptor cells, and the nose contains olfactory receptor cells (Fig. 28.2*b,c*). Receptor proteins for chemicals are located on the microvilli of taste receptor cells and on the cilia of olfactory receptor cells. When molecules bind to these receptor proteins, nerve impulses are generated in sensory nerve fibers that go to the brain. When they reach the appropriate cortical areas, they are interpreted as taste and smell, respectively.

There are at least five primary types of tastes: bitter, sour, salty, sweet, and umami (Japanese for "savory" or "delicious"). Foods rich in certain amino acids. such as the common seasoning monosodium glutamate (MSG), as well as certain flavors of cheese, beef broth, and some seafood, produce the taste of umami. Taste buds for each primary taste are located throughout the tongue but may be concentrated in particular regions. A particular food can stimulate more than one of these types of taste buds. In this way, the response of taste buds can result in a range of sweet, sour, salty, umami, and bitter tastes. The brain appears to survey the overall pattern of incoming sensory impulses and take a "weighted average" of their taste messages as the perceived taste. Similarly, an odor contains many odor molecules, which activate a characteristic combination of receptor proteins. When this complex information is communicated to the cerebral cortex, we know we have smelled a rose—or an onion!

Animation
Taste

An important function of taste and smell is to trigger reflexes that start the digestive juices flowing. A revolting or repulsive substance in the mouth can initiate the gag reflex or even vomiting. Smell is even more important to our survival. Smells associated with danger, such as smoke, can trigger the fight-or-flight reflex. Unpleasant smells can cause us to sneeze or choke.

Have you ever noticed that a certain aroma vividly brings to mind a certain person or place? A person's perfume may remind you of someone else, or the smell of boxwood may remind you of your grandfather's farm. The olfactory bulbs have direct connections with the limbic system and its centers for emotions and memory. One investigator showed that, when subjects smelled an orange while viewing a painting, they not only remembered the painting when asked about it later but also had many deep feelings about it.

MP3
Taste and Smell

Hearing and Balance

The human ear has two sensory functions: hearing and balance (equilibrium). The sensory receptors for both of these consist of *hair cells* with long microvilli called stereocilia. These microvilli, unlike those of taste cells, are sensitive to mechanical stimulation. Therefore, they are termed *mechanoreceptors.*

The similarity of the sensory receptors for balance and hearing and their presence in the same organ suggest an evolutionary relationship between them. In fact, the sense organs of the mammalian ear may have evolved from a type of sense organ in fishes.

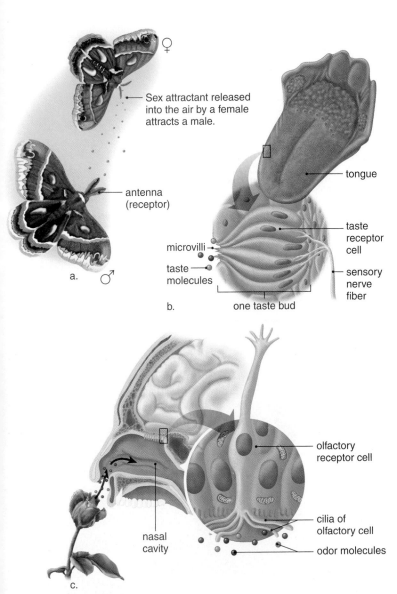

Figure 28.2 **Chemoreceptors.**

Chemoreceptors are common in the animal kingdom. **a.** Male moths have receptor proteins on their antennae for a sex attractant released by the female. **b.** Humans have receptor proteins for taste on the microvilli of taste receptor cells and (**c**) receptor proteins for smell on the cilia of olfactory receptor cells.

Hearing

Most invertebrates cannot hear. Some arthropods, including insects, do have sound receptors but they are quite simple. In insects, the ear consists of a pair of air pockets, each enclosed by a membrane, called the tympanic membrane, that passes sound vibrations to sensory receptors. The human ear has a tympanic membrane also, but it is between the outer and middle ear (**Fig. 28.3***a*). The outer ear collects sound waves that cause the tympanic membrane to move back and forth (vibrate) ever so slightly. Three tiny bones in the middle ear (the ossicles) amplify the sound about 20 times as it moves from one to the other. The last of the ossicles (the stapes) strikes the membrane of the oval window,

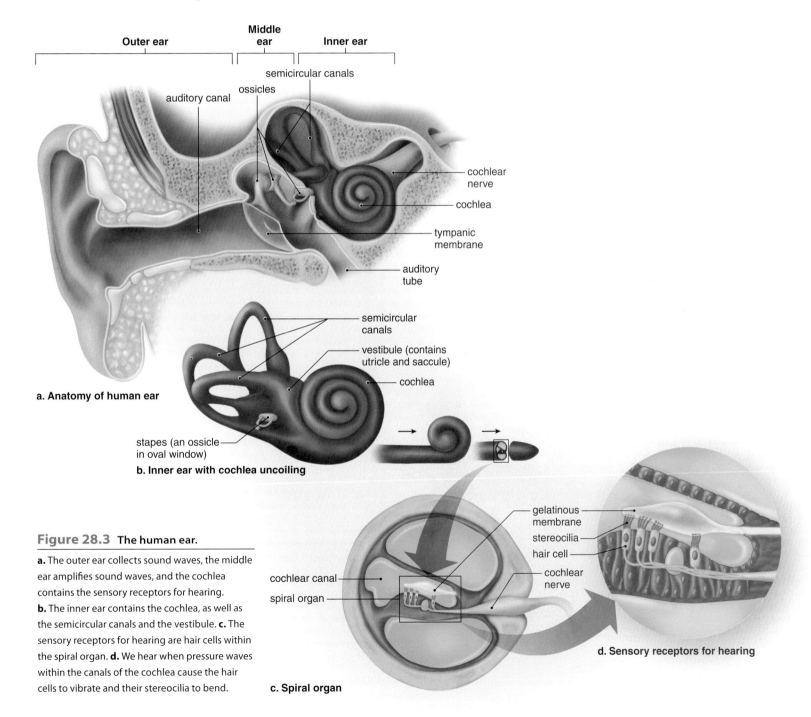

a. Anatomy of human ear

b. Inner ear with cochlea uncoiling

stapes (an ossicle in oval window)

semicircular canals

vestibule (contains utricle and saccule)

cochlea

Figure 28.3 **The human ear.**

a. The outer ear collects sound waves, the middle ear amplifies sound waves, and the cochlea contains the sensory receptors for hearing. **b.** The inner ear contains the cochlea, as well as the semicircular canals and the vestibule. **c.** The sensory receptors for hearing are hair cells within the spiral organ. **d.** We hear when pressure waves within the canals of the cochlea cause the hair cells to vibrate and their stereocilia to bend.

cochlear canal

spiral organ

c. Spiral organ

gelatinous membrane

stereocilia

hair cell

cochlear nerve

d. Sensory receptors for hearing

causing it to vibrate; in this way, the pressure is passed to a fluid within the hearing portion of the inner ear called the cochlea (Fig. 28.3*b*). The term *cochlea* means "snail shell." Specifically, the sensory receptors of hearing are located in the cochlear canal of the cochlea. The sensory receptors for hearing are hair cells whose stereocilia are embedded in a gelatinous membrane. Collectively, they are called the **spiral organ** (organ of Corti; Fig. 28.3*c,d*).

The outer ear and middle ear, which collect and amplify sound waves, are filled with air. The auditory tube relieves pressure in the middle ear. But the inner ear is filled with fluid; therefore, fluid pressure waves actually stimulate the spiral organ. When the stapes strikes the membrane of the oval window, pressure waves cause the hair cells to move up and down, and the stereocilia of the hair cells embedded in the gelatinous membrane bend. The hair cells of the spiral organ synapse with the cochlear nerve, and when their stereocilia bend, nerve impulses begin in the cochlear nerve and travel to the brain stem. When these impulses reach the auditory areas of the cerebral cortex, they are interpreted as sound.

Animation
Hearing

Each part of the spiral organ is sensitive to different wave frequencies, or *pitch*. Near the tip, the spiral organ responds to low pitches, such as a tuba; near the base, it responds to higher pitches, such as a bell or whistle. The nerve fibers from each region along the length of the spiral organ lead to slightly different areas in the brain. The pitch sensation we experience depends on which region of the spiral organ vibrates and which area of the brain is stimulated.

Volume is a function of the amplitude of sound waves. Loud noises cause the spiral organ to vibrate to a greater extent. The brain interprets the resulting increased stimulation as volume. It is believed that the brain interprets the *tone* of a sound based on the distribution of the stimulated hair cells.

Animation
Effect of Sound
Waves on Cochlear
Structures

Hearing Loss Especially when we are young, the middle ear is subject to infections that can lead to hearing impairment. Today, it is quite common for youngsters to have "tubes" put into the tympanic membrane to allow the middle ear to drain, in an effort to prevent this type of hearing loss. The mobility of ossicles decreases with age, and if bone grows over the stapes, the only remedy is implantation of an artificial stapes that can move.

Deafness due to middle ear damage is called **conduction deafness.** Deafness due to spiral organ damage is called **nerve deafness.** In today's society, noise pollution is common, and even city traffic can be loud enough to damage the stereocilia of hair cells (**Fig. 28.4**). It stands to reason, then, that frequently attending rock concerts, constantly playing a stereo loudly, or using earphones at high volume can also damage hearing. The first hint of danger can be temporary hearing loss, a "full" feeling in the ears, muffled hearing, or tinnitus (ringing in the ears). If exposure to noise is unavoidable, specially designed noise-reduction earmuffs are available, and it is possible to purchase earplugs made from compressible, sponge-like material at a drugstore or sporting goods store. These earplugs are not the same as those worn for swimming, and they should not be worn interchangeably. Finally, you should be aware that some medicines may damage the ability to hear. Anyone taking anticancer drugs, such as cisplatin, and certain antibiotics, such as streptomycin, should be especially careful to protect the ears from loud noises.

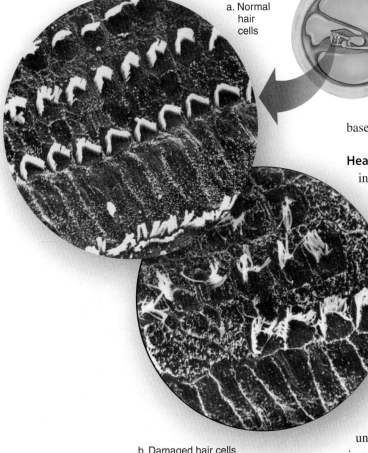

a. Normal hair cells

b. Damaged hair cells

Figure 28.4 **Effect of noise on hearing.**

a. Normal hair cells in the spiral organ of a guinea pig. **b.** Spiral organ damaged by 24-hour exposure to a noise level equivalent to that of a heavy-metal rock concert. Damaged cells cannot be replaced, so hearing is permanently impaired.

Balance

Humans have two senses of balance (equilibrium): rotational and gravitational. We are able to detect the rotational (angular) movement of the head as well as the straight-line movement of the head with respect to gravity.

Rotational equilibrium involves the semicircular canals (see Fig. 28.3*b*). In the base of each canal, hair cells have stereocilia embedded within a gelatinous membrane (**Fig. 28.5***a*). Because there are three semicircular canals, each responds to head movement in a different plane of space. As fluid within a semicircular canal flows over and displaces the gelatinous membrane, the stereocilia of the hair cells bend, and the pattern of impulses carried to the central nervous system (CNS) changes. These data, usually supplemented by vision, tell the brain how the head is moving. *Vertigo* is dizziness and a sensation of rotation. It is possible to bring on a feeling of vertigo by spinning rapidly and stopping suddenly. Now the person feels like the room is spinning because of sudden stimulation of stereocilia in the semicircular canals.

Gravitational equilibrium refers to the position of the head in relation to gravity. It depends on the **utricle** and **saccule,** two membranous sacs located in the inner ear (see Fig. 28.3*b*). Both of these sacs also contain hair cells with stereocilia in a gelatinous membrane (Fig. 28.5*b*). Calcium carbonate ($CaCO_3$) granules (otoliths) rest on this membrane. When the head moves forward or back, up or down, the otoliths are displaced and the membrane moves, bending the stereocilia of the hair cells. This movement alters the frequency of nerve impulses to the CNS. These data, usually supplemented by vision, tell the brain the direction of the movement of the head.

Animation Sense and Balance

MP3 The Sense of Hearing and Equilibrium

Similar Receptors in Other Animals

Gravitational equilibrium organs, called **statocysts,** are found in several types of invertebrates, including cnidarians, molluscs, and crustaceans. These organs give information only about the position of the head; they are not involved in the sensation of movement (**Fig. 28.6***a*). When the head stops moving, a small particle called a statolith stimulates the cilia of the closest hair cells, and these cilia generate impulses, indicating the position of the head.

The **lateral line** system of fishes uses sense organs similar to those in the human inner ear (Fig. 28.6*b*). In bony fishes, the system consists of sense organs located within a canal that has openings to the outside. As you might expect, the sense organ is a collection of hair cells with cilia embedded in a gelatinous membrane. Water currents and pressure waves from nearby objects cause the membrane and the cilia of the hair cells to bend. Thereafter, the hair cells initiate nerve impulses that go to the brain. Fishes use these data not for hearing or balance but to locate other fish, including predators, prey, and mates.

Figure 28.6 Sensory receptors in other animals.

a. Invertebrates use statocysts to determine their position. When the statolith stops moving, cilia of the nearest hair cells are stimulated, telling the position of the head. **b.** The lateral line of fishes is not for hearing or balance. It is for knowing the location of other fishes.

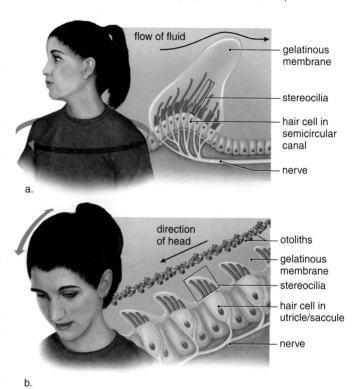

a.

b.

Figure 28.5 **Sense of balance in humans.**

a. Sensory receptors for rotational equilibrium. When the head rotates, the gelatinous membrane is displaced, bending the stereocilia.
b. Sensory receptors for gravitational equilibrium. When the head bends, otoliths are displaced, causing the gelatinous membrane to sag and the stereocilia to bend.

Statocyst

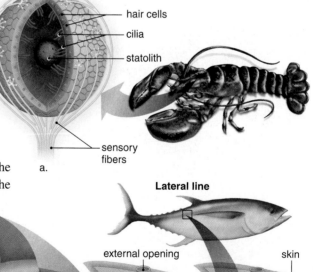

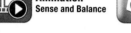

a.

Lateral line

b.

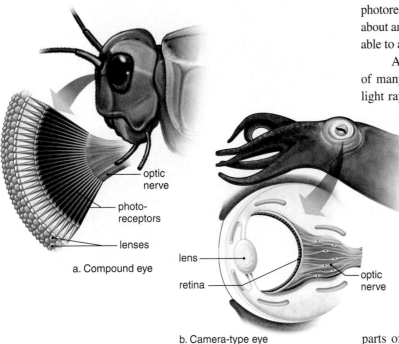

a. Compound eye

b. Camera-type eye

Figure 28.7 Eyes.

a. A compound eye has several visual units. Each visual unit has a lens that focuses light onto photoreceptor cells. **b.** A camera-type eye has one lens that focuses light onto a retina containing many photoreceptors.

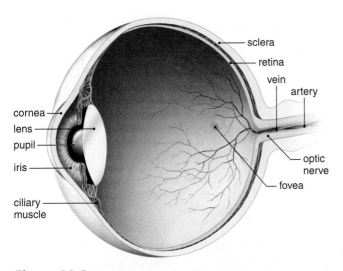

Figure 28.8 The human eye.

Light passes through the cornea and through the pupil (a hole in the iris) and is focused by the lens on the retina, which houses the photoreceptors.

Vision

Sensory receptors that are sensitive to light are called **photoreceptors.** In planarians, eyespots allow these animals to determine only the direction of light. Other photoreceptors form actual images. Image-forming eyes provide information about an object, including how far away it is. Such detailed information is invaluable to an animal.

Among invertebrates, arthropods have **compound eyes** composed of many, independent visual units, each of which possesses a lens to focus light rays on photoreceptors (**Fig. 28.7**a). The image that results from all the stimulated visual units is crude because the small size of compound eyes limits the number of visual units, which still might number as many as 28,000.

Vertebrates (including humans) and certain molluscs, such as the squid and the octopus, have a **camera-type eye** (Fig. 28.7b). A single lens focuses light on photoreceptors, which number in the millions and are closely packed together within a retina. Since molluscs and vertebrates are not closely related, the camera-type eye evolved independently in each group.

Insects have color vision, but they make use of a slightly shorter range of the electromagnetic spectrum compared with humans. However, they can see the longest of the ultraviolet rays, and this enables them to be especially sensitive to the reproductive parts of flowers, which have particular ultraviolet patterns. Some fishes and most reptiles are believed to have color vision, but among mammals, only humans and other primates have expansive color vision. It would seem, then, that this trait was adaptive for a diurnal habit (being active during the day), which accounts for its retention in only a few mammals.

The Human Eye

When looking straight ahead, each of our eyes views the same object from a slightly different angle. This slight displacement of the images permits *binocular vision,* the ability to perceive three-dimensional images and to sense depth. Like the human ear, the human eye has numerous parts, many of which are involved in preparing the stimulus for the sensory receptors. In the case of the ear, sound wave energy is magnified before it reaches the sensory receptors. In the case of the eye, light rays are brought to a focus on the photoreceptors within the **retina.** In **Figure 28.8,** the **cornea** and especially the **lens** are involved in focusing light rays on the photoreceptors. The **iris,** the colored part of the eye, regulates the amount of light that enters the eye by way of the **pupil.** The retina generates nerve impulses, which are sent to the visual part of the cerebral cortex, and from this information, the brain forms an image of the object.

The shape of the lens is controlled by the ciliary muscles. When we view a distant object, the lens remains relatively flat, but when we view a near object, the lens rounds up. With normal aging, the lens loses its ability to accommodate for near objects; therefore, many people need reading glasses when they reach middle age. Aging, or possibly exposure to the sun, also makes the lens subject to cataracts; the lens becomes opaque, incapable of transmitting light rays. Currently, surgery is the only viable treatment for cataracts.

Video
Artificial Eye

Photoreceptors of the Eye

Figure 28.9 illustrates the structure of the photoreceptors in the human eye, which are called **rods** and **cones.** Both types of photoreceptors contain a visual pigment similar to that found in all types of eyes throughout the animal kingdom. The visual pigment in rods is a deep-purple pigment called rhodopsin. **Rhodopsin** is a complex molecule made up of the protein opsin and a light-absorbing molecule called *retinal,* which is a derivative of vitamin A. When a rod absorbs light, rhodopsin splits into opsin and retinal, leading to a cascade of reactions that ends in the generation of nerve impulses. Rods are very sensitive to light and therefore are suited to night vision. (Because carrots are rich in vitamin A, it is true that eating carrots can improve your night vision.) Rod cells are plentiful throughout the retina; therefore, they also provide us with peripheral vision and perception of motion.

The cones, on the other hand, are located primarily in a part of the retina called the **fovea.** Cones are activated by bright light; they allow us to detect the fine detail and color of an object. Color vision depends on three different kinds of cones, which contain pigments called B (blue), G (green), and R (red) pigments. Each pigment is made up of retinal and opsin, but a slight difference in the opsin structure of each accounts for their specific absorption patterns. Various combinations of cones are believed to be stimulated by in-between shades of color.

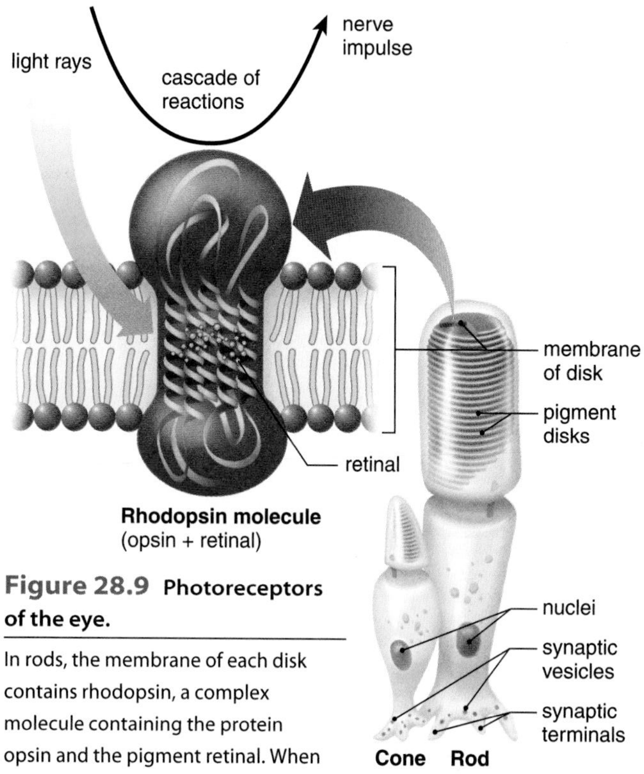

Retina The retina has three layers of cells, and light has to penetrate through the first two layers to reach the photoreceptors (**Fig. 28.10**). The intermediate cells of the middle layer process and relay visual information from the photoreceptors to the ganglion cells that have axons forming the optic nerve. The sensitivity of cones versus rods is mirrored by how directly they connect to ganglion cells. Information from several hundred rods may converge on a single ganglion cell, while cones show very little convergence. As signals pass through the layers of the retina, integration occurs. Integration improves the overall contrast and quality of the information sent to the brain, which uses the information to form an image of the object.

No rods and cones occur where the optic nerve exits the retina. Therefore, no vision is possible in this area. You can prove this to yourself by putting a very small dot to the right of center on a piece of paper. Close your left eye; then use your right hand to move the paper slowly toward your right eye while you look straight ahead. The dot will disappear at one point—this point is your **blind spot.**

Animation Vision

MP3 Sense of Vision

Figure 28.9 Photoreceptors of the eye.

In rods, the membrane of each disk contains rhodopsin, a complex molecule containing the protein opsin and the pigment retinal. When rhodopsin absorbs light energy, it splits, releasing opsin, which sets in motion a cascade of reactions that ends in nerve impulses.

light rays / cascade of reactions / nerve impulse / membrane of disk / pigment disks / retinal

Rhodopsin molecule (opsin + retinal)

nuclei / synaptic vesicles / synaptic terminals / **Cone** **Rod**

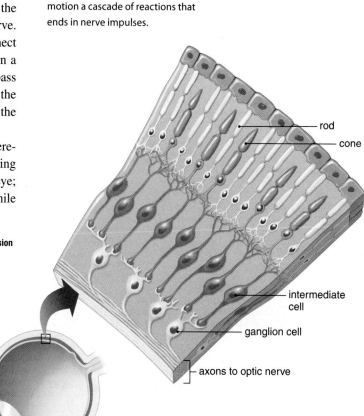

rod / cone / intermediate cell / ganglion cell / axons to optic nerve

Figure 28.10 The retina.

The retina contains a layer of rods and cones, a layer of intermediate cells, and a layer of ganglion cells. Integration of signals occurs at the synapses between the layers, and much processing occurs before nerve impulses are sent to the brain.

Connections and Misconceptions

What is LASIK surgery?

LASIK, which stands for *laser in-situ keratomileusis*, is a quick and painless procedure that involves the use of a laser to permanently change the shape of the cornea. During the LASIK procedure, a small flap of tissue (the conjunctiva) is cut away from the front of the eye. The flap is folded back, exposing the cornea and allowing the surgeon to remove a defined amount of tissue from the cornea. Each pulse of the laser will remove a small amount of corneal tissue, allowing the surgeon to flatten or increase the steepness of the curve of the cornea. After the procedure, the flap of tissue is put back into place and allowed to heal on its own. LASIK patients receive eye drops or medications to help relieve the pain of the procedure. Improvements to vision begin as soon as the day after the surgery but typically take two to three months. Most patients will have vision close to 20/20, but the chances for improved vision are based in part on how good the person's vision was before the surgery.

Seeing in the Dark

Some nocturnal animals rely on vision, but others use sonar (sound waves) to find their way in the dark. Bats flying in a dark room easily avoid obstacles in their path because they echolocate, like submarines. When searching for food, they emit ultrasonic sound (above the range humans can hear) in chirps that bounce off their prey. They are able to determine the distance to dinner by timing the echo's return—a long delay means the prey is far away.

Video **Bat Echolocation**

A bat-inspired sonar walking stick is being perfected to help visually impaired people sense their surroundings. It, too, emits ultrasonic chirps and picks up the echoes from nearby objects. Buttons on the cane's handle vibrate gently to warn a user to dodge a low ceiling or to sidestep objects blocking the path. A fast, strong signal means an obstacle is close by.

Cutaneous Receptors and Proprioceptors

Cutaneous receptors are located in the skin, and **proprioceptors** are located in the muscles and joints. These sensory receptors are connected to the primary sensory area of the cerebral cortex where each part of the body is represented. Here, sensory input is received from the skin, muscles, and joints in each part of the body (**Fig. 28.11**). Thus, these receptors are important to our well-being.

Cutaneous Receptors

Skin, the outermost covering of our body, contains numerous sensory receptors that help us respond to changes in our environment, be aware of dangers, and communicate with others. The sensory receptors in skin are for touch, pressure, pain, and temperature.

Skin has two regions, the epidermis and the dermis (**Fig. 28.12**). The epidermis is packed with cells that become keratinized as they rise to the surface. Among these cells are free nerve endings responsive to cold or to warmth. Cold receptors are far more numerous than warmth receptors, but there are no known structural differences between the two. Also in the epidermis are pain receptors

Figure 28.11 **Sensory input to the primary sensory area of the brain.**

Sensory receptors in the skin, muscles, and joints send nerve impulses to the CNS. Sensation and perception occur when these reach the primary sensory area of the cerebral cortex. A motor response can be initiated by the spinal cord and cerebellum without the involvement of the cerebrum.

sensitive to extremes in temperature or pressure and to chemicals released by damaged tissues. Sometimes, the stimulation of internal receptors is felt as pain in the skin. This is called *referred pain*. For example, pain from the heart may be felt in the left shoulder and arm. This effect most likely happens when nerve impulses from the pain receptors of internal organs travel to the spinal cord and synapse with neurons also receiving impulses from the skin.

Connections and Misconceptions

How does aspirin work?

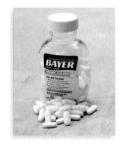

Aspirin is made of a chemical called acetylsalicyclic acid (ASA). When tissue is damaged, it produces large amounts of a type of fatty acid called prostaglandin. Prostaglandin acts as a signal to the pain receptors that tissue damage has occurred, which the brain interprets as pain. Prostaglandins are manufactured in the cell by an enzyme called COX. ASA reduces the capabilities of this enzyme, lowering the amount of prostaglandin produced and the perception of pain.

The dermis of the skin contains sensory receptors for pressure and touch (Fig. 28.12). The Pacinian corpuscles are onion-shaped pressure receptors that lie deep inside the dermis. Several other cutaneous receptors detect touch. A free nerve ending, called a root hair plexus, winds around the base of a hair follicle and produces nerve impulses if the hair is touched. Touch receptors are concentrated in parts of the body essential for sexual stimulation: the fingertips, palms, lips, tongue, nipples, penis, and clitoris.

Proprioceptors

Proprioceptors help the body maintain equilibrium and posture, despite the force of gravity always acting upon the skeleton and muscles. A **muscle spindle** consists of sensory nerve endings wrapped around a few muscle cells within a connective tissue sheath. **Golgi tendon** organs and other sensory receptors are located in the joints. The rapidity of nerve impulses from proprioceptors is proportional to the stretching of the organs they occupy. A motor response results in the contraction of muscle fibers adjoining the proprioceptor.

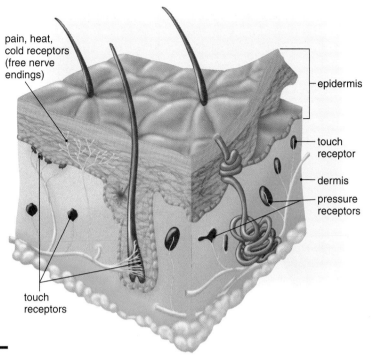

Figure 28.12   Cutaneous receptors.

Numerous receptors are in the skin. Free nerve endings (yellow) in the epidermis detect pain, heat, and cold. Various touch and pressure receptors (red) are in the dermis.

Check Your Progress 28.1

1. Summarize how chemoreceptors, mechanoreceptors, photoreceptors, cutaneous receptors, and proprioreceptors each detect stimuli.

2. Summarize the function of taste receptor cells and list the five primary types of taste.

3. Explain how the ear can distinguish between sounds with different pitches.

4. Distinguish between the structures of the inner ear that are responsible for gravitational equilibrium and those responsible for rotational equilibrium.

5. List, in order, the structures a ray of light encounters on its way to the retina of the human eye.

6. Compare and contrast rod and cone photoreceptors.

7. Explain the role of cutaneous receptors and proprioceptors.

Connecting the Concepts

For more information on the senses, refer to the following discussions.

Section 22.3 describes the relationship among sensory input, negative feedback mechanisms, and homeostasis.

Section 27.1 explains how sensory neurons transfer information to interneurons for integration.

Figure 27.9 illustrates the regions of the cerebral cortex that are involved in the processing of sensory input.

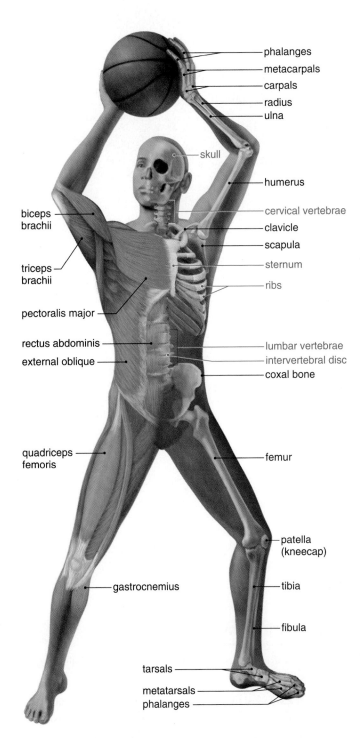

phalanges
metacarpals
carpals
radius
ulna
skull
humerus
biceps brachii
cervical vertebrae
clavicle
scapula
triceps brachii
sternum
ribs
pectoralis major
rectus abdominis
lumbar vertebrae
external oblique
intervertebral disc
coxal bone
quadriceps femoris
femur
patella (kneecap)
gastrocnemius
tibia
fibula
tarsals
metatarsals
phalanges

Figure 28.13 Musculoskeletal system.

Major skeletal muscles and bones of the body are labeled. The human skeleton contains bones that belong to the axial skeleton (red labels) and those that belong to the appendicular skeleton (black labels).

28.2 The Motor Systems

Learning Outcomes

Upon completion of this section, you should be able to

1. Summarize the functions of skeletal muscles and bones.
2. Distinguish among endoskeletons, exoskeletons, and hydrostatic skeletons and give examples of each.
3. Describe the structure of a long bone.
4. Describe the structure and function of a muscle fiber; summarize the process of muscle contraction.
5. List and provide examples of the types of joints in the human body and describe common joint disorders.

In this section, we consider the motor systems of humans—the muscles and the bones—as forming a single system—namely, the musculoskeletal system (**Fig. 28.13**). Combining the two seems appropriate because in many cases the functions of muscles and bones overlap:

Both skeletal muscles and bones support the body and make the movement of body parts possible.

Both skeletal muscles and bones protect internal organs. Skeletal muscles pad the bones that protect the heart and lungs, the brain, and the spinal cord.

Both muscles and bones aid the functioning of other systems. Without the movement of the rib cage, breathing would not occur. As an aid to digestion, the jaws have sockets for teeth; skeletal muscles move the jaws, so that food can be chewed, and smooth muscle moves food along the digestive tract (see Fig. 24.21*b*). Red bone marrow supplies the red blood cells that carry oxygen, and the pumping of cardiac muscle in the wall of the heart moves the blood to the tissues, where exchanges with tissue fluid occur.

In addition to these shared functions, the muscles and bones have individual functions:

Skeletal muscle contraction assists the movement of blood in the veins and lymphatic vessels (see Fig. 23.8). Without the return of lymph to the cardiovascular system and blood to the heart, circulation could not continue.

Skeletal muscles help maintain a constant body temperature (see Fig. 28.17). Skeletal muscle contraction causes ATP to break down, releasing heat that is distributed about the body.

Bones store fat and calcium. Fat is stored in yellow bone marrow (see Fig. 28.15), and the extracellular matrix of bone contains calcium. Calcium ions play a major role in muscle contraction and nerve conduction.

The Human Skeleton

The **endoskeleton** of vertebrates can be contrasted with the **exoskeleton** of arthropods. Both skeletons are jointed, which has helped members of both groups of animals live successfully on land. The endoskeleton of humans is composed of bone, which is living material and capable of growth. Like an exoskeleton, an endoskeleton protects vital internal organs, but unlike an exoskeleton, it need not limit the space available for internal organs because it grows as the animal grows. The soft tissues that surround an endoskeleton

protect it, and injuries to soft tissues are apt to be easier to repair than is the skeleton itself.

The exoskeleton of arthropods is composed of chitin—a strong, flexible, nitrogenous polysaccharide. Besides providing protection against wear and tear and against enemies, an exoskeleton also prevents drying out. Although an arthropod exoskeleton provides support for muscle contractions, it does not grow with the animal, and arthropods molt to rid themselves of an exoskeleton that has become too small (**Fig. 28.14***a*). This process makes them vulnerable to predators.

One other type of skeleton is seen in the animal kingdom. In animals, such as worms, that lack a hard skeleton, a fluid-filled internal cavity can act as a **hydrostatic skeleton** (Fig. 28.14*b*). A hydrostatic skeleton offers support and resistance to the contraction of muscles, so that the animal can move. As an analogy, consider that a garden hose stiffens when filled with water, and that a water-filled balloon changes shape when squeezed at one end. Similarly, an animal with a hydrostatic skeleton can change shape and perform a variety of movements.

Axial and Appendicular Skeletons

The 206 bones of the human skeleton are arranged into an axial skeleton and an appendicular skeleton.

Axial Skeleton   The labels for the bones of the **axial skeleton** are in color in Figure 28.13. The **skull** consists of the **cranium,** which protects the brain, and the facial bones. The most prominent of the **facial bones** are the lower and upper jaws, the cheekbones, and the nasal bones. The **vertebral column** (spine) extends from the skull to the **sacrum** (tailbone). It consists of a series of vertebrae separated by pads of fibrocartilage called the **intervertebral discs.** On occasion, discs can slip or even rupture. A damaged disc pressing against the spinal cord or spinal nerves causes pain. Removal of the disc may be required.

The **rib cage,** composed of the ribs and **sternum** (breastbone), demonstrates how the skeleton can be protective and flexible at the same time. In the United States, automobile accidents are the most common cause of sternum fractures; people with such injuries are typically examined for signs of heart damage. The rib cage protects the heart and lungs but moves when we breathe.

Appendicular Skeleton   The **appendicular skeleton** contains the bones of two girdles and their attached limbs. The **pectoral (shoulder) girdle** and upper limbs are specialized for flexibility; the **pelvic (hip) girdle** and lower limbs are specialized for strength. The pelvic girdle also protects internal organs.

A **clavicle** (collarbone) and a **scapula** (shoulder blade) make up the shoulder girdle. The shoulder pads worn by football players are designed to cushion direct blows that could cause clavicle fractures. The **humerus** of the arm articulates only with the scapula; the joint is stabilized by tendons and ligaments that form a **rotator cuff.** Vigorous circular movements of the arm can lead to rotator cuff injuries. Two bones (the **radius** and **ulna**) contribute to the easy twisting motion of the forearm. The flexible hand contains bones of the wrist, palm, and five fingers.

The pelvic girdle contains two massive **coxal bones**, which form a bowl, called the pelvis, and articulate with the longest and strongest bones of the body, the **femurs** (thighbones). The strength of the femurs allows these massive bones to support the weight of the upper half of the body. The kneecap protects the

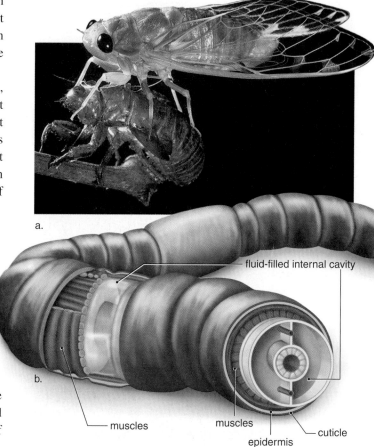

a.

b.

fluid-filled internal cavity

muscles

muscles

epidermis

cuticle

Figure 28.14   Types of skeletons.

a. Arthropods have an exoskeleton that must be shed as they grow.
b. Worms have a hydrostatic skeleton, in which muscle contraction pushes against a fluid-filled internal cavity.

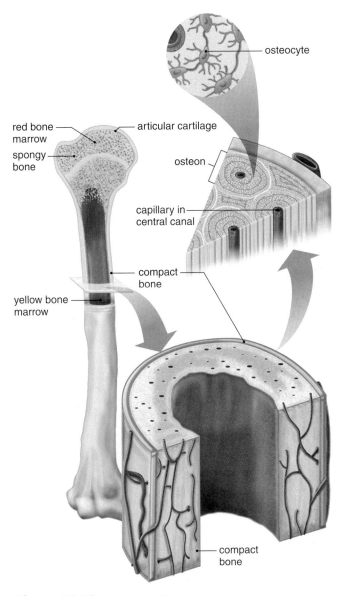

red bone marrow

spongy bone

yellow bone marrow

osteocyte

articular cartilage

osteon

capillary in central canal

compact bone

compact bone

Figure 28.15 Structure of bone.

A bone has a central cavity that is usually filled with yellow bone marrow. Both spongy bone and compact bone are living tissues composed of bone cells within a matrix that contains calcium. The spaces of spongy bone contain red bone marrow.

knee. The **tibia** of the leg is the shinbone. The **fibula** is the more slender bone in the leg. Each foot contains bones of the ankle, instep, and five toes.

Structure of a Bone

When a long bone is split open, as in **Figure 28.15**, a cavity is revealed that is bounded on the sides by compact bone. **Compact bone** contains many **osteons** (Haversian systems), where bone cells called **osteocytes** lie in tiny chambers arranged in concentric circles around central canals. The cells are separated by an extracellular matrix, which contains mineral deposits—primarily, calcium and phosphorus salts. Two other cell types are constantly at work in bones. **Osteoblasts** deposit bone, and **osteoclasts** secrete enzymes that digest the matrix of bone and release calcium into the bloodstream. When a person has **osteoporosis** (weak bones subject to fracture), osteoclasts are working harder than osteoblasts. Intake of high levels of dietary calcium, especially when a person is young and more active, encourages denser bones and lessens the chance of getting osteoporosis later in life.

 Animation Bone Growth in Width

 **Animation** Nano Bones

A long bone has **spongy bone** at each end. Spongy bone has numerous bony bars and plates separated by irregular spaces. The spaces are often filled with **red bone marrow,** a specialized tissue that produces red blood cells. The cavity of a long bone is filled with yellow bone marrow and stores fat. Beyond the spongy bone are a thin shell of compact bone and a layer of hyaline cartilage, which is important to healthy joints.

 Video Bear Bones

MP3 Bone Structure

Skeletal Muscle Structure and Physiology

The three types of muscle—smooth, cardiac, and skeletal—have different structures (see Fig. 22.7). Smooth muscles contain sheets of long, spindle-shaped cells, each with a single nucleus. Cardiac cells are striated (having a striped appearance) and typically possess a single nucleus. Cardiac muscle tissue contains branched chains of cells that interconnect, forming a lattice network. Skeletal muscle cells, called muscle fibers, are also striated. They are quite elongated, and they run the length of a skeletal muscle. Skeletal muscle fibers arise during development when several cells fuse, resulting in one long, multinucleated cell. Skeletal muscle contraction has been studied extensively.

 **Video** Making a Muscle

Connections and Misconceptions

How many muscles are in the human body?

Most experts agree that there are over 600 muscles in the human body. The exact number varies, since some experts lump muscles together under one name, while others split them apart. The smallest of these is the stapedius, a 1.27 millimeter long muscle in the middle ear. The longest muscle is the sartorius, which starts at the hip and extends to the knee. The biggest muscle (in terms of mass) is the gluteus maximus, the muscle that makes up the majority of the buttocks.

biceps brachii (relaxed)

triceps brachii (contracted)

Skeletal Muscle Contraction

When skeletal muscle fibers contract, they shorten. Let's look at details of the process, beginning with the motor axon (**Fig. 28.16**). When nerve impulses travel down a motor axon and arrive at an axon terminal, synaptic vesicles release acetylcholine (ACh) into a synaptic cleft. ACh quickly diffuses across the cleft and binds to receptors in the plasma membrane of a muscle fiber, called the sarcolemma. The sarcolemma generates impulses, which travel along its T tubules to the endoplasmic reticulum, which is called the *sarcoplasmic reticulum* in muscle fibers. The release of calcium from calcium storage sites causes muscle fibers to contract.

Animation
Function of the
Neuromuscular
Junction

The contractile portions of a muscle fiber are many parallel, thread-like **myofibrils** (Fig. 28.16). An electron microscope shows that myofibrils (and therefore skeletal muscle fibers) are striated because of the placement of protein filaments within contractile units called **sarcomeres.** A sarcomere extends between two dark lines called Z lines. Sarcomeres contain thick filaments made up of the protein **myosin** and thin filaments made up of the protein **actin.** As a muscle fiber contracts, the sarcomeres within the myofibrils shorten because actin (thin) filaments slide past the myosin (thick) filaments and approach one another. The movement of actin filaments in relation to myosin filaments is called the **sliding filament model** of muscle contraction. During the sliding process, the sarcomere shortens, even though the filaments themselves remain the same length.

Animation
Sarcomere
Contraction

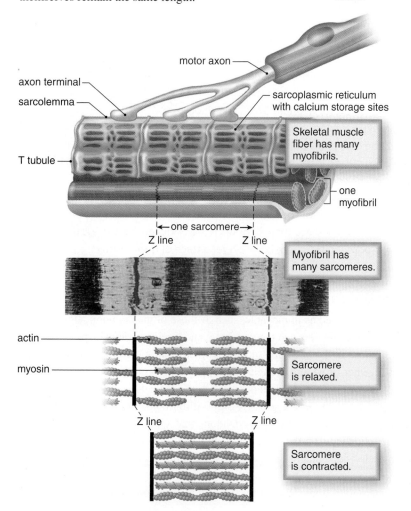

Figure 28.16 Skeletal muscle fiber structure and function.

A muscle fiber contains many myofibrils divided into sarcomeres, which are contractile. When innervated by a motor neuron, the myofibrils contract and the sarcomeres shorten because actin filaments slide past the myosin filaments.

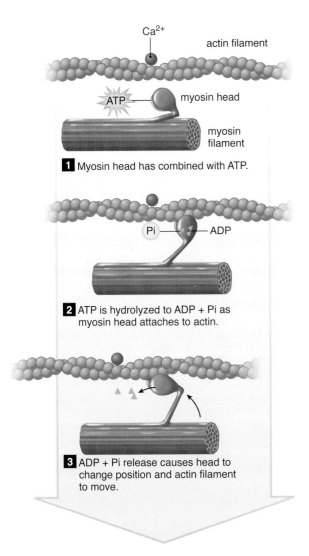

1 Myosin head has combined with ATP.

2 ATP is hydrolyzed to ADP + Pi as myosin head attaches to actin.

3 ADP + Pi release causes head to change position and actin filament to move.

Figure 28.17   Why a muscle shortens when it contracts.
The presence of calcium (Ca^{2+}) sets in motion a chain of events (1–3) that causes myosin heads to attach to and pull an actin filament toward the center of a sarcomere. After binding to other ATP molecules, myosin heads return to their resting position. Then, the chain of events (1–3) occurs again, except that the myosin heads reattach farther along the actin filament.

Why the Filaments Slide   The thick filament is a bundle of myosin molecules, each having a globular head with the capability of attaching to the actin filament when calcium (Ca^{2+}) is present. First, myosin binds to and hydrolyzes ATP. Thus, energized myosin heads attach to an actin filament. The release of ADP + Pi causes myosin to shift its position and pull the actin filaments to the center of the sarcomere. The action is similar to the movement of your hand when you flex your forearm (**Fig. 28.17**).

In the presence of another ATP, a myosin head detaches from actin. Now the heads attach farther along the actin filament. The cycle occurs again and again, and the actin filaments move nearer and nearer the center of the sarcomere each time the cycle is repeated. Contraction continues until nerve impulses cease and calcium ions are returned to their storage sites. The membranes of the sarcoplasmic reticulum contain active transport proteins that pump calcium ions back into the calcium storage sites, and the muscle relaxes. When a person or an animal dies, ATP production ceases. Without ATP, the myosin heads cannot detach from actin, nor can calcium be pumped back into the sarcoplasmic reticulum. As a result, the muscles remain contracted, a phenomenon called *rigor mortis*.

Animation
Breakdown of ATP and Cross-Bridge Movement

Connections and Misconceptions

How do medical examiners use rigor mortis to estimate the time of death?

Body temperature and the presence or absence of rigor mortis allow the time of death to be estimated. For example, the body of someone who has been dead for 3 hours or less will still be warm (close to body temperature, 98.6°F, or 37°C), and rigor mortis will be absent. After approximately 3 hours, the body will be significantly cooler than normal, and rigor mortis will begin to develop. The corpse of an individual dead at 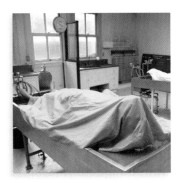 least 8 hours will be in full rigor mortis, and the temperature of the body will be the same as the surroundings. Forensic pathologists know that a person has been dead for more than 24 hours if the body temperature is the same as the environment and there is no longer a trace of rigor mortis.

Skeletal Muscles Move Bones at Joints

Joints are classified as immovable, such as those of the cranium; slightly movable, such as those between the vertebrae; and freely movable (synovial joints), such as those in the knee and hip. In **synovial joints, ligaments** bind the two bones together, providing strength and support and forming a capsule containing lubricating synovial fluid. All of our movements, from those of graceful and agile ballet dancers to those of aggressive and skillful football players, occur because muscles are attached to bones by tendons that span movable joints. Because muscles shorten when they contract, they have to work in antagonistic pairs. If one muscle of an antagonistic pair flexes the joint and raises the limb, the other extends the joint and straightens the limb. **Figure 28.18** illustrates this principle with regard to the movement of the forearm at the elbow joint.

Figure 28.19*b* illustrates the anatomy of a freely movable synovial joint. Note that, in addition to a cavity filled with synovial fluid, a synovial joint may include additional structures—namely, menisci and bursae. Menisci (sing., **meniscus**), are C-shaped pieces of hyaline cartilage between the bones. These give added stability and act as shock absorbers. Fluid-filled sacs called bursae (sing., **bursa**) ease friction between bare areas of bone and overlapping muscles or between skin and tendons.

The **ball-and-socket joints** at the hips and shoulders are synovial joints that allow movement in all planes, even rotational movement (Fig. 28.19*c*). The elbow and knee joints are synovial joints called **hinge joints** because, like a hinged door, they largely permit movement in one direction only (Fig. 28.19*d*).

Joint Disorders   *Sprains* occur when ligaments and tendons are overstretched at a joint. For example, a sprained ankle can result if you turn your ankle too far. Overuse of a joint may cause inflammation of a bursa, called *bursitis*. Tennis elbow is a form of bursitis. A common knee injury is a torn meniscus. Because fragments of menisci can interfere with joint movements, most physicians believe they should be removed. Today, arthroscopic surgery is possible to remove cartilage fragments or to repair ligaments or cartilage. A small instrument bearing a tiny lens and light source is inserted into a joint, as are the surgical instruments. Fluid is then added to distend the joint and allow visualization of its structure. Usually, the surgery is displayed on a monitor, so that the whole operating team can see the operation. Arthroscopy is much less traumatic than surgically opening the knee with long incisions. The benefits of arthroscopy are small incisions, faster healing, a more rapid recovery, and less scarring. Because arthroscopic surgical procedures are often performed on an outpatient basis, the patient is able to return home on the same day.

Rheumatoid arthritis, discussed in Chapter 26, is not as common as osteoarthritis (OA), which is the deterioration of an overworked joint. Constant compression and abrasion continually damage articular cartilage, and eventually it softens, cracks, and wears away entirely in some areas. As the disease progresses, the exposed bone thickens and forms spurs that cause the bone ends to enlarge and restrict joint movement. Weight loss can ease arthritis. Taking off 3 pounds can reduce the load on a hip or knee joint by 9 to 15 pounds. A sensible exercise program helps build up muscles, which stabilize joints. Low-impact activities, such as biking and swimming, are best.

Today, the replacement of damaged joints with a prosthesis (artificial substitute) is often possible. Some people have found glucosamine-chondroitin supplements beneficial as an alternative to joint replacement. Glucosamine, an amino sugar, is thought to promote the formation and repair of cartilage. Chondroitin, a carbohydrate, is a cartilage component that is thought to promote water retention and elasticity and to inhibit enzymes that break down cartilage. Both compounds are naturally produced by the body.

Video
Smart Robo-Knee

Exercise   A sensible exercise program has many benefits. Exercise improves muscular strength, muscular endurance, and flexibility. It improves cardio-respiratory endurance and may lower blood cholesterol levels. People who exercise are less likely to develop various types of cancer. Exercise promotes the activity of osteoblasts; therefore, it helps prevent osteoporosis (see page 550). It helps prevent weight gain, not only because of increased activity but also because as muscle mass increases, the body is less likely to accumulate fat. Exercise even relieves depression and enhances mood.

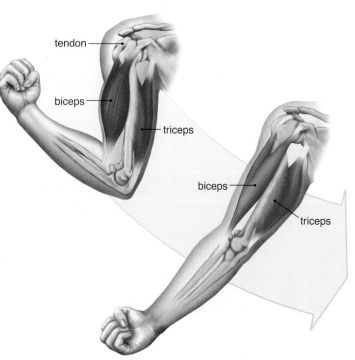

Figure 28.18   Action of muscles.

When muscles contract, they shorten. Therefore, muscles only pull— they cannot push. This causes them to work in antagonistic pairs; each member of the pair pulls on a bone in the opposite direction. For example, (**a**) when the biceps contracts, the forearm flexes (raises), and (**b**) when the triceps contracts, the forearm extends (lowers).

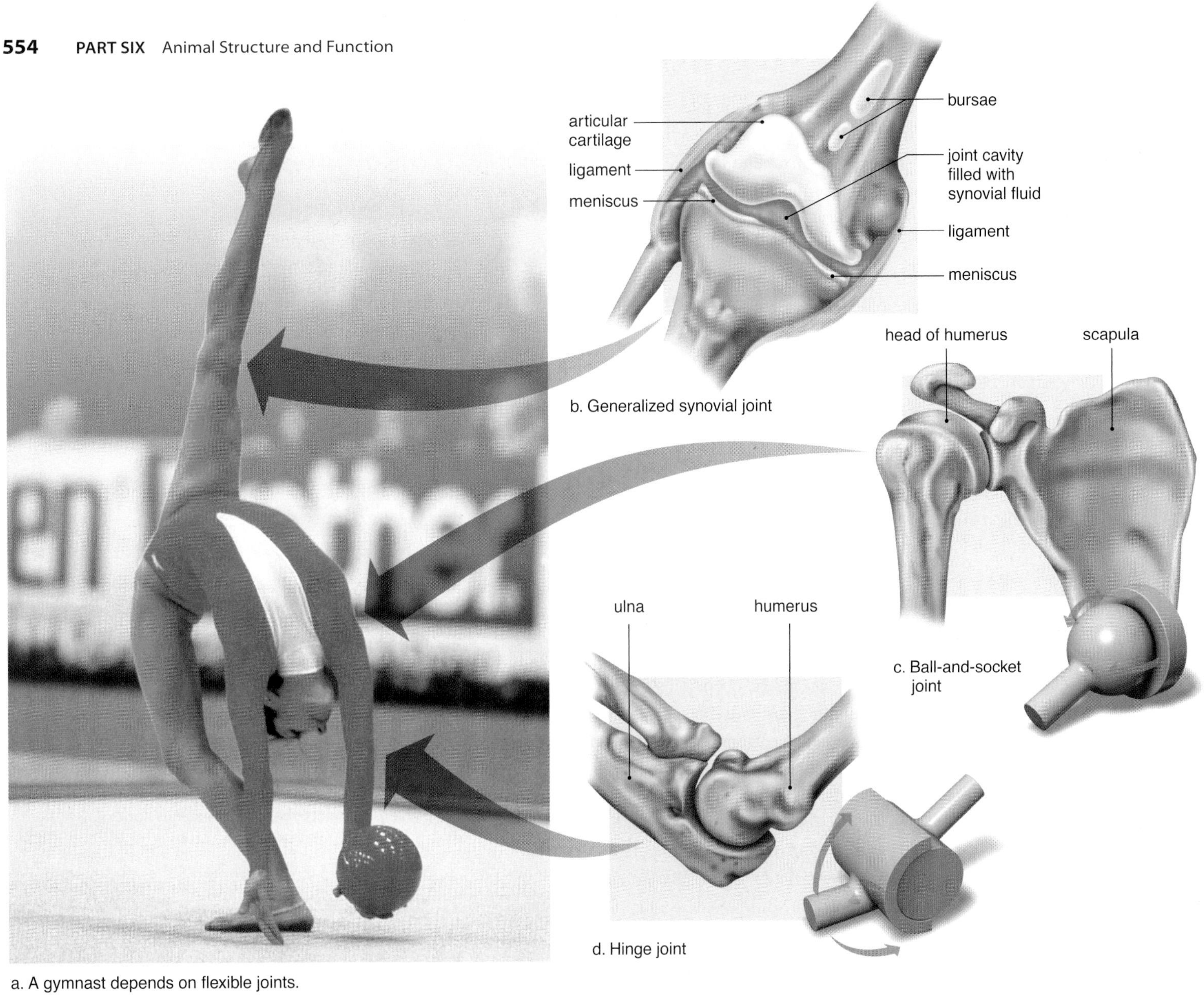

articular cartilage

ligament

meniscus

bursae

joint cavity filled with synovial fluid

ligament

meniscus

b. Generalized synovial joint

head of humerus

scapula

ulna

humerus

c. Ball-and-socket joint

d. Hinge joint

a. A gymnast depends on flexible joints.

Figure 28.19 Synovial joints.

a. The synovial joints of the human skeleton allow the body to be flexible and move with precision even when bearing a weight. **b.** Generalized synovial joint. Problems arise when menisci or ligaments are torn, bursae become inflamed, and articular cartilage wears away. **c.** The shoulder is a ball-and-socket joint that permits movement in three planes. **d.** The elbow is a hinge joint that permits movement in a single plane.

Connecting the Concepts

For more information on the muscular and skeletal systems, refer to the following discussions.

Section 27.1 examines the events that occur within the synapse of a neuromuscular junction.

Section 27.2 explains how the endocrine system maintains blood calcium homeostasis.

Check Your Progress 28.2

1. Summarize the roles of the muscular and skeletal systems.
2. List the three types of muscle, and provide a function for each.
3. List examples of bones in the axial and appendicular skeletons.
4. Contrast the functions of osteoblasts and osteoclasts.
5. Explain why a muscle shortens when it contracts.
6. Explain why a sprain would influence the function of a synovial joint.

Media Study Tools

Muscle Stimulation

The virtual lab "Muscle Stimulation" allows you to visualize how the load on a muscle influences the structures of the muscle at the microscopic level.

The Chapter in Review

Summary

28.1 The Senses

All living things respond to stimuli. In animals, stimuli generate nerve impulses that often go to a CNS, which integrates the information before initiating motor responses:

Stimulus ⟶ Nerve impulse ⟶ CNS ⟶ Motor response

Chemical Senses

Chemoreception is found universally in animals and is therefore believed to be the most primitive sense.

Human taste buds and olfactory receptor cells are chemoreceptors.

- Taste buds have microvilli with receptor proteins that bind to chemicals in food.
- Olfactory receptor cells have cilia with receptor proteins that bind to odor molecules.

Hearing and Balance

Mammals have an ear that may have evolved from the lateral line of fishes.

The sensory receptors for hearing are hair cells with stereocilia that respond to pressure waves.

- Hair cells respond to stimuli that have been received by the outer ear and amplified by the ossicles in the middle ear.
- Hair cells are found in the spiral organ and are located in the cochlear canal of the cochlea. The spiral organ generates nerve impulses that travel to the brain.

The sensory receptors for balance (equilibrium) are also hair cells with stereocilia.

- Hair cells in the base of the semicircular canals provide rotational equilibrium.
- Hair cells in the utricle and saccule provide gravitational equilibrium.

Vision

Arthropods have a compound eye; squids and humans have a camera-type eye. In humans, the photoreceptors

- Respond to light that has been focused by the cornea and lens.

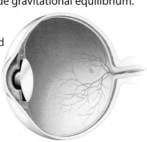

- Consist of two types, rods and cones. In rods, rhodopsin splits into opsin and retinal.
- Communicate with the next layer of cells in the retina. Integration occurs in the three layers of the retina before nerve impulses go to the brain.

Cutaneous Receptors and Proprioceptors

These receptors communicate with the primary sensory area of the brain. They consist of receptors for hot, cold, pain, touch, and pressure (cutaneous) and stretching (proprioceptors).

28.2 The Motor Systems

Together, the muscles and bones

- Support the body and allow parts to move
- Help protect internal organs
- Assist the functioning of other systems

In addition,

- Skeletal muscle contraction assists movement of blood in cardiovascular veins and lymphatic vessels.
- Skeletal muscles provide heat that warms the body.
- Bones are storage areas for calcium and phosphorus salts, as well as sites for blood cell formation.

The Human Skeleton

Arthropods have an exoskeleton, and molting is needed to replace it as the animal grows. Humans have an endoskeleton that grows with them. Worms have a hydrostatic skeleton.

The human skeleton is divided into two parts:

- The axial skeleton is made up of the skull, vertebral column, sternum, and ribs.
- The appendicular skeleton is composed of the girdles and their appendages.

Bone contains

- Compact bone (site of calcium storage) and spongy bone (site of red bone marrow)
- A cavity (site of fat storage)

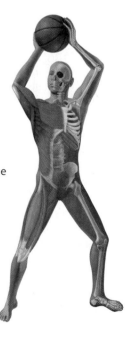

Skeletal Muscle Structure and Physiology

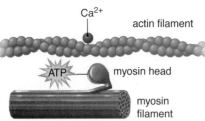

Skeletal muscle is one of the three types of muscles:

- Smooth muscle is composed of spindle-shaped cells that form a sheet.
- Cardiac muscle has striated cells that form a lattice network.
- Skeletal muscle has striated, tubular cells (fibers) that run the length of the muscle.

In a skeletal muscle cell,

- Myofibrils, myosin filaments, and actin filaments are arranged in a sarcomere.
- Myosin filaments pull actin filaments, and sarcomeres shorten.

Skeletal muscles

- Move bones at joints
- Work in antagonistic pairs
- Move bones at synovial joints that are subject to a number of disorders, such as sprains, bursitis, and torn ligaments and cartilage

Key Terms

actin 551
appendicular skeleton 549
axial skeleton 549
ball-and-socket joints 553
blind spot 545
bursa 553
camera-type eye 544
clavicle 549
compact bone 550
compound eye 544
conduction deafness 542
cone 545
cornea 544
coxal bones 549
cranium 549
cutaneous receptor 546
endoskeleton 548
exoskeleton 548
facial bones 549
femur 549
fibula 550
fovea 545
Golgi tendon 547
gravitational equilibrium 543
hinge joint 553
humerus 549
hydrostatic skeleton 549
intervertebral discs 549
iris 544
lateral line 543
lens 544
ligament 552
meniscus 553
muscle spindle 547
myofibril 551
myosin 551

nerve deafness 542
osteoblast 550
osteoclast 550
osteocyte 550
osteon 550
osteoporosis 550
pectoral (shoulder) girdle 549
pelvic (hip) girdle 549
photoreceptor 544
proprioceptor 546
pupil 544
radius 549
red bone marrow 550
retina 544
rhodopsin 545
rib cage 549
rod 545
rotational equilibrium 543
rotator cuff 549
saccule 543
sacrum 549
sarcomere 551
scapula 549
skull 549
sliding filament model 551
spiral organ 542
spongy bone 550
statocyst 543
sternum 549
synovial joint 552
taste buds 540
tibia 550
ulna 549
utricle 543
vertebral column 549

Testing Yourself

Choose the best answer for each question.

1. The most primitive sense is probably the ability to detect
   a. light.   c. chemicals.
   b. sound.   d. pressure.

2. Label the parts of the ear in the following illustration.

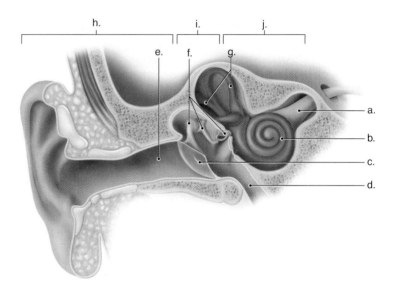

3. The function of the lateral line system in fishes is to
   a. detect sound.
   b. locate other fish.
   c. maintain gravitational equilibrium.
   d. maintain rotational equilibrium.

4. The human eye focuses by
   a. changing the thickness of the lens.
   b. changing the shape of the lens.
   c. opening and closing the pupil.
   d. rotating the lens.

5. Label the parts of the eye in the following illustration.

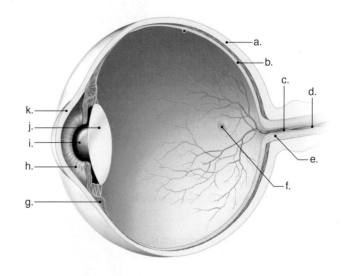

6. Cold receptors differ from warmth receptors in
   a. abundance.   c. structure.
   b. shape.   d. All of these are correct.

7. Unlike an endoskeleton, an exoskeleton
    a. grows with the animal.
    b. is composed of chitin.
    c. is jointed.
    d. protects internal organs.

8. A component of the appendicular skeleton is the
    a. rib cage.   c. femur.
    b. skull.   d. vertebral column.

For questions 9–12, identify the type of muscle in the key that matches the description. Some answers may be used more than once. Some questions may have more than one answer.

Key:

    a. smooth
    b. cardiac
    c. skeletal

9. composed of cells with one nucleus each
10. composed of fibers that result when many cells fuse
11. composed of a lattice network of branched cells
12. striated

13. The deterioration of a joint over time can cause
    a. osteoarthritis.   c. bursitis.
    b. a sprain.   d. tendonitis.

14. The _____ contains the sensory receptors for hearing.
    a. utricle   d. osteon
    b. fovea   e. spiral organ
    c. saccule

15. The thick filaments in a sarcomere are composed mainly of
    a. calcium.   d. rhodopsin.
    b. actin.   e. Both b and c are correct.
    c. myosin.

16. A freely movable, fluid-filled joint is a(n) _____ joint.
    a. synovial   d. otolithic
    b. proprioceptive   e. hydrostatic
    c. Golgi

17. The _____ is the region of the eye that controls the amount of light that enters.
    a. cornea   d. fovea
    b. pupil   e. retina
    c. lens

18. _____ are gravitational equilibrium organs found in several types of invertebrates.
    a. Utricles   d. Statocysts
    b. Rods   e. Spiral organs
    c. Saccules

19. The _____ is the portion of the retina containing most of the cones.
    a. blind spot   d. lens
    b. cornea   e. fovea
    c. iris

20. The contractile portion of a skeletal muscle fiber is the
    a. myofibril.   c. thick filament.
    b. thin filament.   d. sarcoplasmic reticulum.

21. Which cell type deposits bone tissue?
    a. osteocyte   c. osteoblast
    b. osteoclast   d. None of these are correct.

Thinking Scientifically

1. The two leading causes of blindness are age-related macular degeneration and diabetic retinopathy. Both are characterized by the development of an abnormally high number of blood vessels (angiogenesis) in the retina. Why do you suppose angiogenesis impairs vision? Progress in cancer research has led to new strategies for the treatment of these eye diseases. Do you see a connection between the causes of blindness and cancer?

2. Human genome researchers have found a family of approximately 80 genes that encode receptors for bitter-tasting compounds. The genes encode proteins made in taste receptor cells of the tongue. Typically, many types of taste receptors are expressed per cell on the tongue. However, in the olfactory system, each cell expresses only 1 of the 1,000 olfactory receptor genes. Different cells express different genes. How do you suppose this difference affects our ability to taste versus smell different chemicals? Why might the two systems have evolved such different patterns of gene expression?

Bioethical Issue

Deafness: Disability or Cultural Identity?

A lesbian couple, both of whom are deaf, sought out a deaf sperm donor in order to have a deaf child. The women considered deafness to be a cultural identity, rather than a disability. They felt that they would be better parents to a deaf child because they understood the "culture" of deafness. Should this couple have been allowed to choose a father based on his ability to contribute genes for deafness? In a situation like this, whose rights should prevail, the parents' rights to choose their mate or the child's rights to have a chance for normal hearing? Is deafness a disability or a cultural identity?

29

Reproduction and Development

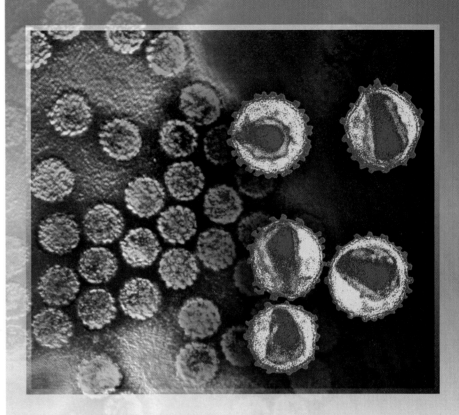

STDs and Infertility

Most people know how sexually transmitted diseases (STDs) are contracted, but many do not understand their treatment or their long-term consequences. Bacterial STDs, such as chlamydia and gonorrhea, are very easy to treat, but when untreated they can have lasting effects, including permanent infertility. Many people are surprised to find that they have become infertile as a result of an STD. If these diseases are easy to treat, why would anyone not be treated and take the risk of becoming infertile? There are many reasons, but one of the most important is that many people with an STD do not have any symptoms of the infection. As long as everything appears normal, they do not seek treatment; meanwhile, they may pass the disease to others and run the risk of long-term complications. This is the primary reason that testing for STDs (even when everything "seems OK") is so important. In addition to the bacterial STDs, several viral STDs are of concern—not only the well-known human immunodeficiency virus (HIV) but also human papillomavirus (HPV). HPV is the primary cause of cervical cancer (the cervix is at the entrance to the uterus). There are usually no noticeable signs of early cervical cancer, but it can be detected by having yearly checkups. Treatment for cervical cancer can result in the inability to have children if the cervix and the uterus must be surgically removed.

Reproduction is a fascinating topic with relevance not only to STDs but also to birth control, reproductive technology, and development. In this chapter, you will learn about reproduction, including some unique issues faced by humans. You will also learn about the portion of human development that occurs before birth and about the birth process itself.

OUTLINE

BEFORE YOU BEGIN

Before beginning this chapter, take a few moments to review the following discussions.

Figure 9.2 What is the role of mitosis and meiosis in the human life cycle?

Section 9.1 What is the role of spermatogenesis and oogenesis?

Section 27.2 Which of the hormones produced by the pituitary glands are associated with sexual reproduction?

29.1 How Animals Reproduce

Learning Outcomes

Upon completion of this section, you should be able to

1. Describe the processes of asexual and sexual reproduction in animals.
2. Explain the adaptations necessary for animals to reproduce on land versus in water.

Animals usually reproduce sexually, but some can reproduce asexually. In asexual reproduction, there is only one parent, and in sexual reproduction, there are two parents.

Asexual Versus Sexual Reproduction

Hydras can reproduce by budding. A new individual arises as an outgrowth (bud) of the parent (**Fig. 29.1**). Many flatworms can constrict into two halves; each half regenerates to become a new individual. Fragmentation, followed by **regeneration,** is also seen among sponges, echinoderms, and corals. Chopping up a sea star does not kill it; instead, each fragment can grow into another animal if a portion of the oral disk is retained with the cut fragment.

   Parthenogenesis is a modification of sexual reproduction in which an unfertilized egg develops into a complete individual. In honeybees, the queen bee can fertilize eggs or allow eggs to pass unfertilized as she lays them. The fertilized eggs become diploid females called workers, and the unfertilized eggs become haploid males called drones. Parthenogenesis is also observed in some fish (including sharks), amphibian, and reptile species.

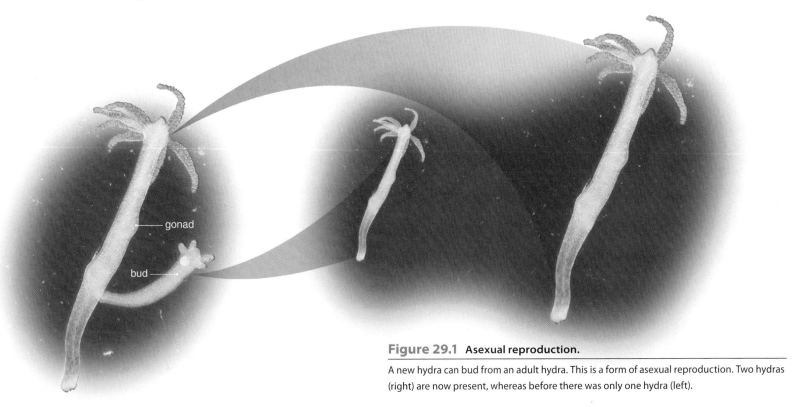

gonad

bud

Figure 29.1 Asexual reproduction.

A new hydra can bud from an adult hydra. This is a form of asexual reproduction. Two hydras (right) are now present, whereas before there was only one hydra (left).

Figure 29.2 Reproducing in water.

Animals that reproduce in water have no need to protect their eggs and embryos from drying out. Here, male and female frogs are mating. They deposit their gametes in the water, where fertilization takes place. The egg contains yolk, which nourishes the embryo until it is a free-swimming larva that can feed for itself.

In sexual reproduction, animals usually produce gametes in specialized organs called **gonads.** The gonads are **testes,** which produce sperm, and **ovaries,** which produce eggs. Eggs or sperm are derived from germ cells that become specialized for this purpose during early development. During sexual reproduction, the egg of one parent is usually fertilized by the sperm of another, and a **zygote** (fertilized egg) results. Even among earthworms, which are **hermaphroditic** (each worm has both male and female sex organs), cross-fertilization occurs. In coral reef fishes called wrasses, a male has a harem of several females. If the male dies, the largest female becomes a male. Many aquatic animals practice **external fertilization**—that is, eggs and sperm join outside the body in the water (**Fig. 29.2**).

Copulation is sexual union to facilitate the reception of sperm by a female. In terrestrial animals, males typically have a penis for depositing sperm into the vagina of females (**Fig. 29.3**). Aquatic animals also have copulatory organs. Lobsters and crayfish have modified swimmerets. Among terrestrial animals, birds lack a penis or vagina. They have a cloaca, a chamber that receives products from the digestive, urinary, and reproductive tracts. A male transfers sperm to a female after placing his cloacal opening against hers.

Reproduction on Land Versus in Water

Many aquatic animals have a larval stage, an immature form capable of feeding. Since the larva has a different lifestyle than the adult, it is able to use a different food source. In sea stars, the bilaterally symmetrical larva undergoes metamorphosis to become a radially symmetrical juvenile. Crayfish, on the other hand, do not have a larval stage; the egg hatches into a tiny juvenile with the same form as the adult. Egg-laying animals are *oviparous,* producing eggs that will hatch after ejection from the body.

Reptiles, particularly birds, provide their eggs with plentiful yolk, a rich nutrient material, and they have no larval stage. Complete development takes place within a shelled egg containing **extraembryonic membranes** to serve the needs of the embryo and prevent it from drying out. The shelled egg frees these animals from the need to reproduce in the water—a significant adaptation to the terrestrial environment.

Birds, in particular, tend their eggs, and newly hatched birds usually have to be fed before they are able to fly away and seek food for themselves. Complex hormones and neural regulation are involved in the reproductive behavior of parental birds. Other animals take a different course. They do not deposit and tend their eggs; instead, they are *ovoviviparous,* meaning that their eggs are retained in the body until they hatch, releasing fully developed offspring that have a way of life like that of the parent. Oysters, which are molluscs, retain their eggs in the mantle cavity, and male sea horses, which are vertebrates, have a brood pouch in which the eggs develop. Garter snakes, water snakes, and pit vipers retain their eggs in their body until they hatch, thus giving birth to living young.

Figure 29.3 Reproducing on land.

Animals that reproduce on land need to protect their gametes and embryos from drying out. Here, the male passes sperm to the female by way of a penis, and the developing embryo/fetus will remain in the female's body until it is capable of living independently.

Finally, most mammals are *viviparous,* meaning that they produce living young. After offspring are born, the mother supplies the nutrients needed for further growth. Viviparity represents the ultimate in caring for the unborn. How did viviparity among certain mammals come about? Some mammals, such as the duckbill platypus and the spiny anteater, lay eggs. In contrast, marsupial offspring are born in a very immature state; they finish their development in a pouch, where they are nourished on milk. Only among primates, including humans, does a single embryo usually develop. The **placenta** is a complex structure derived, in part, from the chorion, the outermost membrane surrounding an embryo, which first appeared in a shelled egg. The evolution of the placenta allowed the developing offspring to exchange materials with the mother internally.

Connections and Misconceptions

For viviparous mammals, what are typical gestation periods?

Gestation periods, the time of development in the uterus from conception to birth, vary greatly between species. The average gestation periods for some common mammals are as follows (listed in days of gestation): hamster, 16; mouse, 21; squirrel, 44; pig, 114; grizzly bear, 220; human, 270; giraffe, 425; killer whale, 500; Indian elephant, 624.

Connecting the Concepts

For more information on the animals mentioned in this section, refer to the following discussions.

Figures 19.4 and **19.5** illustrate the evolutionary relationships of the major groups of animals.

Section 19.5 examines the importance of the amniotic egg in vertebrate evolution.

Check Your Progress 29.1

1. Compare and contrast parthenogenesis with asexual reproduction.
2. Describe the advantages and disadvantages of reproduction on land and in water.
3. Detail the differences among oviparous, ovoviviparous, and viviparous animals.

29.2 Human Reproduction

Learning Outcomes

Upon completion of this section, you should be able to

1. Describe the structures of the male and female reproductive systems and their functions.
2. Explain how hormones regulate the male and female reproductive systems.
3. Evaluate the effectiveness of various means of birth control and explain how they work.
4. Identify the causes of male and female infertility.
5. Describe the assisted reproductive technologies.
6. Identify the causative agents of common sexually transmitted diseases.

In human males and females, the reproductive system consists of two components: (1) the gonads, either testes or ovaries, which produce gametes and sex hormones, and (2) accessory organs that conduct gametes, and in the case of the female, house the embryo/fetus.

Male Reproductive System

The human male reproductive system includes the testes (sing., testis), the epididymis (pl., epididymides), the vas deferens (pl., vasa deferentia), and the urethra (**Fig. 29.4**). (The urethra in males is a part of both the urinary system and the reproductive system.) The paired testes, which produce sperm, are suspended within the scrotum. The testes begin their development inside the abdominal cavity, but they descend into the scrotum as embryonic development proceeds. If the testes do not descend soon after birth—and the male does not receive hormone therapy or undergo surgery to place the testes in the scrotum—sterility (the inability to produce offspring) results. This type of sterility occurs because normal sperm production is inhibited at body temperature; a slightly cooler temperature is required.

Sperm produced by the testes mature within the **epididymis,** a coiled tubule lying just outside each testis. Maturation seems to be required for the sperm to swim to the egg. Once the sperm have matured, they are propelled into the **vas deferens** by muscular contractions. The vasa deferentia are severed or blocked in a surgical form of birth control called a vasectomy (see Table 29.1). Sperm are stored in both the epididymis and the vas deferens. When a male becomes sexually aroused, sperm enter first the ejaculatory duct and then the urethra, part of which is located within the penis.

**Video
Human Sperm**

The **penis** is a cylindrical organ that hangs in front of the scrotum. Three cylindrical columns of spongy, erectile tissue containing distensible blood spaces extend through the shaft of the penis (Fig. 29.4*b*). During sexual arousal, nervous reflexes cause an increase in arterial blood flow to the penis. This increased blood flow fills the blood spaces in the erectile tissue, and the penis, which is normally limp (flaccid), stiffens and increases in size. These changes are called an **erection.** If the penis fails to become erect, the condition is called erectile dysfunction. Drugs such as Viagra, Levitra, and Cialis work

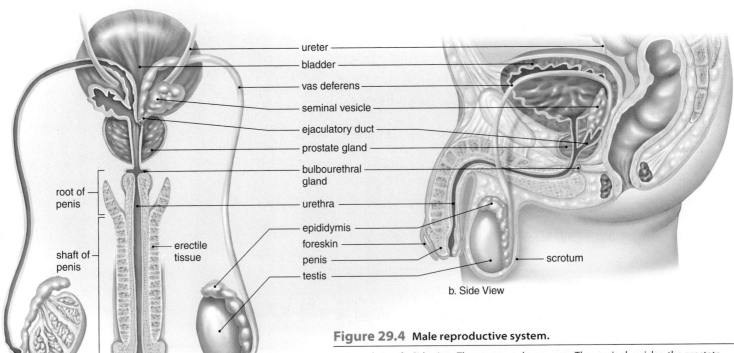

a. Frontal View

b. Side View

Figure 29.4 Male reproductive system.

a. Frontal view. **b.** Side view. The testes produce sperm. The seminal vesicles, the prostate gland, and the bulbourethral glands provide a fluid medium for the sperm. Circumcision is the removal of the foreskin.

by increasing blood flow to the penis, so that when a man is sexually excited, he can achieve and keep an erection.

Semen (seminal fluid) is a thick, whitish fluid that contains sperm and secretions from three glands: the seminal vesicles, prostate gland, and bulbo-urethral glands. The **seminal vesicles** lie at the base of the bladder. Each joins a vas deferens to form an ejaculatory duct that enters the urethra. As sperm pass from the vas deferens into the ejaculatory duct, these vesicles secrete a thick, viscous fluid containing nutrients for use by the sperm. Just below the bladder is the **prostate gland,** which secretes a milky, alkaline fluid believed to activate or increase the motility of the sperm and neutralize the acidity of the urethra from urination. In older men, the prostate gland may become enlarged, thereby constricting the urethra and making urination difficult. Also, prostate cancer is the most common form of cancer in men. Slightly below the prostate gland, one on each side of the urethra, is a pair of small glands called **bulbourethral glands,** which have mucous secretions with a lubricating effect. Notice from Figure 29.4 that the urethra also carries urine from the bladder during urination.

If sexual arousal reaches its peak, ejaculation follows an erection. The first phase of ejaculation is called emission. During emission, the spinal cord sends nerve impulses via appropriate nerve fibers to the epididymides and vasa deferentia. Their muscular walls contract, causing sperm to enter the ejaculatory ducts, whereupon the seminal vesicles, prostate gland, and bulbourethral glands release their secretions. Secretions from the bulbourethral glands occur first and may or may not contain sperm.

During the second phase of ejaculation, called expulsion, rhythmic contractions of muscles at the base of the penis and within the urethral wall expel semen in spurts from the opening of the urethra. These contractions are an example of release from muscle tension. An erection lasts for only a limited amount of time. The penis then returns to its normal, flaccid state. Following ejaculation, a male may typically experience a time, called the refractory period, during which stimulation does not bring about an erection. The contractions that expel semen from the penis are a part of male **orgasm,** the physiological and psychological sensations that occur at the climax of sexual stimulation.

The Testes

A longitudinal section of a testis shows that it is composed of compartments, called lobules, each of which contains one to three tightly coiled **seminiferous tubules (Fig. 29.5a,b).** A microscopic cross section of a seminiferous tubule reveals that it is packed with cells undergoing spermatogenesis, a process that involves reducing the chromosome number from diploid (2n) to haploid (n). Also present are *Sertoli cells,* which support, nourish, and regulate the production of sperm (Fig. 29.5c). A sperm has three distinct parts: a head, a middle piece, and a tail. The head contains a nucleus and is capped by a membrane-bounded acrosome that contains digestive enzymes, so that the sperm can penetrate the outer membrane of an egg. The tail is a flagellum that allows sperm to swim toward the egg, and the middle piece contains energy-producing mitochondria.

The ejaculated semen of a normal human male contains 40 million sperm per milliliter, ensuring an adequate number for fertilization to take place. Fewer than 100 sperm ever reach the vicinity of the egg, however, and only 1 sperm normally enters an egg.

MP3
Male Reproductive Anatomy and Physiology

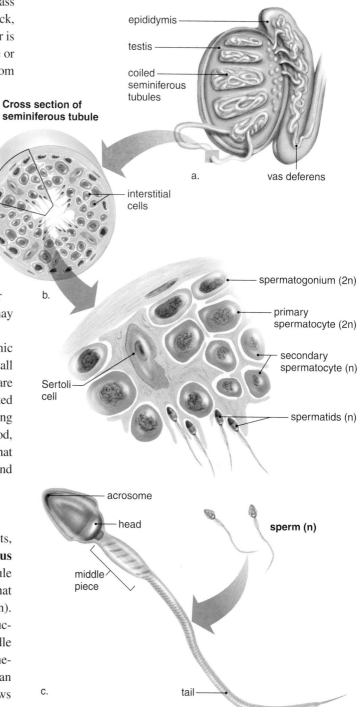

Figure 29.5 Seminiferous tubules.

a. The testes contain seminiferous tubules, where sperm are produced. **b.** Cross section of a tubule. As spermatogenesis occurs, the chromosome number is reduced to the haploid number. **c.** A sperm has a head, a middle piece, and a tail. The nucleus is in the head, which is capped by an acrosome.

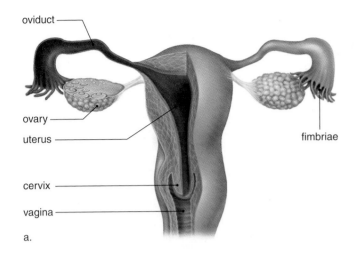

oviduct

ovary

uterus

fimbriae

cervix

vagina

a.

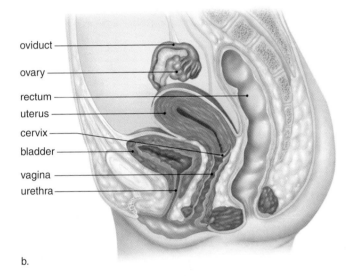

oviduct

ovary

rectum

uterus

cervix

bladder

vagina

urethra

b.

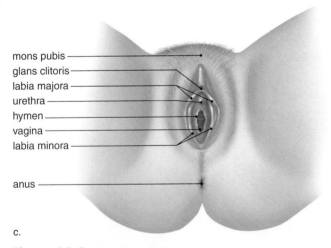

mons pubis
glans clitoris
labia majora
urethra
hymen
vagina
labia minora

anus

c.

Figure 29.6 **Female reproductive system.**

a. Frontal view of the female reproductive system. The ovaries produce one oocyte (egg) per month. Fertilization occurs in the oviduct, and development occurs in the uterus. The vagina is the birth canal and organ of sexual intercourse. **b.** Side view of the female reproductive system plus nearby organs. **c.** Female external genitals, the vulva. At birth, the opening of the vagina is partially occluded by a membrane called the hymen. Physical activities and sexual intercourse disrupt the hymen.

Hormonal Regulation in Males

The hypothalamus has ultimate control of the testes' sexual function because it secretes a hormone called gonadotropin-releasing hormone, or GnRH, that stimulates the anterior pituitary to produce the gonadotropic hormones. Both males and females have two gonadotropic hormones—follicle-stimulating hormone (FSH) and luteinizing hormone (LH). In males, FSH promotes spermatogenesis in the seminiferous tubules. LH in males was formerly called interstitial cell–stimulating hormone (ICSH) because it controls the production of testosterone by the interstitial cells, which are scattered in the spaces between the seminiferous tubules.

Testosterone, the main sex hormone in males, is essential for the normal development and functioning of the sexual organs. Testosterone is also necessary for the maturation of sperm. In addition, testosterone brings about and maintains the male **secondary sex characteristics** that develop at the time of **puberty,** the time of life when sexual maturity is attained. Males are generally taller than females and have broader shoulders and longer legs relative to trunk length. The deeper voice of males compared with females is due to males' having a larger larynx with longer vocal cords. Because the so-called Adam's apple is a part of the larynx, it is usually more prominent in males than in females.

Testosterone causes males to develop noticeable hair on the face, chest, and occasionally other regions of the body, such as the back. Testosterone also leads to the receding hairline and pattern baldness that occur in males. Testosterone is responsible for the greater muscular development in males, which is why steroids are sometimes abused.

Female Reproductive System

The human female reproductive system includes the ovaries, the oviducts, the uterus, and the vagina (**Fig. 29.6***a,b*). The oviducts, also called uterine or fallopian tubes, extend from the ovaries to the uterus; however, the oviducts are not attached to the ovaries. Instead, the oviducts have finger-like projections, called **fimbriae** (sing., fimbria), that sweep over the ovaries. When an oocyte, an immature egg cell, ruptures from an ovary during ovulation, it usually is swept into an oviduct by the combined action of the fimbriae and the beating of cilia that line the oviducts. Fertilization, if it occurs, normally takes place in the first one-third of an oviduct, and the developing embryo is propelled slowly by ciliary movement and tubular muscle contraction to the uterus. The **uterus** is a thick-walled, muscular organ about the size and shape of an inverted pear. The narrow end of the uterus is called the **cervix.** An embryo completes its development after embedding itself in the uterine lining, called the **endometrium.** A small opening at the cervix leads to the vaginal canal. The vagina is a tube at a 45° angle with the body's vertical axis. The mucosal lining of the vagina lies in folds, and therefore the vagina can expand. This ability to expand is especially important when the vagina serves as the birth canal, and it can facilitate sexual intercourse, when the penis is inserted into the vagina.

The external genital organs of a female are known collectively as the vulva (Fig. 29.6*c*). The mons pubis and two folds of skin called labia minora and labia majora are on each side of the urethral and vaginal openings. At the juncture of the labia minora is the clitoris, which is homologous to the penis of males. The clitoris has a shaft of erectile tissue and is capped by a pea-shaped glans. The many sensory receptors of the clitoris allow it to function as a sexually sensitive organ. Most females are born with a thin membrane, called the hymen, which partially obstructs the vaginal opening and has no apparent biological function. The hymen is typically ruptured by physical activities, including sexual intercourse, tampon insertion, and even athletics.

The Ovaries

An oogonium, an undifferentiated germ cell, in the ovary gives rise to an oocyte surrounded by epithelium (**Fig. 29.7**). This is called a primary **follicle.** An ovary contains many primary follicles, each containing an oocyte. At birth, a female has as many as 2 million primary follicles, but the number has been reduced to 300,000–400,000 by the time of puberty. Only a small number of primary follicles (about 400) ever mature and produce a secondary oocyte. When mature, the follicle balloons out on the surface of the ovary and bursts, releasing the secondary oocyte surrounded by follicle cells. The release of a secondary oocyte from a mature follicle is termed **ovulation.** Oogenesis is completed when and if the secondary oocyte is fertilized by a sperm. A follicle that has lost its oocyte develops into a **corpus luteum** and stays inside the ovary. If fertilization and pregnancy do not occur, the corpus luteum begins to degenerate after about 10 days.

Animation
Maturation of the
Follicle and Oocyte

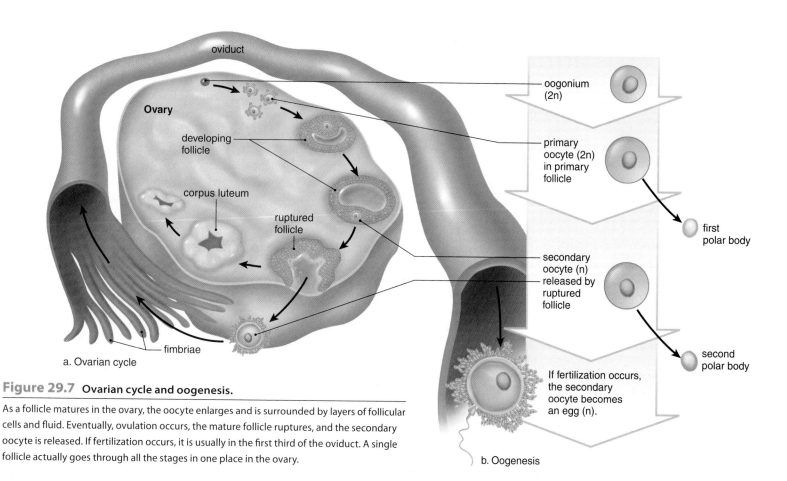

Figure 29.7 Ovarian cycle and oogenesis.

As a follicle matures in the ovary, the oocyte enlarges and is surrounded by layers of follicular cells and fluid. Eventually, ovulation occurs, the mature follicle ruptures, and the secondary oocyte is released. If fertilization occurs, it is usually in the first third of the oviduct. A single follicle actually goes through all the stages in one place in the ovary.

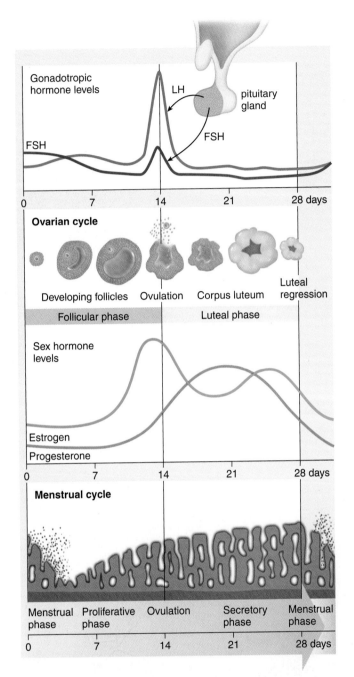

Figure 29.8 Ovarian and menstrual cycles.
During the follicular phase of the ovarian cycle (pink), FSH released by the anterior pituitary promotes the maturation of follicles in the ovary. Ovarian follicles produce increasing levels of estrogen, which cause the endometrium to thicken during the proliferative phase of the menstrual cycle (bottom). After ovulation and during the luteal phase of the ovarian cycle (yellow), LH promotes the development of a corpus luteum. This structure produces increasing levels of progesterone, which cause the endometrium to become secretory. Menstruation begins when progesterone production declines to a low level.

The ovarian cycle is controlled by the gonadotropic hormones FSH and LH from the anterior pituitary gland (**Fig. 29.8**). During the first half, or **follicular phase,** of the cycle (pink in Fig. 29.8), FSH promotes the development of follicles that secrete primarily estrogens, called *estrogen* for the sake of simplicity. As the blood level of estrogen rises, it exerts feedback control over FSH secretion, and ovulation occurs. Ovulation marks the end of the follicular phase. During the second half, or **luteal phase**, of the ovarian cycle (yellow in Fig. 29.8), LH promotes the development of a corpus luteum, which secretes primarily progesterone. As the blood level of progesterone rises, it exerts feedback control over LH secretion, so that the corpus luteum begins to degenerate if fertilization does not occur. As the luteal phase comes to an end, menstruation occurs.

Animation Female Reproductive System

Notice that the female sex hormones estrogen and progesterone affect the endometrium of the uterus, causing the series of events known as the **menstrual cycle** (Fig. 29.8, *bottom*). The 28-day (on average) menstrual cycle in the nonpregnant female is divided as follows:

During *days 1–5,* female sex hormones are at a low level in the body, causing the endometrium to disintegrate and its blood vessels to rupture. A flow of blood, mucus, and degenerating endometrium, known as the **menses,** passes out of the vagina during **menstruation,** also known as the menstrual period.

During *days 6–13,* increased production of estrogen by ovarian follicles causes the endometrium to thicken and become vascular and glandular. This is called the proliferative phase of the menstrual cycle.

Ovulation usually occurs on *day 14* of the 28-day cycle.

During *days 15–28,* increased production of progesterone by the corpus luteum causes the endometrium to double in thickness and the uterine glands to mature, producing a thick, mucoid secretion. This is called the secretory phase of the menstrual cycle. The endometrium now is prepared to receive the developing embryo. But if fertilization does not occur and no embryo embeds itself, the corpus luteum degenerates and the low level of sex hormones in the female body causes the endometrium to break down. Menses begins, marking day 1 of the next cycle. Even while menstruation is occurring, the anterior pituitary begins to increase its production of FSH, and new follicles begin to mature.

MP3 Female Reproductive Anatomy and Physiology

Hormonal Regulation in Females

Estrogen and **progesterone** are the female sex hormones. Estrogen, in particular, is essential for the normal development and functioning of the female reproductive organs. Estrogen is also largely responsible for the secondary sex characteristics in females, including body hair and fat distribution. In general, females have a more rounded appearance than males because of a greater accumulation of fat beneath their skin. Also, the pelvic girdle aligns so that females have wider hips than males, and the thighs converge at a greater angle toward the knees. Both estrogen and progesterone are required for breast development as well.

Menopause, which usually occurs between ages 45 and 55, is the time in a woman's life when the ovarian and menstrual cycles cease. Menopause is not complete until menstruation has been absent for a year.

Connections and Misconceptions

What is hormone replacement therapy?

Hormone replacement therapy after menopause. The fluctuation of hormone levels during menopause can cause symptoms such as hot flashes, mood swings, trouble sleeping, increased abdominal fat, and thinning hair, which HRT can alleviate. The use of HRT has its pros and cons; there is evidence that using HRT after menopause can help prevent bone loss, decrease the risk of colorectal cancer, and decrease certain types of heart disease. Studies also show that some types of HRT in certain patients can increase incidences of stroke and blood clots. Women on HRT should be evaluated every six months by their physician.

Aspects of Reproduction

Important topics related to human reproduction include control of reproduction, infertility, and sexually transmitted diseases.

Control of Reproduction

Table 29.1 lists various means of birth control and gives their rate of effectiveness, assuming consistent and correct usage. **Figure 29.9** illustrates some birth control devices. **Abstinence**—that is, not engaging in sexual intercourse—is very reliable

Table 29.1 Common Birth Control Methods

Name	Procedure	Effectiveness*	Name	Procedure	Effectiveness*
Abstinence	Refrain from sexual intercourse	100%	Diaphragm	Latex cap inserted into vagina to cover cervix before intercourse	With jelly, about 90%

Sterilization
  Vasectomy
  Tubal ligation
	Vas deferens cut and tied
	Oviducts cut and tied
		Almost 100%
		Almost 100%
			Cervical cap	Latex cap held by suction over cervix	Almost 85%

Combined estrogen/ progesterone available as a pill, an injectable, or a vaginal ring and patch	Pill is taken daily; injectable and ring last a month; patch is replaced weekly	About 100%	Male condom	Latex sheath fitted over erect penis	About 85%
			Female condom	Polyurethane liner fitted inside vagina	About 85%
Progesterone only available as a tube implant and an injectable	Implant lasts 3 years; injectable lasts 3 weeks	About 95%	Coitus interruptus	Penis withdrawn before ejaculation	About 75%
			Jellies, creams, foams	Spermicidal products inserted before intercourse	About 75%
Intrauterine device (IUD)	Newest hormone-free device contains progesterone and lasts up to 10 years	More than 90%	Natural family planning	Day of ovulation determined by record keeping; various methods of testing	About 70%
Vaginal sponge	Sponge permeated with spermicide is inserted into vagina	About 90%	Douche	Vagina cleansed after intercourse	Less than 70%

*The percentage of sexually active women per year who will not get pregnant using this method.

and has the added advantage of avoiding sexually transmitted diseases. Oral contraception (the **birth control pill**) often involves taking a combination of estrogen and progesterone on a daily basis. These hormones effectively shut down the pituitary production of both FSH and LH, so that no follicle in the ovary begins to develop in there; because ovulation does not occur, pregnancy cannot take place. Because of possible side effects, including headaches, blurred vision, chest or abdominal pain, and swollen legs, women taking birth control pills should see a physician regularly.

Contraceptive implants use a synthetic progesterone to prevent ovulation by disrupting the ovarian cycle. The older version of the implant consists of six match-sized, time-release capsules that are surgically implanted under the skin of a woman's upper arm. The newest version consists of a single capsule that remains effective for about three years.

Contraceptive injections are available as progesterone only or as a combination of estrogen and progesterone. The length of time between injections can vary from a few weeks to three months.

Interest in barrier methods of birth control has recently increased because they offer some protection against sexually transmitted diseases. A **female condom** consists of a large, polyurethane tube with a flexible ring that fits onto the cervix. The open end of the tube has a ring that covers the external genitals. A **male condom** is most often a latex sheath that fits over the erect penis. The ejaculate is trapped inside the sheath and thus does not enter the vagina. When used in conjunction with a spermicide, the protection is better than with the condom alone. The **diaphragm** is a soft, latex cup with a flexible rim that lodges behind the pubic bone and fits over the cervix. Each woman must be properly fitted by a physician, and the diaphragm can be inserted into the vagina no more than two hours before sexual relations. Also, it must be used with spermicidal jelly or cream and should be left in place for at least six hours after sexual relations. The **cervical cap** is a minidiaphragm.

An **intrauterine device (IUD)** is a small piece of molded plastic that is inserted into the uterus by a physician or another qualified health-care practitioner. IUDs are believed to alter the environment of the uterus and oviducts, so that fertilization probably will not occur—but if it should occur, implantation cannot take place. One type of IUD has copper wire wrapped around the plastic.

Contraceptive vaccines are now being developed. For example, a vaccine intended to immunize women against human chorionic gonadotropin (hCG), a hormone necessary to maintain the implantation of the embryo, has been successful in a limited clinical trial. Since hCG is not normally present in the body, no autoimmune reaction is expected, but the immunization does wear off with time. Other

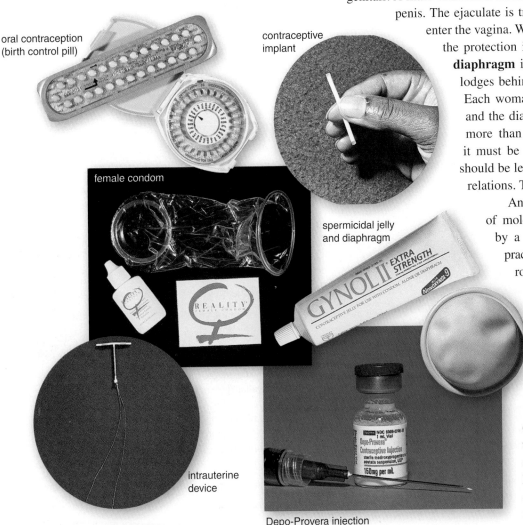

oral contraception
(birth control pill)

contraceptive
implant

female condom

spermicidal jelly
and diaphragm

intrauterine
device

Depo-Provera injection

Figure 29.9 Contraceptive devices.

researchers believe that it would also be possible to develop a safe antisperm vaccine for women.

A morning-after pill, or emergency contraception, is a medication that will prevent pregnancy after unprotected intercourse. One type, a kit called Preven, contains four synthetic progesterone pills; two are taken up to 72 hours after unprotected intercourse, and two more are taken 12 hours later. The medication upsets the normal menstrual cycle, making it difficult for an embryo to implant itself in the endometrium. In a recent study, it was estimated that the medication was 75% effective in preventing unintended pregnancies.

Mifepristone, better known as RU-486, is a pill that is presently used to cause the loss of an implanted embryo by blocking the progesterone receptor proteins of endometrial cells. When taken in conjunction with a prostaglandin to induce uterine contractions, RU-486 is 95% effective. It is possible that someday this medication will also be a "morning-after pill," taken when menstruation is late without evidence that pregnancy has occurred.

Much research today is devoted to developing safe and effective hormonal birth control for males. Implants, pills, patches, and injections are being explored as ways to deliver testosterone and/or progesterone at adequate levels to suppress sperm production. Even the most successful formulations are still in the experimental stage and are unlikely to be available outside of clinical trials for at least a few more years.

Infertility

Control of reproduction does not only mean preventing pregnancy. It also involves methods of treating infertility (**Fig. 29.10**). **Infertility** is the failure of a couple to achieve pregnancy after one year of regular, unprotected intercourse. The American Medical Association estimates that 15% of all couples are infertile. The cause of infertility can be attributed to the male (40%), the female (40%), or both (20%).

The most frequent cause of infertility in males is low sperm count and/or a large proportion of abnormal sperm, which can be due to environmental influences. Physicians advise that a sedentary lifestyle coupled with smoking and alcohol consumption can lead to male infertility. When males spend most of the day driving or sitting in front of a computer or television, the testes' temperature remains too high for adequate sperm production.

In females, body weight is an important fertility factor. Only if a woman is of normal weight do fat cells produce a hormone, called leptin, that stimulates the hypothalamus to release GnRH, so that follicle formation begins in the ovaries. If a woman is overweight, the ovaries may contain many small, ineffective follicles and ovulation does not occur. About 10% of women of childbearing age have an endocrine condition that may be due to an inability to utilize insulin properly. The condition is called polycystic ovary syndrome (PCOS) because the ovaries contain many cysts (small, fluid-filled sacs) but no functioning follicles and ovulation does not occur. An impaired ability to respond to insulin is implicated because many women with PCOS eventually develop type 2 diabetes. Other causes of infertility in females are blocked oviducts due to pelvic inflammatory disease (see page 572) and endometriosis, the presence of uterine tissue outside the uterus, particularly in the oviducts and on the abdominal organs.

Sometimes the causes of infertility can be corrected by medical intervention, so that couples can have children. It is also possible to give females fertility drugs, gonadotropic hormones that stimulate the ovaries and bring about ovulation. Such hormone treatments have been known to cause multiple ovulations and multiple births.

Video New Fertility Boost

Figure 29.10 Control of reproduction.

The medical profession can offer help to control reproduction. The number of births can be decreased by the proper use of birth control, and the number of births can be increased through assisted reproductive technologies.

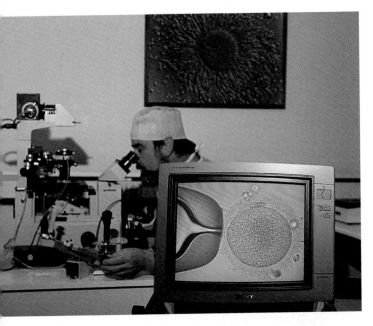

Figure 29.11 Intracytoplasmic sperm injection (ICSI).
A microscope connected to a television screen (at left) is used to carry out ICSI. On the screen, note that a pipette (at left of egg) holds the egg steady while a needle (not visible) introduces the sperm into the egg, ensuring fertilization.

Connections and Misconceptions

How many babies are born as a result of ART in the United States?

The very first IVF baby born in the United States was Elizabeth Carr on December 28, 1981. Since that time, assisted reproductive technologies have improved, along with their  success rates. In 2009, more than 4 million babies were born in the United States, with more than 1% of that total conceived with assisted reproductive technologies.

When reproduction does not occur in the usual manner, many couples adopt a child. Others try the assisted reproductive technologies discussed in the following paragraphs.

Assisted Reproductive Technologies

Assisted reproductive technologies (ART) are techniques used to increase the chances of pregnancy. Often, sperm and/or eggs are retrieved from the testes and ovaries, and fertilization takes place in a clinical or laboratory setting.

Artificial Insemination by Donor (AID)   During artificial insemination, harvested sperm are placed in the vagina by a physician. Sometimes a woman is artificially inseminated by her partner's sperm. This technique is especially helpful if the partner has a low sperm count, because the sperm can be collected over a period of time and concentrated, so that the sperm count is sufficient to result in fertilization. Often, however, a woman is inseminated by sperm acquired from a donor who is a complete stranger to her.

In Vitro Fertilization (IVF) and Intracytoplasmic Sperm Injection (ICSI)   During IVF and ICSI (**Fig. 29.11**), conception occurs outside the body in a laboratory. Ultrasound machines can spot follicles in the ovaries that hold immature eggs; therefore, the latest method is to forgo the administration of fertility drugs and retrieve immature eggs by using a needle. In IVF, the immature eggs are then brought to maturity in glassware before concentrated sperm are added. In ICSI, a single sperm is injected into an egg, usually because a male has severe infertility problems. After about two to four days, embryos are ready to be transferred to the uterus of the woman, who is now in the secretory phase of her menstrual cycle. If desired, embryos can be tested for a genetic disease, and only those found to be free of disease will be used. If implantation is successful and development is normal, pregnancy continues to term.

Video IVF

If several embryos are produced and transferred at a time, multiple births are common. Or there may be excess embryos left over and these may be frozen for transfer later, or they may be donated to other couples or used in research.

Gamete Intrafallopian Transfer (GIFT)   The term **gamete** refers to a sex cell, either a sperm or an egg. Gamete intrafallopian transfer was devised to overcome the low success rate (15–20%) of in vitro fertilization. The method is similar to in vitro fertilization, except the eggs and the sperm are placed in the oviducts immediately after they have been brought together. GIFT has the advantage of being a one-step procedure for the woman—the eggs are removed and reintroduced all in the same time period. A variation on this procedure is to fertilize the eggs in the laboratory and then place the zygotes in the oviducts.

Sexually Transmitted Diseases

Abstinence is the best protection against the spread of sexually transmitted diseases (STDs). For those who are sexually active, a latex condom offers some protection. Among STDs caused by viruses, treatment is available for AIDS and genital herpes, but these conditions are not curable. Only STDs caused by bacteria (e.g., chlamydia, gonorrhea, and syphilis) are curable with antibiotics.

STDs Caused by Viruses **Acquired immunodeficiency syndrome (AIDS)** is caused by a virus called **human immunodeficiency virus (HIV)**. HIV attacks the type of lymphocyte known as helper T cells. Helper T cells, you will recall, stimulate the activities of B lymphocytes, which produce antibodies. After an HIV infection sets in, helper T cells begin to decline in number, and the person becomes debilitated and more susceptible to other types of infections (**Fig. 29.12**). AIDS has three stages of infection, called categories A, B, and C.

During the category A stage, which may last about a year, the individual is an asymptomatic carrier. He or she may exhibit no symptoms but can pass on the infection. Immediately after infection and before the blood test is positive, a large number of infectious viruses are present in the blood and can be passed on to another person. Even after the blood test is positive, the person remains well as long as the body produces sufficient helper T cells to keep the count higher than 500 per mm³. With a combination therapy of several drugs, AIDS patients can remain in this stage indefinitely.

During the category B stage, which may last six to eight years, the lymph nodes swell and the person may experience weight loss, night sweats, fatigue, fever, and diarrhea. Infections such as thrush (white sores on the tongue and in the mouth) and herpes recur.

Finally, the person may progress to category C, which is full-blown AIDS characterized by nervous disorders and the development of an opportunistic disease, such as an unusual type of pneumonia or skin cancer. Opportunistic diseases are those that occur only in individuals who have little or no capability of fighting an infection. Without intensive medical treatment, the AIDS patient dies about seven to nine years after infection.

Genital warts are caused by the human papillomaviruses (HPVs). According to the Centers for Disease Control and Prevention (CDC), genital HPV is the most common STD in the United States; about half of all sexually active individuals will acquire an HPV infection at some point in their lives. Many times, carriers either do not have any sign of warts or merely have flat lesions. When present, the warts commonly are seen on the penis and foreskin of men and near the vaginal opening in women. Genital warts are associated with cancer of the cervix, as well as tumors of the vulva, vagina, anus, and penis.

Genital herpes is characterized by painful blisters on the genitals. Once the blisters rupture, they leave painful ulcers, which may take as long as three weeks or as little as five days to heal. The blisters may be accompanied by fever, pain on urination, swollen lymph nodes in the groin, and in women, a copious discharge. After the ulcers heal, the disease is only latent, and blisters can recur, although usually at less frequent intervals and with milder symptoms. Fever, stress, sunlight, and menstruation are associated with the recurrence of symptoms.

Hepatitis is an infection of the liver and can lead to liver failure, liver cancer, and death. Several types of hepatitis exist, some of which can be transmitted sexually. Each type of hepatitis and the virus that causes it are designated by the same letter.

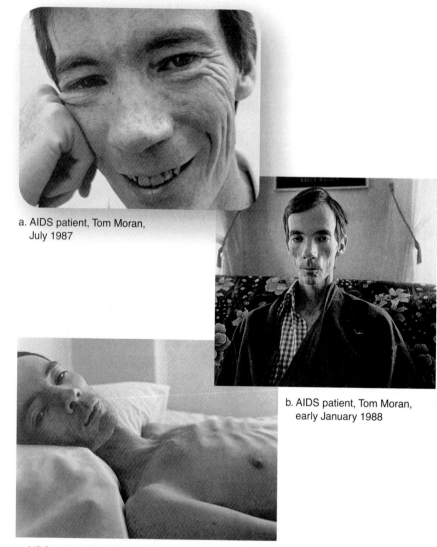

a. AIDS patient, Tom Moran, July 1987

b. AIDS patient, Tom Moran, early January 1988

c. AIDS patient, Tom Moran, late January 1988

Figure 29.12 **The course of AIDS.**

This individual went through all the stages of AIDS before he died. You should protect yourself from sexually transmitted diseases (STDs). If you have an STD, you are more susceptible to getting another one.

Figure 29.13 **Condom use can help reduce the spread of STDs.**

The following are some guidelines for preventing the spread of STDs:

1. Abstain from sexual intercourse or develop a long-term monogamous (always the same person) relationship with a person who is free of STDs.

2. Refrain from having multiple sex partners or a relationship with a person who does have multiple sex partners.

3. Be aware that having relations with an intravenous drug user is risky because the behavior of this group puts them at risk for AIDS and hepatitis B.

4. Avoid anal intercourse because HIV has easy access through the lining of the rectum.

5. Always use a latex condom if your partner has not been free of STDs for the past five years.

6. Avoid oral sex because this may be a means of transmitting AIDS and other STDs.

7. Stop, if possible, the habit of injecting drugs; if you cannot stop, at least always use a sterile needle.

Hepatitis A is usually acquired from sewage-contaminated drinking water, but this infection can also be sexually transmitted through oral/anal contact. Hepatitis B, which is spread in the same manner as AIDS, is even more infectious. Fortunately, a vaccine is available for hepatitis B. Hepatitis C is spread when a person comes in contact with the blood of an infected person.

STDs Caused by Bacteria **Chlamydia** is a bacterial infection of the lower reproductive tract that is usually mild or asymptomatic, especially in women. About 8 to 21 days after infection, men may experience a mild burning sensation upon urination and a mucoid discharge. Women may have a vaginal discharge, along with the symptoms of a urinary tract infection. Chlamydia also causes cervical ulcerations, which increase the risk of acquiring AIDS. If the infection is misdiagnosed or if a woman does not seek medical help, there is a risk of the infection spreading from the cervix to the oviducts, so that **pelvic inflammatory disease (PID)** results. This very painful condition can result in blockage of the oviducts, with the possibility of sterility and infertility.

Gonorrhea is easier to diagnose in males than in females because males are more likely to experience painful urination and a thick, greenish-yellow urethral discharge. In males and females, a latent infection leads to PID, which affects the vasa deferentia or oviducts. As the inflamed tubes heal, they may become partially or completely blocked by scar tissue, resulting in sterility or infertility.

Syphilis has three stages, which are typically separated by latent periods. During the final stage, syphilis may affect the cardiovascular and/or nervous system. An infected person may become mentally retarded, become blind, walk with a shuffle, or show signs of insanity. Gummas, which are large, destructive ulcers, may develop on the skin or in the internal organs. Syphilitic bacteria can cross the placenta, causing birth defects or stillbirth. Syphilis is easily diagnosed with a blood test.

STDs Caused by Other Organisms Females very often have vaginitis, or infection of the vagina, caused by either the flagellated protozoan *Trichomonas vaginalis* or the yeast *Candida albicans*. **Trichomoniasis** is most often acquired through sexual intercourse, and the asymptomatic male is usually the reservoir of infection. *Candida albicans,* however, is an organism normally found in the vagina; its growth simply increases beyond normal under certain circumstances. For example, women taking birth control pills are sometimes prone to yeast infections. Also, the legitimate and indiscriminate use of antibiotics for infections elsewhere in the body can alter the normal balance of organisms in the vagina, so that a yeast infection flares up.

Figure 29.13 lists the precautions everyone should take to avoid sexually transmitted diseases.

Connecting the Concepts

For more information on the material presented in this section, refer to the following discussions.

Figure 17.5 illustrates the reproductive cycle of the HIV virus.

Section 26.5 explains how the HIV virus causes AIDS.

Section 27.2 examines the role of hormones as chemical regulators of the organ systems of the body.

1. List the structures involved in sperm development and ejaculation.

2. Summarize the function of testosterone, LH, and FSH in the male reproductive system.

3. Detail the function of a few contraceptive devices.

4. Describe the process of ovulation in the ovaries.

5. Detail the events of the menstrual cycle in females.

6. Compare and contrast the structure and function of the ovaries and the testes.

7. Categorize STDs according to how they are transmitted (viruses, bacteria, and other causes).

8. Predict what could occur if the estrogen level of a female were deficient.

9. Outline a pro-con argument on assisted reproductive technologies.

29.3 Human Development

Learning Outcomes

Upon completion of this section, you should be able to

1. Describe the processes of fertilization, embryonic development, fetal development, and birth.

2. Identify the stages of embryonic development and explain the significance of gastrulation and neurulation.

3. Describe the structure and function of the placenta.

Development encompasses all the events that occur from the time of fertilization until the animal is fully formed—in this case, a human being. Humans and other mammals have stages similar to those of all animals but with some marked differences, chiefly because of the presence of extraembryonic membranes (membranes outside the embryo). Developing mammalian embryos (and fetuses in placental mammals, such as humans), such as those of reptiles and birds, depend on these membranes to protect and nourish them. **Figure 29.14** shows what these membranes are and what they do in a mammal.

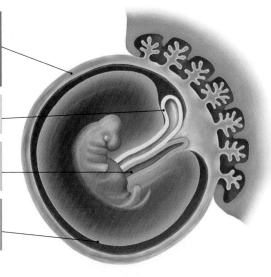

chorion
becomes part of the placenta where the embryo/fetus receives oxygen and nutrient molecules and rids itself of waste molecules

yolk sac
first site of blood cell formation

allantois
its blood vessels become the blood vessels of the umbilical cord

amnion
contains the amniotic fluid, which cushions and protects the embryo

Figure 29.14 The extraembryonic membranes.

Humans, like other animals that reproduce on land, are dependent on these membranes to protect and nourish the embryo (and later the fetus).

Fertilization

Fertilization, which results in a zygote, requires that the sperm and the secondary oocyte interact (**Fig. 29.15**). The plasma membrane of the secondary oocyte is surrounded by an extracellular material termed the zona pellucida. In turn, the zona pellucida is surrounded by a few layers of adhering follicle cells.

During fertilization, a sperm moves past the leftover follicle cells, and acrosomal enzymes released by exocytosis digest a route through the zona pellucida. A sperm binds and fuses to the secondary oocyte's plasma membrane, and its nucleus enters. Only then does the secondary oocyte complete meiosis II and become an egg. Finally, the haploid egg and sperm nuclei fuse. Only one sperm should fertilize an egg, or else the zygote will have too many chromosomes and the resulting zygote will not be viable. Changes in the zona pellucida prevent the binding and penetration of additional sperm (called polyspermy).

Early Embryonic Development

The first two months of development are considered the embryonic period. Approximately six days of development occur in the oviduct before the embryo implants itself in the uterine lining (endometrium; **Fig. 29.16**). During the first

Figure 29.15 Fertilization.

1 The sperm makes its way through the adhering follicle cells.
2 Acrosomal enzymes digest a portion of the zona pellucida. **3** Sperm binds to and fuses with egg plasma membrane. **4** Sperm nucleus enters cytoplasm of egg. **5** Sperm and egg nuclei fuse to produce a zygote.

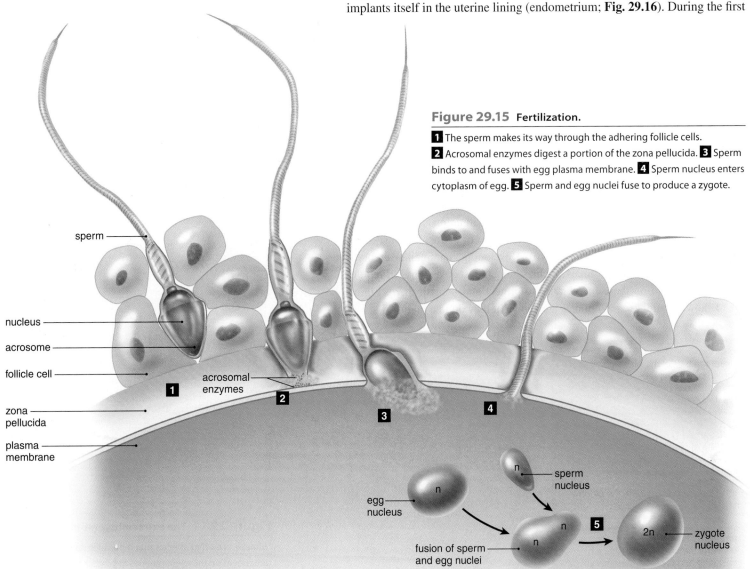

stage of development, the embryo becomes multicellular. Following fertilization, the zygote undergoes **cleavage,** which is cell division without growth. DNA replication and mitotic cell division occur repeatedly, and the cells get smaller with each division. Notice that cleavage only increases the number of cells; it does not change the original volume of the egg cytoplasm. The resulting tightly packed ball of cells is called a **morula.**

The cells of the morula continue to divide, but they also secrete a fluid into the center of a ball of cells. A hollow ball of cells, called the **blastocyst,** is formed, surrounding a fluid-filled cavity called the blastocoel. Within the ball is an inner cell mass that will go on to become the embryo. The outer layer of cells is the first sign of the chorion, the extraembryonic membrane that will contribute to the development of the placenta. As the embryo implants itself in the uterine lining (endometrium), the placenta begins to form and to secrete the hormone human chorionic gonadotropin (hCG). This hormone is the basis for the pregnancy test, and it maintains the corpus luteum past the time it normally disintegrates inside the ovary. Because of hCG, the endometrium is maintained until this function is taken over by estrogen and progesterone, produced by the placenta. Ovulation and menstruation do not normally occur during pregnancy.

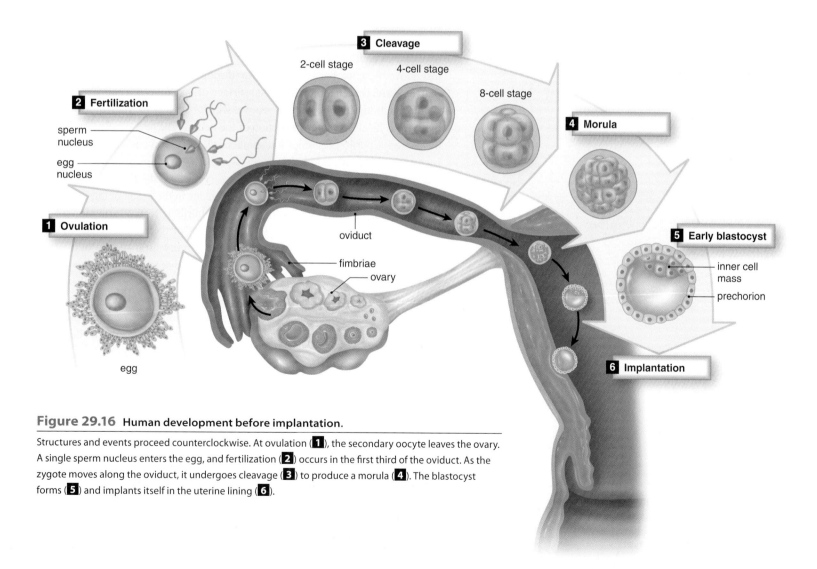

Figure 29.16 Human development before implantation.

Structures and events proceed counterclockwise. At ovulation (**1**), the secondary oocyte leaves the ovary. A single sperm nucleus enters the egg, and fertilization (**2**) occurs in the first third of the oviduct. As the zygote moves along the oviduct, it undergoes cleavage (**3**) to produce a morula (**4**). The blastocyst forms (**5**) and implants itself in the uterine lining (**6**).

Figure 29.17 Gastrulation and neurulation.

a. Gastrulation is the movement of cells to establish the three germ layers: ectoderm, mesoderm, and endoderm. **b.** During neurulation, the neural tube, which becomes the central nervous system, develops right above the notochord. Later, the notochord is replaced by the vertebral column. The neural crests contain ectodermal cells that did not participate in forming the neural tube. They develop into a number of different structures, including the autonomic nervous system.

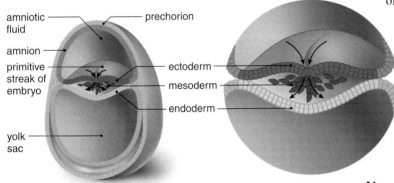

a. Gastrulation

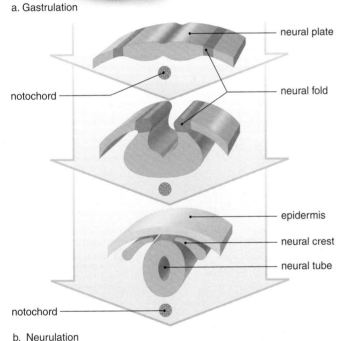

b. Neurulation

Later Embryonic Development

Gastrulation

During the third week of development, a slit called the primitive streak appears. The **gastrula** stage is evident when certain cells begin to push, or invaginate, at the primitive streak, creating three layers of cells (**Fig. 29.17***a*). Also evident is the **amnion,** the extraembryonic membrane that surrounds the developing embryo and later the fetus. The amnion encloses the amniotic fluid, which offers protection from sudden blows and movements, allows freedom of fetal movement, helps maintain a relatively constant temperature, and prevents drying out, which assists in lung development.

Gastrulation is complete when the three layers of cells that will develop into adult organs have been produced. The outer layer is the **ectoderm;** the inner layer is the **endoderm;** and the third, or middle, layer of cells is called the **mesoderm.** Ectoderm, mesoderm, and endoderm are called the embryonic **germ layers.** As shown in **Figure 29.18,** the organs of an animal's body develop from these three germ layers.

Neurulation

The first organs to form are those of the central nervous system (see Fig. 29.17*b*). The newly formed mesoderm cells that lie along the main longitudinal axis of the animal coalesce to form a dorsal supporting rod called the **notochord.** The central nervous system develops from midline ectoderm located just above the notochord. During **neurulation,** thickening of cells, called the **neural plate,** is seen along the dorsal surface of the embryo. Then, neural folds develop on both sides of a neural groove, which becomes the **neural tube** when these folds fuse. The anterior portion of the neural tube becomes the brain, and the posterior portion becomes the spinal cord. Development of the neural tube is an example of **induction,** the process by which one tissue or organ influences the development of another. Induction occurs because the tissue initiating the induction releases a chemical that turns on genes in the tissue being induced.

Midline mesoderm cells that did not contribute to the formation of the notochord now become two longitudinal masses of tissue. These two masses become blocked off into somites, which are serially arranged on both sides along the length of the notochord. Somites give rise to the vertebrae and to muscles associated with the axial skeleton. The sequential order of the vertebrae and the muscles of the trunk testify that chordates are segmented animals. Lateral to the somites, the mesoderm splits and forms the mesodermal lining of the coelom. In addition, the neural crest consists of a band of cells that develops

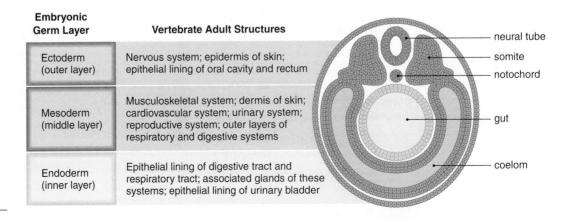

Embryonic Germ Layer	Vertebrate Adult Structures
Ectoderm (outer layer)	Nervous system; epidermis of skin; epithelial lining of oral cavity and rectum
Mesoderm (middle layer)	Musculoskeletal system; dermis of skin; cardiovascular system; urinary system; reproductive system; outer layers of respiratory and digestive systems
Endoderm (inner layer)	Epithelial lining of digestive tract and respiratory tract; associated glands of these systems; epithelial lining of urinary bladder

Figure 29.18 The germ layers and associated organs.

Figure 29.19 Human embryo at the beginning of the fifth week.

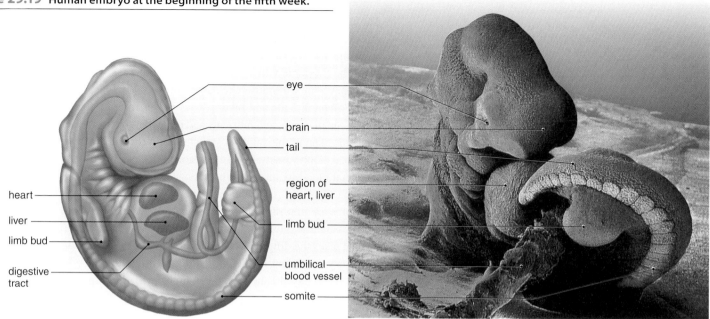

where the neural tube pinches off from the ectoderm. These cells migrate to various locations, where they contribute to the formation of skin and muscles, in addition to the adrenal medulla and the ganglia of the peripheral nervous system.

At the end of the third week, over a dozen somites are evident, and the blood vessels and gut have begun to develop. At this point, the embryo is about 2 millimeters (mm) long.

Organ Formation Continues

A human embryo at five weeks has little flippers called limb buds (**Fig. 29.19**). Later, the arms and legs develop from the limb buds, and even the hands and feet become apparent. During the fifth week, the head enlarges and the sense organs become more prominent. It is possible to make out the developing eyes, ears, and nose.

The umbilical cord has developed from a bridge of mesoderm called the body stalk, which connects the caudal (tail) end of the embryo with the chorion. A fourth extraembryonic membrane, the allantois, is contained within this stalk, and its blood vessels become the umbilical blood vessels. The head and the tail then lift up as the body stalk moves anteriorly by constriction. Once this process is complete, the **umbilical cord,** which connects the developing embryo to the placenta, is fully formed (**Fig. 29.20**).

A remarkable change in external appearance occurs during the sixth through eighth weeks of development—the embryo becomes easily recognized as human. Concurrent with brain development, the neck region develops, making the head distinct from the body. The nervous system is developed well enough to permit reflex actions, such as a startle response to touch. At the end of this period, the embryo is about 38 mm (1.5 inches) long and weighs no more than an aspirin tablet, even though all its organ systems are established.

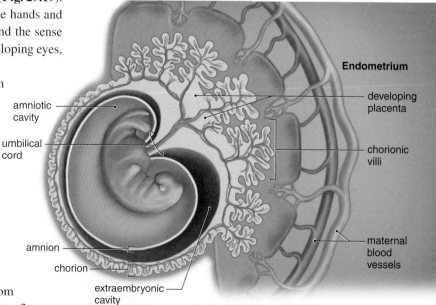

Figure 29.20 Placenta.

Blood vessels within the umbilical cord lead to the placenta, where exchange takes place between fetal blood and maternal blood.

Placenta

The placenta has a fetal side contributed by the chorion, the outermost extra-embryonic membrane, and a maternal side consisting of uterine tissues. Notice in Figure 29.20 that the chorion has tree-like projections, called the **chorionic villi.** The chorionic villi are surrounded by maternal blood, yet maternal and fetal blood never mix, because exchange always takes place across the walls of the villi. Carbon dioxide and other wastes move from the fetal side to the maternal side of the placenta, and nutrients and oxygen move from the maternal side to the fetal side. The umbilical cord stretches between the placenta and the fetus. The umbilical blood vessels are an extension of the fetal circulatory system and simply take fetal blood to and from the placenta. Harmful chemicals can also cross the placenta. This is of particular concern during the embryonic period, when various structures are first forming. Each organ or part seems to have a sensitive period, during which a substance can alter its normal development. For example, if a woman takes the drug thalidomide, a tranquilizer, between days 27 and 40 of her pregnancy, the infant is likely to be born with deformed limbs. After day 40, however, the limbs will develop normally.

Animation
Fetal Development and Risk

Fetal Development and Birth

Fetal development encompasses the third to the ninth months (**Figs. 29.21–29.24**). Fetal development is marked by an extreme increase in size. Weight multiplies 600 times, going from less than 28 grams to 3 kilograms. During this time, too,

Figure 29.21 A 9-week fetus.

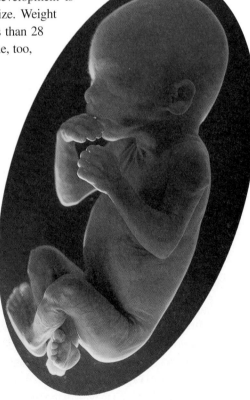

Figure 29.22
A 12- to 16-week fetus.

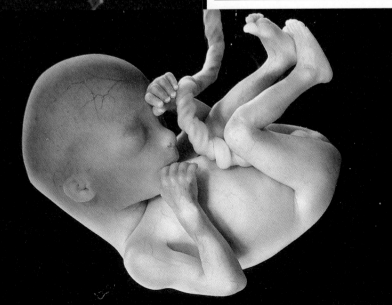

Figure 29.23 A 20- to 28-week fetus.

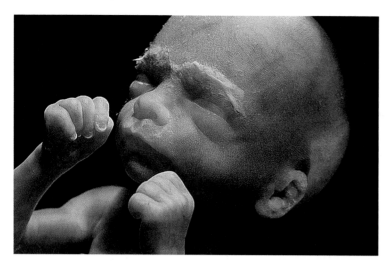

Figure 29.24 A 24-week fetus.

the fetus grows to about 50 centimeters in length. The genitalia appear in the third month, so it is possible to tell if the fetus is male or female.

Soon after the third month, hair, eyebrows, and eyelashes add finishing touches to the face and head. In the same way, fingernails and toenails complete the hands and feet. Later, during the fifth through seventh months, a fine, downy hair (lanugo) covers the limbs and trunk, only to disappear later. The fetus looks very old because the skin is growing so fast that it wrinkles. A waxy, almost cheese-like substance (called vernix caseosa) protects the wrinkly skin from the watery amniotic fluid.

The fetus at first only flexes its limbs and nods its head, but later it can move its limbs vigorously to avoid discomfort. The mother feels these movements from about the fourth month on. The other systems of the body also begin to function. As early as 10 weeks, the fetal heartbeat can be heard through a stethoscope. A fetus born at 22 weeks has a chance of surviving, although the lungs are still immature and often cannot capture oxygen adequately. Weight gain during the last couple of months increases the likelihood of survival.

Connections and Misconceptions

Why is the female gender sometimes referred to as the "default sex"?

The term *default sex* has to do with the presence or absence of the Y chromosome in the fetus. On the Y chromosome is a gene called *sex determining region Y (SRY),* which produces a protein that causes Sertoli cells in the testes to produce Mullerian inhibiting substance (MIS). This causes Leydig cells in the testes to produce testosterone, which signals the development of the male sex organs (vas deferens, epididymis, penis, etc.). Without the *SRY* gene and this hormone cascade effect, female structures (uterus, fallopian tubes, ovaries, etc.) will begin to form.

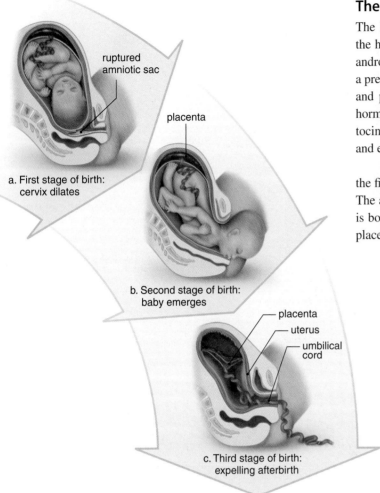

Figure 29.25 Three stages of birth (parturition).

a. Dilation of cervix. **b.** Birth of baby. **c.** Expulsion of afterbirth (placenta).

The Stages of Birth

The latest findings suggest that, when the fetal brain is sufficiently mature, the hypothalamus causes the pituitary to stimulate the adrenal cortex, so that androgens are released into the bloodstream. The placenta uses androgens as a precursor for estrogen, a hormone that stimulates the production of oxytocin and prostaglandin (a molecule produced by many cells that acts as a local hormone). All three of these molecules—estrogen, oxytocin, and prostaglandin—cause the uterus to contract and expel the fetus.

Video
Triggering Birth

The process of birth (parturition) has three stages (**Fig. 29.25**). During the first stage, the cervix dilates to allow passage of the baby's head and body. The amnion usually bursts about this time. During the second stage, the baby is born and the umbilical cord is cut (**Fig. 29.26**). During the third stage, the placenta is delivered.

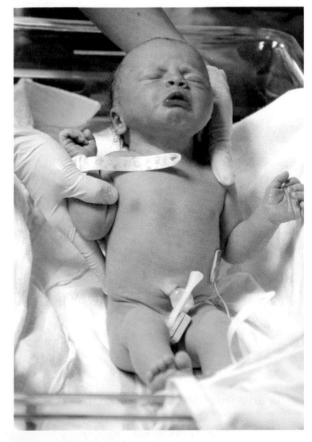

Figure 29.26 Newborn (40 weeks).

Connecting the Concepts

For more information on the topics in this section, refer to the following discussion.

Section 27.2 describes the hormones produced by the hypothalamus, pituitary gland, and adrenal cortex.

Check Your Progress 29.3

1. Explain how a morula is formed.
2. Describe the significance of germ layer formation during gastrulation.
3. Explain how induction occurs.
4. Contrast embryonic development with fetal development.
5. Explain how the placenta is used during development.
6. Predict what might occur if neurulation does not happen properly.

Media Study Tools

 plus+

www.mhhe.com/maderessentials3

Enhance your study of this chapter with study tools and practice tests. Also ask your instructor about the resources available through ConnectPlus, including the media-rich eBook, interactive learning tools, and animations.

The Chapter in Review

Summary

29.1 How Animals Reproduce

Animals usually reproduce sexually, but some can also reproduce asexually.

- In asexual reproduction, a single parent produces offspring.
- In sexual reproduction, gametes produced by gonads join to form a zygote that develops into an offspring.
- Animals adapted to reproducing in water shed their gametes into the water; fertilization and development occur in water. Animals that reproduce on land protect their gametes and embryos from drying out.
- Reptiles and mammals have extraembryonic membranes.

29.2 Human Reproduction

Male Reproductive System

- Sperm are produced in the testis, mature in the epididymis, and are stored in the epididymis and vas deferens.
- Sperm enter the urethra (in the penis) prior to ejaculation, along with seminal fluid (produced by seminal vesicles, the prostate gland, and bulbourethral glands).
- Spermatogenesis occurs in the seminiferous tubules of the testes, which also produce testosterone in interstitial cells.

Hormone Regulation in Males Testosterone brings about the maturation of the sex organs during puberty and promotes the secondary sex characteristics of males.

Female Reproductive System

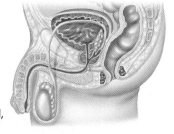

- Oocytes are produced in the ovary and move through the oviduct to the uterus. The uterus opens into the vagina. The external genital area of women includes the clitoris, the labia minora, the labia majora, and the vaginal opening.
- A follicle produces a secondary oocyte and becomes a corpus luteum. The follicle and later the corpus luteum produce estrogen and progesterone.

The ovarian and menstrual cycles last 28 days. The events of these cycles are as follows:

- Days 1–13: Menstruation occurs for five days; the anterior pituitary produces FSH, and follicles produce primarily estrogen. Estrogen causes the endometrium to increase in thickness.
- Day 14: Ovulation occurs.
- Days 15–28: LH from the anterior pituitary causes the corpus luteum to produce progesterone. Progesterone causes the endometrium to become secretory.

Estrogen and Progesterone Estrogen and progesterone bring about the maturation of the sex organs during puberty and promote the secondary sex characteristics of females.

Aspects of Reproduction

Numerous birth control methods and devices are available for those who wish to prevent pregnancy. Infertile couples are increasingly making use of assisted methods of reproduction.

Sexually transmitted diseases include the following:

- AIDS, an epidemic disease that destroys the immune system
- Genital warts, which can lead to cancer of the cervix
- Genital herpes, which repeatedly flares up
- Hepatitis, especially types A and B
- Chlamydia and gonorrhea, which cause pelvic inflammatory disease (PID)
- Syphilis, which leads to cardiovascular and neurological complications if untreated

29.3 Human Development

Development encompasses all the events that occur from fertilization to a fully formed animal, in this case a human being.

Fertilization

The acrosome of a sperm releases enzymes that digest a pathway for the sperm through the zona pellucida. The sperm nucleus enters the egg and fuses with the egg nucleus.

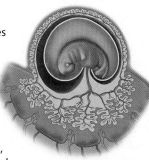

Early Embryonic Development (Months 1 and 2)

- Cleavage, which occurs in the oviduct, is cell division and formation of a morula and blastocyst. The blastocyst implants itself in the endometrium.
- Gastrulation is invagination of cells into the blastocoel, which results in formation of the germ layers. The germ layers are the ectoderm, mesoderm, and endoderm.
- Organ formation can be related to the germ layers. Organ development begins with neural tube formation. Induction helps account for the steady progression of organ formation during embryonic development.
- The placenta has a fetal side and a maternal side. Gases and nutrients are exchanged at the placenta. The umbilical blood vessels are an extension of the embryo/fetal cardiovascular system and carry blood to and from the placenta.

Fetal Development and Birth (Months 3–9)

- During fetal development, the refinement of organ systems occurs and the fetus adds weight.
- Birth has three stages: The cervix dilates, the baby is born, and the placenta is delivered.

Key Terms

abstinence   567
acquired immunodeficiency
   syndrome (AIDS)   571
amnion   576
birth control pill   568
blastocyst   575
bulbourethral glands   563
cervical cap   568
cervix   564
chlamydia   572
chorionic villi   578
cleavage   575
contraceptive implant   568
contraceptive injection   568
contraceptive vaccine   568
copulation   560
corpus luteum   565
diaphragm   568
ectoderm   576
endoderm   576
endometrium   564
epididymis   562
erection   562
estrogen   566
external fertilization   560
extraembryonic membranes   560
female condom   568
fertilization   574
fimbriae (sing., fimbria)   564
follicle   565
follicular phase   566
gamete   570
gastrula   576
gastrulation   576
genital herpes   571
genital warts   571
germ layer   576
gonads   560
gonorrhea   572
hepatitis   571
hermaphroditic   560

human immunodeficiency
   virus (HIV)   571
induction   576
infertility   569
intrauterine device (IUD)   568
luteal phase   566
male condom 568
menopause   566
menses   566
menstrual cycle   566
menstruation   566
mesoderm   576
morula   575
neural plate   576
neural tube   576
neurulation   576
notochord   576
orgasm   563
ovary   560
ovulation   565
parthenogenesis   559
pelvic inflammatory disease
   (PID)   572
penis   562
placenta   561
progesterone   566
prostate gland   563
puberty   564
regeneration   559
secondary sex characteristic   564
semen (seminal fluid)   563
seminal vesicle   563
seminiferous tubule   563
syphilis   572
testis (pl., testes)   560
testosterone   564
trichomoniasis   572
umbilical cord   577
uterus   564
vas deferens   562
zygote   560

Testing Yourself

Choose the best answer for each question.

1. The evolutionary significance of the placenta is that it allowed
   a. animals to retain their eggs until they hatched.
   b. offspring to be born in an immature state and then develop outside the mother.
   c. offspring to exchange materials with the mother while developing inside the mother.
   d. mammals to develop the ability to produce milk.

2. Sperm are never found in the
   a. prostate gland.
   b. epididymides.
   c. urethra.
   d. vasa deferentia.
   e. ejaculatory duct.

3. The function of the prostate gland is to
   a. cause sperm to reproduce.
   b. improve sperm motility.
   c. provide nutrients for sperm.
   d. extend the life of sperm.

4. Spermatogenesis produces cells that are
   a. diploid and genetically identical to each other.
   b. diploid and genetically different from each other.
   c. haploid and genetically identical to each other.
   d. haploid and genetically different from each other.

5. Label the parts of the male reproductive system in the following illustration.

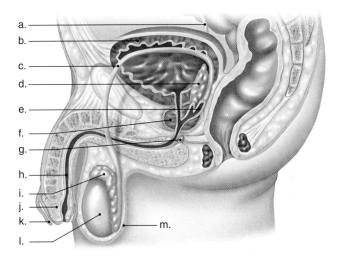

6. Label the parts of the female reproductive system in the following illustration.

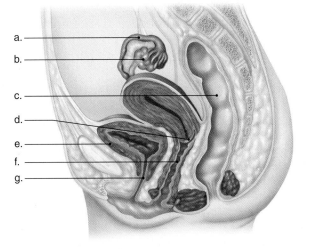

7. Secondary sex characteristics in women are due mainly to the effects of
   a. estrogen.
   b. progesterone.
   c. follicle-stimulating hormone.
   d. luteinizing hormone.
   e. testosterone.

8. It is thought that the process of birth begins when
   a. the fetal heart is sufficiently developed.
   b. the umbilical cord can no longer support the growth of the fetus.
   c. the fetal brain is mature enough to release androgens.
   d. all organ systems have completed development.

For questions 9–12, identify the stage of embryonic development in the key that matches the description.

Key:

   a. cleavage
   b. gastrulation
   c. neurulation
   d. organ formation

9. The notochord is formed.

10. The zygote divides without increasing in size.

11. Ectoderm and endoderm are formed.

12. Reflex actions, such as response to touch, develop.

13. Hepatitis is caused by a
    a. bacterium.
    b. fungus.
    c. protozoan.
    d. virus.

14. In human males, sterility results when the testes are located inside the abdominal cavity instead of the scrotum because
    a. the sperm cannot pass through the vas deferens.
    b. sperm production is inhibited at body temperature.
    c. the sperm cannot travel the extra distance.
    d. digestive juices destroy the sperm.

15. In the human menstrual cycle, ovulation usually occurs
    a. during days 1–5.
    b. during days 6–13.
    c. on day 14.
    d. during days 15–28.

16. In the testes, the _____ produce(s) the sperm.
    a. seminiferous tubules
    b. bulbourethral glands
    c. seminal vesicles
    d. prostate gland

17. Exchange of gases and waste between maternal and fetal blood occurs across the walls of the
    a. corpus luteum.
    b. seminiferous tubules.
    c. chorionic villi.
    d. germ layers.

18. The release of a secondary oocyte from a follicle is
    a. fertilization.
    b. menses.
    c. parturition.
    d. copulation.
    e. ovulation.

19. The assisted reproductive technology (ART) in which harvested sperm are placed in the vagina by a physician is
    a. in vitro fertilization (IVF).
    b. artificial insemination by donor (AID).
    c. gamete intrafallopian transfer (GIFT).
    d. intracytoplasmic sperm injection (ICSI).

20. Which pair is matched incorrectly?
    a. HIV—AIDS
    b. HPVs—genital warts
    c. chlamydia—pelvic inflammatory disease (PID)
    d. gonorrhea—liver failure

21. The process by which one tissue or organ influences the development of another is
    a. induction.
    b. insemination.
    c. neurulation.
    d. cleavage.
    e. gastrulation.

22. Human fetal development encompasses
    a. the third to the ninth months.
    b. the first to the third months.
    c. the first to the ninth months.
    d. the sixth to the ninth months.

23. The _____ is a dorsal supporting rod that forms from mesoderm during embryonic development.
    a. neural plate
    b. neural tube
    c. umbilical cord
    d. notochord
    e. morula

Thinking Scientifically

1. Female athletes who train intensively often stop menstruating. The important factor appears to be the reduction of body fat below a certain level. Give a possible evolutionary explanation for a relationship between body fat in females and reproductive cycles.

2. Human females undergo menopause typically between the ages of 45 and 55. In most other animal species, both males and females maintain their reproductive capacity throughout their lives. Why do you suppose human females lose that ability in midlife?

Bioethical Issue

The HPV Vaccine

In June 2006, the Food and Drug Administration (FDA) licensed Gardasil, a vaccine that protects against four types of human papillomaviruses (HPVs), including the two known to cause 90% of genital warts and 70% of cervical cancers. The HPV vaccine is intended for use before the onset of sexual activity; thus, it is recommended for 11- and 12-year-old girls, although it can be given to any female between the ages of 9 and 26. Gardasil typically has no side effects other than temporary soreness in the injection site. The vaccine is nearly 100% effective against four types of HPVs; however, because there are over 40 types of sexually transmitted HPVs, Gardasil does not provide absolute protection against either genital warts or cervical cancer.

When the HPV vaccine was first made available, some parents were eager to have their daughters vaccinated. But enthusiasm for Gardasil has not been universal. In 2007, the governor of Texas signed an executive order (later overturned by the state legislature) requiring all schoolgirls to receive the HPV vaccine. Controversy ensued, as some parents were upset by the idea of vaccinating their young daughters against an STD. Some opponents argued that requiring Gardasil would encourage promiscuity, perhaps lulling young women into a false sense of security, so they might believe themselves completely immune to genital warts and cervical cancer.

Should parents be required to have their daughters vaccinated with Gardasil? Or should such vaccinations be up to the discretion of the parents? Does the HPV vaccine encourage reckless sexual behavior, or is it a practical means of preventing a potentially life-threatening disease?

30

Ecology and Populations

OUTLINE

BEFORE YOU BEGIN

Before beginning this chapter, take a few moments to review the following discussions.

Section 1.3 What is the relationship between ecosystems and the biosphere?

Section 16.2 What have been the causes of mass extinctions in the past?

Controlling Population Growth

The world's human population is growing at an alarming rate. Some estimates predict it will exceed 9 billion people within 50 years. If this rate of growth continues, the planet will eventually be unable to support the human population. Various countries have found ways to encourage their citizens to limit population growth. One of the most scrutinized strategies is China's one-child policy. This policy allows each Chinese couple to have only one child. Punishments for those who do not follow the policy include imposing fines of as much as eight times the annual household income, destroying a family's home, and even forcing a pregnant woman to terminate her pregnancy. While some officials believe that such a drastic policy is necessary to decrease the growth of the Chinese population, others feel the policy is too harsh and no more effective than the much less restrictive policies, such as family planning and birth control education, that are being practiced in other overpopulated countries. The situation in China poses a major ethical dilemma. Is it more important to control population growth or to respect individuals' reproductive rights?

In this chapter, you will learn about ecology and population dynamics in all sorts of populations, including humans. One important lesson to be learned is that there are no easy solutions when it comes to ecological problems.

30.1 The Scope of Ecology

Learning Outcomes

Upon completion of this section, you should be able to

1. List and describe the hierarchical levels of biological organization.
2. Explain the relationship between ecology and environmental science.

Ecology is the scientific study of the interactions of organisms with one another and with their physical environment. Ecology is one of the two biological sciences of most interest to the public today, the other being genetics. Ecological studies offer information key to the survival of all species, present and future. Understanding ecology will help us make informed decisions, ranging from what kind of car to drive to how to support the preservation of a forested area in our towns. Current ecological decisions will affect not only our lives but also the lives of generations to come.

Ecology involves the study of several levels of biological organization (**Fig. 30.1**). Some ecologists study **organisms,** focusing on their adaptations to a particular environment. For example, organismal ecologists might investigate what attributes of a clownfish in a coral reef allow it to live in warm tropical waters.

Organisms are usually part of a **population,** a group of individuals of the same species in a given location at the same time. At the population level of study, ecologists describe changes in population size over time. For example, a population ecologist might compare the number of clownfishes living in a given location today with data she obtained from the same location 20 years ago.

A **community** consists of all the various populations at a particular locale. A coral reef contains numerous populations of fishes, crustaceans, corals, algae, and so forth. At the community level, ecologists study how the interactions between populations affect the populations' well-being. For example, a community ecologist might study how a decrease in algal populations affects the population sizes of crustaceans and fishes living on the coral reef.

Video
Coral Reef
Ecology

An **ecosystem** consists of a community of living organisms as well as their physical environment. As an example of how the physical environment affects a community, consider that the presence of suspended particles in the water decreases the amount of sunlight reaching algae living on the coral reef. Without solar energy, algae cannot produce the organic nutrients they and the other populations require.

Finally, the **biosphere** is the portion of the Earth's surface—air, water, and land—where living things exist. Having information about the many levels of organization within a coral reef allows ecologists to understand how a coral reef contributes to the biodiversity and dynamics of the biosphere.

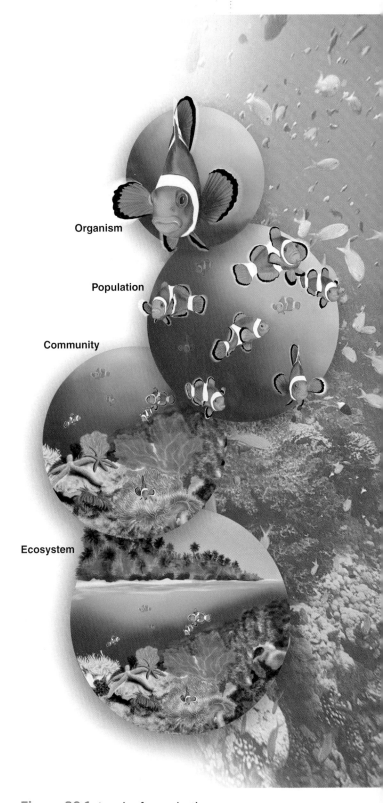

Organism

Population

Community

Ecosystem

Figure 30.1 Levels of organization.

The study of ecology encompasses the organism, population, community, and ecosystem levels of biological organization.

a. After a burn

b. Before

Figure 30.2 **Studying the effects of fire.**

Fire is a natural occurrence in lodgepole pine forests, such as these in Yellowstone National Park. Over time, the recently burned forest (**a**) will mature and look like the unburned forest (**b**).

Ecology: A Biological Science

Ecology began as a part of **natural history,** the discipline of observing and describing organisms in their environment, but today ecology is also an experimental science. A central goal of ecology is to develop models that explain and predict the distribution and abundance of populations based on their interactions within an ecosystem. Achieving such a goal involves testing hypotheses. For example, ecologists might formulate and test hypotheses about the role fire plays in maintaining a lodgepole pine forest (**Fig. 30.2**). To test these hypotheses, ecologists can compare the characteristics of a community before and after a prescribed burn of the area by professionals under controlled conditions. Ultimately, ecologists wish to develop models about the distribution and abundance of ecosystems within the biosphere.

Ecology and Environmental Science

The field of environmental science applies ecological principles to practical human concerns. Comprehension of ecological principles helps us understand why a functioning biosphere is critical to our survival and why our population of 6 billion people and our overconsumption of resources pose threats to the biosphere.

Conservation biology is a discipline of ecology that studies all aspects of biodiversity, with the goal of conserving natural resources, including wildlife, for the benefit of this generation and future generations. Conservation biology recognizes that wildlife species are an integral part of a well-functioning biosphere on which human life depends.

Video
Tallgrass Prairie Ecology

Connecting the Concepts

For more information on the study of ecosystems, refer to the following discussions.

Section 31.2 describes the roles of energy and living organisms in an ecosystem.

Section 31.3 examines some of the major ecosystems on the planet.

Section 32.2 explores how human activity is negatively influencing the health of many ecosystems worldwide.

Check Your Progress 30.1

1. List the levels of biological organization above organ systems.
2. Distinguish an ecosystem from a community.
3. Explain why an environmental scientist is also considered an ecologist.

30.2 The Human Population

Learning Outcomes

Upon completion of this section, you should be able to

1. Explain the relationship among the birthrate, death rate, and annual growth rate of a population.
2. Distinguish between more-developed countries (MDCs) and less-developed countries (LDCs) with regard to population growth.
3. Compare the environmental impact of MDCs and LDCs.

The human population practically covers the face of the Earth. However, on both a large scale and a small scale, the human population has a clumped distribution; 56% of the world's people live in Asia, and most Asian populations live in China and India. Mongolia has a population density of only 0.25 person per square kilometer, while Bangladesh has a density of over 1,000 persons per square kilometer. On every continent, human population densities are highest along the coasts.

Present Population Growth

For most of the twentieth century, the growth of the global human population was relatively slow, but as more individuals achieved reproductive age, the growth rate began to increase. The Industrial Revolution (1800s) brought an increase in the production of food and jobs. At this time, the growth curve for the human population began to increase steeply, so that currently the population is undergoing rapid growth (**Fig. 30.3**).

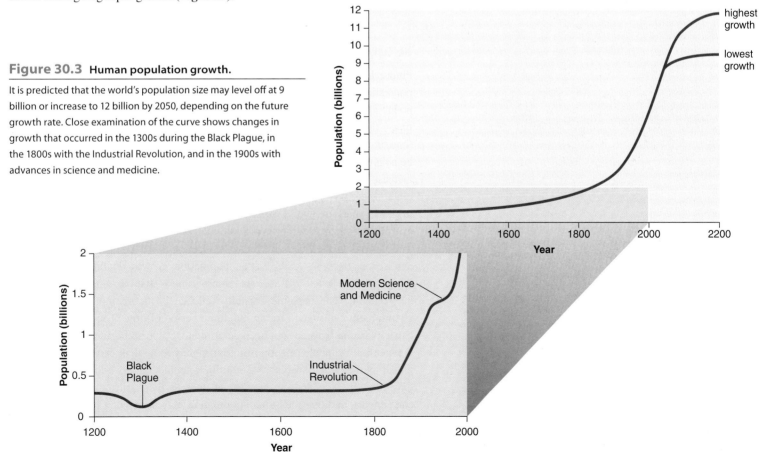

Figure 30.3   Human population growth.

It is predicted that the world's population size may level off at 9 billion or increase to 12 billion by 2050, depending on the future growth rate. Close examination of the curve shows changes in growth that occurred in the 1300s during the Black Plague, in the 1800s with the Industrial Revolution, and in the 1900s with advances in science and medicine.

Just as with populations in nature, the growth rate of the human population is determined by the difference between the number of people born and the number of people who die each year. For example, the current global birthrate is an estimated 19.9 per 1,000 per year, while the global death rate is approximately 8.2 per 1,000 per year. From these numbers, we can calculate that the current annual growth rate of the human population is

$$(19.9 - 8.2)/1,000 = (11.7)/1,000 = 1.17\%$$

Connections and Misconceptions

How do birthrates and death rates vary among countries?

In 2009, the country with the highest birthrate was Niger (51.6%), followed by Mali (49.1%) and Uganda (47.8%), while Japan had the lowest birthrate (7.64%). For death rates, Swaziland had the highest rate (30.8%) and the United Arab Emirates was the lowest (2.1%). For overall annual growth rates, the fastest growing country in 2009 was the United Arab Emirates (3.7%). Several countries, including Montenegro (−0.8%) and Bulgaria (−0.8%) experienced population declines. In comparison, for the United States the 2009 birthrate was 13.82%, the death rate was 8.38%, and the annual growth rate was 0.98%. (Source: www.globalhealthfacts.com.)

Future Population Growth

Future growth of the human population is of great concern. It's possible that population increases will exceed the rate at which resources can be supplied. The potential for dire consequences can be appreciated by considering the *doubling time*—the length of time it takes for a population to double in size. The doubling time of the human population is now estimated to be less than 35 years. Will we be able to meet the extreme demands for resources by such a large population increase within such a short period of time? Already, there are areas across the globe where people have inadequate access to fresh water, food, and shelter, yet in just 35 years, the world would need to double the amount of food, jobs, water, energy, and other resources in order to maintain the present standard of living.

Rapid growth usually begins to decline when resources, such as food and space, become scarce. Then population growth declines to zero, and the population levels off at a size appropriate to the carrying capacity of its environment. The carrying capacity reflects the number of individuals the environment can sustain for an indefinite period of time. The Earth's carrying capacity for humans has not been determined. Some authorities think the Earth is potentially capable of supporting far more people than currently inhabit the planet, perhaps as many as 50–100 billion people. Others believe that we may have already exceeded the number of humans the Earth can support, and that the human population may undergo a catastrophic crash to a much smaller size.

Animation
World Hunger

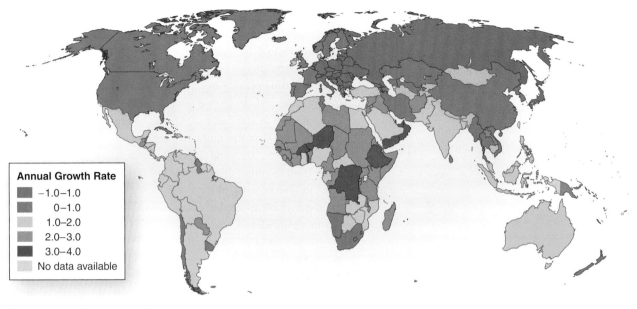

Annual Growth Rate

- −1.0–1.0
- 0–1.0
- 1.0–2.0
- 2.0–3.0
- 3.0–4.0
- No data available

Figure 30.4 **Global population growth rates.**

The highest population growth rates are found in Africa and the Middle East, while the growth rates throughout Europe are much lower.

Source: GlobalHealthFacts.org (2010). Note: Based on 2009 data.

More-Developed Versus Less-Developed Countries

Complicating the issue of future human population growth is the fact that not all countries have the same growth rate (**Fig. 30.4**). The countries of the world today can be divided into two groups (**Fig. 30.5**). In the **more-developed countries (MDCs),** such as those in North America and Europe, population growth is modest and the people enjoy a fairly good standard of living. In the **less-developed countries (LDCs),** such as those in Latin America, Asia, and Africa, population growth is dramatic and the majority of the people live in poverty.

The MDCs

The MDCs did not always have a low growth rate. Between 1850 and 1950, their populations doubled, largely because of the decline in the death rate due to the development of modern medicine and improved socioeconomic conditions. The decline in the death rate was followed shortly thereafter by a decline in birthrate, so that populations in the MDCs have experienced only modest growth since 1950. This sequence of events (i.e., decreased death rate followed by decreased birthrate) is termed a *demographic transition.*

The growth rate for the MDCs as a whole is now about 0.2%, but populations in several countries are not growing at all or are actually decreasing in size. The MDCs are expected to increase by 52 million between 2002 and 2050, but this modest amount will still keep their total population at just about 1.2 billion. In contrast to the other MDCs, population growth in the United States is not leveling off and continues to increase. The United States currently has a growth rate of 0.98%. This higher rate is due to the fact that many people immigrate to the United States each year, and a large number of the women are still of reproductive age. Therefore, it is unlikely that the United States will experience a decline in its growth rate in the near future.

a.

b.

Figure 30.5 **More-developed versus less-developed countries.**

People in the (**a**) more-developed countries have a high standard of living and will contribute the least to world population growth, while people in the (**b**) less-developed countries have a low standard of living and will contribute the most to world population growth.

The LDCs

Death rates began to decline sharply in the LDCs following World War II, with the introduction of modern medicine, but the birthrate remained high. The *collective* growth rate of the LDCs peaked at 2.5% between 1960 and 1965, and since that time the rate has declined to 1.5%. Unfortunately, the growth rates in 46 of the less-developed countries have not declined. Thirty-five of these countries are in sub-Saharan Africa, where women on average are presently bearing more than five children each.

Between 2002 and 2050, the population of the LDCs is expected to jump from 5 billion to at least 8 billion. Africa will not share appreciably in this increase because of the many deaths there due to AIDS. The majority of the increase is expected to occur in Asia. Asia, with 31% of the world's arable (farmable) land, is already home to 59% of the human population. Twelve of the world's most polluted cities are located in Asia. If the human population increases as expected, Asia will experience even more urban pollution, acute water scarcity, and a significant loss of wildlife over the next 50 years.

Comparing Age Structures

An age structure diagram is divided into three groups: prereproductive, reproductive, and postreproductive (**Fig. 30.6**). Many MDCs have a stable age structure, meaning that the number of individuals in each group is just about the same. Therefore, MDC populations are expected to remain about the same size if couples have two children, and to decline if each couple has fewer than two children. This is often called **replacement reproduction,** since if each couple produces only two children, then the population growth should be zero. In reality, due to mortality, the replacement reproduction rate is often slightly greater than two children per couple. In contrast, the age structure diagram of most LDCs has a pyramid shape, with the prereproductive group being the largest. Therefore, LDC populations will continue to expand, even after replacement reproduction is attained. It might seem at first that replacement reproduction would cause an LDC population to undergo zero population growth and therefore no increase in population size. However, if there are more young women entering the reproductive years than there are older women leaving them, replacement reproduction still results in population growth. We will take a closer look at age structure diagrams in section 30.3.

Figure 30.6 Age structure diagrams.

The diagrams predict that (**a**) the LDCs will expand rapidly due to their age distributions, whereas (**b**) the MDCs are approaching stabilization.

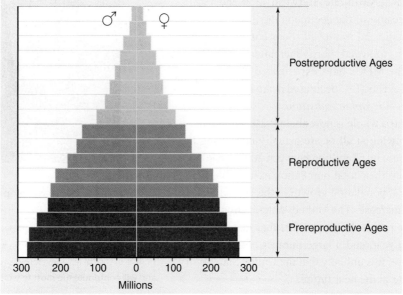

a. Less-developed countries (LDCs)

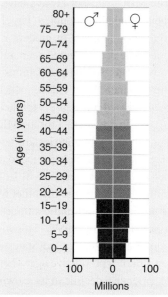

b. More-developed countries (MDCs)

Population Growth and Environmental Impact

Population growth is putting extreme pressure on each country's social organization, the Earth's resources, and the biosphere. Since the population of the LDCs is still growing at a significant rate, it might seem that their population increase will be the main cause of future environmental degradation. But this is not necessarily the case, because the MDCs consume a much larger proportion of the Earth's resources than do the LDCs. This consumption leads to environmental degradation, which is also of the utmost concern.

Environmental Impact

The environmental impact (E.I.) of a population is measured not only in terms of population size but also in terms of resource consumption per capita and the pollution resulting from this consumption. In other words,

$$\text{E.I.} = \text{population size} \times \text{resource consumption per capita}$$
$$= \text{pollution per unit of resource used}$$

Therefore, there are two possible causes of overpopulation: population size and resource consumption. Overpopulation is more obvious in LDCs; resource consumption is more obvious in MDCs, because per capita consumption is so much higher in the MDCs. For example, an average American family, in terms of per capita resource consumption and water requirements, is the equivalent of 30 people in India. We need to realize that only a limited number of people can be sustained anywhere near the standard of living enjoyed in the MDCs.

The comparative environmental impacts of MDCs and LDCs are shown in **Figure 30.7.** The MDCs account for only about one-fourth of the world population. However, compared with the LDCs, MDCs account for 90% of the hazardous waste production, due to their high rate of consumption of such resources as fossil fuel, metal, and paper.

Connecting the Concepts

For more information on the topics presented in this section, refer to the following discussions.

Section 32.1 examines how the increase in human populations generates pollution and strains natural resources.

Section 32.2 explores the relationship between human population growth and a global loss of biodiversity.

Check Your Progress 30.2

1 Calculate the annual growth rate of a population that is experiencing a birthrate of 20.5% and a death rate of 9.8%.

2 Compare the characteristics of an MDC with those of an LDC, and give an example of each.

3 Evaluate an age-structure diagram to determine if the population will experience a population growth or decline in the future.

4 Contrast the environmental impact of an MDC versus an LDC.

LDCs 78%   MDCs 22%

a. Population

LDCs 10%   MDCs 90%

b. Hazardous wastes

LDCs 40%   MDCs 60%

c. Consumption: fossil fuels

LDCs 20%   MDCs 80%

d. Consumption: metals

LDCs 25%   MDCs 75%

e. Consumption: paper

Figure 30.7 Environmental impacts.

a. The combined population of the MDCs is much smaller than that of the LDCs. **b.** MDCs produce most of the world's hazardous wastes. **c.–e.** MDCs consume much more fossil fuel, metals, and paper than do LDCs.

30.3 Characteristics of Populations

Learning Outcomes

Upon completion of this section, you should be able to

1. Distinguish between the types of population distribution.
2. Describe how age structure diagrams are constructed and explain how they can be used to predict future growth.
3. Describe the three types of survivorship curves and give examples of each.
4. Explain biotic potential and list the criteria that influence it.
5. Compare and contrast exponential and logistic growth patterns.
6. Compare and contrast density-dependent and density-independent factors.

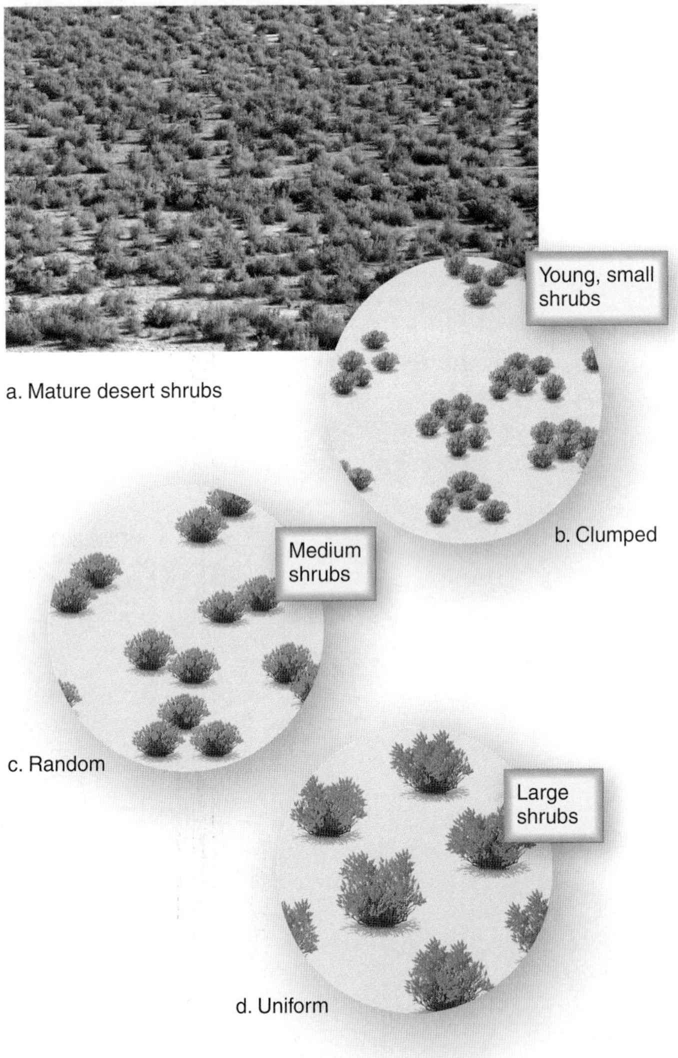

Various characteristics of all populations change over time. Populations are periodically subject to environmental instability. During such times of environmental instability, individuals are under the pressures of natural selection. Individuals better adapted to the environment will leave behind more offspring than those that are not as well adapted. Once the population becomes adapted to the new environment, an increase in size will occur again. So when an ecologist lists the characteristics of a population, this is a snapshot of the characteristics at a particular time and place.

Distribution and Densitys

Resources are the components of an environment that support its organisms. Light, water, space, mates, and food are some important resources for wildlife populations. The availability of resources influences the spatial distribution of individuals in a population. Three terms—**clumped, random,** and **uniform** distributions—are used to describe the observed patterns of distribution. In a study of desert shrubs, it was found that the distribution changes from clumped to random to uniform as the plants mature (**Fig. 30.8**). After sufficient study, competition for belowground resources was found to cause the distribution pattern to become uniform.

Suppose we were to step back and consider the distribution of individuals, not within a single desert or forest or pond but within a species' range. A **range** is that portion of the globe where the species can be found. On this scale, all the members of a species within the range make up the population. Members of a population are clumped within the range because they are located in areas suitable for their growth.

Population density is the number of individuals per unit area or volume. Population density tends to be higher in areas with plentiful resources than in areas with scarce resources. Other factors can be involved, however. In general, population density declines with increasing body size. Consider that tree seedlings live at a much higher population density than do mature trees. Therefore, more than one factor must be taken into account when explaining population densities.

Figure 30.8 Patterns of distribution within a population.

a. A population of mature desert shrubs. **b.** Young, small desert shrubs are clumped. **c.** Medium shrubs are randomly distributed. **d.** Large shrubs are uniformly distributed.

Population Growth

As mentioned on page 588, the growth rate is dependent on the number of individuals born and the number of individuals who die each year. Populations grow when the number of births exceeds the number of deaths. To take another example, if the number of births is 30 per year, and the number of deaths is 10 per year per 1,000 individuals, the growth rate is 2%.

In cases where the number of deaths exceeds the number of births, the value of the growth rate is negative—the population is shrinking.

Demographics and Population Growth

Availability of resources and certain characteristics of a population—called **demographics**—influence its growth rate.

One demographic characteristic of interest for all populations is the **age structure** of a population. It is customary to use bar diagrams to depict the age structure of a population. The population is divided into members that are *prereproductive, reproductive,* or *postreproductive,* based on their age. If the prereproductive and reproductive individuals outnumber those that are postreproductive, the birthrate will exceed the death rate (**Fig. 30.9***a*). In contrast, if the postreproductive individuals exceed those that are prereproductive and/or reproductive, the number of individuals in the population will decrease over time because the death rate will exceed the birthrate (Fig. 30.9*b*). If the numbers of prereproductive, reproductive, and postreproductive individuals are approximately equal, the birthrate and death rate is equal, and the number of individuals in the population remains relatively stable over time (Fig. 30.9*c*).

Ecologists also study patterns of **survivorship**—how age at death influences population size. For example, if members of a population die young, the number of reproductive individuals is lower than otherwise. First, an ecologist constructs a life table, which lists the number of individuals that die or survive at each age. The best way to arrive at a life table is to identify a large number of individuals that are born at about the same time and keep records on them from birth until death.

Connections and Misconceptions

Where are the highest and lowest population densities of humans?

Based on 2009 demographic data, the country with the highest population density is the Principality of Monaco (in Europe), with a density of 35,282 people per square  kilometer. Greenland technically has the lowest population density (0), followed by Mongolia and the Western Sahara, both with an average population density of 2 people per square kilometer.

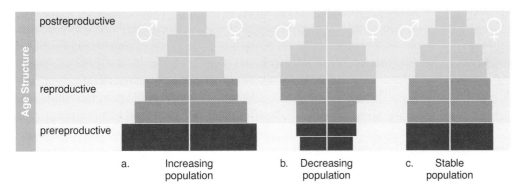

Figure 30.9 Age structure diagrams.

Typical age structure diagrams for hypothetical populations that are (**a**) increasing, (**b**) decreasing, and (**c**) stable. Different numbers of individuals in each age class create these distinctive shapes. In each diagram, the left half represents males, while the right half represents females.

Age (years)	Number of survivors at beginning of year	Number of deaths during year
0–1	1,000	199
1–2	801	12
2–3	789	13
3–4	776	12
4–5	764	30
5–6	734	46
6–7	688	48
7–8	640	69
8–9	571	132
9–10	439	187
10–11	252	156
11–12	96	90
12–13	6	3
13–14	3	3
14–15	0	

1,000 – 199

801 – 12

etc.

a.

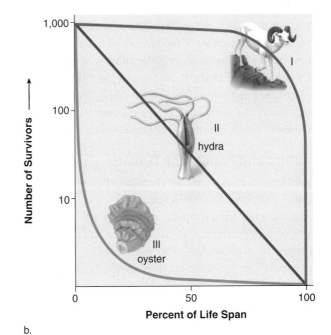

b.

Figure 30.10   Life table and survivorship curves.

a. A life table for the Dall sheep. **b.** Three typical survivorship curves. Among Dall sheep, with a type I curve, most individuals survive until old age, when they gradually die off. Among hydras, with a type II curve, there is an equal chance of death at all ages. Among oysters, with a type III curve, most die when they are young, and few become adults that are able to survive until old age.

A life table for Dall sheep is given in **Figure 30.10a**. This table was constructed by gathering a large number of skulls and counting the growth rings on the horns to estimate each sheep's age at death, given that one growth ring is produced annually. Plotting the number of survivors per 1,000 births against age produces a *survivorship curve* (Fig. 30.10b). While each species has its own pattern, three types of survivorship curves are common. In a type I curve, typical of Dall sheep and humans, survival is high until old age, when deaths increase due to illness. In a type II curve, the possibility of death is equally likely over all age groups. In a type III curve, death is likely among the young, and few individuals reach old age.

Biotic Potential   How quickly a population increases over time depends on its **biotic potential,** the highest possible rate of increase for a population when resources are unlimited (**Fig. 30.11**). Whether the biotic potential is high or low depends on the demographic characteristics of the population, such as the following:

1. Abundance of resources
2. Usual number of offspring per reproduction
3. Chances of survival until age of reproduction
4. How often each individual reproduces
5. Age at which reproduction begins

Patterns of Population Growth

The patterns of population growth are dependent on (1) the biotic potential of the species combined with other demographics and (2) the availability of resources. The two fundamental patterns of population growth are exponential growth and logistic growth.

a.

Figure 30.11   Biotic potential.

A population's maximum growth rate under ideal conditions—that is, its biotic potential—is greatly influenced by the number of offspring produced in each reproductive event. **a.** Mice, which produce many offspring that quickly mature to produce more offspring, have a much higher biotic potential than (**b**) the rhinoceros, which produces only one or two offspring per infrequent reproductive events.

b.

Exponential Growth

Suppose ecologists are studying the growth of a population of insects that are capable of infesting and taking over an area. Under these circumstances, **exponential growth** is expected. An exponential pattern of population growth results in a J-shaped curve (**Fig. 30.12**). This growth pattern can be likened to compound interest at a bank: The more your money increases, the more interest you will get. If the insect population has 2,000 individuals and the per capita rate of increase is 20% per month, there will be 2,400 insects after one month, 2,880 after two months, 3,456 after three months, and so forth.

Notice that a J-shaped curve has two phases:

Lag phase. Growth is slow because the number of individuals in the population is small.

Exponential growth phase. Growth is accelerating.

Usually, exponential growth can continue only as long as resources in the environment are unlimited. When the number of individuals in the population approaches the number that can be supported by available resources, competition for resources among individuals increases.

Video Exponential Growth

Logistic Growth

As resources decrease, population growth levels off, and a pattern of population growth called **logistic growth** is expected. Logistic growth results in an S-shaped growth curve (**Fig. 30.13**).

Notice that an S-shaped curve has four phases:

Lag phase. Growth is slow because the number of individuals in the population is small.

Exponential growth phase. Growth is accelerating due to biotic potential (see Fig. 30.11).

Deceleration phase. The rate of population growth slows because of increased competition among individuals for available resources.

Stable equilibrium phase. Little, if any, growth takes place because births and deaths are about equal.

The stable equilibrium phase is said to occur at the carrying capacity of the environment. The **carrying capacity** is the total number of individuals that the resources can support. This number is not a constant and varies with the circumstances. For example, a large island can support a larger population of penguins than a smaller island, because the smaller island has a limited amount of space for nesting sites.

Applications   Our knowledge of logistic growth has practical applications. The model predicts that exponential growth will occur only when population size is much lower than the carrying capacity. So if humans are using a fish population as a continuous food source, it is best to maintain that population size in the exponential phase of growth when biotic potential is having its full effect and the birthrate is the highest. If people overfish, the fish population will sink into the lag phase, and it may be years before exponential growth recurs (**Fig. 30.14**). On the other hand, if people are trying to limit the growth of a pest, it is best to reduce the carrying capacity, rather than the population size. Reducing the population size only encourages exponential growth to begin once again. Farmers can reduce the carrying capacity for a pest by alternating rows of different crops instead of growing one type of crop throughout the entire field.

Video Big Fish, Small Fish

Video Ocean Fishing Ban

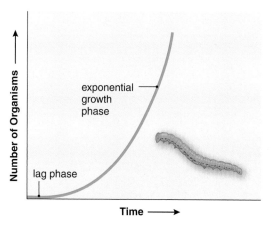

Figure 30.12 Exponential growth.

Exponential growth results in a J-shaped growth curve because the growth rate is positive.

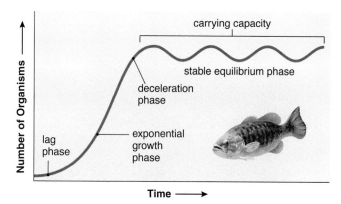

Figure 30.13 Logistic growth. Logistic growth produces an S-shaped growth curve because competition for resources increases as the population approaches the carrying capacity.

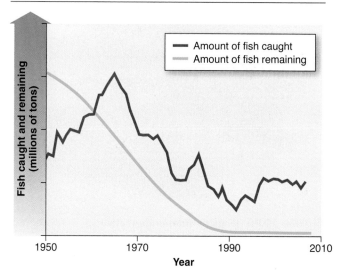

Figure 30.14 Sustainable population density.

As the amount of fish caught increased over the 1950s and 1960s, the remaining amount decreased. Because fishing still continues, fish population growth is maintained in its lag phase. Fishing must be reduced to allow the fish population to enter an exponential growth phase. Now a balance between fish population growth and fish caught could be maintained regularly at a higher amount.

Factors That Regulate Population Growth

Ecologists have long recognized that the environment contains both biotic (living) and abiotic (nonliving) components and that these two components play an important role in regulating population size in nature.

Density-Independent Factors

Abiotic factors, such as weather and natural disasters, are typically **density-independent factors.** Abiotic factors can cause sudden and catastrophic reductions in population size. However, density-independent factors cannot in and of themselves regulate population size because the effect is not influenced by the number of individuals in the population. In other words, the intensity of the effect does not increase with increased population size. For example, the proportion of a population killed in a flash flood is independent of density—floods don't necessarily kill a larger percentage of a dense population than of a less dense population. If a mouse population has only 5 members and 3 drown in a flash flood, the death rate is 60% (**Fig. 30.15a**). If a mouse population has 20 mice and 12 drown, the death rate is still 60% (Fig. 30.15b).

Density-Dependent Factors

Biotic factors are considered **density-dependent factors** because the percentage of the population affected does increase as the density of the population increases. Competition, predation, and parasitism are all biotic factors that increase in intensity as the density increases. We will discuss these interactions between populations again in Chapter 31 because they influence community composition and diversity.

Competition   **Competition** can occur when members of the same species attempt to use resources (such as light, food, or space) that are in limited supply but are necessary for survival. As a result, not all members of the population can have access to the resource to the degree necessary to ensure survival, reproduction, or some other aspect of their life cycle.

Video
Booby Chick Competition

Let's consider a woodpecker population in which members have to compete for nesting sites. Each pair of birds requires a tree hole in order to raise offspring. If there are more holes than breeding pairs, each pair can have a hole in which to lay eggs and rear young birds (**Fig. 30.16a**). But if there are fewer holes than breeding pairs, each pair must compete to acquire a nesting site (Fig. 30.16b). Pairs that fail to gain access to holes will be unable to contribute new members to the population.

A well-known example of competition for food concerns the reindeer on St. Paul Island. Overpopulation led to an insufficiency of food to sustain all members, causing the population to crash—that is, it became drastically reduced.

Video
Hungry Reindeer

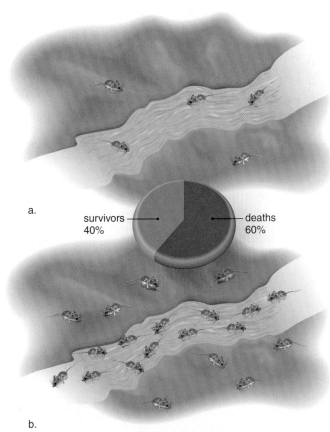

a.

survivors 40%   deaths 60%

b.

Figure 30.15   **Density-independent effects.**

The impact of a density-independent factor, such as weather or a natural disaster, is not influenced by population density. Two populations of field mice are in the path of a flash flood. **a.** In the low-density population, 3 out of the 5 mice drown, a 60% death rate. **b.** In the high-density population, 12 out of 20 mice drown—also a 60% death rate.

Figure 30.16 **Density-dependent effects—competition.**

The impact of competition is directly proportional to the density of a population. **a.** When density is low, every member of the population has access to the resource. **b.** When density is high, members of the population must compete to gain access to available resources, and some fail to do so.

Resource partitioning among different age groups is a way to reduce competition for food. During the life cycle of butterflies, the caterpillar stage requires different food than the adult stage. Caterpillars graze on leaves, while adult butterflies feed on nectar produced by flowers. Therefore, the parents are less apt to compete with their offspring for food.

Predation   **Predation** occurs when one organism, the predator, eats another, the prey. In the broadest sense, predation can include not only animals such as lions that kill zebras but also filter-feeding blue whales that strain krill from the ocean waters, and even herbivorous deer that browse on trees and bushes.

The effect of predation generally increases as the prey population increases in density, because prey are easier to find when hiding places are limited. Consider a field inhabited by a population of mice (**Fig. 30.17**). Each mouse must have a hole in which to hide to avoid being eaten by a hawk. If there are 102 mice, but only 100 holes, 2 mice will be left out in the open. It might be hard for the hawk to find only 2 mice in the field. If neither mouse is caught, the predation rate is $0/2 = 0\%$. However, if there are 200 mice and only 100 holes, there is a greater chance that the hawk will be able to find some of the 100 mice without holes. If half of the exposed mice are caught, the predation rate is $50/100 = 50\%$. Therefore, increasing the density of the available prey has increased the proportion of the population preyed upon.

Predator-Prey Population Cycles   Rather than remaining steady, some predator and prey populations experience an increase, and then a decrease, in population size. A famous example of such a cycle occurs between the snowshoe hare and the Canada lynx, a type of small cat (**Fig. 30.18**). The snowshoe hare is a common herbivore in the coniferous forests of North America, where it feeds on terminal twigs of various shrubs and small trees. Investigators first assumed that the predatory lynx brings about a decrease in the hare population, and this decrease in prey brings about a subsequent decrease in the lynx population. Once the hare population recovers, so does the lynx population,

a.

b.

Figure 30.17 **Density-dependent effects—predation.**

The impact of predation on a population is directly proportional to the density of the population. **a.** In a low-density population, the chances of a predator finding the prey are low, resulting in little predation (mortality rate of 0/2, or 0%). **b.** In a higher-density population, there is a greater likelihood of the predator locating potential prey, resulting in a greater predation rate (mortality rate of 50/100, or 50%).

and the result is a boom–bust cycle. But some biologists noted that the decline in snowshoe hare abundance is accompanied by low growth and reproductive rates, which could be signs of food shortage. A field experiment showed that, if the lynx predator was denied access to the hare population, the cycling of the hare population still occurred based on food availability. The results suggested that a hare–food cycle and a predator–hare cycle have combined to produce the pattern observed in Figure 30.18.

Connecting the Concepts

For more information on populations, refer to the following discussions.

Section 30.4 examines how a species responds to the carrying capacity of an environment.

Section 31.1 explores how competition shapes community organization.

Check Your Progress 30.3

1. Explain how an age structure diagram may be used to predict the future growth of a population.

2. Describe how reproductive strategies differ between species that exhibit a type I and type III survivorship curve.

3. Compare and contrast an exponential growth curve and a logistic growth curve.

4. List the four demographic characteristics of a population that determine its biotic potential.

5. List examples of density-independent and density-dependent factors that regulate population growth.

6. Explain why competition and predation are considered density-dependent factors regulating population growth.

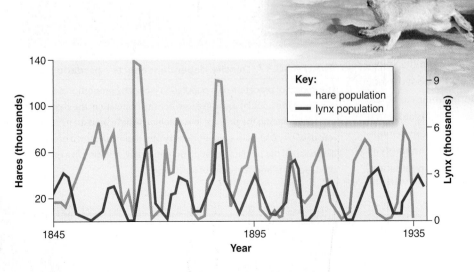

Figure 30.18 Predator-prey cycling of a lynx and a snowshoe hare.

The number of pelts received yearly by the Hudson Bay Company for almost 100 years shows a pattern of 10-year cycles in population densities. The snowshoe hare (prey) population reaches a peak before that of the lynx (predator) by a year or more.

Data from D. A. MacLuiich, *Fluctuations in the Numbers of the Varying Hare* (Lepus americanus), University of Toronto Press, Toronto, 1937, reprinted 1974, p. 543.

30.4 Life History Patterns and Extinction

Learning Outcomes

Upon completion of this section, you should be able to

1. Distinguish between opportunistic and equilibrium populations.
2. Describe factors that can lead to the extinction of a species.

A **life history** consists of a particular mix of the characteristics we have been discussing. After studying many types of populations, from mayflies to humans, ecologists have discovered two fundamental and contrasting patterns: the opportunistic life history pattern and the equilibrium life history pattern.

Opportunistic species tend to exhibit exponential growth. The members of the population are small in size, mature early, have a short life span, and provide limited parental care for a great number of offspring (**Fig. 30.19***a*). Density-independent factors tend to regulate the population size, which is large enough to survive an event that threatens to annihilate it. A population has a high dispersal capacity. Various types of insects and weeds provide the best examples of opportunistic species.

Equilibrium species exhibit logistic growth, and the population size remains close to or at the carrying capacity (Fig. 30.19*b*). Resources are relatively scarce, and the individuals best able to compete—those with phenotypes best suited to the environment—have the largest number of offspring. They allocate energy to their own growth and survival and to the growth and survival of their few offspring. Therefore, they are fairly large, are slow to mature, and have a fairly long life span. The growth of equilibrium species tends to be regulated by density-dependent factors. Various birds and mammals are the best examples of equilibrium species.

Animation
Life History Strategies

Extinction

Extinction is the total disappearance of a species or higher group. Which species shown in Figure 30.19, the dandelion or the mountain gorilla, is apt to become extinct? Because the dandelion matures quickly, produces many offspring at one time, and has seeds dispersed widely by wind, it can withstand a local decimation more easily than the mountain gorilla.

A study of equilibrium species shows that three other factors—size of geographic range, degree of habitat tolerance, and size of local populations—can help determine whether an equilibrium species is in danger of extinction. **Figure 30.20** compares several equilibrium species on the basis of these three factors. The mountain gorilla has a restricted geographic range, narrow habitat tolerance (few preferred places to live), and small local population. This combination of characteristics makes the mountain gorilla very vulnerable to extinction. The possibility of extinction increases depending on whether a species is similar to the gorilla in one, two, or three ways. Such population studies can assist conservationists and others trying to preserve the biosphere's biodiversity.

Figure 30.19 Life history patterns.

a. Dandelions are an opportunistic species, whereas (**b**) mountain gorillas are an equilibrium species.

Opportunistic Pattern

Small individuals
Short life span
Fast to mature
Many offspring
Little or no care of offspring

a.

Equilibrium Pattern

Large individuals
Long life span
Slow to mature
Few offspring
Much care of offspring

b.

Most common

	Species	
Restricted geographic range Broad habitat tolerance Large local population	Galápagos medium ground finch	Monterey pine
Extensive geographic range Narrow habitat tolerance Large local population	California grey whale	Fremont cottonwood
Extensive geographic range Broad habitat tolerance Small local population	Tiger	Grassfern
Restricted geographic range Narrow habitat tolerance Large local population	Fish crow	Haleakala silversword
Restricted geographic range Broad habitat tolerance Small local population	Tasmanian devil	*Welwitschia*
Extensive geographic range Narrow habitat tolerance Small local population	Northern spotted owl	Pacific yew
Restricted geographic range Narrow habitat tolerance Small local population	Mountain gorilla	*Pritchardia kaalae*

Rarest

Figure 30.20 Vulnerability to extinction.

Vulnerability is particularly tied to range, habitat, and size of population. The amount of white-highlighted text in the boxes indicates the chances and causes of extinction.

Connections and Misconceptions

How are humans influencing the rates of extinction?

Many conservation biologists now believe that we are in the midst of one of the largest periods of mass extinctions since the end of the Cretaceous 65 MYA (see Table 16.1). The major causes of this "Sixth Mass Extinction" are human activity (population growth and resource use) and global climate change. While extinction is a normal aspect of life, the current extinction rates are between 100 and 1,000 times the background levels of one to five species per year. Some estimates suggest that as much as 38% of all species on the planet, including most primates, birds, and amphibians, may be in danger of extinction before the end of this century.

Connecting the Concepts

For more information on the material in this section, refer to the following discussions.

Table 16.1 outlines the major extinctions in the history of life.

Section 32.2 examines the direct and indirect benefits of maintaining high levels of biodiversity.

Check Your Progress 30.4

1. Contrast opportunistic populations with equilibrium populations.
2. Define *extinction*.
3. List five factors that can determine whether an equilibrium population is in danger of extinction.

Media Study Tools

www.mhhe.com/maderessentials3

Enhance your study of this chapter with study tools and practice tests. Also ask your instructor about the resources available through ConnectPlus, including the media-rich eBook, interactive learning tools, and animations.

Population Biology

The virtual lab "Population Biology" provides an interactive look at the factors that influence population size.

The Chapter in Review

Summary

30.1 The Scope of Ecology

Ecology is an experimental science that studies the interactions of organisms with each other and with the physical environment. The levels of biological organization studied by ecologists are

- Organisms
- Population
- Community
- Ecosystem
- Biosphere

Environmental scientists and conservation biologists both utilize the principles of ecology in their work.

30.2 The Human Population

The human population exhibits clumped distribution (on both a large scale and a small scale) and is undergoing rapid growth.

Future Population Growth

At the present growth rate, the doubling time for the human population is estimated to be 35 years. This rate of growth will put extreme demands on resources, and growth will decline due to resource scarcity. Eventually, the population will most likely level off at its carrying capacity.

More-Developed Countries (MDCs) The growth rate for MDCs is about 0.1%. The total MDC population between 2002 and 2050 will remain at around 1.2 billion.

Less-Developed Countries (LDCs) The growth rate for LDCs is about 1.6%. The LDC population between 2002 and 2050 is expected to increase from 5 billion to 8 billion.

Comparing Age Structures

The age structure of populations is divided into three age groups: prereproductive, reproductive, and postreproductive. The MDCs are approaching stabilization, with just about equal numbers in each group. The LDCs will continue to expand because their prereproductive group is the largest.

Population Growth and Environmental Impact

Two types of environmental impact can occur: The LDCs put stress on the biosphere due to population growth, and the MDCs put stress on the biosphere due to resource consumption and waste production.

30.3 Characteristics of Populations

Distribution patterns and population density are both dependent on resource availability.

Distribution Patterns Clumped, random, uniform.

Population Density The number of individuals per unit area is higher in areas with abundant resources and lower when resources are limited.

Demographics and Population Growth Population growth can be calculated based on the annual birthrate and death rate. Population growth is determined by

- Resource availability
- Demographics (age structures, survivorship, and biotic potential—the highest rate of increase possible)

Patterns of Population Growth

The two patterns of population growth are exponential growth and logistic growth.

Exponential Growth Exponential growth results in a J-shaped curve. The two phases of exponential growth are lag phase (slow growth) and exponential growth phase (accelerating growth).

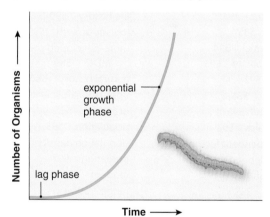

Logistic Growth Logistic growth results in an S-shaped curve. The four phases of logistic growth are lag phase (slow growth), exponential growth phase (accelerating growth), deceleration phase (slowing growth), and stable equilibrium phase (relatively no growth).

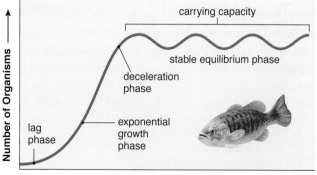

Factors That Regulate Population Growth

The two factors that regulate population growth are density-independent and density-dependent factors.

Density-independent factors include abiotic factors, such as weather and natural disasters. The effect of the factor is not dependent on density.

Density-dependent factors include biotic factors, such as competition and predation. The effect of the factor is dependent on density.

30.4 Life History Patterns and Extinction

The two fundamental life history patterns are exhibited by opportunistic populations and equilibrium populations.

Opportunistic populations are characterized by individuals that have short life spans, mature quickly, produce many offspring, have strong dispersal ability, provide little or no care of offspring, and exhibit exponential growth.

Equilibrium populations are characterized by individuals that have long life spans, mature slowly, produce few offspring, provide much care of offspring, and exhibit logistic growth.

Extinction

Extinction is the total disappearance of a species or higher group. Opportunistic populations are less likely than equilibrium populations to become extinct. Three factors, in particular, influence the vulnerability of equilibrium populations to extinction: size of geographic range, degree of habitat tolerance, and size of local populations.

Key Terms

age structure   593
biosphere   585
biotic potential   594
carrying capacity   595
clumped   592
community   585
competition   596
demographics   593
density-dependent factor   596
density-independent factor   596
ecology   585
ecosystem   585
equilibrium species   599
exponential growth   595
extinction   599
less-developed country (LDC)   589

life history   599
logistic growth   595
more-developed country (MDC)   589
natural history   586
opportunistic species   599
organism   585
population   585
population density   592
predation   597
random   592
range   592
replacement reproduction   590
resource   592
survivorship   593
uniform   592

Testing Yourself

Choose the best answer for each question.

1. Decreased death rate followed by decreased birthrate has occurred in
   a. MDCs.
   b. LDCs.
   c. MDCs and LDCs.
   d. neither MDCs nor LDCs.

2. Human societies at present are characterized by
   a. ever-increasing population growth.
   b. unsustainable practices.
   c. concentrations along coastlines.
   d. All of these are correct.

3. A population's maximum growth rate is also called its
   a. carrying capacity.
   b. biotic potential.
   c. growth curve.
   d. replacement rate.

4. Which of these levels of ecological study involves both abiotic and biotic components?
   a. organisms
   b. populations
   c. communities
   d. ecosystem
   e. All of these are correct.

5. The biological level of organization subject to evolution by natural selection is called
   a. an organism.
   b. a population.
   c. a community.
   d. an ecosystem.
   e. the biosphere.

6. Globally, if a population lives only along a lake's shoreline, members of this population exhibit which type of spatial distribution?
   a. variable
   b. clumped
   c. random
   d. uniform

7. Which of the following are likely to have the highest population density?
   a. zebras on the African savanna during the dry season
   b. mice in the frozen tundra
   c. moose in a Canadian forest
   d. earthworms in an organically rich soil

8. Calculate the growth rate of a population of 500 individuals in which the birthrate is 10 per year and the death rate is 5 per year.
   a. 5%
   b. −5%
   c. 1%
   d. −1%

9. If the human birthrate were reduced to 15 per 1,000 per year and the death rate remained the same (9 per 1,000), what would be the growth rate?
   a. 9%
   b. 6%
   c. 10%
   d. 0.6%
   e. 15%

10. Replacement reproduction in a population with a pyramid-shaped age structure diagram results in
    a. no population growth.
    b. population growth.
    c. a decline in the population.

11. Label the following age structure diagrams to indicate whether the population is stable, increasing, or decreasing.

a. _____   b. _____   c. _____

12. If the number of prereproductive and reproductive members of a population exceeds the number of postreproductive members, the population will
    a. grow.
    b. remain stable.
    c. decline.

13. An S-shaped growth curve indicates
    a. logistic growth.
    b. logarithmic growth.
    c. exponential growth.
    d. additive growth.

14. Exponential growth occurs
    a. when population size is increasing.
    b. at the carrying capacity of the environment.
    c. when people are poor and don't have enough to eat.

For statements 15–18, indicate the interaction in the key exemplified by the scenario.

Key:

    a. density-independent factor
    b. competition
    c. predation
    d. predator-prey cycle

15. A severe drought destroys the entire food supply of a herd of gazelle.

16. A population of feral cats increases in size as the mouse population increases and then crashes regularly after the mouse population enters periods of decline.

17. Swift coyotes are able to catch rabbits and live to reproduce. Other coyotes lack a good food source and are not strong enough to reproduce.

18. Deer prefer to feed on a dense thicket of oak saplings rather than more widely spaced young oak trees.

19. Which of the following is not an adaptive feature of an opportunistic life history pattern?
    a. many offspring
    b. little or no care of offspring
    c. long life span
    d. small individuals
    e. individuals that mature quickly

20. When the carrying capacity of the environment is exceeded, the population typically
    a. increases, but at a slower rate.
    b. reverts to an opportunistic life history strategy.
    c. decreases to below the carrying capacity.
    d. experiences extinction immediately.

21. A pyramid-shaped age distribution means that
    a. the prereproductive group is the largest group.
    b. the population will grow for some time in the future.
    c. the country is more likely an LDC than an MDC.
    d. fewer women are leaving the reproductive years than entering them.
    e. All of these are correct.

22. Which of these is a density-independent factor?
    a. competition   c. weather
    b. predation   d. resource availability

23. Which of the following would not limit the biotic potential of an organism?
    a. a decrease in predators of this population
    b. an increase in parasites of this population
    c. a consistently high degree of competition between this population and other species
    d. a decrease in natural resources available to this population
    e. All of these would limit a population's biotic potential.

Thinking Scientifically

1. In Sri Lanka, the death rate is 6 per 1,000, while the birthrate is 19 per 1,000. Calculate the current population growth rate. The formula for the doubling time of a population is

   $t = 0.69/r$, where t is the doubling time and r is the growth rate.

   Determine the doubling time (the number of years it will take to double the population size) in Sri Lanka. Be sure to convert your growth rate from a percentage (e.g., 2.1%) to a decimal value (0.021). If the birthrate drops to 10 per 1,000, what is the doubling time?

2. Assume that a population of dandelions grows exponentially, so that it doubles in size every week. The population (beginning with one plant) expands to fill a field in 20 weeks. How many weeks will it take to fill one-quarter of the field? How long will it take to fill half the field?

3. What type data would you collect to show that a population is undergoing stabilizing selection, as in Figure 15.2?

Bioethical Issue

Saving Species from Extinction

Species that are more prone to certain risk factors are more likely than others to become extinct. For example, species with a unique lineage, such as the giant panda, are likely to be at severe risk of extinction. Should our limited resources for species protection be focused on species that are at the highest risk of extinction? Some argue that high-risk species are less successful products of evolution and should not receive extraordinary protection; consequently, all species at risk of extinction should be equally protected. Which camp are you in? Support your position.

31

Communities and Ecosystems

BEFORE YOU BEGIN

Before beginning this chapter, take a few moments to review the following discussions.

Figure 1.12 What are the roles of producers, consumers, and decomposers in an ecosystem?

Section 6.1 How do photosynthetic organisms convert CO_2 and water to sugar?

Section 30.2 What is the environmental impact of human population growth?

Chemical Cycling

Essential elements are elements necessary for all living things. Nitrogen is one of these elements, and even though it is the most abundant gas in Earth's atmosphere, it is not biologically available to plants until it has been "fixed." This "fixing" process is part of the nitrogen cycle, and much of it is performed by soil bacteria. Once this process is complete, plants can absorb the nitrogen from the soil in the form of ammonium and nitrate ions, and other organisms receive this nitrogen when they feed on plants. Carbon dioxide, another atmospheric gas, is the source of all carbon used in life's organic molecules. Plants and other autotrophs convert carbon dioxide into usable organic molecules through the process of photosynthesis. Animals that then feed on plants acquire the necessary organic molecules, which get broken down through the process of cellular respiration. Carbon dioxide, a waste product of cellular respiration, is then exhaled back into the atmosphere, thus completing the carbon cycle.

In order for nutrients to cycle and energy to flow through both the environment and its organisms, many levels of interactions among species and their habitats must exist. In this chapter, you will learn about communities; types of interactions among species that influence natural selection; characteristics of aquatic and terrestrial ecosystems; and chemical cycling, including the nitrogen and carbon cycles.

31.1 Ecology of Communities

Learning Outcomes

Upon completion of this section, you should be able to

1. Define community and coevolution and provide an example of each.
2. Distinguish between species richness and species diversity.
3. Explain the process of ecological succession.
4. Compare and contrast primary and secondary succession.
5. Describe and provide examples of the types of interactions with a community.
6. Distinguish between a habitat and a niche.

A **community** is an assemblage of populations of multiple species, interacting with one another within a single environment. For example, the various species living in and on a fallen log, such as plants, fungi, worms, and insects, interact with one another and form a community. The fungi break down the log and provide food for the earthworms and insects living in and on the log. Those insects may feed on one another, too. If birds flying throughout the forest feed on the insects and worms living in and on the log, they are also part of the larger forest community.

Communities come in different sizes, and it is sometimes difficult to decide where one community ends and another one begins. The relationships and interactions between species in a community form over time. Some of the relationships between species are products of **coevolution,** by which an evolutionary change in one species results in an evolutionary change in another. **Figure 31.1** gives examples of how flowering plants and their pollinators are

a.

b.

c.

Figure 31.1 Coevolution.

Flowers and pollinators have evolved to be suited to one another. **a.** Hummingbird-pollinated flowers are usually red, a color these birds can see, and the petals are recurved to allow the stamens to dust the birds' heads. **b.** The reward offered by the flower is not always food. This orchid looks and smells like the female of this wasp's species. The male tries to copulate with flower after flower and in the process transfers pollen. **c.** Bats are nocturnal, and the flowers they pollinate are white or light-colored, making them visible in moonlight. The flowers smell like bats and are large and sturdy, enabling them to withstand insertion of the bat's head as it uses its long, bristly tongue to lap up nectar and pollen.

co-adapted. The Australian orchid, *Chiloglottis trapeziformis,* resembles the body of a wasp, and its odor mimics the pheromones of a female wasp. Therefore, male wasps are attracted to the flower, and when they attempt to mate with it, they become covered with pollen, which they transfer to the next orchid flower (Fig. 31.1*b*). The orchid is dependent on wasps for pollination because neither wind nor other insects pollinate these flowers.

Video Pollinators

All species in a community possess adaptations suitable to the conditions of the physical environment. An **ecosystem** consists of these species interacting with each other and with the physical environment. If the physical environment of an ecosystem changes, corresponding changes will most likely occur in the species that make up the community and in the relationships among these species. Extinction of species can occur when environmental change is too rapid for suitable adaptations to evolve.

Rapid environmental changes can be detrimental to humans, too, even though technology increases our ability to adapt. Sometimes the economy of an area is dependent on the climate and, in some cases, the species composition of an aquatic or terrestrial ecosystem. Therefore, human activities that negatively impact the community of the area can also negatively affect the economy of that area. Knowledge of community and ecosystem ecology will help you better understand how humans can, in the end, be detrimental to ourselves.

Community Composition and Diversity

Two characteristics of communities—species composition and diversity—allow us to compare communities. The species composition of a community, also known as **species richness,** is simply a listing of the various species found in that community. The **diversity** of a community includes both species richness and species distribution.

Species Composition

It is apparent by comparing the photographs in **Figure 31.2** that a coniferous forest has a different species composition than a tropical rain forest. Narrow-leaved evergreen tree species are prominent in the coniferous forest, whereas broad-leaved evergreen tree species are numerous in the tropical rain forest. As the list of mammals demonstrates, a coniferous community and a tropical rain forest community contain different species of mammals. Ecologists comparing these two communities would also find differences in the number of other plants and animal species. In the end, ecologists would conclude that not only are the species compositions of these two communities different but the tropical rain forest supports more species and therefore higher species richness.

Diversity

The diversity of a community goes beyond species richness to include distribution and relative abundance, the number of individuals of each species per unit area. For example, suppose a deciduous forest in West Virginia has, among individuals of other species, 76 poplar trees but only 1 American elm. If you were simply walking through this forest, you could miss the lone American elm. If, instead, the forest had 36 poplar trees and 41 American elm trees, the forest would seem more diverse to you, and indeed would have a higher diversity value, although both would have the same species richness value. The greater the species richness and the more even the distribution of the species in the community, the greater the diversity.

moose
lynx
snowshoe hare
bear
red fox
wolf

a.

monkey
sloth
anteater
kinkajou
jaguar
tapir
bat

b.

Figure 31.2 Community species composition.

Communities differ in their species composition, as exemplified by the predominant plants and animals in (**a**) a coniferous forest and (**b**) a tropical rain forest. Some mammals found in coniferous forests and in tropical rain forests are listed.

Ecological Succession

Community species composition and diversity do change over time, although they may seem to remain static because it can take decades—often longer than the human life span—for noticeable changes to occur. Natural forces, such as glaciers, volcanic eruptions, lightning-ignited forest fires, hurricanes, tornadoes, and floods are considered disturbances and bring about community changes. Communities also change because of human activities, such as logging, road building, sedimentation, and farming. A more or less orderly process of community change is known as **ecological succession.**

Ecologists have developed models to explain why succession occurs and to predict patterns of succession following a disturbance. The climax-pattern model says that the climate of an area always leads to the same stable **climax community,** a specific assemblage of bacterial, fungal, plant, and animal species (**Fig. 31.3**). For example, a coniferous forest community is expected in northern latitudes, a deciduous forest in temperate zones, and a tropical forest in areas with a tropical climate. Now that scientists know that disturbances influence community composition and diversity and that, despite the climate, the composition of a community is not always the same, the climax model of succession is being modified.

Two Types of Succession

Ecologists define two types of ecological succession: primary and secondary (**Fig. 31.4**). **Primary succession** starts where soil has not yet formed. For example, hardened lava flows and the scraped bedrock that remains following a glacial retreat are subject to primary succession. **Secondary succession** begins, for example, in a cultivated field that is no longer farmed, where soil is already present. With both primary and secondary succession, a progression of species occurs over time, as first spores of fungi and vascular seedless plants and then seeds of nonvascular plants, followed by seeds of gymnosperms and/or angiosperms, are carried

a.

b.

Figure 31.3 Climax communities.

Does succession in a particular area always lead to the same climax community? For example, (**a**) temperate rain forests occur only where there is adequate rainfall, and (**b**) deserts occur where rainfall is minimal. Even so, exactly the same mix of plants and animals may not always arise because the assemblage of organisms depends on which organisms, by chance, migrate to the area.

Figure 31.4 Primary and secondary succession.

Primary succession begins on areas of bare rock. Secondary succession begins in areas where soil remains following natural or human-caused disturbance.

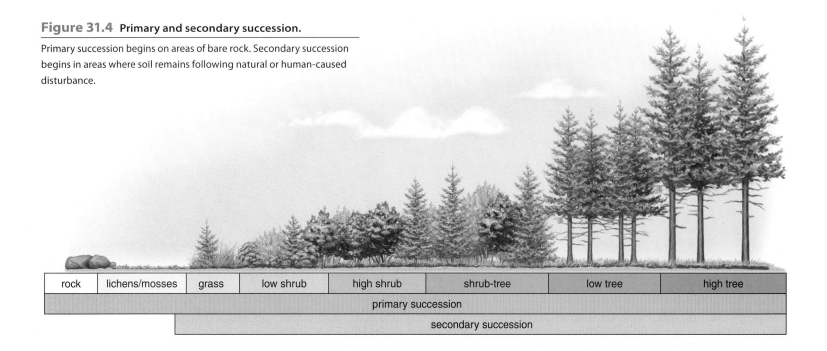

rock	lichens/mosses	grass	low shrub	high shrub	shrub-tree	low tree	high tree
primary succession								
secondary succession								

into the area by wind, water, or animals from the surrounding regions (**Fig. 31.5**; see also Fig. 31.4).

The first species to appear in an area undergoing either primary or secondary succession are called opportunistic pioneer species. These species are small in stature, short-lived, and quick to mature, and they produce numerous offspring per reproductive event. The first pioneer species to arrive are photosynthetic organisms, such as lichens and mosses. Lichens are two organisms from different kingdoms, living as one. The fungal partners of lichens play a critical role by breaking down rock or lava into usable mineral nutrition, not only for their algal partners but also for pioneer plants. The mycorrhizal fungal partners of plants (see Chapter 16) pass minerals directly to plants, so that they can grow successfully in poor soil. Pioneer plant species that become established in an area are accompanied by pioneer herbivore species (e.g., insects) and then carnivore species (e.g., small mammals). As the community continues to change, equilibrium species become established in the area. Equilibrium species, such as deer, wolves, and bears, are larger in size, long-lived, and slow to mature, and they produce few offspring per reproductive event.

a. During the first year, only the remains of corn plants are seen.

b. During the second year, wild grasses have invaded the area.

c. By the fifth year, the grasses look more mature, and sedges have joined them.

d. After twenty years, the juniper trees are mature, and there are also birch and maple trees in addition to the blackberry shrubs.

Figure 31.5 Secondary succession.

Secondary successional changes in a western Pennsylvania field: (**a**) first year, (**b**) second year, (**c**) fifth year, and (**d**) after 20 years.

Interactions in Communities

Species interactions, especially competition for resources, fashion a community into a dynamic system of interspecies relationships. In **Table 31.1,** the plus and minus signs tell how the relationship affects the abundance of the two interacting species. **Competition** between two species for limited resources has a negative effect on the abundance of both species. In **predation,** one animal, the predator, feeds on another, the prey (see Fig. 30.18); in **parasitism,** one species obtains nutrients from another species, called the host, but usually does not kill the host. **Commensalism** is a relationship in which one species benefits, while the second species is not harmed. Commensalism often occurs when one species provides a home or transportation for another. In **mutualism,** two species interact in such a way that both benefit.

Ecological Niche

Each species occupies a particular position in the community, in both a spatial and a functional sense. Spatially, species live in a particular area of the community, or **habitat,** such as underground, in the trees, or in shallow water. Functionally, each species plays a role, such as a photosynthesizer, predator, prey, parasite, or decomposer. The **ecological niche** of a species incorporates the role the species plays in its community, its habitat, and its interactions with other species. The niche includes the living and nonliving resources that individuals in the population need to meet their energy, nutrient, and survival demands. The habitat of an insect called a backswimmer is a pond or lake, where it eats other insects (**Fig. 31.6**). The pond or lake must contain vegetation where the backswimmer can hide from its predators, including fish and birds. The pond water must be clear enough for the backswimmer to see its prey and warm enough for it to maintain a good metabolic rate. Since it is difficult to describe and measure the total niche of a species in a community, ecologists often focus on a certain aspect of a species' niche, as with the birds featured in **Figure 31.7.**

Table 31.1 Species Interactions

Interaction	Expected Outcome
Competition (− −)	Abundance of both species decreases.
Predation (+ −)	Abundance of predator increases, and abundance of prey decreases.
Parasitism (+ −)	Abundance of parasite increases, and abundance of host decreases.
Commensalism (+ 0)	Abundance of one species increases, and the other is not affected.
Mutualism (+ +)	Abundance of both species increases.

Figure 31.6 Niche of a backswimmer.

Backswimmers require warm, clear pond water containing insects that they can eat and vegetation where they can hide from predators.

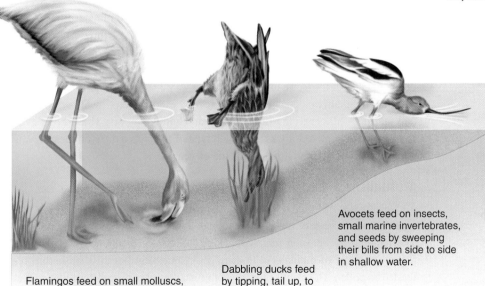

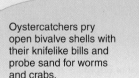

Oystercatchers pry open bivalve shells with their knifelike bills and probe sand for worms and crabs.

Plovers dart around on beaches and grasslands hunting for insects and small invertebrates.

Avocets feed on insects, small marine invertebrates, and seeds by sweeping their bills from side to side in shallow water.

Flamingos feed on small molluscs, crustaceans, and vegetable matter strained from mud pumped through their bills by their powerful tongues.

Dabbling ducks feed by tipping, tail up, to reach aquatic plants, seeds, snails, and insects.

Figure 31.7 Feeding niches for wading birds.

Flamingos feed in deeper water by filter feeding; dabbling ducks feed in shallower areas by upending; avocets feed by sifting. Oystercatchers and plovers have adaptations, such as shorter legs, for feeding in shallows.

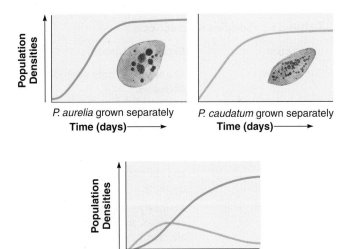

Figure 31.8 Competitive exclusion principle demonstrated by *Paramecium*.

The competitive exclusion principle states that no two species occupy exactly the same niche. When grown separately, *Paramecium caudatum* and *Paramecium aurelia* exhibit logistic growth. When grown together, *P. aurelia* excludes *P. caudatum*.

Data from G. F. Gause, *The Struggle for Existence*, 1934, Williams & Wilkins Company, Baltimore, MD, p. 557.

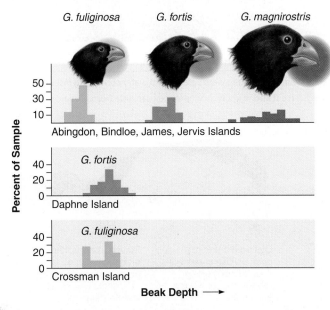

Figure 31.9 Character displacement in finches on the Galápagos Islands.

When *Geospiza fuliginosa*, *G. fortis*, and *G. magnirostris* coexist on the same island, their beak sizes are appropriate for eating small, medium-size, and large seeds, respectively. When *G. fortis* and *G. fuliginosa* are on separate islands, their beaks have the same intermediate size, which allows them to eat seeds of various sizes. Character displacement is evidence that resource partitioning has occurred.

Competition

Competition for resources, such as light, space, or nutrients, contributes to the niche of each species and helps structure the community. Laboratory experiments helped ecologists formulate the **competitive exclusion principle,** which states that no two species can occupy the same niche at the same time. In the 1930s, G. F. Gause grew two species of *Paramecium* in one test tube containing a fixed amount of bacterial food. Although populations of each species survived when grown in separate test tubes, only one species, *Paramecium aurelia,* survived when the two species were grown together (**Fig. 31.8**). *P. aurelia* acquired more of the food resource and had a higher population growth rate than did *P. caudatum*. Eventually, as the *P. aurelia* population grew and obtained an increasingly greater proportion of the food resource, the number of *P. caudatum* individuals decreased and the population died out.

Niche Specialization   Competition for resources does not always lead to localized extinction of a species. Multiple species coexist in communities by partitioning, or sharing, resources. In another laboratory experiment using other species of *Paramecium,* Gause found that the two species could survive in the same test tube if one species consumed bacteria at the bottom of the tube and the other ate bacteria suspended in solution. This **resource partitioning** decreased competition between the two species, leading to increased niche specialization. What could have been one niche became two, more specialized niches due to species differences in feeding behavior.

**Video
Cichlid
Specialization**

When three species of ground finches of the Galápagos Islands live on the same islands, their beak sizes differ, and each feeds on a different-sized seed (**Fig. 31.9**). When the finches live on separate islands, their beaks tend to be the same intermediate size, enabling each to feed on a wider range of seeds. Such **character displacement** often is viewed as evidence that competition and resource partitioning have taken place.

**Video
Finches
Adaptive
Radiation**

The niche specialization that permits the coexistence of multiple species can be very subtle. Species of warblers that live in North American forests are all about the same size, and all feed on budworms, a type of caterpillar found on spruce trees. Robert MacArthur recorded the length of time each warbler species spent in different regions of spruce canopies to determine where each species did most of its feeding. He discovered that each species primarily used different parts of the tree canopy and, in that way, had a more specialized niche. As another example, consider that three types of birds—swallows, swifts, and martins—all eat flying insects and parachuting spiders. These birds even frequently fly in mixed flocks. But each type of bird has a different nesting site and migrates at a slightly different time of year.

Mutualism

Mutualism, a symbiotic relationship in which both members benefit, is now recognized to be at least as important as competition in shaping community structure. The relationship between plants and their pollinators mentioned previously is a good example of mutualism. Perhaps the relationship began when herbivores feasted on pollen. The plant's provision of nectar may have spared the pollen and, at the same time, allowed the animal to become an agent of pollination. By now, pollinator mouthparts have become adapted to gathering the nectar of a particular plant species, and this species is dependent on the pollinator for dispersing pollen. As also mentioned previously, lichens can grow on rocks because the fungal member leaches minerals, which are provided to the algal partner. The algal partner, in turn, photosynthesizes and provides organic food for both members of the relationship.

In tropical America, ants form mutualistic relationships with certain plants. The bullhorn acacia tree is adapted to provide a home for ants of the species *Pseudomyrmex ferruginea* (**Fig. 31.10**). Unlike other acacias, this species has swollen thorns with a hollow interior, where ant larvae can grow and develop. In addition to housing the ants, acacias provide them with food. The ants feed from nectaries at the base of the leaves and eat fat- and protein-containing nodules called Beltian bodies, which are found at the tips of the leaves. For 24 hours a day, ants protect the tree from herbivores that would like to feed on it. The ants are so critical to the trees' survival that, when the ants on experimental trees were poisoned, the trees died.

Video — Thorn Tree-Ant Mutualism

The outcome of mutualism is an intricate web of species interdependencies critical to the community. For example, in areas of the western United States, the branches and cones of whitebark pine are turned upward, meaning that the seeds do not fall to the ground when the cones open. Birds called Clark's nutcrackers eat the seeds of whitebark pine trees and store them in the ground (**Fig. 31.11**).

Figure 31.10 Mutualism.

The bullhorn acacia tree is adapted to provide nourishment for a mutualistic ant species. **a.** The thorns are hollow, and the ants live inside. **b.** The bases of the leaves have nectaries (openings) where ants can feed. **c.** The tips of the leaves of the bullhorn acacia have Beltian bodies, which ants harvest for larval food.

Figure 31.11 Interdependence of species.

(**a**) The Clark's nutcracker feeds on the seeds of (**b**) the white bark pine. The nutcracker stores the seeds in the ground, and these seeds may be found and eaten by (**c**) the grizzly bear. Clark's nutcrackers and grizzly bears disperse the seed of the pine.

Connections and Misconceptions

Do humans form mutualistic relationships with other organisms?

One of the best examples of mutualism between humans and another species involves the *Escherichia coli* that live in our large intestine. While some strains of *E. coli* have a bad reputation for causing intestinal problems, the majority of the *E. coli* in our intestine are beneficial. In exchange for a  warm environment, and plenty of food, the intestinal *E. coli* produce vitamin K and assist in the breakdown of fiber into glucose.

Therefore, Clark's nutcrackers are critical seed dispersers for the trees. Also, grizzly bears find the stored seeds and consume them. Whitebark pine seeds do not germinate unless their seed coats are exposed to fire. When natural forest fires in the area are suppressed, whitebark pine trees decline in number, and so do Clark's nutcrackers and grizzly bears. When lightning-ignited fires are allowed to burn, or prescribed burning is used in the area, the whitebark pine populations increase, as do the populations of Clark's nutcrackers and grizzly bears.

 Video Fruit Bat Seed Dispersal

 Video Ant-Caterpillar Mutualism

Community Stability

As witnessed by the discussion of succession, community stability is fragile. However, some communities have one species that stabilizes the community, helps maintain its characteristics, and essentially helps hold its web of interactions together. Such a species is known as a **keystone species,** or species on which the existence of a large number of other species in the ecosystem depends. The term *keystone* comes from the name for the center stone of an arch that holds the other stones in place, so that the arch can keep its shape.

Keystone species are not always the most numerous in the community. However, the loss of a keystone species can lead to the extinction of other species and a loss of diversity. For example, bats are designated as keystone species in tropical forests. Bats are pollinators, and they disperse the seeds of certain tropical trees. When bats are killed off or their roosts are destroyed, these trees fail to reproduce. The grizzly bear is a keystone species in the northwestern United States and Canada. Grizzly bears disperse as many as 7,000 berry seeds in one dung pile. Grizzly bears also kill the young of many herbivorous mammals, such as deer, and thereby keep their populations under control.

The sea otter is a keystone species of a kelp forest ecosystem. Kelp forests, created by large, brown seaweeds, provide a home for a vast assortment of organisms. The kelp forests occur just off the coast and protect coastline ecosystems from damaging wave action. Among other species, sea otters eat sea urchins, keeping their population size in check. Otherwise, sea urchins feed on the kelp, causing the kelp forest and its associated species to decline severely. Fishermen don't like sea otters because they also prey on abalone, a mollusc prized for its commercial value. Many do not realize that, without the otters, abalone and many other species would not be around because their natural habitat, a kelp forest, would no longer exist.

Native Versus Exotic Species

Native species are species indigenous to an area. They colonize an area without intentional or accidental human assistance. For example, you naturally find maple trees in Vermont and many other states of the eastern United States. The introduction of **exotic species,** or nonnative species, into a community greatly disrupts normal interactions, and therefore a community's web of species. Populations of exotic species may grow exponentially because they are often better competitors, or because their population size is not controlled by their natural predators or diseases. The unique assemblage of native species on an island often cannot compete well against an exotic species. For example, myrtle trees, introduced into the Hawaiian Islands from the Canary Islands, are mutualistic with a type of bacterium capable of fixing atmospheric nitrogen. The bacterium provides the plant with a form of nitrogen it can use. This feature allows myrtle trees to become established on nitrogen-poor volcanic soil, a distinct advantage in Hawaii. Once established, myrtle trees halt the normal succession of native plants on volcanic soil.

Exotic species disrupt communities in continental areas as well. The red fox was deliberately imported into Australia to prey on the previously introduced European rabbit, but instead the red fox has now reduced the populations of native small mammals (**Fig. 31.12***a*). The brown tree snake was accidentally introduced onto a number of islands in the Pacific Ocean (Fig. 31.12*b*). The brown tree snake eats eggs, nestlings, and adult birds. On Guam, the snakes have reduced 10 native bird species to the point of extinction. On the Galápagos Islands, black rats accidentally carried to the islands by ships have reduced populations of the giant tortoise. Goats and feral pigs have changed the vegetation on the islands from highland forest to pampas-like grasslands and destroyed stands of cactuses. In the United States, gypsy moths, zebra mussels, the Chestnut blight fungus, fire ants, and African bees are well-known exotic species that have killed native species. At least two species, fire ants and African bees, have attacked humans, with serious consequences.

Video
Alien Invasion

a.

b.

Figure 31.12 Exotic species.

Human introduction of exotic species, such as (**a**) the red fox and (**b**) the brown tree snake, has disrupted communities in Australia and Guam, respectively.

Connecting the Concepts

For more information on the material presented in this section, refer to the following discussions.

Section 14.1 explores how both variation and fitness play a role in the evolutionary processes that drive the coevolution of species.

Section 21.3 examines how seed dispersal in some species of flowering plants has coevolved with animals.

Check Your Progress 31.1

1. Describe an example of coevolution.
2. Contrast species richness with diversity.
3. Contrast primary succession with secondary succession.
4. List the five major types of species interactions in a community.
5. Distinguish between a habitat and an ecological niche.
6. Describe the role of a keystone species in an ecosystem.

Figure 31.13   Producers.

Green plants and algae are photoautotrophs.

a. Herbivores

b. Carnivores

Figure 31.14   Consumers.

a. Caterpillars and giraffes are herbivores. **b.** A praying mantis and an osprey are carnivores.

31.2 Ecology of Ecosystems

Learning Outcomes

Upon completion of this section, you should be able to

1. Define an ecosystem and provide examples.
2. Distinguish between autotrophs and heterotrophs.
3. Compare how energy and chemicals interact with an ecosystem.
4. Explain the different trophic levels and the formation of ecological pyramids.
5. Summarize the biogeochemical cycles and state how human activity is influencing each cycle.

An ecosystem is more inclusive than a community because community ecology considers only how species interact with one another. When studying ecosystem ecology, interactions with the physical environment are also considered. For example, one important aspect of ecological niche is how an organism acquires food. It is obvious that autotrophs interact with the physical environment, but so do heterotrophs.

Autotrophs

Autotrophs take in only inorganic nutrients (e.g., CO_2 and minerals) and an outside energy source to produce organic nutrients for their own use and for all the other members of a community. They are called **producers** because they produce food. *Photoautotrophs,* often called photosynthetic organisms, produce most of the organic nutrients for the biosphere (**Fig. 31.13**). Algae of all types possess chlorophyll and carry on photosynthesis in freshwater and marine habitats. Algae make up the phytoplankton, which are photosynthesizing organisms suspended in water. Green plants are the dominant photosynthesizers on land. All photosynthesizing organisms release O_2 into the atmosphere.

    Some bacteria are *chemoautotrophs*. They obtain energy by oxidizing inorganic compounds, such as ammonia, nitrites, and sulfides, and they use this energy to synthesize organic compounds. Chemoautotrophs have been found to support communities in some caves and at hydrothermal vents along deep-sea oceanic ridges.

Video Plants

Heterotrophs

Heterotrophs need a source of preformed organic nutrients, and they release CO_2 into the atmosphere. They are called **consumers** because they consume food. **Herbivores** are animals that graze directly on algae or plants (**Fig. 31.14***a*). In aquatic habitats, zooplankton, such as protozoans, are herbivores; in terrestrial habitats, insects, including caterpillars, play that role. Among larger herbivores, giraffes browse on trees. **Carnivores** eat other animals; for example, a praying mantis catches and eats caterpillars, and an osprey preys on and eats fish (Fig. 31.14*b*). These examples illustrate that there are primary consumers (e.g., giraffe), secondary consumers (e.g., praying mantis), and tertiary consumers (e.g., osprey). Sometimes tertiary consumers are called top predators. **Omnivores** are animals that eat both plants and animals. As you likely know, most humans are omnivores.

Video Snake Eating

Connections and Misconceptions

How do chemoautotrophs on the ocean floor produce food?

The chemoautotrophs on the ocean floor use chemical energy instead of sunlight to make food. Volcanoes at the bottom of the ocean floor release hydrogen sulfide gas through cracks called hydrothermal vents. (Hydrogen sulfide is the nasty-smelling gas we associate with rotten eggs.) Some  chemoautotrophs split hydrogen sulfide to obtain the energy needed to link carbon atoms together to form glucose. The glucose contained in these chemoautotrophs sustains a variety of bizarre organisms, such as giant tube worms, anglerfish, and giant clams.

The **decomposers** are heterotrophic bacteria and fungi, such as molds and mushrooms, that break down dead organic matter, including animal wastes (**Fig. 31.15**). They perform a very valuable service, because they release inorganic nutrients (CO_2 and minerals), which are then taken up by plants once more. Otherwise, plants would rely on minerals to be slowly released from rocks. **Detritus** is composed of the remains of dead organisms plus the bacteria and fungi that aid in decay. Fanworms feed on detritus floating in marine waters, while clams take detritus from the sea bottom. Earthworms and some beetles, termites, and maggots are soil detritus feeders.

Figure 31.15 Decomposers.

Fungi and bacteria are decomposers.

 Video
Decomposers

Energy Flow and Chemical Cycling

The living components of ecosystems process energy and chemicals. Energy flow through an ecosystem begins when producers absorb solar energy, and chemical cycling begins when producers take in inorganic nutrients from the physical environment (**Fig. 31.16**). Thereafter, via photosynthesis, producers convert the solar energy and inorganic nutrients into chemical energy in the form of organic nutrients, such as carbohydrates. Producers synthesize organic nutrients directly for themselves and indirectly for the heterotrophic components of the ecosystem. Energy flows through an ecosystem because as organic nutrients pass from one component of the ecosystem to another, as when an herbivore eats a plant or a carnivore eats an herbivore, a portion is used as an energy source. Eventually, the energy dissipates into the environment as heat. Therefore, the vast majority of ecosystems cannot exist without a continual supply of solar energy.

Only a portion of the organic nutrients made by producers is passed on to consumers because plants use organic molecules to fuel their own cellular respiration. Similarly, only a small percentage of nutrients consumed by lower-level consumers, such as herbivores, is available to higher-level consumers, or carnivores. As **Figure 31.17** demonstrates, a certain amount of the food eaten by an herbivore is never digested and is eliminated as feces. Metabolic wastes are excreted as urine. Of the assimilated energy, a large portion is used during cellular respiration for the production of ATP and thereafter becomes heat. Only the remaining energy, which is converted into increased body weight or additional offspring, becomes available to carnivores.

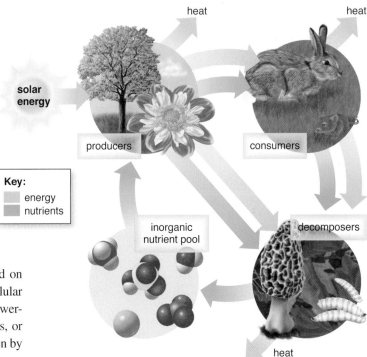

Figure 31.16 Chemical cycling and energy flow.

Chemicals cycle within, but energy flows through, an ecosystem. As energy is repeatedly passed from one component to another, all the chemical energy derived from solar energy dissipates as heat.

Figure 31.17 Energy balances.

Only about 10% of the nutrients and energy taken in by an herbivore is passed on to carnivores. A large portion goes to detritus feeders. Another large portion is used for cellular respiration.

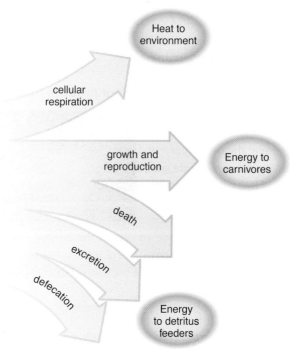

The elimination of feces and urine by a heterotroph, and indeed the death of all organisms, does not mean that organic nutrients are lost to an ecosystem. Instead, they represent the organic nutrients made available to decomposers. Decomposers convert the organic nutrients back into inorganic chemicals and release them to the soil or atmosphere. Chemicals complete their cycle within an ecosystem when producers absorb inorganic chemicals from the atmosphere or soil.

Video
Salmon Farming

Energy Flow

Applying the principles discussed so far to a temperate deciduous forest, ecologists can draw a **food web** to represent the interconnecting paths of energy flow between the component species of the ecosystem. In **Figure 31.18**, the green arrows are part of a **grazing food web** because the web begins with plants, such as the oak trees depicted. A **detrital food web** (orange arrows) begins with bacteria and fungi. In the grazing food web, caterpillars and other herbivorous insects feed on the leaves of the trees, while other herbivores, including mice, rabbits, and deer, feed on leaves at or near the ground. Birds, chipmunks, and mice feed on fruits and nuts of the trees, but, in fact, they are omnivores because they also feed on caterpillars and other insects. These herbivores and omnivores all provide food for a number of different carnivores. In the detrital food web, detritus, which includes smaller decomposers (such as bacteria and fungi), is food for larger organisms. Because some of these organisms, such as shrews and salamanders, become food for aboveground animals, the detrital and grazing food webs are joined.

We tend to think that the aboveground parts of trees are the largest storage form of organic matter and energy, but this is not necessarily the case. In temperate deciduous forests, the organic matter lying on the forest floor and mixed into the soil, along with the underground roots of the trees, contains over twice the energy of the leaves, branches, and trunks of living trees combined. Therefore, more energy and matter in a forest may be stored in or funneled through the detrital food web than the grazing food web.

Trophic Levels and Ecological Pyramids   The arrangement of component species in **Figure 31.19** suggests that organisms are linked to one another in a straight line according to feeding relationships, or who eats whom. Diagrams that

Key:
→ grazing food web
→ detrital food web

fruits and nuts

birds

hawks

chipmunks

owls

mice

leaves

leaf-eating insects

snakes

rabbits

fishers

old leaves, dead twigs

deer

skunks

foxes

shrews

salamanders

bacteria and fungi

invertebrates

carnivorous invertebrates

Figure 31.18 **Food webs.**

The grazing and detrital food webs of ecosystems are linked.

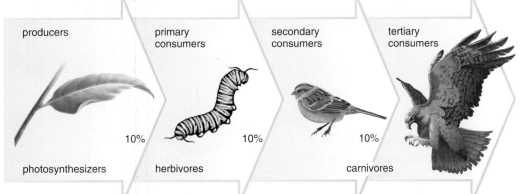

producers

primary consumers

secondary consumers

tertiary consumers

photosynthesizers

10%

herbivores

10%

10%

carnivores

Figure 31.19 **Food chain.**

A food chain diagrams a single path of energy flow in an ecosystem. Ten percent of the energy goes toward growth and reproduction on each step in the path.

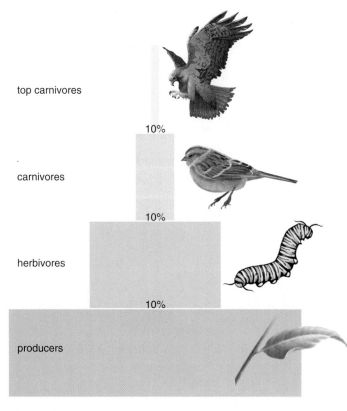

Figure 31.20 Ecological pyramid.

An ecological pyramid depicts the loss of nutrients and energy from one trophic level to the next.

show a single path of energy flow in an ecosystem are called **food chains** (Fig. 31.19). A **trophic level** is a level of nourishment within a food web or chain. In the grazing food web (see Fig. 31.18), from left to right: The trees are producers (the first trophic level), the first series of animals are herbivores (the second trophic level), and many of the animals in the next series are carnivores (the third and possibly fourth trophic levels). Food chains are short because energy is lost between trophic levels. In general, only about 10% of the energy of one trophic level is available to the next trophic level. Therefore, if an herbivore population consumes 1,000 kg of plant material, only about 100 kg is converted to herbivore tissue, 10 kg to first-level carnivores, and 1 kg to second-level carnivores. The so-called 10% rule of thumb explains why few carnivores can be supported in a food web. The flow of energy with large losses between successive trophic levels is sometimes depicted as an **ecological pyramid** (**Fig. 31.20**).

A pyramid based on the number of organisms can run into problems because, for example, one tree can support many herbivores. Pyramids of **biomass,** which is the number of organisms multiplied by their weight, eliminate size as a factor. Even then, apparent inconsistencies can arise. In aquatic ecosystems, such as lakes and open seas where algae are the only producers, the herbivores at some point in time may have a greater biomass than the producers. Why? Even though the algae reproduce rapidly, they are also consumed at a high rate.

Chemical Cycling

The pathways by which chemicals cycle within ecosystems involve both living components (producers, consumers, decomposers) and nonliving components (rock, inorganic nutrients, atmosphere) and therefore are known as **biogeochemical cycles.** Biogeochemical cycles can be sedimentary or gaseous. In a *sedimentary cycle,* such as the phosphorus cycle, the chemical is absorbed from the sediment by plant roots, passed to heterotrophs, and eventually returned to the soil by decomposers, usually in the same general area. In a *gaseous cycle,* such as the nitrogen and carbon cycles, the element returns to and is withdrawn from the atmosphere as a gas.

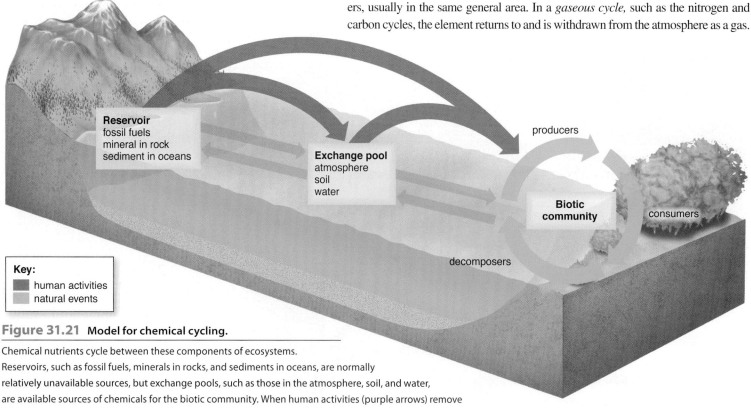

Figure 31.21 Model for chemical cycling.

Chemical nutrients cycle between these components of ecosystems.
Reservoirs, such as fossil fuels, minerals in rocks, and sediments in oceans, are normally relatively unavailable sources, but exchange pools, such as those in the atmosphere, soil, and water, are available sources of chemicals for the biotic community. When human activities (purple arrows) remove chemicals from reservoirs and make them available to the biotic community, pollution can result.

Chemical cycling of an element may involve **reservoirs** and exchange pools as well as the biotic community (**Fig. 31.21**). A *reservoir* is a source normally unavailable to organisms. For example, much carbon is found in calcium carbonate shells in ocean bottom sediments. An *exchange pool* is a source from which organisms generally take elements. For example, photosynthesizers can use carbon dioxide in the atmosphere for their carbon needs. The *biotic community* consists of the autotrophic and heterotrophic species of an ecosystem that feed on one another. Human activities, such as mining or burning fossil fuels, increase the amounts of chemical elements removed from reservoirs and cycling within ecosystems. As a result, the physical environment of the ecosystem contains excess chemicals that, in turn, may alter the species composition and diversity of the biotic community.

Phosphorus Cycle

On land, the very slow weathering of rocks fostered by fungi adds phosphates (PO_4^{2-} and HPO_4^{2-}) to the soil, some of which become available to terrestrial plants for uptake (**Fig. 31.22**). Phosphates made available by weathering also run off into aquatic ecosystems, where algae absorb the phosphates from the water before they become trapped in sediments. Phosphates in sediments become available again only when a geological upheaval exposes sedimentary rocks to weathering once more.

Producers use phosphates in a variety of molecules, including phospholipids, ATP, and the nucleotides that become a part of DNA and RNA. Animals consume producers and incorporate some of the phosphates into teeth, bones, and shells, which take many years to decompose. Decomposition of dead plant and animal material and animal wastes does, however, make phosphates available to producers faster than weathering. Because most of the available phosphates are used in food chains, phosphate is usually a limiting inorganic nutrient for ecosystems. In other words, the finite supply of phosphates limits plant growth, and therefore primary productivity.

The importance of phosphates and calcium to population growth is demonstrated by considering the fate of lemmings every four years (**Fig. 31.23**). You may have heard that lemmings dash mindlessly over cliffs into the sea; ecologists tell us that these lemmings are actually migrating to find food. What happened? Every four years or so, grasses and sedges of the tundra (see Fig. 31.28) become rich in minerals, and the lemming population starts to explode. Once the lemmings number in the millions, the grasses and sedges of the tundra suffer a decline caused by a lack of minerals. Now the lemming population suffers a crash, but it takes about four years before the animals decompose in this cold region and minerals return to the producers. Then the cycle begins again.

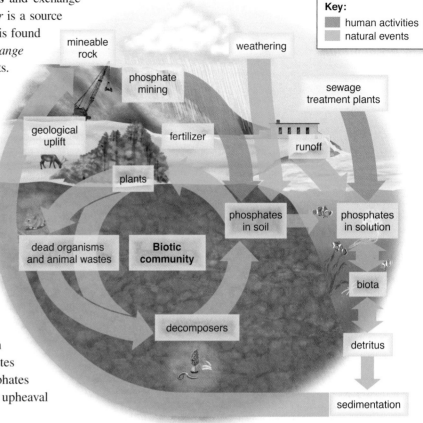

Figure 31.22 The phosphorus cycle.

The phosphorus cycle is a sedimentary biogeochemical cycle. Globally, phosphates flow into large bodies of water and become a part of sedimentary rocks. Thousands or millions of years later, the seafloor can rise; the phosphates are then exposed to weathering and become available. Locally, phosphates cycle within a community when plants on land and algae in the water take them up. Animals gain phosphates when they feed on plants or algae. Decomposers return phosphates to plants or algae, and the cycle within the community begins again.

Figure 31.23 Lemmings.

Lemmings, small rodents, feed on plant species in an Arctic community called the tundra.

Human Activities   A **transfer rate** is the amount of a nutrient that moves from one component of the environment to another within a specified period of time. Human activities affect the dynamics of a community by altering transfer rates. For example, humans mine phosphate ores and use them to make fertilizers, animal feed supplements, and detergents. Phosphate ores are slightly radioactive; therefore, mining phosphate poses a health threat to all organisms, including the miners. Animal wastes from livestock feedlots, fertilizers from lawns and cropland, and untreated and treated sewage discharged from cities all add excess phosphates to nearby waters. The result is **eutrophication,** or overenrichment, of a body of water, which causes an algal overpopulation called an algal bloom. When the algae die and decay, oxygen is consumed, causing fish kills. In the mid-1970s, Lake Erie was dying because of eutrophication. Control of nutrient phosphates, particularly in sewage effluent and household detergents, reversed the situation.

Nitrogen Cycle

Nitrogen, in the form of nitrogen gas (N_2), comprises about 78% of the atmosphere by volume. But plants cannot make use of nitrogen gas. Instead, plants rely on various types of bacteria to make nitrogen available to them. Therefore, nitrogen, like phosphorus, is a limiting inorganic nutrient of producers in ecosystems.

Plants can take up both ammonium (NH_4^+) and nitrate (NO_3^-) from the soil and incorporate the nitrogen into amino acids and nucleic acids. Two processes, nitrogen fixation and nitrification, convert nitrogen gas, N_2, into NH_4^+ and NO_3^-, respectively (**Fig. 31.24**). *Nitrogen fixation* occurs when nitrogen gas is converted to ammonium. Some cyanobacteria in aquatic ecosystems and some free-living, nitrogen-fixing bacteria in soil are able to fix nitrogen in this way. Other nitrogen-fixing bacteria live in nodules on the roots of legumes, plants such as peas, beans, and alfalfa. They make organic compounds containing nitrogen available to the host plant.

Nitrification is the production of nitrates. Ammonium in the soil is converted to nitrate by certain nitrifying soil bacteria in a two-step process. First, nitrite-producing bacteria convert ammonium to nitrite (NO_2^-), and then nitrate-producing bacteria convert nitrite to nitrate. In Figure 31.24, notice that the biotic community subcycle in the nitrogen cycle does not depend on the presence of nitrogen gas.

Denitrification is the conversion of nitrate to nitrogen gas, which enters the atmosphere. Denitrifying bacteria are chemoautotrophs, living in the anaerobic mud of lakes, bogs, and estuaries, that carry out this process as a part of their own metabolism. In the nitrogen cycle, denitrification counterbalanced nitrogen fixation until humans started making fertilizer.

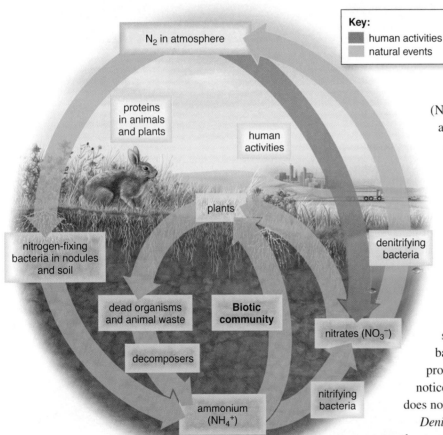

Figure 31.24   The nitrogen cycle.

The nitrogen cycle is a gaseous biogeochemical cycle normally maintained by the work of several populations of soil bacteria.

Video
Dung Beetles

Human Activities Human activities significantly alter the transfer rates in the nitrogen cycle by producing fertilizers from N_2—in fact, humans nearly double the fixation rate. The nitrate in fertilizers, just like phosphate, can leach out of agricultural soils into surface waters, leading to eutrophication. Deforestation by humans also causes a loss of nitrogen to groundwater and makes regrowth of the forest difficult. The underground water supplies in farming areas today are more apt to contain excess nitrate. A high concentration of nitrates interferes with blood oxygen levels, and infants below the age of six months who drink water containing excessive amounts of nitrate can become seriously ill and, if untreated, may die.

To cut back on fertilizer use, it might be possible to genetically engineer soil bacteria with increased nitrogen fixation rates. Also, farmers could grow legumes that increase the nitrogen content of the soil (**Fig. 31.25**). In one study, the rotation of legumes and winter wheat produced a better yield than fertilizers after several years.

Carbon Cycle

In the carbon cycle, organisms in both terrestrial and aquatic ecosystems exchange carbon dioxide with the atmosphere (**Fig. 31.26**). On land, plants take up carbon dioxide from the air, and through photosynthesis, they incorporate carbon into organic nutrients, which are used by autotrophs and heterotrophs alike. When aerobic organisms respire, a portion of this carbon is returned to the atmosphere as carbon dioxide, a waste product of cellular respiration.

In aquatic ecosystems, the exchange of carbon dioxide with the atmosphere is indirect. Carbon dioxide from the air combines with water to produce bicarbonate ion (HCO_3^-), a source of carbon for algae that produce food for themselves and for heterotrophs. Similarly, when aquatic organisms respire, the carbon dioxide they give off becomes bicarbonate ion. The amount of bicarbonate in the water is in equilibrium with the amount of carbon dioxide in the air.

Living and dead organisms contain organic carbon and serve as one of the reservoirs for the carbon cycle. The world's biotic components, particularly trees, contain billions of tons of organic carbon, and additional tons are estimated to be held in the remains of plants and animals in the soil. If dead plant and animal remains fail to decompose, they are subjected to extremely slow physical processes that transform them into coal, oil, and natural gas, the **fossil fuels.** Most of the fossil fuels were formed during the Carboniferous period, 286–360 MYA, when an exceptionally large amount of organic matter was buried before decomposing. Another reservoir is the calcium carbonate ($CaCO_3$) that accumulates in limestone and shells. Many marine organisms have calcium carbonate shells that remain in bottom sediments long after the organisms have died. Geological forces change these sediments into limestone.

Figure 31.25 Root nodules.

Bacteria that live in nodules on the roots of plants in the legume family, such as pea plants, convert nitrogen in the air to a form that plants can use to make proteins.

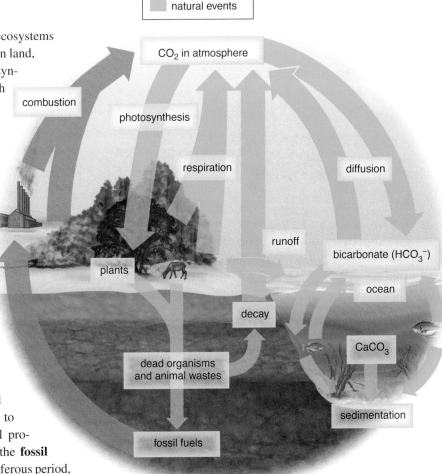

Figure 31.26 The carbon cycle.

The carbon cycle is a gaseous biogeochemical cycle. Producers take in carbon dioxide from the atmosphere and convert it to the organic molecules that feed all organisms. Fossil fuels arise when organisms die but do not decompose. The burning of fossil fuels releases carbon dioxide and causes environmental pollution.

Human Activities   The transfer rates of carbon dioxide due to photosynthesis and cellular respiration are just about even. However, more carbon dioxide is being deposited in the atmosphere than is being removed. This increase is largely due to the burning of fossil fuels and the destruction of forests to make way for farmland and pasture. Because of fossil fuel consumption and deforestation, CO_2 in the atmosphere is predicted to rise from 0.36% today to 0.75% by 2100. CO_2 concentrations in the atmosphere would rise higher except for the fact that the oceans take up CO_2, and so far they have been taking up more CO_2 than they vent.

Connections and Misconceptions

What are some other greenhouse gases?

In addition to carbon dioxide, these gases also play a role in the greenhouse effect:

- Methane (CH_4). A single molecule of methane has 21 times the warming potential of a molecule of carbon dioxide, making it a powerful greenhouse gas. Methane is a natural by-product of the decay of organic material, but it is also released by landfills and the production of coal, natural gas, and oil.

- Nitrous oxide (N_2O). Nitrous oxide is released from the combustion of fossil fuels and as gaseous waste from many industrial activities.

- Hydrofluorocarbons. These were initially produced to reduce the levels of ozone-depleting compounds in the upper atmosphere. Unfortunately, although present in very small quantities, they are potent greenhouse gases.

The increased amount of carbon dioxide (and other gases) in the atmosphere is predicted to cause a rise in temperature called **global warming.** These gases allow the sun's rays to pass through, but they absorb and reradiate heat back to the Earth, a phenomenon called the **greenhouse effect.** Scientists predict that, if the ice at the poles melts and sea levels rise as a result, many of the world's most populous cities will be flooded. Furthermore, weather pattern changes might cause the American Midwest to become a dust bowl.

 Video
Green Concrete

 Video
Global Warming

 Video
Karoo Global
Warming

Connecting the Concepts

For more information on the material presented in this section, refer to the following discussions.

Section 20.6 explores the role of mycorrhizas in providing plants with nitrogen.

Section .1 examines the importance of water as a natural resource.

Section 32.1 explores the relationship between fossil fuel use and global climate change.

Check Your Progress 31.2

1. Compare how autotrophs and heterotrophs acquire energy.
2. Describe the differences between energy flow and chemical cycling in an ecosystem.
3. Compare an ecological pyramid with a food chain.
4. Summarize the biogeochemical cycling of phosphorus, nitrogen, and carbon.
5. Explain how human activity is influencing the phosphorus and nitrogen cycles.
6. Summarize the relationship between carbon and global warming.

31.3 Ecology of Major Ecosystems

Learning Outcomes

Upon completion of this section, you should be able to

1. List the two major types of ecosystems that make up the biosphere.
2. List the major types of terrestrial ecosystems.
3. Explain primary productivity and relate this to the different types of aquatic and terrestrial ecosystems.

The **biosphere,** which encompasses all the ecosystems on Earth, is the final level of biological organization. **Aquatic ecosystems** are divided into those composed of fresh water and those composed of salt water (marine ecosystems; **Fig. 31.27**). The ocean, a marine ecosystem, covers 70% of the Earth's surface. Two types of freshwater ecosystems are those with standing water, such as lakes and ponds, and those with running water, such as rivers and streams. The richest marine ecosystems lie near the coasts. Coral reefs are located offshore, while marshes occur where rivers meet the sea.

Figure 31.27 The major aquatic ecosystems.

Aquatic ecosystems are divided into those that have salt water, such as (**a**) the ocean, and (**b**) those that have fresh water, such as a river. Saltwater, or marine, ecosystems also include (**c**) coral reefs and (**d**) marshes.

Figure 31.28 The major terrestrial ecosystems.

The tundra is the northernmost terrestrial ecosystem and has the lowest average temperature of all the terrestrial ecosystems, with minimal to moderate rainfall. The taiga, a coniferous forest that encircles the globe, also has a low average temperature, but moderate rainfall. Temperate forests have moderate temperatures and occur where rainfall is moderate yet sufficient to support trees. A tropical grassland (savanna) has high temperatures and moderate/seasonal rainfall. A temperate grassland (prairie) has low to high temperatures with low annual rainfall. Deserts have extreme and changeable temperatures with minimal rainfall. Tropical rain forests, which generally occur near the equator, have a high average temperature and the greatest amount of rainfall of all the terrestrial ecosystems.

Scientists recognize several distinctive major types of **terrestrial ecosystems,** also called *biomes* (**Fig. 31.28**). Temperature and rainfall define the biomes, which contain communities adapted to the regional climate. The tropical rain forests, which occur at the equator, have a high average temperature and the greatest amount of rainfall of all the biomes. They are dominated by large, evergreen, broad-leaved trees. The savanna is a tropical grassland with a high temperature and alternating wet and dry seasons. Temperate grasslands receive less rainfall than temperate forests (in which trees lose their leaves during the winter) and more water than deserts, which lack trees. The taiga is a very cold northern coniferous forest, and the tundra, which borders the North Pole, is also very cold, with long winters and a short growing season. In the tundra, a permafrost persists, even during the summer, and prevents large plants from becoming established.

 Animation Biomes

 Video Tallgrass Prairie Ecology

Primary Productivity

One way to compare ecosystems is based on **primary productivity,** the rate at which producers capture and store energy as organic nutrients over a certain length of time. Temperature and moisture, and secondarily the nature of the soil, influence the primary productivity and, as already discussed, the assemblage of species in an ecosystem. In terrestrial ecosystems, primary productivity is generally lowest in high-latitude tundras and deserts, and it is highest at the equator where tropical forests occur (**Fig. 31.29**). The high productivity of tropical rain forests provides varied niches and much food for consumers. The number and diversity of species in tropical rain forests are the highest of all the terrestrial ecosystems. Therefore, conservation biologists are interested in preserving as much of this biome as possible.

The primary productivity of aquatic communities is largely dependent on the availability of inorganic nutrients. Estuaries, swamps, and marshes are rich in organic nutrients and in decomposers that convert those organic

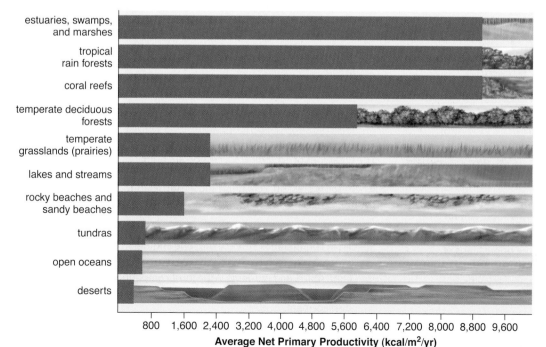

estuaries, swamps, and marshes
tropical rain forests
coral reefs
temperate deciduous forests
temperate grasslands (prairies)
lakes and streams
rocky beaches and sandy beaches
tundras
open oceans
deserts

800 1,600 2,400 3,200 4,000 4,800 5,600 6,400 7,200 8,000 8,800 9,600
Average Net Primary Productivity (kcal/m²/yr)

Figure 31.29 Primary productivity.

Ecologists can compare ecosystems based on primary productivity, the rate at which producers convert and store solar energy as chemical energy.

Connecting the Concepts

For more information on the topics in this section, refer to the following discussions.

Section 32.1 examines the importance of water as a natural resource.

Section 32.2 explores the direct and indirect benefits of the biodiversity found in aquatic and terrestrial ecosystems.

Check Your Progress 31.3

1. Describe the two major types of ecosystems of the biosphere.
2. List the terrestrial ecosystems of the world.
3. Explain why swamps have higher levels of primary productivity than open oceans.

---

nutrients into their inorganic chemical components. Estuaries, swamps, and marshes also contain a large number of varied species, particularly in the early stages of their development before they venture forth into the sea. Therefore, all of these coastal regions are in great need of preservation. The open ocean has a productivity between that of a desert and that of the tundra because it lacks a concentrated supply of inorganic nutrients. Coral reefs exist near the coasts in warm tropical waters where currents and waves bring nutrients and sunlight penetrates to the ocean floor. Coral reefs are areas of remarkable biological abundance, equivalent to that of tropical rain forests.

**Video
Coral Reef
Ecosystems**

Connections and Misconceptions

What is an estuary?

An estuary is a region where fresh water, usually from a river, meets the ocean. The water in an estuary zone is often called "brackish" because it is a mixture of salt and fresh water. Estuaries are considered to be some of the most productive ecosystems in the world. Not only do they provide food sources, such as shrimp, clams and other seafood, but they also play an important role in the chemical cycling of nitrogen and phosphorus. Estuaries are fragile ecosystems; pollution, oil spills (such as the one that threatened Louisiana's Bartaria Bay in 2010), and human activity can easily reduce the productivity of an estuary.

Media Study Tools

www.mhhe.com/maderessentials3

Enhance your study of this chapter with study tools and practice tests. Also ask your instructor about the resources available through ConnectPlus, including the media-rich eBook, interactive learning tools, and animations.

Model Ecosystems

The virtual lab "Model Ecosystems" provides an interactive look at some of the factors that influence the structure of an ecosystem.

The Chapter in Review

Summary

31.1 Ecology of Communities

Knowledge of community and ecosystem ecology is important for understanding the impacts of human alterations to the environment.

Community A community is an assemblage of the populations of different species interacting with each other in a given area.

Ecosystem An ecosystem consists of species interacting with one another and with the physical environment.

Ecological Succession

bare rock → lichens/mosses → grasses → shrubs → trees

The two types of ecological succession are primary succession (begins on bare rock) and secondary succession (following a disturbance; begins where soil is present). Ecological succession leads to a stable climax community.

Interactions in Communities

Species in communities interact with one another in the following ways:

- **Competition** Species vie with one another for resources, such as light, space, and nutrients. Aspects of competition are the competitive exclusion principle, resource partitioning, and character displacement.
- **Predation** One species (predator) eats another species (prey).

- **Parasitism** One species (parasite) obtains nutrients from another species (host) but does not kill the host species.
- **Commensalism** One species benefits from the relationship, while the other species is not harmed.
- **Mutualism** Two species interact in a way that benefits both.

Ecological Niche The ecological niche of a species is defined by the role it plays in its community, the habitat, and its interactions with other species.

Keystone Species The interactions of a keystone species in the community hold the community and its species together. Removal of a keystone species can lead to species extinctions and loss of diversity. An example of a keystone species is the grizzly bear.

Native Versus Exotic Species Native species are indigenous to a given area and thrive without assistance. Exotic species are introduced into an area and may disrupt the balance and interactions among native species in that area's community.

31.2 Ecology of Ecosystems

In the food chain of an ecosystem, some populations are autotrophs and some are heterotrophs.

Autotrophs

The autotrophs are the producers. They require only inorganic nutrients (e.g., CO_2 and minerals) and an outside energy source to produce organic nutrients for their own use and for the use of other members of the community. Examples of autotrophs are algae, cyanobacteria, and plants.

Heterotrophs

The heterotrophs are the consumers. They require a preformed source of organic nutrients and give off CO_2. Examples of heterotrophs are herbivores (feed on plants), carnivores (feed on animals), and omnivores (feed on both plants and animals). Other heterotrophs are the decomposers (the bacteria and fungi that aid in decay).

Energy Flow and Chemical Cycling

Energy flows through an ecosystem, while chemicals cycle within an ecosystem.

Food Webs and Food Chains Energy flows in an ecosystem through food chains and detrital and grazing food webs.

Trophic Level A trophic level is a level of nourishment in a food web or chain.

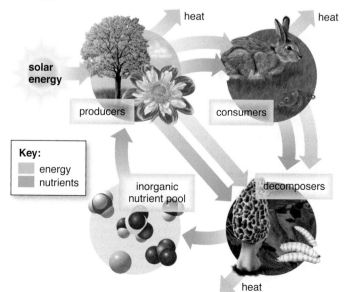

Ecological Pyramid An ecological pyramid illustrates the energy losses that occur between trophic levels.
- Only about 10% of the energy of one trophic level is available to the next trophic level.
- Top carnivores occupy the last and smallest trophic level.

Biogeochemical Cycle Chemicals cycle within an ecosystem through various biogeochemical cycles, such as the phosphorus cycle, the nitrogen cycle, and the carbon cycle. Human activities significantly alter the transfer rates in these cycles.

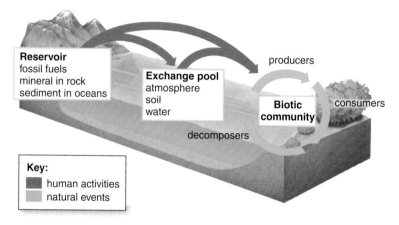

31.3 Ecology of Major Ecosystems

The biosphere encompasses all the major ecosystems of the Earth.

Aquatic Ecosystems The aquatic ecosystems are classified as freshwater ecosystems (rivers, streams, lakes, ponds) and marine ecosystems (oceans, coral reefs, saltwater marshes).

Terrestrial Ecosystems The terrestrial ecosystems are called biomes. The major biomes are
- Tundra
- Taiga
- Temperate forest
- Tropical grassland (savanna)
- Temperate grassland (prairie)
- Desert
- Tropical rain forest

Primary Productivity

Primary productivity is the rate at which producers capture solar energy and convert it to chemical energy over a specified length of time. The number of species in an ecosystem is positively related to its primary productivity.

▬▬ Key Terms ▬▬▬▬▬▬▬▬▬

aquatic ecosystem   623
autotroph   614
biogeochemical cycle   618
biomass   618
biosphere   623
carnivore   614
character displacement   610
climax community   607
coevolution   605
commensalism   609
community   605
competition   609

competitive exclusion
   principle   610
consumer   614
decomposer   615
detrital food web   616
detritus   615
diversity   606
ecological niche   609
ecological pyramid   618
ecological succession   607
ecosystem   606
eutrophication   620

continued on next page

Testing Yourself

Choose the best answer for each question.

1. As diversity increases,
   a. species richness increases and the distribution of species becomes more even.
   b. species richness decreases and the distribution of species becomes more even.
   c. species richness increases and the distribution of species becomes less even.
   d. species richness decreases and the distribution of species becomes less even.

2. Which is not an adaptive trait of an opportunistic pioneer species?
   a. long life span   c. small size
   b. short time to maturity   d. high reproductive output

For statements 3–7, indicate the type of interaction described in each scenario.

Key:
   a. competition
   b. predation
   c. parasitism
   d. commensalism
   e. mutualism

3. An alfalfa plant gains fixed nitrogen from the bacterial species *Rhizobium* in its root system, while *Rhizobium* gains carbohydrates from the plant.

4. Both foxes and coyotes in an area feed primarily on a limited supply of rabbits.

5. Roundworms live and reproduce within a cat's digestive tract.

6. A fungus captures nematodes as a food source.

7. An orchid plant lives in the treetops, gaining access to sun and pollinators but not harming the trees.

8. The abundance of both species is expected to increase as a result of which type of interaction?
   a. predation   d. competition
   b. commensalism   e. parasitism
   c. mutualism

9. According to the competitive exclusion principle,
   a. one species is always more competitive than another for a particular food source.
   b. competition excludes multiple species from using the same food source.
   c. no two species can occupy the same niche at the same time.
   d. competition limits the reproductive capacity of species.

10. In the following diagram, fill in the components of chemical cycling and nutrient flow.

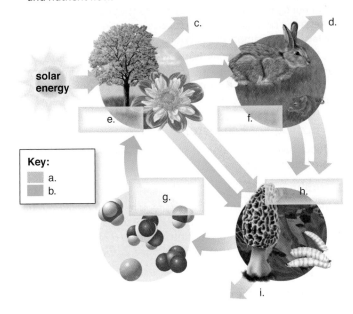

11. An ecological pyramid depicts the amount of _____ in various trophic levels.
    a. food   d. nutrients
    b. organisms   e. Both b and c are correct.
    c. energy

12. Which of the following would be a primary consumer in a vegetable garden?
    a. aphid sucking sap from cucumber leaves
    b. lady beetle eating aphids
    c. songbird eating lady beetles
    d. fox eating songbirds
    e. All of these are correct.

13. Detritus always contains
    a. bacteria and fungi.   c. decaying logs.
    b. leaf litter.   d. animal carcasses.

14. The first trophic level in a food web is occupied by the
    a. producers.   c. secondary consumers.
    b. primary consumers.   d. tertiary consumers.

15. Which of the following represents a grazing food chain?
    a. leaves → detritus feeders → deer → owls
    b. birds → mice → snakes
    c. nuts → leaf-eating insects → chipmunks → hawks
    d. leaves → leaf-eating insects → mice → snakes

16. In a grazing food web, carnivores that eat herbivores are
    a. producers.
    b. primary consumers.
    c. secondary consumers.
    d. tertiary consumers.

17. Identify the components of the ecological pyramid in the following diagram.

a. _____

b. _____

c. _____

d. _____

18. Which of the following is a sedimentary biogeochemical cycle?
    a. carbon
    b. nitrogen
    c. phosphorus

19. Underground oil is an example of a carbon
    a. cycle.
    b. pathway.
    c. reservoir.
    d. exchange pool.

For questions 20–22, match the description to the process in the key.

Key:
        a. nitrogen fixation
        b. nitrification
        c. denitrification

20. nitrate to nitrogen gas
21. ammonium to nitrate
22. nitrogen gas to ammonium

23. Which of the following is not a component of the nitrogen cycle?
    a. proteins
    b. ammonium
    c. decomposers
    d. photosynthesis
    e. bacteria in root nodules

24. Which biome is characterized by a coniferous forest with low average temperature and moderate rainfall?
    a. taiga
    b. savanna
    c. tundra
    d. tropical rain forest
    e. temperate forest

25. Which biome has the lowest primary productivity?
    a. tundra
    b. lake
    c. sandy beach
    d. prairie
    e. temperate forest

26. Ecosystems include which of the following components that communities do not?
    a. energy transfer between members
    b. interaction with physical environment
    c. intricate food webs
    d. herbivores and carnivores

27. Character displacement evolves in response to
    a. predation.
    b. competiton among species for a single niche.
    c. loss of keystone species.
    d. primary succession.

Thinking Scientifically

1. One of the most striking examples of coevolution is between insects and flowers. The earliest angiosperms produced wind-pollinated flowers, which released large quantities of pollen. The ovules were partially exposed and exuded tiny droplets of sugary sap to catch passing pollen. Outline a course of events that could have resulted in the coevolution we observe today between a flower and its pollinator.

2. Over 200 wildlife species have been associated with prairie dog colonies in the Great Plains of the United States. The prairie dog, which burrows, forages, and feeds in the area, acts as a keystone species. If the prairie dogs are destroyed, many other species will suffer as well. How do you think the activities of the prairie dogs influence the survival of so many other species?

Bioethical Issue

Protection of Native Species

Many exotic species, such as zebra mussels and sea lampreys, are so obviously troublesome that most people do not object to programs aimed at controlling their populations. However, some exotic species' eradication programs meet with more resistance. For example, the mute swan, one of the world's largest flying birds, is beautiful and graceful, and it has an impressive presence. However, it is very aggressive and territorial. The mute swan was introduced to the United States from Asia and Europe in the nineteenth century as an ornamental bird but has since established a large wild population. The birds consume large amounts of aquatic vegetation and displace native birds from feeding and nesting areas. The U.S. Fish and Wildlife Service explains that it will be necessary to kill 3,000 mute swans in Maryland in the next two years in order to protect native bird populations. Attempts to limit the size of the mute swan populations in Maryland and other states have been met with opposition by citizens who find the birds beautiful.

Do you feel that native populations need not be protected as long as the exotic species serves a suitable human purpose? Or do you feel that native species should be protected regardless?

32

Human Impact on the Biosphere

OUTLINE

BEFORE YOU BEGIN

Before beginning this chapter, take a few moments to review the following discussions.

Section 11.3 What is a transgenic organism?

Figure 30.7 How do the environmental impacts of a more-developed country (MDC) compare with those of a less-developed country (LDC)?

Figure 31.20 How much energy is transferred up each level of an ecological pyramid?

The Value of Biodiversity

The demands made on the Earth's natural resources by the growth of the human population have created a biodiversity crisis through habitat destruction, the introduction of exotic species, pollution, overexploitation, and the spread of disease.

Biodiversity has both direct and indirect value to humans. Billions of dollars worth of medicines have come from plant, animal, and fungal products. Plants and animals also provide valuable resources, such as timber and food, and they control pests that threaten our crops. Indirect values of biodiversity, often referred to as "ecosystem services," are estimated to be more valuable than the total gross national product of the entire world! For example, preserving biogeochemical cycles, such as the water cycle, provides clean drinking water, and conserving topsoil provides billions of dollars worth of indirect value for crop and erosion control.

Conservation biology is directed at protecting biodiversity and natural resources for the good of all living things, including humans. In this chapter, you will learn not only how humans negatively impact ecosystems but how this is resulting in a tremendous loss of biodiversity across the planet. We will also explore some of the potential solutions that will help us move toward a sustainable society.

32.1 Resources and Pollution

Human beings have certain basic needs, and they use resources to meet these needs. Land, water, food, energy, and minerals are the maximally used resources (**Fig. 32.1**).

Some resources are nonrenewable, and some are renewable. **Nonrenewable resources** are limited in supply. For example, the amount of land, fossil fuels, and minerals is finite and can be exhausted. Better extraction methods and efficient use can make the supply last longer, but eventually these resources will run out. **Renewable resources** are not limited in supply. We can use water and certain forms of energy (e.g., solar energy) or harvest plants and animals for food, and more will always be forthcoming. However, even though these resources are renewable, we must be careful not to squander them. Consider, for example, that most species have a population threshold below which they cannot recover or sustain their numbers, as when the huge herds of buffalo that once roamed the western United States disappeared after being overexploited.

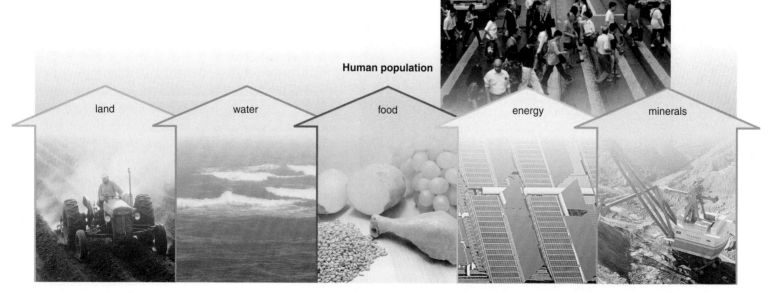

Figure 32.1 Resources.

Humans use land, water, food, energy, and minerals to meet their needs, including a place to live, food to eat, and products that make their lives easier.

Unfortunately, a side effect of resource consumption can be pollution. **Pollution** is any alteration of the environment in an undesirable way. Pollution is often caused by human activities. The human impact on the environment is proportional to the size of the population. As the population grows, so does the need for resources and the amount of pollution generated by using these resources. Consider that six people adding waste to the ocean might not be alarming, but 6 billion people doing so would certainly affect its cleanliness. In modern times, the consumption of mineral and energy resources has grown faster than population size, most likely because people in the less-developed countries (LDCs) have increased their use of them.

 **Video** Ocean Garbage

 Video Inherited Pollution

Land

People require a physical place to live. Worldwide, there are currently more than 32 people for each km² (83 people per mi²) of all available land, including Antarctica, mountain ranges, jungles, and deserts. Naturally, land is also needed for a variety of uses aside from homes, such as agriculture, electric power plants, manufacturing plants, highways, hospitals, and schools.

Beaches and Human Habitation

At least 40% of the world population lives within 100 km (60 mi) of a coastline, and this number is expected to increase. In the United States, over half of the population lives within 80 km (50 mi) of the coasts (including the Great Lakes). Living right on the coast is an unfortunate choice because it accelerates natural beach erosion and loss of habitat for marine organisms.

Beach Erosion An estimated 70% of the world's beaches are eroding; **Figure 32.2** shows how extensive this issue is in the United States. The seas have been rising for the past 12,000 years, ever since the climate turned warmer after the last Ice Age, and consequently the forces associated with waves, tides, and wind have been amplified.

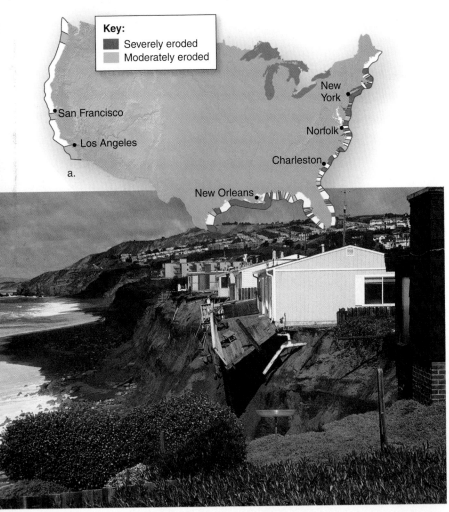

a.

b.

Figure 32.2 Beach erosion.

a. Most of the U.S. coastline is subject to beach erosion. **b.** Therefore, people who choose to live near the coast may eventually lose their homes.

Video Prehistoric Hurricanes

Humans often participate in activities that divert more water to the oceans, contributing to rising seas and beach erosion. For example, humans have filled in coastal wetlands, such as mangrove swamps in the southern United States and saltwater marshes in the northern United States formerly considered "wastelands." With growing recognition of the services provided by wetlands, wetland conservation has been growing in the United States during the past 40 years, but it is just starting in South America, where a project to straighten the Paraná River will drain the world's largest wetland. Reasons to protect coastal wetlands are that they serve as spawning areas for fish and other forms of marine life as well as buffer against hurricane storm surges and protect against shoreline erosion. They are also habitats for certain terrestrial species, including many types of birds.

Humans often try to stabilize beaches by building groynes (structures that extend from the beach into the water) and seawalls. Groynes trap sand and build beaches on one side as well as slowing the longshore currents, but erosion is increased on the other side. Seawalls, in the end, also increase erosion because ocean waves remove sand from in front of and to the side of the seawalls. Importing sand is a better solution, but it is very costly and can disturb plant and animal populations. It's estimated that the U.S. shoreline loses 40% more sediment than it receives, because the building of dams prevents sediment from reaching the coast.

Coastal Pollution   The coasts are particularly subject to pollution because toxic substances placed in freshwater lakes, rivers, and streams may eventually find their way to the coast. Oil spills at sea cause localized harmful effects also.

Semiarid Lands and Human Habitation

Forty percent of the Earth's lands are already deserts, and any land adjacent to a desert is in danger of becoming desert if humans manage it improperly. **Desertification** is the conversion of semiarid land to desert-like conditions (**Fig. 32.3**).

Quite often, desertification is thought to begin when humans allow animals to overgraze the land. The soil can no longer hold rainwater, which runs off instead of keeping the remaining plants alive or recharging wells. Humans then remove whatever vegetation they can find to use as fuel or fodder for their animals. The end result is a desert unable to support agriculture, which is then abandoned as people move on to continue the process someplace else. Some estimate that nearly three-quarters of all rangelands worldwide are in danger of desertification. The famine in Ethiopia during 2002–2003 was due, at least in part, to degradation of the land to the point that it could no longer support human beings and their livestock.

Tropical Rain Forest and Human Habitation

Deforestation, the removal of trees, has long allowed humans to live in areas where forests once covered the land. The concern recently has been that people are settling in tropical rain forests, such as the Amazon, following the building of roads (**Fig. 32.4**). This land, too, is subject to desertification. Soil in the tropics is often thin and nutrient-poor because all the nutrients are tied up in the trees and other vegetation. When the trees are felled and removed and the land is used for agriculture or grazing, it quickly loses its fertility and becomes subject to desertification.

Other consequences of deforestation are a loss of biodiversity and an increase in atmospheric carbon dioxide because trees that once took in carbon dioxide for photosynthesis have been removed. The destruction of tropical rain forests kills many unique plants and stresses the animals that live there because their supply of food is reduced and they no longer have homes to live in.

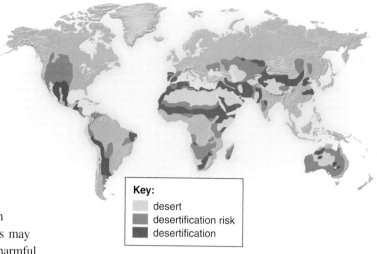

Key:
- desert
- desertification risk
- desertification

Figure 32.3 **Desertification.**

Desertification is a worldwide occurrence that reduces the amount of land suitable for human habitation.

Source: Data from A. Goudie and I. Wilkinson, *The Warm Desert Environment.* Copyright 1977 by Cambridge University Press, New York.

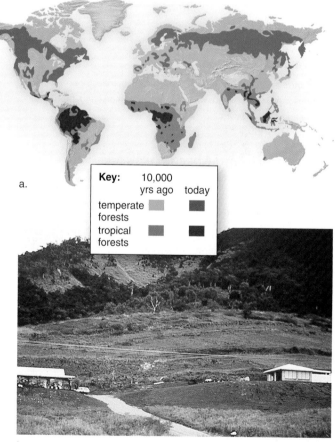

Key: 10,000 yrs ago today

temperate forests

tropical forests

a.

b.

Figure 32.4 **Deforestation.**

a. Nearly half of the world's forest lands have been cleared for farming, logging, and urbanization. **b.** The soil of tropical rain forests is not suitable for long-term farming.

Sources: United Nations Environment Program; World Resources Institute.

a. Agriculture uses most of the fresh water consumed.

b. Industrial use of water is about half that of agricultural use.

c. Domestic use of water is about half that of industrial use.

Figure 32.5 Global water use.

a. Agriculture uses water primarily for irrigation. **b.** Industry uses water in various ways. **c.** People use water for drinking, showering, flushing toilets, and watering lawns.

Connections and Misconceptions

How much water is required to produce your food?

Growing a single serving of lettuce takes about 6 gallons of water. Producing an 8-oz glass of milk requires 49 gallons. That includes the amount of water the cow drinks, the water used to grow the cow's food, and the water needed to process the milk. Producing a single serving of steak consumes more than 2,600 gallons of water.

Water

Access to clean drinking water is considered a human right, but actually most fresh water is used by agriculture and industry (**Fig. 32.5**). Worldwide, 70% of all fresh water is used to irrigate crops! Much of a recent surge in demand for water stems from increased industrial activity and irrigation-intensive agriculture, the type of agriculture that now supplies about 40% of the world's food crops. Domestically, in the more-developed countries (MDCs), more water is usually used for bathing, flushing toilets, and watering lawns than for drinking and cooking. In the water-poor areas of the world, people may not have ready access to drinking water, and if they do, the water may be unclean.

Increasing Water Supplies

Although the needs of the human population overall do not exceed the renewable supply, this is not the case in certain regions of the world. As illustrated in Figure 32.3, about 40% of the world's land is desert, and deserts are bordered by semiarid land. When necessary to support human population growth, the supply of fresh water is increased by damming rivers and withdrawing water from aquifers.

Dams The world's 45,000 large dams catch 14% of all precipitation runoff, provide water for up to 40% of irrigated land, and give some 65 countries more than half their electricity. Damming of certain rivers has been so extensive that they no longer flow as they once did. The Yellow River in China fails to reach the sea most years; the Colorado River barely makes it to the Gulf of California; and even the Rio Grande dries up before it can merge with the Gulf of Mexico. The first and third longest rivers on Earth, the Nile in Egypt and the Ganges in India, respectively, are also so overexploited that, at some times of the year, they hardly make it to the ocean.

Dams have additional drawbacks: (1) Reservoirs behind the dam lose water due to evaporation and seepage into underlying rock beds. The amount of water lost sometimes equals the amount made available! (2) The salt left behind by evaporation and agricultural runoff can make a river's water unusable farther downstream. (3) Sediment buildup causes dams to hold back less water; with time, dams may become useless for storing water. (4) The reduced amount of water below the dam has a negative impact on the native wildlife.

Aquifers To meet their freshwater needs, people are pumping vast amounts of water from **aquifers,** which are natural reservoirs found just below or as much as 1 km below the surface. Aquifers hold about 1,000 times the amount of water that falls on land as precipitation each year. This water accumulates from rain that fell in far-off regions as many as hundreds of thousands of years ago. In the past 50 years, groundwater depletion has become a problem in many areas of the world. In substantial portions of the High Plains Aquifer, which stretches from South Dakota to the Texas Panhandle, much water is pumped out for irrigation or for household water use. In the 1950s, India had 100,000 motorized pumps in operation; today, India has 20 million pumps, a huge increase in groundwater pumping.

Environmental Consequences Removal of water is causing land **subsidence,** settling of the soil as it dries out. In California's San Joaquin Valley, an area of more than 13,000 square km has subsided, and in the worst spot, the surface has dropped more than 9 meters (m)! In some parts of Gujarat, India, the water table has dropped as much as 7 m. Subsidence damages canals, buildings, and underground pipes. Withdrawal of groundwater can cause sinkholes, as when an underground cavern collapses because water no longer holds up its roof.

Saltwater intrusion is another consequence of groundwater depletion. The flow of water from streams and aquifers usually keeps them fairly free of seawater. But as water is withdrawn, the water table can lower to the point that seawater backs up into streams and aquifers. Saltwater intrusion reduces the supply of fresh water along the coast.

Conservation of Water

By 2025, two-thirds of the world's population may be living in countries that are facing serious water shortages. Some solutions for expanding water supplies have been suggested. Using drip irrigation delivers more water to crops and saves about 50% over traditional methods while increasing crop yields (**Fig. 32.6**). Although the first drip systems were developed in 1960, they are used on less than 1% of irrigated land. Most governments subsidize irrigation so heavily that farmers have little incentive to invest in drip systems or other water-saving methods. Planting drought- and salt-tolerant crops would decrease the water required for agriculture. Recycling water and adopting conservation measures could help the world's industries cut their water demands by more than half.

Video
Thames River

Food

In 1950, the human population numbered 2.5 billion, and there was only enough food to provide less than 2,000 calories per person per day; now, with over 6.7 billion people on Earth, the world food supply provides more than 2,500 calories per person per day. Generally speaking, food comes from three activities: growing crops, raising animals, and fishing the seas. Unfortunately, modern farming methods, which have increased the food supply, include some harmful practices:

Animation
World Hunger

1. *Planting of a few genetic varieties.* The majority of farmers practice monoculture, meaning that they plant a single type (strain) of a crop throughout their fields. Unfortunately, the resulting lack of genetic diversity means that a single type of parasite or disease would infect the entire monoculture. This scenario does not take into account any genetic variants that may exhibit disease resistance.

2. *Heavy use of fertilizers, pesticides, and herbicides.* Fertilizer production is energy-intensive, and fertilizer runoff contributes to water pollution. Pesticides reduce soil fertility because they kill beneficial soil organisms as well as pests, and some pesticides and herbicides are linked to the development of cancer.

3. *Generous irrigation.* As already discussed, water is sometimes taken from aquifers whose water content may in the future become so reduced that it could be too expensive to pump out any more.

4. *Excessive fuel consumption.* Irrigation pumps remove water from aquifers, and large farming machines are used to spread fertilizers, pesticides, and herbicides, as well as to sow and harvest the crops. In effect, modern farming methods transform fossil fuel energy into food energy.

a. Drought-resistant plants

tubing

b. Drip irrigation

Figure 32.6 **Conservation measures to save water.**

a. Planting drought-resistant plants in parks and gardens and drought-resistant crops in the field cuts down on the need to irrigate. **b.** Drip irrigation delivers water directly to the roots.

a. Polyculture

b. Contour with no-till farming

Figure 32.7

Conservation methods.

a. Polyculture reduces the ability of one parasite to wipe out an entire crop and reduces the need to use an herbicide to kill weeds. This farmer has planted alfalfa between strips of corn, which also replenishes the nitrogen content of the soil (instead of adding fertilizers). **b.** Contour farming with no-till conserves topsoil because water has less tendency to run off. **c.** Instead of pesticides, it is sometimes possible to use a natural predator. Here ladybugs are eating cottony-cushions scale insects on citrus trees.

c. Biological pest control

Figure 32.7 shows ways to minimize the harmful effects of modern farming practices.

Soil Loss and Degradation

Land suitable for farming and for grazing animals is being degraded worldwide. Topsoil, the topmost portion of the soil, is the richest in organic matter and the most capable of supporting grass and crops. When bare soil is acted on by water and wind, soil erosion occurs and topsoil is lost. As a result, marginal rangeland becomes desert, and farmland loses its productivity.

The custom of planting the same crop in straight rows that facilitate the use of large farming machines has caused the United States and Canada to have one of the highest rates of soil erosion in the world. Conserving the nutrients now being lost could save farmers $20 billion annually in fertilizer costs. Much of the eroded sediment ends up in lakes and streams, where it reduces the ability of aquatic species to survive.

Almost all water contains dissolved salts, and these salts are left behind either when plants take up the water or when a dry climate causes the water to evaporate. Between 25% and 35% of the irrigated western croplands are thought to have undergone **salinization,** an accumulation of mineral salts generated through irrigation. Salinization makes the land unsuitable for growing crops.

Green Revolutions

About 50 years ago, research scientists began to breed tropical wheat and rice varieties specifically for farmers in the LDCs. The dramatic increase in yield due to the introduction of these new varieties around the world was called "the green revolution." These plants helped the world food supply keep pace with the rapid increase in world population. Most green revolution plants are called "high responders" because they need high levels of fertilizer, water, and pesticides in order to produce a high yield. In other words, they require the same subsidies and create the same ecological problems as do modern farming methods.

Genetic Engineering Genetic engineering can produce transgenic plants with new and different traits, among them resistance to both insects and herbicides. When herbicide-resistant crops are planted, weeds are easily controlled, less tillage is needed, and soil erosion is minimized. Researchers also want to produce crops that tolerate salt, drought, and cold. In addition, some progress has been made in increasing the food quality of crops, so that they will supply more of the proteins, vitamins, and minerals people need. Genetically engineered crops are resulting in another green revolution. Nevertheless, some citizens are opposed to the use of genetically engineered crops, fearing that they will damage the environment and lead to health problems in humans.

 Video **Pesticide Plants**

 Video **GM Food Safety**

Domestic Livestock

A low-protein, high-carbohydrate diet consisting only of grains, such as wheat, rice, or corn, can lead to malnutrition. In the LDCs, kwashiorkor, a condition caused by a severe protein deficiency, is seen in infants and children ages 1–3, usually after a new baby arrives in the family and the older children are no longer breast-fed. Such children are lethargic and irritable, and they have bloated abdomens. Mental retardation is expected.

Figure 32.8 Crowding of livestock: hogs milling in a feedlot pen.

b. Drag net fishing

In the MDCs, many people tend to have more than enough protein in their diet. Almost two-thirds of U.S. cropland is devoted to producing livestock feed. This means that a large percentage of the fossil fuel, fertilizer, water, herbicides, and pesticides we use are actually for the purpose of raising livestock. Typically, cattle are range-fed for about four months, and then they are taken to crowded feedlots, where they receive growth hormone and antibiotics while feeding on grain. Most pigs and chickens spend their entire lives in crowded pens and cages (**Fig. 32.8**).

Just as livestock eat a large proportion of the crops in the United States, raising livestock accounts for much of the pollution associated with farming. Consider also that presently fossil fuel energy is needed not just to produce herbicides and pesticides and to grow food but also to make the food available to the livestock. Raising livestock is extremely energy-intensive in the MDCs. In addition, water is used to wash livestock wastes into nearby bodies of water, where they add significantly to water pollution. Whereas human wastes are sent to sewage treatment plants, raw animal wastes are not.

For these reasons, it is prudent to recall the ecological energy pyramid (see Fig. 31.20), which shows that, as you move up the food chain, energy is lost. As a rule of thumb, for every 10 calories of energy from a plant, only 1 calorie is available for the production of animal tissue in an herbivore. In other words, it is possible to feed 10 times as many people on grain as on meat.

a. World fish catch

Fishing

Since 2000, the world fish catch has been on the decline (**Fig. 32.9***a*). Worldwide, between 1970 and 1990, the number of large boats devoted to fishing doubled to 1.2 million. The U.S. fishing fleet participated in this growth due to the availability of federal loans for building fishing boats. The new boats have sonar and depth recorders, and their computers remember the sites of previous catches, so that the boats can go there again. Helicopters, planes, and even satellite data are used to help find fish. The number of North Atlantic swordfish caught in the United States declined 70% from 1980 to 1990, and the average weight fell from 115 to 60 pounds. The Atlantic bluefin tuna is so overfished that it may never recover and instead become extinct.

Modern fishing practices negatively impact biodiversity because a large number of marine animals are caught by chance in the huge nets some

c. Aquafarming

Figure 32.9 Fishing.

a. The world fish catch has declined in recent years (**b**) because modern fishing methods are overexploiting fisheries. **c.** Aquaculture, the farming of aquatic organisms, is decreasing the demand from the oceans.

Source: Data from U.S. Marine Fisheries Service.

fishing boats use (Fig. 32.9b). These animals are discarded. The world's shrimp fishery has an annual catch of 1.8 million tons, but the other animals caught and discarded in the process amount to 9.5 million tons. Raising tuna, shrimp, and other aquatic organisms in controlled settings, called aquafarming, has reduced the fishing pressures on wild populations (Fig. 32.9c). In fact, over 90% of the shrimp available in the United States is produced through aquaculture.

Video Ocean Fishing Ban

Video Big Fish, Small Fish

Energy

About 6% of the world's energy supply comes from nuclear power, and 75% comes from fossil fuels; both of these are finite, nonrenewable sources. Although it was once predicted that the nuclear power industry would fulfill a significant portion of the world's energy needs, this has not happened for two reasons: (1) People are very concerned about nuclear power dangers, such as the meltdown that occurred in 1986 at the Chernobyl nuclear power plant in Russia. (2) Radioactive wastes from nuclear power plants remain a threat to the environment for thousands of years, and we still have not determined the best way to store them safely.

As you learned in Chapter 31, oil, natural gas, and coal are **fossil fuels**—the compressed remains of organisms that died many thousands of years ago. The MDCs presently consume more than twice as much fossil fuel as the LDCs, yet there are many more people in the LDCs than in the MDCs. It has been estimated that each person in the MDCs uses approximately as much energy in one day as a person in an LDC does in one year.

Among the fossil fuels, oil burns more cleanly than coal, which may contain a considerable amount of sulfur. So despite the fact that the United States has a good supply of coal, imported oil is our preferred fossil fuel. Even so, the burning of any fossil fuel contributes to environmental problems because, as it burns, pollutants are emitted into the air.

Fossil Fuels and Global Climate Change

The percentage of carbon dioxide in the atmosphere is 0.03%. However, the level of CO_2 in the atmosphere in 1850 was about 280 parts per million (ppm), and today it is about 350 ppm. This increase is largely due to the burning of fossil fuels and the burning and clearing of forests to make way for farmland and pasture. Human activities are causing the emission of other gases as well. For example, the amount of methane given off by oil and gas wells, rice paddies, and all sorts of organisms, including domesticated cows, is increasing by about 1% a year. These gases are known as **greenhouse gases** because, just like the panes of a greenhouse, they allow solar radiation to pass through but hinder the escape of infrared heat back into space.

Today, data collected around the world show a steady rise in the concentration of the various greenhouse gases. These data are used to generate computer models, which predict that the Earth may warm to temperatures never before experienced by living things. The global climate has already warmed about 0.6°C since the Industrial Revolution. Computer models are unable to consider all possible variables, but the Earth's temperature may rise 1.5°–4.5°C by 2060 if greenhouse emissions continue at the current rates (**Fig. 32.10**).

Some models predict that, as the oceans warm due to global warming, temperatures in the polar regions will rise to a greater degree than in other regions. If so, glaciers will melt and sea levels will rise, not only due to this melting but also because water expands as it warms. Water evaporation will increase, and most likely

a.

b.

Figure 32.10 Global warming.

a. Mean global temperature is expected to rise due to the introduction of greenhouse gases into the atmosphere. Global warming has the potential to significantly affect the world's biodiversity. **b.** A temperature rise of only a few degrees causes coral reefs to "bleach" and become lifeless. As the oceans warm and land recedes, coral reefs can move northward.

precipitation will increase along the coasts, while conditions inland become drier. The occurrence of droughts will reduce agricultural yields and will also cause trees to die off. Coastal agricultural lands, such as the deltas of China and Bangladesh, India, will be inundated, and billions of dollars will have to be spent to keep U.S. coastal cities, such as New York, Boston, Miami, and Galveston, from disappearing into the sea.

 Video Global Warming

 Video Warming Hurts Rice

Renewable Energy Sources

Renewable types of energy include wind power, hydropower, geothermal energy, and solar energy (**Fig. 32.11**).

a.

Figure 32.11 Renewable energy sources.

a. Hydropower dams provide a clean form of energy but can be ecologically disastrous in other ways. **b.** Wind power requires land on which to place enough windmills to generate energy. **c.** Photovoltaic cells on rooftops and (**d**) sun-tracking mirrors on land can now collect diffuse solar energy more cheaply than in the past.

b.

c.

d.

a.

b.

c.

Figure 32.12 Solar-hydrogen revolution.

a. Hydrogen fuel cells. **b.** This bus is powered by hydrogen fuel. **c.** The use of fuel-cell hybrid vehicles, such as this prototype, will reduce air pollution and dependence on fossil fuels.

Wind Power Wind power is expected to account for a significant percentage of our energy needs in the future. Despite the common belief that a huge amount of land is required for the "wind farms" that produce commercial electricity, the actual amount of space for a wind farm compares favorably with the amount of land required by a coal-fired power plant or a solar thermal energy system.

A community that generates its own electricity by using wind power can solve the problem of uneven energy production by selling electricity to a local public utility when an excess is available and buying electricity from the same facility when wind power is in short supply.

Hydropower Hydroelectric plants convert the energy of falling water into electricity. Hydropower accounts for about 10% of the electric power generated in the United States and almost 98% of the total renewable energy used. Brazil, New Zealand, and Switzerland produce at least 75% of their electricity with water power, but Canada is the world's leading hydropower producer. Worldwide, hydropower presently generates 19% of all electricity utilized, but this percentage is expected to rise due to increased use in certain countries. For example, Iceland has an ambitious hydropower project underway because it uses only 10% of its potential capacity.

Much of the hydropower development in recent years has been due to the construction of enormous dams, which have detrimental environmental effects. An alternative choice may be small-scale dams that generate less power per dam but do not have the same environmental impact.

Geothermal Energy Elements such as uranium, thorium, radium, and plutonium undergo radioactive decay below the Earth's surface and then heat the surrounding rocks to hundreds of degrees Centigrade. When the rocks are in contact with underground streams or lakes, huge amounts of steam and hot water are produced. This steam can be piped up to the surface to supply hot water for home heating or to run steam-driven turbogenerators. The California's Geysers project is the world's largest geothermal electricity-generating complex.

Energy and the Solar-Hydrogen Revolution Solar energy is diffuse energy that must be (1) collected, (2) converted to another form, and (3) stored if it is to compete with other available forms of energy. Passive solar heating of a house is successful when its windows face the sun, the building is well insulated, and heat can be stored in water tanks, rocks, bricks, or some other suitable material.

In a **photovoltaic (solar) cell,** a wafer of an electron-emitting metal is in contact with another metal that collects the electrons and passes them along into wires in a steady stream. Ever since the oil shocks of the 1970s, the U.S. government has been supporting the development of photovoltaics. As a result, the price of buying a system has dropped from about $100 per watt to around $4. The photovoltaic cells placed on roofs, for example, generate electricity that can be used inside a building and/or sold back to a power company.

Several types of solar power plants are now operational across the country. In one type, huge reflectors focus sunlight on a pipe containing oil. The heated pipes boil water, generating steam that drives a conventional turbogenerator. In another type, 1,800 sun-tracking mirrors focus sunlight onto a molten salt receiver mounted on a tower. The hot salt generates steam that drives a turbogenerator.

Scientists are working on the possibility of using solar energy to extract hydrogen from water via electrolysis. The hydrogen can then be used as a clean-burning fuel; when it burns, water is produced. Presently, cars have internal combustion engines that run on gasoline. In the future, vehicles are expected to be powered by fuel cells, which use hydrogen to produce electricity (**Fig. 32.12**), and many gasoline–fuel cell hybrid vehicles now exist. The electricity runs a motor that propels the vehicle. Fuel cells are powering buses in Vancouver and Chicago, and more buses are planned.

Hydrogen fuel can be produced locally or in central locations, using energy from photovoltaic cells. If produced in central locations, hydrogen can be piped to filling stations using the natural gas pipes—already plentiful in the United States. The advantages of a solar-hydrogen revolution would be at least twofold: (1) The world would no longer be dependent on oil, and (2) environmental problems, such as acid rain and smog, would begin to lessen.

Video
Spinach Battery

Minerals

Minerals are nonrenewable raw materials in the Earth's crust that can be mined (extracted) and used by humans. Nonrenewable minerals include fossil fuels; nonmetallic raw materials, such as sand, gravel, and phosphate; and metals, such as aluminum, copper, iron, lead, and gold.

Nonrenewable resources are subject to depletion—that is, the supply will eventually run out. A depletion curve is dependent on how fast the resource is used, whether new reserves can be found, and whether recycling and reuse are possible. We can extend our supply of fossil fuels if we conserve our use and if we find new reserves. In addition to these possibilities, metals can be recycled.

Most of the metals mined each year are consumed by the United States, Japan, and Europe, despite the fact that they are primarily mined in South America, South Africa, and countries of the former Soviet Union. One of the greatest threats to the maintenance of ecosystems and biodiversity is surface mining, called strip mining. In the United States, huge machines can go as far as removing mountaintops in order to reach a mineral (**Fig. 32.13**). The land, devoid of vegetation, takes on a surreal appearance, and runoff from rain washes toxic waste deposits into nearby streams and rivers. Legislation now requires that strip miners restore the land to its original condition, a process that can take years to complete.

The most dangerous metals to human health are the heavy metals: lead, mercury, arsenic, cadmium, tin, chromium, zinc, and copper. They are used to produce batteries, electronics, pesticides, medicines, paints, inks, and dyes. In the ionic form, they enter the body and inhibit vital enzymes. That's why these items should be discarded carefully and taken to hazardous waste sites.

Other Sources of Pollution

Synthetic organic compounds and wastes are also pollutants of concern.

Synthetic Organic Compounds

In addition to metals, synthetic organic compounds are of considerable ecological concern due to their detrimental effects on the health of living things, including humans. Synthetic organic compounds play a role in the production of plastics, pesticides, herbicides, cosmetics, coatings, solvents, wood preservatives, and hundreds of other products.

Synthetic organic compounds include halogenated hydrocarbons, in which halogens (chlorine, bromine, fluorine) have replaced certain hydrogens. **Chlorofluorocarbons (CFCs)** are a type of halogenated hydrocarbon in which

Figure 32.13 Modern mining capabilities.
Giant mining machines—some as tall as a 20-story building—can remove an enormous amount of the Earth's crust in one scoop in order to mine for coal or a metal ore.

Connections and Misconceptions

What is methylmercury and why is it dangerous?

Methylmercury is a form of the element mercury that has been bound to a methyl (CH_3) group. Because of this methyl group, methylmercury easily accumulates in the food chain by biological magnification, or the concentration of chemicals in a food chain. Methylmercury is released into the environment by the burning of coal, the mining of certain metals, and the incineration of medical waste. Methylmercury is a powerful neurotoxin that also inhibits the activity of the immune system. Because of this, the Food and Drug Administration (FDA) and the Environmental Protection Agency (EPA) recommend that pregnant women and small children not eat shark, swordfish, tilefish, or king mackerel and limit the amount of albacore tuna to less than 6 ounces per week. Most states have also posted warnings on eating local fish that have been caught from mercury-contaminated waters. For more information, visit the EPA website, www.epa.gov/waterscience/fish.

both chlorine and fluorine atoms replace some of the hydrogen atoms. CFCs have brought about a thinning of the Earth's ozone shield, which protects terrestrial life from the dangerous effects of ultraviolet radiation. In most MDCs, legislation has been passed to prevent the production of any more CFCs. Hydrofluorocarbons, which contain no chlorine, are expected to take their place in coolants and other products. The ozone shield is predicted to recover by 2050; in the meantime, many more cases of skin cancer are expected to occur.

Other synthetic organic chemicals pose a direct and serious threat to the health of all living things. Rachel Carson's book *Silent Spring,* published in 1962, made the public aware of the deleterious effects of pesticides.

Wastes

Every year, the countries of the world discard billions of tons of solid wastes, some on land and some in fresh and marine waters. Solid wastes are visible wastes, some of which are hazardous to our health.

Industrial Wastes   Industrial wastes are generated during the mining and production of a product. Clean-water and clean-air legislation in the early 1970s prevented venting industrial wastes into the atmosphere and flushing them into waterways. Industry turned to land disposal, which was unregulated at the time. The use of deep-well injection, pits with plastic liners, and landfills led to much water pollution and human illness, including cancer. An estimated 5 billion metric tons of highly toxic chemicals were improperly discarded in the United States between 1950 and 1975. The public's concern was so great that the EPA came into existence. Using an allocation of monies called the superfund, the EPA oversees the cleanup of hazardous waste disposal sites in the United States.

Among the most commonly found contaminants are heavy metals (lead, arsenic, cadmium, chromium) and organic compounds (trichloroethylene, toluene, benzene, polychlorinated biphenyls [PCBs], and chloroform). Some of the chemicals used in pesticides, herbicides, plastics, food additives, and personal hygiene products are classified as endocrine disrupters. These products can affect the endocrine system and interfere with reproduction. In the environment, they occur at a level 1,000 times greater than the hormone levels in human blood.

Decomposers are unable to break down these wastes. They enter and remain in the bodies of organisms because they are not excreted. Therefore, they become more concentrated as they pass along a food chain, a process termed **biological magnification (Fig. 32.14).** This effect is most apt to occur in aquatic food chains, which have more links than do terrestrial food chains. Humans are one of the final consumers in both types of food chains, and in some areas, human milk contains detectable amounts of the polychlorinated hydrocarbons DDT and PCBs.

Sometimes industrial wastes accumulate in the mud of deltas and estuaries of highly polluted rivers and cause environmental problems if disturbed. Industrial pollution is being addressed in many MDCs, but it usually has low priority in LDCs.

Sewage   Raw sewage can contribute to oxygen depletion in lakes and rivers. Sewage serves as a fertilizer for plants, which can lead to eutrophication; first there is an algal bloom, and then when the algae use up all the nutrients, the algae die. Decomposition of dead algae robs the water of oxygen, which can result in a massive fish kill. Also, human feces can contain pathogenic

Video
Lead Soil Solution

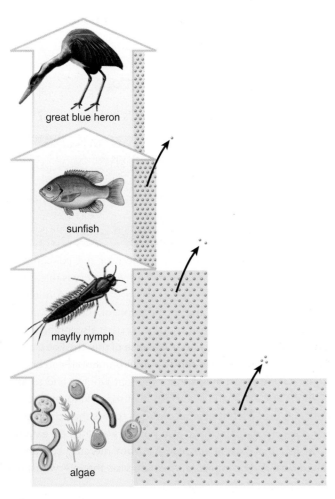

Figure 32.14 Biological magnification.

A poison, such as DDT, that is excreted in relatively small amounts (arrows) becomes maximally concentrated as it passes along a food chain due to the reduced size of the trophic levels.

microorganisms that cause cholera, typhoid fever, and dysentery. In regions of the LDCs where sewage treatment is practically nonexistent, many children die each year from these diseases. Sewage treatment plants use bacteria to break down organic matter to inorganic nutrients, such as nitrates and phosphates, which then enter surface waters.

Check Your Progress 32.1

1. Distinguish between renewable and nonrenewable resources.
2. Describe two problems related to human habitation near oceans.
3. Explain the causes of deforestation and desertification, and list one environmental problem associated with each.
4. List the harmful effects of modern agricultural practices.
5. Summarize the influences of the two green revolutions.
6. Summarize the potential effects of global warming.
7. List four types of renewable energy, and explain the limits of each.

Connecting the Concepts

For more information on the topics presented in this section, refer to the following discussions.

Section 11.3 explores how transgenic organisms are created.

Section 30.3 examines the concept of carrying capacity as it relates to natural resources.

Section 31.2 examines the relationship between the carbon cycle and global warming.

32.2 Biodiversity

Learning Outcomes

Upon completion of this section, you should be able to

1. Define *biodiversity* and briefly list some of the threats to it.
2. List the direct benefits of biodiversity and explain how they are beneficial to the human population.
3. List the indirect benefits of biodiversity and briefly describe their economic benefits.

Biodiversity can be defined as the variety of life on Earth, described in terms of the number of different species. We are presently in a biodiversity crisis—the number of extinctions (loss of species) expected to occur in the near future will, for the first time, be attributable to human activities. According to the U.S. Fish and Wildlife Service (FWS), as of 2010 there are over 530 animal species, and 795 plant species, in the United States that are in danger of extinction. The majority of these species (85%) are threatened by habitat loss (**Fig. 32.15a**), usually associated with the sprawl of urban areas. Other factors that are contributing to the biodiversity crisis are the introduction of exotic species (50%), water and air pollution (24%), and the overexploitation of natural resources (17%). In many cases, endangered species are threatened by multiple factors. For example, the Canada lynx (*Lynx canadensis*) lives in the northern forests of the United States. Naturally, it is a relatively rare species that prefers dense forests. It is currently listed as a threatened species due to habitat loss caused by the creation of roads for snowmobiling and skiing, as well as increased foresting of timber in their natural habitats.

Conservation biology strives to reverse the trend toward the possible extinction of thousands of plants and animals. To bring this about, it is necessary to make all people aware that biodiversity is a resource of immense value—both direct and indirect.

Figure 32.15   Habitat loss.

a. In a study examining records of imperiled U.S. plants and animals, habitat loss emerged as the greatest threat to wildlife. **b.** The Canada lynx is threatened by the loss of habitat and the overexploitation of forests for wood.

Wild species, like the rosy periwinkle, are sources of many medicines.

Wild species, like the nine-banded armadillo, play a role in medical research.

Figure 32.16   **Medicinal value of biodiversity.**

Many wildlife species are sources of medicines for today's ills.

Direct Values of Biodiversity

Figure 32.16 depicts one of the direct values of biodiversity, which include medicines, foods, and other products that benefit human beings.

Medicinal Value

Most of the prescription drugs used in the United States were originally derived from organisms. The rosy periwinkle from Madagascar is an excellent example of a tropical plant that has provided us with useful medicines (Fig. 32.16*a*). Potent chemicals from this plant are now used to treat two forms of cancer: leukemia and Hodgkin disease. Due to these drugs, the survival rate for childhood leukemia has gone from 10% to 90%, and Hodgkin disease is now usually curable. Although the value of saving a life cannot be calculated, it is still sometimes easier for us to appreciate the worth of a resource if it is explained in monetary terms. Thus, researchers tell us that, judging from the success rate in the past, an additional 328 types of drugs are yet to be found in tropical rain forests, and the value of this resource to society is probably $147 billion.

The popular antibiotic penicillin is derived from a fungus, and certain species of bacteria produce the antibiotics tetracycline and streptomycin. These drugs have proven to be indispensable in the treatment of diseases, including certain sexually transmitted diseases.

Leprosy is among the diseases for which there is, as yet, no cure. The bacterium that causes leprosy will not grow in the laboratory, but scientists discovered that it grows naturally in the nine-banded armadillo (Fig. 32.16*b*). Having a source for the bacterium may make it possible to find a potential cure for leprosy. The blood of horseshoe crabs contains a substance called limulus amoebocyte lysate, which is used to ensure that medical devices, such as pacemakers, surgical implants, and prostheses, are free of bacteria. Blood is taken from 250,000 horseshoe crabs a year, and then they are returned to the sea unharmed.

 Video Indigenous Medicine   **Video** Good Poison

Agricultural Value

Crops such as wheat, corn, and rice are derived from wild plants that have been modified to be high producers. The same high-yield, genetically similar strains tend to be grown worldwide. When cultivated rice crops in Africa were being devastated by a virus, researchers grew wild rice plants from thousands of seed samples until they found one that contained a gene for resistance to the virus. These wild plants were then used in a breeding program to transfer the gene into high-yield rice plants. If this variety of wild rice had become extinct before its resistance could be discovered, rice cultivation in Africa might have collapsed.

Biological pest controls—specifically, natural predators and parasites—are often preferable to chemical pesticides (**Fig. 32.17***a*). When a rice pest called the brown planthopper became resistant to pesticides, farmers began to use natural enemies of the brown planthopper instead. The economic savings were calculated at well over $1 billion. Similarly, cotton growers in Cañete Valley, Peru, found that pesticides were no longer working against the cotton aphid due to resistance. Research identified natural predators, which cotton farmers are now using to an even greater degree.

Most flowering plants are pollinated by animals, such as bees, wasps, butterflies, beetles, birds, and bats (Fig. 32.17*b*). The honeybee has been domesticated, and it pollinates almost $10 billion worth of food crops annually in the United States. The danger of this dependency on a single species is exemplified by mites that have wiped out more than 20% of the commercial honeybee population in the United States. Where can we get resistant bees? From the wild, of course; the value of wild pollinators to the U.S. agricultural economy has been calculated at $4.1 to $6.7 billion a year.

Consumptive Use Value

We have had much success in cultivating crops, keeping domesticated animals, growing trees on plantations, and so on. However, the environment provides all sorts of other products that are sold in marketplaces worldwide, including wild fruits and vegetables, skins, fibers, beeswax, and seaweed. Also, some people obtain their meat directly from the environment. In one study, researchers calculated that the economic value of wild pig in the diet of native hunters in Sarawak, East Malaysia, was about $40 million per year.

Similarly, many trees are still felled in the natural environment for their wood. Researchers have calculated that a species-rich forest in the Peruvian Amazon is worth far more if the forest is used for fruit and rubber production than for timber production (Fig. 32.17*d*). Fruit and the latex needed to produce rubber can be brought to market for an unlimited number of years, whereas once the trees are gone, no more timber can be harvested.

Figure 32.17   Agriculture and the consumptive value of wildlife.

Many wildlife species help account for our bountiful harvests and are sources of food for the world.

Wild species, like ladybugs, play a role in biological control of agricultural pests.

Wild species, like the long-nosed bat, are pollinators of agricultural and other plants.

Wild species, like many marine species, provide us with food.

Wild species, like rubber trees, can provide a product indefinitely if the forest is not destroyed.

biogeochemical cycles
waste disposal
provision of fresh water
prevention of soil erosion
regulation of climate
ecotourism

Figure 32.18 Indirect value of ecosystems.

Forests and the oceans perform many of the functions listed as the indirect value of ecosystems.

Indirect Values of Biodiversity

All wild species play important roles in the ecosystems to which they belong. If we want to preserve them, it is more economical to save the ecosystems than the individual species. Ecosystems perform many indirect services for modern humans, who increasingly live in cities. These services are said to be indirect because they are wide-ranging and not easily perceptible (**Fig. 32.18**).

Biogeochemical Cycles

Recall from Chapter 31 that ecosystems are characterized by energy flow and chemical cycling. The biodiversity within ecosystems contributes to the workings of the water, carbon, nitrogen, phosphorus, and other biogeochemical cycles. We are dependent on these cycles for fresh water, removal of carbon dioxide from the atmosphere, uptake of excess soil nitrogen, and provision of phosphate. When human activities upset one aspect of biogeochemical cycles, other parts within the cycles will also be affected. Technology is unable to artificially contribute to or create any of the biogeochemical cycles.

> **Video**
> Dung Beetles

Waste Disposal

Decomposers break down dead organic matter and other types of wastes to inorganic nutrients that are used by the producers within ecosystems. This function aids humans immensely because we dump millions of tons of waste material into natural ecosystems each year. If not for decomposition, waste would soon cover the entire surface of our planet. We can build sewage treatment plants, but they are expensive, and few of them break down solid wastes completely to inorganic nutrients. It is less expensive and more efficient to water plants and trees with partially treated wastewater and let soil bacteria cleanse it completely.

> **Video**
> Decomposers

Biological communities are also capable of breaking down and immobilizing pollutants, such as heavy metals and pesticides that humans release into the environment. A review of wetland functions in Canada assigned a value of $50,000 per hectare (2.471 acres, or 10,000 square meters) per year to the ability of natural areas to purify water and take up pollutants.

> **Video**
> Good Tobacco

Provision of Fresh Water

Few terrestrial organisms are adapted to living in a salty environment—they need fresh water. The water cycle continually supplies fresh water to terrestrial ecosystems. Humans use fresh water in innumerable ways, including drinking it and irrigating their crops. Freshwater ecosystems, such as rivers and lakes, also provide fish and other types of organisms for food.

Unlike other commodities, there is no substitute for fresh water. We can remove salt from seawater to obtain fresh water, but the cost of desalination is about four to eight times the average cost of fresh water acquired via the water cycle.

Forests and other natural ecosystems exert a "sponge effect." They soak up water and then release it at a regular rate. When rain falls in a natural area, plant foliage and dead leaves lessen its impact, and the soil slowly absorbs it, especially if the soil has been aerated by organisms. The water-holding capacity of forests reduces the possibility and degree of flooding. The value of a marshland outside Boston, Massachusetts, has been estimated at $72,000 per hectare per year solely on its ability to reduce floods. Forests release water slowly for days or weeks after the rains have ceased. Comparing rivers in West African coffee plantations, those flowing through forests release twice as much water halfway through the dry season and three to five times as much at the end of the dry season. These data show the water-retaining ability of forests.

Prevention of Soil Erosion

Intact terrestrial ecosystems naturally retain soil and prevent soil erosion. The importance of this ecosystem attribute is especially noticeable following deforestation. In Pakistan, the world's largest dam, the Tarbela Dam, is losing its storage capacity of 12 billion cubic meters many years sooner than expected because silt is building up behind the dam due to deforestation of areas upriver. At one time, the Philippines were exporting $100 million worth of oysters, mussels, clams, and cockles each year. Now, silt carried down rivers following deforestation is smothering the mangrove ecosystem that serves as a nursery for many marine species. Most coastal ecosystems are not as bountiful as they once were because of deforestation elsewhere and its consequences.

Regulation of Climate

At the local level, trees provide shade, block drying winds, and reduce the need for fans and air conditioners during the summer.

Globally, forests regulate the climate because they take up carbon dioxide, a greenhouse gas. When the leaves of trees photosynthesize, they use carbon dioxide, which is stored as the wood of the tree, When trees are cut and burned, carbon is no longer released into the soil through natural decomposition; instead, carbon dioxide is released into the atmosphere. Subsequently, the reduction of forests worldwide reduces the amount of carbon dioxide removed from the atmosphere, amplifying the issue. Deforestation often removes soil nutrients needed by future tree generations, which in turn limits reforestation.

Ecotourism

Almost everyone prefers to vacation in the natural beauty of an ecosystem (**Fig. 32.19**). In the United States, nearly 100 million people enjoy vacationing in a natural setting. To do so, they spend $5 billion each year on fees, travel, lodging, and food. Many tourists want to go sport fishing, whale watching, boat riding, hiking, birdwatching, and the like. Some merely want to immerse themselves in the beauty of a natural environment. Some less-developed countries are realizing that there is more economic potential in ecotourism than in cutting forests for timber.

Connecting the Concepts

For more information on these topics, refer to the following discussions.

Section 18.2 explains the importance of pollinators in the evolution of the flowering plants.

Section 18.3 explores the biology of the fungi.

Section 31.2 provides illustrations of the nitrogen, phosphorus, and carbon cycles.

Check Your Progress 32.2

1. Summarize the direct and indirect values of biodiversity.

2. Explain how modern medicine and agriculture are dependent on biodiversity.

3. Hypothesize how the preservation of biodiversity can help reduce the problems associated with overexplotation of water, land, food, or energy resources.

Figure 32.19 Ecotourism.

Whale watchers experience the thrill of seeing an orca surfacing off the coast of Washington State.

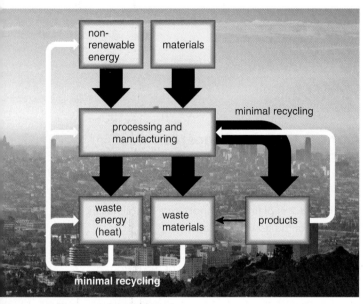

a. Human society at present

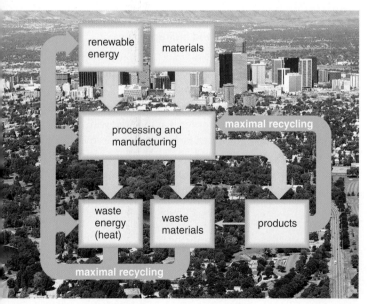

b. Sustainable society

Figure 32.20 Current human society versus a sustainable society.

a. Our "throw-away" society is characterized by high input of energy and raw material, large output of waste materials and energy in the form of heat, and minimal recycling (white arrows). **b.** A sustainable society would be characterized by the use of only renewable energy sources, reuse of heat and waste materials, and maximal recycling of products (blue arrows).

32.3 Working Toward a Sustainable Society

Learning Outcomes

Upon completion of this section, you should be able to

1. Explain why human society is unsustainable in its current form.
2. List some activities that may help make rural societies more sustainable.
3. List some activities that may help make urban societies more sustainable.

A **sustainable society** would be able to provide the same goods and services for future generations of human beings as it does now. At the same time, biodiversity would be preserved.

Today's Society

The following evidence indicates that at present, human society is most likely not sustainable (**Fig. 32.20***a*):

- A considerable proportion of land, and therefore natural ecosystems, is being used for human purposes (homes, agriculture, factories, etc.).
- Agriculture requires large inputs of nonrenewable fossil fuel energy, fertilizer, and pesticides, which create much pollution. More fresh water is used for agriculture than in homes.
- At least half of the agricultural yield in the United States goes toward feeding animals. According to the 10-to-1 rule of thumb, it takes 10 pounds of grain to grow 1 pound of meat. Therefore, it is wasteful for citizens in MDCs to eat as much meat as they do. Also, animal sewage pollutes water.
- Even though fresh water is a renewable resource, we are running out of the available supply.
- Our society uses primarily nonrenewable fossil fuel energy, which leads to acid precipitation, smog, and various other pollutants entering the ecosystems.
- Minerals are nonrenewable, and the mining, processing, and use of minerals are responsible for much environmental pollution.

Characteristics of a Sustainable Society

A natural ecosystem can offer clues as to what a sustainable human society would be like. A natural ecosystem is characterized by the use of renewable solar energy, and its materials cycle through the various populations back to the producers once again. If we want to develop a sustainable society, we, too, should use renewable energy sources and recycle materials (Fig. 32.20*b*).

We can apply these principles to both rural areas and urban areas, as suggested in the next two sections.

Rural Sustainability

In rural areas, we should put the emphasis on preservation. We should preserve ecosystems, including terrestrial ecosystems—such as forests and prairies—and aquatic ecosystems—both freshwater and brackish ones along the coast. We should also preserve agricultural land, groves of fruit trees, and other areas that provide us with renewable resources.

Meiosis II- Prophase II- no pairing of chromosomes. Anaphase II- haploid number of dyads at spindle equator. Anaphase II- sister chromatids separate, become daughter chromosomes that move to the poles. Telophase II- four haploid daughter cells, different from each other genetically and from the parent cell; **2.** Mitosis- all cells except germ cells for growth and repair. Meiosis- in germ cells only for formation of gametes; **3.** During metaphase I, the pairs of homologues are attached to the spindle with either homologue facing either spindle pole. During metaphase II (and mitosis) dyads are attached to the spindle with sister chromatids facing opposite poles. **(9.4) 1.** If nondisjunction occurs during meiosis I or II, a gamete, and therefore a zygote, could have an abnormal chromosome number; **2.** Monosomy results when an individual is missing one chromosome. Trisomy results when an individual has one extra chromosome; **3.** Chromosome abnormalities in the sex chromosomes are from abnormalities in chromosome X or Y only. In the autosome, they can occur in any of the other 22 pairs of chromosomes.

Testing Yourself

1. a; **2.** d; **3.** d; **4.** a; **5.** c; **6.** c; **7.** a; **8.** d; **9.** a; **10.** f; **11.** h; **12.** c; **13.** g; **14.** b; **15.** e; **16. a.** chromosomes condense and do not pair, homologues condense and undergo synapsis; **b.** dyads at spindle equator, tetrads at spindle equator; **c.** sister chromatids separate, homologues separate; **d.** two daughter cells following mitosis, (top) two daughter cells result from meiosis I, (bottom) four daughter cells result from meiosis II; see Fig. 9.8, p. 148, in text; **17.** d; **18.** c; **19.** c; **20.** c; **21.** b

Thinking Scientifically

1. A mosaic such as (46,XY/47,XXY) may result from nondisjunction occurring during mitosis, especially at a very early stage of development. A fertile Klinefelter mosaic may arise from an individual in which karyotypically normal (46,XY) embryonic cells gave rise to the testes. **2.** Hypotheses to explain the relationship between the increase in Down syndrome with advancing maternal age include: (1) The older a woman is, the longer her oocytes have been arrested in meiosis, and the higher the risk that these oocytes have been exposed to mutagens which might cause nondisjunction; (2) A woman may have a pool of oocytes resulting from nondisjunction which take longer to mature, and thus are more commonly released as she ages; (3) Estrogen levels (which control the rate of meiosis in developing oocytes) drop with advancing maternal age and may slow down the rate of meiosis. This might allow nondisjunction to occur more frequently in older women. **3.** Large amounts of genetic variability in a population allow the population to quickly adapt to changing conditions. For example, if a population is suddenly exposed to a new pathogen, it is likely that at least a few individuals will be able to survive the pathogen and the species would continue.

CHAPTER 10

Check Your Progress

(10.1) 1. Mendel selected seven traits (stem length, pod shape, seed shape, seed color, flower position, flower color, pod color) and studied offspring through successive generations to learn about transmission of traits. He determined the law of segregation and law of independent assortment; **2.** Genotype is the genetic make-up of an individual, while phenotype is the physical appearance of the individual; **3.** Because a dominant phenotype could be either heterozygous or homozygous dominant; **4.** Genotypic ratio 1:2:1, phenotypic ratio 3:1; **5.** Heterozygous for both traits; **6.** a) All long winged, gray bodied flies b) Either 1 long wings, gray body: 1 long wings, black body or 1 long wings, gray body: 1 short wings, gray body. **(10.2) 1.** Incomplete dominance is exhibited when the heterozygote has an intermediate phenotype between that of either homozygote; an example is pink four-o'clock flowers; **2.** Codominance is exhibited when both alleles are fully expressed. Possible genotypes: I^AI^A, I^Ai, I^BI^B, I^Bi, ii, I^AI^B; **3.** Polygenic inheritance occurs when a trait is governed by two or more sets of alleles. Multifactorial traits are ones that are controlled by polygenes subjected to environmental influences. Pleiotropy occurs when a single gene has more than one effect. **(10.3) 1.** Since males are XY, some sperm will contain an X chromosome and others will contain a Y; **2.** Since sex-linked disorders are mainly on the X chromosome, the mother is the only one who can pass on those traits to her son; **3.** Since males only have one X chromosome, they would be affected, not carriers. Only females can be carriers, if one of their XX chromosomes is affected. **(10.4) 1.** Crossing-over causes certain daughter chromosomes to have a new combination of alleles. The gametes that contain these daughter chromosomes are called recombinant gametes; **2.** Mendel's law of independent assortment because linked alleles normally stay together in the gametes; **3.** The percentage of crossing-over indicates the distance between alleles. When many crosses are done, the collective data can be used to tell the sequence of the gene loci involved.

Testing Yourself

1. c; **2.** d; **3.** d; **4.** d; **5.** a; **6.** e; **7. a.** TG; **b.** tg; **c.** $TtGg$; **d.** $TTGG$; **e.** $TtGg$; **f.** $TTgg$; **g.** $Ttgg$; **h.** $TtGG$; **i.** $ttGG$; **j.** $TtGg$; **k.** $ttGg$; **l.** $ttgg$; see Fig. 10.6, p. 160, in text; **8.** e; **9.** c; **10.** c; **11.** d; **12.** a; **13.** e; **14.** c; **15.** d; **16.** e; **17.** b; 18. e; **19.** d; **20.** a,b; **21.** a,b,c,d; **22.** c,d; **23.** a, b, d; **24.** a; **25.** b

Thinking Scientifically

1. a. $CcPp$ x $CcPp$ would produce 9/16 purple ($C\_P\_$) and 7/16 white offspring (C_pp, $ccP\_$, or $ccpp$). **2.** The chromosomal segment contained the *SRY* gene. An XX man has an *SRY* gene on an X chromosome, and would show the male phenotype. XY women have a Y chromosome that is missing its *SRY,* so they would have the female phenotype. **3.** In a changing environment, a particular combination of alleles may occur to produce an advantageous polygenic trait. The individuals with this trait would be selected to reproduce more than other individuals.

CHAPTER 11

Check Your Progress

(11.1) 1. Series of nucleotides containing a deoxyribose sugar, a phosphate, and a nitrogen containing base of either cytosine, thymine, adenine or guanine double stranded twisted into a helix; **2.** DNA contains the sugar deoxyribose, RNA has the sugar ribose. DNA has the base thymine, RNA has the base uracil. DNA is double stranded, RNA is single stranded; **3.** Unwinding of the double helix, complementary base pairing, then joining of the new nucleotides. **(11.2) 1.** Using a DNA strand as a template, RNA polymerase adds complementary RNA nucleotides to a growing RNA strand; **2.** mRNA strand is made in transcription, rRNA composes the ribosome subunits which attach to the mRNA strand, and tRNA molecules bring the amino acids to the translation complex; **3.** A point mutation changes a single nucleotide, likely due to a mutagen or error in replication, and can change the amino acid at that location in the gene; A frameshift mutation, caused by extra or missing nucleotides due to a mutagen, error in replication, or transposon, changes the entire reading frame of the protein made from that point in the gene on. **(11.3) 1.** Restriction enzymes are used to cut strands of DNA at certain sequences useful in cloning; **2.** Transgenic animals can produce milk with antibiotics or vaccines. Transgenic plants can be produced that are pesticide resistant; **3.** PCR is used to make numerous copies of sequences of DNA useful for testing. If there is a limited amount of DNA, for instance from a crime scene, making mass copies is needed for diagnostic testing. **(11.4) 1.** Genomics research attempts to understand how genes direct growth and development in order to control the activities of cells and organisms. Genomics also strives to understand the function of the intergenic sequences within the genome; **2.** To understand evolutionary history and diversity; **3.** Proteomics is crucial in the development of new drugs to treat diseases.

Testing Yourself

1. a. sugar-phosphate backbone; **b.** base; **c.** hydrogen bonds; **d.** sugar; **e.** nucleotide; see Fig. 11.5, p. 178, in text; **2.** a; **3.** a; **4.** d; **5.** b; **6.** b; **7.** a; **8.** c; **9.** b; **10.** b; **11.** c; **12.** 1) TTAGCCATA 2) UUAGCCAUA 3) Leu, Ala, Ile; **13.** b; **14.** c; **15.** c; **16.** c; **17.** a; **18.** c; **19.** d

Thinking Scientifically

1. DNA from the heat-killed virulent strain contains genes that make them virulent. Somehow, the living non-virulent bacteria obtained some of this DNA, converting them into a virulent strain. **2.** The principles of evolution tell us that it would be too disadvantageous to maintain this DNA if it didn't have a necessary function in these animals.

CHAPTER 12

Check Your Progress

(12.1) 1. Reproductive cloning creates an individual that is genetically identical to the 2n nucleus donor. Therapeutic cloning provides specialized cells for medical procedures; **2.** In reproductive cloning the nucleus is removed from a G_0 animal cell where it is fused with an egg cell with its nucleus removed. The cell is cultured into an embryo, implanted into a surrogate mother; **3.** In therapeutic cloning, the nucleus is removed from a G_0 cell and fused with an egg cell with its nucleus removed. The cell is cultured into an embryo and specialized tissue cells are produced for use. **(12.2) 1.** In prokaryotes, gene

photosynthesis and vision are adapted to utilizing this portion of the electronic spectrum, which is the most abundant high-energy radiation source available.

CHAPTER 7

Check Your Progress

(7.1) 1. Breathing takes in oxygen needed for cellular respiration and rids the body of carbon dioxide, a waste product of cellular respiration. **2.** Glycolysis-splitting sugar; the preparatory reaction- which prepares pyruvate for citric acid cycle; citric acid cycle- oxidizes acetyl CoA to produce ATP; electron transport chain- transfers energy from high to low states to produce ATP. **3.** Carbon dioxide is released during the preparatory reaction and citric acid cycle. Water is the end product of the electron transport chain. **(7.2) 1.** Glycolysis occurs in the cytoplasm of the cell. **2.** During the energy-investment step, ATP breakdown provides the phosphate groups that activate substrates. During the energy-harvesting steps, NADH and ATP are produced. **3.** 2 ATP molecules. **(7.3) 1.** 4 carbon dioxide; 6 NADH; 2 FADH2; 2 ATP. **2.** As electrons move from one carrier to another in the cristae, energy is released and this energy is used to pump hydrogen ions from the matrix into the intermembrane space. The flow of hydrogen ions back down the concentration gradient into the matrix drives the synthesis of ATP by ATP synthase. **3.** Each fat or oil molecule contains three long-chain fatty acids and one glycerol. Many more acetyl-CoA result from the breakdown of the three fatty acids than from the breakdown of one glucose. Therefore, many more ATP result from the breakdown of a fat than from glucose. **(7.4) 1.** Cellular respiration uses oxygen and results in a higher yield of ATP than fermentation, which does not require oxygen. **2.** Humans produce lactate; yeast produces ethyl alcohol and carbon dioxide. **3.** Since humans utilize oxygen and need a constant high yield of ATP, cellular respiration is a more efficient process than fermentation.

Testing Yourself

1. a; **2.** c; **3.** a; **4.** a; **5.** c; **6.** a; **7.** a; **8.** c; **9.** b; **10. a.** Pyruvate is broken down to an acetyl group; **b.** Acetyl group is taken up and a C_6 molecule results; **c.** Oxidation results in NADH and CO_2; **d.** ATP is produced by substrate-level ATP synthesis; **e.** Oxidation produces more NADH and $FADH_2$; **11.** c; **12.** a, c, d; **13.** a; **14.** a, b, c; **15.** d; **16.** b, c; **17.** a; **18.** b, c, d; **19.** c; **20.** d; **21.** c; **22.** b; **23.** b; **24.** a; **25.** c; **26.** c; **27.** a; **28. a.** citric acid cycle; **b.** anaerobic; **c.** pyruvate; **d.** fermentation; **29.** c

Thinking Scientifically

1. Malonate is a product of one enzymatic reaction in the citric acid cycle and a substrate of the next enzyme in the cycle. Due to a feedback mechanism, excessive amounts of malonate will inhibit the enzyme that produced it, thereby stopping the citric acid cycle. **2.** As we age, our mitochondria lose efficiency and do not break down fats as quickly. This causes fats to accumulate in cells. Since fats and oils enter the cellular respiration pathway as acetyl CoA, there are plenty of citric acid cycle substrates to create energy, and more glucose is not needed.

Since the cells already have plenty of energy, they ignore the insulin signal and do not readily take up glucose. Thus, fatty acid accumulation can lead to insulin resistance. **3.** The inner membrane of mitochondria has invaginations and this membrane would have been the plasma membrane of a free-living prokaryote cell. The inner membrane of mitochondria bears a metabolic pathway.

CHAPTER 8

Check Your Progress

(8.1) 1. Mature organisms have to replenish worn-out cells and repair injuries. Unicellular organisms divide in order to reproduce, and multicellular organisms develop from a single cell; **2.** Growth and cell division normally occur; **3.** Chromatin consists of strands of DNA that are organized around histone proteins. Chromatin forms a zigzag that is folded into loops. When cellular reproduction occurs, the chromatin is further condensed into large loops that produce a greatly compacted chromosome. **(8.2) 1.** The cell cycle is the orderly arrangement that duplication follows to ensure it is done accurately; **2.** Interphase- the cell is performing its normal function and getting ready for G1, S, G2 phases. G1-organelles are duplicated. S- DNA is synthesized. G2- all proteins needed for cell division are made; **3.** No, since interphase includes G1, S, G2 to prepare the cell to divide, without it, the cell would not divide. **(8.3) 1.** Prophase- chromosomes condense and consist of two sister chromatids. Metaphase-chromosome align down the spindle equator. Anaphase- sister chromatids separate becoming daughter chromosomes. Telophase- nuclear envelope forms around the daughter chromosomes; **2.** Spindle has spindle fibers made of microtubules that attach to chromosomes at the centromere in order to move them around the dividing cell. Without the spindle, the chromosomes would not move correctly; **3.** Daughter cells would not contain the proper amount of organelles or cytosol and would not be able to function properly. **(8.4) 1.** It is important for each step of the cell cycle to be completed correctly before the next one begins, and checkpoints allow the cell to make sure that happens; **2.** A signal is a molecule that promotes or inhibits an event. Internal and external signals control the cell cycle; **3.** If apoptosis is not regulated, cells could terminate prematurely or not at all and cause malfunction in physiology. **(8.5) 1.** Cancer occurs when the cell cycle is not regulated properly; **2.** Cancer cells lose their normal functions, have abnormal nuclei, form tumors, do not undergo apoptosis, undergo metastasis, and promote angiogenesis. Avoid tobacco use, reduce sun and radon exposure, avoid heavy drinking and obesity, consume food rich in vitamins A and C, avoid chemically preserved foods, and include vegetables from the cabbage family in your diet; **3.** Radiation and chemotherapy damage DNA or otherwise interfere with the completion of mitosis. Hormonal therapy interferes with cell's reception of an external signal to divide.

Testing Yourself

1. d; **2.** b; **3.** a; **4.** c; **5.** a; **6.** b; **7. a.** G_1; **b.** S; **c.** G_2; **d.** prophase; **e.** metaphase; **f.** anaphase; **g.** telophase; **h.** cytokinesis; see Fig. 8.3, p. 126, in text; **8.** a; **9.** d;

10. c; **11.** c; **12.** b; **13.** a, b; **14.** b; **15.** b; **16.** c; **17.** b; Each chromosome has one chromatid; therefore, the answer is 42; **18.** d; Each chromosome has two chromatids; therefore, the answer is 84; **19.** a; **20.** a; **21.** c; **22.** e; **23.** e; **24.** d; **25.** b;

Thinking Scientifically

1. Cancer results from a series of mutations ("hits") that accumulate in cells over a person's lifetime. People exposed to high levels of radiation are likely to have a higher number of somatic mutations than the general population. They would then require fewer additional random mutations to occur before they develop cancer. A reasonable explanation for the appearances of different cancers at different times is that some cancers require a larger number of mutations to occur before they develop. Cancers that require more mutations to occur would more likely occur later in life. **2.** Sister chromatids of chromosomes (inside the nucleus) must link to microtubules produced by centrosomes (outside the nucleus) in order to be pulled apart during mitosis. Therefore, microtubules cannot attach to and separate chromatids if the nuclear membrane is present. The current hypothesis is that cytoplasmic dynein, a molecular motor protein, tears the nuclear envelope and transports pieces of the membrane away from the area along microtubules. Once pieces are torn away, the nuclear envelope loses shape and cytoplasmic proteins flow into the nuclear region. The rush of cytoplasmic proteins is believed to stimulate chromosome condensation and spindle formation. So, dismantling the nuclear membrane is coordinated with chromosome behavior during mitosis. **3.** Checkpoints ensure that when the cell cycle is finished, the chromosomal content is normal. If the chromosomal content is not normal, then cancer can become prevalent and the species may become extinct.

CHAPTER 9

Check Your Progress

(9.1) 1. Homologous chromosome have the same size, shape and constrictions and occur in pairs, one from each parent; **2.** During meiosis I, homologous chromosomes separate and during meiosis II, sister chromatids separate; **3.** Crossing-over alters the types of alleles on the chromosomes and daughter cells have different combinations of the haploid number of chromosomes. Genetic variation is the basis of evolution. **(9.2) 1.** Tetrad formation allows crossing-over to occur and prepares homologues for separation during anaphase I; **2.** The parent cell is diploid and each chromosome has two chromatids. A daughter cell is haploid and each chromosome has one chromatid; **3.** The daughter cells would not have the correct number of chromosomes. **(9.3) 1.** Mitosis- Prophase- no pairing of chromosomes; no crossing over. Metaphase- dyads at spindle equator. Anaphase- sister chromatids separate, becoming daughter chromosomes that move to the poles. Telophase- two identical diploid cells. Meiosis-Prophase I- pairing of homologous chromosomes, crossing over. Metaphase I- tetrads at spindle equator. Anaphase I- homologues of each tetrad separate and dyads move to poles. Telophase I- two haploid daughter cells not identical to parent.

adequate surface area for exchange of waste and nutrients. (4.2) 1. Prokaryotes- lack a nucleus, small, not complex. Eukaryotes- nucleus, large, complex. 2. Harmful- disease causing; Beneficial- decompose dead material. 3. Flagella- tail-like for propulsion; fimbriae- small bristle-like fibers for surface attachment; conjugation pili- rigid tubular structures used to pass DNA between cells. All are made of protein. (4.3) 1. Phospholipid bilayer with embedded proteins. 2. Proteins embedded in the bilayer have a pattern and the entire bilayer is semi-fluid. 3. Channel- allow passage of certain molecules; Transport- passage of ions and molecules; Cell recognition- distinguish between foreign and own cells; Receptor- binds signal molecules; Enzymatic- participate in metabolic reactions; Junction- form cell junctions between cells. (4.4) 1. Nucleus; chromosomes- store DNA; nucleolus- produces ribosomal RNA; Ribosomes- small and large subunits; aid in production of proteins. 2. Endoplasmic reticulum- Rough ER- synthesize, modifies proteins; Smooth ER- synthesize lipids; Golgi apparatus- transfer station and modifies proteins; Lysosomes- digests proteins. 3. Vacuoles store substances. Vacuoles in some protists rid the cell of water. Vacuoles in plants store pigments as well as toxic substances for protection. 4. Both specialize in energy conversion. Mitochondria- small, double membrane organelle that produces ATP; Chloroplasts- Large, double membrane with interior thylakoid stacks of granum found in plants where photosynthesis occurs. 5. Cytoskeleton- Microtubules- small, hollow cylinders of 13 long chains of tubulin dimers; maintain shape of cell and act as tracks for organelle movement; Intermediate fibers- intermediate in size ropelike assemblies of proteins; supports nucleus and plasma membrane; Actin filaments- two chains of globular actin monomers twisted to form long filament; supports the cell and microvilli. Motor Proteins- Myosin- contracts muscle cells; Kinesin and Dynein- move organelles within the cell. (4.5) 1. Primary- contains cellulose fibrils and noncellulose substances. Secondary- forms inside primary wall containing cellulose and lignin. 2. Fibrous proteins and polysaccharides. 3. Adhesion- internal cytoplasmic plaques attached to the cytoskeleton within each cell to prevent overstretching; Tight- plasma membrane proteins attached to each other forming a barrier; Gap- two identical channel proteins are lined by six plasma protein allowing the junction to open and close to move materials between cells.

Testing Yourself

1. d; 2. a; 3. b; 4. b; 5. c; 6. d; 7. a; 8. b; 9. d; 10. e; 11. c; 12. b; 13. a; 14. b; 15. c; 16. d; 17. c; 18. a. vesicle; b. centrioles; c. mitochondrion; d. rough endoplasmic reticulum; e. smooth endoplasmic reticulum; f. lysosome; g. Golgi apparatus; h. nucleus; 19. a

Thinking Scientifically

1. a. A dietician may suggest a diet that would not include the offending long-chain fatty acids. b. Lysosomes are similar to peroxisomes in that they are single-membrane organelles containing enzymes that break down cellular compounds. 2. The plastid must be necessary to the life of the

parasite because the parasite dies if the plastid no longer functions. 3. Just as a house cannot store its contents, so a cell cannot be different from its environment without a boundary.

CHAPTER 5
Check Your Progress

(5.1) 1. Potential energy is stored energy; kinetic energy is energy of motion. Pulling an arrow on a bow is potential energy, and when the arrow is shot from the bow, it turns into kinetic energy. 2. First Law- Energy cannot be created or destroyed but it can be changed from one form to another. Second Law- Energy cannot be changed from one form to another without loss of usable energy. 3. Cells carry out many energy requiring processes and reactions to maintain their organization. (5.2) 1. ATP is produced through a process called cellular respiration. ATP is not stored because it is unstable. 2. Myosin combines with ATP prior to its breakdown. Release of ADP + Ⓟ causes myosin to change position and pull on the actin filament. 3. Energy flows from the sun to the Earth through all living things. This flow maintains the levels of biological organization from molecules to organisms to ecosystems. (5.3) 1. Various reactions of a pathway produce a product that is usable in one or more other ways. Also, the reactions of a metabolic pathway allow energy to be used or generated in some increments. 2. A slight change in shape of the active site of an enzyme allows the substrate to fit more easily into that active site. 3. It ensures that the enzyme is only active when product levels are low and more product is needed. (5.4) 1. All substances move from an area of higher concentration to an area of lower concentration and no energy is required. 2. The size of the cell is affected, whether it swells (in a hypotonic solution) or shrivels (in a hypertonic solution). 3. Exocytosis transports materials out of a cell; endocytosis transports materials into the cell; phagocytosis transports large materials into the cell, normally food; pinocytosis transports large amounts of liquid into the cell; receptor-mediated endocytosis forms a vesicle around molecules and their receptors.

Testing Yourself

1. d; 2. e; 3. d; 4. d; 5. d; 6. a, d; 7. b, c; 8. b, c; 9. a, d; 10. a. without enzyme; b. with enzyme; see Fig. 5.10, p. 85, in text; 11. c; 12. a, b, c; 13. b, c, d; 14. d; 15. d; 16. a; 17. b; 18. c; 19. b; 20. e; 21. a; 22. d; 23. c; 24. b; 25. a; 26. b; 27. b; 28. c; 29. b

Thinking Scientifically

1. Using GTP instead of ATP as an energy source for a subset of enzymes allows the cell to maintain separate "pools" of energy available for these reactions. For example, using GTP to fuel protein synthesis allows a cell to continue making proteins even when energy levels (the ATP pool) are low. It also allows protein synthesis to continue when oxidative phosphorylation, the process by which most of a cell's ATP is produced, becomes blocked. Since GTP is produced directly from the citric acid cycle, energy for protein synthesis is still available. 2. A normal channel protein allows chloride to leave the cells. When chloride leaves, water follows by osmosis, keeping the outer surfaces hydrated. Since

the defective cystic fibrosis chloride channel does not allow chloride to leave the cell, water remains inside, and the cell surface will be covered with a thick mucus. 3. The composition of the archaeal plasma membrane is different from that of bacteria and eukaryotes. Furthermore, archaeal enzymes have evolved to withstand this high temperature.

CHAPTER 6
Check Your Progress

(6.1) 1. Photosynthesis produces carbohydrates in plants and most food chains lead back to plants. A by-product of photosynthesis is oxygen which is needed to sustain most life on Earth. 2. Stomata- small holes in leaf surface for gas exchange; chloroplasts- organelles that carry on photosynthesis; chlorophyll- pigment in chloroplasts that absorb energy. 3. Reduction- when a molecule gains hydrogen atoms; Oxidation- when a molecule gives up hydrogen atoms. (6.2) 1. During light reactions the pigments in the thylakoid membrane absorb solar energy and store the energy in the form of a hydrogen ion gradient. 2. Electrons are removed from water as it is split by PS II. They are passed from PS II to PS I through an electron transport chain and are transferred to $NADP^+$ as it becomes NADPH to capture energy. 3. The thylakoid membrane is structured to include PSII, PSI, ATP synthase complex, and an electron transport chain; this structure allows the light reactions to occur. (6.3) 1. Carbon dioxide fixation, carbon dioxide reduction, and regeneration of RuBP. 2. The light reactions occur in the membrane of the thylakoid and capture and store the sun's energy as a hydrogen ion gradient. Calvin cycle reactions occur in the stroma of the chloroplasts and produce carbohydrates. 3. Since plants are limited in their ability to obtain nutrients, they must make everything they need from the nutrients they have. (6.4) 1. C_3 plants include columbines, carbon dioxide is taken up by the Calvin cycle directly into the mesophyll cells and the first detectable molecule is a C_3 molecule. In C_4 plants, like corn, carbon dioxide fixation results in C_4 molecules in the mesophyll cells. The C_4 molecule releases carbon dioxide to the Calvin cycle in the bundle sheath cells. 2. The mesophyll cells are exposed to carbon dioxide and the bundle sheath cells, where the Calvin cycle occurs, are not. 3. It allows these plants to live in warm, arid regions.

Testing Yourself

1. c; 2. a; 3. d; 4. d; 5. a; 6. a; 7. b; 8. d; 9. a. stroma; b. electron transport chain; c. ATP; d. NADPH; e. ATP synthase complex; f. thylakoid membrane; g. thylakoid space; see Fig. 6.7, p. 100, in text; 10. b; 11. c; 12. a, b, c; 13. b; 14. b; 15. c; 16. d; 17. a; 18. d; 19. d; 20. a; 21. d

Thinking Scientifically

1. The alga was producing O_2 as a byproduct of photosynthesis wherever its chlorophyll was able to absorb the needed light energy. The bacteria were congregated wherever this O_2 was produced, and were using it for cellular respiration. 2. Both

Appendix A   Answer Key

CHAPTER 1

Check Your Progress

(1.1) 1. From smallest to largest the levels of biological organization are: small molecule, large molecules, cells, tissues, organs, organ systems, complex organism. The basic characteristics of life are: living things are organized, they acquire materials and energy, they maintain an internal environment, they respond to stimuli, they can reproduce and develop, and they have adaptations. **2.** Homeostasis- the maintenance of internal conditions within certain boundaries. It is important for metabolic processes to continue in order for living organisms to survive. **3.** Living organisms having adaptations to their environments and being able to reproduce and share new traits with offspring are two examples representing evolution in living organisms. **(1.2) 1.** Species, genus, family, order, class, phylum, kingdom, domain. **2.** Natural selection is the process in which populations adapt to the environment through variation in their DNA and production of more offspring than resources can support leading to competition for resources. Natural selection is an important biological concept that explains population changes, like those seen in antibiotic resistant bacteria. **3.** Descent with modification shows common ancestry and diversity, both important to the study of evolution. **(1.3) 1.** Communities are populations interacting with each other; an ecosystem is the community and its interactions with the physical environment. **2.** Energy flows one way from the sun through populations; chemicals are recycled through populations. **3.** Pro: Advances in technology and farming improve crops; Con: Pollution. **(1.4) 1.** Observation, hypothesis, experiment/observations, conclusion, scientific theory. **2.** Actual experiments may vary. Controlled studies are important to determine if the results are valid by having a comparison of a variable. **3.** The conclusion of one experiment can often lead to the beginning of another experiment; improvements on past studies. **(1.5) 1.** Branch of ethics concerned with development and consequences of biological technology, which has an impact on everyone. **2.** Biodiversity leads to variation which is needed for adaptation and evolution. **3.** The extinction of species will lead to an overall loss of biodiversity.

Testing Yourself

1. d; **2. a.** small molecules; **b.** large molecules; **c.** cells; **d.** tissues; **e.** organs; **f.** organ systems; **g.** complex organisms; **3.** b; **4.** d; **5.** c; **6.** c; **7.** e; **8.** d; **9.** b; **10.** b; **11.** d; **12.** c; **13.** b; **14.** c; **15.** c; **16.** c; **17.** a; **18.** b

Thinking Scientifically

1. Domain Archaea gave rise to Eukarya. Kingdom Protista gave rise to plants, animals, and fungi. **2.** Nasal mucus contains an antibacterial chemical. Tears also contain an antibacterial chemical.

CHAPTER 2

Check Your Progress

(2.1) 1. Nucleus with protons (+) and neutrons (neutral) surrounded by shells of electrons (–). **2.** Atoms are stable, isotopes can be unstable due to varying numbers of neutrons. Radioactive isotopes can be used in PET scans and to kill cancer cells. **3.** Both are types of chemical bonds. Covalent bonds share pairs of elections; ionic bonds have one atom giving an electron to another atom. **4.** If an atom has two or more shells, the outer shell (the valence shell) is most stable when it has eight electrons. The valence shell determines reactivity. **5.** Reactants are molecules that participate in a reaction, products are molecules formed by reactions. Balancing means the same number of each type of atom occurs on both sides of the arrow. **(2.2) 1.** Water is a solvent, its molecules are adhesive and cohesive, it has a high surface tension, it has a high heat capacity, it is less dense than ice. **2.** Water is a polar molecule and its special properties come from its ability to form hydrogen bonds. **3.** Plant: Adhesion and cohesion allow water to travel from the roots to the leaves; Human: high heat capacity allows sweat to cool us; Fish: Water less dense than ice allows freezing from top down, protecting fish in winter. **(2.3) 1.** pH scale is from 0-14 going from more acidic to neutral to more basic. **2.** pH scale indicates the number of hydrogen ions in a solution. **3.** Buffers resist pH changes by taking up excess hydrogen or hydroxide ions.

Testing Yourself

1. a; **2.** c; **3.** d; **4.** d; **5.** d; **6.** c; **7.** a; **8.** a; **9.** b; **10.** c; **11.** b; **12.** c

Thinking Scientifically

1. When the bird enters the water, surface tension produced by hydrogen bonding keeps the surface smooth and continuous. When the bird flies out of the water, the cohesiveness of water molecules causes drops of water to be pulled out with the bird. **2.** Silicon, with one extra shell, is too big to form the same varied molecules as carbon.

CHAPTER 3

Check Your Progress

(3.1) 1. Organic molecules contain carbon and hydrogen; these are molecules that make up cells and organisms. **2.** Bonded atoms with the same chemical properties always reacting in the same way regardless of the carbon skeleton to which it is attached; they determine chemical properties. **3.** Carbon can make up to 4 other bonds leading to diversity, and when bonded to another carbon, is very stable. **(3.2) 1.** Monosaccharides- glucose; disaccharides-maltose; polysaccharides- glycogen **2.** Primary-order of amino acids; secondary- alpha helix and pleated sheets; tertiary- globular shape; quaternary-

more than one polypeptide. **3.** Carbohydrates are ringed structures used for energy and structure; lipids have varied structures and can function in energy storage and insulation. **4.** Hydrophilic head groups arrange themselves to interact with water while hydrophobic tail groups crowd inward, away from the water. This creates a bilayer surrounding the cell essential to its structure and function. **5.** Fats and oils are normally chains while steroids are four fused rings. **6.** They are used for the same function but in different manners by different cells due to their structure. **7.** By varying the number and sequence of amino acids. **8.** DNA- phosphate, deoxyribose sugar, nitrogen containing base (cytosine, thymine, adenine, guanine) double stranded; stores genetic information. RNA- phosphate, ribose sugar, nitrogen containing base (cytosine, uracil, adenine, guanine) single stranded; aids in production of proteins. Protein-chain of amino acids; support, metabolism, transport, defense, regulation, motion.

Testing Yourself

1. c; **2.** c; **3.** a; **4.** d; **5.** b; **6.** a; **7.** a; **8.** a; **9.** c; **10.** b; **11.** d; **12.** b; **13.** d; **14.** c; **15.** a; **16.** a; **17.** c; **18.** d; **19.** e; **20.** b; **21.** a; **22.** d; **23.** c; **24.** d; **25.** d; **26.** d; **27.** c; **28.** a; **29.** a; **30.** a; **31.** e; **32.** d; **33.** a

Thinking Scientifically

1. a. The plasma membrane. **b.** It increases, because the plant improves its ability to function at lower temperatures by increasing the proportion of linolenic acid in its plasma membrane. Chilling the plants results in injury due to reduced membrane function because the lipids in the plasma membrane solidify slightly. This does not allow the embedded proteins to float within the membrane as rapidly and impairs the membrane's function. **c.** Canola oil remains liquid, while butter (the saturated fat) is a solid. **d.** The fatty acids within the plasma membrane of cold-tolerant plants are mostly unsaturated, whereas the membrane fatty acids of cold-intolerant plants are more saturated and tend to solidify at lower temperatures. This explains why cold-intolerant plants, such as tropical plants, are often damaged by exposure to low temperatures. **2.** A protein's ability to function correctly is dependent upon its shape. In this case, the amino acid with the polar *R* group may interact with other polar *R* groups within the protein and change the enzyme's shape. **3.** Mutations help drive evolution so that the DNA codes for a different sequence of amino acids in a protein. Different mutations occur as two species evolve separately.

CHAPTER 4

Check Your Progress

(4.1) 1. Microscopes allow us to see cells that are not visible to the naked eye, which includes most cells. **2.** Frog egg- light or eye; animal cell- light or electron; amino acid- electron; chloroplast- light or electron; plant cell- light or electron. **3.** To have

For questions 8–12, indicate the type of energy associated with each statement. Some questions may have more than one answer.

Key:

    a. fossil fuels
    b. hydropower
    c. geothermal energy
    d. wind power
    e. solar energy

8. may require the building of dams
9. minimal environmental impact
10. may lead to inland droughts
11. must be stored at times
12. may be used to extract hydrogen from water

13. A major negative effect of the dumping of raw sewage into lakes and rivers is
    a. oxygen depletion.
    b. the buildup of carbon.
    c. a reduction of light penetration into the water.
    d. an increase in populations of small fish.

14. Show how the following diagram must change in order to develop a sustainable society.

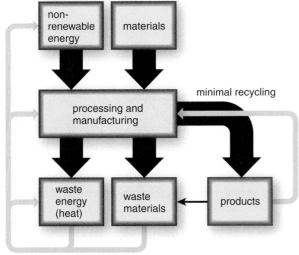

minimal recycling

15. The preservation of ecosystems indirectly provides fresh water because
    a. trees produce water as a result of photosynthesis.
    b. animals excrete water-based products.
    c. forests soak up water and release it slowly.
    d. ecosystems promote the growth of bacteria that release water into the environment.

16. A renewable energy source is
    a. hydropower.   **d.** solar energy.
    b. natural gas.   **e.** Both a and d are correct.
    c. coal.

17. A nonrenewable energy source is
    a. wind.   **c.** solar energy.
    b. geothermal energy.   **d.** fossil fuel.

18. Most fresh water is held in
    a. groundwater.   **d.** ice and snow.
    b. lakes.   **e.** oceans.
    c. rivers.

19. Evaporation of excess water from the irrigation of farmland causes
    a. salinization.
    b. accumulation of mineral salts.
    c. subsidence.
    d. loss of topsoil.
    e. Both a and b are correct.

20. Which of these is an indirect value of species?
    a. participation in biogeochemical cycles
    b. participation in waste disposal
    c. provision of fresh water
    d. prevention of soil erosion
    e. All of these are indirect values.

21. Which of the following is not a result of deforestation?
    a. siltation behind dams
    b. increase in soil nutrients
    c. decrease in local oxygen output
    d. increase in local carbon dioxide

Thinking Scientifically

1. In the early 1960s, Soviet officials built dams on two large Siberian rivers to support a developing agricultural economy based on cotton, a crop with a high demand for water. Since that time, the Aral Sea (which is fed by those two rivers) in central Asia has shrunk to 25% of its original size. Now plans are being discussed to build a canal to divert water from two other Siberian rivers into the dammed ones in order to keep up with agricultural water demands. What might be the positive and negative environmental impacts of connecting the two river systems for the purpose of increasing the supply of water for cotton production? Another option is to remove the dams from the first two rivers and let them flow freely again. What might be the positive and negative impacts of this strategy?

2. Every time a new species becomes extinct, we lose a potential source of human medicine. Since we do not know which species are likely to provide valuable medicines, it is difficult to know where to focus conservation efforts. If you were in charge of determining high priorities for species conservation in a country, with the goal of identifying plants with medicinal value, what factors would you consider in making your decisions?

3. Knowing that organisms are closely adapted to their environment, why might global climate change cause the extinction of mammals even living in the tropics?

Bioethical Issue

Control of Beach Erosion

Beach erosion is a serious problem that is difficult to address. One control strategy has been to build hard structures, such as seawalls and goynes, to reduce the impact of waves. However, these structures are expensive to construct and actually accelerate beach erosion by concentrating the power of waves. Construction of new barriers has recently been banned in several states. The opposite approach to saving beaches is "managed retreat." This method allows coastal erosion to occur naturally by removing human-made beach structures. Several parks have adopted this strategy by removing all hard structures. Should private property owners on beaches be encouraged or forced to remove physical barriers as well? If so, should they be compensated for the loss of beach that will occur as a result of natural erosion? Should incentives be put in place to encourage the planting of native vegetation to reduce erosion?

- Modern farming methods increase the food supply, but some methods harm the land, pollute the water, and consume fossil fuels excessively.
- Transgenic plants can increase the food supply and reduce the need for chemicals.
- Raising livestock contributes to water pollution and uses fossil fuel energy.
- The increased number and high efficiency of fishing boats have caused the world fish catch to decline.

Energy

Most of the world's energy is supplied by the burning of fossil fuels, a nonrenewable resource. The burning of fossil fuels causes pollutants and gases to enter the air.

- Burning fossil fuels and burning to clear land for farming adds CO_2, a greenhouse gas, to the atmosphere. Other gases are emitted by various means. Greenhouse gases allow solar radiation to pass through the atmosphere to the Earth, but infrared heat cannot escape back into space.
- Renewable energy sources include wind power, hydropower, geothermal energy, and solar energy.

Minerals

Minerals are a nonrenewable resource that can be mined. These raw materials include sand, gravel, phosphate, and metals. The act of mining causes destruction of the land by erosion, loss of vegetation, and toxic runoff into bodies of water. Some metals are dangerous to human health. Land ruined by mining can take many years to recover.

Synthetic Organic Compounds

Compounds such as chlorofluorocarbons (CFCs) are detrimental to the ozone layer and to the health of living things, including humans.

Wastes Raw sewage and industrial wastes can pollute land and bodies of water. Some industrial wastes cause biological magnification, by which toxins become more concentrated as they pass along a food chain.

32.2 Biodiversity

Biodiversity is the variety of life on Earth. It has both direct values and indirect values.

Direct Values of Biodiversity

The direct values of biodiversity are

- Medicinal value (medicines derived from living organisms)
- Agricultural value (crops derived from wild plants)
- Biological pest controls and animal pollinators
- Consumptive use values (food production)

Indirect Values of Biodiversity

Biodiversity in ecosystems contributes to

- The function of biogeochemical cycles (water, carbon, nitrogen, phosphorus, and others)
- Waste disposal (through the action of decomposers and the ability of natural communities to purify water and take up pollutants)
- Freshwater provision through the water biogeochemical cycle
- Prevention of soil erosion, which occurs naturally in intact ecosystems
- Climate regulation (plants take up carbon dioxide)
- Ecotourism (human enjoyment of a beautiful ecosystem)

32.3 Working Toward a Sustainable Society

A sustainable society would use only renewable energy sources, would reuse heat and waste materials, and would recycle almost everything. It would also provide the same goods and services presently provided and would preserve biodiversity.

Key Terms

aquifer 634	nonrenewable resources 631
biodiversity 643	photovoltaic (solar) cell 640
biological magnification 642	pollution 632
chlorofluorocarbons (CFCs) 641	renewable resources 631
deforestation 633	salinization 636
desertification 633	saltwater intrusion 635
fossil fuel 638	subsidence 634
greenhouse gases 638	sustainable society 648
mineral 641	

Testing Yourself

Choose the best answer for each question.

1. Beach erosion is caused by
   a. rising sea levels.
   b. mining for minerals.
   c. human development.
   d. preserving coastal wetlands.
   e. Both a and c are correct.

2. Desertification typically happens because
   a. deserts naturally expand in size.
   b. humans allow overgrazing.
   c. desert animals wander into adjacent areas for food.
   d. humans tap into limited water supplies for their water needs in the nearby desert.

3. Soils in tropical rain forests are typically nutrient-poor because
   a. they receive more water than other ecosystems.
   b. they are sandy.
   c. nitrogen-fixing bacteria are absent.
   d. nutrients are tied up in vegetation.

4. Most of the fresh water in the world is used for
   a. drinking.   d. cooking.
   b. supporting industry.   e. bathing.
   c. irrigating crops.

5. Which of the following will help conserve fresh water?
   a. saltwater intrusion
   b. drought-tolerant crops
   c. increased irrigation
   d. use of water contained in aquifers

6. The first green revolution resulted from the development of
   a. high-responder wheat and rice varieties.
   b. effective irrigation systems.
   c. transgenic crop plants.
   d. crops designed for animal feed.

7. The preferred fossil fuel in the United States is
   a. coal because it produces less pollution than oil.
   b. oil because it produces less pollution than coal.
   c. coal because it is more abundant than oil.
   d. oil because it is more abundant than coal.

Connections and Misconceptions

What are some simple things YOU can do to conserve energy and/or water and help solve environmental problems, such as global warming?

The following are a few things you could easily do:

- Change the light bulbs in your home to compact fluorescent bulbs. They use 75–80% less electricity than incandescent bulbs.
- Walk, ride your bike, carpool, or use mass transit. It will save you a lot of money, too!
- Get cloth or mesh bags for groceries and other purchases. Plastic bags may take 10–20 years to degrade. They're also dangerous to wildlife if mistaken for food and consumed.
- Turn off the water while you brush your teeth. If you don't finish a bottle of water, use it to water your plants. In dry climates, plant native plants that won't require frequent watering.

- Improve storm-water management by using sediment traps for storm drains, artificial wetlands, and holding ponds. Increase the use of porous surfaces for walking paths, parking lots, and roads. These surfaces reflect less heat while soaking up rainwater runoff.
- Instead of traditional grasses, plant native species that attract bees and butterflies and require less water and fertilizers.
- Create *greenbelts* that suit the particular urban setting. Include plentiful walking and bicycle paths.
- Revitalize old sections of a city before developing new sections.
- Use lighting fixtures that hug the walls or ground and send light down; control noise levels by designing quiet motors.
- Promote sustainability by encouraging the recycling of business equipment; use low-maintenance building materials, rather than wood.

Video
Fishing for Trouble

Connecting the Concepts

For more information on these topics, refer to the following discussion.

Section 30.2 examines the environmental impacts of human population growth.

Check Your Progress 32.3

1 List the characteristics of a sustainable society.

2 Summarize the steps that could increase rural sustainability.

3 Summarize the steps that could increase urban sustainability.

Media Study Tools

www.mhhe.com/maderessentials3

Enhance your study of this chapter with study tools and practice tests. Also ask your instructor about the resources available through ConnectPlus, including the media-rich eBook, interactive learning tools, and animations.

The Chapter in Review

Summary

32.1 Resources and Pollution

Five resources are maximally used by humans:

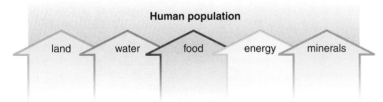

Resources are either nonrenewable or renewable.

Nonrenewable resources are limited in supply:
- Land
- Fossil fuel
- Minerals

Renewable resources are not limited in supply:
- Water
- Solar energy
- Food

Land

Human activities, such as habitation, farming, and mining, contribute to erosion, pollution, desertification, deforestation, and loss of biodiversity.

Water

Industry and agriculture use most of the freshwater supply. Water supplies are increased by damming rivers and drawing from aquifers. As aquifers are depleted, subsidence, sinkhole formation, and saltwater intrusion can occur. If used by industries, water conservation methods could cut world water consumption by half.

Food

Food comes from growing crops, raising animals, and fishing.

It is imperative that we take all possible steps to preserve what remains of our topsoil and replant areas with native grasses, as necessary. Native grasses stabilize the soil and rebuild soil nutrients while serving as a source of renewable biofuel. Trees can be planted to break the wind and protect the soil from erosion while providing a product. Creative solutions to today's ecological problems are very much needed.

Here are some other possible ways to help make rural areas sustainable:

- Plant *cover crops,* which often are a mixture of legumes and grasses, to stabilize the soil between rows of cash crops or between seasonal plantings of cash crops.
- Practice *multiuse farming* by planting a variety of crops and use a variety of farming techniques to increase the amount of organic matter in the soil.
- Replenish soil nutrients through composting, organic gardening, or other self-renewable methods.
- Use low-flow or trickle irrigation, retention ponds, and other water-conserving methods.
- Increase the planting of *cultivars* (plants propagated vegetatively), which are resistant to blight, rust, insect damage, salt, drought, and encroachment by noxious weeds.
- Use *precision farming (PF)* techniques that rely on accumulated knowledge to reduce habitat destruction while improving crop yields.
- Use *integrated pest management (IPM),* which encourages the growth of competitive beneficial insects and uses biological controls to reduce the abundance of a pest.
- Plant a variety of species, including native plants, to reduce our dependence on traditional crops.
- Plant *multipurpose trees*—trees with the ability to provide numerous products and perform a variety of functions, in addition to serving as windbreaks (**Fig. 32.21**). Remember that mature trees can provide many types of products: Mature rubber trees provide rubber, and tagua nuts are an excellent substitute for ivory, for example.
- Maintain and restore wetlands, especially in hurricane- or tsunami-prone areas. Protect deltas from storm damage. By protecting wetlands, we protect the spawning ground for many valuable fish nurseries.
- Use renewable forms of energy, such as wind and biofuel.
- Support local farmers, fishermen, and feed stores by buying food products produced close to home.

Urban Sustainability

More and more people are moving to cities. Much thought needs to be given to serving the needs of new arrivals, without overexpansion of the city. Here are some other ways to help make a city sustainable:

- Design an energy-efficient transportation system to rapidly move people about.
- Use solar or geothermal energy to heat buildings; cool them with an air-conditioning system that uses seawater; and in general, use conservation methods to regulate the temperature of buildings.

  Video
  Solar Cooking

- Utilize *green roofs*—a wild garden of grasses, herbs, and vegetables on the tops of buildings—to assist in temperature control, supply food, reduce the amount of rainwater runoff, and create visual appeal (**Fig. 32.22**).

Figure 32.21   Rural sustainability.

Trees planted by a farmer to break the wind and prevent soil erosion can also have other purposes, such as supplying fruits and nuts.

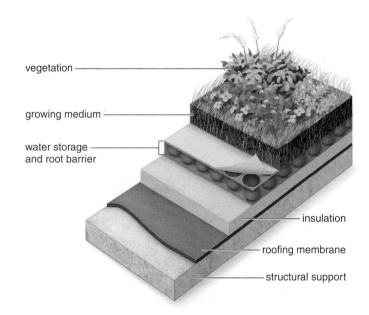

vegetation

growing medium

water storage and root barrier

insulation

roofing membrane

structural support

Figure 32.22   Urban sustainability.

Green roofs are ecologically sound and can be visually beautiful.

expression is controlled by use of operons with a promoter region to control expression of genes and repressors to inhibit gene expression. In eukaryotes, each gene has its own promoter and a variety of mechanisms are used to control gene expression; **2.** At the transcription level of gene expression regulation occurs by chromosome condensation, DNA transcription, and mRNA processing. At the translation level of gene expression regulation occurs by mRNA translation and protein activity; **3.** The signal causes a receptor to initiate a signal transduction pathway, the end result of which affects the metabolism of the cell, such as alteration in gene expression. **(12.3) 1.** Cancer results from an accumulation of mutations over time; **2.** Influences that contribute to cancer development include proto-oncogenes becoming oncogenes like the Ras oncogene that is found in 20-30% of human cancers. Tumor suppressor genes can become inactive as in retinoblastoma. Absence of telomere shortening can occur allowing a cell to continue to divide when it should undergo apoptosis. Chromosomal rearrangements can occur as in the Philadelphia chromosome. Inheritance of cancer causing alleles can occur as in RB and RET genes; **3.** When proto-oncogenes are active, the cell cycle occurs normally and apoptosis is restrained. When tumor suppressor genes are active, the cell cycle is restrained and apoptosis is promoted.

Testing Yourself

1. b; **2.** c; **3.** d; **4. a.** regulatory gene; **b.** promoter; **c.** operator; **d.** lactose metabolizing genes; **e.** mRNA; **f.** active repressor; **g.** RNA polymerase; see Fig. 12.5, p. 201, in text; **5.** b; **6.** d; **7.** c; **8.** a; **9.** b; **10.** c; **11.** d; **12.** c; **13.** c; **14.** c; **15.** e; **16.** e; **17.** d; **18.** c; **19.** a

Thinking Scientifically

1. If a mutation in cyclin D leads to its expression outside of the G_1 stage, or prevents it from being destroyed during the S stage, the tumor suppressor RB would be inhibited at other points during the cell cycle. This would tilt the balance of cell cycle regulation towards promotion of the cell cycle, and rapid growth might result. There are other possible explanations as well. **2.** With alternative mRNA splicing, one gene can serve as the coding sequence for a number of different proteins.

CHAPTER 13

Check Your Progress

(13.1) 1. During amniocentesis, a sample of amniotic fluid is collected. Cells from this fluid are cultured and used for karyotyping; **2.** Chorionic villus sampling can be performed earlier in pregnancy and does not require a cell culture period; however CVS carries a higher risk of miscarriage than amniocentesis; **3.** The chromosomes are the most condensed during metaphase. **(13.2) 1.** A duplication chromosome contains a repeated segment while a deletion chromosome is missing a segment. A translocation is an exchange of chromosome segments between two nonhomologous chromosomes while an inversion occurs when a chromosome segment is turned 180 degrees within the same chromosome. **2.** As long as the translocated chromosomes are inherited together and control of gene expression is not disrupted, the individual still has the normal genotype and would

appear normal; **3.** In an autosomal dominant disorder a child can be unaffected even when the parents are affected (heterozygous). In an autosomal recessive disorder a child can be affected even if the parents are not, as in cystic fibrosis and sickle cell disease. In an X-linked recessive disorder, like color blindness, more males than females have the disorder. In X-linked dominant disorders daughters of affected males have 100% chance of having the disorder and females can pass equally to both sons and daughters. **(13.3) 1.** When testing for a protein, the test looks for a proper protein made from a gene. When testing the DNA, genetic markers show differences in the DNA due to presence of the abnormal allele. Another DNA test is genetic profiling that looks at an individual's entire genome to find abnormal genes; **2.** Ultrasound images help doctors evaluate the fetal anatomy only, not the genome of the fetus. Amniocentesis uses a long needle to withdraw amniotic fluid containing fetal cells. It can be performed between the 14th to 17th week of pregnancy only and has a 0.3% risk of spontaneous abortion. In chorionic villus sampling (CVS) a suction tube is used to remove cells from the chorion where the placenta develops. This procedure can be performed as early as 5 weeks pregnancy and it carries a 0.8% spontaneous abortion rate; **3.** Advantage- Both procedures implant only a healthy embryo into the mother. Disadvantage- Both procedures require in vitro fertilization. Testing the eggs means that only healthy eggs will be fertilized. This has an advantage over testing the embryo in which embryos with genetic defects are currently discarded; **4.** Fetal blood cells still have nuclei, allowing them to be separated from maternal blood cells by using a cell sorter; **5.** The change in base sequence of a mutant allele may cause restriction enzymes to cleave DNA at unexpected places altering the normal fragmentation pattern; **6.** If a woman is heterozygous for a genetic disorder, the first polar body receives one allele while the egg receives the other. If the polar body contains the mutant allele, the egg will probably be genetically normal. **(13.4) 1.** Bone marrow stem cells are capable of dividing after genetic alteration to produce cells carrying the alteration; **2.** Genes may be delivered by spraying them into the nose, or by injecting them into particular organs. Liposomes and adenoviruses may also be used to carry the genes into a cell and organs; **3.** Adenoviruses containing an anti-inflammatory gene are injected into the affected joint to alleviate pain.

Testing Yourself

1. a; **2.** c; **3.** d; **4. a.** *aa*; **b.** *A_*; **c.** *Aa*; **d.** *Aa*; **e.** *Aa*; **f.** *Aa*; **g.** *aa*; **h.** *aa*; **i.** *A_*; see Fig. 13.8, p. 220, in text; **5. a.** *Aa*; **b.** *Aa*; **c.** *aa*; **d.** *Aa*; **e.** *aa*; **f.** *Aa*; **g.** *Aa*; **h.** *aa*; see Fig. 13.9, p. 220, in text; **6. a.** X^bY; **b.** X^BX^b; **c.** X^BY; **d.** X^BY; **e.** X^bY; see Fig. 13.10, p. 221, in text; **7.** b; **8.** a, e; **9.** b, d, e; **10.** a, c; **11.** b, d; **12.** e; **13.** e, f; **14.** a, b, c, d; **15.** f; **16.** h; **17.** b; **18.** g, h; **19.** a; **20.** d; **21.** Parents: *Aa* and *Aa*; child: *aa*; **22.** Yes, the son may inherit the father's recessive allele; **23.** Child: $X^{cb}X^{cb}$; Mother: $X^{CB}X^{cb}$ or $X^{cb}X^{cb}$; Father: $X^{cb}Y$; **24.** b; **25.** b; **26.** d; **27.** c

Thinking Scientifically

1. In many cases, gene dosage effects may cause disease. For example, you have learned that some genetic disorders are dominant because having a single mutant allele of the gene is enough to alter the

phenotype. In these cases, the mutant allele may have an additional function, not possessed by the normal allele, which causes the disorder. Likewise, having a third or fourth copy of a gene may cause too much of the protein to be made, which may lead to a genetic disorder if the increased enzyme activity is harmful. Some genes are already present in multiple copies, for example, rRNA genes. Having more of them would not be expected to cause problems. **2.** The adenovirus may have inserted itself into or nearby a gene that is involved in controlling the cell cycle, such as a proto-oncogene or a tumor suppressor gene. In these cases, disrupting the gene may have inactivated it. In the case cited, the virus was found to have inserted itself near a proto-oncogene called *LMO-2*. **3.** When diseases are homozygous recessive, the recessive allele can remain hidden in a portion of the populace at all times. Only when two carriers mate is there a chance that the offspring will have the disease. Sometimes the recessive trait confers an evolutionary advantage. For example, carriers of sickle cell disease are more resistant to malaria. See Table 15.2, p. 255, in text.

CHAPTER 14

Check Your Progress

(14.1) 1. Prior to Darwin, most people believed the Earth was only a few thousand years old and species had remained exactly the same; **2.** Catastrophists proposed that previous species were eliminated by a catastrophe and new ones replaced them from surrounding areas or they were created anew; **3.** Cuvier founded the science of paleontology. He said that fossils are found in strata, and a change in species can be observed over time; **4.** Use and disuse of structures can produce heritable changes; **5.** Since the most adapted individuals to an environment survive and reproduce, natural selection of traits that allow survival in an environment lead to adaptation; **6.** Without diversity and variation in a gene pool, individuals will not be able to adapt to changing environments and if the environment changes, the species may die out. **(14.2) 1.** The finches show that a common ancestor can give rise to other species through adaptation to unique environments- in this case, the environment on different islands; **2.** These structures show that present day organisms have an ancestor from the past. Vestigial structures are useless today but they were functional in the ancestor; **3.** Homologous structures share an anatomical similarity that reflects a common ancestry. Analogous structures have a common function but do not serve as evidence of a common ancestry.

Testing Yourself

1. b; **2.** d; **3.** e; **4.** c; **5.** b; **6.** b; **7. a.** Originally, giraffes had short necks. **b.** Giraffes stretched their necks in order to reach food. **c.** Today, most giraffes have long necks. **d.** Originally, giraffe neck length varied. **e.** Struggle to exist caused long-necked giraffes to have the most offspring. **f.** Today, most giraffes have long necks; **8.** e; **9.** d; **10.** d; **11.** c; **12.** b; **13.** a; **14.** c; **15.** b; **16.** a; **17.** c; **18.** e; **19.** e; **20.** e; **21.** e; **22.** d; **23.** a

Thinking Scientifically

1. Humans and other animals share this common structure because we share a common ancestor that had such a structure. The dietary history of humans suggests that humans previously ate more plants than they currently do. **2.** Different genes are under varying levels of evolutionary constraint. **3.** Islands have populations that are both isolated and with different environments from a mainland population.

CHAPTER 15

Check Your Progress

(15.1) 1. Directional selection occurs when one extreme phenotype is favored; stabilizing selection occurs when the intermediate phenotype is favored; **2.** Since the intermediate phenotype is selected against, two distinctly different phenotypes are seen in the population; **3.** Mutation, recombination, gene flow, disruptive selection, and diploidy; **4.** Sickle cell disease occurs at a higher incidence in populations subject to malaria because the heterozygote is favored by natural selection in these regions. **(15.2) 1.** The various alleles at all the gene loci in all individuals make up the gene pool of the population. The gene pool is described in terms of allele frequency; **2.** The Hardy-Weinberg principle states that an equilibrium of genotype frequencies in a gene pool will remain in effect in each succeeding generation of a sexually reproducing population as long as five conditions are met. Microevolution can be detected and measured by noting the amount of deviation from a Hardy-Weinberg equilibrium of genotype frequencies in the gene pool of a population; **3.** Mutations, gene flow, nonrandom mating, genetic drift, and natural selection; **4.** Mutations are ultimately responsible for all the genetic variability upon which evolution depends; **5.** Gene flow allows migration into or out of populations allowing microevolution to occur. Nonrandom mating has individuals paired according to their genotypes and phenotypes leading to microevolution; **6.** The chance that a rare genotype might not or might participate in passing alleles to the next generation is greater. The result would be a decrease or increase in the frequency of this genotype.

Testing Yourself

1. a; **2.** d; **3.** c; **4.** a; **5.** a; **6.** c; **7.** d; **8.** b; **9.** a; **10.** c; **11.** b; **12.** b; **13.** c; **14.** c; **15.** c; **16.** c; **17.** c; **18.** b; **19.** a; **20.** a; **21.** Curve is shifted to one extreme or the other; **22.** Curve narrows, but mean remains the same; **23.** The one curve splits into two curves, each curve having its mean at a value that is now seen as an extreme value.

Thinking Scientifically

1. a. The most fit individuals of a population are the ones that reproduce more than other individuals. Therefore, it is adaptive for the male to be aggressive during nest construction because mating hasn't taken place yet. Once the egg is laid, the male has the assurance that the offspring is his and energy should be conserved. After the offspring has hatched, it would be disadvantageous to spend energy on aggression. **b.** If genetic drift of alleles was causing the male to be aggressive, he would be aggressive all the time without regard to the reproductive cycle.

CHAPTER 16

Check Your Progress

(16.1) 1. A biological species is a group of organisms that can interbreed. The members of this species are reproductively isolated from members of other species; **2.** The biological species concept cannot be applied to asexually reproducing organisms. In addition, reproductive isolation may not be complete, so some breeding between species may occur; **3.** Prezygotic isolating mechanisms do not allow zygotes to be formed, as in habitat, temporal, behavioral, mechanical, and gametic isolation. Postzygotic mechanisms allow zygotes to be formed, but the zygotes either die or become offspring that are infertile, as in zygote mortality, hybrid sterility, and F_2 fitness; **4.** Both result in reproductive isolation. Allopatric speciation occurs when a geographic barrier stops gene flow, while sympatric speciation produces reproductively isolated groups without geographic isolation; **5.** One mainland finch species was probably ancestral to all existing species on the Galapagos Islands. Each new species evolved in response to the unique environment on the island it inhabits. **(16.2) 1.** Fossils allow you to see changes occurring over time; **2.** The fossil record shows times of equilibrium (no change) being interrupted by relatively sudden, rapid changes; **3.** Continental drift has contributed to several extinctions, so has loss of habitat, and the crash of a meteorite into the Earth; **4.** During the Precambian era Earth forms. During the Paleozoic era jawed fishes diversify and seedless vascular plants appear. During the Mesozoic era the first mammals and dinosaurs appear. During the Cenozoic era the first hominids appear. **(16.3) 1.** Species, genus, family, order, class, phylum, kingdom, domain; **2.** The fossil records, anatomical homologies, and molecular data are used; **3.** Homologous structures are similar in anatomy because they are derived from a common ancestor. Analogous structures have the same function in different groups and do not share a common ancestry; **4.** Because rRNA molecules are found in ribosomes that carry out protein synthesis, their nucleotide sequence changes very little over time. Therefore, similarities in rRNA sequence can be reliably used to determine who is related to whom; **5.** Linnaean classification sometimes fails to place common ancestors and all their descendants in the same group even though they all share the same derived traits; **6.** C is a fish because it shares a common ancestor with B and D.

Testing Yourself

1. c; **2.** c; **3.** b; **4.** f; **5.** a; **6.** g; **7.** d; **8.** e; **9.** h; **10. a.** species 1; **b.** geographic barrier; **c.** genetic changes; **d.** species 2; **e.** genetic changes; **f.** species 3; **11.** b; **12.** a; **13.** c; **14.** a, b, c; **15.** d; **16.** b; **17.** e; **18.** e; **19. a.** 3, different colors; **b.** vertebrae, amniotic egg and internal fertilization; **c.** snake and lizard, they share the most derived traits; **20.** b; **21.** a; **22.** c

Thinking Scientifically

1. a. ferns; **b.** produce seed; **c.** naked seeds; **d.** conifers, needle-like leaves; **e.** Ginkgos, fan-shaped leaves; **f.** enclosed seeds; **g.** monocots, one embryonic leaf; **h.** eudicots, two embryonic leaves. **2.** Species get stranded on islands that are far away from a mainland and adapt to the island's environment, often different from the mainland environment from which they came. The Florida Keys are too close together and too close to the Florida mainland. Species are able to move among the islands and thus do not speciate.

CHAPTER 17

Check Your Progress

(17.1) 1. Viruses are noncellular structures composed of an outer capsid of protein subunits and an inner core of DNA or RNA; **2.** Viruses attach to their host cell and enter the cell. In a lytic lifecycle, the virus takes over the metabolic machinery of the cell and begins to make copies of the viral genome and capsid. It assembles progeny viruses that escape from the cell, either by bursting the cell or budding through the cell membrane. In a lysogenic lifecycle, the virus does not begin replication, but it may do so in the future. The virus is said to be latent; **3.** Emerging viruses can occur by being transported from one place to another, due to high mutation rates, or change in mode of transmission. **(17.2) 1.** A viroid is composed of naked RNA, while a prion is composed of protein; **2.** Kuru is a prion disorder causing a fatal infection and a neurodegenerative disorder. Mad cow disease is another prion disorder resulting in a neurodegenerative disorder; **3.** Prions are derived from normal proteins that have become prions due to a change in shape. **(17.3) 1.** They were anaerobes and did not use oxygen; **2.** Cocci (spheres), bacilli (rods), and spirilla or spirochetes (spirals); **3.** Conjugation (transfer of DNA from cell to cell), transformation (transfer of DNA from environment to cell), and transduction (transfer of DNA using a virus); **4.** Bacterial DNA is a closed circle of double stranded DNA whereas eukaryotic DNA is a linear double strand; **5.** A cell-like structure with an outer membrane, most likely formed through the abiotic synthesis of component macromolecules; **6.** The plasma membrane and cell wall of archaea are unusual which allows them to survive in extreme environments. They also have unusual metabolic characteristics. Because of these, they do not compete with bacteria. **(17.4) 1.** Double membranes and distinct genomes; **2.** Both algae and protozoans are protists and belong to domain Eukarya. Algae are photoautotrophs; protozoans are chemoheterotrophs; **3.** Water molds have a cell wall containing cellulose while slime molds do not have a cell wall. Water molds form flagellated spores while slime mold spores are windblown.

Testing Yourself

1. c; **2. a.** capsid; **b.** protein subunit; **c.** DNA/ nucleic acid; **d.** spike; **3.** d; **4.** b; **5.** a; **6.** b; **7.** d; **8.** c; **9.** c; **10.** c; **11.** c; **12.** a; **13.** b; **14.** a; **15.** d; **16.** c; **17.** c; **18.** d; **19.** a; **20. a**; **21. a.** contractile vacuole (partially full); **b.** pellicle; **c.** cilia; **d.** food vacuole; **e.** oral groove; **f.** macronucleus; **g.** micronucleus; **h.** gullet; **i.** anal pore; **j.** contractile vacuole (full)

Thinking Scientifically

1. As members of the domain Eukarya, protists are more closely related to archaea than to bacteria. Archaea and eukarya have certain ribosomal proteins in common that bacteria do not share; they initiate transcription in the same fashion, and have similar tRNAs. Viruses are not shown on the tree because they are acellular and are not considered to belong to any of the three domains of life. **2.** Bacteriophages could infect bacteria in a patient's body and initiate a lytic cycle that would kill the bacteria. The most important benefit would be the ability to treat infections of antibiotic-resistant bacteria against which drugs are ineffective. The main shortcoming is that bacteria would almost certainly evolve resistance to bacteriophage infection.

CHAPTER 18

Check Your Progress

(18.1) 1. Both charophytes and plants contain chlorophylls *a* and *b* plus accessory pigments, store carbohydrates as starch, have cellulose in their cell walls, and form structures that protect and nourish the zygote; **2.** Protection of the embryo, the presence of vascular tissue, presence of megaphylls, seed production, and the presence of flowers; **3.** Meiosis forms a haploid spore from a diploid sporangium, which undergoes mitosis and forms gametes. Following fertilization, they become a diploid zygote that develops into a sporophyte. **(18.2) 1.** Nonvascular plants do not have true roots, stems, or leaves like vascular plants do; **2.** Plants provide food, shelter, and habitats to organisms in the ecosystem as well as provide oxygen to the atmosphere. Forests are important to the water and carbon cycle; **3.** During pollination, the pollen grain germinates and a nonflagellated sperm travels in a pollen tube to the egg produced by the female gametophyte. Following fertilization the zygote becomes the sporophyte embryo; tissue within the ovule becomes the stored food and the ovule becomes the seed coat; **4.** Sporophyte produces spores by meiosis; microspores become pollen grain and megaspore becomes embryo sac; double fertilization produces fertilized egg and 3n endosperm nucleus; ovule becomes seed where endosperm is taken up by cotyledon; ovary becomes fruit; **5.** Nonvascular plants need to be close to a water source not only to be able to get water for photosynthesis (since they cannot get it from the soil, because they do not have roots) but also for reproduction; **6.** Without gymnosperms and angiosperms, food, chemicals, medicines, and building materials would be economic hardships. **(18.3) 1.** In black bread mold, meiosis produces spores in sporangia. In a mushroom, meiosis produces spores in clublike structures attached to the gills; **2.** When fungi secrete enzymes that digest the remains of plants and animals in the environment, inorganic nutrients are returned to plants; **3.** A lichen is composed of a fungus and a cyanobacterium or a green alga. Each partner gives to the other. The fungus provides protection, water, and minerals, and the photosynthesizer provides organic nutrients.

Testing Yourself

1. a; **2.** e; **3.** e; **4.** b; **5. a.** sporophyte (2n); **b.** meiosis; **c.** spore (n); **d.** gametophyte (n); **e.** gametes (n); **f.** zygote (2n); **6.** c, d, e; **7.** a, b, c, d, e; **8.** d, e; **9.** a; **10.** d; **11.** a, b, c; **12.** e; **13.** a, b, c, d, e; **14. a.** anther; **b.** filament; **c.** stigma; **d.** style; **e.** ovary; **f.** ovule; **g.** sepals (calyx); **h.** petals (corolla); **15.** e; **16.** b; **17.** b; **18.** e; **19.** e; **20.** c; **21.** e; **22.** a; **23.** b; **24.** d; **25.** e

Thinking Scientifically

1. The original soil contains mycorrhizal fungi, which have mutualistic relationships with plant roots and enhance the growth of plants. **2.** Protists, plants, and fungi all belong to Domain Eukarya. The protist ancestors of plants resembled certain green algae, possessing chlorophylls *a* and *b* and accessory pigments for photosynthesis, storing carbohydrates as starch, having cellulose in their cell walls, and utilizing an alternation of generations life cycle. The protist ancestors of fungi probably resembled modern chytrids in having flagellated gametes and spores, a feature that has since been lost in the evolution of the other fungal groups. Cell walls may have contained chitin, a feature shared by all modern fungi, including chytrids. Fungal ancestors may have had an alternation of generations life cycle, as do some chytrids.

CHAPTER 19

Check Your Progress

(19.1) 1. Differences in symmetry (bilateral and radial), cephalization (localization of brain and specialized sensory organs), and segmentation (repetition of body parts along the length of the body); **2.** Protostomes are characterized by spiral cell division when the embryo first forms and a blastopore that becomes the mouth. If coelom is present, it is formed by the splitting of the mesoderm. Deuterostomes are characterized by radial cell division when the embryo first forms and a blastopore that becomes the anus. The primitive gut outpockets to form the coelom; **3.** Without bilateral symmetry, animals would not develop with definite right and left sides and all animals would have radial symmetry. **(19.2) 1.** Sponges have cellular level of organization and cnidarians have tissue level of organization; **2.** Both are invertebrates. Sponges have sac-like bodies perforated by many holes, water enters through pores and circulates past collar cells before entering the osculum. Cnidarians are invertebrates with radial symmetry and tentacles with cnidocytes in either a polyp or medusa body style; **3.** Cnidarians probably cannot regenerate since they do not have radial symmetry. **(19.3) 1.** The *lopho* portion of their name is derived from a tentacle-like feeding structure called a lophophore. The *trocho* portion of the name refers to a larval stage, called a trochophore, that is characterized by a distinct band of cilia; **2.** Flatworms have bilateral symmetry and are acoelomates. Annelids are segmented and have a well-developed coelom; **3.** In gastropods the foot is ventrally flattened, in cephalopods the foot has evolved into tentacles about the head, and in bivalves the muscular foot projects ventrally from the two-parted shell. **(19.4) 1.** Roundworms secrete a nonliving cuticle and undergo molting; **2.** Jointed appendages, exoskeleton, segmentation, well developed nervous system, variety of respiratory organs, and reduced competition through metamorphosis; **3.** Insects may be so diverse on Earth because most show remarkable behavior adaptations. **(19.5) 1.** Echinoderms have radial symmetry as adults and a water vascular system with tube feet for locomotion; **2.** Mammals are amniotes and share an ancestor with reptiles; **3.** Cartilaginous fish have a skeleton made of cartilage instead of bone, as seen in bony fish; **4.** The chordates have a dorsal supporting rod, dorsal tubular nerve cord, pharyngeal pouches, and a tail; **5.** Amphibians have jointed limbs, eyelids for keeping their eyes moist, ears for picking up sound waves, a sound producing larynx, and lungs in adults; **6.** Mammals have adapted to most of the environments on Earth. **(19.6) 1.** Several groups of extinct humanlike hominins preceded the evolution of modern humans. These groups have been divided into the early hominins, the australopithecines, and Homo; **2.** The multiregional continuity hypothesis proposes that evolution to modern humans was essentially similar in several different places. The out of Africa hypothesis proposes that *H. sapiens* evolved from *H. erectus* only in Africa then *H. sapiens* migrated to Europe; **3.** There could possibly have been "missing links" between australopithecines, *Homo habilis*, or *Homo erectus* that displayed many intermediate traits and behaviors.

Testing Yourself

1. b; **2. a.** ectoderm; **b.** mesoderm; **c.** endoderm; **d.** ectoderm; **e.** mesoderm; **f.** endoderm; **g.** pseudocoelom; **h.** ectoderm; **i.** mesentery; **j.** endoderm; **k.** coelom; **l.** mesoderm; see Fig. 19.8, p. 339, in text; **3.** d; **4.** g; **5.** a, b, c, d, e, g; **6.** a, b, c, f; **7.** a; **8.** e; **9.** c; **10.** f; **11.** b; **12.** e; **13.** c; **14.** a; **15.** a; **16.** a; **17.** d; **18.** a; **19.** b; **20.** e; **21.** a; **22.** e; **23.** a; **24.** b; **25.** d; **26.** d; **27.** b; **28.** b; **29.** a; **30.** a.

Thinking Scientifically

1. The recent findings support the idea that a constant lineage exists in Africa from primitive to modern humans. However, it does not prove that this lineage exists only in Africa. This discovery of similar fossils in Asia and Europe would provide support for the multiregional continuity hypothesis. **2.** Animals that are sessile tend to be radially symmetrical because their food comes to them from all directions. There is no need to have anterior and posterior body regions. Animals that move through their environment are bilaterally symmetrical, with the anterior portion containing sensory organs. This allows the animal to sense and respond to the environment as it travels through it.

CHAPTER 20

Check Your Progress

(20.1) 1. Vegetative organs include the leaves, stem, and root; **2.** Some leaves (tendrils) attach to objects, some leaves capture insects, and others (spines) protect the plant from being eaten; **3.** Monocots; xylem and phloem in a ring in the root; scattered vascular bundles in the stem; parallel leaf veins. Eudicots; phloem located between arms of xylem in the root; vascular bundles form a ring in the stem; netted leaf veins. **(20.2) 1.** Epidermal tissue; epidermal cells; ground tissue; parenchyma, collenchyma, and sclerenchyma cells; vascular tissue: xylem (vessel elements and tracheids) and phloem (sieve tube members and companion cells); **2.** Xylem transports water and minerals from root to leaves. Phloem transports organic compounds throughout the plant; **3.** Without ground tissue, plants could not properly function; without epidermal tissue the plant would not have a protective covering. **(20.3) 1.** Most of the epidermis is composed of tightly packed epidermal cells covered by a waxy cuticle. Many stomata and their guard cells are also present. Sometimes hairs or glands play a protective role; **2.** Simple leaf has one blade off a petiole. Compound leaf has several leaflets off a blade off a petiole. A twice compound leaf has multiple blades with multiple leaflets off a petiole; **3.** The leaf contains stoma for gas exchange and veins that carry nutrients to and from the leaf. The mesophyll contains large amounts of surface area for gas exchange and water loss. (**20.4**) **1.** Nonwoody stems only experience primary growth and contain epidermis, cortex, and pith. Woody stems experience both primary and secondary growth and contain bark, wood, and pith; **2.** Bark is composed of cork, cork cambium, cortex, and phloem; **3.** The new secondary xylem produced by the vascular cambium every growing season builds up, forming the annual rings of wood. An annual ring often consists of spring wood followed by summer wood. In the spring, when water is plentiful, wood contains wide vessels with thin walls. In the summer, when conditions are dry, wood contains numerous thick walled tracheids and also fibers. (**20.5**) **1.** The root apical meristem is located at the tip of the root and is covered by the root cap; **2.** Cortex: food storage; endodermis: control of mineral uptake; pericycle: transports water, minerals and organic nutrients; epidermis: protection and may contain root hairs; **3.** Both roots and stems have apical meristems that allow them to increase in length. They are capable of branching; stems branch at lateral buds, while root branches originate in the pericycle. (**20.6**) **1.** Carbon, hydrogen, oxygen, calcium, phosphorus, potassium, nitrogen, sulfur,and magnesium; **2.** Bacteria in soil or root nodules fix aerial nitrogen. Those in the soil form nitrate or ammonium that plant roots absorb. Those in root nodules form nitrogen compounds that are passed to the plant; **3.** Without mutualistic fungi, plants would not be able to obtain necessary inorganic nutrients. (**20.7**) **1.** Transpiration, evaporation of water at the leaves, causes water to move upward in xylem; **2.** Because of cohesion and adhesion, a continuous water column exists in xylem. Without a continuous sequence of water molecules, transpiration would be unable to pull water from roots to the leaves; **3.** Sugar is transported into phloem at a source, when water follows by osmosis, a pressure builds that causes phloem sap to flow toward a sink, where sugar is transported out of phloem.

Testing Yourself

1. c; **2.** b; **3.** c; **4.** c; **5. a.** terminal bud; **b.** blade (leaf); **c.** lateral bud; **d.** node; **e.** shoot system; **f.** vascular tissues; **g.** root system; see Fig. 20.1, p. 371, in text; **6.** b; **7.** a; **8.** a, b, c, d; **9.** b; **10.** a, b, c; **11.** b; **12.** c; **13.** a; **14.** d; **15. a.** cuticle; **b.** upper epidermis; **c.** palisade mesophyll; **d.** leaf vein; **e.** spongy mesophyll; **f.** lower epidermis; **g.** guard cell; see Fig. 20.9, p. 377, in text; **16. a.** epidermis; **b.** cortex; **c.** vascular bundle; **d.** xylem; **e.** phloem; **f.** pith; see Fig. 20.11*a*, p. 378, in text; **17. a.** endodermis; **b.** pericycle; **c.** xylem; **d.** phloem; **e.** cortex; **f.** epidermis; **g.** root hair; **h.** apical meristem; **i.** root cap; **j.** zone of cell division; **k.** zone of elongation; **l.** zone of maturation; see Fig. 20.13, p. 381, in text.

Thinking Scientifically

1. There are unique cell recognition proteins on the surface of the nitrogen-fixing bacteria that the plant recognizes. **2.** When the stomata open at night, both CO_2 and water in the fog are available.

CHAPTER 21

Check Your Progress

(**21.1**) **1.** Auxins, gibberellins, cytokinins, abscisic acid, and ethylene; **2.** Auxins- plant growth and development; gibberellins- elongation of cells; cytokinins- influence plant growth and prevent aging in plants; abscisic acid- closes stomata; ethylene-leaf, fruit, and flower drop promotion; **3.** They are probably found in different areas of the plant to regulate those areas. (**21.2**) **1.** Both are differential growth responses to external stimuli. Whereas positive tropisms cause plant growth toward a stimulus, negative tropisms cause plant growth away from a stimulus; **2.** Short-day plants flower when the night is longer than a critical length; long-day plants flower when the night is shorter than a critical length; day-neutral plants do not depend on day/night length to stimulate flowering; **3.** Without phytochrome plants would not be able to function in detection of photoperiod stimulation of seed germination or development of normal stems and leaves. (**21.3**) **1.** Sepals protect the petals; petals attract pollinators; stamen consists of anther and filament and are the "male" portion; carpel has stigma, style, ovary, and ovules and is the "female" portion; **2.** After fertilization of haploid microspore and megaspore, the seed goes through mitosis, a diploid sporophyte develops to produce haploid microspores and megaspores by meiosis; **3.** Seed structure is important so the gamete will be nourished and will not desiccate. These characteristics allow seeds to be dispersed widely and stay viable. (**21.4**) **1.** Plants with seedless fruits are propagated asexually, usually by cuttings; **2.** Both are ways of producing many plant embryos from a single plant. Plants propagated in tissue culture can differ when mutations arise, but the plant embryos derived from meristem culture are identical; **3.** Pros- GMPs may be more tolerant of high salt, drought, and cold; they may be more disease and pest resistant; they may have improved nutritional quality; and they may also be modified in order to make products of medical importance to human. Cons- people might be allergic to GMPs and GMPs could also harm the environment.

Testing Yourself

1. d; **2.** a; **3.** e; **4.** a; **5.** a; **6.** e; **7.** b; **8.** c; **9.** a; **10.** b; **11. a.** sporophyte; **b.** meiosis; **c.** microspore; **d.** megaspore; **e.** male gametophyte; **f.** female gametophyte; **g.** egg; **h.** sperm; **i.** zygote; **12.** d; **13.** d; **14.** b; **15.** d; **16.** b; **17.** a; **18.** b; **19.** a; **20.** e; **21. a.** stigma; **b.** style; **c.** carpel; **d.** ovary; **e.** ovule; **f.** receptacle; **g.** peduncle; **h.** sepal; **i.** petal; **j.** stamen; **k.** filament; **l.** anther; **22.** a; **23.** b; **24.** c; **25.** d; **26.** e; **27.** a; **28.** d; **29. a.** seed coat; **b.** embryo; **c.** cotyledon; **d.** seed coat; **e.** first true leaves; **f.** cotyledons (two); **g.** seed coat

Thinking Scientifically

1. When plants are shipped long distances, they produce ethylene as a result of the stress. Ethylene builds up within the shipping container, and leaf abscission is stimulated. The plastic sleeves promote ethylene buildup near the plant, amplifying the abscission response. **2.** You could determine whether flowers or stems are responsible by cutting the flowers off the stems and looking to see whether the stems still exhibit heliotropism. If so, then the stems rather than the flowers are responsible. You could determine which portion of the stem is responsible for heliotropism by shading different portions of the stem and determining which treatment blocks the response. **3.** Although there are few pollinators available, plants that bloom at this time of year generally do not have to compete with other species to attract pollinators that are available. The flowers of early-blooming species tend to be low-growing to protect them from cold snaps that may occur, often bear hairs that will keep frost away from the plant, and sometimes close up at night to preserve warmth.

CHAPTER 22

Check Your Progress

(**22.1**) **1.** Epithelial- covering and lining; connective-binds and supports; muscular- locomotion; nervous-receives stimuli and conducts nerve impulses; **2.** Epithelial cells can be squamous, cuboidal or columnar and layers can be simple (one layer of cells), striated (2 or more layers), or pseudostratified (one layer, but looks like stratified); **3.** Epithelial covers and lines organs because of the high amount of wear and tear in these areas and the ease with which epithelial tissue can be repaired; **4.** Both types can be further subdivided. Epithelial tissue contains compacted cells in layers; connective tissue is composed of many different types of cells within a matrix of fibers and ground substance; **5.** Matrix of the different connective tissues relates to its functions- bone has a hard matrix for stability, blood has a liquid matrix for circulation, and cartilage has a semi-fluid matrix for flexibility; **6.** Nervous tissue receives stimuli from receptors in muscle tissue and sends nerve impulses to the muscle tissue in response. (**22.2**) **1.** Both are transport systems. The cardiovascular system carries blood throughout the body, while the lymphatic system absorbs fat from the digestive system and collects excess fluid and returns it to the blood; **2.** Both the nervous and endocrine systems coordinate and regulate the functions of other systems; **3.** Cardiovascular-transports nutrients and gases, removes wastes; Lymphatic and immune- collects excess fluids; Respiratory- exchanges gas with blood; Urinary- rids blood of wastes; Digestive- receives food and digests into nutrient molecules for entry into the blood; Nervous- receives stimuli and conducts impulses; Endocrine- secretes hormones; Integumentary-protection; Skeletal- protection, movement; Muscular- protection, movement; Reproductive-production of gametes. (**22.3**) **1.** External environment is the area surrounding the organism, internal environment is inside the organism; **2.** A sensor detects a change in the internal environment; the control center initiates an effect that brings conditions back to normal and the sensor is no longer activated; **3.** First the control center brings about constriction of the blood vessels of the skin. If the temperature continues to drop, shivering begins in order to generate heat. Blood vessels of the skin dilate and sweat glands are activated in response to a rise of internal temperature.

Testing Yourself

1. a. cell; **b.** tissue; **c.** organ; **d.** organ system; **e.** organism; **2.** c; **3.** b; **4.** a, b; **5.** a; **6.** a, b, c, d; **7.** a, b, c; **8.** c, d; **9.** c; **10. a.** white blood cell; **b.** plasma; **c.** red blood cells; **d.** platelets; **11.** b; **12.** a; **13. a.** dendrite; **b.** cell body; **c.** nucleus; **d.** nucleus of Schwann cell; **e.** axon; **14.** e; **15.** c; **16.** d; **17.** d; **18.** d; **19.** b; **20.** a; **21.** e; **22.** b; **23.** e; **24.** c; **25.** b; **26.** d; **27.** b; **28.** a; **29.** a; **30.** c; **31.** c

Thinking Scientifically

1. Even though the heart is made of cardiac muscle instead of skeletal muscle, the probable link between the two conditions is that types of muscle tissue are involved. The cardiac muscle may deteriorate in a similar way to that of skeletal muscle. **2.** There is interaction between the internal and external environments because the body has the nervous system for protection, keeping homeostasis of the internal environment no matter what occurs in the external. It would not be possible; loss of interaction would possibly result in injury and death. **3.** All cells need to carry out exchanges with the external environment either directly or indirectly. With an increase in size, specialized organs and organ systems carry out exchanges with the external environment so that cells are able to carry out exchanges with the internal environment, blood, and tissue fluid. Just as a large company uses specialized rooms to carry out its functions so does a large animal use specialized organs that work together within a system.

CHAPTER 23

Check Your Progress

(23.1) 1. Both use a heart to pump fluid. An open system pumps hemolymph through channels and cavities. The hemolymph eventually drains back to the heart. A closed system pumps blood through vessels that carry blood both away from and back to the heart; **2.** In fishes, the one-circuit pathway utilizes a heart with one atrium and one ventricle to send blood to the gill capillaries and then to the systemic capillaries in a single loop. The two-circuit pathway pumps blood to both the pulmonary and systemic capillaries simultaneously; **3.** Both have two atria. The amphibian heart is a three-chambered heart, with only one ventricle. This causes oxygen-poor and oxygen-rich blood to mix. The mammalian heart is a four-chambered heart, the right side receiving oxygen-poor blood and pumping it to the lungs where it picks up oxygen. The left side receives oxygen-rich blood and sends it out to the body. **(23.2) 1.** A heartbeat is a cycle of events. Phase 1 (atrial systole): the atria contract and pass blood to the ventricles. Phase 2 (ventricular systole): the ventricles contract and blood moves into the attached arteries. Phase 3 (atrial and ventricular diastole): both the atria and ventricles relax while the heart fills with blood; **2.** Arteries have very well developed walls with a thick middle layer of elastic fibers and smooth muscle. Capillary walls are composed of an epithelium only one cell thick. Veins have flabby walls because the middle layer is not as thick as in arteries. Capillaries are used for gas exchange because diffusion can be done quickest with the thinnest walls of all the vessels; **3.** Homeostasis is maintained in blood volume, pH balance, and chemical concentration in the cardiovascular system because the lymphatic system adds fluid and certain substances back to the blood. **(23.3.) 1.** All are involved in the immune response. Granular leukocytes: Neutrophils- phagocytize pathogens and cellular debris. Eosinophils- use granular contents to digest large pathogens, such as worms, and reduce inflammation. Basophils- promote blood flow to injured tissues and the inflammatory response. Agranular leukocytes: Lymphocytes- responsible for specific immunity; B cells produce antibodies; T cells destroy cancer and virus infected cells. Monocytes- become macrophages that phagocytize pathogens and cellular debris; **2.** Hypertension is high blood pressure and can be treated with medication and changes in diet and exercise. Atherosclerosis is an accumulation of soft masses of fatty materials beneath the lining of the arteries. Treatments include medication, surgery, changes in diet and exercise; **3.** Without platelets, blood clotting would be inefficient or nonexistent and lead to hemorrhage.

Testing Yourself

1. c; **2. a.** pulmonary vein; **b.** aorta; **c.** renal artery; **d.** lymphatic vessel; **e.** pulmonary artery; **f.** superior vena cava; **g.** inferior vena cava; **h.** hepatic vein; **i.** hepatic portal vein; **j.** renal vein; **3.** a; **4. a.** superior vena cava; **b.** aortic semilunar valve; **c.** pulmonary semilunar valve; **d.** right atrium; **e.** tricuspid valve; **f.** right ventricle; **g.** inferior vena cava; **h.** aorta; **i.** pulmonary trunk; **j.** pulmonary arteries; **k.** pulmonary veins; **l.** left atrium; **m.** bicuspid valve; **n.** septum; **o.** left ventricle; **5.** a; **6.** d; **7.** b; **8.** c; **9.** e; **10.** a; **11.** d; **12.** b; **13.** a; **14.** a; **15.** a; **16.** b; **17.** c,d; **18.** b; **19.** a; **20.** d; **21.** a; **22.** b; **23.** c; **24.** b; **25.** c; **26.** b; **27.** a; **28.** e; **29.** a; **30.** c; **31.** d; **32.** e; **33.** c; **34. a.** blood pressure; **b.** osmotic pressure; **c.** blood pressure; **d.** osmotic pressure

Thinking Scientifically

1. The four-chambered heart keeps O_2-poor and O_2-rich blood separate. This allows the maximum amount of oxygen to be delivered to the cells. This in turn allows for the greatest amount of energy to be produced. In other words, a four-chambered heart supports the high metabolism of an endothermic animal. **2.** Lower oxygen levels found at higher altitudes stimulate an increased production of red blood cells and hemoglobin. Therefore, if an athlete trains at higher altitudes long enough for this to take affect, he/she will have the ability to carry a higher amount of oxygen in his/her bloodstream. This effect will continue even at lower altitudes for a few months after the high altitude training has stopped. More oxygen means more energy through aerobic cellular respiration for muscle contraction.

CHAPTER 24

Check Your Progress

(24.1) 1. The three steps of respiration are breathing, external exchange, and internal exchange; **2.** Upper respiratory tract includes nose, pharynx and larynx. The lower respiratory tract includes the trachea, bronchi, bronchioles and alveoli; **3.** Oxygen follows its concentration gradient; there is more oxygen in the lungs than in the blood and there is more oxygen in the blood than in tissue fluid. Also, environmental conditions in the tissues cause hemoglobin to give up oxygen. Carbon dioxide combines with water to form carbonic acid. Carbonic acid breaks down to the bicarbonate ion and H^+, which combines with hemoglobin. **(24.2) 1.**

The kidney functions are excretion of nitrogenous wastes, maintenance of the water-salt balance of blood, and maintenance of the acid-base balance of blood; **2.** A nephron consists of the nephron capsule and proximal tubule which function in filtration, the nephron loop which functions in reabsorption, and the distal tubule which functions in secretion; **3.** Water-salt balance and pH balance is important to homeostasis to maintain blood volume and composition to maintain circulation and proper gas and waste exchange. **(24.3) 1.** Ingest and break down food, absorb nutrients, and eliminate indigestible material; **2.** Peristalsis is a rhythmic contraction of smooth muscle that moves materials through the digestive tract; **3.** The stomachs of ruminants are divided into several chambers, including the rumen. The rumen houses bacteria and protozoans which secrete enzymes that digest cellulose; **4.** The small intestine digests all types of food and it produces enzymes that complete the breakdown of food to small molecules that can be absorbed across its villi. The villi are folds in the wall of the small intestine that are lined with cells that have microvilli formed by their plasma membranes. The small intestine is relatively long and skinny. No digestion occurs in the large intestine, but it does store digestive remains until they are eliminated. The large intestine also absorbs water, salts, and some vitamins. The large intestine lacks microvilli and villi. It is shorter in length and larger in diameter than the small intestine; **5.** The liver and pancreas secrete bile and pancreatic juice, respectively, into the duodenum. These substances contribute to the digestion and absorption of nutrients; **6.** Salivary amylase breaks down starch; pepsin breaks down proteins; trypsin breaks down proteins; peptidase breaks down peptides; lipase breaks down fats.

Testing Yourself

1. a. nostril; **b.** nasal cavity; **c.** pharynx; **d.** epiglottis; **e.** glottis; **f.** larynx; **g.** trachea; **h.** bronchus; **i.** bronchiole; **j.** alveoli; **k.** lung; **l.** diaphragm; **2.** c; **3.** d; **4.** b; **5.** a; **6.** d; **7.** c; **8. a.** kidney; **b.** ureter; **c.** bladder; **d.** urethra; **9.** c; **10.** d; **11.** b; **12.** a; **13.** d; **14.** a; **15.** d; **16.** b; **17.** a; **18.** c; **19.** b; **20.** d; **21.** c; **22.** c; **23.** d; **24. a.** salivary glands; **b.** mouth; **c.** liver; **d.** gallbladder; **e.** duodenum; **f.** cecum; **g.** appendix; **h.** anus; **i.** pharynx; **j.** esophagus; **k.** diaphragm; **l.** stomach; **m.** pancreas; **n.** large intestine; **o.** small intestine; **p.** rectum; **25.** a; **26.** b; **27.** a; **28.** d; **29.** b; **30.** a; **31.** b

Thinking Scientifically

1. Protein digestion begins in the stomach. The pepsin produced by glands in the stomach would break down the insulin, since it is a protein. The insulin would not make it to the small intestine where most substances are absorbed. **2.** Without normal amounts and concentrations of bile, the individual will have difficulty digesting and absorbing fats. Fats that remain in the intestines may cause abdominal discomfort. Individuals may also suffer from fat-soluble vitamin deficiencies due to the difficulty they will have in absorbing them. **3.** Grasshoppers have a tracheal system consisting of air tubes to carry out gas exchange and their exoskeleton is both air- and waterproof. They excrete uric acid which requires little water to excrete and their Malpighian tubules absorb most of the water back into their bodies.

CHAPTER 25

Check Your Progress

(25.1) 1. Vitamins, minerals, and water do not provide energy; **2.** So-called "junk" foods do contain nutrients; **3.** Empty-calorie foods that are rich sources of simple sugars and/or fat include sugar-sweetened soft drinks, pastries, and cookies. These are called empty-calorie foods because they provide little else besides calories. **(25.2) 1.** Carbohydrates include sugars and starches. Lipids are oils and fats. Proteins are composed of 20 amino acids. Minerals include calcium and iron. Vitamins include vitamins A and C. Water; **2.** Overindulgence in carbohydrates can lead to obesity; in lipids can lead to elevated cholesterol levels and increased risk of cardiovascular disease; in proteins can lead to increased calcium loss in urine, formation of kidney stones, and obesity; **3.** Without vitamins and minerals, proper organ development and functioning would not occur, as well as proper protein synthesis and function. **(25.3) 1.** Daily caloric needs are based on metabolic functions; physical activity; and digestion, absorption, and processing of food; **2.** Disorders linked to obesity include cardiovascular disease due to increased blood lipid levels and plaque buildup and diabetes, the improper functioning of pancreas producing enough insulin to regulate blood sugar levels; **3.** It is important to understand your daily energy needs so you bring in enough calories to cover those needs without overindulging or underindulging. **(25.4) 1.** A balanced meal provides all the needed nutrients and energy for the body's output; **2.** Grains are an important source of carbohydrates; vegetables and fruits are a source of carbohydrates and fiber; oils are a source of lipids; milk is a source of carbohydrates and oils; meats and beans are a good source of proteins. **3.** Contains specific dietary information about the product and the nutrients it provides; Serving size, calories, fat, cholesterol, sodium, total carbohydrate, dietary fiber, sugar, protein, vitamins A and C, calcium, iron, and % daily value for each.

Testing Yourself

1. b; **2.** a ; **3.** f ; **4.** a; **5.** b; **6.** e; **7.** c; **8.** b; **9.** d; **10.** e; **11.** d; **12.** a; **13. a.** grains; **b.** 6 oz.; **c.** vegetables; **d.** 2.5 cups; **e.** fruits; **f.** 2 cups; **g.** oils; **h.** 1–2 Tbsp; **i.** milk; **j.** 2 cups kids, 3 cups if older; **k.** meats and beans; **l.** 5.5 oz; **14.** c; **15.** d; **16.** a; **17.** d; **18.** c; **19.** b; **20.** a; **21.** a; **22.** b; **23.** a; **24.** c;

Thinking Scientifically

1. See table below. **2.** All nine essential amino acids are easily obtainable from eating meat. However, our dentition and also the length of our digestive tracts suggests that we evolved as omnivores. Normally, we need both plant matter (for fiber) and animal matter (for amino acids) in our diet. A vegetarian diet takes careful planning.

CHAPTER 26

Check Your Progress

(26.1) 1. Antigens are the agents that cause a response from and activate the immune system; **2.** Red bone marrow; produces blood cells including white blood cells; thymus; aids in maturation of T lymphocytes and testing of their ability to function; spleen; filters blood of pathogens and debris; lymph nodes; filter lymph of pathogens and debris; **3.** Macrophages- phagocytocize pathogens, inflammatory response and specific immunity; Mast cells- release histamines, inflammatory response; Neutrophils- phagocytize pathogens; inflammatory response; Natural killer cells- kill virus-infected and tumor cells; Lymphocytes- specific immunity; B cells- produce plasma and memory cells; Plasma cells- produce specific antibodies; Memory cells- ready to produce B and T cells in the future; T cells- regulate immune response, produce cytotoxic T cells and helper T cells; Cytotoxic T cells- kill virus-infected cells and cancer cells; Helper T cells- regulate immunity. **(26.2) 1.** Damaged cells and mast cells release chemicals that cause capillaries to dilate and become more permeable. Neutrophils and macrophages migrate to the site of injury and phagocytize pathogens; **2.** The complement system produces proteins that complement other forms of immunity. Certain complement proteins also poke holes in invading pathogens, causing them to burst;

3. Natural killer cells attack cells that have lost their ability to produce self proteins, often because of viral infection. **(26.3) 1.** Antibodies produced by plasma cells bind to specific antigens. Such antibody-antigen complexes mark the antigen for destruction; **2.** Both are involved with specific defenses against disease. B cells are produced and mature in the bone marrow and give rise to plasma cells that produce antibodies. T cells are produced in the bone marrow and mature in the thymus as cytotoxic or helper T cells; **3.** Antibody response uses proteins called antibodies produced by B cells to fight pathogens. Cellular response uses entire cells, T cells, to fight pathogens; **4.** MHC markers are unique to each individual, except identical twins, and allow the immune system to identify "self" from "non-self" cells; they are the markers that make tissue rejection occur. **(26.4) 1.** Vaccinations only work if acquired immunity is possible. People cannot develop an immunity to most sexually transmitted diseases, and so a vaccine cannot be developed; **2.** To increase the antibody count to a concentration high enough to fight pathogens; **3.** Active immunity involves being exposed to the pathogen and actively fighting it off as well as producing memory cells for the pathogen. Passive immunity is temporary and involves giving a patient antibodies that they did not produce against a pathogen. **(26.5) 1.** Allergies are hypersensitive responses to antigens called allergens; **2.** An immediate allergic response occurs within seconds of an exposure to an allergen and is caused by chemicals released by mast cells. Delayed allergic responses take longer to develop and are probably initiated by memory T cells; **3.** The HIV virus lives in and destroys helper T cells, which are necessary for the activity of all other immune system cells.

Testing Yourself

1. b; **2. a.** red bone marrow; **b.** thymus; **c.** lymph nodes; **d.** spleen; **e.** lymph vessels; **3.** d; **4.** a; **5.** c; **6.** a; **7.** b; **8.** d; **9.** e; **10.** b; **11.** c; **12.** b; **13.** a; **14.** b; **15.** c; **16.** e; **17.** d; **18.** a; **19.** e; **20.** b; **21.** b; **22.** b; **23.** e; **24.** b; **25.** d; **26.** d; **27.** d; **28. a.** hair shaft; **b.** oil gland; **c.** sweat glands; **d.** dermis; **e.** epidermis; **29.** e; **30.** a; **31.** a; **32.** e; **33.** c; **34.** b; **35.** d; **36.** a; **37.** b; **38.** a; **39.** c; **40.** a

Thinking Scientifically

1. The tonsils are composed of lymph tissue and help to capture pathogens as they enter the digestive and respiratory tracts. Because of their location and function, they often contain bacteria and that bacteria can overwhelm the tissue of the tonsils and cause an infection. Either through the body's usual defenses or through help from medication, individuals can usually overcome these infections. The tonsils, once healthy, will once again aid the body in the fight against infection. Only if they become chronically infected should they be removed. At that point, they no longer perform their function. **2.** Immunosuppressive drugs dampen the body's response not only to the transplanted tissue but to other foreign cells as well. Therefore, the body's ability to fight infections is weakened and the transplant patients become more susceptible to all types of infections. **3.** Our immune system recognizes pathogens by their antigens. If a pathogen changes its antigens then the immune system has a reduced ability to recognize that pathogen.

Ch. 25.1 Thinking Scientifically

	Soda then 6.5 oz	Soda Now 20 oz	Calorie Increase
Fat (g)	0	0	
Carbohydrate (g)	21	65	
Protein (g)	0	0	
Total Calories	84	260	176
	Fries then 2.6 oz	Fries Now 6 oz	Calorie Increase
Fat (g)	11	25	
Carbohydrate (g)	28	65	
Protein (g)	3	7	
Total Calories	223	513	290
	Bagels then 3 oz	Bagels Now 6 oz	Calorie Increase
Fat (g)	0.5	1.0	
Carbohydrate (g)	40	80	
Protein (g)	6	12	
Total Calories	188.5	377	188.5

Therefore, our immune system is a selective agent for pathogens that can change their antigens and thereby escape recognition. Also, the greater the ability of a pathogen to change its antigens, the greater difficulty it is for scientists to make an effective vaccine.

CHAPTER 27

Check Your Progress

(27.1) 1. Sensory neurons take nerve impulses from sensory receptors to the central nervous system. Interneurons carry nerves impulses between parts of the central nervous system. Motor neurons carry nerve impulses from the central nervous system to muscles or glands; **2.** Cell body contains the nucleus and other organelles. Dendrites receive signals from sensory receptors or other neurons. An axon is a long single projection that conducts nerve impulses; **3.** First: Na^+ gates open and Na^+ moves to the inside; the inside of the axon becomes positive. Second: K^+ moves to the outside; the inside of the axon becomes negative again; **4.** Nerve impulses at the axon terminal cause synaptic vesicles to release a neurotransmitter into the synaptic cleft. The neurotransmitters then diffuse across the cleft and bind to receptor proteins; **5.** Central nervous system consists of the brain (cerebrum, diencephalon, cerebellum, brain stem) and spinal cord; **6.** Our awareness of a reflex is dependent on impulses that travel in tracts from the spinal cord to the brain or from sense organs to the brain, where awareness of both the stimulus and response occurs; **7.** The parasympathetic division keeps the "house" in order by maintaining the internal responses associated with a relaxed state, such as pupil contraction, food digestion, and slow heartbeat. **(27.2) 1.** Steroid hormones enter the cell and stimulate genes to produce enzymes that alter cell activity. Peptide hormones induce an enzyme cascade that alters cell activity; **2.** The hypothalamus helps regulate the internal environment and stimulates the pituitary, which produces hormones that stimulate other glands and control other responses; **3.** Both hormones regulate blood calcium levels. Calcitonin is produced by the thyroid gland, and reduces blood calcium. PTH is produced by the parathyroid glands, and increases blood calcium; **4.** Insulin lowers blood sugar levels by stimulating the uptake of glucose by cells. Glucagon raises blood sugar levels by stimulating the breakdown of glycogen to glucose; **5.** People with type 1 diabetes do not produce insulin, so they require daily insulin injection. People with type 2 diabetes produce insulin, but the liver and muscle cells do not respond to it as they should; **6.** A hormone controls the level of a product. When that level varies from normal, the amount of hormone is altered until the product's level returns to normal.

Testing Yourself

1. a; **2.** c; **3.** d; **4.** c; **5.** a; **6.** c; **7. a.** cerebrum; **b.** skull; **c.** corpus callosum; **d.** diencephalon; **e.** pineal gland; **f.** pituitary gland; **g.** cerebellum; **h.** spinal cord; **i.** medulla oblongata; **j.** pons; **k.** midbrain; **l.** brain stem; see Fig. 27.8, p. 522, in text; **8.** b; **9.** a, c; **10.** a; **11.** d; **12.** a; **13.** d; **14.** d; **15.** b; **16. a.** pineal gland; **b.** pituitary gland

(hypophysis); **c.** hypothalamus; **d.** parathyroid gland; **e.** thyroid gland; **f.** thymus gland; **g.** adrenal gland; **h.** pancreas; see Fig. 27.14, p. 527, in text; **17.** d; **18.** a; **19.** d; **20.** a; **21.** c; **22.** d; **23.** e; **24.** a; **25.** c; **26.** b; **27.** b; **28.** c

Thinking Scientifically

1. Parkinson's patients do not exhibit a constriction of blood vessels, so their blood pressure slowly decreases. The loss of function of the sympathetic division prevents the body from regulating blood pressure, so rapid changes in blood pressure can occur, leading to orthostatic hypotension. **2.** Diabetes type 1 is thought to result when an environmental agent, such as a virus, causes T cells to destroy the pancreatic islets. Vaccines that use a live virus may introduce that agent. For example, the mumps and rubella viruses may remain in the body for years after vaccination and can infect pancreatic islet cells. Infection lowers the levels of insulin secreted by these cells. In addition, exposure to the viruses may lead to an autoimmune disorder that causes the destruction of pancreatic islet cells. **3.** The brain is an energetically-expensive organ, accounting for a substantial proportion of our daily calorie intake. The selective pressure in an environment with scarce food, like that on Borneo, would be to sacrifice brain size to conserve energy. In an environment with plentiful food like Sumatra, the orangutans can afford to spend extra energy on nervous tissue, and thus reap the advantage of larger brain size.

CHAPTER 28

Check Your Progress

(28.1) 1. Chemoreceptors detect chemicals; mechanoreceptors detect mechanical stimulation; photoreceptors detect light; cutaneous receptors detect touch, pain, temperature, and pressure; proprioceptors detect changes in positions of joints, muscles, and bones; **2.** After chemicals bind to the receptor proteins of taste receptor cells, nerve impulses go to the brain, which interprets the nerve impulses as tastes. Bitter, sour, salty, sweet, and umami are the five types of taste; **3.** Each part of the spiral organ is sensitive to a different pitch. The nerve fibers in each part connect to and stimulate a different part of the brain; **4.** The utricle and saccule are responsible for gravitational equilibrium, and the semicircular canals for rotational equilibrium; **5.** Cornea, pupil, lens, retina; **6.** Both contain a visual pigment consisting of retinal and opsin. All rods have the same visual pigment rhodopsin. The opsin portion of the pigment varies slightly from one type of cone to the next. Rods are distributed throughout the retina, while cones are concentrated in the fovea. Rods are sensitive to low levels of light, and cones are activated by bright light. Rods enable us to see in dim light, have peripheral vision, and detect motion; and cones provide us with detailed color vision; **7.** Cutaneous receptors are found in the skin and respond to touch, pressure, pain, and temperature. Proprioceptors are found in skeletal muscles and joints where they help maintain equilibrium and posture. **(28.2) 1.** Bones and muscles support the body and allow body parts to move, protect internal organs, and help other systems to function; **2.** Smooth muscle is involuntary and found in organs

and vessels; cardiac muscle is involuntary and found in the walls of the heart; skeletal muscle is voluntary and is used to move the skeleton; **3.** The axial skeleton is composed of the skull, vertebral column, ribs, and sternum. The appendicular skeleton is composed of the shoulder girdle, the pelvic girdle, and their attached limbs; **4.** Osteoblasts deposit bone, and osteoclasts break down bone; **5.** The globular heads of myosin molecules attach to actin filaments and pull them toward the center of the sarcomere. Then ATP allows the heads to be released, and they reattach at a new location along the actin filament, resulting in contraction; **6.** A sprain occurs when ligaments and tendons are overstretched and can result in a torn meniscus in the synovial joint.

Testing Yourself

1. c; **2. a.** cochlear nerve; **b.** cochlea; **c.** tympanic membrane; **d.** auditory tube; **e.** auditory canal; **f.** ossicles; **g.** semicircular canals; **h.** outer ear; **i.** middle ear; **j.** inner ear; ; **3.** b; **4.** b; **5. a.** sclera; **b.** retina; **c.** vein; **d.** artery; **e.** optic nerve; **f.** fovea; **g.** ciliary muscle; **h.** iris; **i.** pupil; **j.** cornea; **k.** lens; **6.** a; **7.** b; **8.** c; **9.** a; **10.** c; **11.** b; **12.** b,c; **13.** a; **14.** e,; **15.** c; **16.** a; **17.** b; **18.** d; **19.** e; **20.** a; **21.** c

Thinking Scientifically

1. The blood vessels interfere with the transmission of light to the back of the eye, consequently impairing the signal sent to the brain. Both cancer and eye diseases involve the growth of new blood vessels. Anti-angiogenesis drugs that have been developed for cancer treatment are currently being studied for the treatment of eye diseases. **2.** Our olfactory system allows us to recognize and discriminate between many different odors. Therefore, we can detect the difference between a dangerous odor, such as smoke, and a harmless one, such as flower fragrance. On the other hand, every natural toxin tastes bitter, so it is more important to be able to sense bitterness than to discriminate between bitter tastes. Therefore, humans can identify bitter compounds by taste, but cannot distinguish between different compounds.

CHAPTER 29

Check Your Progress

(29.1) 1. Both are forms of reproduction that do not require fertilization. Parthenogenesis is the development of an unfertilized egg into an adult, while asexual reproduction is the development of part of the body into an adult; **2.** Animals that reproduce in the water have no need to protect their eggs and embryos from drying out, but they do have to worry more about predators destroying the zygotes. Animals that reproduce on land need to protect their gametes and embryos from drying out; **3.** Egg-laying animals are oviparous. Animals that retain their eggs in the body until hatched are called ovoviviparous. Mammals that produce living young are viviparous. **(29.2) 1.** Sperm are produced in the seminiferous tubules in the testis, mature in the epididymis, are propelled into the vas deferens and out the penis; **2.** Testosterone is the main sex hormone and is essential for the normal development and functioning of male sex organs. FSH functions in sperm production, and LH controls the production of testosterone; **3.** Condoms trap male ejaculate inside the sheath. The birth control pill regulates the hormones needed for proper ovulation in females. IUD fits in the uterus and prevents sperm from fertilizing the egg in the fallopian tube; **4.** During the follicular phase of the ovarian cycle, FSH released by the anterior pituitary promotes the maturation of follicles in the ovary. Ovarian follicles produce increasing levels of estrogen, which exert feedback control over FSH secretion and result in ovulation; **5.** After ovulation and during the luteal phase of the ovarian cycle, LH promotes the development of a corpus luteum. This structure produces increasing levels of progesterone, which causes the endometrium to become secretory. Menstruation begins when progesterone production declines to a low level. During the follicular phase, FSH promotes the development of ovarian follicles which produce estrogens that cause the endometrium to thicken; **6.** An ovary contains many primary follicles, each containing an oocyte, and is the female gonad. The testes, the male gonads, are suspended in the scrotum and contain seminiferous tubules; **7.** STDs caused by bacteria include chlamydia, gonorrhea, and syphilis. STDs caused by a virus include HIV, genital warts, genital herpes and hepatitis. Trichimonas is caused by a parasitic protozoan; **8.** If a female had low estrogen levels, ovulation and menstruation would not occur correctly; **9.** PRO- help couples with fertility problems become pregnant; CON- expensive, low success rate in some cases, medically invasive in some cases. **(29.3) 1.** The zygote divides without increasing in size, forming a ball of cells called a morula; **2.** Each of the germ layers gives rise to particular structures and organs; **3.** One tissue releases a chemical that turns on genes in another tissue, influencing its development; **4.** Embryonic development occurs during months 0-2 and is characterized by the establishment of organ systems. Fetal development occurs during months 3-9 and is characterized by an increase in size; **5.** The afterbirth contains the placenta, the region of exchange with the mother's bloodstream; **6.** Without neuralization, formation of the central nervous system would not occur.

Testing Yourself

1. c; **2.** a; **3.** b; **4.** d; **5. a.** ureter; **b.** bladder; **c.** vas deferens; **d.** seminal vesicle; **e.** ejaculatory duct; **f.** prostate gland; **g.** bulbourethral gland; **h.** urethra; **i.** epididymis; **j.** penis; **k.** foreskin; **l.** testis; **m.** scrotum; **6. a.** oviduct; **b.** ovary; **c.** rectum; **d.** cervix; **e.** bladder; **f.** vagina; **g.** urethra; ; **7.** a; **8.** d; **9.** c; **10.** a; **11.** b; **12.** d; **13.** d; **14.** b; **15.** c; **16.** a; **17.** c; **18.** e; **19.** b; **20.** d; **21.** a; **22.** a; **23.** d

Thinking Scientifically

1. Fetuses take a high caloric toll on their mothers. In modern times food supply is usually sufficient for our needs, but earlier in human history the food supply may not have been as dependable. Mothers with low percentages of body fat might starve during pregnancy or the prolonged period of breast-feeding that early human infants presumably required, and thus would not be able to reproduce at all. By stopping the reproductive cycle when the body senses insufficient food reserves, the woman is perhaps better able to survive to a time when food

is more plentiful. **2.** The most logical (although controversial) explanation is that menopause in women developed early in human evolution. It allowed women to channel their efforts into caring for their existing children, increasing the survival rate of these children. This would give the woman the best chance to pass her genes through generations, so it increased her fitness. Since men were not the main caregivers for their children, their fitness would not be increased if they evolved the ability to undergo menopause.

CHAPTER 30

Check Your Progress

(30.1) 1. Organisms, populations, communities, ecosystems, and the biosphere; **2.** Ecosystem consists of a community of living organisms as well as their physical environment. Community consists of all the various populations at a particular locale; **3.** An environmental scientist also studies biodiversity and conservation biology; topics in ecology. **(30.2) 1.** 10.7%; **2.** MDCs such as those in Europe and North America have modest population growth and a good standard of living. LDCs such as those in Latin America, Asia, and Africa have a fast rate of population growth, and most of the people live in poverty; **3.** An age structure diagram shows different numbers of individuals in each age class, and the diagrams have different shapes for increasing, decreasing, or stable populations; **4.** MDCs strain available resources due to high resource consumption while LDCs strain available resources due to large population sizes. **(30.3) 1.** Type I populations tend to have fewer offspring since most of those offspring live to reproduce. Type III populations have large numbers of offspring since most of them will die before they reproduce; **2.** Type I has survival likely to old age, type III has few survivals to old age, most dying at a young age; **3.** Both plot number of organisms vs. time. Exponential growth produces a J shaped curve and occurs when resources are unlimited. Logistic growth produces an S shaped curve and occurs when resources are limited; **4.** Usual numbers of offspring per reproduction; chances of survival until age of reproduction; how often each individual reproduces; age at which reproduction begins; **5.** Independent- weather and natural disasters; Dependent- competition, predation, and parasitism; **6.** With both factors, the percentage of the population affected increases as the density of the population increases. **(30.4) 1.** Opportunistic populations exhibit exponential growth and are regulated by density-independent effects. Equilibrium populations exhibit logistic growth and remain near carrying capacity; **2.** Extinction is the total disappearance of a species or higher group; **3.** Time to maturity; number of offspring produced; size of geographic range; degree of habitat tolerance; and size of local populations.

Testing Yourself

1. a; **2.** d; **3.** b; **4.** d; **5.** b; **6.** b; **7.** d; **8.** c; **9.** d; **10.** b; **11. a.** increasing; **b.** decreasing; **c.** stable; **12.** a; **13.** a; **14.** a; **15.** a; **16.** d; **17.** b; **18.** c; **19.** c; **20.** c; **21.** e; **22.** c; **23.** a;

Thinking Scientifically

1. 53 years; 172.5 years. **2.** ¼ filled takes 18 weeks; ½ filled takes 19 weeks. **3.** The number of surviving individuals and the variable being studied. In Fig. 15.2, that variable would be clutch size.

CHAPTER 31

Check Your Progress

(31.1) 1. Flowers and pollinators have evolved to be suited to one another; **2.** Species richness is the list of species in a community (species composition) while diversity encompasses composition, distribution, and relative abundance of each species; **3.** Primary succession occurs where soil has not yet been formed, while secondary succession occurs where soil is present; **4.** Competition, predation, parasitism, commensalism, and mutualism; **5.** A habitat is the special location of a species. It is one component of an ecological niche, which also includes the role the species plays in the community and its interactions with other species; **6.** A keystone species stabilizes the community, helps maintain its characteristics, and helps hold the web of interactions together. **(31.2) 1.** Autotrophs take in only inorganic nutrients and acquire energy from an outside source, often the sun. Heterotrophs acquire energy from consuming autotrophs; **2.** Energy flows from the sun to the Earth through living organisms until it is all lost as heat. Chemicals cycle throughout the environment through the biotic and abiotic; **3.** A food chain depicts a single path of energy flow in an ecosystem, while an ecological pyramid depicts the entire flow of energy between trophic levels; **4.** The phosphorous cycle is a sedimentary biogeochemical cycle and the nitrogen and carbon cycles are gaseous biogeochemical cycles; **5.** Through mining and eutrophication by excessive fertilizers; **6.** The burning of fossil fuels increases the amount of carbon dioxide in the atmosphere, contributing to global warming. **(31.3) 1.** The biosphere is divided into aquatic ecosystems and terrestrial ecosystems; **2.** Taiga, savanna, prairie, temperate forest, desert, tropical rain forest, and tundra; **3.** Swamps have a more concentrated supply of nutrients and decomposers.

Testing Yourself

1. a; **2.** a; **3.** e; **4.** a; **5.** c; **6.** b; **7.** d; **8.** c; **9.** c; **10. a.** energy; **b.** nutrients; **c.** heat; **d.** heat; **e.** producer; **f.** consumer; **g.** inorganic nutrients; **h.** decomposers; **11.** e; **12.** a; **13.** a; **14.** a; **15.** d; **16.** c; **17. a.** tertiary consumer; **b.** secondary consumer; **c.** primary consumer; **d.** producer; **18.** c; **19.** c; **20.** c; **21.** b; **22.** a; **23.** d; **24.** a; **25.** a; **26.** b; **27.** b

Thinking Scientifically

1. Insects that are best adapted for obtaining the most food from flowers that are most accessible to them reproduce more. Those best adapted flowers for attracting insect pollinators produce more offspring and thus the next generation, and likewise those insects that obtain the most food produce more offspring than others. This process continues throughout time, leading to very specific adaptations. **2.** The prairie dog is a keystone species because, for example, other species such as coyotes, hawks, and snakes all rely on prairie dogs for food. The black-footed ferret cannot survive unless the prairie dog remains. Other animals such as burrowing white-tailed rabbits live in abandoned prairie dog burrows, and bison, pronghorn antelope, and cattle prefer to graze near their towns because prairie dog foraging increases seed dispersal and the quality of grassland plants.

CHAPTER 32

Check Your Progress

(32.1) **1.** Renewable resources are not limited in supply, while nonrenewable resources are; **2.** Beach erosion occurs due to land development on the shore, destruction of wetlands and coastal pollution occur due to human activities in the vicinity of the water; **3.** Desertification occurs when humans allow animals to overgraze in semiarid regions. The loss of vegetation prevents soil from holding water and so it runs off. Deforestation occurs when trees are removed leading to a loss of biodiversity; **4.** Limited genetic diversity, heavy use of chemicals, frequent irrigation and excessive fuel consumption; **5.** The first occurred in the 1950s and resulted from the breeding of high-yielding rice and wheat varieties. The second is under way now and is the result of the creation of transgenic crops; **6.** Global warming will cause sea levels to rise, evaporation to increase, and precipitation to increase near coastlines and decrease inland; **7.** Hydropower-large dams have adverse environmental effects, geothermal- must pipe the steam to the surface, wind- requires appropriate land for the placement of windmills, and solar- must be collected, converted to another form, and stored. (32.2) **1.** Direct values include medicine, agriculture, and human consumption. Indirect values include contributions to biogeochemical cycles, waste disposal, provision of fresh water, prevention of soil erosion, regulation of climate, and ecotourism; **2.** Most modern medicines are derived from organisms so the more biodiversity available, the increased potential of more medicines. Same with agriculture: biodiversity will lead to variation in crops; **3.** Biodiversity gives options, so if any resource is becoming limited, another source can be substituted. (32.3) **1.** Use of renewable energy resources, reuse of heat and waste materials, and maximal recycling; **2.** Plant cover crops, practice multiuse farming, replenish soil nutrients, conserve water, plant cultivars, use precision farming, use integrated pest management, plant a variety of species, plant multipurpose trees, maintain and restore wetlands, use renewable energy, and buy locally; **3.** Energy-efficient transportation system, utilize green roofs, create greenbelts, use solar or geothermal energy to heat buildings, improve storm-water management, plant native species, revitalize old sections of the city before building new, change lighting, reduce noise levels, and promote sustainable building practices.

Testing Yourself

1. e; **2.** b; **3.** d; **4.** c; **5.** b; **6.** a; **7.** b; **8.** b; **9.** c, d, e; **10.** a; **11.** a, d, e; **12.** e; **13.** a; **14.** Use more renewable resources than non-renewable resources, increase recycling, refine processing to reduce waste; **15.** c; **16.** e; **17.** d; **18.** a; **19.** e; **20.** e; **21.** b

Thinking Scientifically

1. Diverting the two rivers into one would create a large lake that could serve as habitat for animal and plant species. However, downstream from these two rivers, the aquatic habitat would be damaged, likely reducing the number of species. Removing the dams would replenish life downstream but the water level is still 25% of the original volume. **2.** How rare the species is, how difficult it will be to preserve it versus the cost of preserving it, the potential benefit to human medicine. **3.** The internal temperature of mammals must stay within the normal range. If the external temperature is too high for homeostatic mechanisms to control and their internal temperature rises above normal, they will die off.

Appendix B Metric System

Unit and Abbreviation	Metric Equivalent	Approximate English-to-Metric Equivalents	Units of Temperature
Length			
nanometer (nm)	$= 10^{-9}$ m $(10^{-3}$ μm)		
micrometer (μm)	$= 10^{-6}$ m $(10^{-3}$ mm)		
millimeter (mm)	$= 0.001 (10^{-3})$ m		
centimeter (cm)	$= 0.01 (10^{-2})$ m	1 inch = 2.54 cm 1 foot = 30.5 cm	
meter (m)	$= 100 (10^{2})$ cm $= 1{,}000$ mm	1 foot = 0.30 m 1 yard = 0.91 m	
kilometer (km)	$= 1{,}000 (10^{3})$ m	1 mi = 1.6 km	
Weight (mass)			
nanogram (ng)	$= 10^{-9}$ g		
microgram (μg)	$= 10^{-6}$ g		
milligram (mg)	$= 10^{-3}$ g		
gram (g)	$= 1{,}000$ mg	1 ounce = 28.3 g 1 pound = 454 g	
kilogram (kg)	$= 1{,}000 (10^{3})$ g	= 0.45 kg	
metric ton (t)	$= 1{,}000$ kg	1 ton = 0.91 t	
Volume			
microliter (μl)	$= 10^{-6}$ l $(10^{-3}$ ml)		
milliliter (ml)	$= 10^{-3}$ l $= 1$ cm^3 (cc) $= 1{,}000$ mm^3	1 tsp = 5 ml 1 fl oz = 30 ml	
liter (l)	$= 1{,}000$ ml	1 pint = 0.47 l 1 quart = 0.95 l 1 gallon = 3.79 l	
kiloliter (kl)	$= 1{,}000$ l		

°C	°F	
100	212	Water boils at standard temperature and pressure.
71	160	Flash pasteurization of milk
57	134	Highest recorded temperature in the United States, Death Valley, July 10, 1913
41	105.8	Average body temperature of a marathon runner in hot weather
37	98.6	Human body temperature
13.7	56.66	Human survival is still possible at this temperature.
0	32.0	Water freezes at standard temperature and pressure.

To convert temperature scales:

$$°C = \frac{(°F - 32)}{1.8}$$

$$°F = 1.8(°C) + 32$$

Glossary

A

abdominal cavity Located inferior to the thoracic cavity and separated from it by the diaphragm; it contains most other internal organs except the heart and lungs. 454

abscisic acid Plant hormone that causes stomata to close and initiates and maintains dormancy. 394

abscission Dropping of leaves, fruits, or flowers from a plant. 394

abstinence Method of birth control; the practice of not engaging in sexual intercourse. 567

acetylcholine (ACh) Neurotransmitter active in both the peripheral and central nervous systems. 520

acetylcholinesterase (AChE) Enzyme that breaks down acetylcholine within a synapse. 520

acetyl-CoA Molecule made up of a 2-carbon acetyl group attached to coenzyme A. During cellular respiration, the acetyl group enters the citric acid cycle for further breakdown. 113

acid Molecules tending to raise the hydrogen ion concentration in a solution and to lower its pH numerically. 31

acoelomate Animal without a coelom, as in flatworms. 339

acquired immunodeficiency syndrome (AIDS) Disease caused by a retrovirus and transmitted via body fluids; characterized by failure of the immune system. 509, 571

actin Muscle protein making up the thin filaments in a sarcomere. Its movement shortens the sarcomere, yielding muscle contraction. Actin filaments play a role in the movement of the cell and its organelles. 551

actin filament Cytoskeletal filament of eukaryotic cells composed of the protein actin; also refers to the thin filaments of muscle cells. 70

action potential Electrochemical changes that take place across the axomembrane; the nerve impulse. 518

active immunity Resistance to disease due to the immune system's response to a microorganism or a vaccine. 506

active site Region on the surface of an enzyme where the substrate binds and where the reaction occurs. 84

active transport Use of a plasma membrane carrier protein to move a molecule or an ion from a region of lower concentration to one of higher concentration. It opposes equilibrium and requires energy. 88

adaptation Organism's modification in structure, function, or behavior suitable to the environment. 4, 238

adaptive radiation Rapid evolution of several species from a common ancestor into new ecological or geographical zones. 271

Addison disease Condition resulting from a deficiency of adrenal cortex hormones; characterized by low blood glucose, weight loss, and weakness. 532

adenine (A) One of four nitrogen-containing bases in nucleotides composing the structure of DNA and RNA. 176

adenosine triphosphate (ATP) Nucleotide with three phosphate groups. The breakdown of ATP into ADP + Ⓟ makes energy available for energy-requiring processes in cells. 68

adhesion junction Junction between cells in which the adjacent plasma membranes do not touch but are held together by intercellular filaments attached to button-like thickenings. 73

adrenal cortex Located in the adrenal gland; produces the glucocorticoid and mineralocorticoid hormones. 531

adrenal gland Gland that lies atop a kidney. The *adrenal medulla* produces the hormones epinephrine and norepinephrine, and the *adrenal cortex* produces the glucocorticoid and mineralocorticoid hormones. 531

adrenal medulla Inner portion of the adrenal gland; secretes the hormones epinephrine and norepinephrine. 531

adrenocorticotropic hormone (ACTH) Hormone secreted by the anterior lobe of the pituitary gland that stimulates activity in the adrenal cortex. 529

adult stem cell Found in many organs of an adult's body, can become like an embryonic stem cell by adding four genes. 200

age structure In demographics, a display of the age groups of a population. A growing population has a pyramid-shaped diagram. 593

agglutination Clumping of red blood cells due to a reaction between antigens on red blood cell plasma membranes and antibodies in the plasma. 503

algae (sing., alga) Type of protist that carries on photosynthesis. Unicellular forms are a part of phytoplankton, and multicellular forms are called seaweed. 301

alkaptonuria Rare autosomal genetic disorder in which the individual is unable to metabolize homogentisic acid. 222

allele Alternative form of a gene. Alleles occur at the same locus on homologous chromosomes. 142

allergen Foreign substance capable of stimulating an allergic response. 508

allergy Immune response to substances that usually are not recognized as foreign. 508

allopatric speciation Origin of new species between populations that are separated geographically. 270

alternation of generations Life cycle, typical of plants, in which a diploid sporophyte alternates with a haploid gametophyte. 312, 398

alternative mRNA processing Lets cells produce multiple proteins from the same gene by changing the way exons are joined. 204

alveolus (pl., alveoli) In humans, terminal, microscopic, grape-like air sac in the lungs. 453

Alzheimer disease (AD) Brain disorder characterized by a general loss of mental abilities. 524

amino acid Organic molecule composed of an amino group and an acid group; covalently bonds to produce peptide molecules. 46

amniocentesis Procedure for removing amniotic fluid surrounding the developing fetus for testing of the fluid or cells in the fluid. 216

amnion Extraembryonic membrane of reptiles (including birds) and mammals that forms an enclosing, fluid-filled sac. 576

amoeboid Cell that moves and engulfs debris with pseudopods. 303

amphibian Member of a class of vertebrates that includes frogs, toads, and salamanders. They are still tied to a watery environment for reproduction. 355

analogous structure Structure that has a similar function in separate lineages but differs in anatomy and ancestry. 245, 289

analogy Similarity of characteristics due to convergent evolution. 279

anaphase Mitotic phase during which daughter chromosomes move toward the poles of the spindle. 128

angiogenesis Formation of new blood vessels; one mechanism by which cancer spreads. 134

angiosperm Flowering plant. The seeds are borne within a fruit. 312

animal Multicellular, heterotrophic organism belonging to the animal kingdom. 8

annelid Segmented worms, such as the earthworm and the clam worm. 344

annual ring Layer of wood (secondary xylem) usually produced during one growing season. 380

anorexia nervosa Eating disorder characterized by a morbid fear of gaining weight. 489

anterior pituitary Portion of the pituitary gland that is controlled by the hypothalamus. It produces six types of hormones, some of which control other endocrine glands. 529

anther In flowering plants, pollen-bearing portion of the stamen. 319, 399

anthropoid Group of primates that includes monkeys, apes, and humans. 360

antibody Protein produced in response to the presence of an antigen. Each antibody combines with a specific antigen. 443, 503

anticodon Three-base sequence in a transfer RNA molecule base that pairs with a complementary codon in mRNA. 184

antidiuretic hormone (ADH) Hormone, secreted by the posterior pituitary, that increases the permeability of the collecting ducts in a kidney. 530

antigen Foreign substance, usually a protein or a polysaccharide, that stimulates the immune system to react, such as to produce antibodies. 443, 497

antigen-presenting cell (APC) Cell that displays an antigen to certain cells of the immune system, so that they can defend the body against that antigen. 503

antioxidant Substance, such as vitamins C, E, and A, that defends the body against free radicals. 482

anus Outlet of the digestive tube. 467

aorta In humans, the major systemic artery that takes blood from the heart to the tissues. 434

aortic body Structure located in the walls of the aorta; contains chemoreceptors sensitive to hydrogen ion and carbon dioxide concentrations in the blood. 454

apical dominance Influence of a terminal bud in suppressing the growth of lateral buds. 393

apical meristem In vascular plants, masses of cells in the root and shoot that reproduce and elongate as primary growth occurs. 374

apoptosis Programmed cell death involving a cascade of specific cellular events, leading to death and destruction of the cell. 132

appendicular skeleton Part of the vertebrate skeleton forming the appendages, shoulder girdle, and hip girdle. 549

appendix In humans, small, tubular appendage containing lymphatic tissue, that extends outward from the cecum of the large intestine; a part of the immune system. 467, 497

aquaporin Membrane channel through which water moves by simple diffusion. 86

aquatic ecosystem Freshwater ecosystem (river, stream, lake, pond) or saltwater (marine) ecosystem (ocean, coral reef, saltwater marsh). 623

aquifer Rock layers that contain water and release it in appreciable quantities to wells or springs. 634

arachnid Group of arthropods that contains spiders, scorpions, and ticks. 348

archaea Prokaryotic members of the domain Archaea. Many live in unique habitats, such as swamps, highly saline environments, and submarine thermal vents. 299

arteriole Vessel that takes blood from an artery to capillaries. 436

artery Blood vessel that transports blood away from the heart. 435, 436

arthropod Invertebrate with an exoskeleton and jointed appendages, such as crustaceans and insects. 347

artificial selection Breeders selecting individuals with particular traits, or combinations of traits, to reproduce over others. 240

assortative mating Individuals tend to mate with those that have the same phenotype with respect to certain characteristics. 260

aster Short, radiating fibers produced by the centrosomes in animal cells. 128

atom Smallest particle of an element that displays the properties of the element. 21

atomic mass Mass of an atom equal to the number of protons plus the number of neutrons within the nucleus. 21

atomic number Number of protons within the nucleus of an atom. 21

atomic symbol One or two letters that represent the name of an element—e.g., H stands for a hydrogen atom, and Na stands for a sodium atom. 21

ATP synthase Enzyme that is part of an *ATP synthase complex* and functions in the production of ATP in chloroplasts and mitochondria. 100

atrioventricular valve Heart valve located between an atrium and a ventricle. 434

atrium (pl., atria) Chamber; particularly an upper chamber of the heart lying above a ventricle. 434

australopithecine One of several species of *Australopithecus,* a genus that contains the first generally recognized hominins. 362

autoimmune disease Disease that results when the immune system mistakenly attacks the body's own tissues. 509

autonomic system Portion of the peripheral nervous system that regulates internal organs. 526

autosomal trait A characteristic or trait that is determined by a gene located on one of the 22 sets of autosomal chromosomes. 220

autosome Any chromosome other than the sex-determining pair. 141

autotroph Organism that can capture energy and synthesize organic molecules from inorganic nutrients. 614

auxin Plant hormone regulating growth, particularly cell elongation; also called indoleacetic acid (IAA). 392

AV (atrioventricular) node Small region of neuromuscular tissue that transmits impulses received from the SA node to the ventricular walls. 436

axial skeleton Part of the vertebrate skeleton forming the vertical support or axis, including the skull, rib cage, and vertebral column. 549

axon Elongated portion of a neuron that conducts nerve impulses, typically from the cell body to the synapse. 517

B

bacteria Prokaryotic members of the domain Bacteria; most diverse and prevalent organisms on Earth. 294

bacteriophage Virus that infects bacteria. 289

ball-and-socket joint Synovial joint (hip and shoulder joints) that allows movement in all directions. 553

bark External part of a tree, containing cork, cork cambium, and phloem. 379

Barr body Dark-staining body (discovered by M. Barr) in the nuclei of female mammals that contains a condensed, inactive X chromosome. 151

base Molecules tending to lower the hydrogen ion concentration in a solution and raise the pH numerically. 31

B-cell receptor (BCR) Complex on the surface of a B cell that binds an antigen and stimulates the B cell. 502

benign tumor Tumor that does not invade adjacent tissue and stays at the site of origin. 134

bicarbonate ion Ion that participates in buffering the blood, and the form in which carbon dioxide is transported in the bloodstream. 456

bilateral symmetry Body plan having two corresponding or complementary halves. 336

bile Secretion of the liver that is temporarily stored and concentrated in the gallbladder before being released into the small intestine, where it emulsifies fat. 466

binary fission Splitting of a parent cell into two daughter cells; serves as an asexual form of reproduction in bacteria. 124, 296

binge-eating disorder Condition characterized by overeating episodes that are not followed by purging. 489

binomial name Scientific name of an organism, the first part of which designates the genus and the second part of which designates the specific epithet. 6

biodiversity Total number of species, the variability of their genes, and the communities in which they live. 15, 643

bioethics The branch of ethics concerned with the development and implementation of biological technology. 15

biogeochemical cycle Circulating pathway of elements, such as carbon and nitrogen, involving exchange pools, storage areas, and biotic communities. 618

biogeography Study of the geographical distribution of organisms. 236

bioinformatics Computer technologies used to study the genome and other molecular data. 193

biological magnification Process by which substances become more concentrated in organisms in the higher trophic levels of a food web. 642

biological molecule Organic molecule in cells: carbohydrates, lipids, proteins, and nucleic acids. 38

biology Scientific study of life. 11

biomass Weight of one or more organisms. 618

bioremediation Cleanup of the environment using bacteria to break down pollutants, such as oil spills and Agent Orange. 297

biosphere Zone of air, land, and water at the surface of the Earth in which organisms are found. 10, 585, 623

biotechnology Genetic engineering and other techniques that make use of natural biological systems to create a product or achieve a particular result desired by humans. 189

biotic potential Maximum population growth rate under ideal conditions. 594

bipedalism Walking erect on two feet. 362

bird Endothermic reptile that has feathers and wings, is often adapted for flight, and lays hard-shelled eggs. 356

birth control pill Oral contraception containing estrogen and progesterone. 568

bivalve Type of mollusc with a shell composed of two valves; includes clams, oysters, and scallops. 344

blade Broad, expanded portion of a plant leaf that may be single or compound leaflets. 376

blastocyst Early stage of animal embryonic development that consists of a hollow, fluid-filled ball of cells. 575

blind spot Region of the retina, lacking rods or cones, where the optic nerve exits the retina. 545

blood Fluid circulated by the heart through a closed system of vessels. 418, 441

blood pressure Force of blood pushing against the inside wall of blood vessels. 436

B lymphocyte (B cell) Lymphocyte that matures in the bone marrow and, when stimulated by the presence of a specific antigen, gives rise to antibody-producing plasma cells. 498

body mass index (BMI) Calculation used to determine whether or not a person is overweight or obese. 485

bolus Mass of chewed food mixed with saliva. 463

bone Connective tissue having protein fibers and a hard matrix of inorganic salts, notably calcium salts. 418

bony fishes Fishes with a bony rather than cartilaginous skeleton. 355

bottleneck effect Evolutionary event in which a significant percentage of a population is prevented from reproducing, therefore reducing variation. Population bottlenecks increase the influence of *genetic drift*. 261

brain stem In mammals, portion of the brain consisting of the medulla oblongata, pons, and midbrain. 524

bronchiole In terrestrial vertebrates, small tube that conducts air from a bronchus to the alveoli. 453

bronchus (pl., bronchi) In terrestrial vertebrates, branch of the trachea that leads to the lungs. 453

brown algae Marine photosynthetic protists with a notable abundance of xanthophyll pigments; includes well-known seaweeds of northern rocky shores. 302

bryophyte Nonvascular plant—the mosses, liverworts, and hornworts. These plants have no vascular tissue and occur in moist locations. 312

buffer Substance or group of substances that tends to resist pH changes of a solution, thus stabilizing its relative acidity and basicity. 32

bulbourethral glands Two small structures located below the prostate gland in males; adds secretions to semen. 563

bulimia nervosa Eating disorder characterized by binge eating followed by purging via self-induced vomiting or use of a laxative. 489

bursa Sac-like, fluid-filled structure, lined with synovial membrane, near a synovial joint. 553

C

C₃ plant Plant that fixes carbon dioxide via the Calvin cycle. The first stable product of C_3 photosynthesis is a 3-carbon compound. 103

C₄ plant Plant that fixes carbon dioxide to produce a C_4 molecule that releases carbon dioxide to the Calvin cycle. 103

calcitonin Hormone secreted by the thyroid gland that increases the blood calcium level. 530

calorie Amount of heat energy required to raise the temperature of 1 gram of water 1°C. 78

Calvin cycle reactions Portion of photosynthesis that takes place in the stroma of chloroplasts and can occur in the dark. It uses the products of the light reactions to reduce CO_2 to a carbohydrate. 96

calyx The sepals collectively; the outermost flower whorl. 319

CAM Crassulacean-acid metabolism. A plant fixes carbon dioxide at night to produce a C_4 molecule that releases carbon dioxide to the Calvin cycle during the day. 104

camera-type eye Type of eye found in vertebrates and certain molluscs. A single lens focuses an image on closely packed photoreceptors. 544

cancer Malignant tumor whose nondifferentiated cells exhibit loss of contact inhibition, uncontrolled growth, and the ability to invade tissue and metastasize. 134

capillary Microscopic blood vessel. Gases and other substances are exchanged across the walls of a capillary between blood and tissue fluid. 436

capsid Protective protein container of the genetic material of a virus. 28

capsule Gelatinous layer surrounding the cells of green algae and certain bacteria. 58

carbohydrate Class of organic compounds that includes monosaccharides, disaccharides, and polysaccharides; present in food as sugars, starch, and fiber. 39, 478

carbon dioxide (CO_2) fixation Photosynthetic reaction in which carbon dioxide is attached to an organic compound. 101

carcinogenesis Development of cancer. 134

cardiac cycle One complete cycle of systole and diastole for all heart chambers. 435

cardiac muscle Striated, involuntary muscle tissue found only in the heart. 419

cardiovascular system Organ system in which blood vessels distribute blood under the pumping action of the heart. 422, 433

carnivore Consumer in a food chain that eats other animals. 463, 614

carotenoid Yellow or orange pigment that serves as an accessory to chlorophyll in photosynthesis. 98

carotid body Structure located at the branching of the carotid arteries; contains chemoreceptors sensitive to hydrogen ion and carbon dioxide concentrations in blood. 454

carpel Ovule-bearing unit that is a part of a pistil. 319, 399

carrier Heterozygous individual who has no apparent abnormality but can pass on an allele for a recessively inherited genetic disorder. 168

carrying capacity Largest number of organisms of a particular species that can be maintained indefinitely by a given environment. 595

cartilage Connective tissue in which the cells lie within lacunae embedded in a flexible, proteinaceous matrix. 418

cartilaginous fishes Fishes with a cartilaginous rather than bony skeleton; includes sharks, rays, and skates. 354

cell Smallest unit that displays the properties of life; composed of cytoplasm surrounded by a plasma membrane. 2, 55

cell body Portion of a neuron that contains a nucleus and from which dendrites and an axon extend. 517

cell cycle Repeating sequence of events in eukaryotes that involves cell growth and nuclear division; consists of the stages G_1, S, G_2, and M. 126

cell plate Structure across a dividing plant cell that signals the location of new plasma membranes and cell walls. 131

cell-signaling The process by which cells communicate with other cells and detect changes in their environment. 205

cell theory One of the major theories of biology; states that all organisms are made up of cells. Cells are capable of self-reproduction and come only from preexisting cells. 57

cellular respiration Metabolic reaction that uses the energy from carbohydrates, fatty acid, or amino acid breakdown to produce ATP molecules. 69

cellulose Polysaccharide that is the major complex carbohydrate in plant cell walls. 40

cell wall Structure that surrounds a plant, protistan, fungal, or bacterial cell and maintains the cell's shape and rigidity. 58, 72

central nervous system (CNS) Portion of the nervous system consisting of the brain and spinal cord. 517

centriole Cell organelle, existing in pairs, that occurs in the centrosome and may help organize a mitotic spindle for chromosome movement during animal cell division. 70, 128

centromere Constricted region on a chromosome joining two sister chromatids. 125

centrosome Central microtubule organizing center of cells. In animal cells, it contains two centrioles. 70, 128

cephalization Having a well-recognized anterior head with a brain and sensory receptors. 38

cephalopod Type of mollusc in which a modified foot develops into the head region; includes squids, cuttlefish, octopuses, and nautiluses. 344

cerebellum In terrestrial vertebrates, portion of the brain that coordinates skeletal muscles to produce smooth, graceful motions. 523

cerebral cortex Outer layer of cerebral hemispheres; receives sensory information and controls motor activities. 522

cervical cap Birth control device made of latex or rubber in the shape of a cup, which covers the cervix; considered a minidiaphragm. 568

cervix Narrow end of the uterus leading into the vagina. 564

character displacement Tendency for characteristics to be more divergent when similar species belong to the same community than when they are isolated from one another. 610

charophytes A group of freshwater green algae thought to be the closest relatives of land plants. 310

checkpoint In the cell cycle, one of several points where the cell cycle can stop or continue on, depending on the internal signal it receives; ensures that each step of the cell cycle is completed before the next one begins. 131

chemoautotroph Organism able to synthesize organic molecules by using carbon dioxide as the carbon source and the oxidation of an inorganic substance (such as hydrogen sulfide) as the energy source. 296

chemoheterotroph Organism that is unable to reproduce its own organic molecules and therefore requires organic nutrients in its diet. 297

chitin Strong but flexible nitrogenous polysaccharide in the exoskeleton of arthropods. 41, 347

chlamydia Bacterial STD of the lower reproductive tract that can result in pelvic inflammatory disease if not treated. 572

chlorofluorocarbons (CFCs) Organic compounds containing carbon, chlorine, and fluorine atoms. CFCs, such as freon, can deplete the ozone shield by releasing chlorine atoms in the upper atmosphere. 641

chlorophyll Green pigment that absorbs solar energy and is important in algal and plant photosynthesis; occurs as chlorophyll *a* and chlorophyll *b*. 95, 98

chloroplast Membrane-bounded organelle in algae and plants with chlorophyll-containing membranous thylakoids; where photosynthesis takes place. 67, 95

cholesterol One of the major lipids found in animal plasma membranes; makes the membrane impermeable to many molecules. 44, 480

chordate Animal that has a dorsal tubular nerve cord, a notochord, pharyngeal pouches, and a postanal tail at some point in its life cycle. 352

chorionic villi Tree-like extensions of the chorion (fetal portion of the placenta) that project into the maternal portion of the placenta. 578

chorionic villus sampling (CVS) Prenatal test in which a sample of chorionic villi cells is removed for diagnostic purposes. 216

chromatid Single DNA strand of a chromosome. Chromosomes may consist of a pair of sister chromatids. 126

chromatin Network of fibrils consisting of DNA and associated proteins observed within a nucleus that is not dividing. 64, 125

chromosome Structure, consisting of DNA complexed with proteins, that transmits genetic information from the previous generation of cells and organisms to the next generation. 64, 124

chromosome map Sequence that shows the relative distance between gene loci on a chromosome. 170

chyme Thick, semiliquid food material that passes from the stomach to the small intestine. 465

ciliate Complex, unicellular protist that moves by means of cilia and digests food in food vacuoles. 303

cilium (pl., cilia) Short, hair-like projection from the plasma membrane, occurring usually in larger numbers. 71

citric acid cycle Cycle of reactions in mitochondria that begins with citric acid. It breaks down an acetyl group and produces CO_2, ATP, NADH, and $FADH_2$; also called the Krebs cycle. 110

clade A group of organisms that includes a common ancestor and all of its descendants and that has its own particular derived traits. 280

cladistics School of systematics that uses derived characters to determine monophyletic groups and construct cladograms. 280

cladogram Branching diagrammatic tree used to depict the evolutionary history of a group of organisms. 280

class One of the categories, or taxa, used by taxonomists to group species; the taxon above the order level. 6, 278

clavicle Collarbone. 549

cleavage Cell division without cytoplasmic addition or enlargement; occurs during the first stage of animal development. 575

climax community In ecology, the community that results when succession has come to an end. 607

clonal selection model States that an antigen selects which lymphocyte will undergo clonal expansion and produce more lymphocytes bearing the same type of receptor. 502

closed circulatory system In all vertebrates and some invertebrates, cardiovascular system composed of a muscular heart and blood vessels. 433

club mosses Lycophyges, a group of seedless vascular plants also known as ground pines. 314

clumped Spatial distribution of individuals in a population in which individuals are more dense in one area than in another. 592

cnidarian Invertebrates existing as either a polyp or a medusa with two tissue layers and radial symmetry. 341

codominance Inheritance pattern in which both alleles of a gene are equally expressed in a heterozygote. 164

codon Three base sequence in messenger RNA that causes the addition of a particular amino acid into a protein, or termination of translation. 182

coelom Body cavity, lying between the digestive tract and body wall, that is completely lined by mesoderm. 339, 463

coenzyme Nonprotein organic molecule that aids the action of the enzyme to which it is loosely bound. 100, 110

coenzyme A (CoA) Molecule that helps oxidate pyruvate in the preparatory (prep) reaction during cellular respiration. 110

coevolution Joint evolution in which one species exerts selective pressure on the other species. 605

cohesion-tension model Explanation for upward transport of water in xylem based on transpiration-created tension and the cohesive properties of water molecules. 385

collecting duct Duct in the kidneys that receives fluid from several nephrons. The reabsorption of water occurs here. 458

collenchyma Plant tissue composed of cells with unevenly thickened walls; supports growth of stems and petioles. 374

color blindness Deficiency in one or more of the three kinds of cone cells responsible for color vision; a common X-linked recessive disorder. 223

columnar epithelium Type of epithelial tissue in which the cells are rectangular, with the nuclei typically located at the bottom of each cell. 415

commensalism Symbiotic relationship in which one species is benefited and the other is neither harmed nor benefited. 609

common ancestor Ancestor held in common by at least two lines of descent. 268

community Assemblage of populations interacting with one another in the same environment. 10, 585, 605

compact bone Type of bone that contains osteons consisting of concentric layers of matrix and osteocytes in lacunae. 418, 550

competition Interaction between two organisms in which both require the same limited resource, which results in harm to both. 596, 609

competitive exclusion principle Theory that no two species can occupy the same niche. 610

complementary base pairing Hydrogen bonding between particular purines and pyrimidines in DNA. 177

complement system Series of proteins in plasma that form a nonspecific defense mechanism against a pathogen invasion; complements the antigen-antibody reaction. 500

compound Substance having two or more different elements united chemically in a fixed ratio. 24

compound eye Type of eye found in arthropods. It is composed of many independent visual units. 544

conclusion Statement made following an experiment as to whether or not the results support the hypothesis. 12

conduction deafness Deafness due to middle ear damage. 542

cone In vertebrate eyes, photoreceptor cell that responds to bright light and makes color vision possible. In conifers, structure that bears either pollen (male gametophyte) or seeds (female gametophyte). The sporangia-bearing structures of certain seedless vascular plants are also termed cones. 318, 545

conifer Member of a group of cone-bearing gymnosperm plants, including pine, cedar, and spruce trees. 318

conjugation Transfer of genetic material from one cell to another. 296

conjugation pili Tubular, rigid bacterial structures that enable the transmission of DNA from cell to cell. 59

connective tissue Type of animal tissue that binds structures together, provides support and protection, fills spaces, stores fat, and forms blood cells. Adipose tissue, cartilage, bone, and blood are types of connective tissue. 417

consumer Organism that feeds on another organism in a food chain; primary consumers eat plants, and secondary consumers eat animals. 614

contact inhibition In cell culture, the point where cells stop dividing when they become a one-cell-thick sheet. 132

continuous ambulatory peritoneal dialysis (CAPD) Dialysis that takes place inside the body using the peritoneum, the natural lining of the abdomen, as the dialysis membrane. 460

contraceptive implant Birth control method utilizing synthetic progesterone; prevents ovulation by disrupting the ovarian cycle. 567

contraceptive injection Birth control method utilizing progesterone or estrogen and progesterone; prevents ovulation by disrupting the ovarian cycle. 567

contraceptive vaccine Under development, this birth control method immunizes against the hormone hCG, crucial to maintaining implantation of the embryo. 567

control group Sample that goes through all the steps of an experiment but does not contain the experimental variable being tested; a standard against which the results of an experiment are checked. 12

convergent evolution Similarity in structure in distantly related groups due to adaptation to the environment. 279

copulation Sexual union between a male and a female. 560

cork cambium Lateral meristem that produces cork. 379

cork cell Dead cell in the outer covering of the bark of trees; may be sloughed off. 379

cornea Transparent, anterior portion of the outer layer of the eyeball. 544

corolla Petals, collectively; usually the conspicuously colored flower whorl. 319

corpus luteum Follicle that has released an egg and increases its secretion of progesterone. 565

cortex In plants, ground tissue bounded by the epidermis and vascular tissue in stems and roots; in animals, outer layer of an organ, such as the cortex of the kidneys or adrenal gland. 378

cotyledon Seed leaf of an embryo of a flowering plant; provides nutrient molecules for the developing plant before photosynthesis begins. 321, 373, 402

coupled reaction Reaction that occurs simultaneously. One is an exergonic reaction that releases energy, and the other is an endergonic reaction that requires an input of energy in order to occur. 81

covalent bond Chemical bond in which atoms share one pair of electrons. 25

coxal bones Two massive bones that form the pelvis and articulate with the femurs (thighbones). 549

cranial nerve Nerve that arises from the brain. 525

cranium Portion of the skull that protects the brain. 549

cristae Short, finger-like projections formed by the folding of the inner membrane of mitochondria. 69

Cro-Magnon Common name for the first fossils to be designated *Homo sapiens*. 364

crossing-over Exchange of segments between nonsister chromatids of a tetrad during meiosis. 144

crustacean Member of a group of marine arthropods that contains, among others, shrimps, crabs, crayfish, and lobsters. 348

cuboidal epithelium Type of epithelial tissue in wich the cells are cube-shaped. 415

culture Total pattern of human behavior; includes technology and the arts and is dependent on the capacity to speak and transmit knowledge. 363

Cushing syndrome Condition resulting from hypersecretion of glucocorticoids; characterized by thin arms and legs and a "moon face," and accompanied by high blood glucose and sodium levels. 532

cutaneous receptor Sensory receptor for pressure and touch; found in the dermis of the skin. 546

cuticle Waxy layer covering the epidermis of plants; protects the plant against water loss and disease-causing organisms. 374

cyanobacteria Photosynthetic bacteria that contain chlorophyll and release O_2; formerly called blue-green algae. 296

cyclin Protein that cycles in quantity as the cell cycle progresses; combines with and activates the kinases that promote the events of the cycle. 132

cystic fibrosis Generalized, autosomal recessive disorder of infants and children in which there is widespread dysfunction of the exocrine glands. 222

cytokine A chemical messenger, secreted by cells of the immune system, that stimulates other cells to perform their various functions. 502

cytokinesis Division of the cytoplasm following mitosis and meiosis. 127

cytokinin Plant hormone that promotes cell division; often works in combination with auxin during organ development in plant embryos. 394

cytoplasm Contents of a cell between the nucleus (nucleoid region of bacteria) and the plasma membrane. 57

cytosine (C) One of four nitrogen-containing bases in the nucleotides composing the structure of DNA and RNA; pairs with guanine. 176

cytoskeleton Internal framework of the cell, consisting of microtubules, actin filaments, and intermediate filaments. 63, 69

cytotoxic T cell (T_c cell) T cell that attacks and kills antigen-bearing cells. 504

D

data (sing., datum) Facts, or information, collected through observation and/or experimentation. 12

day-neutral plant Plant whose flowering is not dependent on day length—e.g., tomato and cucumber. 397

deciduous Shedding leaves annually. 371

decomposer Organism, usually a bacterium or fungus, that breaks down organic matter into inorganic nutrients that can be recycled in the environment. 615

deficiency disorder Disorder caused by a lack of certain vitamins or minerals in the diet. Examples are bleeding gums, rickets (vitamin deficiency), osteoporosis, and anemia (mineral deficiency). 476

deforestation Removal of trees from a forest in a way that reduces the size of the forest. 633

dehydration synthesis reaction Chemical reaction resulting in a covalent bond with the accompanying loss of a water molecule. 39

delayed allergic response Allergic response initiated at the site of the allergen by sensitized T cells, involving macrophages and regulated by cytokines. 508

deletion Change in chromosome structure in which the end of a chromosome breaks off or two simultaneous breaks lead to the loss of an internal segment; often causes abnormalities—e.g., cri du chat syndrome. 217

demographics Study of human populations, their characteristics, and their changes. 593

denatured Loss of an enzyme's normal shape, so that it no longer functions; caused by a less than optimal pH and temperature. 48

dendrite Part of a neuron that sends signals toward the cell body. 517

dense fibrous connective tissue Type of connective tissue containing many collagen fibers packed together; found in tendons and ligaments, for example. 418

density-dependent factor Biotic factor, such as disease or competition, that affects population size according to the population's density. 596

density-independent factor Abiotic factor, such as fire or flood, that affects population size independent of the population's density. 596

deoxyribose In DNA, pentose sugar that has one less hydroxyl group than ribose. 39

desertification Denuding and degrading a once fertile land, initiating a desert-producing cycle that feeds on itself and causes long-term changes in the soil, climate, and biota of an area. 633

detrital food web Complex pattern of interlocking and crisscrossing food chains that begins with a population of detritivores. 616

detritus Organic matter produced by decomposition of substances, such as tissues and animal wastes. 615

deuterostome Coelomate animals in which the second embryonic opening is associated with the mouth. The first embryonic opening, the blastopore, is associated with the anus. 339

diabetes mellitus Condition characterized by a high blood glucose level and the appearance of glucose in the urine, due to a deficiency of insulin production and failure of cells to take up glucose. 532

dialysate Material that passes through the membrane in dialysis. 460

diaphragm In mammals, dome-shaped muscularized sheet separating the thoracic cavity from the abdominal cavity important in inhalation. Also, a birth control device consisting of a soft rubber or latex cup that fits over the cervix. 454, 568

diarrhea Excessively frequent and watery bowel movements. 468

diastole Relaxation period of a heart chamber during the cardiac cycle. 435

diet A person's typical food choices. A balanced diet contains all the nutrients in the right proportions to maintain a healthy body. 477

dietary supplement Nutrient or plant product (e.g., herbal teas, protein supplements) used to enhance health. They do not undergo the same safety and effectiveness testing required for prescription drugs. 491

digestive system Organ system that includes the mouth, esophagus, stomach, small intestine, and large intestine (colon). It receives and digests food into nutrient molecules and has associated organs: teeth, tongue, salivary glands, liver, gallbladder, and pancreas. 422

dihybrid Individual that is heterozygous for two traits; shows the phenotype governed by the dominant alleles but carries the recessive alleles. 161

diploid (2n) number Cell condition in which two of each type of chromosome are present. 142

directional selection Outcome of natural selection in which an extreme phenotype at one end of a population distribution is favored over all other phenotypes, leading to one distinct form. 261

disaccharide Sugar that contains two units of a monosaccharide. One example is maltose. 39

disruptive selection Outcome of natural selection in which both extreme phenotypes at the ends of a population distribution are favored over the average phenotype, leading to more than one distinct form. 253

distal tubule Final portion of a nephron that joins with a collecting duct; associated with tubular secretion. 458

diversity Amount of each specific species in a community. 606

DNA (deoxyribonucleic acid) Nucleic acid polymer produced from covalent bonding of nucleotide monomers that contain the sugar deoxyribose; the genetic material of nearly all organisms. 48

DNA fingerprinting Use of DNA fragment lengths resulting from restriction enzyme cleavage to identify individuals. 191

DNA ligase Enzyme that links DNA fragments; used during production of recombinant DNA to join foreign DNA to vector DNA. 179

DNA microarray Thousands of different single-stranded DNA fragments arranged in an array (grid) on a glass slide; used to detect and measure gene expression. 225

DNA polymerase During replication, enzyme that joins the nucleotides complementary to a DNA template. 179

DNA replication Synthesis of a new DNA double helix prior to mitosis or meiosis in eukaryotic cells and during prokaryotic fission in prokaryotic cells. 125, 179

domain Largest of the categories, or taxa, used by taxonomists to group species. The three domains are Archaea, Bacteria, and Eukarya. 6, 278

domain Archaea One of the three domains of life; contains prokaryotic cells that often live in extreme habitats and have unique genetic, biochemical, and physiological characteristics; its members are sometimes referred to as *archaea*. 7, 283

domain Bacteria One of the three domains of life; contains prokaryotic cells that differ from archaea because they have unique genetic, biochemical, and physiological characteristics. 7, 283

domain Eukarya One of the three domains of life, consisting of organisms with eukaryotic cells and further classified into the kingdoms Protista, Fungi, Plantae, and Animalia. 7, 283

dominant allele Allele that exerts its phenotypic effect in the heterozygote. It masks the expression of the recessive allele. 159

dormancy In plants, cessation of growth under conditions that seem appropriate for growth. 393

dorsal root ganglion Mass of sensory neuron cell bodies located in the dorsal root of a spinal nerve. 525

double fertilization In flowering plants, one sperm nucleus unites with the egg nucleus, and a second sperm nucleus unites with the polar nuclei of an embryo sac. 320, 401

Down syndrome Trisomy 21. The individual has three copies of chromosome 21 and the following characteristics: mental disabilities of varying degree, short stature, an eyelid fold, stubby fingers, and a palm crease. 150

Duchenne muscular dystrophy Chronic, progressive disease affecting the shoulder and hip (pelvic) girdles, commencing in early childhood; characterized by increasing weakness of the muscles followed by atrophy and a peculiar swaying gait with the legs kept wide apart. It is transmitted as an X-linked trait; affected individuals, predominantly males, rarely survive to maturity. Death is usually due to respiratory weakness or heart failure. 223

duodenum First part of the small intestine, where chyme enters from the stomach. 468

duplication Change in chromosome structure in which a particular segment is present more than once in the same chromosome. 218

dyad Chromosome composed of two sister chromatids. 144

E

echinoderm Marine invertebrate, such as a sea star, sea urchin, and sand dollar; characterized by radial symmetry and a water vascular system. 351

ecological niche Role an organism plays in its community, including its habitat and its interactions with other organisms. 609

ecological pyramid Pictorial graph based on the biomass, number of organisms, or energy content of various trophic levels in a food web—from the producer to the final consumer populations. 618

ecological succession Gradual replacement of communities in an area following a disturbance (secondary succession) or the creation of new soil (primary succession). 607

ecology Study of the interactions of organisms with other organisms and with the physical and chemical environment. 585

ecosystem Biological community together with the associated abiotic environment; characterized by a flow of energy and a cycling of inorganic nutrients. 10, 585, 606

ectoderm Outermost primary tissue layer of an animal embryo; gives rise to the nervous system and the outer layer of the integument. 576

ectothermic Having a body temperature that varies according to the environmental temperature. 356

edema Swelling due to tissue fluid accumulation in the intercellular spaces. 460

electrocardiogram (ECG) Recording of the electrical activity associated with the heartbeat. 436

electron Negative subatomic particle moving about in an energy level around the nucleus of an atom. 21

electronegativity Ability of an atom to attract electrons toward itself in a chemical bond. 27

electron shell Concentric energy levels in which electrons orbit. 23

electron transport chain Passage of electrons along a series of electron carriers from a higher to lower energy levels; the energy released is used to synthesize ATP. 99, 110

element Substance that cannot be broken down into substances with different properties; composed of only one type of atom. 20

embryo sac Female gametophyte of flowering plants. 401

embryonic stem cell Undifferentiated cell, obtained from an embryo, that can be manipulated to become a specialized cell type, such as a muscle, nerve, or red blood cell. 200

emerging virus Causative agent of a disease that is new or is demonstrating increased prevalence, such as the viruses that cause AIDS, SARS, and avian influenza. 291

emulsification Breaking up of fat globules into smaller droplets by the action of bile salts or any other emulsifier. 466

endocrine gland Ductless organ that secretes hormone(s) into the bloodstream. 468, 527

endocrine system Organ system involved in the coordination of body activities; uses hormones as chemical signals secreted into the bloodstream. 422

endocytosis Process by which substances are moved into the cell from the environment by phagocytosis (cellular eating) or pinocytosis (cellular drinking); includes receptor-mediated endocytosis. 89

endoderm Innermost primary tissue layer of an animal embryo that gives rise to the linings of the digestive tract and associated structures. 576

endodermis Internal plant root tissue forming a boundary between the cortex and the vascular cylinder. 382

endomembrane system Collection of membranous structures involved in transport within the cell. 66

endometrium Mucous membrane lining the interior surface of the uterus. 564

endoplasmic reticulum (ER) System of membranous saccules and channels in the cytoplasm, often with attached ribosomes. 65

endoskeleton Protective internal skeleton, as in vertebrates. 548

endosperm In flowering plants, nutritive storage tissue derived from the union of a sperm nucleus and polar nuclei in the embryo sac. 320

endospore Spore formed within a cell; formed by certain bacteria. 296

endosymbiotic theory Explanation of the evolution of eukaryotic organelles by phagocytosis of prokaryotes. 300

endotherm Animal that generates its own internal heat. 357

energy Capacity to do work and bring about change; occurs in a variety of forms. 3, 78

energy laws Two laws explaining energy and its relationships and exchanges. The first, also called the "law of conservation," says that energy cannot be created or destroyed but can only be changed from one form to another; the second says that energy cannot be changed from one form to another without a loss of usable energy. 79

energy of activation (E$_a$) Energy that must be added in order for molecules to react with one another. 85

enhancer Element that stimulates transcription of nearby genes; functions by acting as a binding site for transcription factors. 204

entropy Measure of disorder or randomness. 79

enzyme Organic catalyst, usually a protein, that speeds a reaction in cells due to its shape. 45, 83

enzyme inhibition Means by which cells regulate enzyme activity; may be competitive or noncompetitive inhibition. 84

epidermal tissue (epidermis) In plants, tissue that covers roots, leaves, and stems of nonwoody organisms. 374

epididymis Coiled tubule next to the testes, where sperm mature and may be stored for a short time. 562

epiglottis Structure that covers the glottis and closes off the air tract during the process of swallowing. 464

epinephrine Adrenaline; hormone secreted by the adrenal medulla in times of stress. 531

epithelial tissue Tissue that lines hollow organs and covers surfaces. 415

equilibrium species Species demonstrating a life history pattern in which members exhibit logistic population growth and the population size remains at or near the carrying capacity. Its members are large and slow to mature, have a long life span and few offspring, and provide much care to offspring (e.g., bears, lions). 599

erection Increase in blood flow to the penis during sexual arousal, causing the penis to stiffen and become erect. 562

esophagus Muscular tube for moving swallowed food from the pharynx to the stomach. 464

essential nutrient Substance in the diet that contributes to good health; must be supplied by the diet because the body either cannot synthesize it or makes insufficient amounts to meet the body's needs. 477

estrogen Female sex hormone that helps maintain sexual organs and secondary sex characteristics. 566

ethylene Plant hormone that causes ripening of fruit and is involved in abscission. 395

euchromatin Chromatin that is extended and accessible for transcription. 203

eudicot Eudicotyledon; flowering plant group. Members have two embryonic leaves (cotyledons), net-veined leaves, vascular bundles in a ring, and flower parts in fours or fives and their multiples. 373

eugenics Gene manipulation to control desired traits in offspring. 227

eukaryotic cell Type of cell that has a membrane-bounded nucleus and membranous organelles; found in organisms within the domain Eukarya. 57

eutrophication Enrichment of water by inorganic nutrients used by phytoplankton. Often, overenrichment caused by human activities leads to excessive bacterial growth and oxygen depletion. 620

evergreen Land plant that sheds leaves over a long period, so that some leaves are always present. 371

evolution Descent of organisms from common ancestors, with the development of genetic and phenotypic changes over time that make them more suited to the environment. 5, 233

evolutionary tree Diagram that shows the evolutionary history of groups of organisms. 274

exocrine gland Gland that discharges its secretion into ducts. The pancreas is an exocrine gland when it secretes pancreatic juice into the duodenum. 468

exocytosis Process in which an intracellular vesicle fuses with the plasma membrane, so that the vesicle's contents are released outside the cell. 89

exophthalmos Abnormal protrusion of the eyes, often a sign of hyperthyroidism (Graves disease). 530

exoskeleton Protective external skeleton, as in arthropods. 548

exotic species Species that is new to a community (nonnative). 613

experiment Series of actions undertaken to collect data with which to test a hypothesis. 12

experimental design Artificial situation devised to test a hypothesis. 12

experimental variable In a scientific experiment, a condition of the experiment that is deliberately changed. 12

exponential growth Growth, particularly of a population, in which the increase occurs in the same manner as compound interest. 595

external fertilization Fertilization of an egg by sperm that occurs outside the body, as in many aquatic animals. 560

extinction Total disappearance of a species or higher group. 15, 599

extracellular matrix (ECM) Meshwork of polysaccharides and proteins that provides support for an animal cell and affects its behavior. 72

extraembryonic membranes Membranes that are not a part of an embryo but are necessary to the continued existence and health of the embryo. 560

ex vivo gene therapy Gene therapy in which infected tissue is removed from an organism, injected with normal genes, and then returned to the organism. The normal genes divide, producing normal genes instead of infected ones. 228

F

facial bone Axial skeleton bone, including the lower and upper jawbones, cheekbones, and nasal bones. 549

facilitated diffusion Passive transfer of a substance into or out of a cell along a concentration gradient by a process that requires a carrier. 86

family One of the categories, or taxa, used by taxonomists to group species; the taxon above the genus level. 6, 278

fat Organic molecule that contains glycerol and fatty acids and is found in adipose tissue of vertebrates. 41, 479

fatty acid Molecule that contains a hydrocarbon chain and ends with an acid group. 42

feedback inhibition Mechanism for regulating metabolic pathways in which the concentration of the product is kept within a certain range until binding shuts down the pathway and no more product is produced. 84

female condom Birth control method that blocks the entrance of sperm to the uterus and prevents STDs. 568

female gametophyte In seed plants, the gametophyte that produces an egg; in flowering plants, an embryo sac. 401

femur Thighbone, which is the longest, strongest bone of the body. 549

fermentation Anaerobic breakdown of glucose that results in a gain of 2 ATP and end products, such as alcohol and lactate. 118

fern Member of a group of plants that have large fronds. In the sexual life cycle, the independent gametophyte produces flagellated sperm, and the vascular sporophyte produces windblown spores. 315

fertilization Fusion of sperm and egg nuclei, producing a zygote that develops into a new individual. 141, 574

fiber Structure resembling a thread; also plant material that is nondigestible. 479

fibroblast Connective tissue cell that synthesizes fibers and ground substance. 417

fibrous protein Structural protein with only a secondary structure (e.g., keratin, silk, collagen). 48

fibula Outer, more slender bone in a human's lower leg. 550

filament End-to-end chain of cells that forms as cell division occurs in only one plane; in plants, the elongated stalk of a stamen. 319, 399

filter feeder Animal that obtains nourishment by straining minute organic particles from the water in a way that deposits them in the animal's digestive tract. 340

filtration Movement of small molecules from a blood capillary into the nephron capsule due to the action of blood pressure. 458

fimbriae (sing., fimbria) Small, bristle-like fibers on the surface of bacterial cells that enable them to adhere to surfaces. 59, 564

fitness The reproductive success of an individual relative to other members of a population. 239, 254

five-kingdom system System of classification that contains the kingdoms Monera, Protista, Plantae, Animalia, and Fungi. 282

flagellum (pl., flagella) Long, slender extension used for locomotion by some bacteria, protozoans, and sperm. 58, 71, 295

flatworm Invertebrate, such as a planarian and tapeworm, that has a thin body, three-branched gastrovascular cavity, and ladder-type nervous system. 342

flower Reproductive organ of a flowering plant, consisting of several kinds of modified leaves arranged in concentric rings and attached to a modified stem called a receptacle. 311, 399

follicle In the ovary of animals, structure that contains an oocyte; site of oocyte production. 565

follicular phase First half of the ovarian cycle, during which the follicle matures and much estrogen (and some progesterone) is produced. 566

food chain Order in which one population feeds on another in an ecosystem, thereby showing the flow of energy from a detritivore (detrital food chain) or a producer (grazing food chain) to the final consumer. 618

food web In ecosystems, a complex pattern of linked and crisscrossing food chains. 616

foraminiferan Protozoan; marine amoeba having a calcium carbonate skeleton, many of which make limestone formations, such as the White Cliffs of Dover. 303

formed elements Constituents of blood that are either cellular (red blood cells and white blood cells) or at least cellular in origin (platelets). 441

fossil Evidence of an organism that has been preserved in the Earth's crust. 235, 272

fossil fuel Fuel such as oil, coal, and natural gas that is the result of partial decomposition of plants and animals coupled with exposure to heat and pressure for millions of years. 621, 638

fossil record History of life recorded from remains from the past. 242

founder effect Effect of establishing a new population by a small number of individuals, carrying only a small fraction of the original population's genetic variation. The alleles carried by these individuals often, by chance, do not occur in the same frequency as the parent population; an example of genetic drift. 261

fovea Region of the retina consisting of densely packed cones; responsible for the greatest visual acuity. 545

frameshift mutation Alteration in a gene due to deletion of a base, so that the reading "frame" is shifted; can result in a nonfunctional protein. 187

frond Fern leaf. 315

fruit In flowering plants, the structure that forms from an ovary and associated tissues and encloses seeds. 321, 402

functional group Specific cluster of atoms attached to the carbon skeleton of organic molecules that enters into reactions and behaves in a predictable way. 37

fungus (pl., fungi) Saprotrophic decomposer. The body is made up of filaments, called hyphae, that form a mass called a mycelium. 7, 323

G

gallbladder Organ attached to the liver that stores and concentrates bile. 466

gamete Haploid sex cell: egg or sperm. 570

gametophyte Haploid generation of the alternation of generations life cycle of a plant; produces gametes that unite to form a diploid zygote. 312, 398

ganglion Collection of neuron cell bodies usually outside the central nervous system. 525

gap junction Junction between cells formed by the joining of two adjacent plasma membranes; lends strength and allows ions, sugars, and small molecules to pass between cells. 73

gastropod Mollusc with a broad, flat foot for crawling (e.g., snails and slugs). 344

gastrula Stage of animal development during which the germ layers form, at least in part, by invagination. 576

gastrulation Formation of a gastrula from a blastula; characterized by an invagination of the cell layers to form a cap-like structure. 576

gene Unit of heredity existing as alleles on the chromosomes. In diploid organisms, typically two alleles are inherited—one from each parent. 4

gene flow Sharing of alleles among multiple populations through interbreeding. 259

gene linkage Existence of several alleles on the same chromosome. 170

gene locus Specific location of a particular gene on homologous chromosomes. 159

gene mutation Change in the sequence of bases in a gene. 187

gene pool Total of all the genes of all the individuals in a population. 256

gene therapy Correction of a detrimental mutation by the addition of new DNA and its insertion in a genome. 228

genetically modified The process by which an organism's genetic material is changed, usually by using recombinant DNA technology. 189

genetic counseling Prospective parents consult a counselor who determines the genotype of each and whether an unborn child will have a genetic disorder. 215

genetic drift Mechanism of evolution due to random changes in the allelic frequencies of a population; more likely to occur in small populations or when only a few individuals of a large population reproduce. 260

genetic engineering Alteration of genomes for medical or industrial purposes. 188

genetic marker Abnormality in the sequence of a base at a particular location on a chromosome signifying a disorder. 225

genital herpes Viral STD characterized by painful blisters on the genitals. 571

genital warts Viral STD caused by the human papillomaviruses; associated with cervical cancer and penile tumors; most common STD in the United States. 571

genomics Study of genomes. 192

genotype Alleles of an organism for a particular trait or traits; often designated by letters—for example, *BB* or *Aa*. 159

genus One of the categories, or taxa, used by taxonomists to group species; contains those species that are most closely related through evolution. 6, 278

germinate Beginning of growth of a seed, spore, or zygote, especially after a period of dormancy. 404

germ layer Primary tissue layer of a vertebrate embryo—namely, ectoderm, mesoderm, or endoderm. 338, 576

gibberellin Plant hormone promoting increased stem growth; also involved in flowering and seed germination. 393

global warming Predicted increase in the Earth's temperature due to human activities that promote the greenhouse effect. 622

globular protein Protein (i.e., enzymes) whose polypeptides give it a tertiary structure and a globular shape. 48

glottis Opening for airflow in the larynx. 453

glucagon Hormone, secreted by the pancreas, that causes the liver to break down glycogen and raises the blood glucose level. 468, 532

glucocorticoid Type of hormone, secreted by the adrenal cortex, that influences carbohydrate, fat, and protein metabolism. 531

glucose Six-carbon sugar that organisms degrade as a source of energy during cellular respiration. 39

glycerol Three-carbon carbohydrate with three hydroxyl groups attached; a component of fats and oils. 42

glycogen Storage polysaccharide found in animals; composed of glucose molecules joined in a linear fashion but having numerous branches. 40

glycolysis Anaerobic breakdown of glucose that results in a gain of 2 ATP and the end product pyruvate. 110

golden-brown algae Marine photosynthetic protist with elaborate shells of silica; represented by diatoms. 302

Golgi apparatus Organelle, consisting of saccules and vesicles, that processes, packages, and distributes molecules about or from the cell. 66

Golgi tendon Proprioceptive sensory receptor located where skeletal muscle inserts into tendons; sensitive to changes in tendon tension. 547

gonadotropic hormones (FSH and LH) Substances, secreted by the anterior pituitary, that regulate the activity of the ovaries and testes. 529

gonads Organs that produce gametes. The ovary produces eggs, and the testis produces sperm. 560

gonorrhea Bacterial STD that can lead to sterility or infertility. 572

granum (pl., grana) Stack of chlorophyll-containing thylakoids in a chloroplast. 68, 95

gravitational equilibrium Maintenance of balance when the head and body are motionless. 543

gravitropism Growth response of plant roots and stems to Earth's gravity. Roots demonstrate positive gravitropism, and stems demonstrate negative gravitropism. 396

grazing food web Complex pattern of linked and crisscrossing food chains that begins with a population of photosynthesizers serving as producers. 616

green algae Members of a diverse group of photosynthetic protists that contain chlorophylls *a* and *b* and have other biochemical characteristics like those of plants. 302

greenhouse effect Reradiation of solar heat toward the Earth, caused by gases such as carbon dioxide, methane, nitrous oxide, water vapor, ozone, and nitrous oxide in the atmosphere. 622

greenhouse gases Gases in the atmosphere, such as carbon dioxide, methane, nitrous oxide, water vapor, ozone, and nitrous oxide, that (like the panes of a greenhouse) allow the sun's rays to pass through but trap the heat. 638

ground tissue Tissue that constitutes most of the body of a plant; consists of parenchyma, collenchyma, and sclerenchyma cells that function in storage, basic metabolism, and support. 374

growth factor Chemical signal that regulates mitosis and differentiation of cells that have receptors for it; important in such processes as fetal development, tissue maintenance and repair, and hematopoiesis; sometimes a contributing factor in cancer. 208

growth hormone (GH) Substance secreted by the anterior pituitary; controls the size of an individual by promoting cell division, protein synthesis, and bone growth. 530

guanine (G) One of four nitrogen-containing bases in nucleotides composing the structure of DNA and RNA; pairs with cytosine. 176

gymnosperm Type of woody seed plant in which the seeds are not enclosed by fruit and are usually borne in cones, such as those of the conifers. 312

H

habitat Place where an organism lives and is able to survive and reproduce. 609

halophile Type of archaea that lives in extremely salty habitats. 299

haploid (n) number Cell condition in which only one of each type of chromosome is present. 142

heart Muscular organ whose contraction causes blood to circulate in the body of an animal. 432

heart attack Myocardial infarction; damage to the myocardium due to blocked circulation in the coronary arteries. 445

heat Type of kinetic energy. Captured solar energy eventually dissipates as heat in the environment. 79

helicase The enzyme in DNA replication that separates DNA strands by breaking the hydrogen bonds between the strands. 179

helper T cell (T_H cell) Cell that secretes cytokines, which stimulate all kinds of immune cells. 504

heme Iron-containing group found in hemoglobin. 456

hemocoel Body cavity in arthropods where exchange between hemolymph and tissues occurs. 432

hemodialysis Cleansing of blood by using an artificial membrane that causes substances to diffuse from blood into a dialysis fluid. 460

hemoglobin Iron-containing respiratory pigment occurring in vertebrate red blood cells and in the blood plasma of some invertebrates. 45, 441

hepatitis Viral infection of the liver; can be transmitted sexually and can lead to liver failure and liver cancer. 571

herbaceous Plants that have nonwoody stems. 378

herbivore Primary consumer in a grazing food chain; plant eater. 463, 614

hermaphroditic Having both male and female sex organs in an animal. 343, 560

heterochromatin Highly compacted chromatin that is not accessible for transcription. 202

heterotroph Organism that cannot synthesize organic compounds from inorganic substances and therefore must take in organic food. 614

heterozygous Possessing unlike alleles for a particular trait. 160

hinge joint Synovial joint (elbow and knee joints) that permits movement in one direction only. 553

histamine Substance, produced by basophils in blood and mast cells in connective tissue, that causes capillaries to dilate. 500

histone Protein molecule responsible for packing chromatin. 125

homeostasis Maintenance of normal internal conditions in a cell or an organism by means of self-regulating mechanisms. 3, 424

hominid Family designation for both humans and African apes. 360

hominin Designation that includes humans and extinct species very closely related to humans. 361

Homo erectus Early *Homo* who used fire and migrated out of Africa to Europe and Asia. 363

Homo habilis Early *Homo*, dated between 2.0 and 1.9 million years ago, who is believed to have been the first tool user. 362

homologous chromosome Homologue; member of a pair of chromosomes that are alike and come together in synapsis during prophase of the first meiotic division. 142

homologous structure In evolution, a structure that is similar in different types of organisms because these organisms are derived from a common ancestor. 245, 278

homologue Homologous chromosome; member of a pair of chromosomes that are alike and come together in synapsis during prophase of the first meiotic division. 142

Homo sapiens Modern humans. 363

homozygous Possessing two identical alleles for a particular trait. 159

human immunodeficiency virus (HIV) Virus responsible for AIDS. 509, 571

humerus Long bone of the arm that articulates with the scapula. 549

Huntington disease Genetic disease marked by progressive deterioration of the nervous system and resulting in neuromuscular disorders. 223

hyaline cartilage Cartilage whose cells lie in lacunae separated by a white, translucent matrix containing very fine collagen fibers. 418

hybridization Crossing of different species of organisms or different varieties of plants. 406

hydrogen (H) bond Weak bond that arises between a slightly positive hydrogen atom of one molecule and a slightly negative atom of another molecule or between parts of the same molecule. 27

hydrogen ion (H⁻) Hydrogen atom that has lost its electron and therefore bears a positive charge. 30

hydrolysis reaction Splitting of a compound by the addition of water, with the H⁺ being incorporated into one molecule and the OH⁻ into the other. 39

hydrophilic Type of molecule that interacts with water by dissolving in water and/or by forming hydrogen bonds with water molecules. 28, 37

hydrophobic Type of molecule that does not interact with water because it is nonpolar. 28, 37

hydrostatic skeleton Fluid-filled body compartment that provides support for muscle contraction resulting in movement; seen in cnidarians, flatworms, roundworms, and segmented worms. 549

hydroxide ion (OH⁻) One of two ions that result when a water molecule dissociates. It has gained an electron and therefore bears a negative charge (OH⁻). 30

hypertension Elevated blood pressure, particularly the diastolic pressure. 444

hypertonic solution Higher solute concentration (less water) than the cytoplasm of a cell; causes cells to lose water by osmosis. 88

hyphae Filaments of the vegetative body of a fungus. 324

hypothalamic-releasing hormone One of many hormones produced by the hypothalamus that stimulates the secretion of an anterior pituitary hormone. 529

hypothalamus In vertebrates, part of the brain that helps regulate the internal environment of the body—for example, heart rate, body temperature, and water balance. 523

hypothesis Supposition established by reasoning after consideration of available evidence. It can be tested by obtaining more data, often by experimentation. 11

hypotonic solution Lower solute (more water) concentration than the cytoplasm of a cell; causes cells to gain water by osmosis. 88

I

immediate allergic response Allergic response that occurs within seconds of contact with an allergen; caused by the attachment of the allergen to receptors on the surface of mast cells. 508

immune system All the cells in the body that protect the body against foreign organisms and substances as well as cancerous cells. 422, 497

immunity Ability of the body to protect itself from foreign substances and cells, including disease-causing agents. 497

incomplete dominance Inheritance pattern in which the offspring has an intermediate phenotype, as when a red-flowered plant and a white-flowered plant produce pink-flowered offspring. 163

induced fit model Change in the shape of an enzyme's active site that enhances the fit between the active site and its substrate(s). 84

induction Ability of a chemical or tissue to influence the development of another tissue. 576

inductive reasoning Using specific observations and the process of logic and reasoning to arrive at a hypothesis. 11

industrial melanism Increased frequency of darkly pigmented (melanic) forms in a population when soot and pollution make lightly pigmented forms easier for predators to see against a pigmented background. 258

infertility Inability to have as many children as desired. 569

inflammatory response Tissue response to injury that is characterized by redness, swelling, pain, and heat. 500

innate immunity A function of the immune system that protects the body against pathogens in a nonspecific manner. 499

inorganic chemistry Study of compounds not having a carbon basis; chemistry of the nonliving world. 36

insect Type of arthropod. The head has antennae, compound eyes, and simple eyes; the thorax has three pairs of legs and often wings; and the abdomen has internal organs. 350

insulin Hormone, secreted by the pancreas, that lowers blood glucose level by promoting the uptake of glucose by cells and the conversion of glucose to glycogen by the liver and skeletal muscles. 468, 532

integration Summing up of excitatory and inhibitory signals by a neuron or a part of the brain. 520

integumentary system Organ system consisting of skin and various organs, such as hair, in the skin. 423

intercalated disks Region that holds adjacent cardiac muscle cells together. The disks appear as dense bands at right angles to the muscle striations. 419

intergenic DNA The sequence of nuecleotides that lie between genes on a chromosome. 192

interkinesis Period of time between meiosis I and meiosis II during which no DNA replication takes place. 147

intermembrane space Space that occurs between the outer and inner membranes of a mitochondrion. 116

interneuron In the central nervous system, a neuron that conveys messages between parts of the central nervous system. 518

internode In vascular plants, the region of a stem between two successive nodes. 372

interphase Stages of the cell cycle (G₁, S, G₂) during which growth and DNA synthesis occur when the nucleus is not actively dividing. 126

intervertebral discs Pads of cartilage that separate vertebrae. 549

intestinal enzyme Enzyme, produced by the epithelial cells on the surface of villi, which functions in the digestion of small organic molecules. 466

intrauterine device (IUD) Birth control device consisting of a small piece of molded plastic inserted into the uterus; believed to alter the uterine environment, so that fertilization does not occur. 568

inversion Change in chromosome structure in which a segment of a chromosome is turned around 180°. This reversed sequence of genes can lead to altered gene activity and abnormalities. 219

in vivo gene therapy Gene therapy in which normal genes are injected directly into an organism for growth. 228

ion Charged particle that carries a negative or positive charge. 24

ionic bond Chemical bond in which ions are attracted to one another by opposite charges. 24

iris Muscular ring that surrounds the pupil and regulates the passage of light through this opening. 544

isomer Molecule with the same molecular formula as another but having a different structure and therefore a different shape. 37

isotonic solution Solution that is equal in solute concentration to that of the cytoplasm of a cell; causes a cell to neither lose nor gain water by osmosis. 87

isotope Atom of the same element having the same atomic number but a different mass number due to the number of neutrons. 22

J

jawless fishes Type of fish that have no jaws; includes hagfishes and lampreys. 354

K

karyotype Chromosomes arranged by pairs according to their size, shape, and general appearance in mitotic metaphase. 141, 216

keystone species Species whose activities significantly affect community structure. 612

kidneys Paired organs of the vertebrate urinary system that regulate the chemical composition of the blood and produce the waste product urine. 457

kilocalorie Caloric value of food; 1,000 calories. 78

kinase Enzyme that activates another enzyme by adding a phosphate group. 132

kinetic energy Energy associated with motion. 78

kingdom One of the categories, or taxa, used by taxonomists to group species; the taxon above phylum. 6, 278

Klinefelter syndrome Condition caused by the inheritance of XXY chromosomes. 151

L

lacteal Lymphatic vessel in an intestinal villus; aids in the absorption of fats. 467

lacuna Small pit or hollow cavity, as in bone or cartilage, where a cell or cells are located. 418

lancelet Invertebrate chordate with a body that resembles a lancet and has the four chordate characteristics as an adult. 353

large intestine In vertebrates, portion of the digestive tract that follows the small intestine; in humans, consists of the cecum, colon, rectum, and anal canal. 467

larynx Voice box; cartilaginous organ located between the pharynx and the trachea; in humans, contains the vocal cords. 452

lateral bud Site on a stem where lateral branches grow. 372

lateral line Canal system containing sensory receptors that allow fishes and amphibians to detect water currents and pressure waves from nearby objects. 543

law Principle. 13

law of independent assortment Alleles of unlinked genes assort independently of each other during meiosis, so that the gametes contain all possible combinations of alleles. 160

law of segregation Separation of alleles from each other during meiosis, so that the gametes contain one from each pair. Each resulting gamete has an equal chance of receiving either allele. 159

leaf Lateral appendage of a stem, highly variable in structure, often containing cells that carry out photosynthesis. 371

leaf vein Vascular tissue in a leaf. 373

leech Blood-sucking annelid, usually found in fresh water, with a sucker at each end of a segmented body. 345

lens In the vertebrate eye, a clear, membrane-like structure behind the iris; brings objects into focus. 544

less-developed country (LDC) Country that is becoming industrialized. Typically, population growth is expanding rapidly and the majority of people live in poverty. 589

lichen Symbiotic relationship between certain fungi and algae, in which the fungi possibly provide inorganic food or water and the algae provide organic food. 326

life cycle Recurring pattern of genetically programmed events by which individuals grow, develop, maintain themselves, and reproduce. 142

life history Adaptations in characteristics that influence an organism's biology, such as how many offspring it produces, its survival, and factors (such as age and size) that determine its reproductive maturity. 599

ligament Tough cord or band of dense fibrous tissue that binds bone to bone at a joint. 418, 552

light reactions Portion of photosynthesis that captures solar energy and takes place in thylakoid membranes of chloroplasts; produces ATP and NADPH. 96

lignin Chemical that hardens the cell walls of plants. 314

limbic system In humans, functional association of various brain centers, including the amygdala and hippocampus; governs learning and memory and various emotions, such as pleasure, fear, and happiness. 524

lineage Evolutionary line of descent. 360

lipase Fat-digesting enzyme secreted by the pancreas. 466

lipid Class of organic compounds that tends to be soluble in nonpolar solvents; includes fats and oils. 41

liver Large, dark red internal organ that produces urea and bile, detoxifies the blood, stores glycogen, and produces the plasma proteins, among other functions. 466

lobe-finned fishes Type of fish with limb-like fins. 355

logistic growth Population increase that results in an S-shaped curve. Growth is slow at first, steepens, and then levels off due to environmental resistance. 595

long-day plant/short-night plant Plant that flowers when day length is longer and night is shorter than a critical length—e.g., wheat, barley, clover, and spinach. 397

loose fibrous connective tissue Tissue composed mainly of fibroblasts widely separated by a matrix containing collagen and elastic fibers. 417

lumen Cavity of a tubular organ. 415

lung Internal respiratory organ containing moist surfaces for gas exchange. 355, 452

luteal phase Second half of the ovarian cycle, during which the corpus luteum develops and much progesterone (and some estrogen) is produced. 566

lycophyte Club mosses; seedless plant with microphylls and well-developed vascular tissue in roots, stems, and leaves. 314

lymph Fluid, derived from tissue fluid, that is carried in lymphatic vessels. 439

lymphatic organs Organs other than a lymphatic vessel that are part of the lymphatic system; lymph nodes, tonsils, spleen, thymus gland, and bone marrow. 497

lymphatic system Organ system consisting of lymphatic vessels and lymphatic organs; transports lymph and lipids and aids the immune system. 422, 438

lymphatic vessel Vessel that carries lymph. 439

lymph node Mass of lymphatic tissue located along the course of a lymphatic vessel. 498

lymphocyte Specialized white blood cell that functions in specific defense; occurs in two forms—T lymphocytes and B lymphocytes. 443

lysogenic cycle Bacteriophage life cycle in which the virus incorporates its DNA into that of a bacterium; occurs preliminary to the lytic cycle. 289

lysosome Membrane-bounded vesicle that contains hydrolytic enzymes for digesting macromolecules. 66

lytic cycle Bacteriophage life cycle in which a virus takes over the operation of a bacterium immediately upon entering it and subsequently destroys the bacterium. 289

M

macroevolution Large-scale evolutionary change, such as the formation of new species. 266

macronutrient Essential element needed in large amounts by plants and humans. In plants, nitrogen, calcium, and sulfur are needed for plant growth; in humans, carbohydrates, lipids, and proteins supply the body's energy needs. 477

macrophage In vertebrates, large, phagocytic cell derived from a monocyte that ingests microbes and debris. 443, 498

major histocompatability complex (MHC) Cell surface proteins that recognize "self" and bind antigens for presentation to a T cell. 503

major mineral Essential inorganic nutrient (such as calcium, potassium, phosphorus, sodium, chloride, magnesium, or sulfur) required daily by humans to regulate metabolic activities and maintain health. 481

male condom Sheath used to cover the penis during sexual intercourse; used as a contraceptive and, if latex, to minimize the risk of transmitting infection. 568

male gametophyte In seed plants, the gametophyte that produces sperm; a pollen grain. 401

malignant Invasive tumor that may spread. 134

Malpighian tubule Blind, thread-like excretory tubule near the anterior end of an insect's hindgut. 459

maltase Enzyme, produced in the small intestine, that breaks down maltose to two glucose molecules. 470

mammal Endothermic vertebrate characterized especially by the presence of hair and mammary glands. 358

Marfan syndrome Congenital disorder of connective tissue characterized by abnormally long extremities. 223

marsupial Member of a group of mammals bearing immature young nursed in a marsupium, or pouch—for example, kangaroo and opossum. 358

mass extinction Episode of large-scale extinction in which large numbers of species disappear in a few million years or less. 275

mast cell Connective tissue cell that releases histamine in allergic reactions. 500

matrix Unstructured, semifluid substance that fills the space between cells in connective tissues and inside organelles. 69, 417

matter Anything that takes up space and has mass. 20

medulla oblongata In vertebrates, part of the brain stem that is continuous with the spinal cord; controls heartbeat, blood pressure, breathing, and other vital functions. 524

megaphylls Large leaves with complex networks of veins. 311

megaspore One of two types of spores produced by seed plants; develops into a female gametophyte (embryo sac). 398

meiosis, meiosis I, meiosis II Type of nuclear division that occurs as part of sexual reproduction in which the daughter cells receive the haploid number of chromosomes in varied combinations. 141, 143

memory Capacity of the brain to store and retrieve information about past sensations and perceptions; essential to learning. 524

meniscus Cartilaginous wedges that separate the surfaces of bones in synovial joints. 553

menopause Termination of the ovarian and menstrual cycles in older women. 566

menses Flow of blood during menstruation. 566

menstrual cycle Cycle that runs concurrently with the ovarian cycle. It prepares the uterus to receive a developing zygote. 566

menstruation Periodic shedding of tissue and blood from the inner lining of the uterus in primates. 566

meristem Undifferentiated embryonic tissue in the active growth regions of plants. 374

mesoderm Middle primary tissue layer of an animal embryo that gives rise to muscle, several internal organs, and connective tissue layers. 576

mesophyll Inner, thickest layer of a leaf consisting of palisade and spongy mesophyll; the site of most of photosynthesis. 377

mesophyll cells The primary site of photosynthesis in a plant. 103

messenger RNA (mRNA) Type of RNA formed from a DNA template and bearing coded information for the amino acid sequence of a polypeptide. 180

metabolic pathway Series of linked reactions, beginning with a particular reactant and terminating with an end product. 83

metabolism All of the chemical reactions that occur in a cell during growth and repair. 3

metaphase Mitotic phase during which chromosomes are aligned at the spindle equator. 128

metastasis Spread of cancer from the place of origin throughout the body; caused by the ability of cancer cells to migrate and invade tissues. 134

methanogen Type of archaea that lives in oxygen-free habitats, such as swamps, and releases methane gas. 299

methemoglobinemia Disorder resulting from the accumulation of methemoglobin in the blood, causing the skin of those affected to appear bluish-purple. 221

microevolution Small, measurable evolutionary changes in a population from generation to generation; change in allele frequencies within a population over time. 256

micronutrient Essential element needed in small amounts by plants and humans. In plants, boron, copper, and zinc are needed for plant growth; in humans, vitamins and minerals help regulate metabolism and physiological development. 477

microphylls Small, narrow leaves with single, unbranched veins. 311

microspore One of two types of spores produced by seed plants; develops into a male gametophyte (pollen grain). 398

microtubule Small, cylindrical organelle composed of tubulin protein around an empty central core; present in the cytoplasm, centrioles, cilia, and flagella. 70

microvillus (pl., microvilli) Cylindrical process that extends from an epithelial cell of a villus; increases the surface area of the cell. 467

midbrain Part of the brain stem located between the diencephalon and the pons. 524

mineral Naturally occurring inorganic substance containing one or more elements. Certain minerals are needed in the diet. As raw materials in the Earth, they are nonrenewable resources. 481, 641

mineralocorticoid Hormones secreted by the adrenal cortex that regulate salt and water balance, leading to increases in blood volume and blood pressure. 531

mitochondrion Membrane-bounded organelle in which ATP molecules are produced during the process of cellular respiration. 67

mitosis Process in which a parent nucleus produces two daughter nuclei, each having the same number and kinds of chromosomes as the parent nucleus. 127

model Simulation of a process; aids conceptual understanding until the process can be studied firsthand; a hypothesis that describes how a particular process might be carried out. 12

molecule Union of two or more atoms of the same element; also, the smallest part of a compound that retains the properties of the compound. 24

molting Periodic shedding of the exoskeleton in arthropods or the cuticle in roundworms. 347

monocot Monocotyledon; flowering plant group. Members have one embryonic leaf (cotyledon), parallel-veined leaves, scattered vascular bundles, and flower parts in threes or multiples of three. 373

monocyte Type of granular leukocyte that functions as a phagocyte, particularly after it becomes a macrophage. 442

monomer Small molecule that is a subunit of a polymer—e.g., glucose is a monomer of starch. 39

monosaccharide Simple sugar; a carbohydrate that cannot be decomposed by hydrolysis—e.g., glucose. 39

monotreme Egg-laying mammal—e.g., duckbill platypus and spiny anteater. 358

more-developed country (MDC) Country that is industrialized. Typically, population growth is slow and the people enjoy a good standard of living. 589

morula Spherical mass of cells resulting from cleavage during animal development prior to the blastula stage. 575

mosaic evolution Concept that human characteristics did not evolve at the same rate. For example, some body parts are more human-like than others in early hominins. 362

motor neuron Nerve cell that conducts nerve impulses away from the central nervous system and innervates effectors (muscle and glands). 518

motor protein Protein associated with the cytoskeleton; allows movement of transport vesicles and organelles. 69

mouth In humans, organ of the digestive tract where food is chewed and mixed with saliva. 463

mRNA transcript Complementary copy of the sequence of bases in template. 183

multicellular Organism composed of many cells; usually has organized tissues, organs, and organ systems. 2

multifactorial trait Trait controlled by polygenes subject to environmental influences. Each dominant allele contributes to the phenotype in an additive and like manner. 164

multipotent A term used to describe a stem cell that has the potential to form a limited number of cell types. 200

multiregional continuity hypothesis Proposal that modern humans evolved separately in at least three places: Asia, Africa, and Europe. 363

muscle dysmorphia Mental state in which a person thinks his or her body is underdeveloped and becomes preoccupied with body-building and diet; affects more men than women. 489

muscle spindle Proprioceptor wrapped around a few muscle cells within a connective tissue sheath; can respond to changes in muscle length. 547

muscular system System of muscles that produces movement, both movement in the body and movement of its limbs. The principal components are skeletal muscle, smooth muscle, and cardiac muscle. 423

muscular tissue Type of animal tissue composed of fibers that shorten and lengthen to produce movements. 418

mutagen Agent, such as radiation or a chemical, that brings about a mutation in DNA. 187

mutation Change made in the DNA base-pair sequences in a gene due to a copying error or external sources called mutagens. Such changes generate variation in a gene pool. 259

mutualism Symbiotic relationship in which both species benefit in terms of growth and reproduction. 609

mycelium Tangled mass of hyphal filaments composing the vegetative body of a fungus. 324

mycorrhizal association Mutualistic relationship between fungal hyphae and roots of vascular plants. 327, 384

myelin sheath White, fatty material—derived from the membranes of tightly spiraled cells—that forms a covering for nerve fibers. 517

myofibril Specific muscle cell organelle containing a linear arrangement of sarcomeres, which shorten to produce muscle contraction. 551

myosin Muscle protein making up the thick filaments in a sarcomere. It pulls actin to shorten the sarcomere, yielding muscle contraction. 551

N

nasal cavity One of two canals in the nose, separated by a septum. 452

native species Indigenous species that colonize an area without human assistance. 613

natural history Study of how organisms are influenced by factors such as climate, predation, competition, and evolution; uses field observations instead of experimentation. 586

natural killer (NK) cell Lymphocyte that causes an infected or cancerous cell to burst. 501

natural selection Mechanism of evolution caused by environmental selection of organisms most fit to reproduce; results in adaptation to the environment. 8, 238, 251

Neandertal Later *Homo* with a sturdy build that lived during the last Ice Age in Europe and the Middle East; hunted large game and left evidence of being culturally advanced. 364

negative feedback Mechanism of homeostatic response by which the output of a system suppresses or inhibits activity of the system. 425, 530

nematocyst In cnidarians, a capsule that contains a thread-like fiber, the release of which aids in the capture of prey. 341

nephridium Segmentally arranged, paired excretory tubules of many invertebrates, as in the earthworm. 344

nephron Microscopic kidney unit that regulates blood composition by filtration, reabsorption, and secretion. 458

nephron capsule Cup-like structure that is the initial portion of a nephron. 458

nephron loop Portion of a nephron between the proximal and distal tubules; functions in water reabsorption. 458

nerve Bundle of long axons outside the central nervous system. 420, 525

nerve deafness Deafness due to spiral organ damage. 542

nervous system Organ system, consisting of the brain, spinal cord, and associated nerves, that coordinates the other organ systems of the body. 422

nervous tissue Tissue that contains nerve cells (neurons), which conduct impulses, and neuroglia, which support, protect, and provide nutrients to neurons. 420

neural plate Region of the dorsal surface of the chordate embryo that marks the future location of the neural tube. 576

neural tube Tube formed by closure of the neural groove during development. In vertebrates, the neural tube develops into the spinal cord and brain. 576

neuroglia Nonconducting nerve cells that are intimately associated with neurons and function in a supportive capacity. 420

neuron Nerve cell that characteristically has three parts: dendrites, a cell body, and an axon. 420, 517

neurotransmitter Chemical stored at the ends of axons; responsible for transmission across a synapse. 520

neurulation Development of the central nervous system organs in an embryo. 576

neutron Neutral subatomic particle, located in the nucleus and assigned one atomic mass unit. 21

neutrophil Granular leukocyte that is the most abundant of the white blood cells; first to respond to infection. 442, 500

node In plants, the place where one or more leaves attach to a stem. 372

nodes of Ranvier In the peripheral nervous system, gaps in the myelin sheath encasing long axons. 518

nondisjunction Failure of homologous chromosomes or daughter chromosomes to separate during meiosis I and meiosis II, respectively. 150

nonrandom mating Mating among individuals on the basis of their phenotypic similarities or differences, rather than mating on a random basis. 260

nonrenewable resources Minerals, fossil fuels, and other materials present in essentially fixed amounts (within human time scale) in our environment. 631

nonsister chromatid A tetrad consists of four chromatids. Only the chromatids belonging to a homologue are sister chromatids and the others, belonging to the other homologue, are nonsister chromatids. 144

nonvascular plant Bryophyte, such as mosses, that has no vascular tissue and either occurs in moist locations or has adaptations for living in dry locations. 313

norepinephrine (NE) Neurotransmitter of the postganglionic fibers in the sympathetic division of the autonomic system; also, a hormone produced by the adrenal medulla. 520

notochord Cartilaginous-like, supportive dorsal rod in all chordates sometime in their life cycle; replaced by vertebrae in vertebrates. 352, 576

nuclear envelope Double membrane that surrounds the nucleus in eukaryotic cells and is connected to the endoplasmic reticulum; has pores that allow substances to pass between the nucleus and the cytoplasm. 65

nuclear pore Opening in the nuclear envelope that permits the passage of proteins into the nucleus and ribosomal subunits out of the nucleus. 65

nuclease Enzyme that catalyzes decomposition of nucleic acids. 466

nucleic acid Polymer of nucleotides. Both DNA and RNA are nucleic acids. 48

nucleoid Region of prokaryotic cells where DNA is located. It is not bounded by a nuclear envelope. 58, 295

nucleolus In the nucleus, a dark-staining, spherical body that produces ribosomal subunits. 64

nucleosome In the nucleus of a eukaryotic cell, a unit composed of DNA wound around a core of eight histone proteins, giving the appearance of a string of beads. 125

nucleotide Monomer of DNA and RNA consisting of a 5-carbon sugar bonded to a nitrogenous base and a phosphate group. 48

nucleus Center of an atom, in which protons and neutrons are found; membrane-bounded organelle within a eukaryotic cell that contains chromosomes and controls the structure and function of the cell. 21, 62

nutrient Chemical substance in foods that is essential to the diet and contributes to good health. 476

nutrition According to the Council on Food and Nutrition of the American Medical Association, "the science of food; the nutrients and the substances therein; their action, interaction, and balance in relation to health and disease; and the process by which the organism (i.e., body) ingests, digests, absorbs, transports, utilizes, and excretes food substances." 476

O

obesity Excess adipose tissue; exceeding desirable weight by more than 20%. 479

observation Step in the scientific method by which data are collected before a conclusion is drawn. 11

octet rule States that an atom other than hydrogen tends to form bonds until it has eight electrons in its outer shell. An atom that already has eight electrons in its outer shell does not react and is inert. 24

oil Triglyceride, usually of plant origin, that is composed of glycerol and three fatty acids and is liquid in consistency due to many unsaturated bonds in the hydrocarbon chains of the fatty acids. 41

omnivore Organism in a food chain that feeds on both plants and animals. 463, 614

oncogene Cancer-causing gene. 208

oogenesis Production of eggs in females by the process of meiosis and maturation. 143

open circulatory system Circulatory system, such as that found in a grasshopper, in which a tubular heart pumps hemolymph through channels and body cavities. 432

operon Group of structural and regulating genes that function as a single unit. 201

opportunistic species Population demonstrating a life history pattern in which members exhibit exponential population growth. Its members are small in size, mature early, have a short life span, produce many offspring, and provide little or no care to offspring (e.g., dandelions). 599

order One of the categories, or taxa, used by taxonomists to group species; the taxon above the family level. 6, 278

organ Combination of two or more different tissues performing a common function. 3, 413

organelle Small, often membranous structure in the cytoplasm having a specific structure and function. 62

organic Molecule that always contains carbon and hydrogen and often contains oxygen as well. Organic molecules are associated with living things. 36

organic chemistry Study of carbon compounds; chemistry of the living world. 36

organism Individual living thing. 2, 585

organ system Group of related organs working together. 3

orgasm Physiological and psychological sensations that occur at the climax of sexual stimulation. 563

osmosis Diffusion of water through a semipermeable membrane. 87

osteoblast Bone-forming cell. 550

osteoclast Cell that causes erosion of bone. 550

osteocyte Branched bone cell embedded in a calcium-containing extracellular matrix. 550

osteon Haversian system; cylindrical unit containing bone cells that surround an osteonic canal. 550

osteoporosis Condition in which bones break easily because calcium is removed from them faster than it is replaced. 550

ostium (pl., ostia) In the heart of an animal with an open circulatory system, an opening that allows blood to fill the relaxed heart. 432

out-of-Africa hypothesis Proposal that modern humans originated only in Africa; then they migrated and supplanted populations of *Homo* in Asia and Europe about 100,000 years ago (see also *replacement model*). 363

ovary In animals, the female gonad that produces an egg and female sex hormones; in flowering plants, the enlarged, ovule-bearing portion of the carpel that develops into a fruit. 319, 399, 560

ovulation Bursting of a follicle when a secondary oocyte is released from the ovary. If fertilization occurs, the secondary oocyte becomes an egg. 565

ovule In seed plants, a structure that contains the female gametophyte and has the potential to develop into a seed. 317, 399

oxidation Loss of one or more electrons from an atom or a molecule; in biological systems, generally the loss of hydrogen atoms. 96

oxygen deficit Amount of oxygen needed to metabolize lactate, a compound that accumulates during vigorous exercise. 118

oxytocin Hormone, released by the posterior pituitary, that causes contraction of the uterus and milk letdown. 530

P

paleontology Study of fossils, which yields knowledge about the history of life. 272

palisade mesophyll In a plant leaf, layer of tissue containing elongated cells with many chloroplasts. 377

pancreas Internal organ that produces digestive enzymes and the hormones insulin and glucagon. 466, 532

pancreatic amylase Enzyme that digests starch to maltose. 466

pancreatic islets Masses of cells that constitute the endocrine portion of the pancreas. 532

parasitism Symbiotic relationship in which one species (the *parasite*) obtains nutrients from another species (the *host*) but does not usually kill the host. 609

parasympathetic division Division of the autonomic system that is active under normal conditions; uses acetylcholine as a neurotransmitter. 526

parathyroid gland Gland embedded in the posterior surface of the thyroid gland; produces parathyroid hormone. 531

parathyroid hormone (PTH) Hormone, secreted by the four parathyroid glands, that increases the blood calcium level and decreases the phosphate level. 531

parenchyma Plant tissue composed of the least specialized of all plant cells; found in all organs of a plant. 374

parthenogenesis Development of an egg cell into a whole organism without fertilization. 559

passive immunity Protection against infection acquired by transfer of antibodies to a susceptible individual. 507

pathogen Disease-causing agent, such as viruses, parasitic bacteria, fungi, and animals. 298

pectoral (shoulder) girdle Portion of the vertebrate skeleton that provides support and attachment for the upper (fore-) limbs; consists of the scapula and clavicle on each side of the body. 549

pedigree Chart showing the relationships of relatives and which ones have a particular trait. 220

pelvic (hip) girdle Portion of the vertebrate skeleton to which the lower (hind-) limbs are attached; coxal bones. 549

pelvic inflammatory disease (PID) Latent infection of gonorrhea or chlamydia in the vasa deferentia or uterine tubes. 572

penis Male copulatory organ; in humans, the male organ of sexual intercourse. 562

pepsin Enzyme, secreted by gastric glands, that digests proteins to peptides. 464

peptidase Intestinal enzyme that breaks down short chains of amino acids to individual amino acids that are absorbed across the intestinal wall. 470

peptide Two or more amino acids joined together by covalent bonding. 46

peptide bond Type of covalent bond that joins two amino acids. 46

peptide hormone Type of hormone that is a protein or a peptide or is derived from an amino acid. 528

peptidoglycan Unique molecule found in bacterial cell walls. 295

perennial plant Flowering plant that lives more than one growing season because the underground parts regrow each season. 372

pericycle Layer of cells surrounding the vascular tissue of roots; produces branch roots. 382

peripheral nervous system (PNS) Nerves and ganglia that lie outside the central nervous system. 517

peristalsis Wave-like contractions that propel substances along a tubular structure, such as the esophagus. 464

petal Flower part just inside the sepals; often conspicuously colored to attract pollinators. 319, 399

petiole Part of a plant leaf that connects the blade to the stem. 376

pH A logarithmic measure of the hydrogen ion concentration. 32

phagocytosis Process by which amoeboid cells engulf large substances, forming an intracellular vacuole. 89

pharynx In vertebrates, common passageway for both food intake and air movement; located between the mouth and the esophagus. 452

phenotype Visible expression of a genotype—e.g., brown eyes or attached earlobes. 160

phloem Vascular tissue that conducts organic solutes in plants; contains sieve-tube members and companion cells. 314, 373

phospholipid Molecule that forms the *phospholipid bilayer* of plasma membranes. It has a polar, hydrophilic head bonded to two nonpolar, hydrophobic tails. 43

photoautotroph Organism able to synthesize organic molecules by using carbon dioxide as a carbon source and sunlight as an energy source. 296

photoperiod Relative lengths of daylight and darkness that affect the physiology and behavior of an organism. 396

photoreceptor Sensory receptor that responds to light stimuli. 544

photosynthesis Process occurring usually within chloroplasts whereby chlorophyll-containing organelles trap solar energy to reduce carbon dioxide to carbohydrate. 3, 94

photosystem Photosynthetic unit in which solar energy is absorbed and high-energy electrons are generated; contains a pigment complex and an electron acceptor; occurs as PS (photosystem) I and PS II. 98

phototropism Growth response of plant stems to light; stems demonstrate positive phototropism. 396

photovoltaic (solar) cell Energy-conversion device that captures solar energy and directly converts it to electrical current. 640

pH scale Measurement scale for hydrogen ion concentration. 32

phylogenetic tree Diagram that indicates common ancestors and lines of descent among groups of organisms. 278

phylogeny Evolutionary history of a group of organisms. 277

phylum One of the categories, or taxa, used by taxonomists to group species; the taxon above the class level. 6, 278

phytochrome Photoreversible plant pigment involved in photoperiodism and other responses of plants, such as etiolation. 397

pinocytosis Process by which vesicle formation takes macromolecules into the cell. 89

pith Parenchyma tissue in the center of some stems and roots. 379

pituitary gland Small gland that lies just inferior to the hypothalamus; consists of the anterior and posterior pituitary, both of which produce hormones. 528

placebo Treatment that contains no medication but appears to be the same treatment as that administered to other test groups in a controlled study. 14

placenta Organ formed during the development of placental mammals from the chorion and the uterine wall; allows the embryo, and then the fetus, to acquire nutrients and rid itself of wastes; produces hormones that regulate pregnancy. 561

placental mammal Mammal characterized by the presence of a placenta during the development of an offspring. 359

planarian Free-living flatworm with a ladder-like nervous system. 342

plant Multicellular, usually photosynthetic, organism belonging to the plant kingdom. 7

plaque Accumulation of soft masses of fatty material, particularly cholesterol, beneath the inner linings of the arteries. 444

plasma In vertebrates, the liquid portion of blood; contains nutrients, wastes, salts, and proteins. 441

plasma membrane Membrane surrounding the cytoplasm; consists of a phospholipid bilayer with embedded proteins. It regulates the entrance and exit of molecules from a cell. 59

plasmid Self-duplicating ring of accessory DNA in the cytoplasm of bacteria. 295

plasmodesmata In plants, cytoplasmic strands that extend through pores in the cell wall and connect the cytoplasm of two adjacent cells. 72

plasmolysis Contraction of the cell contents due to the loss of water. 88

platelet In blood, formed element that is necessary to blood clotting. 418, 443

pleiotropy Inheritance pattern in which one gene affects many phenotypic characteristics of the individual. 166

point mutation Alteration in a gene due to a change in a single nucleotide. The results of this mutation vary. 187

polar In chemistry, bond in which the sharing of electrons between atoms is unequal. 27

pollen grain In seed plants, structure that is derived from a microspore and develops into a male gametophyte. 317, 401

pollen tube In seed plants, tube that forms when a pollen grain lands on the stigma and germinates. The tube grows, passing between the cells of the stigma and the style to reach the egg inside an ovule, where fertilization occurs. 320

pollination In gymnosperms, the transfer of pollen from pollen cone to seed cone; in angiosperms, the transfer of pollen from anther to stigma. 317, 401

pollution Any environmental change that adversely affects the lives and health of living things. 632

polygenic The contribution of two or more genes to a phenotype. 251

polygenic inheritance Pattern of inheritance in which a trait is controlled by several allelic pairs. Each dominant allele contributes to the phenotype in an additive and like manner. 164

polymer Macromolecule consisting of covalently bonded monomers. For example, a polypeptide is a polymer of monomers called amino acids. 39

polymerase chain reaction (PCR) Technique that uses the enzyme DNA polymerase to produce millions of copies of a particular piece of DNA. 190

polyp Small, abnormal growth that arises from the epithelial lining. 468

polypeptide Polymer of many amino acids linked by peptide bonds. 46

polyribosome String of ribosomes simultaneously translating regions of the same mRNA strand during protein synthesis. 185

polysaccharide Polymer made from sugar monomers. The polysaccharides starch and glycogen are polymers of glucose monomers. 40

pons Part of the brain stem located between the midbrain and the medulla oblongata. 524

population Group of interbreeding individuals of the same species occupying the same area at the same time. 2, 10, 256, 585

population density Number of individuals per unit area or volume living in a particular habitat. 592

population genetics Study of gene frequencies and their changes within a population. 256

portal system Pathway of blood flow that begins and ends in capillaries, such as the portal system between the small intestine and the liver. 438

posterior pituitary Portion of the pituitary gland that stores and secretes oxytocin and antidiuretic hormone produced by the hypothalamus. 530

postzygotic isolating mechanism Anatomical or physiological difference between two species that prevents successful reproduction after mating has taken place. 269

potential energy Stored energy as a result of location or spatial arrangement. 78

predation Interaction in which one organism (the *predator*) uses another (the *prey*) as a food source. 597, 609

prefrontal area In the frontal lobe, association area that receives information from other association areas and uses it to reason and plan actions. 523

preparatory (prep) reaction Reaction that oxidizes pyruvate with the release of carbon dioxide; results in acetyl-CoA and connects glycolysis to the citric acid cycle. 110

pressure-flow model Explanation for phloem transport. Osmotic pressure following active transport of sugar into phloem brings a flow of sap from a source to a sink. 386

prezygotic isolating mechanism Anatomical or behavioral difference between two species that prevents the possibility of mating. 268

primary motor area In the frontal lobe, area where voluntary commands begin. Each section controls a part of the body. 523

primary productivity Amount of biomass produced primarily by photosynthesizers. 625

primary sensory area In the parietal lobe, area where sensory information arrives from the skin and skeletal muscles. 523

primary succession Stage in ecological succession, which involves the creation of new soil. 607

principle Law; theory that is generally accepted by an overwhelming number of scientists. 13

prion Infectious particle consisting of protein only and no nucleic acid. 292

producer At the start of a grazing food chain, a photosynthetic organism that makes its own food—e.g., green plants on land and algae in water. 614

product Substance that forms as a result of a reaction. 26

progesterone Female sex hormone that helps maintain sexual organs and secondary sex characteristics. 566

prokaryote Organism that lacks the membrane-bounded nucleus and membranous organelles typical of eukaryotes. 293

prokaryotic cell Cell lacking a membrane-bounded nucleus and organelles; the cell type within the domains Bacteria and Archaea. 57

prolactin (PRL) Hormone, secreted by the anterior pituitary, that stimulates the production of milk from the mammary glands. 530

promoter In an operon, a sequence of DNA in which RNA polymerase binds prior to transcription. 183, 201

prophase Mitotic phase during which chromatin condenses, so that chromosomes appear. Chromosomes are scattered. 128

proprioceptor Sensory receptor that helps the brain determine the position of the limbs. 546

prosimian Group of primates that includes lemurs and tarsiers and may resemble the first primates to have evolved. 360

prostate gland Gland located around the male urethra below the urinary bladder; adds secretions to semen. 563

protein Molecule consisting of one or more polypeptides; a macronutrient in the diet that is digested to amino acids used by cells to synthesize cellular proteins. 45, 480

proteome Collection of proteins resulting from the translation of genes into proteins. 193

proteomics Study of the structure, function, and interaction of proteins. 193

protist Member of the kingdom Protista. 7, 300

protocell Cell-like structure with an outer membrane, thought to have preceded true cells. 294

proton Positive subatomic particle located in the nucleus and assigned one atomic mass unit. 21

proto-oncogene Normal gene that can become an oncogene through mutation. 208

protostome Group of coelomate animals in which the first embryonic opening (the blastopore) is associated with the mouth. 339

protozoan Chemoheterotrophic, unicellular protist that moves by flagella, cilia, or pseudopodia or is immobile. 301

proximal tubule Portion of a nephron following the nephron capsule where reabsorption of filtrate occurs. 458

pseudocoelomate Animal, such as a roundworm, with a body cavity incompletely lined by mesoderm. 339

pseudopod Cytoplasmic extension of amoeboid protists; used for locomotion and engulfing food. 303

puberty Period of life when secondary sex changes occur in humans; marked by the onset of menses in females and sperm production in males. 564

pulmonary artery Blood vessel that takes blood away from the heart to the lungs. 434

pulmonary circuit Circulatory pathway between the lungs and the heart. 433, 438

pulmonary trunk Large blood vessel that divides into the pulmonary arteries; takes blood from the heart to the lungs. 434

pulmonary vein Blood vessel that takes blood to the heart from the lungs. 434

pulse Vibration felt in arterial walls due to expansion of the aorta following ventricle contraction. 435

Punnett square Grid used to calculate the expected results of simple genetic crosses. 158

pupil Opening in the center of the iris of the vertebrate eye. 544

R

radial symmetry Body plan in which similar parts are arranged around a central axis, like spokes of a wheel. 336

radiolarian Protozoan; marine amoeba having a calcium carbonate skeleton, many of which make limestone formations, such as the White Cliffs of Dover. 303

radius Thicker and shorter forearm bone that contributes to the easy twisting motion of the forearm. 549

random Spatial distribution of individuals in a population in which individuals have an equal chance of living anywhere within an area. 592

range Portion of the globe where a certain species can be found. 592

ray-finned fishes Group of bony fishes with fins supported by parallel bony rays connected by webs of thin tissue. 355

reabsorption Movement of primarily nutrient molecules and water from the contents of the nephron into blood at the proximal tubule. 458

reactant Substance that participates in a reaction. 26

receptacle Area where a flower attaches to a floral stalk. 399

receptor-mediated endocytosis Selective uptake of molecules into a cell by vacuole formation after they bind to specific receptor proteins in the plasma membrane. 89

recessive allele Allele that exerts its phenotypic effect only in the homozygote. Its expression is masked by a dominant allele. 159

recombinant DNA (rDNA) DNA that contains genes from more than one source. 188

recombinant gamete New combination of alleles incorporated into a gamete during crossing-over. 169

red algae Marine photosynthetic protists with a notable abundance of phycobilin pigments; includes coralline algae of coral reefs. 302

red blood cell Erythrocyte; contains hemoglobin and carries oxygen from the lungs or gills to the tissues in vertebrates. 418, 441

red bone marrow Vascularized, modified connective tissue that is sometimes found in the cavities of spongy bone within certain bones; site of blood cell formation. 498, 550

redox reaction Oxidation-reduction reaction; one molecule loses electrons (oxidation) while another molecule gains electrons (reduction). 96

reduction Gain of electrons by an atom or a molecule with a concurrent storage of energy. In biological systems, the electrons are accompanied by hydrogen ions. 96

reflex Automatic, involuntary response of an organism to a stimulus. 525

regeneration Ability of sponges and echinoderms to "generate" into another organism if separated into pieces. 559

regulatory gene In an operon, a gene that codes for a protein that regulates the expression of other genes. 201

release factor Protein complex that binds to an mRNA codon at the A site of the ribosome, causing termination of the transcription. 186

renal cortex Outer portion of the kidney that appears granular. 458

renal medulla Inner portion of the kidney that consists of renal pyramids. 458

renal pelvis Hollow chamber in the kidney that lies inside the renal medulla and receives freshly prepared urine from the collecting ducts. 458

renewable resources Resources normally replaced or replenished by natural processes; resources not depleted by moderate use. Examples include solar energy, biological resources (such as forests and fisheries), biological organisms, and some biogeochemical cycles. 631

replacement model Proposal that modern humans originated only in Africa, then migrated and supplanted populations of *Homo* in Asia and Europe about 100,000 years ago (see also *out-of-Africa hypothesis*). 363

replacement reproduction A measure of population dynamics in which each person is replaced by only one child in the next generation. 590

repressor In an operon, a protein molecule that binds to an operator, preventing transcription of structural genes. 201

reproduce To produce a new individual of the same kind. 4

reproductive cloning The process that creates a new individual that is genetically identical to the original individual. 198

reproductive system Organ system that contains male or female organs and specializes in the production of offspring. 423

reptile Terrestrial vertebrate with internal fertilization, scaly skin, and a shelled egg; includes snakes, lizards, turtles, crocodiles, and birds. 356

reservoir In chemical cycling, a source (e.g., fossil fuels, minerals in rocks, and ocean sediments) that is normally unavailable to organisms. 619

resource A necessity for an organism to live and reproduce; includes food, water, shelter, and nesting sites. 592

resource partitioning Mechanism that increases the number of niches by apportioning the supply of a resource, such as food or living space, between species. 610

respiration Sequence of events that results in gas exchange between the cells of the body and the environment. 451

respiratory system Organ system consisting of the lungs and tubes; brings oxygen into the lungs and takes carbon dioxide out. 422

restriction enzyme Bacterial enzyme that stops viral reproduction by cleaving viral DNA; used to cut DNA at specific points during production of recombinant DNA. 188

retina Innermost layer of the vertebrate eyeball, containing the photoreceptors—rods and cones. 544

retrovirus RNA virus containing the enzyme reverse transcriptase, which carries out RNA/DNA transcription. 291

rhodopsin In rods and cones, light-absorbing molecule that contains a pigment and the protein opsin. 545

rib cage Flexible skeletal structure, composed of the ribs and sternum, that protects the heart and lungs and moves when we breathe. 549

ribose Pentose sugar found in RNA. 39

ribosomal RNA (rRNA) Type of RNA found in ribosomes that translate messenger RNAs to produce proteins. 180

ribosome RNA and protein in two subunits; site of protein synthesis in the cytoplasm. 58, 180

RNA (ribonucleic acid) Nucleic acid produced from covalent bonding of nucleotide monomers that contain the sugar ribose. Three major forms are messenger RNA, ribosomal RNA, and transfer RNA. 48, 180

RNA polymerase During transcription, an enzyme that joins nucleotides complementary to a DNA template. 183

rod In vertebrate eyes, photoreceptor that responds to dim light. 545

root cap Protective cover of a plant root tip, whose cells are constantly replaced as they are ground off when the root pushes through rough soil particles. 380

root hair Extension of a root epidermal cell that collectively increases the surface area for the absorption of water and minerals. 372

root nodule Plant root structure that contains nitrogen-fixing bacteria. 384

root system Main plant root and all of its lateral (side) branches. 371

rotational equilibrium Maintenance of balance when the head and body are suddenly moved or rotated. 543

rotator cuff Tendons and ligaments that stabilize the shoulder joint. 549

rough ER Membranous system of tubules, vesicles, and sacs in cells; has attached ribosomes. 66

roundworm Invertebrate with a cylindrical body covered by a cuticle that molts; also has a complete digestive tract and a pseudocoelom. Some forms are free-living in water and soil, and many are parasitic. 346.

RuBP carboxylase (rubisco) Enzyme required for carbon dioxide fixation (atmospheric CO_2 attaches to RuBP) in the Calvin cycle. 101

rule of multiplication The chance of two (or more) independent events' occurring together is the product of their chance of occurring separately. 161

S

saccule Sac-like cavity in the vestibule of the vertebrate inner ear; contains sensory receptors for gravitational equilibrium. 543

sacrum Large, triangular bone at the base of the vertebral column; consists of five vertebrae that are fused together. 549

salinization Process in which mineral salts accumulate in the soil, killing plants; occurs when soils in dry climates are irrigated profusely. 636

salivary amylase In humans, enzyme in saliva that digests starch to maltose. 463

salivary gland In humans, gland associated with the mouth that secretes saliva. 463

salt Compound produced by a reaction between an acid and a base. 25

saltatory conduction Movement of nerve impulses from one node of Ranvier to another along a myelinated axon. 519

saltwater intrusion Movement of salt water into freshwater aquifers in coastal areas where groundwater is withdrawn faster than it is replenished. 635

SA (sinoatrial) node Pacemaker; small region of neuromuscular tissue that initiates the heartbeat. 435

saprotroph Organism that secretes digestive enzymes and absorbs the resulting nutrients back across the plasma membrane; fungus or bacterium that decomposes the remains of plants, animals, and microbes in the soil. 297, 326

sarcomere One of many identical units, arranged linearly in a myofibril, whose contraction produces muscle contraction. 551

saturated fatty acid Fatty acid molecule that lacks double bonds between the carbons of its hydrocarbon chain. The chain bears the maximum number of hydrogens possible. 42

scapula Shoulder blade. 549

scientific theory Concept supported by a broad range of observations, experiments, and data. 13

sclerenchyma Plant tissue composed of cells with heavily lignified cell walls; functions in support. 375

secondary sex characteristic Trait that is sometimes helpful but not necessary for reproduction and is maintained by the sex hormones in males and females. 564

secondary succession In ecological succession, the stage at which there is gradual replacement of communities in an area following a disturbance. 607

secretion In the cell, release of a substance by exocytosis from a cell that may be a gland or part of a gland; in the urinary system, movement of certain molecules from blood into the distal tubule of a nephron, so that they are added to urine. 66, 459

seed Mature ovule that contains an embryo, with stored food enclosed in a protective coat. 311, 402

segmentation Repetition of body units, as in the earthworm. 340

semen (seminal fluid) Thick, whitish fluid consisting of sperm and secretions from several glands of the male reproductive tract. 563

semiconservative Duplication of DNA resulting in two double helix molecules, each having one parental and one new strand. 179

semilunar valve Valve resembling a half-moon located between the ventricles and their attached vessels. 434

seminal vesicle Convoluted, sac-like structure attached to the vas deferens near the base of the urinary bladder in males; adds secretions to semen. 563

seminiferous tubule Long, coiled structure contained in chambers of the testis; where sperm are produced. 563

senescence Sum of the processes involving aging, decline, and eventual death of a plant or plant part. 394

sensory neuron Nerve cell that transmits nerve impulses to the central nervous system after a sensory receptor has been stimulated. 518

sepal Outermost, leaf-like covering of a flower; usually green. 319, 399

septum Partition that divides two areas. The septum in the heart separates the right half from the left half. 434

serotonin In the human central nervous system, neurotransmitter that affects mood, sleep, emotion, and appetite. 520

sex chromosome Chromosome that determines the sex of an individual. In humans, females have two X chromosomes, and males have an X and a Y chromosome. 141

sex-linked trait A characteristic or trait that is determined by a gene located on either the X or the Y chromosome. 220

sexual selection Changes in males and females, often due to male competition and female selectivity, leading to increased fitness. 260

shoot system Aboveground portion of a plant consisting of the stem, leaves, and flowers. 371

short-day plant/long-night plant Plant that flowers when day length is shorter and the night is longer than a critical length—e.g., cocklebur, poinsettia, and chrysanthemum. 397

sickle cell disease Hereditary disease in which red blood cells are narrow and curved, so that they are unable to pass through capillaries and are destroyed; causes chronic anemia. 222

sieve-tube member Member that joins with others in the phloem tissue of plants as a means of transport for nutrient sap. 376

signal Molecule that stimulates or inhibits an event in the cell cycle. 132

signal transduction pathway Activation and inhibition of intracellular targets after binding of growth factors. 205

simple diffusion Movement of molecules or ions from a region of higher to lower concentration; requires no energy and tends to lead to an equal distribution. 86

simple goiter Condition in which an enlarged thyroid produces low levels of thyroxine. 530

sinus Cavity into which hemolymph flows and baths the organs in an open circulatory system; also, air-filled spaces in nasal cavities. 432, 452

sister chromatid One of two genetically identical chromosomal units that are the result of DNA replication and are attached to each other at the centromere. 125

skeletal muscle Striated, voluntary muscle tissue that comprises skeletal muscles; also called striated muscle. 419

skeletal system System of bones, cartilage, and ligaments that works with the muscular system to protect the body and provide support for locomotion and movement. 423

skull Protective skeletal structure consisting of the cranium and the facial bones. 549

sliding filament model Explanation for muscle contraction based on the movement of actin filaments in relation to myosin filaments. 551

slime mold Protists that decompose dead material and feed on bacteria by phagocytosis; vegetative state is mobile and amoeboid. 304

small intestine In vertebrates, portion of the digestive tract that precedes the large intestine; in humans, the duodenum, jejunum, and ileum; responsible for most digestion and absorption. 465

smooth ER Membranous system of tubules, vesicles, and sacs in eukaryotic cells; lacks attached ribosomes. 66

smooth muscle Nonstriated, involuntary muscle in the walls of internal organs. 419

sodium-potassium pump Carrier protein in the plasma membrane that moves sodium ions out of and potassium ions into animal cells; important in nerve and muscle cells. 88

solute Substance dissolved in a solvent, forming a solution. 86

solution Fluid (the solvent) that contains a dissolved solid (the solute). 86

solvent Liquid portion of a solution that dissolves a solute. 86

somatic cell Body cell; excludes cells that undergo meiosis and become sperm or eggs. 133

somatic system Portion of the peripheral nervous system containing motor neurons that control skeletal muscles. 525

sori (sing., sorus) Clusters of sporangia on the underside of a fern's frond. 315

speciation Origin of new species due to the evolutionary process of descent with modification. 238, 266

species Group of similarly constructed organisms capable of interbreeding and producing fertile offspring; organisms that share a gene pool; the taxon at the lowest level of classification. 6, 278

species richness List of different species in a community. 606

spermatogenesis Production of sperm in males by the process of meiosis and maturation. 143

spinal cord In vertebrates, the nerve cord that is continuous with the base of the brain and housed within the vertebral column. 522

spinal nerve Nerve that arises from the spinal cord. 525

spindle Microtubule structure that brings about chromosomal movement during nuclear division. 127

spindle equator Disk formed during metaphase, in which all of a cell's chromosomes lie in a single plane at right angles to the spindle fibers. 128

spiral organ Organ of Corti; in the vertebrate inner ear, the structure that contains auditory receptors. 542

spleen Large, glandular, lymphatic organ in the upper left region of the abdomen; stores and purifies blood. 498

sponge Invertebrate animal of the phylum Porifera; pore-bearing filter feeder whose inner body wall is lined by collar cells. 340

spongy bone Type of bone that has an irregular, mesh-like arrangement of thin plates of bone. 550

spongy mesophyll In a plant leaf, layer of tissue containing loosely packed cells, increasing the amount of surface area for gas exchange. 377

sporangium Spore-producing plant or fungal structure. 313

spore Asexual reproductive or resting cell capable of developing into a new organism without fusion with another cell, in contrast to a gamete. 312

sporophyte Diploid generation of the alternation of generations life cycle of a plant; produces haploid spores that develop into the haploid generation. 312, 398

sporozoan Spore-forming protist that has no means of locomotion and is typically a parasite with a complex life cycle having both sexual and asexual phases. 303

squamous epithelium Type of epithelial tissue in which the cells are flattened. 415

stabilizing selection Outcome of natural selection in which extreme phenotypes are eliminated and the average phenotype is conserved. 253

stamen In flowering plants, portion of the flower that consists of a filament and an anther containing pollen sacs where pollen is produced. 319, 399

starch In plants, storage polysaccharide composed of glucose molecules joined in a linear fashion with few side chains. 40

statocyst Gravitational equilibrium organ found in some invertebrates. It provides information about head position. 543

stem Usually the upright, vertical portion of a plant that transports substances to and from the leaves. 372

sternum Breastbone. 549

steroid Type of lipid molecule having a complex of four carbon rings—e.g., cholesterol, estrogen, progesterone, and testosterone. 44

steroid hormone Type of hormone that has the same complex of four carbon rings, but each one has different side chains. 528

stigma In flowering plants, portion of the carpel where pollen grains adhere and germinate before fertilization can occur. 319, 399

stoma (pl., stomata) Small opening between two guard cells on the underside of leaf epidermis through which gases pass. 95, 374

stomach In vertebrates, muscular sac that mixes food with gastric juices to form chyme, which enters the small intestine. 464

strata Layers of rock or sedimentary material, often containing fossils. 235

striated Having bands; in cardiac and skeletal muscle, alternating light and dark bands produced by the distribution of contractile proteins. 419

stroke Condition resulting when an arteriole in the brain bursts or becomes blocked by an embolism; cerebrovascular accident. 444

stroma Fluid within a chloroplast that contains enzymes involved in the synthesis of carbohydrates during photosynthesis. 68, 95

style Elongated, central portion of the carpel between the ovary and stigma. 319, 399

subsidence Gradual settlement of a portion of the Earth's surface. 634

subspecies Taxonomic designation of populations within a species, each with somewhat distinctive genotypic and phenotypic characteristics; readily interbreed with other such subspecies. 260

substrate Reactant in a reaction controlled by an enzyme. 84

surface-area-to-volume ratio Ratio of a cell's outside area to its internal volume. 56

survivorship Probability of newborn individuals of a cohort surviving to particular ages. 593

sustainable society Ability of a society or an ecosystem to maintain itself while providing services to human beings. 648

symbiotic Type of relationship that occurs when two different species live together in a unique way. It may be beneficial, neutral, or detrimental to one or both species. 297

sympathetic division Division of the autonomic system that is active when an organism is under stress; uses norepinephrine as a neurotransmitter. 527

sympatric speciation Origin of new species in populations that overlap geographically. 270

synapse Junction between neurons consisting of the axon membrane, the synaptic cleft, and a dendrite. 519

synapsis Pairing of homologous chromosomes during meiosis I. 144

synaptic cleft Small gap between membranes of a synapse. 519

synovial joint Freely moving joint in which two bones are separated by a cavity. 552

syphilis Bacterial STD with three stages separated by latent periods; may result in blindness, cause birth defects or stillbirth, and affect the cardiovascular and/or nervous system. 572

systematics Study of the diversity of organisms to classify them and determine their evolutionary relationships. 6, 277

systemic circuit Circulatory pathway of blood flow between the tissues and the heart. 433, 438

systole Contraction period of the heart during the cardiac cycle. 435

T

taste buds In the vertebrate mouth, structures containing sensory receptors for taste. In humans, most taste buds are on the tongue. 540

taxon (pl., taxa) Group of organisms that fills a particular classification category. 277

taxonomy Branch of biology concerned with identifying, describing, and naming organisms. 6, 277

T-cell receptor (TCR) On the T-cell surface, receptor consisting of two antigen-binding peptide chains; associated with a large number of other glycoproteins. Binding of antigen to the TCR, usually in association with MHC, activates the T cell. 503

technology Application of scientific knowledge for a practical purpose. 15

telomere Long, repeating DNA sequence at the ends of chromosomes; functions as a cap and keeps chromosomes from fusing with each other. 133, 209

telophase Mitotic phase during which daughter cells are located at each pole. 128

template Parental strand of DNA that serves as a guide for the complementary daughter strand produced during DNA replication. 179

tendon Strap of fibrous connective tissue that connects skeletal muscle to bone. 418

terminal bud Bud that develops at the apex of a shoot. 371

terrestrial ecosystem Biome; tundra, taiga, temperate forests, tropical grasslands (savanna), temperate grasslands (prairie), deserts, and tropical rain forests. 625

testcross Cross between an individual with the dominant phenotype and an individual with the recessive phenotype. The resulting phenotypic ratio indicates whether the dominant phenotype is homozygous or heterozygous. 158

test group In a controlled experiment, the group subjected to the experimental variable. 12

testis (pl., testes) Male gonad that produces sperm and the male sex hormones. 560

testosterone Male sex hormone that helps maintain sexual organs and secondary sex characteristics. 564

tetany Severe twitching caused by involuntary contraction of the skeletal muscles due to a calcium imbalance. 531

tetrad Homologous chromosomes, each having sister chromatids that are joined during meiosis; also called bivalent. 144

thalamus In vertebrates, portion of the diencephalon that passes on selected sensory information to the cerebrum. 523

therapeutic cloning Cloning done to create mature cells of various types; also used to learn about specialization of cells and provide cells and tissue to treat human illnesses. 199

thermoacidophile Type of archaea that lives in hot, acidic, aquatic habitats (such as hot springs) or near hydrothermal vents. 299

thoracic cavity Cavity located in the ventral cavity above the abdominal cavity and separated from the abdominal cavity by the diaphragm; contains the heart and lungs. 454

three-domain system System of classification that recognizes three domains: Bacteria, Archaea, and Eukarya. 283

thrombin Enzyme that converts fibrinogen to fibrin threads during blood clotting. 444

thylakoid Flattened sac within a granum whose membrane contains chlorophyll and where the light reactions of photosynthesis occur. 68, 95

thymine (T) One of four nitrogen-containing bases in nucleotides composing the structure of DNA; pairs with adenine. 176

thymus gland Lymphatic organ involved in the development and functioning of the immune system. T cells mature in the thymus gland. 498

thyroid gland Large gland in the neck; produces several important hormones, including thyroxine, triiodothyronine, and calcitonin. 530

thyroid-stimulating hormone (TSH) Substance, produced by the anterior pituitary, that causes the thyroid to secrete thyroxine and triiodothyronine. 529

thyroxine (T_4) Hormone, secreted from the thyroid gland, that promotes growth and development. In general, it increases the metabolic rate in cells. 530

tibia Shinbone. 550

tight junction Junction between cells when adjacent plasma membrane proteins join to form an impermeable barrier. 73

tissue Group of similar cells combined to perform a common function. 3, 413

tissue culture Process of growing tissue artificially, usually in a liquid medium in laboratory glassware. 405

tissue fluid Fluid that surrounds the body's cells; consists of dissolved substances that leave the blood capillaries by filtration and diffusion. 422

T lymphocyte (T cell) Lymphocyte that matures in the thymus and exists in four varieties, one of which kills antigen-bearing cells outright. 498

tonsils Partially encapsulated lymphatic nodules in the pharynx. 497

totipotent Cell that has the full genetic potential of the organism, including the potential to develop into a complete organism. 200, 405

tracer Substance having an attached radioactive isotope that allows a researcher to track its whereabouts in a biological system. 22

trace mineral Essential inorganic nutrient (such as zinc, iron, copper, iodine, or selenium) needed daily by humans to regulate metabolic activities and maintain good health. 481

trachea (pl., tracheae) In tetrapod vertebrates, air tube (windpipe) that runs between the larynx and the bronchi. 453

tracheid In vascular plants, type of cell in xylem that has tapered ends and pits through which water and minerals flow. 376

tract Bundles of myelinated axons in the central nervous system. 522

trans fat A form of unsaturated fatty acid in which the hydrogen atoms of the carbon chain are on opposite sides of the double bond. 43

transcription Process whereby a DNA strand serves as a template for the formation of mRNA. 181

transcription activator Protein that speeds transcription. 204

transcription factor In eukaryotes, protein required for the initiation of transcription by RNA polymerase. 203

transduction Exchange of DNA between bacteria by means of a bacteriophage. 296

transfer rate Amount of a substance that moves from one component of the environment to another within a specified period of time. 620

transfer RNA (tRNA) Type of RNA that transfers a particular amino acid to a ribosome during protein synthesis. At one end it binds to the amino acid, and at the other end it has an anticodon that binds to an mRNA codon. 180

transformation Taking up of extraneous genetic material from the environment by bacteria. 296

transgenic organism In the environment, free-living organism that has had a foreign gene inserted into it. 188

transgenic plant Genetically modified plant; plant that carries the genes of another organism as a result of DNA technology. 407

translation Process whereby ribosomes use the sequence of codons in mRNA to produce a polypeptide with a particular sequence of amino acids. 181

translocation Movement of a chromosomal segment from one chromosome to another nonhomologous chromosome, leading to abnormalities—e.g., Down syndrome. 185, 210, 218

transpiration Plant's loss of water to the atmosphere, mainly through evaporation at leaf stomata. 385

transport vesicles Small, membranous sacs that carry products between endomembrane system organelles and prevent the products from entering the cytoplasm. 63

transposon DNA sequence capable of randomly moving from one site to another in the genome. 187

trichomoniasis Sexually transmitted disease caused by the parasitic protozoan *Trichomonas vaginalis*. 572

triglyceride Neutral fat composed of glycerol and three fatty acids. 42, 479

triplet code During gene expression, each sequence of three nucleotide bases stands for a particular amino acid. 182

trophic level Feeding level of one or more populations in a food web. 618

tropism In plants, growth response toward or away from a directional stimulus. 396

trypanosome Parasitic zooflagellate that causes severe disease in humans and domestic animals, including a condition called sleeping sickness. 303

trypsin Protein-digesting enzyme secreted by the pancreas. 466

tumor Group of cells derived from a single, mutated cell that has repeatedly undergone cell division. Benign tumors remain at their site of origin, while malignant tumors metastasize. 134

tumor suppressor gene Gene that codes for a protein that ordinarily suppresses cell division. Inactivity can lead to a tumor. 208

tunicate Type of primitive invertebrate chordate. 352

Turner syndrome Condition caused by the inheritance of a single X chromosome. 151

U

ulna Larger forearm bone that, together with the radius, contributes to the easy twisting motion of the forearm. 549

umbilical cord Cord connecting the fetus to the placenta, through which blood vessels pass. 577

unicellular Made up of a single cell, as in bacteria. 2

uniform Spatial distribution of individuals in a population in which individuals are dispersed uniformly through the area. 592

uniformitarianism Belief, espoused by Charles Lyell, that geological forces act at a continuous, uniform rate. 236

unsaturated fatty acid Fatty acid molecule that has one or more double bonds between the carbons of its hydrocarbon chain. The chain bears fewer hydrogens than the maximum number possible. 42

uracil (U) Pyrimidine base that replaces thymine in RNA; pairs with adenine. 180

ureter Tubular structure conducting urine from the kidney to the urinary bladder. 457

urethra Tubular structure that receives urine from the bladder and carries it to the outside of the body. 457

urinary bladder Organ where urine is stored. 457

urinary system Organ system consisting of the kidneys and urinary bladder; rids the body of nitrogenous wastes and helps regulate the water-salt balance of the blood. 422

urine Liquid waste product made by the nephrons of the vertebrate kidney through the processes of filtration, reabsorption, and secretion. 457

uterus In mammals, expanded portion of the female reproductive tract through which eggs pass to the environment or in which an embryo develops and is nourished before birth. 564

utricle Sac-like cavity in the vestibule of the vertebrate inner ear; contains sensory receptors for gravitational equilibrium. 543

V

vaccine Antigens prepared in such a way that they can promote active immunity without causing disease. 506

vacuole Membrane-bounded sac, larger than a vesicle; usually functions in storage and can contain a variety of substances. In plants, the central vacuole fills much of the interior of the cell. 67

valence shell Outer shell of an atom. 24

vascular bundle In plants, primary phloem and primary xylem enclosed by a bundle sheath. 373

vascular cambium In plants, lateral meristem that produces secondary phloem and secondary xylem. 374

vascular cylinder In eudicots, tissues in the middle of a root, consisting of the pericycle and vascular tissues. 380

vascular plant Plant that has vascular tissue (xylem and phloem); includes seedless vascular plants (e.g., ferns) and seed plants (gymnosperms and angiosperms). 314

vascular tissue Transport tissue in plants, consisting of xylem and phloem. 314, 374

vas deferens Tube that leads from the epididymis to the urethra in males. 562

vector Piece of DNA that can have a foreign DNA attached to it. A common vector is a plasmid. 188

vein Blood vessel that arises from venules and transports blood toward the heart. 435

vena cava (pl., venae cavae) Large, systemic vein that returns blood to the right atrium of the heart in tetrapods; either the superior or inferior vena cava. 434

ventricle Cavity in an organ, such as a lower chamber of the heart or the ventricles of the brain. 434

venule Vessel that takes blood from capillaries to a vein. 437

vertebral column Portion of the vertebrate endoskeleton that houses the spinal cord; consists of many vertebrae separated by intervertebral discs. 549

vertebrate Chordate in which the notochord is replaced by a vertebral column. 340

vesicle Small, membrane-bounded sac that stores substances in a cell. 66

vessel element Cell that joins with others to form a major conducting tube in xylem. 376

vestigial structure Remains of a structure that was functional in an ancestor but is no longer functional in the present organism. 244

villus (pl., villi) Small, finger-like projection of the inner small intestinal wall. 466

viroid Infectious strand of RNA devoid of a capsid and much smaller than a virus. 292

virus Noncellular, parasitic agent consisting of an outer capsid and an inner core of nucleic acid. 288

vitamin Essential requirement in the diet, needed in small amounts. Vitamins are often part of coenzymes. 482

vocal cord In the human larynx, folds of tissue that create vocal sounds when they vibrate. 453

W

water mold Filamentous organisms having cell walls made of cellulose. Most are decomposers of dead freshwater organisms, but some are parasites of aquatic or terrestrial organisms. 304

white blood cell Leukocyte, of which there are several types, each having a specific function in protecting the body from invasion by foreign substances and organisms. 418, 442

wild-type Allele or phenotype that is the most common for a certain gene in a population. 161

wood Secondary xylem that builds up year after year in woody plants, becoming annual rings. 379

X

X-linked Allele that is located on an X chromosome but may control a trait that has nothing to do with the sex characteristics of an animal. 167

xylem Vascular tissue that transports water and mineral solutes upward through the plant body. It contains vessel elements and tracheids. 314, 373

Z

zooflagellate Nonphotosynthetic protist that moves by flagella. Typically, zooflagellates enter into symbiotic relationships, and some are parasitic. 303

zygote Diploid cell formed by the union of two gametes; the product of fertilization. 143, 560

Credits

Photographs

Front Matter

Page vi (nerve cells): © Dennis Kunkel Microscopy, Inc./Visuals Unlimited/Corbis; p. vii (bacteriophage): © Lee Simon/Photo Researchers, Inc.; p. viii (bathroom scale): © Photodisc/Getty RF; p. x (paramecium): © Dr. David Patterson/SPL/Photo Researchers, Inc.; p. x (leaf): © Jeremy Burgess/SPL/Photo Researchers, Inc.; p. xiii (lab manual cover): © Minden Pictures/Masterfile; p. xiv (crab): © Brand X Pictures/PunchStock RF; p. xv (butterfly): © Wally Eberhart/Getty Images; p. xix (tortoise): © Digital Vision/PunchStock RF; p. xx (gecko): Kevin Hanley/Animals Animals; p. xxi (egg and sperm): © Pascal Goetgheluck/SPL/Photo Researchers, Inc.; p. xxii (blood cells): © Eye of Science/Photo Researchers, Inc.; p. xxiii (snake): © Joseph Collins/Photo Researchers, Inc.; p. xxv (lady bug): © Anthony Mercieca/Photo Researchers, Inc.; p. xxvii (frog): © Kevin Schaefer and Martha Hill.

Chapter 1

Opener: © David Scharf/SPL/Photo Researchers, Inc.; 1.1(forest): © Steven P. Lynch; p. 3(butterfly): © Stockbyte RF; p. 3(toucan): © IT Stock/PunchStock RF; p. 3(frog): © MedioImages/SuperStock RF; p. 3(orchid): © Max and Bea Hunn/Visuals Unlimited; 1.2a: © age fotostock/SuperStock; 1.2b: © Ed Young/Corbis; 1.3(family): © Vol. 113/PhotoDisc/Getty RF; 1.3(DNA): © David Mack/SPL/Photo Researchers, Inc.; 1.5(main): © Ralph Robinson/Visuals Unlimited; 1.6(main): © A.B. Dowsett/SPL/Photo Researchers, Inc.; 1.7: © Tom Adams/Visuals Unlimited; 1.8: © Corbis RF; 1.9: © Brenda Tharp/Photo Researchers, Inc.; 1.10: © PhotoDisc/PunchStock RF; 1.13a: © Kathy Sloane/Photo Researchers, Inc.; 1.13b: © Getty Images/Digital Vision RF; 1.13c: © Luiz C. Marigo/Peter Arnold/Photolibrary/Photolibrary; 1.15(students): © image100 Ltd RF; 1.15(surgeon): © Phanie/Photo Researchers, Inc.; 1.16: © The McGraw-Hill Companies, Inc./Mark Dierker, photographer.

Chapter 2

Opener: © Steve Smith; 2.1: © Tim Pannell/Corbis RF; p. 20(supernova): © NASA/JPL-Caltech/STSci; 2.4(both): Courtesy National Institutes of Health; 2.5a: © Natasja Weitsz/Getty Images; 2.5b: © Geoff Tompkinson/SPL/Photo Researchers, Inc.; 2.7b(crystals): © Evelyn Jo Johnson; 2.9: © Fritz Polking/Peter Arnold/Photolibrary; 2.10: © Corbis RF; p. 29(water strider): © Martin Shields/Alamy; 2.11a: © The McGraw-Hill Companies, Inc./Jill Braaten, photographer; 2.11b: © Amanda Langford/SuperStock RF; p. 31(steel girders): © Corbis RF; 2.17a: © Frederica Georgia/Photo Researchers, Inc.; 2.17b: © Digital Vision/PunchStock RF.

Chapter 3

Opener: © Cole Group/Getty RF; 3.1a: © Lary Lefever/Grant Heilman Photography; 3.1b: © PhotoDisc/Getty RF; 3.1c: © H. Pol/CNRI SPL/Photo Researchers, Inc.; p. 37(foods): © Corbis RF; 3.4–3.6: © The McGraw-Hill Companies, Inc./John Thoeming, photographer; p. 40(cola): © The McGraw-Hill Companies, Inc./Richard Hutchings, photographer; p. 41(fiber): © The McGraw-Hill Companies, Inc./John Thoeming, photographer; 3.10a: © Jeremy Burgess/SPL/Photo Researchers, Inc.; 3.10b: © Don W. Fawcett/Photo Researchers, Inc.; 3.10c: © J.D. Litvay/Visuals Unlimited; 3.11: © Lightwave Photography, Inc./Animals Animals; 3.15b–c: © Corbis RF; 3.16a(woman): © Reuters/Corbis; 3.16a(web): © OSF/Animals Animals; 3.16b: © P. Motta & S. Correr/Photo Researchers, Inc.; 3.16c: © Duomo/Corbis; p. 46(curly hair): © Getty RF; 3.21: © Eye of Science/Photo Researchers, Inc.

Chapter 4

Opener: © Ed Reschke/Peter Arnold/Photolibrary; 4.1(TEM, lymphocyte): © Dr. Gopal Murti/Visuals Unlimited; 4.1(electron microscope): © Inga Spence/Visuals Unlimited; 4.1(light microscope): © Corbis Images/Jupiter Images RF; 4.1(Euglena): © Tom Adams/Visuals Unlimited; 4.1(SEM, spider): © Science Photo Library. Getty RF; p. 55(squid): © Karen Gowlett-Holmes/Getty Images; 4.4: © Ralph A. Slepecky/Visuals Unlimited; 4.7a: © Dr. Dennis Kunkel/Visuals Unlimited; 4.8a: © Newcomb/Wergin/Biological Photo Service; 4.9b: Courtesy E.G. Pollock; 4.11: © R. Bolender and D. Fawcett/Visuals Unlimited; 4.13a: © Roland/Birke/Peter Arnold/Photolibrary; 4.13b: © Newcomb/Wergin/Biological Photo Service; 4.13c: © The McGraw Hill Companies, Inc./Al Telser, photographer; 4.14a: © Dr. George Chapman/Visuals Unlimited; 4.15b: Courtesy Keith Porter; 4.19: © Don W. Fawcett/Photo Researchers, Inc.; 4.20a(cilia): © Manfred Kage/Peter Arnold/Photolibrary; 4.20a(sperm): © David M. Phillips/Photo Researchers, Inc.; 4.20b (both): © William L. Dentler/Biological Photo Service; 4.21: © E.H. Newcomb/Biological Photo Service; 4.23a:From Douglas E. Kelly, Journal of Cell Biology, 28 (1966:51). Reproduced by permission of The Rockefeller University Press.

Chapter 5

Opener: © age fotostock/SuperStock; 5.1(woman): © GoodShoot/SuperStock RF; p. 79(energy for life): © Vol. 41/PhotoDisc/Getty RF; 5.2(both): © Keith Eng, 2008; p. 81(creatine): © The McGraw-Hill Companies, Inc./Mark Dierker, photographer; 5.6(leaves): © Comstock/PunchStock RF; 5.6(woman): © Getty RF; 5.8(antibiotic): © Tom Pantages/ Phototake; 5.8(bacteria): © Oliver Meckes/Ottawa/Photo Researchers, Inc.; 5.16(both): Courtesy Mark Bretscher.

Chapter 6

Opener: © Stone/Getty RF; 6.1(kelp): © Chuck Davis/Stone/Getty Images; 6.1(diatoms): © Ed Reschke/Peter Arnold/Photolibrary; 6.1(Euglena): © Tom Adams/Visuals Unlimited; 6.1(sunflower): © Corbis RF; 6.1(moss): © Steven P. Lynch; 6.1(cyanobacteria): © Sherman Thomas/Visuals Unlimited; 6.2(mesophyll cell): © Manfred Kage/Peter Arnold/Photolibrary; 6.2(chloroplast): © Dr. George Chapman/Visuals Unlimited; p. 96(diatoms): © Carolina Biological Supply/Phototake; p. 98(grow light): Courtesy Hydrofarm Horticultural Products; 6.5(both): © Digital Vision/Getty RF; p. 102(forest): PhotoLink/Getty RF; 6.10: © Brand X Pictures/PunchStock RF; 6.12: Courtesy USDA/Doug Wilson, photographer; 6.13: © PhotoLink/Getty RF.

Chapter 7

Opener: © PhotoDisc/Getty RF; 7.1: © Corbis RF; 7.5: Courtesy Dr. Keith Porter; p. 116(exhaust): © Getty RF; p. 117(chromium picolinate): © The McGraw-Hill Companies, Inc./Mark Dierker, photographer; 7.10: © C Squared Studios/Getty RF; p. 118(lactic acid): © Bill Robbins/Index Stock Imagery; 7.11(runner): © Rubber Ball Productions/RF Index Stock Imagery, Inc.; 7.11(wine, bread): © C Squared Studios/Getty RF.

Chapter 8

Opener: © Simon Fraser/Royal Victoria Infirmary, Newcastle Upon Tyne/SPL/Photo Researchers, Inc.; 8.1a: © Corbis RF; 8.1b(top): © McDonald Wildlife Photography/Animals Animals; 8.1b(bottom): © Kevin Hanley/Animals Animals; 8.1c: © Biophoto Associates/Photo Researchers, Inc.; 8.1d(left): © Anatomical Travelogue/Photo Researchers, Inc.; 8.1d(right): © Brand X Pictures/PunchStock RF; p. 125(ferns): © 2008 Robbin Moran; 8.2(chomatin): Courtesy O.L. Miller, Jr. and Steve L. McKnight; 8.5(all): Courtesy Dr. Andrew S. Bajer, University of Oregon; 8.6(all): © Ed Reschke; 8.7: © Thomas Deerinck/Visuals Unlimited; 8.7: © SPL/Getty RF; 8.8: © B.A. Palevitz and E.H. Newcomb/BPS/Tom Stack & Associates; p. 132(patient): © Antonia Reeve/SPL/Photo Researchers, Inc.; 8.10(blebs): Courtesy Klaus Hahn, Cecelia Subauste, and Gary Bikoch from Science Vol. 280, April 3, 1998 p. 32; 8.10(apoptotic cell): Courtesy Douglas R. Green, LaJolla Institute for Allergy and Immunology, San Diego; 8.11: © Nancy Kedersha/Immunogen/SPL/Photo Researchers, Inc.; p. 134(cells): © VVG/SPL/Photo Researchers, Inc.; 8.12: © Breast Scanning Unit, Kings College Hospital, London/SPL/Photo Researchers, Inc.; 8.13(cabbage): © PhotoLink/Getty RF; 8.13(broccoli): © Corbis RF; 8.13(berries): © Cole Group/Getty RF; 8.13(oranges): © Corbis RF; 8.13(chard): © Roy Morsch.

© Dwight Kuhn; 18.11: © Phototake; 18.12 (both), 18.13a (forest): © Steven P. Lynch; 18.13a(pollen cones): © Runk/Schoenberger/Grant Heilman Photography; 18.13b: © Ed Reschke/Peter Arnold/Photolibrary; 18.13c(tree): © T. Daniel/Bruce Coleman/Photoshot; 18.13c(berries): © Evelyn Jo Johnson; 18.14: Courtesy Stephen McCabe/Arboretum at University of California Santa Cruz; 18.17a: © IT Stock Free/Alamy RF; 18.17b: © H. Eisenbeiss/Photo Researchers, Inc.; 18.17c: © Anthony Mercieca/Photo Researchers, Inc.; 18.17d: © Merlin D. Tuttle/Bat Conservation International; 18.18(all): © Walther H. Hodge/Peter Arnold/Photolibrary; 18.19(left): © Imagesource/JupiterImages RF; 18.19(center): © Brand X Pictures/Alamy RF; 18.19(right): © Thinkstock Images/JupiterImages RF; 18.21a: © Gary R. Robinson/Visuals Unlimited; 18.21b: © Corbis RF; 18.21c: © Miguel Champi/Alamy; 18.22a: © Gary T. Cole/Biological Photo Service; 18.23(bread): © Precision Graphics/Kim Brucker, Photographer; 18.23(inset): © James W. Richardson/Visuals Unlimited; 18.24b: © Glenn Oliver/Visuals Unlimited; 18.25:Reproduced with permission of the Freshwater Biological Association on behalf of The Estate of Dr Hilda Canter-Lund. (c) The Freshwater Biological Association.; 18.26: © L. West/Photo Researchers, Inc.; 18.27b(crustose): © Digital Vision/Getty RF; 18.27b (fruticose, foliose): © Steven P. Lynch; 18.28: © R. Roncandore/Visuals Unlimited; 18.29: © The McGraw-Hill Companies, Inc./John Thoeming, photographer; 18.30a: © Kingsley Stern; p. 329(*Amanita* mushroom): © Holmes Garden Photos/Alamy; 18.31a: © Everett S. Beneke/Visuals Unlimited; 18.31b: © John Hadfield/SPL/Photo Researchers, Inc.; 18.31c: © P. Marazzi/SPL/Photo Researchers, Inc.

Chapter 19

Opener: © D. Parer & E. Parer-Cook/Ardea; 19.1(adult frog): © Dwight Kuhn; 19.1(top row, all): © CABISCO/Phototake; 19.1(bottom row, all): © Dwight Kuhn; 19.9a: © Andrew J. Martinez/Photo Researchers, Inc.; 19.10a: © CABISCO/Visuals Unlimited; 19.10b: © CABISCO/Phototake; 19.11a: © Runk/Schoenberger/Grant Heilman Photography; 19.11b: © Ron Taylor/Bruce Coleman/Photoshot; 19.13a: © James Webb/Phototake; 19.13b: © NIBSC/SPL/Photo Researchers, Inc.; 19.15(snail): © Bob Coyle; 19.15(nudibranch): © Kenneth W. Fink/Bruce Coleman/Photoshot; 19.15(octopus): © Alex Kerstitch/Bruce Coleman/Photoshot; 19.15(nautilus): © Douglas Faulkner/Photo Researchers, Inc.; 19.15(scallop): Courtesy Larry S. Roberts; 19.15(mussels): © Rick Harbo; 19.17a: © Heather Angel/Natural Visions; 19.17b: © James Carmichael/Nature Photographics; 19.17c: © St. Bartholomews Hospital/Photo Researchers, Inc.; 19.18c: © John Burbidge/SPL/Photo Researchers, Inc.; 19.20(all): © John Shaw/Tom Stack & Associates; 19.21a: © Kim Taylor/Bruce Coleman/Photoshot; 19.21b: © James Carmichael/Nature Photographics; 19.21c: © NHPA/Photoshot; 19.21d: © Kjell Sandved/Butterfly Alphabet; 19.22a: © C. Allan Morgan/Peter Arnold/Photolibrary; 19.22b: © John Bell/Bruce Coleman/Photoshot; 19.22c(ventral): © Daniel Lyons/Bruce Coleman/Photoshot; 19.22c(dorsal): © Ken Lucas/Ardea; 19.22d: © Geof de Feu/Imagestate; 19.22e: © David M. Dennis/Animals Animals; 19.23(scale): © Science VU/Visuals Unlimited;

19.23(beetle): © Kjell Sandved/Bruce Coleman/Photoshot; 19.23(housefly): © L. West/Bruce Coleman/Photoshot; 19.23(lacewing): © Glenn Oliver/Visuals Unlimited; 19.23(walking stick): © Art Wolfe; 19.23(honeybee): Courtesy of The National Human Genome Research Institute; 19.23(flea): © Edward S. Ross; 19.23(leaf hopper): © Farley Bridges; 19.23(dragonfly): © Creatas Images/PictureQuest RF; 19.23(grasshopper): © Gary Meszaros/Visuals Unlimited; 19.23(moth): © Dr. Cleveland P. Hickman; 19.25a(sea lily): © P. Danna/Peter Arnold/Photolibrary; 19.25a(feather star): © Carl Roessler/Sea Images, Inc.; 19.25b: © Alex Kerstitch/Visuals Unlimited; 19.25c: © Jim Greenfield/Imagequestmarine.com; 19.25d(sea urchins): © Randy Morse/Animals Animals; 19.25d(sand dollar): © Roger Steene/Imagequestmarine.com; 19.28a: © Corbis RF; 19.28b, 19.31a: © Heather Angel/Natural Visions; 19.31b: © Ingram Publishing/Alamy RF; 19.32b: © John Dudak/Phototake; 19.33: © Zig Leszczynski/Animals Animals; 19.34a: © Martin Harvey/Gallo Images/Corbis; 19.35: © Tony Camacho/SPL/Photo Researchers, Inc.; 19.36a: © Kirtley Perkins/Visuals Unlimited; 19.36b: © Thomas Kitchin/Tom Stack & Associates; 19.36c: © Brian Parker/Tom Stack & Associates; 19.37a: © B.G. Thompson/Photo Researchers, Inc.; 19.37b: © Leonard Lee Rue/Photo Researchers; 19.37c: © Fritz Prenzel/Animals Animals; 19.38a: © Leonard Rue; 19.38b: © Gallo Images-Gerald Hinde/Getty Images; 19.38c: © Denise Tackett; 19.38d: © Mike Bacon; 19.41a: © Dan Dreyfus and Associates; 19.41b: © John Reader/Photo Researchers, Inc.; p. 364(*Homo floresiensis*): © Mauricio Anton/Photo Researchers, Inc.; 19.43: Courtesy American Museum of Natural History. Neg #608; p. 365(Neandertals): © The Field Museum #A102513c.

Chapter 20

Opener: © John & Barbara Gerlach/Visuals Unlimited/Corbis; p. 371(redwoods): © Corbis RF; 20.2a: © Michael Gadomski/Photo Researchers, Inc.; 20.2b: © Steven P. Lynch; 20.2c: © Nature Picture Library; 20.2d: © The McGraw Hill Companies, Inc./Carlyn Iverson, photographer; 20.3a: © Jonathan Buckley/Getty Images; 20.3b: © Evelyn Jo Johnson; 20.3c: © David Newman/Visuals Unlimited; 20.5a: © Runk/Schoenberger/Grant Heilman Photography; 20.5b: © J. Robert Waaland/Biological Photo Service; 20.5c, 20.6a–c: © Biophoto Associates/Photo Researchers, Inc.; p. 375(paper): © Comstock Images/Jupiterimages; 20.7a(right): Courtesy Wilfred A. Cote, SUNY College of Environmental Forestry; 20.7b(left): © Randy Moore/Visuals Unlimited; 20.9: © Jeremy Burgess/SPL/Photo Researchers, Inc.; 20.10: © J. Robert Waaland/Biological Photo Service; 20.11a: Courtesy Ray F. Evert, University of Wisconsin, Madison; 20.11b: © CABISCO/Phototake; p. 379(cork): © Thomas Dressler/Getty Images; 20.12(bottom, circular): © Ed Reschke/Peter Arnold/Photolibrary; 20.13(top left): © CABISCO/Phototake; 20.13(lower right): © Visuals Unlimited/Corbis; 20.14: © John D. Cunningham/Visuals Unlimited; 20.15(both): © Ken Wagner/Phototake; 20.16: Courtesy Mary E. Doohan; p. 383(fertilizer): © Evelyn Jo Johnson; 20.17a: © Dwight Kuhn; 20.17b: © Runk/Schoenberger/Grant Heilman Photography; p. 390(stem): Courtesy Ray F. Evert, University of Wisconsin, Madison.

Chapter 21

Opener: © Carla Gottgens/Bloomberg via Getty Images; 21.2(both): Courtesy Prof. Malcolm B. Wilkins; 21.3: © Robert E. Lyons/Visuals Unlimited; 21.4a(right): Courtesy John Selier/Virginia Tech Forestry Department; 21.4b-2(stoma): © Jeremy Burgess/SPL/Photo Researchers, Inc.; 21.5(both): © Kingsley Stern; 21.5c: © SuperStock RF; 21.6: © Runk/Schoenberger/Grant Heilman Photography; 21.7: © Kingsley Stern; 21.9: © BananaStock/PunchStock RF; 21.12a: Courtesy Timothy J. Fehr; 21.12b: © Steven P. Lynch; 21.13a: © Arthur C. Smith/Grant Heilman Photography; 21.13b: © Grant Heilman Photography; 21.14(pollen grain): Courtesy Graham Kent; 21.14(embryo sac): © Ed Reschke; 21.15a: © Vol. 7/PhotoDisc/Getty RF; 21.15b: © Simko/Visuals Unlimited; 21.15c: © Dwight Kuhn; 21.16f: © Jack Bostrack/Visuals Unlimited; 21.17(walnut half): © Jack Star/PhotoLink/Getty RF; 21.17(whole walnut): © Getty RF; 21.17(peas): © Andrew Thomas/Getty Images; 21.17(peach half): © Alamy RF; 21.17(whole peach): © Jupiterimages/Image Source RF; 21.17(apples): © Photolink/Getty RF; p. 403(tomatoes): © Getty RF; 21.18a: © Marie Read/Animals Animals; 21.18b: © Scott Camazine/Photo Researchers, Inc.; 21.18c: © Getty RF; 21.18d: © James Mauseth; 21.22(all): Courtesy Prof. Dr. Hans-Ulrich Koop, from Plant Cell Reports, 17:601-604; 21.23: © Kingsley Stern; 21.24(all): Courtesy Monsanto; 21.25: © David Hoffman Photo Library/Alamy; 21.26b(both): Courtesy Dr. Eduardo Blumwald; 21.26c: © Getty RF; p. 408(organic foods): © Brand X Pictures/Getty RF.

Chapter 22

Figure 22.3(squamous, cuboidal, pseudostratified): © Ed Reschke; 22.3(columnar): © Ed Reschke/Peter Arnold/Photolibrary/Photolibrary; 22.4(left): © Jim Merli/Visuals Unlimited; 22.4(right): © Ed Reschke; p. 416(skinned elbow): © Eric Bean/Getty Images; 22.5(dense fibrous): © The McGraw-Hill Companies, Inc.; 22.5(hyaline, adipose, bone): © Ed Reschke; 22.7–22.8: © Ed Reschke; p. 420(neurons): © Steve Gschmeissner/Photo Researchers, Inc.

Chapter 23

Opener: © Du Cane Medical Imaging, Ltd/Photo Researchers, Inc.; p. 435(heart valves): © Biophoto Associates/Photo Researchers, Inc.; 23.6b, c (both): © Ed Reschke; p. 437(varicose veins): © BSIP/Photo Researchers, Inc.; p. 442(blood cells): © Prof P.M. Motta, and S. Correr/SPL/Photo Researchers, Inc.; 23.15d: © Manfred Kage/Peter Arnold/Photolibrary/Photolibrary; 23.16(right): © Biophoto Associates/Photo Researchers, Inc.; 23.17: © Pascal Goethgheluck/SPL/Photo Researchers, Inc.

Chapter 24

Opener: © Ansel Horn/Phototake; 24.3: © Photo Insolite Realite/SPL/Photo Researchers, Inc.; p. 453(runny nose): © Franc Lucas/Getty Images; 24.4b: © Ed Reschke; 24.22: © Ardea London Ltd; p. 465(woman): © Hill Creek Pictures/Getty RF; 24.24c: © Fawcett, Hirokawa, Heuser/SPL/Photo Researchers, Inc.

Chapter 25

Opener: © Reuters/Corbis; 25.1: © Getty RF; 25.2a: © Dr. Ken Greer/Visuals Unlimited, Inc.; 25.2b: © ISM/

Phototake; 25.3: © Precision Graphics/Kim Brucker, photographer; 25.4: © The McGraw-Hill Companies, Inc., John Thoeming, photographer; 25.5: © Cole Group/Getty RF; 25.6: © Volume 20/PhotoDisc/Getty RF; 25.7: © Lilli Day/Getty RF; 25.9a: © Corbis RF; 25.9b: © Lester Lefkowitz/The Image Bank/Getty Images; 25.9c: © Sean Bagshaw/Photo Researchers, Inc.; p. 487(pedometer): © Chris Kerrigan RF; 25.11: © Getty RF; 25.12: © Corbis RF; 25.13a: © Tony Freeman/PhotoEdit; 25.13b: © PhotoLink/Getty RF; p. 492(tea): © Getty RF; p. 493(bread): © Precision Graphics/Kim Brucker, photographer.

Chapter 26

Opener: © AFP/Getty Images; p. 498(biopsy): © National Cancer Institute-digital version copyright Science Faction/Science Faction/Corbis; 26.2: © Comstock/PunchStock RF; p. 500(antihistimines): © The McGraw-Hill Companies, Inc., Jill Braaten, photographer; p. 501(fever): © Vol. 41/Getty RF; 26.4a: © Dennis Kunkel/Phototake; 26.6b: Courtesy Dr. Arthur J. Olson, Scripps Institute; 26.8a: © Steve Gschmeissner/Photo Researchers, Inc.; p. 507(MMR vaccine): © Saturn Stills/Photo Researchers, Inc.; 26.10(pollen): © David Scharf/SPL/Photo Researchers, Inc.; 26.10(girl): © Damien Lovegrove/SPL/Photo Researchers, Inc.; 26.11: © Southern Illinois University/Photo Researchers, Inc.; 26.12a: © NIBSC/Photo Researchers, Inc.

Chapter 27

Opener left(normal brain): © SPL/Getty RF; opener right(half brain): © Neil Bordon/Photo Researchers, Inc.; p. 520(nerve cells): © Dennis Kunkel Microscopy, Inc./Visuals Unlimited/Corbis; 27.7a(both): © Science VU/Visuals Unlimited; p. 521(Marinol): © Time & Life Pictures/Getty Images; 27.10: © Cole Group/Getty RF; 27.13: © Tom McHugh/Photo Researchers, Inc.; p. 530(pregnant woman): © Owen Franken/Corbis; 27.18: © Biophoto Associates/Photo Researchers, Inc.; 27.19: © Dr. P. Marazzi/SPL/Photo Researchers, Inc.; 27.20a(both): *Atlas of Pediatric Physical Diagnosis*, 2nd edition by Zitelli and Davis, 1992. Mosby-Wolfe Europe LTd., London, U.K.; p. 533(woman): © Getty RF.

Chapter 28

Opener: © Corbis RF; 28.4a(both): Courtesy Dr. Yeohash Raphael, the University of Michigan, Ann Arbor; p. 546(LASIK surgery): © Pascal Goetgheluck/Photo Researchers, Inc.; p. 547(aspirin): © The McGraw-Hill Companies, Inc./Photo by Eric Misko, Elite Images Photography; 28.14a: © Michael Fogden/OSF/Animals Animals; 28.16(myofibril): Courtesy Hugh Huxley; p. 552(rigor mortis): © Michael Matthews/Alamy; 28.19a: © Gerard Vandystadt/Photo Researchers, Inc.

Chapter 29

Opener left (small, green-colored virus, HPV): © Phototake Inc./Phototake; opener right (larger virus, HIV): © Dr. Hans Gelderblom/Visuals Unlimited/

Corbis; 29.1: © Dennis Kunkel/Visuals Unlimited; 29.2: © Hans Pfletschinger/Peter Arnold/Photolibrary/Photolibrary; 29.3: © Leonard Lee Rue, Jr./Photo Researchers, Inc.; p. 561(gestation): © Gemstone Images/Corbis; p. 567(HRT): © Philippe Garo/Photo Researchers, Inc.; 29.9(pill): © The McGraw-Hill Companies, Inc./Bob Coyle, photographer; 29.9(implant): © Population Council/Karen Tweedy-Holmes; 29.9(female condom, jelly & diaphragm, Depo-Provera): © The McGraw-Hill Companies, Inc./Bob Coyle, photographer; 29.9(IUD): © The McGraw-Hill Companies, Inc./Vincent Ho, photographer; 29.10: © Myrleen Ferguson Cate/PhotoEdit; 29.11: © CC Studio/SPL/Photo Researchers, Inc.; p. 570(newborns): © PhotoDisc/Getty RF; 29.12(all): © Nicholas Nixon; 29.13: © PhotoAlto/PunchStock RF; 29.19, 29.21: © Lennart Nilsson, *A Child is Born*, Dell Publishing Company; 29.22: © John Watney/Photo Researchers, Inc.; 29.23: © James Stevenson/SPL/Photo Researchers, Inc.; 29.24: © Lennart Nilsson, *A Child is Born*, 1990 Delacorte Press, Pg. 111(top frame), p. 579(ultrasound/default sex): © Nic Cleave Photography/Alamy; 29.26: © Dennis MacDonald/PhotoEdit.

Chapter 30

Opener: © Photo Link/Getty RF; 30.1: © Getty RF; 30.2a: © Jeff Henry/National Park Service; 30.2b: © William P. Cunningham; p. 588(babies): © ER Productions/Corbis; 30.5a: © Jeff Greenberg/Peter Arnold/Photolibrary/Photolibrary; 30.5b: © Ben Osborne/OSB/Animals Animals; 30.7(MDC, all): © Comstock/Getty RF; 30.7(LDC, all): © Alasdair Drysdale RF; 30.8a: © Evelyn Jo Johnson; p. 593(Monaco): © PhotoDisc/PunchStock RF; 30.11a: © E. R. Degginger/Photo Researchers, Inc.; 30.11b: © Corbis RF; 30.18: © Creatas/PunchStock RF; 30.19a: © Ted Levin/Animals Animals; 30.19b: © Martin Harvey/Gallo Images/Corbis; p. 600(lemur/extinction): © Corbis RF; 30.20a: © Jeff Henry/National Park Service; 30.20b: © William P. Cunningham.

Chapter 31

Opener: © Stock Trek/Getty RF; 31.1a: © Anthony Mercieca/Photo Researchers, Inc.; 31.1b: Courtesy Dr. Rod Peakall, School of Botany and Zoology, Australian National University; 31.1c: © Merlin D. Tuttle/Bat Conservation International; 31.2a: © Charlie Ott/Photo Researchers, Inc.; 31.2b: © Michael Graybill and Jan Hodder/Biological Photo Service; 31.3a–b: © Vol. 63/Corbis RF; 31.5(all): © Breck P. Kent/Animals Animals; 31.6: © Steve Austin/Papilio/Corbis; 31.10(all): Courtesy Daniel Janzen; 31.11a: © George Armistead/VIREO; 31.11b: © Jack Wilburn/Animals Animals; 31.11c: © Eastcott/Momatiuk/Animals Animals; p. 612(bacteria): © Science Photo Library/Getty RF; 31.12a: © Stockbyte RF; 31.12b: © John Mithcell/Photo Researchers, Inc.; 31.13(diatoms): © Ed Reschke/Peter Arnold/Photolibrary/Photolibrary; 31.13(tree): © Authors Image/PunchStock RF; 31.14a(caterpillar): © Corbis RF; 31.14a(giraffes): © George W. Cox; 31.14b(insect): © Scott Camazine;

31.14b(osprey): © Joe McDonald/Visuals Unlimited; p. 615(chemoautotroph/fish): © Oxford Scientific/Getty Images; 31.15(mushrooms): © IT Stock/age fotostock RF; 31.15(bacteria): © David M. Phillips/Visuals Unlimited; 31.17: © Sylvia S. Mader; 31.23: © Tom McHugh/Photo Researchers, Inc.; 31.25: © Biophoto Associates/Photo Researchers, Inc.; p. 622(smoke stacks): © Corbis RF; 31.27a, d: © Vol. 121/Corbis RF; 31.27b: © The McGraw-Hill Companies, Inc./Carlyn Iverson, photographer; 31.27c: © David Hall/Photo Researchers, Inc.; 31.28(taiga, desert, temperate): © Vol. 63/Corbis RF; 31.28(prairie): © Vol. 86/Corbis RF; 31.28(tropical rain forest, tundra): © Vol. 121/Corbis RF; 31.28(savanna): © Gregory Dimijan/Photo Researchers, Inc.; p. 626(estuary): © Cameron Davidson/Getty Images.

Chapter 32

Opener: © L. Hobbs/PhotoLink/Getty RF; 32.1(land): © Vol. 39/PhotoDisc/Getty RF; 32.1(water): © Evelyn Jo Johnson; 32.1(food): © The McGraw-Hill Companies, Inc., John Thoeming, photographer; 32.1(energy): © Gerald and Buff Corsi/Visuals Unlimited; 32.1(minerals): © James P. Blair/National Geographic Image Collection; 32.1(population): © Photo Link/Getty RF; 32.2b: © Melvin Zucker/Visuals Unlimited; 32.4b: © Courtesy Carla Montgomery; 32.5a: © Corbis RF; 32.5b: © Jeremy Samuelson/FoodPix/Getty Images; 32.5c: © Stockbyte/PunchStock RF; p. 634(milk): © PhotoSpin, Inc/Alamy RF; 32.6a: © Jodi Jacobson/Peter Arnold/Photolibrary; 32.6b: © Inga Spence/Visuals Unlimited; 32.7a: © Laish Bristo/Visuals Unlimited; 32.7b: © Inga Spence/Visuals Unlimited; 32.7c: Courtesy V. Jane Windsor, Division of Plant Industry, Florida Department of Agriculture and Consumer Services; 32.8: © Corbis RF; 32.9b: © Shane Moore/Animals Animals; 32.9c: © Michael Pole/Corbis; 32.10b: Courtesy Walther C. Japp, Florida Fish and Wildlife Conservation Commission; 32.11a: © David L. Pearson/Visuals Unlimited; 32.11b: © S.K. Patrick/Visuals Unlimited; 32.11c: © Argus Foto Archiv/Peter Arnold/Photolibrary/Photolibrary; 32.11d: © Gerald and Buff Corsi/Visuals Unlimited; 32.12a: © Matt Stiveson/DOE/NREL; 32.12b: © Courtesy DaimlerChrysler; 32.12c: © Warren Gretz/DOE/NREL; 32.13: © James P. Blair, National Geographic Image Collection; p. 641(swordfish): © Norbert Wu/Minden Pictures; 32.15b: © Daniel Cox/Getty Images; 32.16(periwinkle): © Kevin Schaefer/Peter Arnold/Photolibrary/Photolibrary; 32.16(armadillo): © John Cancalosi/Peter Arnold/Photolibrary/Photolibrary; 32.17a: © Anthony Mercieca/Photo Researchers, Inc.; 32.17b: © Merlin D. Tuttle/Bat Conservation International; 32.17c: © Herve Donnezan/Photo Researchers, Inc.; 32.17d: © Bryn Campbell/Stone/Getty; 32.18(mountains): © William Smithey Jr./Getty Images; 32.18(seashore): © Vol. 121/Corbis RF; 32.19: © Stuart Westmorland/Corbis; 32.20a: © PhotoLink/Getty RF; 32.20b: © Scenics of America/PhotoLink/Getty RF; 32.21: © Dan McCoy/Rainbow; p. 650: © Getty RF.

Index

Note: Page numbers followed by *f* and *t* indicate figures and tables, respectively.

Thymus gland, 498, 527f
Thyroid gland, 527f, 530–531
Thyroid-stimulating hormone (TSH), 529
Thyroxine, 530
Tibia, 548f, 550
Tight junctions, 73, 73f
Tinea, 329f, 330
Tissue, 2f, 3
    capillary exchange in, 440, 440f
    connective, 414, 414f, 417–418
    defined, 413
    epithelial, 414, 414f, 415–417
    of immune system, 497–499
    muscular, 414, 414f, 419
    nervous, 414, 414f, 420
    plant, 374–376
    structure and function, 414, 421
Tissue culture, 405
Tissue fluid, 422
Tissue rejection, 505
Titin, 46
T lymphocytes (T cells), 498, 498t
    activation of, 503–504, 504f
    cellular response and, 503–505
    characteristics of, 504t
Tobacco mosaic virus, 290, 290f
Tonsillitis, 452
Totipotent, 200, 405
Toxins, 298
Trace minerals, 481–482, 481t
Tracer, 22
Trachea, 452f, 453
Transcription, 181, 182–183, 183f
    gene expression in eukaryotes, 202–204
    gene expression in prokaryotes, 201–202
Transcription activators, 204, 204f
Transcription factors, 203–204, 204f
Transduction, 296
Trans fat, 43, 43f, 475, 480
Transfer rate, 620
Transfer RNA, 65, 180, 184, 184f
Transformation, 296
Transgenic animals, 174, 188, 189–190, 189f
Transgenic bacteria, 189, 189f
Transgenic plants, 189–190, 189f, 407, 407f
Transitional epithelium, 416
Translation, 184–186, 184f, 185f
Translocation, 185, 210
    chromosomes, 218
Transpiration, 385
Transport proteins, 61, 61f
Transport system of body, 422, 422f
Transposon, 187, 187f
Trees
    annual ring, 380
    bark of, 379
    wood of, 378–379
    world's tallest, 371, 371f
Treponema pallidum, 295, 295f
Triassic period, 273t
Trichinosis, 347
Trichomoniasis, 572

Tricuspid valve, 434
Triglycerides, 42, 479
Triplet code, 182, 182f
Trisomy, 150
Trisomy 21, 150–151
Tropical level, 618
Tropism, 396, 396f
True-breeding plants, 157
Trypanosome, 303
Trypsin, 466
Tuberculosis, cause of, 287
Tumor, 134
Tunicates, 352
Turgid, 88
Turner syndrome, 151, 151f
Two-trait inheritance, 160–161
Type 1 diabetes, 488, 532–533
Type 2 diabetes, 479, 488, 533
Typhoid fever, 299

U

Ulcer, 464
Ulna, 548f, 549
Ultrasound, 226, 226f
Umbilical cord, 577
Uncoating, of virus, 290f
Uniform, 592
Uniformitarianism, 236
Unsaturated fatty acids, 42, 43f
Upper respiratory tract, 452–453, 452f
Uracil (U), 48f, 49, 180
Urban sustainability, 649–650
Ureter, 457, 457f
Urethra, 457–458, 457f
Urinary bladder, 457, 457f
Urinary system, 413, 413f, 422, 422f, 457–461
    functions of kidneys, 457
    kidney problems, 460–461
    urine formation, 458–460
Urine, 457
    formation of, 458–460
Uterus, 564, 564f
Utricle, 543

V

Vaccines
    contraceptive, 568–569
    historical perspective on, 496
    immunity and, 506
    MMR vaccine, 507
Vacuoles, 67, 67f
Vagina, 564, 564f
Vaginitis, 572
Valence shell, 23–24, 23f, 24f
Valine (Val), 49
Valium, 520
Varicose veins, 437, 437f
Vascular bundles, 373
Vascular cambium, 374
Vascular cylinder, 380, 381f

Vascular plants, 314–317
Vascular tissue
    of plants, 374, 375–376, 375f
    of roots, 380–382
Vas deferens, 562
Vector, 188, 188f
Veins, 435, 437, 437f
Venae cavae, 434, 434f
Ventricles, 434, 434f
Ventricular fibrillation, 436
Venules, 437
Vertebral column, 549
Vertebrate animals, 340
    circulatory pathways in, 433
Vertebrate tissue, 414, 414f
Vertigo, 543
Vessel elements, 376
Vestigial structures, 244–245
Villi, 466, 466f, 467
Viroids, 292
Viruses, 288–292
    animal viruses, 290–292
    bacteriophage reproduction, 289
    cancer and, 197
    defined, 288
    drug control of, 292
    emerging, 291
    H1N1, 291
    parts of, 288, 288f
    plant viruses, 290
    reproduction of, 288–289
    retroviruses, 291
    sexually transmitted diseases,
        571–572
Visceral muscle, 419
Visible light, 97, 97f
Vision, 544–546
    in the dark, 546
    LASIK surgery, 546
    photoreceptors in eye, 545
    types of eyes, 544
Vitamin deficiencies, 482–483, 483t
Vitamins, 482–483, 483t
    A, 483t
    B₁, 483t
    B₂, 483t
    C, 476, 482–483, 483t
    D, 482–483, 483t, 531
    E, 483t
    K, 467, 483t
Viviparous, 561
Vocal cords, 453
Volvox, 302
Vulva, 564, 564f

W

Wallace, Alfred Russel, 240–242
Waste
    biodiversity and disposal of, 646
    industrial waste, 642
    sewage, 642